Foundations of Optical System Analysis and Design

Foundations of Optical System Analysis and Design

Lakshminarayan Hazra

(Emeritus Professor, University of Calcutta,
West Bengal, India)

CRC Press
Taylor & Francis Group
Boca Raton London New York

CRC Press is an imprint of the
Taylor & Francis Group, an informa business

First edition published 2022
by CRC Press
6000 Broken Sound Parkway NW, Suite 300, Boca Raton, FL 33487-2742

and by CRC Press
2 Park Square, Milton Park, Abingdon, Oxon, OX14 4RN

CRC Press is an imprint of Taylor & Francis Group, LLC

Library of Congress Cataloguing-in-Publication Data
Names: Hazra, Lakshminarayan (Lakshminarayan N.), author.
Title: Foundations of optical system analysis and design / Lakshminarayan Hazra.
Description: First edition. | Boca Raton: CRC Press, 2021. |
Includes bibliographical references and index. |
Summary: "Analysis and design of optical/photonic systems call for applications of
principles of optics of varying degrees of approximation at different stages.
This book is designed to provide adequate material for 'self-learning' and also act
as a handy reference for the students of related Masters and Doctoral courses" – Provided by publisher.
Identifiers: LCCN 2021012295 (print) | LCCN 2021012296 (ebook) |
ISBN 9781498744928 (hardback) | ISBN 9781032037080 (paperback) |
ISBN 9780429154812 (ebook)
Subjects: LCSH: Optical instruments–Design and construction. | Optics–Mathematics. | Aberration.
Classification: LCC QC372.2.D4 H397 2021 (print) |
LCC QC372.2.D4 (ebook) | DDC 621.36–dc23
LC record available at https://lccn.loc.gov/2021012295
LC ebook record available at https://lccn.loc.gov/2021012296

ISBN: 978-1-4987-4492-8 (hbk)
ISBN: 978-1-0320-3708-0 (pbk)
ISBN: 978-0-429-15481-2 (ebk)

DOI: 10.1201/9780429154812

Typeset in Times
by Newgen Publishing UK

In loving memory of my parents
Mother: Shrimati Bidyutlata Hazra
Father: Shri Kashi Nath Hazra

Contents

Preface

Optical systems have played a key role in the sustenance and development of human life and living since the days of antiquity. Observational sciences commenced their journey in the seventeenth century on the basis of experimental observations with the help of two optical systems, namely the telescope and the microscope. The evolution of modern science and technology during the next four centuries is associated with concomitant transformations of these systems in terms of scope, range, and sensitivity, among other qualities.

In the recent past, the emergence of novel means for the generation, manipulation, detection, processing, and display of light and electromagnetic waves of the neighbouring wavebands has not only opened up new frontiers for optical systems, but has also given rise to a metamorphosis of them, often rendering them beyond recognition. However, despite this change, the core functions continue to be performed by an optical system. This book aims to lay a foundation for analysis and synthesis/design of this optical system.

Many excellent books, monographs, book chapters, and journal publications exist on topics like 'Geometrical Optics', 'Applied Optics', 'Ray Optics', 'Theories of Image Formation', 'Aberration Theory', 'Optical System Design' and 'Lens Design', etc., and, in general, they serve the purpose they sought after. However, during my investigations in different areas of this field, I came across some interesting features. Textbooks published anywhere in the world on topics like 'Geometrical Optics', 'Physical Optics', 'Optical Instrumentation', etc. are more or less similar. However, books or book chapters on topics like 'Theory of Aberrations', 'Optical Design', 'Applied Optics', etc. appearing from different parts of the USA, the UK, the European continent, Russia, Japan, Australia, etc. often look different, either due to differences in basic treatment or differences in notation, and often each of them make no cognizance of research work carried out by others. In the field of journal publications, this is manifested by papers reporting more or less similar works carried out independently. On the other hand, excellent publications by some individuals, who incidentally do not appear to belong to any proactive group, simply disappear into oblivion. The problem is aggravated if the publication is in a language other than English. Even now, English translations of many significant publications in other languages are not readily available.

Owing to the current proliferation of areas related to optical science and technology, there is a growing demand of new types of optical systems. Many types of powerful commercial and in-house software are available to cater to these needs. Nevertheless, an understanding of the foundations of analysis and design of optical systems is a sine qua non for effective utilization of the software in order to achieve one's desired goals.

This book aims to alleviate some of these problems. In order to inculcate a rational scientific approach, necessary historical background is provided for each topic dealt with in this book. The references detailed therein provide a veritable knowledge bank for the inquisitive reader. Despite covering material from a wide range of sources, the mainstream of the book follows the same notation and representation throughout the book for the sake of ease in understanding. The main approach adopted for each topic is now followed more or less universally.

A few years back, a tentative proposal for 'A Course on Foundations of Optical System Analysis and Design (FOSAD)' was presented in the Twelfth International Topical Meeting on Education and Training in Optics and Photonics (ETOP2013), organized by the International Commission for Optics at Porto, Portugal [Proc. SPIE.9289, doi: 10.1117/12.2070381 (2013)]. A lot of the comments and feedback from the participants, and subsequently from readers of the publication, led to some modifications in the course content, ushering in the present structure of the book.

Chapter 1 presents an early history of major developments in optical systems and the related principles of optics from the days of antiquity to the end of the seventeenth century. The metamorphosis of optical systems, engendered by the proliferation of novel optical devices, components, and elements is noted. The current definition of 'Optics' per the International Commission for Optics (ICO) is described. A list of types of optical systems and their constituents is given to underscore the wide range of current optical systems. Different schemes for classification of optical systems are noted. For the sake of convenient analysis and synthesis, several theories of light with varying degrees of approximation are utilized in practice. A flowchart demonstrating the interrelationship between the theories is presented. In practice, Maxwellian electrodynamics provide a powerful theory encompassing almost all phenomena in classical optics and related systems. However, as shown in the flowchart, at least six different theories with varying levels of approximation are routinely used for the analysis and design of optical systems. The foundations of each theory are dealt with in Chapter 2.

Prefaced with a statement on the raison d'être for paraxial optics in analysis and synthesis of optical systems, paraxial treatments for both on-axis and off-axial imaging in a refracting spherical interface are presented in Chapter 3, highlighting the degrees of approximation in aperture and field parameters in each case. This is followed by the derivation of a general object-image relationship for imaging by an axisymmetric system. The tricky interrelationship between vergence and power is underscored with ray diagrams for nine different cases. The geometric nature of image formation by an ideal Gaussian system is described. The inevitable distortion in paraxial images of a line or a plane inclined with the axis by such a system is demonstrated. The Scheimpflug principle is explained, as is the occurrence of keystone distortion in the paraxial image of a square object placed on an incline with the axis in the object space.

A paraxial analysis of the roles of different types of stops, such as aperture stop, field stop, and glare stop, is presented in Chapter 4. The phenomenon of vignetting in extra-axial imaging is explained, and the significant role of field lens for the matching of pupils in cascaded lens systems is elucidated. The optical imaging system of the human eye and the optical requirements for proper functioning of visual optical instruments are noted. Different forms of paraxial optical invariants are presented. Suitable choices for paraxial ray variables for obtaining best possible estimates for the real ray imaging performance in the final image space are described, and the appropriate definition for angular magnification in afocal systems is given. Exact definitions for F-number and numerical aperture are presented. Paraxial expressions for depth of focus and depth of field are derived. Different types of telecentric stops are illustrated. Examples of the use of stops in illumination systems are given.

Several techniques for facilitating paraxial treatment of optical systems have been proposed. Most of them have the common goal of enhancing analyticity in the paraxial treatment of multicomponent optical systems of increasing complexity. Some of these techniques have dedicated areas of application. The major techniques are noted in Chapter 5, and three specific approaches, the matrix method, the Gaussian brackets, and the Delano diagram, are enunciated in this chapter.

All optical systems accomplish their functions by transport of energy. For most practical purposes, a relatively simple model, as provided in 'radiometry' is adequate for analyzing energy considerations in source and receiver/detector and propagation of electromagnetic waves. For visual systems with the human eye as detector, 'photometry' has emerged as a special form of radiometry. The interrelationship between the two is deliberated upon in Chapter 6 along with the basic assumptions inherent in this model. The conservation relations for radiance/illuminance in optical imaging systems are derived from first principles. It is shown that the light flux collected by the entrance pupil of a lens system is proportional to the square of the paraxial optical invariant. The cosine-to-the-fourth law of illumination in images of extraaxial points of extended objects is derived, and indications on methods to circumvent this limitation are also given.

The imaging characteristics of a real optical system differ from the predictions obtained by using a paraxial model of the system. For a specific system, these differences can be determined by tracing real rays through the system using finite ray tracing formulae. An alternative approach to understand the characteristics of real optical systems makes use of propagation laws of real rays. The latter approach is known as the Hamiltonian treatment, whereas the former one is called the Lagrangian method. Chapter 7 commences with a description of the basic concepts of Hamiltonian optics and their relationship with Lagrangian optics. This is followed by a detailed exposition of the feasibility of perfect imaging by real rays for objects of different dimensions. Different types of Cartesian ovals are derived. The requirements of ideal imaging by Maxwellian perfect instrument are discussed. The interesting possibility of perfect imaging by inhomogeneous lenses, e.g. Maxwell's fish-eye lens, Luneburg lens, etc., is highlighted. The case of aplanatic points and surfaces is discussed. The optical cosine rule for stigmatic imaging of two neighbouring points is derived, and Abbe's sine condition and Herschel's condition follow as special cases.

The same chapter contains a discussion on real ray invariants, which include derivations of different forms of skew ray invariant as well as the generalized Lagrange invariant. The concept of astigmatism of surfaces is discussed beginning with the fundamentals. The S and T ray tracing formulae are derived. The sagittal invariant in axisymmetric imaging is underscored. Chapter 7 concludes with a broad classification of aberrations of optical systems.

Chapter 8 commences with a brief early history of optical aberrations, and deals with monochromatic aberrations. Different measures of aberration used in treatment of different types of optical imaging systems are discussed, along with their interrelationships. The relations between wave aberration, ray aberration, and Hamilton-Bruns' eikonal are given. The nature of the phenomenon of caustic near focus is also discussed. A power series expansion of the wave aberration function is carried out. Detailed analysis of aberrations of various orders in case of axially symmetric imaging systems is done, and each term up to quaternary order is identified. The phenomena of aplanatism and isoplanatism are discussed. The asymmetric component of wave aberration is identified, and it is shown that total linear coma can be determined from properties of axial pencil alone. Two important topics, the 'Offence against sine condition' and the 'Staeble-Lihotzky condition', are covered in this chapter.

Chapter 9, which concerns chromatic aberrations, deals with paraxial chromatism, i.e. the changes in paraxial characteristics of dioptric or catadioptric systems occurring with changes in working wavelength. In a brief account of the dispersion of optical materials, a set of practically useful interpolation formulae for determination of refractive index of optical glasses at any wavelength over a given range is explored. The role of matching relative partial dispersions of optical materials of the constituent elements for reducing secondary spectrum in achromatic doublets is described. The use of apochromats for reducing tertiary spectrum and of superachromats for reducing chromatic spectra of higher orders are discussed. Harting's criterion for complete or total achromatization within a given wavelength range is described. The use of a dialyte, i.e. an air-spaced doublet, in a one-glass achromatic Schupmann doublet is explained. The feasibility of multicomponent air-spaced achromats with normal glasses, and in some cases with only one type of optical material, is underscored. Exact conditions for achromatism in thick lenses or compound lens systems are elucidated.

Chapter 10 deals with evaluation of finite or total wave aberration and ray aberrations in axially symmetric imaging systems. The formulae are specifically directed towards the determination of wave aberration of the image forming wavefront passing through the real exit pupil, so that the numerical results obtained thereby can be directly utilized for the purposes of image quality evaluation on the basis of diffraction.

In the latter half of Chapter 10, different measures for nonparaxial chromatism that involve variation in monochromatic aberrations with changes in working wavelength are discussed. Two different approaches for representing the phenomenon of spherochromatism are explained. It is shown that the widely used Conrady $(d - D)$ chromatic aberration formula describes the chromatic aberration of image space associated rays. A method for evaluating the exact chromatic aberration of object space associated rays is presented.

A generalized set of extended paraxial coordinates, called 'canonical coordinates' by H.H. Hopkins, serves to extend the linear rules of paraxial optics to arbitrarily large fields and apertures. This leads to a consistent definition of pupils and oblique magnifications for a real optical system. The failure of an optical system to meet these required relations leads to aberrations and is closely related to the establishment of isoplanatism. What's more, use of these canonical coordinates in the definition of wavefront aberration facilitates proper formulation and accurate evaluation of diffraction-based image evaluation criteria for both axial and extra-axial imaging. The basics of Hopkins' canonical coordinates and variables in the treatment of image formation are highlighted in Chapter 11.

The relations between the Seidel aberration coefficients and the coefficients of primary wave aberration as obtained via power series expansion of the wave aberration function are given in Chapter 12. The primary aberrations can be conveniently determined from data of paraxial ray tracing with suitably chosen ray variables. Further, primary aberrations of a cascaded system can be obtained by a summation of primary aberrations contributed by the constituent refraction surfaces, and the validity of these assertions is proven in this chapter. The formulae (based on Abbe's refraction invariant) for computation of the five monochromatic Seidel aberrations in a few typical systems are given. Convenient formulae for computing the primary wavefront aberrations corresponding to the axial colour and the transverse colour are derived.

Chapter 13 starts with a brief review of the diverse approaches adopted for evaluation of aberrations of higher orders. For the sake of aberrational analysis in optical systems incorporating necessary higher order aberration coefficients, a special treatment is reported. The method reduces the computational load significantly.

Thin lens approximation in modelling lens systems is very useful in deriving the general principles of lens structures and in obtaining useful results in systems in which the field and the aperture are not too large. Expressions for primary aberrations of a thin lens with different object and image space media are presented in Chapter 14. They are given in terms of modified Coddington variables. Practical use of thin lens aberration theory in the structural design of lens systems are enunciated. Suitable procedures for transition from thin lens to thick lens and vice versa are discussed. A short note on thin lens modelling of diffractive lenses is provided.

Chapter 15 deals with the effects of stop shift and conjugate shift on axi-symmetric imaging systems. The relationships between the Seidel pupil aberrations and Seidel image aberrations are derived. Conjugate shift implies an axial shift of the object plane with the concomitant axial shift of the image plane. Effects of conjugate shift on the coefficients of Seidel aberrations in object imagery is enunciated. The use of the 'symmetrical principle' in tackling asymmetrical aberrations is explained, along with a brief description of the Bow-Sutton conditions.

In the case of optical systems with small values of residual aberrations, the ray-optical theory fails to provide correct interpretation for experimental observations. It is necessary to incorporate the diffraction effects arising out of the finite size of the aperture of the imaging system in the analysis for proper explanation of the observations. Chapter 16 describes the Huygens-Fresnel principle for propagation of light waves, and uses it for determining the diffraction image of a point object with an aberration-free axi-symmetric lens system. The concepts of confocal pupil sphere and image sphere are discussed, and the consequences of omission of quadratic phase term in evaluation of complex amplitude on the image plane are discussed. The anamorphic stretching of the extraaxial point spread function (PSF) is described. The Airy pattern and distribution of energy in the 'factor of encircled energy' in aberration-free imaging systems are illustrated. Detailed physical interpretations for different criteria for two-point resolution, including Rayleigh criterion, Sparrow criterion, Dawes' criterion, etc., are provided. We will then underscore the basic principles of four techniques for breaking the diffraction limit of resolution: (i) use of phase shifting mask, (ii) use of mixed/phase filters for superresolution over a restricted field of view, (iii) confocal scanning, and (iv) near field super resolving aperture scanning.

The combined effects of diffraction and residual aberrations of an imaging system on the quality of image of a bright point object on a dark background are reported in Chapter 17. Aberration tolerances as per the Rayleigh quarter wavelength rule and the Strehl criterion are described. The relationship between the Strehl ratio and the variance of wave aberration is derived. We will show that use of local variances of the wavefront can enhance the range of validity of the interrelationship between the two. The technique of aberration balancing for relaxing stringent tolerances is described with illustrations, e.g., the tolerance on secondary spherical aberration with optimum values for primary spherical aberration and the defect of focus. We present an approach for fast evaluation of variance of wave aberration from ray trace data, and a criterion based on the factor of encircled energy is reported for aberration balancing and tolerancing applicable for moderately corrected systems. The chapter describes the Zernike circle polynomials and explains their important role in aberration analysis. The role of the Fresnel number in focusing/imaging is discussed. The requirements for special treatment needed for tackling problems of imaging/focusing in optical systems with large numerical aperture are also briefly noted.

Chapter 18 presents a system theoretic viewpoint in optical image formation. The process of optical image formation is modelled as a linear two-port system with the object as input and the image as output of the system. Conditions for validity of this model are explained. A physical interpretation of the kernel of Fourier transform is provided for underscoring the usefulness of Fourier analysis in the treatment of extended object imagery. The image quality evaluation parameters, including the optical transfer function (OTF), the modulation transfer function (MTF), and the phase transfer function (PTF), are defined, and the Abbe Theory of coherent image formation is explained.

We then present the relations between the transfer function, the point spread function, and the pupil function separately for the two cases of coherent illumination and incoherent illumination. The effects of residual aberrations and of nonuniform amplitude over the pupil function on the transfer function are described.

Aberration tolerances based on the OTF are defined in terms of variance of the wave aberration difference function. A method for fast evaluation of variance of the wave aberration difference function from finite ray trace data is described. The concept of through-focus MTF is explained. Different interrelationships between PSF, LSF, ESF, BSF, and OTF are presented. We emphasize the importance of taking into account the effects of anamorphic imagery in the off axis region on OTF analysis, and provide the correct procedure for evaluating transfer function in cascaded optical systems. The appropriate definitions for image evaluation parameters, such as polychromatic PSF and polychromatic OTF, are given, and we highlight the use of information-theoretic concepts in image evaluation.

Chapter 19 deals with the basics of lens design. Steps of the iterative procedure adopted in practical lens design methodology are explained, and common approaches for tackling aberrations are described. The structures and working principles of some special types of lens systems are given. Lenses with working wavelength beyond the visible range are described. A brief description of some unconventional lenses and lenses using unconventional optical elements is given. A broad classification of lenses based on aperture, field of view, axial location of aperture stop, image quality, and generic type of lens elements provides a link between disparate types of lenses and underscores the inherent unity in diversity of the large types of optical imaging lenses. Some well-known structures of lenses used in infinity conjugate systems are identified, and their basic optical layouts are given, and we then elucidate on the important topic of manufacturing tolerances.

For successful implementation in practice, the iterative procedure of lens design is formulated mathematically as a problem of nonlinear optimization in a constrained multivariate hyperspace of the design variables. Chapter 20 starts with brief history of developments in lens design optimization since its conception. We provide the mathematical preliminaries of nonlinear optimization, and underscore the advantages of using a formulation based on nonlinear least squares. We then discuss techniques for the handling of constraints. The chapter presents the details of a damped least squares (DLS) method for lens design optimization. Methods for evaluation of aberration derivatives are also described.

The nonlinear optimization methods mentioned above can only provide the local optimum in the immediate neighbourhood of the given starting point of a multimodal objective function. Chapter 21 discusses different approaches for overcoming this limitation. Usually, the number of local optima continues increasing exponentially with the increase in the number of variables of this function. The nature of optimality of these optima is not similar, and determination of the global optimum, or even a 'good'/quasi-global optimum, depends on the ability to choose a suitable starting point. Several heuristic approaches to circumvent this problem in case of lens design optimization have been proposed. However, the solution obtained therefrom cannot be called a global solution.

A number of stochastic methods, such as simulated annealing, evolutionary algorithms, neural networks, fuzzy logic, deep learning, swarm optimization, and different combinations of the above hold better promise in this regard. We provide a glimpse of the large number of nature-inspired and bio-inspired algorithms that are currently being explored to tackle global optimization problems, and describe a prophylactic approach for global or quasi-global synthesis of practical lens systems. Although the lens design optimization problem is usually formulated as a single-objective optimization problem, in reality, it is essentially a multi-objective optimization (MO) problem. In this chapter, we explore the concept of pareto-optimality in dealing with an MO problem.

Hopefully, the material presented in these twenty-one chapters covers the foundations of analysis and design of optical systems and provides a sourcebook for the inquisitive student or scholar in this field. Few topics, as mentioned in the epilogue, could not be accommodated, but proper references on them are noted. Although on some topics the number of publications being referred to appears large, it should be noted that no attempt has been made to refer to all relevant publications. However, conscious attempts are made to include the original and the important works in a chronological order, published from anywhere in the world. Omission of any particular reference on a topic may have occurred because I did not come across it. In any case, it should not be interpreted as my lack of regard for its merit.

Lakshminarayan Hazra

Acknowledgments

This book is based on my unique perspective for optical systems. It has had a long period of gestation, and many academics from around the globe played important roles in this. At the outset, I express my sincere gratitude to my mentors, Professor Manoranjan De who induced my interest in the field of applied optics and optical engineering during my graduation at the University of Calcutta, and Professor Harold Horace Hopkins who introduced me to the practice of analysis and design of optical systems during my tenure of Nuffield Foundation fellowship at the University of Reading, UK.

For the greater part of my professional career, I worked as a member of the faculty at the University of Calcutta. One of the major privileges of the position was that I suffered practically no interference from higher-ups on fixing my academic choices and goals. On the other hand, I had to suffer perennially from limited access to knowledge banks of all conceivable forms because state universities located at a provincial capital in a third world country have very limited facilities in this regard. The recent proliferation of internet facilities has marginally improved the situation. On my part, generous cooperation from my friends and students working abroad kept me afloat; also, I had been assisted profusely by my hosts during my sojourns abroad in collecting rare references. I take this opportunity to put on record my sincere appreciation of their help.

Dr. John Macdonald, Dr. Steve Dobson, and Dr. Azmi Kadkly of the University of Reading in England made my first exposure to computer aided design of lens systems highly enjoyable and illuminating. My hosts, Professor Claude A. Delisle and Professor Roger A. Lessard at Centre d'Optique, Photonique et Lasers at Laval University, Quebec, Canada, during my visiting professorship in the nineties, provided a congenial atmosphere for assimilation of fast changes in the field of classical optics. The seeds of the current enterprise were sown during my visiting professorship under the Erasmus Mundus program of the European Commission at Warsaw Technological University, Poland. Prof. Margarita Kujawinska provided a perfect ambience for me to formulate and deliver a course of lectures on optical system design. Prof. Marcin Lesniewski provided me with first-hand knowledge on the fast changes occurring in optical engineering in Eastern Europe. A few years later, I was invited to be a JSPS fellow at the Center for Optics Research and Education at Utsunomiya University, Japan. With support from my hosts, Professor Toyohiko Yatagai and Professor Yukitoshi Otani, I delivered a course of lectures on 'Foundations of Optical System Analysis and Design'. The course was attended by both students of the centre and engineers from the optics industry of the region. Later on, these courses were delivered at a few other places, most notably at the Indian Institute of Space Science and Technology, Thiruvananthapuram, India. It was organized by my host, Professor C.S. Narayanamurthy, during the formation of their new M. Tech. program in Optical Engineering. Parts of the course were also delivered as parts of my lecture course in my university, and in several invited lectures in different parts of India and abroad. Interaction with participants of these lectures, as well as feedback from them, led to many modifications in the course content.

Over the years, interactions with many individuals have played distinct roles in framing my perspective on this topic. I would like to put on record my sincere appreciation for Professor Adolf Lohmann, Professor Olaf Bryngdahl, Dr. K.J. Rosenbruch, Professor Pierre Chavel, Professor Christopher Dainty, Dr. Florian Bociort, Professor José Sasián, Professor Virendra N. Mahajan, Prof. Yunlong Sheng, Prof. Vasilios Sarafis, Prof. Yontian Wang, Prof. Jorge Ojeda-Castañeda, Professor James Bradford Cole, Professor Jari Turunen, and Prof. Irina Livshits from abroad, and Professor Kehar Singh, Professor Rajpal Singh Sirohi, Professor Bishnu P. Pal, Professor Anurag Sharma, Dr. Rafiz Hradaynath, Dr. Arun Kumar Gupta, and Dr Nimai C. Das from India.

In the beginning of my professional life, I worked in an optics factory, and then in an R&D institution. First-hand knowledge gained therefrom on the design, fabrication, and testing of optical systems provided me a special insight into their operations. I express my gratitude to the individuals who helped me in gaining this knowledge: Mr. Amal Ghosal, Mr. Samad, Mr. B.D. Pande, Mr. Vasant Bande, Mr. Raj Kumar Seth, Dr. E.K. Murthy, and Dr. Jagdish Prasad.

My colleagues in the Applied Optics and Photonics department of the University of Calcutta deserve special mention in shaping my professional journey. They are: Professor Subodh Chandra Som, Professor Ajay Kumar Chakraborty, Professor Amitabha Basuray, Professor Samir Kumar Sarkar, Professor Ajay Ghosh, Professor Kallol Bhattacharya, Professor Rajib Chakraborty, Dr. Mina Ray, and Dr. Kanik Palodhi. Incidentally, some of them were also my students.

Students have always been a source of inspiration and encouragement. Often they had to put up with my experiments with 'Teaching of Optics'. To express my gratitude for their indirect contribution in the realization of this book, I mention some of them whose undergraduate/graduate level dissertations were supervised by me on related topics: Akshyay Banerjee, Paresh Kundu, Partha Pratim Goswami, Sourangsu Bandopadhyay, Anirban Guha, Anup Kumar Samui, Kallol Bhattacharya, Yiping Han, Otman Filali Meknassi, Jaya Basu, Saswati Banerjee, Maumita Chakraborti, Sanghamitra Chatterjee, Kanik Palodhi, Ujjwal Dutta, Santa Sircar, Sourav Pal, Nasrin Reza, Pubali Mukhopadhyay, Somparna Mukhopadhyay, Atri Haldar, and Indrani Bhattacharya, among others.

Professor Ajay Ghatak, ex-Indian Institute of Technology, Delhi and Professor Swapan Saha, ex-Indian Institute of Astrophysics, Bangalore deserve special mention for encouraging me continually to write a book on a similar theme. I wish I could fulfil their expectations.

I would like to thank Debalina Bhattacharya and Arpita Chakraborty for their help with some of the drawings, Atanu Dutta Khan for his help in sorting out few critical MS Office features, and Dr. Soma Chakraborty for her help in locating the three Vaiśeşika Sûtras of Kanâda quoted in the Introduction. Sri Kalyan Kumar Banerjee and Sri Haraprasad Bhattacharyya extended constant support and encouragement during the long journey of completing this book.

Aastha Sharma and Shikha Garg of CRC Press deserve special mention for their cooperation and patience with me over a prolonged period during compilation of the book. On the other hand, I am indebted to Suba Ramya and Bronwyn Hemus for their excellent performance during the stage of production of the book.

Finally, I would like to express my appreciation for my wife, Sukla and my son, Arnab for being a constant source of inspiration during the difficult days of working with this book.

Lakshminarayan Hazra

Author Brief Biography

Prof. Lakshminarayan Hazra obtained his M. Tech. and Ph.D. degrees from the University of Calcutta, and after several years in M/s National Instruments Limited, Calcutta and the Central Scientific Instruments Organisation (C.S.I.O.), Chandigarh, he joined the faculty of the Department of Applied Physics of the University of Calcutta in 1979.

Early in his career, he devised new techniques for imaging in telescopes, working in turbulent medium, and in collaboration with Prof. H. H. Hopkins of University of Reading, U.K., he designed a remote access zoom objective for monitoring open-heart surgery in 1983. Prof. Hazra developed new methods for structural design of optical systems, as well as the optical design software 'Ray Analysis Package' (RAP). He pioneered the use of Walsh functions in image analysis and synthesis, and, in association with researchers at Laval University in Canada, investigated the exact surface relief profile of diffractive lenses in the non-paraxial region.

He took the lead in establishing the Department of Applied Optics and Photonics at the University of Calcutta in 2005. He is currently associated as an Emeritus Professor with this department. He is a Distinguished Fellow of the Optical Society of India (OSI), where he was General Secretary for more than 15 years, and currently he is the Editor-in-Chief of the OSI '*Journal of Optics*', published in collaboration with M/s Springer. He is a fellow of both the Optical Society of America and the SPIE. Prof. Hazra is the representative for the Indian Territory in the International Commission for Optics. He was conferred the 'Eminent Teacher Award' by the University of Calcutta in 2019.

Prof. Hazra has published more than 100 papers in archival journals, and a few research monographs. He has delivered more than 250 invited talks on his research in different parts of India and in many countries of the world, namely the USA, the UK, Canada, Japan, Austria, Poland, France, Germany, the Netherlands, Italy, Slovenia, Malaysia, China, Finland, and Russia, among others. He provides consultancy services to the industry on optical instrumentation, and regularly works for the promotion of applied and modern optics. He was a Nuffield Foundation Fellow (1982-83) in the UK, a visiting Professor (1991 – 97) at the Centre d'Optique, Photonique et Lasers, Laval University in Quebec, Canada, an Erasmus Mundus Visiting Professor in the OpSciTech program of the European Commission in 2008, and a Visiting (Invitation) Professor of the Japan Society for Promotion of Science in 2012.

His areas of research interest include Optical System Design, Zoom Lens Design, Diffractive Optics, Optical and Ophthalmic Instrumentation, Aberration Theory, Theories of Image Formation, Super Resolution, Global Optimization, Laser Beam Shaping, Optical Tweezers, Fractal Optics, Metamaterials, Pareto-optimality, and Solar Concentrator Optics. He has undertaken major research projects (funded by government agencies and private industries) on many of these topics. He has supervised the theses of more than twenty doctoral students and more than a hundred Bachelor and Master level students both in India and abroad.

1

Introduction

1.1 An Early History of Optical Systems and Optics

The history of optical systems can be traced back to the Palaeolithic Age, or old Stone Age, starting around 40,000 BCE, when, it can only be surmised, human beings would be surprised and intrigued by observing the images formed by some naturally occurring surfaces.

1.1.1 Early History of Mirrors

In the Palaeolithic Age (around 40,000-10,000 BCE) and the Mesolithic Age (10,000–8000 BCE), the mirrors used by humans were most likely surfaces of lakes or pools of still water. In the early Neolithic age (8000–3000 BCE), the surface of water collected in a primitive vessel of some sort might have acted as a mirror, but historical evidence suggests that the formal manufacture of mirrors took place sometime in that age. The earliest manufactured mirrors were pieces of polished stone, e.g., obsidian, a naturally occurring volcanic glass. Some of the obsidian mirrors excavated by archaeologists at Çatal Höyük, located in Anatolia within the Konya Plain (modern day Turkey), have been dated to around 6000 BCE [1–3].

Mirrors of polished copper were crafted in Mesopotamia from 4000 BCE, and in ancient Egypt from around 3000 BCE. Polished bronze mirrors were made by the Egyptians from 2900 BCE.

On the Indian subcontinent, manufacture of bronze mirrors goes back to the time between 2800 BCE and 2500 BCE, during the Indus valley civilization [4–5]. Bronze mirrors, in general, are of low tin content (usually less than 10 percent). However, historical evidence exists of use of bronze mirrors with a much higher content of tin in different parts of India from an early period. A high-tin metal alloy mirror of high reflectance, known as 'Aranmula' mirror, manufactured in the state of Kerala in South India, continues to be made even now [6–7].

Some of the earliest examples of Chinese copper and bronze mirrors belonged to the late Neolithic Qijia culture from around 2000 BCE [8]. In Europe, bronze mirrors from the Bronze Age have been discovered from various places, including Britain and Italy. Polished stone and pyrite mirrors from different parts of central and south America have been found, dating from 2000 BCE [9]. Interestingly, the oldest mirrors from the Inca period (in Peru) predate the Olmec mirrors (in Mexico) by about 800 years [10].

1.1.2 Early History of Lenses

In comparison with mirrors, the known history of lenses is shorter by several millennia. The earliest lenses, discovered in Egypt, are dated between 2600–2400 BCE [11]. These lenses were found in the eyes of Egyptian statues. A set of lenses of larger diameter has been discovered in Troy (now Western Turkey) dating from 2200 BCE, whilst another set of lenses of the Minoan era were discovered in Crete and dates from 1500 BCE. The very good optical quality of these lenses prompted some scholars to argue that there was widespread use of lenses in antiquity, spanning several millennia [12–13]. Recently discovered ancient artificial eyes have been dated to 3000 BCE (for one discovered in Iran), and 5000 BCE (for one discovered in Spain) [14].

DOI: 10.1201/9780429154812-1

The 'Nimrud' lens, also known as 'Layard' lens, in memory of Austin Henry Layard who discovered it in 1850 CE, was unearthed at the Assyrian palace at Nivedeh, near the river Tigris in modern day Iraq [15–17]. It is dated around 1000 BCE, and credited by many as the earliest known manufactured lens for use either as a magnifying lens or as a burning lens to start fire by concentrating sunlight. It should be noted that all these early lenses were made of rock crystal, which is pure transparent crystalline quartz (SiO_2), i.e., silica.

1.1.3 Early History of Glass Making

The history of glass making dates back to at least 3600 BCE in central north Syria, Mesopotamia and Egypt, but manufacturing of glass was not very widespread for a long time. In Mesopotamia, it was revived in 700 BCE, and in Egypt in 500 BCE. For the next 500 years, Egypt, Syria and other countries along the coast of the Mediterranean Sea were centres for glass manufacturing. In the first century BCE, Syrian craftsmen invented the blow pipe, and this made the production of glass easier, faster and cheaper. Glass production flourished in the Roman empire, and spread from Italy to all countries under its rule. From the first century CE onwards, use of magnifying glasses increased, and quality lenses were made from glass throughout the Roman empire [18–21]. In India, development of glass technology began in 1700 BCE [22]. In ancient China, glass making had a later start during the Warring States period (475–220 BCE).

The brief history of optical systems of the ancient period, as enunciated above, forms an integral part of the history of optics.

1.1.4 Ancient History of Optics in Europe, India and China

In all major civilizations, the ancient history of optics concerned the nature of light [23], and the visual perception [24]. In fifth century BCE, the pre-Socratic philosopher, Empedocles argued that vision occurred when light issued from the eyes. This was what is known as the 'Emission Theory', or the 'Extramission Theory'. Plato held this theory, as did Hero, Euclid and Ptolemy in the Hellenistic age. Later, Aristotle, around 350 BCE, advocated for a 'theory of intromission' by which the eye received rays rather than directed them outward.

A few other theories involved different combinations or modifications of these two approaches. On the nature of light, in ancient India, the philosophical schools of Sāmkhya and Vaiśeşika developed theories of light in the period of the sixth to the fifth century BCE [25–27]. The concept of a ray of light was developed in these schools, and the theory of vision in Vaiśeşika school of philosophy comprised of a combination of the theories of extramission and intromission.

For example, below are three Vaiśeşika Sûtras of Kanâda. The first one defines the nature of light, whilst the next two define the nature of darkness. The numbers after each statement refer to the identifying number of the Sûtra in Vaiśeşika philosophy:

(i) तेजो रुपस्पर्शबत् । २/१/३ [Tejo rūpasparśavat//2/1/3//]
'Fire or light is identifiable by feel and sight'.
(ii) द्रब्यगुणकर्मनिष्पत्तिबैधर्म्यादभावस्तमः । ५/२/१९ [Dravyaguṇakarmmaniṣpattivaidharāmmyādabhāvastmamaḥ //5/2/19//]
'Darkness is non-existence, because it is different in its production from substance, attribute and action'.
(iii) तेजसोद्रब्यान्तरेणावरणाच्च । ५/२/२० [Tejaso dravyāntareṇāvaraṇācca //5/2/20]
'(Darkness is non-existence), also because (it is produced) from the obscuration of light by another substance'.

It is interesting to note that the concept of a ray as a direction of propagation of light was also developed in the early Hellenic period. A work entitled *Optics* by Euclid of Alexandria (300 BCE), often referred to as the founder of geometry, is the earliest surviving Greek treatise on 'Geometrical Optics'. It deals with the rectilinear propagation of light, shadows, perspective, parallax, etc [28]. Hero CE of Alexandria (10–70 CE), also known as Heron of Alexandria, is considered the greatest experimenter of antiquity. It is surprising to note that he derived the laws of reflection of light by invoking the stationarity principle. His treatise entitled *Catoptrica* deals with the progression of light, reflection, and the use of mirrors [29]. Ptolemy (100–170 CE), also of Alexandria, wrote a treatise called *Optics* that survives in a poor Arabic translation. He wrote about the properties of light, including reflection, refraction, and colour. His works influenced the subsequent investigators significantly [30–31].

A book entitled *Catoptrics* used to be attributed to Euclid. The book covers mathematical theory of mirrors, particularly the images formed by plane and spherical concave mirrors. However, the authorship of the available version of the book is disputed, and it is argued that it might have been compiled by the fourth century CE mathematician, Theon of Alexandria [32].

It is also noteworthy that, during the Warring States period in China, Mo Zi (385 BCE), who established the Mohism school dealing with natural sciences and engineering, deliberated on many phenomena of light in his book [33–34].

1.1.5 Optics Activities in the Middle East in 10th Century CE

After the fourth century CE, there was a lull in investigations in optics and related areas for a few centuries, as evidenced by the absence of any significant report. A revival of activities took place in the middle east in the tenth century CE. Ibn Sahl, a Persian mathematician, developed geometrical treatments for burning mirrors and lenses. Although he did not put forward any formal law of refraction, his analysis was mostly correct [35]. Another Persian mathematician, Al Quhi, developed geometrical treatments for mirrors from different conic sections [36–37]. The well-known Arab scholar, Ibn al-Haytham, also known as Alhazen or Alhacen (965–1040 CE), made important contributions in optics, and visual perception in particular [38–39]. He would take recourse to scientific experimentation methods during his investigations. He decisively upheld the theory of intromission in visual perception. He also investigated human eye, dispersion of light, refraction, camera obscura, etc. The principles underlying the camera obscura were first correctly determined by Alhacen [40]. He realized that whatever the shape of the aperture of a pinhole, the image would exhibit the shape of the luminous source. Further, he understood that the image is a composite formed by the superposition of numerous overlapping images cast by individual points of the luminous source. He produced many publications in different areas of optics, his most influential work being a seven volume treatise '*Kitāb al-Manāẓir*' ('*Book of Optics*') published around 1011–1021 CE [41–44]. It is significant to note that he even dealt with binocular vision in his analyses [45].

Around the late twelfth or early thirteenth century CE, *Kitāb al-Manāẓir*, the treatise of Ibn al-Haytham, was rendered from Arabic into Latin under the title *De Aspectibus*, or *De Perspectiva*. Along with Ptolemy's optics (translated from the available Arabic version to Latin), the two treatises set the tune for the revival of investigations on optics in Europe for the next few centuries.

1.1.6 Practical Optical Systems of the Early Days

During the period of antiquity (3000–476 BCE), the major optical systems consisted of burning mirrors and burning lenses. Scattered evidence exists for use of visual aid devices in these times, the most prominent one being the use of an emerald by emperor Nero, as mentioned by Pliny the Elder (23–79 CE). Another discovery was an optical magnifier in the form of a water filled spherical glass container—when placed on small sized letters or figures, they are magnified. Both Pliny and Seneca of Rome described this effect in first century CE.

The use of mirrors and lenses as magnifying devices started picking up in 1000 CE, with the growing availability of colourless, transparent glass of good quality. Glass production flourished in many regions surrounding the Mediterranean Sea, so much so that Alexandria emerged as a major centre of glass manufacturing around 1000 CE. By the time of Crusades, northern Italy, and Venice in particular, had turned into a glass manufacturing hub in the western world.

1.1.7 Reading stones and Discovery of Eyeglasses

The regular use of reading stone (an approximately hemispherical lens that could be placed on top of letters so that people with presbyopia could read more easily) began around 1000 CE. Initially it used to be made by cutting a glass sphere in half, but later on plano-convex lenses of different apertures were used. The reading stones or reading glasses were precursors to eyeglasses, or spectacles, popularly, the 'glasses' The development of first eyeglasses took place in northern Italy in the second half of the thirteenth century, by about 1290 CE. Fortunately, the first Italian spectacle makers had access to the well-established glass works around Venice, an industry handed down to the Venetians by the Byzantine traders. There are rival claimants for this invention from the neighbouring cities of Pisa, Florence and Venice [46]. However, two Italians are usually quoted as inventors of spectacles—Alexandro della Spina, a Dominican monk of Pisa, and Salvino d'Armati of Florence. It is interesting to note that, for environmental considerations, all glass making equipment was transferred to the island of Murano in 1291 CE. Later, centres for making of eyeglasses developed in Netherlands and Germany.

1.1.8 Revival of Investigations on Optics in Europe by Roger Bacon in the 13th century CE

As mentioned earlier, in the thirteenth century, a revival of investigations on optics in Europe was based on Ptolemy's *Optics* and Alhacen's *De Perspectiva*. In part five of his treatise, *Opus Major*, Roger Bacon put forward an account of available knowledge in optics. He clearly identified the specific roles of plano-convex and plano-concave lenses. The Polish natural philosopher, Witelo, or Vitello, wrote a treatise entitled *Perspectiva* that gave a synthetic description of different aspects of light and vision. This is the largest treatise on optics in the middle ages. A third treatise, *Perspectiva Communis*, was written by John Pecham, or Peckham, towards the end of the century. The three treatises, written within a period of two decades, influenced the subsequent investigators during the next few centuries [47–49].

1.1.9 Optics during Renaissance in Europe

Leonardo da Vinci (1452–1519 CE), the Italian polymath of the High Renaissance, and, one of the greatest painters of all times, studied the eye anatomically and made a model of it, showing where he supposed the rays to cross twice inside to form an upright image. However, in his description of the camera obscura, he correctly showed the image reversal. Leonardo embarked upon the concept of contact lens for correction of eye defects. He advocated extramission theory of vision in 1480 CE, but reversed his position to the theory of intromission in 1490 CE. Leonardo, a practical man, also discussed methods for making eyeglasses. Significantly enough, certain passages of his notebook can be interpreted as instructions for making telescopes, which only appeared a century later [50–51]. Unfortunately, he never organized his notes into publishable form, so they were practically unknown until Venturi published them in 1797 CE [52]. About half a century later, Francesco Maurolico (or Maurolycus) (1494–1577 CE), a mathematician and astronomer from Sicily, carried out significant investigations into optics. He solved a problem that bothered earlier thinkers: how the image of the sun is round even if formed by a square pinhole? Maurolico asserted that the eye acted as a double convex lens, and described appropriate treatment of myopia and hypermetropia of the eye by using contact lenses. He noted that the rays from an object point give rise to a surface of intense light concentration. This surface is now known as the caustic surface. He also explained colours of rainbow by multiple reflections in water drops. Francesco compiled a book entitled *Photismi de Lumine* in 1567 CE based on his investigations in optics, but the book was only published long after, in 1611 CE, as a consequence of the revival of interest in optics engendered by Galileo's discoveries using a telescope (further discussion of which below).

One of the most interesting books in popular science was *Magia Naturalis* (*Natural Magic*) by Giambattista della Porta, also known as Giovanni Battista della Porta. The first edition of the book, published in 1558 CE, consisted of four parts and it grew in size with subsequent editions, the largest one, published in 1589, consisting of 20 parts. Of interest, the seventeenth part of the book dealt with magic involving optics, where he laid down the exact theory of images formed by multiple mirrors. Della

Porta compared the eye and its pupil with the camera obscura and its stop. He wrote a book called *De Refractione* in 1593, where he put forward his optical observations in a scientific manner, including his theory of imaging by lenses. Recently, a reassessment of the optics of Porta has been published by various scholars of the history of science [54–56].

1.1.10 Invention of Telescope and Microscope

After the Renaissance Period, Europe ushered in the Age of Enlightenment, or the Age of Reason, in the seventeenth century. Unprecedented developments took place in optics during this century. Novel optical systems that emerged that not only opened new frontiers in observational sciences, but, to a large extent, also set the course of developments in different areas of science, e.g., astronomy, biology, defence etc. in the centuries that followed.

From the days of antiquity until the sixteenth century, the optical systems used in practice, such as the burning mirror or lens, magnifying mirror or lens, the reading stone and the eyeglass or the spectacle, all were a single component system. The early seventeenth century witnessed the arrival of optical systems consisting of more than one air-spaced lens and/or mirrors with novel imaging characteristics. The foremost among them is the telescope.

> Most historians make Holland the country of the telescope's origin and 1608 as the year of its birth ... Yet there is still doubt as to the name of the inventor, for the idea seems to have germinated in several minds at once. Many writers follow tradition and hand the merit unconditionally to Hans Lippershey, an obscure spectacle maker of Middleburg in Zeeland.
>
> [57]

However, there were two other claimants for the invention. James Metius, more correctly, Jacob Adriaanzoon, a native of Alkamaar, claimed he had made a telescope, whilst the third was Zacharias Jansen, another Middleburg spectacle maker. Zacharias' son, Johannes said his father invented the telescope in 1590 and used it to look at the moon and stars, but Zacharias' daughter, Sara, was uncertain about the year of invention, and gave either 1611 or 1619 as date of the invention [58–61].

'Owing to the clamours of Metius and Jansen and to the fact that "many other persons had a knowledge of the invention", the States-General finally declined to give a patent to Lippershey' [57].

Galileo first heard of Lippershey's invention in May 1609, when he was in Padua, Italy. The telescopes made by the spectacle makers in Holland had approximately 3X magnification. Galileo grasped the general principles of image formation with the help of two separated lenses, one convex and the other concave, and also developed his own lens grinding and polishing facility. He crafted several versions of telescopes with gradually increasing values of magnification, the largest magnification he could achieve with his telescope was approximately 30X. Often Galileo has been referred to as the inventor of the telescope, but he himself would always point out that the credit was due to the 'Dutchman'.

In the early days, this optical device consisting of two spectacle lenses and used to be called 'spyglasses'. In 1611 CE, Giovanni Demisiani, a Greek court mathematician, first coined the name 'teleskopos' ('tele': far, 'skopein': to look or see) for the particular instrument used by Galileo for his astronomical observations, and Prince Frederico Cesi, founder and President of the Academia dei Lincei, an Italian Science Academy located in Rome, agreed to this name in a banquet held in honour of Galileo. On some occasions, Galileo expressed his desire to write the principles of optics adopted by him in the synthesis of his optical instruments, but unfortunately it was never realized. Few scholars of the history of science have reported on this aspect of Galileo's accomplishments [62–63].

The invention of the telescope and that of the microscope are interwoven. The same spectacle makers of Holland, who first explored optical systems consisting of more than a single lens, looked at both types of configuration—lenses that could magnify both far and near objects. For the obvious possibility of utilization for military purposes, the telescopes immediately received much more attention in comparison with the microscopes. Historically, the father-son duo Hans and Zacharias Janssen claimed to have made the first compound microscope in 1590 CE, and are often recognized as such [64]. However, initially, the poor quality of the lenses and the low magnification obtained within the compound microscope limited its practical use.

After his astounding success in the use of telescopes for astronomical observations, it was Galileo who used an improved two component design for the compound microscope and rekindled interest in these systems, which he would call the 'occhiolino' or 'little eye'. Giovanni Faber coined the term 'microscope' for the occhiolino, submitted by Galileo to the Academia dei Lincei in 1625 CE [65–66].

After a few decades, in Delft, Holland, Antonij van Leeuwenhoek (1632–1723 CE) improved the quality of microscopes significantly, and used them to carry out many scientific observations in biological objects, so much so that he is known as the father of microbiology. He did not publish any book, but a series of his correspondences with members of the Royal Society, which he began writing in 1673, contain highlights of his research in microscopy and the development of microbiology [67]. Meanwhile, in England, Robert Hooke carried out investigations into biology by using his microscopes, and his well-known treatise *Micrographia* was published by the Royal Society in 1665 CE [68]. An epistemology of Robert Hooke's microscope has been carried out recently [69].

1.1.11 Investigations on Optics and Optical Systems by Johannes Kepler

In the early seventeenth century, the great mathematician and astronomer, Johannes Kepler wrote two books on optics. The first one, *Ad Vitellionem Paralipomena quibus astronomie pars optica traditur,* was written as a supplement to Vitello's treatise [70]. It dealt with problems encountered when using different optical methods for astronomical measurements, and their possible remedies. He realized that it is necessary to know the correct mathematical relation between the paths of incident and refracted rays in refraction of light. It seems he arrived at the correct relation, but he abandoned it because the results obtained by it did not agree with the values that Vitello gave, relating the angles of refraction to angles of incidence. In fact, Vitello's results were erroneous, and thus Kepler failed to discover the law of refraction. He did, however, determine the correct role of the eye lens in human vision, and confirmed that the retinal image is inverted. He identified the different types defects of the human eye, and corresponding visual aids for rectifying them. He also looked into the effects on the image of changes in variables in the camera obscura.

Kepler's second book, the *Dioptrice*, was published in 1611, soon after the astronomical discoveries of Galileo [71]. Assuming that for small angles, the angle of refraction is proportional to the angle of incidence, Kepler developed the laws of first order optics of image formation by a thin lens, and a system of thin lenses at finite separations. He gave a design of a telescope with two convex lenses for improving the small field of view encountered in Galilean telescopes. This type of telescope is now known as Keplerian telescope. Incidentally, Kepler is the name behind the word 'focus'. A burning mirror or lens produces a convergent beam out of an incident parallel beam of light, usually sunlight. The point of convergence used to be called a burning point. Kepler named it a 'fireplace', which, in Latinized form, became 'focus' [72].

1.1.12 Discovery of the Laws of Refraction

René Descartes first published the sine law of refraction in his *La Dioptrique* in 1636, nearly 1,600 years after the discovery of the law of reflection by Hero of Alexandria [73]. In fact, the discovery of the law of refraction has an interesting history. From the time of Ptolemy in the Hellenistic period, an approximate form of the law of refraction valid for small angles of incidence was already known [31]. It is known from his treatise on optics written in 984 CE by the great Persian mathematician and physicist, Ibn Sahl, that he was aware of a technique for determining the correct refracted ray corresponding to an incident ray. While projecting the pioneering role of Ibn Sahl in anaclastics (dioptrics), the French scholar Roshdi Rashed commented on the contribution of Ibn Sahl in relation to laws of refraction:

> Even if he did not state the law explicitly, it underlies all of his researches on lenses, and his contribution is of utmost importance. Ibn Sahl's discretion in this matter is apparently

> not fortuitous. It would seem due to the absence of inquiry into the physical causes of refraction—that is to say, to the lack of any attempt to justify this mode of propagating light. [35]

This may be the reason why Alhacen did not make any mention in his *Book of Optics* of this contribution by Ibn Sahl.

It is said that the English mathematician and astronomer Thomas Harriot knew the law of refraction as early as in 1602 CE, but he never published it. When Kepler asked him for the law, Herriott sent him precisely computed tables of data, instead of sending the law itself [74]. Again, in 1621 CE, the Dutch astronomer and mathematician, Willebrord Snellius (known in the English-speaking world as Snell) derived the law in a manuscript, but never published it. A few scientists, including Christian Huygens, became aware of this manuscript, and, although indirectly, through them Snell's contribution became known. A bitter argument arose on the priority of the discovery by Snell and Descartes, but it is now accepted that they derived the law independently. Descartes' presentation of the law of refraction had several faults in its reasoning, amended by Pierre de Fermat who applied his stationarity principle (initially it was presented by Fermat as 'the principle of least path') [75]. In his *La Dioptrique*, Descartes presented results of his investigations in imaging by the eye and human vision, including the binocular vision. He worked out the surfaces required for stigmatic imaging of a point, and in his memory they are now called Cartesian surfaces.

1.1.13 Discovery of the phenomena of Diffraction, Interference and Double Refraction

In the 1660s, a few significant discoveries involving light were reported. The Italian mathematician and physicist, Francisco Maria Grimaldi (1618–1663 CE) discovered the diffraction of light while carrying out experiments with very narrow beams of light. Grimaldi himself coined the name 'diffraction' for this phenomenon. His voluminous book on light, *Physico-Mathesis de Lumine, Coloribus and Iride* was published in 1665 CE [76]. The formation of fringe pattern by superposition of light obtained by reflection of light from two surfaces of a thin transparent film was first described by Robert Hooke in his book *Micrographia* published in 1664 CE [68]. The Danish naturalist Rasmus Bartholin (Latinized name Erasmus Bartholinus, 1625–1698 CE) discovered the phenomenon of double refraction produced by calcite or Iceland spar, and published his discovery in 1669 CE [77].

1.1.14 Newton's Contributions in Optics and Rømer's Discovery of Finite Speed of Light

In 1671 CE, Isaac Newton (1642–1727 CE) reported his famous experiment on the separation of white light into the colours of the spectrum by means of prisms [78]. His lectures on different aspects of optics, e.g. geometrical treatment of image formation, minimum deviation of prisms and elements of spherical aberration and chromatic aberration, were delivered in Latin during 1669 CE in public schools of the University of Cambridge and were later compiled and translated into English [79]. In 1675 CE, the Danish astronomer, Ole Rømer, the son-in-law of Rasmus Bartholin, discovered that the speed of light is finite from his observations of the eclipses of satellites of the planet Jupiter [80].

Opticks, the well-known book of optics by Newton, was published in 1704 CE [81] and consists of three books. Book I deals with basic definitions, reflection, refraction, dispersion, and analysis and synthesis of colours. Book II deals with the study and interpretation of the fringes observed in thin films (as noted by Hooke in his book). These fringes are now known as Newton's fringes. (Note that the name 'interference' for this phenomenon was given nearly a century later by Thomas Young). Book III deals with the phenomenon of diffraction of light. Newton propounded the corpuscular theory of light, and put forward his explanation of the different phenomena of light in terms of 'corpuscles' of light. On account of the availability of very few types of optical glass, and limited ability for accurately measuring dispersion characteristics of glasses, Newton concluded wrongly that all glass materials were of same dispersion, so that it would be impossible to develop achromatic refracting systems. To get rid of chromatic aberration, he went for reflecting systems. The Newtonian reflecting telescopes are well known, and he also proposed reflecting microscopes.

1.1.15 Contributions by Christian Huygens in Instrumental Optics and in Development of the Wave Theory of Light

The Dutch polymath, Christian Huygens (1629–1695 CE), started his scientific investigations following in the footsteps of Kepler and Galileo. He extended the theoretical analysis of image formation initiated by Kepler by developing treatments of paraxial optics for lens systems with finite thicknesses. He also developed the conjugate distance formula with the foci as references before Newton. To carry out his astronomical pursuits, he collaborated with his brother Constantin in the development of lens grinding and polishing procedures for the sake of fabricating lenses of good optical quality. They started making their own telescopes in 1652 CE, followed by telescopes of larger focal length to reduce chromatic defects in the image, and of larger aperture to collect more light. The largest one developed by the Huygens brothers had a focal length of 123 feet, and an aperture of 7½ inches. This telescope was commissioned in 1686, and it was donated to the Royal Society in London. For setting up very long telescopes, Huygens developed an open-air arrangement for aligning the objective and the eyepiece, obviating the need for the long metallic barrels to hold the telescope. He developed an eyepiece that today still bears his name. Huygens nearly completed his first book, *Dioptrica,* in 1653, but it was published posthumously in 1703 CE [82].

In the general scientific community, Huygens is more well-known for his wave theory of light. He put forward the theory of the envelope of secondary wavelets for determining the propagation of waves. He derived not only the laws of reflection and refraction from his wave theory, but also the law of refraction in doubly refracting crystals. Huygens realized that if the velocity of light varied with the direction, the spherical wavefront would deform to ellipsoids, and this explains the double refraction effect in Iceland spar. By using a pair of these elements, Huygens discovered the phenomenon of polarization of light, although the term 'polarization' was coined long after by Étienne-Louis Malus in 1811. The book *Traité de la lumière* was completed by Huygens in 1678, but it was only published in 1690 CE [83–84].

From the days of antiquity till the end of sixteenth century, optical imaging systems used to be developed mostly by a trial-and-error approach, without any reliable scientific basis, aside from a few cases involving plane and curved mirrors. The phenomenal developments of two key tools of observational sciences, the telescope and the microscope, in the seventeenth century could only be possible with concomitant developments in the theory of image formation, fabrication of precision optical elements, and optical testing by the great opticists of the period, Galileo, Kepler, Descartes, Grimaldi, Hooke, Newton and Huygens. Not only the exact law of refraction, but a large number of optical phenomena such as diffraction, interference, double refraction and polarization of light were discovered in this century. On the nature of light, two contending theories, the corpuscular theory of Newton and wave theory of Huygens were proposed.

These developments heralded the beginning of a new era in optics and optical systems. The pace of developments in related areas accelerated in the succeeding period, and continues unabated, so much so that, like many other branches of physics, distinct disciplines of optical sciences and engineering have emerged to carry forward light (in the extended sense) based technologies. A chronicle of the developments in optical physics is given in many books, e.g. the 'Historical Introduction' provided by Max Born and Emil Wolf in their classic *Principles of Optics* [85]. Several theoretical treatments for tackling the problems of analysis and synthesis of optical systems emerged over the years, and novel approaches continue to be reported. The pertinent history of these treatments will be presented in related chapters in this book.

1.2 Metamorphosis of Optical Systems in the Late Twentieth Century

The emergence of electro-optic display, in the 1930s, forged a link between optics, electronics and communication in a composite system. From the mid-50s, developments in the field of solid-state technology, and semiconductors in particular, brought forward a sea change in the components involved in generation, transmission, manipulation and detection of light. In the last six decades, the optical systems have gradually undergone a metamorphosis with the advent of: new sources of light, e.g. lasers, light emitting diodes (LED); new types of photon detectors and detector arrays like charged coupled devices (CCD) and complementary metal oxide semiconductor (CMOS) devices; optical fibres; light manipulators, e.g. spatial light modulators (SLM); micro-opto-electro-mechanical-systems (MOEMS); image intensifiers; new types of electro-optic and liquid crystal display devices; novel optical elements and components; and last, but not least, the integration of digital computers with their phenomenal number crunching ability, as a dedicated component of the optical systems. The range and scope of applications of optics-related technologies have increased by leaps and bounds in the recent past, so much so that they have now become indispensable for advances in almost all areas of human life, from health, industry, agriculture and entertainment to defence and space exploration, to name just a few.

The current burgeoning of interest in optics and related fields has been accentuated by the growing realization that the key to sustained growth and further development in frontier technologies of the twenty first century is held by concomitant developments in this area. A new phrase 'photonics' has been coined to identify these 'new age optics'. 'Optics', the science of light is now amalgamated with 'photonics', the science and technology of photons.

1.3 Current Definition of Optics

The word 'optics' is derived from the medieval Latin 'opticus' (meaning 'of seeing'), which in turn is derived from ancient Greek 'ὀπτικός'. Traditionally, 'optics' used to be defined as the science of visible light, but it is now known that visible light constitutes a very small part of the electromagnetic spectrum (Figure 1.1), and electromagnetic waves with wavelengths both shorter than the visible wavelengths, e.g. ultraviolet, X-rays and gamma rays, and longer than the visible wavelengths, e.g. infrared, terahertz and microwaves, have properties identical with the visible wavelengths. As per the statutes of the International Commission for Optics:

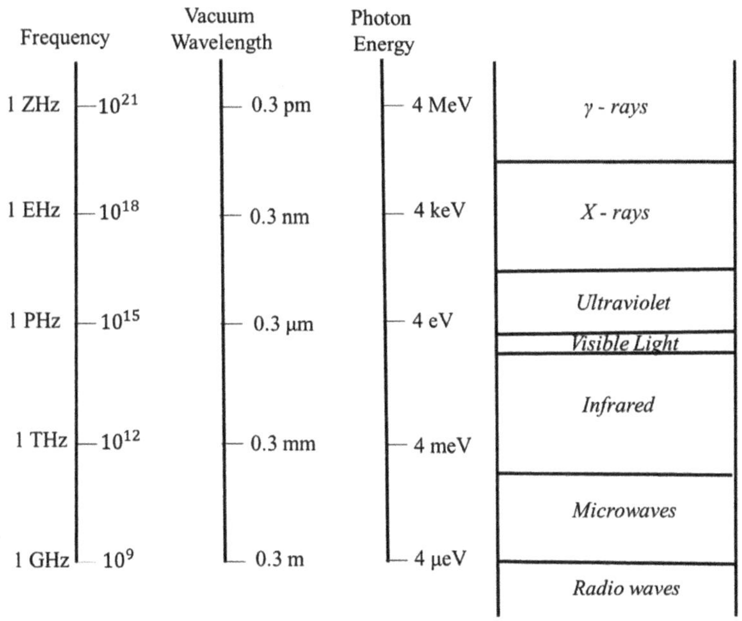

FIGURE 1.1 The electromagnetic-photon spectrum.

> Optics and photonics are defined as the fields of science and engineering encompassing the physical phenomena and technologies associated with the generation, transmission, manipulation, detection, and utilization of light. It extends on both sides of the visible part of the electromagnetic spectrum as far as the same concepts apply.
>
> [86]

1.4 Types of Optical Systems

All physical systems forming images are usually categorized as optical systems. Of course, the optical system most widely used by humans is the eye—approximately 80 per cent of information about the outside world is carried to the brain by images formed on retina of the eye. Spectacles, or more commonly, glasses, have been in use for nearly a millennium, so much so that the average person is prone to associate the word 'optics' with glasses. All systems and devices used for restoring the loss of vision occurring from diseases like myopia, hypermetropia, presbyopia, glaucoma, etc., are of paramount importance for humans. Ophthalmic optics, rightfully, continues to remain on the forefront of research and development.

On the other hand, the two optical instruments, namely the telescope and the microscope, continue to play a vital role in development of modern science and technology. Whereas the telescope makes a distant object appear to be near, so that the details of the object are more clearly visible, the microscope makes minute details, which are invisible to unaided eye, visible. Over the years, both these instruments have undergone major changes and modifications to cater to the growing needs of observational sciences. For the sake of analysis and design, many other optical systems that apparently do not have a direct link with either a telescope or a microscope can often be treated as a derivative from one or the other. The case of spectroscope that uses two telescopes in opposite mode is a glaring example.

Non-imaging optical systems constitute another class of optical systems. In general, analysis and synthesis of these systems are carried out by techniques which are somewhat different from the ones adopted for imaging systems. However, often the treatments for non-imaging systems can make effective use of approaches and tools developed for analysis and synthesis of imaging optical systems with minor modifications, if necessary.

Currently optical systems are found in all areas of human venture, from astronomy, biology, communication, industry, health sciences, space sciences, defence, entertainment and environmental sciences to many gadgets used by humans in daily living. Some of the well-known optical systems are listed below:

- Telescopic systems;
- Microscopic systems;
- Microlithographic systems;
- Photographic systems;
- Projection systems (still, cine, TV, multimedia, etc.);
- Metrological systems;
- Relay systems;
- Range and/or direction finding systems;
- Guidance and navigation systems;
- Holographic/image processing systems;
- Remote vision systems;
- Machine vision systems;
- Aerial/space reconnaissance systems;
- Radiation collection systems;
- Spectroscopic systems;
- Spectrophotometric/spectroradiometric systems;
- Scanning systems;
- Bar code readers;
- Alignment systems;

- Systems for virtual/augmented reality;
- Experimental setups using optical/photonic components;
- Remote control systems;
- Medical endoscopes (rigid and flexible);
- Periscopes/borescopes;
- Fibre-optic communication systems;
- Free space optical communication systems;
- Ring laser/fibre optic gyroscopes;
- Colorimeters;
- Optical pyrometers;
- Heliometers and coronagraphs;
- Tomographic systems;
- Laser cutting, drilling and engraving systems;
- Fundus camera;
- Ophthalmic auto refraction systems; and
- Laser assisted ophthalmic surgical machines

These optical systems consist of a multitude of optical components and devices, including just some listed below:

- Lenses;
- Mirrors;
- Prisms;
- Polarizers;
- Beam splitters;
- Spectral filters;
- Gratings;
- Gradient index (GRIN) elements;
- Thermal sources;
- Lasers;
- Fibres/faceplates;
- Image intensifiers;
- Light emitting diode (LED) display;
- Liquid crystal display (LCD);
- Spatial light modulators (SLM);
- Thermal detectors;
- Photomultipliers;
- Photon detector/detector array;
- Charged coupled device (CCD) detector/detector array;
- Complementary metal oxide (CMOS) detector/detector array;
- Micro-electro-mechanical-systems (MEMS) devices;
- Micro-optic devices;
- Micro-opto-electro-mechanical-systems (MOEMS) devices;

Of course, every single optical system does not involve all these elements. An optical system consists of a set of select elements from the above list. Different permutations and combinations of components of the above list go into the formation of the wide range of optical systems noted. Physical dimensions of optical systems vary over a large range. At one extreme are the micro-optical systems, and the very large 30 or 40 meter telescopes at the other extreme. The demands of the constituent modules or elements are significantly different in the two cases and, of course, there are significant variations for the in-between cases. However, a great deal of similarity in operating principles of these systems, irrespective of the large variation in their dimensions, should not be overlooked.

1.5 Classification of Optical Systems

Optical systems in practical use can be classified from different points of view, e.g. size, functionality, optical characteristics, etc. Some examples are given below.

(a) Size of the elements and devices: large optics – bulk optics – small optics – micro optics;
(b) Working wavelengths: monochromatic – broadband – polychromatic – hyperchromatic;
(c) Working wavelength range: microwaves – terahertz (THz) – far infrared (IR) – mid IR – Near IR – visible – ultraviolet (UV) – extreme UV – x- rays – γ rays;
(d) Illumination levels: low light level – high light level – variable light level;
(e) Coherence in illumination: fully coherent – incoherent – partially coherent;
(f) Nature of object illumination: self-luminous – trans illuminated;
(g) Size of object: point like object – extended object;
(h) Location of object/image: (i) object at infinity, image at a finite distance; (ii) object at a finite distance, image at infinity; (iii) both object and image at a finite distance; (iv) both object and image at infinity;
(i) Location of entrance pupil (Ent. P.) / exit pupil (Ex. P.): (i) Ent. P. at infinity, Ex. P. at a finite distance; (ii) Ent. P. at a finite distance, Ex. P. at infinity; (iii) Both Ent. P. and Ex. P. at a finite distance; (iv) both Ent. P. and Ex. P. at infinity;
(j) Nature of detector: visual (human eye) – thermal detector – photon detector;
(k) Nature of optics: catoptric – dioptric – catadioptric;
(l) Nature of operation: imaging – nonimaging;
(m) Mode of operation: scanning – non-scanning; and
(n) Quality of imaging: diffraction limited – detector limited – user-specified.

The list is long, but not exhaustive. Depending on the nature of optical elements used in a system, other classifications—namely fixed focus – variable focus; aspheric – spherical optics; homogeneous – gradient index; regular – anamorphic; smooth – diffractive etc.—can also be made.

1.6 Performance Assessment of Optical Systems

The desired performance goals of an optical system are usually characterized in terms of a set of specifications in accordance with user specified requirements. This calls for the appropriate modelling of the optical system with approximations of various orders. Many analytical and semi-analytical approaches exist for tackling this problem. Approaches based on approximations of lower orders, namely the paraxial/Gaussian model, and the primary/Seidel aberration model, provide identical or equivalent descriptions of the system.

For approximations with higher orders, and in cases of the evaluation of total aberration in particular, this is no longer true, and the results are highly susceptible to the actual implementation of the approximation procedure in practice.

In cases of imaging systems, the evaluation of diffraction-based parameters of image quality from the constructional parameters of the system can only be done by using a set of total aberration values obtained for specific points on the exit pupil for each desired field point. The process involves the numerical evaluation of oscillatory integrals—a highly involved procedure [87]. Due to the possibility of variation in the procedure adopted at one or more of these stages, the final results for image evaluation parameters calculated by different groups often do not match well, and the mismatch may be striking at larger values of field of view. A correct assessment for performance of an optical imaging system at a large field of view requires special measures, e.g., accurate determination of the real pupil, proper selection of ray grid pattern and suitable choice of aberration polynomial etc.

1.7 Optical Design: System Design and Lens Design

The design of an optical system involves the determination of its constituent elements, components, and devices. Traditionally, the term 'optical design' implies the design of optical elements, components, and devices that constitute the optical system. On a broader level, 'optical system design' includes the generation and/or detection of light used in the system [88]. Not only the characteristics of the generator and/or detector of light, but also the mode of their generation and/or detection significantly affect the conceptualization and practical implementation of the optical systems. Therefore, the appropriate generator and/or detector of light to cater to the needs of a specific optical system must be known or decided upon at the outset of the process of optical design.

Most optical systems can be modelled as provider of a one-to-one correspondence, or a one-to-many correspondence, or a many-to-one correspondence between a prespecified input and its output. These models can be related to one or the other type of imaging optical systems that are constituted of a single or a multitude of lens elements. In the present context, the term 'lens' includes the reflecting mirrors, the refracting prisms, the anamorphic attachments, the diffractive lenses, the inhomogeneous lenses, etc., as well as the traditional refracting lens elements. A selected list of important publications over the last 125 years in optical design, and lens design in particular, is the references [89–114]. Traditionally, the terms 'optical design' and 'lens design' were often used interchangeably. However, in view of the current proliferation of a large number of optical systems transcending the boundaries of conventional optical systems, often in conjunction with digital techniques for overcoming the limitations of all-optical methods, at times 'optical lens design' is often reckoned as an integral subset of 'optical system design', the other subsets being the 'optomechanical design' and 'optoelectronic design'. In this work, the primary thrust is on 'optical lens design' as enunciated above.

1.8 Theories of Light

Optics provides a veritable playground for theoretical modelling of physical phenomena. Using varying degrees of approximations and assumptions, a multitude of theories of light has emerged over the years for explaining phenomena relating to light and light-matter interactions [115].

At one end, the theory of quantum electrodynamics involving the quantization of both the field and the energy is, in principle, considered to be the all-encompassing theory in this venture of modelling. Application of this theory in practical problems of engineering optics is, however, limited. The theory of Maxwellian electrodynamics, that provides the backbone of practical engineering optics, may be considered as a special case of quantum electrodynamics when the quantization effects are ignored.

For tackling light propagation problems in linear dielectric media, an approximate theory called 'vector wave optics' is sufficient. In the case of propagation in isotropic media, a further simplification gives rise to 'scalar wave optics'. In the limit of working wavelength of illumination tending to zero, wave optics reduces to 'ray optics', which is traditionally called 'geometrical optics'. Practical treatment of optical systems with components in arbitrary orientation involves characterization of arbitrary rays in the system with respect to a base ray. With respect to the base ray as a reference, an arbitrary ray can be specified in terms of two parameters, one of them being an angle parameter θ representing the angle between the arbitrary ray and the base ray, and a distance parameter d representing the distance between the arbitrary ray and the base ray. Both θ and d change as an arbitrary ray propagates through the system. 'Ray optics' take a simplified form called 'linear optics' or 'parabasal optics' in the limit of both θ and d tending to zero. For rotationally symmetric optical systems, the base ray is the optical axis, and 'linear optics' reduces to 'paraxial optics' or 'Gaussian optics', in memory of Carl Friedrich Gauss who first worked out this model. Last, in most optical systems, the axial thicknesses t of the constituent lens elements tend to be much smaller than both the radii R of the refracting interfaces and the semi-diameters of their openings ρ, i.e. $t \ll R$, $t \ll \rho$. 'Paraxial optics' reduces to a simpler model called 'thin lens optics' in the limit $t \to 0$. The 'thin lens optics' model provides significant simplification in analytical treatments for the conceptual

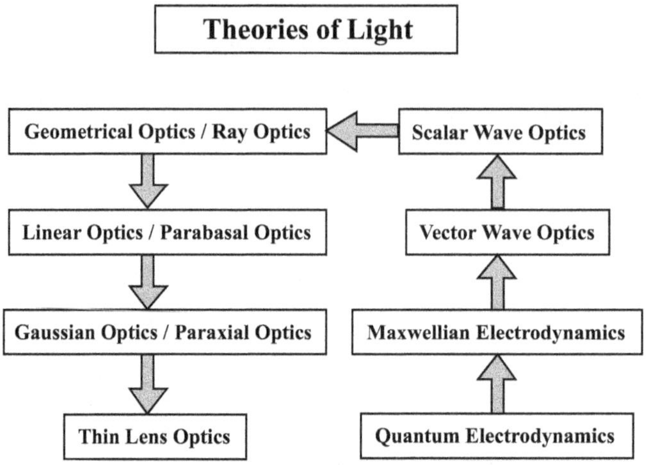

FIGURE 1.2 Interrelationship between theories of light.

development of the structural design of optical systems. An interrelationship between the theories of optics, as mentioned above, is shown in Figure 1.2. It is important to note that each of these models has its own niche in practical usage.

REFERENCES

1 H.C. Fioratti, 'The Origins of the Mirrors and their use in the Ancient World', 'Mirrors in the Medieval Period' and 'Mirrors during the Renaissance Period', *L'Antiquaire & The Connoisseur Inc., New York* (2008).
2 J.M. Enoch, 'History of Mirrors Dating Back 8000 years', *Optometry and Vision Science*, Vol. 83 (2006) pp. 775–781.
3 J.M. Enoch, 'Archeological optics: the very first known mirrors and lenses', *J. Mod. Opt.*, Vol. 54 (2007) pp. 1221–1239.
4 V. Shinde, Chapter 9 on 'Current Perspectives on the Harappan Civilization', in *A Companion to South Asia in the Past*, Eds., G.R. Schug and S.R. Walimbe, Wiley-Blackwell, Hoboken, New Jersey (2016) pp. 125–144.
5 S. Srinivasan, 'Mirrors: Metal mirrors from India', in *Encyclopaedia of the History of Science, Technology and Medicine in Non-Western Cultures', Vol. 2*, Ed. H. Selin, Springer, Berlin (2008) pp. 1699–1704.
6 S. Srinivasan and I. Glover, 'Skilled mirror craft of intermetallic delta high-tin bronze ($Cu_{31}Sn_8$,32.8%) from Aranmula, *Kerala' Current Science*, Vol. 93 (2007) pp. 35–40.
7 M.S. Swapna, V.P.N. Nampoori, and S. Sankararaman, 'Photoacoustics: a non-destructive evaluation technique for thermal and optical characterization of metal mirrors', *J. Opt. (India)* Vol. 47 (2018) pp. 405–411.
8 Z. Shoukang and H. Tangkun, 'Studies of ancient Chinese mirrors and other bronze artefacts', in Metal Plating and Patination: Cultural, Technical and Historical developments, Eds. S.L. Niece and P. Craddock, Elsevier (1993) 50–62.
9 J.J. Lunazzi, 'On the quality of the Olmec mirrors and its utilization', *Proc. SPIE*, Vol. 2730 (1996) pp. 2–7.
10 J. J. Lunazzi, 'Olmec mirrors: an example of archaeological American mirrors', arXiv.physics/0701328v1 (2007).
11 J.M. Enoch, 'First known lenses originating in Egypt about 4600 years ago', *Documenta Ophthalmologica*, Vol. 99 (1999) pp. 303–314.
12 G. Sines and Y.A. Sakellarakis, 'Lenses in Antiquity', *American Journal of Archaeology*, Vol. 91 (1987) pp. 191–196.

13 D. Plantzos, 'Crystals and Lenses in the Graeco-Roman World', *Amer. J. Archaeology*, Vol. 101 (1997) pp. 451–464.
14 J.M. Enoch, 'The fascinating early history of optics! Archaeological Optics, 2009; our knowledge of the early history of lenses, mirrors and artificial eyes!', *Proc. SPIE*, Vol. 7428 (2009) 742803-1–742803-42.
15 M.G. Tomilin, O.Y. Neverov, and G. Sines, 'The first lenses of the ancients', *J. Opt. Technol.*, Vol. 64 (1997) pp. 1066–1072.
16 K.M. Paasch, 'The history of optics: From ancient times to the middle ages', (1999) pp. 1–5. www.researchgate.net/publication/318708288.
17 B. Laufer, 'Optical Lenses. I. Burning-Lenses in China and India', *T'oung Pao, Second Series*, Vol. 16, No. 2 (1915) pp. 169–228.
18 H.C. King, 'Glass and Lenses in Antiquity', *The Optician*, Vol. 136 (1958) pp. 221–224.
19 C. Fryer, 'Glass and Lenses in Ancient Times', *The Optician*, Vol. 195 (1988) pp. 22–33.
20 G. Di Pasquale, 'Scientific and technological use of glass in Graeco-Roman world', in *When Glass Matters*, Ed. M. Beretta, L.S. Olschki, Firenze (2004) pp. 31–76.
21 H. Hanna, 'The Use of Magnifying Lenses in the Classical World', in *Greek Art & Architecture*, A. Gill (2010) pp. 1–17.
22 A.K. Kanungo and R.H. Brill, 'Kopia, India's First Glassmaking Site: Dating and Chemical Analysis', *J. Glass Studies*, Vol. 51 (2009) pp. 11–25.
23 V. Ronchi, *The Nature of Light: An Historical Survey*, Translator, V. Barocas, Heinemann, London (1970) [originally published in Italian as *Storie della Luce* (1939)].
24 P. Thibodeau, Chapter 8 entitled 'Ancient Optics: Theories and Problems of Vision', in *A Companion to Science, Technology, and Medicine in Ancient Greece and Rome,* Ed. G.L. Irby, J. Wiley, New Jersey (2016).
25 P. Chakravarti, *Origin and Development of the Sāmkhya Systems of Thought*, Metropolitan, Calcutta (1951).
26 D. Chattopadhyaya, *Indian Philosophy: A Popular Introduction*, People's Publishing House, New Delhi (1986).
27 S. Kak, 'The Nature of Light in Indian Epistemology', *Proc. SPIE*, Vol. 8121 (2011) 81211G-1–81211G-8.
28 Euclid, *Optics. Omnibus V.10*, Cactus, Athens (2004); also, in H.E. Burton, Translator, 'The Optics of Euclid', *J. Opt. Soc. Am.*, Vol. 35 (1945) pp. 357–372.
29 L. Nix and W. Schmidt, *Heronis Alexandrini opera quae supersunt omnia, Vol. II, Fasc. I, Mechanica et Catoptrica*, B. G. Teubner, Leipzig (1900).
30 A. Mark Smith, 'Ptolemy's Theory of Visual Perception: An English Translation of the "Optics" with Introduction and Commentary' *Trans. Amer. Philosoph. Soc., New Series*, Vol. 86, No. 2 (1996) pp. 1–61, 63–261, 263–269, 279–300.
31 A. Mark Smith. 'Ptolemy and the Foundations of Mathematical Optics: A Source based guided study', *Trans. Amer. Philosoph. Soc., New Series*, Vol. 89, No. 3 (1999) pp. 1, 3–49, 51–77, 79–147, 149–172.
32 J.J. O'Connor and E.F. Robertson, 'Theon of Alexandria', MacTutor History of Mathematics Archive, University of St. Andrews.
33 L.A. Wu, G.L Long, Q. Gong, and G.C. Guo, 'Optics in Ancient China', *AAPPS Bulletin*, Vol. 25 (2015) pp. 6–10.
34 J. Needham, 'Light(optics)' in *Science and Civilisation in China, Volume 4 Physics and Physical Technology, Part I: Physics*, Cambridge University Press, Cambridge (1962) pp. 78–125.
35 R. Rashed, 'A Pioneer in Anaclastics: Ibn Sahl on Burning Mirrors and Lenses', *ISIS, Chicago*, Vol. 81 (1990) pp. 464–491.
36 R. Rashed, *Géométrie et dioptrique au Xe siècle: Ibn Sahl, al-Quhi et Ibn al-Haytham*, Les Belles Lettres, Paris (1993).
37 M. Zghal, H.-E. Bouali, Z.B. Lakhdar, and H. Hamam, 'The first steps for learning optics: Ibn Sahl's, Al-Haytham's and Young's works on refraction as typical examples', *Proc. SPIE*, Vol. 9665 (2007) 966509-1 to 966509-7.
38 N. El-Bizri, 'A Philosophical Perspective on Alhazen's *Optics*', *Arabic Sciences and Philosophy*, Vol. 15 (2005) pp. 189–218.
39 A. Tbakhi and S.S. Amr, 'Ibn Al-Haytham: Father of Modern Optics', *Ann. Saudi Med.*, Vol. 27 (2007) pp. 464–467.

40 D.C. Lindberg, 'The Theory of Pinhole images from antiquity to the thirteenth century', *Arch. Hist. Exact Sci.*, Vol. 5 (1968) pp. 154–176.
41 A. Mark Smith, 'Alhacen's Theory of Visual Perception', Books I-II-III of Alhacen's *De Aspectibus*, Trans. *Amer. Philosophical Soc., New Series*, Vol. 91, No5, Section 2 (2001) pp. 339–819.
42 A. Mark Smith, 'Alhacen on the Principles of Reflection', Books IV-V of Alhacen's *De Aspectibus, Trans. Amer. Philosophical Soc., New Series,* Vol. 96, No. 3, Section 2 (2006) pp. 289–697.
43 A. Mark Smith, 'Alhacen on Image Formation and Distortion in mirrors', Book VI of Alhacen's De Aspectibus, *Trans. Amer. Philosophical Soc., New Series*, Vol. 98, No.1, Section 2 (2008) pp. 155–327.
44 A. Mark Smith, 'Alhacen on Refraction', Book VII of Alhacen's *De Aspectibus, Trans. Amer. Philosophical Soc., New Series*, Vol. 100, No. 3, Section 2 (2010) pp. 213–439.
45 D. Raynaud, 'Chapter 5. Ibn al-Haytham on Binocular Vision', in *Studies on Binocular Vision, Springer, Switzerland* (2016) pp. 71–93.
46 E. Rosen, 'The Invention of eyeglasses', *J. Hist. Medicine and Allied Sciences*, Vol. 11 (1956), Part I. 13–53, Pt II, pp. 183–218.
47 R. Bacon, 'Part Five: Optical Science' in *The Opus Major of Roger Bacon,* Vol. II, (A translation by R.B. Burke), Russell & Russell, New York (1962) pp. 419–582 [original Latin version was completed in 1267 CE].
48 D. Lindberg, 'Lines of Influence in Thirteenth-Century Optics: Bacon, Witelo and Pecham', *Speculum*, Vol. 46 (1971) pp. 66–83.
49 A.W. Fryczkowski, L. Bieganowski, and C.N. Nye, 'Witelo – Polish Vision Scientist of the Middle Ages: Father of Physiological Optics', *Survey of Ophthalmology*, Vol. 41 (1996) pp. 255–260.
50 D. Raynaud, 'La Perspective Aérienne de Léonard de Vinci et ses origines dans l'optique d'Ibn Al-Haytham (*De Aspectibus*, III, 7), *Arabic Sciences and Philosophy*, Vol. 19 (2009) pp. 225–246.
51 F. Fiorani, 'Leonardo's Optics in the 1470s', in *Leonardo da Vinci and Optics: Theory and Pictorial Practice*, Eds., F. Fiorani and A. Nova, Marsilio, Venice (2015) pp. 265–292.
52 J.B. Venturi, *Essai sur les ouvrages Physico-Mathématiques de Léonard De Vinci*, DuPrat, Paris (1797).
53 F. Maurolycus, *Photismi de Lumine*, Tarquinij Longi, Neapoli (1611). English Translation: Ed. H. Crew, *The Photismi de Lumine of Maurolycus: A Chapter in Late Medieval Opyics*, Macmillan, New York (1940).
54 A. Borelli, 'Thinking with Optical Objects: Glass Spheres, Lenses and Refraction in Giovani attista Della Porta's Optical Writings', *Journal of Early Modern Studies*, Vol. 3 (2014) pp. 39–61.
55 A. Borrelli, G. Hon, and Y. Zik, Eds., *The Optics of Giambattista Della Porta (ca. 1535–1615): A Reassessment*, Springer International Publishing AG (2017).
56 A. Mark Smith, 'Optical Magic in the Late Renaissance: Giambattista Della Porta's De Refractione of 1593', *Trans. Amer. Philosophical Soc.*, Vol. 107, Pt 1 (2019).
57 H.B. King, *The History of the Telescope*, Dover, New York (1979), p. 30, 32 [originally published by Charles Griffin, England (1955)].
58 L. Bell, *The Telescope*, Dover Publications, New York (1981) [originally published by Mc-Graw Hill, New York (1922)].
59 A. Van Helden. 'The invention of the telescope', *Trans. Amer. Philosophical Soc.*, Vol. 67, No. 4 (1977) pp. 1–67.
60 V. Ilardi, *Renaissance Vision from Spectacles to Telescopes*, American Philosophical Society, Philadelphia (2007).
61 A. Van Helden, S. Dupré, R.V. Gent and H. Zuidervaart, Eds., *The origins of the telescope*, KNAW Press, Amsterdam (2010).
62 Y. Zik and G. Hon, 'Magnifcation: how to turn a spyglass into an astronomical telescope', *Arch, Hist. Exact Sciences*, Vol. 66 (2012) pp. 439–464.
63 Y. Zik and G. Hon, 'Galileo's Knowledge of Optics and the Functioning of the Telescope—Revised', *Physics, arXiv, History and Philosophy of Physics* (2012).
64 D. Bardell, The Invention of the Microscope, *BIOS*, Vol 75 (2004) 78–84.
65 S.I. Vavilov, 'Galileo in the history of optics', *Sov. Phys. Usp.*, Vol. 7 (1965) pp. 596–616.
66 M. Herzberger, 'Optics from Euclid to Huygens' *Appl. Opt.*, Vol. 5 (1966) pp. 1383–1393.
67 C. Dobell, *Antony van Leeuwenhoek and his "little animals"*, Harcourt, Brace, New York (1932); Reprinted Dover, New York (1960).

68 R. Hooke, *Micrographia,* Dover Publications, New York (1961) [originally published by the Royal Society, London (1667)].
69 I. Lawson, *Robert Hooke's Microscope: The Epistemology of an instrument*, Ph. D. Thesis University of Sydney (2015).
70 J. Kepler, *Optics: Paralipomena to Witelo & Optical Part of Astronomy* (Translated by W.H. Donahue) Green Lion Press, Santa Fe (2000) [original Latin book '*Ad Vitellionem Paralipomena quibus astronomie pars optica traditur*' was published in Frankfurt (1604)].
71 J. Kepler, *Dioptrice*, Franc, Augsberg (1611) (in Latin).
72 J. Marek, 'Kepler and Optics', in *Vistas in Astronomy*, Vol. 18 (1975) pp. 849–854.
73 R. Descartes, 'La Dioptrique' (1636), in *Discours de la method*, Garnier-Flammarion, Paris (1966) pp. 97–162.
74 A. Kwan, J. Dudley, and E. Lantz, 'Who really discovered Snell's law?', *Physics World*, Vol. 15, No. 4 (2002) pp. 64.
75 P. Tannery and C. Henry (eds.) *Œuvres de Fermat*, Gauthier-Villars, Paris (1891–1922).
76 F.M. Grimaldi, *Physico-Mathesis de Lumine, Coloribus and Iride*, Haredis Victorij Benati, Bologna, Italy (1665) [in Latin].
77 R. Bartholin, *Experimenta Crystalli Islandici Disdiaclastici Quibus Mira Et Insolite Refractio Detegitur*, Daniel Paulli, Copenhagen ('Hafniæ'), Denmark (1669). English Translation: *Experiments with the double refracting Iceland crystal which led to the discovery of a marvelous and strange refraction*, Translator, Werner Brandt, Westtown, Pennsylvania (1959).
78 I. Newton, 'New Theory about Light and Colors', *Phil. Trans. Royal Soc., London*, Vol. 6, Issue 80 (1671) pp. 3075–3087.
79 I. Newton, *Optical Lectures read in the Public Schools of the University of Cambridge Anno Domini 1669*, (translated into English out of the original Latin), Francis Fayram, South Entrance of Royal Exchange, London (1728).
80 O. Rømer, 'A Demonstration concerning the motion of Light', *Philos. Trans. Royal Soc. London*, Vol. 12 (1677) pp. 893–894. [original article 'Demonstration Touchant le mouvement de la lumiere', *J. des Scavans* (1676) pp. 233–236].
81 I. Newton, *Opticks: A Treatise of the Reflexions, Refractions and Colours of Light*, Royal Society, London (1704); Dover, New York (1952).
82 C. Huygens, *Dioptrica*, C. Boutesteyn, Leiden (1703) [in Latin]. Also, *Dioptrique*, in *Œuvres Complètes de Christian Huygens, Tome XIII*, Société Hollandaise des Sciences, Fascicule I and Fascicule II, Martinus Nijhoff, La Haye (1916).
83 C. Huygens, Traité de la Lumière, and P. van der Aa, Leyden (1690). *English Translation, Treatise on Light*, Translator, S.P. Thomson, Macmillan, London (1912); Dover, New York (1962).
84 F. Jan Dijksterhuis, *Lenses and Waves: Christiaan Huygens and the Mathematical Science of Optics in the Seventeenth Century*, Ph. D. Thesis, University of Twente (1999). Published by Springer, Dordrecht (2004).
85 M. Born and E. Wolf, 'Historical Introduction' in *Principles of Optics*, Cambridge University Press, Cambridge (1999) pp. xxv–xxxiii [first published by Pergamon, London in 1959].
86 Vide ICO Newsletter, No. 46, January 2001; also 'Article 1 Objective' in *Statutes of the International Commission for Optics*. [e-ico.org/about/statutes]
87 H.H. Hopkins, 'The Development of Image Assessment Methods', *Proc. SPIE*, Vol. 46 (1975) pp. 2–18.
88 R. Kingslake, *Optical System Design*, Academic, Orlando (1983)
89 S. Czapski, *Grundzüge der Theorie der Optischen Instrumente nach Abbe*, Second Edition, J.A. Barth, Leipzig (1904). [originally published in 1893].
90 H. Dennis Taylor, *A System of Applied Optics*, Macmillan, London (1906).
91 E.T. Whittaker, *The Theory of Optical Instruments*, Hafner, New York (1907).
92 É. Turrière, *Optique Industrielle*, Delagrave, Paris (1920).
93 M. von Rohr, Ed., *Geometrical Investigation of the Formation of Images in Optical Instruments*, (Translated from the original German version by R. Kanthack), His Majesty's Stationery Office, London (1920).
94 I.C. Gardner, 'Application of algebraic aberration equations to optical design', *Scientific Papers of the Bureau of Standards*, Vol. 22 (1926) pp. 73–203.

95 A.E. Conrady, *Applied Optics and Optical Design, Part One*, Dover (1957), [originally published by Oxford University Press (1929)].
96 M. Berek, *Grundlagen der praktischen Optik: Analyse und Synthese optischer Systeme*, Walter de Gruyter, Berlin (1930).
97 H. Chrétien, *Calcul des Conbinaisons Optiques*, Masson, Paris (1980) [first published in 1938].
98 L.C. Martin, *Technical Optics*, Vol. I (1948), Vol. II (1950), Pitman, London.
99 H.H. Hopkins, *Wave Theory of Aberrations*, Oxford University Press, Oxford (1950).
100 D. Argentieri, *Ottica Industriale*, Ulrico Hoepli, Milano (1954). [in Italian]
101 A.E. Conrady, *Applied Optics and Optical Design, Part Two*, Dover (1960).
102 A. Cox, *A system of Optical Design*, The Focal Press, London (1964).
103 W. Brouwer and A. Walther, 'Design of Optical Instruments' in *Advanced Optical Techniques*, Ed. A.C.S. van Heel, North Holland, Amsterdam (1967) pp. 571–631.
104 R. Kingslake, *Lens Design Fundamentals*, Academic, New York (1978).
105 O.N. Stavroudis, *Modular Optical Design*, Springer, Berlin (1982).
106 G.G. Slyussarev, *Aberration and Optical Design Theory*, Adam Hilger, Bristol (1984) [first published in 1969 as *Metodi Rascheta Opticheskikh Sistem* (in Russian) by Mashinostroenie Press, Leningrad].
107 D.C. O'Shea, *Elements of Modern Optical Design*, John Wiley, New York (1985).
108 W.T. Welford, *Aberrations of Optical Systems*, Academic, New York (1986).
109 R.R. Shannon, *The Art and Science of Optical Design*, Cambridge University Press, Cambridge (1997).
110 M. Laikin, *Lens Design*, CRC Press, Taylor & Francis, Boca Raton (2006).
111 R.E. Fischer, B. Tadic-Galeb, and P.R. *Yoder, Optical System Design*, McGraw Hill, New York (2008).
112 M.J. Riedl, *Optical Design: Applying the Fundamentals*, Vol TT84, SPIE Press, Bellingham, Washington (2009).
113 C. Velzel, *A Course in Lens Design*, Springer, Dordrecht (2014).
114 J. Sasian, *Introduction to Lens design*, Cambridge University Press, Cambridge (2019).
115 V. Guillemin and S. Sternberg, *Symplectic techniques in physics*, Cambridge University Press, Cambridge (1984).

2

From Maxwell's Equations to Thin Lens Optics

Maxwell's equations hold the key for understanding different phenomena in optics. All theories of optics used in practice can be derived from these equations. In Section 1.8 of the last chapter, we discussed the interrelationship between theories of optics. At different stages of analysis and synthesis of the various optical systems in practice, a variety of approximations can often be invoked to obtain a useful model that can facilitate significant treatment of the problem. The model or models are based on one or the other theories of light mentioned earlier. Starting with a brief description of Maxwell's equations, the various assumptions that are often made during the course of development of these models are briefly discussed in this chapter.

2.1 Maxwell's Equations

Maxwell's equations [1] summarize the classical properties of an electromagnetic field—the electric vector **E**, the electric displacement vector **D**, the magnetic vector **H**, the magnetic flux density (or the magnetic induction) vector **B**, the electric current density vector **J**, and charge density ρ—in a set of four simple equations. We denote a vector by a letter in bold font. The four equations in vectorial form are:

$$\nabla \times \mathbf{E} = -\frac{\partial \mathbf{B}}{\partial t} \tag{2.1}$$

$$\nabla \times \mathbf{H} = -\frac{\partial \boldsymbol{D}}{\partial t} + \mathbf{J} \tag{2.2}$$

$$\nabla . \mathbf{D} = \rho \tag{2.3}$$

$$\nabla . \mathbf{B} = 0 \tag{2.4}$$

The symbols × and **.** represent a vector cross product and a vector dot product, respectively, while ∇, pronounced 'del', is a vector function space operator, which, when expressed in differential terms, can be defined in terms of the partial derivatives with respect to space coordinates. In Cartesian coordinate system

$$\nabla = \tilde{\mathbf{i}}(\partial / \partial x) + \tilde{\mathbf{j}}(\partial / \partial y) + \tilde{\mathbf{k}}(\partial / \partial z) \tag{2.5}$$

$\tilde{\mathbf{i}}$, $\tilde{\mathbf{j}}$, and $\tilde{\mathbf{k}}$ are the unit vectors along x, y, and z coordinates, respectively. The divergence and curl of a vector **A** with components A_x, A_y, and A_z are given by dot product and cross product respectively of the operator ∇ and **A**.

$$\nabla . \mathbf{A} = \frac{\partial A_x}{\partial x} + \frac{\partial A_y}{\partial y} + \frac{\partial A_z}{\partial z} \tag{2.6}$$

DOI: 10.1201/9780429154812-2

$$\nabla\times\mathbf{A}=\begin{vmatrix}\tilde{\mathbf{i}} & \tilde{\mathbf{j}} & \tilde{\mathbf{k}}\\ \frac{\partial}{\partial x} & \frac{\partial}{\partial y} & \frac{\partial}{\partial z}\\ A_x & A_y & A_z\end{vmatrix}=\left(\frac{\partial A_z}{\partial y}-\frac{\partial A_y}{\partial z}\right)\tilde{\mathbf{i}}+\left(\frac{\partial A_x}{\partial z}-\frac{\partial A_z}{\partial x}\right)\tilde{\mathbf{j}}+\left(\frac{\partial A_y}{\partial x}-\frac{\partial A_x}{\partial y}\right)\tilde{\mathbf{k}} \tag{2.7}$$

In practical optical systems we are usually concerned with that part of the electromagnetic field which contains no charges or currents, i.e.

$$\mathbf{J}=0 \text{ and } \rho=0 \tag{2.8}$$

Then the four equations (2.1) – (2.4) reduce to

$$\nabla\times\mathbf{E}=-\frac{\partial\mathbf{B}}{\partial t} \tag{2.9}$$

$$\nabla\times\mathbf{H}=\frac{\partial\mathbf{D}}{\partial t} \tag{2.10}$$

$$\nabla.\mathbf{D}=0 \tag{2.11}$$

$$\nabla.\mathbf{B}=0 \tag{2.12}$$

In general, the relationship between the two electric vectors **D** and **E** can be complicated, being tensorial or nonlinear in nature. However, for linear, isotropic media, the two electric vectors have a simple linear relationship:

$$\mathbf{D}=\epsilon\mathbf{E} \tag{2.13}$$

The constant ϵ is the dielectric permittivity of the medium. Note that in vacuum **D** and **E** are identical fields. But the value of permittivity ϵ_0 in vacuum has a non-unity value since the two electric vectors are measured in different units. In SI units, $\epsilon_0 = 8.854 \times 10^{-12}\ C^2N^{-1}m^{-2}$. The ratio (ϵ/ϵ_0) is called relative permittivity, or the dielectric constant.

The relation between the two magnetic vectors, **B** and **H** is similarly given by

$$\mathbf{B}=\mu\mathbf{H} \tag{2.14}$$

The constant μ is known as the magnetic permeability. The value of permeability in a vacuum is μ_0. In SI units, $\mu_0 = 4\pi \times 10^{-7}\ Ns^2C^{-2}$. The ratio (μ/μ_0) is called the relative magnetic permeability μ_r. In optical wavelengths, for most optical materials, $\mu_r = 1$, i.e., $\mu = \mu_0$. Thus the magnetic properties of materials tend to play a much less significant role in optics compared to radio waves or microwaves. More details on different aspects of electromagnetic theory can be found in references [2–9].

2.2 The Wave Equation

Maxwell's equations relate the field vectors by simultaneous differential equations. By mathematical manipulation, differential equations are obtained that each of the field vectors must satisfy separately. Analytical treatment of the optics phenomena can be facilitated in many cases by using these equations which are commonly called wave equations.

Substituting from (2.14) in (2.9), we obtain

$$\frac{1}{\mu}(\nabla \times \mathbf{E}) = -\frac{\partial \mathbf{H}}{\partial t} \tag{2.15}$$

Taking curl of both sides

$$\nabla \times \left[\frac{1}{\mu}(\nabla \times \mathbf{E})\right] = -\frac{\partial}{\partial t}[\nabla \times \mathbf{H}] \tag{2.16}$$

Using the identity

$$\nabla \times (u\mathbf{V}) = u(\nabla \times \mathbf{V}) + (\nabla u) \times \mathbf{V} \tag{2.17}$$

and substituting from (2.10) we obtain

$$\frac{1}{\mu}[\nabla \times \nabla \times \mathbf{E}] + \nabla\left(\frac{1}{\mu}\right) \times (\nabla \times \mathbf{E}) = -\frac{\partial^2 \mathbf{D}}{\partial t^2} \tag{2.18}$$

Using the identity

$$\nabla \times \nabla \times \mathbf{V} = \nabla(\nabla.\mathbf{V}) - \nabla^2 \mathbf{V} \tag{2.19}$$

and substituting for **D** from (2.13), (2.18) reduces to

$$\nabla(\nabla.\mathbf{E}) - \nabla^2 \mathbf{E} + \mu \nabla\left(\frac{1}{\mu}\right) \times (\nabla \times \mathbf{E}) = -\epsilon\mu \frac{\partial^2 \mathbf{E}}{\partial t^2} \tag{2.20}$$

In the above, the operator ∇^2 is called the Laplacian, and is expressed in Cartesian coordinate system as

$$\nabla^2 = \nabla \cdot \nabla = \frac{\partial^2}{\partial x^2} + \frac{\partial^2}{\partial y^2} + \frac{\partial^2}{\partial z^2} \tag{2.21}$$

Applying the identity

$$\nabla.(u\mathbf{V}) = u(\nabla.\mathbf{V}) + \mathbf{V}.\nabla u \tag{2.22}$$

we obtain from (2.11) and (2.13)

$$\epsilon(\nabla.\mathbf{E}) + \mathbf{E}.\nabla\epsilon = 0 \tag{2.23}$$

so that

$$\nabla.\mathbf{E} = -\mathbf{E}.\nabla(\ln\epsilon) \tag{2.24}$$

Also

$$\mu \nabla\left(\frac{1}{\mu}\right) = -\frac{\nabla\mu}{\mu} = -\nabla(\ln \mu) \tag{2.25}$$

Substituting from (2.24) and (2.25), (2.20) can be written as

$$\nabla^2\mathbf{E} + \nabla\left[\mathbf{E}.\nabla(\ln\epsilon)\right] + \left[\nabla(\ln\mu)\right]\times(\nabla\times\mathbf{E}) = \epsilon\mu\frac{\partial^2\mathbf{E}}{\partial t^2} \tag{2.26}$$

This is a second order differential equation involving the electric vector **E**. A similar equation for the magnetic vector **H** can be written as

$$\nabla^2\mathbf{H} + \nabla\left[\mathbf{H}.\nabla(\ln\epsilon)\right] + \left[\nabla(\ln\mu)\right]\times(\nabla\times\mathbf{H}) = \epsilon\mu\frac{\partial^2\mathbf{H}}{\partial t^2} \tag{2.27}$$

Although vectors **E** and **H** appear singly in equations (2.26) and (2.27) respectively, it is important to note that the two vectors are not independent, and they are related by equations (2.9) and (2.10) with (2.13) and (2.14).

As mentioned earlier, for almost all optical media of interest in instrumental optics, at optical frequencies, $\mu = \mu_0$. Therefore

$$\nabla(\ln\mu) = 0 \tag{2.28}$$

Also

$$\nabla(\ln\epsilon) = \frac{\nabla\epsilon}{\epsilon} \tag{2.29}$$

The general wave equations (2.26) and (2.27) reduce to

$$\nabla^2\mathbf{E} + \nabla\left[\mathbf{E}.\frac{\nabla\epsilon}{\epsilon}\right] = \epsilon\mu\frac{\partial^2\mathbf{E}}{\partial t^2} \tag{2.30}$$

$$\nabla^2\mathbf{H} + \nabla\left[\mathbf{H}.\frac{\nabla\epsilon}{\epsilon}\right] = \epsilon\mu\frac{\partial^2\mathbf{H}}{\partial t^2} \tag{2.31}$$

If the optical medium is homogeneous

$$\nabla\epsilon = 0 \tag{2.32}$$

so that the wave equations (2.30) and (2.31) reduce to the well-known forms

$$\nabla^2\mathbf{E} = \epsilon\mu\frac{\partial^2\mathbf{E}}{\partial t^2} \tag{2.33}$$

$$\nabla^2\mathbf{H} = \epsilon\mu\frac{\partial^2\mathbf{H}}{\partial t^2} \tag{2.34}$$

These are standard wave equations, and they suggest the existence of electromagnetic waves propagating with a velocity v given by

$$v = 1/(\epsilon\mu)^{\frac{1}{2}} \tag{2.35}$$

It should be noted that the concept of a velocity of an electromagnetic wave actually has an unambiguous meaning only in connection with very simplistic waves, e.g. plane waves. That v does not represent the velocity of propagation of an arbitrary solution of the wave equation is obvious if we bear in mind that these equations also admit standing waves as solutions [4].

The two vector equations (2.33) and (2.34) actually represent six scalar equations. In Cartesian coordinates (2.33) represents

$$\frac{\partial^2 E_x}{\partial x^2}+\frac{\partial^2 E_x}{\partial y^2}+\frac{\partial^2 E_x}{\partial z^2}=\epsilon\mu\frac{\partial^2 E_x}{\partial t^2} \tag{2.36}$$

$$\frac{\partial^2 E_y}{\partial x^2}+\frac{\partial^2 E_y}{\partial y^2}+\frac{\partial^2 E_y}{\partial z^2}=\epsilon\mu\frac{\partial^2 E_y}{\partial t^2} \tag{2.37}$$

$$\frac{\partial^2 E_z}{\partial x^2}+\frac{\partial^2 E_z}{\partial y^2}+\frac{\partial^2 E_z}{\partial z^2}=\epsilon\mu\frac{\partial^2 E_z}{\partial t^2} \tag{2.38}$$

and (2.34) represents

$$\frac{\partial^2 H_x}{\partial x^2}+\frac{\partial^2 H_x}{\partial y^2}+\frac{\partial^2 H_x}{\partial z^2}=\epsilon\mu\frac{\partial^2 H_x}{\partial t^2} \tag{2.39}$$

$$\frac{\partial^2 H_y}{\partial x^2}+\frac{\partial^2 H_y}{\partial y^2}+\frac{\partial^2 H_y}{\partial z^2}=\epsilon\mu\frac{\partial^2 H_y}{\partial t^2} \tag{2.40}$$

$$\frac{\partial^2 H_z}{\partial x^2}+\frac{\partial^2 H_z}{\partial y^2}+\frac{\partial^2 H_z}{\partial z^2}=\epsilon\mu\frac{\partial^2 H_z}{\partial t^2} \tag{2.41}$$

This shows that each of the components of the electric vector **E** and the magnetic vector **H** satisfies the scalar wave equation [10]

$$\nabla^2\psi=\frac{1}{v^2}\frac{\partial^2\psi}{\partial t^2} \tag{2.42}$$

The physical significance of $v=(\epsilon\mu)^{-\frac{1}{2}}$ is that it is the velocity of an electromagnetic wave in a medium of dielectric constant (ϵ/ϵ_0). The velocity of an electromagnetic wave in vacuum is

$$c=(\epsilon_0\mu_0)^{-\frac{1}{2}}=2.99792458\times10^8\,\mathrm{ms}^{-1} \tag{2.43}$$

The ratio (c/v) is called the refractive index n of the medium. For optical materials at optical wavelengths

$$n=(c/v)=(\epsilon/\epsilon_0)^{\frac{1}{2}} \tag{2.44}$$

2.3 Characteristics of the Harmonic Plane Wave Solution of the Scalar Wave Equation

The scalar wave equation is a linear, homogeneous, second order differential equation. Any solution of (2.40) of the form

$$\psi(\mathbf{r},t) = Af(\mathbf{s}.\mathbf{r} - vt) \tag{2.45}$$

represents a plane wave propagating with a velocity v along the direction of the unit vector **s** in the three-dimensional space. **r** is a position vector, and

$$\mathbf{r}.\mathbf{s} = \text{constant} \tag{2.46}$$

represents planes perpendicular to the unit vector **s** (Figure 2.1). **r** and **s** are expressed as

$$\mathbf{r} = \tilde{\mathbf{i}}x + \tilde{\mathbf{j}}y + \tilde{\mathbf{k}}z \tag{2.47}$$

$$\mathbf{s} = \tilde{\mathbf{i}}s_x + \tilde{\mathbf{j}}s_y + \tilde{\mathbf{k}}s_z \tag{2.48}$$

where (x, y, z) are coordinates of a point P on the plane, and s_x, s_y and s_z are the three components of the vector **s**. The three components are the three direction cosines of the unit vector along **s**.

In general, the function f in (2.45) is arbitrary; the only condition it must satisfy is that f is twice differentiable with respect to both space and time. Of special importance are the set of plane waves called harmonic plane waves. This is the case primarily because, by Fourier analysis, any arbitrary three-dimensional wave can be represented as a superposition of a set of harmonic plane waves, each of the latter having specific values for frequency, amplitude, and direction of propagation. It may be noted that the concept of a plane wave is a mathematical abstraction. A 'perfect' plane wave cannot be generated in practice. Nevertheless, plane waves play indispensable roles in the mathematical models developed for the analysis of propagation of electromagnetic waves. Often approximate plane waves that are generated experimentally are sufficient in practical applications.

A harmonic plane wave solution of the scalar wave equation is

$$\psi(\mathbf{r},t) = A\cos\left[k(\mathbf{s}.\mathbf{r} - vt)\right] \tag{2.49}$$

Note that the multiplier k is introduced to make the argument of the cosine function dimensionless. Substituting $\boldsymbol{k} = k\boldsymbol{s}$ and $\omega = kv$ in (2.45) we get

$$\psi(\mathbf{r},t) = A\cos\left[\boldsymbol{k}.\boldsymbol{r} - \omega t\right] \tag{2.50}$$

Or, in complex notation

$$\psi(\mathbf{r},t) = Ae^{i(\boldsymbol{k}.\boldsymbol{r} - \omega t)} \tag{2.51}$$

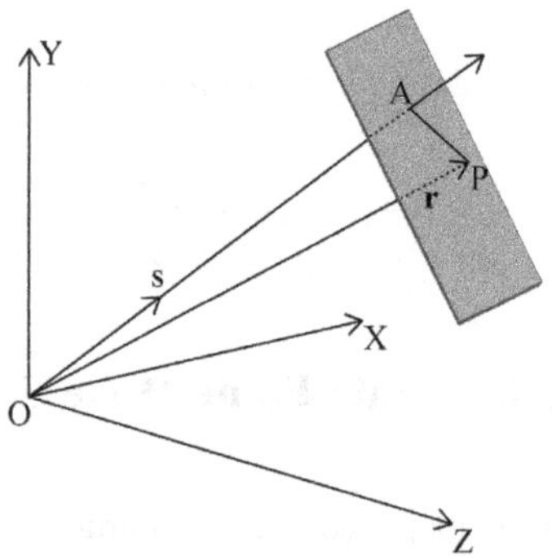

FIGURE 2.1 A plane wave propagating in the direction of unit vector ***S***.

The right-hand side of the equation (2.50) represents a harmonic plane wave propagating along the direction of vector $\boldsymbol{k}$, and $\boldsymbol{r}$ is the position vector of a point in the three-dimensional space. This plane wave corresponds with the real part of the right-hand side of the equation (2.51). A fixed value of $\boldsymbol{k.r}$ represents a plane perpendicular to the direction of propagation. Using (2.47) and (2.48), $\boldsymbol{k.r}$ is given by

$$\boldsymbol{k.r} = k\boldsymbol{s.r} = k\left(s_x x + s_y y + s_z z\right) \tag{2.52}$$

The wave is periodic in both space and time. Holding either **r** or t fixed in place results in a sinusoidal disturbance in time or space, respectively.

The spatial period is known as wavelength λ, the length of a full period harmonic wave. For a fixed value of t, the spatially repetitive nature of these harmonic functions can be expressed as

$$\psi(\mathbf{r}) = \psi(\mathbf{r} + \lambda \boldsymbol{s}) \tag{2.53}$$

where **s** is a unit vector along the direction of $\boldsymbol{k}$, and $\mathbf{s} = \boldsymbol{k} / \mathrm{k}$. In exponential form

$$Ae^{i\boldsymbol{k.r}} = Ae^{i\boldsymbol{k}.(\boldsymbol{r}+\lambda \boldsymbol{s})} = Ae^{i\boldsymbol{k.r}}e^{i\lambda k} \tag{2.54}$$

It follows

$$e^{i\lambda k} = 1 = e^{i2\pi} \tag{2.55}$$

Therefore, k = 2π/λ. The vector $\boldsymbol{k}$, whose magnitude is the propagation constant k, is called the propagation vector.

The phase velocity of a plane wave given by equation (2.42) is equal to the propagation velocity of the wavefront. Let the scalar component of **r** in the direction of **k** be r_k. The disturbance on a wavefront is constant; so, if the front moves along **k** a distance dr_k, after a time dt we have

$$\mathrm{Ae}^{\mathrm{i}(\mathrm{kr_k} - \omega \mathrm{t})} = \mathrm{Ae}^{\mathrm{i}(\mathrm{kr_k} + \mathrm{kdr_k} - \omega \mathrm{t} - \omega \mathrm{dt})} \tag{2.56}$$

and so

$$kdr_k = \omega d\mathrm{t} \tag{2.57}$$

The magnitude of the wave velocity v is given by

$$\mathrm{v} = \frac{dr_k}{d\mathrm{t}} = \frac{\omega}{k} \tag{2.58}$$

The temporal period T, the amount of time taken for a full period wave to pass a stationary observer at a point **r** can be obtained as

$$\psi(\mathbf{r},\mathrm{t}) = \psi(\mathbf{r},\mathrm{t} + \mathrm{T}) \tag{2.59}$$

Using (2.51) we get

$$\mathrm{Ae}^{-\mathrm{i}\omega\mathrm{t}} = \mathrm{Ae}^{-\mathrm{i}\omega\mathrm{t}}e^{-i\omega\mathrm{T}} \tag{2.60}$$

so that $\omega\mathrm{T} = 2\pi$. From (2.58) we obtain $\mathrm{kv}T = 2\pi$. Thus, the temporal period T is given by

$$\mathrm{T} = \lambda / \mathrm{v} \tag{2.61}$$

The inverse of T is the temporal frequency ν, the number of full period oscillations per unit of time. Thus $\nu = \frac{v}{\lambda}$, and $v = \nu\lambda$.

Also ω = 2π/T = 2πν. ω is called the angular temporal frequency. Note that the dimensions of k and ω are $[L]^{-1}$ and $[T]^{-1}$ respectively, so that the argument of the wave function (2.50) becomes dimensionless.

When an electromagnetic wave passes through materials of different wave speeds, its frequency remains unchanged, but its wavelength changes. The vacuum wavelength λ_0 of the wave is related with the material wavelength λ by $\lambda_0 = cT = (c/v)\lambda = n\lambda$. The dielectric constant of an optical material is dependent on the wavelength of electromagnetic radiation, and this implies that refractive index n is a function of wavelength n (λ). This phenomenon is called dispersion, and this is the reason for the occurrence of chromatic aberration in refractive optical systems.

2.3.1 Inhomogeneous Waves

A general time harmonic, real, scalar wave of frequency ω may be defined as a solution of the scalar wave equation (2.42). It may be expressed as

$$\psi(\mathbf{r},t) = A(\mathbf{r})\cos\left[g(\mathbf{r}) - \omega t\right] \tag{2.62}$$

where g is a real scalar function of position. In complex notation, the general solution may be expressed as

$$\psi(\mathbf{r},t) = A(\mathbf{r})e^{i\left[g(\mathbf{r}) - \omega t\right]} \tag{2.63}$$

This may be rewritten as

$$\psi(\mathbf{r},t) = U(\mathbf{r})e^{-i\omega t} \tag{2.64}$$

where

$$U(\mathbf{r}) = A(\mathbf{r})e^{ig(\mathbf{r})} \tag{2.65}$$

At a given time, the surfaces given by

$$g(\mathbf{r}) = \text{constant} \tag{2.66}$$

are called cophasal or equi-phasal surfaces. They are also known as phase fronts or wavefronts. On the other hand, surfaces of constant amplitude are given by

$$A(\mathbf{r}) = \text{constant} \tag{2.67}$$

For a plane wave, amplitude A is independent of **r**, and so surfaces of equal amplitude and surfaces of equal phase are coincident. Nevertheless, for the general case, the two sets of surfaces may not be coincident; also, the amplitude may not be the same on all points of the wavefront. This type of wave is called an inhomogeneous wave. Common examples are light waves emerging from lasers and light waves used in optical systems with very large numerical apertures. Note that these inhomogeneous waves may exist in homogeneous media.

2.4 Wave Equation for Propagation of Light in Inhomogeneous Media

In inhomogeneous media, the refractive index n, or the dielectric permittivity ϵ is not constant in space, and therefore $\nabla\epsilon \neq 0$. The well-known wave equations (2.33) and (2.34) are valid only in the case of homogeneous media, and, strictly speaking, rigorous treatment for propagation of light in inhomogeneous media requires the use of the differential equations given in (2.30) and (2.31). However, tackling the latter equations is considerably difficult compared to the wave equations (2.33) – (2.34). Therefore, for most practical problems, ingenious approaches are developed to obviate the need for more difficult wave equations.

An order-of-magnitude analysis of the terms appearing in equation (2.30) may be carried out to obtain a rule of thumb for neglect of the second term in the left-hand side of equation (2.30) [11]. The harmonic electromagnetic plane wave is represented as

$$\mathbf{E} = \mathbf{E}_0 e^{i(k.r-\omega t)} \tag{2.68}$$

The direction of propagation of this wave is along the vector $\boldsymbol{k}$. It may be noted that the order of magnitude of the first term in the left-hand side of (2.30) and the term on the right-hand side of (2.30) are equal, so that

$$\left|\nabla^2\mathbf{E}\right| \approx \left|\epsilon\mu\frac{\partial^2\mathbf{E}}{\partial t^2}\right| = \omega^2\epsilon\mu\mathbf{E} = \frac{\omega^2}{v^2}\mathbf{E} = \left(\frac{2\pi}{\lambda}\right)^2\mathbf{E} \tag{2.69}$$

An estimate of the order of magnitude of the second term in the left-hand side of (2.30) may be obtained by replacing the operator with a derivative with respect to some direction S in space. Thus

$$\nabla\left(\mathbf{E}.\frac{\nabla\epsilon}{\epsilon}\right) \approx \frac{\partial}{\partial S}\left(\mathbf{E}.\frac{\nabla\epsilon}{\epsilon}\right) = R(1+B) \tag{2.70}$$

In the above equation

$$\mathrm{R} = \left(\frac{2\pi}{\lambda}\right)\mathrm{E}.\frac{\nabla\epsilon}{\epsilon} = \left(\frac{2\pi}{\lambda}\right)\mathrm{E}.\frac{\epsilon_2-\epsilon_1}{\epsilon\Delta\mathrm{S}} \tag{2.71}$$

and

$$B = \frac{\mathbf{E}.\dfrac{\partial}{\partial S}\left(\dfrac{\nabla\epsilon}{\epsilon}\right)}{\dfrac{2\pi}{\lambda}\mathbf{E}.\dfrac{\nabla\epsilon}{\epsilon}} \cong \frac{1}{2\pi}\lambda\frac{\left(\dfrac{|\nabla\epsilon|}{\epsilon}\right)_2 - \left(\dfrac{|\nabla\epsilon|}{\epsilon}\right)_1}{\dfrac{|\nabla\epsilon|}{\epsilon}\Delta S} \tag{2.72}$$

For $\Delta S = \lambda$, R and B reduce to

$$R = \left(\frac{2\pi}{\lambda}\right)\mathbf{E}\frac{\left[(\epsilon_2-\epsilon_1)/\epsilon\right]}{\lambda} \tag{2.73}$$

$$B = \left(\frac{1}{2\pi}\right)\frac{|\nabla\epsilon|_2 - |\nabla\epsilon|_1}{|\nabla\epsilon|} \tag{2.74}$$

From (2.69) and (2.70), an estimate for the ratio T of the second term in the left-hand side of equation (2.30) with the first term in the left-hand side of the equation can be obtained. Using equations (2.73) and (2.74) T can be expressed as

$$T \approx \frac{R(1+B)}{\left(\frac{2\pi}{\lambda}\right)^2 \mathbf{E}} = \left(\frac{1}{2\pi}\right)\left\{\frac{\epsilon_2 - \epsilon_1}{\epsilon}\right\}\left[1 + \left(\frac{1}{2\pi}\right)\frac{|\nabla\epsilon|_2 - |\nabla\epsilon|_1}{|\nabla\epsilon|}\right] \tag{2.75}$$

Thus, if the relative change in dielectric permittivity $\nabla\epsilon$ over the distance of one wavelength is less than unity, and the change in the gradient of dielectric permittivity $\nabla\epsilon$ over the distance of one wavelength is less than the gradient itself, the ratio $T \approx 0$. For most optical media with inhomogeneous but continuous dielectric constant, e.g. gradient index (GRIN) elements, this condition is usually satisfied, and therefore the relatively simpler wave equation (2.33) can be used in place of the more difficult equation (2.30).

Most optical instruments consist of one or more optical elements embedded in a medium, e.g., air which is of uniform dielectric constant. The optical elements, e.g. lenses or prisms, consist of uniform dielectric constants of different values. At the interfaces of the optical elements and air, the value of T is likely to be large. The total region between, say, the object and the image, is inhomogeneous, and it is incorrect to use wave equations (2.33) and (2.34) for studying the propagation of light from the object to the image. However, the problem is circumvented by taking recourse to an analysis based on the 'piecewise constant' approach. The wave equations (2.33) – (2.34) are rigorously valid in each subregion of the optical system. Subregions are the regions between two consecutive interfaces, as well as the regions between the object/image and the neighbouring interface. The optical medium in each subregion is usually homogeneous with uniform dielectric constant. Even in the special case of inhomogeneous GRIN elements, the inhomogeneity is such that the ratio $T \approx 0$, thereby validating the use of (2.33) in the subregion consisting of the GRIN material. The transition from one subregion to the next across an interface is tackled by means of suitably formulated boundary conditions. Therefore, in practice, use of wave equations (2.33)–(2.34) is sufficient for studying the light propagation problems in optical instruments.

2.4.1 Boundary Conditions

The interfaces referred to above are surfaces of discontinuity across which the physical properties of the medium change abruptly. The interface is a boundary between two media, in which both the dielectric permittivity ϵ and magnetic permeability μ are smooth and continuous.

In Figure 2.2, U is a surface of discontinuity across which the physical properties of the optical medium change abruptly from ϵ_1, μ_1 to ε_2, μ_2. Let us replace part of the interface by a thin transition layer in the form of a 'coin'. Let the thickness of the coin be δh. Within the coin, ϵ and μ vary rapidly but continuously from their values from one flat side to the other flat side. Let the area of the flat side of the coin be δA, and let $\boldsymbol{n}_1$, $\boldsymbol{n}_2$ be unit outward normals on the flat sides of the coin towards the first and the second medium respectively. If $\mathbf{n}$ is the unit normal pointing from the first to the second medium, $\boldsymbol{n}_1 = -\mathbf{n}$ and $\boldsymbol{n}_2 = \mathbf{n}$.

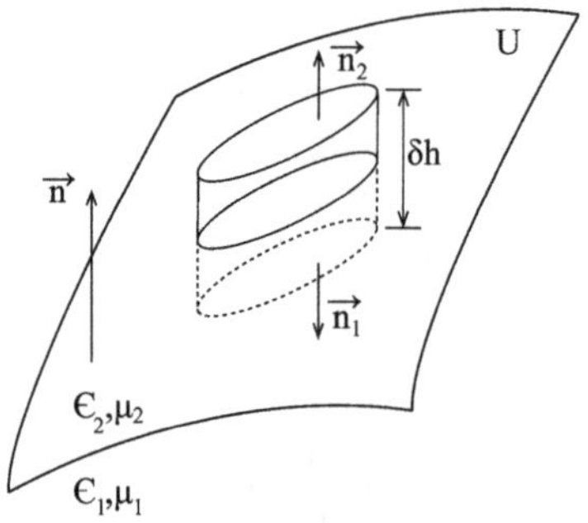

FIGURE 2.2 A coin in the infinitesimally thin transition layer replacing a surface of discontinuity U.

We assume

- the magnetic induction **B** and the electric displacement **D** and their derivatives to be continuous throughout the volume of the coin
- δA is so small that values of $\mathbf{B}^{(1)}, \mathbf{B}^{(2)}$ and $\mathbf{D}^{(1)}, \mathbf{D}^{(2)}$ are constant on respective sides of the coin
- $\delta h \to 0$, so that the two flat sides of the coin coalesce, and contribution from the curved surface of the coin can be neglected

Consequently, an application of Gauss divergence theorem, as shown below,

$$\int \nabla . \mathbf{F}\, dV = \int \mathbf{F} . \mathbf{n}\, dS \tag{2.76}$$

to both sides of equations (2.11) and (2.12) yields

$$\mathbf{n}.\left(\mathbf{B}^{(2)} - \mathbf{B}^{(1)}\right) = 0 \tag{2.77}$$

$$\mathbf{n}.\left(\mathbf{D}^{(2)} - \mathbf{D}^{(1)}\right) = 0 \tag{2.78}$$

Normal components of magnetic induction and electric displacement in the two media can be written as

$$\mathrm{B}_{\mathrm{n}}^{(2)} = \mathbf{n}.\mathbf{B}^{(2)}, \mathrm{B}_{\mathrm{n}}^{(1)} = \mathbf{n}.\mathbf{B}^{(1)}, \mathrm{D}_{\mathrm{n}}^{(2)} = \mathbf{n}.\mathbf{D}^{(2)}, \mathrm{D}_{\mathrm{n}}^{(1)} = \mathbf{n}.\mathbf{D}^{(1)} \tag{2.79}$$

where the subscript n indicates normal component of the corresponding term.

From (2.77) – (2.79) it follows

$$\mathrm{B}_{\mathrm{n}}^{(2)} = \mathrm{B}_{\mathrm{n}}^{(1)}, \quad \text{and} \quad \mathrm{D}_{\mathrm{n}}^{(2)} = \mathrm{D}_{\mathrm{n}}^{(1)} \tag{2.80}$$

Therefore, the normal components of magnetic induction and electric displacement are continuous across the surface of discontinuity. Note that the use of (2.11) in this derivation automatically assumes the absence of any surface charge.

Let a small rectangular loop $A_1B_1B_2A_2$ replace the 'coin' in the transition layer (Figure 2.3). The length and the width of the rectangle are δl and δw respectively. Let $\mathbf{n}, \mathbf{b}$, and $\mathbf{t}$ represent a unit normal pointing from the first to the second medium, a unit vector perpendicular to the plane of the rectangle, and a unit tangent along the surface respectively. The unit vectors $\mathbf{t}_1$ along P_1Q_1 and $\mathbf{t}_2$ along P_2Q_2 are given by

$$\mathbf{t}_1 = -\mathbf{t} = -(\mathbf{b} \times \mathbf{n}) \qquad \mathbf{t}_1 = \mathbf{t} = (\mathbf{b} \times \mathbf{n}) \tag{2.81}$$

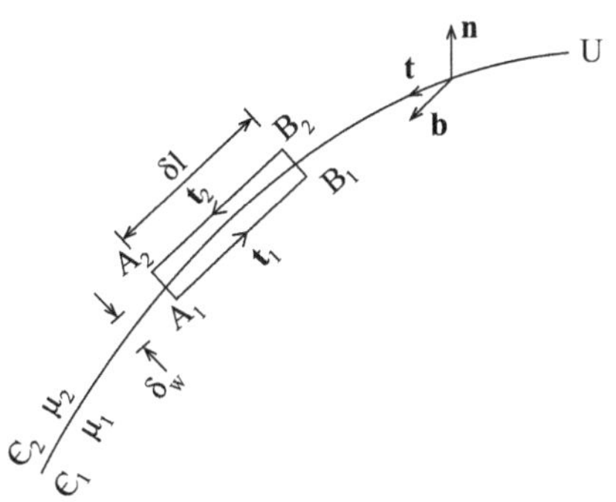

FIGURE 2.3 A rectangular loop in the infinitesimally thin transition layer replacing a surface of discontinuity U.

We assume

- The length δl of the rectangle is small, so that $\mathbf{E}^{(1)}$, $\mathbf{E}^{(2)}$ and $\mathbf{H}^{(1)}$, $\mathbf{H}^{(2)}$ have constant values along corresponding segments
- Both $\mathbf{E}$ and $\frac{\partial \mathbf{B}}{\partial t}$ remain finite over the small rectangle
- As the width δw of the rectangle tends to zero, the line segments A_1B_1 and A_2B_2 coalesce, and the contribution from the end faces are neglected.

An application of Stokes theorem, as given below,

$$\int (\nabla \times \mathbf{F}).d\mathbf{S} = \int \mathbf{F}.d\mathbf{l} \tag{2.82}$$

to both sides of (2.9) and (2.10) yields

$$\mathbf{n} \times \left(\mathbf{E}^{(2)} - \mathbf{E}^{(1)}\right) = 0 \tag{2.83}$$

$$\mathbf{n} \times \left(\mathbf{H}^{(2)} - \mathbf{H}^{(1)}\right) = 0 \tag{2.84}$$

Tangential components of the electric vector and the magnetic vector in the two media can be written as

$$\mathbf{E}_t^{(2)} = \mathbf{n} \times \mathbf{E}^{(2)},\ \mathbf{E}_t^{(1)} = \mathbf{n} \times \mathbf{E}^{(1)},\ \mathbf{H}_t^{(2)} = \mathbf{n} \times \mathbf{H}^{(2)},\ \mathbf{H}_t^{(1)} = \mathbf{n} \times \mathbf{H}^{(1)} \tag{2.85}$$

where the subscript t indicates the tangential component of the corresponding term. From (2.83) – (2.85), it follows

$$\mathbf{E}_t^{(2)} = \mathbf{E}_t^{(1)} \text{ and } \mathbf{H}_t^{(1)} = \mathbf{H}_t^{(1)} \tag{2.86}$$

Therefore, the tangential components of the electric vector and the magnetic vector are continuous across the surface of discontinuity. Note that the use of (2.10) in this derivation automatically assumes the absence of any surface current.

2.5 Vector Waves and Polarization

The general solution of the wave equation (2.33) can be written in the form

$$\mathbf{E}(\mathbf{r},t) = \mathbf{E}_0(\mathbf{r},t)e^{i\varphi(\mathbf{r},t)} \tag{2.87}$$

where $\mathbf{E}_0(\mathbf{r},t)$ and $\varphi(\mathbf{r},t)$ are the amplitude and phase, respectively, of the wave at the point with position vector $\mathbf{r}$ and time t. The harmonic electromagnetic plane wave propagating along the direction $\boldsymbol{k}$ can be represented as a special case of the general solution as

$$\mathbf{E}(\mathbf{r},t) = \mathbf{E}_0 \cos(\boldsymbol{k}.\boldsymbol{r} - \omega t) = \mathbf{E}_0 e^{i(\boldsymbol{k}.\boldsymbol{r} - \omega t)} \tag{2.88}$$

Note that the amplitude $\mathbf{E}_0$ of the plane wave is independent of position $\boldsymbol{r}$ and time t. Similarly, the magnetic vector of the electromagnetic plane wave is represented as

$$\mathbf{H}(\mathbf{r},t) = \mathbf{H}_0 \cos(\boldsymbol{k}.\boldsymbol{r} - \omega t) = \mathbf{H}_0 e^{i(\boldsymbol{k}.\boldsymbol{r} - \omega t)} \tag{2.89}$$

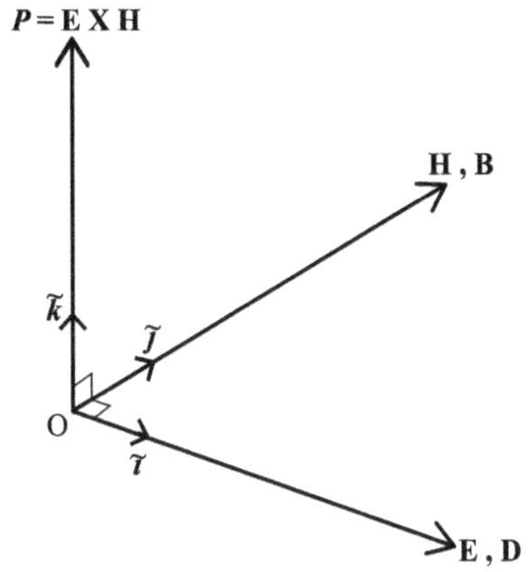

FIGURE 2.4 Orthogonal triad of vectors.

It may be noted that

$$\nabla e^{ik.r} = ike^{ik.r} \text{ and and } \frac{\partial\left(e^{-i\omega t}\right)}{\partial t} = -i\omega e^{-i\omega t} \tag{2.90}$$

Therefore, substituting the plane wave solution in Maxwell's equations implies replacing the operators ∇ and $\left(\frac{\partial}{\partial t}\right)$ in equations (2.9)-(2.12) by ik and $-i\omega$ respectively. This leads to the following equations

$$\boldsymbol{k} \times \mathbf{E} = \omega\mu\mathbf{H} = \omega\mathbf{B} \tag{2.91}$$

$$\boldsymbol{k} \times \mathbf{H} = -\omega\mathbf{D} = -\omega\varepsilon\mathbf{E} \tag{2.92}$$

$$\boldsymbol{k}.\mathbf{E} = 0 \tag{2.93}$$

$$\boldsymbol{k}.\mathbf{H} = 0 \tag{2.94}$$

It follows from equations (2.93) and (2.94) that the vectors **E** and **H** of a plane electromagnetic wave are perpendicular to the direction of propagation $\boldsymbol{k}$. Relations (2.91) and (2.92) show that vectors **E** and **H** are mutually perpendicular. These relations express transversality of the electromagnetic field, i.e. the electric and magnetic field vectors lie in planes normal to the direction of propagation. It should be noted that the three vectors **D**, **B**, and $\boldsymbol{k}$ are mutually orthogonal because of isotropy of the medium. The orthogonal triad of vectors formed by **E, H,** $\boldsymbol{k}$ and **D**, **B**, $\boldsymbol{k}$ in an isotropic medium are shown in Figure 2.4.

The amount of energy transported by this electromagnetic wave is represented by the Poynting vector $\mathcal{P}$ given by

$$\mathcal{P} = \mathbf{E} \times \mathbf{H} \tag{2.95}$$

This vector has dimensions of energy per unit time per unit area, and its units are watt/m^2. The vector $\mathcal{P}$ lies parallel to $\boldsymbol{k}$ in an isotropic medium; it is also called the intensity of the wave.

2.5.1 Polarization of Light Waves

For a uniform plane electromagnetic wave, at a given point in space, the electric vector **E** and the magnetic vector **H** are mutually orthogonal. They lie on a plane that is transverse or perpendicular to the

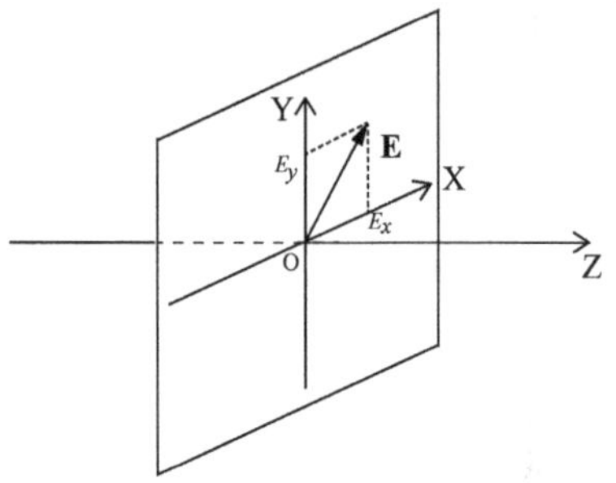

FIGURE 2.5 Electric vector **E** at point O.

direction of the propagation vector $\boldsymbol{k}$. This definition does not imply any specific or fixed direction per se for the vector **E** (or the vector **H**) on the transverse plane. In general, the direction of the vector **E** may change over time. In what follows, a separate description of the magnetic vector **H** is not necessary, for the direction of **H** is always orthogonal to the direction of **E**.

Polarization of a uniform plane wave describes the time-varying behaviour of the electric vector **E** at a fixed point in space. Without loss of generality, the direction of propagation of the wave is taken to be along the Z direction of a Cartesian system of coordinates. Coordinates X and Y are located on the transverse plane at the point O; the electric vector at point O lies on the transverse plane (Figure 2.5).

In the (X, Y, Z) system, the electric vector **E** can be expressed as

$$\mathbf{E} = \tilde{\mathbf{i}}E_x + \tilde{\mathbf{j}}E_y + \tilde{\mathbf{k}}E_z \tag{2.96}$$

On account of transversality of the electromagnetic field,

$$E_z = 0, E_x \neq 0, E_y \neq 0 \tag{2.97}$$

Therefore,

$$\mathbf{E} = \tilde{\mathbf{i}}E_x + \tilde{\mathbf{j}}E_y \tag{2.98}$$

This may be interpreted as the electric vector **E** of the plane wave being consisted of two electric vectors $\mathbf{E}_1$ and $\mathbf{E}_2$, which are given by

$$\mathbf{E}_1 = \tilde{\mathbf{i}}E_x \quad \mathbf{E}_2 = \tilde{\mathbf{j}}E_y \tag{2.99}$$

The electric vectors $\mathbf{E}_1$ and $\mathbf{E}_2$ are fixed on the X-axis and Y-axis respectively. Corresponding plane waves are said to be polarized in the directions of the X-axis and Y-axis respectively. E_x and E_y are real numbers denoting amplitudes of the two plane waves. Direct superposition of these two plane waves, which are linearly polarized in orthogonal directions, gives rise to the plane wave whose electric vector is **E**. It may be noted from (2.98) that the constituent plane waves with electric vectors $\mathbf{E}_1$ and $\mathbf{E}_2$ are in phase. For given values of z and t, the instantaneous expression for **E** is

$$\mathbf{E}(z,t) = \left(\tilde{\mathbf{i}}E_x + \tilde{\mathbf{j}}E_y\right) = \left[\tilde{\mathbf{i}}E_{x_0} + \tilde{\mathbf{j}}E_{y_0}\right]e^{-i\omega t}e^{ikz} = \tilde{s}E_0\cos(kz - \omega t) \tag{2.100}$$

The direction $\tilde{s}$ of the resultant vector makes an angle φ with the X-axis. φ is given by

$$\varphi = \tan^{-1}\left(\frac{\mathrm{E}_{y_0}}{\mathrm{E}_{x_0}}\right) \tag{2.101}$$

It is obvious that for all values of z and t, the direction of the electric vector remains the same. This plane wave is said to be linearly polarized along the direction $\tilde{s}$.

If the two constituent electric vectors $\mathbf{E}_1$ and $\mathbf{E}_2$ are not in phase, i.e., if they reach their maximum/minimum at different instants of time, both the direction and the magnitude of the resultant electric vector varies with time. Let the phase difference between $\mathbf{E}_1$ and $\mathbf{E}_2$ be δ. The instantaneous expressions for $\mathbf{E}_1$ and $\mathbf{E}_2$ are

$$\mathbf{E}_1(z,t) = \tilde{\mathbf{i}}E_x = \tilde{\mathbf{i}}E_{x_0}\cos(kz - \omega t) \tag{2.102}$$

$$\mathbf{E}_2(z,t) = \tilde{\mathbf{j}}E_y = \tilde{\mathbf{j}}E_{y_0}\cos(kz - \omega t - \delta) \tag{2.103}$$

From (2.102) and (2.103)

$$\left(\frac{\mathrm{E_x}}{\mathrm{E}_{x_0}}\right) = \cos(kz - \omega t)$$

$$\left(\frac{\mathrm{E_y}}{\mathrm{E}_{y_0}}\right) = \cos(kz - \omega t)\cos\delta + \sin(kz - \omega t)\sin\delta \tag{2.105}$$

By algebraic manipulation, we have

$$\left(\frac{E_y}{E_{y_0}}\right)^2 - 2\left(\frac{E_y}{E_{y_0}}\right)\left(\frac{E_x}{E_{x_0}}\right)\cos\delta + \left(\frac{E_x}{E_{x_0}}\right)^2 = \sin^2\delta \tag{2.106}$$

This is the equation of an ellipse whose major axis makes an angle α with the X-axis such that

$$\tan(2\alpha) = \frac{2E_{x_0}E_{y_0}\cos\delta}{\left(E_{x_0}{}^2 - E_{y_0}{}^2\right)} \tag{2.107}$$

The ellipse is the locus of the tip of the electric vector with time (Figure 2.6). The plane wave with such electric vector is said to be elliptically polarized. For $\delta = \pm\frac{\pi}{2}, \pm\frac{3\pi}{2}, \pm\frac{5\pi}{2}, \ldots$, $\alpha = 0$. The equation (2.106) reduces to

$$\left(\frac{E_y}{E_{y_0}}\right)^2 + \left(\frac{E_x}{E_{x_0}}\right)^2 = 1 \tag{2.108}$$

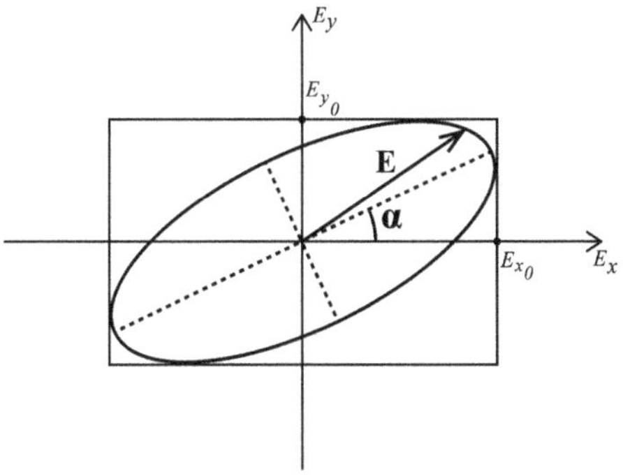

FIGURE 2.6 Polarization ellipse.

It is the more familiar form of regular ellipse with the axes coinciding with the X and Y axes. Further, $E_{x_0} = E_{y_0} = E_0$, and (2.108) reduces to

$$E_y^2 + E_x^2 = E_0^2 \tag{2.109}$$

This is the equation of a circle, and the plane wave with such an electric vector is said to be circularly polarized. For phase difference $\delta = 0, \pm 2\pi, \pm 4\pi, \ldots$, (2.106) reduces to

$$E_y = \frac{E_{y0}}{E_{x0}} E_x \tag{2.110}$$

For phase difference $\delta = \pm\pi, \pm 3\pi, \pm 5\pi \ldots$, one obtains

$$E_y = -\frac{E_{y0}}{E_{x0}} E_x \tag{2.111}$$

(2.110) and (2.111) represent two straight lines with slope $+\left(\frac{E_{y0}}{E_{x0}}\right)$ and $-\left(\frac{E_{y_0}}{E_{x_0}}\right)$ respectively. Plane waves with such electric vectors represent linearly polarized light.

It should be noted that a perfectly monochromatic plane wave is an abstract concept, for to be 'perfect' the wave needs to exist in the time domain from $-\infty$ to $+\infty$, an impossible proposition. Nevertheless, this abstract model facilitates our understanding and treatment of practical problems. By definition, a perfectly monochromatic plane wave has to be perfectly polarized.

In general, the term 'polarization' refers to the behaviour of an electric field at a particular point in an electromagnetic field. In general, different points of a wave may be in different states of polarization.

2.6 Propagation of Light in Absorbing/Semi-Absorbing Media

The propagation of electromagnetic waves considered so far in this chapter assumes the medium to be non-absorbing, at least for the range of operating wavelengths. The assumption has been incorporated in the analysis by considering both the electric current density vector **J** and the charge density ρ over the region of the medium under consideration as zero, already mentioned in expression (2.8). For absorbing/semi-absorbing media, neither **J** nor ρ can be taken as zero. In an electromagnetic field, the current density vector **J** is related with the electric vector **E** and the magnetic flux density vector **B** by

$$\mathbf{J} = \sigma(\mathbf{E} + \mathbf{v} \times \mathbf{B}) \tag{2.112}$$

where σ is the electrical conductivity of the medium [12]. Normally the velocity, **v**, of the charges is so small that the second term can be ignored, so that (2.112) reduces to

$$\mathbf{J} = \sigma\mathbf{E} \tag{2.113}$$

This relation is a form of Ohm's law.

Thus, Maxwell's relations appropriate to this case are given by

$$\nabla \times \mathbf{E} = -\mu \frac{\partial \mathbf{H}}{\partial t} \tag{2.114}$$

$$\nabla \times \mathbf{H} = \epsilon \frac{\partial \mathbf{E}}{\partial t} + \sigma\mathbf{E} \tag{2.115}$$

$$\nabla.\mathbf{E} = \frac{\rho}{\epsilon} \tag{2.116}$$

$$\nabla.\mathbf{H} = 0 \tag{2.117}$$

Differentiation of (2.116) with respect to t gives

$$\nabla.\frac{\partial \mathbf{E}}{\partial t} = \frac{1}{\epsilon}\frac{\partial \rho}{\partial t} \tag{2.118}$$

Taking divergence of (2.115), we get

$$\nabla.(\nabla \times \mathbf{H}) = \epsilon \nabla.\frac{\partial \mathbf{E}}{\partial t} + \sigma \nabla.\mathbf{E} = 0 \tag{2.119}$$

Using (2.116) and (2.118), (2.119) reduces to

$$\frac{\partial \rho}{\partial t} = -\frac{\sigma}{\epsilon}\rho \tag{2.120}$$

On integration one obtains

$$\rho(t) = \rho(0)e^{-t/\tau} \tag{2.121}$$

where relaxation time $\tau = \left(\frac{\epsilon}{\sigma}\right)$. This shows that initial charge density $\rho(0)$ falls off exponentially with a very small relaxation time, since the value of σ is very high. For metals, τ is typically of the order of 10^{-18} sec. The following treatment for the propagation of e m wave in absorbing media is valid when the accumulated charge, if any, has vanished, and the charge density ρ can be considered as zero. Therefore, out of the four equations (2.114) – (2.117), the third equation (2.116) reduces to

$$\nabla.\mathbf{E} = 0 \tag{2.122}$$

and the other three equations remain the same.

Taking curl of both sides of equation (2.114), and substituting from (2.115) and (2.122), the modified wave equation for **E** is obtained as

$$\nabla^2\mathbf{E} = \mu\epsilon\frac{\partial^2 \mathbf{E}}{\partial t^2} + \mu\sigma\frac{\partial \mathbf{E}}{\partial t} \tag{2.123}$$

Similarly, from (2.115), (2.114), and (2.117), the modified wave equation for **H** is given by

$$\nabla^2\mathbf{H} = \mu\epsilon\frac{\partial^2 \mathbf{H}}{\partial t^2} + \mu\sigma\frac{\partial \mathbf{H}}{\partial t} \tag{2.124}$$

Assuming a strictly monochromatic field of angular frequency ω, e.g.

$$\mathbf{E} = \mathbf{E}_0 e^{-i\omega t} \text{ and } \mathbf{H} = \mathbf{H}_0 e^{-i\omega t} \tag{2.125}$$

and noting that $\partial / \partial t \equiv -i\omega$, by substitution, (2.123) and (2.124) become

$$\nabla^2 \mathbf{E} + \omega^2 \mu \left(\epsilon + i\frac{\sigma}{\omega} \right) \mathbf{E} = 0 \tag{2.126}$$

$$\nabla^2 \mathbf{H} + \omega^2 \mu \left(\epsilon + i\frac{\sigma}{\omega} \right) \mathbf{H} = 0 \tag{2.127}$$

By the same substitution in (2.114) and (2.115), and rewriting (2.122) and (2.117), Maxwell's equations for the absorbing medium become

$$\nabla \times \mathbf{E} - i\omega\mu \mathbf{H} = 0 \tag{2.128}$$

$$\nabla \times \mathbf{H} + i\omega \left(\epsilon + i\frac{\sigma}{\omega} \right) \mathbf{E} = 0 \tag{2.129}$$

$$\nabla . \mathbf{E} = 0 \tag{2.130}$$

$$\nabla . \mathbf{H} = 0 \tag{2.131}$$

For the non-absorbing case, the wave equations are

$$\nabla^2 \mathbf{E} + \omega^2 \mu \epsilon \mathbf{E} = 0 \tag{2.132}$$

$$\nabla^2 \mathbf{H} + \omega^2 \mu \epsilon \mathbf{H} = 0 \tag{2.133}$$

Three of the four Maxwell's equations, (2.128), (2.130), and (2.131) are the same for both the absorbing and the non-absorbing case. Equation (2.129) for the absorbing case becomes

$$\nabla \times \mathbf{H} + i\omega\epsilon \mathbf{E} = 0 \tag{2.134}$$

in the non-absorbing case.

From the analogy of the expressions for the absorbing and non-absorbing medium, it is obvious that the formal relations in the absorbing or semi-absorbing case can be obtained from the corresponding relations of the non-absorbing case if the permittivity in these latter expressions are replaced by the complex permittivity given by

$$\tilde{\epsilon} = \left(\epsilon + i\frac{\sigma}{\omega} \right) \tag{2.135}$$

Similarly, the modified wave equations can be derived from the normal wave equations for the non-absorbing case by replacing the wave number k with a complex wave number $\tilde{k}$ given by

$$\tilde{k}^2 = \omega^2 \mu \left(\epsilon + i\frac{\sigma}{\omega} \right) \tag{2.136}$$

The analogy can be made closer still by introducing complex phase velocity $\tilde{v}$ and complex refractive index $\tilde{n}$ and defining them respectively by

$$\tilde{v} = \frac{c}{\sqrt{\mu \tilde{\epsilon}}} \tag{2.137}$$

and

$$\tilde{n} = \frac{c}{\tilde{v}} = \sqrt{\mu\tilde{\epsilon}} = \frac{c}{\omega}\tilde{k} \tag{2.138}$$

Let

$$\tilde{n} = n(1 + ia) \tag{2.139}$$

where n and a are real, and a is called the attenuation index, or the extinction coefficient. Squaring (2.139), we get

$$\tilde{n}^2 = n^2(1 - a^2 + i2a) \tag{2.140}$$

From (2.138) and (2.135),

$$\tilde{n}^2 = \mu\tilde{\epsilon} = \mu\epsilon + i\frac{\mu\sigma}{\omega} \tag{2.141}$$

Equating the real and imaginary parts of (2.140) and (2,141),

$$n^2(1 - a^2) = \mu\epsilon \tag{2.142}$$

and

$$n^2 a = \frac{\mu\sigma}{2\omega} \tag{2.143}$$

The quantities n and a are related with the material constants ϵ and σ by the equations (2.142) and (2.143). However, it should be noted that the permittivity ϵ and conductivity σ of the medium is a function of the frequency, and so n and a are also functions of frequency. As in the case of a non-absorbing medium, the simplest solution of the wave equation in the case of an absorbing medium is a plane time-harmonic wave given by

$$\mathbf{E} = \mathbf{E}_0 e^{i\{\tilde{k}(\mathbf{r}.\mathbf{s}) - \omega t\}} \tag{2.144}$$

From (2.138) and (2.139), we get

$$\tilde{k} = \frac{\omega\tilde{n}}{c} = \frac{\omega n(1 + ia)}{c} \tag{2.145}$$

Substituting for $\tilde{k}$ in (2.144) the plane wave solution takes the form

$$\mathbf{E} = \mathbf{E}_0 e^{-\frac{\omega}{c}na(\mathbf{r}.\mathbf{s})} e^{i\omega\left\{\frac{n}{c}(\mathbf{r}.\mathbf{s}) - \omega t\right\}} \tag{2.146}$$

This is a plane wave with wavelength $\lambda = 2\pi\left(\frac{c}{\omega n}\right)$, propagating along the direction of the unit vector **s**. The amplitude of the plane wave decreases exponentially during its propagation, i.e. the wave is absorbed as it propagates [13].

2.7 Transition to Scalar Theory

It has been shown earlier that in a linear, isotropic, homogeneous, and nondispersive medium, each of the three rectangular components of both vectors **E** and **H**, namely $E_x, E_y, E_z, H_x, H_y, H_z$ satisfy the scalar wave equation

$$\nabla^2 \psi(\mathbf{r},t) = \frac{1}{v^2}\frac{\partial^2 \psi(\mathbf{r},t)}{\partial t^2} = \frac{n^2}{c^2}\frac{\partial^2 \psi(\mathbf{r},t)}{\partial t^2} \tag{2.147}$$

where the dependence on position $\boldsymbol{r}$ and time t are explicitly introduced. In practice, a vacuum is the only perfect nondispersive medium. Almost all dielectric mediums are dispersive. The problem of propagation of an electromagnetic field in a dispersive medium is usually solved for monochromatic waves, and the property of linearity of wave propagation phenomenon is invoked to tackle the corresponding problems for nonmonochromatic waves.

A strictly monochromatic scalar wave field may be written as

$$\psi(\mathbf{r},t) = A(\mathbf{r})\cos\left[\varnothing(\mathbf{r}) - \omega t\right] \tag{2.148}$$

where $A(\mathbf{r})$ and $\varnothing(\mathbf{r})$ are the amplitude and phase, respectively, of the wave at the position **r**. The temporal frequency of the monochromatic wave is $\nu(= \omega / 2\pi)$. In phasor representation, the equation (2.148) can be rewritten as

$$\psi(\mathbf{r},t) = \mathrm{Re}\left[U(\mathbf{r})e^{-i\omega t}\right] \tag{2.149}$$

where Re represents 'real part of'; $U(\mathbf{r})$, called a phasor, is a complex function of position **r**, and is given by

$$U(\mathbf{r}) = A(\mathbf{r})e^{i\varnothing(\mathbf{r})} \tag{2.150}$$

$U(\mathbf{r})$ is also called complex amplitude of the wave.

Substituting for $\psi(\mathbf{r},t)$ from (2.149) in the wave equation (2.147) yields

$$\left(\nabla^2 + k^2\right)U(\mathbf{r}) = 0 \tag{2.151}$$

where $k(= 2\pi / \lambda)$ is called the wave number. Note that λ is the wavelength in the dielectric medium and is given by

$$\lambda = \left(c / n\nu\right) \tag{2.152}$$

This wave equation (2.151) is called the 'reduced' wave equation, since it is 'time independent' and usually a wave equation contains both space and time dependent terms [2]. It is also known as the Helmholtz equation. In analyses based on scalar theory, complex amplitude of any propagating wave must satisfy the single equation (2.151) instead of six different equations (2.36) – (2.41) corresponding to six rectangular components of the **E** and **H** vectors. Scalar wave theory has been very widely used in the study of diffraction problems. Several theories with varying degrees of accuracy exist. In principle, so long as propagation of light in homogeneous or inhomogeneous media does not create significant coupling between the components of the vectors **E** and **H**, analysis based on scalar theory can provide useful results. The results will be accurate so long as the size of the diffracting structures is large compared with the wavelength light. In optical instruments, diffraction effects arise at the boundaries of the different apertures or diaphragms of the system elements. The effect of coupling remains significant around a distance of about one or two wavelengths from the physical boundary. Except in the case of very large aperture systems requiring tight focusing or very high frequency gratings, scalar theory of diffraction is adequate for most practical purposes. As a rule of thumb, the minimum diffracting structure should be more than five, or say, ten times the wavelength of light. This minimum value is to be decided by the order of accuracy demanded in specific application.

2.8 'Ray Optics' Under Small Wavelength Approximation

The electromagnetic field encountered in optical systems is characterized by very rapid oscillations. For visible light, the frequency of oscillation is of the order of 10^{14} s^{-1}, corresponding to wavelength of the order of 1 micron (μm). Even for the extended spectral domain of optics as defined currently, the wavelength varies approximately from 10^{-2} m to 10^{-14} m. A simple model for propagation of electromagnetic waves in optical systems is obtained by substituting the value of the wavelength with zero in the actual formulae. This model is called 'Geometrical optics', for it allows formulation of optical laws in the language of geometry [14–15]. An alternative, and probably more appropriate nomenclature is 'Ray optics', as used in some German optics literature, which calls it 'Strahlenoptik' ('Strahl' in German means 'ray' in English) [16].

2.8.1 The Eikonal Equation

A monochromatic wave, travelling in an optical medium where the refractive index is fixed or is changing very slowly over a spatial distance of the order of the wavelength can be represented as

$$U(\mathbf{r}) = A(\mathbf{r})e^{ik_0 S(\mathbf{r})} \tag{2.153}$$

where $A(\mathbf{r})$ is the amplitude and $k_0 S(\mathbf{r})$ is the phase of the wave. k_0 is equal to the vacuum wave number $(2\pi/\lambda_0)$. $S(\mathbf{r})$ is called the 'Eikonal' function, and is a function of the refractive index $n(\mathbf{r})$ of the medium.

Substituting $U(r)$ in the Helmholtz equation (2.151), we obtain

$$k_0^2\left[n^2 - |\nabla S|^2\right]A + \nabla^2 A - ik_0\left[2\nabla S.\nabla A + A\nabla^2 S\right] = 0 \tag{2.154}$$

The real and imaginary parts on the left-hand side of this equation must vanish independently.

Thus we have

$$k_0^2\left[n^2 - |\nabla S|^2\right]A + \nabla^2 A = 0 \tag{2.155}$$

and

$$k_0\left[2\nabla S.\nabla A + A\nabla^2 S\right] = 0 \tag{2.156}$$

From equation (2.155)

$$|\nabla S|^2 = n^2 + \left(\frac{\lambda_0}{2\pi}\right)^2 \frac{\nabla^2 A}{A} \tag{2.157}$$

As $\lambda_0 \to 0$, the second term on the right-hand side of the above equation, vanishes, leading to

$$|\nabla S(\mathbf{r})|^2 = n(\mathbf{r})^2 \tag{2.158}$$

In rectangular coordinates it can be written explicitly as

$$\left(\frac{\partial S}{\partial x}\right)^2+\left(\frac{\partial S}{\partial y}\right)^2+\left(\frac{\partial S}{\partial z}\right)^2=n^2(x,y,z) \tag{2.159}$$

Equation (2.159) is known as the eikonal equation, or the Hamilton-Jacobi equation. The equation may be interpreted as the limit of the Helmholtz equation as the wavelength approaches zero. This is the fundamental equation of geometrical optics, for the wave propagation under geometrical optics approximation is uniquely specified by the eikonal equation. The surfaces

$$S(r)=\text{constant} \tag{2.160}$$

are surfaces of equal phase, and they are called the geometrical wavefronts of the electromagnetic disturbance. Note that for known $S(\mathbf{r})$, amplitude $A(\mathbf{r})$ can be determined from equation (2.156).

2.8.2 Equation for Light Rays

In geometrical or ray optics, a ray is defined as an orthogonal trajectory to the geometrical wavefronts [17–18]. This implies that a ray passing through a point on a wavefront is perpendicular to the wavefront at that point, and remains so for successive wavefronts at every point of its journey through the system (see Figure 2.7).

Figure 2.8 shows two close points, P and Q, with position vectors $\mathbf{r}$ and $(\mathbf{r}+d\mathbf{r})$ respectively on a ray in a rectangular Cartesian coordinate system with the origin at O. Let ds be the distance PQ along the ray. A unit ray vector $\mathbf{u}(\mathbf{r})$ is defined as

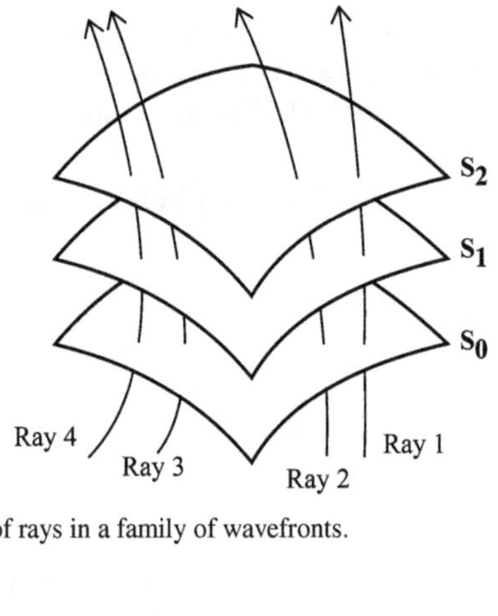

FIGURE 2.7 Orthogonal trajectory of rays in a family of wavefronts.

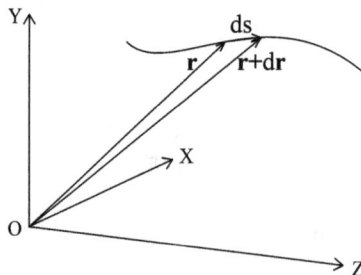

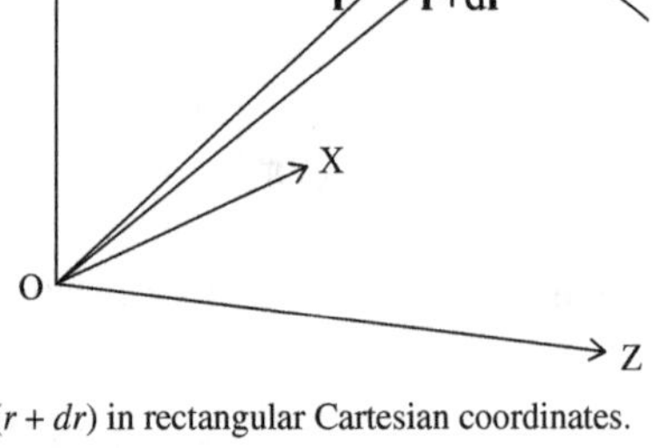

FIGURE 2.8 Position vectors *r* and (*r* + *dr*) in rectangular Cartesian coordinates.

$$\mathbf{u}(\mathbf{r}) = \frac{d\mathbf{r}}{ds} \tag{2.161}$$

By definition, the unit ray vector is tangential to the ray and perpendicular to the wavefronts given by (2.160). A vector perpendicular to the wavefronts can be defined as

$$\mathbf{v}(\mathbf{r}) = \nabla S \tag{2.162}$$

From the eikonal equation (2.158), magnitude of the vector **v** is given by

$$|\mathbf{v}| = n \tag{2.163}$$

The vectors **u** and **v** are parallel to each other, and so, $\mathbf{u} = \mathbf{v} / n$. From (2.161) – (2.163) we get

$$n\frac{d\mathbf{r}}{ds} = \nabla S \tag{2.164}$$

Note that the differentiation with respect to s can be expressed by a dot product of the unit vector given by (2.161) and the operator ∇, as given below.

$$\frac{d}{ds} = \left(\frac{dx}{ds}\frac{\partial}{\partial x} + \frac{dy}{ds}\frac{\partial}{\partial y} + \frac{dz}{ds}\frac{\partial}{\partial z}\right) = \left(\frac{d\mathbf{r}}{ds}\right).\nabla \tag{2.165}$$

Differentiating (2.164) with respect to s and using (2.165), we obtain

$$\frac{d}{ds}\left(n\frac{d\mathbf{r}}{ds}\right) = \frac{d}{ds}(\nabla S) = \left(\frac{d\mathbf{r}}{ds}\right).\nabla(\nabla S) = \frac{1}{n}\nabla S.\nabla(\nabla S) = \frac{1}{2n}\nabla\left[(\nabla S)^2\right] \tag{2.166}$$

Substituting from the eikonal equation, one obtains the vector form of the differential equation for light rays as

$$\frac{d}{ds}\left(n\frac{d\mathbf{r}}{ds}\right) = \frac{1}{2n}\nabla\left(n^2\right) = \nabla n \tag{2.167}$$

In a homogeneous medium *n* is a constant; therefore $\nabla n = 0$. The ray equation reduces to

$$\frac{d^2\mathbf{r}}{ds^2} = 0 \tag{2.168}$$

The solution of this equation is

$$\mathbf{r} = \mathbf{a}s + \mathbf{b} \tag{2.169}$$

This is the vector equation of a straight line in the direction of the vector **a**, and passing through the point **r** = **b.** Thus, in a homogeneous medium, the path of a ray is a straight line (see Figure 2.9).

Ray paths in inhomogeneous media are obtained from the equation (2.167). The ray equation and the eikonal equation are two alternate but equivalent descriptions for propagation of electromagnetic waves as per the model of geometrical optics.

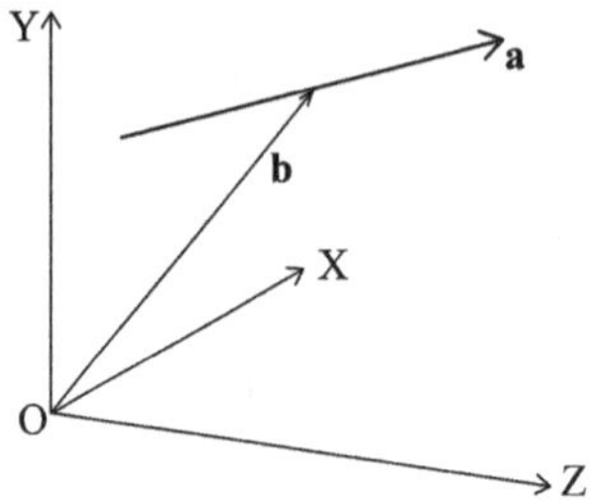

FIGURE 2.9 Straight-line ray path in a homogeneous medium.

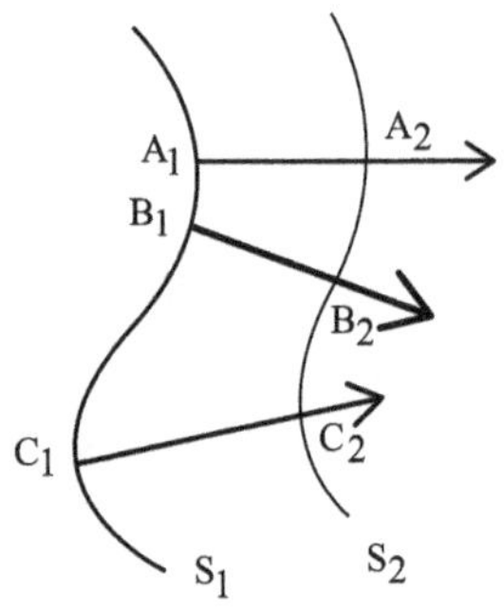

FIGURE 2.10 OPLs for rays between two wavefronts are equal.

2.9 Basic Principles of Ray Optics

Principles of ray optics play a predominant role in conceptualization, analysis, and synthesis of optical systems. Ray optics is also called geometrical optics, for the basic similarity of abstraction involved in the definition of a ray in optics and a line or curve in geometry; in both cases the definition strictly holds in the limit of corresponding thickness tending to zero. Nevertheless, ray geometry of an optical system can provide a clear visualization of propagation and manipulation of light in the three-dimensional space of the system.

All principles of ray optics in isotropic media can be derived from the basic proposition that rays are wave normals, and a wavefront represents a surface of equal phase. Thus, if $A_1B_1C_1$ and $A_2B_2C_2$ are two successive wavefronts (Figure 2.10), having A_1A_2, B_1B_2 and C_1C_2 as rays, i.e. wave normals, the wave elements at A_1, B_1 and C_1 may be thought of as travelling along the rays with phase velocity v, and the times of flight $t_{A_1A_2}$, $t_{B_1B_2}$, and $t_{C_1C_2}$ are all equal. It follows also that the intercepts along rays between two wavefronts are all equal. Thus in Figure 2.10, let

$$A_1A_2 = B_1B_2 = C_1C_2 = D \tag{2.170}$$

The time taken by the wave to cover this distance is given by

$$t = \frac{D}{v} = \frac{1}{c}(nD) \tag{2.171}$$

where (nD), the product of the refractive index and the geometrical length along the ray, is known as the optical path length.

Since the times of flight along the rays between the two wavefronts are equal, it follows that, for the wavefronts in Figure 2.10, for example,

$$[A_1A_2] = [B_1B_2] = [C_1C_2] \tag{2.172}$$

where the square brackets are used to denote optical path lengths. The optical path length (OPL) along a ray between two points, P and Q, in an inhomogeneous medium is given by the line integral, shown below.

$$\mathrm{OPL} = \int_{P}^{Q} n(s)\,ds \qquad (2.173)$$

where the integration is taken along the ray path from P to Q, and ds is an element of the path. If the medium between the two points P and Q consists of a set of M homogeneous media placed one after another, the ray path changes discontinuously at each of the (M-1) refracting interfaces, and it consists of straight ray segments Δs_m in each medium of refractive index n_m. The OPL is then given by

$$\mathrm{OPL} = \sum_{m=1}^{M} n_m \Delta s_m \qquad (2.174)$$

The 'ray optics' model of light propagation provides useful tools for tackling many difficult problems of reflection, refraction, diffraction, and scattering encountered in instrumental optics. This is not always straightforward, and to circumvent the occurrence of misinterpretation or gross errors in analysis, suitable strategies in adoption of the ray optics model need to be incorporated. In some other cases, however, analyses based on ray optics need to be supplemented by treatments based on 'scalar wave optics', or 'vector wave optics', or 'electromagnetic theory'. An example for each of these cases is given below:

- The correspondence between the geometrical wavefronts and iso-phase (equal phase) fronts of wave optics is strictly valid in regions far away from the foci or edges of shadow regions. In addition, there is no correspondence between the two when the geometrical wavefronts are almost plane. Therefore, practical approaches for analysis and synthesis of optical systems involve adoption of suitable methods that retain the validity of assuming coincidence of geometrical wavefronts with corresponding phase fronts. A specific example is the determination of wave aberration from the deformation of an actual wavefront in comparison with a reference surface. However, the results can be useful if and only if the measurement is carried out in regions far away from the focal region [19].
- At any refracting interface, a part of the incident light is refracted, and the remaining part is reflected back. The paths of the refracted and reflected rays, or the refracted and reflected wavefronts, can be determined by analyses based solely on ray optics. But, to obtain quantitative results on the amounts of reflected and refracted light corresponding to a given amount of light incident at a given direction on the interface, it is necessary to invoke Fresnel equations derived from the electromagnetic theory by using suitable boundary conditions. This is shown in Section 2.10 below.

2.9.1 The Laws of Refraction

Analysis of optical systems involves tracking the changes in shape and orientation of wavefronts, or, equivalently, of the wave normals or rays, during their propagation through various constituent optical elements of the system. This calls for suitable methods for taking into account the effects of a multitude of refractions taking place at different planes or curved interfaces in the course of their propagation. Out of a large number of laws or principles developed over a period of several centuries, we mention four of them below. The latter have direct implications in treatments of instrumental optics. Indeed, they are not independent characteristics—any one of them can be derived from the other, the difference lying in relative ease with which one or the other can tackle specific phenomena. Note that the case of reflection can be treated as a special case of refraction, as shown later in Section 2.9.4. The interrelationship between rays and wavefronts is intimately linked with the normals on plane curves and surfaces. A few characteristics of the latter are put forward below for the sake of a better understanding of what follows next.

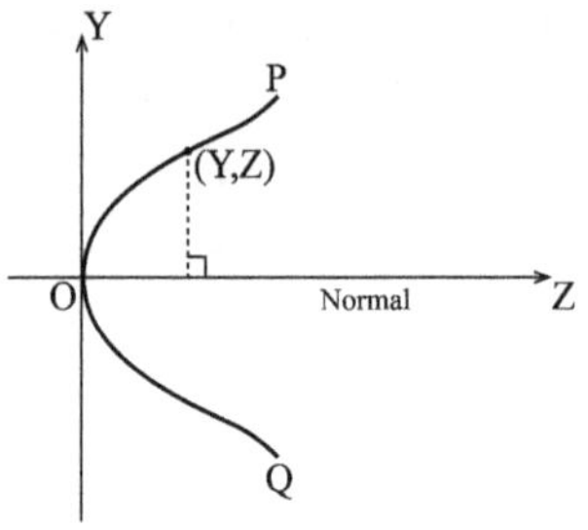

FIGURE 2.11 A continuous plane curve POQ in two-dimensional space.

2.9.1.1 A Plane Curve in the Neighbourhood of a Point on It

In Figure 2.11, POQ is a continuous plane curve in two-dimensional space. Let O be the origin of a rectangular coordinate system with Z-axis along the normal at O, and Y-axis along the tangent at O. The equation of the curve in the neighbourhood of O may be represented in the form

$$Z = f(Y) \tag{2.175}$$

For points near to O, (2.175) this may be expanded by Maclaurin's theorem, to

$$Z = a_0 + a_1 Y + a_2 Y^2 + a_3 Y^3 + \ldots \tag{2.176}$$

At Y = 0,

$$Z = 0 \text{ and } \frac{dZ}{dY} = 0 \tag{2.177}$$

Therefore, $a_0 = a_1 = 0$, and (2.176) reduces to

$$Z = a_2 Y^2 + a_3 Y^3 + \ldots = a_2 Y^2 + O(3) \tag{2.178}$$

where $O(n)$ represents a sum of terms in degree n and higher in Y, and the value of $O(3)$ can be made as insignificantly small compared to $a_2 Y^2$, as desired by choosing a small enough value of Y. In this notation, Z can be represented as $Z = O(2)$.

2.9.1.2 A Continuous Surface in the Neighbourhood of a Point on It

By similar treatment, the equation of a continuous surface in the neighbourhood of a point O on it can be represented in the form

$$Z = f(X, Y) \tag{2.179}$$

Let the point O on the surface be considered an origin of a rectangular coordinate system. The normal to the surface at O is taken as the Z axis, and the (X, Y) axes lie on the tangent surface at O. Any plane containing the normal OZ cuts the surface in a plane curve. Such a curve is known as a normal section of the surface. Figure 2.12 shows a normal section POQ of the continuous surface.

In the neighbourhood of the point O, equation (2.167) may be expanded as

$$Z = a_0 + a_1 X + b_1 Y + a_2 X^2 + b_2 Y^2 + c_2 XY + O(3) \tag{2.180}$$

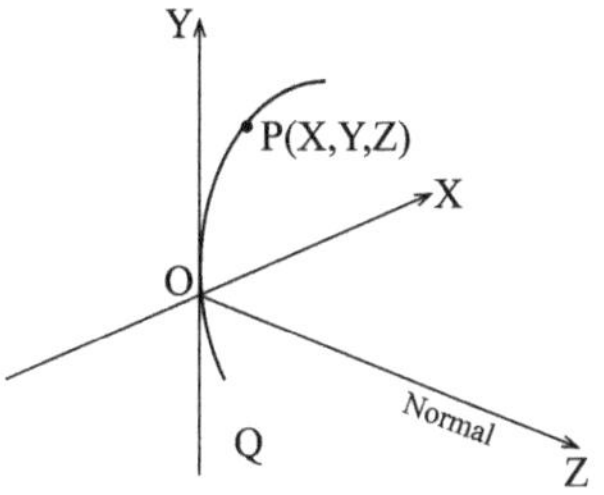

FIGURE 2.12 A normal section POQ through O of a smooth surface in three-dimensional space.

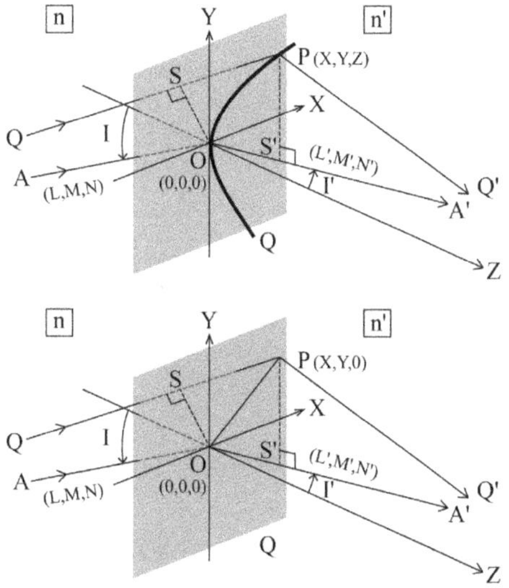

FIGURE 2.13 Refraction of wave element OS at (i) a curved refracting interface POQ; (ii) a plane refracting interface at XY plane.

At the origin O, i.e. at X =Y =0,

$$Z = 0, \frac{dZ}{dX} = 0, \frac{dZ}{dY} = 0 \tag{2.181}$$

Therefore (2.180) reduces to

$$Z = a_2X^2 + b_2Y^2 + c_2XY + O(3) \tag{2.182}$$

By suitable rotation of the (X, Y) axes about the OZ axis, the product term vanishes, and the form of the equation of the surface in the neighbourhood of O is given by

$$Z = a_2X^2 + b_2Y^2 + O(3) = O(2) \tag{2.183}$$

2.9.1.3 Snell's Laws of Refraction

A normal section POQ of a continuous refracting interface between two media of refractive indices *n* and *n'* is shown in Figure 2.13. AO is a ray incident at O on the continuous refracting surface. The point of incidence O is taken as the origin of the rectangular coordinate system. (*X, Y*) axes are located on the tangent plane at O. Z-axis is taken along the normal to the refracting interface at point O. Let the direction cosines of the incident ray AO and the refracted ray OA′ be (L, M, N), and (L′, M′, N′), respectively. QSP is a neighbouring ray of the incident pencil, the incident at the point P with coordinates (X, Y, Z). SO is the wavefront to which QSP and AO are normals, and PS′ is the wavefront in the second medium. The direction of the refracted ray OSB is determined by the condition that it is normal to the wave element PS′.

Since SO and PS′ are surfaces of constant phase, the phase difference between P and S must be equal to that between S′ and O. Thus, by (2.172), the optical path lengths [SP] and [OS′] must be equal. Since PS′ is normal to the ray OS′B, [OS′] is given by

$$[OS'] = n'(L'X + M'Y + N'Z) \tag{2.184}$$

By (2.183), *Z* is found to be of second degree in (*X, Y*), so that

$$[OS'] = n'(L'X + M'Y) + O(2) \tag{2.185}$$

Similarly, [SP] is given by

$$[SP] = n(LX + MY) + O(2) \tag{2.186}$$

Equating (2.185) and (2.186) gives

$$(n'L' - nL)X + (n'M' - nM)Y + O(2) = 0 \tag{2.187}$$

which must hold for all rays of the incident pencil near to AO. It may be noted that, in the special case when the refracting interface is a plane at the (X, Y) plane, as shown in Figure 2.13(ii), the term *O* (2) will be absent in the right-hand side of the expressions (2.185) – (2.187).

It is obvious that (2.187) can be valid for all values of L and M, if and only if the coefficients of X and Y are separately zero. Thus, we get

$$n'L' = nL \quad \text{and} \quad n'M' = nM \tag{2.188}$$

Suppose now that the (X, Y) axes are chosen such that Y lies in, and X is perpendicular to the plane of incidence, i.e. the plane containing the normal OZ and the incident ray AO. In this case, L = 0, and so $L' = 0$. Also, M = sin I, and M′ = sin I′, where I and I′ are the angles of incidence and refraction. Thus

$$n' \sin I' = n \sin I \tag{2.189}$$

The laws of refraction of rays, i.e. wave normals, contained in (2.187) are thus that the refracted ray is coplanar with the incident ray and the surface normal at the point of incidence, and that the angles of incidence and refraction satisfy the refraction law (2.189). These laws are well-known as Snell's laws of refraction. Snell's laws hold good on all points of a smooth refracting interface.

The two laws of refraction can be compactly written in vector form. Let **r** and **r′** be unit vectors along the incident and refracted rays along the direction of propagation of light (see Figure 2.14).

If **p** is a unit vector along the normal to the refracting surface at O, the equation

$$n'(\boldsymbol{r}' \times \boldsymbol{p}) = n(\boldsymbol{r} \times \boldsymbol{p}) \tag{2.190}$$

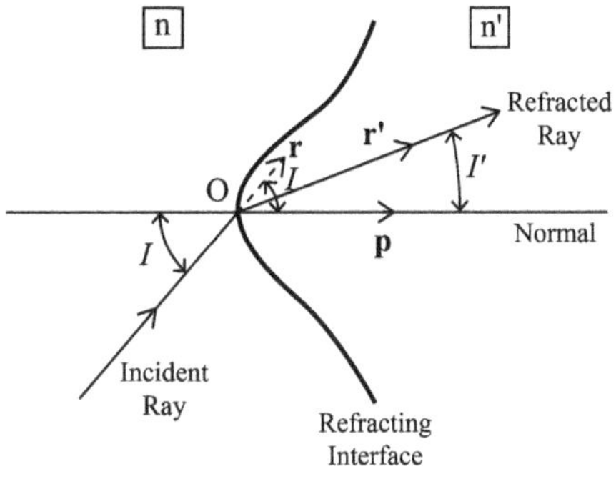

FIGURE 2.14 Vector form of Snell's laws of refraction.

where × denotes a vector product, expresses both laws of refraction [19–21]. The direction of the vector **p** is chosen to make the angle (**r**, **p**) acute. The normal to the plane containing **r**′ and **p** is the same as that of the plane containing **r** and **p**. The relation (2.189) is also satisfied since I and I′ are the acute angles (**r**, **p**) and (**r**′, **p**), respectively. It is important to note the conventions defining the senses of the vectors **r**, **r**′ and **p**.

Equating vector products of **p** with the two sides of equation (2.190), we get

$$n'\left[\mathbf{p}\times(\mathbf{r}'\times\mathbf{p})\right]=n\left[\mathbf{p}\times(\mathbf{r}\times\mathbf{p})\right] \tag{2.191}$$

Expanding triple vector products on both sides, this relation gives

$$n'\left[(\mathbf{p}\cdot\mathbf{p})\mathbf{r}'-(\mathbf{p}\cdot\mathbf{r}')\mathbf{p}\right]=n\left[(\mathbf{p}\cdot\mathbf{p})\mathbf{r}-(\mathbf{p}\cdot\mathbf{r})\mathbf{p}\right] \tag{2.192}$$

Since **r**, **r**′ and **p** are all unit vectors, and $(\mathbf{p}\cdot\mathbf{r})$ and $(\mathbf{p}\cdot\mathbf{r}')$ are the acute angles I and I′, this reduces to

$$n'\mathbf{r}'-n\mathbf{r}=\mathbf{p}\{n'\cos I'-n\cos I\} \tag{2.193}$$

which gives **r**′ explicitly. Let (L, M, N), (L′, M′, N′) and (i,j,k) be the direction cosines of the incident ray, the refracted ray, and the normal at the point of incidence, respectively. This gives

$$\mathbf{r}=\tilde{\mathbf{i}}L+\tilde{\mathbf{j}}M+\tilde{\mathbf{k}}N \tag{2.194}$$

$$\mathbf{r}'=\tilde{\mathbf{i}}L'+\tilde{\mathbf{j}}M'+\tilde{\mathbf{k}}N' \tag{2.195}$$

$$\mathbf{p}=\tilde{\mathbf{i}}i+\tilde{\mathbf{j}}j+\tilde{\mathbf{k}}k \tag{2.196}$$

where $(\tilde{i},\tilde{j},\tilde{k})$ are the unit vectors along the three orthogonal directions.

Substituting **r**, **r**′ and **p** from (2.194) – (2.196) in (2.193), and writing it in terms of rectangular components, we get

$$n'L'-nL=i\Delta(n\cos I) \tag{2.197}$$

$$n'M'-nM=j\Delta(n\cos I) \tag{2.198}$$

$$n'N'-nN=k\Delta(n\cos I) \tag{2.199}$$

where

$$\Delta(n\cos I) = \{n'\cos I' - n\cos I\} \tag{2.200}$$

and the signs of the cosines are determined by the senses ascribed to the vectors **r**, **r′** and **p**. They are both positive. The formulae (2.197) – (2.200) are the basic refraction equations for tracing rays through optical systems.

2.9.2 Refraction in a Medium of Negative Refractive Index

All naturally occurring optical materials have a positive refractive index. Artificially structured materials, also called metamaterials, may be developed to exhibit negative refractive indices [22–23]. Refraction in a medium of a negative refractive index can also be treated by means of Snell's laws with proper interpretation of ray vectors [24].

In Figure 2.15, AO is a ray incident on a point O of a smooth refracting interface between two media of refractive index n and n'. The normal at O on the surface is shown in the figure. Equation (2.176) may be rewritten as

$$n\sin I = n'\sin I' \tag{2.201}$$

Let **r**, $\tilde{\mathbf{r}}'$, and **p** represent unit vectors along the incident ray, the refracted ray, and the normal. The angle of incidence I = Angle (**p**, **r**). Let $n' = \mu$, where $\mu > 0$, and let the corresponding angle of refraction be $\tilde{I}'$ where $\tilde{I}'$ = Angle $(\mathbf{p}, \tilde{\mathbf{r}}')$. From (2.189)

$$\sin\tilde{I}' = \frac{n}{\mu}\sin I \tag{2.202}$$

For $n' = \mu$, where $\mu < 0$, let the angle of refraction be $\hat{I}'$. From (2.189)

$$\sin\hat{I}' = -\frac{n}{\mu}\sin I \tag{2.203}$$

Therefore $\hat{I}' = -\tilde{I}'$. For the same value of the angle of incidence I, in the same incident medium of refractive index n, the direction of the refracted ray in the case of refraction in a medium of negative refractive index does not lie in the same quadrant as in the case of refraction in a medium of positive refractive index. The angles of refraction are of opposite sign; however, their magnitude will be the same if the magnitude of refractive index is the same for both cases.

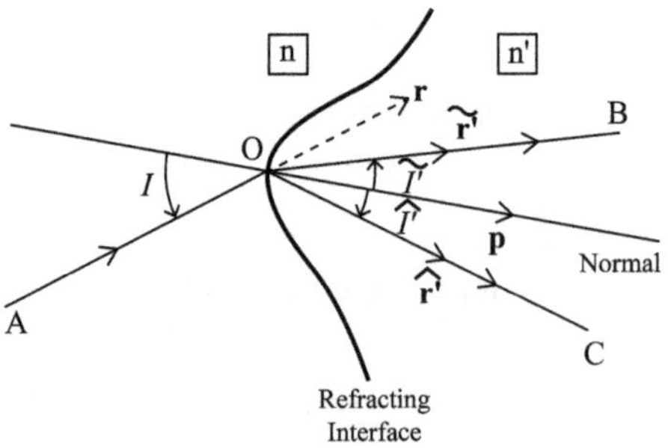

FIGURE 2.15 Case of refraction: Incident ray: **r**; Angle of incidence $I = (\mathbf{p}, \mathbf{r})$;
Refracted ray (i) $\tilde{r}'$ when $n' > 0$; Angle of refraction $\tilde{I}' = (\mathbf{p}, \tilde{\mathbf{r}}')$
(ii) $\hat{r}'$ when $n' < 0$; Angle of refraction $\hat{I}' = (\mathbf{p}, \hat{\mathbf{r}}')$

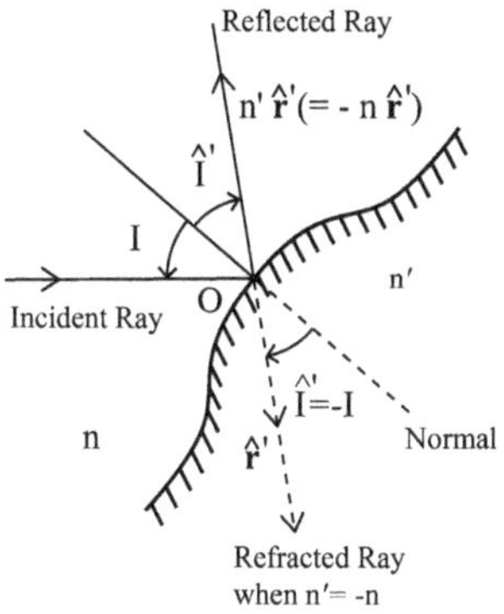

FIGURE 2.16 Case of reflection: Incident ray *r*; Angle of incidence I
When n′ = −n, Refracted ray $\hat{\mathbf{r}}'$; Reflected ray $n'\hat{r}' = (-n\hat{r}')$;
Angle of reflection $\hat{I}' = -I$.

2.9.3 The Case of Reflection

Reflection may be considered as a special case of refraction in a medium whose refractive index is the negative of that of the incident medium. In Figure 2.16, a ray incident at the point O on a smooth refracting interface between two media of refractive index n and n′ respectively. The path of the refracted ray when n′ = −n is shown with dashed lines. Let **r**, **p** and $\tilde{\mathbf{r}}'$ be unit vectors along the incident ray AO, the normal to the surface at O and the refracted ray respectively. The reflected ray at O is in the direction opposite to that of the refracted ray; the unit vector along the reflected ray is $\tilde{\mathbf{r}}'$, and in vector notation the reflected ray is given by n′ $\tilde{\mathbf{r}}'$ (=−n$\tilde{\mathbf{r}}'$). It may be noted that the magnitudes of the direction cosines of the unit vector along the reflected ray are the same as those of the refracted ray with a real refractive index of the refraction medium having a value that is negative of the refractive index of the medium of incidence. The exact values of direction cosines incorporating both magnitude and sense call for a multiplication of these magnitudes by −n.

2.9.3.1 Total Internal Reflection

If n′ <n, the angle of refraction is larger than the angle of incidence. The angle of incidence that corresponds to the value 90° for the angle of refraction is called the critical angle I_c. From (2.189), I_c is given by

$$I_c = \sin^{-1}(n'/n) \tag{2.204}$$

All rays with angles of incidence in the range $90° \geq I \geq I_c$ are not refracted, and they are reflected back in the incident medium. This phenomenon is known as 'Total internal reflection', and it is often utilized in prisms used in optical systems for image inversion or reversion.

2.9.4 Fermat's Principle

Pierre de Fermat enunciated a principle that states that the path of a ray between any two point, say, A and A′, is such that the time taken for light to travel from A to A′ along the ray is less than the time that would be taken along any other path in the neighbourhood of the ray. This principle of least time implies that the optical path length from A to A′ is minimum along the ray itself. It was subsequently observed that, more strictly stated, the condition is that the optical path is stationary for small displacements of the ray path.

In Figure 2.17, AO is a ray incident in a medium of refractive index *n* on a refracting interface PO, and refracted as OA′ into a medium of refractive index *n′*. Let BPB′ be any nearby path, not necessarily itself a ray path. The Z-axis is along the normal to the interface at O, and (X, Y) axes lie in the tangent plane at O. The point O is considered to be the origin of the rectangular coordinate system. The direction

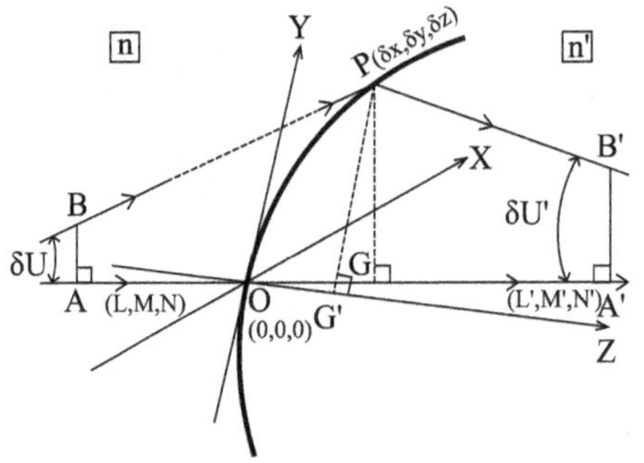

FIGURE 2.17 Fermat's principle; AOA′ is a ray path; BPB′ is a neighbouring path.

cosines of the incident and refracted portions of the ray APA′ are (L, M, N) and (L′, M′, N′) respectively. The segments BP and PB′ of the nearby path BPB′ will have direction cosines $(L+\delta L, M+\delta M, N+\delta N)$ and $(L'+\delta L', M'+\delta M', N'+\delta N')$, these lines being only slightly inclined to AO and OA′, respectively. Similarly, the point P will have the coordinates $(\delta X, \delta Y, \delta Z)$. However, the nearby path will be fully specified by changes in the two direction cosines $(\delta L, \delta M)$, and the two coordinates $(\delta X, \delta Y)$ of the point P, for δN and δZ are determined by the relation $L^2+M^2+N^2=1$ and the equation of the surface respectively.

Let AB and A′B′ be two perpendiculars drawn at point A on the segment AO and at point A′ on the segment OA′ respectively of the ray AOA′; the perpendiculars intersect the nearby path BPB′ at B and B′ respectively. The optical path difference $[BPB']-[AOA']$ is a function of $(\delta L, \delta M)$ and $(\delta X, \delta Y)$.

Let PG and PG′ be two perpendiculars drawn on AO (extended) and OA′ respectively. Then

$$[BPB'] = n(BP) + n'(PB') = n\left(\frac{AG}{\cos\delta U}\right) + n'\left(\frac{G'A'}{\cos\delta U'}\right) \tag{2.205}$$

where $\delta U, \delta U'$ are the small inclinations of BP to AO and of PB′ to OA′. Thus

$$\begin{aligned}[BPB'] &= n(AG) + n'(GA') + O(2)\\ &= n(AO) + n(OG) + n'(OA') - n'(OG') + O(2)\end{aligned} \tag{2.206}$$

so that

$$[BPB'] - [AOA'] = -\{n'(OG) - n(OG')\} + O(2) \tag{2.207}$$

Projection of OP on the incident and refracted parts of the ray APA′ yields

$$OG = L\delta X + M\delta Y + N\delta Z = L\delta X + M\delta Y + O(2) \tag{2.208}$$

$$OG' = L'\delta X + M'\delta Y + N'\delta Z = L'\delta X + M'\delta Y + O(2) \tag{2.209}$$

Thus

$$\{n'(OG) - n(OG')\} = \delta X\Delta(nL) + \delta Y\Delta(nM) + O(2) \tag{2.210}$$

Since direction cosines of the normal OZ are (0, 0, 1), from (2.197) and (2.198) we get

$$\Delta(nL) = \Delta(nM) = 0 \tag{2.211}$$

From (2.207), (2.210), and (2.211,) it follows

$$[BPB'] - [AOA'] = O(2) \tag{2.212}$$

This is Fermat's principle. Although the derivation is shown above for the case of a single refraction between two homogeneous media, the principle is valid for the path of a ray passing through any number of reflections and refractions. Indeed, the implications of this principle go far beyond the limited domain of ray optics. A brief interpretation of Fermat's principle is given below in the context of ray optics.

Let $[BPB'] = P$, and $[AOA'] = P_0$ The difference δP between these two optical path lengths may involve squares or products of the variables $\delta L, \delta M, \delta X$ and δY, but no term of the first power. Thus, δP is of the form

$$\delta P = A\delta L^2 + B\delta M^2 + C\delta X^2 + D\delta Y^2 + \alpha\delta L\delta M + \beta\delta X\delta Y + E\delta L^3 + F\delta M^3 + \ldots \tag{2.213}$$

It follows that

$$\frac{\partial P}{\partial L} = \frac{\partial P}{\partial M} = \frac{\partial P}{\partial X} = \frac{\partial P}{\partial Y} = 0 \tag{2.214}$$

The stationary character of an optical ray path against small displacements implies that the actual ray path in an optical system between two points is either a minimum or a maximum, or a point of inflexion among the paths available in its immediate neighbourhood.

2.9.5 The Path Differential Theorem

APP′A′ is a ray in a general optical system (Figure 2.18). Let $\mathbf{s} \equiv (L, M, N)$ be the unit vector for the direction of the ray in the initial space of refractive index n, and $\mathbf{s}' \equiv (L', M', N')$ be the unit vector for the ray direction in the final space of refractive index n'. In the initial space, a point B is shifted from the point A by an infinitesimal amount $\mathbf{dr}$. Similarly, in the final space, a point B′ is shifted from the point A′ by an infinitesimal amount $d\,\mathbf{r}'$. Let BQQ′B″ be the ray from point B to B′.

The optical path lengths along the two rays are:

$$E = [APP'A'] \text{ and } E + dE = [BQQ'B'] \tag{2.215}$$

A point P, lying far away from either A or B, is chosen on the ray AA′ in the initial space so that the angle ∠BPA is infinitesimally small, and BP ≅ DP, where BD is perpendicular to AP. In the final space, a point P′ is similarly chosen where the angle ∠B′P′A′ is infinitesimally small, and P′B′ ≅ P′D′, where B′D′ ⊥ P′A′(extended). By Fermat's principle, the actual optical path length $[BQ\ldots Q'B']$ of the ray BB′ is equal to the optical path length $[BP\ldots P'B']$ along the neighbouring path. Thus

$$dE = [BQ\ldots Q'B'] - [AP\ldots P'A'] = [BP\ldots P'B'] - [AP\ldots P'A'] = [A'D'] - [AD] \tag{2.216}$$

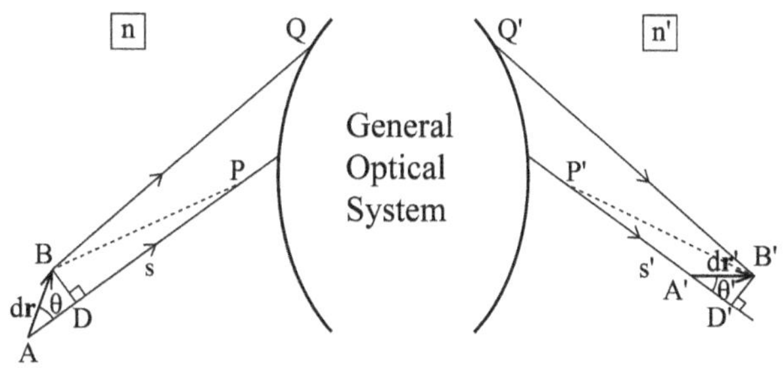

FIGURE 2.18 Optical path differential {[BB′] – [AA′]} in a general optical system.

AD and A′D′ are the projections of the shifts AB and A′B′ on the ray APP′A′ in the initial and final space, respectively, and thus can be represented as inner products. Thus, the path differential dE is given by

$$dE = n'\mathbf{s}'.\mathbf{dr}' - n\mathbf{s}.\mathbf{dr} \tag{2.217}$$

The path differential theorem given by (2.217) has many applications in ray optics, particularly in aberration theory. It should be iterated that the path differential theorem involves the difference between two neighbouring ray paths, whereas in Fermat's theorem, the ray path between two points is defined as a path that is stationary among all possible paths between the two points; the latter paths are not necessarily ray paths. Hopkins [25] has shown that the path differential $dE = O(3)$.

2.9.6 Malus-Dupin Theorem

In a homogeneous medium, the normals to a wavefront form a family of straight lines; this family is called a congruence. An alternative term for the same family of rays is 'orthotomic system of rays'. [26] In a three-dimensional space this is a two parameter congruence, since each line can be specified in terms of two coordinates of the point on the surface at which the given line is the normal. Note that the third coordinate of the point is determined by the equation of the surface. The family of rays, which are the normals to a wavefront, is a 'normal congruence', since there exists a surface, namely the wavefront, which cuts each of the rays orthogonally. It is clearly possible to have a family of straight lines for which no such surface exists; such family is called a 'skew congruence'. The theorem of Malus and Dupin posits, 'A normal congruence of rays remains a normal congruence after any number of refractions and reflections'. Alternatively, the theorem states, 'An orthotomic system of rays remains so after any number of refractions or reflections'. An important consequence of this theorem is that the wavefronts after refraction, or reflection, are uniquely specified if the ray configuration is known and conversely, the ray configuration in the incident and in either the refraction medium or the reflection medium can be worked out when the wavefronts in the corresponding media are known.

First, a proof of the proposition that 'the wavefronts in an isotropic medium are all orthogonal to the rays' is given below. In Figure 2.19, AB is a given wavefront with normals AA′ and BB′. A second wavefront A′B′ is found, as established in (2.172), by taking equal distances along the normals to AB. Two perpendiculars, AB_0 and $A'B_0'$, are drawn to the ray AA′. Since AB_0 is tangent to the wavefront AB at A, the distance BB_0 is of the order of $(AB_0)^2$, i.e. O (2). Conversely, if $B'B_0'$ is O (2), the line $A'B_0'$ will be tangent to the wavefront A′B′ at A′.

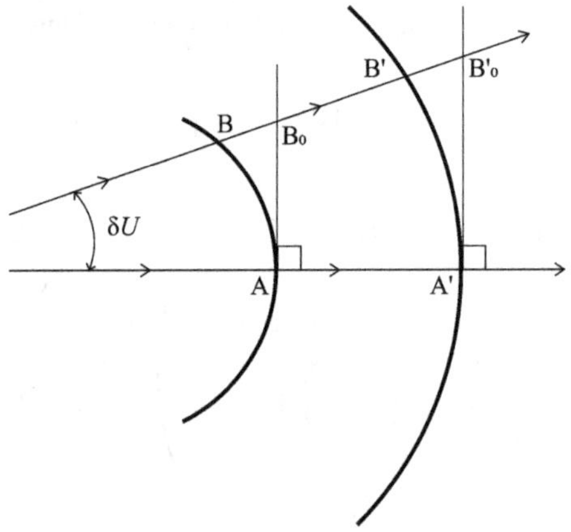

FIGURE 2.19 Wavefronts in an isotropic medium are orthogonal to the rays.

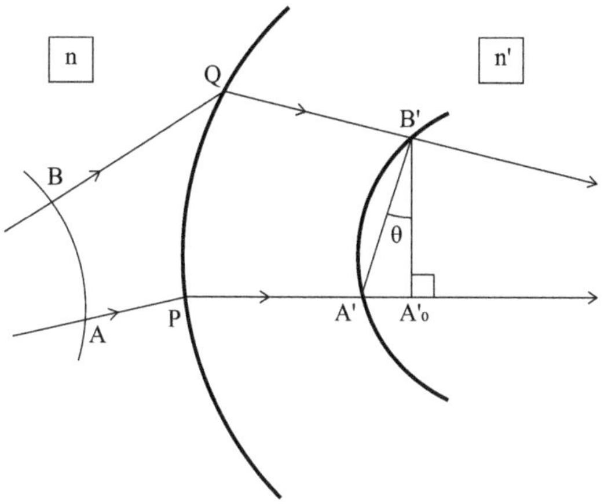

FIGURE 2.20 The theorem of Malus and Dupin.

Since $BB_0 = O(2)$, and $BB' = AA'$,

$$B'B_0' = B_0B_0' - BB' + BB_0$$

$$= \frac{AA'}{\cos \delta U} - AA' + O(2) = O(2) \tag{2.218}$$

where δU is the angle between AA′ and BB′. Thus, $A'B_0'$ is tangent to the wavefront at A′. Hence the wavefront A′B′ cuts the ray AA′ orthogonally at A′. This will be true for all rays normal to AB, and for all wavefronts constructed by taking equal intercepts from AB along the rays. Consequently, a normal congruence of rays remains so during propagation in an isotropic medium.

Figure 2.20 shows a refracting interface PQ between two media of refractive indices n and n'. In the medium of incidence, AP and BQ are two rays normal to the wavefront AB. These rays are refracted as PA′ and QB′ respectively. A′B′ is a wavefront after refraction; A′ and B′ are two points on the wavefront where the refracted rays intersect the wavefront. By definition,

$$[BQB'] = [APA'] \tag{2.219}$$

A perpendicular $B'A_0'$ is drawn from the point B′ on the refracted ray path PA′. By Fermat's principle,

$$[BQB'] = [APA_0'] + O(2) \tag{2.220}$$

The quantity $O(2)$ being of at least to the second degree in the variables specifying the inclination and lateral displacement of BQ relative to AP. Thus

$$[A'A_0'] = \sigma \tan\theta = O(2) \tag{2.221}$$

where, $\sigma = B'A_0'$, and $\theta = \angle A'B'A_0'$. Clearly, $\sigma = O(1)$, and, therefore $\tan\theta = O(1)$. Thus, θ tends to zero as the ray BQB′ is taken limitingly close to APA′. It follows that the wavefront B′A′ cuts the ray PA′ orthogonally at A′. This will be true for any ray of the family of rays, and for any other wavefront in the medium of refraction, and hence the normals to any given wavefront remain at a normal congruence after refraction. It is obvious the result holds good for any number of susccessive refractions. As explained earlier, the case of reflection can be considered a special case of refraction.

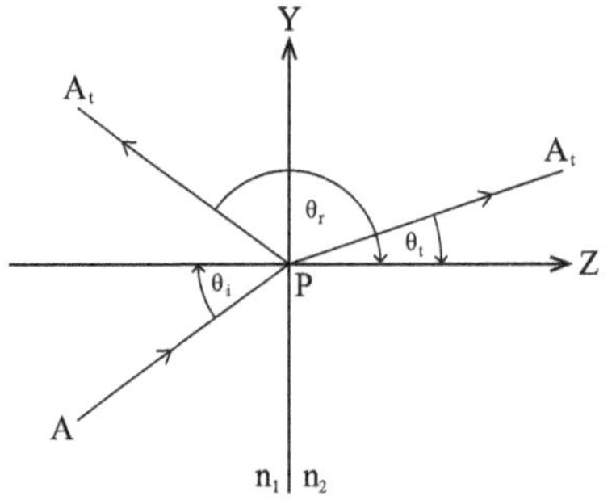

FIGURE 2.21 A plane wave along AP is reflected along PA_r, and is refracted along PA_t by a plane surface of discontinuity between two homogeneous and isotropic optical media.

2.10 Division of Energy of a Light Wave Incident on a Surface of Discontinuity

The 'ray optics' model, as enunciated above, is highly useful in analysis and synthesis of optical systems, and provides directions of the reflected ray and refracted ray corresponding to a ray incident on a surface of discontinuity. Nevertheless, this model cannot provide answers to the question how the incident light energy is distributed among the reflected and the transmitted rays. Indeed, this distribution depends on the angle of incidence and the state of polarization of the incident wave, and can be determined by the appropriate use of boundary conditions described in Section 2.4.1.

Figure 2.21 shows a plane surface of discontinuity between two homogeneous and isotropic media. Both media are of zero conductivity, and of magnetic permeability $\mu_1, \mu_2 \approx 1$. Refractive indices of the two media are $n_1 = \sqrt{\epsilon_1}$ and $n_2 = \sqrt{\epsilon_2}$. The surface normal at P is the Z-axis. A plane electromagnetic wave is incident along AP on the surface at P. The plane of incidence is the YZ plane. The reflected wave is along PA_r, and the transmitted wave is along PA_t. The rays AP, PA_r, and PA_t make angles θ_i, θ_r and θ_t with the Z-axis, respectively. Without loss of generality, we can consider two special cases of linear polarization; in one case, denoted by ||, the electric vector of the incident wave lies parallel to the plane of incidence, and in the case of the other, denoted by ⊥, the electric vector is normal to this plane. Any other state of polarization, plane or otherwise, of the incident wave can be considered a linear superposition of the two special cases mentioned above. Other names for '||' polarization are 'parallel', 'horizontal', 'TM' (Transverse Magnetic), 'π', or 'p', etc.; similarly, other names for '⊥' polarization are 'perpendicular', 'vertical', 'TE' (Transverse Electric), 'σ', or 's', etc. The letter 's' for perpendicular polarization originates from the German word 'senkrecht' meaning 'perpendicular'. Using the boundary conditions given in Section 2.4.1, the amplitude reflection and transmission coefficients for the two cases of polarization can be obtained as

$$r_{\parallel} = \frac{-n_2 \cos\theta_i + n_1 \cos\theta_t}{n_2 \cos\theta_i + n_1 \cos\theta_t} = \frac{\tan(\theta_t - \theta_i)}{\tan(\theta_i + \theta_t)} \tag{2.222}$$

$$r_{\perp} = \frac{n_1 \cos\theta_i - n_2 \cos\theta_t}{n_1 \cos\theta_i + n_2 \cos\theta_t} = -\frac{\sin(\theta_i - \theta_t)}{\sin(\theta_i + \theta_t)} \tag{2.223}$$

$$t_{\parallel} = \frac{2 n_1 \cos\theta_i}{n_2 \cos\theta_i + n_1 \cos\theta_t} = \frac{2\cos\theta_i \sin\theta_t}{\sin(\theta_i + \theta_t)\cos(\theta_i - \theta_t)} \tag{2.224}$$

$$t_{\perp} = \frac{2 n_1 \cos\theta_i}{n_1 \cos\theta_i + n_2 \cos\theta_t} = \frac{2\cos\theta_i \sin\theta_t}{\sin(\theta_i + \theta_t)} \tag{2.225}$$

[4, 10, 27–29]

Equations (2.222) – (2.225) are called Fresnel formulae, as Augustine Fresnel first derived them in 1823 CE [30].

The ratios of the energy in the reflected wave and the energy in the transmitted wave to the energy in the incident wave are called reflectivity $\mathcal{R}$ and transmissivity $\mathcal{T}$, respectively. These are different for the waves with two different states of polarization. They are given by

$$\mathcal{R}_{\parallel} = \left|r_{\parallel}\right|^2 \tag{2.226}$$

$$\mathcal{R}_{\perp} = \left|r_{\perp}\right|^2 \tag{2.227}$$

$$\mathcal{T}_{\parallel} = \left(\frac{n_2}{n_1}\right)\left(\frac{\cos\theta_t}{\cos\theta_i}\right)\left|t_{\parallel}\right|^2 \tag{2.228}$$

$$\mathcal{T}_{\perp} = \left(\frac{n_2}{n_1}\right)\left(\frac{\cos\theta_t}{\cos\theta_i}\right)\left|t_{\perp}\right|^2 \tag{2.229}$$

Note that in some literature, there is a reversal of the sign in the expression for $r_{\parallel}$. This occurs due to choice of reference in the case of reflection. However, it does not affect the expressions for reflectivity and transmissivity.

For normal incidence, the difference in reflectivity and transmissivity for two states of polarization disappears, and we get

$$\mathcal{R} = \mathcal{R}_{\parallel} = \mathcal{R}_{\perp} = \left(\frac{n_2 - n_1}{n_2 + n_1}\right)^2 = \left(\frac{n-1}{n+1}\right)^2 \tag{2.230}$$

$$\mathcal{T} = \mathcal{T}_{\parallel} = \mathcal{T}_{\perp} = \frac{4n_2 n_1}{\left(n_2 + n_1\right)^2} = \frac{4n}{\left(n+1\right)^2} \tag{2.231}$$

where $n = \left(n_2 / n_1\right)$. Therefore, for normal incidence at an air glass interface, $n \approx 1.5$, and so $\mathcal{R} = 0.04$, and $\mathcal{T} = 0.96$.

2.10.1 Phase Changes in Reflected and Transmitted Waves

From the Fresnel formula, as given in (2.222) – (2.225), it is seen that in the case of a transmitted wave, both $t_{\parallel}$ and $t_{\perp}$ are positive, irrespective of the values of the angle of incidence θ_i, or the angle of refraction θ_t, i.e. irrespective of whether $n_1 > n_2$, or $n_1 < n_2$. Thus, both the components of the transmitted wave are in phase with the incident wave. However, in the case of the reflected wave, both $r_{\parallel}$ and $r_{\perp}$ change sign depending on the relative magnitudes of angles θ_i and θ_t [4, 31].

For $n_2 < n_1, \theta_t > \theta_i$, so $r_{\perp}$ is positive. For $n_2 > n_1, \theta_t < \theta_i$, so $r_{\perp}$ is negative. The positive value implies that the corresponding reflected wave is in phase with the incident wave, whereas the negative value implies that the phase of the corresponding reflected wave differs from that of the incident wave by π.

However, for $n_2 < n_1$, $\theta_t > \theta_i$, the numerator $\tan\left(\theta_i - \theta_t\right)$ in the expression for $r_{\parallel}$ is negative, but the sign of the denominator $\tan\left(\theta_i + \theta_t\right)$ is positive if $\left(\theta_i + \theta_t\right) < \pi/2$, and negative if $\left(\theta_i + \theta_t\right) > \pi/2$. Therefore, for $\left(\theta_i + \theta_t\right) < \pi/2$, the phase of the corresponding reflected wave differs from that of the incident wave by π; and for $\left(\theta_i + \theta_t\right) > \pi/2$, the incident wave and the corresponding reflected wave are in phase.

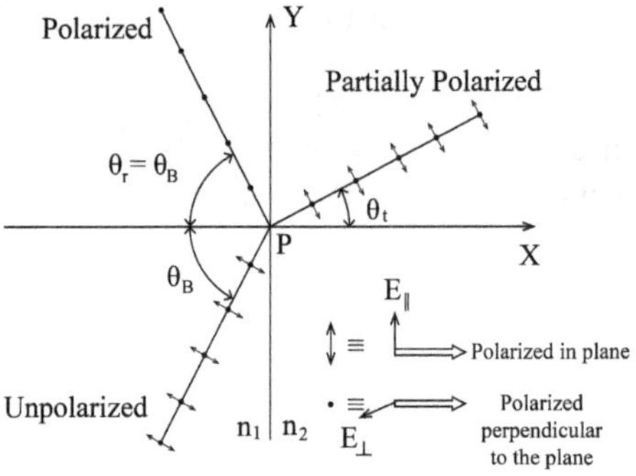

FIGURE 2.22 Polarization by reflection at Brewster angle of incidence.

2.10.2 Brewster's Law

From (2.222) we note that, when $(\theta_i + \theta_t) = \pi/2$, $\tan(\theta_i + \theta_t) = \infty$, so that $\mathcal{R}_{\parallel} = r_{\parallel} = 0$. In this case the directions of the transmitted ray and the reflected ray are perpendicular to each other (Figure 2.22). Using Snell's law, we get

$$n_1 \sin\theta_i = n_2 \sin\theta_t = n_2 \sin\left(\frac{\pi}{2} - \theta_i\right) = n_2 \cos\theta_i$$

or,

$$\tan\theta_i = \frac{n_2}{n_1} \tag{2.233}$$

This particular angle of incidence, given by (2.233), is called the polarizing angle or Brewster angle θ_B [32]. If unpolarized light is incident at this angle, the electric vector of the reflected light has no component in the plane of incidence. For an air-glass interface, say, $n_1 = 1$, and $n_2 = 1.5$, the Brewster angle $\theta_B = \tan^{-1}(1.5) = 56.3°$.

2.11 From General Ray Optics to Parabasal Optics, Paraxial Optics and Thin Lens Optics

The principles of ray optics enunciated in the earlier sections are sufficient to determine the wavefront or ray geometry in the final or in any intermediate space of an optical system corresponding to an incident wavefront in the initial space by taking recourse to suitable numerical procedures. On the other hand, extensive theoretical analyses on the characteristics of deformation of the wavefront or changes in orientation of the corresponding family of rays have been carried out in the theory of aberrations by making use of symmetry considerations. Nevertheless, all attempts to obtain analytical relations between actual wavefront deformation or changes in corresponding ray geometry and the constructional parameters of a real optical system have floundered, and have not been successful in practice. This has led to the emergence of approximate models providing the required analyticity for facilitating analysis and synthesis of optical systems.

Figure 2.23 shows the passage of a bundle of rays through a general optical system. The qualifier 'general' of the optical system implies that its refracting interfaces can be 'freeform', i.e. they need not necessarily have to comply with some kind of symmetric shape, and obviously such optical systems are characterized by non-existence of any unique axis of rotational symmetry. Characterization of changes in the shape of the wavefront or the corresponding ray configuration in such systems is done by measuring

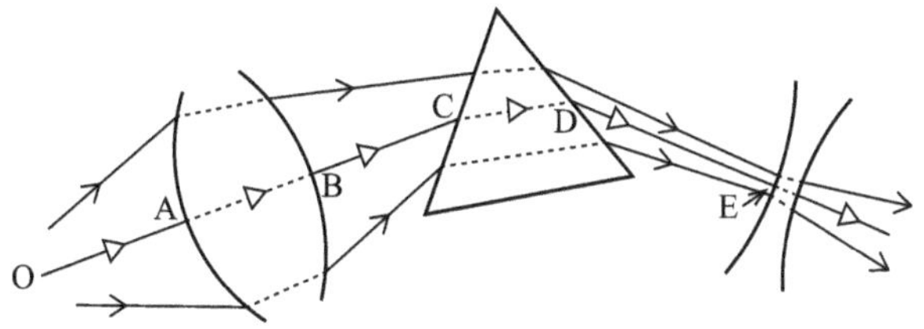

FIGURE 2.23 Base ray OABCO′ of a four parameter family of rays in a general optical system.

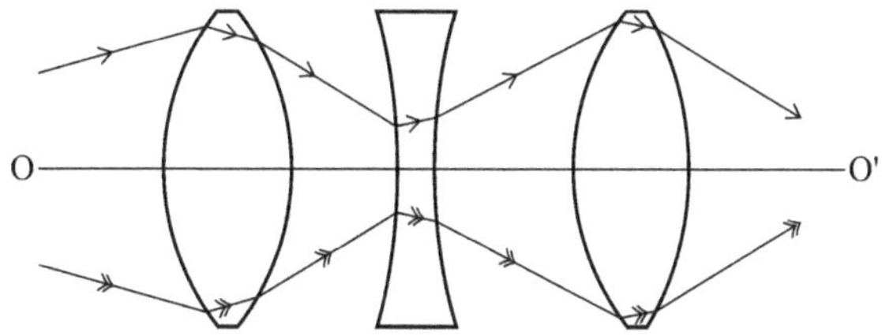

FIGURE 2.24 Optical axis is the base ray for optical systems with rotational symmetry around the axis.

these changes with reference to parameters provided by a suitably chosen central ray of the bundle. OABCO' is one such ray for the ray bundle shown in the figure. This ray is called the *base ray*. In a general optical system, the base ray changes its direction after every refraction. For an axially symmetric optical system, the optical axis plays the role of this 'base ray' so that the direction of the base ray remains unchanged after refractions. (Figure 2.24).

Approximate analytical treatment for image formation by axially symmetric systems is obtained by considering the behaviour of rays that lie in the neighbourhood of this optical axis. The treatment is called '*Paraxial Optics*', where the prefix 'par', derived from Latin, implies 'equal', or 'of the same level'. In memory of Carl Friedrich Gauss, who first carried out a systematic treatment for this type of analysis, the treatment is also called '*Gaussian Optics*'. For general optical systems, similar types of approximate analytical treatments can be developed around the base ray, and such analyses constitute '*Parabasal Optics*'. Indeed, even for axially symmetric optical systems, a better insight on the quality of imaging of off-axial points can be obtained by undertaking an '*extended*' paraxial analysis around the base ray of the oblique bundle of rays involved in the imaging of off-axial points. By symmetry considerations, the obvious choice of base ray in case of a ray bundle originating from an off-axial point is an oblique meridional ray. An orthogonal ray system around the base ray is used to carry out the 'extended paraxial analysis'. The model obtained by this type of analysis is also known as '*First order optics*'. Thomas Young initiated this type of approximate analysis for axially symmetric systems [33–35]. First order optics can also be developed when the base ray is a skew ray [36–39]. It is obvious that the analytical treatment obtained thereby is identical with the parabasal treatment of an asymmetric or general optical system mentioned earlier [40–41]. Allvar Gullstrand first developed a complete theory of first order optics of image formation by general optical systems [42–43]. An extensive review of 'Geometrical Optics' is presented by D.S. Goodman in *OSA Handbook of Optics* [44].

For a conceptual understanding of the behaviour of optical lens systems, a further level of simplification in analysis may be obtained by neglecting the axial thicknesses of individual lens elements in comparison with the radii of curvature of the refracting interfaces and with the diameter of the effective aperture of the lens. Paraxial expressions for the 'thin lens model' of a lens system are obtained from the corresponding paraxial expressions of the real lens system by substituting zero for all axial thicknesses of the real thick lens elements.

Consequently, the change in ray height from one surface to the next of the individual lens elements is also neglected, and the lens element is simulated as a plane surface with optical power to deviate an incident ray path—the angle of deviation being linearly proportional to the height of incidence and power of the lens (Figure 2.25). '*Thin Lens Optics*' provides practical, useful analytical relations even for complex

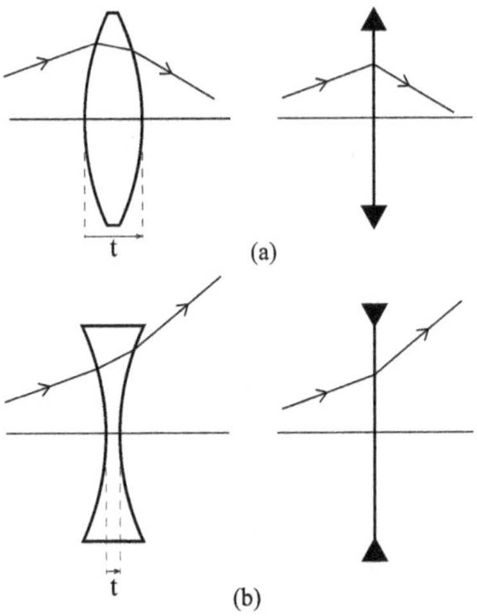

FIGURE 2.25 Ray paths in thin lens model (on the right) of a single thick lens (on the left). (a) a converging lens (b) a diverging lens

lens systems; these relations facilitate understanding of the characteristics of different types of lenses and lens systems, and play an important role in the structural design of optical lens systems.

REFERENCES

1 J.C. Maxwell, 'A dynamical theory of the electromagnetic field', *Phil. Trans. R. Soc. London*, Vol. 155 (1865) pp. 459–512.
2 A. Sommerfeld, *Optics*, Academic Press, New York (1954).
3 R.K. Luneburg, *Mathematical Theory of Optics*, University of California Press, Berkeley (1964).
4 M. Born and E. Wolf, *Principles of Optics*, Cambridge University Press, Cambridge (1999).
5 J.A. Stratton, *Electromagnetic Theory*, McGraw Hill (1941).
6 J.D. Jackson, *Classical Electrodynamics*, Wiley (1998).
7 E.C. Jordan and K.G. Balmain, *Electromagnetic Waves & Radiating Systems*, Prentice-Hall, New Jersey (1968).
8 D.K. Cheng, *Field and Wave Electromagnetics*, Addison Wesley Longman, Singapore (1989).
9 S. Ramo, J.R. Whinnery, and T.V. Duzer, *Fields and Waves in Communication Electronics*, John Wiley, Singapore (1997).
10 E. Hecht, *Optics*, Pearson Education, Delhi (2002).
11 D. Marcuse, *Light Transmission Optics*, Van Nostrand Reinhold, New York (1982).
12 D.J. Griffiths, *Introduction to Electrodynamics*, Prentice-Hall, New Jersey (1989).
13 K.K. Sharma, *Optics*, Elsevier, Amsterdam (2006).
14 M. Kline and I.W. Kay, *Electromagnetic Theory and Geometrical Optics*, Interscience, New York (1965).
15 A.R. Mickelson, *Physical Optics*, Van Nostrand Reinhold, New York (1992).
16 M. Herzberger, *Strahlenoptik*, Springer Verlag, Berlin (1931) [in German].
17 S. Solimeno, B. Crosignani, and P. DiPorto, *Guiding, Diffraction, and Confinement of Optical Radiation*, Academic Press, Orlando (1986).
18 R. Jozwicki, Optyka Instrumentalna, Wydawnictwa Naukowo-techniczne, Warszawa (1970) [in Polish].
19 W.T. Welford, *Aberrations of Optical Systems*, Adam Hilger, Bristol (1986).
20 M. Herzberger, *Modern Geometrical Optics*, Interscience, New York, (1958).
21 S.G. Lipson, H. Lipson, and D.S. Tannhauser, *Optical Physics*, Cambridge University Press, Cambridge (1995).
22 V.G. Veselago, 'The electrodynamics of substances with simultaneously negative values of ε and μ', *Sov. Phys. Usp*. Vol. 10 (1968) pp. 509–514.

23 J.B. Pedry, 'Negative refraction makes a perfect lens', *Phys. Rev. Letts.*, Vol. 85 (2000) pp. 3966–3969.
24 P. Tassin, I. Veretennicoff, and G. Van der Sande, 'Veselago's lens consisting of left-handed materials with arbitrary index of refraction', *Opt. Commun.*, Vol. 264 (2000) pp. 130–134.
25 H.H. Hopkins, 'An extension of Fermat's theorem', *Opt. Acta*, Vol. 17 (1970) pp. 223–225.
26 O.N. Stavroudis, *The Optics of Rays, Wavefronts and Caustics*, Academic, New York (1972).
27 K.D. Möller, *Optics*, University Science Books, California (1988).
28 A.N. Matveev, *Optics*, Mir Publishers, Moscow (1988).
29 C.A. DiMarzio, *Optics for Engineers*, CRC Press, Florida (2012).
30 A. Fresnel, 'Mémoire sur la loi des modifications que la reflexion imprime a la lumière polarisée', l'Académie des sciences, 7 Janvier 1823, reprinted as part of 'Théorie de la lumière – deuxième section', in *Œuvres completes d'Augustine Fresnel*, Tome Premier, Imprimerie Impériale (1866).
31 A. Pramanik, *Electromagnetism: Theory and Applications*, Prentice-Hall, New Delhi (2003).
32 D. Brewster, 'On the laws which regulate the polarization of light by reflection from transparent bodies', *Phil. Trans. Roy. Soc. London*, Vol. 105 (1815) pp. 125–159.
33 T. Young, 'On the mechanism of the eye', *Phil. Trans.* Vol. 91 (1801) pp. 23–88; also in, *Miscellaneous Works of the late Thomas Young*, Vol. I, Ed. G. Peacock, John Murray, London (1855) pp. 12–63.
34 T. Smith, 'The contribution of Thomas Young to geometrical optics, and their application to present-day questions', *Proc. Phys. Soc. B*, Vol. 62 (1949) pp. 619–629.
35 R. Kingslake, *Lens Design Fundamentals*, Academic Press, New York (1978) pp. 190–191.
36 Ref. 3 pp. 234–243.
37 Ref. 16 pp. 77–99.
38 R.A. Sampson, 'A Continuation of Gauss's Dioptrische Untersuchungen', *Proc. London Math. Soc.*, Vol. XXIX, (1897) pp. 33–52.
39 T. Smith, 'Imagery around a skew ray', *Trans. Opt. Soc. (London)*, Vol. 31 (1929–30) pp. 131–156.
40 H.A. Buchdahl, An Introduction to Hamiltonian Optics, Cambridge Univ. Press, London (1970) pp. 26–31.
41 A.E. Siegman, *Lasers*, Univ. Sci. Books, California (1986) pp. 616–623.
42 A. Gullstrand, 'Appendices to Part I. Optical Imagery', in *Helmholtz's Treatise on Physiological Optics*, Ed. J.P.C. Southall, Dover, New York (1962) pp. 261–300.
43 A. Gullstrand, 'Tatsachen und Fiktionen in der Lehre von der optischen Abbildung', *Archive für Opt.* Vol. 1 (1907) pp. 1–41, 81–97.
44 D.S. Goodman, 'General Principles of Geometrical Optics', in *Handbook of Optics*, Vol. I, Optical Society of America, Eds., M. Bass, E.W.V. Stryland, D.R. Williams, and W.L. Wolfe, McGraw-Hill, New York (1995).

3

Paraxial Optics

The prefix 'par' in the word 'paraxial' is derived from Latin, and it means 'equal' or 'near'. Paraxial optics is a gross approximation of real optics, and this model is valid for an infinitesimally small region surrounding the axis of an axially symmetric optical system. During its passage through an axially symmetric optical system, not only is the convergence angle of a paraxial ray—i.e. the angle the ray makes with the optical axis—infinitesimally small, but the perpendicular distance of any point of the ray from the optical axis is also vanishingly small.

Optical imaging means creating a magnified/minified real or virtual reproduction of an object (also either real or virtual) by optical means. In the jargon of optics, the adjective 'real' for a point, line, object, image, etc. implies actual physical existence of the latter, so that it can be directly accessed. On the other hand, when the adjective 'virtual' is applied, it indicates an apparent existence so that the pertinent item appears to exist, but it cannot be directly accessed physically. To make it directly accessible, it is required to use an additional imaging system. Everyday experience of virtual/real images of real/virtual objects as formed by plane mirrors should facilitate clarification of the concepts of 'reality' and 'virtuality' used in practical optics. An optical imaging system transforms the three-dimensional (3-D) object space into a 3-D image space. Both the object space and the image space pervade the whole three-dimensional space within which the imaging system is located. Where light is travelling from left to right, the 3-D space lying on the left of the system is real object space, and the same lying on the right of the imaging system is virtual object space. For the same system, the 3-D space lying on the left of the system is virtual image space, and the same lying on the right of the imaging system is real image space.

3.1 *Raison d'Être* for Paraxial Analysis

A faithful reproduction of an object calls for a one-to-one correspondence between an object point P and the corresponding image point P′. The points P and P′ are called conjugate points. As mentioned above, in general, both/any one of the two points can be real or virtual. From a ray optical point of view, this implies that the role of the imaging system is to ensure that all rays originating from the object point P and passing through the system should converge at the corresponding image point P′. Let C′ be the curve described by the point P′ in the image space corresponding to a curve C described by the point P in the object space. Fidelity of reproduction demands that the curve C′ be geometrically similar to curve C. In the most general case, goals of imaging may be different from exact similarity in geometrical shapes of C and C′; in such a case the fidelity in reproduction of image needs to be based on a suitable measure M, e.g.

$$M = \left| C'_{target} - C'_{obtained} \right| \tag{3.1}$$

where C'_{target} is the target or desired value for C′, and $C'_{obtained}$ is the value obtained for C′ by the imaging system. Additionally, in the case of imaging systems operating over an extended range of wavelengths, the relative spectral composition of the object should remain unaltered in the reproduced image. Again, in general, a prespecified relative spectral composition may be a desired goal in the image corresponding to a given relative spectral composition in the object.

DOI: 10.1201/9780429154812-3

Indeed, formation of high-fidelity images, as enunciated above, is a tall order for an optical imaging system, except for in special cases. This is so because not all rays originating from an object point are taking part in image formation. Firstly, because of the finite size of the aperture of a real imaging system, a truncated part of the incident pencil of rays is transmitted by the system to the image plane. Next, in general, all rays of this finite pencil do not all converge to a single point that can be unequivocally called an image point corresponding to the object point P. Indeed, what one obtains in practice is convergence of the image forming pencil of rays to a small 3-D region in the image space. It is obvious that the choice of a specific point of this region as the image point cannot be a unique one. Consequently, the choice of a specific plane as image plane can also not be made uniquely. One of the primary tasks of image analysis and synthesis is to develop suitable measures for characterizing these image defects.

On the other hand, from the point of view of a practical user, it is necessary to develop a set of specifications pertaining to the imaging functions to be carried out by a lens system. These specifications need to be unique for a particular imaging function; obviously, all lens systems contending for the job need to satisfy primarily these specifications as best as possible. Usually, at the stage of laying out the specifications, imaging is assumed to be 'perfect'. In practice, imperfection in imaging performed by the lens systems is treated separately in the next level of analysis and synthesis.

Paraxial analysis of real lens systems provides the much-sought correspondence with the unique specifications mentioned above. Specifications like size/location of object/image, object-to-image throw, magnification, relative aperture in terms of F-no., or numerical aperture are used to indicate particular imaging functions to be carried out by the system, and paraxial treatment of real lens systems provides, in a relatively simple manner, these characteristic specifications for them.

For the sake of conformity with publications by H.H. Hopkins and his coworkers, in what follows we have mostly used similar notations for different optical quantities. However, in sign convention for paraxial angles, we adopt the current practice of following the convention of analytical geometry. This is opposite of the convention for paraxial angles followed earlier by Smith, Conrady, Hopkins, Wynne, and many others. The approximation analysis for paraxial imagery follows the treatment of Hopkins [1–2].

3.2 Imaging by a Single Spherical Interface

In Figure 3.1, O is an object point, and AQ is a spherical refracting interface between two optical media of refractive index n and n′. The axis ray OACO′ passes undeviated through the centre of curvature C. The radius of curvature of the spherical surface AQ is r. A ray OP is incident on the point P of the interface, and the refracted ray PV′ intersects the axis at V′. PQ is perpendicular to P on the axis. $PQ = \hat{Y}$, and AQ = s. As the point P is taken increasingly closer to the vertex A of the surface AQ, the point of intersection V' tends to a limiting position O'.

This limiting point O is said to be the paraxial image of the object point O. Alternatively, the point O′ is sometimes called the paraxial focus of the refracted pencil of rays. Since AV′ and PV′ are normals to the

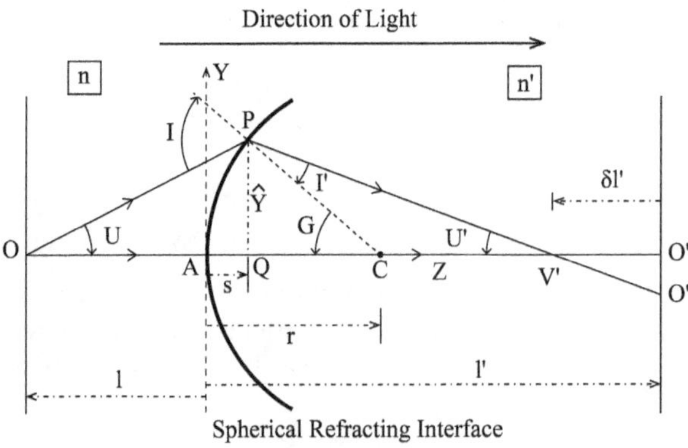

FIGURE 3.1 Image of a point on the axis by a single refraction; Meridional YZ section.

refracted wavefront, O′ is the point of intersection of two limitingly close normals around the axis, and so O′ is the centre of curvature of the part of the refracted wavefront in the immediate neighbourhood of the axis.

A ray at finite aperture will, in general, intersect the axis ray at a point, such as V′, and not at O′. This ray will intersect the transverse plane at O′ at a point O″. For different positions of the point P on the refracting interface, the incident ray OP will have different angles of convergence, and so the corresponding refracted rays PV′ will follow different paths, so that the location of the point V′ on the axis will be different. The deviation $\delta l''$ of the ray intersection point V′ from the paraxial image point O′ along the longitudinal axis is a measure of the non-paraxial behaviour of the rays at finite apertures. $\delta l'$ is a function of the convergence angle of the incident ray U, or equivalently of the ray height on the interface, $\hat{Y}$. Similarly, the deviation O′O″ along the transverse direction can be used as a measure of the non-paraxial behaviour of the finite rays.

As noted earlier, the concept of a paraxial ray is a mathematical abstraction; *a paraxial ray cannot exist in practice*. This is so because no matter how close P is taken to A, the ray PV′ will not actually pass through the point O′ unless P is made coincident with A.

3.3 Sign Convention

So far, no universally followed sign convention exists in ray optics. Over the years, different investigators, at times upholding their justification for a particular choice, followed different conventions, but often it was a matter of personal discretion. A certain degree of caution on sign convention is warranted while following the equations of ray optics given in optics literature. Currently there is a growing trend to follow a convention, which is in conformity with coordinate geometry. In this book, the same is adopted. The convention followed is:

(i) The direction of the travel of light is from left to right, unless otherwise stated.

(ii) The point of intersection of the refracting interface and the axis is taken as the local origin of a coordinate system. In Figure 3.1, O is the local origin with rectangular coordinates (0, 0, 0). The Z-axis is along the optical axis; the XY plane is tangential to the interface at the origin O. Figure 3.1 shows the ray geometry in the meridional YZ plane. Looking from the object side, a right-handed coordinate system is adopted for determining the unique direction of the X-axis.

(iii) All distances are to be measured from O as directed segments. Distances to the top, or to the right of O are taken as positive. For example, in Figure 3.1, object distance OA (= l) is negative; the height of the point of incidence P on the interface (= Y), and the image distance OA′ (= l′) are positive. The radius of curvature AC (= r) is positive.

(iv) The sign of the angle of convergence of a ray, or a normal with respect to the axis, e.g., angles U, U′, and G, follow the convention of coordinate geometry. If a ray or a normal can be made coincident with the axis by a clockwise rotation of less than 90^0, the angle is taken as positive. Accordingly, in Figure 3.1, the angle U is positive, and the angles U′ and G are negative. A useful mnemonic for remembering the sign convention is *'RANA'*, from Ray to Axis, and from Normal to Axis. *'RANA'* is the Latin word for 'frog', and it implies a 'king' in many Indo-European languages.

(v) The angle of incidence I or the angle of refraction I′ is taken as positive, if the incident ray or the refracted ray can be made coincident with the normal by a clockwise rotation of less than 90^0.

Alternatively, the signs of I and I′ may be taken to be those defined by the relations

$$I = U - G \tag{3.2}$$

$$I' = U' - G \tag{3.3}$$

The relations (3.2) and (3.3) may be derived from the geometry of triangles OPC and CPV′ in Figure 3.1, and using the signs of U, U′, and G. Accordingly, in Figure 3.1, the angles I and I′ are positive. A useful mnemonic for this sign convention is *'RON'*, from Ray to Normal.

3.4 Paraxial Approximation

3.4.1 On-Axis Imaging

The paraxial formula relating the object distance l and image distance l′ to the radius of curvature r, and the refractive indices n and n′ are obtained below. The law of refraction for the ray OPV′ gives

$$n' \sin I' = n \sin I \tag{3.4}$$

$\sin x$ can be expanded as

$$\sin x = x - \frac{x^3}{3!} + \frac{x^5}{5!} - \ldots = x + O(3) \tag{3.5}$$

where $O(3)$ denotes a quantity containing only terms of degree 3 and above in x. Using (3.5), (3.4) can be expressed as

$$n'I' = nI + A(3) \tag{3.6}$$

where $A(3)$ denotes a quantity containing only terms of degree ≥ 3 in aperture variable, e.g. U, $\hat{Y}$, or G, etc.

Using (3.2) and (3.3), this relation becomes

$$n'(U' - G) = n(U - G) + O(3) \tag{3.7}$$

From Δ QPC in Figure 3.1, we get

$$\sin G = -\frac{Y}{r} \tag{3.8}$$

since $\hat{Y}$ and r are positive, and G is negative.

Expanding $\sin G$ we get

$$G = -\frac{\hat{Y}}{r} + A(3) \tag{3.9}$$

Similarly, from Δ OPQ of Figure 3.1

$$\tan U = \frac{PQ}{OQ} = \frac{\hat{Y}}{-1+s} = -\frac{\hat{Y}}{1+s} \tag{3.10}$$

since U and $\hat{Y}$ are positive, and l is negative.

From ΔQPV′ of Figure 3.1

$$\tan U' = \frac{PQ}{QV'} = \frac{Y}{l' + \delta l' - s} \tag{3.11}$$

since l' and $\hat{Y}$ are positive, and U′ is negative.

From the relations (3.10) and (3.11), it follows

$$U = -\frac{\hat{Y}}{l - s} + A(3) \tag{3.12}$$

$$U' = -\frac{\hat{Y}}{l' + \delta l' - s} + A(3) \tag{3.13}$$

(3.7) may be written

$$n'\left(\frac{1}{l' + \delta l' - s} - \frac{1}{r}\right) = n\left(\frac{1}{l - s} - \frac{1}{r}\right) + A(3) \tag{3.14}$$

The sag s for the point P on the spherical surface AQ can be expressed as

$$s = \frac{1}{2r}\hat{Y}^2 + A(4) = A(2) \tag{3.15}$$

$\delta l'$ can be expressed in a power series of the aperture variable $\hat{Y}$ as

$$\delta l'(\hat{Y}) = a_0 + a_1\hat{Y} + a_2\hat{Y}^2 + A(3) \tag{3.16}$$

By definition

$$\delta l' = 0 \text{ when } \hat{Y} = 0 \tag{3.17}$$

Also, by symmetry,

$$\delta l'(\hat{Y}) = \delta l'(-\hat{Y}) \tag{3.18}$$

The relations (3.17) and (3.18) imply that in (3.16)

$$a_0 = a_1 = 0 \qquad (3.18)$$

Therefore,

$$\delta l' = a_2\hat{Y}^2 + A(3) = A(2) \tag{3.19}$$

In view of (3.15) and (3.19), the relation (3.14) reduces to

$$n'\left(\frac{1}{l'} - \frac{1}{r}\right) = n\left(\frac{1}{l} - \frac{1}{r}\right) + A(2) \tag{3.20}$$

The required paraxial formula is, therefore,

$$\frac{n'}{l'} - \frac{n}{l} = \frac{(n' - n)}{r} \tag{3.21}$$

since, in paraxial approximation, terms involving powers of aperture higher than the first are neglected. The above relation can be looked at two different ways. Regarded as an equation relating the axial locations of the centres of curvature O and O′ of the incident and refracted wavefronts, respectively, the relation (3.21) is exact. On the other hand, regarded as an approximate formula for the focus of a finite ray, (3.21) has errors of the order of square of the aperture.

3.4.1.1 Power and Focal Length of a Single Surface

The right-hand side of the expression (3.21) gives power K of a single refracting surface.

$$K = (n' - n)c \tag{3.22}$$

Power K is positive when both c and (n′ - n) is of the same sign. In Figure 3.2(a) and Figure 3.2(d), power of the surface is positive, whereas in Figure 3.2(b) and Figure 3.2(c), the power of the surface is negative.

In the former case, the surface is converging, and in the latter case, it is diverging. Note that the dimension of power is $[\text{Length}]^{-1}$. In instrumental or ophthalmic optics, the unit of power is usually given as $[\text{m}]^{-1}$. The latter is given a special name called diopter, denoted by D or Δ.

(3.21) can be rewritten as

$$\frac{n'}{l'} - \frac{n}{l} = K \tag{3.21}$$

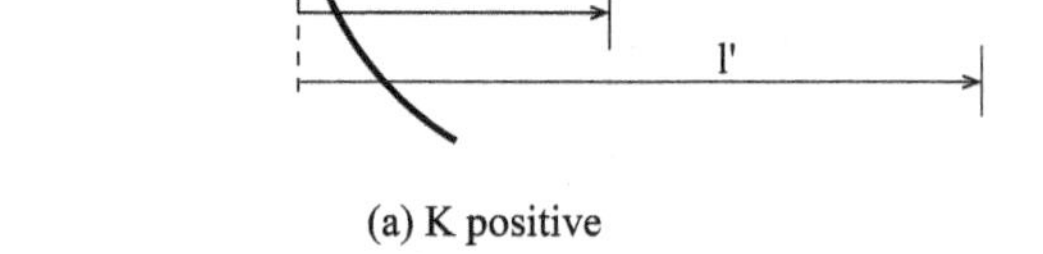

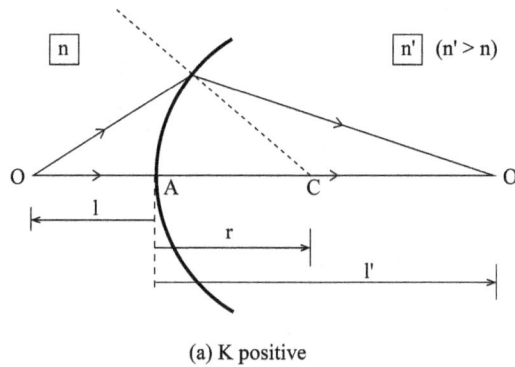

FIGURE 3.2(a) A surface of positive power.

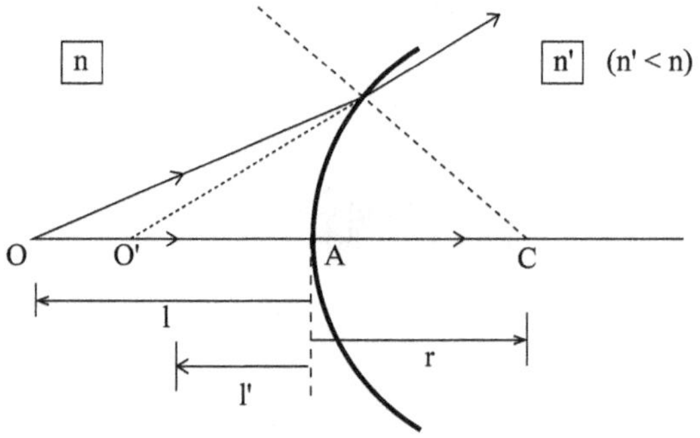

FIGURE 3.2(b) A surface of negative power.

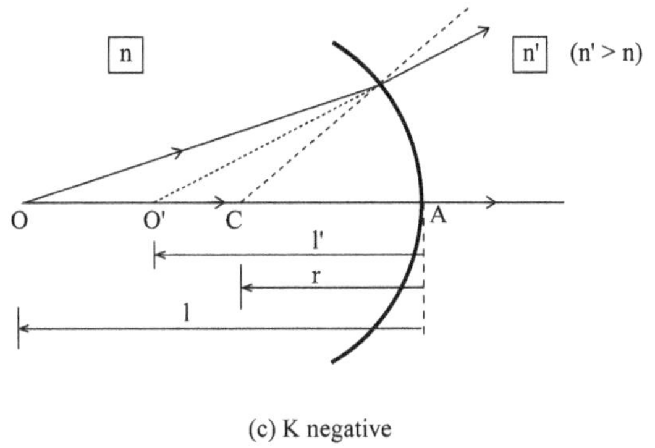

FIGURE 3.2(c) A surface of negative power.

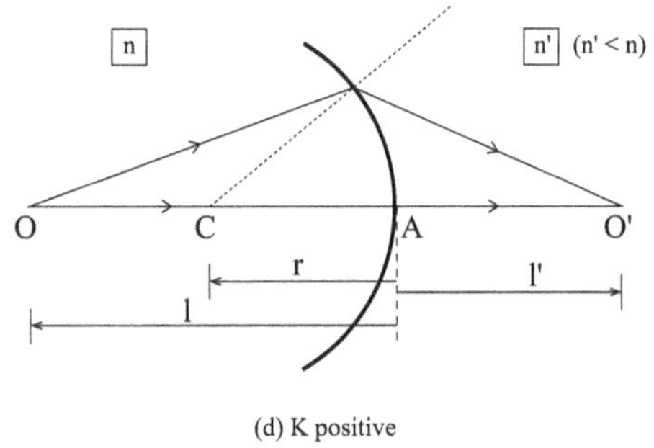

FIGURE 3.2(d) A surface of positive power.

The reciprocal of power K is called the equivalent focal length F, i.e.

$$F = \frac{1}{K} \tag{3.22}$$

Image space focal length, or second focal length f′, is defined as the image distance corresponding to an axial object point at infinity, i.e. $l = \infty$. From (3.21), f′ for a single spherical refracting interface (Figure 3.3(a)) is given by

$$f' = \frac{n'}{K} = \frac{n'}{c(n' - n)} = n'F \tag{3.23}$$

Similarly, the object space focal length, or first focal length f, is equal to the object distance corresponding to the image distance $l' = \infty$ (Figure 3.3 (b)). It is given by

$$f = -\frac{n}{K} = -\frac{n}{c(n' - n)} = -nF \tag{3.24}$$

Note that for a single refracting interface, $|f'| \neq |f|$ since $n \neq n'$.

3.4.2 Extra-Axial Imaging

In Figure 3.4, Q is a point on the object plane OQ perpendicular to the axis through O. The spherical refracting interface AP, between optical media of refractive index n and n′, has its centre at C, and its

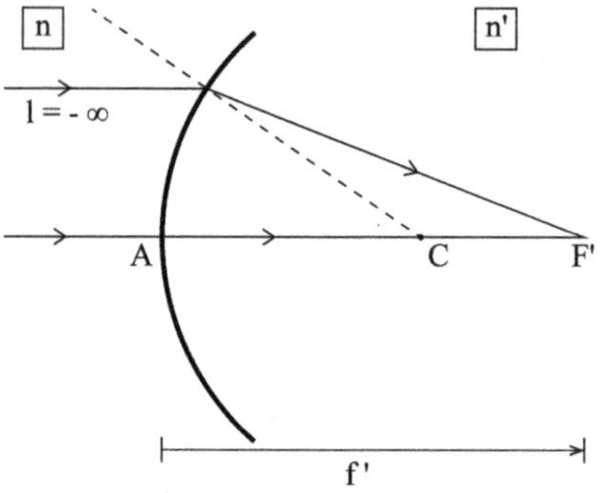

FIGURE 3.3(a) Image space focal length of a single refracting surface, $f' = n'F = n'\left(\frac{1}{K}\right)$.

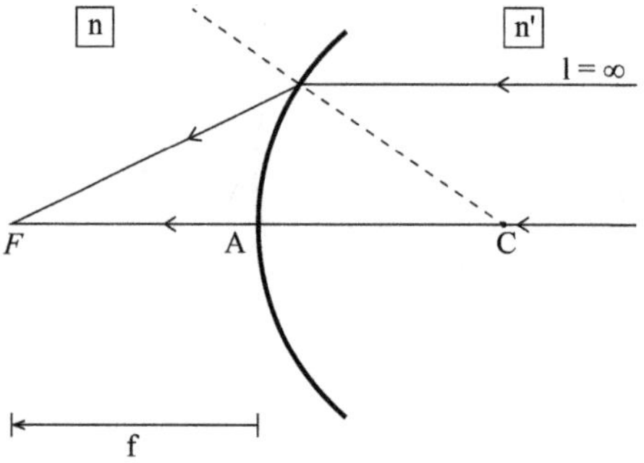

FIGURE 3.3(b) Object space focal length of a single refracting surface, $f = -nF = -n\left(\frac{1}{K}\right)$.

vertex on the axis at A. O′ is the paraxial image of the axial object point O. The paraxial image plane is taken as the plane perpendicular to the optical axis at O′.

The ray Q C Q_c', passing through the centre of curvature C, is undeviated and intersects the paraxial image plane at Q_c'. Let OQ = η, and O′ $Q_c' = \eta_c'$. From the similar triangles Δ OCQ and Δ O′ C Q_c', it is seen that

$$\frac{\eta_c'}{\eta'} = \frac{(l'-r)}{(l-r)} = \frac{nl'\left\{n'\left(\frac{1}{r}-\frac{1}{l'}\right)\right\}}{n'l\left\{n\left(\frac{1}{r}-\frac{1}{l}\right)\right\}} \tag{3.25}$$

which, by (3.21) above, gives

$$\eta_c' = \left(\frac{nl'}{n'l}\right). \tag{3.26}$$

Consider the ray QAQ_A', which passes through the vertex A of the refracting interface. For this ray, G = 0, and therefore, $I = U_0$, and $I' = U_0'$. This ray intersects the paraxial image plane at a point Q_A', which is different from Q_C'. Let $O'Q_A' = \eta_A'$.

From Δ OAQ and Δ $O'AQ_A'$ in the diagram,

$$\tan U_0 = \frac{\eta}{l} \quad \text{and} \quad \tan U_0' = \frac{\eta_A'}{l'} \tag{3.27}$$

For an infinitesimally small value of η,

$$I_0 = U_0 = \frac{\eta}{l} + F(3) \tag{3.28}$$

$$I_0' = U_0' = \frac{\eta_A'}{l'} + F(3) \tag{3.29}$$

where $F(3)$ denotes a quantity containing only terms of degree ≥ 3 in field variable, e.g. η, U_0, etc. By applying the law of refraction, it follows

$$\frac{n'\eta_A'}{l'} = \frac{n\eta}{l} + F(3) \tag{3.30}$$

so that

$$\eta_A' = \left(\frac{nl'}{n'l}\right)\eta + F(3) \tag{3.31}$$

gives the height of intersection at the paraxial image plane of the ray refracted at A. For limitingly small values of η, the two rays considered above will focus more and more nearly on the image plane with the same image height given by

$$\eta_A' \approx \eta_C' \approx \eta' = \left(\frac{nl'}{n'l}\right)\eta \tag{3.32}$$

However, η' as given above defines an exact value for the paraxial image height corresponding to the object point Q. The corresponding paraxial magnification is

$$M = \frac{\eta'}{\eta} = \frac{nl'}{n'l} = \left(\frac{n}{n'}\right)\left(\frac{l'}{l}\right) \tag{3.33}$$

This is the transverse magnification for an infinitesimally small element of object lying normal to the optical axis.

It should be noted that the actual rays will not, in general, focus on the paraxial image plane. On the paraxial image plane,

$$OQ_A' = \eta_A' = OQ_C' + Q_C'Q_A' = \eta_C' + \delta\eta' = \eta' + \delta\eta' \tag{3.34}$$

In Figure 3.4, it is seen that the refracted ray AQ_A' intersects the auxiliary axis QCQ_C' at a point T′. $\delta\eta'$ can be approximately given by

$$\delta\eta' \cong (T'Q_C')\delta U' \tag{3.35}$$

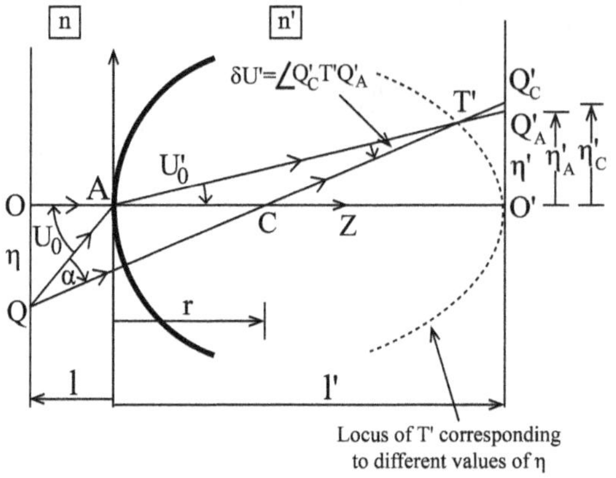

FIGURE 3.4 Imaging of an extra-axial object point Q by a spherical refracting interface.

Expressing the aperture angle $\angle AQC = \alpha$, it may be noted that $\delta U'$ is a function of α. By symmetry,

$$\delta U'(+\alpha) = -\delta U'(-\alpha) \tag{3.36}$$

Therefore

$$\delta U' = A(1) \tag{3.37}$$

$T'Q'_C$ is a function of η, and from consideration of symmetry, it follows

$$T'Q'_C(+\eta) = T'Q'_C(-\eta) \tag{3.38}$$

Also, it is seen that $T'Q'_C = 0$, when $\eta = 0$. Therefore

$$(T'Q'_C) = F(2) \tag{3.39}$$

Using (3.34), (3.35), (3.37), and (3.39) it follows

$$\eta'_A = \eta' + F(2)A(1) \tag{3.40}$$

where $[F(2)A(1)]$ denotes a quantity that is of at least degree 2 in the object or image height, and of not less than degree 1 in the aperture.

3.4.3 Paraxial Ray Diagram

In Figure 3.5, a paraxial ray diagram for the case of imaging by a single spherical surface is presented.

O′ is the paraxial image of the point O on the axis, and Q′ is the paraxial image of the extra-axial point Q. Note that Q′ is the same as Q_C' of Figure 3.4. The subscript is dropped as all paraxial rays from Q are supposed to meet at Q′. OQ and O′Q′ lie on two planes transverse to the optical axis OO′ at O and O′, respectively. As elucidated above, this paraxial model of imaging is valid if and only if it is assumed that all angles U, U′, G, I, and $I' \rightarrow 0$, and heights η and $\eta' \rightarrow 0$. Indeed, the paraxial model gives an exact description of imaging if only the rays lying near the axis are taken into consideration. However, for the sake of illustration, in paraxial ray diagrams, the ray paths have to be shown such that they have large angles with the axis, or the distance of the ray path from the axis is not negligible. Often, to emphasize

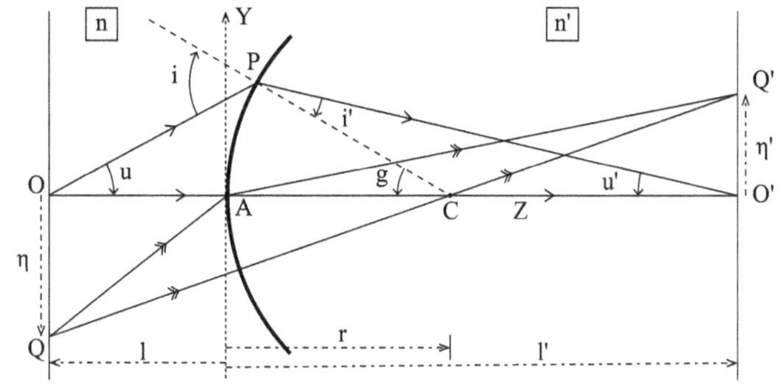

FIGURE 3.5 Paraxial Ray Diagram for imaging of object OQ at O′Q′ by a single spherical refracting interface.

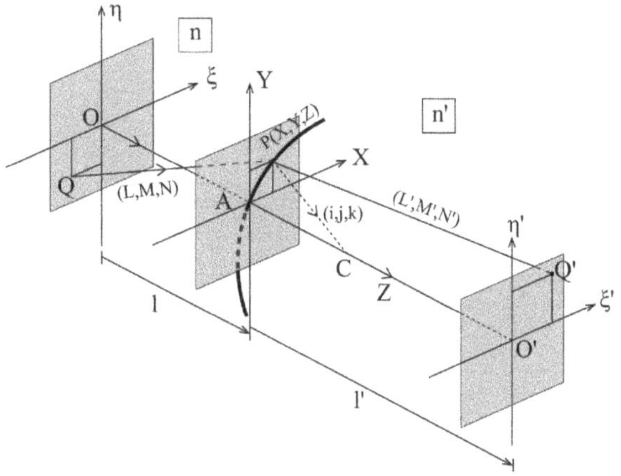

FIGURE 3.6 Paraxial imaging by a smooth surface of revolution about the axis.

the 'paraxial' nature of these ray diagrams, it is customary to use corresponding lowercase letters for the convergence angles or the angles of incidence and refraction for underscoring the paraxial nature of these angles.

For the sake of the analysis and synthesis of optical systems, paraxial ray diagrams, if suitably formulated, can provide practically useful estimates for the actual ray paths. Comparing Figure 3.1 with Figure 3.5, it may be noted that for the incident ray OP, the paraxial ray path PO′ for the refracted ray is a close estimate for the actual ray path PV′. Similarly, from Figure 3.4 and Figure 3.5, it is seen that for the incident ray QA, the paraxial ray path AQ′ is a close estimate for the actual ray path $AT'Q_A'$ of the refracted ray.

3.4.4 Paraxial Imaging by a Smooth Surface of Revolution about the Axis

In Figure 3.6, Q is a point on a plane OQ, which is normal to the axis AO of a smooth surface of revolution of which AP is a normal section. The vertex A of the surface AP is taken as the origin of the right-handed rectangular coordinate system, with the Z-axis along the positive direction of the axis, and the X- and Y-axes lying on the tangent plane at A. The corresponding rectangular axes on the plane OQ are ξ and η, respectively. Let the axial distance AO be equal to l. QP is a ray from the point Q, intersecting the refracting surface at P. Let (X, Y, Z) be the coordinates of P, (ξ, η, l) be the coordinates of Q, and (L, M, N) be the direction cosines of QP. The equation of the line QP is

$$\frac{\xi - X}{L} = \frac{\eta - Y}{M} = \frac{l - Z}{N} \tag{3.41}$$

Considering L, M, and N as functions of (ξ, η, X, Y), the following conclusions can be drawn from geometrical considerations:

$$L(0,0,0,0) = 0 \tag{3.42}$$

$$M(0,0,0,0) = 0 \tag{3.43}$$

$$N(0,0,0,0) = 1 \tag{3.44}$$

$$L(\xi,\eta,X,Y) = -L(-\xi,-\eta,-X,-Y) \tag{3.45}$$

$$M(\xi,\eta,X,Y) = -M(-\xi,-\eta,-X,-Y) \tag{3.46}$$

$$N(\xi,\eta,X,Y) = N(-\xi,-\eta,-X,-Y) \tag{3.47}$$

Therefore, it follows

$$L(\xi,\eta,X,Y) = D(1) \tag{3.48}$$

$$M(\xi,\eta,X,Y) = D(1) \tag{3.49}$$

$$N(\xi,\eta,X,Y) = 1 + D(2) \tag{3.50}$$

$$\frac{L}{N} = \frac{M}{N} = D(1) \tag{3.51}$$

where $D(n)$ denotes any quantity containing only terms of total degree ≥ n in any, or all of the variables (ξ,η,X,Y).

The surface of revolution AP can be represented by an equation of the form

$$Z = f(X,Y) \tag{3.52}$$

From Figure 3.6 it is apparent that at the origin A,

$$Z = f(0.0) = 0 \tag{3.53}$$

Also, by symmetry,

$$Z(X,Y) = Z(-X,-Y) \tag{3.54}$$

From (3.52) and (3.53) it follows

$$Z = D(2) \tag{3.55}$$

Using (3.50), (3.51), and (3.55), from (3.41) one obtains

$$\xi = X + lL\left(\frac{1}{N}\right) - Z\left(\frac{L}{N}\right) = X + lL - lD(1)D(2) + D(2)D(1) \tag{3.56}$$

Thus

$$\xi = X + lL + D(3) \tag{3.57}$$

Similarly, η can be obtained as

$$\eta = Y + lM + D(3) \tag{3.58}$$

Let the refracted ray PQ′ cut the plane O′Q′ at a distance l′ from A, at the point Q′. Coordinates of the point Q′ are (ξ', η', l'). If (L′, M′, N′) are the direction cosines of the refracted ray PQ′, ξ' and η' can be expressed as

$$\xi' = X + l'L' + D(3) \tag{3.59}$$

$$\eta' = Y + l'M' + D(3) \tag{3.60}$$

Let the refractive index of the object space be n, and that of the image space be n′. Multiplying both sides of equation (3.57) by (n/l), and of equation (3.59) by (n′/l′), and subtracting the former from the latter, one obtains

$$\frac{n'\xi'}{l'} - \frac{n\xi}{l} = X\left(\frac{n'}{l'} - \frac{n}{l}\right) + n'L' - nL + D(3) \tag{3.61}$$

Similarly,

$$\frac{n'\eta'}{l'} - \frac{n\eta}{l} = Y\left(\frac{n'}{l'} - \frac{n}{l}\right) + n'M' - nM + D(3) \tag{3.62}$$

Relations (3.61) – (3.62) are valid for any plane O′Q′.

The laws of refraction (2.186) – (2.187) of Chapter 2 give

$$n'L' - nL = i\Delta(n\cos I) \tag{3.63}$$

$$n'M' - nM = j\Delta(n\cos I) \tag{3.64}$$

where (i, j, k) are the direction cosines of the normal PC to the refracting surface at P. From consideration of symmetry, the equation of the surface of revolution AP in the neighbourhood of A is of the form

$$Z = \frac{1}{2}c\left(X^2 + Y^2\right) + D(4) \tag{3.65}$$

where c is the curvature of the surface at vertex A. The corresponding radius of curvature is $r = (1/c)$. The equation of the surface AP can, therefore, be written in the form

$$F(X,Y,Z) = Z - \frac{1}{2}c\left(X^2 + Y^2\right) + D(4) = 0 \tag{3.66}$$

The direction cosines (i, j) of the normal PC are given by

$$i = \frac{\partial F}{\partial x} = -cX + D(3) \tag{3.67}$$

$$j = \frac{\partial F}{\partial Y} = -cY + D(3) \tag{3.68}$$

From (3.63) and (3.64), it follows

$$n'L' - nL = i(n'\cos I' - n\cos I) = i(n' - n) + D(2) \tag{3.69}$$

$$n'M' - nM = j(n'\cos I' - n\cos I) = j(n' - n) + D(2) \tag{3.70}$$

since $\cos I$ and $\cos I$, are of the form [1+ D(2)]. Using relations (3.67) – (3.70), equations (3.61) – (3.62) give

$$\frac{n'\xi'}{l'} - \frac{n\xi}{l} = X\left\{\left(\frac{n'}{l'} - \frac{n}{l}\right) - c(n' - n)\right\} + D(3) \tag{3.71}$$

$$\frac{n'\eta'}{l'} - \frac{n\eta}{l} = Y\left\{\left(\frac{n'}{l'} - \frac{n}{l}\right) - c(n' - n)\right\} + D(3) \tag{3.72}$$

for $X = D(1), Y = D(1)$, and $D(1)D(2) = D(3)$.

For any ray QPQ′ lying limitingly close to the axis, terms of D (3) may be neglected. If the image plane is chosen such that

$$\left(\frac{n'}{l'} - \frac{n}{l}\right) = c(n' - n) = \frac{(n' - n)}{r} \tag{3.73}$$

the coordinates (ξ', η') of Q′ are given by

$$\xi' = \left(\frac{nl'}{n'l}\right)\xi \qquad \eta' = \left(\frac{nl'}{n'l}\right)\eta \tag{3.74}$$

It is to be noted that, under this approximation, all rays from Q (ξ, η) meet at the point Q′ (ξ', η'). The image plane chosen by the condition (3.73) is the paraxial image plane corresponding to a spherical refracting interface with the same vertex curvature, as shown in (3.21). An element of object at O is imaged at O′ as a geometrically similar figure with a magnification

$$M = \frac{\eta'}{\eta} = \frac{\xi'}{\xi} = \frac{nl'}{n'l} \tag{3.75}$$

as shown earlier in (3.33) for a spherical refracting interface.

Thus, for the smooth surface of revolution, the paraxial imaging characteristics are similar to those of a spherical refracting interface with the same vertex curvature. The smooth surfaces of revolution considered above include those generated by the rotation of the conic sections—namely, ellipses, hyperbolae, and parabola—about their axes. However, they do not include surfaces with cusps, so that a cone with axis along AO is not included in this treatment. Optical elements like axicons need to be treated separately.

3.5 Paraxial Imaging by Axially Symmetric System of Surfaces

Almost all optical systems consist of a number of refracting and/or reflecting surfaces. In most cases, the latter are smooth surfaces of revolution, and they are spaced along a common axis so that the optical system is axially symmetric. In the case of spherical interfaces, it is sufficient to have the centres of the spheres lie on a straight line that becomes an optical axis for the system. For other surfaces of revolution, the axes of the surfaces need to be coincident, and this common axis becomes the optical axis of the axially symmetric system.

In order to find the location and magnification of the image of an object formed by an axially symmetric system, the formulae (3.21) and (3.33) may be applied at successive surfaces.

3.5.1 Notation

In order to forestall occurrence of misleading ambiguity, it is necessary to use a consistent notation for dealing with a system of several surfaces. The key features of the adopted notation are:

(i) Vertices of the surfaces of a system consisting of k surfaces are labelled as A_s, $s = 1,\ldots,k$.
(ii) Every quantity representing a characteristic of the system, e.g., curvature, refractive index, or separation, etc., is given a subscript to denote the surface number with which it is associated.
(iii) Each quantity corresponding to the paraxial ray, e.g., convergence angle, height on a surface, or distance of the object/image corresponding to refraction at the specific surface is given a subscript that denotes the associated surface.
(iv) For each surface, the image space quantities are primed.

Since the image space for surface s becomes object space for surface (s+1), some of the quantities can be denoted in more than one way, e.g. refractive index of the medium after the s-th surface can be denoted by $n'_s \equiv n_{s+1}$. Similarly, $u'_{s-1} = u_s$, $d'_s = -d_{s+1}$, etc. Note that $d'_s = \overrightarrow{A_sA_{s+1}}$, and $d_{s+1} = \overrightarrow{A_{s+1}A_s}$. Figure 3.7 illustrates this scheme for notation.

3.5.2 Paraxial Ray Tracing

Figure 3.8 shows the passage of a paraxial ray originating from an axial object point O_1 through the first two surfaces of an axially symmetric system. The first refracting surface with axial radius of curvature r_1 forms an image O'_1 of the object point O_1. $l_1 = A_1O_1$ is the object distance. The image distance $l'_1 = A_1O'_1$ is determined by using the refraction relation (3.21) or (3.73) as

$$\frac{n'_1}{l'_1} - \frac{n_1}{l_1} = \frac{(n'_1 - n_1)}{r_1} \tag{3.76}$$

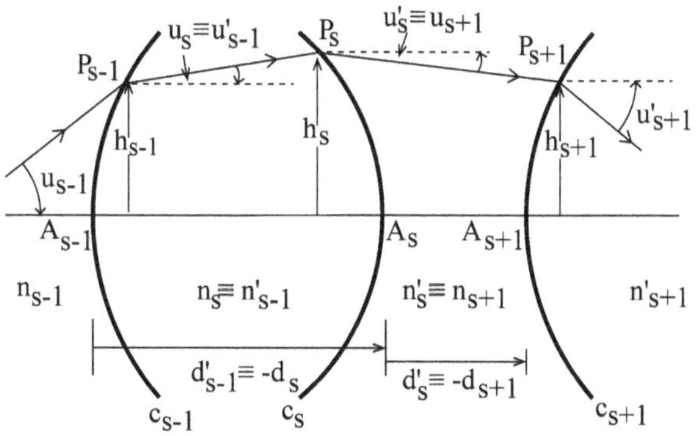

FIGURE 3.7 Notations for paraxial parameters in imaging by axially symmeric system of surfaces.

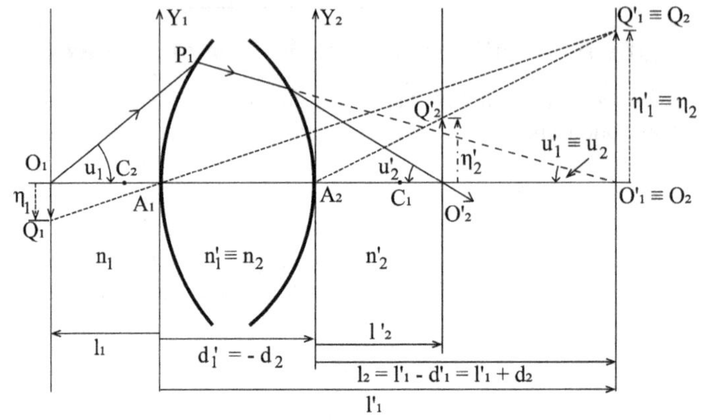

FIGURE 3.8 Passage of a paraxial ray through an axially symmetric system of surfaces.

For the second refracting surface at A_2, the object is at $O_1' \equiv O_2$, and the object distance l_2 is given by

$$l_2 = l_1' - d_1' \tag{3.77}$$

The second refracting surface forms the image of O_2 at O_2' . The corresponding image distance l_2' is given by

$$\frac{n_2'}{l_2'} - \frac{n_2}{l_2} = \frac{(n_2' - n_2)}{r_2} \tag{3.78}$$

By dropping the subscripts, the refraction and transfer formulae can be written conveniently as

$$\frac{n'}{l'} - \frac{n}{l} = \frac{(n' - n)}{r} \tag{3.79}$$

$$l_+' = l' - d' \tag{3.80}$$

using the subscript + to denote the quantity relating to the next surface.

Successive use of these formulae to successive surfaces of the axially symmetric system consisting of k surfaces gives the distance $l_k' = A_k O_k'$, where O_k' is the image formed by the system of k surfaces. The magnification M of a given object $O_1 Q_1 = \eta_1$, produced by the system of k surfaces, will be given by

$$M = \frac{\eta_k'}{\eta_1} = \left[\frac{\eta_1'}{\eta_1} \frac{\eta_2'}{\eta_2} \cdots\cdots\cdots \frac{\eta_{k-1}'}{\eta_{k-1}} \frac{\eta_k'}{\eta_k} \right] \tag{3.81}$$

Note that $\eta_1' = \eta_2, \eta_2' = \eta_3, \ldots\ldots\ldots \eta_{k-1}' = \eta_k$.

Using (3.75) at each of the k surfaces, (3.81) may be written as

$$M = \left(\frac{n_1}{n_k'} \right) \left\{ \frac{l_1'}{l_1} \frac{l_2'}{l_2} \cdots\cdots\cdots \frac{l_{k-1}'}{l_{k-1}} \frac{l_k'}{l_k} \right\} \tag{3.82}$$

since $n_1' = n_2, n_2' = n_3, \ldots\ldots\ldots n_{k-1}' = n_k$.

As it is, the paraxial refraction, transfer, and magnification formulae, given above in (3.79), (3.80), and (3.82), respectively, become indeterminate whenever l_1 or any of $l'_s, s = 1,\ldots,k$ takes values tending to ∞ or zero.

This problem is circumvented by using paraxial convergence angle variable u and paraxial height variable h, instead of the object/image distance variables l and l′. Note that the paraxial refraction equation (3.79) is an exact relation between l, the distance of the centre of curvature of the axial element of the incident wavefront from the vertex of the refracting surface, and l′, the distance of the centre of curvature of the axial element of the refracted wavefront. These centres of curvature are the axial object and image points. The transfer formula (3.80) is also exact; it merely expresses a shift of origin of the coordinate a distance d′ along the optical axis.

These formulae remain exact even if a constant multiplier, say h, multiplies both sides of a relation. Following the sign convention adopted for paraxial angles, a paraxial angle variable u is defined as

$$u = -\frac{h}{l} \tag{3.83}$$

u′ is given by

$$u' = -\frac{h}{l'} \tag{3.84}$$

Multiplying both sides of (3.79) by h, and substituting from (3.83) and (3.84), we get the new form of paraxial refraction formula

$$n'u' - nu = -hc(n' - n) = -hK \tag{3.85}$$

where curvature $c\left[=\left(\frac{1}{r}\right)\right]$ is used instead of the radius of curvature r. K is the power of the surface, and has been defined earlier in (3.22) as $K = (n' - n)c$. The quantity h can be considered as the paraxial height variable (Figure 3.9).

Multiplying both sides of (3.80) by $u'(= u_+)$, the new form of paraxial transfer formula is obtained as

$$h_+ = h + d'u' \tag{3.86}$$

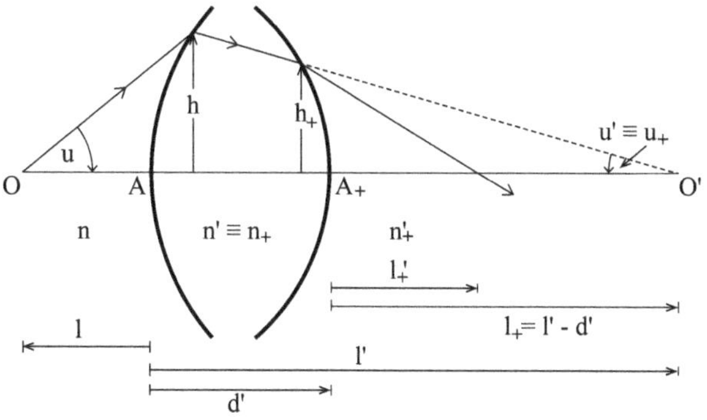

FIGURE 3.9 $(nu - h)$ scheme for paraxial ray tracing.

(3.85) and (3.86) are in the form of recurrence relations, and can be used to determine h_k and u'_k for an axially symmetric system consisting of *k* surfaces. The axial distance of the final image $O' \equiv O'_k$ from the vertex of the k-th surface, l'_k is given by

$$l'_k = -\frac{h_k}{u'_k} \tag{3.87}$$

In terms of paraxial angle variables, the magnification formula (3.81) and (3.82) become

$$M = M_1 M_2 \dots M_s \dots M_k = \frac{n_1 u_1}{n'_1 u'_1} \frac{n_2 u_2}{n'_2 u'_2} \dots \frac{n_s u_s}{n'_s u'_s} \dots \frac{n_k u_k}{n'_k u'_k} \tag{3.88}$$

Since $n_{s+1} u_{s+1} = n'_s u'_s$, $s = 1, \dots, (k-1)$, (3.88) reduces to

$$M = \frac{n_1 u_1}{n'_k u'_k} \tag{3.89}$$

Note that the use of paraxial angle variables has considerably simplified the expression for overall magnification of the system; since the final expression involves only the object space and the final image space quantities, no special measures need to be adopted, even if any of the intermediate paraxial angle variables become zero or infinity.

3.5.3 The Paraxial Invariant

The magnification *M* of the image of an object formed by an axially symmetric system consisting of k surfaces is

$$M = \frac{\eta'_k}{\eta_1} \tag{3.90}$$

where η_1 and η'_k are the height of the object in the initial object space and the height of the image in the last image space, respectively. From (3.89) and (3.90) it follows

$$n_1 u_1 \eta_1 = n'_k u'_k \eta'_k \tag{3.91}$$

The magnification corresponding to imaging at the s-th surface is

$$M_s = \frac{\eta'_s}{\eta_s} = \frac{n_s u_s}{n'_s u'_s} \tag{3.92}$$

Therefore

$$n_s u_s \eta_s = n'_s u'_s \eta'_s \tag{3.93}$$

This relation is valid for all values of s = 1, 2…, k. Recalling the notations

$$n'_s = n_{s+1}, u'_s = u_{s+1}, \eta'_s = \eta_{s+1} \tag{3.94}$$

it follows

$$n_1u_1\eta_1 = n_2u_2\eta_2 = \ldots = n_ku_k\eta_k = n_{k+1}u_{k+1}\eta_{k+1} \tag{3.95}$$

Thus, the product of the refractive index, the paraxial angle variable, and the height of the object in any space of an axially symmetric system consisting of several surfaces has the same value in all spaces, including the object, image, and intermediate spaces. This invariant relationship leads to an optical invariant of paramount importance in the analysis and synthesis of optical systems. It will be deliberated upon in the next chapter.

3.6 Paraxial Imaging by a Single Mirror

In Figure 3.10, AP is a reflecting surface of revolution of vertex curvature $c = (1/r)$, where $r = AC$.

A is the vertex of the surface on the optical axis, and C is the centre of curvature of the surface around the vertex. OQ is a transverse object located at the axial point O, where $AO = l$.

Let $OQ = \eta$. To determine the paraxial image corresponding to the object OQ, two rays originating from the axial point O, OA, and OP are taken as incident rays on the surface AP. After reflection, corresponding reflected rays are PO″ and AO. The latter rays intersect each other at O′ in case (b) where the reflecting surface is a concave mirror. In case (a), where the reflecting surface is a convex mirror, the extended part of the ray PO″ intersects the extended part of ray AO at point O′. Thus in case (a), O′ is a virtual image, and in case (b), O′ is a real image. The image distance $AO' = l'$. The incident ray QA is reflected as AQ″. In figure (b), the ray AQ″ intersects the perpendicular on the optical axis at O′, forming a real inverted image O′Q′ of the object OQ.

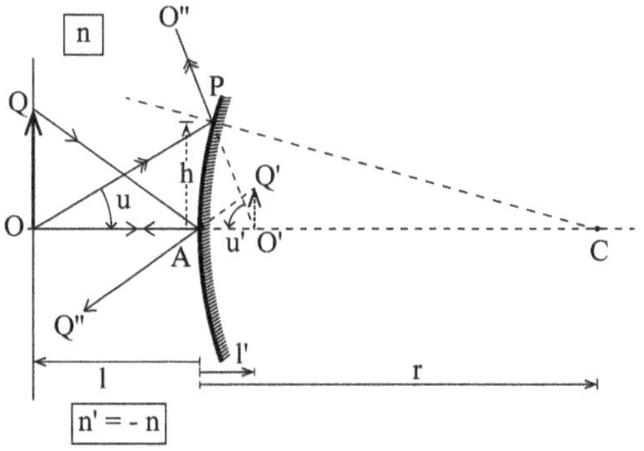

FIGURE 3.10(a) Paraxial imaging by a single convex mirror when light is incident from the left.

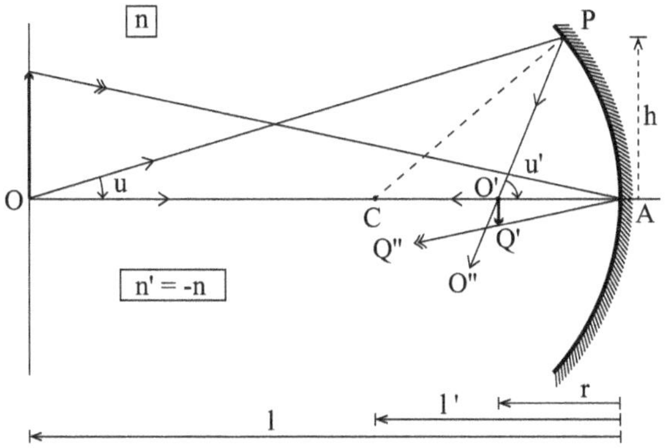

FIGURE 3.10(b) Paraxial imaging by a single concave mirror when light is incident from the left.

Let $O'Q' = \eta'$. In Figure 3.10(a), the extended part of the reflected ray AQ″ intersects the perpendicular on the optical axis at O′ at the point Q′, forming an erect virtual image O′Q′ of the object OQ. The paraxial convergence angle of the incident ray OP is u, and of the reflected ray PO″ is u′. As mentioned earlier in Section 2.9.4, the effect of reflection can be taken care of in refraction equation, if the refractive index of the medium after reflection is taken as negative, i.e., $n' = -n$.

We recall the paraxial refraction equation from (3.85) is

$$n'u' - nu = -hc(n' - n) \tag{3.96}$$

So, for reflection in the single mirror, the paraxial angle variables u and u′ are related by

$$u' + u = -2hc \tag{3.97}$$

Substituting $h = -lu = -l'u'$, the relation between the object distance l and image distance l′ is

$$\frac{1}{l'} + \frac{1}{l} = 2c \tag{3.98}$$

The magnification M is given by

$$M = \frac{\eta'}{\eta} = \left(\frac{n}{n'}\right)\left(\frac{u}{u'}\right) = \left(\frac{n}{n'}\right)\left(\frac{l'}{l}\right) \tag{3.99}$$

Substituting $n' = -n$, the expression for M reduces to

$$M = -\left(\frac{u}{u'}\right) = -\left(\frac{l'}{l}\right) \tag{3.100}$$

Note that (3.100) gives positive and negative magnification, respectively, for the cases shown in Figures 3.10(a) and (b).

Power K of a single reflecting mirror is obtained by substituting $n' = -n$ in (3.22) as

$$K = -2nc \tag{3.101}$$

For a mirror in air $n = 1$, and so $K = -2c$. Thus, the concave mirror in Figure 3.10(b) is of positive power, and the convex mirror of Figure 3.10(a) is of negative power.

Figure 3.11(a) and Figure 3.11(b) show formation of paraxial images by a convex mirror and a concave mirror, respectively, when light is incident from the right. It should be noted that for applying general formulae to the case of mirrors, the indices n and n′ need to be taken as positive or negative according to whether the light travels from left to right or from right to left. Therefore, if the mirrors of Figure 3.11 are in air, $n = -1$ and the power of the mirror in each case is given by $K = +2c$. Thus the convex mirror of Figure 3.11(a) is of negative power, and the concave mirror of Figure 3.11(b) is of positive power. Comparing the corresponding cases of Figure 3.10, it is seen that the concave mirror is of positive power and the convex mirror is of negative power in both cases. Thus, as defined here, power is an inherent property of the mirror surface. In addition, it is seen that a mirror of positive power increases the convergence of light, and the light is diverged by a mirror of negative power.

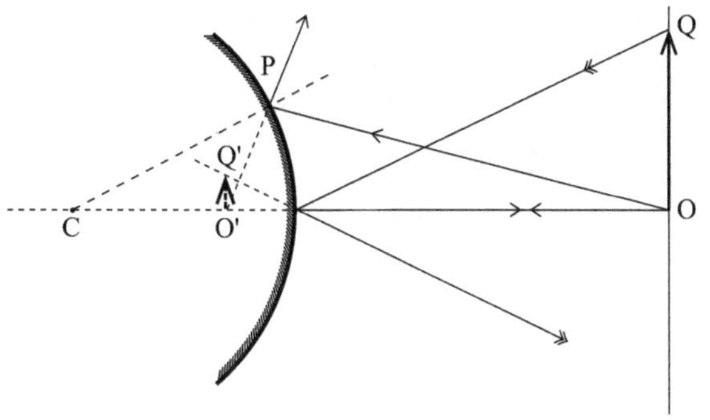

FIGURE 3.11 (a) Paraxial imaging by a single convex mirror when light is incident from the right.

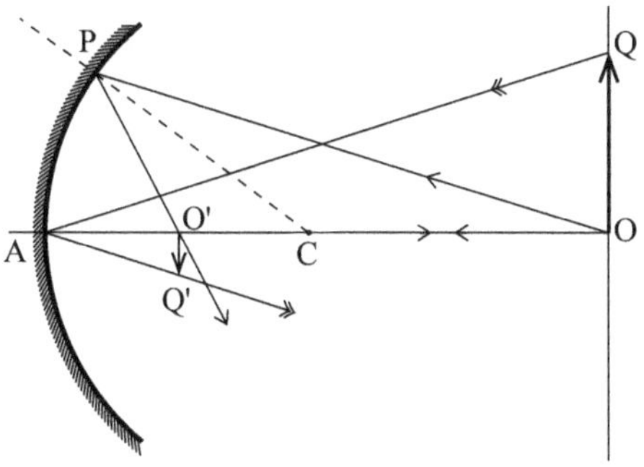

FIGURE 3.11 (b) Paraxial imaging by a single concave mirror when light is incident from the right.

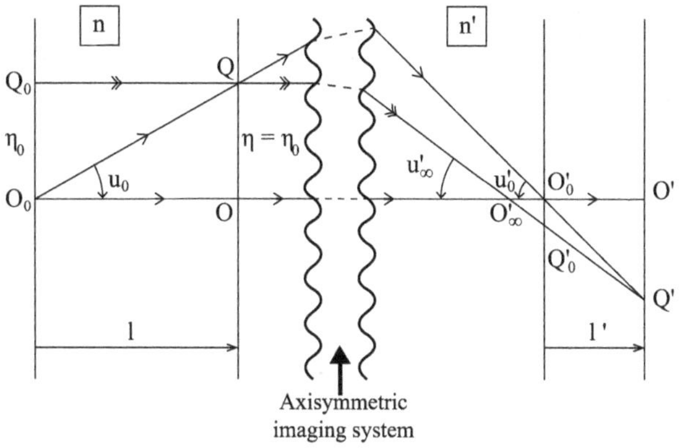

FIGURE 3.12 Effects of an axial shift of the object plane (transverse size of object unchanged) on the axial location of the image plane and on the size of the transverse image.

3.7 The General Object-Image Relation for an Axisymmetric System

$O_0O \ldots O'_0\, O'$ is the optical axis of an axisymmetric imaging system. (Figure 3.12) $O_0Q_0 (= \eta_0)$ is a transverse object located at O_0 on the axis in the object space of refractive index *n*. Let two paraxial rays from O_0, O_0O and O_0Q be incident on the system, and emerge as $O'_0\, O'$ and $O'_0\, Q'$ in the image space of refractive index n'. O'_0 is the paraxial image of O_0, and the transverse plane at O'_0 is the conjugate image plane corresponding to the transverse object O_0Q_0.

Let $OQ(=\eta)$ be a transverse object located at O on the axis where $O_0O = l$. Assume $\eta = \eta_0$. A paraxial ray Q_0Q,, parallel to the axis, is incident on the system, and emerges as $O'_\infty O'_0 Q'$ in the image space. O'_∞ is the point of intersection of the emergent ray with the optical axis in the image space. It is designated so because the corresponding incident ray in the object space is parallel to the axis, i.e. its convergence angle is zero, and so it may be considered to have originated from negative infinity in the object space. The ray $Q_0Q\ldots O'_\infty Q'_0 Q'$ intersects the transverse plane at Q'_0 at the point Q'_0, so that Q'_0 is the paraxial image of Q_0. Let M_0 be the magnification between the conjugate planes O_0Q_0 and $O'_0Q'_0(=\eta'_0)$. Thus $M_0 = (\eta'_0/\eta_0)$.

Note that the two rays $Q\ldots Q'_\infty Q'_0 Q'$ and $Q\ldots O'_0 Q'$ originating from Q in the object space are intersecting at Q' in the image space. Therefore, Q' is the image of the point Q. The foot O' of the perpendicular from Q' on the axis is the image of the axial object point O. Let M be the magnification between the conjugate planes $O'Q'(=\eta')$ and OQ. Thus $M = (\eta'/\eta) = (\eta'/\eta_0)$.

Also, the longitudinal shift of the object plane by $O_0O = l$ in the object space corresponds to a longitudinal shift $O'_0O' = l'$ in the image space.

From (3.85), the expression for paraxial refraction relation for a single refracting surface can be written as

$$n'u' - nu = -hK \tag{3.102}$$

by substituting power $K = (n' - n)c$ from (3.22). For a ray incident from infinity $(u = 0)$ at incidence height $h = \eta_0$, the above relation reduces to

$$n'u'_\infty = -\eta_0 K \tag{3.103}$$

In Figure 3.12, Q_0Q is a ray incident from infinity on the system, and in the image space the emergent ray has a convergence angle u'_∞. By analogy with a single surface, relation (3.103) can be used to define a quantity K characterizing the system. As it is, (3.103) implies u'_∞ is proportional to incidence height η_0, and K is the constant of proportionality. Note that the dimension of K is $[L]^{-1}$, and it may be called power of the system. The reciprocal of K, $F\left[=\left(\frac{1}{K}\right)\right]$ is called the equivalent focal length.

Expressing u'_∞ in terms of η, η_0 and l' with the help of Figure 3.12, and using the sign convention, it follows

$$\frac{n'(\eta' - \eta'_0)}{l'} = -\eta_0 K \tag{3.104}$$

In terms of magnifications M_0 and M, the relation becomes

$$\frac{n'M_0}{l'} - \frac{n'M}{l'} = \frac{n'(M_0 - M)}{l'} = K \tag{3.105}$$

Using the paraxial invariant relationship for the conjugates O_0, O'_0, we get

$$n'u'_0\,\eta'_0 = nu_0\eta_0 \tag{3.106}$$

Substituting

$$u'_0 = \frac{\eta'}{l'}, u_0 = \frac{\eta}{l} \tag{3.107}$$

it follows

$$\frac{n'\eta'\eta'_0}{l'} = \frac{n\eta\eta_0}{l} \tag{3.108}$$

An important general formula for longitudinal magnification

$$M_L = O'_0O'/O_0O = l'/l \tag{3.109}$$

is obtained as

$$\frac{l'}{l} = \left(\frac{n'}{n}\right) M_0 M \tag{3.110}$$

Using (3.110), (3.105) can be rewritten as

$$\frac{n'M_0}{l'} - \frac{n}{lM_0} = K = \frac{1}{F} \tag{3.111}$$

This equation is one of the most important relations of Gaussian optics. It indicates how the power K of an axisymmetric system can be conveniently determined experimentally. It shows that if the location of one set of conjugate points and the magnification between them is known, this can be used as a reference for tackling the problem of location of other paraxial conjugate points.

If, in Figure 3.12, a ray Q'_0Q' is taken parallel to the axis in the image space and is used to define a paraxial angle u_∞ in the object space, the quantity K will be defined by $nu_\infty = \eta'_0K$ in place of (3.103). Similar considerations as used above will lead to the same relation (3.111), showing that the quantity K is the same in the two cases. The equivalent power K, and the equivalent focal length F are intrinsic properties of the system.

From (3.105), the distance l′, of the image O′ from O'_0 is given by

$$l' = n'(M_0 - M)F \tag{3.112}$$

Using (3.110), the distance l, of the object O from O_0 is given by

$$l = n\left(\frac{1}{M} - \frac{1}{M_0}\right)F \tag{3.113}$$

Formulae (3.112) and (3.113) assume that $K \neq 0$, for otherwise $F = \infty$. This is the case for afocal or telescopic systems, where an object at infinity is imaged at infinity. This special case will be dealt with in the next chapter.

Relations (3.112) and (3.113) show that in any nonzero power system, there exist conjugate planes for any stipulated magnification. Out of them, one set of conjugate planes will have magnification equal to +1. These planes are called principal or unit planes of the system. When object and image distances are measured from the respective principal planes, $M_0 = +1$ and the formulae (3.111) – (3.113) reduce to

$$\frac{n'}{l'} - \frac{n}{l} = \frac{1}{F} = K \tag{3.114}$$

$$l = n\left(\frac{1}{M} - 1\right)F \tag{3.115}$$

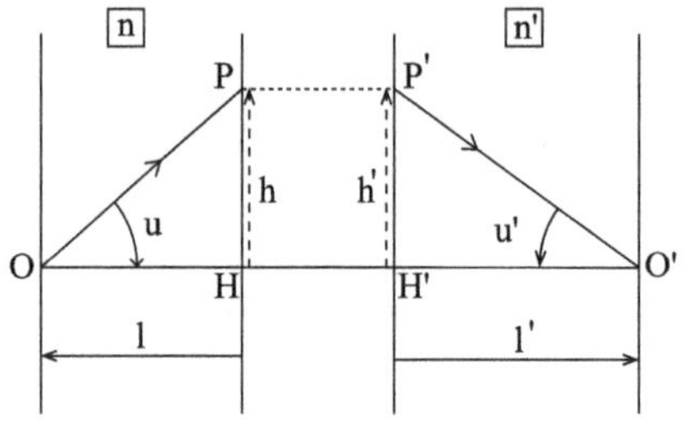

FIGURE 3.13 Object and Image distances l and l′ measured from the principal points H and H′.

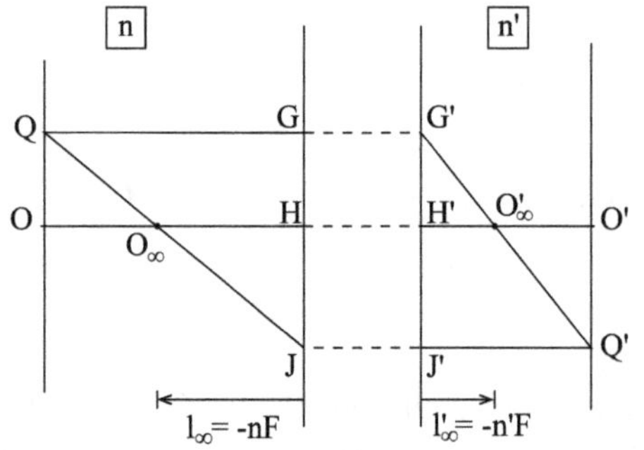

FIGURE 3.14 Geometrical construction for finding the paraxial image.

$$l' = n'(1-M)F \tag{3.116}$$

When object and image distances refer to the principal planes, the general conjugate relation (3.114) has exactly the same form (3.21) as that for a surface of power K. With reference to Figure 3.13, suppose a paraxial ray OP with convergence angle u is incident on the first principal plane at a height $h = HP = -lu$. The paraxial image of P is at P′ on the second principal plane, such that $H'P' = -l'u' = h' = h$. Multiplying both sides of (3.114) by h, we get

$$n'u' - nu = -hK \tag{3.117}$$

The magnification between the transverse planes at O and O′ is given by

$$M = \frac{\eta'}{\eta} = \frac{nu}{n'u'} = \frac{nl'}{n'l} \tag{3.118}$$

This expression is the same as (3.33), derived earlier for single surface imaging. Thus, any axially symmetric system behaves exactly as if it were a simple system of power K when the principal planes are used as reference planes.

3.7.1 A Geometrical Construction for Finding the Paraxial Image

If the positions of the principal points H and H' of any axially symmetric systems are known together with the equivalent focal length F, the formulae presented above can be used to determine the location of the paraxial image, as well as its magnification. However, for quick visualization of the paraxial image, a geometrical construction is useful. This is illustrated in Figure 3.14. The points O_∞ and O'_∞ on the axis are the axial locations of an image at infinity, and of an object at infinity, respectively. The point O_∞ is known as the first principal focus and the point O'_∞ as the second principal focus. Often they are denoted by F and F′ respectively. A caveat is called for on the use of these notations. Typically, image space quantities are denoted by primed symbols, which are the images of the corresponding unprimed symbols, located in the object space. But, F′ or O'_∞ are not the image of F or O_∞. Here, the prime symbol implies that the corresponding focus lies in the image space. The distances $HO_\infty (= l_\infty = f)$ and $H'O'_\infty (= l'_\infty = f')$ are called the first and the second principal focal length of the system. They are related with the equivalent focal length F by

$$l_\infty = f = -nF, \quad l'_\infty = f' = n'F \tag{3.119}$$

Note that the equivalent focal length $F = l'_\infty = f'$, if the image space refractive index $n' = 1$; also, $F = -l_\infty = -f$, if the object space refractive index $n = 1$.

For determining the paraxial image of the object OQ, consider two rays from the extra-axial point Q. One of them, QG, is parallel to the axis and is incident on the first principal plane at point G. Since the transverse magnification between the principal planes is +1, the paraxial image of point G on the first principal plane is G′ on the second principal plane, such that H′G′ = HG. The emergent ray from G′ must pass through the second principal focus O'_∞, and it is shown as $G'O'_\infty Q'$.

The second ray from Q, $QO_\infty J$, is taken such that it passes through the first principal focus O_∞ in the object space, and it is incident on the first principal plane at point J. The corresponding ray from the second principal plane emerges from point J′, such that H′J′ = H J. This emergent ray J′Q′ must be parallel to the axis. The two rays in the image space, namely G′Q′ and J′Q′, corresponding to the two object space rays originating from the point Q, QG and QJ, intersect at point Q′. Thus, Q′ is the paraxial image of Q. Draw a perpendicular Q′O′ on the optical axis. O′Q′ is the paraxial image of OQ.

3.7.2 Paraxial Imaging and Projective Transformation (Collineation)

In the paraxial model of imaging, a unique image point corresponds with an object point. Also, corresponding to a ray passing through the object point, there is a unique ray passing through the image point. 'This unique point-to-point correspondence by means of rectilinear rays between image and object is called "Collineation" – a term introduced by Möbius in his great work entitled *Der barycentrische Calcul* (Leipzig, 1827)' [3]. The latter led to new developments in the field of projective geometry [4]. An authoritative contemporary account of the approach is presented by Born and Wolf [5].

3.8 Cardinal Points in Gaussian Optics

It has been shown above that, under paraxial approximation, any axially symmetric optical system of non-zero power can be uniquely represented by the axial location of the principal or unit points and the equivalent focal length. Carl Friedrich Gauss first made this observation in his *Dioptrische Untersuchungen*, where he first worked out the theory of the refraction of paraxial rays through a centred system of lenses. The paraxial treatment of optical systems is also called Gaussian optics [6]. However, Lord Rayleigh noted that about a century before Gauss' report, considerable progress along similar lines were made by Roger Cotes, the editor of the second edition of Newton's *Principia* [7]. Due to his premature death, Cotes' works were never published. Fortunately, Robert Smith reported them in his book on optics [8].

It may be noted that the axial locations of the first and second principal foci can be ascertained from the above data if the refractive indices of the object and image spaces are known.

It is sometimes useful to define a pair of conjugate axial points such that any ray incident on one of them emerges parallel to itself from the other point. This implies that in the paraxial invariance relation

$$nu\eta = n'u'\eta' \tag{3.120}$$

u = u′, and so transverse magnification between these two conjugates is

$$M = \left(\frac{\eta'}{\eta}\right) = \left(\frac{n}{n'}\right) \tag{3.121}$$

These two axial conjugate points are called nodal points, and are denoted by N and N′ (Figure 3.15). Using (3.113) and (3.114), axial locations of N and N′, with regard to the principal points H and H′ in the object space and image space, respectively, are given by

$$HN = \tilde{l} = (n' - n)F \tag{3.122}$$

$$H'N' = \tilde{l}' = (n' - n)F \tag{3.123}$$

When n′ = n, the values of $\tilde{l}$ and $\tilde{l}'$ are zero, and the nodal points coincide with the principal or unit points.

In Gaussian optics, the six points—namely the first and the second principal foci F and F′, the first and the second principal or unit points H and H', and the first and the second nodal points N and N′—are called the cardinal points of an axisymmetric imaging system. For paraxial analysis, it is sufficient to know the location of the four cardinal points, which include the two foci, and either the two principal points or the two nodal points. The conjugate planes at the nodal points are often called the nodal planes. However, it should be noted that, except for the axial points, the equality of the convergence angles of an incident and the corresponding emergent ray does not hold good for paraxial rays passing through other points of the nodal planes. A well-known practical application of the concept of nodal points is the method of measurement of equivalent focal length of a lens system by nodal slide method.

3.8.1 Determining Location of Cardinal Points from System Data

Axial locations of four of the cardinal points are already given above with respect to principal points in the corresponding object or image space (Figure 3.15). They are given as:

$$HF = f = -nF;\ H'F' = f' = n'F;\ HN = H'N' = (n' - n)F \tag{3.124}$$

It remains to be determined what the equivalent focal length F is, and to specify the axial location of the principal points H and H′ with respect to a suitable reference point in the system. Since the first and last vertices of an axisymmetric system are usually accessible, H and H′ are usually specified with respect to

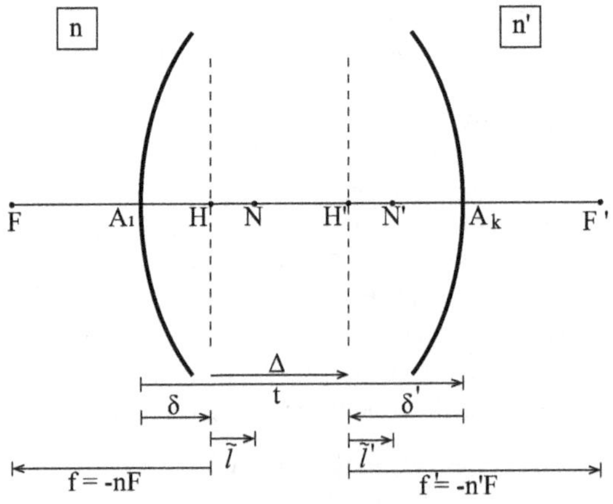

FIGURE 3.15 Axial location of the cardinal points and the Gaussian constants

the first vertex A_1 and the last vertex A_k, where k is the total number of refracting interfaces in the system. The notations for these specifications are:

$$A_kH = \delta, \quad A_kH' = \delta' \tag{3.125}$$

The equivalent focal length F and the quantities δ and δ′ for any given system can be determined as follows. A ray, parallel to the axis (i.e., $u_1 = 0$), is traced from left, with any convenient incidence height h_1 at the first surface, through all k refracting interfaces of the system successively, using the paraxial refraction and transfer formulae (3.85) and (3.86). The paraxial variables for this ray in the final space are h_k and u'_k. From Figure 3.16, it is seen that the second principal focal length f′ and the back focal length f_b are given by

$$f' = -\frac{h_1}{u'_k} \qquad f_b = -\frac{h_k}{u'_k} \tag{3.126}$$

Note that f′ and f_b are the distance of the second principal focus F′ from the second principal point H′ and the vertex of the last surface, A_k, of the system, respectively. The equivalent focal length F of the system is related to f′ with the relation $F = \left(\frac{f'}{n}\right)$. From the diagram, it may be noted that

$$\delta' = A_kH' = -H'A_k = -(f' - f_b) = \frac{(h_1 - h_k)}{u'_k} \tag{3.127}$$

For determining the Gaussian constants of the system in the object space, a ray is traced backwards from the image space (Figure 3.17). This ray is parallel to the axis in the image space, and is incident

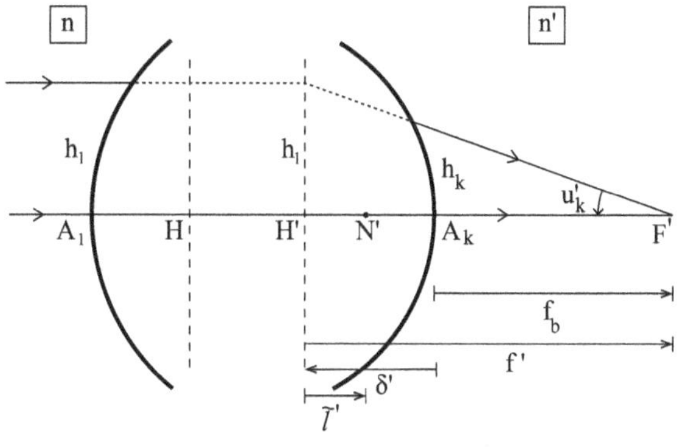

FIGURE 3.16 Forward ray tracing for determining the Gaussian constants of the system in the image space.

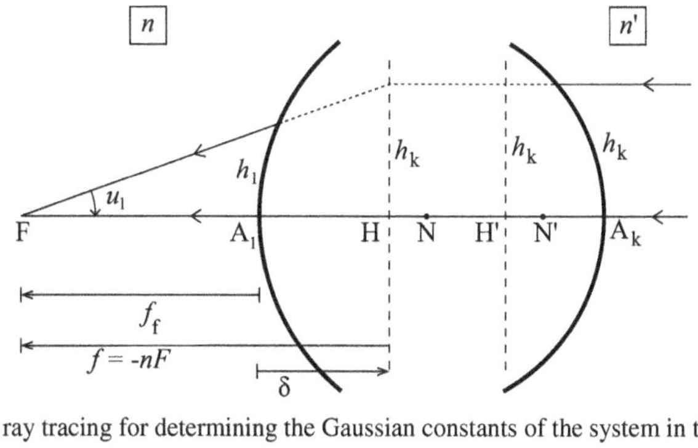

FIGURE 3.17 Backward ray tracing for determining the Gaussian constants of the system in the object space.

on the last surface at the height h_k. Note that, for computational purposes, in ray tracing, the propagation of the ray from right to left can be taken into account by changing not only the sign of the refractive index of the medium from positive to negative, but also the axial distance between the vertices of the interfaces enclosing the medium, which is to be taken as negative. In catadioptric systems, these two changes in system parameters need to be incorporated every time the ray changes its direction of travel from left to right and vice versa. On completion of the backward ray tracing, the paraxial variables of the ray in the object space, h_1 and u_1 are obtained. From Figure 3.17, it is seen that the first principal focal length $(= f)$, i.e. the distance of the first principal focus F from the first principal point H, and the front focal length $(= f_f)$, i.e. the distance of the first principal focus F from the first vertex A_1, are given by

$$f = -\frac{h_k}{u_1} \qquad f_f = -\frac{h_1}{u_1} \tag{3.128}$$

Note that the equivalent focal length F of the system is related with the first principal focal length by the relation $F = -\left(\frac{f}{n}\right)$. From Figure 3.17 it is seen that

$$\delta = A_1H = -HA_1 = -\left(f - f_f\right) = -\frac{\left(h_1 - h_k\right)}{u_1} \tag{3.129}$$

The distance $HH'$of the second principal point H′ from the first principal point H is called the Gaussian thickness of the system. In earlier literature, this distance was sometimes referred to as 'interstitium' (meaning 'interspace'). From Figure 3.14, with due regard to sign of the quantities, it is seen that

$$HH' = \Delta = t + \delta' - \delta \tag{3.130}$$

where $t = A_1A_k$. Note that, for a general axisymmetric system consisting of many refracting the Gaussian thickness Δ can be positive or negative depending on whether H′ is located on the right or left of H.

3.8.2 Illustrative Cases

3.8.2.1 A Single Refracting Surface

Figure 3.18 shows the cardinal points of a single refracting surface of axial curvature $c = \left(\frac{1}{r}\right)$, where r is the corresponding radius of curvature, and C is the centre of curvature. The vertex of the surface is located

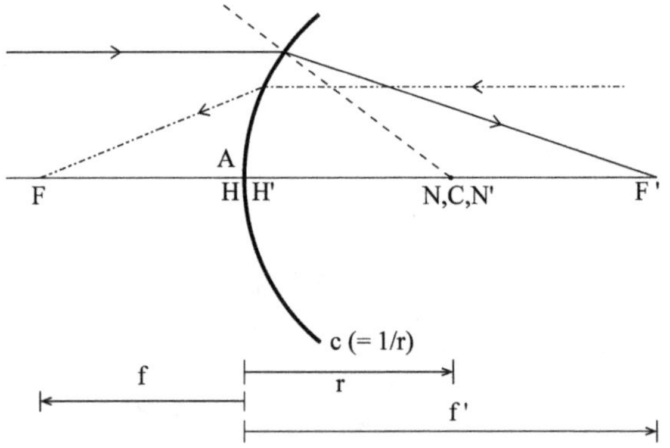

FIGURE 3.18 Cardinal points of a single refracting surface of axial curvature c.

on the axis at point A. The refractive indices of the optical media on the object and image spaces of the surface are n and n′, respectively. From (3.22), the power of the surface is $K = (n' - n)c$. The equivalent focal length of the single refracting surface is

$$F = \left(\frac{1}{K}\right) = \left[r/(n' - n)\right] \tag{3.131}$$

A paraxial ray, parallel to the axis and originating in the object space, is traced to determine the cardinal points in the image space. The emergent ray intersects the optical axis at F′, which is the image space focus or the second principal focus. A solid line in the diagram shows this ray.

From (3.127), $\delta' = 0$, since $h_1 = h_k$. The second principal point H′ is, therefore, located at the vertex A. The second principal focal length, or the image space focal length is given by

$$H'F' = AF' = n'F = f' \tag{3.132}$$

From (3.124), the distance of the second nodal point N′ from the second principal point H′ is $H'N' = \tilde{l}' = (n' - n)F$. Substituting for F from (3.131), $\tilde{l}' = r$. The second nodal point or the image space nodal point is located at the centre of the curvature of the refracting surface.

For determining cardinal points in the object space, a paraxial ray, parallel to the axis, is traced backwards from the image space, and the point of intersection of this ray with the optical axis in the object space gives the object space focus, or the first principal focus F. Chain lines in the diagram denote this ray. Since $h_1 = h_k$, from (3.129), $\delta = 0$; this implies that the first principal point H is also located at the vertex A of the surface. The first principal focal length or the object space focal length is

$$HF = AF = -nF = f \tag{3.133}$$

Also, $HN = \tilde{l} = (n' - n)F = r$. So, the first nodal point, or the object space nodal point, is also located at the centre of curvature C.

3.8.2.2 A System of Two Separated Components

Consider an axisymmetric system consisting of two components separated along the axis by a distance d. Each of the two components is an axisymmetric system specified by the location of its principal planes and its power. Let the Gaussian thickness and power of the two components be Δ_1, K_1 and Δ_2, K_2. The refractive indices in the initial, final, and intermediate spaces are n_1, n_2', and μ respectively. The axial separation d between the two components is the distance between the second principal point of the first component and the first principal point of the second component, i.e. $H_1'H_2 = d$ (Figure 3.19).

Tracing a ray, parallel to the axis in the initial space (i.e. $u_1 = 0$) and at height h_1, successively through the two components, we get

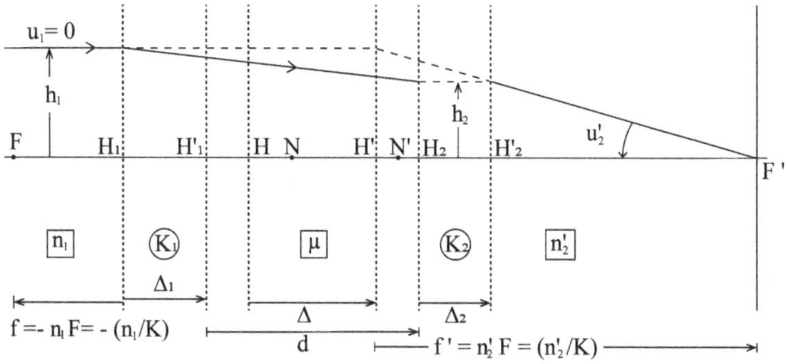

FIGURE 3.19 Cardinal points of an axially symmetric system consisting of two axially separated components.

$$\mu u_1' = -h_1 K_1 \tag{3.134}$$

$$h_2 = h_1 + d u_1' = h_1 \left(1 - \frac{dK_1}{\mu}\right) \tag{3.135}$$

$$n_2' u_2' - \mu u_1' = -h_2 K_2 = -h_1 K_2 \left(1 - \frac{dK_1}{\mu}\right) \tag{3.136}$$

Adding (3.134) and (3.136), it follows

$$n_2' u_2' = -h_1 \left[K_1 + K_2 - \frac{d}{\mu} K_1 K_2\right] \tag{3.137}$$

Following (3.103), power K of the system is given by

$$K = -\left(n_2' u_2' / h_1\right) \tag{3.138}$$

so that the expression for power of the composite system reduces to

$$K = K_1 + K_2 - \frac{d}{\mu} K_1 K_2 \tag{3.139}$$

The equivalent focal length F of the system is given by the reciprocal of K.

The axial location of the second principal point H′ of the composite system can be specified in terms of $\delta' = H_2'H'$. Using (3.127), (3.135), and (3.139) we get

$$\delta' = \frac{(h_1 - h_2)}{u_2'} = \frac{dK_1 / \mu}{-K / n_2'} = -\frac{n_2' dK_1}{\mu K} \tag{3.141}$$

Similarly, the axial location of the first principal point of the composite system can be specified in terms of $\delta = H_1 H$ as

$$\delta = \frac{n_1 dK_2}{\mu K} \tag{3.142}$$

The axial location of nodal points N and N′ of the composite system with respect to its principal points can be specified by using relation (3.124)

$$HN = H'N' = \frac{\left(n_2' - n_1\right)}{K} \tag{3.143}$$

where K of the composite system is given by relation (3.140). The object space or the first principal focus, and the image space or the second principal focus are F and F′, respectively, and the corresponding focal lengths are:

$$HF = f = -n_1 F = -\left(n_1 / K\right) \qquad H'F' = f' = n_2' F = \left(n_2 / K\right) \tag{3.144}$$

with K given by (3.140).

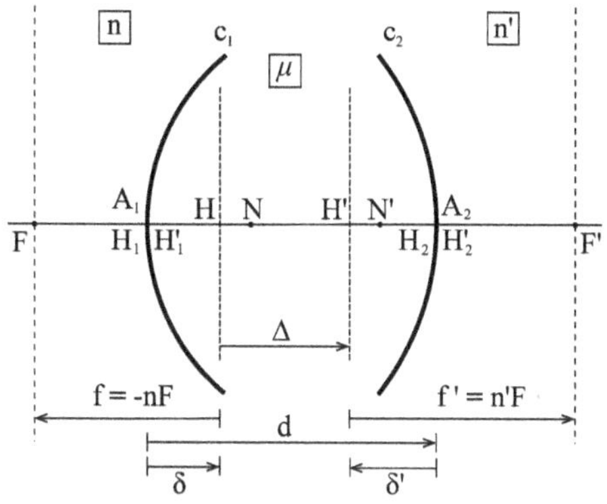

FIGURE 3.20 Cardinal points of a single thick lens with different refractive indices for the object and image spaces.

3.8.2.3 A Thick Lens with Different Refractive Indices for the Object and the Image Spaces

The formulae for power and location of the cardinal points developed above can be applied for studying the same characteristics of a thick lens. The latter is considered a system consisting of two components separated by the axial thickness *d* of the thick lens. The components are the two refracting interfaces with axial curvatures c_1 and c_2. Let the refractive index of the optical material of the thick lens be μ. Assume the object space and the image space are unequal, with refractive indices n and n′, respectively (Figure 3.20).

Powers K_1 and K_2 of the two refracting interfaces are:

$$K_1 = (\mu - n)c_1 \tag{3.145}$$

$$K_2 = (n' - \mu)c_2 \tag{3.146}$$

Using (3.140), power K of the thick lens is given by

$$K = (\mu - n)c_1 + (n' - \mu)c_2 - \frac{d}{\mu}(\mu - n)(n' - \mu)c_1c_2 \tag{3.147}$$

The axial locations of the principal points H and H′ of the thick lens are obtained from (3.141) and (3.142) as

$$A_2H' = \delta' = -\left(\frac{n'd}{\mu}\right)\left(\frac{K_1}{K}\right) \tag{3.148}$$

$$A_1H = \delta = \left(\frac{nd}{\mu}\right)\left(\frac{K_2}{K}\right) \tag{3.149}$$

The ratio of δ′ and δ is given by

$$\frac{\delta'}{\delta} = -\left(\frac{n'}{n}\right)\left(\frac{K_1}{K_2}\right) = \left(\frac{n'}{n}\right)\left[\frac{(\mu - n)}{(\mu - n')}\right]\left(\frac{c_1}{c_2}\right) \tag{3.150}$$

The axial locations of the nodal points N and N′ are specified by

$$HN = H'N' = (n' - n)F \tag{3.151}$$

where the equivalent focal length F is given by the reciprocal of power K of the thick lens as given by (3.147). The axial location of the object space and image space foci (F and F′) of the thick lens are specified by

$$H'F' = f' = n'F \qquad HF = f = -nF \tag{3.152}$$

When the refractive indices of the object and the image spaces are equal, i.e. $n = n'$, expressions for K_1, K_2, and K reduce to

$$K_1 = (\mu - n)c_1 \qquad K_2 = -(\mu - n)c_2 \tag{3.153}$$

$$K = (\mu - n)(c_1 - c_2) + \frac{d}{\mu}(\mu - n)^2 c_1 c_2 \tag{3.154}$$

Also δ' and δ become

$$\delta' = -\left(\frac{nd}{\mu}\right)\left(\frac{K_1}{K}\right) \qquad \delta = \left(\frac{nd}{\mu}\right)\left(\frac{K_2}{K}\right) \tag{3.155}$$

From (3.153) and (3.155), it follows

$$\frac{\delta'}{\delta} = \frac{A_2H'}{A_1H} = \frac{c_1}{c_2} \tag{3.156}$$

In this case, since $n = n'$, the nodal points N and N′ coincide with the principal points H and H′. The axial locations of the object space and image space foci, F and F′, of the thick lens with the same refractive index n in both the object and the image space are specified by $A_1F = f = -nF$, and $A_2F' = f' = nF$.

For a thick lens in air, $n = n' = 1$. Note that, for most applications in optical engineering, the refractive index of air is taken as 1.00, since the refractive indices of optical materials are specified with respect to air, and not vacuum. The refractive index of air at ordinary temperatures and pressures is 1.0003, and this value may be used in some demanding applications [9].

The relations (3.153) – (3.155) reduce to

$$K_1 = (\mu - 1)c_1 \qquad K_2 = -(\mu - 1)c_2 \tag{3.157}$$

$$K = \frac{1}{F} = (\mu - 1)(c_1 - c_2) + \frac{d}{\mu}(\mu - 1)^2 c_1 c_2 \tag{3.158}$$

$$\delta' = -\left(\frac{d}{\mu}\right)\left(\frac{K_1}{K}\right) \qquad \delta = \left(\frac{d}{\mu}\right)\left(\frac{K_2}{K}\right) \tag{3.159}$$

Note that the equivalent focal length F of the thick lens is equal to the second principal focal length $f'(= H'F')$, when $n' = 1$. The expression for power of the thick lens, as given in (3.158), consists of two terms. The first term is independent of thickness d of the lens. The second term becomes zero when $d = 0$, or when either of the two curvatures c_1 or c_2 is zero. Note that when either c_1 or c_2 is zero, power K of the thick lens is independent of the thickness d of the lens, and it becomes equal to K_2 or K_1, respectively. When $c_1 = c_2 (= c)$, the first term is zero, and the power of the thick lens is directly proportional to the thickness d, and to the square of the curvature, c^2.

The common notion that a single lens that is thicker in the middle is of positive power, and that a lens that is thicker around the edge is of negative power, is true for lenses of moderate thicknesses, but does not necessarily hold good for very thick lenses [10]. Let a thick lens have $c_1 > 0$, and $c_2 < 0$. The first term in the expression (3.158) for power is positive, and the second term is negative. For smaller values of thickness d, the magnitude of the first term is larger than that of the second term; however, for very large values of thickness d, the magnitude of the second term will be larger than that of the first term, resulting in negative value for power of the lens.

The location of the principal planes H and H′ are determined by δ and δ'. From (3.159) and (3.157), it follows

$$\frac{\delta'}{\delta} = -\frac{K_1}{K_2} = \frac{c_1}{c_2} \tag{3.160}$$

For an equi-convex or an equi-concave lens, $c_2 = -c_1$, so that $\delta' = \delta$. For a plano-vex or plano-cave lens with $c_2 = 0$, $K_2 = 0$, and, therefore, $\delta = 0$. This implies that the first principal point H is located at the first vertex A_1 of the thick lens. If $c_1 = 0$, then $K_1 = 0$, and $\delta' = 0$, i.e. the second principal point H′ is located at the second vertex A_2 of the thick lens. For meniscus lenses, both the principal points may be located outside the lens.

For a single thick lens, the 'interstitium' $HH'(=\Delta)$ is given by

$$\Delta = d + \delta' - \delta = d\left\{1 - \frac{1}{\mu} + \left(\frac{\mu - 1}{\mu}\right)^2 dc_1c_2\right\} \tag{3.161}$$

For lenses with thickness $d \ll \{R_1, R_2\}$ where $R_1 = (1/c_1)$, and $R_2 = (1/c_2)$, an approximate value for Δ is

$$\Delta \sim d\left(1 - \frac{1}{\mu}\right) \tag{3.162}$$

At visible wavelengths, $\mu \sim 1.5$, and so the magnitude of Δ is approximately equal to $\frac{d}{3}$. Figures 3.21(a)–(d) demonstrate approximate locations of the principal planes of single thick lenses of different shapes; the lenses are of positive power. Figures 3.22(a)–(d) show the same for four single thick lenses of different shapes, but of negative power. Note that in each case of Figure 3.21 and of Figure 3.22, since the lens is in air, the nodal points coincide with the principal points, and the locations of the focal points F and F′ are specified by $HF = f = -F$, and $H'F' = f' = F$.

Figure 3.23(a) shows a ball lens in air. It consists of a glass sphere of radius r and refractive index μ. Using (3.157) – (3.158), power of the ball lens is given by

$$K = \frac{2(\mu - 1)}{\mu r} \tag{3.163}$$

From (3.159), locations of the principal points of the ball lens are obtained as

$$\delta = +r \qquad \delta' = -r \tag{3.164}$$

This implies that both the principal points H and H′ are located at the centre C of the ball lens. From (3.163), the equivalent focal length, H′F′, of the ball lens is obtained as

$$F = f' = H'F' = \frac{\mu r}{2(\mu - 1)} \tag{3.165}$$

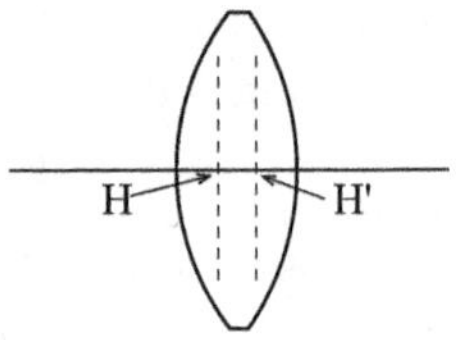

FIGURE 3.21(a) Principal planes of a thick bi-convex lens – positive power.

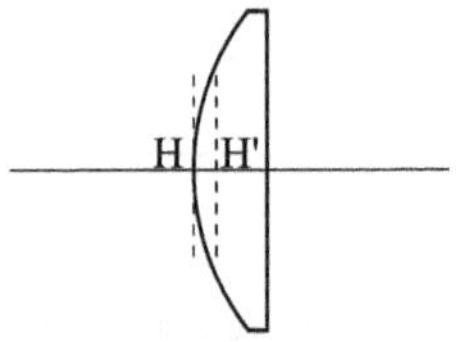

FIGURE 3.21(b) Principal planes of a thick plano-convex lens with plane surface on the right – positive power.

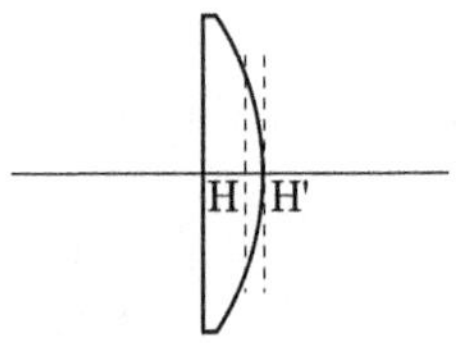

FIGURE 3.21(c) Principal planes of a thick plano-convex lens with plane surface on the left – positive power.

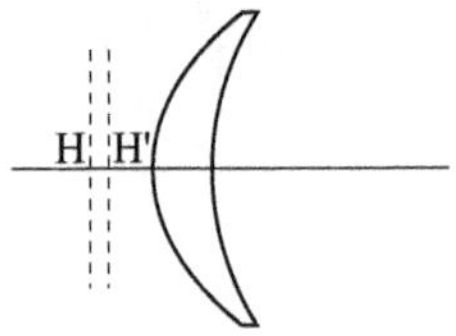

FIGURE 3.21(d) Principal planes of a thick meniscus lens curved towards right – positive power.

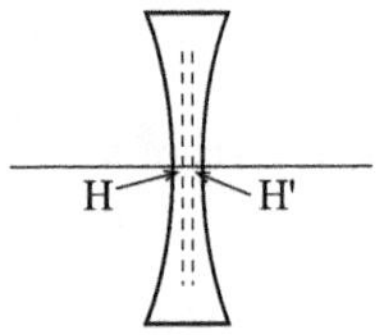

FIGURE 3.22(a) Principal planes of a thick bi-concave lens – negative power.

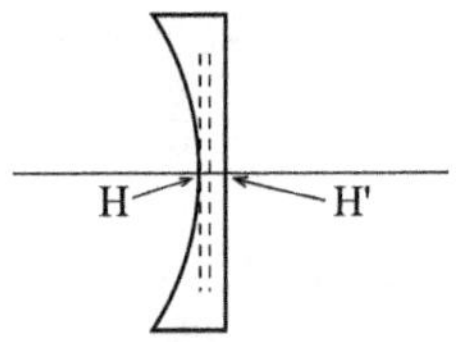

FIGURE 3.22(b) Principal planes of a thick plano-concave lens with plane surface on the right – negative power.

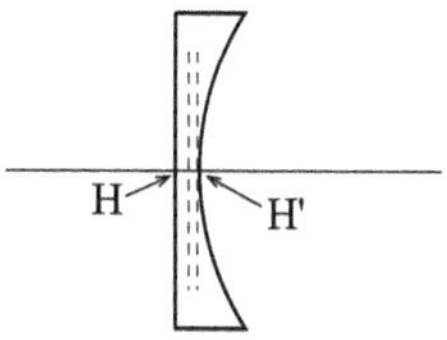

FIGURE 3.22(c) Principal planes of a thick plano-concave lens with plane surface on the left – negative power.

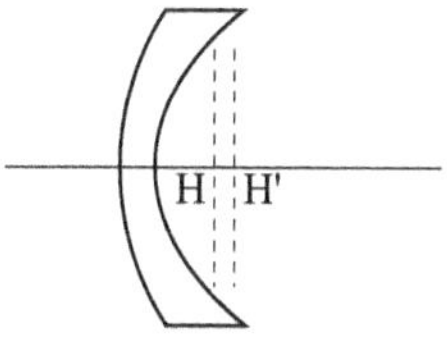

FIGURE 3.22(d) Principal planes of a thick meniscus lens curved towards right – negative power.

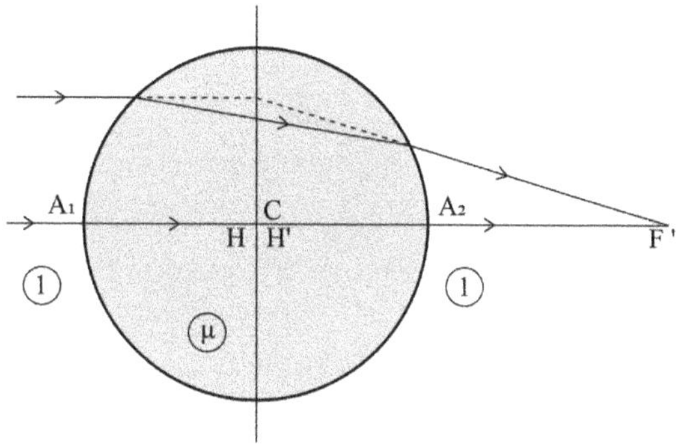

FIGURE 3.23(a) Principal planes of a ball lens in air.

The back focal length, A_2F', of the ball lens can be determined by paraxial ray tracing, and it is given by

$$f_b = \frac{(2-\mu)r}{2(\mu-1)} \tag{3.166}$$

Assuming $\mu \sim 1.5$, effective focal length and back focal length of a ball lens is approximately equal to 1.5 r and 0.5 r, respectively.

Figure 3.23(b) shows a concentric meniscus lens of axial thickness d. So, $R_1 = R_2 + d$, where $R_1 = (1/c_1)$ and $R_2 = (1/c_2)$, c_1 and c_2 being the axial curvatures of the first and second surfaces, respectively. From (3.156), by substitution, the power of the concentric meniscus lens is seen to be equal to $\left[-\frac{(\mu-1)}{\mu}dc_1c_2\right]$. Note that the power and the equivalent focal length of a concentric meniscus lens is negative. Using (3.157) and (3.159), we get

$$A_1H = \delta = R_1 \qquad A_2H' = \delta' = R_2 \tag{3.167}$$

Thus, both the principal points H and H′ are located at the common centre C of the two surfaces of the concentric meniscus lens.

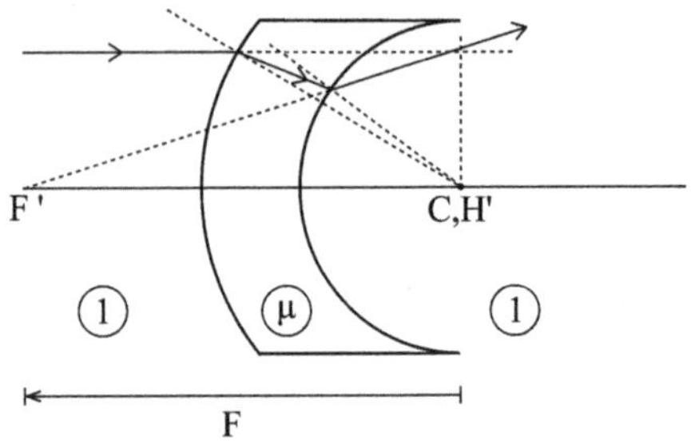

FIGURE 3.23(b) Principal planes of a concentric meniscus lens.

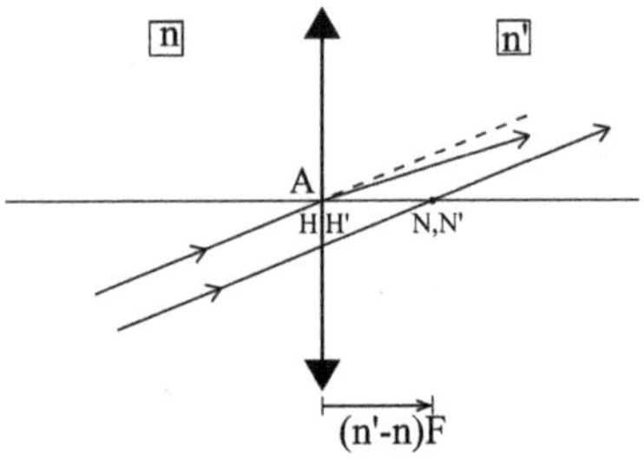

FIGURE 3.24 Principal planes of a single thin lens with different refractive indices for the object and the image spaces.

3.8.2.4 A Thin Lens with Different Refractive Indices for the Object and the Image Spaces

Usually, the axial thickness of a single lens is much smaller compared to both the radii of curvature of its refracting surfaces and the diameter of its aperture. Relatively simpler expressions for the optical characteristics of the lens can be obtained by neglecting its thickness and by assuming that the axial thickness $d \sim 0$. This approximate model of 'thin lens optics' is particularly useful at the stage of conceptualization and structural design of lens systems.

First, we consider the general case of a single thin lens of refractive index μ; the object space and the image space are assumed to have refractive index n and n′, respectively. Let the axial curvatures of the two refracting surfaces be c_1 and c_2. From (3.147), by substituting $d = 0$, we get the value of power K of the thin lens as

$$K = (\mu - n)c_1 + (n' - \mu)c_2 \tag{3.168}$$

Similarly, from (3.148) – (3.149), values of both $\delta'(= AH')$ and $\delta(= AH)$ are seen to be zero, i.e. both principal planes are coincident on the thin lens.

Since the refractive index of the object space and that of the image space are different, an oblique ray incident at the centre of the lens will always be deviated (Figure 3.24). From (3.151) it may be noted that the nodal points N and N′ are coincident with each other, and they are axially shifted from the lens by

$$AN = AN' = (n' - n)F = \frac{(n' - n)}{K} \tag{3.169}$$

An oblique ray incident on the lens along PN with a paraxial convergence angle u emerges undeviated along NP′ with a paraxial convergence angle $u' = u$. This may be checked by using the paraxial refraction equation

$$n'u' - nu = -hK \tag{3.170}$$

An incident ray towards N with $u = -hK/(n' - n)$ corresponds to an emergent ray with convergence angle $u' = -hK/(n' - n)$.

When $n' = n$, the nodal points N and N′ coincide with the principal points H and H′, and the quadruple is located at the axial point A of the thin lens. Then, any ray incident on A is refracted with no change in angle of convergence. The expression for power reduces to

$$K = (\mu - n)(c_1 - c_2) \tag{3.171}$$

The same conclusions hold good in the common case of a thin lens in air, with a further reduction in the expression for power as

$$K = (\mu - 1)(c_1 - c_2) \tag{3.172}$$

3.8.2.5 Two Separated Thin Lenses in Air

Figure 3.25 shows different axisymmetric systems consisting of two thin lenses separated by an axial distance d. It is assumed that the refractive index of the object space, the space between the two lenses, and the image space is air. For each of the two lenses, its nodal points coincide with the principal points. The two principal planes of a thin lens are located on the thin lens. Let the powers of the two thin lenses be K_1 and K_2, the equivalent focal lengths of the thin lenses being $F_1 (= 1/K_1)$ and $F_2 (= 1/K_2)$, respectively. Substituting $\mu = 1$ in (3.140), power K of the composite system is expressed as

$$K = K_1 + K_2 - dK_1K_2 \tag{3.173}$$

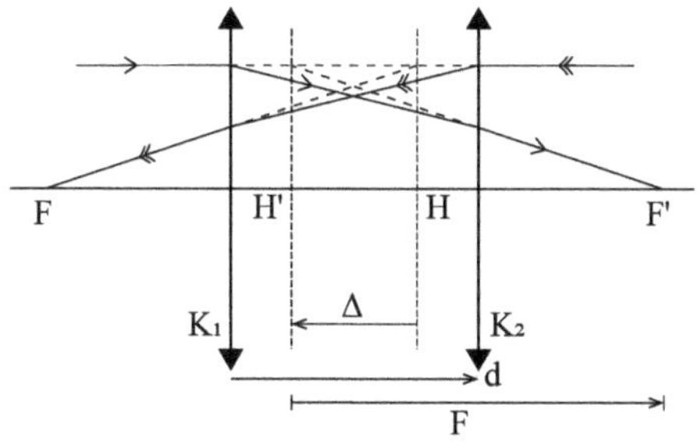

FIGURE 3.25(a) Two thin lenses (of positive powers K_1 and K_2) are axially separated by distance $d < \left\{ \left(\frac{1}{K_1}\right) + \left(\frac{1}{K_2}\right) \right\}$.

For the composite system, the equivalent power is postive and the Gaussian thickness Δ is negative.

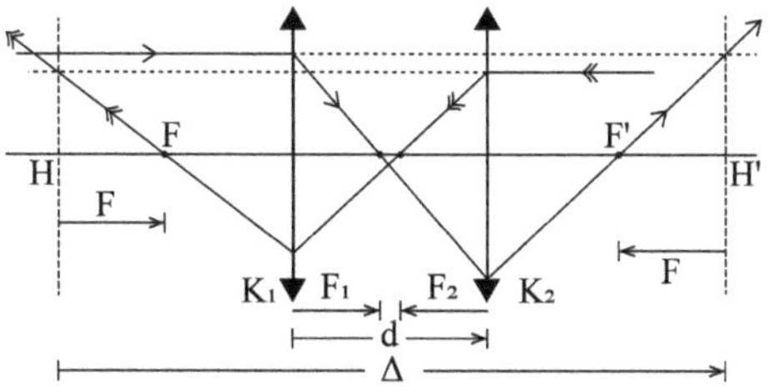

FIGURE 3.25(b) Two thin lenses (of positive powers K_1 and K_2) are axially separated by distance $d > \left\{ \left(\frac{1}{K_1}\right) + \left(\frac{1}{K_2}\right) \right\}$.

For the composite system, the equivalent power is negative and the Gaussian thickness Δ is positive.

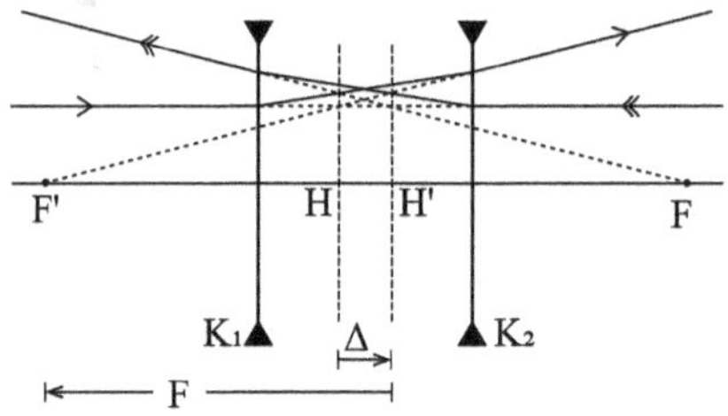

FIGURE 3.25(c) Two thin lenses (of negative powers K_1 and K_2) are axially separated by distance d. For any value of d, the equivalent power is negative and the Gaussian thickness Δ is positive for the composite system.

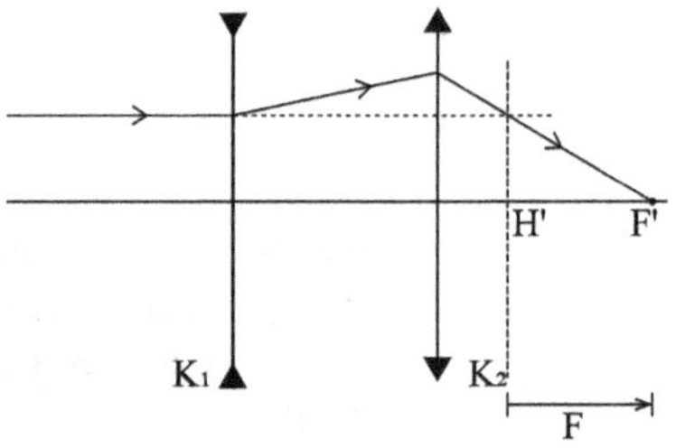

FIGURE 3.25(e) Two axially separated thin lenses of negative power K_1 and positive power K_2 constitute a wide angle or retrofocus lens. The equivalent focal length F of the composite system is smaller than its back focal length.

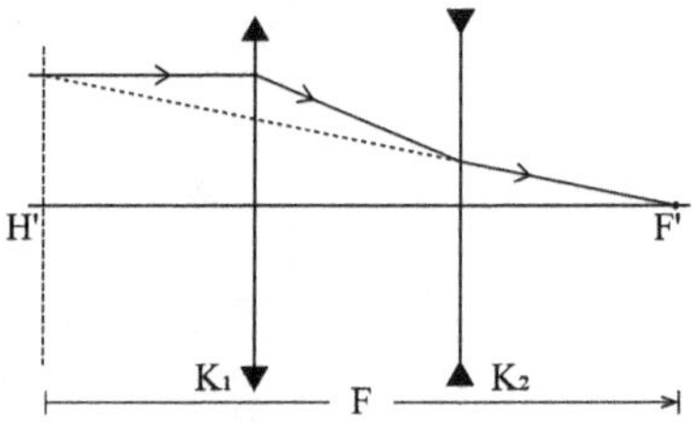

FIGURE 3.25(d) Two axially separated thin lenses of positive power K_1 and negative power K_2 constitute a telephoto lens. The equivalent focal length F of the composite system is l arger than the length of the system from the first lens to the focus.

Axial locations H and H′ of the first principal point and the second principal point are specified by $\delta = A_1H$ and $\delta' = A_2H'$. Substituting $n_2' = 1$ in (3.141), and $n_1 = 1$ in (3.142), expressions for δ' and δ are obtained as

$$\delta'' = -\frac{dK_1}{\mu K} \qquad \delta = \frac{dK_2}{\mu K} \tag{3.174}$$

Gaussian thickness of the composite system, or the distance between the principal points $HH' = \Delta$ is given by

$$HH' = \Delta = -d^2 \frac{K_1 K_2}{K} \tag{3.175}$$

Figures 3.25(a)–(b) show the case of positive values for both K_1 and K_2. For values of $d < (F_1 + F_2)$, equivalent power K is positive, and Gaussian thickness Δ is negative. However, for values of $d > (F_1 + F_2)$,

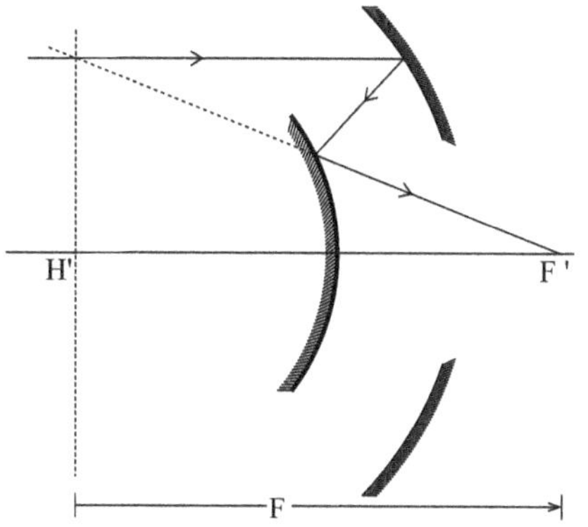

FIGURE 3.26 A 'mirror lens' consisting of two axially separated catoptric or reflecting surfaces.

equivalent power K becomes negative, and Gaussian thickness Δ becomes positive. Figure 3.25(c) shows the case of negative values for K_1 and K_2. In this case, for any value of d, equivalent power K is negative, and Gaussian thickness Δ is positive.

Figure 3.25(d) shows the case of positive value for K_1, and negative value for K_2. Note that the equivalent focal length F of the composite system is larger than the length of the system from the first component to the focus. For observing finer detail in very distant scenes, a lens system with a very long equivalent focal length is needed. These are called telephoto lenses. All refractive telephoto lenses providing large equivalent focal lengths are based on this type of composite structure.

Figure 3.25(e) shows the case of negative value for K_1, and positive value for K_2. This structure provides a value of equivalent focal length that is not only smaller than the system length, but is also smaller than the back focal length. This principle is utilized in the design of wide angle or retro focus lenses providing very small focal lengths for the lens systems.

Incidentally, it may be noted that telephoto effects that are larger than what can be achieved by dioptric (i.e. refracting) optics can be obtained by using catoptric (i.e. reflecting) systems consisting of two reflecting surfaces (Figure 3.26). In the parlance of photography, these systems are sometimes called 'mirror lenses'.

3.9 The Object and Image Positions for Systems of Finite Power

In Figure 3.27(a), a thin lens of finite power K is located at A on the optical axis. The equivalent focal length *F* of the lens is equal to $\left(\frac{1}{K}\right)$. An object at O on the axis is imaged by the lens at O′ with a magnification $M = \left(\frac{l'}{l}\right)$. The object and image spaces are assumed to be air with refractive index unity. By substituting n = n′ = 1, we obtain from (3.115)–(3.116) the following expressions for the object distance l and image distance l′

$$l' = (1 - M)F \qquad l = \left(\frac{1}{M} - 1\right)F \tag{3.176}$$

The throw T is the distance from the object to the image and for the thin lens it is given by

$$T = (l' - l) = \left\{(1 - M) - \left(\frac{1}{M} - 1\right)\right\}F \tag{3.177}$$

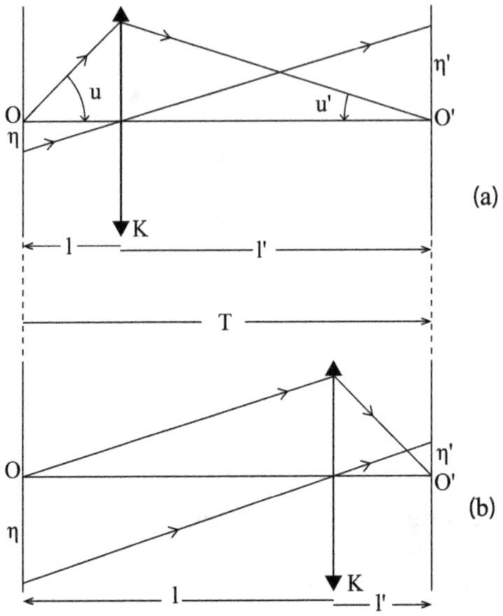

FIGURE 3.27 A lens of finite power produces image for two different conjugates with same object-to-image throw. (a) Magnification M (b) Magnification (1/M).

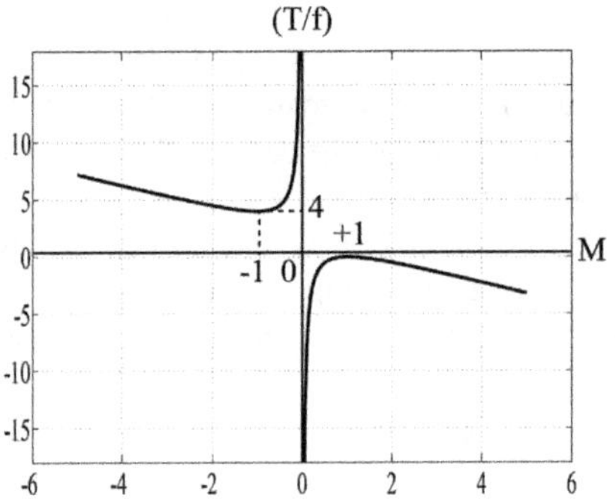

FIGURE 3.28 Throw – Magnification diagram.

Note that this expression for T can be rewritten as

$$T = \left\{ \left(\frac{1}{M} - 1 \right) - (M - 1) \right\} F \tag{3.178}$$

This implies the existence of another axial location of the thin lens that forms an image of the object O at the same location of the image, O′, i.e. the same value of throw T, but with a different magnification $\left(\frac{1}{M} \right)$. This is illustrated in Figure 3.27(b). Thus we get

$$T = \left(2 - M - \frac{1}{M} \right) F = -\frac{(M-1)^2}{M} F \tag{3.170}$$

Figure 3.28 shows the variation of (T/F) against magnification M. From the figure, it is obvious that a lens cannot form an image for which $0 < |T| < |4F|$. The curve has a stationary value when $M = \pm 1$. This

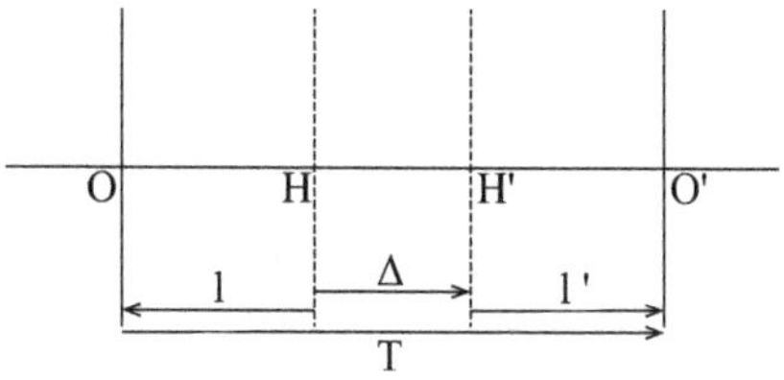

FIGURE 3.29 Object-to-image throw T and Gaussian thickness Δ.

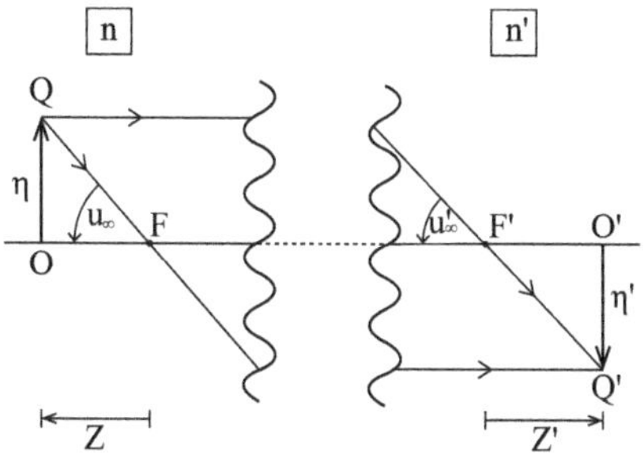

FIGURE 3.30 Newton's form of the conjugate equation.

characteristic is often utilized in the design of zoom systems. Normally, a change in magnification is associated with a change in location of the image plane. The latter is significantly reduced when changes in magnification take place around the stationary points.

In the case of a thin lens, the Gaussian thickness is zero. The observations made above for a thin lens remain valid for a general axisymmetric system, when the object distance l and the image distance l′ are measured with respect to the first principal focus H and the second principal focus H′, respectively, and the object-to- image throw T is taken as (Figure 3.29)

$$T = (l' - l) + \Delta \tag{3.180}$$

The sign and magnitude of the Gaussian thickness Δ can bring about major departures from the predictions based on the assumption of thin lens, particularly in the case of compound systems with large inter-lens separations. Nevertheless, even in such cases, the thin lens approach facilitates conceptualization and visualization of the problem at hand.

3.10 Newton's Form of the Conjugate Equation

For an axisymmetric imaging system of finite power *K*, the conjugate equation takes a simple form if the object and the image distances are measured with respect to the first and the second principal focus. Let the first principal focus in the object space of refractive index n be located at F, and the second principal focus in the image space of refractive index *n′* be located at F′ on the axis. An object of height $OQ\,(=\eta)$ is located at O on the axis, and the corresponding image $O'Q'\,(=\eta')$ is located at O′ on the axis. Let FO = z, and F′O′= z′ (Figure 3.30).

A ray QG, parallel to the axis, is incident on the system at height η. It emerges in the image space as F′Q′. The convergence angle of this ray in the image space is u'_∞. By using paraxial refraction relation

$$n'u'_\infty = -\eta K = -\frac{\eta}{F} \tag{3.181}$$

From $\Delta F'O'Q'$, $u'_{\infty} = \frac{\eta'}{z'}$. Therefore

$$z' = -\frac{\eta'}{\eta} n'F = -n'MF \quad (3.182)$$

Similarly, considering the ray path Q′…FQ, we get

$$z = nF\frac{\eta}{\eta'} = \frac{nF}{M} \quad (3.183)$$

Eliminating M from (3.180) and (3.181)

$$zz' = -nn'F^2 \quad (3.184)$$

This is Newton's conjugate formula. However, this formula is applicable only for systems of non-zero finite power, i.e. when $K \neq 0$.

3.11 Afocal Systems

After derivation of the general object-image relation in Section 3.7, all axisymmetric systems considered so far are assumed to be of power $K \neq 0$. For an imaging system with zero power, the image of an object at infinity is formed also at infinity. This implies that the principal foci of the system are located at infinity. By substituting $K = 0$ in (3.105), we get $M = M_0$, i.e. the transverse magnification is the same for all positions of the object. Thus, principal planes with unity positive magnification do not exist in these systems.

It is obvious that an afocal system cannot be formed with a single lens element, and it has to be constituted with at least two or more lens elements. Figure 3.31 shows an afocal system formed by two positive lenses.

In order to develop object-image conjugate relation for an afocal system it is necessary to determine the axial location for one object-image conjugate. For example, in the above system, an object O_1 located at front focus F_1 of the first lens component will be imaged as O_1' at the back focus F_2' of the second lens component. Once the axial locations of the two points are known, relation (3.110), rewritten in the form shown below

$$l' = \left(\frac{n'}{n}\right) M_0^2 l \quad (3.185)$$

can be used as object-image conjugate relation, where l is the axial distance of the object with respect to the axial point O_1, and l′ is the distance of the corresponding image from the axial point O_1'. Figure 3.32 illustrates this conjugate relation, which is valid for afocal systems.

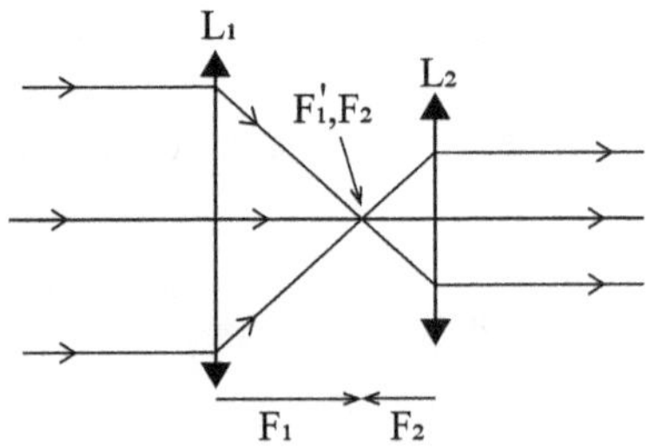

FIGURE 3.31 An afocal system consisting of two lenses, both of positive focal length.

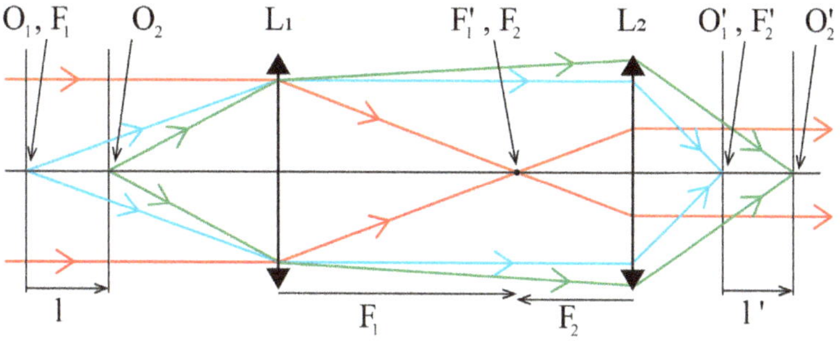

FIGURE 3.32 Object-image conjugate relation for finite conjugate imaging in afocal systems [Pseudo-colour used for better visualization].

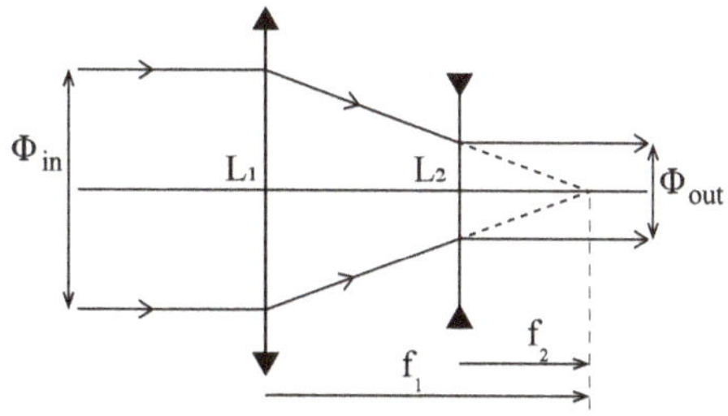

FIGURE 3.33(a) Beam Compressor: A lens of negative focal length follows a lens of positive focal length.

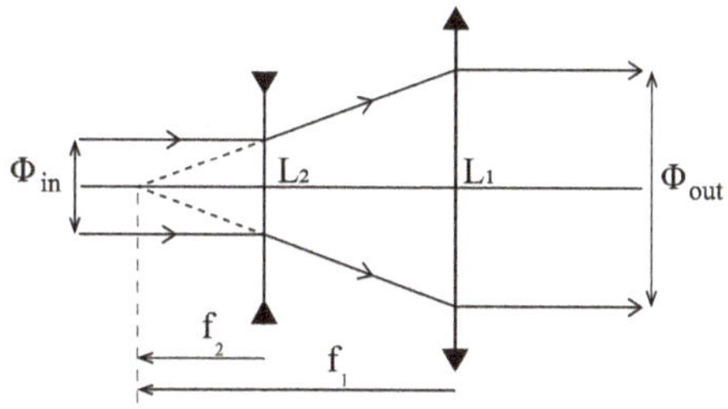

FIGURE 3.33(b) Beam Expander: A lens of positive focal length follows a lens of negative focal length.

Note that sometimes, afocal systems are also referred to as telescopic systems. They are characterized by transverse magnification, and not the power, which is zero for any such system. For the case shown in Figure 3.30, where the afocal system consists of two lenses separated by the algebraic sum of their focal lengths, the transverse magnification is given by $M = -(F_2 / F_1)$. M is negative as shown in the figure, for both F_1 and F_2 are taken as positive. It is obvious that the positive value of M is obtained when the two lenses have focal lengths of the opposite sign.

An interesting special case emerges when $F_1 = F_2$, and $n' = n$. Both the transverse and the longitudinal magnification become unity for all object distances. The axial separation between the object and the image is four times the focal length of the lenses. For a near object, the axial position and the size of the image remain constant while the optical system is moved along the axis.

Since transverse magnification is the same for all object positions, afocal systems are widely used in optical metrological applications. It should also be noted that afocal systems can provide large changes in object-to-image throw, whilst the transverse magnification remains constant. Other major applications of afocal systems include beam expanders and beam compressors (Figure 3.33).

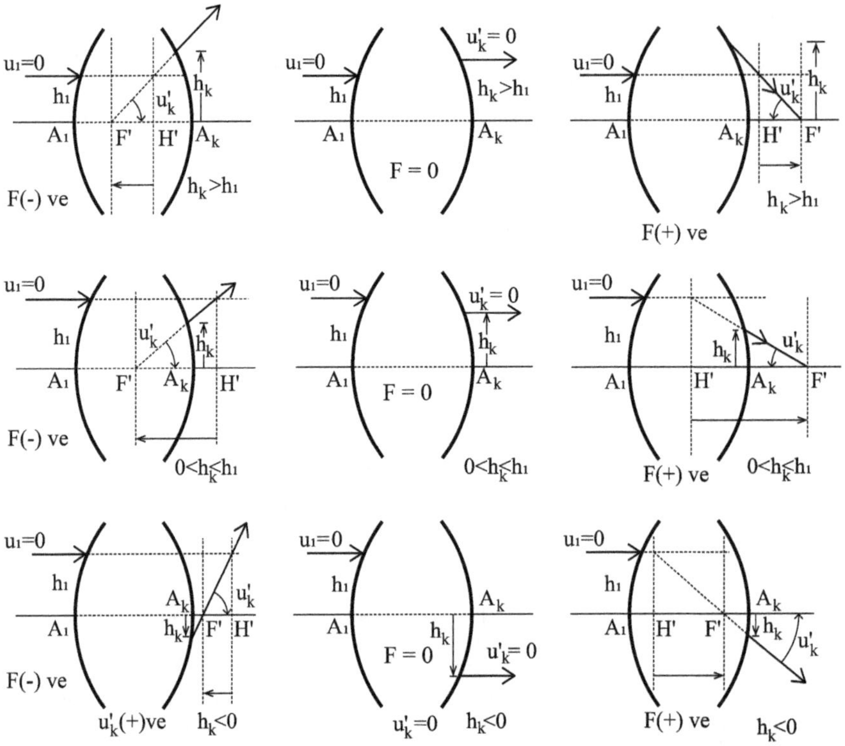

FIGURE 3.34 Nine imaging configurations shown in a 3 x 3 format. Each of the three systems in first column has F < 0, in second column has F = 0, and in third column has F > 0. Each of the three systems in first row has $h_k \geq h_1$, in second row has $0 < h_k < h_1$, and in third row has $h_k \leq 0$.

3.12 Vergence and Power

Power of a lens is usually associated with the vergence of the emergent beam corresponding to an incident beam parallel to the axis. If the emergent beam is convergent, it is presumed that the lens is of positive power, or that its equivalent focal length is positive. On the other hand, if the emergent beam is divergent, the lens is thought to be of negative power, or its equivalent focal length is thought to be negative. This type of conjecture on the sign of power of a lens holds good for a single lens element, but in the case of a lens system consisting of multiple lens elements, such conjecture may turn out to be incorrect in many practical situations.

The power K of an axisymmetric lens system, consisting of k number of refracting interfaces and placed in air, is either positive or negative if the second principal focus F′ of the system is located on the right or left of the second principal or unit point H', i.e. according to the sign of H′F′. On the other hand, vergence of the lens is usually associated with the sign of the convergence angle u'_k of the ray emerging from the k-th surface for a paraxial ray incident on the first surface at height h_1 and convergence angle $u_1 = 0$. The emergent beam is divergent or convergent if u'_k is positive or negative. For a single thin lens, $h_1 = h_k$, and the principal planes are located on the lens, so that the observations made above are correct in general. But for a general multi-element lens system, in general, $h_k \neq h_1$, and the distance δ' of the second principal focus from the last vertex A_k of the system may be positive, zero, or negative, opening up new possibilities that are not achievable with single thin lenses. Also, it becomes feasible to have imaging systems with zero power, i.e. afocal systems, in the case of multi-element lens systems.

Figure 3.34 shows nine imaging configurations in a 3X3 format. The first column shows three configurations where the equivalent focal length F is negative, i.e. power K is also negative. In each case, the incident ray is taken at height h_1 on the first surface (vertex at A_1) and convergence angle $u_1 = 0$. The

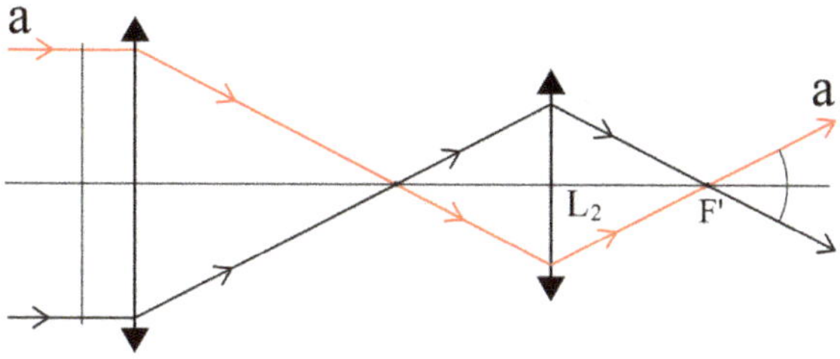

FIGURE 3.35 The equivalent focal length of the composite system is negative. The emergent beam is convergent between L_2 and F′, and divergent beyond F′ [Pseudo-colour used for better visualization].

corresponding ray emerging from the k-th surface (vertex at A_k) of the system at height h_k has a convergence angle u'_k. The axial locations of the second principal point H′ and the second principal focus F′ are also indicated in each of the three diagrams. The third column depicts three configurations where the equivalent focal length F is positive, i.e. positive power K. Details of its depiction are for similar quantities as in column one. The second column presents three configurations where power $K = 0$. In eachconfiguration, the incident ray on the first surface has the specifications, height h_1, and convergence angle $u_1 = 0$. The emergent ray from the k-th surface is parallel to the axis, i.e. $u'_k = 0$, in all three cases, but the height h_k of the emergent ray is different in the three cases.

Row-wise disposition of the configurations underscores the effect of relative magnitudes of h_1 and h_k on vergence of the emergent beams. The three configurations shown in the first row correspond to the situation, $h_k \geq h_1$. The second row presents the three cases corresponding to $0 < h_k < h_1$. The third row depicts the ray geometry for the three cases in which $h_k \leq 0$. Note that in case of both the first and the third element of the first row, the second principal or unit point falls on the last vertex, i.e., $H' \equiv A_k$, when $h_k = h_1$. Similarly, for both the first and the third element of the third row, the second principal focus lies on the last vertex, i.e., $F' \equiv A_k$, when $h_k = 0$.

The first system and the third system of the third row present two interesting cases. In the former case, the emergent beam of the lens' system of negative power is convergent in the image space lying between A_k and F′. In the latter case, the emergent beam of the lens' system of positive power is divergent in the image space beyond A_k, the vertex of the last surface. Figure 3.35 illustrates an implementation of the former case using two separate thin lenses.

3.13 Geometrical Nature of Image Formation by an Ideal Gaussian System

The Gaussian properties of an axially symmetric optical system are valid in the case of limitingly small object and aperture size. Under this paraxial approximation, an array of points on a transverse object plane is imaged stigmatically as a geometrically similar array of points on the transverse image plane. 'An ideal Gaussian system' is defined to be a system where this characteristic holds for all object and aperture sizes, and for all axial positions of the object plane. This model of image formation defines a one-to-one correspondence between all points in the object space and their respective image points.

3.13.1 Imaging of a Two-Dimensional Object on a Transverse Plane

Figure 3.36 shows an ideal imaging system where an object on the transverse object plane at O is imaged with transverse magnification M on the transverse image plane at O′. Let the four points on the ξ-η plane at O, namely $A(\xi_1,\eta_1), B(\xi_2,\eta_2), C(\xi_3,\eta_3)$ and $D(\xi_4,\eta_4)$ be imaged on the ξ'-η' plane at O′ as $A'(\xi'_1,\eta'_1), B(\xi'_2,\eta'_2), C(\xi'_3,\eta'_3)$ and $D(\xi'_4,\eta'_4)$, respectively. The slopes m_1, m_2 of the two lines AB and CD are given by

$$m_1 = \frac{\eta_1 - \eta_2}{\xi_1 - \xi_2} \qquad m_2 = \frac{\eta_3 - \eta_4}{\xi_3 - \xi_4} \tag{3.186}$$

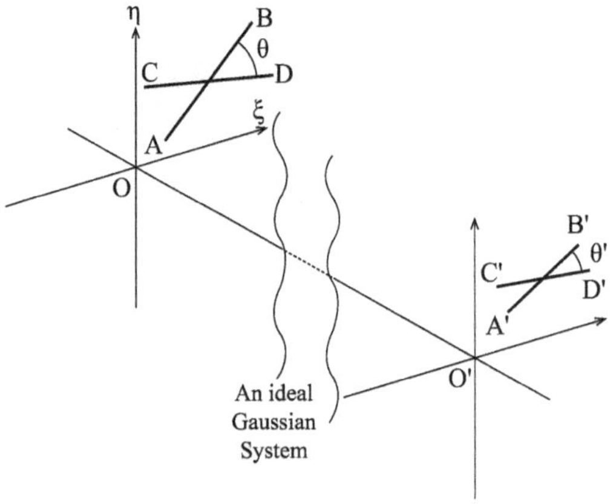

FIGURE 3.36 The angle of intersection between two lines on a transverse image plane remains the same as the angle between them on the transverse object plane.

The angle of intersection θ between AB and CD is given by

$$\tan\theta = \frac{m_1 - m_2}{1 + m_1 m_2} \tag{3.187}$$

The slopes m_1', m_2' of the two lines A′B′ and C′D′ can be expressed as

$$m_1' = \frac{\eta_1' - \eta_2'}{\xi_1' - \xi_2'} \qquad m_2' = \frac{\eta_3' - \eta_4'}{\xi_3' - \xi_4'} \tag{3.188}$$

The angle of intersection θ′ between A′B′ and C′D′ is

$$\tan\theta' = \frac{m_1' - m_2'}{1 + m_1' m_2'} \tag{3.189}$$

Since $\xi_i' = M\xi, \eta_i' = M\eta$ for $i = 1, \ldots, 4$, it follows $m_1' = m_1, m_2' = m_2$, and $\theta' = \theta$. The angle of intersection between two lines on a transverse image plane remains the same as the angle between them on the transverse object plane. Since transverse magnification is the same irrespective of the location of the object on the transverse object plane, any curved line is also imaged as a geometrically similar curved line. It follows that, per the model of an ideal Gaussian system, in an axially symmetric system, the shape of the image formed on a transverse image plane is identical with the shape of the corresponding object on the transverse object plane.

3.13.2 Imaging of Any Line in the Object Space

Figure 3.37 shows an axisymmetric imaging system with its axis $O_0 O O_0' O'$. A point Q_0 is on the transverse plane at the axial point O_0. Objects on this plane are imaged with magnification M_0 on the transverse plane at O_0'. Let Q_0' be the image of the point Q_0. Parallel Cartesian coordinates, centred on the respective axial points of the planes, specify coordinates of points on the transverse planes. Thus coordinates of Q_0 and Q_0' are (ξ_0, η_0) and (ξ_0', η_0'), respectively. Let Q be an arbitrary point located on the line Q_0Q. The point Q is the point of intersection of the line Q_0Q with the transverse plane located at the point O on the axis. The distance $O_0O = l$. Coordinates of the point Q are (ξ, η). The point Q is imaged

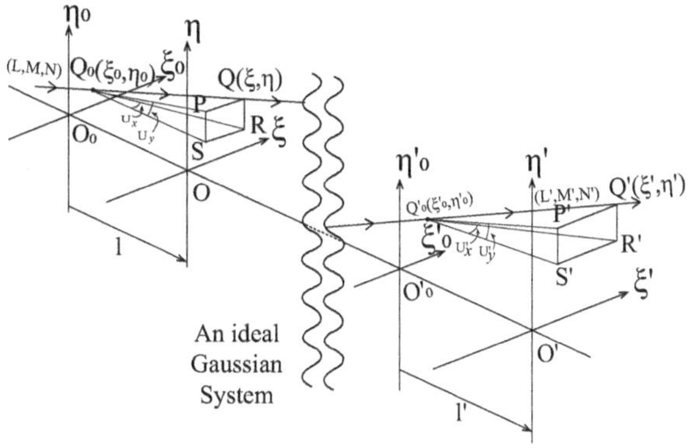

FIGURE 3.37 Ideal Gaussian model of imaging of a straight line Q_0Q by an axisymmetric system.

at the point Q′, located on the transverse plane at O′. Objects on the plane at O are imaged on the plane at O′ with transverse magnification M. Coordinates of the point Q′ are (ξ', η').

Let the direction cosines of the lines Q_0Q and $Q_0'Q'$ be (L, M, N) and (L', M', N'), respectively. The directions of the two lines Q_0Q and $Q_0'Q'$ are specified in terms of the angles U_x, U_y and U_x', U_y' respectively. The angles are indicated in the diagram (Figure 3.37). The signs of the angles are determined from the relations

$$\frac{L}{N} = \tan U_x \quad \frac{M}{N} = \tan U_y \quad \frac{L'}{N'} = \tan U_x'' \quad \frac{M'}{N'} = \tan U_y' \tag{3.190}$$

With the notations indicated in Figure 3.37,

$$\tan U_x = \frac{\xi - \xi_0}{l} \quad \tan U_y = \frac{\eta - \eta_0}{l} \quad \tan U_x' = \frac{\xi' - \xi_0'}{l} \quad \tan U_y' = \frac{\eta' - \eta_0'}{l'} \tag{3.191}$$

Substituting $\xi_0' = M_0\xi_0$, and $\xi' = M\xi$, and using (3.110), it follows

$$\tan U_x' = \left(\frac{n}{n'}\right)\frac{[M\xi - M_0\xi_0]}{lM_0M} \tag{3.192}$$

From (3.113) we get

$$\frac{n}{M} = \frac{n}{M_0} + \frac{l}{F} \tag{3.193}$$

From (3.192) and (3.193), by algebraic manipulation it follows

$$n'\tan U_x' = \frac{n\tan U_x}{M_0} - \frac{\xi_0}{F} \tag{3.194}$$

Similarly, it is also found that

$$n'\tan U_y' = \frac{n\tan U_y}{M_0} - \frac{\eta_0}{F} \tag{3.195}$$

For all object points on the line Q_0Q of Figure 3.35, the values of $\tan U_x, \tan U_y, \xi_0, \eta_0$ are the same. The images of these points lie on the straight line $Q_0'Q'$, whose direction is specified by the angles U_x' and U_y' as given by the equations (3.194) and (3.195), respectively. It is, therefore, proven that the image of any straight line by an ideal Gaussian system is a straight line. An immediate consequence is that the images of all planes in the object space are planes in the image space.

3.13.3 Suitable Values for Paraxial Angle and Height Variables in an Ideal Gaussian System

In Figure 3.37, if the points Q_0 and Q_0' are taken to be on the optical axis, i.e. at O_0 and O_0', respectively, then $\xi_0 = \eta_0 = 0$. From (3.194) – (3.195) the transverse magnification between these points is

$$M_0 = \frac{n \tan U_x}{n' \tan U_x'} = \frac{n \tan U_y}{n' \tan U_y'} \tag{3.196}$$

When the points O_0 and O_0' are chosen to be at the principal or unit points, $M_0 = +1$. The equations (3.194) – (3.195) reduce to

$$n' \tan U_x' = n \tan U_x - \xi_0 K \tag{3.197}$$

$$n' \tan U_y' = n \tan U_y - \eta_0 K \tag{3.198}$$

Note that, in the above equations, (ξ_0, η_0) are coordinates of the point Q_0 where the ray in the object space intersects the first principal plane. Comparison of (3.196) – (3.198) with (3.117) – (3.118) shows that, for an ideal Gaussian system, the paraxial angle variables have to be replaced by tangents of the angles of rays. The paraxial height variables then become the heights of the ray intersection points in planes perpendicular to the optical axis. Consequently, it follows that in Gaussian optics, the angular magnification for any ray is to be defined as the ratio of the tangent of the ray angle in the image space to the tangent of the corresponding ray angle in the object space, and not the ratio of the angles.

3.14 Gaussian Image of a Line Inclined with the Axis

Figure 3.38 shows an axially symmetric system of non-zero power. The first and the second principal planes of the system are located respectively at points H and H' on the axis. The refractive indices of the object and the image spaces are n and n′, respectively. A ray OG, inclined at an angle θ with the axis, is

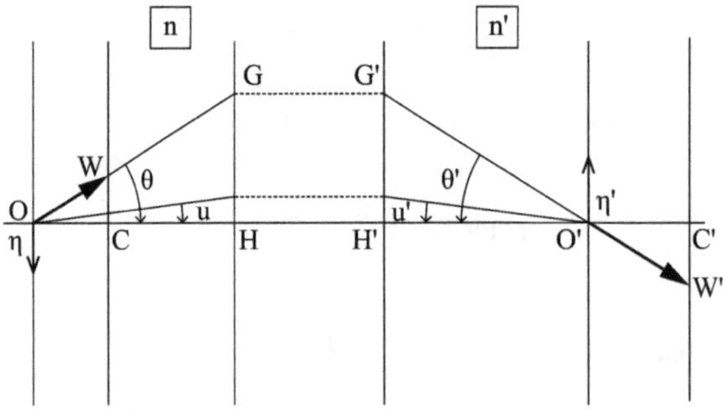

FIGURE 3.38 Gaussian image of a line inclined with the axis.

incident on the first principal plane at G. It emerges from the point G′ of the second principal plane and intersects the optical axis at the point O′. By definition of unit planes, the points G and G′ are such that GH = G′H′. Thus O′ is the image of O. Let HO = l, and H′O′ = l′. In the object space, OW is a line inclined at an angle θ with the axis. From the tip W of the line OW, WC is drawn perpendicular on the axis. Let the image plane corresponding to the object plane WC be at W′C′. The ray BGG′O′ intersects the plane W′C′ at W′. Thus, W′ is the image of W, and the line O′W′ is the image of the line OW. Let the angle of inclination of the line O′W′ with the axis be θ′.

Considering the points O and O′ as the local origin of coordinates, and following our sign convention, we get from the triangles OGH and O′G′H′,

$$l \tan\theta = l' \tan\theta' \tag{3.199}$$

since GH = G′H′. Recalling the paraxial invariant in relation to the imaging of the transverse object plane at O on the transverse image plane at O′, we write

$$nu\eta = n'u'\eta' \tag{3.200}$$

Using paraxial angle and height variables (shown in the diagram) we get

$$h = -lu = h' = -l'u' \tag{3.201}$$

Therefore

$$\frac{l}{l'} = \frac{u'}{u} = \frac{n\eta}{n'\eta'} \tag{3.202}$$

and from (3.199) and (3.202) it follows

$$n\eta \tan\theta = n'\eta' \tan\theta' \tag{3.203}$$

So, the product $n\eta\tan\theta$ is an invariant. The angle of inclination, θ′, of the line image is given by

$$\tan\theta' = \left(\frac{n}{n'}\right)\left(\frac{\eta}{\eta'}\right)\tan\theta = \left(\frac{n}{n'}\right)\left(\frac{1}{M}\right)\tan\theta \tag{3.204}$$

Since the principal planes do not exist in the case of afocal systems, the above derivation does not hold good when the imaging system is afocal or telescopic. For the latter systems, from (3.185), by substituting $M_0 = M$,

$$l' = \left(\frac{n'}{n}\right)M^2 l \tag{3.205}$$

Since $y' = My$, we get

$$\tan\theta' = \frac{y'}{l'} = \left(\frac{n}{n'}\right)\left(\frac{1}{M}\right)\tan\theta \tag{3.206}$$

same as (3.204).

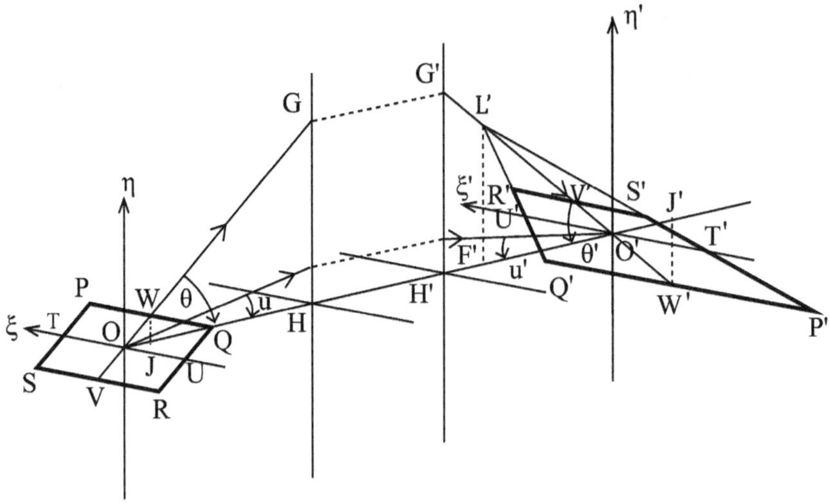

FIGURE 3.39 Keystone distortion in the Gaussian image of a tilted plane.

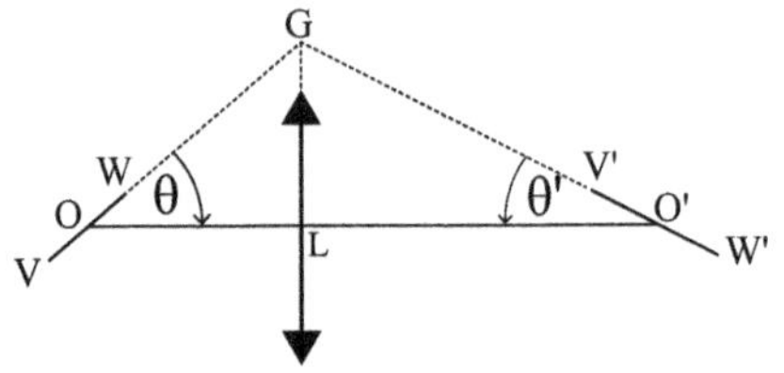

FIGURE 3.40 An illustration of the Scheimpflug principle for the special case of a thin lens in air.

3.15 Gaussian Image of a Tilted Plane: The Scheimpflug Principle

Figure 3.39 shows the same axisymmetric system as in Figure 3.38. A square object PQRS has its centre at O on the axis. Along ξ axis the square extends from U to T. The square is inclined with respect to the optical axis by an angle $\theta = \angle WOH = \angle GOH$. The axial point O is imaged at O′, and the line TOU is imaged as T′O′U′ along the ξ' axis. The transverse magnification between the planes at O and O′ is M. From the diagram, it is seen that the inclined object plane contains the line TOU and the point G. Correspondingly, the inclined image plane contains the line T′O′U′ and the point G′, and is thus uniquely determined. Let the inclination of the image plane with respect to the optical axis be $\theta' = \angle G'O'H'$. Indeed, Figure 3.38 shows the meridional section of the system shown in Figure 3.39, and the angle of inclination θ' is given by expression (3.204).

In the early years of the last century, Captain Theodor Scheimpflug of the Austrian army devised a systematic method for correcting perspective distortion in aerial photographs, and embarked upon a simple geometrical construction for determining the image plane when the object plane and the lens plane are not parallel to each other. This construction is well known as 'Scheimpflug principle' [11]. In his patent, Scheimpflug carried out extensive analysis for correction of this distortion [12]. Merklinger has recently noted that, although the principle is associated with the name of Scheimpflug, primacy on the discovery of the principle belongs to Jules Carpentier, and even Scheimpflug noted the same in his patent [13–14].

Figure 3.40 illustrates the special case of a thin lens in air. The transverse magnification between the conjugates O, O′ is M. From (3.204), by substituting $n = n' = 1$, the inclination of the image plane at O′, corresponding to an object plane, inclined by an angle θ with respect to the axis at O, is given by

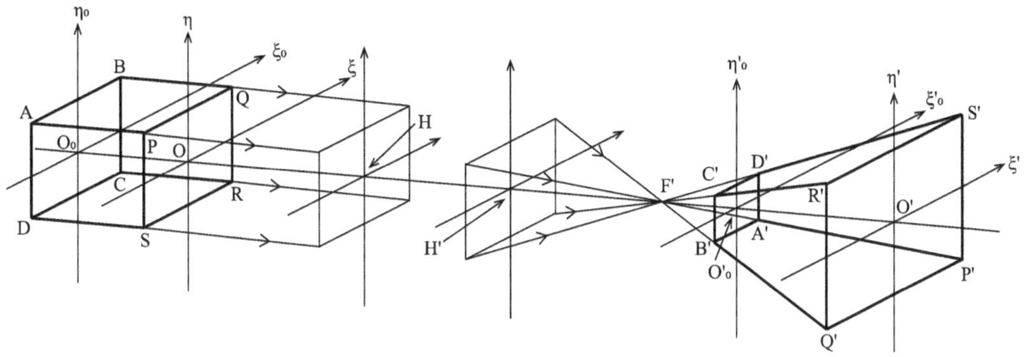

FIGURE 3.41 Gaussian mage of a cube placed symmetrically on the optical axis of an axisymmetric imaging system.

$$\theta' = \tan^{-1}\left\{\left(\frac{1}{M}\right)\tan\theta\right\} \tag{3.207}$$

Geometrically, the central line VOW (extended) of the object plane intersects the principal plane of the thin lens at G. The line GO′ lies on the image plane that is inclined to the optical axis by an angle $\theta'=\angle GO'L$. As shown in the last section, in case of afocal or telescopic systems, the angle θ' is given by the same relation (3.207), although the derivation is to be done differently.

3.15.1 Shape of the Image

In Figure 3.39, the lines PQ and SR in the square object are parallel to the axis O ξ, and, therefore their images P′Q′ and S′R′ are parallel to the axis O′ ξ'. But the line PQ is nearer to the principal plane at H, and the line SR is further from it. So, in general, magnifications of the corresponding images are unequal, and the image P′Q′R′S′ is in the shape of a keystone. The shape of this keystone can be determined by following three successive steps.

Step I. Let the second principal focus of the system be at F′, and F′L′ be a perpendicular drawn on the axis at F′. This perpendicular is on the meridional plane, and it intersects the ray G′O′ at L′. Each point on the perpendicular F′L′ is the point of intersection on the image space focal plane of a set of parallel rays in the object space making the same angle with the axis. The point L′ is, thus, the point of intersection of all rays in the object space parallel to the ray VOW that is inclined at angle θ with the optical axis. Such rays are STP, RUQ, etc.

Step II. The line TU lying on the axis O ξ is imaged as line T′U′ lying on the O′ ξ' axis. Joining L′T′ and L′U′, give the paths for the rays S′T′P′ and R′U′Q′ corresponding to the object space rays STP and RUQ respectively.

Step III. Let two perpendiculars WJ and W′J′ be drawn from W in the object space and W′ in the image space on the optical axis. The distances of the feet of perpendiculars J and J′ from O and O′ are OJ= $\tilde{l}$. O′J′ = $\tilde{l}'$ Since $\angle WOJ = \theta$, and $\angle W'O'J' = \theta'$, we get

$$\tilde{l} = \tilde{\eta}\cos\theta \quad \tilde{l}' = \tilde{\eta}'\cos\theta' \tag{3.208}$$

where OW = η, and O′W′ = $\tilde{\eta}'$. Taking the conjugates O and O′, between whom the transverse magnification is M, as references, it follows from (3.111)

$$\frac{M\eta'}{\tilde{\eta}'\cos\theta'} - \frac{n}{M\tilde{\eta}\cos\theta} = K = \frac{1}{F} \tag{3.209}$$

where K and F are the power and equivalent focal length of the system, respectively. All other quantities being known, the value of $\tilde{\eta}' = O'W'$ can be determined from (3.209). In order to determine the position of V′, or the distance O′V′, it is necessary to replace $\tilde{\eta}$ by $-\tilde{\eta}$ in (3.209).

It should be noted that this construction fails in the case of afocal or telescopic systems of zero power. However, it is obvious that, due to the difference in transverse and longitudinal magnifications, as given by (3.185), the image of a square on the tilted object plane of an afocal system is a rectangle on the corresponding tilted image plane.

3.16 Gaussian Image of a Cube

ABCD and PQRS are the two opposite vertical faces of a cube. The central points of the two vertical faces are located on the optical axis at O_0 and O respectively. H and H' are the principal points in the object space and the image space respectively. F′ is the image space principal focus. The refractive indices of the object and the image spaces are n and n′, respectively. The end faces at O_0 and O are imaged as squares at O'_0 and O′. Rays along the edges of the object cube, i.e. rays along AP, BQ, DS, and CR are parallel to the axis. Their points of incidence on the principal plane at H constitute a square. As shown in the diagram (Figure 3.41), the four rays emerge from the four corners of an exactly identical square on the principal plane at H' and pass through the second principal focus F′. The image of the cube is thus a truncated pyramid.

If the transverse magnifications for the conjugate positions O_0, O'_0, and O, O′ are M_0 and M, respectively, from (3.110) we get

$$l' = \left(\frac{n'}{n}\right) M_0 M l \tag{3.210}$$

where $l = O_0O$ and $l' = O'_0O'$ are the axial lengths of the object and the image, respectively. For the case when l is very small, $M_0 \approx M$, and (3.210) reduces to

$$l' = \left(\frac{n'}{n}\right) M^2 l \tag{3.211}$$

This implies that the image can be considered a rectangular parallelepiped when the size of the cube is very small. It should be noted that this is a gross approximation, and can lead to serious errors in many analyses, e.g. the action of stereoscopic microscope.

Although the construction given above is not valid in the case of afocal systems, since the transverse magnification of an afocal system is the same for any axial location of the object plane, the image of a cube is, in general, a rectangular parallelepiped whose length to height ratio is equal to $\left(\frac{n'}{n}\right)M$, and the image is another identical cube when $n' = n = 1$, and $M = \pm 1$.

REFERENCES

1 H.H. Hopkins, *Geometrical Theory of Image Formation*, University of Reading, U.K., personal communication (1982).
2 H.H. Hopkins, 'The nature of the Paraxial Approximation', *J. Mod. Opt.*, Vol. 38, (1991) pp. 427–445.
3 J.P.C. Southall, *The Principles and Methods of Geometrical Optics: Especially as applied to the Theory of Optical Instruments*, Macmillan, New York (1910), p. 201.
4 A.F. Möbius, Der barycentrische Calcul: ein neues Hülfsmittel zur analytischen Behandlung der Geometrie, Johann Ambrosius Barth, Leipzig (1827).
5 M. Born and E. Wolf, 'Projective transformation (collineation) with axial symmetry' in *Principles of Optics*, Cambridge University Press, Cambridge (1999) pp. 160–167. [First published by Pergamon, London in 1959].

6 C.F. Gauss, *Dioptrische Untersuchungen*, Lecture delivered on 10 Dec 1840. Published in Abhandlungen der Mathematischen Classe der Königlichen Gesellscaft der Wissenschaften zu Göttingen, Erster Band von den Jahren 1838–1841 (1843) pp. 1–34.
7 Lord Rayleigh (J. W. Strutt), 'Notes, chiefly Historical, on some Fundamental Propositions in Optics', *Phil. Mag.*, (5) Vol. 21 (1886) 466–476.
8 R. Smith, Chapter V in *A Compleat System of Opticks*, Book II, Cambridge (1738).
9 W.T. Welford, *Aberrations of Optical Systems*, Adam Hilger, Bristol (1986) p. 37.
10 P. Mouroulis and J. Macdonald, *Geometrical Optics and Optical Design*, Oxford University Press, London (1996) p. 88.
11 R. Kingslake, *Optical System design*, Academic Press, Orlando, 1983, p. 58.
12 T. Scheimpflug, 'Improved Method and Apparatus for the Systematic Alteration or Distortion of Plane Pictures and Images by Means of Lenses and Mirrors for Photography and for other purposes', *British Patent, GB* 1196 (1904).
13 H.M. Merklinger, 'Scheimpflug's Patent', *Photo Techniques*, Vol. 17, No. 6, Nov/Dec 1996.
14 J. Carpentier, 'Improvements in Enlarging or like Cameras', *British Patent, GB* 1139 (1901).

4

Paraxial Analysis of the Role of Stops

In the paraxial analysis of axisymmetric optical imaging systems carried out so far, we have not needed to take into account explicitly the size of elements or components of any real system in the transverse direction. So far, the analysis was based upon the premises that there is a one-to-one correspondence between the points in the object space, and their paraxial images in the image space. For a given object point, any two paraxial rays need to be traced from the point through the system for determining the paraxial image, which is given by the point of intersection of the two paraxial rays in the image space. On the other hand, if any two points on a paraxial ray are known in the object space, the straight line joining the paraxial images of the two points (assuming the image space to be homogeneous) gives the corresponding paraxial ray in the image space. The inherent linearity of the paraxial ray tracing formulae ensures that the axial distances predicted by them remain unaltered if both the paraxial angle variable u and the paraxial height variable h are multiplied by the same factor. Similarly, in Gaussian analysis, a parameter like transverse magnification $M = (\eta' / \eta)$ is not dependent on the absolute value of transverse size of the object, η. The effect of multiplying η by a scaling factor is that the size of the corresponding transverse image η' is multiplied by the same scaling factor, keeping M unchanged. Thus, in paraxial simulation of imaging by axisymmetric systems, the choice of these variables, and the corresponding paraxial rays, need not be unique, and multiple alternative options are available. It may so happen that many rays used in paraxial analysis are providing useful results, as sought for, but the rays are not physically realizable.

In general, the one-to-one correspondence between points in the object space and their paraxial images in the image space does not hold good for points in the object space and their real images. In real systems, the rays originating from an object point do not all intersect at the same point in the image space, thereby ruling out the existence of any single point that can be unambiguously defined as a point image corresponding to the object point. Indeed, a multitude of rays, originating from the point object, traverses through the system, and the rays then intersect each other within a small region in the three-dimensional image space. The shape and size of this region can be controlled by tailoring the pencil of rays taking part in the image formation. Also, at times it becomes necessary to have control over the light intensity in the images. Often, an iris diaphragm with an adjustable diameter is used for this control in some optical imaging systems, e.g., the camera. This is called an aperture stop. Even in the absence of a physical aperture stop, one of the elements or the components of the imaging lens system acts as an effective aperture stop. Correct analysis of the effects of limitations in the transverse extent of elements and components of an imaging system calls for real ray analysis. Nevertheless, suitably formulated paraxial analysis provides a useful approximation, and these aspects will be dwelt upon in subsequent sections of this chapter.

4.1 Aperture Stop and the Pupils

Many optical imaging systems have to operate under varying conditions of illumination. To regulate the amount of light reaching the image of a point object, usually an iris diaphragm, with an opening whose size can be controlled, is used. Keeping in view the analogous role of 'iris' in the human eye, Ernst Abbe introduced the term for the actual diaphragm that sets the size of the regular cone of light forming the image of an axial object point. This diaphragm, which is actually an 'aperture stop', is placed symmetrically with its centre on the axis. The lower the radius of the circular opening of the aperture, the less light reaches the image point, and vice versa.

Again, in real lens systems, rays in the image space that correspond to a point object do not all pass exactly through the image point, giving rise to loss of clarity in the image. In order to obtain sharper

DOI: 10.1201/9780429154812-4

images, it becomes necessary to limit the size of the image-forming pencils by using an aperture stop. In order to improve the quality of extra-axial image points, the pencils of rays forming the images of extra-axial object points are frequently made to pass eccentrically through some components of the system, and this is accomplished in practice with the help of an appropriate aperture stop.

The prime purposes of aperture stops in imaging lens systems are mentioned above. It should be noted that there are many lens systems where no explicit diaphragm is used to function as an aperture stop. In such cases, the rim of one of the surfaces of the lens elements or components, of any mirror, or of any other opening of the imaging system acts as the aperture stop, limiting the size of the cones of light forming the image points. For the sake of understanding the aberrational characteristics of any lens system, it is mandatory to know the size and axial location of the aperture stop of the imaging lens system.

The image of the 'aperture stop' in the object space, which is formed by part of the imaging system lying between the object and the aperture stop, is called the entrance pupil. If the aperture stop is located in the object space, the aperture stop itself is also the entrance pupil. Similarly, the image of the aperture stop in the image space, formed by the part of the imaging system that lies between the aperture stop and the image, is called the exit pupil. If the aperture stop is located in the real image space, the aperture stop itself is also the exit pupil. Similarly, if the aperture stop is located in the real object space, the aperture stop itself is also the entrance pupil. By definition, as above, the entrance pupil and the exit pupil are images of each other.

Figure 4.1(a) shows a thin lens at L on the axis. The lens is of semi-diameter $L\hat{A}$. An aperture stop S is located at the lens; it is of semi-diameter $SA = LA \leq L\hat{A}$. The image-forming pencil of rays emerging from the axial object point O forms the right circular cone of base AB and semi-vertex angle U in the object space, and it is transformed into the pencil of rays converging to the image point O′. In the image space, the pencil forms the right circular cone of base AB and semi-vertex angle U′. The aperture stop can be located away from the lens. Figure 4.1(b) shows the imaging geometry when the aperture stop is located in the object space between the object point O and the lens L. Note that the diameter of the stop is to be proportionately changed to keep the image-forming cone of light coming from the axial object point O unaltered. Figure 4.1(c) demonstrates the geometry when the aperture stop is located in the image

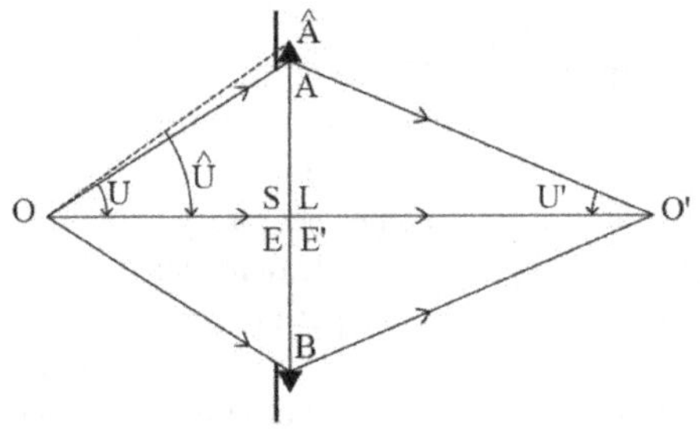

FIGURE 4.1(a) Aperture stop S located on the thin lens.

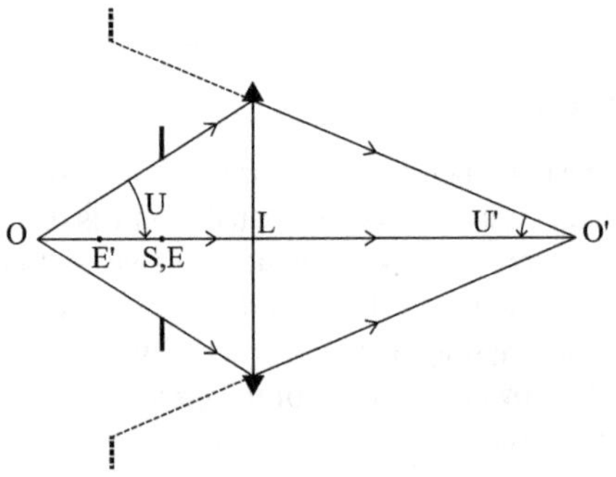

FIGURE 4.1(b) Aperture stop S located in the object space between the axial point O and the thin lens L.

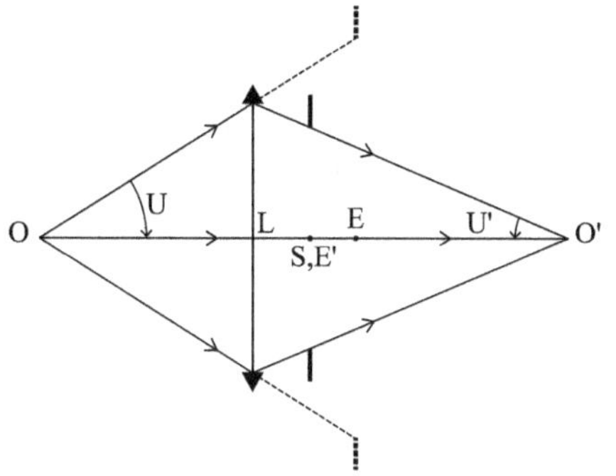

FIGURE 4.1(c) Aperture stop S located in the image space between the thin lens L and the image point O′.

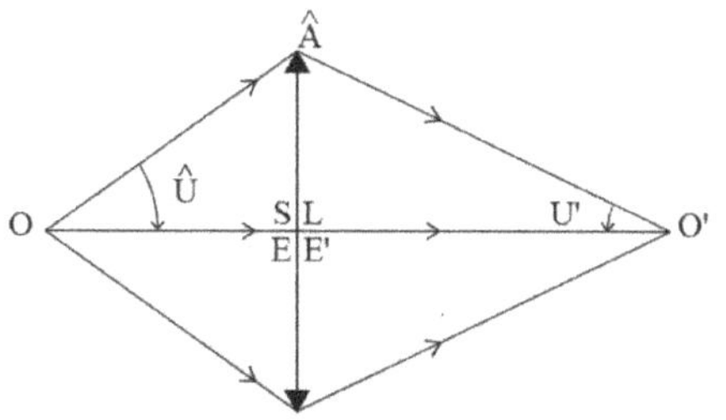

FIGURE 4.1(d) Rim of the lens as the aperture stop S.

space between the lens L and the image point O′. In the absence of any separate aperture stop, the rim of the lens acts as the aperture stop (Figure 4.1 (d)), but the imaging geometry is not necessarily the same. The right circular cone of light rays in the object space has the base $S\hat{A} = L\hat{A}$ and the semi-vertex angle $\hat{U}$. Corresponding changes take place in the right circular cone of the pencil of rays in the image space. The axial locations of the entrance pupil and the exit pupil are indicated in Figures 4.1(a)–(d) by E and E′, respectively.

Figure 4.2(a) shows the imaging geometry for the case of two separated thin lenses, L_1 and L_2, with an aperture stop S located between them. Figure 4.2(b) illustrates the situation for a Cooke triplet lens consisting of three separated thin lenses, L_1, L_2, and L_3, with an aperture stop located at the second lens. In each case, tentative locations for the entrance pupil and the exit pupil of the multi-element imaging system are shown in the diagrams. Figure 4.2(c) shows how the pupils of an axisymmetric lens system can often be seen. From the object side, if one looks into the lens against a bright background, the entrance pupil is visible. Similarly, the exit pupil is visible when one looks into the lens from the image side against a bright background. Note that in this particular case, the aperture stop is located inside the lens system, and both the entrance pupil and the exit pupil are virtual. Also, the sizes of the two pupils are not the same, and neither of them are of the same size as the aperture stop.

The first comprehensive report on the theory of stops was given by von Rohr [1]. He based his report on the early work of Petzval, [2] followed by the systematic investigations in the field by Ernst Abbe [3].

4.1.1 Conjugate Location and Aperture Stop

In axisymmetric imaging systems consisting of separated elements (a separate aperture stop, if any, is included in the list of elements), each with finite transverse extent, any one of the elements is a potential aperture stop, and, in general, no single element can act as aperture stop for all positions of the axial point in the object space. Therefore, it becomes imperative to determine the effective aperture stop for a specific conjugate location [4].

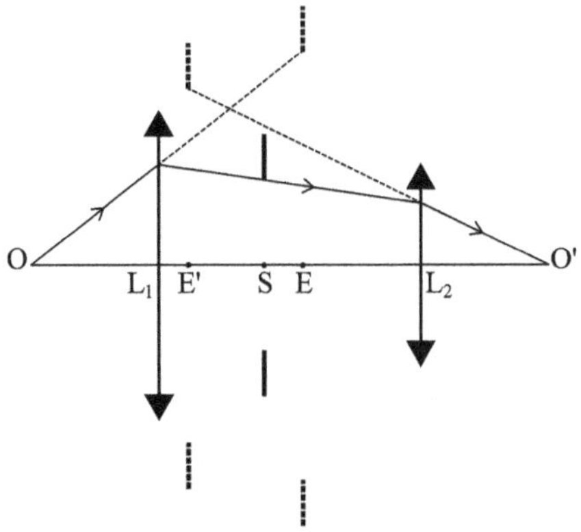

FIGURE 4.2(a) Imaging of axial point O by two separated thin lenses L_1 and L_2 with an aperture stop S located between them. The entrance pupil and the exit pupil are located at E and E′, respectively.

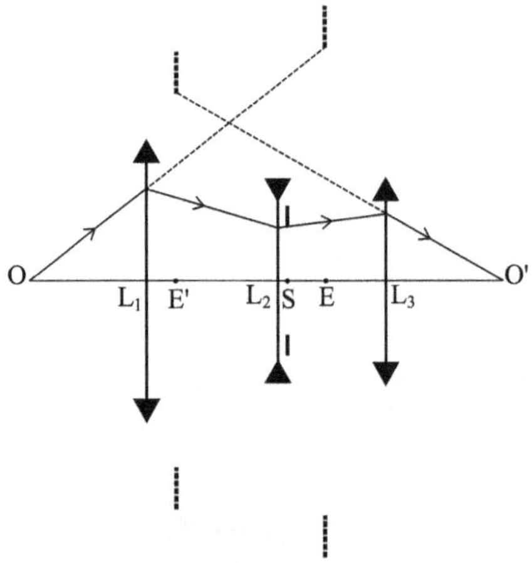

FIGURE 4.2(b) Location of aperture stop S in a Cooke triplet lens. O: Axial object; O′: Axial image; E: Entrance pupil; E′: Exit pupil.

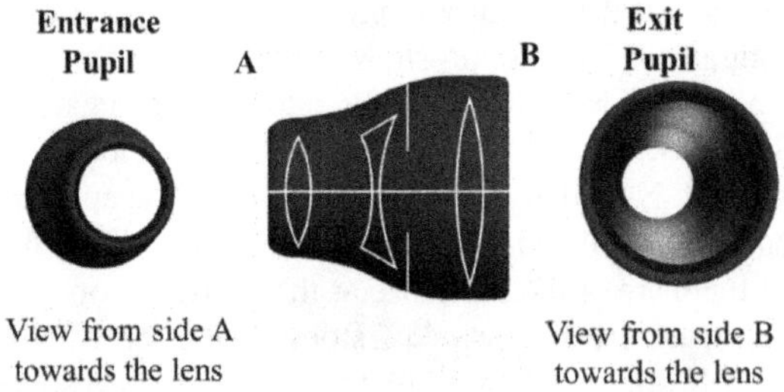

FIGURE 4.2(c) Views of the entrance pupil and exit pupil by looking normally through the lens system against a bright background from the object side and the image side, respectively.

Figure 4.3(a) shows the meridional section of an imaging system consisting of a single thin lens L with a remote stop S on the axis. $\hat{A}$ and $\hat{S}$ are two points on the rim of the lens and the rim of the opening of the aperture stop, respectively. The straight line $\hat{A}\hat{S}$ intersects the optical axis at point $\hat{Z}$. It can be checked from the diagram that for all axial object points lying left of point $\hat{Z}$, the diaphragm at S acts as the aperture stop. But for all axial object points lying to the right of $\hat{Z}$, the rim of the lens aperture at L acts as the aperture stop.

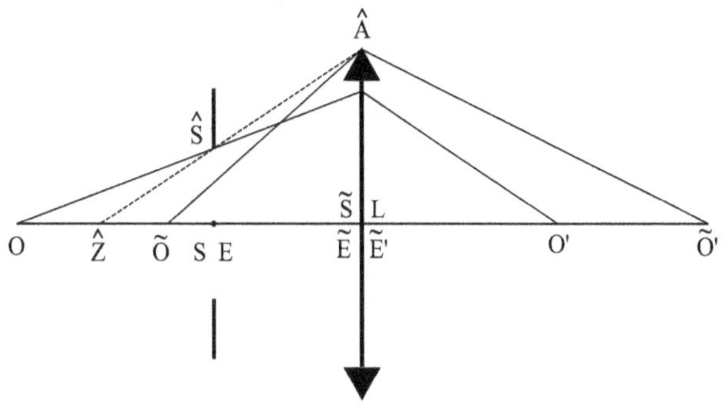

FIGURE 4.3(a) For all object points lying left of the axial point $\hat{Z}$, the diaphragm S is the aperture stop. For all object points lying on the right of $\hat{Z}$, the rim of the lens L is the aperture stop.

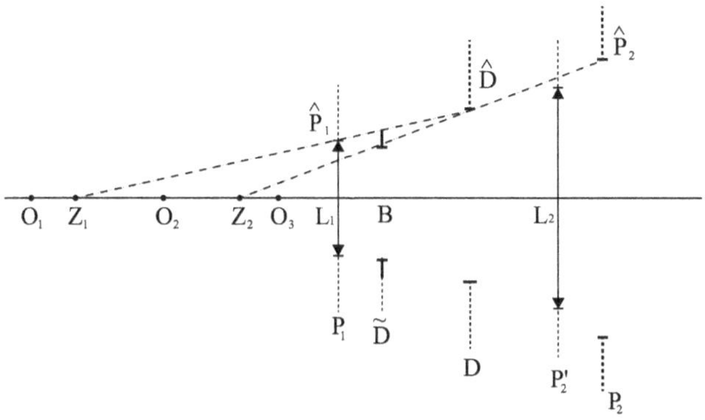

FIGURE 4.3(b) Different openings act as stop and pupils for different axial locations of the object.

Figure 4.3(b) shows the meridional section of an imaging system consisting of two separated lenses, L_1 and L_2, with a stop located at a point B on the axis. Each one of the three openings P_1, P_2', and $\tilde{D}$, which were provided by apertures of the two lenses and of the stop, is a potential aperture stop. Let the images of P_2' and $\tilde{D}$ formed by the lens L_1 in the object space be P_2 and D respectively. $\hat{P}_1$, $\hat{P}_2$, and $\hat{D}$ are the points on the rim of the apertures P_1, P_2, and D respectively. The straight line $\hat{D}\hat{P}_1$ intersects the optical axis at the point Z_1, and the straight line $\hat{P}_2\hat{D}$ intersects the axis at Z_2. From the diagram it is seen that, for any axial point O_1 lying on the left of the point Z_1,

$$\angle \hat{P}_1 O_1 L_1 < \angle \hat{D} O_1 L_1 < \angle \hat{P}_2 O_1 L_1 \tag{4.1}$$

so that P_1 is the aperture stop as well as the entrance pupil, and its image at P_1' (not shown in Figure 4.3(b)) formed by lens L_2 is the exit pupil. For axial points O_2 lying between Z_1 and Z_2,

$$\angle \hat{D} O_2 L_1 < \angle \hat{P}_2 O_2 L_1 < \angle \hat{P}_1 O_2 L_1 \tag{4.2}$$

Therefore, for axial point O_2, $\tilde{D}$ is the aperture stop, D is the entrance pupil, and D′ is the exit pupil. For axial points O_3 lying on the right of Z_2,

$$\angle \hat{P}_2 O_3 L_1 < \angle \hat{D} O_3 L_1 < \angle \hat{P}_1 O_3 L_1 \tag{4.3}$$

so that P_2 is the entrance pupil, and P_2' is both the aperture stop and the exit pupil.

4.2 Extra-Axial Imagery and Vignetting

The definition of relative aperture, e.g. F-number or numerical aperture, etc., of any axi-symmetric imaging system is based on the configuration of axial imagery. The shape of the actual pencil of rays forming the image of the axial object point changes as it traverses the image forming system. In the object space, this pencil of rays diverging from the object point constitutes a right circular cone with the entrance pupil as its base and the axial object point as its vertex. In the image space, the corresponding converging bundle of rays constitutes a right circular cone with the exit pupil as its base, and the axial image point as its vertex. In intermediate spaces, the image-forming pencil of rays constitutes truncated right circular cones. The cross-section of the image-forming pencil of rays on the aperture stop must exactly match the axially symmetric opening in the aperture stop, and all refracting interfaces must have their diameters large enough to allow unhindered passage of the axial bundle of rays. Accordingly, the clear openings in all surfaces of the system need to be larger than, or at least equal to the footprint of the axial beam on the corresponding surface. The quality of axial imagery is independent of the actual transverse extent of a surface so long as the inequality constraint is satisfied.

However, the transverse extent of the surfaces plays a major role in the imaging of extra-axial points. Figure 4.4(a) shows an imaging system consisting of a single thin lens and a remote front stop. It may be noted that the diameter of the lens is slightly larger than the footprint formed on the lens by the cone of light coming from the axial point O. For an extra-axial object point O_1, not far from the axis, the cone of light originating from O_1 and passing through the aperture stop is incident obliquely on the lens, and passes unhindered to form the image at O_1' (Figure 4.4(b)). For extra-axial points like O_2, lying further away from the axis (Figure 4.4(c)), it is noted that the oblique cone of light, originating from O_2 and passing by the aperture stop, is not fully admitted by the aperture of the lens system. Only a part of the cone of light originating from the object point takes part in formation of the image at the point O_2'. Such limitation of an extra-axial pencil by the rim of a lens is termed 'vignetting'. Figure 4.4(d) shows

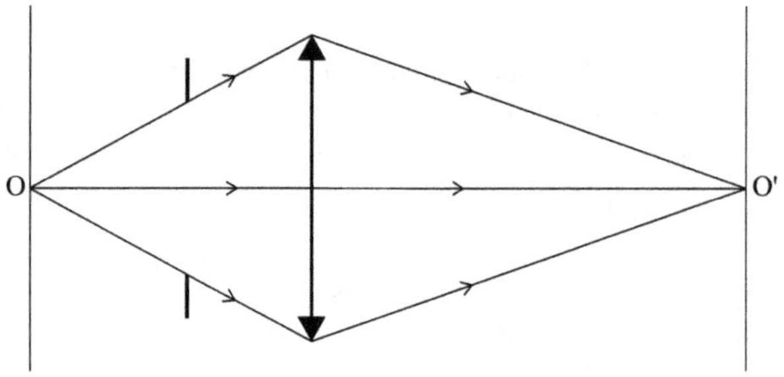

FIGURE 4.4(a) Pencil of light forming the image of an axial point O by a system consisting of a single thin lens and a remote front stop.

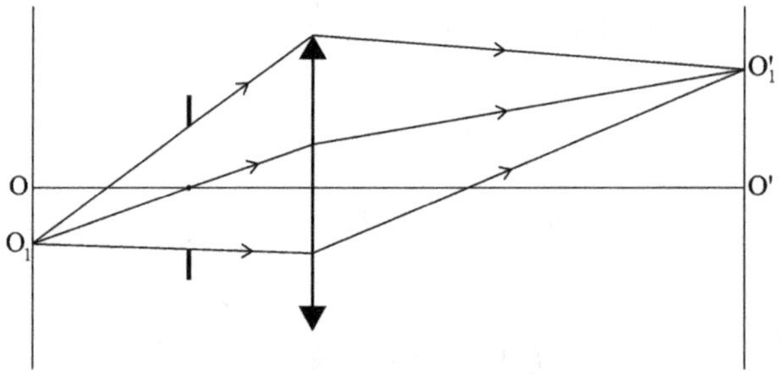

FIGURE 4.4(b) Pencil of light forming the image of an extraaxial point O_1, not far from the axis, by the same system as Figure 4.4(a).

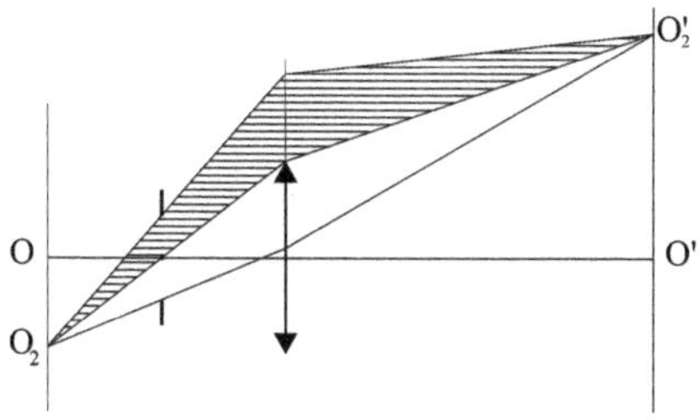

FIGURE 4.4(c) Pencil of light forming the image of an extraaxial point O_2, lying far away from the axis, by the same system as Figure 4.4(a); a part of the pencil (shown by hatched lines) admitted by the entrance pupil/stop is cut off by the lens aperture.

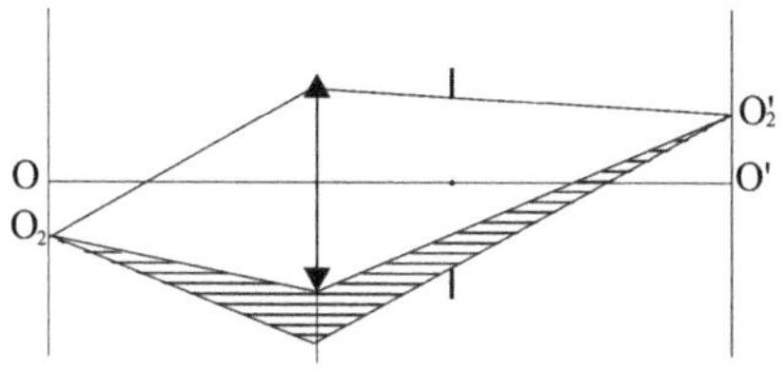

FIGURE 4.4(d) Pencil of light forming the image of an extraaxial point O_2, lying far away from the axis, by the same lens as Figure 4.4(a) but with the stop in the image space; a part of the pencil (shown by hatched lines) that could have passed unhindered through the stop/exit pupil is not being admitted on account of the finite size of the aperture of the lens.

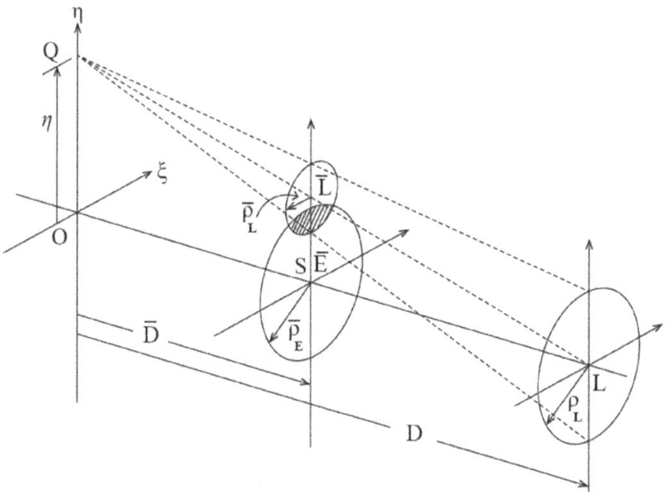

FIGURE 4.5 The shaded overlap region in the vignetted entrance pupil forms the base of the image forming pencil of rays originating from the extraaxial point Q in an imaging system consisting of a single lens with a remote front stop.

vignetting in the case of imaging of an extra-axial point O_3 by a single thin lens with a remote stop in the rear.

In both cases of front and rear remote stop of a single lens, the full aperture is not transmitted by the lens system for extra-axial object points that lie away from the axis. The effect of the vignetting of the shading regions in Figure 4.4(c) and Figure 4.4(d) is to produce a cat's eye shaped pupil. Figure 4.5 shows the effect for the case of a single lens with a remote front stop located at points L and S on the axis, respectively. The transverse object OQ of height $OQ = \eta$ is located at O on the axis. In this case, the aperture stop is also the entrance pupil, so that S and $\bar{E}$ is the same point. Let the semi-diameters of the entrance pupil and the lens be $\bar{\rho}_E$ and ρ_L, respectively. The axial distances $O\bar{E} = \bar{D}$, and $OL = D$. Any

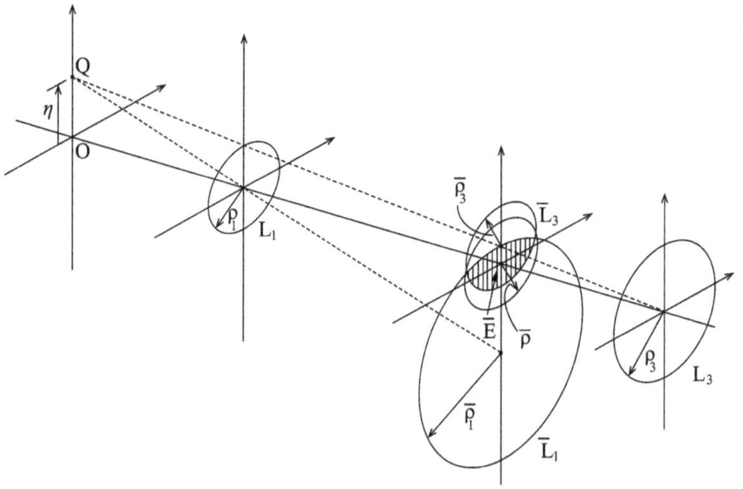

FIGURE 4.6 Cat's eye diagram demonstrating the vignetting effect in the object space of the triplet system of Figure 4.2(b).

ray from an object point Q passes unimpeded through the system, if it passes through both the apertures located at $\bar{E}$ and L. For the object point Q, let a projection be drawn on the plane of the entrance pupil at $\bar{E}$, using Q as the centre of projection. The projection is a circle with its centre $\bar{L}$ and radius $\bar{\rho}_L$ given by

$$\overline{E}\overline{L} = \frac{(D - \bar{D})}{D}\eta \tag{4.4}$$

$$\bar{\rho}_L = \frac{\bar{D}}{D}\rho_L \tag{4.5}$$

Under paraxial approximation, the shaded overlap region of the two circles is the vignetted entrance pupil corresponding to the extra-axial point Q. A similar procedure adopted in the image space will yield the vignetted exit pupil.

Figure 4.6 shows the cat's eye diagram in the object space for the triplet lens of Figure 4.2(b). The first lens at L_1 is located in the object space. Its semi-diameter is ρ_1. In the triplet lens of Figure 4.2(b) the aperture stop is located on the second lens. The image of the second lens (and the aperture stop) in the object space is located at $\bar{E}$, and its semi-diameter is $\bar{\rho}$. By definition, it is the entrance pupil of the system. L_3 is the axial location of the image of the third lens in the object space. The semi-diameter of this image is ρ_3. Therefore, in the object space there are three potential diaphragms for limiting the size of the image-forming beams corresponding to different object points. For the sake of analyzing the imaging of the extra-axial point Q at height $OQ = \eta$, projections of the circles at L_1 and L_3 are drawn on the entrance pupil plane at $\bar{E}$, with Q as the centre of projection. Any ray originating from Q will pass through the system if it passes through the common area of overlap of the three circles of semi-diameters $\bar{\rho}_1$, $\bar{\rho}$ and $\bar{\rho}_3$ with their centres at $\bar{L}_1\bar{E}$, and $\bar{L}_3$, respectively, on the entrance pupil plane. The shaded region in the figure is this common area. Note that, in the case of the axial point O, corresponding points $\bar{E}, \bar{L}_3$, and $\bar{L}_1$ coincide with each other on the entrance pupil plane, and $\bar{\rho}_1, \bar{\rho}_3 \geq \bar{\rho}$.

4.2.1 Vignetting Stop

In complex optical systems consisting of many lens elements, the vignetted cat's eye diagram can take complicated shapes [5–6]. Nevertheless, in most practical applications an ellipse can approximate it [7]. However, the rate at which vignetting takes effect in such systems with increasing heights is different for

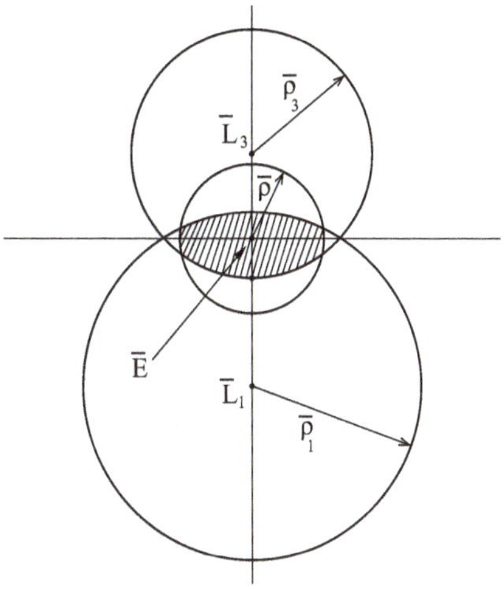

FIGURE 4.7 A side view of the cat's eye diagram of the vignetted pupil for the triplet system of Figure 4.2(b).

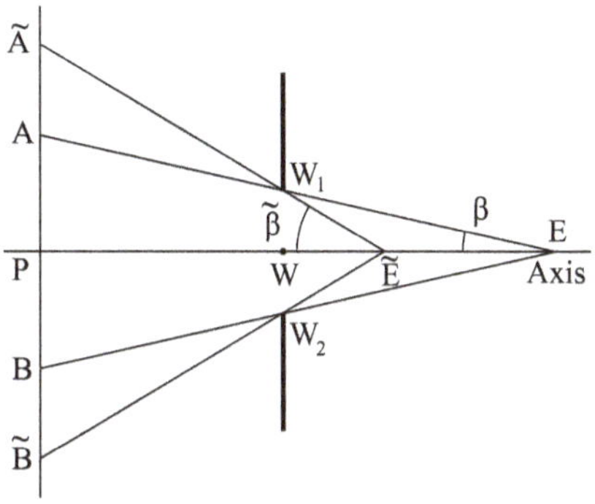

FIGURE 4.8 Effects of axial shift from E to $\tilde{E}$ in the location of an observer looking through a fixed field stop/entrance window W_1W_2 on an angular field of view (for objects at either a finite distance or an infinite distance) and linear field of view (in the case of objects at a linear distance only).

different surfaces. In such cases a side view of the cat's eye diagram is often helpful to identify the surface at which vignetting can occur most effectively. Figure 4.7 shows a side view of the cat's eye diagram for the triplet system already discussed. Occasionally, if no surface is found to be located suitably for undertaking the task, a vignetting stop in addition to the usual aperture stop is used in practice for stopping out the badly aberrated outer part of oblique pencils.

4.3 Field Stop and the Windows [8]

4.3.1 Field of View

In Figure 4.8, W is the midpoint of the opening W_1W_2 of a window. PWE is a line perpendicular to W_1WW_2. Let the eye of an observer be located at E. Assume a transverse object APB located at a finite distance from the window. At this position of the eye, only the part AB of the object will be visible to the eye. With the given positions of the window, the eye, and the object, the semi field of view of the observer in linear measure is equal to PA. The semi angular field of view of the observer is $\angle AEP = \beta$. If the eye of the observer is advanced towards the window to a position $\tilde{E}$, the linear semi field of view increases to $P\tilde{A}$; correspondingly the angular semi field of view increases to $\tilde{\beta}$. For an object at infinity or very far from the window, the linear measure cannot be used, and the observable space is specified by the angular field of view.

4.3.2 Field Stop, Entrance, and Exit Windows

It is seen from Figure 4.8 that, for a specific position of the observer's eye on the axis, the field of view of the observer is determined by the window. Therefore, W_1W_2 is called the 'field stop'. The image of the field stop in the object space is called the entrance window, and its image in the image space is called the exit window. In Figure 4.8, the imaging element is the eye lens. The field stop is in the object space, so it is also the entrance window. The exit window is the image of the field stop formed by the eye lens.

4.3.2.1 Looking at an Image Formed by a Plane Mirror

Figure 4.9 shows the field of view, the field stop, and the windows when an observer looks at a plane mirror W_1WW_2. The eye lens of the observer is located at E′ on the axis. The mirror forms a virtual image of the eye lens at E. Ray paths forming the image Q′ of the object point Q are shown in the figure to illustrate that the entrance pupil of the imaging system is at E, and the exit pupil is at E′. It is important to note that although the plane mirror forms a perfect three-dimensional image of the three-dimensional object space, no ray from points beyond a restricted region of the object space can reach the exit pupil after reflection from the mirror due to the finite size of the mirror. In Figure 4.9, the observable image points Q′ lie within the region $T_1'W_1WW_2T_2'$, corresponding to object points Q lying within $T_1W_1WW_2T_2$. The observer cannot observe images $\hat{Q}'$ of object points $\hat{Q}$. The border of the mirror is, therefore, the field stop of the system. As shown in the figure, the semi angular field of view β is the same in both the object space and the image space, and the field stop is also the entrance window and the exit window.

4.3.2.2 Looking at Image Formed by a Convex Spherical Mirror

In Figure 4.10, W_1WW_2 is a convex spherical mirror with the centre at C on the axis E′WC. The eye of the observer is located at E′; its virtual image by the mirror is at E on the axis. Thus the entrance and exit pupils of the imaging system are located at E and E′ on the axis. Paraxial ray paths from the object point Q to the mirror, and then to the eye E′ of the observer are shown. The rays, incident on the eye, appear to come from point Q′, the image of point Q by the mirror. As before, the image points Q′ located within the

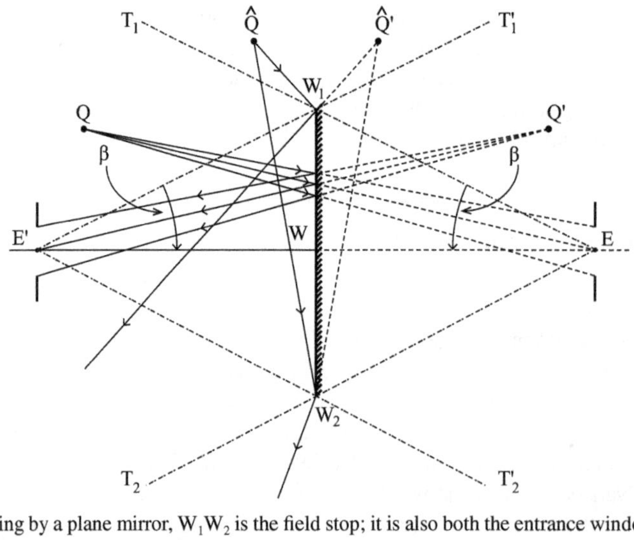

FIGURE 4.9 In imaging by a plane mirror, W_1W_2 is the field stop; it is also both the entrance window and the exit window. Angular field of view is the same for both the object space and the image space.

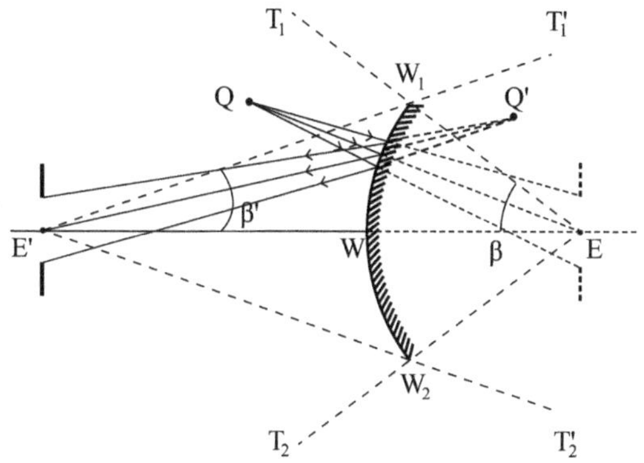

FIGURE 4.10 In imaging by a convex spherical mirror, W_1W_2 is the field stop; it is also both the entrance window and the exit window. Semi-angular field of view in object space: β, and in the image space: β′.$(\beta \neq \beta')$.

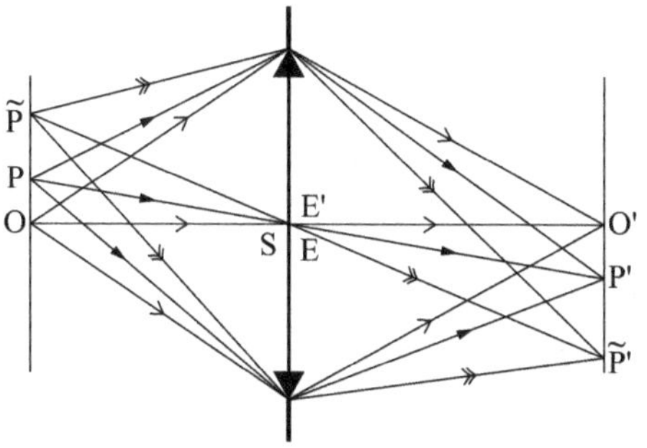

FIGURE 4.11 Imaging by a single lens with an aperture stop on it; with increasing distance of the extra-axial point from the axis, the obliquity of its image forming a cone of light increases – leading to loss of illumination at the corresponding image point.

region $T_1'W_1WW_2T_2'$ and corresponding to object points Q lying within $T_1W_1WW_2T_2$ are observable by the viewer. The semi angular field of view in the object space and the image space are unequal, and they are β and β′, respectively. In this imaging, the border of the mirror W_1W_2 is the field stop; it is also the entrance window and the exit window, since it may be considered to be located in both the object space and the image space.

4.3.2.3 Imaging by a Single Lens with a Stop on It

A transverse object OP located at point O on the axis is imaged by a single lens with a stop on it; the transverse image O′P′ is located at O′ on the axis. The entrance pupil E and the exit pupil E′ coincide with the stop S on the lens (Figure 4.11). As it is, the stop S acts as an aperture stop, and it does not affect the field covered by the lens. With increasing distance of the point P from the axis, the obliquity of the image-forming cone of light increases, and consequently the illumination at the image point decreases. Therefore, any definition of an effective field of view of this imaging system has to be based on acceptable levels of illumination in extra-axial points.

4.3.2.4 Imaging by a Single Lens with an Aperture Stop on it and a Remote Diaphragm in the Front

Figure 4.12 shows a single lens with an aperture stop on it; the lens images the transverse object OP at O′P′. A remote circular diaphragm is located at the point F on the axis. It is seen that this diaphragm limits the field being imaged by the lens. The semi-angular field of view is β (=∠PEO), and the semi field of

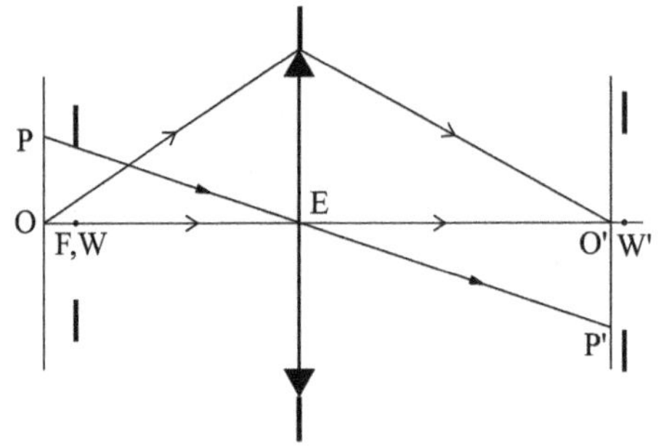

FIGURE 4.12 Imaging by a single lens with an aperture stop on it and a remote diaphragm in the front. The latter acts as a field stop/entrance window.

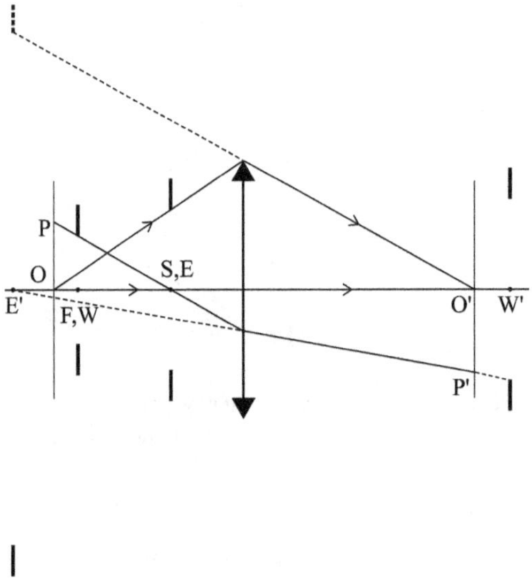

FIGURE 4.13 Imaging by a single lens with two limiting diaphragms at S and W on the axis in front of the lens; axial locations of S and W are chosen arbitrarily, and they are assigned roles of aperture stop and field stop respectively.

view in linear measure is η (= PO). Note that the image space field of view is equal to the object space field of view β. The diaphragm at W is the field stop, and since it is located in the object space, it is also the entrance window. The image of the field stop in the image space is located at W′ on the axis, and this image is the exit window.

4.3.2.5 Appropriate Positioning of Aperture Stop and Field Stop

Aperture stop and field stop are supposed to play two distinctly different roles in an optical system. Changing the diameter of an aperture stop should not affect the field size, and should affect only the size of the axial cone of light rays forming the image. On the other hand, changing the diameter of a field stop should affect only the size of the object/image field, and should by no means affect the level of illumination in the image. This calls for appropriate positioning on the axis of the iris diaphragms that are supposed to play the role of a field stop and an aperture stop.

For the sake of illustration, Figure 4.13 shows a single lens L with an aperture stop S and a field stop F located arbitrarily on the axis in front of the lens. Figure 4.14(a) shows that the semi-angular aperture and semi-angular field of view of the imaging system are α and β, respectively. When the diaphragm at

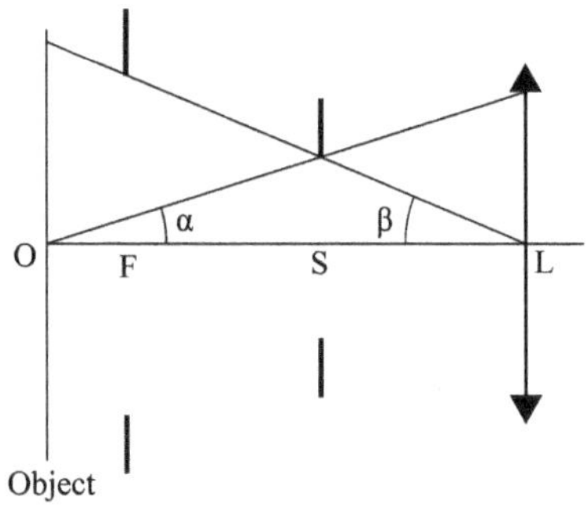

FIGURE 4.14(a) Semi-angular aperture α and semi-field of view β in the system of Figure 4.13.

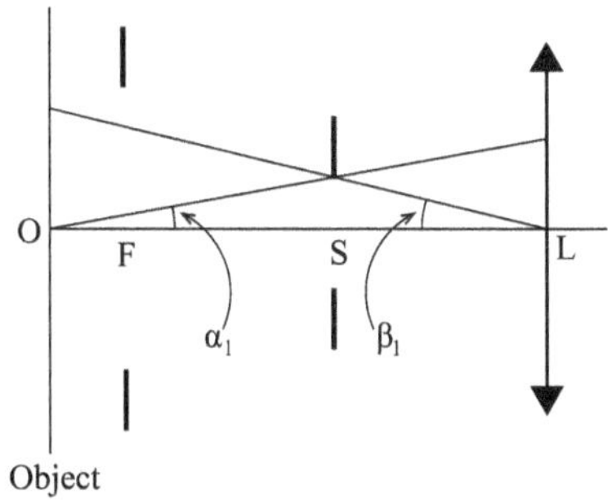

FIGURE 4.14(b) When the diaphragm at S in the system of Figure 4.13 is stopped down, both semi-angular aperture and semi-field of view are decreased to α_1 and β_1, respectively.

S is stopped down, as shown in Figure 4.14(b), the semi-angular aperture and the semi-angular field of view become u_1 and β_1, respectively, where $\alpha_1 < \alpha$ and $\beta_1 < \beta$. Thus the role of the aperture stop and the field stop is messed up by inappropriate locations of the stops. Figure 4.15 demonstrates the appropriate locations for the two types of stops in this case. An iris diaphragm with controllable semi-diameter ρ_{FS} (where $S_1A_1 < \rho_{FS} < S_1B_1$) and located at a point S_1 on the axis (where S_1 lies between the axial object point O and the point S_2) will act as a field stop over a substantial working range determined by exact location of S_1 on the axis. Similarly, an iris diaphragm with controllable semi-diameter ρ_{AS} (where $S_3A_3 < \rho_{AS} < S_3B_3$) and located at a point S_3 on the axis (where S_3 lies between L and S_2) will act as an aperture stop over a working range decided by the axial location of the point S_3.

4.3.2.6 Aperture Stop and Field Stop in Imaging Lenses with No Dedicated Physical Stop

Many imaging lenses operating under more or less the same level of illumination usually do not have dedicated physical diaphragms to act as either aperture stop or field stop or both. Rims of one of the lenses or other optical elements determine the effective aperture stop and the effective field stop. Often it is possible to identify the surface playing the predominate role in effective aperture stop by tracing a paraxial ray from the axial object point to the corresponding image point, and by determining the heights $h_i, i = 1,\ldots,k$ for all k surfaces. Let $\rho_i, i = 1,\ldots,k$ be the corresponding semi-diameters of the surfaces. The surface with the highest value for the ratio $\left|\left(\frac{h_i}{\rho_i}\right)\right|$ is the nearest neighbour estimate for the effective aperture stop. For some systems it becomes possible to use the effect of vignetting in the search for effective

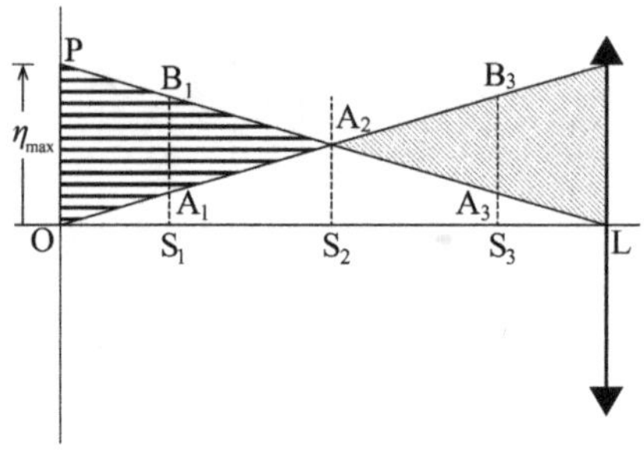

FIGURE 4.15 Appropriate axial locations for aperture stop and field stop in imaging by a single lens.

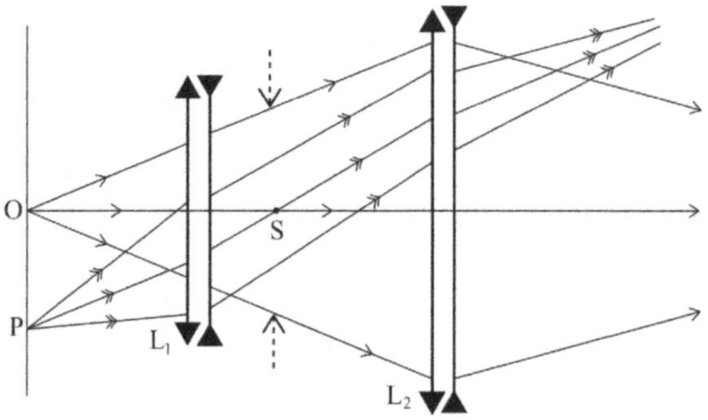

FIGURE 4.16 Axial location of the effective aperture stop for imaging of an extra-axial point P in a Petzval type lens, which does not contain any physical diaphragm to act as an aperture stop.

aperture stop. Figure 4.16 shows a Petzval type projection lens consisting of two cemented doublets placed apart by a distance. As shown in the figure, the lower part of the image-forming pencil from the off-axial point P is vignetted by lens L_1, and the second lens, L_2, vignettes the upper part of the same cone. For an extraaxial point P, the point of intersection of the central ray of its actual forming pencil with the optical axis of the system is the location of the effective aperture stop on the axis. This approach for determining the axial location of an effective aperture stop can be used in practice for systems with a small field of view. For points lying in the outer region of large field of view systems, the axial location of the midpoint of the vignetted bundle of image-forming rays can be significantly different from S. Indeed, the location of effective aperture stop becomes different for different extra-axial points.

The case for the field stop is somewhat similar. Once the location of the effective aperture stop is known, the location of the entrance pupil can be determined by paraxial ray tracing. A paraxial ray from the centre of the entrance pupil is traced through the system up to the exit pupil, and the height of the ray at all surfaces $\bar{h}_i, i = 1,\ldots,k$ is noted. The surface with the highest value for the ratio $\left|\frac{\bar{h}_i}{\rho_i}\right|, i = 1,\ldots,k$ is the nearest neighbour to the effective field stop.

4.3.2.7 *Imaging by a Multicomponent Lens System*

Figure 4.17 shows imaging by an axisymmetric system consisting of *k* refracting interfaces. A_1 is the vertex of the first interface, and A_k is the vertex of the last interface. The object POQ is imaged as P′O′Q′. E and E′ are the axial locations of the entrance pupil and the exit pupil. The semi-angular aperture of the system in the object space and the image space are α_0 and α_i, respectively. W and W′ are the axial locations of the entrance window and the exit window. The corresponding semi-angular field of view in the object space and the image space are β_o and β_i, respectively. Note that, in general,

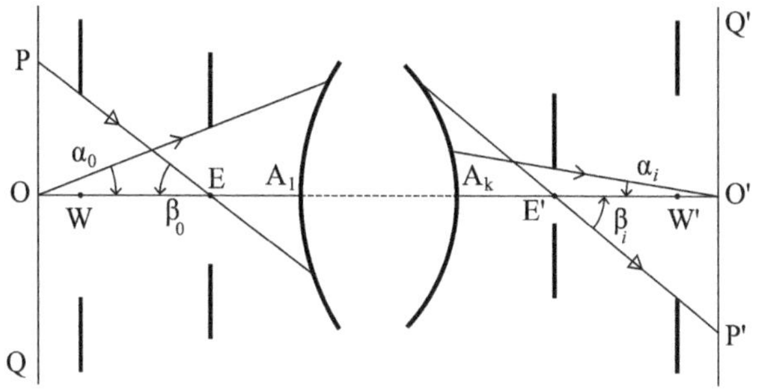

FIGURE 4.17 In general, in an axisymmetric imaging system, the semi-angular aperture and the semi-field of view in the object space and the image spaces are not equal. $|\alpha_0| \neq |\alpha_i|$ and $|\beta_o| \neq |\beta_i|$.

$$|\alpha_o| \neq |\alpha_i| \qquad |\beta_o| \neq |\beta_i| \tag{4.6}$$

In the absence of a dedicated physical field stop, the finite contour of the detector is the effective field stop. However, in multicomponent systems, a physical field stop is often placed near a plane where a reasonably sharp intermediate image is formed. One of the reasons for this is to cut off highly vignetted or ill-defined outer parts of the image. In long narrow relay systems, e.g. submarine periscopes, medical endoscopes, etc., it is advisable to use a field stop at all locations of the intermediate images.

But the duplicate field stops should have a diameter that is slightly in excess of the principal field stop, the latter often being the one located near the detector. This will prevent the occurrence of undesirable non-circular fields of view in case of slight misalignment in the system. In order to eliminate the possibility of carrying forward unwanted blemishes of the intermediate images, the field stops are usually placed near, and not exactly on, the intermediate images.

4.3.2.8 Paraxial Marginal Ray and Paraxial Pupil Ray

In paraxial analysis of imaging by optical systems, two rays play major roles. The first one is the paraxial marginal ray (PMR), which originates from the axial object point and passes through the margin or edge of the entrance pupil. In paraxial treatment, the PMR passes through the margin of both the aperture stop and the exit pupil. The other ray is the paraxial pupil ray (PPR). It originates from the edge of the field and passes through the centre of the entrance pupil. The pupil ray is also known as the principal ray, the central ray, or the chief ray. The PPR passes through the edge of the field stop, the entrance window, and the exit window.

4.4 Glare Stop, Baffles, and the Like [9]

The quality of imaging by an optical system does not only depend on how faithfully the system is reproducing the object in the image space, but, to a large extent, it is also dependent on how efficiently the system is preventing stray radiation from appearing on the image. The use of a single field stop alone at the detector is not adequate for blocking radiation from points outside the desirable field of view. The primary role of hoods in front of camera lenses is to stop light from bright points lying outside the field of view by blocking (Figure 4.18). In absence of the hood, the unwanted light enters the system through the camera lens and affects the quality of image formation by giving rise to many unpredictable effects, e.g. ghost imaging, contrast reduction, uneven illumination, etc. However, external hoods can play a limited role, for they are required to ensure that light from the useful field of view is not blocked.

If a real exit pupil is available, a diaphragm can be used there to arrest propagation of stray radiation in the final image. This diaphragm is called a 'glare stop'. Figure 4.19 shows the use of such a glare stop in an erecting telescope. The aperture stop S and the entrance pupil E are located at the primary objective, and the exit pupil is at E′. The glare stop is located at E′. Stray light entering through the objective lens

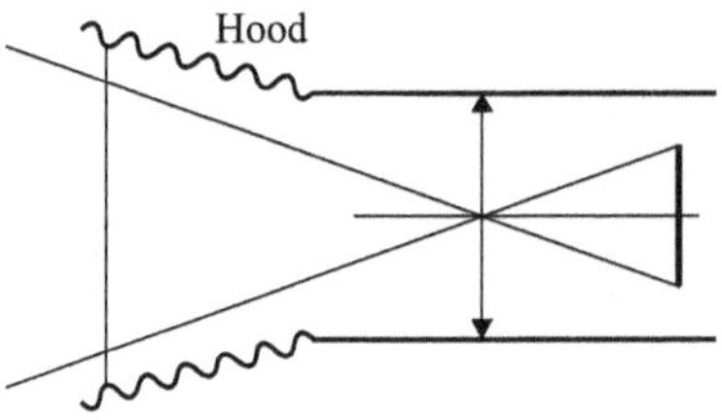

FIGURE 4.18 In a camera lens, a hood is often used to obstruct stray light from bright points lying outside the field of view.

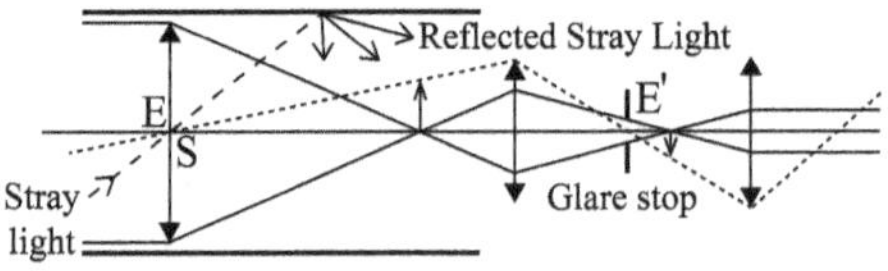

FIGURE 4.19 A glare stop located at the exit pupil E′ to obstruct stray light.

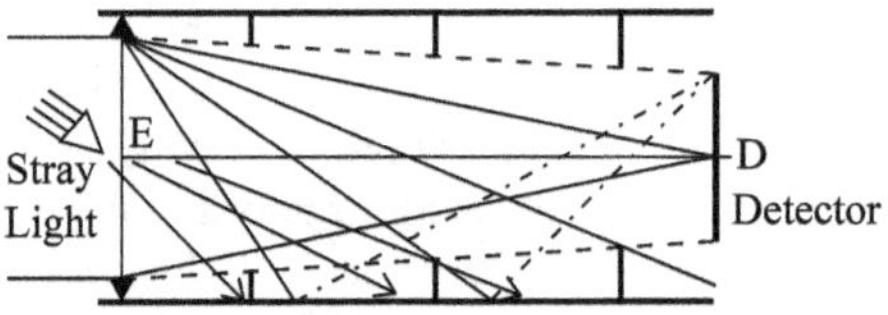

FIGURE 4.20 To obstruct stray light, often baffles are used in the inside of a barrel containing the imaging lens and the detector.

will be reflected from the inside wall of the barrel or other elements of the system, if there are any, and appears as though it is coming from the outer part of the entrance pupil, so that it will be blocked by the glare stop at E′. A paraxial analysis of the role of glare stops is often adequate in practice [10].

In the case of erecting telescopes, the glare stop is also known as the 'erector stop'. In the field of coronagraphs, the glare stop is known as 'Lyot stop', in memory of B. Lyot who pioneered the use of such stops in coronagraphs. In the field of long wavelength thermal imaging, the wall itself is a source of unwanted thermal radiation, and the latter is prevented from reaching the final image by using a suitable glare stop on the exit pupil of the system. In infrared engineering, the glare stops are also known as 'cold stops'.

In some systems, no suitable location of the real exit pupil is available where a glare stop can be located. In such situations, a remedy is sought by using 'baffles' in the inside of the barrel containing the imaging lens and the detector. The clear diameters of the baffles are such that a central truncated zone with the entrance pupil E as the base and the detector D on the top allow propagation of unhindered radiation from the entrance pupil to the detector (Figure 4.20). Stray radiation, after one or more reflection from the inside surface of the barrel, cannot reach the detector D.

4.5 Pupil Matching in Compound Systems

Figure 4.21 shows two lens systems in tandem or cascade. System 1 images the axial point O_1 at O_1', which acts as object O_2 for system 2 and images it at O_2'. The entrance and exit pupils of systems 1 and 2 are shown at E_1E_1' and E_2, E_2', respectively. The whole pencil of light from O_1 passing through system 1 will pass through system 2 if the diameter of the entrance pupil of system 2 is sufficiently large to accept the full aperture of the pencil of rays forming the image at O_1'.

The entrance pupil E_2 increasingly vignettes the pencils of light from the exit pupil E_1', proceeding to form extra-axial image points until points such as O_1' in Figure 4.21, when the pencil of light misses the entrance pupil of the second system altogether. The field of view of the combined system will thus be extremely limited.

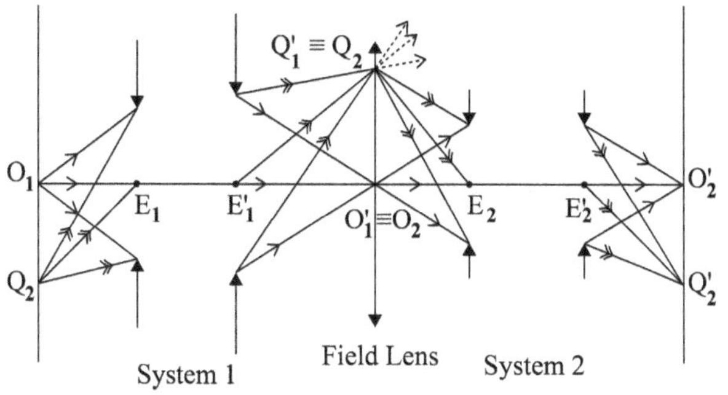

FIGURE 4.21 Pupil matching of optical systems in cascade or tandem by field lenses.

In order to increase the size of field of the composite system, a lens is placed at, or near, the image plane $O_1'Q_1'$. The focal length of the lens is to be such that the exit pupil E_1' of system 1 is imaged at the entrance pupil E_2 of the system 2. Within the paraxial approximation, all rays emerging from the exit pupil of system 1 also pass through the entrance pupil of the following system 2. Such a lens is called a field lens. Its function is to increase the field of view by matching the exit pupil of one system to the entrance pupil of the following system.

Field lenses play a very important role in all but the simplest form of visual instruments. For the sake of obtaining a useful field of view, all eyepieces use a field lens. In relay systems, e.g. submarine periscopes or medical endoscopes, a field lens is to be used with each intermediate image. In general, a field lens is used with the intermediate image of an imaging system to obtain a useful field of view.

In cascaded systems, the field lens does not directly take part in the formation of images of objects, yet it can affect the image quality indirectly. A field lens, as shown schematically in Figure 4.21, will not usually image the points of the entrance pupil E_1' stigmatically at corresponding points in the entrance pupil E_2, and when this effect is severe, it will cause non-uniform illumination in the image formed at O_2'. It is also seen that the field lens shown here is a short focus lens of positive power, which has the effect of giving a large curvature of the image. Consequently, the image is no longer sharply focused on the image plane but on a surface curved concave to the direction of propagation of light. These effects of field lenses have to be taken into account in the overall design of compound optical systems.

4.6 Optical Imaging System of the Human Eye [11]

Figure 4.22 presents a schematic diagram of the human eye. The sclera is the hard outer part that gives the eyeball its shape. The cornea is a transparent part of the sclera through which light passes. It is about 0.5 mm thick; on outer side of it is air, and inside is a transparent liquid called aqueous humour of refractive index 1.336. The cornea provides about 40 diopters, or approximately two-thirds, of the total refractive power of the eye system. The eye lens, of mean refractive index 1.413 provides the remaining one-third of refractive power. The space between the lens and the retina is filled with a transparent liquid called vitreous humour of refractive index 1.336. Ciliary muscles can change the shape of the eye lens by changing the curvatures of the two surfaces; the resulting effect is to change focal length, or the power of the eye lens for 'accommodation', i.e. to bring an object, irrespective of its distance from the eye, into sharp focus. The range of power of the eye lens is approximately from 20 to 30 diopters. The iris controls the diameter of the eye pupil. It is the aperture stop of the system. Depending on the level of illumination, the diameter of the pupil varies from 1.5 to 8 mm.

The retina consists of layers of blood vessels, photoreceptors, and nerve fibres. The location where the nerve fibres leave the eyeball and proceed towards the brain is shown as the optic nerve in the diagram. A small spot on the retina in that region is called a blind spot, for there is no photoreceptor there. There are two types of photoreceptors, namely cones and rods. There are about seven million cones. They are active during daylight condition and provide photopic vision and colour sensation. There are three types

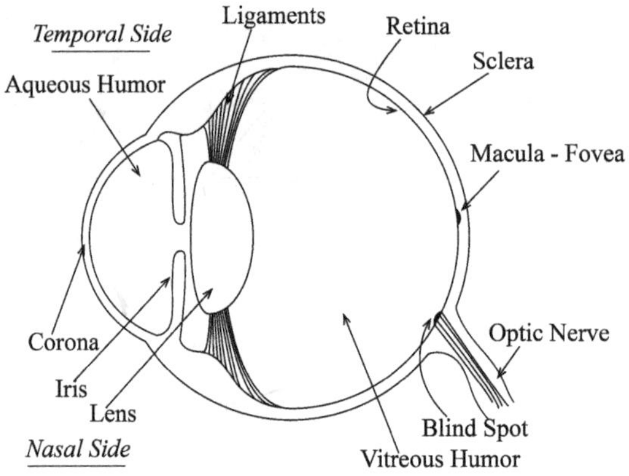

FIGURE 4.22 A schematic diagram of the human eye.

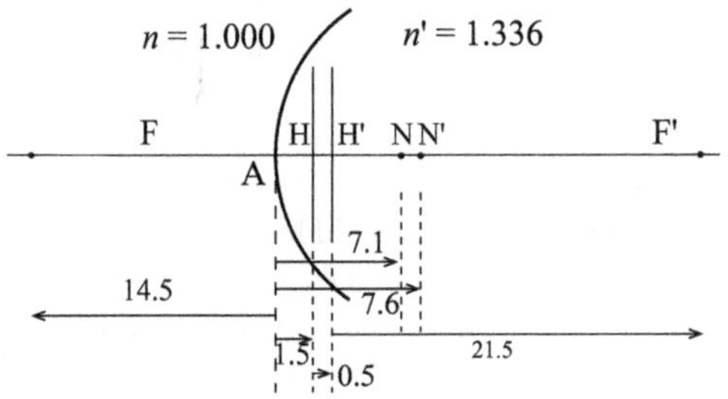

FIGURE 4.23 Location of the paraxial cardinal points of a normal or emmetropic eye under relaxed condition.

of cones—red-sensitive, green-sensitive, and blue-sensitive—for providing the sensation of colour. At low light level, the cones are inactive, and rods act as photoreceptors at night-time and provide scotopic vision. Rods are of only one type, and they have no colour sensation. There are about 125 million rods in a human eye. They are spread all over the retina except the fovea, where rods are absent. Compared to other regions of the retina, the cones are densely packed in the fovea so that the image formed over the fovea is much sharper than the images formed at other regions of the retina. In order to see more detail of a particular part of an object, the eye is rotated around its axis of rotation to form an image of that part on the fovea.

The optical system of the human eye differs considerably with age and individual. The refracting interfaces of the eye lens are, in general, aspheric, and the crystalline lens is not constituted of perfectly homogeneous medium. The optical medium of the eye lens has a higher refractive index at the axis compared to the outer region, like the GRIN (gradient index) lens. Modelling of the optical imaging system is quite tricky. There are different models with varying degrees of approximation, e.g. Gullstrand number 1, Gullstrand number 2, Le Grand, Emsley, etc. Each of them has two forms, one for the relaxed eye, and the other for the accommodated eye.

4.6.1 Paraxial Cardinal Points of the Human Eye

For the sake of design and analysis of visual optical instruments, Figure 4.23 presents the location of cardinal points of the human eye under a relaxed condition. A is the vertex of the anterior surface of the cornea. F and F′ are the first and the second principal foci. Note that F′ is on the retina. H and H′ are the first and the second principal points, and N and N′ are the nodal points. Note that the object space medium is taken as air ($n = 1.000$), and the image space medium is vitreous humour of refractive index 1.336. All distances are in millimetres and give average values. The equivalent focal length is 16mm.

4.6.1.1 Correction of Defective Vision by Spectacles

Accommodation of the eye, as mentioned above, calls for smooth changes in the power of the eye lens so that the image formed on the fixed image plane, i.e. the retina, continues to be at clear focus even when the object distance is changed. The ability of accommodation of a normal or *emmetropic* eye is such that it can bring an object lying at any distance between infinity and a 'near point'. The location of this near point is not the same for all humans, and depends on the age and background of the individual. By convention, it is taken at a distance 250mm from the eye, although young eyes can often accommodate closer. A *myopic* or 'near-sighted' eye can bring to focus on the retina objects lying at distances shorter than 250 mm, but it cannot bring to focus distant objects on the retina; distant objects are focused on a plane nearer to the eye lens, and, therefore, the corresponding image formed on the retina is blurred. A lens of suitable negative power used in conjunction with the myopic eye lens can restore clear focus on the retina. The opposite of myopia is hypermetropia or hyperopia. An eye with such characteristics is called a 'far-sighted' eye and it can bring distant objects into perfect focus on the retina, but the near point of such an eye is larger than 250 mm. The remedy is to use a lens of suitable positive power in conjunction with the defective eye.

4.6.1.2 Position of Spectacle Lens with Respect to Eye Lens

In case of short or long sightedness, spectacle lenses are often used to restore proper vision. The correct axial location of the spectacle lens should be such that its second principal plane is coincident with the first principal focus of the eye. This can be checked as follows. Let K_{eye} and $K_{spectacle}$ be powers of the eye lens and the spectacle lens, respectively. If the separation between the second principal point of the spectacle and the first principal point of the eye lens is d, then the power K of the spectacle-eye combination is given by

$$K = K_{eye} + K_{spectacle} - dK_{eye}K_{spectacle} \tag{4.7}$$

If the second principal point of the spectacle coincides with the first principal focus of the eye lens, $d = F_{eye} = \left(1/K_{eye}\right)$, and consequently, $K = K_{eye}$. The equivalent focal length of the combination is thus the same as that of the eye alone, and consequently, the effect of the spectacle lens is to correct the focus of the eye without changing the size of the retinal image.

Thus, an observer wearing spectacles requires 8 to 10 mm additional clear distance from the exit surface of a visual instrument compared with those who do not need spectacles. For comfortable viewing, the distance should be more than 10 mm.

4.6.2 Pupils and Centre of Rotation of the Human Eye

Figure 4.24 shows the pupils E, E′ of the eye. From the anterior surface of the cornea the distance of the entrance pupil E is 3.0 mm, and the distance of the exit pupil is 3.5 mm. The distance from E to the centre of rotation R of the eye is 11.5 mm. The magnification between the iris and the entrance pupil is (9/8). Thus, when the iris has a diameter of 4 mm, the diameter of the entrance pupil is 4.5 mm.

4.6.2.1 Position of Exit Pupil in Visual Instruments

Following the principle of pupil matching in cascaded systems, the exit pupil of a visual instrument should coincide with the entrance pupil of the eye so that the whole field of view is visible. In order to give sufficient clearance for eyelashes, a minimum distance of about 7 mm is needed between the last surface of the instrument and the anterior surface of the cornea. As the distance of the entrance pupil of the eye from the anterior surface of cornea is approximately 3 mm, the distance of the exit pupil of the visual instrument from its last surface should be not less than 10 mm. This distance is called the 'eye clearance'

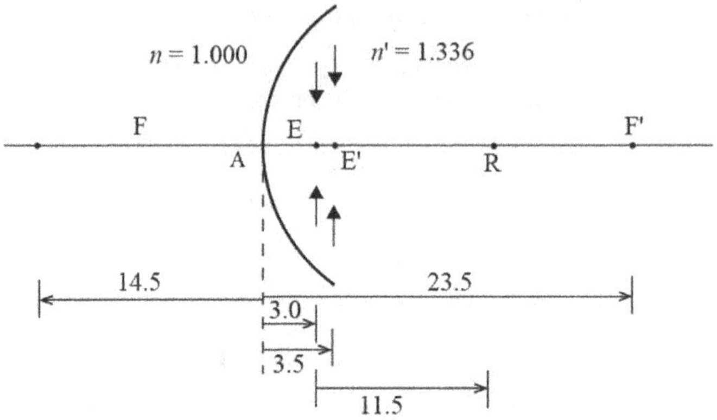

FIGURE 4.24 Location of the entrance pupil E and the exit pupil E′ of the human eye. R is the position of centre of rotation of eye in the socket.

of the visual instrument. Eye clearance with less than 10 mm leads to discomfort of the observer, for the observer's head has to move to see outer parts of the field of view. Whenever possible, it is advantageous to make the eye clearance a few mm more than 10mm. It should also be noted that for observers with spectacles, since the appropriate location of the spectacle should be such that its second principal point coincides with the first principal focus of the eye lens, about 8 mm more eye clearance compared to observers without spectacles is required for viewing comfortably in visual instruments.

The observation made above assumes that the diameter ρ'_{sys} of the exit pupil of the visual instrument is much smaller than the diameter ρ_{eye} of the entrance pupil of the observer's eye. The field of view of the human eye is approximately 150^0 high by 210^0 wide. However, visual acuity is low over the whole field except for the region called the fovea on the retina. This region of high visual acuity of the human retina covers an angular field of view of approximately 1^0. To see any part of the field of view more clearly, the eye rotates around a centre of rotation so that the image of that part is formed over the fovea of the retina. A typical example is a microscope that usually has an exit pupil of 1 mm diameter. Assume the diameter of the observer's eye pupil to be 4 mm. The distance of the centre of rotation from the entrance pupil of the eye is 11.5 mm. Note that in all figures, the eye pupil is denoted by bold triangular heads, and the pupil of the optical system is denoted by bold arrowheads. Figure 4.25(a) shows that the field of view of the overall system is uniformly illuminated on the retina when the exit pupil E'_{sys} of the instrument and the entrance pupil E_{eye} of the eye are coincident. Figure 4.25(b) shows the ray geometry when the eye rotates to bring the region around the point A focused on the fovea. Note that the entrance pupil of the eye is vignetting the exit pupil of the instrument. This will happen when the angle of rotation is more than $10^0\left[=\tan^{-1}\left(2/11.5\right)\right]$. The rapid drop in illumination with increasing rotation limits the usefulness of this process to a small rotation of the eye, forcing the observer to rotate the eye in order to bring objects in the outer region of the field into sharp focus. The vignetting effect occurs more when the exit pupil of the instrument has a diameter comparable to diameter of the eye. This problem can be circumvented by making the centre E'_{sys} of the exit pupil of the visual instrument coincident with the centre of rotation of the eye.

Figure 4.25(a) shows that the field of view is uniformly illuminated when the diameter of the exit pupil of the visual instrument is smaller than the diameter of the entrance pupil, and the two pupils are coincident. However, when the eye is rotated to bring the point of interest into focus on the fovea, the field of view is still uniformly illuminated, but vignetting effects may lead to a decrease of illumination on the fovea (Figure 4.25(b)). On the other hand, when the diameters of the two pupils are comparable, and the exit pupil of the instrument is coincident with the centre of rotation, the field of view is non-uniformly illuminated (Figure 4.26(a)). Vignetting effects in the case of even moderate rotation is discernible [12]. Nevertheless, full illumination is obtained for the neighbourhood of the point imaged on the fovea even when the eye is rotated (Figure 4.26(b)).

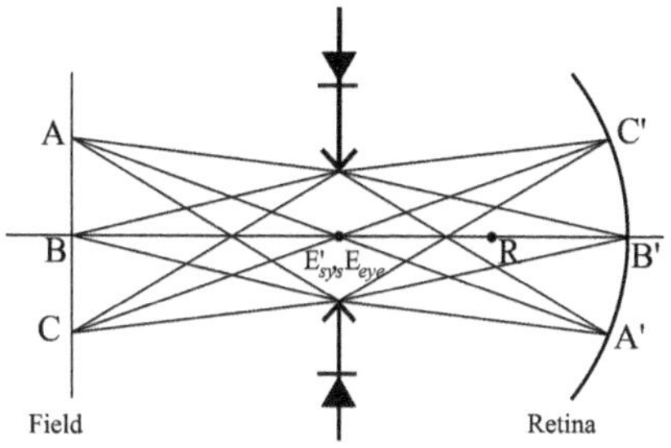

FIGURE 4.25(a) $\rho'_{sys} \ll \rho_{eye}$. E'_{sys} is coincident with E_{eye}. Field of view is illuminated uniformly.

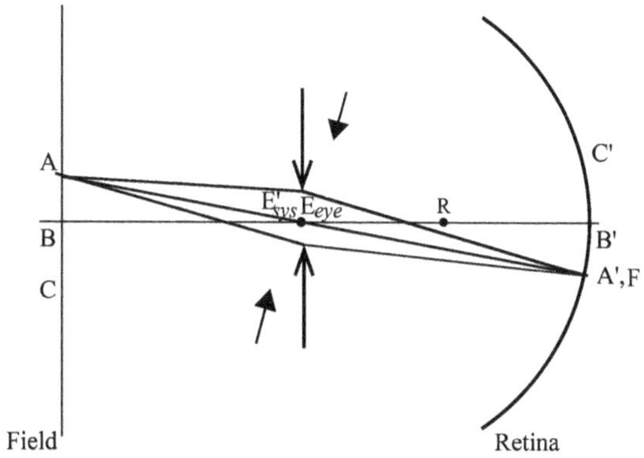

FIGURE 4.25(b) $\rho'_{sys} \ll \rho_{eye}$. E'_{sys} is coincident with E_{eye}. Eye is rotated across R on the axis of rotation to bring image of the point A focused on the fovea at F. Field of view is illuminated uniformly. When $\rho'_{sys} \geq \rho_{eye}$, vignetting effects will reduce level of illumination.

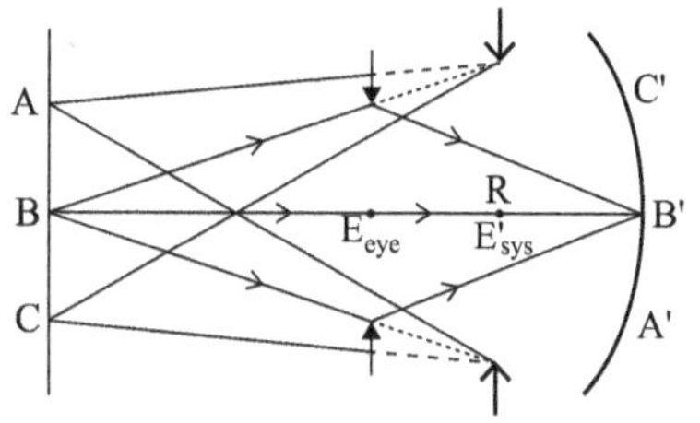

FIGURE 4.26(a) ρ'_{sys} is comparable with ρ_{eye}. E'_{sys} is coincident with R. Field of view is always seen non-uniformly illuminated because of vignetting by the eye pupil.

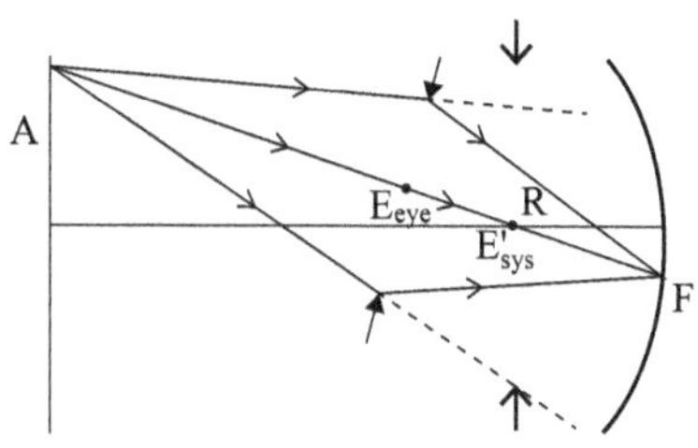

FIGURE 4.26(b) Ray diagram for the same imaging set up as Figure 4.26(a) when the eye is rotated for imaging of point A on fovea F.

4.6.3 Visual Magnification of an Eyepiece or Magnifier

When an object is brought nearer to the eye along its optical axis, the size of the retinal image of the same object goes on increasingly concurrently, thereby allowing more detail of the object to become visible (Figure 4.27). This increase in detail that can be observed is offset by a loss of sharpness due to focus error when the object is brought nearer than the least distance of distinct vision. This is the minimum object distance from the eye where the finest detail of an object can be seen by unaided eye. This distance is usually taken to be equal to 250 mm for a normal eye. The size of the retinal image, s_{250}, for an object of height η placed at a distance of 250 mm from the eye is (Figure 4.28)

$$s_{250} = f_{eye} \tan \beta_{250} \tag{4.8}$$

A magnifier or eyepiece of focal length f_m is placed in front of the eye. The distance of the magnifier from the eye is d (Figure 4.29), and it forms a virtual image η' of the object of height η placed at a distance l from the magnifier. The virtual image of the object is formed on the plane of the least distance of distinct vision. The height of the corresponding retinal image is s′. Magnification obtained by using the eyepiece or magnifier is given by the ratio of the corresponding retinal image to the size of the retinal image of the same object placed at the least distance of distinct vision, and can be written as

$$M = \frac{s'}{s_{250}} = \frac{f_{eye} \tan \beta'}{f_{eye} \tan \beta_{250}} = \frac{\eta' / (-250)}{\eta / (-250)} = \frac{\eta'}{\eta} \tag{4.9}$$

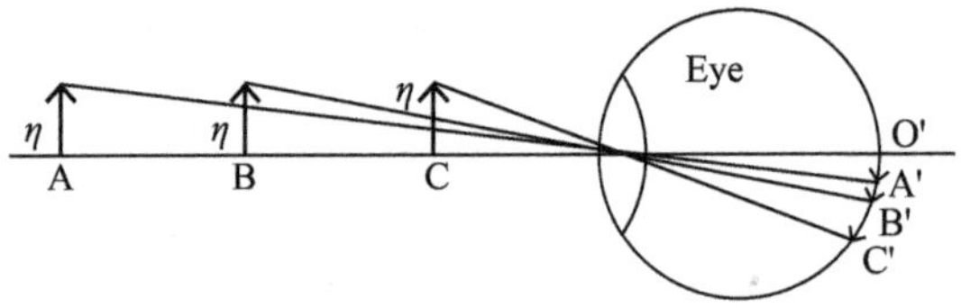

FIGURE 4.27 Size of the retinal image increases as the object is brought nearer to the eye.

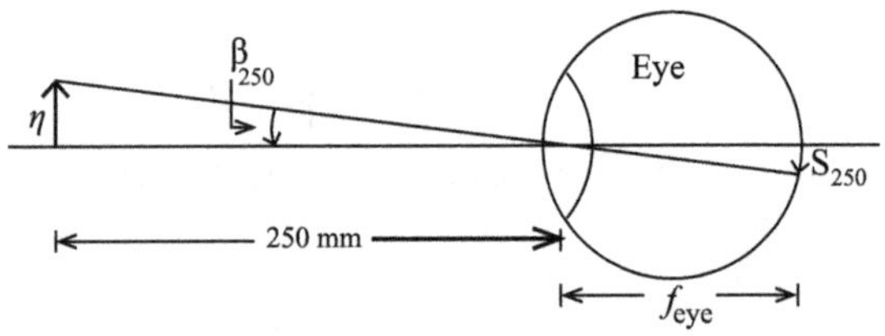

FIGURE 4.28 Least distance of distinct vision is 250 mm for a normal eye.

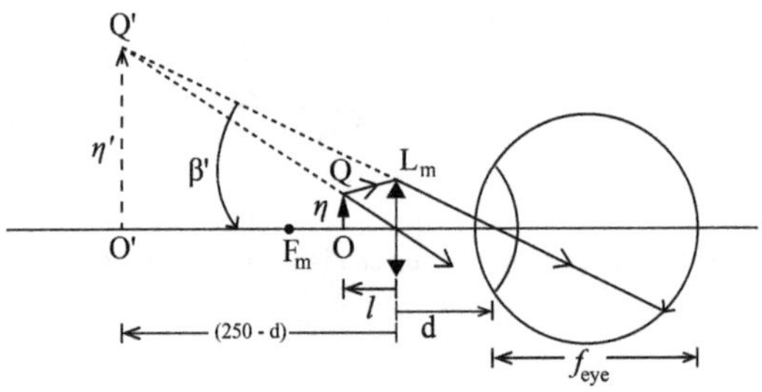

FIGURE 4.29 Visual magnification by using a magnifier.

From Figure 4.29, using the conjugate equation for the lens L_m, the above relation reduces to

$$M = \frac{\eta'}{\eta} = \frac{(250 - d)}{l} = 1 + \frac{(250 - d)}{f_m} \tag{4.10}$$

When the object is at the front focus of the lens L_m, the image is formed at infinity. This is called the normal magnification for a relaxed eye. The corresponding magnification is given by

$$M_{normal} = \frac{[\eta/(-f_m)]}{[\eta/(-250)]} = \frac{250}{f_m} \tag{4.11}$$

Note that M_{normal} is independent of d, the axial location of the magnifier with respect to the eye. It may be noted that the maximum value of magnification obtainable with a magnifier of focal length f_m is

$$M_{maximum} = 1 + \frac{250}{f_m} \tag{4.12}$$

This is obtained when d = 0, i.e., the magnifier is on the eye, and the image is formed on the plane of least distance of distinct vision.

Note that the definition of visual magnification given above is conventional. Different observers will focus an eyepiece at different distances from the eye. In order to give unambiguous meaning to the magnification of an eyepiece or magnifier, it is necessary to adopt a standard image distance, and the average value 250 mm for the least distance of distinct vision is taken as the standard.

4.7 Optical (Paraxial) Invariant: Paraxial Variables and Real Finite Rays

In Section 3.5.3 of the earlier chapter, it is shown that the paraxial magnification formula can be rewritten in a form that provides a paraxial invariant for propagation of a paraxial ray through an axisymmetric imaging system consisting of multiple refracting interfaces. Indicating parameters of the object space with subscript 1, and those of the image space with subscript $(k+1)$, the invariant relationship is expressed as

$$n_1u_1\eta_1 = n_2u_2\eta_2 = \ldots = n_ju_j\eta_j = \ldots = n_ku_k\eta_k = n_{k+1}u_{k+1}\eta_{k+1} \tag{4.13}$$

where n_j, u_j, and η_j are the refractive index, convergence angle variable of a paraxial ray from the axial object point, and the height of the object in the j-th intermediate space.

The importance of this paraxial invariant in ray-optical treatment of optical systems cannot be overemphasized. Lord Rayleigh noted that an early form of this relationship was reported by Robert Smith in Book II of his 'A Compleat System of Optics' in 1738 [13–14]. It is apparent that he obtained the cue from 'Cotes theorem', an unpublished work of Roger Cotes. Later on, Lagrange deliberated further on this relationship in two seminal papers published in 1778 and 1803 [15–16]. However, the treatments of both Smith and Lagrange were restricted to certain specified points only. Later on, Helmholtz presented the relation in its current form, and his treatment could be applied for any pair of conjugate points [17]. This paraxial invariant relationship is noted differently as the Helmholtz equation, the Smith-Helmholtz equation, the Lagrange-Helmholtz relation, or the Lagrange invariant, etc. In view of its widespread application in different areas of optics, it is often called the '*Optical Invariant*'.

The optical invariant acquires greater importance when the paraxial variables are defined appropriately in relation to corresponding real finite rays, i.e. the aperture rays from axial object point at finite apertures and the pupil rays from extra-axial object points at finite field angles. In optical systems engineering the optical invariant,commonly noted as *H*—provides practically useful cues in tackling many problems. A few of them are noted below:

- Illumination in an image/Light throughput (Étendue) in an optical system is directly proportional to H^2.
- Information content in an image is directly proportional to H.
- In general, aberrational complexity in an imaging system rises fast with an increase in H; the dependence is highly nonlinear.
- In diffraction theory of image formation, the normalized diffraction variables for points on the object and the image plane are proportional to $\left(\frac{H}{\lambda}\right)$, where λ is the working wavelength.
- Derivations in theory of aberrations are based on the invariance of H. It is more conspicuous in Buchdahl aberration theory [18].

4.7.1 Different Forms of Paraxial Invariant

4.7.1.1 Paraxial Invariant in Star Space

In a multicomponent axisymmetric optical imaging system, the object space, the image space, or any of the intermediate spaces become a star space, if the image (real or virtual) formed in the space lies at infinity, or very far away from the neighbourhood. Note that each of these spaces are nominally extended to the whole three-dimensional space, although the corresponding real space, flanked by two refracting interfaces, is limited in extent. If the j-th space is the star space, $u_j \approx 0$ and $\eta_j \to \infty$, and therefore, the form of H as given by (4.14) becomes indeterminate for this case. To circumvent this problem, it becomes necessary to use the variables of the paraxial pupil ray.

Figure 4.30 shows the paraxial marginal ray $P_{j-1}O_jP_j$ and the paraxial pupil ray $\bar{P}_{j-1}E_j\bar{P}_j$ in the j-th space of refractive index $n_j\left(=n'_{j-1}\right)$. The vertices on the axis of the (j-1)-th and the j-th refracting interfaces are at A_{j-1} and A_j, respectively. The intermediate image and the effective aperture stop at the j-th space are located on the axis at O_j and E_j respectively. Note that in each space there are paraxial images of the object and the aperture stop of the system. They are shown diagrammatically in Figure 4.30. The image height $O_jQ_j = \eta_j$, and the semi-aperture $S_jE_j = h_{E_j}$. A useful convention is to choose the size of the object height η, and consequently that of the angle u so that H is always of positive sign.

Let $A_jO_j = l_j$, $A_jE_j = \bar{l}_j$ and $E_jO_j = R_j$. In the diagram, the angle u_j is positive, and the angle $\bar{u}_j$is negative. With the sign convention in view,

$$u_j = -\frac{h_{E_j}}{\left(l_j - \bar{l}_j\right)} = -\frac{h_{E_j}}{R_j} \tag{4.14}$$

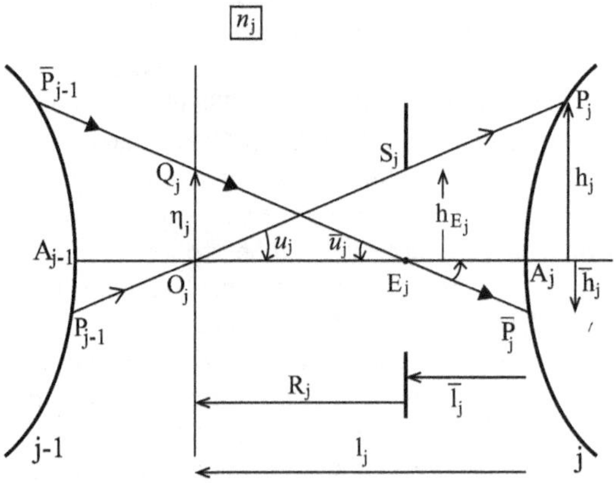

FIGURE 4.30 Paraxial/Optical invariant H.

$$\bar{u}_j = \frac{\eta_j}{\left(l_j - \bar{l}_j\right)} = \frac{\eta_j}{R_j} \tag{4.15}$$

Therefore, the expression for H in the j-th space can be written as

$$H = n_j u_j \eta_j = n_j \left(-\frac{h_{E_j}}{R_j} \right) \eta_j = -n_j h_{E_j} \bar{u}_j \tag{4.16}$$

If the object is at infinity in the first space j = 1, and $H = -n_1 h_{E_1} \bar{u}_1$. Similarly, if the image is at infinity in the last space $j = k$, in that space $H = -n_k h_{E_k} \bar{u}_k$.

Formula (4.17) for H is exact, and remain determinate for the cases of an object or image at infinity. It can also be used for determining H in case of other object or image distances. However, the formula becomes indeterminate when the effective aperture stop in the j-th space is located at infinity, or at a large distance away. For, in such case, $\bar{u}_j \approx 0$, and $h_{E_j} \to \infty$. Of course, the other formula using the paraxial marginal ray can also be used in such a case.

4.7.1.2 A Generalized Formula for Paraxial Invariant H

In Figure 4.30, let the height of the PMR on the j-th refracting interface be h_j, and the height of the PPR on the same surface be $\bar{h}_j$. With the sign convention in view,

$$\bar{h}_j = -\bar{u}_j \bar{l}_j \tag{4.17}$$

$$h_j = -u_j l_j \tag{4.18}$$

$$\eta_j = \bar{u}_j R_j = \bar{u}_j l_j - \bar{u}_j \bar{l}_j \tag{4.19}$$

Using (4.18) – (4.20), H can be written as

$$H = n_j u_j \eta_j = n_j u_j \bar{u}_j l_j - n_j u_j u_j \bar{l}_j = n_j \left(u_j \bar{h}_j - \bar{u}_j h_j \right) \tag{4.20}$$

It is important to note that the paraxial ray heights h_j and $\bar{h}_j$ can be chosen at any transverse surface of the given space of the system, not necessarily a refracting interface. If the image plane is chosen as the surface, $h_j = 0$, and $\bar{h}_j = \eta_j$. Then (4.21) reduces to $H = n_j u_j \eta_j$. On the other hand, if the effective stop surface is chosen as the surface where $h_j = h_{E_j}$ and $\bar{h}_j = 0$, H is then given by $H = -n_j h_{E_j} \bar{u}_j$, as shown in (4.17) above.

4.7.1.3 An Expression for Power K in Terms of H and Angle Variables of the PMR and the PPR

H has the dimension of length. Using its reciprocal, a simple formula for equivalent power of the imaging system can be derived from the angle variables of two paraxial rays in the initial object space and the final image space. Figure 4.31 shows the PMR and the PPR in the object and the image spaces of refractive index n and n′, respectively, of an axisymmetric system of finite power K. H, H′ are the axial locations of the conjugate principal or unit planes.

The convergence angle of the PMR is u in the object space, and u′ in the image space; the same for the PPR is $\bar{u}$ in the object space, and $\bar{u}'$ in the image space. The height of the PMR and the PPR on the principal planes is $h = h'$, and $\bar{h} = \bar{h}'$, respectively. Using paraxial refraction law for the PMR and the PPR, one can write

$$n'u' - nu = hK \tag{4.21}$$

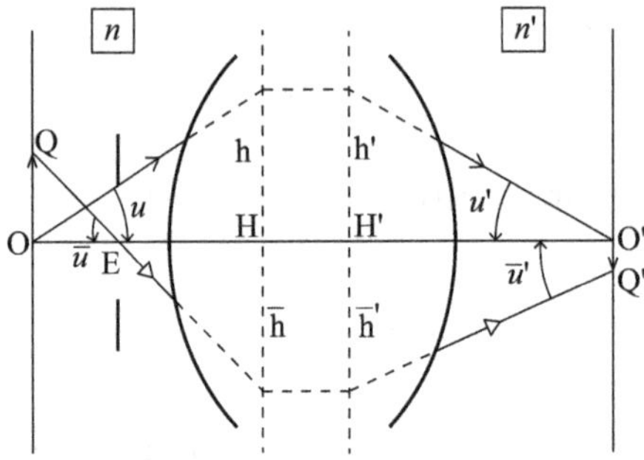

FIGURE 4.31 Passage of PMR and PPR through an axisymmetric imaging system of non-zero power K.

$$n'\bar{u}' - n\bar{u} = \bar{h}K \tag{4.22}$$

Substituting for $\bar{h}$ and h in the generalized form of the optical invariant H presented above,

$$H = n\left(\bar{h}u - h\bar{u}\right) = \frac{nn'}{K}\left(u'\bar{u} - u\bar{u}'\right) \tag{4.23}$$

Power K of the system is given by

$$K = \frac{nn'\left(u'\bar{u} - u\bar{u}'\right)}{H} \tag{4.24}$$

Thus, when H is known, the power of a finite conjugate system can be determined from the ray trace data of the PMR and the PPR, obviating the need for tracing a separate ray from infinity for determining the power of the system. Note that the expression remains determinate even when the object or the image is at infinity.

4.7.2 Paraxial Ray Variables and Real Finite Rays

So far, the invariant H has been considered solely in terms of paraxial variables. By definition, the values of these paraxial variables, e.g. u and h, are infinitesimally small quantities. On account of linearity of paraxial ray tracing equations, multiplication of both the angle and the height variable by the same factor does not change the output paraxial parameters, e.g. focal length, magnification, etc., that come as a ratio of two paraxial variables. Otherwise, the paraxial model obtained by assigning arbitrary finite values to paraxial parameters often fails to provide even a rough glimpse of the behaviour of the real system. The appropriate choice of the paraxial angle and height variables plays an important role in creating useful paraxial models of real optical systems. Use of finite values for angle and height variables in paraxial ray tracing gives rise to 'quasi-paraxial optics', and these fictitious rays are called 'zero rays' in some Russian texts. 'A "zero ray" is a fictitious ray refracting (reflecting) in the same manner as a *paraxial ray* and meeting the optical axis at the same distance as the paraxial ray, but traversing the principal planes at actual, i.e., non-Gaussian, heights above the axis' [19].

4.7.2.1 Paraxial Ray Variables in an Ideal Gaussian System

In an ideal Gaussian system, all points in the object space have their stigmatic image points in the image space. Not only are all points on the object imaged stigmatically on the image, but all points on the aperture stop are also imaged stigmatically on the entrance pupil in the object space, and on the exit pupil in the image space; thus all points on the entrance pupil are effectively imaged stigmatically on points on the exit pupil. It has been shown earlier in Section 3.13.3 that for an ideal Gaussian system, the paraxial angles are replaced by the tangents of the angles that the finite rays make with the axis. Figure 4.32 shows two Gaussian rays, OP…P′O′ and QE…E′Q′ in the initial object space and the final image space of an axisymmetric imaging system. The object and image planes are located at O and O′ on the axis. The entrance and the exit pupils are located at E and E′ respectively on the axis. The intersection heights of the ray OP…P′O′ at the entrance pupil is PE = Y, and at the exit pupil is P′E′ = Y′. In the object space, $\angle$POE = U, and $\angle$QEO = $\bar{U}$, and in the image space, $\angle$P′O′E′ = U′, and $\angle$Q′E′O′ = $\bar{U}'$.

For this system, let the paraxial angles in the object space be defined as

$$u = \tan U \qquad \bar{u} = \tan \bar{U} \tag{4.25}$$

The corresponding angles and the paraxial angle variables in the image space are

$$u' = \tan U' \qquad \bar{u}' = \tan \bar{U}'$$

The paraxial height variables at the entrance and the exit pupil are

$$h = -Ru = -R\tan U = Y \qquad h' = -R'u' = -R'\tan U' = Y' \tag{4.26}$$

The Gaussian object and image heights η and η' are given by

$$\eta = R\tan\bar{U} = R\bar{U} \qquad \eta' = R'\tan\bar{U}' = R'\bar{U}' \tag{4.27}$$

The paraxial formula for the invariant H becomes

$$H = n\tan U\eta = -nY\tan\bar{U} \tag{4.28}$$

in the initial object space. In the final image space, the expression for H takes the form

$$H = n'\tan U'\eta' = -n'Y'\tan\bar{U}' \tag{4.29}$$

4.7.2.2 Paraxial Ray Variables in a Real Optical System

Unlike the ideal Gaussian systems, a real optical system does not produce simultaneously stigmatic images of points lying on all planes in the object space. At best, a real system can produce on an image plane stigmatic images of given points on a single object plane. For points on other object planes, the images will not in general be stigmatic. Thus, in the case of a corrected real system corresponding to the paraxial imaging system meant for the object-image conjugate lying at *O, O′*, as shown in Figure 4.32, the real marginal ray OP originating from the axial point O in the object space will pass through the paraxial

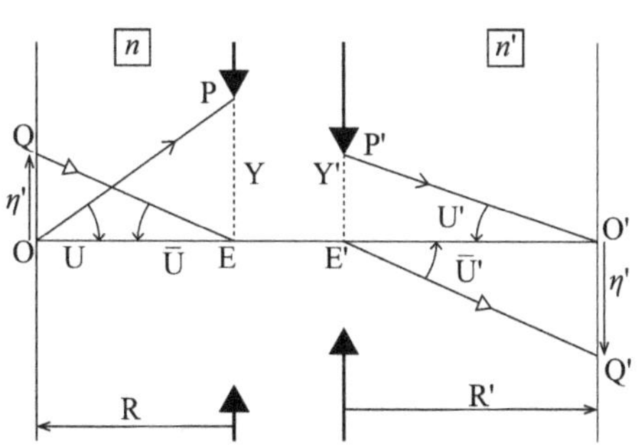

FIGURE 4.32 The optical invariant for an ideal Gaussian system.

image point O′ in the image space. However, this ray will not, in general, pass through the point P', the paraxial image of the point P. Similarly, the real pupil ray QE, originating from the point Q on the object plane, will pass through the paraxial image point Q′. But this ray will not, in general, pass through E′, the paraxial image of the point E. The entrance pupil is assumed here to be the aperture stop. In general, for a separate aperture stop, the aperture ray, passing through the margin of the aperture stop, will not pass through the margin of the entrance pupil or that of the exit pupil. Similarly, the pupil ray, passing through the centre of the aperture stop, will not pass through the centre of the entrance pupil or that of the exit pupil. Thus, in corrected optical systems, both the Gaussian rays and the actual rays pass through the same points on the object plane and the image plane, but not necessarily through the same points in the entrance pupil or the exit pupil.

4.7.2.3 Choice of Appropriate Values for Paraxial Angles u and $\bar{u}$

Figure 4.33 shows an aperture ray coming from the axial object point O in the initial object space. The ray OBB_1P meets the first refracting surface at P, and the tangent surface at vertex A of the first surface at the point B_1. AB is a part of a sphere with O as centre and OA as radius. The ray intersects this sphere at B. BD is perpendicular from B on the axis.

If $U(=\angle BOA)$ is the aperture angle of the finite ray OBP, possible choices for the paraxial angle u are the radian value of U, $\sin U$, or $\tan U$. In fact, any function of U is permissible, which for small values of U can be approximated by the angle U with an error of smaller order of magnitude. Use of $u = U$ implies that the paraxial ray height on the first surface is $h = -lu = -lU$ =length of the arc BA. For the choice $u = \sin U$, the paraxial ray height on the first surface becomes $h = -lu = -l\sin U$ =BD, whereas for the choice $u = \tan U$, the paraxial ray height on the first surface, $h = -lu = -l\tan U = B_1A$. Thus, the paraxial ray height h corresponds to three different quantities of the finite ray OBB_1 for the three different choices for u in the object space. Obviously, it would be advantageous if a choice on paraxial angle variable in the object space was made so that the value of the same quantity of the same finite ray in the final image space of a corrected optical system was also given by the paraxial angle variable in that space.

Figure 4.34 shows the passage of an axial marginal ray and a pupil ray in the initial object space of refractive index n, and the final image space of refractive index n′ of a nominally corrected axi-symmetric imaging system. The object OQ and its paraxial image O′Q′ have heights η and η′ respectively. EP and E′P′ are the paraxial entrance and exit pupils. EB and E'B' are sections of spheres with centres at O and O′, and radii $R(=EO)$ and $R'(=E'O')$, respectively. An edge ray of the entering pencil of rays from the axial point O is shown as OB. This ray passes through a point at the edge of the aperture stop. The convergence angle of this ray with the axis $\angle BOE = \alpha$. If the paraxial entrance pupil is also the aperture stop, the edge ray will pass through point P lying at the edge of the pupil. In general, this ray does not necessarily pass through the edge of the paraxial entrance pupil. The general notation used for the convergence angle of this marginal finite ray in the object space is U. In Figure 4.34, the edge ray is taken as the finite ray entering the system. Therefore, in the object space, $U = \alpha$. The finite ray OBB′O″ from the axial object point intersects the optical axis at O″ in the final image space, and the sphere E′B′ at the point B′. The

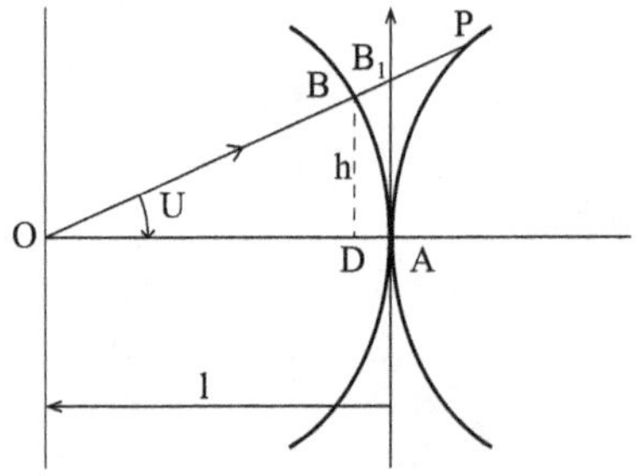

FIGURE 4.33 An aperture ray OBB_1P from the axial object point O making finite angle U with the axis.

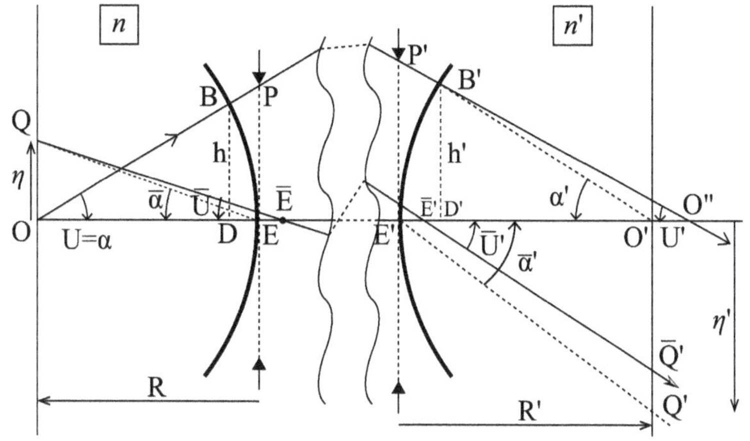

FIGURE 4.34 The axial marginal ray and the pupil ray: actual rays (solid lines) and corresponding paraxial rays (dashed lines) for a nominally corrected system.

convergence angle of the finite ray in the image space, $\angle B'O''E' = U'$. The angle $\angle B'O'E' = \alpha'$. If the system is properly corrected, $O'O'' = 0$, and, consequently $U' = \alpha'$.

Out of the three available choices mentioned above for paraxial angle variable, for the paraxial ray from the axial object point O, as shown in Figure 4.33 and Figure 4.34, the preferred choice is

$$u = \sin U \tag{4.30}$$

The paraxial incidence height at the first surface is then equal to the height BD shown in the diagrams. This preference stems from the 'sine condition' that is satisfied by the axial pencil of rays of any perfectly corrected optical system. This condition, proved later in Chapter 7 (Section 7.2.4.1), states that

$$n' \sin \alpha' \eta' = n \sin \alpha \eta \tag{4.31}$$

Thus, for a corrected optical system, the choice

$$u = \sin U = \sin \alpha \tag{4.32}$$

for paraxial convergence angle in the initial object space leads to the exact relation

$$u' = \sin \alpha' = \sin U' \tag{4.33}$$

When the sine condition is not satisfied, (4.28) is to be replaced, in general, by

$$u' = \sin \alpha' \neq \sin U' \tag{4.34}$$

However, even for nominally corrected systems, the relation reduces to

$$u' = \sin \alpha' \cong \sin U' \tag{4.35}$$

Errors in approximating U' by α' are small in all cases of practically useful imaging systems.

An actual pupil ray $Q\bar{E}\bar{E}'\bar{Q}'$, coming from the extra-axial object point Q is shown in Figure 4.34. It intersects the paraxial image plane at $\bar{Q}'$. Note that this ray will not normally pass through the centres E and E' of the paraxial entrance and exit pupils. If the image point corresponding to object point Q is free from aberration, and the system is corrected for distortion, the pupil ray $\bar{E}'Q'$ and all other rays of the image-forming pencil pass through the point Q', and $\bar{Q}'Q' = 0$. In this case, there will be a finite ray of

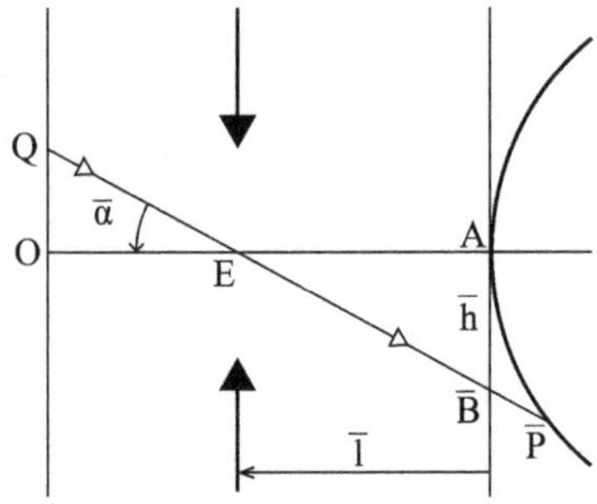

FIGURE 4.35 A pupil ray $QE\bar{B}\bar{P}$ with paraxial convergence angle $\bar{\alpha}$.

the image-forming pencil which passes through the points E′ and Q′. Similarly, in the object space, there will be a finite ray QE, not necessarily the same ray, which passes through the two points Q and E. These two rays will coincide with the object space and image space parts of the paraxial pupil ray QEE′Q′.

For the paraxial pupil ray $QE\bar{B}\bar{P}$ from an extra-axial object point Q, a useful choice for the paraxial angle variable (Figure 4.35) is

$$\bar{u} = \tan\bar{\alpha} \tag{4.36}$$

Where $\bar{\alpha} = \angle QEO$.

The height of the paraxial pupil ray on the first surface is then equal to the intersection height $\bar{h}\left(=\bar{B}A\right)$ of the same ray on the tangent plane to the first surface at the vertex A. Tracing the paraxial pupil ray QEE′Q′ with these values of $\bar{u}$ and $\bar{h}$ in the object space will yield the pupil ray angle in the final image space as

$$\bar{u}' = \tan\bar{\alpha}' \tag{4.37}$$

where $\bar{\alpha}' = \angle Q'E'O'$. Note that, with this choice for the paraxial pupil angle, even if distortion is not exactly zero, $\bar{u}$ and $\bar{u}'$ satisfy the relations

$$\bar{u} = \tan\bar{\alpha} \cong \tan\bar{U} \tag{4.38}$$

$$\bar{u}' = \tan\bar{\alpha}' \cong \tan\bar{U}' \tag{4.39}$$

$\bar{U}\left(=\angle\bar{Q}\bar{E}O\right)$ and $\bar{U}'\left(=\angle\bar{Q}'\bar{E}'O'\right)$ are the convergence angles of the real pupil ray $Q\bar{E}\bar{E}'\bar{Q}'$ in the object space and the image space, respectively. Thus, the choice of suitable values for convergence angles of the paraxial marginal ray and paraxial pupil ray can render useful information about the imaging characteristics of corresponding finite rays. However, it should be emphasized that this correspondence between the paraxial rays and corresponding finite rays does not hold good in the intermediate spaces of imaging optical systems. Since there is no requirement on the quality of intermediate images, normally the images are not even nominally aberration free, and often the magnitude of such aberrations is so large that the paraxial variables provide very poor approximations for the corresponding finite ray parameters.

4.8 Angular Magnification in Afocal Systems

In Section 3.11 of last chapter, it was shown that an afocal or telescopic system is characterized by transverse magnification, which is the same for all finite conjugates of the system. An afocal system forms the image of an object at infinity also at infinity. Figure 4.36 shows an afocal system with its entrance pupil of semi-diameter h and the exit pupil of semi-diameter h′ at E and E′ on the axis. The transverse

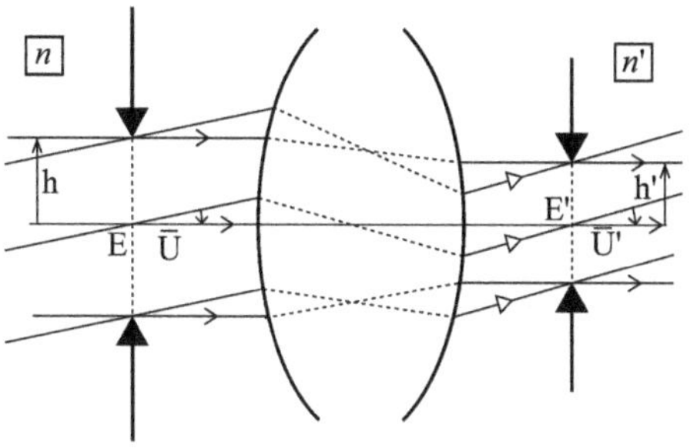

FIGURE 4.36 Angular magnification of an afocal or telescopic system.

magnification of the system is $M = (h'/h)$. The refractive indices of the object space and the image space are n and n′, respectively. The pupil ray from the extra-axial object point at infinity makes an angle $\bar{U}$ with the optical axis. The same ray makes an angle $\bar{U}'$ with the axis in the image space. The optical invariant H of the afocal system is

$$H = nh\tan\bar{U} = n'h'\tan\bar{U}' \tag{4.40}$$

The angular magnification M_A is defined by the ratio of the angular size of the image to that of the object. Therefore, from (4.41)

$$M_A = \frac{\tan\bar{U}'}{\tan\bar{U}} = \frac{nh}{n'h'} = \left(\frac{n}{n'}\right)\frac{1}{M} \tag{4.41}$$

Sometimes it is suggested to define angular magnification in terms of the angles in the object and image space, i.e., to define it as $\left(\frac{\bar{U}'}{\bar{U}}\right)$. However, for observing an angular object lying at infinity, it is necessary to produce a real image by the observing system. If the observing system, e.g. the human eye or a camera, is of focal length, f, the radius of the real image of the angular object of semi-angle $\bar{U}$ becomes $f\tan\bar{U}$, and the size of the angular object is perceived as such. If the angular object is observed through the telescopic system of Figure 4.36, the radius of the image is $f\tan\bar{U}'$, which is then perceived as the size of the distant angular object. The factor $\left(\frac{\tan\bar{U}'}{\tan\bar{U}}\right)$ provides a measure of the change in size of the angular object when the latter is viewed through the telescopic system. Therefore, it seems more reasonable to define angular magnification in terms of tangents of the angles rather than the angles themselves.

4.9 F-number and Numerical Aperture

F-number, also called the relative aperture, F- No., F#, or F/N (N indicating value of the parameter), and numerical aperture are two metrics for the light gathering ability of optical systems. Usually, F-number is used for infinite conjugate imagery with object at infinity and image at a finite distance. Numerical aperture is used in the case of finite conjugate imagery, where both the object and the image lie at a finite distance from the system. Figure 4.37 shows the infinite conjugate imaging by a thin lens with a stop on it, and the passage of the paraxial marginal ray from the axial object point at infinity up to the second principal focus. The clear semi-diameter of the lens is h, and its second principal focal length $f' = n'F$. Figure 4.38 shows the passage of the paraxial marginal ray from an axial object point at infinity in case of an axisymmetric imaging system with the entrance and the exit pupils located at E, E′, and the first and the second principal points located at H, H′, respectively.

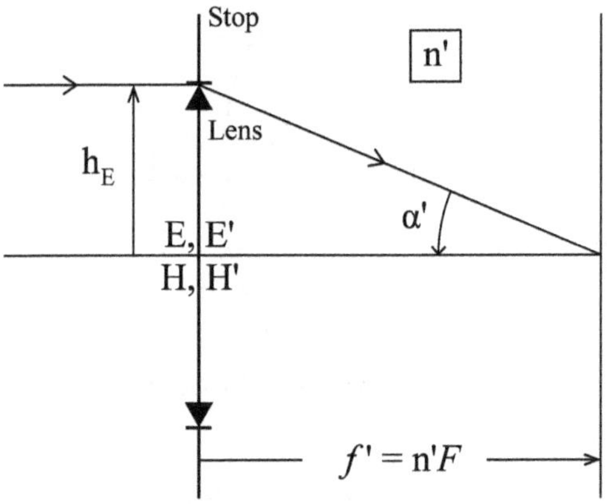

FIGURE 4.37 Infinite conjugate imaging by a thin lens with a stop on it.

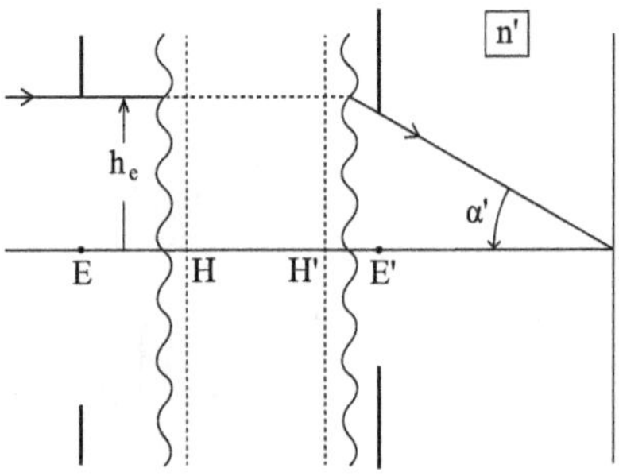

FIGURE 4.38 Infinite conjugate imaging by an axisymmetric imaging lens.

In Figure 4.37, the four points E, E', H, H' are coincident on the axis at the lens. The refractive index of the image space is n' and the semi-angular aperture is α' in both cases. The image space numerical aperture is defined as

$$NA = n' \sin \alpha' \tag{4.42}$$

Note that the angle α', as shown in Figure 4.37 or Figure 4.38, is negative. But the numerical aperture NA is always taken as positive by convention, so that the sign of this angle is ignored.

In paraxial approximation

$$\sin \alpha' \cong \tan \alpha' = \frac{h_e}{n'F} \tag{4.43}$$

so that

$$NA = \frac{h_e}{F} \tag{4.44}$$

where h_e is the height of the paraxial marginal ray at the entrance pupil. The relative aperture or $F_\#$ is defined as the ratio of the equivalent focal length to the diameter of the entrance pupil

$$F_\# = \frac{F}{2h_e} = \frac{1}{2(NA)} \tag{4.45}$$

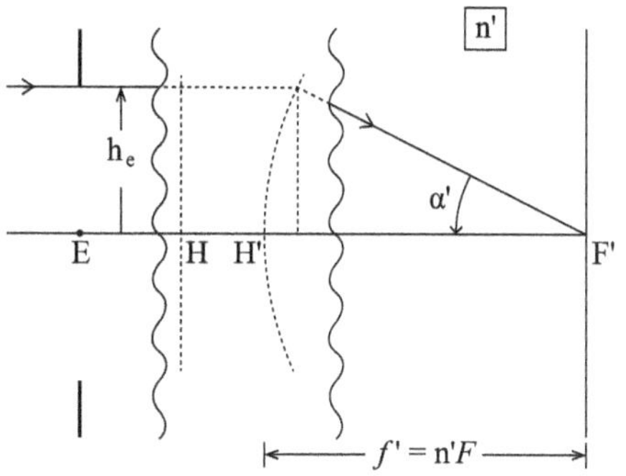

FIGURE 4.39 Second principal surface in infinite conjugate imaging.

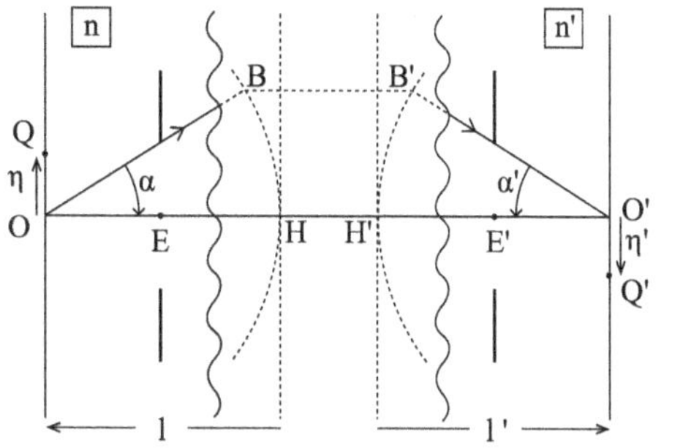

FIGURE 4.40 Meridional section of the object space principal sphere HB and the image space principal sphere H′B′.

The concept of principal planes is strictly valid under paraxial approximation. It has been mentioned above that corrected optical imaging systems satisfy the Abbe sine condition, and this leads to a useful paraxial modelling of real optical systems; it suggests the use of the sine of convergence angle of the marginal ray as the paraxial angle variable. This model dispenses with the principal planes, and replaces them with principal surfaces. This model can be applied even in the case of nominally corrected systems, because the error involved thereby can be neglected in most practical cases. In Figure 4.39, a meridional section of the second principal surface is shown. This spherical surface has its centre at F′, and radius $H'F' = f'$. Note that, in this model, the point of intersection of the emergent ray in the image space with the second principal surface is at a height h_e from the axis. Therefore, $\sin\alpha' = (h_e / f')$ directly, and there is no need for the approximation shown in equation (4.44). In the object space, the first principal surface becomes the principal plane, as shown in the figure, for the axial object is at infinity. Figure 4.40 illustrates the principal surfaces in the case of finite conjugate imagery. In this case, there is also an object space numerical aperture, $(n\sin\alpha)$, where n is the refractive index of the object space, and α is the semi-angular aperture of the marginal ray in the object space. The two numerical apertures are related by the Abbe sine condition

$$(n\sin\alpha)\eta = (n'\sin\alpha')\eta' \tag{4.46}$$

where η and η′ are object and image heights, respectively. Thus, the ratio of the object space and the image space to the numerical apertures is equal to magnification $M = (\eta' / \eta)$.

Measures of relative aperture, like $F_{\#}$ or NA, are, in fact, characteristics of the cone of light forming image of the axial object point. Associating these measures with a lens implies the largest angular size of the axial bundle that can pass through it, e.g. the largest cone of light that can pass through an F/4 lens in infinite conjugate imagery is F/4. Any beam whose $F_{\#}$ is less than 4 will be truncated while passing through the lens, and all beams whose $F_{\#}$ is larger than 4 will pass through the lens system. Evidently, it is the opposite in the case of numerical aperture. The largest NA of the axial cone of light that can pass through an objective with NA, say 0.6, is 0.6; all axial cones with NA<0.6 will pass through the system; but all axial cones with NA>0.6 will be truncated while passing through the lens.

For the sake of convenience in control of exposure, the iris diaphragm or the aperture stop is calibrated numerically to indicate light transmission. The approved $F_{\#}$ scale is given in the sequence 0.5, 1.0, 1.4, 2.0, 2.8, 4.0, 5.6, 8.0, 11.0 22.0,…, 64.0. It can be seen from (4.46)

$$\left(F_{\#}\right)_{min} = \frac{1}{2(NA)_{max}} = 0.5 \tag{4.47}$$

because $(NA)_{max} = 1$, when $n' = 1$. The ratio of two neighbouring members of the ascending sequence is approximately equal to $\sqrt{2}$, implying a reduction of the diameter of the iris diaphragm by $\sqrt{2}$, and consequently reducing the light gathered by the system by $\left(\sqrt{2}\right)^2 = 2$ times. The light gathering ability is inversely proportional to the square of the F-number. Note that nonimaging systems, e.g. concentrators are not required to satisfy Abbe sine condition for ensuring better quality. Therefore, values of F-number lower than 0.5 are admissible. Of course, practical feasibility of the same is determined by manufacturing limitations.

Another measure called T-No. or T-number is used in this context. It takes into account the transmission of the lens. The relation between F-No., T-No. and transmission is

$$T - \text{Number} = \frac{F - \text{Number}}{\sqrt{\text{transmission}}} \tag{4.48}$$

4.10 Depth of Focus and Depth of Field

4.10.1 Expressions for Depth of Focus, Depth of Field, and Hyperfocal Distance for a Single Thin Lens with Stop on It

Figure 4.41 shows the paraxial image O′ of an axial object point O formed by a single thin lens with stop on it. The object and image distances are $LO = l,\ LO' = l'$. For the sake of simplicity, it is assumed that $n = n' = 1$. When the image plane is shifted either in-focus to O_1' or out-of-focus to O_2', the cone of rays forming the image point O′ will intersect these planes on a blur circle of diameter ρ. On other planes located between O_1' and O_2', the diameter of the blur circle is less than ρ. If the least resolvable distance of a detector (e.g. the grain size in photographic film or the pixel size in a CCD, etc.) placed on the image plane is greater than or equal to ρ, the image quality will remain the same irrespective of exact axial location of the image plane within the range $O_1'O_2'$. $\delta z' \left(= O_1' O_2'\right)$ is called the depth of focus for imaging at O′.

Assume the diameter of clear aperture of the lens as D. Using the properties of similar triangles, it follows

$$\frac{D}{l'} = \frac{\rho}{\delta z'/2}, \text{ or } \delta z' = \frac{2\rho l'}{D} \tag{4.49}$$

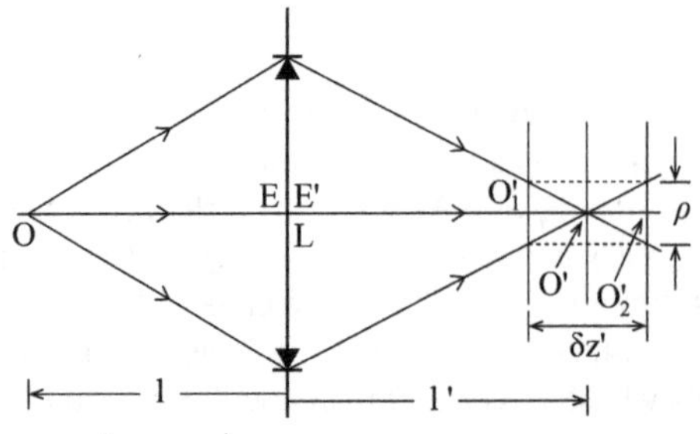

FIGURE 4.41 Axial depth of focus $\delta Z' \left(= O_1' O_2'\right)$ of the axial image point O′ formed by a thin lens with a stop on it.

Assuming $n = n' = 1$ for the sake of simplicity, and substituting $l' = (1 - M)F$ from (3.176) we get

$$\delta z' = \frac{2\rho F(1-M)}{D} = 2\rho(1-M)F_{\#} \tag{4.50}$$

In the above, F is the equivalent focal length, M is the transverse magnification for the conjugates O.O′ and $F_{\#}$ is the F-number of the lens. For an object at infinity, $M = 0$, and the corresponding depth of focus reduces to

$$(\delta z')_{inf} = 2\rho F_{\#} \tag{4.51}$$

Sometimes, particularly in the field of photography, a working F-number is defined to obtain identical expression for depth of focus in the case of finite conjugate imaging. The working F-number is defined as

$$F_{\#}^{*} = \frac{l'}{D} = (1-M)F_{\#} \tag{4.52}$$

In terms of the working $F_{\#}^{*}$, the depth of focus in finite conjugate imagery is defined as

$$\delta z' = 2\rho F_{\#}^{*} \tag{4.53}$$

The image of the axial segment $O_1'O_2'$ in the object space i.e., $O_1O_2 (= \delta z)$ is called the depth of field (Figure 4.42). If the least resolvable transverse distance of the recording device placed at O′ is ρ, an object located anywhere inside the depth of field O_1O_2 will appear to be in focus on a recording device at O′. Note that the word 'field' used in the term 'depth of field' refers to the longitudinal field or depth of a three-dimensional object, and is not to be confused with the transverse field. The depth of field δz and depth of focus δz′ are related, approximately, by the longitudinal magnification M_L, as

$$\delta z' \cong M_L \delta z \cong M^2 \delta z \tag{4.54}$$

A better estimate for the parameters of interest in depth of field is presented below. In Figure 4.42, the imaging system is a thin lens of clear diameter D with a stop on it. The object plane at O is focused at the image plane at O′. The depth of focus of imaging at O′ is $\delta z' = O_1'O_2'$, corresponding to ρ, the maximum acceptable diameter for the blur circle. In the object space, the axial points O_1 and O_2 are the objects corresponding to O_1' and O_2', respectively. By symmetry, the maximum acceptable diameter ρ_0 for the circle of confusion in the object space is $\rho_0 = \rho / M$, where M is the transverse magnification between the

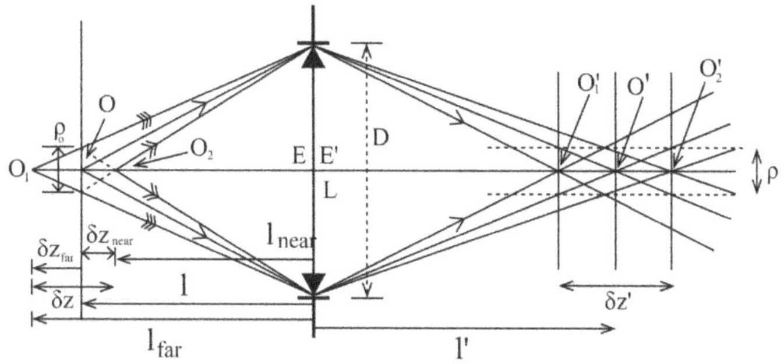

FIGURE 4.42 Depth of field δZ corresponding to depth of focus δZ′ of thin lens with a stop on it.

conjugates O, O′. The terminal points of the depth of field, namely O_1 and O_2, are called the 'far point' and 'near point', respectively. Accordingly, the far depth of field $\delta z_{far} = OO_1$, and the near depth of field $\delta z_{near} = OO_2$ The total depth of field δz corresponding to depth of focus $\delta z'$ is given by

$$\delta z = \delta z_{far} + \delta z_{near} \tag{4.55}$$

From similar triangles in the object space, we get

$$\frac{\rho_0}{D} = \frac{l_{far} - l}{l_{far}} = \frac{l - l_{near}}{l_{near}} \tag{4.56}$$

It follows

$$l_{far} = \frac{lD}{D - \rho_0} \qquad l_{near} = \frac{lD}{D + \rho_0} \tag{4.57}$$

and

$$\delta z_{far} = \left| \frac{l\rho_0}{D - \rho_0} \right| \qquad \delta z_{near} = \left| \frac{l\rho_0}{D + \rho_0} \right| \tag{4.58}$$

Note that the far depth of field and the near depth of field are not equal and, $\delta z_{far} > \delta z_{near}$. The location of the far point and the near point varies with object distance l. It is also seen that when $\rho_0 = D$, $\delta z_{far} = \infty$, i.e., the far point is at infinity. The object distance for which the far point of the depth of field is at infinity is called 'the hyperfocal distance'. When the lens is focused at this distance, the depth of field is maximum. Figure 4.43 shows the ray geometry for determining the hyperfocal distance l_{hyp}. The far point is imaged at F′, the near point is imaged at O'_{near}, and the point O_{hyp} is imaged at O'_{hyp}. The depth of focus $\delta z' = 2\rho l' / D$.

Let $LO'_{hyp} = l'$ and $LO_{hyp} = l_{hyp}$. It follows

$$l' = F + \rho \frac{l'}{D} \tag{4.59}$$

Using lens conjugate formula, we get

$$l_{hyp} = -\frac{FD}{\rho} \text{ and } l_{near} = -\frac{FD}{2\rho} = \frac{l_{hyp}}{2} \tag{4.60}$$

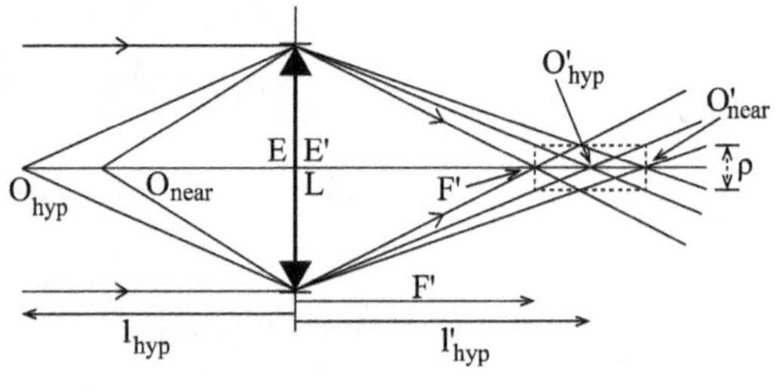

FIGURE 4.43 Hyperfocal distance l_{hyp}.

4.10.2 General Expressions for Depth of Focus, Depth of Field, and Hyperfocal Distance for an Axisymmetric Imaging System

The expressions for the depth of focus, the depth of field, and the hyperfocal distance given in the earlier section are valid only for the case of a thin lens with a stop on the lens, and needs to be modified in the case of a general axisymmetric system.

Figure 4.44 shows a general axisymmetric optical imaging system with its object space and image space principal points located at H,H′, respectively. For the sake of simplicity, it is assumed that $n = n' = 1.0$. The entrance pupil and the exit pupil of the system are located at E,E′ on the axis. Their semi-diameters are $\bar{h}_e$ and $\bar{h}'_e$, respectively. Let the second principal focus of the system be located at F′. The axial object point O is imaged at O′. The maximum diameter of the blur circle is ρ on the transverse planes at O'_1 and O'_2 lying on in-focus and out-focus sides of the image point O′. For any transverse plane inside the range $O'_1O'_2$, the maximum diameter of the blur circle is less than ρ. The distances

$$O'_1O'_2 = \delta z' \quad H'O' = l' \quad H'F' = F \quad H'E' = \bar{l}' \tag{4.61}$$

From similar triangles ΔPF′H′ and ΔB′F′E′, we get

$$\frac{\bar{h}_e}{F} = \frac{\bar{h}'_e}{F - \bar{l}'} \tag{4.62}$$

On algebraic manipulation, it follows

$$\bar{l}' = F(1 - \bar{M}) \tag{4.63}$$

where pupil magnification $\bar{M} = \frac{\bar{h}'_e}{\bar{h}_e}$. Let $\angle B'O'E' = \alpha'$. From ΔB′O′E′, by substituting $l' = (1 - M)F$ where M is the transverse magnification for the conjugates O, O′, we get

$$\tan\alpha' = -\frac{\bar{h}'_e}{l' - \bar{l}'} = \frac{\bar{h}_e}{F}\frac{1}{\left(\frac{M}{\bar{M}} - 1\right)} \tag{4.64}$$

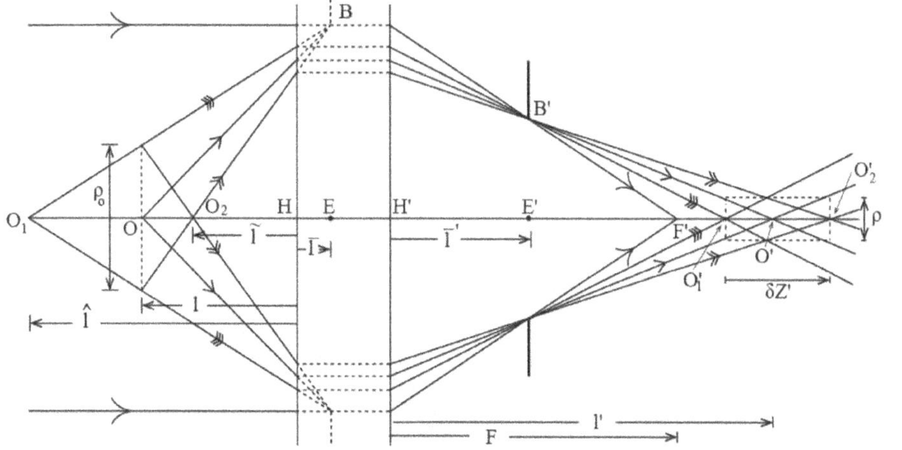

FIGURE 4.44 Depth of focus and depth of field of a general axisymmetric imaging system.

The depth of focus $\delta z'$ is given by

$$\delta z' = -\frac{\rho}{\tan\alpha'} = 2\rho\left(1-\frac{M}{\bar{M}}\right)F_{\#} \qquad (4.65)$$

Note that when $\bar{M}$ is unity, the expression (4.66) for depth of focus reduces to expression (4.50), derived earlier.

The maximum acceptable diameter for the circle of confusion on the focused plane is $\rho_0 = (\rho / M)$, where M is the transverse magnification between the conjugates O, O'. Let the far point and near point of the depth of field at O be O_1 and O_2, respectively, and the far depth of field $\delta z_{far} = OO_1$ and the near depth of field $\delta z_{near} = OO_2$. The distances

$$HO_1 = \hat{l} \qquad HO_2 = \tilde{l} \qquad HE = \bar{l} \qquad (4.66)$$

From similar triangles in the object space

$$\frac{\rho_0}{2\bar{h}_e} = \frac{-\delta z_{far}}{-\hat{l}+\bar{l}} = \frac{-\delta z_{far}}{-\delta z_{far} + (\bar{l} - l)} \qquad (4.67)$$

Using the relations

$$l = \left(\frac{1}{M} - 1\right)F \qquad \bar{l} = \left(\frac{1}{\bar{M}} - 1\right)F \qquad (4.68)$$

We get from (4.68)

$$\delta z_{far} = \frac{\rho_0 F\left(1-\frac{M}{\bar{M}}\right)}{M\left(2\bar{h}_e - \rho_0\right)} \qquad (4.69)$$

Using similar triangles, it also follows

$$\frac{\rho_0}{\delta z_{near}} = \frac{2\bar{h}_e}{-\tilde{l}+\bar{l}} = \frac{2\bar{h}_e}{-\delta z_{near} + (\bar{l} - l)} \qquad (4.70)$$

Thus, we get

$$\delta z_{near} = -\frac{\rho_0 F\left(1-\frac{M}{\bar{M}}\right)}{M\left(2\bar{h}_e + \rho_0\right)} \qquad (4.71)$$

The magnitudes of the far depth of field and the near depth of field are thus given by

$$\delta z_{far} = \left|\frac{\rho_0 F\left(1-\frac{M}{\bar{M}}\right)}{M\left(2\bar{h}_e - \rho_0\right)}\right| \qquad \delta z_{near} = \left|-\frac{\rho_0 F\left(1-\frac{M}{\bar{M}}\right)}{M\left(2\bar{h}_e + \rho_0\right)}\right| \qquad (4.72)$$

In case of $\bar{M} = +1$, the expression reduces to the expression (4.59).

When the far point of the depth of field of the axial object point O is at infinity, the far point is imaged at F′. The image distance l′ of the axial image point O′ corresponding to the object point O is given by

$$l' = F + \rho\left(1 - \frac{M}{\bar{M}}\right)\frac{F}{2\bar{h}_e} \tag{4.73}$$

Using the conjugate distance formula for imaging between O, O′, the hyperfocal distance is given by

$$l_{hyp} = -\frac{F}{\rho}\left[\frac{2\bar{h}_e}{\left(1-\frac{M}{\bar{M}}\right)} + \rho\right] \cong -\frac{F}{\rho}\left[\frac{2\bar{h}_e}{\left(1-\frac{M}{\bar{M}}\right)}\right] = -\frac{F\left(2\bar{h}_e\right)}{\rho}\left[\frac{\bar{M}}{\bar{M}-M}\right] \tag{4.74}$$

For the case $\bar{M} = +1$, and $|M| \ll +1$, this expression for hyperfocal distance reduces to the simpler expression (4.67) given above.

4.11 Telecentric Stops

If the aperture stop or the entrance pupil of an imaging lens coincides with the first principal focus of the lens, the exit pupil in the image space is located at infinity. This imaging system is said to be telecentric in the image space. Figure 4.45(a) shows an image space telecentric system imaging an object OQ of height η as O′Q′ of height η'. The paraxial principal ray QS is parallel to the axis in the image space. If the image plane at O′ is shifted by a small distance Δ, the image point Q′ turns into a blur circle. However, the centre of the blur circle appears as the location of the image on the new image plane. Since the centre is on the pupil ray, it is at a height η' from the axis, and thus the image height remains unchanged. By contrast, Figure 4.45(b) shows the same lens with the aperture stop located at a non-telecentric position. The paraxial pupil ray is inclined to the axis, and, therefore, the height of the image is changed by $\delta\eta'$ when the image plane is shifted by Δ.

Figure 4.46 shows an object space telecentric system realized with the help of an aperture stop/exit pupil located at the second principal focus F′ of the lens system. By locating the aperture stop at the common focus (the second principal focus of the first lens coinciding with the first principal focus of the second lens) of two lenses L_1 and L_2, an imaging system that is telecentric in both the object space and the image space can be realized (Figure 4.47). Note that the last system is also an afocal system that provides fixed transverse magnification for all conjugate positions.

Telecentric stops are widely used in optical measuring instruments, e.g., profile or contour projectors, machine vision systems, etc., and optical systems with strict dimensional tolerances, e.g., orthographic camera systems, microlithography, etc.

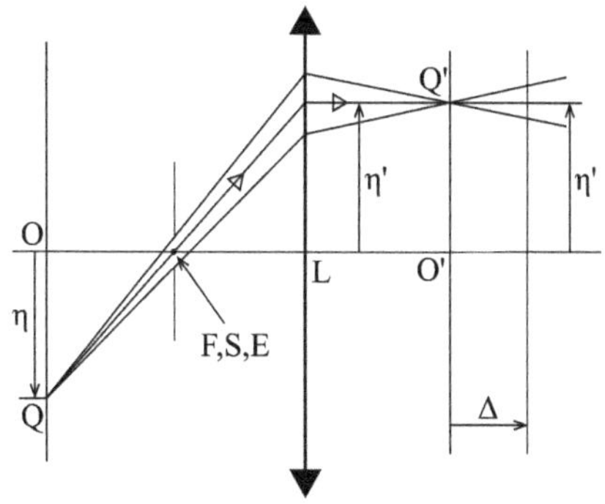

FIGURE 4.45(a) Image space telecentric system. Stop S and entrance pupil E is coincident with the object space focus F.

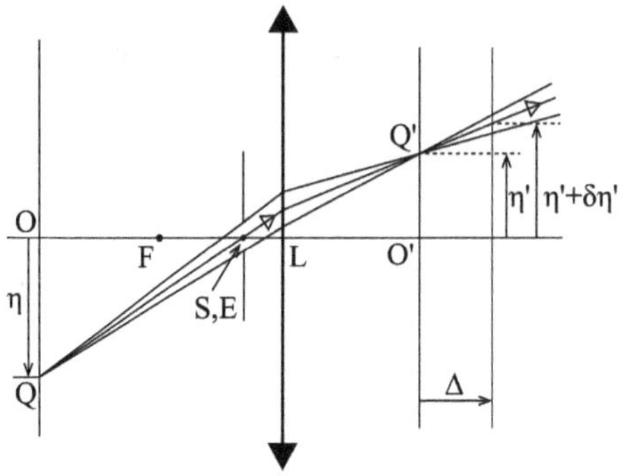

FIGURE 4.45(b) General imaging by a non-telecentric set up.

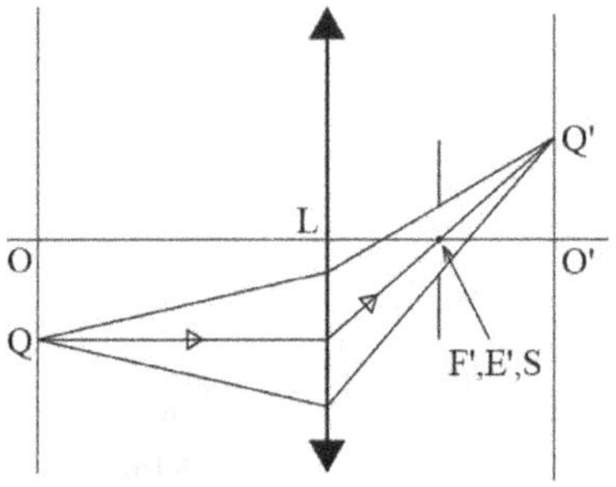

FIGURE 4.46 An object space telecentric system. Stop S and exit pupil E′ is coincident with image space focus F′.

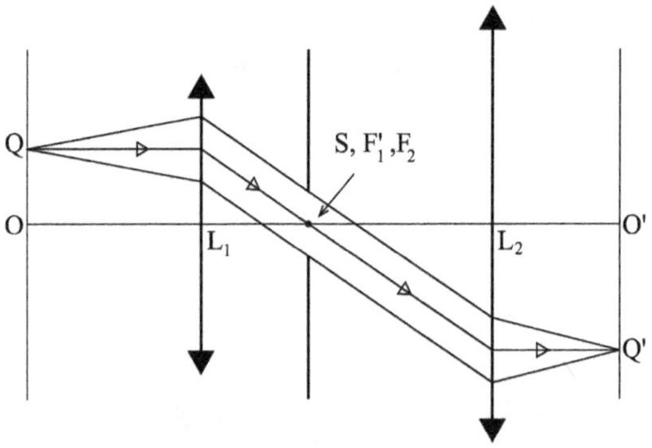

FIGURE 4.47 Both object space and image space telecentricity. Aperture stop S is at common focus $F_1' \equiv F_2$.

4.12 Stops in Illumination Systems

Illumination systems are integral components of projection systems or microscopic systems, where the object needs to be suitably illuminated for obtaining images of the required brightness. In addition, pupils of the illuminating system and of the projection/microscope objective should match so that the non-uniformity of illumination in the final image is minimized.

4.12.1 Slide Projector

Figure 4.48 shows the optical layout of a slide projector. The entrance and exit pupils of the condenser lens and the projection lens are at E_C, E_c', and E_p, E_p' respectively. The slide is placed on the gate AB

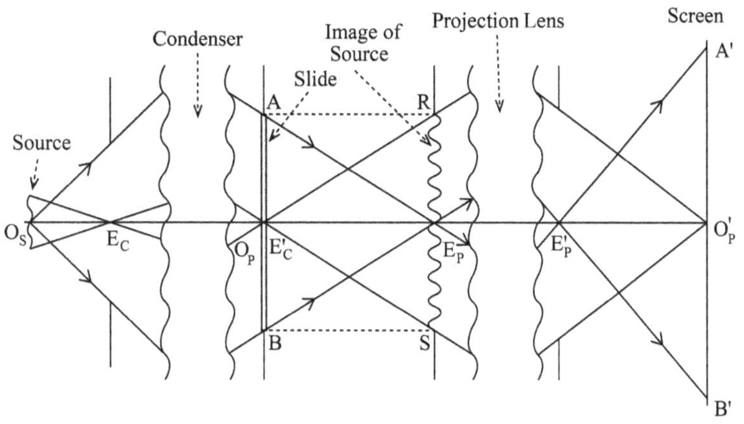

FIGURE 4.48 Optical layout of a slide projector.

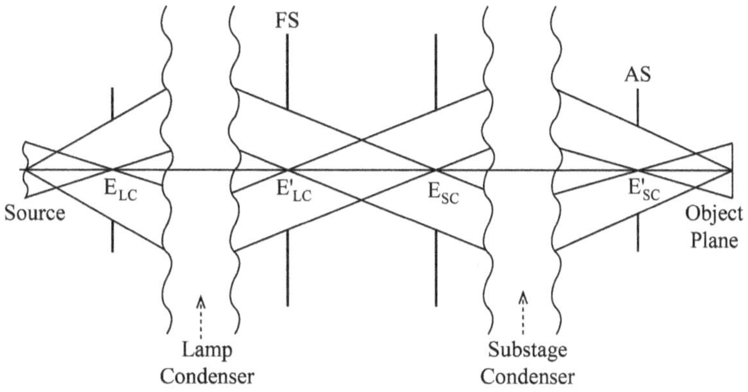

FIGURE 4.49 Optical layout of the Kohler illumination system.

located at the exit pupil E'_C. The latter acts as an aperture stop for the condenser lens, and as a field stop for the projection lens. The projection lens images the slide AB as A′B′ on the screen at O'_P. The condenser lens images the source on the entrance pupil of the projection lens. Note that the roles of the aperture stop and the field stop are interchanged for the two lens systems, namely the condenser lens and the projection lens. The images of the source and of the slide are widely separated to forestall the occurrence of undesirable non-uniformity in illumination on the screen.

The contrast in the final image may be seriously impaired by stray light that can arise from scattering at the edges of the lenses or parts of the mounts holding different lens elements. Occurrence of stray light can be prevented by slight under illumination. The lines AR and BS joining the periphery of the gate to the extremities of the image RS of the source filament define the light cone in the space between the gate and the projection lens. The magnification of the source by the condenser is usefully made small enough for this light cone to intercept the first component of the projection lens at slightly less than the full diameter. This limits the amount of light, which falls on the lens edges and mounts, and consequently an excessive amount of stray light does not reach the final image plane.

4.12.2 The Köhler Illumination System in Microscopes

The Köhler illumination system consists of two condensers in cascade lying in between the source and the object plane (Figure 4.49). The lamp condenser images the lamp source on the entrance pupil E_{SC} of the substage condenser, and the substage condenser images the exit pupil E'_{LC} of the lamp condenser on the object plane. An iris diaphragm is placed at the exit pupil E'_{LC} of the lamp condenser, and it acts as a field stop for the substage condenser. Another iris diaphragm placed on the exit pupil of the substage condenser acts as its aperture stop. By this optical layout, the different sizes of object fields corresponding to different microscope objectives can be catered to by changing the diameter of the field stop FS. The necessary level of illumination can be controlled independently by means of the aperture stop AS.

From the two examples considered above, it is clear that prevention of any occurrence of non-uniformity of illumination in the final observation plane calls for the suitable matching of appropriate stops in the illumination system to the positions of object and pupils of the observing system.

REFERENCES

1 M.V. Rohr, 'The limitation of rays in optical systems (Theory of stops), Chapter IX in *Geometrical Investigation of the Formation of Images in Optical Instruments*, Ed. M. Von Rohr, His Majesty's Stationary Office, London, (1920) pp. 473–514.
2 J. Petzval, 'Bericht über Dioptrische Untersuchungen', *Wien Bericht*, Vol. 26 (1857) pp. 33–90.
3 E. Abbe, 'Über die Bestimmung der Lichtstärke optischer Instrumente. Mit besonderer Berücksichtigung der Mikroskops und der Apparate zur Lichtconcentration', *Jena Zeits. f. Med. u. Naturw.*, Vol. 6, (1871) pp. 263–291.
4 F. Jenkins and H.E. White, *Fundamentals of Optics*, McGraw-Hill, New York (1937) pp.121–122.
5 C.G. Wynne, 'Vignetting', *Proc. Phys. Soc.*, Vol. 56 (1944) pp. 366–371.
6 M. Herzberger, *Modern Geometrical Optics*, Interscience, New York (1958) pp. 101–107.
7 W.B. King, 'The Approximation of Vignetted Pupil Shape by an Ellipse', *Appl. Opt.*, Vol. 7 (1968) pp. 197–201.
8 W.J. Smith, *Modern Optical Engineering*, McGraw-Hill, New York (2000) pp. 141–147.
9 W.J. Smith, ibid., pp. 147–150.
10 G. Smith, 'Veiling Glare due to Reflections from Component Surfaces: The Paraxial Approximation', *Opt. Acta*, Vol. 18 (1971) pp. 815–827.
11 G. Smith and D.A. Atchison, The eye and visual optical instruments, Cambridge University Press, Cambridge (1997).
12 M. Rosete-Aguilar and J.L Rayces, 'Eye rotation and vignetting in visual instruments', *Appl. Opt.*, Vol. 41 (2002) pp. 6593–6602.
13 J.W. Strutt, Lord Rayleigh, 'Notes, chiefly Historical, on some Fundamental Propositions in Optics', Phil. Mag., (5) Vol. 21 (1886) pp. 466–476.
14 R. Smith, *A Compleat System of Optics*, in four books, viz. a popular, a mathematical, a mechanical and a philosophical treatise, Cambridge (1738).
15 J.L. de Lagrange, 'Sur la théorie des lunettes', Nouveaux Mémoires de l'Académie Royale des Sciences et Belles-Lettres de Berlin (1778), in *Œuvres de Lagrange, tome IV*, Gauthier-Villars, Paris (1869).
16 J.L. de Lagrange, 'Sur une loi générale d'Optique', Nouveaux Mémoires de l'Academie Royale des Sciences et belles-lettres de Berlin (1803), in *Œuvres de Lagrange, tome V*, Gauthier-Villars, Paris (1870).
17 H. von Helmholtz, 'Optical Imagery in a system of spherical refracting surfaces', in *Helmholtz's Treatise on Physiological Optics*, Vol. I, Ed. J.P.C. Southall, The Optical Society of America (1924) p.74. The original German book *Handbuch der Physiologischen Optik* by Helmholtz was published in 1866.
18 H.A. Buchdahl, Optical Aberration Coefficients, Dover, New York (1968). [Originally published by Oxford Univ. Press (1954).
19 B.N. Begunov, N.P. Zakaznov, S.I. Kiryushin, and V.I. Kuzichev, *Optical Instrumentation, Theory and Design*, Translated from Russian by M. Edelev, MIR Publishers, Moscow (1988) pp. 52–56.

5

Towards Facilitating Paraxial Treatment

Paraxial treatment of optical systems essentially involves tracing selected paraxial rays through the interfaces of the system from the object space to the image space. This involves the repeated use of the paraxial refraction formula and paraxial transfer formula, presented earlier as equations (3.85) and (3.86) in Chapter 3. The equations are reproduced below:

$$n'u' - nu = -hK \tag{5.1}$$

$$h_{+} = h + d'u' \tag{5.2}$$

where $K = (n' - n)c$.

In this $(nu - h)$ paraxial ray trace, u is the paraxial angle variable, and h is the paraxial height variable. Numerical implementation of paraxial ray tracing equations is quite straightforward, and even in case of very complex optical systems, desired results can be obtained. However, any direct algebraic approach to obtain analytical expressions for the paraxial characteristics of the system, either in terms of the object space variables or in terms of the constructional parameters of the system, e.g., the curvatures of the interfaces and their axial separations, leads to unwieldy expressions involving continued fractions even for moderately complex optical systems.

In order to circumvent this limitation, and to facilitate analytical treatments in paraxial optics pertaining to different applications, investigators have proposed many interesting approaches from the early days of developments in this field. Some of them are listed below:

- Matrix method [1–27]
- Gaussian Bracket [28–42]
- Delano diagram [43–62]
- Use of Wigner distribution function [63–64]
- Symplectic and Lie algebraic approach [65–68]
- Canonical operator approach [69–72]

Most of these approaches have found their niche in specific practical applications. So far, use of the first three of the list has been quite significant, and we describe below their salient features.

5.1 Matrix Treatment of Paraxial Optics

R.A. Sampson [1] used matrices for parabasal analysis in general optical systems. A few decades later, T. Smith [2] presented a matrix treatment specifically for paraxial optics. Harrold [3], O'Neill [4], Halbach [5], and Richards [6] followed his work. Later on Brouwer [7], Brouwer and Walther [8], Blaker [9], and Gerrard and Burch [10] popularized this approach. Sinclair proposed specifications of optical systems by paraxial transfer matrices [11]. A spurt in practical use of the matrix treatment occurred in connection with the developments in non-symmetrical systems, beam propagation problems, and resonator design [12–27].

DOI: 10.1201/9780429154812-5

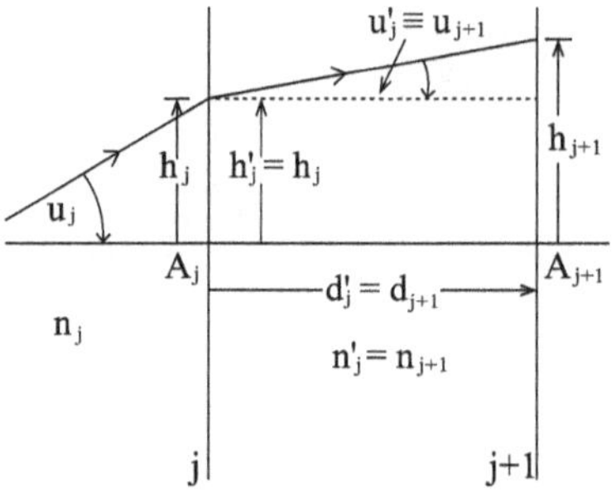

FIGURE 5.1 Refraction at the j-th surface followed by translation/transfer to the (j+1)-th surface.

5.1.1 The Refraction Matrix and the Translation/Transfer Matrix

Figure 5.1 shows the refraction of a paraxial ray at the j-th surface of a system, followed by transfer to the (j+1)-th surface. The power of the j-th surface is K_j, the axial separation between the j-th and the (j+1)-th surfaces is $d'_j = d_{j+1}$, and the convergence angle of the paraxial ray incident on the j-th surface is u_j. The convergence angle of the refracted ray at the j-th surface is $u'_j = u_{j+1}$. Since this is a paraxial analysis, the j-th and the (j+1)-th refracting surfaces are shown as plane surfaces that are perpendicular to the optical axis at the corresponding vertices A_j and A_{j+1}, respectively. The heights of the paraxial ray at the j-th and (j+1)-th surfaces are h_j and h_{j+1}. Let us define reduced separation $\tilde{d}$ and reduced paraxial angle variable $\tilde{u}$ as

$$\tilde{d} = \frac{d}{n} \qquad \tilde{u} = nu \tag{5.3}$$

The paraxial refraction equation for the j-th surface takes the form

$$\tilde{u}'_j - \tilde{u}_j = -h_j K_j \tag{5.4}$$

Distances normal to the axis, such as the incidence height h, are not affected by the use of the reduced variables. At the j-th surface, the heights of the incident ray and the refracted ray is the same, so that

$$h'_j = h_j \tag{5.5}$$

The subsequent transfer equation for transfer to the $(j+1)$-th surface reduces to

$$h_{j+1} = h'_j + \tilde{d}_{j+1}\tilde{u}'_j \tag{5.6}$$

The paraxial ray refracted from the j-th surface is the same paraxial ray incident on the $(j+1)$-th surface, and, therefore, the convergence angle is same for both of them. Thus,

$$\tilde{u}'_j = u_{j+1} \tag{5.7}$$

Equations (5.4) – (5.5) can be rewritten as

$$h'_j = 1.h_j + 0.\tilde{u}_j \tag{5.8}$$

$$\tilde{u}'_j = -K_j h_j + 1.\tilde{u}_j \tag{5.9}$$

Similarly, (5.6) – (5.7) can be rewritten as

$$h_{j+1} = 1.h'_j + \tilde{d}_{j+1}\tilde{u}'_j \tag{5.10}$$

$$\tilde{u}_{j+1} = 0.h'_j + 1.\tilde{u}'_j \tag{5.11}$$

The set of equations (5.8) – (5.9) and (5.10) – (5.11) may be rewritten in matrix form as given below.

$$\begin{bmatrix} h'_j \\ \tilde{u}'_j \end{bmatrix} = \begin{bmatrix} 1 & 0 \\ -K_j & 1 \end{bmatrix} \begin{bmatrix} h_j \\ \tilde{u}_j \end{bmatrix} = \mathcal{R}_j \begin{bmatrix} h'_j \\ \tilde{u}'_j \end{bmatrix} \tag{5.12}$$

$$\begin{bmatrix} h_{j+1} \\ \tilde{u}_{j+1} \end{bmatrix} = \begin{bmatrix} 1 & \tilde{d}_{j+1} \\ 0 & 1 \end{bmatrix} \begin{bmatrix} h'_j \\ \tilde{u}'_j \end{bmatrix} = \mathcal{T}_{j+1} \begin{bmatrix} h'_j \\ \tilde{u}'_j \end{bmatrix} \tag{5.13}$$

The refraction matrix $\mathcal{R}_j$ is a 2×2 unimodular matrix; the subscript j indicates refraction of the ray at the j-th surface. The translation or transfer matrix $\mathcal{T}_{j+1}$ is also a 2×2 unimodular matrix; the subscript j+1 indicates translation of the ray from the j-th surface to the (j+1)-th surface. From (5.12) and (5.13) we get

$$\begin{bmatrix} h_{j+1} \\ \tilde{u}_{j+1} \end{bmatrix} = \mathcal{T}_{j+1} \begin{bmatrix} h'_j \\ \tilde{u}'_j \end{bmatrix} = \mathcal{T}_{j+1}\mathcal{R}_j \begin{bmatrix} h_j \\ \tilde{u}_j \end{bmatrix} \tag{5.14}$$

5.1.2 The System Matrix

Since, in general, matrix multiplication is not commutative, so the order of multiplication is essential to obtain correct results. Extending to the case of many, say k, refracting surfaces, values of h and ũ in the $(k+1)$-th space are related with the same in the 1st space by

$$\begin{bmatrix} h_k \\ \tilde{u}'_k \end{bmatrix} = \mathcal{R}_k \mathcal{T}_k \mathcal{R}_{k-1} \ldots\ldots \mathcal{T}_2 \mathcal{R}_1 \begin{bmatrix} h_1 \\ \tilde{u}_1 \end{bmatrix} \tag{5.15}$$

The product $\mathcal{R}_k\, \mathcal{T}_k\, \mathcal{R}_{k-1} \ldots\ldots \mathcal{T}_2\, \mathcal{R}_1 \equiv \mathcal{S}$ is also a unimodular matrix; it is called the 'System matrix'. The matrix $\mathcal{S}$ can be written as

$$\mathcal{S} = \begin{bmatrix} A & B \\ C & D \end{bmatrix} \tag{5.16}$$

The elements A, B, C, and D of the matrix $\mathcal{S}$ are functions of powers of the surfaces and the reduced inter surface separations. Figure 5.2 shows a single thick lens of reduced axial thickness $\tilde{d}_2$. The powers of the first and the second surfaces of the lens are K_1, and K_2, respectively. The system matrix of this lens is known as lens matrix, and it is given by

$$\mathcal{S}_{(\text{thick lens})} = \mathcal{R}_2\mathcal{T}_2\mathcal{R}_1 = \begin{bmatrix} 1 & 0 \\ -K_2 & 1 \end{bmatrix} \begin{bmatrix} 1 & \tilde{d}_2 \\ 0 & 1 \end{bmatrix} \begin{bmatrix} 1 & 0 \\ -K_1 & 1 \end{bmatrix}$$

$$= \begin{bmatrix} 1 - K_1\tilde{d}_2 & \tilde{d}_2 \\ -K_1 - K_2 + K_1K_2\tilde{d}_2 & -K_2\tilde{d}_2 + 1 \end{bmatrix} \tag{5.17}$$

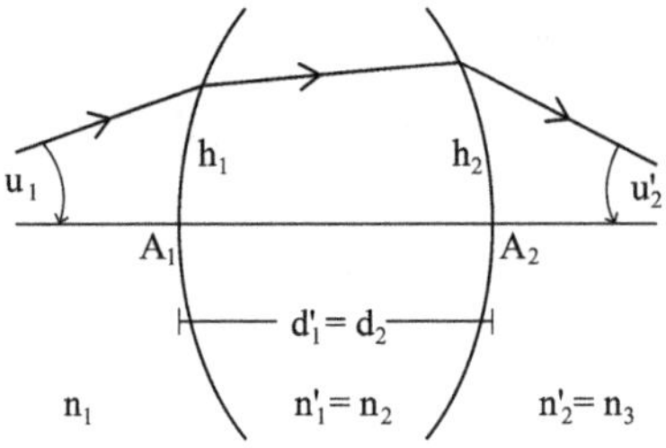

FIGURE 5.2 Passage of a paraxial ray though a single thick lens of reduced axial thickness $\tilde{d}_2 = n_2 d_2$.

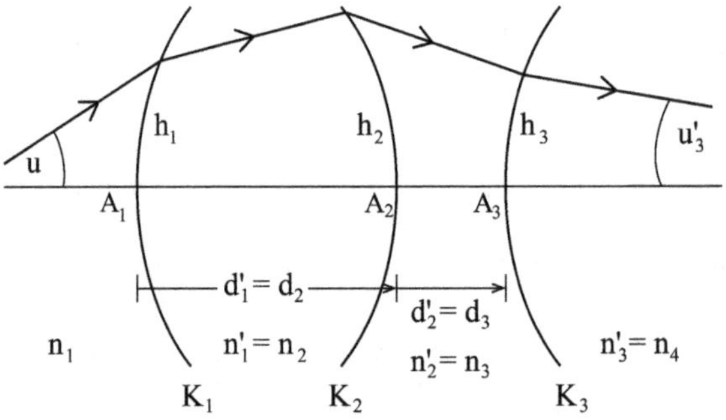

FIGURE 5.3 Passage of a paraxial ray through a cemented doublet lens.

and the ray vector in the image space of the thick lens is given in terms of the same in the object space as

$$\begin{bmatrix} h_2 \\ \tilde{u}_2' \end{bmatrix} = \mathcal{S}_{(\text{thick lens})} \begin{bmatrix} h_1 \\ \tilde{u}_1 \end{bmatrix} \tag{5.18}$$

For a thin lens, $\tilde{d}_2 = 0$, and the lens matrix simplifies to

$$\mathcal{S}_{(\text{thin lens})} = \begin{bmatrix} 1 & 0 \\ -(K_1 + K_2) & 1 \end{bmatrix} \tag{5.19}$$

If there is one more refracting interface of power K_3 in front of the above thick lens, and its reduced distance from the vertex of the second surface is $\tilde{d}_3$, the thick lens becomes a doublet (Figure 5.3), and its system matrix is

$$\mathcal{S}_{(\text{doublet})} = \mathcal{R}_3 \mathcal{T}_3 \mathcal{R}_2 \mathcal{T}_2 \mathcal{R}_1 = \mathcal{R}_3 \mathcal{T}_3 \begin{bmatrix} 1 - K_1 \tilde{d}_2 & \tilde{d}_2 \\ -K_1 - K_2 + K_1 K_2 \tilde{d}_2 & -K_2 \tilde{d}_2 + 1 \end{bmatrix}$$

$$= \begin{bmatrix} 1 & 0 \\ -K_3 & 1 \end{bmatrix} \begin{bmatrix} 1 & \tilde{d}_3 \\ 0 & 1 \end{bmatrix} \begin{bmatrix} 1 - K_1 \tilde{d}_2 & \tilde{d}_2 \\ -K_1 - K_2 + K_1 K_2 \tilde{d}_2 & -K_2 \tilde{d}_2 + 1 \end{bmatrix}$$

$$= \begin{bmatrix} \left\{1 - K_2 \tilde{d}_3 - K_1 \varnothing\right\} & \varnothing \\ \left\{-(K_1 + K_2 + K_3) + K_2 \left(K_1 \tilde{d}_2 + K_3 \tilde{d}_3\right) + K_1 K_3 \varnothing\right\} & \left\{1 - K_2 \tilde{d}_2 - K_3 \varnothing\right\} \end{bmatrix} \tag{5.20}$$

where

$$\emptyset = \left(\tilde{d}_2 + \tilde{d}_3 - K_2\tilde{d}_2\tilde{d}_3\right) \tag{5.21}$$

The ray vector in the image space of the thick lens is given in terms of the same in the object space as

$$\begin{bmatrix} h_3 \\ \tilde{u}_3' \end{bmatrix} = \mathcal{S}_{(\text{doublet})} \begin{bmatrix} h_1 \\ \tilde{u}_1 \end{bmatrix} \tag{5.22}$$

5.1.3 The Conjugate Matrix

In an axi-symmetric optical system consisting of k number of refracting interfaces, the ray vector in the $(k+1)$-th space is related with the ray vector in the first space by means of the system matrix, as shown in (5.15) above. The object in the object space and the image in the image space can be incorporated in the framework by pre- and post-multiplication of the system matrix $\mathbb{S}$ by suitable translation matrices. Let the object and the image heights η and η' be represented as h_0 and h_{k+1} (Figure 5.4). The object surface is considered as the zero-th surface, and the final image surface is taken as the $(k+1)$-th surface. The refractive index is the same on both sides of the object surface, and on both sides of the image surface. Therefore, $n_0' = n_1$, and $n_{k+1}' = n_{k+1}$. The conjugate matrix $\mathcal{C}$ for the axi-symmetric imaging system consisting of k number of refracting interfaces is given by

$$\mathcal{C} = \mathcal{T}_{k+1}\ \mathcal{S}\mathcal{T}_1 \tag{5.23}$$

and we have

$$\begin{bmatrix} h_{k+1} \\ \tilde{u}_{k+1}' \end{bmatrix} = \mathcal{C}\begin{bmatrix} h_0 \\ \tilde{u}_0 \end{bmatrix} = \begin{bmatrix} c_{11} & c_{12} \\ c_{21} & c_{22} \end{bmatrix}\begin{bmatrix} h_0 \\ \tilde{u}_0 \end{bmatrix} \tag{5.24}$$

In the case of finite conjugate imaging, not any arbitrary 2×2 matrix can be a conjugate matrix. By definition, Q_k' with height $\eta'(=h_{k+1})$ can be a conjugate of the object point Q with height $\eta(=h_0)$, if and only if all ray vectors with the same value for h_0 and different values for $\tilde{u}_0$ in the object space lead to the same value for h_{k+1} of the ray vector in the final image space. From (5.24) we note that this is only possible if $c_{12} = 0$, for then we get

$$h_{k+1} = c_{11}h_0 + c_{12}\tilde{u}_0 = c_{11}h_0 \tag{5.25}$$

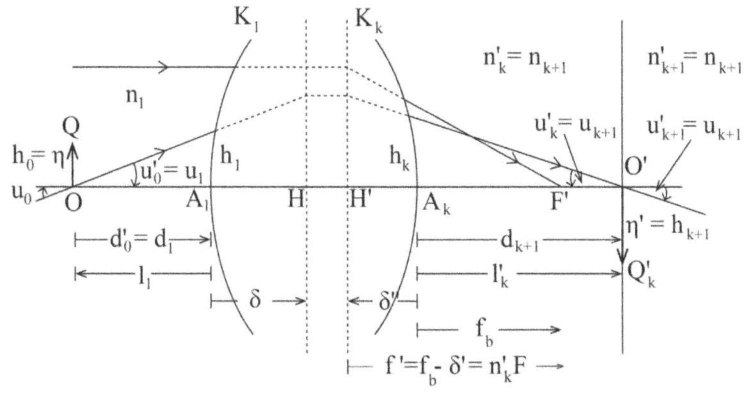

FIGURE 5.4 An axisymmetric imaging system consisting of k refracting interfaces images an object OQ of height $\eta(=h_0)$ as $O_k'Q_k'(=\eta'=h_{k+1})$.

i.e. h_{k+1} is independent of $\tilde{u}_0$, and so all ray vectors with the same value of h_0 lead to same value of h_{k+1} in the image space. Thus, in finite conjugate imaging, $c_{12} = 0$, and the conjugate matrix $\mathcal{C}$ is to be of the form

$$\mathcal{C} = \begin{bmatrix} c_{11} & 0 \\ c_{21} & c_{22} \end{bmatrix} \tag{5.26}$$

Three other special cases are:

(i) $c_{11} = 0$. From (5.24) we get $h_{k+1} = c_{11}h_0 + c_{12}\tilde{u}_0 = c_{12}\tilde{u}_0$. Thus h_{k+1} is the same for all ray vectors of the object space, which have a fixed value of $\tilde{u}_0$ but different values for h_0. This implies the special case of the object at infinity, and the image on the focal plane in the image space.

(ii) $c_{21} = 0$. From (5.24) it follows $\tilde{u}'_{k+1} = c_{21}h_0 + c_{22}\tilde{u}_0 = c_{22}\tilde{u}_0$. Therefore, the direction u'_{k+1} of the emergent ray in the image space is directly proportional to the direction u_0 of the incident ray in the object. This implies an afocal or telescopic imaging system where a parallel bundle of incident rays emerges parallel in the image space.

(iii) $c_{22} = 0$. From (5.24), we get $\tilde{u}'_{k+1} = c_{21}h_0 + c_{22}\tilde{u}_0 = c_{21}h_0$. Thus, ray vectors with the same value for h_0 but different values for $\tilde{u}_0$ in the object space give rise to ray vectors in the image space that have the same value for $\tilde{u}'_{k+1}$, irrespective of their heights h_{k+1}. This implies the special case of the image at infinity and the object on the focal plane in the object space.

5.1.4 Detailed Form of the Conjugate Matrix in the Case of Finite Conjugate Imaging

By definition, the conjugate matrix $\mathcal{C}$ is related with the system matrix $\mathbb{S}$ by (as shown in (5.23))

$$\mathcal{C} = \mathcal{T}_{k+1}\,\mathcal{S}\mathcal{T}_1 \tag{5.27}$$

In Figure 5.4, the object distance is $A_1O = l_1$, and the image distance $A_kO'_k = l'_k$. It follows

$$d'_0 = d_1 = OA_1 = -l_1 \qquad d_{k+1} = A_kO'_k = l'_k \tag{5.28}$$

Therefore, the translation matrices $\mathcal{T}_1$ and $\mathcal{T}_{k+1}$ are given by

$$T_1 = \begin{bmatrix} 1 & -\tilde{l}_1 \\ 0 & 1 \end{bmatrix} \text{ and } T_{k+1} = \begin{bmatrix} 1 & \tilde{l}'_k \\ 0 & 1 \end{bmatrix} \tag{5.29}$$

where the reduced distances $\tilde{l}_1 = \dfrac{l_1}{n_1}$ and $\tilde{l}'_k = \dfrac{l'_k}{n'_k}$ are used. Substituting these forms for $\mathcal{T}_1$ and $\mathcal{T}_{k+1,}$ and for $\mathcal{S}$ from (5.16), we get from (5.27) the expression for the conjugate matrix as

$$\mathcal{C} = \begin{bmatrix} \left\{A + C\tilde{l}'_k\right\} & \left\{\tilde{l}'_k\left(D - C\tilde{l}_1\right) - A\tilde{l}_1 + B\right\} \\ C & \left\{D - C\tilde{l}_1\right\} \end{bmatrix} \tag{5.30}$$

In the case of finite conjugate imaging,

$$c_{12} = \left\{\tilde{l}'_k\left(D - C\tilde{l}_1\right) - A\tilde{l}_1 + B\right\} = 0 \tag{5.31}$$

It follows

$$\tilde{l}'_k = \frac{A\tilde{l}_1 - B}{-C\tilde{l}_1 + D} \tag{5.32}$$

$$\tilde{l}_1 = \frac{B + D\tilde{l}'_k}{A + C\tilde{l}'_k} \tag{5.33}$$

The location of the image plane corresponding to reduced object distance $\tilde{l}_1$, and similarly, the location of the object plane corresponding to reduced image distance $\tilde{l}'_k$ can thus be determined from (5.32) and (5.33), respectively, if the elements A, B, C, and D of the system matrix $\mathcal{S}$ are known. The conjugate matrix reduces to

$$\mathcal{C}_{(\text{finite conjugate})} = \begin{bmatrix} \{A + C\tilde{l}'_k\} & 0 \\ C & \{D - C\tilde{l}_1\} \end{bmatrix} \tag{5.34}$$

From (5.24), since $c_{12} = 0$, it follows

$$h_{k+1} = c_{11} h_0 \tag{5.35}$$

Substituting for n_{11} from (5.34), the transverse magnification M is given by

$$M = \frac{h_{k+1}}{h_0} = c_{11} = \{A + C\tilde{l}'_k\} \tag{5.36}$$

5.1.4.1 Location of the Cardinal Points of the System: Equivalent Focal Length and Power

Principal points or unit points are conjugate points of magnification M = +1. From (5.36), the distance $\delta' (= A_k H')$ of the image space principal or unit point H′ from the vertex A_k of the k-th refracting interface can be obtained as

$$\delta' = n'_k \left(\frac{1 - A}{C} \right) \tag{5.37}$$

Since the determinant of the unimodular system matrix $\mathbb{S}$ is one, we have

$$AD - BC = 1 \tag{5.38}$$

The distance $\delta (= A_1 H)$ of the object space principal or unit point H from the vertex A_1 of the 1st surface can be obtained from (5.33) by using (5.37) – (5.38) as

$$\delta = n_1 \left(\frac{D - 1}{C} \right) \tag{5.39}$$

It has been shown earlier that if the element c_{11} of the conjugate matrix is zero, the image is on the second or image space focal plane. From (5.34), we note

$$c_{11} = A + C\tilde{l}'_k \tag{5.40}$$

Substituting $c_{11} = 0$, we get the back focal distance $f_b \left(= A_k F'\right)$ represented in terms of elements of the system matrix as

$$f_b = -n'_k \left(\frac{A}{C}\right) \tag{5.41}$$

The secondary focal length f' is equal to the distance of F' from the second principal point H'. It is given by

$$f' = f_b - \delta' = -n'_k \left(\frac{A}{C}\right) - n'_k \left(\frac{1-A}{C}\right) = -\frac{n'_k}{C} \tag{5.42}$$

The expression for equivalent focal length F of the system reduces to

$$F = \frac{f'}{n'_k} = -\frac{1}{C} \tag{5.43}$$

The power K of the system is represented as

$$K = \left(\frac{1}{F}\right) = -C \tag{5.44}$$

From (5.15) and (5.16), using the system matrix, we get the parameters of the ray in the image space in terms of the same in the object space as

$$\begin{bmatrix} h_k \\ \tilde{u}'_k \end{bmatrix} = \begin{bmatrix} A & B \\ C & D \end{bmatrix} \begin{bmatrix} h_1 \\ \tilde{u}_1 \end{bmatrix} \tag{5.45}$$

By definition, the paraxial convergence angle u_1 of a ray PN in object space is equal to the paraxial convergence angle $\tilde{u}'_k$ of the emergent ray N′P′ in the image space (Figure 5.5).

From (5.45) we get

$$n'_k u'_k = C h_1 + D n_1 u_1 \tag{5.46}$$

Let

$$u_1 = u'_k = \hat{u} \tag{5.47}$$

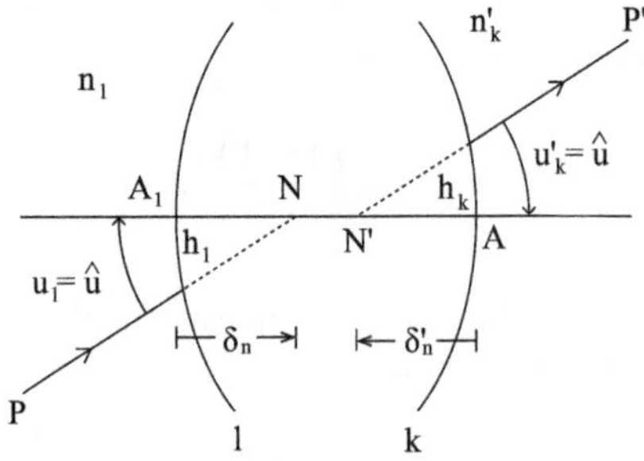

FIGURE 5.5 Nodal points N, N' of an axisymmetric imaging lens system; $\angle PNA_1 = \angle P'N'A_k$.

It follows

$$h_1 = \frac{\left(n'_k - Dn_1\right)}{C}\hat{u} \tag{5.48}$$

The distance δ_n of the object-space nodal point N from the vertex A_1 of the first refracting interface is given by

$$\delta_n = -\frac{h_1}{\hat{u}} = \left(\frac{Dn_1 - n'_k}{C}\right) \tag{5.49}$$

From (5.45)

$$h_k = Ah_1 + Bn_1u_1 \tag{5.50}$$

Substituting for h_1 from (5.48), and using relation (5.38), we get

$$h_k = \frac{An'_k - n_1}{C}\hat{u} \tag{5.51}$$

The distance δ'_n of the image space nodal point N′ from the vertex A_k of the k-th refracting interface is given by

$$\delta'_n = -\frac{h_k}{u'_k} = -\frac{h_k}{\hat{u}} = \left(\frac{n_1 - An'_k}{C}\right) \tag{5.52}$$

From (5.37), (5.39), (5.49), and (5.52), we note that for $n_1 = n'_k$,

$$\delta_n = \delta \text{ and } \delta'_n = \delta' \tag{5.53}$$

i.e., the nodal points N and N′ coincide with the principal points H and H′, when the refractive index of the object space and that of the image space are the same.

5.2 Gaussian Brackets in Paraxial Optics

Carl Friedrich Gauss [28] republished a set of recursion formulae developed earlier by Euler, among other things, to solve the problem of finding the common denominator of two numbers. Gauss found this algorithm to be useful in solving a linear Diophantine equation.

More than a century later, Herzberger [29] rekindled interest on this approach by showing the inherent compatibility of the symbols, developed by Gauss in his treatment, for finding out the dependence of the paraxial characteristics, e.g. focal length, magnification, etc., with the constructional parameters of an optical imaging system. These symbols are now known as *Gaussian brackets.* Besides Herzberger [29–32], Zimmer [33] used these brackets in treatments of paraxial optics. Bergstein [34–37] proposed a modified version while dealing with the general theory of optically compensated varifocal systems. Tanaka [38] made extensive use of Gaussian brackets in the analysis and synthesis of different types of optical systems. Other users of Gaussian brackets include Pegis and Peck [39], Minami [40], and Smirnov [41]. Bhattacharya and Hazra used the Gaussian brackets to obtain analytical derivatives of some paraxial characteristics of optical systems with respect to constructional parameters [42].

5.2.1 Gaussian Brackets: Definition

Let $a_1, a_2, \ldots, a_n$ be an ordered set of quantities (real numbers or real functions), the symbol $[a_1, a_2, \ldots, a_n]$ is defined as the Gaussian bracket with the help of a recursion formula as follows. An empty bracket is defined as having a value of 1. A bracket with one element is the element itself. The recursion relation is:

$$[a_1, a_2, \ldots, a_n] = [a_1, a_2, \ldots, a_{n-1}]a_n + [a_1, a_2, \ldots, a_{n-2}] \tag{5.54}$$

The first few brackets are:

$$[\;] = 1$$

$$[a_1] = a_1$$

$$[a_1, a_2] = a_1 a_2 + 1$$

$$[a_1, a_2, a_3] = a_1 a_2 a_3 + a_1 + a_2$$

$$[a_1, a_2, a_3, a_4] = a_1 a_2 a_3 a_4 + a_1 a_2 + a_1 a_4 + a_3 a_4 + 1$$

$$[a_1, a_2, a_3, a_4, a_5] = a_1 a_2 a_3 a_4 a_5 + a_1 a_2 a_3 + a_1 a_2 a_5 + a_1 a_4 a_5 + a_3 a_4 a_5 + a_1 + a_3 + a_5$$

$$\begin{aligned}[a_1, a_2, a_3, a_4, a_5, a_6] = {} & a_1 a_2 a_3 a_4 a_5 a_6 + a_1 a_2 a_3 a_4 + a_1 a_2 a_3 a_6 + a_1 a_2 a_5 a_6 + a_1 a_4 a_5 a_6 \\ & + a_3 a_4 a_5 a_6 + a_1 a_2 + a_1 a_4 + a_1 a_6 + a_3 a_4 + a_3 a_6 + a_5 a_6 + 1\end{aligned} \tag{5.55}$$

The recursion relation (5.54) implies that the bracket containing n elements is equal to the sum of:

(a) the product of all n elements,
(b) all products of (n - 2) elements, which can be formed with indices increasing from left to right starting with an odd index and with alternating odd and even indices, and
(c) all products of (n - 4) elements under the same conditions, and so on.

When n is even, a last term of 1 is included, whereas, when n is odd, the addition ends with the sum of the elements with odd indices [29].

5.2.2 Few Pertinent Theorems of Gaussian Brackets

Theorem 1. A Gaussian bracket is a linear function of any of its elements a_m. Therefore,

$$[a_1, a_2, \ldots, a_n] = Ca_m + D \tag{5.56}$$

where

$$\begin{aligned} C &= [a_2, \ldots, a_n] && m = 1 \\ &= [a_1, a_2, \ldots, a_{m-1}][a_{m+1}, \ldots, a_n] && 1 < m < n \\ &= [a_1, a_2, \ldots, a_{n-1}] && m = n \end{aligned} \tag{5.57}$$

and

$$D = [a_3, \ldots, a_n] \qquad m = 1$$

$$= [a_1, a_2, \ldots, a_{m-2}, (a_{m-1} + a_{m+1}), a_{m+2}, \ldots, a_n] \qquad 1 < m < n$$

$$= [a_1, a_2, \ldots, a_{n-2}] \qquad m = n \tag{5.58}$$

Corollary:

A Gaussian bracket can be easily differentiated with respect to its elements, e.g. from (5.56), it follows directly

$$\frac{\partial}{\partial a_m}[a_1, a_2, \ldots, a_n] = C \tag{5.59}$$

Theorem 2. A Gaussian bracket is symmetric.

$$[a_1, a_2, \ldots, a_n] = [a_n, a_{n-1}, \ldots, a_1] \tag{5.60}$$

Theorem 3. A Gaussian bracket is a linear function of any of its partial brackets.

$$[a_1, \ldots, a_n] = [a_1, \ldots, a_m][a_{m+1}, \ldots, a_n] + [a_1, \ldots, a_{m-1}][a_{m+2}, \ldots, a_n] \tag{5.61}$$

5.2.3 Elements of System Matrix in Terms of Gaussian Brackets

In Section 5.1.2, it has been shown that pre-multiplication of the ray vector, in the initial space of an axi-symmetric system consisting of k refracting interfaces and by the unimodular system matrix , $\mathcal{S}$ yields the corresponding ray vector in the $(k+1)$th space. Using the same notation used above we have

$$\begin{bmatrix} h_k \\ \tilde{u}'_k \end{bmatrix} = \mathcal{S}\begin{bmatrix} h_1 \\ u_1 \end{bmatrix} = \begin{bmatrix} A & B \\ C & D \end{bmatrix}\begin{bmatrix} h_1 \\ u_1 \end{bmatrix} \tag{5.62}$$

The system matrix is obtained from the refraction matrices of the interfaces and the transfer matrices from one interface to the next by matrix multiplication as shown below.

$$\mathcal{S} = \begin{bmatrix} A & B \\ C & D \end{bmatrix} = \mathcal{R}_k \mathcal{T}_k \mathcal{R}_{r-1} \cdots \cdots \mathcal{T}_2 \mathcal{R}_i \tag{5.63}$$

The four elements A, B, C, and D of the system matrix can be conveniently expressed in terms of Gaussian brackets. For a system consisting of two refracting surfaces of power K_1 and K_2, respectively, and separated axially by reduced distance $\tilde{d}_2$, the system matrix, derived in (5.17), is given by

$$\mathcal{S}_{(\text{thick lens})} = \begin{bmatrix} 1 - K_1\tilde{d}_2 & \tilde{d}_2 \\ -K_1 - K_2 + K_1 K_2 \tilde{d}_2 & -K_2\tilde{d}_2 + 1 \end{bmatrix} \tag{5.64}$$

The four elements of this system matrix can be expressed in terms of Gaussian brackets as

$$A = 1 - K_1\tilde{d}_2 = [-K_1, \tilde{d}_2] \tag{5.65}$$

$$B = \tilde{d}_2 = [\tilde{d}_2] \tag{5.66}$$

$$C = K_1K_2\tilde{d}_2 - K_1 - K_2 = [-K_1,\tilde{d}_2,-K_2] \tag{5.67}$$

$$D = -K_2\tilde{d}_2 + 1 = [\tilde{d}_2,-K_2] \tag{5.68}$$

For a system consisting of three refracting surfaces of power K_1, K_2, and K_3 with inter surface reduced separations $\tilde{d}_2$ and $\tilde{d}_3$, respectively, the system matrix, as shown in (5.20), is given by

$$\mathcal{S}_{(\text{doublet})} = \begin{bmatrix} \{1 - K_2\tilde{d}_3 - K_1\varnothing\} & \varnothing \\ \{-(K_1 + K_2 + K_3) + K_2(K_1\tilde{d}_2 + K_3\tilde{d}_3) + K_1K_3\varnothing\} & \{1 - K_2\tilde{d}_2 - K_3\varnothing\} \end{bmatrix} \tag{5.69}$$

where

$$\varnothing = (\tilde{d}_2 + \tilde{d}_3 - K_2\tilde{d}_2\tilde{d}_3) \tag{5.70}$$

The four elements of the system matrix of this system are

$$\begin{aligned} A &= 1 - K_2\tilde{d}_3 - K_1\tilde{d}_2 - K_1\tilde{d}_3 + K_1K_2\tilde{d}_2\tilde{d}_3 \\ &= (-K_1)\tilde{d}_2(-K_2)\tilde{d}_3 + (-K_1)\tilde{d}_2 + (-K_1)\tilde{d}_3 + (-K_2)\tilde{d}_3 + 1 \\ &= [-K_1,\tilde{d}_2,-K_2,\tilde{d}_3] \end{aligned} \tag{5.71}$$

$$\begin{aligned} B &= \tilde{d}_2 + \tilde{d}_3 - K_2\tilde{d}_2\tilde{d}_3 = \tilde{d}_2(-K_2)\tilde{d}_3 + \tilde{d}_2 + \tilde{d}_3 \\ &= [\tilde{d}_2,-K_2,\tilde{d}_3] \end{aligned} \tag{5.72}$$

$$\begin{aligned} C &= -(K_1 + K_2 + K_3) + K_2(K_1\tilde{d}_2 + K_3\tilde{d}_3) + K_1K_3(\tilde{d}_2 + \tilde{d}_3 - K_2\tilde{d}_2\tilde{d}_3) \\ &= (-K_1)\tilde{d}_2(-K_2)\tilde{d}_3(-K_3) + (-K_1)\tilde{d}_2(-K_2) + (-K_1)\tilde{d}_2(-K_3) \\ &\quad + (-K_1)\tilde{d}_3(-K_3) + (-K_2)\tilde{d}_3(-K_3) + (-K_1) + (-K_2) + (-K_3) \\ &= [-K_1,\tilde{d}_2,-K_2,\tilde{d}_3,-K_3] \end{aligned} \tag{5.73}$$

$$\begin{aligned} D &= 1 - K_2\tilde{d}_2 - K_3(\tilde{d}_2 + \tilde{d}_3 - K_2\tilde{d}_2\tilde{d}_3) \\ &= \tilde{d}_2(-K_2)\tilde{d}_3(-K_3) + \tilde{d}_2(-K_2) + \tilde{d}_2(-K_3) + \tilde{d}_3(-K_3) + 1 \\ &= [\tilde{d}_2,-K_2,\tilde{d}_3,-K_3] \end{aligned} \tag{5.74}$$

Table 5.1 shows the expressions for elements A, B, C, and D of the system matrix in terms of Gaussian brackets for axi-symmetric systems with two and three refracting interfaces. From this table, by induction, the expressions for the elements A, B, C, and D of the system matrix for an axi-symmetric system with k refracting interfaces can be derived as

$$A = [-K_1,\tilde{d}_2,-K_2,\tilde{d}_3,\ldots\ldots,-K_{k-1},\tilde{d}_k] \tag{5.75}$$

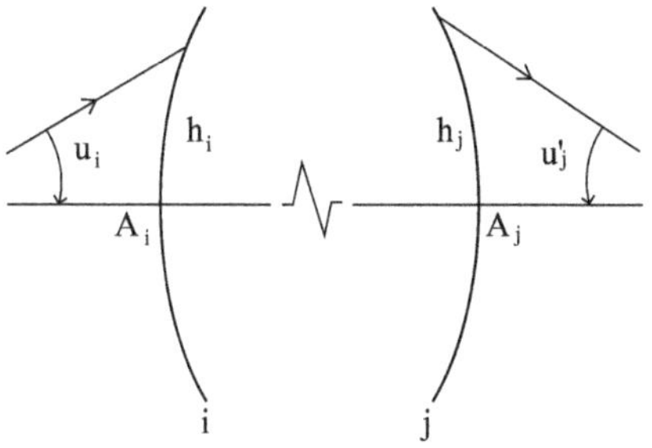

FIGURE 5.6 A subsystem consisting of surfaces i to j of an axisymmetric system of k interfaces. The paraxial ray incident on the k-th surface and the corresponding emergent ray from the j-th surface are shown.

$$B = \left[\tilde{d}_2, -K_2, \tilde{d}_3, -K_3 \ldots\ldots, -K_{k-1}, \tilde{d}_k\right] \tag{5.76}$$

$$C = \left[-K_1, \tilde{d}_2, -K_2, \tilde{d}_3, \ldots\ldots, -K_{k-1}, \tilde{d}_k, -K_k\right] \tag{5.77}$$

$$D = \left[\tilde{d}_2, -K_2, \tilde{d}_3, -K_3, \ldots\ldots, -K_{k-1}, \tilde{d}_k, -K_k\right] \tag{5.78}$$

The elements A, B, C, and D as defined above corresponds to the total system matrix for a system consisting of k refracting interfaces. Similar matrices can be defined for subsystems consisting of a set of consecutive refracting interfaces. For example, let us consider a subsystem (Figure 5.6) consisting of the consecutive refracting interfaces $i, i+1, i+2, \ldots, j-1, j$, where these interfaces form a part of the overall system consisting of k refracting interfaces, so that the sequence of surfaces in the overall system of k interfaces is:

$$1, 2, \ldots, i, i+1, \ldots, j-1, j, \ldots, k-1, k \tag{5.79}$$

Ray vector in the $(j+1)$-th space can be expressed in terms of the ray vector in the i-th space with the help of 'sub-system' matrix ${}^{i}_{j}\mathcal{S}$ as

$$\begin{bmatrix} h_j \\ \tilde{u}'_j \end{bmatrix} = {}^{i}_{j}\mathcal{S} \begin{bmatrix} h_i \\ \tilde{u}_i \end{bmatrix} = \begin{bmatrix} {}^{i}_{j}A & {}^{i}_{j}C \\ {}^{i}_{j}B & {}^{i}_{j}D \end{bmatrix} \begin{bmatrix} h_i \\ \tilde{u}_i \end{bmatrix} \tag{5.80}$$

Elements of the 'sub-system' matrix ${}^{i}_{j}\mathcal{S}$ can be expressed in terms of Gaussian brackets as shown below.

$${}^{i}_{j}A = \left[-K_i, \tilde{d}_{i+1}, -K_{i+1}, \tilde{d}_{i+2}, \ldots, -K_{j-1}, \tilde{d}_j\right] \tag{5.81}$$

$${}^{i}_{j}B = \left[d_{i+1}, -K_{i+1}, \tilde{d}_{i+2}, -K_{i+2}, \ldots, -K_{j-1}, \tilde{d}_j\right] \tag{5.82}$$

$${}^{i}_{j}C = \left[-K_i, \tilde{d}_{i+1}, -K_{i+1}, \tilde{d}_{i+2}, \ldots, -K_{j-1}, \tilde{d}_j, -K_j\right] \tag{5.83}$$

$${}^{i}_{j}D = \left[\tilde{d}_{i+1}, -K_{i+1}, \tilde{d}_{i+2}, -K_{i+2}, \ldots, -K_{j-1}, \tilde{d}_j, -K_j\right] \tag{5.84}$$

5.3 Delano Diagram in Paraxial Design of Optical Systems

Taking a cue from T. Smith's graphical construction [43–44] for paraxial treatment of periscopic systems, Delano devised a generalized graphical approach [45] for facilitating paraxial synthesis of optical systems.

As such, the Delano diagram of an optical system does not provide any more information than a sketch of the paraxial marginal ray and paraxial pupil ray in the usual paraxial ray diagram. Nevertheless, in many practical cases, this succinct and different way of presenting paraxial requirements and constraints of an optical system can facilitate solving the inverse problem of design by providing visualization of effective paraxial layouts for the purpose. Several authors have deliberated on different aspects of Delano diagram in their publications [46–62].

The notations y and $\bar{y}$ of the Delano y,$\bar{y}$ diagram represent the heights of the paraxial marginal ray and of the paraxial pupil or principal ray, respectively. In this book, notations used for the corresponding parameters are h and $\bar{h}$. However, in conformity with the literature on the Delano diagram, in this section we shall use the notations y and $\bar{y}$ for the paraxial heights.

5.3.1 A Paraxial Skew Ray and the y,$\bar{y}$ Diagram

Because of the linear nature of paraxial optics, it is possible to obtain information about the paraxial marginal ray (PMR) and the paraxial pupil ray (PPR) from the ray trace data of a suitably chosen paraxial skew ray. Figure 5.7 shows the object space of an axially symmetric imaging system. The object plane and the entrance pupil are located at the axial points O and E, respectively. The transverse object is assumed to be circular with its radius $\eta = OS$. The radius of the entrance pupil is y_E. A right-handed XYZ coordinate system is followed with the Z-axis along the optical axis, and the mutually perpendicular X- and Y-axes lying on a transverse plane.

ST is a skew ray from the point S lying at the edge of the object along the X-axis, and passing through point T lying at the edge of the entrance pupil along the Y-axis. The projection of the skew ray ST on the YZ plane is the paraxial marginal ray OT, and the projection of ST on the XZ plane is the paraxial pupil

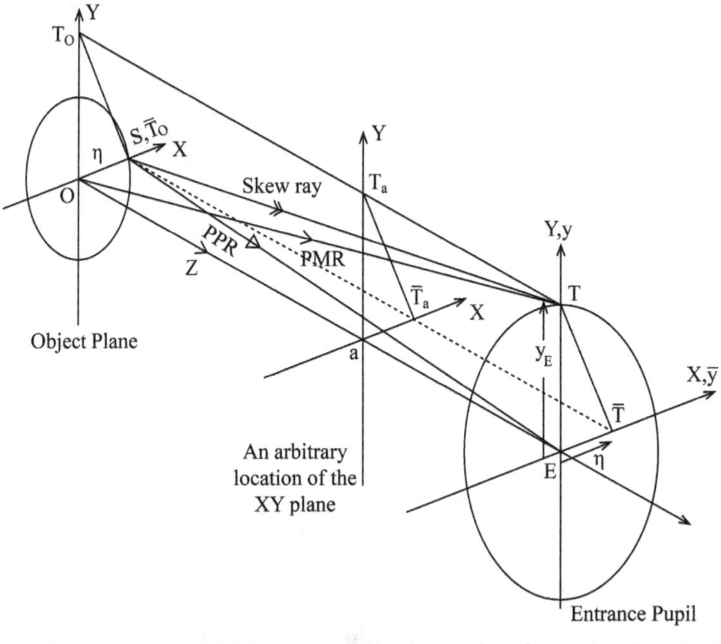

FIGURE 5.7 A special paraxial skew ray ST, from the point S lying at edge of the object along the X-axis and passing through point T lying at the edge of the pupil ray along the Y-axis. The projection of the skew ray on the meridional (YZ) plane is the PMR, and on the sagittal (XZ) plane is the PPR. Projection of the skew ray on the (XY) plane is $T\bar{T}$. Note that the latter projection is invariant with the axial location of the (XY) plane.

$$T\bar{T} = T_a\bar{T}_a = T_0\bar{T}_0$$

$$E\bar{T} = \bar{y} = OS \quad ET = y$$

ray SE. The projection of the skew ray on the XY plane located at E is $T\bar{T}$. It is important to note that this projection is invariant with changes in axial location of the transverse XY plane. The projection of the skew ray on the XY plane located at O is $T_0\bar{T}_o$, and the projection on the XY plane located at an arbitrary point a on the axis is $T_a\bar{T}_a$. It may be checked that

$$T_a\bar{T}_a = T_0\bar{T}_0 = T\bar{T} \tag{5.85}$$

Thus, the single line $T\bar{T}$ contains height information of both the PMR and the PPR during their passage from the plane at O to the plane at E. The (X, Y) coordinates of a point on the line $T\bar{T}$ correspond to the heights of the PPR and PMR, respectively. In order to underscore this, Delano renamed the Y- and X-axes of the side view as y and $\bar{y}$, and called it the $y, \bar{y}$ diagram.

The constitution of the $y, \bar{y}$ diagram may be looked upon in a different manner. The functional relationship between the height y and the distance z along the axis of a PMR can be expressed as $y(z)$; similarly, the height $\bar{y}$ and the distance z along the axis of a PPR can be expressed as $\bar{y}(z)$. Plotting y versus $\bar{y}$, treating z as a free parameter yields the corresponding $y, \bar{y}$ diagram.

5.3.2 Illustrative $y, \bar{y}$ Diagrams

Figure 5.8(a) shows the paraxial ray diagram of finite conjugate imaging by a single thin lens with a remote stop. The corresponding $y, \bar{y}$ diagram is given in Figure 5.8(b). The PMR and the PPR are shown in the paraxial ray diagram. The points A, E, and A′ in the $y, \bar{y}$ diagram correspond with the object plane, the entrance pupil/stop plane, and the image plane. Note that at point A, $y = 0$, and $\bar{y} = \eta = AP$; at point E, $y = h_E = EE_m$, $\bar{y} = 0$; and at point A′, $y = 0$, and $\bar{y} = \eta' = A'P'$. In this case, the entrance pupil E is also the aperture stop S. Therefore, point E is coincident with the point S. The coordinates of point L in the $y, \bar{y}$ diagram are $\bar{y} = \bar{h}_L$, $y = h_L$, and L represents the single thin lens. The passage of paraxial rays from the object plane at A to the image plane at A′ is represented in the $y, \bar{y}$ diagram by two line segments AL and LA′. The line segments in the object and the image spaces are called object line and image line. The exit pupil E′ is located at the point of intersection of the image line LA′ with the y-axis. Since it is

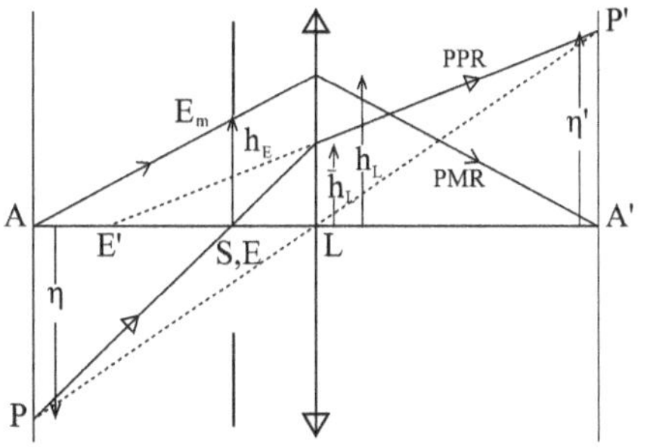

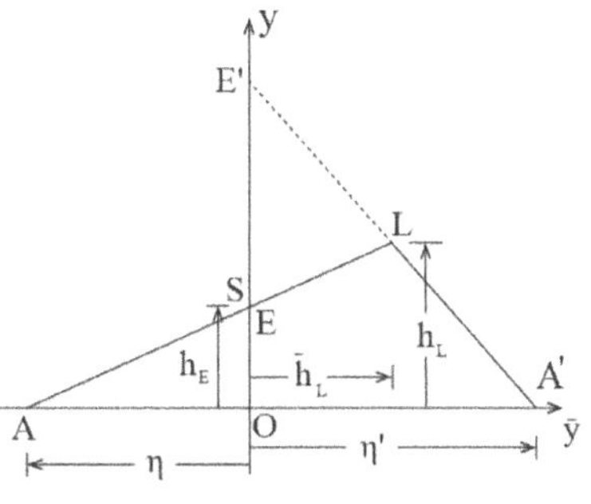

FIGURE 5.8(a) A paraxial ray diagram of finite conjugate imaging by a single thin lens with remote stop.

FIGURE 5.8(b) Corresponding $y-\bar{y}$ diagram.

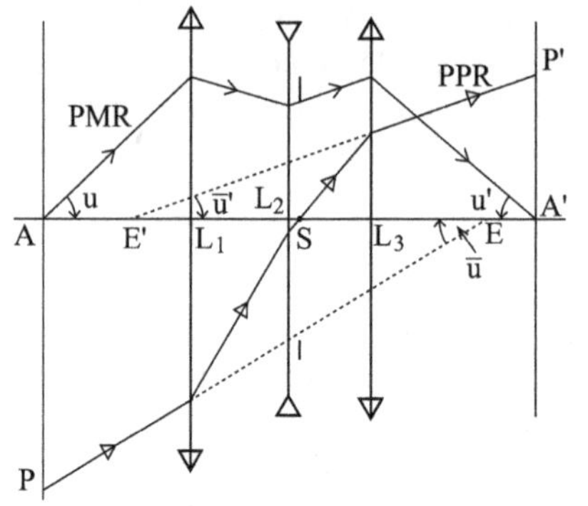

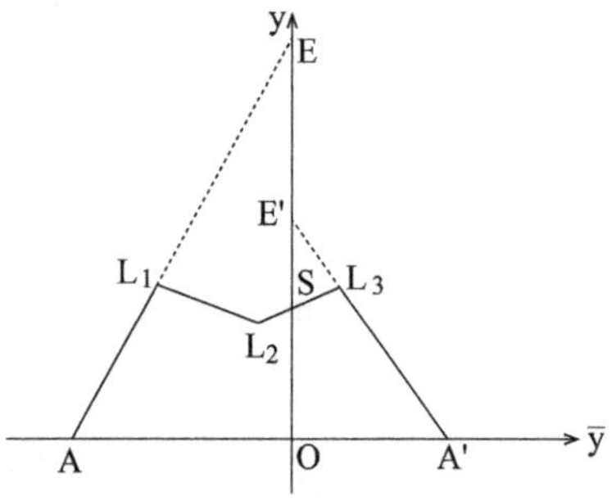

FIGURE 5.9(a) Paraxial Marginal Ray (PMR) and Paraxial Pupil Ray (PPR) in a Cooke (Taylor) triplet lens.
FIGURE 5.9(b) $y-\bar{y}$ diagram corresponding to the triplet lens of Figure 5.9(a).

necessary to extend the line segment in order to obtain the intersection point, it is a virtual exit pupil. The semi-diameter of the exit pupil is equal to OE′.

Figure 5.9(a) shows the paraxial ray diagram of finite conjugate imaging by a Cooke triplet lens consisting of three thin lenses. The corresponding $y,\bar{y}$ diagram is given in Figure 5.9(b). Note that the $y-\bar{y}$ line consists of broken line segments. The points L_1, L_2, and L_3 represent the three lenses of the triplet; each of them is at the point of intersection of two neighbouring line segments belonging to the incident and emergent space of the lens. The object line and image line, as defined above, are AL_1 and L_3A'. The point of intersection of the line segment L_2L_3 with the y-axis represents the aperture stop S. The points E and E′, which are the points of intersection of AL_1 (extended) and L_3A' (extended) with the y-axis, represent the entrance pupil and the exit pupil, respectively; both of them are virtual.

In general, the point of intersection of a line segment with the $\bar{y}$-axis represents an object/image plane. If the line segment needs to be extended to obtain the point of intersection, the latter represents a virtual object/image plane; otherwise, the intersection point represents a real object/image plane. For the two special line segments, namely the object line and the image line, the intersection points represent the object plane and the image plane of the imaging system, respectively. All other intersection points represent intermediate image planes.

Similarly, the point of intersection of a line segment with the y-axis represents the plane of an entrance/exit pupil. If the line segment needs to be extended to obtain the point of intersection, the latter represents a virtual entrance/exit pupil plane; otherwise, the intersection point represents a real entrance/exit pupil plane. The latter is the aperture stop of the system. For the two special line segments, namely the object line and the image line, the intersection points represent the entrance pupil plane and the exit pupil plane of the imaging system, respectively. All other intersection points represent intermediate pupil planes.

5.3.3 Axial Distances

One of the intriguing features of the Delano $y-\bar{y}$ diagram is the representation of axial distances. Figure 5.10(a) shows the propagation of the PMR and the PPR from the plane at axial point A_1 to the

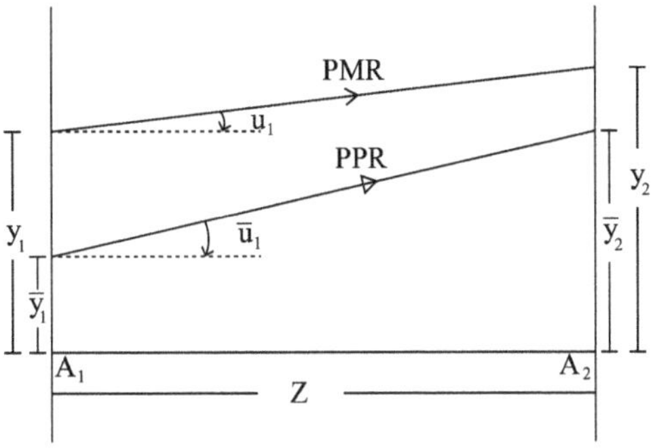

FIGURE 5.10(a) Propagation of PMR and PPR from plane at A_1 to plane at A_2.

$A_1A_2 = z$

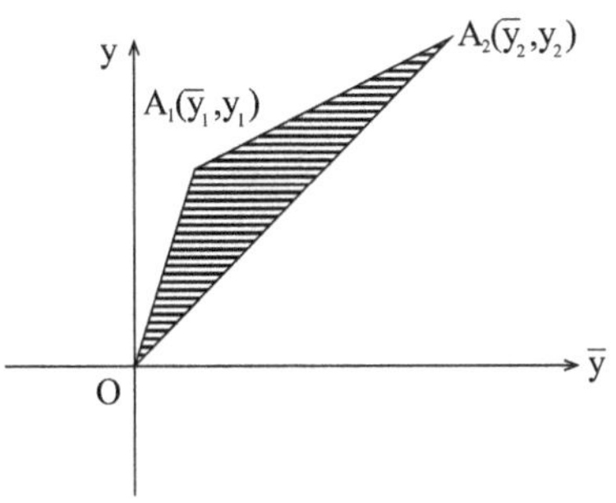

FIGURE 5.10(b) Corresponding $y - \bar{y}$ diagram. Area of the shaded triangle OA_1A_2 is directly proportional to the axial distance z.

plane at axial point A_2. The distance $A_1A_2 = z$. Let the heights of the PMR at the two planes be y_1, y_2, and that of the PPR be $\bar{y}_1, \bar{y}_2$. The convergence angles of the PMR and the PMR in the medium of refractive index n are u_1 and $\bar{u}_1$, respectively. Using transfer equation (5.2), we get

$$y_2 = y_1 + zu_1 \tag{5.86}$$

$$\bar{y}_2 = \bar{y}_1 + z\bar{u}_1 \tag{5.87}$$

From (4.24), the optical invariant H can be written as

$$H = n(\bar{y}_1 u_1 - y_1 \bar{u}_1) \tag{5.88}$$

The coordinates of the vertices O, A_1, and A_2 are $(0,0)$, $(\bar{y}_1, y_1)$, and $(\bar{y}_2, y_2)$, respectively. The area A of the triangle OA_1A_2 in the $y, \bar{y}$ diagram (Figure 5.10(b)) is equal to

$$A = \frac{(\bar{y}_1 y_2 - y_1 \bar{y}_2)}{2} \tag{5.89}$$

Using (5.86) – (5.89), we get

$$z = \frac{2nA}{H} \tag{5.90}$$

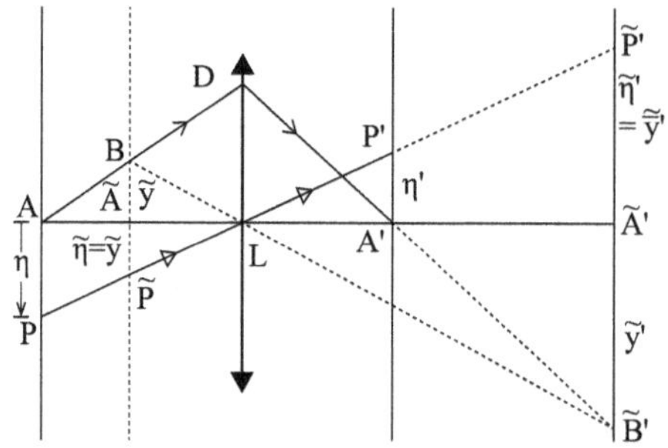

FIGURE 5.11 Finite conjugate imaging by a single thin lens with stop on it. AP is imaged at A′P′ with magnification $m_0 = (\eta' / \eta)$. $B\tilde{A}\tilde{P}$ is imaged at $\tilde{B}'\tilde{A}'\tilde{P}'$ with magnification $m = (\tilde{\eta}' / \tilde{\eta})$.

5.3.4 Conjugate Lines

Figure 5.11 shows the paraxial ray diagram for finite conjugate imaging by a single thin lens L with a stop on it. The transverse object $AP(=\eta)$ located at A is imaged as $A'P'(=\eta')$ at A'. The transverse magnification $m_0 = (\eta' / \eta)$. The PMR is ABDA′, and the PPR is $P\bar{P}LP'$. For a transverse object located at $\tilde{A}$ on the axis, the ray ABDA′ intersects the transverse plane $\tilde{A}B$ at B. Let $\tilde{A}B = \tilde{y}$. Similarly, the ray $P\tilde{P}LP'$ intersects the transverse plane at $\tilde{P}$. Let $\tilde{A}\tilde{P} = \tilde{\eta}$. The lines BL (extended) and DA′ (extended) intersect at the point $\tilde{B}'$. Therefore, $\tilde{B}'$ is the image of B. The perpendicular $\tilde{B}'\tilde{A}'\tilde{P}'$ is the image of $B\tilde{A}\tilde{P}$. The line $P\tilde{P}L\tilde{P}'$ (extended) intersects the line $\tilde{B}'\tilde{A}'\tilde{P}'$ at $\tilde{P}'$. Thus, $\tilde{P}'$ is the image of $\tilde{P}$. Let $\tilde{A}'\tilde{P}' = \tilde{\eta}'$, and $\tilde{A}'\tilde{B}' = \tilde{y}'$. Using properties of similar triangles, we can write

$$\frac{\tilde{y}'}{\tilde{y}} = \frac{\tilde{\eta}'}{\tilde{\eta}} = \frac{L\tilde{A}'}{L\tilde{A}} = m, (\text{say}) \tag{5.91}$$

For the plane at $\tilde{A}$, $y = \tilde{y}$, and $\bar{y} = \tilde{\eta}$. Similarly, for the plane at $\tilde{A}'$, $y = \tilde{y}'$, and $\bar{y} = \tilde{\eta}'$. Denoting $\tilde{\eta}$, by $\tilde{\bar{y}}$, we get, from (5.91)

$$\frac{\tilde{y}'}{\tilde{y}} = \frac{\tilde{\bar{y}}'}{\tilde{\bar{y}}} = m \tag{5.92}$$

Rearranging terms, it follows

$$\frac{\tilde{y}}{\tilde{\bar{y}}} = \frac{\tilde{y}'}{\tilde{\bar{y}}'} \tag{5.93}$$

Note that the object plane at $\tilde{A}$ is represented by the point $\tilde{A}$ on the object line, and the corresponding image plane at $\tilde{A}'$ is represented by the point $\tilde{A}'$ on the image line in the $y,\bar{y}$ diagram (Figure 5.12). Equation (5.93) implies that the line $\tilde{A}\tilde{A}'$ passes through the origin O. $\tilde{A}\tilde{A}'$ is called a conjugate line; it corresponds to magnification m. The conjugate line AA′ lies on the $\bar{y}$-axis. It corresponds to magnification $m_0 = (\eta' / \eta)$. The y-axis is a conjugate line that represents pupil imagery between the entrance and the exit pupils. In Figure 5.11, the stop is on the lens, and, therefore both the entrance pupil and the exit pupil are on the lens; magnification between the pupils is unity. This is represented by L, the common point of intersection of the object line, image line, and the y-axis.

The concept of conjugate lines, as enunciated above for the case of a singlet, holds good for any axi-symmetric imaging system.

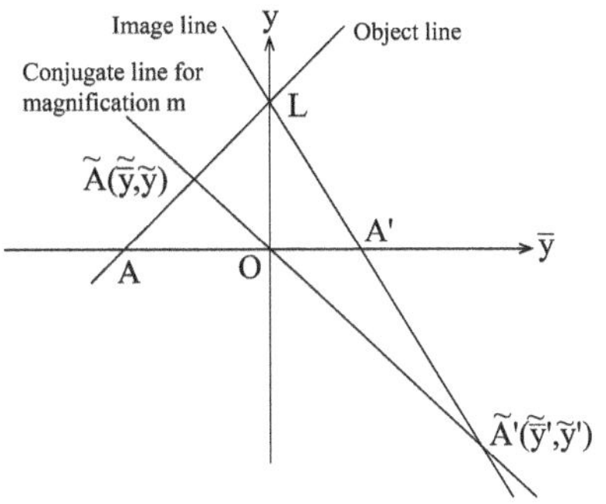

FIGURE 5.12 $\tilde{A}\tilde{A}'$ is a conjugate line of magnification m in the $y-\bar{y}$ diagram for the imaging system of Figure 5.11.

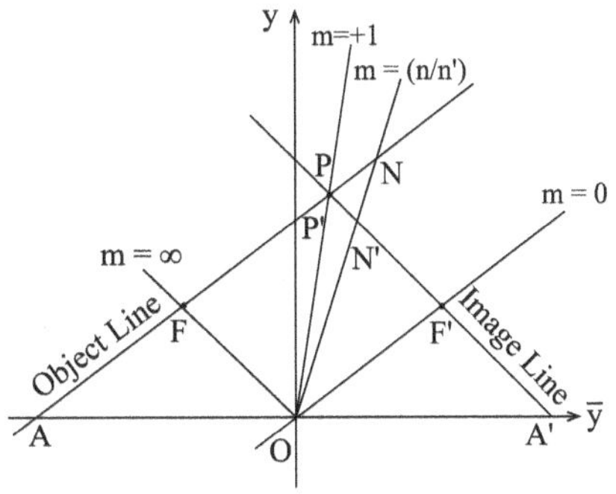

FIGURE 5.13 Cardinal points on a $y-\bar{y}$ diagram.

5.3.5 Cardinal Points

Figure 5.13 shows the object line and the image line of the $y-\bar{y}$ diagram representation of an axisymmetric lens system. Locations of the points on the $y-\bar{y}$ diagram corresponding to planes containing the different cardinal points are obtained by invoking conjugate lines with special magnifications.

The principal or unit planes are determined by means of conjugate line with magnification +1. Location of the first principal plane in the diagram is given by P, the point of intersection of the object line with the conjugate line corresponding to $m=+1$. Similarly, the location of the second principal plane in the diagram is given by P′, the point of intersection of the image line with the conjugate line corresponding to $m=+1$. By definition, for the principal planes, $y=y'$, and $\bar{y}=\bar{y}'$. Therefore, the points P and P′ coincide in the $y-\bar{y}$ diagram. Note that this coincidence does not necessarily imply coincidence of the two points in the real space of the system.

If the refractive indices of the object space and the image space are n and n′ respectively, the magnification between the nodal planes is (n/n'). Thus, the locations of the object space and the image space nodal planes in the $y-\bar{y}$ diagram are obtained by the points of intersection of the conjugate line corresponding to $m=(n/n')$ with the object line and the image line. In Figure 5.12 the two nodal planes are represented by the points N and N′. In the special case when $n=n'$, the points N and N′ coincide with the points P and P′.

The location of the second principal focal plane is represented in the $y-\bar{y}$ diagram by the point F′; this is the point of intersection of the conjugate line with magnification $m=0$, i.e. a line parallel to the object line and passing through the origin O, with the image line. Similarly, the location of the first principal focal plane is represented in the $y-\bar{y}$ diagram by the point F; it is the point of intersection of the conjugate line with magnification $m=\infty$, i.e., a line.

The six points P, P′, N, N′, and F, F′ in the $y-\bar{y}$ diagram in Figure 5.13 indicate location of the planes containing the six cardinal points of the system. More details on the characteristics and applications of the Delano $y-\bar{y}$ diagram can be found in references [46–62].

TABLE 5.1

Elements A, B, C and D of the system matrix in terms of Gaussian brackets for axisymmetric systems with two and three refracting interfaces

Element of System Matrix	Two refracting interfaces	Three refracting interfaces
A	$\left[-K_1, \tilde{d}_2\right]$	$\left[-K_1, \tilde{d}_2, -K_2, \tilde{d}_3\right]$
B	$\left[\tilde{d}_2\right]$	$\left[\tilde{d}_2, -K_2, \tilde{d}_3\right]$
C	$\left[-K_1, \tilde{d}_2, -K_2\right]$	$\left[-K_1, \tilde{d}_2, -K_2, \tilde{d}_3, -K_3\right]$
D	$\left[\tilde{d}_2, -K_2\right]$	$\left[\tilde{d}_2, -K_2, \tilde{d}_3, -K_3\right]$

REFERENCES

1 R.A. Sampson, 'A Continuation of Gauss's Dioptrische Untersuchungen', *Proc. London Math. Soc.*, Vol. XXIX, (1897) pp. 33–52.
2 T. Smith, 'On tracing rays through an optical system', *Proc. Phys. Soc. (London)*, Vol. 57, (1945) pp. 286–297.
3 J.H. Harrold, 'Matrix algebra for ideal lens problems', *J. Opt. Soc. Am.* Vol. 44, (1954) pp. 254–254.
4 E.L. O'Neill, *Introduction to Statistical Optics*, Addison-Wesley, Reading, Massachusetts (1963).
5 K. Halbach, 'Matrix representation of Gaussian optics', *Am. J. Phys.* Vol. 32 (1964) pp. 90–108.
6 P.I. Richards, 'Conventions in matrix optics', *Am. J. Phys.* Vol. 32 (1964) pp. 890–897.
7 W. Brouwer, *Matrix methods in optical instrument design*, Benjamin, New York (1964).
8 W. Brouwer and A. Walther, 'Geometrical Optics', Chapter 16 in *Advanced Optical Techniques*, Ed. A.C.S. Van Heel, North-Holland, Amsterdam (1967) pp. 519–533.
9 J.W. Blaker, *Geometric Optics: The Matrix Theory*, Marcel Dekker, New York (1971).
10 A. Gerrard and J.M. Burch, *Introduction to matrix methods in optics*, John Wiley, London (1975).
11 D.C. Sinclair, 'The specifications of optical systems by paraxial transfer matrices', in *Applications of Geometrical Optics II, Proc. SPIE* 0039 (1974) pp. 141–150.
12 H.H. Arsenault, 'A matrix representation for non-symmetrical optical systems', *J. Opt. (Paris)*, Vol. 11 (1980) pp. 87–91.
13 S.A. Rodionov, 'Matrix technique of Gaussian optics near an arbitrary ray', *Opt. Spectrosc.* Vol. 50 (1981) pp. 531–535.
14 H.H. Arsenault and B. Macukow, 'Factorization of the transfer matrix for symmetrical optical systems', *J. Opt. Soc. Am.* Vol. 73 (1983) pp. 1350–1357.
15 H. Kogelnik and T. Li, 'Laser beams and resonators', *Appl. Opt.*, Vol. 5 (1966) pp. 1550–1567.
16 L.W. Casperson, 'Synthesis of Gaussian beam optical systems', *Appl. Opt.*, Vol. 20 (1981) pp. 2243–2249.
17 C. Fog, 'Synthesis of optical systems', *Appl. Opt.*, Vol. 21, (1982) pp. 1530–1531.
18 J.A. Arnaud, 'Hamiltonian theory of beam mode propagation', in *Progress in Optics, Vol. XI*, ed. E. Wolf, North-Holland, Amsterdam (1973) pp. 247–304.
19 G. Wooters and E.W. Silvertooth, 'Optically compensated zoom lens', *J. Opt. Soc. Am.*, Vol. 55, (1965) pp. 347–351.
20 E.C.G. Sudarshan, N. Mukunda, and R. Simon, 'Realization of first order optical systems using thin lenses', *Opt. Acta*, Vol. 32 (1985) pp. 855–872.
21 W. Shaomin, 'Matrix methods in treating decentered optical systems', *Opt. Quantum Electron.*, Vol. 17 (1985) pp. 1–36.
22 A.E. Siegman, *Lasers*, Univ. Sci. Books, California (1986).
23 E. Elbaz and F. Roux, *Optique matricielle*, ellipses, Paris (1989).
24 A. Nussbaum, *Optical System Design*, Prentice Hall PTR, NJ (1998).
25 R. Ditteon, *Modern Geometrical Optics*, Wiley, New York (1998).
26 H. Gross, *Handbook of Optical Systems,* Vol. 1 Wiley-VCH, Weinheim (2005) pp. 41–58.
27 G. Kloos, Matrix Methods for Optical Layout, Tutorial Text, SPIE, Bellingham, Washington (2007).

28 C.F. Gauss, *Disquisitiones arithmeticae*, Fleischer, Lipsiae (1801) [ebook from eod, Humboldt-Universität zu Berlin].
29 M. Herzberger, 'Gaussian optics and Gaussian brackets', *J. Opt. Soc. Am.* Vol. 33 (1943) pp. 651–655.
30 M. Herzberger, 'Precalculation of optical systems', *J. Opt. Soc. Am.* Vol. 42 (1952) pp. 637–640.
31 M. Herzberger, *Modern Geometrical Optics*, Interscience, New York, (1958).
32 M. Herzberger, 'Geometrical Optics' in *Handbook of Physics*, Ed. Condon and Odishaw, McGraw Hill, New York (1958) pp. 6–46.
33 H.G. Zimmer, *Geometrical Optics*, Springer, New York (1970) [translated by R. N. Wilson from *Geometrische Optik*, Springer, Berlin, (1967)] pp.45–52.
34 L. Bergstein, 'General theory of optically compensated varifocal systems', *J. Opt. Soc. Am.*, Vol. 48 (1958) pp. 154–171.
35 L. Bergstein and L. Motz, 'Two-component optically compensated varifocal system', *J. Opt. Soc. Am.*, Vol. 52 (1962) pp. 353–362.
36 L. Bergstein and L. Motz, 'Three-component optically compensated varifocal system', *J. Opt. Soc. Am.*, Vol. 52 (1962) pp. 363–375.
37 L. Bergstein and L. Motz, 'Four-component optically compensated varifocal system', *J. Opt. Soc. Am.*, Vol. 52 (1962) pp. 376–388.
38 K. Tanaka, 'Paraxial Theory in optical design in terms of Gaussian brackets', in Progress in Optics, Vol. XXIII, Ed. E. Wolf, Elsevier, (1986) pp. 63–111.
39 R.J. Pegis and W.J. Peck, 'First-order design theory for linearly compensated zoom systems', *J. Opt. Soc. Am.*, Vol. 52 (1962) pp. 905–911.
40 S. Minami, 'Thin lens theory of zoom systems', *Kogaku (Japan J. Opt.)* Vol. 1 (1972) 329 (in Japanese).
41 S.E. Smirnov, 'Calculation of a system of two thin components having a specified value of image curvature', *Sov. J. Opt. Technol.*, Vol. 50 (1983) pp. 26–29.
42 K. Bhattacharya and L.N. Hazra, 'Analytical derivatives for optical system analysis: use of Gaussian brackets', *J. Opt. (India)*, Vol. 18, (1989) pp. 57–67.
43 T. Smith, 'A projective treatment of submarine periscope', *Trans. Opt. Soc., (London)* Vol. 23 (1921–22) pp. 217–219.
44 T. Smith, 'Periscopes' in *Dictionary of Applied Physics*, Vol. IV, Ed. R. Glazebrook, Macmillan, London (1923).
45 E. Delano, 'First-order design and the $y, \bar{y}$ diagram, *Appl. Opt.*, Vol. 2 (1963) pp. 1251–1256.
46 R.J. Pegis, T.P. Vogl, A.K. Rigler, and R. Walters, 'Semiautomatic generation of optical prototypes', *Appl. Opt.*, Vol. 6 (1967) pp. 969–972.
47 Ref. 26 pp. 52–74.
48 F.J. López-López, 'Normalization of the Delano diagram', *Appl. Opt.*, Vol. 9 (1970) pp. 2485–2488.
49 H. Slevogt, 'Hilfsmittel für die anschauliche analyse von optischen systemen', *Optik*, Vol. 30 (1970) pp. 431–433.
50 R.V. Shack, 'Analytic system design with pencil and ruler – the advantages of the $y - \bar{y}$ diagram', in *Applications of Geometrical Optics II, Proc. SPIE* 0039 (1974) pp. 127–140.
51 F.J. López-López, 'Analytical aspects of the $y - \bar{y}$ diagram', in *Applications of Geometrical Optics II, Proc. SPIE* 0039 (1974) pp. 151–164.
52 W. Besenmatter, 'Das Delano-Diagramm des Vario-Glaukar-Objectivs', *Optik*, Vol. 47 (1977) pp. 153–166.
53 W. Besenmatter, 'Analyse des Vignettierungsverhaltens der Vario-Objektive mit Hilfe des Delano-Diagramms', *Optik*, Vol. 48 (1977) pp. 289–304.
54 W. Besenmatter, 'Analyse der primären Wirkung asphahärischer Flächen mit Hilfe des Delano-Diagramms', *Optik*, Vol. 51 (1978) pp. 385–396.
55 J.S. Loomis, 'The Use of Paraxial Solves in Lens Design', *Proc. SPIE*, Vol. 147 (1978) pp. 6–11.
56 W. Besenmatter, 'Designing zoom lenses aided by the Delano diagram', *Proc. SPIE*, Vol. 237 (1980) pp. 242–25
57 M. Takeda, 'New diagram for nomographic design and analysis of paraxial optical systems, *J. Opt. Soc. Am.*, Vol. 70 (1980) pp. 236–242.
58 O.N. Stavroudis, *Modular Optical Design*, Springer Verlag, Berlin (1982) pp. 15–32.
59 M.E. Harrigan, R.P. Loce, and J. Rogers, 'Use of the $y - \bar{y}$ diagram in GRIN rod design, *Appl. Opt.*, Vol. 27 (1988) pp. 459–464.

60 D.M. Brown, 'Global optimization using the y-ybar diagram', in *Current Developments in Optical Design and Optical Engineering*, Eds. R.E. Fisher and W.J. Smith, Proc. SPIE, Vol. 1527 (1991) pp. 19–25.
61 D. Kessler and R. V. Shack, $Y - \bar{Y}$ Diagram, a powerful optical design method for laser systems, *Appl. Opt.*, Vol. 31 (1992) pp. 2692–2707.
62 R. Ditteon, Modern Geometrical Optics, Wiley-Interscience, New York (2003), Chapter 8, pp. 278–299.
63 M.J. Bastiaans, The Wigner distribution function applied to optical signals and systems, *Opt. Commun.*, Vol. 25 (1978) pp. 26–30.
64 M.J. Bastiaans, Wigner distribution function and its application to first-order optics, *J. Opt. Soc. Am.*, Vol. 69 (1979) pp. 1710–1716.
65 O.N. Stavroudis, *The Optics of Rays, Wavefronts and Caustics*, Academic, New York (1972) pp. 281–297.
66 A.J. Dragt, 'Lie algebraic theory of geometrical optics and optical aberrations', *J. Opt. Soc. Am.*, Vol. 72 (1982) pp. 372–379.
67 A.J. Dragt, E. Forest, and K.B. Wolf, 'Foundations of a Lie algebraic theory of geometrical optics', in Lie Methods in Optics, Eds. J. Sánchez Mondragón, K.B. Wolf, Lecture Notes in Physics, Vol. 250, Springer, Berlin (1986) pp. 105–157.
68 J.F. Cariñena, C. López, and J. Nasarre, 'Symplectic and Lie algebraic techniques in geometric optics', *arXiv.physics/9708035v1 [physics.optics]* 29 August 1997.
69 J. Shamir, 'Cylindrical lens systems described by operator algebra', *Appl. Opt.*, Vol. 18 (1979) pp. 4195–4202.
70 M. Nazarathy and J. Shamir, 'First-order optics – a canonical operator representation: lossless systems', *J. Opt. Soc. Am.*, Vol. 72 (1982) pp. 356–364.
71 M. Nazarathy and J. Shamir, 'First-order optics: operator representation for systems with loss or gain', *J. Opt. Soc. Am.*, Vol. 72 (1982) pp. 1398–1408.
72 M. Nazarathy, A. Hardy, and J. Shamir, 'Generalized mode propagation in first-order optical systems with loss or gain', *J. Opt. Soc. Am.*, Vol. 72 (1982) pp. 1409–1420.

6

The Photometry and Radiometry of Optical Systems

An optical system accomplishes its functions, e.g. imaging, by the transport of light energy. For most practical purposes, a relatively simple model, as provided in 'radiometry', is adequate for analyzing energy considerations during propagation of light, or electromagnetic waves in general. This model is based on the following assumptions:

- Light or electromagnetic energy is transported along the trajectories of geometrical rays in the system,
- Sizes of the source or the apertures are such that the diffraction effects can be ignored, and
- The radiation field is assumed completely incoherent, so that interference effects are not taken into account.

Mandel and Wolf have discussed the foundations of classical radiometry [1–2]. It needs to be reiterated that in classical treatments of photometry and radiometry, the radiation is assumed to be incoherent. Therefore, the ray optical treatment for transport of energy is adequate. However, in some applications, such as laser illumination, this approximation is not valid, and so the treatment needs to incorporate the diffraction effects [3–4].

Like many other topics of practical engineering, over the years a plethora of units and conventions has emerged in the fields of radiometry and photometry, so much so that often they turn out to be more confusing than they are illuminating. In this treatise, we shall mostly use the units specified by Système Internationale d'Unités, or the SI units, and a few non-SI units that are still widely in vogue in the field.

6.1 Radiometry and Photometry: Interrelationship

Radiometry deals with the measurement of radiant energy of electromagnetic (e.m.) radiation. Optical radiometry pertains to any part of the range of wavelengths of e.m. radiation that is currently considered as optical wavelength (as shown in Figure 1.1 in Chapter 1). Practical radiometry in optical systems has to take into account the variation in the characteristics of its components with wavelength, e.g. the spectral variation in emission from the source, in atmospheric transmission, and in transmission by the optics, as well as the spectral variation in response of the detector. Photometry implies the human eye as the detector, and therefore it is concerned with visible part of the e.m. spectrum. The wavelength sensitivity of the human eye over the visible range provides the weighting factor in correlating the radiometric and corresponding photometric quantities. For details on photometry and radiometry we refer the books by Walsh [5], Boyd [6], and McCluney [7], and a recent publication by Gross [8].

The unit of radiant power or flux is watt, while the unit of luminous power or flux is lumen. 'Candela', the base SI unit of luminous intensity, is defined as

$$1 \text{ candela} = 1 \text{ lumen per steradian} \tag{6.1}$$

where steradian is a measure of solid angle in three-dimensional space.

The concept of a solid angle in three dimensions is similar to the concept of the angle in two dimensions. In two dimensions, an angle, projected on to a circle with its centre in the vertex of the angle, forms a circular arc. The angle is described in radians by the ratio of the length of the arc and the radius of the

DOI: 10.1201/9780429154812-6

circle. Similarly, in three dimensions, a solid angle, projected onto a sphere, gives an area on the spherical surface. The solid angle is denoted in steradian by the ratio of this area and square of the radius of the sphere. Obviously, a sphere forms a solid angle of 4π steradians. The word steradian is derived from the Greek, στερεός (stereos) meaning solid + radian. Steradian is a dimensionless unit, and it is identified by the symbol 'sr'. Like radian, steradian is also considered an SI derived unit from 1995.

The relation (6.1) can be expressed succinctly as

$$1\ \mathrm{cd} = 1\mathrm{lm}\ \mathrm{sr}^{-1} \tag{6.2}$$

Currently candela, one of the seven base SI units, is defined as 'the luminous intensity, in a given direction, of a source that emits monochromatic radiation of 540 THz $\left(=540\times10^{12}\,\mathrm{Hertz}\right)$ and that has a radiant intensity in that direction of $\left(1/683\right)$ watt per steradian' [7]. The factor $\left(1/683\right)$ arises out of the choice to make luminous intensity of candela as near as possible with that of the earlier standard of luminous intensity, namely candlepower, which it superseded. The wavelength of electromagnetic radiation of frequency 540 THz is 555nm. Thus, 1 watt of radiant power (flux) at 555nm gives rise to luminous power (flux) of 683 lumens. In photopic vision, the peak sensitivity of the eye occurs at this wavelength. The sensitivity of the eye gradually falls from this peak value at 555nm on both sides of this wavelength. The spectral luminous efficiency for monochromatic radiant energy of wavelength λ is the ratio $K(\lambda)$ of the luminous to radiant power. The latter is expressed in terms of lumen per watt. The relative spectral luminous efficiency function, commonly called eye sensitivity function, $V(\lambda)$, is the value of $K(\lambda)$ normalized by its peak value at 555nm, so that

$$K(\lambda) = 683\times V(\lambda) \tag{6.3}$$

For polychromatic sources, the luminosity of the source is defined as the ratio of total luminous power to total radiant power. If $P(\lambda)$ represents the power spectral density, i.e. the radiant power emitted per unit wavelength by the source, its luminosity, or luminous efficacy is given by

$$K = 683\times\frac{\int V(\lambda)P(\lambda)d\lambda}{\int P(\lambda)d\lambda}\ \mathrm{lm/W} \tag{6.4}$$

The product of the luminous efficacy and the electrical-to-optical power conversion efficiency gives luminous efficiency of a polychromatic source. It is also expressed in terms of lumen per watt. A degree of caution is warranted, as these definitions are not followed universally, e.g. in the lighting community, the luminous efficiency is often referred to as luminous efficacy.

As the visual response of human beings is not uniform, it is obvious that different observers will have different $V(\lambda)$ curves. Therefore, it is necessary to specify agreed data for the 'standard observer'. The 'International Commission for Illumination' (Commission Internationale d'Eclairage, CIE) introduced the photopic eye sensitivity function $V(\lambda)$. Tabulated values for $V(\lambda)$ are given over (360nm, 825nm) at an interval of 5nm. It is known as the CIE 1931 $V(\lambda)$ function, and has been accepted widely as acceptable data for a standard observer. Later on, it was found that this data underestimated the human eye sensitivity for the spectral region below 460nm. To rectify this, the CIE introduced a modified $V(\lambda)$ function, and this is known as the CIE 1978 $V(\lambda)$ function.

The scotopic eye sensitivity function $V'(\lambda)$ is different from this photopic eye sensitivity function. The peak sensitivity in the scotopic vision regime occurs at 507nm. This phenomenon of shift of the maximum is known as the Purkinje effect. The CIE provides numerical values for $V'(\lambda)$, and the latter is known as the CIE 1951 $V'(\lambda)$ function. Table 6.1 gives values for $V(\lambda)$ and $V'(\lambda)$ over a useful working range at an interval of 10nm. These are values of the photopic and scotopic eye sensitivity functions normalized by their maximum values at 555nm and 507nm, respectively [9].

TABLE 6.1

Tabulated values of CIE 1931 $V(\lambda)$ photopic eye sensitivity function, CIE 1978 $V(\lambda)$ photopic eye sensitivity function and CIE 1951 $V'(\lambda)$ scotopic eye sensitivity function

<table>
<tr><th>λ (nm)</th><th>CIE 1931 V(λ)</th><th>CIE 1978 V(λ)</th><th>CIE 1951 V′(λ)</th></tr>
<tr><td>380</td><td>0.0000</td><td>0.0002</td><td>0.0006</td></tr>
<tr><td>390</td><td>0.0001</td><td>0.0008</td><td>0.0022</td></tr>
<tr><td>400</td><td>0.0004</td><td>0.0028</td><td>0.0093</td></tr>
<tr><td>410</td><td>0.0012</td><td>0.0074</td><td>0.0348</td></tr>
<tr><td>420</td><td>0.0040</td><td>0.0175</td><td>0.0966</td></tr>
<tr><td>430</td><td>0.0116</td><td>0.0273</td><td>0.1998</td></tr>
<tr><td>440</td><td>0.0230</td><td>0.0379</td><td>0.3281</td></tr>
<tr><td>450</td><td>0.0380</td><td>0.0468</td><td>0.4550</td></tr>
<tr><td>460</td><td colspan="2">0.0600</td><td>0.5670</td></tr>
<tr><td>470</td><td colspan="2">0.0910</td><td>0.6760</td></tr>
<tr><td>480</td><td colspan="2">0.1390</td><td>0.7930</td></tr>
<tr><td>490</td><td colspan="2">0.2080</td><td>0.9040</td></tr>
<tr><td>500</td><td colspan="2">0.3230</td><td>0.9820</td></tr>
<tr><td>510</td><td colspan="2">0.5030</td><td>0.9970</td></tr>
<tr><td>520</td><td colspan="2">0.7100</td><td>0.9350</td></tr>
<tr><td>530</td><td colspan="2">0.8620</td><td>0.8110</td></tr>
<tr><td>540</td><td colspan="2">0.9540</td><td>0.6500</td></tr>
<tr><td>550</td><td colspan="2">0.9950</td><td>0.4810</td></tr>
<tr><td>560</td><td colspan="2">0.9950</td><td>0.3288</td></tr>
<tr><td>570</td><td colspan="2">0.9520</td><td>0.2076</td></tr>
<tr><td>580</td><td colspan="2">0.8700</td><td>0.1212</td></tr>
<tr><td>590</td><td colspan="2">0.7570</td><td>0.0655</td></tr>
<tr><td>600</td><td colspan="2">0.6310</td><td>0.0332</td></tr>
<tr><td>610</td><td colspan="2">0.5030</td><td>0.0159</td></tr>
<tr><td>620</td><td colspan="2">0.3810</td><td>0.0074</td></tr>
<tr><td>630</td><td colspan="2">0.2650</td><td>0.0033</td></tr>
<tr><td>640</td><td colspan="2">0.1750</td><td>0.0015</td></tr>
<tr><td>650</td><td colspan="2">0.1070</td><td>0.0007</td></tr>
<tr><td>660</td><td colspan="2">0.0610</td><td>0.0003</td></tr>
<tr><td>670</td><td colspan="2">0.0320</td><td>0.0001</td></tr>
<tr><td>680</td><td colspan="2">0.0170</td><td>0.0000</td></tr>
<tr><td>690</td><td colspan="2">0.0082</td><td>0.0000</td></tr>
<tr><td>700</td><td colspan="2">0.0041</td><td>0.0000</td></tr>
<tr><td>710</td><td colspan="2">0.0021</td><td>0.0000</td></tr>
<tr><td>720</td><td colspan="2">0.0010</td><td>0.0000</td></tr>
<tr><td>730</td><td colspan="2">0.0005</td><td>0.0000</td></tr>
<tr><td>740</td><td colspan="2">0.0002</td><td>0.0000</td></tr>
<tr><td>750</td><td colspan="2">0.0001</td><td>0.0000</td></tr>
</table>

It may be noted that the two maximum values are not the same. Whereas in photopic vision, the peak luminous flux 683 lumen is obtained per unit watt of radiant power at 555nm, in dark-adapted scotopic vision the peak luminous flux 1700 lumen per watt is obtained at 507nm. Note that in scotopic vision, at 555nm, the luminous flux obtained per unit watt of radiant power is 683 lumens, the same value as obtained in photopic vision.

TABLE 6.2

Radiometric and photometric quantities and units

Radiometry Quantity	Unit	Photometry Quantity	Unit
Radiant Energy	Joule (J)	Luminous Energy	Lumen-second (lm s), or Talbot
Radiant Flux (Power)	Watt (W)	Luminous Flux (Power)	Lumen (lm)
Radiant intensity	Watt per steradian $\left(Wsr^{-1}\right)$	Luminous intensity	Lumen per steradian $\left(lmsr^{-1}\right)$, or Candela (cd)
Radiant exitance (emittance)	Watt per square meter $\left(Wm^{-2}\right)$	Luminous exitance (emittance)	Lumen per square meter $\left(lmm^{-2}\right)$, or lux
Radiance	Watt per square meter per steradian $\left(Wm^{-2}sr^{-1}\right)$	Luminance (Brightness)	Lumen per square meter per steradian $\left(lmm^{-2}sr^{-1}\right)$, or nit
Irradiance	Watt per square meter $\left(Wm^{-2}\right)$	Illuminance (Illumination)	Lumen per square meter $\left(lmm^{-2}\right)$, or lux

6.2 Fundamental Radiometric and Photometric Quantities

Like many other engineering disciplines, there was hardly any standardization in the early period of development in the field of radiometry and photometry. Different units for the same quantity were in existence, often creating confusion in practice. The history of their development is colourful and interesting. Nevertheless, many of them were found to be redundant and illogical [10]. Currently, the SI (Système Internationale) units are universally accepted and used; a few derived non-SI units are still in vogue [11–12]. Table 6.2 presents a list of radiometric and photometric quantities discussed here and their units.

6.2.1 Radiant or Luminous Flux (Power)

Light flux is a measure of the rate of flow of light energy. If one has radiant flux or power e_λ of wavelength λ, the luminous flux or power is, $e_\lambda V(\lambda)$, where $V(\lambda)$ is the eye sensitivity function.

The unit of radiant power or flux is watt (w), and the unit for luminous flux or power is lumen (lm). In a beam, whose spectral power density is $P(\lambda)$, the radiant flux or power Φ_r is given by

$$\Phi_r = \int_0^\infty P(\lambda)d\lambda \text{ watt}(w) \tag{6.5}$$

The luminous flux or power Φ_l is given by

$$\Phi_l = 683 \times \int_0^\infty P(\lambda)V(\lambda)d\lambda \text{ lumen}(lm) \tag{6.6}$$

Since $V(\lambda)$ is effectively zero for $\lambda > 760nm$ and $\lambda < 380nm$, the limits of integration may be changed, and the expression for luminous flux Φ reduces to

$$\Phi_l = 683 \times \int_{380}^{760} P(\lambda)V(\lambda)d\lambda \text{ lumen}(lm) \tag{6.7}$$

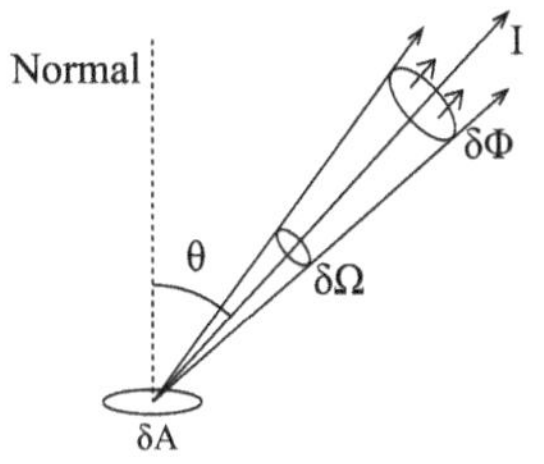

FIGURE 6.1 Intensity I along the direction θ from an elementary source δA.

6.2.2 Radiant or Luminous Intensity of a Source

Radiant or luminous flux emitted by a source is, in general, not the same in all directions. Often it becomes important to know the corresponding flux density in given directions. Intensity is a measure of this flux density, and is defined as the ratio of the flux contained in an infinitesimally narrow cone of solid angle $(\delta\Omega)$ in a given direction (Figure 6.1).

Radiant intensity is given by

$$I_r = \frac{\delta\Phi_r}{\delta\Omega} \tag{6.8}$$

Luminous intensity is defined as

$$I_l = \frac{\delta\Phi_l}{\delta\Omega} \tag{6.9}$$

The unit of radiant intensity is watt per steradian (wsr^{-1}), and the unit of luminous intensity is lumen per steradian ($lmsr^{-1}$). The latter is called candela (cd), one of the base SI units. The intensity of a 100W incandescent lamp is approximately double the intensity of a 50W lamp, if the energy conversion efficiency is the same in both of them.

For practical purposes, many sources may be regarded as point sources. A point source is called a uniform point source when its intensity is the same in all directions. For a uniform point source, the total emitted flux is given by

$$\Phi = \int I d\Omega = I \int d\Omega = 4\pi I \tag{6.10}$$

$$\text{so that } I = \frac{\Phi}{4\pi} \tag{6.11}$$

Note that we use Φ, I, etc. without any subscript in some relations where the relation holds good for both radiometry and photometry. In order to obtain radiometric relations, Φ, I in (6.10) and (6.11) are to be replaced by Φ_r, I_r; the photometric relations are obtained by replacing Φ, I by Φ_l, I_l.

For a non-uniform point source emitting total luminous flux Φ_l, the average value of its luminous intensity over all directions is given by $\left(\Phi_l/4\pi\right)$, which is called the 'mean spherical candle power' of the source, a term widely used in illumination engineering for characterizing sources.

6.2.3 Radiant (Luminous) Emittance or Exitance of a Source

Radiant emittance or radiant exitance M_r of a source is the radiant power emitted in all directions per unit surface area of the source. M_r is given by

$$M_r = \frac{\delta\Phi_r}{\delta A} \tag{6.12}$$

Corresponding luminous emittance or exitance M_l of the source is given by

$$M_l = \frac{\delta\Phi_l}{\delta A} \tag{6.13}$$

For a polychromatic beam, Φ_r and Φ_l in (6.12) and (6.13) are given by (6.5) and (6.7), respectively. The unit for radiant emittance or exitance is watt per square meter $\left(W/m^2\right)$. For luminous emittance or exitance, the unit is lumen per square meter $\left(lm/m^2\right)$. Since the additional light output is obtained by an increase in filament area, the emittance or exitance of a 100 W bulb is approximately the same of a 50 W bulb.

6.2.4 Radiance (Luminance) of a Source

Radiance L_r of a source in a direction is the radiant intensity in that direction per unit projected area of the source along the same direction (Figure 6.2). Using (6.8) L_r can be expressed as

$$L_r = \frac{\delta I_r}{\delta A \cos\theta} = \frac{\delta^2 \Phi_r}{\delta\Omega \delta A \cos\theta} \tag{6.14}$$

The corresponding expression for luminance L_l of a source is

$$L_l = \frac{\delta I_l}{\delta A \cos\theta} = \frac{\delta^2 \Phi_l}{\delta\Omega \delta A \cos\theta} \tag{6.15}$$

In the above, θ is the angle between the normal to the source and the direction of observation (Figure 6.2). The unit for radiance is watt per steradian per square meter, $\left(W sr^{-1} m^{-2}\right)$, and the unit for luminance is lumen per steradian per square meter, $\left(lm sr^{-1} m^{-2}\right)$ or candela per square meter, $\left(cd/m^2\right)$. The radiance or luminance of a 100 W incandescent bulb is about the same of a 50 W incandescent bulb. In photometry, this unit of luminance is often abbreviated as 'nit'. Note that luminance is an objective measure of the intrinsic brightness of a source.

6.2.4.1 *Lambertian Source*

Radiance L_r or Luminance L_l of a source is, in general, not the same along the different directions of observation. Similarly, the radiant or luminous intensity of a source is also a function of the direction of observation. However, if the latter variation is like

$$I(\theta) = I_0 \cos\theta \tag{6.16}$$

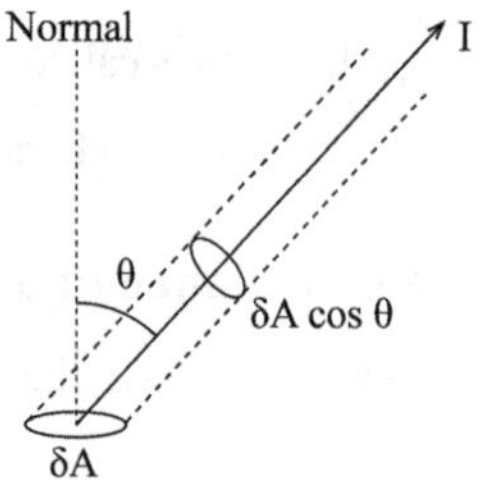

FIGURE 6.2 Projected area $\delta A \cos\theta$ of the elementary source δA along the direction of observation of intensity.

where I_0 is the luminous intensity along the normal to the source, (6.15) shows that L is independent of the direction of observation. Relation (6.16) is known as Lambert's cosine law of emission. A source surface, which obeys this law, is called a uniformly diffusing surface, or a uniform diffuser. Such surface is also known as a Lambert radiator. The concept of a Lambert radiator is also used in radiometry, although it was initially used in photometric measurements based on brightness. Strictly speaking, in practice there is no perfectly Lambertian source covering all angles of observation, but some diffusing surfaces display approximate Lambertian behaviour over a large, albeit restricted region around the normal to the surface. The best Lambertian source is a blackbody radiator. Lasers are extremely poor Lambertian sources, since the output forms a strongly directional beam.

6.2.5 Irradiance (Illuminance/Illumination) of a Receiving Surface

If a radiant flux is incident on a surface, the latter is said to be irradiated, and the flux received per unit area at any point on the surface is called the irradiance. Irradiance E_r is defined as

$$E_r = \frac{\delta\Phi_r}{\delta A} \tag{6.17}$$

In photometry, the corresponding term is called illuminance or illumination E_l, and the same is represented as

$$E_l = \frac{\delta\Phi_l}{\delta A} \tag{6.18}$$

Figure 6.3 shows a point source P irradiating (illuminating) an elemental area δa located around a point Q. The normal to the area δa is inclined to the line PQ by an angle θ. The area δa constitutes a solid angle $\delta\Omega\left(= \delta a \cos\theta / r^2\right)$ at the point P. The flux radiated into the solid angle δΩ is

$$\delta\Phi = I\delta\Omega = \frac{I\delta a \cos\theta}{r^2} \tag{6.19}$$

where I is the intensity of P in the direction $\overline{PQ}$. The illumination at δa is given by

$$E = \frac{\delta\Phi}{\delta a} = \frac{I\cos\theta}{r^2} \tag{6.20}$$

Equation (6.20) embodies two fundamental laws of radiometry and photometry, namely the inverse square law and the cosine law of irradiance/illumination.

The unit for irradiance is Watt per square meter $\left(Wm^{-2}\right)$, and the unit for illuminance or illumination is lumen per square meter $\left(lmm^{-2}\right)$. The latter has a special name 'lux' that is used exclusively for illuminance, and it helps to avoid any confusion arising out of the use of the same unit lumen per square meter (lmm^{-2}) for luminous emittance or exitance.

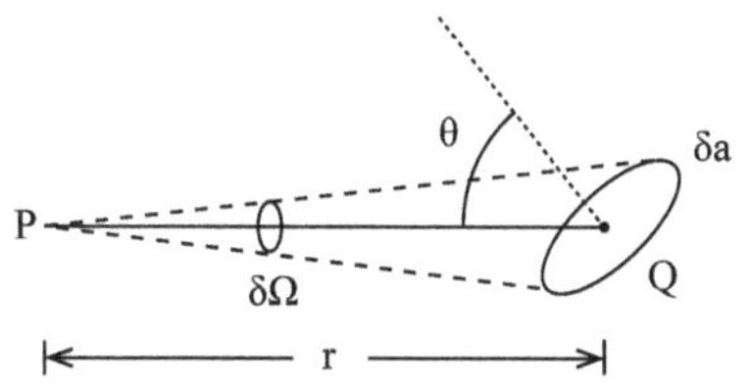

FIGURE 6.3 Illumination of a surface element δs by a point source P.

6.3 Conservation of Radiance/Luminance (Brightness) in Optical Imaging Systems

Mouroulis and Macdonald [13] have given a simplified treatment for deriving the conservation relations for radiance/luminance in optical imaging systems. Their treatment is based on the paraxial optical invariant. We provide below a similar derivation. Figure 6.4 shows a paraxial ray diagram for imaging a circular object of semi-diameter η by an axi-symmetric imaging system.

The object and the image are located at O and O′, respectively. The image is circular with a semi-diameter η'. The entrance and the exit pupils are located at E and E′; both pupils are circular with radii $h_E, h_{E'}$, respectively. The paraxial marginal ray (PMR) and the paraxial pupil ray (PPR) are shown., indicated by an arrow and an empty triangle in the diagram. The refractive indices of the object space and the image space are n and n′ respectively.

The paraxial optical invariant H is

$$H = nu\eta = -nh_E\bar{u} = -n'h'_E\bar{u}' = n'u'\eta' \tag{6.21}$$

By squaring and then multiplying by π^2, we get

$$n^2\pi u^2\pi\eta^2 = n^2\pi h_E^2\pi\bar{u}^2 = n'^2 h'_E{}^2\bar{u}'^2 = n^2\pi u'^2\pi\eta'^2 \tag{6.22}$$

Considering all the paraxial convergence angles $u, u', \bar{u}, \bar{u}'$ as small,

$\pi u^2 \cong \delta\Omega_O$ = Solid angle subtended by the entrance pupil at the object point O
$\pi\bar{u}^2 \cong \delta\Omega_E$ = Solid angle subtended by the object at E, the centre of the entrance pupil
$\pi u'^2 \cong \delta\Omega_{O'}$ = Solid angle subtended by the exit pupil at the mage point O′
$\pi\bar{u}'^2 \cong \delta\Omega_{E'}$ = Solid angle subtended by the image at E′, the centre of the exit pupil

Areas of the object, image, entrance pupil, and exit pupil are:

$\delta s = \pi\eta^2$ = Area of the object
$\delta a = \pi h_E^2$ = Area of the entrance pupil
$\delta a' = \pi h_{E'}^2$ = Area of the exit pupil
$\delta s' = \pi\eta'^2$ = Area of the image

Substituting these values (6.22) reduces to

$$n^2\delta\Omega_O\delta s = n^2\delta\Omega_E\delta a = n'^2\delta\Omega_{E'}\delta a' = n'^2\delta\Omega_{O'}\delta s' \tag{6.23}$$

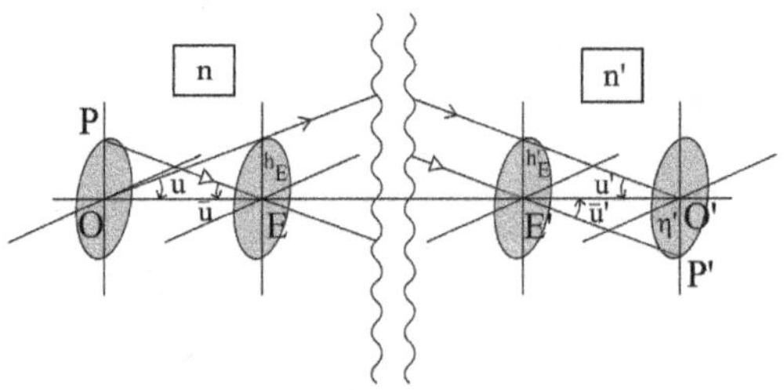

FIGURE 6.4 PPR and PMR in an axi-symmetric imaging system.

Flux $\Delta\Phi$ emitted by the object and admitted by the entrance pupil will be conserved in absence of any loss during propagation through the system. If $L_O, L_E, L_{E'}, L_{O'}$ are the radiance (luminance/brightness) of the object, the entrance pupil, the exit pupil, and the image, respectively, we have

$$\Delta\Phi = L_O \delta\Omega_O \delta s = L_E \delta\Omega_E \delta a = L_{E'} \delta\Omega_{E'} \delta a' = L_{O'} \delta\Omega_{O'} \delta s' \tag{6.24}$$

From (6.23) and (6.24) we get

$$\frac{L_O}{n^2} = \frac{L_E}{n^2} = \frac{L_{E'}}{n'^2} = \frac{L_{O'}}{n'^2} \tag{6.25}$$

For n = n′, (6.25) reduces to

$$L_O = L_E = L_{E'} = L_{O'} \tag{6.26}$$

Therefore, in absence of any loss by absorption, scattering, or vignetting, etc., radiance or luminance (brightness) is conserved in an optical imaging system. This implies that the image cannot be brighter than the object, whatever the configuration is. The image considered above is an aerial image, and the conclusion on brightness refers to the same.

6.4 Flux Radiated Into a Cone by a Small Circular Lambertian Source

By definition (5.9), the flux $\delta\Phi$ radiated into a cone with its axis lying along any direction from a point source is equal to $I\delta\Omega$, where I is the intensity of the source and $\delta\Omega$ is the solid angle of the cone with the point source at its vertex. It is assumed here that the intensity of the point source is the same along all directions from the point source. In practice, no source is actually a point source, and even any real source of small size has a finite surface area. The intensity of radiation from such a source decreases at increasing angles from the normal, as per Lambert's cosine law of intensity. Therefore, the actual flux radiated into a regular cone of finite size, say of solid angle Ω, vertex at O and axis along the normal to the source at O, will be less than $I\Omega$ where I is the luminous intensity of the source along the normal at O.

Figure 6.5 shows a small Lambertian source S_1S_2 of area δs and luminance L, radiating into a cone of semi-vertical angle $\alpha\,(= \angle QOP)$. OPQ is a section of a sphere with O as centre and OP (= r) as its radius. Two co-axial cones Q_1OP and Q_2OP with their axes along OP have semi-vertical angles θ and $(\theta + \delta\theta)$, respectively. The solid angle of the conical shell formed by the two cones is equal to the area intercepted on the surface of the sphere divided by the square of the radius of the sphere, and it can be expressed as

$$\delta\Omega = \frac{(2\pi r \sin\theta) r\delta\theta}{r^2} \tag{6.27}$$

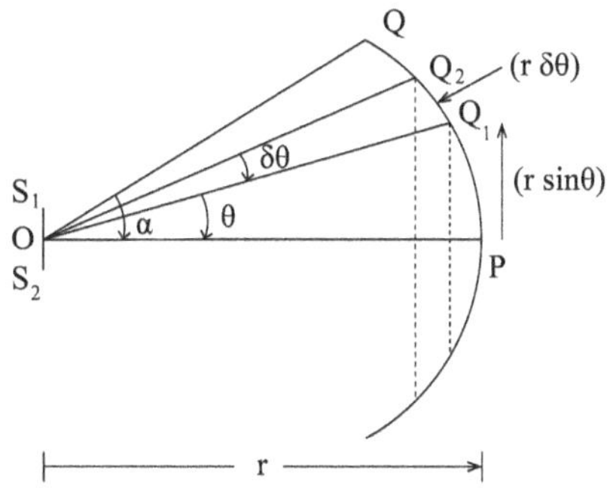

FIGURE 6.5 Flux radiated into a cone from a small plane source S_1S_2 of area δs.

The flux radiated into the shell is given by

$$\delta\Phi = (L\delta s\cos\theta)\delta\Omega = (\pi L\delta s)\sin 2\theta\delta\theta \tag{6.28}$$

Integrating over the semi-vertical angle of the cone, we obtain the expression for flux $\delta\Phi$ in the cone as

$$\delta\Phi = (\pi L\delta s)\int_0^{\alpha} \sin 2\theta \, d\theta = (\pi L\delta s)\sin^2\alpha \tag{6.29}$$

Assuming no emission in other directions, the total power $\widehat{\delta\Phi}$ emitted by the source S_1S_2 is equal to the power emitted in the forward hemisphere. $\widehat{\delta\Phi}$ is obtained by putting $\alpha = \pi/2$ in (6.29) as

$$\widehat{\delta\Phi} = \pi L\delta s \tag{6.30}$$

The radiant emittance or exitance M of the source is

$$M = \frac{\widehat{\delta\Phi}}{\delta s} = \pi L \tag{6.31}$$

6.5 Flux Collected by Entrance Pupil of a Lens System

Figure 6.6 shows a small circular Lambertian source of area δs and radiance L centred on the optical axis. The semi-diameter of the circular entrance pupil at E on the axis subtends an angle α at the source. Using (6.29), the flux radiated into the solid angle defined by the rim of the entrance pupil is approximately given by

$$\Delta\Phi = \pi L\delta s \sin^2\alpha \tag{6.32}$$

Assuming the source to be a circular disk of semi-diameter η, $\delta s = \pi\eta^2$. (6.32) reduces to

$$\Delta\Phi = \pi^2 L(\eta\sin\alpha)^2 \tag{6.33}$$

If the angle variable u of the paraxial marginal ray is taken as $u = \sin\alpha$ (6.33) can be expressed in terms of the paraxial optical invariant H as

$$\Delta\Phi = \left(\frac{\pi}{n}\right)^2 LH^2 \tag{6.34}$$

where n is the refractive index of the optical medium of the source space, and $H = nu\eta$. Thus, the flux collected by the entrance pupil of a lens system is directly proportional to the square of the paraxial optical invariant H.

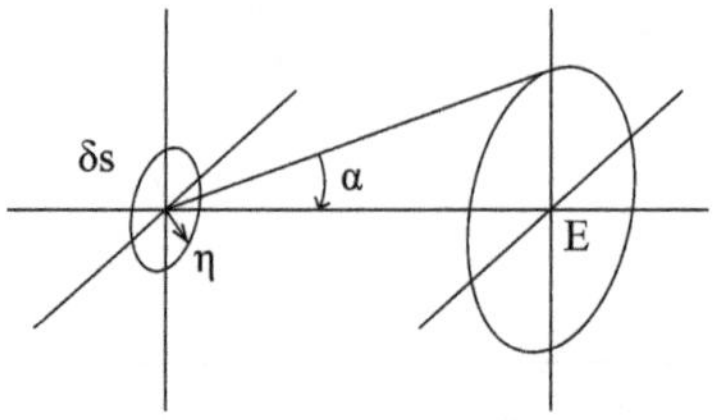

FIGURE 6.6 Flux collected by the entrance pupil of a lens system.

6.6 Irradiance of an Image

Assuming the object to be a small circular Lambertian source of area δs, semi-diameter η and luminance L, the flux $\Delta\Phi$ entering the entrance pupil of a lens is given by (6.33) as $\pi^2 L(\eta \sin\alpha)^2$. Assuming τ as the transmission factor of the lens, the total flux reaching the image is $\tau\Delta\Phi$. Let the semi-diameter of the circular image be η', and the magnification be $M = (\eta'/\eta)$. The image irradiance E is equal to

$$E = \frac{\tau\Delta\Phi}{\pi\eta'^2} = \pi\tau L \frac{(\eta \sin\alpha)^2}{\eta'^2} \tag{6.35}$$

Let the semi-angular aperture of the paraxial marginal ray in the image space be α', and the refractive index of the image space be n'. Using the relation for paraxial optical invariant H

$$H = n \sin\alpha\eta = n' \sin\alpha'\eta' \tag{6.36}$$

(6.35) reduces to

$$E = \pi\tau L \left(\frac{n'}{n}\right)^2 \sin^2\alpha' \tag{6.37}$$

For an object at infinity

$$F_{\#} = \frac{1}{(2n' \sin\alpha')} \tag{6.38}$$

Therefore, for an object at infinity, the expression for irradiance $\bar{E}$ reduces to

$$\breve{E} = \left(\frac{\pi\tau L}{n^2}\right)\frac{1}{4F_{\#}^2} \tag{6.39}$$

The image irradiance is directly proportional to the transmission function of the lens, the scene radiance is inversely proportional to the square of $F_{\#}$ of the imaging lens system. In infrared imaging, the ratio of scene radiance to image irradiance is called G number of the imaging system. For $n = 1$, from (6.39) G number can be expressed as

$$G - \text{number} = \frac{L}{\breve{E}} = \frac{4F_{\#}^2}{\pi\tau} \tag{6.40}$$

In the case of finite conjugate imagery, the irradiance of the image can be expressed as

$$E = \pi\tau L \left(\frac{H}{n\eta'}\right)^2 = \pi\tau L \left(\frac{\sin\alpha}{M}\right)^2 \tag{6.41}$$

Note that the image irradiance in both finite conjugate and infinite conjugate imagery is directly proportional to the square of the paraxial optical invariant. The image irradiance is also proportional to object radiance. Therefore, the exposure for photographing the filament of a 100 W bulb should be about the same as that for a 50 W bulb, since the radiance is the same for both bulbs.

6.7 Off-Axial Irradiance/Illuminance

The drastic fall in illumination in the outer part of the field of view came to notice as soon as optical systems with increased field of view were implemented in practice. One of the early reports of scientific investigations on this aspect of optical imaging was made more than a century ago [14–15]. Figure 6.7 shows the imaging of an object OP by an axi-symmetric imaging system. The axial object point O is imaged at O′, and an elemental object of area δs at the off-axial point P is imaged as δs′ at the off-axial image point P′. The circular entrance and exit pupils of area a,a′ are located at E and E′, respectively. The convergence angles of the paraxial pupil ray PE…E′P′ in the object and image spaces are θ, θ'. The distances $OE = \bar{l}, O'E' = \bar{l}'$. The semi-angular aperture of the entrance pupil with respect to the axial object point O is α. It is assumed that α is small.

The solid angle $\Delta\Omega_0$ subtended by the entrance pupil at the axial object point O is

$$\Delta\Omega_0 \cong \frac{a}{\bar{l}^2} \tag{6.42}$$

The solid angle $\Delta\Omega_P$ subtended by the entrance pupil at the off-axial object point P is

$$\Delta\Omega_P \cong \frac{a}{\bar{l}^2}\cos^3\theta \tag{6.43}$$

Flux $(\Delta\Phi)_P$ collected by the entrance pupil from the off-axial source element δs at P is

$$(\Delta\Phi)_P = L\Delta\Omega_P\delta s\cos\theta = \frac{LA\delta s}{\bar{l}^2}\cos^4\theta \tag{6.44}$$

Flux $(\Delta\Phi)_0$ collected by the entrance pupil from the axial object point O is

$$(\Delta\Phi)_0 = \frac{LA\delta s}{\bar{l}^2} \tag{6.45}$$

Therefore

$$(\Delta\Phi)_P = (\Delta\Phi)_0\cos^4\theta \tag{6.46}$$

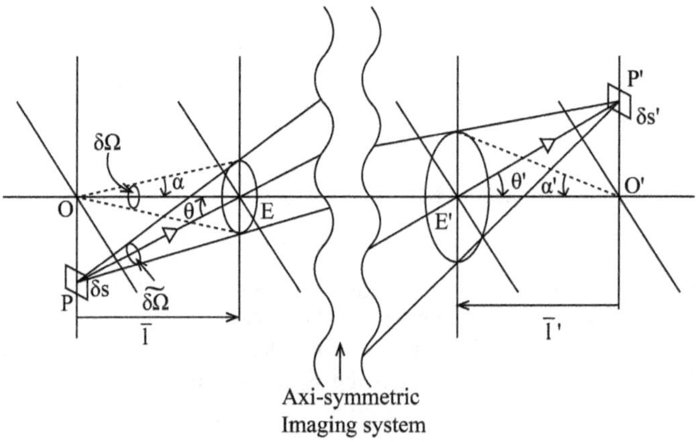

FIGURE 6.7 Irradiance/Illuminance at off-axial image point P′ in an axisymmetric imaging system.

Assuming no loss in the system, this flux will pass through the exit pupil, and will be distributed over $\delta s'$ at P′ on the image plane, so that $(\Delta\Phi)_{P'} = (\Delta\Phi)_{P}$. If M is the transverse magnification $M = (\eta'/\eta)$, we get

$$\delta s' = M^2 \delta s \tag{6.47}$$

Irradiance (illuminance/illumination) at P′ is

$$E_{P'} = \frac{(\Delta\Phi)_{P'}}{\delta s'} = \frac{(\Delta\Phi)_{P}}{\delta s'} \tag{6.48}$$

Substituting from (6.46) and (6.47), (6.48) reduces to

$$E_{P'} = E_{O'} \cos^4\theta \tag{6.49}$$

where

$$E_{O'} = \frac{La}{(M\bar{l})^2} \tag{6.50}$$

Note that the variation of irradiance is given in terms of object space semi-field of view θ in (6.49).

When E and E′ are coincident with the nodal points N and N′ of the system, $\theta = \theta'$. Relation (6.49) becomes

$$E_{P'} = E_{O'} \cos^4\theta' \tag{6.51}$$

In general, $\theta' \neq \theta$, and therefore (6.51) does not follow directly from (6.49). However, the dependence of $E_{P'}$ on θ′ can be determined directly from the image space geometry of Figure 6.7. Assuming radiance or luminance of the exit pupil as $L_{E'}$, the flux reaching the elemental area $\delta s'$ at P′ on the image plane is

$$\delta\Phi_{P'} = \frac{L_{E'} a' \delta s'}{\bar{l}'^2} \cos^4\theta' \tag{6.52}$$

The irradiance $E_{P'}$ at the point P′ is

$$E_{P'} = \frac{\delta\Phi_{P'}}{\delta s'} = \frac{L_{E'} a'}{\bar{l}'^2} \cos^4\theta' \tag{6.53}$$

At $\theta' = 0, E_{O'} = \frac{L_{E'} a'}{\bar{l}'^2}$. Rewriting $E_{P'}$ as $E_{P'}(\theta')$ to emphasize its dependence on θ′, it follows

$$E_{P'}(\theta') = E_{O'} \cos^4\theta' \tag{6.54}$$

This represents the famous 'cosine-to-the-fourth law of illumination'. The derivation does not only assume absence of any loss because of absorption or scattering, but it also assumes the system free from aberrations and vignetting. More specifically, validity of the law assumes

- The entrance/exit pupils subtend small angles α / α' at the axial object/image.
- The angular pupil size remains independent of field of view.
- The local area magnification $(\delta s' / \delta s)$ is independent of field of view.

Indeed, by introduction of selective distortion in off-axial imaging, or equivalently by introducing a calculated amount of pupil coma, the limitations set by the 'cosine-to-the-fourth law of illumination' can be overcome. A recent publication deals with relative illumination in more detail using an irradiance transport equation. The problem has been retackled by various investigators from different standpoints and with varying degrees of approximation [16–27].

6.8 Irradiance/Illuminance From a Large Circular Lambertian Source

Strictly speaking, the sources considered in earlier sections of this chapter were small Lambertian sources, but they were assumed infinitesimally small while deriving the formulae. Therefore, the derived formulae are approximate. We now extend the treatment to the case of large circular Lambertian sources.

Figure 6.8 shows a large circular Lambertian source with its centre at O. It is irradiating an elemental area δA located at G on the optical axis. Both the source and the receiving surface are assumed perpendicular to the axis OG. The distance $OG = Z$. The semi-angular aperture of the circular source is α. An infinitesimally small concentric annular shell of the source has inner and outer radii r and $(r + dr)$ respectively. B is the central point of a source element on the annular shell of area $ds = r dr d\psi$. The distance $BG = R = (Z / \cos\theta)$, where $\theta = \angle BGO$. The elemental solid angle $\widehat{\delta\Omega}$ subtended at the point B on the source element ds by the area δA is

$$\widehat{\delta\Omega} = \frac{\delta A \cos\theta}{(Z / \cos\theta)^2} = \frac{\delta A}{Z^2}\cos^3\theta \tag{6.55}$$

If L is the radiance/luminance of the source, the flux $\delta\Phi$ at δA due to δs is

$$\delta\Phi = L\delta s \cos\theta \delta\Omega = L\delta A \frac{\delta s}{Z^2}\cos^4\theta \tag{6.56}$$

The irradiance/illuminance at G for the source element δs is

$$\delta E = \frac{\delta\Phi}{\delta A} = L\frac{\delta s}{Z^2}\cos^4\theta \tag{6.57}$$

Substituting $\delta s = r dr d\psi$, $r = Z\tan\theta$, $dr = Z\sec^2\theta d\theta$, (6.57) reduces to

$$\delta E = \frac{1}{2} L \sin 2\theta d\theta d\psi \tag{6.58}$$

The total irradiance/illuminance at G is given by

$$E = \frac{1}{2} L \int_0^{2\pi}\int_0^{\alpha} \sin 2\theta d\theta d\psi = \pi L \sin^2\alpha \tag{6.59}$$

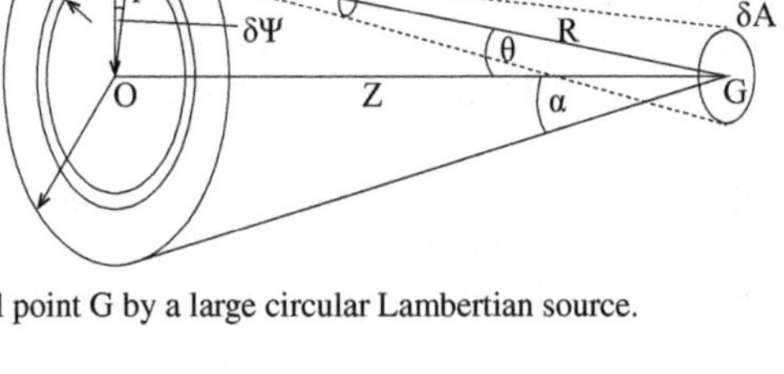

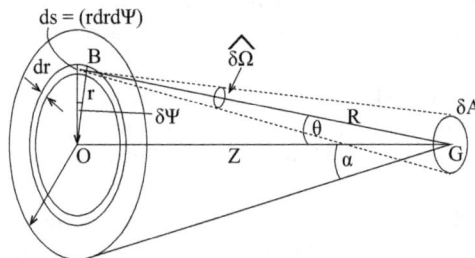

FIGURE 6.8 Irradiance at axial point G by a large circular Lambertian source.

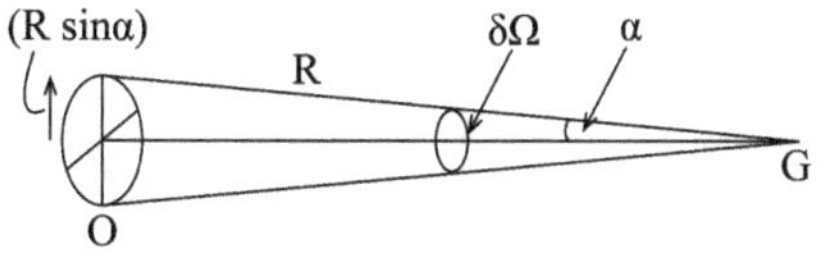

FIGURE 6.9 Solid angle δΩ subtended at G by a distant source at O.

Therefore, for a large circular Lambertian source, the irradiance/illuminance at a point lying on the axis is directly proportional to the radiance/luminance of the source; it is also directly proportional to the square of sine of the semi-angular aperture of the source.

6.8.1 Radiance (Luminance) of a Distant Source

The radiance or luminance of a source in a direction is defined as the intensity in that direction per unit projected area of the source along the same direction. The expressions (6.14) or (6.15), given earlier for radiance or luminance, become indeterminate in the case of distant objects like, say, the sun or the sky, for the lateral size of such a source tends to infinity, and for all practical purposes its size needs to be specified by its angular extent. Figure 6.9 shows a distant source located at a distance R from the point of observation G. The semi-angular extent of the source from the point of observation is α.

As $R \to \infty$, the solid angle $\delta\Omega$ subtended by the source at G is

$$\delta\Omega = \frac{\pi(R\sin\alpha)^2}{R^2} = \pi\sin^2\alpha \tag{6.60}$$

Indeed, for a distant source, there are only two measurable quantities, namely the irradiance (illuminance) at the point of observation and the angular size of the source. Using (6.60) a determinate expression for L in the case of a distant object is obtained from (6.59) as

$$L = \frac{\delta E}{\delta\Omega} \tag{6.61}$$

Thus, the radiance or luminance of a distant source like the sun is given by the ratio of the irradiance or illuminance (illumination) at a point of observation to the solid angle subtended by the distant source at the point of observation.

REFERENCES

1 L. Mandel and E. Wolf, *Optical Coherence and Quantum Optics*, Cambridge University Press, Cambridge (1995), p. 292.
2 E. Wolf, 'Coherence and radiometry', *J. Opt. Soc. Am.*, Vol. 68, No. 1(1978) pp. 6–17.
3 W.H. Steel, M. De, and J.A. Bell, 'Diffraction Corrections in Radiometry', *J. Opt. Soc. Am.*, Vol. 62 (1972) pp. 1099–1103.
4 D.J. Butler, R. Koehler, and G.W. Forbes, 'Diffraction effects in the radiometry of coherent beams', *Appl. Opt.*, Vol. 35 (1996) pp. 2162–2166.
5 J.W.T. Walsh, *Photometry*, Constable, London (1953).
6 R.W. Boyd, *Radiometry and the Detection of Optical Radiation*, J. Wiley, New York (1983).
7 W.R. McCluney, Introduction to Radiometry and Photometry, Artech, Boston (1994).
8 H. Gross, 'Radiometry' Chapter 6 in *Handbook of Optical Systems, Vol. 1. Fundamentals of Technical Optics*, Wiley-VCH, Weinheim (2005) pp. 229–267.
9 CIE 1931 and 1978 data for photopic eye sensitivity function and 1951 data for scotopic eye sensitivity function are available at www.cvrl.org .
10 J.R. Meyer-Arendt, 'Radiometry and Photometry: Units and Conversion Factors', *Appl. Opt.*, Vol. 7, No.10, pp. 2081–2084 (1968).

11 G. Smith and D.A. Atchison, '*The eye and visual optical instruments*', Cambridge University Press, Cambridge (1997), pp. 271–316.
12 A. DiMarzio, '*Optics for Engineers*', CRC Press, New York, (2012), pp. 361–410.
13 P. Mouroulis and J. Macdonald, *Geometrical Optics and Optical Design*, Oxford University Press, Oxford (1997) pp. 127–128.
14 M. v. Rohr, 'Ueber die Lichtvertheilung in der Brennebene photographischer objective mit besonderer Berücksichtigung der bei einfachen Landschaftslinsen und symmetrischen Konstruktionen auftretenden Unterschiede', Vol. 18, No. 1, (1898) pp. 171–180, 197–205.
15 M. v. Rohr, 'Intensity of rays transmitted through optical systems (Photometry of optical instruments)', chapter X in *Geometrical Investigation of the formation of images in optical instruments*, ed. M. von Rohr, His Majesty's Stationary Office, London (1920) pp. 515–554.
Translated by R. Kanthack from *Die Bilderzeugung in Optischen Instrumenten vom Standpunkte der geometrischen Optik*, J. Springer, Berlin (1904).
16 M. Reiss, 'The Cos^4 Law of Illumination', *J. Opt. Soc. Am.*, Vol. 35, No. 1, (1945) pp. 283–288.
17 M. Reiss, 'Notes on the Cos^4 Law of Illumination', *J. Opt. Soc. Am.*, Vol. 38, No. 11, (1948) pp. 980–986.
18 G. Slussareff, 'L'Eclairement de l'image formée par les objectifs photographiques grand-angulaires', *J. Phys. USSR*, Vol. 4, (1941) pp. 537–545.
19 G. Slussareff, 'A Reply to Max Reiss', *J. Opt. Soc. Am.*, Vol. 36, No. 12, (1946), p. 707.
20 I.C. Gardner, 'Validity of the cosine-fourth power law of illumination', *J. Res. Nat. Bur. Standards*, Vol. 39 (1947) pp. 213–219.
21 D.G. Burkhard and D.L. Shealy, 'Simplified formula for the illuminance in an optical system', *Appl. Opt.*, Vol. 20 (1981) pp. 897–909.
22 M.P. Rimmer, 'Relative Illumination calculations', *Proc. SPIE*, Vol. 655 (1986) pp. 99–104.
23 D.G. Koch, 'Simplified irradiance/illuminance calculations in optical systems', *Proc. SPIE*, Vol. 1780 (1992) pp. 226–240.
24 S.J. Dobson and K. Lu, 'Accurate calculation of the irradiance of optical images', *Opt. Eng.* Vol. 37 (1998) pp. 2768–2771.
25 D. Reshidko and J. Sasian, 'Role of aberrations in the relative illumination of a lens system', *Opt. Eng.*, Vol. 55, No. 11, 115105–1/12 (2016).
26 R. Siew, 'Relative illumination and image distortion', *Opt. Eng.*, Vol. 56, No. 4, 049701–1/4 (2017).
27 T.P. Johnson and J. Sasian, 'Image distortion, pupil coma and relative illumination', *Appl. Opt.*, Vol. 59, G19–G23 (2020).

7

Optical Imaging by Real Rays

In Chapters 3–6, the treatment of optical imaging was based on the paraxial model of image formation. The paraxial model of any optical system assumes that both the aperture and the field of the system is infinitesimally small. All real optical systems have finite values of aperture and field, and, therefore, their real imaging characteristics differ from the predictions obtained by using a paraxial model of the system. Often these differences are greater with the increase in size of the aperture and/or the field of the system. Assessment of the quality of image formation by real systems is intimately linked with proper understanding and correct analysis of these differences. For a specific system, these differences are determined by tracing real rays through the system using finite ray tracing formulae. 'Tracing real rays' implies use of exact ray tracing formulae, e.g., use of Snell's laws as given in Section 2.9.1.3, for determining ray segments through the cascade of interfaces of an optical system. This is a topic of paramount importance in optical instrumentation, and it has been receiving the attention of scientists from the early days of observational sciences in the seventeenth century.

An alternative approach to understanding the characteristics of real optical systems makes use of propagation laws of real rays. Whereas the earlier approach is more prevalent in the practical analysis and synthesis of real optical systems, the latter approach is directed towards development of analytical tools for understanding the behaviour of optical systems, and particularly the relationship between characteristics of optical systems and their structure. In the next section, the major highlights of this approach will be underscored.

7.1 Rudiments of Hamiltonian Optics

The importance of Hamiltonian optics in the context of analysis and synthesis of optical systems can be readily appreciated from the following description of correspondence between rays in the initial object space and the final image space. Figure 7.1 depicts a general optical system. Without loss of generality, the initial and final spaces are assumed to be homogeneous, and of refractive indices n and n', respectively. For the sake of complete generality, unrelated rectangular Cartesian coordinate systems (X,Y,Z) and (X',Y',Z'), with their origins at O and O′, are used to specify points in the two spaces. In general, for any two arbitrary points P and Q in the initial and the final spaces, respectively, only one ray APQB passes through both of them. This implies that the points P and Q are not conjugates. Let the point of intersection of the ray segment AP with the XY plane in the initial space be P with coordinates (X, Y, 0). Similarly, in the final space, Q is the point of intersection of the ray segment QB with the X′Y′ plane. The coordinates of the point Q are $(X',Y',0)$. Let the direction cosines of the ray segment AP be (L, M, N), and those of the ray segment QB be (L',M',N'). The direction cosines satisfy the relations

$$L^2 + M^2 + N^2 = 1 \quad \text{and} \quad L'^2 + M'^2 + N'^2 = 1 \tag{7.1}$$

Out of the three direction cosines L, M and, N of a ray, only two are independent; this is also true in the final space. Therefore, the ray AP in the initial space can be uniquely specified by four parameters (X, Y, L, M), and similarly, the ray QB in the final space is specified by (X', Y', L', M'). A change in any one of the four parameters X, Y, L, and M of the ray AP of the initial space gives rise to changes in all four parameters X',Y',L', and M' of the corresponding ray QB in the final space. This observation tends to

DOI: 10.1201/9780429154812-7

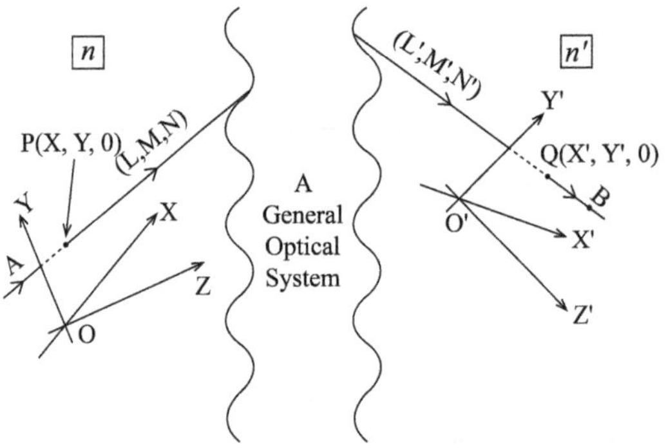

FIGURE 7.1 A real ray APQB passing through a point P on the XY plane in the object space, and a point Q on the X′Y′ plane in the image space.

suggest that a unique correspondence between a real ray in the final space with its counterpart in the initial space calls for knowledge of four functions $f_i, i = 1,\ldots,4$, as given below:

$$X' = f_1(X, Y, L, M) \tag{7.2}$$

$$Y' = f_2(X, Y, L, M) \tag{7.3}$$

$$L' = f_3(X, Y, L, M) \tag{7.4}$$

$$M' = f_4(X, Y, L, M) \tag{7.5}$$

It is important to note that this inference on the necessity of four functions $f_i, i = 1,\ldots,4$ is wrong, for it takes no cognizance of the fact that not any path can be the path of a real ray. The optical path of a real ray has to satisfy the extremal characteristics as laid down in Fermat's principle. Consequently, only one suitably formulated function of four variables is sufficient for the purpose.

The next section will elucidate this point of view.

In the first half of the nineteenth century, Hamilton embarked upon this problem, and gave examples of such functions. He noted that 'the geometrical properties of an optical system of rays, whether straight or curved, whether ordinary or extraordinary, may be deduced by analytic methods, from one fundamental formula, and one characteristic function' [1]. Hamilton's characteristic function represents the optical path length between two points; one of the points is in the initial object space, and the other point is in the final image space. The characteristic function retains its usefulness in characterizing the ray optical characteristics of the optical system so long as only one ray passes between these two points. By no means can the two points be conjugate points because, by definition, two points are conjugate points only when many rays pass between them. For analyzing characteristics of a system in the neighbourhood of conjugate points, different types of characteristic functions need to be used. Hamilton's seminal paper on ray optics, 'Theory of systems of rays' [2] was published in 1828, but unlike his papers on classical mechanics, his papers on ray optics did not receive much attention immediately after their publication. Nevertheless, few reports on the use of Hamilton's characteristic functions appeared in the second half of the nineteenth century. They include the publications by Maxwell [3–4] and Thiesen [5], and the book by Heath [6]. However, there was a resurgence of interest on this approach after Heinrich Bruns [7] dwelt upon this topic again in 1895, and used functions similar to the characteristic functions proposed by Hamilton. Bruns coined the term 'eikonal' for representing these functions. The word 'eikonal' is the German form of the Greek word 'εικών' meaning image, icon, or likeness. Subsequently, in optics literature the word 'eikonal' has become a generic term for all types of characteristic functions. However, for the same concept, some also use the word 'characteristic', as in 'characteristic function'. Note that the scalar function, $S(\boldsymbol{r})$, introduced in Section 2.8.1 for characterizing a light field within the approximations of geometrical optics, is also called the eikonal function; it is closely related with the characteristic function or the eikonal function proposed by Hamilton and Bruns.

In the early days of twentieth century, Klein [8] and Schwarzschild [9] followed up the Bruns' eikonal. On the other hand, Lord Rayleigh, Southall, Prange, Steward, T. Smith, and others reported on the use of Hamilton's characteristic functions in the analysis of optical systems [10–15]. Initially, Hamilton's characteristic functions and Bruns' eikonal were sometimes misconstrued as two independent concepts. After publication of the collected works of Hamilton by Conway and Synge [16], and the interesting polemic between Herzberger and Synge on the interrelation between the two approaches in the open literature, the inherent similarity between the two concepts has been unfolded, so much so that often the term 'Hamilton-Bruns eikonal' is now used for these characteristics [17–36]. The concept of characteristic function or eikonal provides a highly useful analytical tool for investigating the nature of defects, e.g. aberrations occurring in optical systems and the effects of different types of structural symmetry of the systems on the defects [37].

7.1.1 Hamilton's Point Characteristic Function

A is an arbitrary point in the initial space of a general optical system (Figure 7.2). Similarly, B is an arbitrary point in the final space of the system. The two points are not conjugate points, and it is assumed that one and only one ray passes through the two points. With respect to the unrelated rectangular Cartesian coordinate systems (X, Y, Z) and (X',Y',Z') in the initial and final spaces with origins at O and O′, respectively, let the coordinates of A be (X,Y,Z) and that of B be (X',Y',Z'). Hamilton's point characteristic function E is defined as the optical path length between the two points A and B along the ray A,…,B. Therefore,

$$E(X,Y,Z;X',Y',Z') = [A\ldots B] \tag{7.6}$$

This function has already been introduced in Section 2.9.5 while discussing the path differential theorem. From (2.217) we note that, in vector notation, the path differential dE is given by

$$dE = n'\boldsymbol{s}'.\boldsymbol{dr}' - n\boldsymbol{s}.\boldsymbol{dr} \tag{7.7}$$

Assuming direction cosines of the ray segments in the object space and the image space as (L,M,N) and (L',M',N'), respectively, (7.7) can be expressed as

$$dE(X,Y,Z;X',Y',Z') = n'(L'dX' + M'dY' + N'dZ') - n(LdX + MdY + NdZ) \tag{7.8}$$

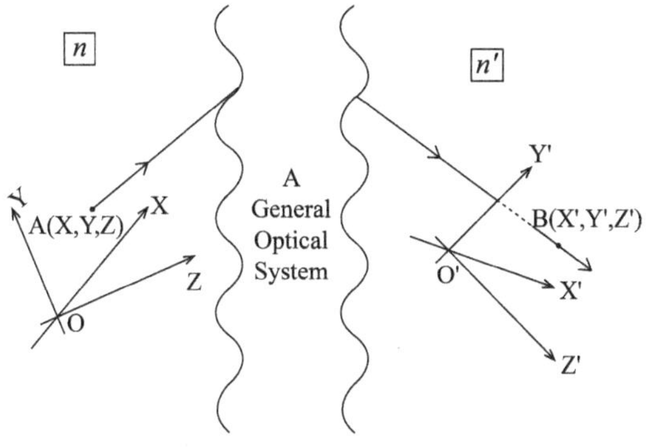

FIGURE 7.2 Hamilton's point characteristic function $E(X,Y,Z;X',Y',Z') = E(A,B) = [A\ldots B]$.

From the total differential of E, as given above, it follows

$$\frac{\partial E}{\partial X} = -nL \quad \frac{\partial E}{\partial Y} = -nM \quad \frac{\partial E}{\partial Z} = -nN \tag{7.9}$$

$$\frac{\partial E}{\partial X'} = -n'L' \quad \frac{\partial E}{\partial Y'} = -n'M' \quad \frac{\partial E}{\partial Z'} = -n'N' \tag{7.10}$$

From (7.9) and (7.10), we note that Hamilton's point characteristic function E has to satisfy the following two partial differential equations

$$\left(\frac{\partial E}{\partial X}\right)^2 + \left(\frac{\partial E}{\partial Y}\right)^2 + \left(\frac{\partial E}{\partial Z}\right)^2 = n^2 \tag{7.11}$$

$$\left(\frac{\partial E}{\partial X'}\right)^2 + \left(\frac{\partial E}{\partial Y'}\right)^2 + \left(\frac{\partial E}{\partial Z'}\right)^2 = n'^2 \tag{7.13}$$

This implies that not every function of the six variables X, Y, Z, X', Y', Z' can be the point characteristic of an optical system—only those functions that satisfy the pair of partial differential equations given above can.

7.1.1.1 Hamilton-Bruns' Point Eikonal

Hamilton's point characteristic function $E(A,B)$ is a six-dimensional function, since an arbitrary point A in the three-dimensional initial space is characterized by three coordinates (X, Y, Z), and similarly the arbitrary point B in the final space also requires three coordinates (X', Y', Z') for its characterization. The number of variables can be conveniently reduced to four, if the points A and B for identifying the ray are constrained to lie on the XY plane in the initial object space, and on the $X'Y'$ plane in the final image space, respectively. Accordingly, let us consider a ray that passes through a point P on the XY plane in the object space, and a point Q on the $X'Y'$ plane in the image space (Figure 7.3). The choice of the coordinate systems (X, Y, Z) and (X', Y', Z') in the initial and the final spaces is, in general, arbitrary. However, with a specific choice for the coordinate systems, any ray from the initial space to the final space can be uniquely specified by the four parameters X, Y, X', Y'.

Hamilton-Bruns' point eikonal S is defined as the optical path length between the points P and Q along the ray passing through them. Therefore,

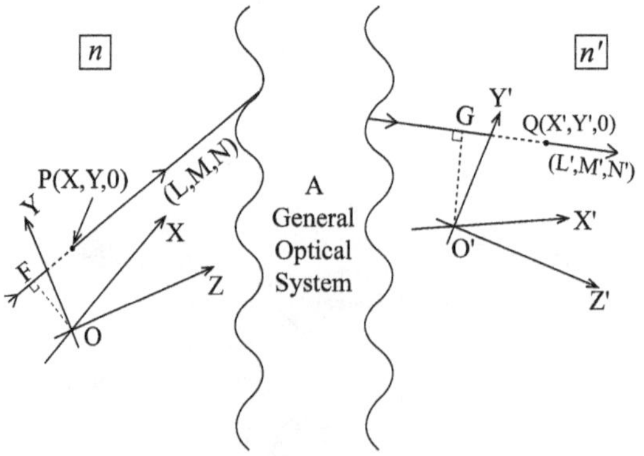

FIGURE 7.3 Four eikonals of Hamilton-Bruns: (i) Point eikonal $S(X,Y;X',Y') = E(X,Y,0;X',Y',0) = [P\ldots Q]$ (ii) Point Angle eikonal $V(X,Y;L',M') = [P\ldots G]$ (iii) Angle Point eikonal $U(L,M;X',Y') = [F\ldots Q]$ (iv) Angle eikonal $W(L,M;L',M') = [F\ldots G]$.

$$S(P,Q) \equiv S(X,Y;X',Y') \equiv E(X,Y,0;X',Y',0) = [P...Q] \tag{7.14}$$

It is to be noted that, apart from the restriction that the two planes, the XY plane and the X′Y′ plane are not conjugate planes, the choice of these two planes is arbitrary, since the coordinate systems (X, Y, Z) and (X',Y',Z') are unrelated. The total differential of Hamilton-Bruns' point eikonal can be obtained from (7.8) by substituting

$$Z = 0,\ Z' = 0,\ dZ = 0,\ dZ' = 0 \tag{7.15}$$

It follows

$$dS(X,Y;X',Y') = n'(L'dX' + M'dY') - n(LdX + MdY) \tag{7.16}$$

Therefore,

$$\frac{\partial S(X,Y;X',Y')}{\partial X} = -nL \tag{7.17}$$

$$\frac{\partial S(X,Y;X',Y')}{\partial Y} = -nM \tag{7.18}$$

$$\frac{\partial S(X,Y;X',Y')}{\partial X'} = n'L' \tag{7.19}$$

$$\frac{\partial S(X,Y;X',Y')}{\partial Y'} = n'M' \tag{7.20}$$

These four equations contain implicit relations between the ray parameters (X, Y, L, M) in the object space and the ray parameters (X',Y',L',M') in the image space.

Looking back to the question raised in Section 7.1, it should be apparent that if the single function $S(X,Y;X',Y')$ is known, then for a given input (X,Y,L,M), the output (X',Y',L',M') could be determined in two steps:

(i) Step 1. Solve the two equations (7.17) – (7.18) to determine X′ and Y′ for the given input (X, Y, L, M).
(ii) Step 2. Use the two equations (7.19) and (7.20) to determine L′ and M′.

Thus, for a given set of input ray parameters, the knowledge of a single function S is sufficient to determine the output ray parameters, and vice versa.

In fact, any other properly formulated characteristic function/eikonal can also be used for this purpose. For example, if we use Hamilton's point characteristic function $E(X,Y,Z;X',Y',Z')$, the ray segment in the initial space is specified by a set of values for (X, Y, Z;L, M, N), The corresponding parameters for the ray segment in the final space can be obtained by using the single function E in two steps:

(i) For the input (X, Y, Z;L, M, N) corresponding to the ray in the initial space, the three equations given in (7.9) can be solved to obtain the three parameters X',Y',Z'.
(ii) Using the input and the values of X',Y',Z', the three equations given in (7.10) can be solved to obtain the three parameters L',M',N'.

The set $(X',Y',Z';L',M',N')$ corresponds to the ray segment in the final space.

7.1.2 Point Angle Eikonal

It is important to note that neither the Hamilton's point characteristic function E, nor the Hamilton-Bruns' point eikonal S can be used when the chosen points in the initial and the final image spaces are conjugates, i.e. when the points are imaged onto each other. This is because, when two points are conjugates, many rays pass through them, and so the basic assumption of only one ray passing through the two points is violated, and the differential relations become indeterminate.

This difficulty is overcome by introducing another function V, obtained from the point eikonal S by Legendre transformation [38]

$$V = S(X,Y;X',Y') - n'(X'L' + Y'M') \tag{7.21}$$

The total differential of V is given by

$$dV = dS - n'(X'dL' + Y'dM') - n'(L'dX' + M'dY') \tag{7.22}$$

Substituting for dS from (7.16), and after algebraic manipulation we get

$$dV = -n'(L'dX' + M'dY') - n(LdX + MdY) \tag{7.23}$$

The above relation shows that it is convenient to consider V as a function of the coordinates x and y in the object space, and the direction cosines L' and M' in the image space. Note that the process of the Legendre transformation leads to a new set of variables. From (7.23) it follows

$$\frac{\partial V}{\partial X} = -nL \qquad \frac{\partial V}{\partial Y} = -nM \tag{7.24}$$

$$\frac{\partial V}{\partial L'} = -n'X' \qquad \frac{\partial V}{\partial M'} = -n'Y' \tag{7.25}$$

The function $V(X,Y,L',M')$ is called the point angle eikonal. The point angle eikonal V is equal to the optical path length $[P\ldots G]$ along the ray from the point P to the point G (Figure 7.3). The point G is the foot of the perpendicular from the origin O' on the ray segment GQ in the image space. Note that the point eikonal S is equal to $[P\ldots Q]$, and the projection of $O'Q$ on the ray segment in the image space is $GQ = (X'L' + Y'M')$. Geometrically, it is evident that $[P\ldots G] = [P\ldots Q] - [GQ]$. The point angle eikonal V can be used when the four variables X, Y, L', M' uniquely specify a ray. Problems crop up when the point (X, Y) alone determines L', M', irrespective of the direction cosines L, M of the ray in the object space. More specifically, if the XY plane, on which the object space point P lies, is the object space focal plane of the system, the differential relations for V become indeterminate, and it cannot be used. The remedy may be sought via the use of angle point eikonal defined below.

7.1.3 Angle Point Eikonal

The Legendre transformation carried out on the point eikonal led to the point angle eikonal where the variables in the image space were changed from point variables to angle variables, leaving the object space variables unchanged. In case of the angle point eikonal U, a Legendre transformation of the point eikonal is carried out such that the object space variables are changed from the point variables to direction variables, leaving the image space variables unchanged. U is defined as

$$U = S(X,Y;X',Y') + n(XL + YM) \tag{7.26}$$

The total differential of U is given by

$$dU = dS + n(XdL + YdM) + n(LdX + MdY) \tag{7.27}$$

Substituting for dS from (7.16), we get

$$dU = n(XdL + YdM) + n'(L'dX' + M'dY') \tag{7.28}$$

From the above differential relation, it is apparent that the angle point eikonal function U can be conveniently expressed in terms of angle variables L, M in the object space and point variables x', y' in the image space. From (7.28) it follows

$$\frac{\partial U}{\partial L} = nX \qquad \frac{\partial U}{\partial M} = nY \tag{7.29}$$

$$\frac{\partial U}{\partial X'} = n'L' \qquad \frac{\partial U}{\partial Y'} = n'M' \tag{7.30}$$

The function $U(L,M,X',Y')$ is called the angle point eikonal. It is equal to the optical path length $[F\ldots Q]$ from F to Q along the ray from the object space to the image space (Figure 7.3). F is the foot of the perpendicular from O on the ray segment in the object space. Note that $S = [P\ldots Q]$, and FP is the projection of OP on the ray in the object space. Therefore, $[FP] = n(LX + MY)$. It is obvious that the angle point eikonal can be used when the four parameters (L,M,X',Y') uniquely specify a ray. The location of the point P on the object space focal plane is no hindrance to its use, but it is obvious that the angle point eikonal cannot be used when the $X'Y'$ plane, on which point Q is located, is the image space focal plane.

Note that, at times, the point angle eikonal V and the angle point eikonal U are referred to as the mixed eikonal of the first kind and the mixed eikonal of the second kind, respectively, in the published literature.

7.1.4 Angle Eikonal

Legendre transformations can be carried out on the point eikonal so that the point variables of both the object space and the image space are changed to direction variables in both spaces. The angle eikonal W is defined as

$$\begin{aligned} W &= S(X,Y;X',Y') - n'(X'L' + Y'M') + n(XL + YM) \\ &\equiv V(X,Y;L',M') + n(XL + YM) \end{aligned} \tag{7.31}$$

The total differential of the angle eikonal is given by

$$dW = dV + n(XdL + YdM) + n(LdX + MdY) \tag{7.32}$$

Substituting for dV from (7.23), and after simplification we get

$$dW = n(XdL + YdM) - n'(X'dL' + Y'dM') \tag{7.33}$$

The above relation shows that the angle eikonal W is a function of the angle variables (L, M) of the object space, and (L', M') of the image space. From (7.33) we get

$$\frac{\partial W(L,M,L',M')}{\partial L} = nX \qquad \frac{\partial W(L,M,L',M')}{\partial M} = nY \tag{7.34}$$

$$\frac{\partial W(L,M,L',M')}{\partial L'} = -n'X' \qquad \frac{\partial W(L,M,L',M')}{\partial M'} = -n'Y' \tag{7.35}$$

Geometrically, the angle eikonal $W(L,M,L',M')$ is equal to the optical path length $[F\ldots G]$ along the ray. F is the foot of the perpendicular from O on the ray segment in the object space, and G is the foot of the perpendicular from O' on the ray segment in the image space. In Figure 7.3, the angle eikonal W is represented as

$$W \equiv [F\ldots G] = [P\ldots Q] - [GQ] + [FP] \tag{7.36}$$

The angle eikonal W can be used whenever the four variables L, M, L', M' uniquely specify a ray. However, it cannot be used in systems where L, M determines L', M'. Therefore, the angle eikonal cannot be used in treatment of afocal or telescopic systems. Of course, any of the eikonals described earlier can be sed for this purpose.

7.1.5 Eikonals and their Uses

The four eikonals, enunciated in the last section, do not exhaust the list of possible eikonals. In fact, any two out of the four variables X, Y, L, M in the initial space, together with any two out of the four variables X', Y', L', M' in the final space can be used to constitute an eikonal. Luneburg [27] and Buchdahl [37] underscored this possibility. Buchdahl also noted that the points (X, Y) and (X', Y') need not necessarily be located on an anterior base plane in the initial space and a posterior base plane in the final spaces. Any suitable base surface can be used in either or both of the initial and final spaces [39].

Eikonals are the building blocks of the beautiful edifice of classical aberration theory. This theory has been extensively developed for axially symmetric optical systems [40–41]. The nature and types of defects (aberrations), their dependence on the aperture and field parameters, and the effects of different types of structural symmetry in the optical systems on the aberrations have been worked out analytically by using eikonals. However, eikonals have been rarely used in practical design of optical systems. In the published literature of the first half of the twentieth century, we could locate only one instance [42]. One of the main reasons for this is the non-availability of suitable analytical expressions for the eikonals in terms of the system parameters, except for in trivial cases. Even for moderately complex optical systems, derivation of these analytical expressions involves elimination of intermediate variables at multiple levels; the process is neither simple nor straightforward. After the widespread availability of digital computers towards the end of the twentieth century, investigators are again looking into the problem with renewed vigour. Walther has put forward the prospects opened up by digital computers in this regard; in particular, he advocated the use of computer algebra for undertaking semi-analytical tasks [43–44]. Forbes et al. and Velzel et al. looked into the bits and pieces of hurdles encountered in practical utilization of the eikonals in optical design [45–53]. De Meijere [54] reported on the design of a triplet derivative using eikonal coefficients.

Asymmetric optical systems display characteristics that are distinctly different from axisymmetric systems. Eikonal theory provides the necessary tools for tackling the problems of analysis of these systems. Forbes and his coworkers have reported on some aspects of them [55–59].

7.1.6 Lagrangian Optics

Traditionally, Hamiltonian Optics has been used extensively in understanding the behaviour of optical systems in general. The building blocks of the edifice of ray optical theory of image formation, e.g., paraxial optics and theory of aberrations, etc., are derived from considerations based on Hamiltonian optics. Notwithstanding the great role played by Hamiltonian Optics in this context, the inherent analytical complexity, as deliberated in the last section, has compelled adoption of a different approach, based on ray tracing with varying degrees of approximation, for analysis and synthesis of optical systems in practice. By analogy with corresponding treatments in the field of dynamics, this approach goes by the name of 'Lagrangian Optics' [60]. Buchdahl has explained the interrelationship between the two approaches in a lucid way [61].

7.2 Perfect Imaging with Real Rays

Let P be a point object located in the object space of an optical imaging system. Out of the total number of rays emerging from point P, only a part of them traverses the system unhindered, and subsequently takes part in image formation. In general, not all rays exiting from the system intersect at a single point. In the unusual case when all such rays intersect at a single point, say P′, the point P′ is said to be a *stigmatic* or *sharp* image of the point P.

In an *ideal* optical instrument, every point P of a three-dimensional object space gives rise to a corresponding stigmatic image P′ in the image space. The points P and P′ in the two spaces are called conjugate points. An optical imaging system that images all points of a three-dimensional object space stigmatically is called an *absolute* instrument.

In an ideal imaging system, when P describes a curve $\mathcal{C}$ in the object space, P' describes a curve $\mathcal{C}'$ in the image space. In general, the two curves are not geometrically similar to each other. If, and only if, every curve $\mathcal{C}$ in the object space is *geometrically similar* to the corresponding curve $\mathcal{C}'$ in the image space, the imaging is said to be *perfect* [62]. A *perfect image* of an object implies not only that the image of each point of the object is *stigmatic*, but that the image of the object is also *free from any defect of curvature and distortion.*

7.2.1 Stigmatic Imaging of a Point

In Figure 7.4, P and P' are two conjugate points of a general optical system. (X, Y, Z) and (X',Y',Z') are rectangular coordinate systems in the object space and image space of the system. The points P and P′ are assumed to lie on the XY plane and the $X'Y'$ plane respectively, so that the coordinates of P and P′ are (X, Y, 0) and $(X',Y',0)$. Let the direction cosines of the central ray of the image-forming pencil of rays be (L, M, N) and (L',M',N').

Obviously, two of the three direction cosines are independent, and so the ray segments in the object space and the image space are specified by (X, Y, L, M) and (X',Y',L',M'). Let us consider the point

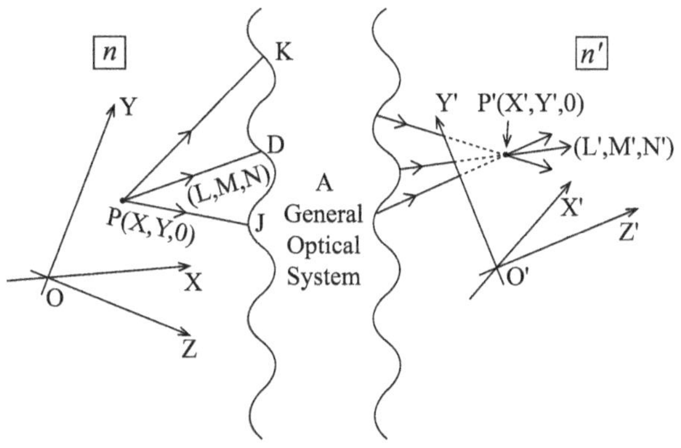

FIGURE 7.4 Conjugate points $P(X,Y,0)$ and $P'(X',Y',0)$.

angle eikonal V (X,Y,L',M'). For a fixed point P(X, Y), V is a function of L′ and M′. From (7.25), the coordinates of the point $P'(X',Y')$ are given by

$$\frac{\partial V}{\partial L'} = -n'X' \qquad \frac{\partial V}{\partial M'} = -n'Y' \tag{7.37}$$

Since P and P′ are assumed to be conjugate points, X′ and Y′ are constants for the image-forming pencil of rays; the image-forming pencil originates from the object point P(X, Y) in the object space and has direction cosines (L, M) within the range

$$L - \Delta L \le L \le L + \Delta L, \qquad M - \Delta M \le M \le M + \Delta M \tag{7.38}$$

Correspondingly, the direction cosines of the image-forming pencil in the image space lie in the range

$$L' - \Delta L' \le L' \le L' + \Delta L', \qquad M' - \Delta M' \le M' \le M' + \Delta M' \tag{7.39}$$

From (7.37) we get

$$V = -n'(X'L' + Y'M') + C \tag{7.39}$$

where C is a constant with respect to L′ and M′. Using (7.21) the constant C can be written as

$$C = V + n'(X'L' + Y'M') = S(X,Y;X',Y') \tag{7.40}$$

The point eikonal $S(X,Y;X',Y')$ is equal to the optical path length of the ray from the point P to the point P′. From (7.39) and (7.40) it follows that 'if P and P′ are perfect conjugate points then all rays through these points have the same optical path length' [63]. For example, the optical path lengths along the three rays (shown in Figure 7.4) from P to P′ are equal.

$$[PK\ldots K'P'] = [PD\ldots D'P'] = [PJ\ldots J'P'] \tag{7.41}$$

7.2.2 Cartesian Oval[64]

7.2.2.1 Finite Conjugate Points

Consider an optical imaging system consisting of two homogeneous media of refractive index *n* and *n′* separated by a surface

$$\mathcal{S}(X,Y,Z) = 0 \tag{7.42}$$

O, the point of intersection of the surface S with the Z-axis, is the origin of the XYZ coordinate system. Let us consider the special case when the surface S is a surface of revolution obtained by rotating a curve

$$\mathcal{C}(Y,Z) = 0 \tag{7.43}$$

around the Z-axis. The XY plane is tangential to the surface at the origin O. Without loss of generality, we assume that a point object A with coordinates $(0,0,l)$ is imaged stigmatically at the point $A'(0,0,l')$. In Figure 7.5, we consider the YZ section. Two rays, APA′ and AOA′, passing through the conjugate

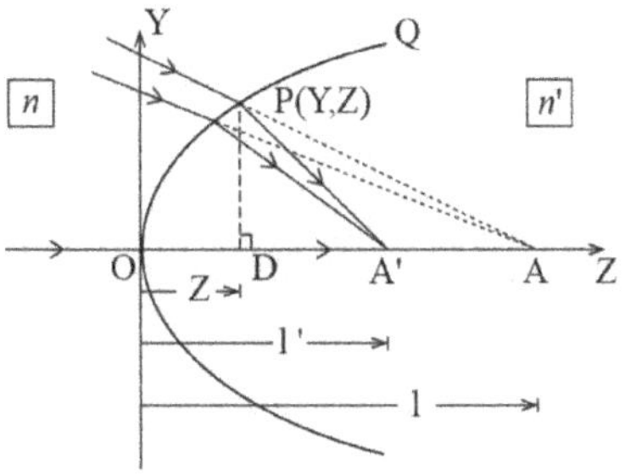

FIGURE 7.5 An arbitrary point $P(Y, Z)$ on a Cartesian oval forming a stigmatic point image A' of an axial point A.

points are shown. P (Y, Z) is an arbitrary point on the curve C. From the condition of equality of optical path lengths, as enunciated in the earlier section, we get

$$[AP]+[PA']=[AO]+[OA'] \tag{7.44}$$

In terms of geometrical path lengths, we write

$$n.AP + n'.PA' = n.AO + n'.OA' \tag{7.45}$$

With due regard for the sign convention,

$$AP = -PA \quad \text{and} \quad AO = -OA = -l \tag{7.46}$$

Let PD be perpendicular from P on the Z-axis. So, PD = Y, and OD = Z.

Therefore, we get

$$-n\sqrt{Y^2 + (l - Z)^2} + n'\sqrt{Y^2 + (l' - Z)^2} = -nl + n'l' \tag{7.47}$$

By squaring twice in succession, we can eliminate the radicals and obtain an algebraic curve of degree four. This curve is known as the Cartesian oval, in memory of René Descartes who first described it in 1637 [64–66]. The corresponding surface of revolution is known as the Cartesian surface [67]. From (7.47), on algebraic manipulation, the fourth degree equation of the generating curve for the Cartesian surface is given by

$$A\left(Y^2 + Z^2\right)^2 + BZ\left(Y^2 + Z^2\right) + C\left(Y^2 + Z^2\right) + DZ^2 + EZ = 0 \tag{7.48}$$

The coefficients are

$$A = \left\{n'^2 - n^2\right\}^2 \tag{7.49}$$

$$B = -4\left\{n'^2 - n^2\right\}\left\{n'^2 l' - n^2 l\right\} \tag{7.50}$$

$$C = 4n'n\left(n'l - nl'\right)\left(n'l' - nl\right) \tag{7.51}$$

$$D = 4\left\{n'^2 l' - n^2 l\right\}^2 \tag{7.52}$$

$$E = -8n'n(n'-n)l'l(n'l'-nl) \tag{7.53}$$

Figure 7.6 shows a Cartesian oval producing a stigmatic image of point A in the homogeneous medium of the refractive index n at point A′ in the homogeneous medium of refractive index n′. Both A and A′ are located on the Z-axis. Obviously, the actual shape of the curve depends on the values of n, n' and l, l'. The fourth degree equation representing the Cartesian oval is a closed curve, as shown in Figure 7.6. However, only part of the curve, from the vertex A to the point where the incident/emergent ray is tangential to the curve, takes part in image formation. The other side of the Cartesian oval comes into play in the case of virtual conjugates. Figure 7.7 illustrates a case where the object point is real and the corresponding image point is virtual.

7.2.2.1.1 Real Image of an Axial Object Point at Infinity

Dividing both sides of (7.48) by $(ll')^2$, we get

$$\left\{\frac{A}{(ll')^2}\right\}(Y^2+Z^2)^2 + \left\{\frac{B}{(ll')^2}\right\}Z(Y^2+Z^2) + \left\{\frac{C}{(ll')^2}\right\}(Y^2+Z^2) + \left\{\frac{D}{(ll')^2}\right\}Z^2 + \left\{\frac{E}{(ll')^2}\right\}Z = 0 \tag{7.54}$$

When the object is at infinity, the limiting forms of the coefficients of the equation (7.54) can be obtained from (7.49) – (7.53) as

$$\lim_{l\to\pm\infty}\left\{\frac{A}{(ll')^2}\right\} = 0 \tag{7.55}$$

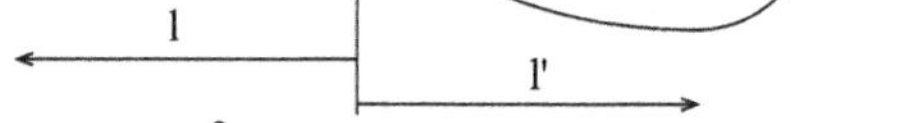

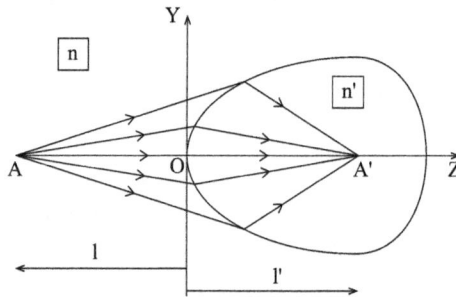

FIGURE 7.6 Cartesian oval with $l' = -l; n' = \frac{3}{2}, n = 1$.

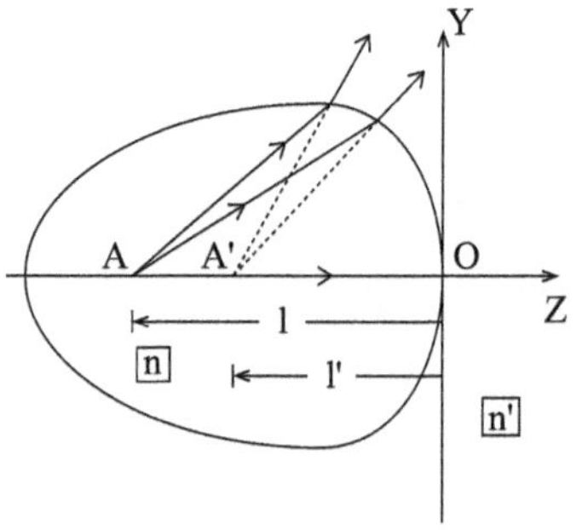

FIGURE 7.7 The other side of the Cartesian oval of Figure 7.6 provides stigmatic imaging of a real point A on the axis as virtual image point A′. $n' = \frac{3}{2}n, n = 1$.

$$\lim_{l\to\pm\infty}\left\{\frac{B}{(ll')^2}\right\}=0 \tag{7.56}$$

$$\lim_{l\to\pm\infty}\left\{\frac{C}{(ll')^2}\right\}=-4\frac{n'^2n^2}{l'^2} \tag{7.57}$$

$$\lim_{l\to\pm\infty}\left\{\frac{D}{(ll')^2}\right\}=4\frac{n^4}{l'^2} \tag{7.58}$$

$$\lim_{l\to\pm\infty}\left\{\frac{E}{(ll')^2}\right\}=8nn'(n'-n)\frac{n}{l'} \tag{7.59}$$

Substituting the limiting values of the coefficients from (7.55) – (7.59) in (7.54), the equation of the corresponding Cartesian oval reduces to a second degree equation as given below

$$n'^2Y^2+\left(n'^2-n^2\right)Z^2-2n'(n'-n)l'Z=0 \tag{7.60}$$

On algebraic manipulation, (7.60) reduces to

$$\frac{\left(Z-l'\dfrac{n'}{n'+n}\right)^2}{l'^2\left(\dfrac{n'}{(n'+n)}\right)^2}+\frac{Y^2}{l'^2\dfrac{(n'-n)}{(n'+n)}}=1 \tag{7.61}$$

Two cases may arise:

Case I. $n'>n$

The equation (7.61) represents an ellipse of the form

$$\frac{\left(Z-\bar{Z}\right)^2}{A^2}+\frac{Y^2}{B^2}=1 \tag{7.62}$$

The centre of the ellipse is at $(0,\bar{z})$. Its semi-major axis A and its semi-minor axis is B. The three parameters $\bar{Z}$, A and B are given by

$$\bar{Z}=l'\frac{n'}{n'+n}\quad A^2=l'^2\left\{\frac{n'}{(n'+n)}\right\}^2\quad B^2=l'^2\frac{(n'-n)}{(n'+n)} \tag{7.63}$$

Figure 7.8 shows the ellipse with its centre at C and the two foci S_1 and S_2. Using (7.63), the eccentricity e of the ellipse can be expressed as

$$e^2=1-\frac{B^2}{A^2}=\left(\frac{n}{n'}\right)^2 \tag{7.64}$$

So that $e=\left(\dfrac{n}{n'}\right)$. The distance OS_2 is given by

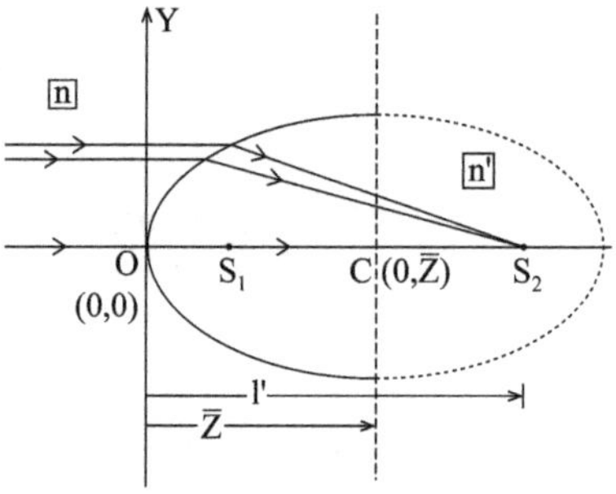

FIGURE 7.8 A stigmatic real image at S_2 of an axial object at infinity formed by one part of a Cartesian oval, which is an ellipse with centre at C and second focus at S_2. $(n' > n)$.

$$OS_2 = OC + CS_2 = A + eA = l' \tag{7.65}$$

Therefore, the stigmatic image of the axial object point at infinity is formed at the second focus S_2 of the ellipse.

Case II.$n' < n$

In order to make the denominator of the second term in (7.61) positive, the equation is rearranged as

$$\frac{\left(Z - l'\dfrac{n'}{n'+n}\right)^2}{l'^2\left(\dfrac{n'}{(n'+n)}\right)^2} - \frac{Y^2}{l'^2\dfrac{(n-n')}{(n'+n)}} = 1 \tag{7.66}$$

(7.66) represents the equation of a hyperbola of the form

$$\frac{(Z-\bar{Z})^2}{A^2} - \frac{Y^2}{B^2} = 1 \tag{7.67}$$

The three parameters $\bar{Z}$, A and B are given by

$$\bar{Z} = l'\frac{n'}{n'+n} \quad A^2 = l'^2\left\{\frac{n'}{(n'+n)}\right\}^2 \quad B^2 = l'^2\frac{(n-n')}{(n'+n)} \tag{7.68}$$

Figure 7.9 shows the hyperbola with its centre at C and the two foci S_1 and S_2. Using (7.68), the eccentricity e of the hyperbola can be expressed as

$$e^2 = 1 + \frac{B^2}{A^2} = \left(\frac{n}{n'}\right)^2 \tag{7.69}$$

So that $e = \left(\dfrac{n}{n'}\right)$. The distance OS_2 is given by

$$OS_2 = OC + CS_2 = A + eA = l' \tag{7.70}$$

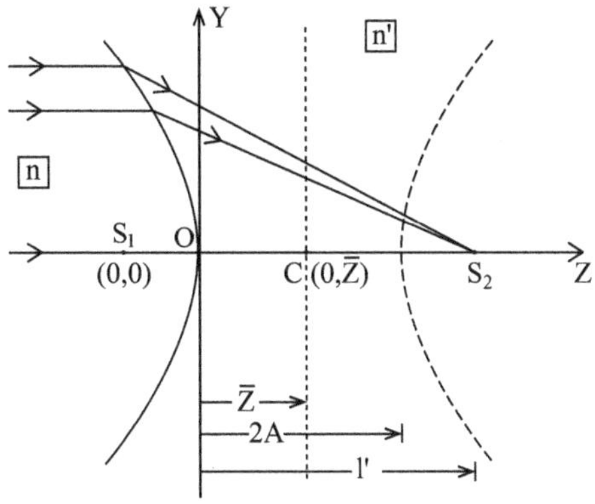

FIGURE 7.9 A stigmatic real image at S_2 of an axial object at infinity formed by one part of a Cartesian oval, which is a hyperbola with centre at C and second focus at S_2. $(n' < n)$.

Therefore, the stigmatic image of the axial object point at infinity is formed at the second focus S_2 of the hyperbola.

7.2.2.1.2 Virtual Image of an Axial Object Point at Infinity

The general equation for the conic section for this case can be obtained from (7.61) by substituting $-l'$ for l', and it is given by

$$\frac{\left(Z + l'\frac{n'}{n'+n}\right)^2}{l'^2\left(\frac{n'}{(n'+n)}\right)^2} + \frac{Y^2}{l'^2\frac{(n'-n)}{(n'+n)}} = 1 \tag{7.71}$$

The centre C of the conic section discussed in the earlier section is located on the right of the vertex (also the origin O of the coordinate system) of the refracting interface so that

$$OC = \bar{Z} = l'\frac{n'}{n'+n} \tag{7.72}$$

In the present case, the centre C of the conic section is located on the left of the vertex, so that

$$OC = -\bar{Z} = -l'\frac{n'}{n'+n} \tag{7.73}$$

However, the values of A^2, B^2 and eccentricity e remain the same as given earlier. The other part of the ellipse/hyperbola becomes the refracting interface, and the stigmatic virtual image of the axial object point at infinity is formed at the first focus S_1. The two cases corresponding to (i) $n' > n$, and (ii) $n' < n$ are illustrated in Figure 7.10 and Figure 7.11, respectively.

7.2.2.2 Cartesian Mirror for Stigmatic Imaging of Finite Conjugate Points

Figure 7.12 shows a Cartesian mirror forming stigmatic virtual image A' of a virtual object point at A. The vertex O of the mirror lies on the axis $A'A$, and it is considered the origin of a rectangular coordinate system. The Z-axis lies along OA. The XY plane is tangential to the axially symmetric mirror surface at

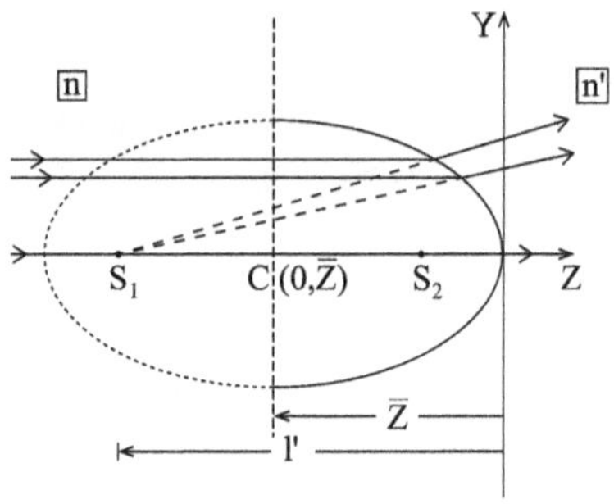

FIGURE 7.10 The other part of the Cartesian oval of Figure 7.8 forms a stigmatic virtual image at S_1 of an axial object point at infinity. S_1 is the first focus of the ellipse. $(n' > n)$.

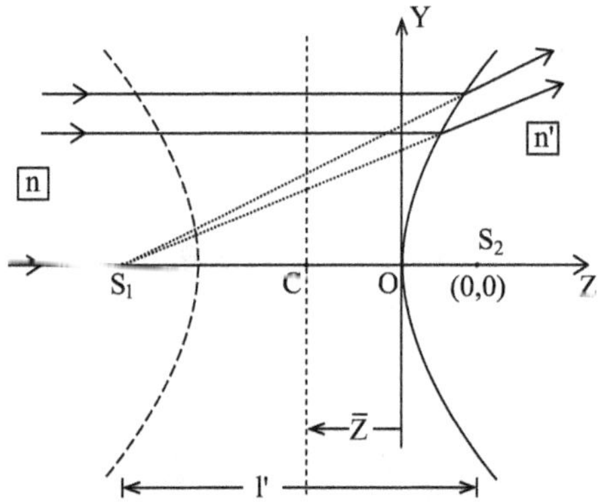

FIGURE 7.11 The other part of the Cartesian oval of Figure 7.9 forms a stigmatic virtual image at S_1 of an axial object point at infinity. S_1 is the first focus of the hyperbola. $(n' < n)$.

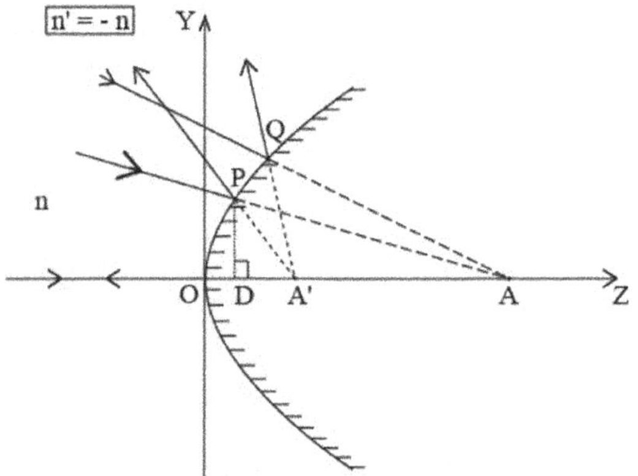

FIGURE 7.12 A Cartesian mirror forming a stigmatic virtual image A′ of a virtual object point A on the axis.

O. The distance of the point A in the object space of refractive index n from the origin O is l. The image point A′ is at a distance l′ from O. OPQ is the YZ section of the Cartesian mirror. The refractive index of the image space is $n' = -n$. Let P (Y, Z) be an arbitrary point on the YZ section of the mirror. PD is perpendicular from P on the axis OA.

By the condition of equality of optical path lengths along all rays between an object point and its stigmatic image point,

$$[APA'] = [AOA'] \tag{7.74}$$

It follows

$$[AP] + [PA'] = [AO] + [OA'] \tag{7.75}$$

With due regard for sign convention

$$AP = -PA \quad AO = -OA = -l \tag{7.76}$$

Substituting from (7.76), we get from (7.75)

$$PA + PA' = OA + OA' \tag{7.77}$$

since $n' = -n$.

Using geometry, (7.72) is given by

$$\sqrt{Y^2 + (l - Z)^2} + \sqrt{Y^2 + (l' - Z)^2} = l + l' \tag{7.78}$$

By squaring both sides and after algebraic manipulation, we obtain the following equation of second degree representing the Cartesian oval for the mirror

$$(l' + l)^2 Y^2 + 4l'lZ^2 - 4(l' + l)l'lZ = 0 \tag{7.79}$$

Alternatively, the general equation (7.48) for a refracting Cartesian interface may be utilized to obtain the equation for a Cartesian mirror by substituting $n' = -n$ in the expressions for the coefficients given in (7.49) – (7.53). It follows, for the case of reflection,

$$A = B = 0 \tag{7.80}$$

$$C = -4n^4 (l' + l)^2 \tag{7.81}$$

$$D = 4n^4 (l' - l)^2 \tag{7.82}$$

$$E = 16n^4 l'l(l' + l) \tag{7.83}$$

Substituting these values for the coefficients A, B, C, D, and E in (7.48), the same result as (7.79) will follow.

7.2.2.2.1 Parabolic Mirror for Object/Image at Infinity

Dividing both sides of (7.79) by $(l'l)^2$ we get

$$\left(\frac{1}{l} + \frac{1}{l'}\right)^2 Y^2 + 4\left(\frac{1}{l'l}\right)Z^2 - 4\left(\frac{1}{l} + \frac{1}{l'}\right)Z = 0 \tag{7.84}$$

When the object is at infinity, $l \rightarrow \pm\infty$, and the expression for the Cartesian mirror reduces to

$$Y^2 = 4l'Z \tag{7.85}$$

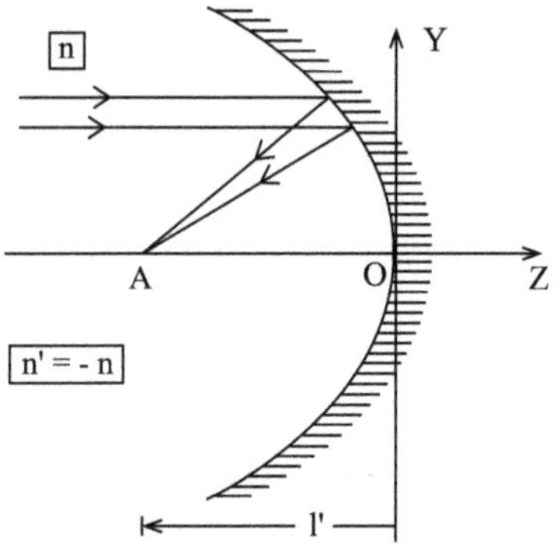

FIGURE 7.13 For an axial object point at infinity, the Cartesian mirror is a paraboloid. A stigmatic image is formed at the focus of the parabola.

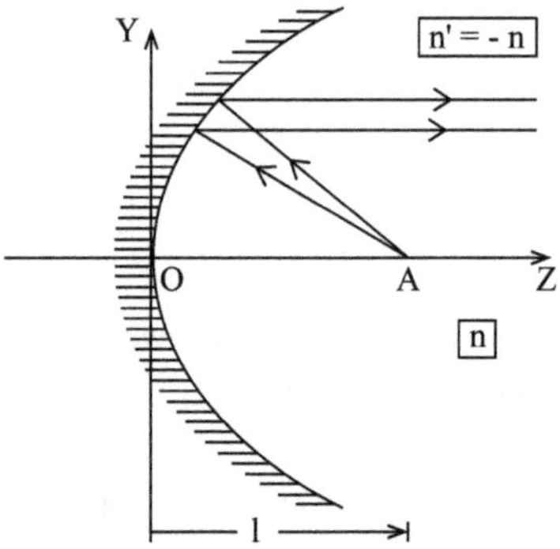

FIGURE 7.14 A point object at the focus of a parabolic Cartesian mirror is imaged stigmatically as an axial image point at infinity.

so that the stigmatic image is formed at the focus A' of a parabola of focal length $l' = OA'$ as shown in Figure 7.13.

When the image is at infinity, $l' \to \pm\infty$, and the expression for the Cartesian mirror is

$$Y^2 = 4lZ \tag{7.86}$$

Figure 7.14 illustrates the case.

7.2.2.2.2 Perfect Imaging of 3-D Object Space by Plane Mirror

For the special case of imaging by reflection, where for any value of l, $l' = -l$, we may obtain the corresponding Cartesian surface from (7.84) as

$$Z = 0 \tag{7.87}$$

Therefore, a plane mirror produces a stigmatic image of the whole object space (Figure 7.15). However, the image of a real object is virtual, and vice versa.

7.2.3 Perfect Imaging of Three-Dimensional Domain

Notwithstanding the great ability of a plane mirror to provide a perfect image of any three-dimensional object, it is important to note that this astonishing feat has two major limitations. By using any number of

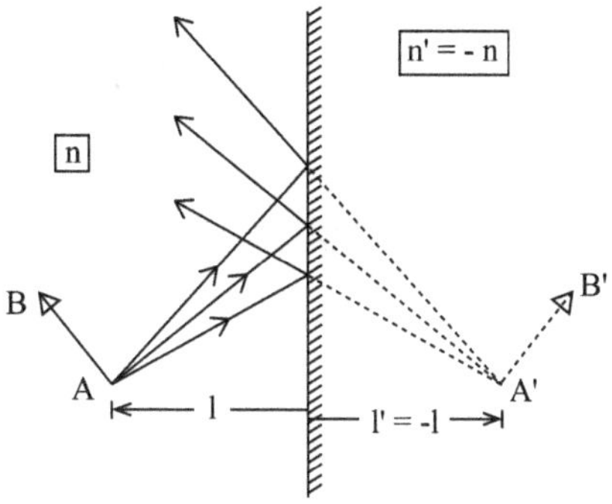

FIGURE 7.15 A plane mirror is a Cartesian oval of degree zero. A three-dimensional real object is stigmatically imaged as a virtual object, and vice versa.

reflections by plane mirrors, the absolute value of magnification cannot be altered from unity, and it is not possible to form a real image of a real object. Attempts to obtain perfect images of three-dimensional objects by a single optical component or any combination of them have also not been successful. In 1858, Maxwell addressed the fundamental question of feasibility of formation of perfect imagery by axisymmetric optical systems and their characteristics in his paper entitled 'On the general laws of optical instruments' [68]. For the first time, this analysis of optical image formation did not invoke any physical image-forming mechanism, and was based on geometrical considerations alone. He gave nine propositions, which may be classified into two groups. The first seven propositions laid down the conditions for obtaining a 'perfect' instrument for optical imaging, and the determination of its characteristics. The approach turned out to be identical with the paraxial model, developed a few years earlier by Carl Friedrich Gauss [69]. The eighth and the ninth propositions dealt with the practical infeasibility of perfect imaging in general, by any kind of optical system. The following two subsections deal with these aspects in more detail.

7.2.3.1 Sufficiency Requirements for Ideal Imaging by Maxwellian 'Perfect' Instrument

Maxwell hypothesized the imaging by an axially symmetric optical instrument to be 'perfect' or 'ideal' when certain conditions, as specified below, are fulfilled.

'A centred optical system is perfect if objects in any plane at right angles with its optical axis are depicted sharply and without distortion in a conjugate image plane also at right angles with the optical axis' [68].

For these 'perfect' or 'ideal' imaging systems, Maxwell proposed a theorem which was rephrased by Conrady [70] and can be stated as:

> A sufficient condition for an optical imaging system to be perfect for all object distances is that it satisfies the above condition of perfect imaging for objects at only two different planes at right angles with the optical axis.

The value of this theorem has also been duly emphasized by Conrady. While in absence of this theorem it would be an endless task to establish that a given system satisfies the criteria of perfect imaging for all conjugate locations, acceptance of the theorem enables checking universal perfection of a given system by testing it for only two object distances. The paraxial or Gaussian model of image formation provides perfect or ideal image formation in the sense defined here. But it is important to note that this model is valid in the limit of both the aperture and field of the imaging system tending to zero. As shown earlier in Chapter 3, the paraxial characteristics of an optical system are uniquely determined from the imaging properties for two pairs of conjugate planes. Following Welford [71], we present a simple derivation of Maxwell's theorem on perfect imaging.

Let two transverse object planes be located at O_1 and O_2 on the optical axis in the object space of an axially symmetric imaging system (Figure 7.16). The corresponding images are located on the transverse

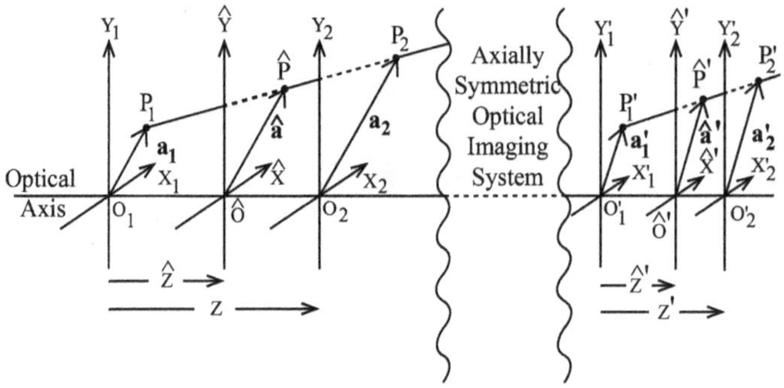

FIGURE 7.16 Three transverse object planes and corresponding image planes in an axially symmetric optical imaging system.

planes at O_1' and O_2', respectively, in the image space. It is assumed that the imaging for these two pairs of conjugate planes is perfect in the sense defined above. Let the position vector of an arbitrary point P_1 on the object plane at O_1 be given by $\mathbf{a}_1 \equiv (X_1, Y_1)$, and let the position vector of its image P_1' on the image plane at O_1' be $\mathbf{a}_1' \equiv (X_1', Y_1')$. The two vectors are related by

$$\mathbf{a}_1' = m_1 \mathbf{a}_1 \tag{7.88}$$

where m_1 is the transverse magnification between the object and image planes at O_1 and O_1', respectively.

Similarly, let the position vectors for an arbitrary point P_2 on the object plane at O_2 be $\mathbf{a}_2 \equiv (X_2, Y_2)$, and its image P_2' on the image plane at O_2' be $\mathbf{a}_2' \equiv (X_2', Y_2')$. Let m_2 be the transverse magnification between the object and image planes at O_2 and O_2', respectively. The two vectors are related by

$$\mathbf{a}_2' = m_2 \mathbf{a}_2 \tag{7.89}$$

The assumption of perfect image formation between the two pairs of conjugate planes implies that any ray passing through P_1 in the object space passes through the point P_1' in the image space. Similarly, any ray passing through the point P_2 in the object space passes through the point P_2' in the image space. Thus, the ray P_1P_2 of the object space transforms into the ray $P_1'P_2'$ in the image space.

Let us consider a third transverse plane located at point $\hat{O}$ on the axis in the object space. The ray P_1P_2 intersects this plane at the point $\hat{P}$. The position vector $\hat{\mathbf{a}}$ of the point $\hat{P}$ is given by

$$\hat{\mathbf{a}} = \left(\frac{\hat{Z}}{Z}\right)(\mathbf{a}_2 - \mathbf{a}_1) + \mathbf{a}_1 \tag{7.90}$$

where $O_1O_2 = Z$, and $\mathbf{O}_1\hat{O} = \hat{Z}$. If all rays passing through this point $\hat{P}$ in the object space also pass through a single point $\hat{P}'$ in the image space, and this point $\hat{P}'$ lies on a fixed transverse plane for a particular choice of $\hat{Z}$ in the object space, then the position vector $\hat{a}'$ of the point $\hat{P}'$ must also be of the form

$$\hat{\mathbf{a}}' = \left(\frac{\hat{Z}'}{Z'}\right)(\mathbf{a}_2' - \mathbf{a}_1') + \mathbf{a}_1' \tag{7.91}$$

where $O_1'\hat{O}' = \hat{Z}'$, and $O_1'O_2' = Z'$. Also, the two position vectors $\hat{\mathbf{a}}$ and $\hat{\mathbf{a}}'$ must be related by a magnification formula similar to (7.88) or (7.89) as

$$\hat{\mathbf{a}}' = \hat{m}\hat{\mathbf{a}} \tag{7.92}$$

$\hat{m}$ defines the transverse magnification between the object plane at $\hat{O}$ and the corresponding image plane at $\hat{O}'$. Equations (7.90) and (7.91) can be rearranged as

$$\hat{\mathbf{a}} = \left(\frac{\hat{Z}}{Z}\right)\mathbf{a}_2 + \left(1 - \frac{\hat{Z}}{Z}\right)\mathbf{a}_1 \tag{7.93}$$

$$\hat{\mathbf{a}} = \left(\frac{m_2\hat{Z}'}{\hat{m}Z'}\right)\mathbf{a}_2 + \left(\frac{m_1}{\hat{m}}\right)\left(1 - \frac{\hat{Z}'}{Z'}\right)\mathbf{a}_1 \tag{7.94}$$

These two equations can be consistent if the following relations hold

$$\hat{m}\left(\frac{Z'}{Z}\right) = m_2\left(\frac{\hat{Z}'}{\hat{Z}}\right) \tag{7.95}$$

$$m_1\left(1 - \frac{\hat{Z}'}{Z'}\right) = \hat{m}\left(1 - \frac{\hat{Z}}{Z}\right) \tag{7.96}$$

The equations (7.95) and (7.96) can be solved to obtain

$$\hat{Z}' = \frac{m_1 Z'\hat{Z}}{\left\{m_1\hat{Z} + m_2\left(Z - \hat{Z}\right)\right\}} \tag{7.97}$$

$$\hat{m} = \frac{m_2 Z}{\left\{m_1\hat{Z} + m_2\left(Z - \hat{Z}\right)\right\}} \tag{7.98}$$

For a set of given values for $Z, Z', \hat{Z}, m_1$, and m_2 equations (7.97) – (7.98) determine the parameters of the perfect image $\hat{P}'$ corresponding to the object point $\hat{P}$. Note that a different value of $\hat{Z}$ corresponds to different values for $\hat{Z}'$, and $\hat{m}$, representing a different pair of conjugate planes. Therefore, it is shown that perfect or ideal image formation for two pairs of conjugate planes implies perfect or ideal image formation for all other planes.

7.2.3.2 Impossibility of Perfect Imaging by Real Rays in Nontrivial Cases

From the propositions of Maxwell, another theorem follows. It states that 'in an absolute optical instrument the optical length along any curve in the object space is equal to the optical length of its image'. 'Optical length' implies the product of geometrical length and refractive index. Maxwell suggested this theorem for the special case when both the object and the image spaces are homogeneous. More rigorous proofs were provided later by Bruns [72], Klein [73], Whittaker [74], and Carathéodory [75–76]. Following Carathéodory, Born and Wolf [77] derived the theorem for heterogeneous object and image spaces. The elegant derivation is not reproduced here, and interested readers may look into ref. [77] for the same.

Figure 7.17 elucidates this theorem. A curve $C_o\left(\equiv A_oB_o\right)$ in the object space is imaged as a curve $C_i\left(\equiv A_iB_i\right)$ in the image space. If the object and image spaces are of heterogeneous refractive index $n_o\left(s_o\right)$ and $n_i\left(s_i\right)$, respectively, according to this theorem it follows that the corresponding optical path lengths are equal.

$$\int_{A_0}^{B_0} n_0\left(s_0\right)ds_0 = \int_{A_i}^{B_i} n_i\left(s_i\right)ds_i \tag{7.99}$$

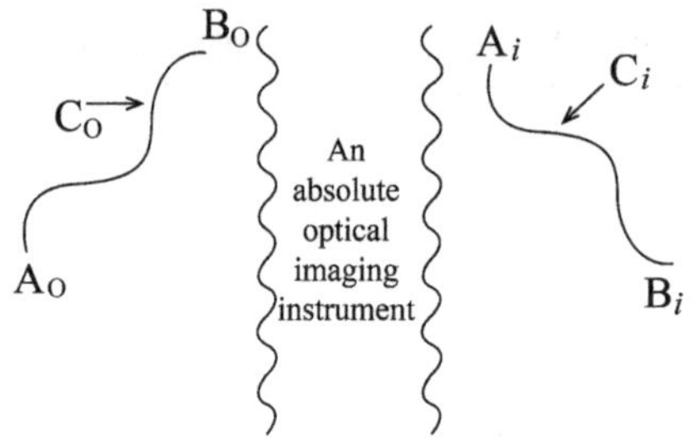

FIGURE 7.17 A curve C_0 in the object space is imaged as a curve C_1 in the image space by an absolute optical imaging instrument.

In the case that both the object space and the image spaces are of homogeneous refractive index n_o and n_i, respectively, i.e. $n_0(s_0) \equiv n_0$ and $n_i(s_i) \equiv n_i$, the optical length relationship reduces to

$$n_o ds_0 = n_i ds_i \tag{7.100}$$

ds_o represents an infinitesimal line element in the object space; ds_i is its image in the image space. It follows that the magnification (ds_i / ds_o) between any two conjugate line elements is equal to (n_o / n_i), and in particular, when $n_o = n_i$, the magnification becomes unity. It follows that '[a] perfect imaging between two homogeneous spaces of equal refractive indices is always trivial in the sense that it produces an image which is congruent with the object' [77]. Therefore, nontrivial imaging by an optical instrument between two homogeneous spaces of equal refractive indices cannot be perfect.

7.2.3.3 Maxwell's 'Fish-Eye' Lens and Luneburg Lens

An absolute instrument providing stigmatic imaging of a three-dimensional domain can be realized in an inhomogeneous medium with spherical symmetry. Maxwell [78] proposed a refractive index function for an inhomogeneous medium with spherical symmetry as

$$n(r) = \frac{1}{1+\left(\frac{r}{a}\right)^2}\hat{n} \tag{7.101}$$

where a and $\hat{n}$ are constants, and r is the distance of the point from a fixed point in the medium.

Figure 7.18 shows the variation of refractive index $n(r)$ with r. It is interesting to note that all rays in this medium follow a circular path, and the circle lies on a plane containing the fixed point, or origin, of the spherically symmetric medium. Proof of this assertion can be derived from equation (2.167) given in Section 2.8.2, and further details can be found in [79–80]. Figure 7.19 illustrates stigmatic imaging of the point P_o at the point P_i in a Maxwell fish-eye medium with the origin at O. Let $OP_o = r_o$ and $OP_i = r_i$. For any circular ray path, we get

$$r_o r_i = a^2 \tag{7.102}$$

Since the medium is of spherical symmetry around the origin at O, it follows that any sphere with radius r_o about the origin O is imaged as a conjugate sphere of radius r_i about the origin with a magnification $\left\{-\frac{r_i}{r_o}\right\}$. The negative sign implies an inversion.

Shafer [81] suggested a modification of the Maxwell fish-eye to obtain a perfect real image of a flat object at a particular wavelength. The image is also flat and the transverse magnification is unity. Note

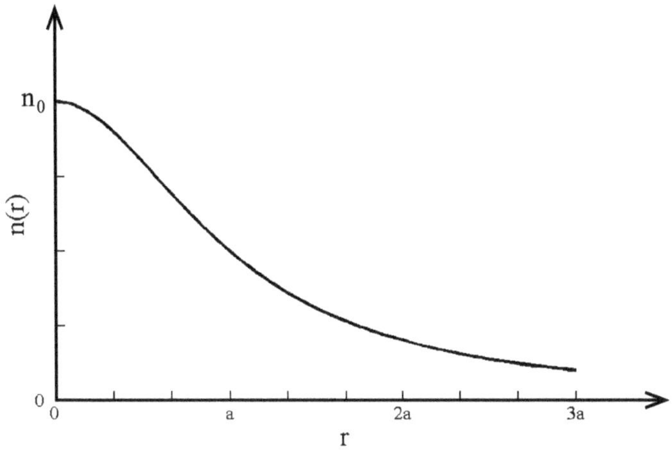

FIGURE 7.18 Variation of a refractive index with a radius in Maxwell's fish-eye lens for $\hat{n} = n_0$.

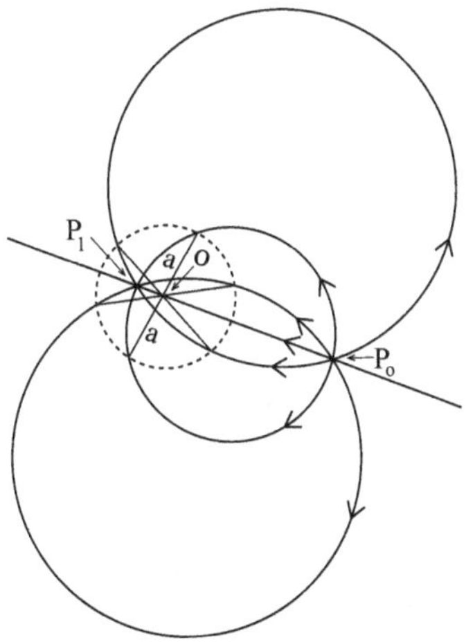

FIGURE 7.19 Few illustrative ray paths for stigmatic imaging of a point P_O at P_i by a Maxwell's fish-eye lens.

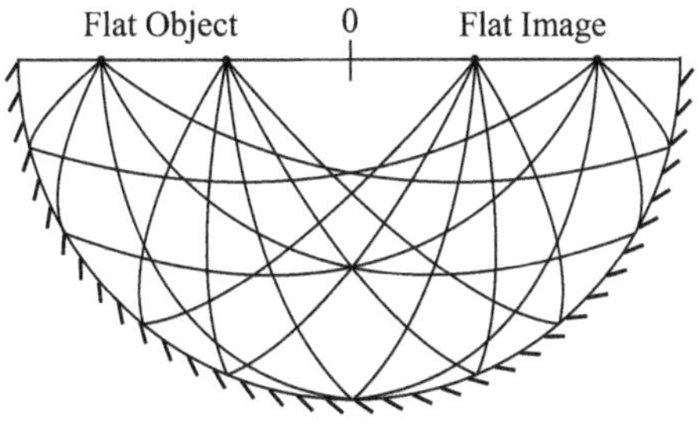

FIGURE 7.20 Perfect imaging of a flat object as a flat image by a Maxwell-Shafer fish-eye lens.

that a plane mirror can also provide a perfect flat image of a flat object with unity magnification, but if the object is real, the image is virtual, and vice versa. The Maxwell-Shafer fish-eye lens consists of one hemisphere of the spherical Maxwell fish-eye lens. A reflecting surface is put on the curved surface of the hemisphere, and the flat object is put on one half of the semi-circular plane surface of the hemisphere; the flat image is formed on the other half of the semicircle (Figure 7.20).

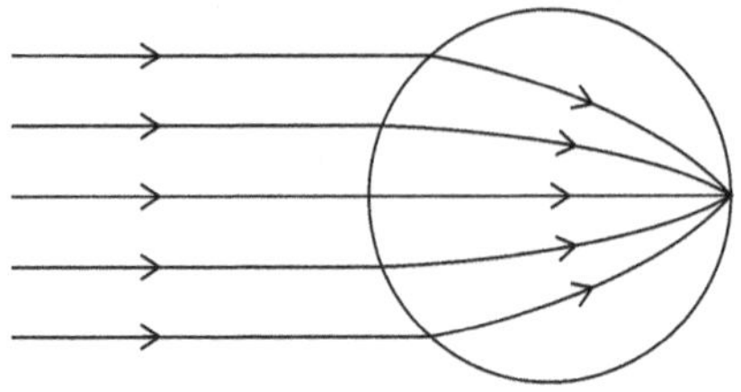

FIGURE 7.21 A spherically symmetric Luneburg lens brings every incident pencil of parallel rays to a perfect focus on the opposite face of the sphere.

Luneburg [82] proposed a generalized form of fish-eye for spherically symmetric medium, where refractive index of the medium is defined as

$$n(r) = \frac{r^{\gamma-1}}{1+\left(\frac{r}{a}\right)^{2\gamma}}\hat{n} \tag{7.103}$$

For $\gamma = 1$, (7.103) reduces to (7.101). It should be noted that in case of the fish-eye, both the object and the image spheres are located inside the inhomogeneous media. Therefore, 'Maxwell fish-eye lens' is also often mentioned as 'Maxwell fish-eye medium'. Luneburg [83] attempted to overcome this limitation, and reported a spherical lens of radius a with refractive index $n(r)$, given by

$$n(r) = \sqrt{2-\left(\frac{r}{a}\right)^2} \tag{7.104}$$

The refractive index of this spherically symmetric inhomogeneous lens is $\sqrt{2}$ at the centre, and decreases gradually to 1, when $r = a$. The lens has the interesting property that every incident pencil of parallel rays is brought to a perfect focus on the opposite face of the sphere (Figure 7.21). Although this lens has not been used so far in bulk optics in visible wavelengths, it has been used in microwave optics and ultrasonic systems [84]. It is also being explored for effective use in integrated optics, and silicon photonics in particular. Several investigations have been carried out to develop 'Generalized Luneburg lenses', which are also known as 'spherically symmetric lenses' in microwave literature. They can also be used for finite conjugate applications. Another significant feature is that they can also circumvent the restriction of image formation strictly on the surface of the lens [85–92]. Inhomogeneous lenses with discontinuities have also been explored. New investigations on the topic are continuing; for more information, see [93–95].

7.2.3.3.1 A Polemic on 'Perfect Imaging' by Maxwell's Fish-Eye Lens

The term 'perfect imaging' used in this chapter follows the prevalent practice in traditional optical engineering, and was defined in Section 7.2 strictly under the realm of ray optics. It is important to note that, in practice, even these perfect images cannot evade the degrading effects of diffraction in real optical systems, so that the ultimate imaging quality of optical systems is 'diffraction limited'. This aspect will be discussed with more detail in Chapter 16. At this point, we note that one of the challenging goals of practical optical imaging is to devise ways for overcoming this limitation. Currently, in optics literature, the term 'perfect imaging' is often being used for defining images whose quality is not degraded by diffraction limitations. Taking a cue from a proposal by Veselago [95] on artificially structured materials with negative refractive indices, Pendry [96] came out with his suggestion for a perfect lens that can produce images that are not subjected to diffraction limitations. For obvious reasons, the possibility of utilizing negative refractive index materials has opened up new frontiers in optics [97–99].

In 2009, Leonhardt [100] claimed that negative refraction is not a mandatory requirement for 'perfect imaging' that overcomes the diffraction limitations, and it can also be achieved with positive refraction. Maxwell's fish-eye provides perfect imaging from a ray optics point of view, and Leonhardt proposed a set up using Maxwell's fish-eye for obtaining perfect images that are not subject to diffraction limitations. This claim has given rise to a polemic that has been brewing for the last few years with evidences and arguments both in favour and against the claim [101–118].

7.2.4 Perfect Imaging of Surfaces

In Section 7.2.3.2, it was shown that the perfect imaging of three-dimensional objects by an optical instrument can only be of the trivial kind, e.g. congruent imaging by a plane mirror, if the object space and the image space are homogeneous. In conjunction with the Maxwellian proposition on sufficiency requirements for the perfect imaging of three-dimensional spaces, this observation on the impossibility of nontrivial perfect imaging of 3-D spaces leads to the conclusion that in an axially symmetric imaging system, two transverse planes cannot be simultaneously imaged perfectly. A corollary implies that, in practice, in axially symmetric systems, perfect imaging is possible for only one set of conjugate planes.

Obviously, this conclusion on conjugate 'planes' cannot be extended in a straightforward manner to conjugate 'surfaces'. Investigations on the feasibility of perfect imaging for more than one set of conjugate surfaces by optical systems in homogeneous object and image space media show that it is indeed possible in certain restricted cases, e.g. when the object and image surfaces are quadratics [119–123].

7.2.4.1 Aplanatic Surfaces and Points

A significant special case of refraction by a spherical interface is the existence of aplanatic spherical surfaces that are perfect images of each other in the object space and the image space.

In Figure 7.22, OP is the YZ section of an axially symmetric refracting interface with n and n′ being refractive indices of the object space and the image space, respectively. The Z-axis is along the axis AO, and XY plane is tangential to the surface at O. Let the axial object point A be imaged as O′. Without the loss of generality, the following treatment is carried out in two dimensions. Let the coordinates of the point P be (y, z). As shown earlier in Section 7.2.2.1, the condition for stigmatic imaging of A at A′ is

$$[\mathrm{APA'}] = [\mathrm{AOA'}] \tag{7.105}$$

This can be rewritten as

$$\mathrm{n.AP} + \mathrm{n'.PA'} = \mathrm{n.AO} + \mathrm{n'.OA'} \tag{7.106}$$

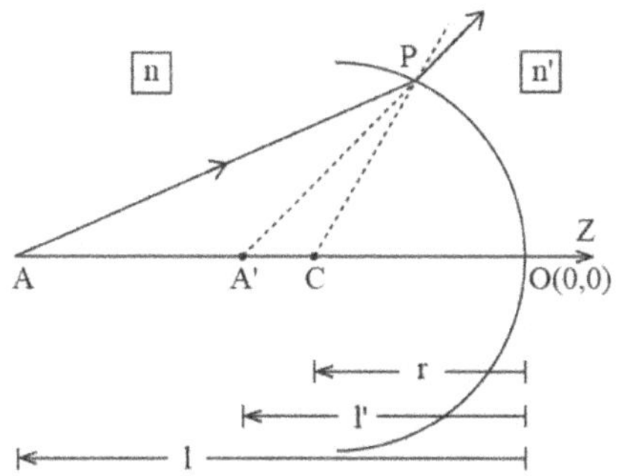

FIGURE 7.22 Aplanatic points A, A′ of a spherical refracting interface: Real A, virtual A′.

Let $OA = l$, and $OA' = l'$. (7.101) reduces to

$$n\sqrt{Y^2 + (Z - l)^2} + n'\sqrt{Y^2 + (Z - l')^2} = -nl + n'l' \tag{7.107}$$

If A and A' are so chosen that

$$n'l' = nl \tag{7.108}$$

then the right-hand side of (7.107) is zero, and we get

$$n\sqrt{Y^2 + (Z - l)^2} = -n'\sqrt{Y^2 + (Z - l')^2} \tag{7.109}$$

The equation of the refracting interface OP for this particular case is obtained by squaring both sides of the equation (7.104) as

$$Y^2 + \left\{Z + \frac{n}{(n' + n)} l\right\}^2 = \left\{\frac{nl}{(n' + n)}\right\}^2 \tag{7.110}$$

This is the equation of a circle of radius $r = \left\{\frac{nl}{(n' + n)}\right\}$ and centre C at $\left\{0, \frac{n}{(n' + n)} l\right\}$. Using (7.108) we get

$$l = \left(\frac{n'}{n}\right) r + r \qquad l' = \left(\frac{n}{n'}\right) r + r \tag{7.111}$$

It follows

$$CA = \left(\frac{n'}{n}\right) r \qquad CA' = \left(\frac{n}{n'}\right) r \tag{7.112}$$

The points A and A′ lie on the axis of the circle OP of centre C. By axial symmetry, OP turns out to be the YZ section of a sphere whose centre is at C, and the points A and A′ lie on the axis OC, passing through the vertex O of the sphere. The points A and A′ are called the aplanatic points of the sphere [124–125]. On account of the symmetry of the sphere, there are not only two aplanatic points but infinitely many, located on two concentric spheres with the centre at C [126]. Figures 7.23 and 7.24 show aplanatic spherical surfaces AR and A′R′ for the two cases $n' > n$, and $n' < n$.

In addition to this perfect imagery between the two surfaces AR and A′R′, the spherical refracting interface is imaged perfectly on itself, and the centre C is also imaged perfectly on itself with large aperture angles. The aplanatic points or surfaces of a spherical refracting interface are used in construction of lens systems of high aperture, e.g. microscope objectives.

Note that in the case of both n and n′ being positive, both the aplanatic points and surfaces cannot be real or virtual. If one of them is real, the other one would be virtual, and vice versa. It is interesting to note that in case of one of the media having a negative refractive index it would be possible to have a real spherical image surface that is aplanatic corresponding to a real spherical object surface [127]. Figures 7.25 and Figure 7.26 present schematic diagrams for the two illustrative cases: (a) $n = 1, n' = -1.5$, and (b) $n = 1, n' = -0.5$.

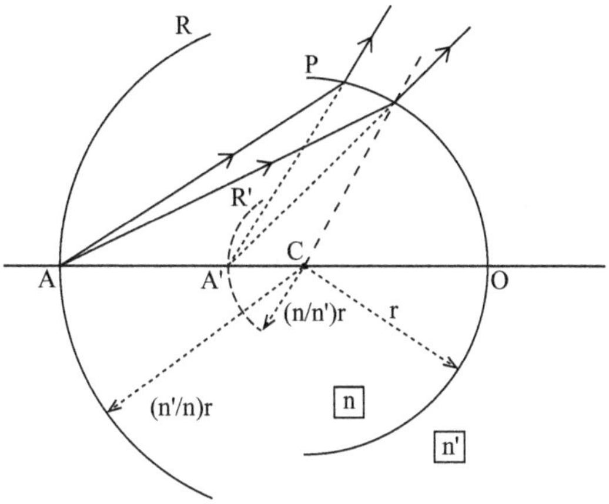

FIGURE 7.23 Aplanatic spherical surfaces AR and A′R′ in the object space and image space of refraction by a spherical interface of radius r. $(n' > n)$.

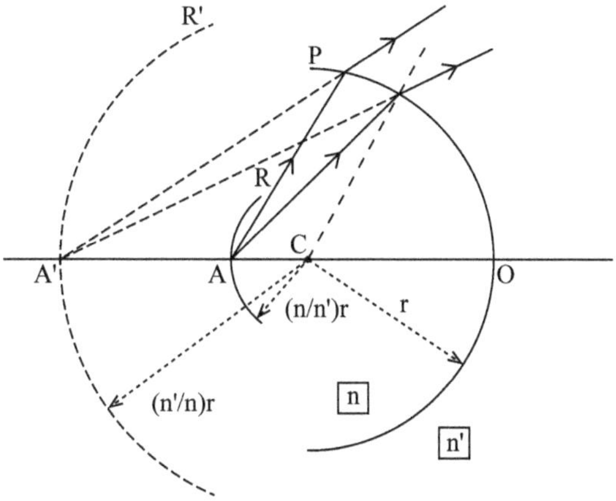

FIGURE 7.24 Aplanatic spherical surfaces AR and A′R′ in the object space and image space of refraction by a spherical interface of radius r. $(n' < n)$.

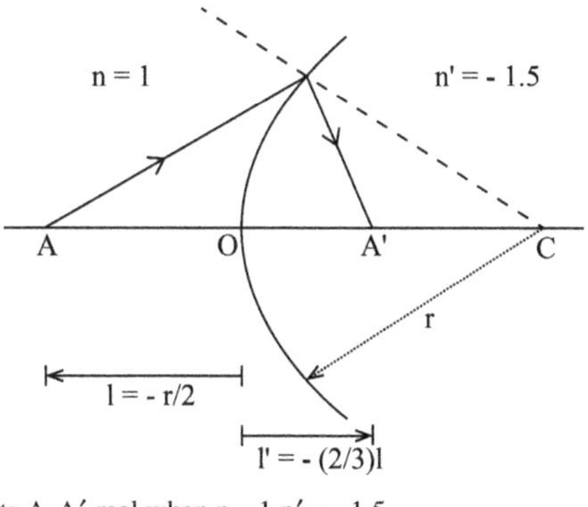

FIGURE 7.25 Both aplanatic points A, A′ real when $n = 1, n' = -1.5$.

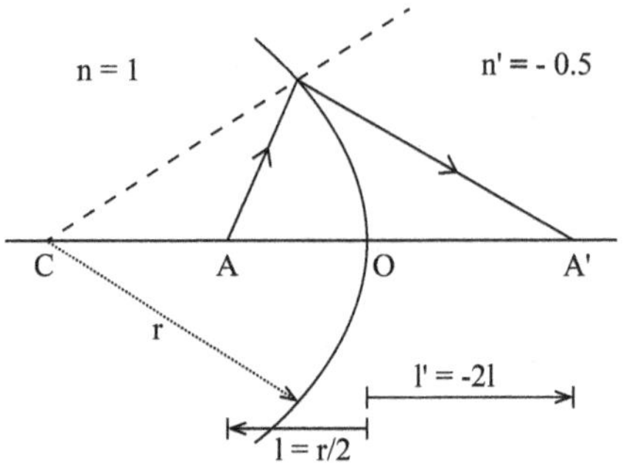

FIGURE 7.26 Both aplanatic points A, A′ real when n = 1, n′ = −0.5

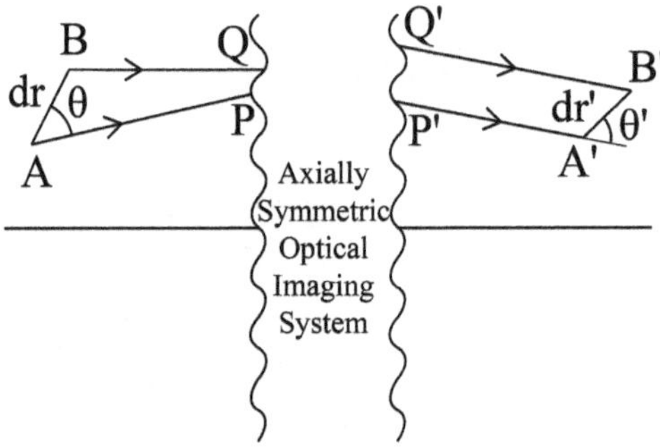

FIGURE 7.27 Perfect imaging of two neighbouring points by an axially symmetric optical imaging system.

7.2.5 Stigmatic Imaging of Two Neighbouring Points: Optical Cosine Rule

Let an object point A be imaged perfectly as an image point A′ (Figure 7.27). Let us assume that another object point B is also imaged perfectly at an image point B′. If the point B is located infinitesimally close to the point A so that AB = **dr**, then the image point B' also lies infinitesimally close to the image point A′, and A′B′ is represented by **dr′**. Let AP…P′A′ and BQ…Q′B′ be two real ray paths from A to A′ and B to B′, respectively. **s** is a unit vector along AP, and **s**' is a unit vector along P′A′. Using the path differential theorem from Section 2.9.5, the difference in the two optical path lengths [BQ…Q′B′] and [AP…P′A′] is given by

$$[BQ\ldots Q'B'] - [AP\ldots P'A'] = n'\mathbf{s}'.\mathbf{dr}' - n\mathbf{s}.\mathbf{dr} = n'dr'\cos\theta' - ndr\cos\theta \tag{7.113}$$

In the above, θ is the angle between the vectors **dr** and **s**, and θ′ is the angle between the vectors **dr′** and **s′**. The refractive indices of the object space and the image space are n and n′, respectively. Consider another arbitrary ray, say, AR…R′A′ from A to A′. The angles θ and θ′ with respect to **dr** and **dr′** will be different for this ray. Nevertheless, the value of the right-hand side in equation (7.113) remains unchanged, for the optical path length from A to its perfect image A′ is the same along all ray paths between the two points. Therefore, the left-hand side of the equation (7.113) remains the same. Defining differential magnification β between the two infinitesimal line elements as

$$\beta = \frac{dr'}{dr} \tag{7.114}$$

It follows from (7.113)

$$n'\beta\cos\theta' - n\cos\theta = C \tag{7.115}$$

C is a constant whose value can be determined by calculating the left-hand side of the above equation for any convenient ray from A to A′. This relation is known as 'The Optical Cosine Law'. This general theorem for perfect imagery in optical systems was put forward by T. Smith [128–131]. Many relations of practical importance can be derived from the optical cosine law [132–135].

7.2.5.1 Abbe's Sine Condition

The optical sine condition was discovered by Ernst Abbe in pursuance of his investigations on improving the quality of microscopic imaging [136]. However, Clausius [137] embarked upon the same condition from thermodynamic considerations before Abbe, and Helmholtz [138] presented a mathematical derivation of the sine condition in this context. The role of sine condition in optical image formation came to the fore since publication of Abbe's next report [139]. Hockin [140] gave a proof using optical path differences. Everett [141] made interesting comments on this proof.

The sine condition can be derived from the optical cosine law. Figure 7.28 shows an axially symmetric optical imaging system with its axis ASS′A′. Let the axial point A in the object space of refractive index be imaged stigmatically at A′ on the axis in the image space of refractive index n′. AB, an infinitesimally small line element, is perpendicular to the axis at A. The condition that the point B is also imaged stigmatically at B′, where B′ lies at the edge of an infinitesimal line element A′B′ that is perpendicular to the axis at A′ in the image space, can be derived by applying the optical cosine law.

APP′A′ is an arbitrary ray from A to A′. The components of this ray in the object space and the image space make large angles U and U′ with the axis.

Let $\theta = \angle BAP, \theta' = \angle B'A'P', AB = d\eta, A'B' = d\eta'$. From (7.115) we can write

$$n'\frac{d\eta'}{d\eta}\cos\theta' - n\cos\theta = C \tag{7.116}$$

For the axial ray ASS′A′, the inclination angles $\angle BAS = \angle B'A'S' = \frac{\pi}{2}$. It follows $C = 0$. Substituting $\theta = \frac{\pi}{2} - U$ and $\theta' = \frac{\pi}{2} - U'$ in (7.116) we get the optical sine condition for finite conjugate imaging as

$$n d\eta \sin U = n' d\eta' \sin U' \tag{7.117}$$

This condition ensures perfect imagery of $d\eta$, an infinitesimally small line element placed perpendicular to the axis, by pencils of large aperture angles.

If the object is at infinity, the object space turns into a star space, and a change of variables is called for by obtaining determinate expressions in the same manner as shown earlier, in Section 4.7.1.1. Let the entrance pupil be located in the object space at E on the axis. Let TS be a section of a sphere with A as

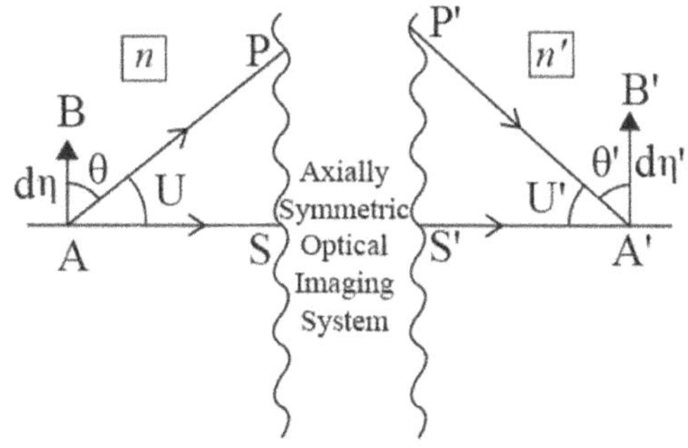

FIGURE 7.28 Abbe sine condition.

centre, and $R(= EA)$ as radius. TD is perpendicular to T on the axis. Let $TD = \rho$, and $\angle BEA = d\beta$. Note that ρ represents a finite semi-aperture of a large semi-diameter, but $d\beta$ represents an angular semi-field of view of infinitesimally small angle. Keeping in view the sign convention,

$$\sin U = -\frac{\rho}{R} \text{ and } d\eta = R \tan d\beta \approx R d\beta \tag{7.118}$$

Changing the variables of the object space from η, U to β, ρ, the left-hand side of (7.104) takes the form

$$n d\eta \sin U = -n d\beta \rho \tag{7.119}$$

Therefore, for the object at infinity, Abbe's sine condition is given by

$$-n d\beta \rho = n' d\eta' \sin U' \tag{7.120}$$

Similarly, if the image is at infinity, Abbe's sine condition takes the form

$$n d\eta \sin U = -n' d\beta' \rho' \tag{7.121}$$

For afocal or telescopic systems, Abbe's sine condition is expressed as

$$n d\beta \rho = n' d\beta' \rho' \tag{7.122}$$

In a paraxial model, correspondence between similar parameters of the object space and the image space is given by the paraxial invariant (see Section 4.7). The paraxial relation corresponding to (7.117) is given by

$$n u \eta = n' u' \eta' \tag{7.123}$$

where u and u′ are the paraxial convergence angles in the object space and the image space, respectively. Note that under paraxial approximation, the object and image heights η and η' are infinitesimally small, and they are identical with the elements $d\eta$ and $d\eta'$ discussed above. By substituting $\eta \equiv d\eta$, and $\eta' \equiv d\eta'$ we get another form of Abbe's sine condition for finite conjugate imaging from (7.117) and (7.123) as

$$\frac{\sin U}{u} = \frac{\sin U'}{u'} \tag{7.124}$$

For the three special cases of imaging considered above, similar expressions for Abbe's sine condition can be obtained.

Case I. Object at infinity, image at a finite distance

The paraxial relation corresponding to relation (7.120) is

$$-n d\beta h = -n' d\eta' u' \tag{7.125}$$

From (7.120) and (7.125), the following form is obtained for Abbe's sine condition:

$$\frac{\rho}{h} = \frac{\sin U'}{u'} \tag{7.126}$$

Case II. Image at infinity, object at a finite distance

The paraxial relation corresponding to relation (7.121) is

$$n d\eta u = -n' d\beta' h' \tag{7.127}$$

From (7.121) and (7.127), Abbe's sine condition takes the form

$$\frac{\sin U}{u} = \frac{\rho'}{h'} \tag{7.128}$$

Case III. Afocal imaging; both object and image at infinity

The paraxial relation corresponding to relation (7.122) is

$$n d\beta h = n' d\beta' h' \tag{7.129}$$

Abbe's sine condition for the afocal case can also be expressed as

$$\frac{\rho}{h} = \frac{\rho'}{h'} \tag{7.130}$$

The practical importance of Abbe's sine condition goes far beyond the discussion made above as a condition that must be satisfied for the perfect imaging of two infinitesimally close points near the axis of an axisymmetric imaging system to occur. In fact, this condition provides a measure for the feasibility of high-quality imaging of objects of small size by large aperture imaging systems. The following observation is highly significant: 'Since the sine condition gives information about the quality of the off-axial image in terms of the properties of axial pencils it is of great importance for optical design' [142]. Different forms of Abbe's sine condition catering to effective use of the sine condition in the presence of residual aberrations of imaging systems are routinely used in practical synthesis of optical imaging systems [143–151].

7.2.5.2 Herschel's Condition

This condition was proposed by William Herschel about fifty years before the proposal on sine condition [152]. Herschel's condition concerns perfect imaging of two infinitesimally close points placed on the optical axis using pencils of large angle in an axially symmetric optical imaging system.

In Figure 7.29, let the axial point A be imaged at the point A′. G is an axial point which is infinitesimally close to A in the object space of refractive index n. Let point G be imaged at the axial point G′ in the image space of refractive index n′. Let AG = dZ, and A′G′ = dZ′ APP′A′ is an arbitrary ray from A

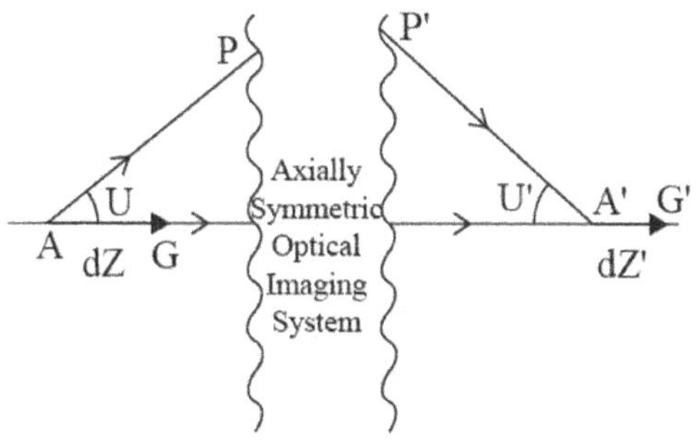

FIGURE 7.29 Herschel condition.

to A′. The components of this ray in the object space and the image space make large angles U and U′ with the axis, respectively.

The sufficient condition for the perfect imaging of the two points A, G at the axial points A′, G′ can be obtained by using the optical cosine law. From (7.115) we can write,

$$n'\frac{dZ'}{dZ}\cos\theta' - n\cos\theta = C \tag{7.131}$$

For the axial ray AGA′G′, $\theta = \theta' = 0$. Therefore, C is given by

$$C = n'\frac{dZ'}{dZ} - n \tag{7.132}$$

From (7.131) and (7.132), for the ray APP′A′ we get the Herschel condition

$$\frac{dZ'}{dZ} = \left(\frac{n}{n'}\right)\left(\frac{\sin\frac{U}{2}}{\sin\frac{U'}{2}}\right)^2 \tag{7.133}$$

This implies that the perfect imaging of the infinitesimally small axial line element dZ as another axial line element dZ' in the image space by pencils of large angles is possible if and only if the finite convergence angles U and U′ satisfy the relation (7.133).

Herschel's condition as enunciated above is concerned with the perfect imaging of axial line elements. Indeed, in tune with the original purpose of the Herschel condition, attempts have been made to extend the condition in a form that enables determining the axial range over which quality of axial imaging remains the same, even though the imaging is not perfect [153–154].

For both conditions, namely the Abbe sine condition and the Herschel condition, the angle U represents a large angle. In the two cases, perfect imaging by wide angle pencil calls for satisfaction of conditions (7.117) or (7.133), respectively, for all values of U in the range (0, U).

7.2.5.3 Incompatibility of Herschel's Condition with Abbe's Sine Condition

From the equation (7.117) expressing Abbe's sine condition, we find that an infinitesimally small transverse line element $d\eta$ placed on the axis in the object space is imaged perfectly as a transverse line element $d\eta'$ on the axis by pencils of large angles if and only if the finite convergence angles U and U′ satisfy the relation

$$\frac{d\eta'}{d\eta} = \left(\frac{n}{n'}\right)\left(\frac{n\sin U}{n'\sin U'}\right) \tag{7.134}$$

On the other hand, the relation (7.133) gives a condition for the perfect imaging of an axial line element dZ in the object space as an axial line element dZ′Type equation here. in the image space.

From (7.133) and (7.134), it is seen that both these conditions can be satisfied simultaneously only in the special case when $U = \pm U'$, implying that absolute values for both the transverse magnification and the axial magnification are equal to $\left(\frac{n}{n'}\right)$. Except for this special case, an axially symmetric optical imaging system that forms a perfect image of a small transverse element by pencils of large angle cannot also form a perfect image of an axial line element concurrently (Figure 7.30).

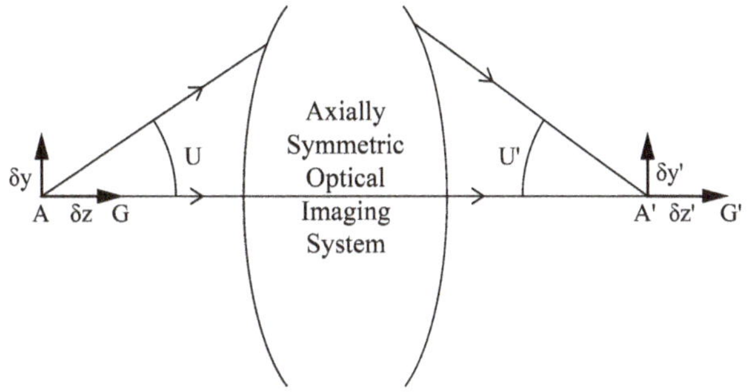

FIGURE 7.30 Abbe sine condition and Herschel condition are mutually incompatible except in the special case when $U = \pm U'$.

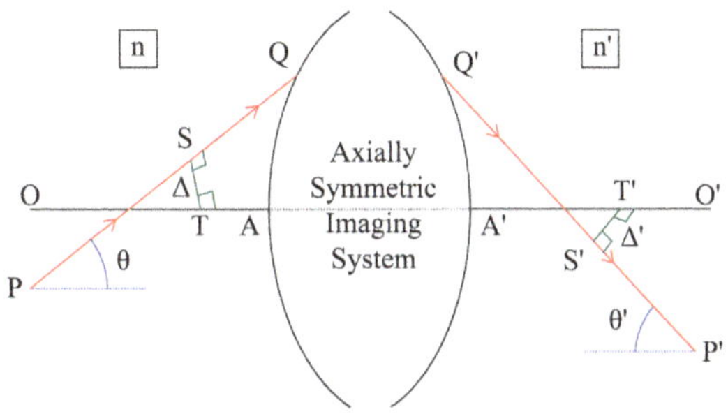

FIGURE 7.31 PQ and P′Q′ are the components of a skew ray PQP′Q′ in the object space and the image space of an axially symmetric optical imaging system. Length of the line of shortest join between the skew lines OA and PQ in the object space is Δ, and the corresponding parameter in image space is Δ'.

7.3 Real Ray Invariants

The optical invariant enunciated in Section 4.7 is a paraxial invariant, i.e., it is an invariant applicable for paraxial rays in an optical system. A similar invariance relationship between components of real rays in the object space, image space, and intermediate spaces of optical imaging systems plays an important role in the analysis and synthesis of these systems. Some of these relations are presented below.

7.3.1 Skew Ray Invariant

Figure 7.31 shows the object and the image spaces of an axially symmetric imaging system. Refractive indices of the two spaces are n and n′, respectively. OAA′O′ is the optical axis of the system. PQQ′P′ is a skew ray with its components PQ and Q′P′ in the object and the image spaces. The optical axis and the skew ray are nonintersecting, and this characteristic of the skew ray is highlighted by using red colour for the skew ray. ST is the line of shortest join between the two skew lines OA and PQ in the object space, and S′T′ is the line of shortest join between the two skew lines A′O′ and Q′P′ in the image space. It should be noted that the line of shortest join between two skew lines is perpendicular to both skew lines. In the object space, the angle between the two skew lines, OA and PQ is θ, and in the image space the angle between the two skew lines A′O′ and Q′P′ is θ'. Let $ST = \Delta$, and $S'T' = \Delta'$.

Using ray tracing formulae for spherical surfaces, T. Smith [155–156] showed that the product $(n\Delta \sin\theta)$ remains invariant in all spaces during propagation of a skew ray from the object to the image space of an axially symmetric optical imaging system. With reference to Figure 7.31, this invariance may be expressed as

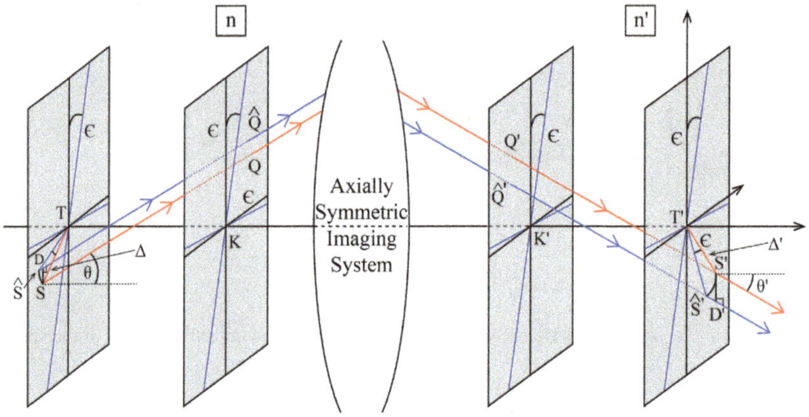

FIGURE 7.32 Derivation of skew ray invariance relationship.

$$n\Delta\sin\theta = \ldots = n_i\Delta_i\sin\theta_i = \ldots = n'\Delta'\sin\theta' \tag{7.135}$$

where the suffix 'i' refers to the corresponding quantity in the i-th intermediate space, and the corresponding unprimed and primed quantities refer to their values in the initial object space and the final image space, respectively.

The product $(n\Delta\sin\theta)$ for a skew ray in an axially symmetric imaging system is called the 'Skew ray invariant', or in short 'Skew invariant'. This invariant provides a measure of the skewness of a skew ray. The 'Skewness' $\mathcal{S}$ of a skew ray can be specified as

$$\mathcal{S} = n\Delta\sin\theta \tag{7.136}$$

The value of $\mathcal{S}$ remains the same in all spaces of an axially symmetric imaging system. Obviously, for any meridional ray $\mathcal{S} = 0$, since $\Delta = 0$ for any meridional ray.

Welford [157–158] derived this invariance for any symmetrical optical system from considerations of symmetry and the implicit use of the Malus-Dupin theorem. A similar derivation is presented below.

7.3.1.1 A Derivation of the Skew Ray Invariance Relationship

TKK′T′ is the optical axis of an axially symmetric imaging system (Figure 7.32). SQQ′S′ is a skew ray. The refractive indices of the object space and the final image space are n and n′, respectively. ST is the line of shortest join between the segment of the skew ray in the object space, i.e. the line SQ and the optical axis. Similarly, S′T′ is the line of shortest join between the optical axis and the segment Q′S′ of the skew ray in the image space. Let ST = Δ, and S′T′ = Δ′.

By definition, $ST \perp SQ$, $ST \perp TK$, $S'T' \perp Q'S'$ and $S'T' \perp K'T'$. If the figure, including the axes, is rotated by an angle ϵ, we obtain the ray $\hat{S}\hat{Q}\hat{Q}'\hat{S}'$. Note that $\hat{S}$ and $\hat{S}'$ are the points of intersection of the latter ray with the transverse planes at T and T′, respectively.

If the angle ϵ is small,

$$[SQ\ldots Q'S'] = \left[\hat{S}\hat{Q}\ldots\hat{Q}'\hat{S}'\right] + D(2) \tag{7.137}$$

where $D(2)$ indicates terms involving degree 2 and above in ϵ. Let D be the foot of perpendicular SD on $\hat{S}\hat{Q}$ from S, and D′ be the foot of perpendicular S′D′ on $\hat{Q}'\hat{S}'$ from S′.

By Malus-Dupin theorem,

$$[SQ\ldots Q'S'] = [D\ldots D'] + R(2) \tag{7.138}$$

where $R(2)$ indicates terms involving degree 2 and above in $(S\hat{S}.\epsilon)$. From (7.124) and (7.125), it follows that

$$[\hat{S}D] = [\hat{S}'D'] \tag{7.139}$$

The angles between the ray segments SQ and Q′S′ with the optical axis are θ and θ', respectively. Therefore, the angles $\angle \hat{S}SD$ and $\angle S'\hat{S}'D'$ are also given by

$$\angle \hat{S}SD = \theta, \text{ and } \angle S'\hat{S}'D' = \theta' \tag{7.140}$$

The optical path lengths $[\hat{S}D]$ and $[\hat{S}'D']$ are given by

$$[\hat{S}D] = n.\hat{S}D = n.S\hat{S}.\sin\theta = n.\Delta.\epsilon.\sin\theta = (n\Delta\sin\theta).\epsilon \tag{7.141}$$

$$[\hat{S}'D'] = n'.\hat{S}'D' = n'.S'\hat{S}'.\sin\theta' = n'.\Delta'.\epsilon.\sin\theta' = (n'\Delta'\sin\theta').\epsilon \tag{7.142}$$

From (7.139), (7.141), and (7.142), we get

$$n\Delta\sin\theta = n'\Delta'\sin\theta' \tag{7.143}$$

7.3.1.2 Cartesian Form of Skew Ray Invariant

In any space of an axially symmetric imaging system, a skew ray is specified by its direction cosines and the coordinates of a point on the ray. In Figure 7.33, the object space of an imaging system is of refractive index n. OA is the optical axis and PQ is a skew ray. The point P lies on the transverse plane at O. Let O be the origin of a rectangular Cartesian coordinate system. Let the coordinates of the point P be $(X, Y, 0)$, and the direction cosines of PQ be (L, M, N). The coordinates (x, y, z) of an arbitrary point R on PQ can be expressed as

$$x = X + \frac{L}{N}z \qquad y = Y + \frac{M}{N}z \tag{7.144}$$

Let RD be perpendicular from R on the axis. The square of the length of RD is given by

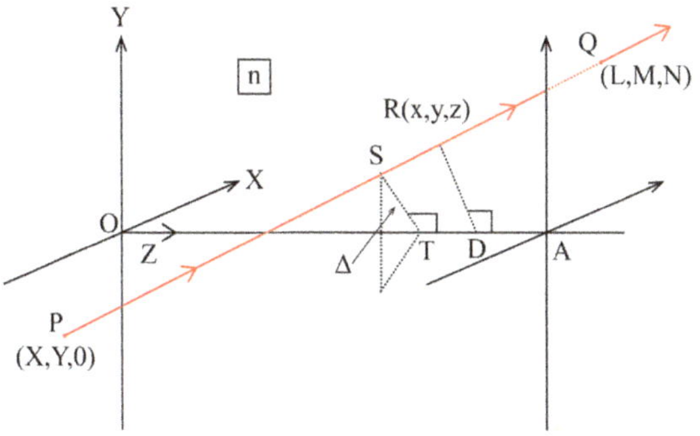

FIGURE 7.33 Parameters of an arbitrary skew ray in the object space of an axially symmetric imaging system.

$$RD^2 = x^2 + y^2 = X^2 + Y^2 + 2\frac{(LX + MY)}{N}z + \frac{(L^2 + M^2)}{N^2}z^2 \tag{7.145}$$

This length varies for different locations of the point R on the line PQ. The value of z, providing minimum value for this length, is obtained from the stationarity condition

$$\frac{d\left[(RD)^2\right]}{dz} = 0 \tag{7.146}$$

It yields

$$z = -\frac{N(LX + MY)}{L^2 + M^2} \tag{7.147}$$

This is the value of z coordinate of the point S where ST is the line of shortest join between the skew ray PQ and the optical axis. The length Δ of this shortest join ST is obtained from (7.145) and (7.147) as

$$\Delta = \frac{(MX - LY)}{\sin\theta} \tag{7.148}$$

Therefore, the Cartesian form of skewness $\mathcal{S}$ of a skew ray in the object space is given by [155]

$$S = n\Delta \sin\theta = n(MX - LY) = n(Mx - Ly) \tag{7.149}$$

Note that x and y are the X and Y coordinates of any point on the skew ray in the object space, where the ray has direction cosines (L, M, N). If the relevant quantities in the i-th intermediate space are represented by the subscript i, and the same in the final image space are represented by primes, from the skew ray invariance relationship we get

$$S = n(Mx - Ly) = \ldots = n_i(M_i x_i - L_i y_i) = \ldots = n'(M'x' - L'y') \tag{7.150}$$

7.3.1.3 *Other Forms of Skew Ray Invariant*

A generalized form of skew invariant was put forward by Herzberger [159–160]. In Figure 7.34, PQ is a skew ray, and OA is the optical axis in the object space of an axially symmetric optical system. Let **k**

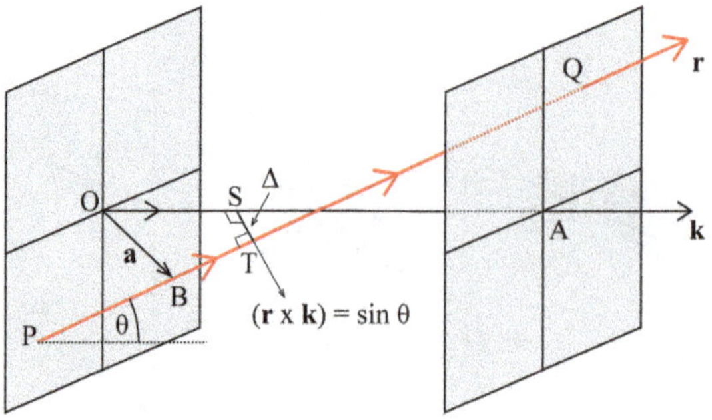

FIGURE 7.34 Generalized skew ray invariant.

be a unit vector along the optical axis OA, and **r** be a unit vector along the skew ray PQ. Also, let **a** be a position vector from any origin on the axis, say O, to any point on the skew ray, say B.

Herzberger stated that the skewness $\mathcal{S}$ of the skew ray is

$$\mathcal{S} = n[\mathbf{a},\mathbf{r},\mathbf{k}] \tag{7.151}$$

where $[\mathbf{a},\mathbf{r},\mathbf{k}]$ represents the scalar triple product of the vectors **a**, **r**, and **k**, and $\mathcal{S}$ remains invariant during propagation through an axially symmetric imaging system. This triple product is also described as the 'optical moment' of the ray around the axis. It may be noted that this skew invariant is equal to the skew invariant proposed by Smith earlier. This may be checked in a straightforward manner.

$(\mathbf{r}\times\mathbf{k})$ is a vector along the common perpendicular ST to the skew ray and the axis; the length of this vector is $\sin\theta$. The projection of the position vector **a** along ST is Δ. Therefore

$$S = n[\mathbf{a},\mathbf{r},\mathbf{k}] = n[\mathbf{a}.(\mathbf{r}\times\mathbf{k})] = n\Delta\sin\theta \tag{7.152}$$

However, the skew invariant proposed by Herzberger is more general, for the choice of the position vector **a** is arbitrary.

From ray tracing equations Buchdahl [161] arrived at the same invariant that he denoted by E_x^*, but did not mention any physical significance of the invariant.

7.3.1.4 Applications of the Skew Invariant

7.3.1.4.1 The Optical Sine Theorem

The optical sine theorem [162–163] provides an invariance relationship for real sagittal rays, and can be derived from the skew ray invariant. Figure 7.35 shows the entrance and the exit pupils at E and E′ on the axis of an axially symmetric imaging system. Let PB…P′B′ be a ray from the axial object point P to the axial image point P′. The ray lies on the XZ plane. The object space segment of this ray, PB, makes a large angle U with the axis, i.e. $\angle BPE = U$. The corresponding angle in the image space is U′. An infinitesimally small object $PQ(=\delta\eta_s)$ lies perpendicular to the XZ plane, and is imaged as $P'Q'(=\delta\eta_s')$. QB…Q′B′ is a skew ray from the point Q in the object space to Q′ in the image space. The refractive indices of the object and the image spaces are n and n′, respectively. Let the skew ray segment QB make an angle θ with the YZ plane in the object space of refractive index n, and the skew ray segment Q′B′ make an angle θ' with the Y′Z′ plane in the image space of refractive index n′. In the object space, let the length of the line of shortest join between the skew ray segment QB and the axis PE be Δ. The corresponding length in the image space is Δ'. By skew ray invariance relation (7.143), it follows

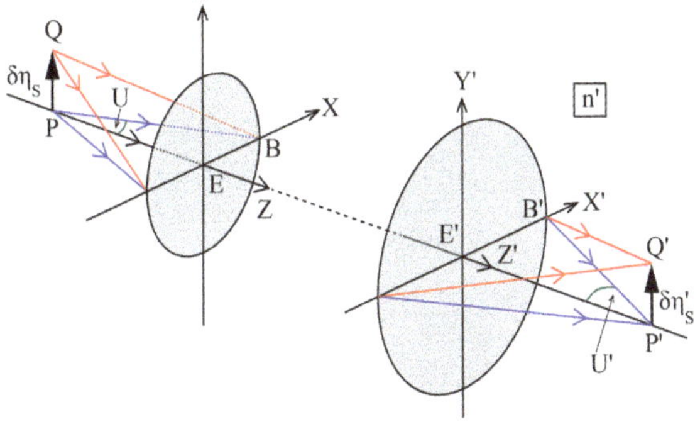

FIGURE 7.35 The optical sine theorem.

$$n\Delta \sin\theta = n'\Delta' \sin\theta' \tag{7.153}$$

For the infinitesimally small object $PQ(=\delta\eta_s)$ placed perpendicular to the XZ plane,

$$\Delta \to \delta\eta_s, \quad \Delta' \to \delta\eta_s', \theta \to U \text{ and } \theta' \to U' \tag{7.154}$$

Therefore, (7.153) reduces to

$$n\delta\eta_s \sin U = n'\delta\eta_s' \sin U' \tag{7.155}$$

This relationship is known as the 'Optical sine theorem'. It gives a magnification for small line objects imaged by rays at large angles in all axially symmetric imaging systems. It is to be noted that, per derivation, the relation is valid only when the object is perpendicular to the plane containing the rays. However, in other cases, the Abbe sine condition, discussed in Section 7.2.5.1, may be utilized.

7.3.1.4.2 Relation between a Point and a Diapoint via a Skew Ray from the Point

In Figure 7.36, YZ plane is the meridian plane of an axially symmetric imaging system. Let the ideal image of a line object OP lying along the Y-axis be O′P′. A skew ray PQ…S′G′ from the off-axis object point P meets the meridian plane Y′Z′ in the image space at the point P_d'. In general, the points P′ and P_d' are different. The point P_d' is called the 'diapoint' corresponding to P via the skew ray PQ…S′G′ [164]. Let PR…T′H′ be another skew ray from P such that on any transverse plane the perpendicular distances of the points of intersection of these two rays from the meridian plane are same, e.g. $|QU| = |RU|$, $|G'V'| = |H'V'|$. Note that, by symmetry, the latter ray also intersects the meridian plane at P_d'. However, it should be noted that a different skew ray from P will intersect the meridian plane at a different point.

Let the direction cosines of the segments of the skew ray PQ…S′G′ in the object and image spaces be (L, M, N) and (L', M', N'), respectively. (X, Y) coordinates of P and P_d' are (0, Y) and $(0, Y_d')$. The refractive indices of the object and image spaces are n and n′, respectively. From (7.149) we get

$$nLY = n'L'Y_d' \tag{7.156}$$

or,

$$nY\cos\alpha = n'Y_d' \cos\alpha' \tag{7.157}$$

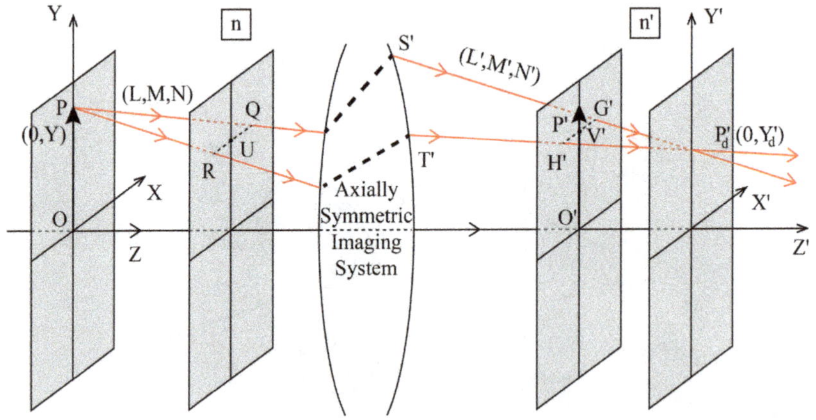

FIGURE 7.36 An object point P and its diapoint P_d' via the skew ray PQ…S′G′.

where α and α′ are the angles made by the segments PQ and S′G′ of the skew ray in the object space and image space with the normal to the meridian plane, i.e. the X-axis. Obviously, the relation can be extended to all diapoints in intermediate spaces.

7.3.1.4.3 Sagittal Magnification Law

Let PE…E′P_s' be the principal ray from the off-axial object point P to the image space of an axially symmetric optical system (Figure 7.37). PS…S′P_s' is a close sagittal ray from the object point to the image space. 'A close sagittal ray' implies that the z-direction cosine of both the sagittal ray and the principal ray is the same, so that the ray segment PE of the principal ray in the object space is an orthogonal projection of PS, the segment of the close sagittal ray in the object space. Similar correspondence exists between the segments of the principal ray and the close sagittal ray in the image space. Considering the principal ray as a base ray, the small angle that the close sagittal ray makes with it is called the parabasal sagittal angle. In Figure 7.37, these angles are u_s and u_s' in the object and the image spaces, respectively. Let OP = η_s, and O′P_s' = η_s'. From (7.149) we get

$$nL\eta_s = n'L'\eta_s' \tag{7.158}$$

L and L′ are the x-direction cosines of the ray segments PS and S′P_s'. Therefore

$$L = \cos\alpha \quad L' = \cos\alpha' \tag{7.159}$$

Since $\alpha = \frac{\pi}{2} - u_s$, and $\alpha' = \frac{\pi}{2} - u_s'$, we get

$$L = \cos\left(\frac{\pi}{2} - u_s\right) = \sin u_s \approx u_s \quad L' = \cos\frac{\pi}{2} - u_s' = \sin u_s' \approx u_s' \tag{7.160}$$

From (7.158), we finally get

$$nu_s\eta_s = n'u_s'\eta_s' \tag{7.161}$$

This is the sagittal magnification law.

7.3.1.4.4 Feasibility of Perfect Imaging of a Pair of Object Planes Simultaneously

In Section 7.2.4, it was noted that in an axially symmetric imaging system the impossibility of nontrivial perfect imaging of 3-D space leads to the conclusion that, in an axially symmetric imaging system, two

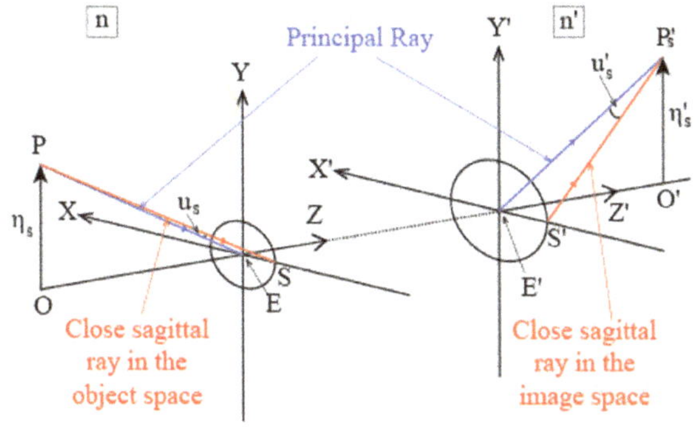

FIGURE 7.37 A sagittal magnification law.

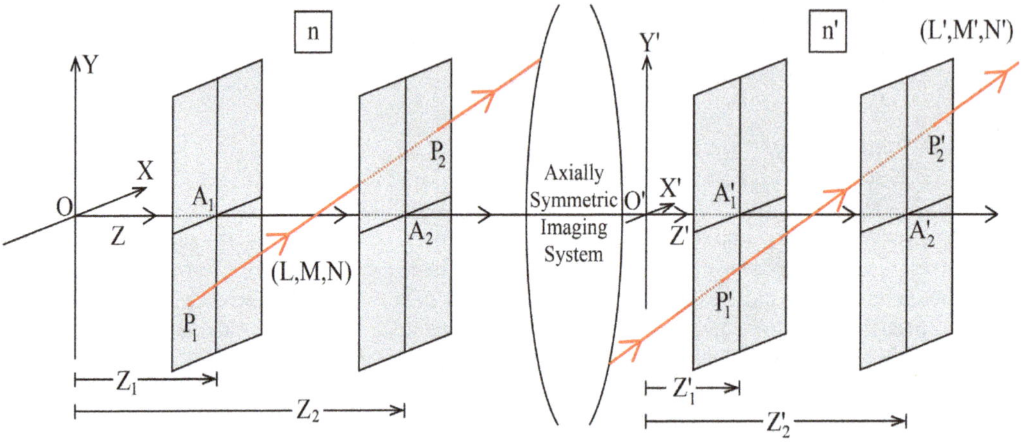

FIGURE 7.38 Impossibility of perfect imaging of two object planes simultaneously.

transverse planes cannot be simultaneously imaged perfectly. The same conclusion follows from the skew ray invariance relationship in a straightforward manner.

Let us postulate that two transverse object planes at A_1 and A_2 on the axis are simultaneously imaged perfectly on transverse planes at A_1' and A_2' with magnifications m_1 and m_2, respectively. The refractive indices of the object space and the image space are n and n′. Point P_1 on the transverse plane at A_1 is imaged perfectly as P_1' on the transverse image plane at A_1'. Similarly, the point P_2 is imaged as P_2' (see Figure 7.38).

With respect to origin O of a rectangular coordinate system in the object space, the coordinates of P_1 and P_2 are (X_1, Y_1, Z_1) and (X_2, Y_2, Z_2), respectively. The coordinates of the points P_1' and P_2' are (m_1X_1, m_1Y_1, Z_1') and (m_2X_2, m_2Y_2, Z_2'), with respect to the origin O′ of a rectangular coordinate system in the image space. Since imaging is assumed to be perfect, the skew ray P_1P_2 in the object space becomes the skew ray $P_1'P_2'$ in the image space. Let the direction cosines of P_1P_2 and $P_1'P_2'$ be (L, M, N) and (L', M', N'), respectively.

The skewness $\mathcal{S}$ of the ray segment P_1P_2 is

$$S = n(MX_1 - LY_1) = \frac{n(X_1Y_2 - X_2Y_1)}{\sqrt{(X_1 - X_2)^2 + (Y_1 - Y_2)^2 + (Z_1 - Z_2)^2}} \tag{7.162}$$

Let $(Z_2 - Z_1) = d$. Then

$$Z_2' - Z_1' = m_1^2 m_2^2 (n'/n)^2 d^2 \tag{7.163}$$

The skewness $\mathcal{S}'$ of the ray segment $P_1'P_2'$ is

$$\mathcal{S}' = n'(M'm_1X_1 - L'm_1Y_1) = n'm_1(M'X_1 - L'Y_1) \tag{7.164}$$

By substitution, this reduces to

$$\mathcal{S}' = \frac{m_1m_2n'(X_1Y_2 - X_2Y_1)}{\sqrt{(m_1X_1 - m_2X_2)^2 + (m_1Y_1 - m_2Y_2)^2 + m_1^2m_2^2(n'/n)^2 d^2}} \tag{7.165}$$

For a lens with nonzero focal length, $m_1 \neq m_2$, so that from (7.162) and (7.165)

$$\mathscr{S} \neq \mathscr{S}' \tag{7.166}$$

This violates the skew ray invariance principle. Therefore, the basic postulate of this analysis is wrong. Perfect imaging of two axially separate object planes by a lens of nonzero focal length is impossible.

For a hypothetical afocal lens, $m_1 = m_2 = m$ (say). The skewness of the skew ray segment in the image space is

$$\mathscr{S}' = \frac{mn'(X_1Y_2 - X_2Y_1)}{\sqrt{(X_1 - X_2)^2 + (Y_1 - Y_2)^2 + m^2(n'/n)^2 d^2}} \tag{7.167}$$

Comparing (7.162) with (7.167), we note $\mathscr{S} \neq \mathscr{S}'$, except for in the special case $m = (n/n')$.

Thus, Maxwell's observations on concurrent perfect imaging of two object planes are vindicated by the use of the skew ray invariance principle. Goodman [165] commented that $(\mathscr{S} - \mathscr{S}')$ may be considered as a measure of the 'degree of impossibility' of perfect imaging of two axially separated object planes by an axially symmetric imaging system.

7.3.2 Generalized Optical Invariant

In Section 4.7, a ray invariant H, which has direct relation with many characteristics of ray propagation through optical systems, was defined. In view of its widespread use in the analysis and synthesis of optical systems, H is often called the 'optical invariant'. However, it should be noted that H is a paraxial invariant, and it can also only be used for axially symmetric optical systems. The 'Generalized Optical invariant', also known as the 'Generalized Lagrange invariant' or the 'Generalized Étendue', circumvents these limitations. This invariant is the ray optics analogue of Liouville's theorem in statistical mechanics [166–170].

Figure 7.39 shows the initial and final spaces of a general optical system. Let P…Q be a real finite ray from the initial to the final spaces. O and O′ are the origins of two unrelated rectangular coordinate systems XYZ and X′Y′Z′ in the two spaces. The points P and Q are chosen to lie on the XY plane in the initial space, and the X′Y′ plane in the final space. The two spaces are of refractive indices n and n′, respectively. The points P and Q have coordinates (X, Y, 0) and $(X', Y', 0)$ in respective coordinate systems. Let the segment of the ray in the initial space have direction cosines (L, M, N). Similarly, (L', M', N') are the direction cosines of the segment of the ray in the final space. If the finite ray is specified by the coordinates

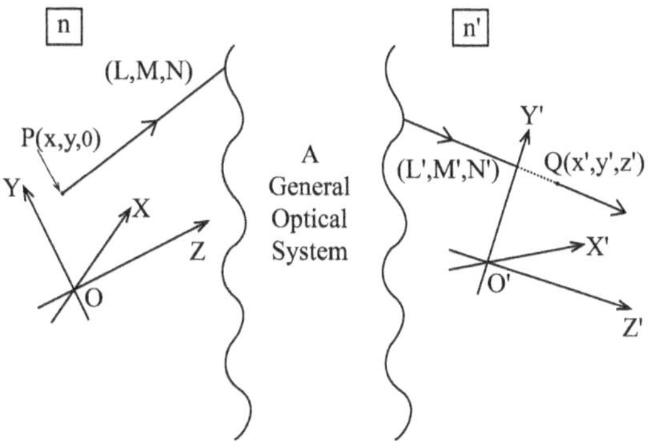

FIGURE 7.39 Initial and final spaces of a general optical system.

of its point of intersection with the XY plane in each space, the ray can be specified with the help of four variables, namely X, Y, L, and M, since only two out of the three direction cosines are independent. Thus, the segment of the ray in the initial space is specified by (X, Y, L, M), and the corresponding segment in the final space is specified by (X', Y', L', M'). A neighbouring ray in the initial space specified by the parameters $(X+dX, Y+dY, L+dL, M+dM)$ corresponds to a ray in the image space specified by $(X'+dX', Y'+dY', L'+dL', M'+dM')$. According to generalized optical invariance relation

$$n^2 dXdYdLdM = n'^2 dX'dY'dL'dM' \tag{7.168}$$

The physical significance of this relation is that the light flux emitted by an element dXdY of a Lambertian source in the XY plane, over a range of solid angle specified by dLdM, remains invariant during propagation through the system. It is assumed that all rays of this infinitesimal pencil get through the system unhindered, and that there are no losses, e.g. absorption or reflection losses, etc. By analogy with similar treatments in other fields, the invariance relation is often recast as

$$dXdYdpdq = dX'dY'dp'dq' \tag{7.169}$$

where the following substitutions have been made:

$$p = nL \quad q = nM \quad p' = n'L' \quad q' = n'M' \tag{7.170}$$

An analogy with Liouville's theorem in statistical mechanics is apparent [168]. In ray optics, the phase space is four-dimensional, optical direction cosines p and q being analogues of generalized momenta, and the spatial coordinates X and Y are the analogues of position coordinates. According to this invariance condition, the volume element dXdYdpdq in phase space that represents this pencil of rays has its volume conserved. It is obvious that the invariance relationship holds good for any of the intermediate spaces of the optical system.

A proof of this relation is given below. The treatment follows the elegant proof given by Welford [166].

7.3.2.1 Derivation of Generalized Optical Invariant

Let us consider the Hamilton-Bruns' point eikonal $S(X, Y, X', Y')$ for the two points P and Q (Figure 7.39). We have, from Section 7.1.1.1,

$$\frac{\partial S(X,Y;X',Y')}{\partial X} = -nL = -p \tag{7.171}$$

$$\frac{\partial S(X,Y;X',Y')}{\partial Y} = -nM = -q \tag{7.172}$$

$$\frac{\partial S(X,Y;X',Y')}{\partial X'} = n'L' = p' \tag{7.173}$$

$$\frac{\partial S(X,Y;X',Y')}{\partial Y'} = n'M' = q' \tag{7.174}$$

The above relations can be represented succinctly as

$$p = -S_X \quad q = -S_Y \quad p' = S_{X'} \quad q' = S_{Y'} \tag{7.175}$$

By differentiating, we get from the above relations

$$dp = -S_{XX'}dX' - S_{XY'}dY' - S_{XX}dX - S_{XY}dY$$

$$dq = -S_{YX'}dX' - S_{YY'}dY' - S_{YX}dX - S_{YY}dY \tag{7.177}$$

$$dp' = +S_{X'X'}dX' + S_{X'Y'}dY' + S_{X'X}dX + S_{X'Y}dY \tag{7.178}$$

$$dq' = +S_{Y'X'}dX' + S_{Y'Y'}dY' + S_{Y'X}dX + S_{Y'Y}dY \tag{7.179}$$

In matrix notation, the relations (7.176) – (7.179) can be represented by the matrix equation

$$\mathbf{Am'} = \mathbf{Bm} \tag{7.180}$$

where **A** and **B** are (4×4) matrices given by

$$\mathbf{A} = \begin{bmatrix} S_{XX'} & S_{XY'} & 0 & 0 \\ S_{YX'} & S_{YY'} & 0 & 0 \\ S_{X'X'} & S_{X'Y'} & -1 & 0 \\ S_{Y'X'} & S_{Y'Y'} & 0 & -1 \end{bmatrix} \quad \mathbf{B} = \begin{bmatrix} -S_{XX} & -S_{XY} & -1 & 0 \\ -S_{YX} & -S_{YY} & 0 & -1 \\ -S_{X'X} & -S_{X'Y} & 0 & 0 \\ -S_{Y'X} & -S_{Y'Y} & 0 & 0 \end{bmatrix} \tag{7.181}$$

m′ and **m** are column vectors as given below:

$$\mathbf{m'} = \begin{bmatrix} dX' \\ dY' \\ dp' \\ dq' \end{bmatrix} \quad \mathbf{m} = \begin{bmatrix} dX \\ dY \\ dp \\ dq \end{bmatrix} \tag{7.182}$$

From (7.180) we get

$$\mathbf{m'} = \left(\mathbf{A}^{-1}\mathbf{B}\right)\mathbf{m} \tag{7.183}$$

This matrix equation can be expanded to

$$dX' = \frac{\partial X'}{\partial X}dX + \frac{\partial X'}{\partial Y}dY + \frac{\partial X'}{\partial p}dp + \frac{\partial X'}{\partial q}dq \tag{7.184}$$

$$dY' = \frac{\partial Y'}{\partial X}dX + \frac{\partial Y'}{\partial Y}dY + \frac{\partial Y'}{\partial p}dp + \frac{\partial Y'}{\partial q}dq \tag{7.185}$$

$$dp' = \frac{\partial p'}{\partial X}dX + \frac{\partial p'}{\partial Y}dY + \frac{\partial p'}{\partial p}dp + \frac{\partial p'}{\partial q}dq \tag{7.186}$$

$$dq' = \frac{\partial q'}{\partial X}dX + \frac{\partial q'}{\partial Y}dY + \frac{\partial q'}{\partial p}dp + \frac{\partial q'}{\partial q}dq \tag{7.187}$$

It is seen that the determinant of the matrix $\left(\mathbf{A}^{-1}\mathbf{B}\right)$ is the Jacobian J

$$\det\left(\mathbf{A}^{-1}\mathbf{B}\right) = J = \frac{\partial\left(X',Y',p',q'\right)}{\partial\left(X,Y,p,q\right)} \tag{7.188}$$

The Jacobian transforms the differential hypervolume element dXdYdpdq of the initial space into the differential hypervolume dX′dY′dp′dq′, so that

$$dX'dY'dp'dq' = \frac{\partial\left(X',Y',p',q'\right)}{\partial\left(X,Y,p,q\right)} dXdYdpdq \tag{7.189}$$

From (7.181) it follows

$$\det\left(\mathbf{A}\right) = \left(S_{XX'}S_{YY'} - S_{XY'}S_{YX'}\right) \tag{7.190}$$

$$\det\left(\mathbf{B}\right) = \left(S_{XX'}S_{YY'} - S_{Y'X}S_{X'Y}\right) \tag{7.191}$$

Since $S_{XY'} = S_{Y'X}$ and $S_{YX'} = S_{X'Y}$, we get

$$\det\left(\mathbf{A}\right) = \det\left(\mathbf{B}\right) \tag{7.192}$$

Therefore, the Jacobian J reduces to

$$J = \det\left(\mathbf{A}^{-1}\mathbf{B}\right) = \det\left(\mathbf{A}^{-1}\right)\det\left(\mathbf{B}\right) = \frac{\det\left(\mathbf{B}\right)}{\det\left(\mathbf{A}\right)} = 1 \tag{7.193}$$

From (7.189) we get

$$dX'dY'dp'dq' = dXdYdpdq \tag{7.194}$$

Thus, the invariance relation is proved. It is important to note that if the Hamilton-Bruns' point eikonal is multivalued between the two points P and Q, the invariant applies separately along each ray between them.

The paraxial invariant H, the optical sine theorem, etc., derived earlier in the context of axially symmetric optical imaging systems can be deduced as special cases of generalized optical invariant. However, the basic importance of the generalized optical invariant stems from the fact that it is applicable for any type of optical system, image-forming or not, and symmetrical or not. Important applications in non-imaging optics and compression of ray bundles can be found in references [170–171].

7.4 Imaging by Rays in the Vicinity of an Arbitrary Ray

It has been elucidated in Chapter 3 that the model of paraxial or Gaussian optics assumes that the rays taking part in image formation lie limitingly close to the optical axis, so much so that the bundle of rays emerging from a point object, or converging to a point image are approximated as homocentric bundles, or equivalently that the corresponding wavefronts are considered as perfectly spherical. Also, in this model, the sag of the surface at the vertex of the refracting/reflecting interface is neglected, and whatever the actual shape of the axially symmetric interface is, the power of the surface is determined by its axial curvature alone. It should be noted that the optical axis is also a ray, and paraxial optics are concerned with rays in the vicinity of this ray. The great success of paraxial optics in rendering effective tools for analysis and synthesis of optical systems prompted exploration of similar models for rays in the vicinity of an arbitrary ray [172–180]. It is obvious that, in general, even a narrow bundle of rays around an arbitrary ray lacks the characteristics of homocentricity or rotational symmetry around that ray, and therefore the local wavefronts become astigmatic.

7.4.1 Elements of Surface Normals and Curvature

In order to describe the phenomenon of astigmatism, the properties of plane curves and continuous surfaces need to be invoked. A brief review of the pertinent theory of surface normals and curvature is presented below for the sake of completeness [181].

7.4.1.1 The Equations of the Normals to a Surface

In rectangular Cartesian coordinate system, a surface is represented by an equation of the form

$$F(X, Y, Z) = 0 \tag{7.195}$$

In principle, this equation can be solved to give Z explicitly in the form

$$Z = f(X, Y) \tag{7.196}$$

This is also another form for representing a surface. Here, Z represents the perpendicular distance of the point (X, Y) on the surface from the XY plane.

In Figure 7.40, let P_0 and P be two neighbouring points on a surface represented by the implicit form (7.195). Let P_0 and P have coordinates (X_0, Y_0, Z_0) and $(X_0 + \delta X, Y_0 + \delta Y, Z_0 + \delta Z)$, and let (L_0, M_0, N_0) be the direction cosines of the normal P_0T at the point P_0 on the surface. The normal P_0T is perpendicular to the line element P_0P. Therefore, the projection of P_0P on the normal must be zero, i. e.

$$L_0\delta X + M_0\delta Y + N_0\delta Z = 0 \tag{7.197}$$

Since both P_0 and P lie on the surface (7.195), we can write

$$\frac{\partial F}{\partial X}\delta X + \frac{\partial F}{\partial Y}\delta Y + \frac{\partial F}{\partial Z}\delta Z = 0 \tag{7.198}$$

Both (7.197) and (7.198) must be satisfied for all values of $(\delta X, \delta Y, \delta Z)$ corresponding to points on the surface in the immediate neighbourhood of the point P_0. This is only possible if

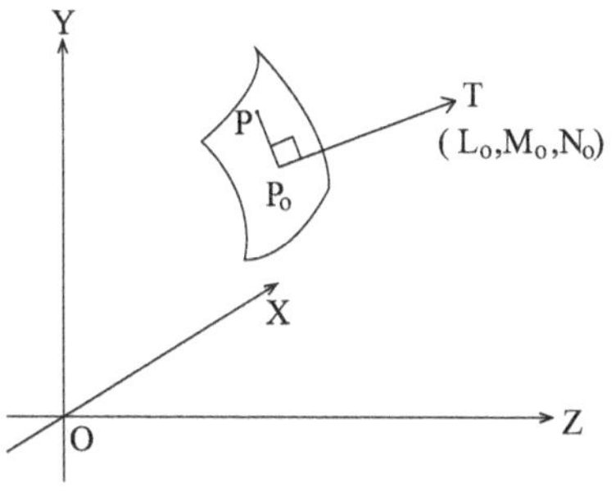

FIGURE 7.40 The normal to a surface.

$$L_0 : M_0 : N_0 = \frac{\partial F}{\partial X} : \frac{\partial F}{\partial Y} : \frac{\partial F}{\partial Z} \tag{7.199}$$

The partial derivatives at P_0 are thus direction ratios of the normal at P_0. The equation of the normal at P_0 on the surface is given by

$$\frac{X - X_0}{\frac{\partial F}{\partial X}} = \frac{Y - Y_0}{\frac{\partial F}{\partial Y}} = \frac{Z - Z_0}{\frac{\partial F}{\partial Z}} \tag{7.200}$$

In two dimensions, say (Y, Z) coordinates, the equation of the normal to a curve at the point (Y_0, Z_0) follows from (7.200) as

$$\frac{Y - Y_0}{\frac{\partial F}{\partial Y}} = \frac{Z - Z_0}{\frac{\partial F}{\partial Z}} \tag{7.201}$$

Note that the partial derivatives specify the direction, but not the sense of the normal. The sense requires a suitable convention to be adopted in any given case.

7.4.1.2 The Curvature of a Plane Curve: Newton's Method

Figure 7.41 shows a plane curve POQ. Suppose it is desired to determine the curvature of the plane curve at O. Let the point O be the origin of a two-dimensional coordinate system where the Y- and the Z-axes are taken along the tangent and the normal to the surface at point O. The equation of the surface can be represented in the form

$$Z = f(Y) \tag{7.202}$$

It has been shown in Section 2.9.1.1 that in the neighbourhood of the point O, this equation may be represented as

$$Z = a_2 Y^2 + O(3) \tag{7.203}$$

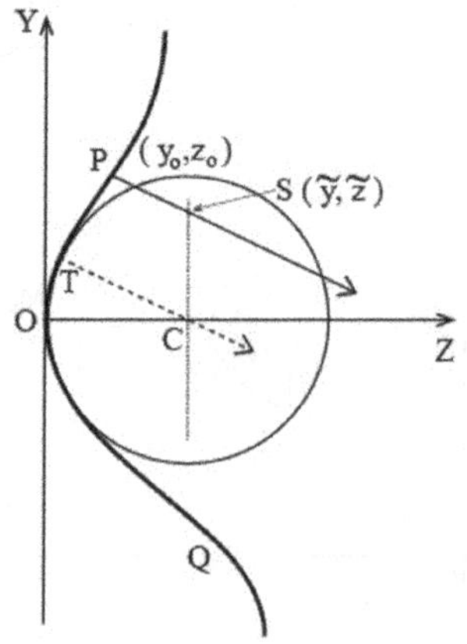

FIGURE 7.41 Curvature of a plane curve.

where $O(3)$ represents terms in degree 3 and above in y.

The equation of a circle of centre C and radius OC = r, as shown in Figure 7.40, is

$$Y^2 + Z^2 - 2rZ = 0 \tag{7.204}$$

This equation may be solved to get

$$Z = \frac{1}{2}cY^2 + \frac{1}{8}c^3Y^4 + \ldots = \frac{1}{2}cY^2 + O(4) \tag{7.205}$$

where $c = (1/r)$ is the curvature of the circle. If c is chosen such that $c = 2a_2$, the circle becomes identical with the curve in the immediate neighbourhood of point O. This is Newton's method for determining the curvature of any plane curve at any point on it. It is required to express the curve in the form (7.203), and then the curvature of the curve at the point is equal to $2a_2$.

The circle defined above is called the circle of curvature at O. All normals to the circle of curvature pass through C, the centre of the circle. Substituting $a_2 = \frac{1}{2}c$ in (7.202), the equation of the curve can be written in parametric form as

$$F(Y,Z) = Z - \frac{1}{2}cY^2 + O(3) \tag{7.206}$$

The partial derivatives of F with respect to Y and Z are given by

$$\frac{\partial F}{\partial Y} = -cY + O(2) \tag{7.207}$$

$$\frac{\partial F}{\partial Z} = 1 + O(2) \tag{7.208}$$

Note that differentiation reduces the quantity of degree n in Y to one of degree $(n-1)$. By substituting the values for the partial derivatives at the point $P(Y_0, Z_0)$, the equation of the normal at any point $P(Y_0, Z_0)$ on the curve POQ can be obtained from (7.201) as

$$\frac{Y - Y_0}{-cY_0 + O(2)} = \frac{Z - Z_0}{1 + O(2)} \tag{7.209}$$

Since $P(Y_0, Z_0)$ is in the neighbourhood of O, from (7.203) it follows that $Z_0 = O(2)$. Therefore, (7.209) reduces to

$$Y = Y_0(1 - cZ) + O(2) \tag{7.210}$$

Let the coordinates of S, the point of intersection of the normal at P with the line parallel to the Y-axis and passing through C be $(\tilde{Y}, \tilde{Z})$. At $\tilde{Z} = r, \tilde{Y} = 0 + O(2) = O(2)$. This implies that normals to the curve in the neighbourhood of O also pass through the centre of the circle of curvature. In this sense, the centre of curvature of a plane curve may be defined to be the point of intersection of limitingly close normals to the curve.

7.4.1.3 The Curvatures of a Surface: Euler's Theorem

In Figure 7.42, O is considered the origin of a rectangular Cartesian coordinate system. The Z-axis lies along OZ, the normal to the surface at the point O, and the X- and the Y-axes lie in the tangent plane

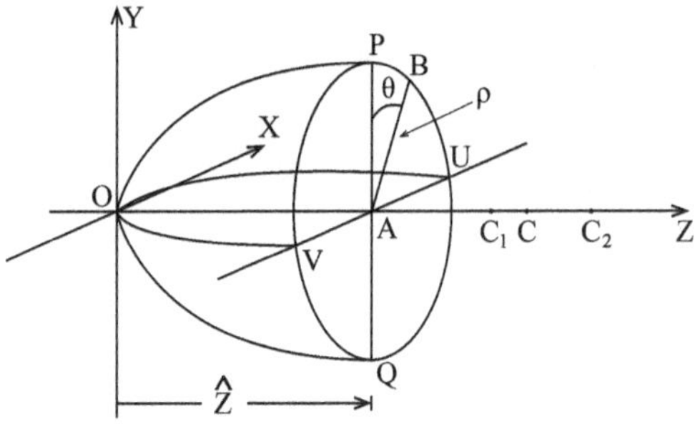

FIGURE 7.42 The curvature of a surface.

at O. Any plane containing the normal OZ intersects the surface in a plane curve, which is known as a normal section of the surface. In Figure 7.42, POQ is a normal section of the surface. This normal section lies in the meridional YZ plane. The centre of curvature of this normal section of the surface is at C_T. If the plane is rotated about OZ into a different azimuth, the surface will have a different normal section which, in general, will have a different centre of curvature. In the figure, it is shown that the curve in the XZ section is UOV. It has a centre of curvature at C_S. In general, the points C_T and C_S are different. The curvature at any point of a surface in a given azimuth is defined to be the curvature at that point of the normal section of the surface in that azimuth.

With reference to the axes shown in Figure 7.42, the equation of a continuous surface in the neighbourhood of O may be written as

$$Z = \alpha X^2 + \beta Y^2 + \gamma XY + O(3) \tag{7.211}$$

The constant term and the linear terms are absent in the series on the right-hand side since

$$Z = \frac{\partial Z}{\partial X} = \frac{\partial Z}{\partial Y} = 0 \quad \text{at } X = Y = 0 \tag{7.212}$$

The product term on the right-hand side of (7.211) can be eliminated by suitable rotation of the X- and Y-axes, and then the equation of the surface takes the form

$$Z = \alpha X^2 + \beta Y^2 + O(3) \tag{7.214}$$

The curve in which the surface cuts the plane $Z = \hat{Z}$ will thus have the equation

$$\alpha X^2 + \beta Y^2 = Z_0 + O(3) \tag{7.215}$$

This curve is known as the indicatrix. For values of Z_0 small enough for the term $O(3)$ to be neglected, (7.215) is the equation of a conic referred to by its principal axes. In Figure 7.42, the X- and Y-axes are chosen such that the principal axes of the indicatrix lie in the planes $X = 0 \text{and} Y = 0$. If c_2 is the curvature at O of the plane curve POQ in the section $X = 0$, and c_1 is the curvature at O of the plane curve UOV in the section $Y = 0$, using Newton's method separately in the two sections we get

$$\alpha = \frac{1}{2} c_1 \quad \text{and} \quad \beta = \frac{1}{2} c_2 \tag{7.216}$$

Thus (7.214) can be rewritten as

$$Z = \frac{1}{2}c_1X^2 + \frac{1}{2}c_2Y^2 + O(3) \tag{7.217}$$

For a point B on the indicatrix having polar coordinates (ρ,θ), where $AB = \rho$, and $\angle PAB = \theta$,

$$X = \rho\sin\theta \quad \text{and} \quad Y = \rho\cos\theta \tag{7.218}$$

In polar coordinates, the equation (7.217) reduces to

$$Z = \frac{1}{2}\left[c_1\sin^2\theta + c_2\cos^2\theta\right]\rho^2 + O(3) \tag{7.219}$$

Regarding Z and ρ as axes, (7.219) is the equation of the normal section of the surface in the azimuth θ. Hence, curvature c in this azimuth is given by

$$c = c_1\sin^2\theta + c_2\cos^2\theta \tag{7.220}$$

In terms of radii of curvature $\frac{1}{c_1} = r_1 = OC_1, \frac{1}{c_2} = r_2 = OC_2$ and $\frac{1}{c} = r = OC$, it follows

$$\frac{1}{r} = \frac{\sin^2\theta}{r_1} + \frac{\cos^2\theta}{r_2} \tag{7.221}$$

The relation (7.221) is known as Euler's formula. It shows that the curvature of a surface in any azimuth θ is uniquely determined by the principal curvatures c_1 and c_2. The azimuths containing the maximum and the minimum curvatures are determined by the condition $\frac{dc}{d\theta} = 0$. From (7.220) we get

$$\frac{dc}{d\theta} = 2(c_1 - c_2)\sin\theta\cos\theta = 0 \tag{7.222}$$

Thus, the maximum and minimum curvatures are along the azimuths $\theta = 0$ and $\theta = \frac{\pi}{2}$. The principal sections are at right angles.

If the principal curvatures are equal, $c_1 = c_2$, the curvature in any section is the same, and the surface element is spherical. The indicatrix is then a circle. If $c_1 \neq c_2$, but both are of the same sign, (7.215) shows that the indicatrix is an ellipse. The concavity of all sections faces the same direction, and the surface is said to be synclastic at the given point. A barrel is an example of a synclastic surface. When c_1 and c_2 are of opposite signs, the indicatrix is a hyperbola. Such a surface is shaped like a saddle, and is said to be anticlastic. In general, any surface element for which $c_1 \neq c_2$ is termed an astigmatic element of surface.

7.4.1.4 The Normals to an Astigmatic Surface

The normals to a spherical surface all pass through a single point, the centre of the sphere, and so the set of these normals constitutes a homocentric, or stigmatic, pencil of lines. By contrast, the normals to an astigmatic surface are not homocentric. However, the normals over a narrow region surrounding a point of the astigmatic surface demonstrate certain discerning features, as described below.

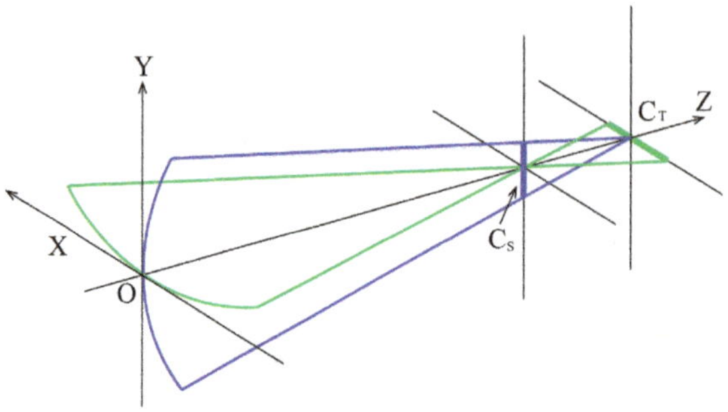

FIGURE 7.43 Normals in the two principal sections, YZ and XZ of an astigmatic surface element.

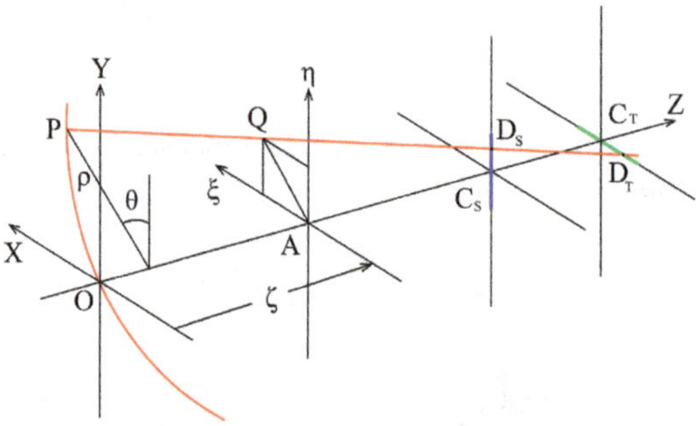

FIGURE 7.44 Normal (in red colour) at point P of the normal section (in red colour) of an astigmatic surface element.

In Figure 7.43, OZ is the normal at O to an astigmatic surface whose principal sections are the planes X = 0, and Y = 0. O is considered to be the origin of a rectangular coordinate system. The X- and Y-axes lie on the tangential plane at O. C_S is the centre of curvature of the normal section in the XZ plane, and C_T is the centre of curvature of the normal section in the YZ plane. Corresponding curvatures in the two sections are c_s and c_t, respectively. These are called the principal curvatures of the astigmatic surface at O. P is a point on the surface in the azimuth θ (Figure 7.44). In parametric form, the equation of the surface may be written as

$$F(X,Y,Z) = Z - \frac{1}{2}\left(c_s X^2 + c_t Y^2\right) + O(3) = 0 \tag{7.223}$$

Let $\left(\tilde{X}, \tilde{Y}, \tilde{Z}\right)$ be the coordinates of the point P. The direction ratios of the normal PQ at P are given by

$$\left(\frac{\partial F}{\partial X}\right)_{\tilde{X}} = -c_s \tilde{X} + O(2) \tag{7.224}$$

$$\left(\frac{\partial F}{\partial Y}\right)_{\tilde{Y}} = -c_t \tilde{Y} + O(2) \tag{7.225}$$

$$\left(\frac{\partial F}{\partial Z}\right)_{\tilde{Z}} = 1 + O(2) \tag{7.226}$$

If (ξ, η, ς) are the coordinates of the point Q, it follows

$$\frac{\xi - \tilde{X}}{-c_s \tilde{X} + O(2)} = \frac{\eta - \tilde{Y}}{-c_t \tilde{Y} + O(2)} = \frac{\varsigma - \tilde{Z}}{1 + O(2)} \tag{7.227}$$

Since the point P is in the neighbourhood of O, $\tilde{Z} = O(2)$. On algebraic manipulation we get from (7.227)

$$\xi = (1 - c_s\varsigma)\tilde{X} + O(2) \cong (1 - c_s\varsigma)\tilde{X} \tag{7.228}$$

$$\eta = (1 - c_t\varsigma)\tilde{Y} + O(2) \cong (1 - c_t\varsigma)\tilde{Y} \tag{7.229}$$

Note that ς is the distance OA, from O to the plane containing the point Q.

For a spherical surface element, say $c_s = c_t = c$. From (7.228) – (7.229), it is seen that for the plane where $\varsigma = \frac{1}{c} = r$, $\xi = \eta = 0$. In other planes where $\varsigma \neq r$, we get

$$\xi^2 + \eta^2 = (1 - c\varsigma)^2 \left(\tilde{X}^2 + \tilde{Y}^2\right) \tag{7.230}$$

This implies that the normals at points lying on the circle $\tilde{X}^2 + \tilde{Y}^2 = \rho^2$ of the spherical surface intersect the plane where $\varsigma \neq r$ in a circle of radius $(1 - c\varsigma)\rho$.

For an astigmatic surface element at O, the two principal centres of curvature are C_S and C_T on the normal OA. The distances of C_S and C_T from O are given by

$$OC_S = r_s = \frac{1}{c_s} \qquad OC_T = r_t = \frac{1}{c_t} \tag{7.231}$$

From (7.228) and (7.229), the point of intersection of the normal at a point P on the astigmatic surface element at O with any plane that is transverse to the normal OA can be determined. On the plane for which $\varsigma = r_s$,

$$\xi = 0 \quad \eta = \left(\frac{r_t - r_s}{r_t}\right)\tilde{Y} \tag{7.232}$$

It implies that the normals from all points of the astigmatic surface on the arc $Y = \tilde{Y}$ intersect the line $C_S D_S$ at a height η along the Y direction. η is proportional to $\tilde{Y}$. The line $C_S D_S$ passes through the principal centre of curvature C_S and is perpendicular to the corresponding principal section. Similarly, on the plane for which $\varsigma = r_t$,

$$\xi = \left(\frac{r_s - r_t}{r_s}\right)\tilde{X} \quad \eta = 0 \tag{7.233}$$

Therefore, the normals to all points on arc $X = \tilde{X}$ intersect the line $C_T D_T$ at a distance ξ from C_T along the X direction. ξ is proportional to $\tilde{X}$. The line $C_T D_T$ passes through the principal centre of curvature C_T, and it is perpendicular to the corresponding principal section. $C_S D_S$ and $C_T D_T$ are called focal lines of the pencil of normals.

From (7.228) – (7.229), we note that normals from points on the surface lying on the circle $\tilde{X}^2 + \tilde{Y}^2 = \rho^2$ intersect the plane $z = \varsigma$ in the ellipse

$$\left(\frac{\xi}{1 - C_s\varsigma}\right)^2 + \left(\frac{\eta}{1 - C_t\varsigma}\right)^2 = \tilde{X}^2 + \tilde{Y}^2 = \rho^2 \tag{7.234}$$

This degenerates to a circle when

$$(1 - c_S\varsigma) = \pm(1 - c_t\varsigma) \tag{7.235}$$

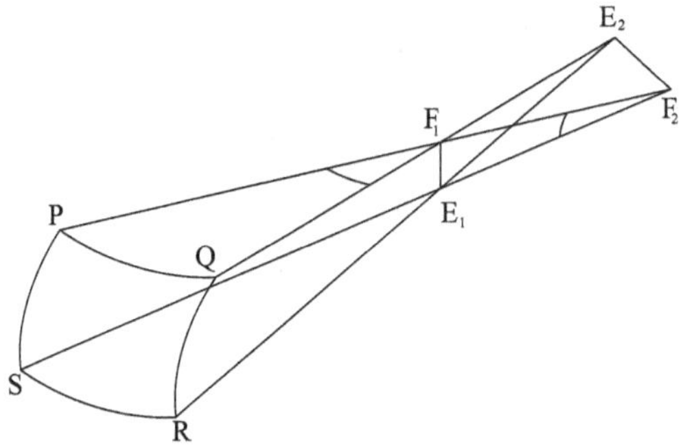

FIGURE 7.45 Focal lines of an astigmatic pencil.

The positive sign corresponds to $\varsigma = 0$, and the negative sign gives

$$\varsigma = \frac{2 r_s r_t}{r_s + r_t} \tag{7.236}$$

Let astigmatic difference $(r_t - r_s) = A$, and mean of curvature radii $\left(\frac{r_t + r_s}{2}\right) = r$. The equation (7.236) can be rewritten as

$$\varsigma = \frac{1}{r}\left(r - \frac{A}{2}\right)\left(r + \frac{A}{2}\right) = r\left[1 - \frac{1}{4}\left(\frac{A}{r}\right)^2\right] \tag{7.237}$$

When $A \ll r$, $\varsigma \cong r$. The given normals then intersect in a circle in the plane midway between C_S and C_T.

Using (7.228) – (7.229), the square of the distance AQ is given by

$$\xi^2 + \eta^2 = (1 - c_s\varsigma)^2 \tilde{X}^2 + (1 - c_t\varsigma)^2 \tilde{Y}^2 \tag{7.238}$$

and this is zero only when (i) $\varsigma = r_s$ and $\tilde{Y} = 0$, and (ii) $\varsigma = r_t$ and $\tilde{X} = 0$. The former are the normals in the principal section of C_S, and they intersect at the point C_S; the latter are the normals in the principal sections of C_T, and they intersect at C_T (Figure 7.43).

It is important to note an important characteristic of an astigmatic surface that although the arc OP of the surface is a circle, with curvature given by Euler's formula, the normal at any point P is skew to the normal at O, unless P lies in one of the two principal sections.

Figure 7.45 illustrates the properties of the focal lines. PQRS is an astigmatic surface element. Normals from points along PQ meet at the point F_1 and intersect at different points along the focal line F_2E_2. If the arc PQ is moved down parallel to itself to the position occupied by SR, the normals rotate about F_2E_2, and their point of intersection describes the focal line F_1E_1. Similarly, the normals from the arc PS focus at F_2. As the arc PS is moved to the position occupied by QR, the normals rotate about F_1E_1, and their point of intersection describes the focal line F_2E_2.

7.4.2 Astigmatism of a Wavefront in General

When a spherical wavefront, originating from an axial object point, is incident normally on a spherical refracting interface, the refracted wavefront is also symmetric around the axis, and it is not astigmatic. This happens on account of axial symmetry of the system. But the situation will not be so when the refracting interface is not symmetric around the axis, e.g. in the case of cylindrical, toroidal, or anamorphic interfaces. The latter surfaces are characterized by their principal sections and corresponding principal curvatures. Although the curvatures of the axially symmetric incident wavefront are the same along all azimuths, the refracted wavefront will become astigmatic on account of the astigmatic nature of the refracting interface. One of the most important optical imaging systems that demonstrates axial astigmatism is the human eye, and Gullstrand [176–178] investigated the phenomenon extensively.

It should be noted that, if the incident wavefront along the axis is astigmatic, the refracted wavefront will, in general, be astigmatic, whether the shape of the refracting interface be spherical or astigmatic. If, at a given refraction, one of the principal sections of the incident wavefront coincides with a principal section of the surface at the point of incidence of the central ray of the wavefront, then, according to Snell's law, this characteristic will remain unchanged for the refracted wavefront. This implies that each principal curvature of the refracted element of the wavefront is a function of only one of the principal curvatures of the surface. This provides a simplified procedure for the analysis of off-axial imagery in axially symmetric imaging systems. This will be enunciated in the following section.

In general, for unsymmetrical or asymmetrical optical systems, or for narrow ray bundles around a skew ray in axi-symmetric optical systems, more involved procedures for analysis are called for [179–196].

7.4.2.1 Rays in the Neighbourhood of a Finite Principal Ray in Axisymmetric Systems

From an axial object point of an imaging system consisting of axially symmetric refracting interfaces, a spherical wavefront is incident normally on the first refracting interface in the object space. Since the incident wavefront is axially symmetric, and all refracting interfaces are also axially symmetric, the wavefront remains axially symmetric during its passage through the system to the final image space. Although the wavefront may change its shape from spherical to aspherical, it remains a surface of revolution in the final image space, and also in all intermediate image spaces. The situation is markedly different in case of an off-axial object point. Due to the oblique incidence of the spherical wavefront from the off-axis object point on the first axially symmetric refracting interface, the refracted wavefront is no longer axially symmetric. It becomes astigmatic in shape. The wavefront remains so in all intermediate spaces of the system during its passage, and emerges as an astigmatic wavefront in the final image space. As mentioned briefly in an earlier section, simplification results when the wavefront in the immediate neighbourhood of the meridian principal ray from an off-axis object point is considered. In this case, one of the principal sections of the astigmatic wavefront is the meridional or tangential plane, and in each space, from the object to the image, the other principal section lies on a plane that contains the segment of the principal ray that is in that space and that is perpendicular to the meridional plane. This plane is called the sagittal plane. Note that the tangential plane is the same for all spaces, from the object to the image, including the intermediate spaces. But the sagittal planes, though always perpendicular to the tangential plane, are different for the spaces. The two centres of curvature of the astigmatic wavefront in each space are called the astigmatic foci along the principal ray in that space. These foci, namely the tangential focus and the sagittal focus, can be determined by tracing close tangential and sagittal rays along the principal ray. The two types of trace can proceed independently in the two principal sections.

This phenomenon of astigmatism was observed long ago—according to Kingslake [197], Isaac Barrow, a teacher of Isaac Newton, first reported on it in 1667. Both Barrow and Newton developed geometrical constructions to study the phenomena.

Convenient algebraic formulae for tracing of T(angential) and S(agittal) rays along a principal ray were reported by Coddington [198]. The equations used for this purpose are often called Coddington equations. However, Young had also reported suitable constructions for determining the astigmatic foci [199–200]. In later years, many reports, some of them with novel methods for derivation of the formulae, appeared in the literature [201–206]. Hopkins [207] introduced the concept of effective 'centres' and 'radii' of curvature to underscore the similarity of S and T ray tracing along a principal ray with paraxial ray tracing. A derivation of S and T ray tracing formulae is given below. We adapt the treatment of Welford [206].

7.4.2.2 Derivation of S and T Ray Tracing Formulae

Figure 7.46 shows a refracting interface of an axially symmetric imaging system with its axis along ACE. The vertex of the interface is at A, and its centre is at C. For convenience, the interface is assumed to be

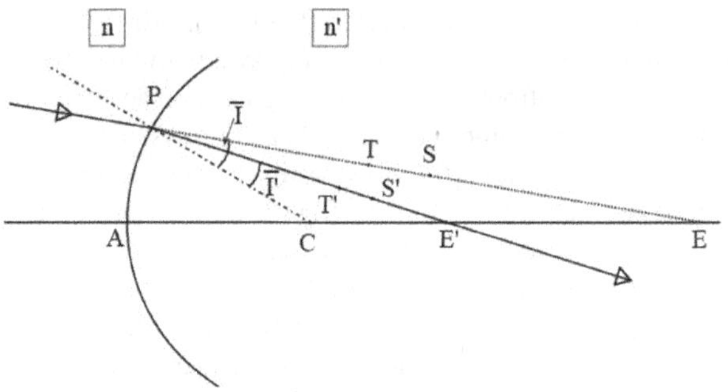

FIGURE 7.46 T and S are the principal centres of curvature of the wavefront element incident at P on the spherical refracting interface AP with centre at C. T′ and S′ are the principal centres of curvature of the refracted wavefront. T, S are called the tangential focus and sagittal focus in the object space. Corresponding foci in the image space are T′ and S′, respectively.

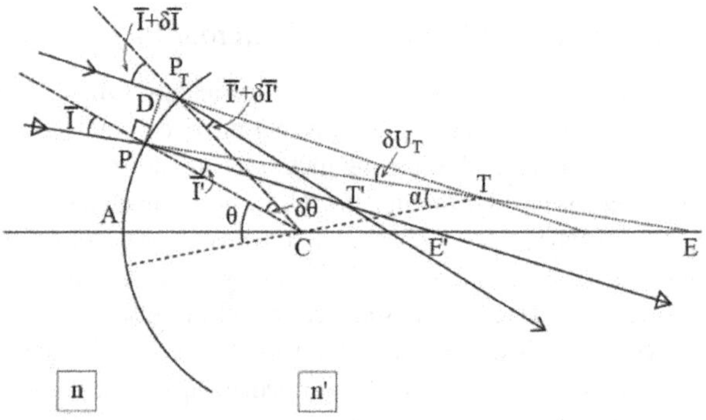

FIGURE 7.47 Derivation of T ray tracing formula.

spherical of curvature c, but the treatment can be extended for any surface of revolution. The refractive indices of the media on the two sides of the interface are n and n′, respectively. PE is a ray (extended) incident on the interface. It makes an angle $\overline{\mathrm{I}}$ with CP, the normal at P. The refracted ray PE′ makes an angle $\overline{\mathrm{I}}$ with the normal CP. T, S are the principal centres of curvature of the incident wavefront element. T is called the tangential focus, and S is called the sagittal focus. The corresponding foci of the refracted wavefront element are T′ and S′.

Figure 7.47 shows the formation of the tangential focus in more detail. $\mathrm{P_T T}$ is a neighbouring ray (extended) of the narrow pencil incident along the principal ray PT. After refraction, the neighbouring ray is refracted as $\mathrm{P_T T'}$. The angles of incidence and refraction of the neighbouring ray are $\left(\overline{\mathrm{I}} + \delta\overline{\mathrm{I}}\right)$ and $\left(\overline{\mathrm{I}}' + \delta\overline{\mathrm{I}}'\right)$, respectively. Let

$$\mathrm{PT} = \mathrm{t} \quad \mathrm{PT}' = \mathrm{t}' \quad \mathrm{PP_T} = \delta\sigma_\mathrm{T} \quad \angle \mathrm{P_T TP} = \delta\mathrm{U_T} \tag{7.239}$$

From $\Delta \mathrm{P_T CT}$, we get

$$\theta + \delta\theta = \overline{\mathrm{I}} + \delta\overline{\mathrm{I}} + \alpha + \delta\mathrm{U_T} \tag{7.240}$$

Also, from ΔPCT,

$$\theta = \overline{\mathrm{I}} + \alpha \tag{7.241}$$

From (7.240) and (7.241), it follows

$$\delta\theta = \delta\bar{I} + \delta U_T \tag{7.242}$$

Let PD be a perpendicular on PT. Then, $\angle P_T PD = \bar{I}$, so that

$$PD = \delta\sigma_T \cos\bar{I} = t\delta U_T \tag{7.243}$$

From the sector $P_T PC$ we get $\delta\theta = c\delta\sigma_T$, where c is the curvature of the spherical interface. From (7.242), we can write

$$\delta\bar{I} = \delta\sigma_T\left(c - \frac{\cos\bar{I}}{t}\right) \tag{7.244}$$

Similarly, $\delta I'$ is given by

$$\delta\bar{I}' = \delta\sigma_T\left(c - \frac{\cos\bar{I}'}{t'}\right) \tag{7.245}$$

Using Snell's law of refraction for the principal ray at P

$$n\sin\bar{I} = n'\sin\bar{I}' \tag{7.246}$$

On differentiation, we get

$$n\cos\bar{I}\delta\bar{I} = n'\cos\bar{I}'\delta\bar{I}' \tag{7.247}$$

By substituting from (7.244) and (7.245), we get

$$n\cos\bar{I}\left(c - \frac{\cos\bar{I}}{t}\right) = n'\cos\bar{I}'\left(c - \frac{\cos\bar{I}'}{t'}\right) \tag{7.248}$$

On algebraic manipulation

$$\frac{n'\cos^2\bar{I}'}{t'} - \frac{n\cos^2\bar{I}}{t} = c\left(n'\cos\bar{I}' - n\cos\bar{I}\right) \tag{7.249}$$

For deriving the relation for sagittal foci, we refer to Figure 7.48. PE (extended) and PE′ are the segments of the principal ray in the incident and refraction spaces. Let S be the sagittal focus in the incident space. This implies that close rays lying on the sagittal plane intersect at point S. The line joining C with S, i.e., CS, may be considered an auxiliary axis. Let the refracted segment of the principal ray intersect CS at S′. A small rotation around CS of the three segments of ray shows that locations of S and S′ are not changed. This implies that close sagittal rays meet at the point S′ after refraction, i.e. S′ is the sagittal image corresponding to S.

Let PS = s and PS′ = s′. From Figure 7.48, we note

$$\alpha = \bar{I} + \angle PSC = \bar{I}' + \angle PS'C \tag{7.250}$$

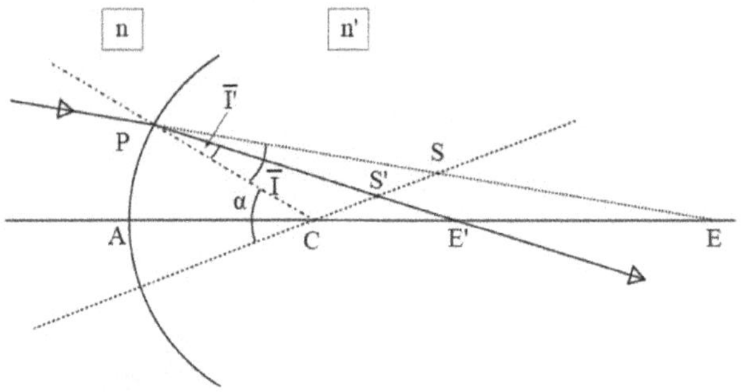

FIGURE 7.48 Derivation of S ray tracing formula.

From ΔPCS

$$\frac{\frac{1}{c}}{\sin\left(\alpha-\bar{I}\right)}=\frac{s}{\sin\left(\pi-\alpha\right)}=\frac{s}{\sin\alpha} \tag{7.251}$$

and from ΔPCS′

$$\frac{\frac{1}{c}}{\sin\left(\alpha-\bar{I}'\right)}=\frac{s'}{\sin\left(\pi-\alpha\right)}=\frac{s'}{\sin\alpha} \tag{7.252}$$

Eliminating α between (7.251) and (7.252), we get

$$\frac{n's}{ns'}=\frac{\left(cs\cos\bar{I}-1\right)}{\left(cs'\cos\bar{I}'-1\right)} \tag{7.253}$$

using (7.246) for the principal ray. On algebraic manipulation, (7.253) reduces to

$$\frac{n'}{s'}-\frac{n}{s}=c\left(n'\cos\bar{I}'-n\cos\bar{I}\right) \tag{7.254}$$

(7.249) and (7.254) are the formulae for tracing T rays and S rays. It may be seen that for an axial ray $I=I'=0$, and both the equations reduce to the paraxial refraction equation.

Hopkins [207] showed that s and t ray tracing along the principal ray can be done in the usual paraxial ray tracing format. For s- ray tracing, let the paraxial sagittal ray be specified by height and angle variables h_S, u_s', where

$$h_S=su_s=s'u_s' \tag{7.255}$$

The refraction equation (7.254) for s ray can be rewritten as

$$n'u_S'-nu_S=h_Sc_S\Delta\left(n\cos\bar{I}\right) \tag{7.256}$$

where

$$\Delta\left(n\cos\bar{I}\right) = n'\cos\bar{I}' - n\cos\bar{I} \tag{7.257}$$

The transfer equation to the next surface is

$$\left(h_S\right)_+ = h_S + u'_S\bar{D}' \tag{7.258}$$

where $\bar{D}'$ is the distance to the next surface along the principal ray.

For t- ray tracing let the paraxial tangential ray be specified by height and angle variables h_T, u'_T, where

$$h_T = \frac{tu_T}{\cos\bar{I}} = \frac{t'u'_T}{\cos\bar{I}'} \tag{7.259}$$

The refraction equation (7.249) for t ray can be rewritten as

$$n'u'_T - nu_T = h_T c_T \Delta\left(n\cos\bar{I}\right) \tag{7.260}$$

The corresponding transfer equation is

$$\left(h_T\right)_+ = h_T + u'_T\bar{D}' \tag{7.261}$$

7.4.2.3 The Sagittal Invariant

In Figure 7.49, $\bar{P}\bar{S}\bar{Q}$ is the pupil ray incident at the point $\bar{P}$ on a refracting interface $A\bar{P}$. The latter images the paraxial object point O at O′. The finite extra-axial object point is at $\bar{Q}$, and its height above the axis is $\bar{\eta}$. The sagittal focus of the incident pencil is at S, and η_S is its height above the optical axis. The refractive index of the object space is n. In the image space of refractive index n′, $\bar{P}S'\bar{Q}'$ is the refracted pupil ray, giving the finite image point at $\bar{Q}'$, and the sagittal focus of the refracted pencil is at S′. The heights of $\bar{Q}'$ and S′ above the optical axis are $\bar{\eta}'$ and η'_S, respectively. Let the direction cosines of the pupil ray in the object space be $\left(0, \bar{M}, \bar{N}\right)$, and the height of the point $\bar{P}$ above the optical axis be $\bar{Y}$. For the paraxial sagittal ray, let u_S, u'_S be the convergence angles with the pupil ray in the object space and image space, respectively, and h_S be its height on the refracting interface. We have

$$h_S = su_S = s'u'_S \tag{7.262}$$

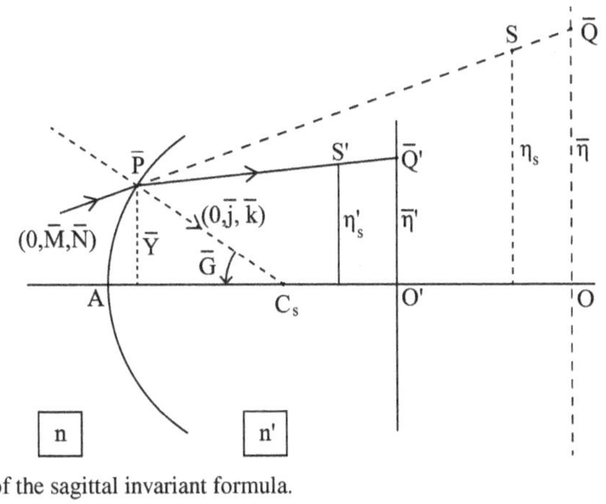

FIGURE 7.49 Derivation of the sagittal invariant formula.

where $S = \bar{P}S$ and $s' = \bar{P}S'$. We define a quantity $\bar{H}_S$ involving only parameters of the sagittal ray in the object space as

$$\bar{H}_S = nu_S\eta_s = nu_S\left(\bar{Y} + \bar{M}s\right) = \bar{Y}\left(nu_S\right) + h_S\left(n\bar{M}\right) \tag{7.263}$$

The change in this quantity on refraction is given by

$$\bar{H}_S' - \bar{H}_S = \bar{Y}\Delta\left(nu_S\right) + h_S\Delta\left(n\bar{M}\right) \tag{7.264}$$

From the relation (7.256), for the law of refraction for the paraxial sagittal ray we get

$$\Delta\left(nu_S\right) = n'u_S' - nu_S = h_S c_S \Delta\left(n\cos\bar{I}\right) \tag{7.265}$$

where $c_S = \frac{1}{r_S} = \frac{1}{\left(\bar{P}C_S\right)}$ is the sagittal curvature at $\bar{P}$.

Using the law of refraction for the finite pupil ray, we get

$$\Delta\left(n\bar{M}\right) = n'\bar{M}' - n\bar{M} = \bar{j}\Delta\left(n\cos\bar{I}\right) \tag{7.266}$$

where $\left(0, \bar{j}, \bar{k}\right)$ are the direction cosines of the surface normal, $\bar{P}C_S$, at $\bar{P}$. Let $\angle\bar{P}C_S A = \bar{G}$. Writing $\bar{j} = -\sin\bar{G} = -\bar{Y}c_S$, (7.266) reduces to

$$\Delta\left(n\bar{M}\right) = -\bar{Y}c_S\Delta\left(n\cos\bar{I}\right) \tag{7.267}$$

Substituting from (7.265) and (7.267), in (7.264) we get

$$\bar{H}_S' - \bar{H}_S = \bar{Y}h_S c_S\Delta\left(n\cos\bar{I}\right) - \bar{Y}h_S c_S\Delta\left(n\cos\bar{I}\right) = 0 \tag{7.268}$$

It follows

$$\bar{H}_S' = \bar{H}_S, \text{i.e. } n'u_S'\eta_S' = nu_S\eta_S \tag{7.269}$$

The product $\bar{H}_S = nu_S\eta_S$ is called the sagittal invariant for any given extra-axial pencil.

Note that there is no similar tangential invariant for an extra-axial pencil. However, a local tangential invariant can be found. It has many important applications in analytical aberration theory. One of them is discussed in Chapter 11.

7.5 Aberrations of Optical Systems

From the discussions in earlier sections on optical imaging by real rays, it is evident that, except for in some special cases, real optical imaging systems cannot produce a perfect image of a three-dimensional object. It is also noted that there is no fundamental limitation barring formation of a perfect image of a plane or curved object on a plane or curved surface by a real optical imaging system. Nevertheless, none of the optical imaging systems used in practice is 'perfect'; the imperfection may be either in the quality of sharpness of the image, or in the shape of the image, or a combination of both. For systems operating in broadband illumination over a range of wavelengths, the imaging performance is affected further by

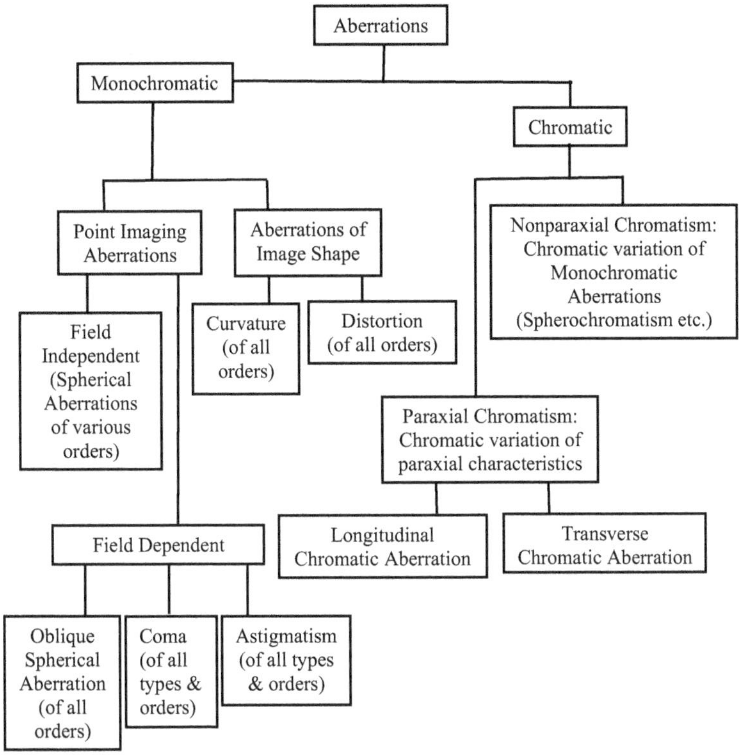

CHART 7.1 A broad classification of optical aberrations in axi-symmetric systems.

dispersion effects arising out of changes in the refractive index of the dioptric elements of the system with a variation in wavelength of light. These chromatic effects can also take place if any of the spaces—the object space, the image space, or any inter element space of the system—consist of dispersive medium. Note that the paraxial characteristics discussed earlier presumed monochromatic illumination. So, both the paraxial characteristics as well as the monochromatic imperfections mentioned above change significantly with changes in wavelength of illumination. All these imperfections are considered as defects in image formation by an optical system, and 'aberration' is a generic term for these imperfections or defects. A broad classification of aberrations of axi-symmetric optical systems is given in Chart 7.1. Each constituent class of aberrations is dealt with in the next chapters.

REFERENCES

1 W.R. Hamilton, 'Supplement to an essay on the theory of systems of rays', *Trans, Roy. Irish Acad.*, Vol. 16, part 1 (1830) pp.1–61.
2 W.R. Hamilton, 'Theory of systems of rays', *Trans. Roy. Irish Acad.*, Vol. 15 (1828) pp. 69–174.
3 J.C. Maxwell, 'On the application of Hamilton's characteristic function to the theory of an optical instrument symmetrical about the axis', *Proc. London Math. Soc.*, Vol. VI (1874–75) pp. 117–122; also article LXXII in *The Scientific papers of James Clark Maxwell, Vol. II*, Ed. W. Niven, Dover, New York (1965) pp. 439–444.
4 J.C. Maxwell, 'On Hamilton's characteristic function for a narrow beam of light', *Proc. London Math. Soc.*, Vol. VI (1874–75) pp. 182–190; also article LXIV in *The Scientific papers of James Clark Maxwell, Vol. II*, Ed. W. Niven, Dover, New York (1965) pp. 381–390.
5 M. Thiesen, 'Beiträge zur Dioptrik', *Berl. Sitzgsber.* Vol. 2 (1890) pp. 799–813.
6 R.S. Heath, *A Treatise on Geometrical Optics*, Cambridge University Press, Cambridge (1887).
7 H. Bruns, 'Das Eikonal', Königl. Sächsischen Ges. D. Wiss. Abhand. Math-phys., Leipz. *Sitzgsber.* Vol. 21 N^0 V (1895) pp. 321–436.

8 F. Klein, 'Über das Brunssche Eikonal', Zeitscrift f. Mathematik u. Physik, Vol. 46 (1901) Gesamm. *Abhand.*, no. XXI, Vol. 2, pp. 601–604.
9 K. Schwarzschild, 'Untersuchungen zur geometrischen Optik I. Einleitung in die Fehlertheorie optischer Instrumente auf Grund des Eikonalbegriffs', *Abhand. d. Königl. Gesell. d. Wissen. z. Göttingen, Math. - Phys. Klasse*, Vol. 4 (1905) pp. 1–31.
10 J.W. Strutt (Lord Rayleigh), 'Hamilton's principle and the five aberrations of von Seidel', *Phil. Mag.*, Vol. XV (1908) pp. 677–68
11 J.P.C. Southall, *The Principles and Methods of Geometrical Optics*, Macmillan, New York (1910).
12 G. Prange, 'W.R. Hamilton's bedeutung für die geom. optik', *Jahresber. D. Dtsch. Math. Vereinigg*, Vol. 30 (1921) pp. 69–82.
13 G. Prange, 'W. R. Hamilton's arbeiten zur strahlenoptik und analytisches mechanik', Nova Acta, Abh. *Der Leopoldina*. Vol. 107, Halle (1923).
14 T. Smith, 'Some uncultivated optical fields', *Trans. Opt. Soc. (London)*, Vol. 23 (1926–27) pp. 225–284.
15 G.C. Steward, *The symmetrical optical system*, Cambridge University Press, London (1928).
16 A.W. Conway and J.L. Synge, Eds., *The Mathematical Papers of Sir William Rowan Hamilton, Vol. I, Geometrical Optics*, Cambridge University Press, London (1931).
17 M. Herzberger, *Strahlenoptik*, Julius Springer, Berlin (1931).
18 A.W. Conway and A.J. McConnell, 'On the determination of Hamilton's principal function', *Proc. Roy. Irish Acad.*, Vol. 41, Sect. A (1932) pp. 18–25.
19 M. Herzberger, 'On the characteristic function of Hamilton, the Eikonal of Bruns and their use in optics', *J. Opt. Soc. Am.*, Vol. 26 (1936) pp. 177–180.
20 J.L. Synge, 'Hamilton's method in geometrical optics', *J. Opt. Soc. Am.*, Vol. 27 (1937) pp. 75–82.
21 M. Herzberger, 'Hamilton's characteristic function and Bruns' eiconal', *J. Opt. Soc. Am.*, Vol. 27 (1937) pp. 133–137.
22 J.L. Synge, 'Hamilton's characteristic function and Bruns' eiconal', *J. Opt. Soc. Am.*, Vol. 27 (1937) pp. 138–144.
23 C. Carathéodory, *Geometrische Optik*, Julius Springer, Berlin (1937).
24 J.L. Synge, *Geometrical Optics*, Cambridge Univ. Press, London (1937).
25 H. Chrétien, Calcul des combinaisons optiques, 5th edition, Masson, Paris (1980) pp. 629–639. (First published in 1938).
26 J. Korringa, *Onderzoekingen op het Gebied der Algebraische Optiek*, Ph. D. thesis, TU, Delft (1942).
27 R.K. Luneburg, *Mathematical Theory of Optics* (Brown University Lectures, 1944), Univ. California Press, Berkeley (1964).
28 W.G. Stephan, *Practische Toepassingen op het Gebied der Algebraische Optica*, Ph. D. Thesis, TU, Delft (1947) [in Dutch].
29 T. Smith, 'Hamilton's characteristic function and the design of optical instruments', in *La Théorie des Images Optiques*, Colloques internationaux du Centre National de la Recherche Scientifique, 14–19 October 1946, Paris (1949).
30 A.C.S. Van Heel, 'Calcul des aberrations du cinquième ordre et projets tenant compte de ces aberrations', in *La Théorie des Images Optiques,* Colloques internationaux du Centre National de la Recherche Scientifique, 14–19 October 1946, Paris (1949).
31 T. Smith, 'Optical Calculations', in *A Dictionary of Applied Physics*, Ed. R. Glazebrook, Peter Smith, New York (1950) pp. 287–315.
32 M. Herzberger, 'Geometrical optics', Chapter 2 in *Handbook of Physics*, Eds. E.U. Condon and H. Odishaw, McGraw Hill, New York (1958) pp. 6–20 to 6–46.
33 R.J. Pegis, 'The modern development of Hamiltonian optics', in *Progress in Optics, Vol. I*, Ed. E. Wolf, North-Holland, Amsterdam (1961) pp. 3–29.
34 W. Brouwer and A. Walther, 'Geometrical Optics', in *Advanced Optical Techniques*, Ed. A.C.S. Van Heel, North-Holland, Amsterdam (1967) pp. 503–570.
35 W.T. Welford, 'Optical calculations and optical instruments, an introduction', in *Handbuch der Physik*, Ed. S. Flugge, Band XXIX Optische Instrumente, Springer, Berlin (1967) pp. 1–42.
36 H. Marx, 'Theorie der geometrisch-optischen bildfehler', in *Handbuch der Physik*, Ed. S. Flugge, Band XXIX Optische Instrumente, Springer, Berlin (1967) pp. 68–191.
37 H.A. Buchdahl, *An Introduction to Hamiltonian Optics*, Cambridge Univ. Press, London (1970).

38 R. Courant and D. Hilbert, Methods of Mathematical Physics, Vol. II, *Interscience, New York* (1953) p. 32.
39 G.W. Forbes, 'New class of characteristic functions in Hamiltonian optics', *J. Opt. Soc. Am.* Vol. 72 (1982) pp. 1698–1701.
40 O.N. Stavroudis, *The Optics of Rays, Wavefronts, and Caustics*, Academic, New York (1972).
41 A. Walther, *The Ray and Wave Theory of Lenses*, Cambridge Univ. Press, New York (1995).
42 C.A.J. Simons, The Application of Eikonal functions in the practice of optical design, Ph. D. thesis, TU Delft (1969).
43 A. Walther, 'Eikonal theory and computer algebra', *J. Opt. Soc. Am. A*, Vol. 13 (1996) pp. 523–531.
44 A. Walther, 'Eikonal Theory and computer algebra II', *J. Opt. Soc. Am. A*, Vol. 13 (1996) pp. 1763–1765.
45 M. Andrews, 'The concatenation of characteristic functions in Hamiltonian optics', *J. Opt. Soc. Am.* Vol. 72 (1982) pp. 1493–1497.
46 G.W. Forbes, 'Order doubling in the determination of characteristic functions', *J. opt. Soc. Am.* Vol. 72 (1982) pp. 1097–1099.
47 G.W. Forbes, 'Concatenation of restricted characteristic functions', *J. Opt. Soc. Am.* Vol.72 (1982) 1702–1706.
48 G.W. Forbes and M. Andrews, 'Concatenation of symmetric systems in Hamiltonian optics', *J. Opt. Soc. Am.* Vol. 73 (1983) pp. 776–781.
49 G.W. Forbes, 'Order doubling in the computation of aberration coefficients', *J. Opt. Soc. Am.* Vol. 73 (1983) pp. 782–788.
50 C.H.F. Velzel, 'Image formation by a general optical system, using Hamilton's method', *J. Opt. Soc. Am. A*, Vol. 4 (1987) pp. 1342–1348.
51 C.H.F. Velzel and J.L.F. de Meijere, 'Characteristic functions and the aberrations of symmetric optical systems. I. Transverse aberrations when the eikonal is given', *J. Opt. Soc. Am. A*, Vol. 5 (1988) pp. 246–250.
52 C.H.F. Velzel and J.L.F. de Meijere, 'Characteristic functions and the aberrations of symmetric optical systems. II. Addition of aberrations', *J. Opt. Soc. Am. A*, Vol. 5 (1988) pp. 251–256.
53 C.H.F. Velzel and J.L.F. de Meijere, 'Characteristic functions and the aberrations of symmetric optical systems. III. Calculation of eikonal coefficients', *J. Opt. Soc. Am. A*, Vol. 5 (1988) pp. 1237–1243.
54 J.L.F. de Meijere, 'The use of characteristic functions in optical design', in *Huygens' Principle 1690-1990: Theory and Applications*, Eds. H. Blok, H.A. Ferwerda and H.K. Kuiken, Elsevier, New York (1992).
55 B.D. Stone and G.W. Forbes, 'Foundations of first order layout for asymmetric systems: an application of Hamilton's methods', *J. Opt. Soc. Am. A*, Vol. 9 (1992) pp. 96–109.
56 B.D. Stone and G.W. Forbes, 'Forms of the characteristic function for asymmetric systems that form sharp images to first order, *J. Opt. Soc. Am. A*, Vol. 9 (1992) pp. 820–831.
57 G.W. Forbes and B.D. Stone, 'Hamilton's angle characteristic in closed form for generally configured conic and toric interfaces, *J. Opt. Soc. Am. A*, Vol. 10 (1993) pp. 1270–1278.
58 G.W. Forbes and B.D. Stone, 'Restricted characteristic functions for general optical configurations', *J. Opt. Soc. Am. A*, Vol. 10 (1993) pp. 1263–1269.
59 M.A. Alonso and G.W. Forbes, 'Generalization of Hamilton's formalism for geometrical optics', *J. Opt. Soc. Am. A*, Vol. 12 (1995) pp. 2744–2752.
60 V. Lakshminarayanan, A. Ghatak, and K. Thyagarajan, *Lagrangian Optics*, Kluwer Academic (2001).
61 Ref. [37] pp. 265–266.
62 M. Born and E. Wolf, *Principles of Optics*, Cambridge University Press, Cambridge (1999).
63 Ref. [27] p. 130.
64 R. Descartes, 'Des figures que doivent avoir les corps transparents pour détourner les rayons par refraction en toutes les façons qui servent a la vue', *La Dioptrique* (Discours huitième), in *Œuvres de Descartes, Vol. 5*, Levrault, Paris (1824) pp. 92–119. [First published in 1637]
65 B. Williamson, 'Genocchi's Theorem on Oval of Descartes', in *An Elementary Treatise on the Integral Calculus*, Appleton, New York (1877) p. 225.
66 B. Williamson, 'Chapter XX. On the Cartesian Oval', in *An Elementary Treatise on the Differential Calculus*, Longmans, London (1884) pp. 375–384.

67 J.P.C. Southall, 'Aplanatic (or Cartesian) Optical Surfaces', *J. Franklin Inst.*, Vol. 193 (1922) pp. 609–626.

68 J.C. Maxwell, 'On the general laws of optical instruments', *Quart. J. Pure Appl. Math.*, Vol.2 (1858) pp. 233–244. Also in, *The Scientific papers of James Clark Maxwell*, Vol. I, Ed. W.D. Niven, Dover, New York (1965) pp. 271–285.

69. C.F. Gauss, 'Dioptrische Untersuchungen' (Dec. 1840), *Königlichen Gesellschaft der Wissenschaften zu Göttingen, Abhandlungen der Mathematischen Classe*, Vol 1, 1838–1841, pp. 1–34.

70 A.E. Conrady, *Applied Optics and Optical Design*, Part I, Dover, New York (1992), originally published by Oxford University Press, London (1929), p. 437.

71 W.T. Welford, *Aberrations of optical systems*, Adam Hilger, Bristol (1986), pp. 3–4.

72 H. Bruns, Ref. 7, p. 370

73 F. Klein, 'Räumliche Kollineation bei optischen instrumenten', *Z. Math. Phys.*, Vol. 46 (1901) pp. 376–382.

74 E.T. Whittaker, The Theory of Optical Instruments, Hafner, New York (1915) pp. 47–48.

75 C. Carathéodory, 'Über den Zusammenhang der Theorie der absoluten optischen Instrumente mit einem Satze der Variationsrechnung', *Sitzgsber. Bayer Akad. Wiss.Math. -naturwiss. Abt.*, Vol. 56 (1926) pp. 1–18.

76 C. Carathéodory, Ref. 23, pp. 70–75.

77 M. Born and E. Wolf, Ref. 62, pp. 153–157.

78 J.C. Maxwell, 'Solutions of Problems', *Cambridge and Dublin Math. J.*, Vol. VIII (1854) p. 188; also, in *The Scientific papers of James Clark Maxwell*, Vol. I, Ed. W.D. Niven, Dover, New York (1965) pp. 76–79.

79 O.N. Stavroudis, Ref. 40 pp. 43–50.

80 E.W. Marchand, *Gradient Index Optics*, Academic, New York (1978) pp. 23–33.

81 D.R. Shafer, 'Doing more with less', *Proc. SPIE*, Vol. 2537 (1995) pp. 2–12.

82 R.K. Luneburg, Ref. 27, p. 179.

83 R.K. Luneburg, Ref. 27, p. 187.

84 S. Cornbleet, *Microwave and Geometrical Optics*, Academic, New York (1994).

85 J. Eaton, 'On spherically symmetric lenses', *IRE Trans. Antennas and Propagation*, Vol. 4 (1952) pp. 66–71.

86 R. Stettler, 'Über die optische Abbildung von Fläschen und Räumen', *Optik*, Vol. 12 (1955) pp. 529–543.

87 S.P. Morgan, 'General solution of the Luneburg lens problem', *J. Appl. Phys.* Vol. 29 (1958) pp. 1358–1368.

88 A.F. Kay, 'Spherically Symmetric Lenses', *IRE Trans. Antennas and Propagation*, Vol. 7 (1959) pp. 32–38.

89 S.P. Morgan, 'Generalizations of Spherically Symmetric Lenses', *IRE Trans. Antennas and Propagation*, Vol. 7 (1959) pp. 342–345.

90 P. Uslenghi, 'Electromagnetic and Optical Behavior of Two Classes of Dielectric Lenses', *IEEE Trans. Antennas and Propagation*, Vol. 17 (1969) pp. 235–236.

91 W.H. Southwell, 'Index profiles for generalized Luneburg lenses and their use in planar optical waveguides', *J. Opt. Soc. Am.*, Vol. 67 (1977) pp. 1010–1014.

92 J. Sochacki, 'Exact analytical solution of the generalized Luneburg lens problem', *J. Opt. Soc. Am.*, Vol. 73 (1983) pp. 789–795.

93 J.C. Miñano, 'Perfect imaging in a homogeneous three-dimensional region', *Opt. Express*, Vol. 14 (2006) pp. 9627–9635.

94 T. Tyc, L. Herzánová, M. Šarbort, and K. Bering, 'Absolute instruments and perfect imaging in geometrical optics', *New J. Phys.*, Vol 13 (2011) 115004.

95 V.G. Veselago, 'The electrodynamics of substances with simultaneously negative values of ε and μ', *Sov. Phys. Usp.*, Vol. 10 (1968) pp. 509–514.

96 J.B. Pendry, 'Negative refraction makes a perfect lens', *Phys. Rev. Lett.*, Vol. 85 (2000) pp. 3966–3069.

97 J.B. Pendry and S. Anantha Ramakrishna, 'Refining the perfect lens', *Physica B*, Vol. 338 (2003) pp. 329–332.

98 J.B. Pendry and D.R. Smith, 'The Quest for the Superlens', *Scientific American*, Vol. 295 (2006) pp. 60–67.

99 X. Zhang and Z. Liu, 'Superlenses to overcome the diffraction limit', *Nature Materials*, Vol. 7 (2008) pp. 435–441.
100 U. Leonhardt, 'Perfect imaging without negative refraction', *New J. Phys.*, Vol. 11 (2009) 093040.
101 R.J. Blaikie, 'Comment on perfect imaging without negative refraction', *New J. Phys.* Vol. 12 (2010) 058001.
102 U. Leonhardt, 'Reply to comment on perfect imaging without negative refraction', *New J. Phys.* Vol. 12 (2010) 058001.
103 R.J. Blaikie, 'Comment on reply to comment on perfect imaging without negative refraction', *New J. Phys.*, Vol 13 (2011) 028001.
104 U. Leonhardt, 'Reply to comment on perfect imaging without negative refraction', *New J. Phys.*, Vol. 13 (2011) 028002.
105 U. Leonhardt and T. Philbin, 'Perfect imaging with positive refraction in three dimensions', *Phys. Rev. A*, Vol. 81 (2010) 011804(R).
106 R. Merlin, 'Comment on perfect imaging with positive refraction in three dimensions', *Phys. Rev. A*, Vol. 82 (2010) 057801.
107 U. Leonhardt and T. Philbin, 'Reply to comment on perfect imaging with positive refraction in three dimensions', *Phys. Rev. A*, Vol. 82 (2010) 057802.
108 F. Sun and S. He, 'Can Maxwell's fish eye lens really give perfect imaging?', *Prog. Electromagn. Res.*, Vol. 108 (2010) pp. 307–322.
109 F. Sun, X.C. Ge, and S. He, 'Can Maxwell's fish eye lens really give perfect imaging? Part II. The case with passive drains', *Prog. Electromagn. Res.*, Vol. 110 (2010) pp. 313–328.
110 Y.G. Ma, S. Sahebdivan, T. Tyc, and U. Leonhardt, 'Evidence for subwavelength imaging with positive refraction', *New J. Phys.*, Vol. 13 (2011) 033016.
111 J.C. Miñano, R. Marqués, J.C. González, P. Benitez, V. Delgrado, D. Grabovičkić, and M. Freire, 'Super-resolution for a point source better than λ/500 using positive refraction, *New J. Phys.*, Vol. 13 (2011) 125009.
112 R. Merlin, 'Maxwell's fish-eye lens and the mirage of perfect imaging', *J. Opt.*, Vol. 13 (2011) 024017.
113 L.A. Pazynin and G.O. Kryvchikova, 'Focusing properties of Maxwell's fish eye medium', *Progr. Electromagn. Res.*, Vol. 131 (2012) pp. 425–440.
114 J.C. Miñano, J. Sánchez-Dehesa, J.C. González, P. Benitez D. Grabovičkić, J. Carbonell, and H. Ahmadpanahi, 'Experimental evidence of superresolution better than λ/105 with positive refraction, *New J. Phys.* Vol. 16 (2014) 033015.
115 T. Tyc and A. Danner, Resolution of Maxwell's fish eye with an optimal active drain, *New J. Phys.*, Vol. 16 (2014) 063001.
116 S. He, F. Sun, S. Guo, S. Zhong, L. Lan, W. Jiang, Y. Ma, and T. Wu, 'Can Maxwell's fish eye lens really give perfect imaging? Part III. A careful reconsideration of the "Evidence for subwavelength imaging with positive refraction"', *Prog. Electromagn. Res.*, Vol. 152 (2015) pp. 1–15.
117 M.A. Alonso, 'Is the Maxwell-Shafer fish eye lens able to form super-resolved images?' *New J. Phys.*, Vol. 17 (2015) 073013.
118 U. Leonhardt and S. Sahebdivan, 'Theory of Maxwell's fish eye with mutually interacting sources and drains', *Phys. Rev. A*, Vol. 92 (2015) 053848.
119 H. Boegehold und M. Herzberger, 'Kann man zwei verschiedene Flächen durch dieselbe Folge von Umdrehungsflächen scharf abbilden?', *Compositio Mathematica*, Vol. 1 (1935) pp. 448–476.
120 M. Herzberger, 'The limitations of optical image formation', *Ann. New York Acad. Sci.*, Vol. 48 (1946) pp. 1–30.
121 T. Smith, 'Secondary conjugate surfaces', *Trans. Opt. Soc. London*, Vol. 32 (1930–31) pp. 129–149.
122 T. Smith, 'On perfect optical instruments', *Proc. Phys. Soc.*, Vol. 60 (1948) pp. 293–304.
123 A. Walther, Ref. 41, pp. 227–237.
124 T. Smith, 'The theory of aplanatic surfaces', *Trans. Opt. Soc. London*, Vol. 29 (1928) pp. 179–186.
125 R.K. Luneburg, Ref. 27, p. 136.
126 O.N. Stavroudis, Ref. 40, pp. 101–102.
127 D. Schurig and D.R. Smith, 'Negative index lens aberrations', *Phys. Rev. E*, Vol. 70 (2004) 065601.
128 T. Smith, 'The Optical Cosine Law', *Trans. Opt. Soc. London*, Vol. 24 (1922) pp. 31–40.
129 H. Boegehold, 'Weitere BemerkungenZum Kosinussatze', *Central-Ztg. F. Opt. u. Mech.*, Vol. 45 (1924) pp. 295–296.

130 H. Boegehold, 'Zum Kosinussatze von A.E. Conrady und T.T. Smith', *Central-Ztg. F. Opt. u. Mech.*, Vol. 45 (1925) pp. 107–108.
131 T. Smith, 'Note on the cosine law', *Trans. Opt. Soc. London*, Vol. 26 (1924–25) pp. 281–286.
132 G.C. Steward, *The Symmetrical Optical System*, Cambridge University Press, London (1928) pp. 57–61.
133 M. Herzberger, *Modern Geometrical Optics*, Interscience, New York (1958) p. 160.
134 W.T. Welford, 'Aplanatism and Isoplanatism', in *Progress in Optics, Vol. XIII*, Ed. E. Wolf, North-Holland, Amsterdam (1976) pp. 284–285.
135 A. Walther, Ref. 41, pp. 51–52.
136 E. Abbe, 'A contribution to the Theory of the microscope, and the nature of microscopic vision', *Proceedings of the Bristol Naturalists' Society, Bristol, New Series*, Vol. I (1874) pp. 200–261; original German article in Schultze's Archiv für mikroscopische Anatomie, Vol. IX (1873) pp. 413–468.
137 R. Clausius, 'Ueber die concentration von wärme und lichtstrahlen und die Gränzen ihrer wirkung, Annalen der Physik und Chemie', Vol. 121 (1864) pp. 1–44.
138 H. v. Helmholtz, 'Die theoretische Grenze für die Leistungsfähigkeit der Mikroskope', *Pogg. Ann. Jubelband* (1874) pp. 557–584. English translation entitled 'On the limits of optical capacity of the microscope', *The Monthly Microscopical Journal: Trans. Roy. Micros. Soc. London, Vol. XVI*, Ed. H. Lawson, (1876) pp. 15–39.
139 E. Abbe 'Ueber die Bedingungen des Aplanatismus der Linsensysteme', *Chapter XI in Gesammelte Abhandlungen von Ernst Abbe*, Vol. I, Gustav Fischer, Jena (1904); also, in Sitzungsberichte der Jenaischen Gesellschaft für Medicin und Naturwissenschaft (1879) pp. 129–142.
140 C. Hockin, 'On the estimation of aperture in the microscope', *J. Roy. Micros. Soc.*, Vol. 4 (1884) pp. 337–347.
141 J.D. Everett, 'Note on Hockin's proof of the sine condition', *Phil. Mag., Ser.6*, Vol. 4 (1902) pp. 170–171.
142 M. Born and E. Wolf, Ref. 62 pp. 178–180.
143 A.E. Conrady, 'The Optical Sine-condition', *Mon. Not, Royal Astron. Soc.*, Vol. 65 (1905) pp. 51–59.
144 A.E. Conrady, Applied Optics and Optical Design, Part 1, Dover, New York (1985) pp. 367–401.(First published Oxford University Press, London, 1929) .
145 H.H. Hopkins, 'The Optical sine condition', *Proc. Phys. Soc.* London (1944) pp. 92–99.
146 J. Picht, Grundlagen der geometrisch-optischen Abbildung, VEB Deutscher Verlag der Wissenschaften, Berlin (1955) pp. 87–101.
147 W.T. Welford, 'Aplanatism and Isoplanatism', in *Progress in Optics, Vol. XIII*, Ed. E. Wolf, North Holland, Amsterdam (1976) pp. 267–292.
148 W.T. Welford, *Aberrations of optical systems*, Adam Hilger, Bristol (1986) pp. 84–87.
149 J. Braat, 'The Abbe sine condition and related imaging conditions in geometrical optics', *Proc. SPIE*, Vol. 3190 (1997) pp. 59–64.
150 H. Gross, Chapter 29 in *Handbook of Optical Systems, Vol.3*, Wiley-VCH, Weinheim (2007) p. 57.
151 M. Mansuripur, 'Abbe's sine condition', in *Classical Optics and its Applications*, Cambridge University Press, Cambridge (2009) pp. 9–21.
152 J.F.W. Herschel, 'XVII. On the aberrations of compound lenses and object glasses', *Phil. Trans. Roy. Soc. London* Vol. 111 (1821) pp. 222–267.
153 H.H. Hopkins, 'Herschel's Condition', *Proc. Phys. Soc. London* (1944) pp. 100–105.
154 H. Gross, Ref. 149, Chapter 29, p. 61.
155 T. Smith, 'On tracing rays through an optical system (Fourth paper)', *Proc. Phys. Soc. London*, Vol. 33 (1921) pp. 174–178.
156 T. Smith, 'Optical Calculations', in *A Dictionary of Applied Physics*, Ed. R. Glazebrook, Peter Smith, New York (1950) pp. 287–315.
157 W.T. Welford, 'A note on the skew invariant of optical systems', *Optica Acta*, Vol. 15 (1968) pp. 621–623.
158 W.T. Welford, Ref. 147, pp. 84–87.
159 M. Herzberger, Ref. 121, p. 30.
160 M. Herzberger, 'Geometrical Optics', Chapter 2 in Handbook of Physics, Eds., E.U. Codon and H. Odishaw, McGraw Hill, New York (1958) pp. 6–20 to 6–46.
161 H.A. Buchdahl, *Optical Aberration Coefficients*, Oxford University Press, London (1954) pp. 3–5.
162 A.E. Conrady, Ref. 70, p. 367.
163 R. Kingslake, *Lens Design Fundamentals*, Academic Press, New York (1978) p. 157.

164 M. Herzberger, Ref. 119, Chapter 7, pp. 64–68.
165 D.S. Goodman, 'Imaging limitations related to the skew invariant', *Proc. SPIE*, Vol. 4442 (2001) pp. 67–77.
166 W.T. Welford, Ref. 147, pp. 87–91.
167 G. Toraldo di Francia, 'Parageometrical optics', *J. Opt. Soc. Am.*, Vol. 40 (1950) pp. 600–602.
168 D.A. McQuarrie, *Statistical Mechanics,* University Science Books, California (2000) pp. 117–121.
169 D. Marcuse, *Light Transmission Optics*, Van Nostrand Reinhold, New York (1972) pp. 112–126.
170 R. Winston, J.C. Miñano and P. Benítez, *Nonimaging Optics*, Elsevier Academic, Amsterdam (2005) pp. 415–420.
171 D. Marcuse, 'Compression of a bundle of light rays', *Appl. Opt.*, Vol. 10 (1971) pp. 494–497.
172 J.C. Maxwell, 'On Hamilton's characteristic function for a narrow beam of light', *Proc. London Math. Soc.*, Vol. 6 (1874) pp. 117–123; also, in *The Scientific papers of James Clark Maxwell*, Vol. I, Ed. W.D. Niven, Dover, New York (1965) pp. 381–390.
173 J. Larmor, 'The characteristics of an asymmetric optical combination', *Proc. London Math. Soc.*, Vol. 20 (1889) pp. 181–194.
174 J. Larmor, 'The simplest specification of a given optical path, and the observations required to determine it', *Proc. London Math. Soc.*, Vol. 23 (1892) pp. 165–173.
175 R.W. Sampson, 'A continuation of Gauss's 'Dioptrische Untersuchungen''', *Proc. London Math. Soc.*, Vol. 29 (1897) pp. 33–83.
176 A. Gullstrand, 'Die reelle optische abbildung', *Svenska Vetensk. Handl.*, Vol. 41 (1906) pp. 1–119.
177 A. Gullstrand, 'Das allgemeine optische abbildungssystem', *Svenska Vetensk. Handl.*, Vol. 55 (1915) pp. 1–139.
178 A. Gullstrand, Appendices in *Helmholtz's treatise on Physiological Optics*, Vol. I, Ed. J.P.C. Southall, Part I. The Dioptrics of the eye, Optical Society of America (1924) pp. 261–482.
179 T. Smith, 'The primordial coefficients of asymmetrical lenses', *Trans. Opt. Soc. London*, Vol. 29 (1928) pp. 167–178.
180 T. Smith, 'Imagery around a skew ray', *Trans. Opt. Soc. London*, Vol. 31 (1929–1930) pp. 131–156.
181 H.H. Hopkins, *Geometrical Theory of image formation*, personal communication, Univ. Reading, UK (1982).
182 R.K. Luneburg, Ref. 27, pp. 234–243.
183 A. Walther, Ref. 41, pp. 238–252.
184 T. Smith, 'Canonical forms in the theory of asymmetrical optical systems', *Trans. Opt. Soc. London*, Vol. 29 (1927/28) pp. 88–98.
185 M. Herzberger, 'First order laws in asymmetrical optical systems Part I', *J. Opt. Soc. Am.*, Vol. 26 (1936) pp. 354–359.
186 M. Herzberger, 'First order laws in asymmetrical optical systems Part II', *J. Opt. Soc. Am.*, Vol. 26 (1936) pp. 389–406.
187 P.J. Sands, 'First order optics of the general optical system', *J. Opt. Soc. Am.*, Vol. 62 (1972) pp. 369–372.
188 S.A. Rodionov, 'Matrix treatment of Gaussian optics near an arbitrary ray', *Opt. Spectrosc. (USSR)*, Vol. 50 (1981) pp. 531–535.
189 H.H. Hopkins, 'Calculation of the aberrations and image assessment for a general optical system', *Opt. Acta*, Vol. 28 (1981) pp. 667–714.
190 H.H. Hopkins, 'Image formation by a general optical system. 1: General Theory', *Appl. Opt.*, Vol. 24 (1985) pp. 2491–2505.
191 J.M. Sasian, 'Review of methods for the design of unsymmetrical optical systems', *Proc. SPIE*, Vol. 1396 (1991) pp. 453–466.
192 B.D. Stone and G.W. Forbes, 'Foundations of first-order layout for asymmetric systems: an application of Hamilton's methods', *J. Opt. Soc. Am. A*, Vol. 9 (1992) pp. 96–109.
193 B.D. Stone and G.W. Forbes, 'Characterization of first-order optical properties for asymmetric systems', *J. Opt. Soc. Am. A*, Vol. 9 (1992) pp. 478–489.
194 B.D. Stone and G.W Forbes, 'Forms of the characteristic function for asymmetric systems that form sharp images to first order', *J. Opt. Soc. Am. A*, Vol. 9 (1992) pp. 820–831.
195 B.D. Stone and G.W. Forbes, 'First-order layout of asymmetric systems: sharp imagery of a single plane object', *J. Opt. Soc. Am. A*, Vol. 9 (1992) pp. 832–843.

196 B.D. Stone and G.W. Forbes, 'Foundations of second-order layout for asymmetric systems', *J. Opt. Soc. Am. A*, Vol. 9 (1992) pp. 2067–2082.
197 R. Kingslake, 'Who? Discovered Coddington's Equations?', *Opt. Phot. News*, Aug. (1994) pp. 20–23.
198 H. Coddington, *A Treatise on the Reflexion and Refraction of Light*, Simpkin and Marshall, London (1829) p. 66.
199 T. Young, 'On the mechanism of the eye', *Phil. Trans. Roy. Soc. London*, Vol. 91 (1801) pp. 23–88.
200 T. Smith, 'The Contributions of Thomas Young to Geometrical Optics, and their Application to Present-day Questions', *Proc. Phys. Soc. B London*, Vol. 62 (1949) pp. 619–629.
201 S. Czapski, *Grundzüge der Theorie der Optischen Instrumente nach Abbe*, 2nd Edition Johann Ambrosius Barth, Leipzig (1904) pp. 92–100.
202 P. Culmann, 'Die Realisierung der Optischen Abbildung', Chap. IV, *Theorie der Optischen Instrumente*, Ed. M. von Rohr, Berlin (1904) p. 167; also, in *Geometrical Investigation of the formation of images in optical instruments*, Trans. R. Kanthack, H. M. Stationery Office, London (1920) pp. 157–170.
203 J.P.C. Southall, *The Principles and Methods of Geometrical Optics*, Macmillan, New York (1910) pp. 340–345.
204 A.E. Conrady, Ref. 70, pp. 409–410.
205 R. Kingslake, *Lens Design Fundamentals*, Academic, New York (1978) pp. 185–187.
206 W.T. Welford, Ref. 147, pp. 187–190.
207 H.H. Hopkins, 'A transformation of known astigmatism formulae', *Proc. Phys. Soc. London*, Vol. 58 (1946) pp, 663–668.

8

Monochromatic Aberrations

8.1 A Journey to the Wonderland of Optical Aberrations: A Brief Early History [1]

It is apparent from the works of Ibn-al-Haytham, also known as Alhazen [2 – 3], and of Roger Bacon [4] that a very elementary conceptual appreciation of aberrations as defects in image formation has been in existence since the time of Hero of Alexandria about two millennia back [5]. But a systematic scientific study on aberrations originated from Kepler [6], who first made an observation on spherical aberration a few years after the advent of the Galilean telescope in 1608 [7]. The cause of this aberration was understood after the discovery of the exact law of refraction by Snell in 1621, and the first mathematical formulation to explain spherical aberration was presented by Descartes in 1637 [8]. In the course of his investigations on the success of the Huygenian eyepiece around 1665, Newton first differentiated between spherical aberration and chromatic aberration, and ascribed the causes to the surfaces and the material, respectively [9]. However, severe limitations in experimental measurements drove him to the wrong conclusion that all glasses have same dispersion characteristics, so that refracting telescopes cannot be made achromatic. A lull set in for about 80 years until Chester Hall and John Dollond developed achromatic doublets in the 1750s [10]. Clairaut investigated the problem of achromatization in more detail in the 1760s [11]. D'Alembert described a triple glass objective in 1764; he was the first to distinguish between the longitudinal and transverse features of spherical aberration and chromatic aberration [12]. Incidentally, another wrong observation by Newton on the correction of spherical aberration of a thin triplet has been noted by Van Heel [13].

In 1762, Clairaut observed extra-axial aberrations and wrongly concluded that they could not be corrected. In the 1780s, the Ramsden eyepiece was developed, and the Ramsden disk provided a demonstration of the concept of the exit pupil [14]. Young [15] and Airy's [16] investigations in the early nineteenth century served as the foundation for Coddington's treatment of astigmatism [17].

A real breakthrough in the treatment and understanding of aberrations took place during 1827–1837 when the seminal works of Hamilton on ray optics were published [18]. We discussed the rudiments of Hamiltonian optics and the relative roles of Hamiltonian optics and Lagrangian optics in analysis and synthesis of optical systems in Section 7.1.

Around 1840, Gauss undertook extensive theoretical investigations on the shape of a wavefront exiting an optical system [19]. Expressing the wavefront as a power series in terms of aperture and field variables, he observed that the first order terms expressed the basic imaging relationships while ignoring the quality of the image. This laid the foundations for paraxial or first order optics, also known as Gaussian optics. He noted that the higher order terms expressed aberrations of the image. Seidel [20] developed systematic formulae for calculation of third order aberrations.

However, Rakich and Wilson [21] have put forward some evidence in support of the primacy of Joseph Petzval in these ventures. It is well-known that, closely on the heels of commercialization of the process of 'Daguerrotype' for photography in 1839, Petzval came forward with his portrait lens [22], which was widely acclaimed as a breakthrough. When Petzval undertook the task, works of Gauss and Seidel were not yet reported. It is presumed that Petzval developed analytical procedures for the treatment of defects in optical imaging, and even developed the theory of higher order aberrations [23–24]. Unfortunately, his extensive manuscript, which was on the verge of publication, was destroyed by thieves, and concurrently, the optics companies that manufactured and marketed the lenses designed by him usurped the total fortune for themselves, depriving Petzval of any share. The shattered Petzval never rewrote the manuscript,

DOI: 10.1201/9780429154812-8

and he moved on to other fields. It was many decades before the topics were revisited by enterprising investigators in optical imaging.

While earlier works in aberration theory were mostly concerned with developments in telescopic and microscopic imagery, in the 1840s rapid developments in photography called for a better understanding of field dependent aberrations in particular and for the development of suitable analytical and numerical tools for studying the combined effects of both aperture and field dependent aberrations.

This topic provided a veritable playground for scientists and mathematicians with a flair for approximation theory, differential geometry, and algebraic analysis. Numerous investigators were involved in the pursuit of these phenomena, and noteworthy contributions were made by Maxwell, Rayleigh, Abbe, Sampson, Schwarzschild, Kerber, Smith, Kohlschütter, Conrady, Turrière, Carathéodory, Berek, Steward, Zernike, Herzberger, Luneburg, Nijboer, Maréchal, Hopkins, Kingslake, Picht, Focke, Wachendorf, Welford, Wynne, and Buchdahl, among others [25–78]. Pitfalls abound this tricky domain so much so that one can witness several polemics that took substantial time and effort to settle.

Allvar Gullstrand, who dealt with the problem of the general neighbourhood of a ray in a normal system, deserves special mention. His work in the field of the dioptric of the eye was crowned with the Nobel prize in 1911 [79–81].

8.2 Monochromatic Aberrations

In this section, we present the theory of monochromatic aberrations.

8.2.1 Measures of Aberration

In the case of aberration free imaging, all rays of the image-forming pencil, corresponding to a homocentric pencil of rays originating from a point in the object space, meet at the desired point in the image space. Equivalently in this case, a spherical wavefront originating from a point in the object space is transformed into a spherical wavefront with its centre at the desired image point in the image space. Any departure from this ideal behaviour is ascribed to aberrations introduced by the imaging system. In the case of general optical systems with no discernible symmetry, not more than two rays meet at a point in the image space, and the multitude of ray intersection points of the image-forming pencil constitutes a close three-dimensional conglomeration or cluster of points instead of the desired point image.

A meaningful measure of aberration needs to be defined in a manner that can unfold the interrelationship between the inherent imaging characteristics of the system with the 'defects' occurring in the image. The relation between the two is neither simple nor straightforward. Nevertheless, various measures of aberration have been proposed from the early days of development of axially symmetric optical systems, and often they have served the purpose they were sought for.

Figure 8.1 shows an optical imaging system symmetric around the axis OO′. Let the desired image point for the axial object point O be O′. E and E′ are the axial locations of the entrance pupil and the exit

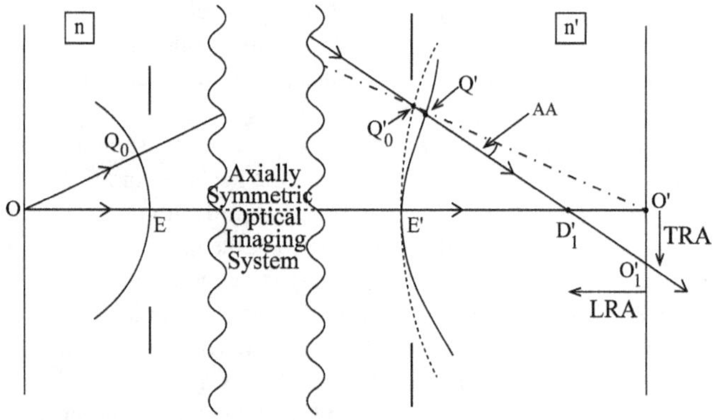

FIGURE 8.1 Finite conjugate axial imaging by an axially symmetric optical imaging system.

pupil. The refractive indices of the object space and the image space are n and n′, respectively. OEE′O′ is the axial ray from the object point O to the image point O′. OQ_0 is an arbitrary ray from the object point O. $E'Q_0'$ is a section of a sphere with O′ as centre and E′O′ as radius. Had the imaging been ideal or aberration free, $Q_0'O'$ would be the image space ray corresponding to the object space ray OQ_0. But the real image space ray is $Q_0'D_1'O_1'$. This ray intersects the optical axis at D_1', and the transverse line at O′ at the point O_1'. The aberration introduced by the ray $OQ_0Q_0'D_1'$ can be measured either by the departure $O'O_1'$ of the ray intersection point O_1' on the transverse plane from the desired image point O′, or by the departure $D_1'O'$ along the optical axis of the point of intersection D_1' of the real ray with the optical axis from the desired image point O′. While $O'O_1'$ is called the transverse ray aberration (TRA), $D_1'O'$ is called the longitudinal ray aberration (LRA).

$E'Q_0'$ is a section of a reference sphere with its centre at O′ and radius E′O′. The point of intersection of the real image space ray $Q_0'D_1'O_1'$ with the reference sphere is Q_0'. The angle $\angle O'Q_0'D_1'$ represents an angular departure of the real image space ray $Q_0'D_1'O_1'$ from the ideal ray path $Q_0'O'$, and this angle can also be considered to be a measure of aberration introduced by the ray. It is called angular aberration (AA), usually denoted by $\delta\psi$. All three measures of aberration are equivalent. However, it may be noted that, while transverse ray aberration and longitudinal ray aberration are invariant to the choice of radius of the reference sphere, angular aberration varies considerably with variation in the radius sphere. Therefore, in the case of finite conjugate imaging, its use is limited. However, the angular aberration provides a highly useful and meaningful measure of aberration for the image at infinity (see Figure 8.2). Both transverse and longitudinal ray aberration are not effective as measures of aberration in this case.

In Figures 8.1 and 8.2, EQ_0 is a section of the spherical wavefront in the object space originating from the axial point O. The corresponding section of the real imaging wavefront is E′Q′. Because of aberrations introduced by the system, the latter is not spherical. In the absence of any aberration introduced by the imaging system, the imaging wavefront passing through E′ would have been perfectly spherical with the centre of the sphere at the desired image point O′. $E'Q_0'$ is a section of this sphere that can be used as a reference for measuring the deformation of the real wavefront E′Q′. This provides another highly useful measure of aberration. Wave aberration (WA) is defined as the optical path length $[Q_0'Q']$ along the ray $Q_0'Q'D_1'O_1'$ [45].

Figure 8.3 shows the paths of the principal ray PEE′P′ and another arbitrary ray $PAA'P_1'$ of the image-forming pupil of rays originating from the off-axial object point P of an axially symmetric imaging system. Transverse object and image planes OP and O′P′ are located respectively at O and O′ on the axis. Entrance and exit pupils are located at E and E′ on the axis. Four parallel rectangular coordinate systems $(\xi,\eta,\varsigma),(\xi',\eta',\varsigma'),(X,Y,Z),$, and (X',Y',Z') with respective origins at O, O′, E, and E′ are used to specify points on and in the neighbourhood of the object, image, entrance pupil, and exit pupil. The off-axial point P is taken on the η-axis without any loss of generality, for a rigid rotation of the coordinate axes around the optical axis of the axially symmetric imaging system can make the η-axis coincident with

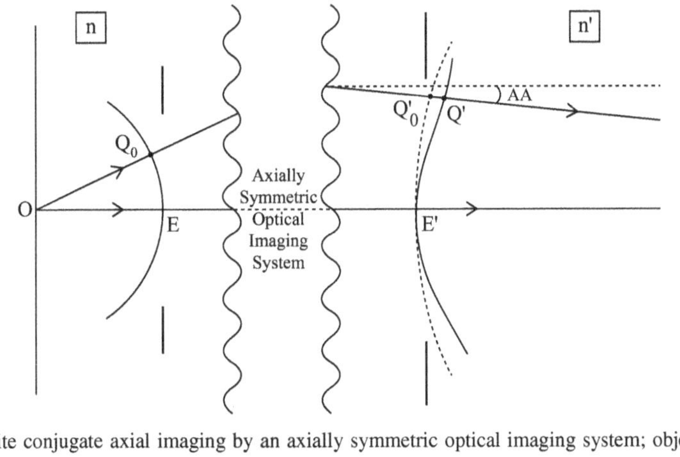

FIGURE 8.2 Infinite conjugate axial imaging by an axially symmetric optical imaging system; object at finite distance, image at infinity.

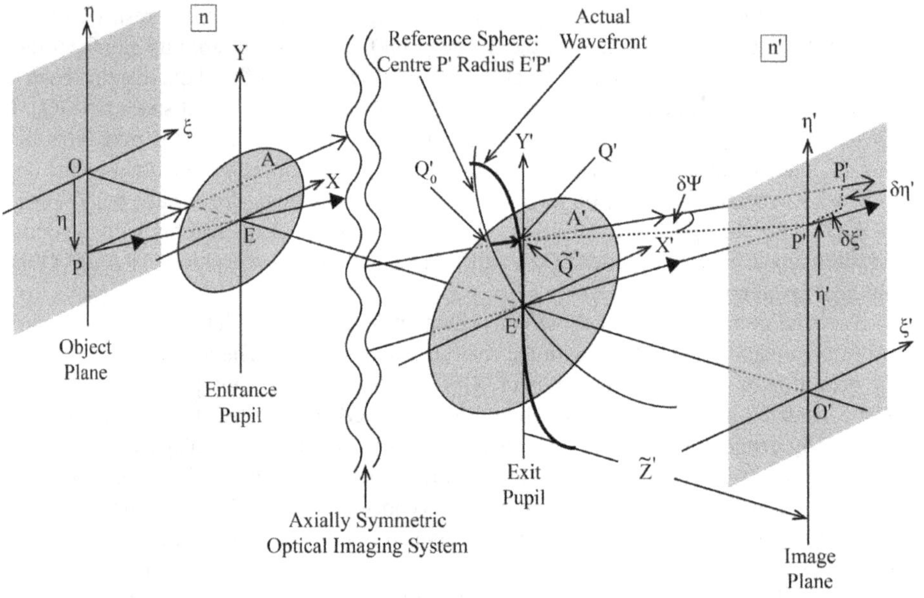

FIGURE 8.3 Imaging of an extraaxial point P by an axially symmetric imaging system.

any arbitrary location for P. Note that, due to symmetry, the principal ray PEE′P′ lies on the meridian plane, but, in general, an arbitrary ray of the image-forming pupil of rays, e.g. $PAA'P'_1$ is a skew ray. It may be reiterated that a skew ray of an axially symmetric system never intersects the axis. Let the skew ray intersect the transverse image plane at P'_1. Local coordinates of the points P, P′, and P'_1 are $(0,\eta,0)$, $(0,\eta',0)$ and $(\delta\xi',\eta'+\delta\eta',0)$, respectively. Two components of the departure of the ray intersection point P'_1 from the desired image point P′ are $\delta\xi'$ and $\delta\eta'$, and these are called components of transverse ray aberration. Note that due to axial symmetry, only one component, say $\delta\eta'$ is required to specify the transverse ray aberration in the case of axial imagery.

In the case of extraaxial imagery, in general, the two ray segments $A'P'_1$ and E′P′ are non-intersecting, and, therefore, it is not possible to define longitudinal aberration in a meaningful way. However, when analysis of extra-axial imagery is restricted to imaging by rays that lie only on the meridional section, longitudinal ray aberration may be defined as the departure of the point of intersection of the aperture ray and of the principal ray from the ideal image point P′ along the principal ray.

A reference sphere with its centre at P′ and $E'P'(=R)$ as radius is used to measure deformation of the actual wavefront E′Q′ in the image space. Note that both the reference sphere and the actual wavefront pass through E′, the centre of the exit pupil. The ray $A'P'_1$ intersects the reference sphere and the actual wavefront at points Q'_0 and Q′respectively. Angular aberration $\delta\psi$ can be defined by the angle $\angle P'Q'_0P'_1$. As in the axial case, it is useful as a measure of aberration when the image is at infinity. Wavefront aberration, or, in short, wave aberration W, is defined by the optical path length $[Q'_0Q']$ along the ray $Q'_0Q'A'P'_1$.

It is important to note that all measures of aberration mentioned above relate to the aberration introduced by a specific ray. In Figure 8.3, that ray is $PAA'P'_1$. Any ray passing through the system is uniquely specified by two points through which the ray is passing. In treatments of aberration theory, one of these points is usually taken as the object point, e.g., P, or the desired image point P′, and the other point is the point of intersection of the ray with either the entrance or the exit pupil plane or with a reference sphere centred on the object/image point and passing through the centre of either the entrance pupil or the exit pupil. In the following treatment, the latter point is taken as the point Q'_0 (Figure 8.3). Let the coordinates of the point Q'_0 be (X',Y',Z'). Note that out of the three coordinates X',Y',Z', only two are independent, since the point Q'_0 lies on a prespecified reference sphere. Considering X',Y' as the two independent coordinates, all measures of aberrations pertaining to a specific object/image point, say P/P′, are functions

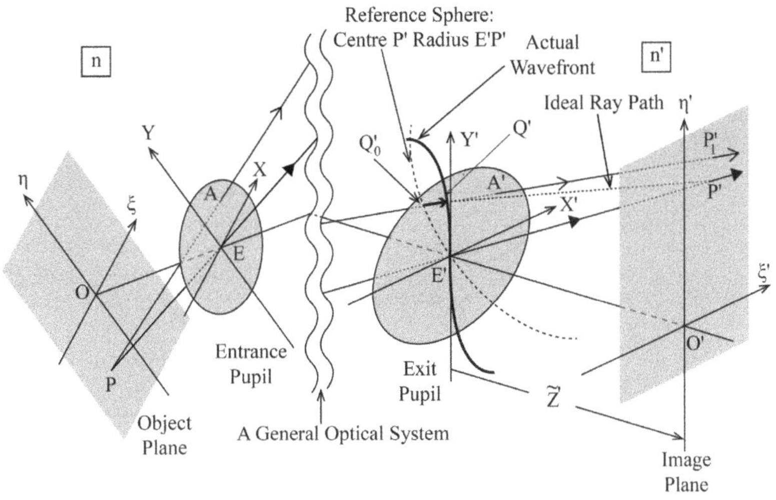

FIGURE 8.4 Imaging of an arbitrary point P in an object by a general optical system.

of (X', Y'), e.g. $W(X', Y')$, $\delta\xi(X', Y')$, $\delta\eta(X', Y')$, $\delta\psi(X', Y')$, etc. Few treatments of aberration theory have considered aberrations in terms of two suitable coordinates of the entrance pupil, although the use of exit pupil coordinates is more widespread.

Figure 8.4 illustrates the situation for a general optical system. A 'general optical system' implies a system with no known structural symmetry, more specifically a system that does not have rotational symmetry around an axis, e.g. a system consisting of tilted or decentred components. These systems, which are also called 'asymmetrical optical systems', may consist of components that have freeform interfaces. Some of the characteristic differences of a general optical system from an axially symmetric system are highlighted in Figure 8.4. Note that an arbitrary ray in a general optical system needs four parameters for its specification; this can by no means be reduced to three coordinates, as is possible in the treatment of axially symmetric systems. Attempts are continuing to be made in order to develop techniques that enable the handling of the general systems in a similar manner as axially symmetric systems [82–84].

8.2.1.1 Undercorrected and Overcorrected Systems

It should be noted that the wave aberration is defined as the optical path length $n'(Q'_0 Q')$, where Q'_0 and Q' are the points of intersection of the aperture ray with the reference sphere and the actual wavefront, respectively. For the sake of simplicity, Figure 8.5(a) and Figure 8.5(b) show the meridian section in the image space of an axially symmetric system in two dimensions. The same notation for the different terms are used. The aperture ray intersects the principal ray at the point $\tilde{P}'$. Note that in Figure 8.5(a), the distance $Q'_0 Q'$ is positive, and consequently, the wave aberration is positive. In this case, the distance $P'\tilde{P}'$ is negative. But in Figure 8.5(b), the distance $Q'_0 Q'$ is negative and consequently, the wave aberration is negative. In this case, the distance $P'\tilde{P}'$ is positive. In the former case, the imaging system is said to be 'undercorrected' for this aberration, whereas the system is said to be 'overcorrected' in the latter case. Presently, a majority of optical engineers adopt this convention.

8.2.2 Ray Aberration and Wave Aberration: Interrelationship

The wave aberration introduced by the ray $PAA'P'_1$ is measured with respect to the reference ray $PEE'P'$, which is the principal ray of the pencil forming the image of the extraaxial point P. It has been defined as the optical path length $\left[Q'_0 Q'\right]$ along the ray $Q'_0 Q' A' P'_1$. Note that Q'_0 and Q' are, respectively, the points of intersection of the ray $Q'_0 Q' A' P'_1$ with the reference sphere and the actual wavefront, both passing through the centre of the exit pupil E′ (Figure 8.3). For a different ray of the pencil originating from P, the

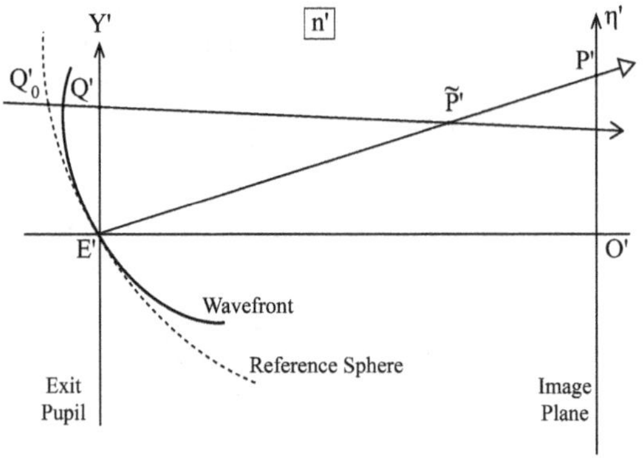

FIGURE 8.5(a) Image space of an undercorrected system.

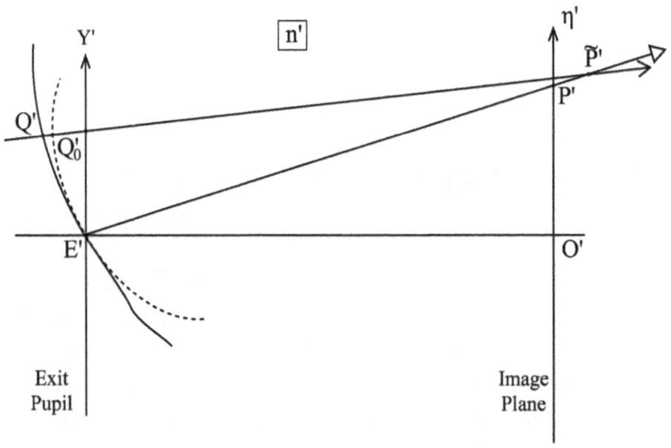

FIGURE 8.5(b) Image space of an overcorrected system.

two points of intersection—and consequently, the wave aberration—will be different. As explained in the earlier section, the wave aberration function for the imaging of object point P is represented by $W(X', Y')$. The treatment presented below is adapted from the one presented by Welford [85]. Using Hamilton's point characteristic function E (see Section 7.1.1), we can write

$$W(X', Y') = n'(Q'_0 Q') = [E(P, Q') - E(P, Q'_0)] = [E(P, E') - E(P, Q'_0)] \tag{8.1}$$

since Q' and E' are on the same wavefront, and the two points are, therefore, at the same optical distance from P. Also, $E(P, E')$ is independent of (X', Y'), for the deformation of the wavefront is measured with respect to the same reference sphere for all values of (X', Y'). Therefore,

$$\frac{\partial W}{\partial X'} = -\frac{\partial E(P, Q'_0)}{\partial X'} - \frac{\partial E(P, Q'_0)}{\partial Z'} \frac{\partial Z'}{\partial X'} \tag{8.2}$$

$$\frac{\partial W}{\partial Y'} = -\frac{\partial E(P, Q'_0)}{\partial Y'} - \frac{\partial E(P, Q'_0)}{\partial Z'} \frac{\partial Z'}{\partial Y'} \tag{8.3}$$

With respect to the origin of the rectangular coordinate system at E', the equation of the reference sphere is given by

$$X'^2 + Y'^2 + Z'^2 - 2\eta' Y' - 2\tilde{Z}' Z' = 0 \tag{8.4}$$

Where $\tilde{Z}' = E'O'$. From (7.1.1) we get

$$\frac{\partial E}{\partial X'} = -n'L' \qquad \frac{\partial E}{\partial Y'} = -n'M' \qquad \frac{\partial E}{\partial Z'} = -n'N' \tag{8.5}$$

From (8.4), by differentiation, it follows

$$\frac{\partial Z'}{\partial X'} = \frac{X'}{(\tilde{Z}' - Z')} \qquad \frac{\partial Z'}{\partial Y'} = \frac{(Y' - \eta')}{(\tilde{Z}' - Z')} \tag{8.6}$$

The direction cosines are given by

$$L' = \frac{(\delta\xi' - X')}{Q_0'P_1'} \qquad M' = \frac{(\eta' + \delta\eta') - Y'}{Q_0'P_1'} \qquad N' = \frac{(\tilde{Z}' - Z')}{Q_0'P_1'} \tag{8.7}$$

Substituting these values in equations (8.2) and (8.3), we get

$$\delta\xi' = -\frac{Q_0'P_1'}{n'}\frac{\partial W}{\partial X'} \tag{8.8}$$

$$\delta\eta' = -\frac{Q_0'P_1'}{n'}\frac{\partial W}{\partial Y'} \tag{8.9}$$

The equations (8.8) and (8.9) relating the wave aberration W with the transverse ray aberrations $\delta\xi', \delta\eta'$ are exact relations, but they involve, on the right-hand side, the distance $Q_0'P_1'$. The latter distance depends on the coordinates of P_1', i.e. on the ray aberration components $\delta\xi', \delta\eta'$. For most practical cases, the angular aberration $\delta\psi$ is so small that the following approximation is valid.

$$Q_0'P' = Q_0'P_1' \cos\delta\psi \approx Q_0'P_1' \tag{8.10}$$

Representing radius $Q_0'P'$ of the reference sphere by R', we obtain from (8.8) – (8.9) the practically useful relation between the wave aberration and the transverse ray aberration as

$$\delta\xi' = -\frac{R'}{n'}\frac{\partial W}{\partial X'} \tag{8.11}$$

$$\delta\eta' = -\frac{R'}{n'}\frac{\partial W}{\partial Y'} \tag{8.12}$$

Note that X', Y' are coordinates of the point Q_0', which is the point of intersection of the ray with the reference sphere, and not of the point of intersection of the ray with the exit pupil plane. The latter point is shown as the point A' in Figure 8.3.

In the literature, sometimes wave aberration W is defined as the optical path length $[Q_0'\tilde{Q}']$, where $\tilde{Q}'$ is the point of intersection of the radial line $Q_0'P'$ with the actual wavefront E′Q′ [86]. Use of this definition for W leads to the following exact relation between the wave aberration and the ray aberration [87–88].

$$\delta\xi' = -\frac{\tilde{Q}'P'}{n'}\frac{\partial W}{\partial X'} = -\frac{R' - W}{n'}\frac{\partial W}{\partial X'} \tag{8.13}$$

$$\delta\eta' = -\frac{\tilde{Q}'P'}{n'}\frac{\partial W}{\partial Y'} = -\frac{R' - W}{n'}\frac{\partial W}{\partial Y'} \tag{8.14}$$

Since for most practical cases, $|W| \ll R'$, the above relations also reduce to (8.11) – (8.12) in practice.

8.2.3 Choice of Reference Sphere and Wave Aberration

In the neighbourhood of the focal region, the aberrated wavefronts forming an image take such complicated shapes that it is not meaningful to use their shapes for a measure of wave aberration. In ray optical treatments of image quality assessment, the shapes of the image-forming wavefronts have been studied at regions far away from the focus. The particular choice for the exit pupil region to study deformation of the wavefront is preferred in cases where image quality assessment needs to take into account the diffraction effects. The ready availability of the shape of the wavefront, or, equivalently, the phase distribution over the wavefront at the exit pupil, can be directly utilized in evaluation of the diffraction pattern, as shown in Chapter 17. On the other hand, the centre of the reference sphere for defining the wave aberration is taken as the point of intersection of the principal ray with the image plane. This is done to develop a formal treatment. However, it must be realized that in aberrated imaging systems, the choice for a suitable centre of a reference sphere is, in general, tricky. An obvious choice for most practical cases is to choose the point that minimizes the effects of aberration on image quality as the centre. This calls for suitable expressions for determining the change in wave aberration when the centre of the reference sphere is changed.

8.2.3.1 Effects of Shift of the Centre of the Reference Sphere on Wave Aberration

Figure 8.6 shows the image space of an axially symmetric imaging system. The transverse image plane and the exit pupil are located respectively at O'_0 and E' on the axis. Let $E'O'_0 = \tilde{Z}'$. (X', Y', Z') and $(\xi', \eta', \varsigma')$ are two parallel rectangular coordinate systems with origins at E' and O'_0. The segment $E'P'_0$ of the principal ray originating from an object point P on the η-axis lies on the meridional plane; it intersects the η'-axis at the point P'_0. An arbitrary aperture ray from P intersects the exit pupil plane at A' and the image plane at P'_1. This ray intersects the wavefront passing through E' at Q'. $E'Q'_0$ is a section of the reference sphere with its centre at P'_0 and radius $E'P'_0 (= R')$. Let the centre of the reference sphere be shifted to the point $\hat{P}'$. $E'\hat{Q}'$ is a section of the new reference sphere with $\hat{P}'$ as the centre, and $E'\hat{P}'$ as the radius. The aperture ray $A'P'_1$ intersects the original reference sphere at point Q'_0 and the new reference sphere at point $\hat{Q}'$. With respect to the origin at E', the coordinates of P'_0, P'_1 and $\hat{P}'$ are $(0, \eta', \tilde{Z}')$, $(\delta\xi', \eta' + \delta\eta', \tilde{Z}')$, and $(\Delta\xi', \eta' + \Delta\eta', \tilde{Z}' + \Delta\varsigma)$. With the same origin coordinates of Q'_0 and $\hat{Q}'$ are (X', Y', Z') and $(X' + \Delta X', Y' + \Delta Y', Z' + \Delta Z')$, respectively.

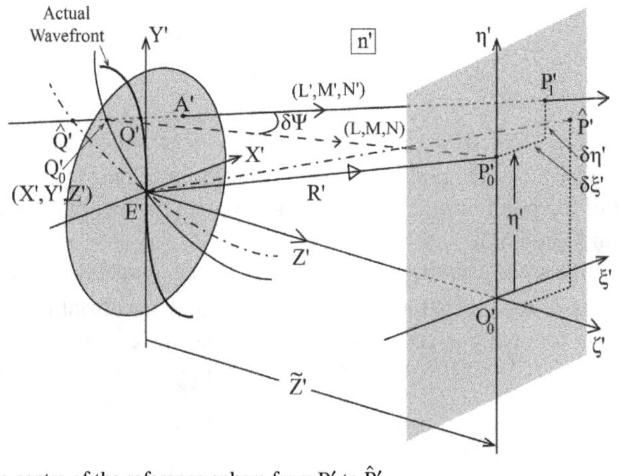

FIGURE 8.6 Shift of the centre of the reference sphere from P'_0 to $\hat{P}'$.

For the specific aperture ray, the wave aberration with respect to the original reference sphere is

$$W = \left[Q_0' Q' \right] \tag{8.15}$$

With respect to the new reference sphere, the wave aberration is given by

$$\left[\hat{Q}' Q' \right] = \left[\hat{Q}' Q_0' \right] + \left[Q_0' Q' \right] = \Delta W + W \tag{8.16}$$

The change in wave aberration due to the shift of the centre of the reference sphere from P_0' to $\hat{P}'$ is equal to

$$\Delta W = \left[\hat{Q}' Q_0' \right] \tag{8.17}$$

The equation of the original reference sphere $E'Q_0'$ is

$$X'^2 + (Y' - \eta')^2 + (Z' - \tilde{Z}')^2 = R'^2 = \eta^2 + \tilde{Z}'^2 \tag{8.18}$$

In the above equation, the coordinates of the centre of the reference sphere is $(0, \eta', \tilde{Z}')$. This implies that the centre is constrained to lie on the η'-axis. In order to incorporate into the analysis the scope for shifting the centre to an arbitrary point, say $(\xi', \eta', \tilde{Z}')$, this restriction needs to be relaxed. Removing this constraint, let the centre of the reference sphere be at coordinates, say, $(\xi', \eta', \tilde{Z}')$. The equation of the reference sphere then becomes

$$X'^2 + Y'^2 + Z'^2 - 2\xi' X' - 2\eta' Y' - 2\tilde{Z}' Z' = 0 \tag{8.19}$$

Differentiating (8.19), we get

$$X'\Delta X' + Y'\Delta Y' + Z'\Delta Z' - \Delta\xi' X' - \xi'\Delta X' - \Delta\eta' Y' - \eta'\Delta Y' - \Delta\tilde{Z}' Z' - \tilde{Z}'\Delta Z' = 0 \tag{8.20}$$

If the centre of the original reference sphere is taken at $P_0'(0, \eta', \tilde{Z}')$ and that of the new reference sphere is taken at $\hat{P}'(\Delta\xi', \eta' + \Delta\eta', \tilde{Z}' + \Delta\varsigma')$, we can substitute $\xi' = 0$ and $\Delta\tilde{Z}' = \Delta\varsigma'$ in (8.20) to obtain

$$X'\Delta X' + Y'\Delta Y' + Z'\Delta Z' - \Delta\xi' X' - \Delta\eta' Y' - \eta'\Delta Y' - \Delta\varsigma' Z' - \tilde{Z}'\Delta Z' = 0 \tag{8.21}$$

If the direction cosines of the aperture ray $\hat{Q}' Q_0' Q' A' P_1'$ are (L', M', N'), using (8.17) it follows

$$\Delta X' = -L' \frac{\Delta W}{n'} \qquad \Delta Y' = -M' \frac{\Delta W}{n'} \qquad \Delta Z' = -N' \frac{\Delta W}{n'} \tag{8.22}$$

From (8.21) and (8.22) we get

$$\Delta W = -\frac{n'(X'\Delta\xi' + Y'\Delta\eta' + Z'\Delta\varsigma')}{L'X' + M'(Y' - \eta') + N'(Z' - \tilde{Z}')} \tag{8.23}$$

The direction cosines (L, M, N) of the radial line $Q_0' P_0'$ are given by

$$L = -\frac{X'} {R'} \qquad M = \frac{(\eta' - Y')}{R'} \qquad N = \frac{(\tilde{Z}' - Z')}{R'} \tag{8.24}$$

With $\angle P_1' Q_0' P' = \delta\psi$ as the angular aberration, it follows

$$\cos\delta\psi = LL' + MM' + NN' = -\frac{L'X' + M'(Y' - \eta') + N'(Z' - \tilde{Z}')}{R'} \tag{8.25}$$

For most practical purposes, the magnitude of $\delta\psi$ is so small that

$$R'\cos\delta\psi \approx R' \tag{8.26}$$

Using (8.23), (8.25), and (8.26), a succinct relation for the change ΔW in wave aberration by a shift of centre of the reference sphere is obtained as

$$\Delta W = \frac{n'}{R'}\left[X'\Delta\xi' + Y'\Delta\eta' + Z'\Delta\varsigma'\right] \tag{8.27}$$

Two special cases of shift of centre of the reference sphere, namely longitudinal shift of the image plane and transverse shift of the chosen image point on the same image plane, are of much practical importance in aberration theory and also in image quality assessment, and they are treated in more detail in the next subsections.

8.2.3.1.1 Longitudinal Shift of Focus

The commonly used expression 'Longitudinal shift of focus' implies a longitudinal shift of the image plane. Figures 8.7 and 8.8 illustrate the effects of this shift with the help of ray geometry in the image space for axial imaging and extra-axial imaging, respectively. For the sake of convenience, the figures are shown in two dimensions. In both diagrams, E' and O_0' are the axial locations of the exit pupil and the original location of the image plane. The axial location of the shifted image plane is at $\hat{O}'$. The longitudinal shift, $O_0'\hat{O}' = R' - R_0' = \Delta\varsigma'$, where $R' = E'\hat{O}'$ and $R_0' = E'O_0'$. The refractive index of the image space n'.

In the extraaxial case (Figure 8.8), $E'P_0'\hat{P}'$ is the image space segment of the principal ray of a pencil of rays originating from an extraaxial object point P lying on the Y-axis. By symmetry, in the axially symmetric imaging system, the principal ray $E'P_0'\hat{P}'$ lies on the meridional plane; it intersects the original image plane at P_0', and the shifted image plane at $\hat{P}'$. The direction cosines of the principal ray $E'P_0'\hat{P}'$ are $\left(0, \bar{M}', \bar{N}'\right)$. $E'Q'$ is the image-forming wavefront passing through the centre E' of the exit pupil. For the location of the image plane at O_0', $E'Q_0'$ is a reference sphere passing through E'; its centre is at P_0', and its radius is $E'P_0' = \bar{R}_0'$. After the longitudinal shift of the image plane to $\hat{O}'$, the effective image point is $\hat{P}'$. $E'\hat{Q}'$ is the new reference sphere passing through E'; its centre is at $\hat{P}'$, and its radius is $E'\hat{P}' = \hat{R}'$. $\hat{Q}'Q_0'Q'P_1'\hat{P}_1'$ is the image space segment of an aperture ray belonging to the same pencil from the object

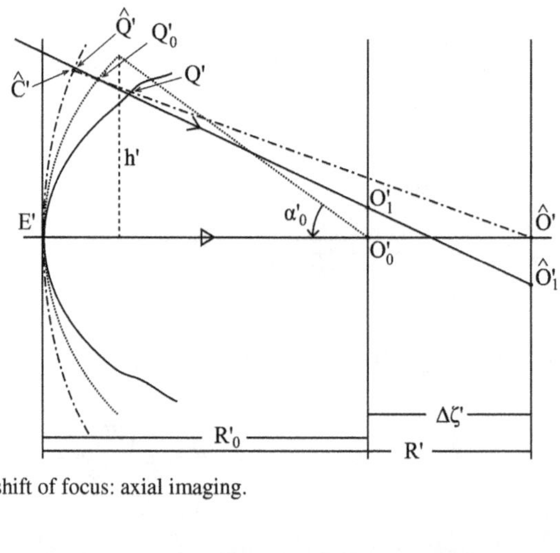

FIGURE 8.7 Longitudinal shift of focus: axial imaging.

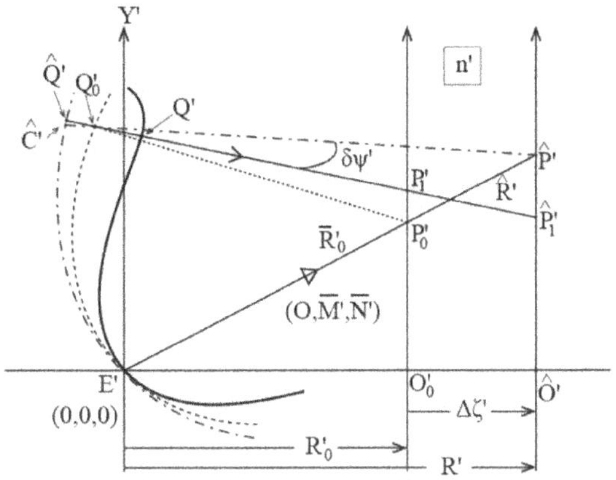

FIGURE 8.8 Longitudinal shift of focus: extraaxial imaging. $E'\hat{P}' = \hat{R}'; E'P_0' = \bar{R}_0'$.

point P. It intersects the imaging wavefront $E'Q'$, the original reference sphere $E'Q_0'$ and the new reference sphere $E'\hat{Q}'$ at Q', Q_0', and $\hat{Q}'$, respectively. The change in wave aberration by the longitudinal shift of the image plane is given by

$$\Delta W_L' = \left[\hat{Q}'Q_0'\right] = \left[\hat{Q}'Q'\right] - \left[Q_0'Q'\right] = \hat{W}' - W_0' \tag{8.28}$$

The angular aberration $\delta\psi'$ of the ray $\hat{Q}'Q_0'Q'P_1'\hat{P}_1'$, with respect to the new reference focus $\hat{P}'$ is

$$\delta\psi' = \angle\hat{P}_1'Q_0'\hat{P}' \tag{8.29}$$

Line $\hat{C}'Q_0'\hat{P}'$ is a radius of the new reference sphere. It may be seen that with negligible error

$$\hat{Q}'Q_0' \approx \frac{\hat{C}'Q_0'}{\cos\delta\psi'} \approx \hat{C}'Q_0' = p'(\text{say}) \tag{8.30}$$

With respect to the origin at $E'(0,0,0)$, the coordinates of $Q_0', \hat{P}'$ and P_0' are (X',Y',Z'), $(0,\bar{M}'\hat{R}',\bar{N}'\hat{R}')$, and $(0, \bar{M}'\bar{R}_0', \bar{N}'\bar{R}_0')$, respectively. The length $Q_0'\hat{P}'$ is given by

$$Q_0'\hat{P}' = \left(\hat{R}' - p'\right) = \left[(X')^2 + \left(Y' - \bar{M}'\hat{R}'\right)^2 + \left(Z' - \bar{N}'\hat{R}'\right)^2\right]^{\frac{1}{2}} \tag{8.31}$$

After algebraic manipulations and neglecting p'^2, we get

$$p' = -\left(\frac{1}{2\hat{R}'}\right)\left(X'^2 + Y'^2 + Z'^2\right) + \left(\bar{M}'Y' + \bar{N}'Z'\right) \tag{8.32}$$

The equation of the original reference sphere is

$$(X')^2 + \left(Y' - \bar{M}'\bar{R}_0'\right)^2 + \left(Z' - \bar{N}'\bar{R}_0'\right)^2 = \bar{R}_0'^2 \tag{8.33}$$

From the above, it follows

$$\left(\bar{M}'Y' + \bar{N}'Z'\right) = \left(\frac{1}{2\bar{R}_0'}\right)\left(X'^2 + Y'^2 + Z'^2\right) \tag{8.34}$$

By substituting from (8.34) in (8.32), we get

$$p' = \frac{1}{2}\left(\frac{1}{\bar{R}_0'} - \frac{1}{\hat{R}'}\right)\left(X'^2 + Y'^2 + Z'^2\right) \tag{8.35}$$

From Figure 8.8, we note

$$\bar{R}_0{}' = \frac{R_0'}{\bar{N}'} \qquad \hat{R}' = \frac{R'}{\bar{N}'} \tag{8.36}$$

Substituting these values, we get from (8.35)

$$p' = \frac{1}{2}\bar{N}'\left(\frac{1}{R_0'} - \frac{1}{R'}\right)\left(X'^2 + Y'^2 + Z'^2\right) \tag{8.37}$$

Normalizing the coordinates X', Y', Z' by the maximum radius h' of the exit pupil, we obtain the reduced coordinates x', y', z', defined by

$$x' = \frac{X'}{h'} \qquad y' = \frac{Y'}{h'} \qquad z' = \frac{Z'}{h'} \tag{8.38}$$

(8.37) can be rewritten as

$$p' = \bar{N}'\frac{1}{2}h'^2\left(\frac{1}{R_0'} - \frac{1}{R'}\right)\left(x'^2 + y'^2 + z'^2\right) \tag{8.39}$$

In the axial case (Figure 8.6), the principal ray $E'O_0'\hat{O}'$ is along the optical axis, so that $\bar{N}'$ is equal to unity. The original and the new reference spheres are centred at O_0' and $\hat{O}'$ with radii R_0' and R', respectively. Therefore,

$$p'_{\text{axial}} = \frac{1}{2}h'^2\left(\frac{1}{R_0'} - \frac{1}{R'}\right)\left(x'^2 + y'^2 + z'^2\right) \tag{8.40}$$

For the edge ray of an axial pencil, $x'^2 + y'^2 + z'^2 = 1$.

The change in wave aberration of the edge ray of an axial pencil for a longitudinal shift of image plane from O_0' to $\hat{O}'$ is given by

$$\left(\Delta W_L'\right)_{\text{axial}}^{\text{edge}} = \frac{1}{2}n'h'^2\left(\frac{1}{R_0'} - \frac{1}{R'}\right) \tag{8.41}$$

From (8.39) and (8.41), the change in wave aberration of the aperture ray in the extraaxial case for a longitudinal shift of image plane from O_0' to $\hat{O}'$ may be expressed as

$$\left(\Delta W_L'\right)_{extraaxial} = \bar{N}'\left(x'^2 + y'^2 + z'^2\right)\left(\Delta W_L'\right)_{axial}^{edge} \tag{8.42}$$

The formulae (8.41) and (8.42) are approximate, but in any practical case the error involved is negligible. The equation (8.41) may be rewritten

$$\left(\Delta W_L'\right)_{axial}^{edge} = \frac{1}{2}n'\left(\frac{h'^2}{R_0'R'}\right)\Delta\varsigma' \tag{8.43}$$

where $\Delta\varsigma' = \left(R' - R_0'\right)$. In most cases, (8.43) may be approximated as

$$\left(\Delta W_L'\right)_{axial}^{edge} = \frac{1}{2}n'\sin^2\alpha_0'\Delta\varsigma' \tag{8.44}$$

α_0' is the semi-angular aperture of the imaging system. However, it is to be noted that for larger values of $\Delta\varsigma'$, the form (8.41) should be used. An approximate expression like (8.44) can also be obtained directly from (8.27), and used in the axial case for smaller apertures. But the treatment in the extraaxial case is more involved than enunciated above.

Hopkins [89] has given a caveat that $(z')^2$ in (8.40) is 'only negligible in the case of the axial image and for smaller apertures. In other case it can lead to very serious error to assume that the effect of a longitudinal shift of image plane is merely to add a term of the form $\left(x'^2 + y'^2\right)$ to the aberration function'.

8.2.3.1.2 Transverse Shift of Focus

Figure 8.9 illustrates the case of transverse shift of focus on a specific image plane located at O_0' on the axis. For the sake of convenience, the diagram is given in two dimensions. E' is the axial location of the exit pupil in the image space of refractive index n'. The distance $E'O_0' = R_0'$. The principal ray $E'P_0'$ intersects the image plane at the point P_0' on the η'-axis. An aperture ray $\tilde{Q}'Q_0'Q'P_1'$ intersects the image plane at the point P_1'. The original reference sphere $E'Q_0'$ through E' is centred on the point P_0', and has the radius $E'P_0' = \bar{R}_0'$. The new reference sphere through E' has its centre at the point $\tilde{P}'$ on the same image plane, and its radius is $E'\tilde{P}' = \tilde{R}'$. The aperture ray intersects the old reference sphere, the new reference sphere, and the wavefront $E'Q'$ at the points Q_0', $\tilde{Q}'$, and Q' respectively.

The change in wave aberration due to transverse shift of the focus from P_0' to $\tilde{P}'$ is given by

$$\Delta W_T' = \tilde{W}' - W_0' = \left[\tilde{Q}'Q' - Q_0'Q'\right] = \left[\tilde{Q}'Q_0'\right] \tag{8.45}$$

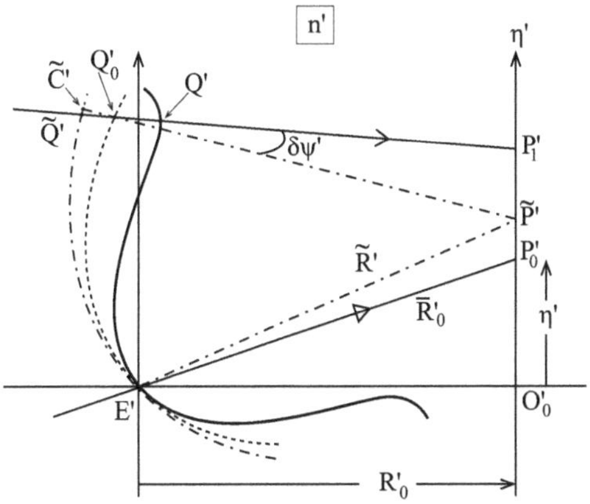

FIGURE 8.9 Transverse shift of focus.

A line $\tilde{C}'Q'_0\,\tilde{P}'$ is drawn to be a radius of the new reference sphere. Angular aberration of the aperture ray with respect to the new focus $\tilde{P}'$ is $\angle P'_1 Q'_0 \tilde{P}' = \delta\psi'$. From Figure 8.9 it is seen that with negligible error we may write

$$\tilde{Q}'Q'_0 = \frac{\tilde{C}'Q'_0}{\cos\delta\psi'} \approx \tilde{C}'Q'_0 = p' \tag{8.46}$$

It follows that

$$Q'_0\tilde{P}' = \tilde{R}' - p' \tag{8.47}$$

The height of the point P'_0, $P'_0O'_0 = \eta'$. With respect to origin at E' of a Cartesian coordinate system, the coordinates of the points Q'_0, P'_0 and $\tilde{P}'$ are (X', Y', Z'), $(0, \eta', R'_0)$ and $\{\Delta\xi', (\eta' + \Delta\eta'), R'_0\}$, respectively. Squaring both sides of (8.47)

$$Q'_0\tilde{P}'^2 = (X' - \Delta\xi')^2 + \{Y' - (\eta' + \Delta\eta')\}^2 + (Z' - R'_0)^2 = (\tilde{R}' - p')^2 \tag{8.48}$$

The equation of the original reference sphere

$$X'^2 + (Y' - \eta')^2 + (Z' - R'_0)^2 = \bar{R}'^2_0 \tag{8.49}$$

The radius $\tilde{R}'$ of the new reference sphere is given by

$$\tilde{R}'^2 = (E'\tilde{P}')^2 = (\Delta\xi')^2 + (\eta' + \delta\eta')^2 + R'^2_0 \tag{8.50}$$

By algebraic manipulation, we get, from (8.48) – (8.50)

$$p'^2 - 2p'\tilde{R}' = -2(X'\Delta\xi' + Y'\Delta\eta') \tag{8.51}$$

Neglecting p'^2 in the above equation, we get the following expression for p' with negligible error

$$p' = \frac{1}{\tilde{R}'}(X'\Delta\xi' + Y'\Delta\eta') \tag{8.52}$$

In practice, for small shifts $\Delta\xi', \Delta\eta'$,

$$\tilde{R}' \cong \bar{R}'_0 = R'_0 / \bar{N}' \tag{8.53}$$

where $(0, \bar{M}', \bar{N}')$ are the direction cosines of the principal ray $E'P'_0$. The change in wave aberration of the aperture ray by a transverse shift of focus from P'_0 to $\tilde{P}'$ is given by

$$\Delta W'_T = n'\bar{N}'\left(\frac{1}{R'_0}\right)(X'\Delta\xi' + Y'\Delta\eta') \tag{8.54}$$

In terms of reduced variables

$$x' = \frac{X'}{h'} \qquad y' = \frac{Y'}{h'} \tag{8.55}$$

where h′ is the radius of the exit pupil. Equation (8.54) may be written as

$$\Delta W'_T = n'\bar{N}'\left(\frac{h'}{R'_0}\right)[x'\Delta\xi' + y'\Delta\eta'] \tag{8.56}$$

In terms of semi-angular aperture α'_0 of the axial pencil for the original object position at O'_0 (Figure 8.7), $\Delta W'_T$ for the aperture ray is given by

$$\Delta W'_T = \bar{N}'(n'\sin\alpha'_0)[x'\Delta\xi' + y'\Delta\eta'] \tag{8.57}$$

8.2.3.2 Effect of Change in Radius of the Reference Sphere on Wave Aberration

The use of small values of radius of the reference sphere is ruled out in measuring aberrations, for the wavefront and corresponding ray patterns take a shape that is usually too complicated for interpretation. This problem can be circumvented by using somewhat larger values for the radius of the reference sphere. The choice of reference sphere passing through the centre of the exit pupil for obtaining a useful measure of aberration can be justified in general for two reasons. First, in any real system the exit pupil cannot lie near the image, and so the problem of interpretation as mentioned above does not arise. The more important factor facilitating this choice is that the wave aberration of the image-forming wavefront obtained thereby can be directly used in the treatment of image quality assessment based on diffraction theory, as discussed in a subsequent chapter.

Figure 8.10 shows image-forming wavefronts $E'_2D'_2$ and $E'_1D'_1$ passing through the points E'_2 and E'_1, respectively on the principal ray $E'_2E'_1M'P'$ in the image space for the chosen image point P′. $B'_2D'_2B'_1D'_1M'$ is an aperture ray of the pencil of rays forming the image at P′. It is assumed that the aperture ray is coplanar with the principal ray and intersects the latter ray at M′. $E'_2D'_2$ and $E'_1D'_1$ are sections of two reference spheres with the centre at P′ and radii $E'_1P'(=R'_1)$ and $E'_2P'(=R'_2)$, respectively. The wave aberration of this aperture ray is $B'_1D'_1(=W'_1)$ when measured with respect to the reference sphere at E'_1, and it is $B'_2D'_2(=W'_2)$ when measured with respect to the reference sphere at E'_2. A radial line $G'_2B'_1P'$ intersects the concentric reference spheres at G'_2 and B'_1. The angular aberration of the aperture ray at B'_1 is $\angle M'B'_1P'(=\delta\psi')$.

The change $\Delta W'(=W'_2 - W'_1)$ in wave aberration by the change $\Delta R'(=R'_2 - R'_1)$ in radius of the reference sphere from R'_1 to R'_2 is given by

$$\Delta W' = B'_2D'_2 - B'_1D'_1 = B'_2B'_1 - D'_2D'_1 \approx \frac{G'_2B'_1}{\cos\delta\psi'} - E'_2E'_1 = \frac{1}{2}\Delta R'\delta\psi'^2 + O(4) \tag{8.58}$$

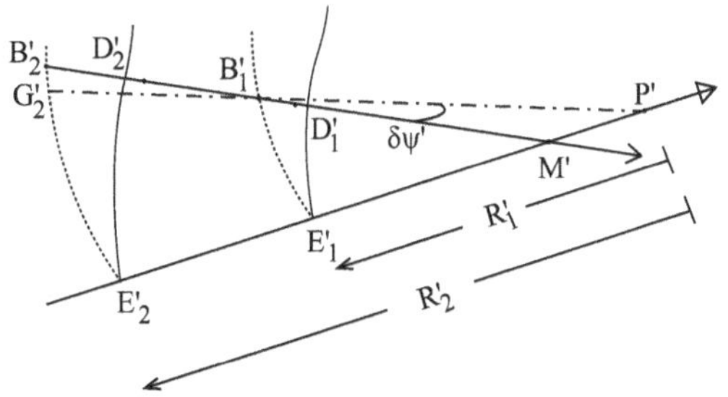

FIGURE 8.10 Effects on wave aberration of a meridional aperture ray $B'_2D'_2B'_1D'_1M'$ by change in radius of reference sphere with centre P′ of the sphere on the principal ray $E'_2E'_1M'P'$ remaining the same.

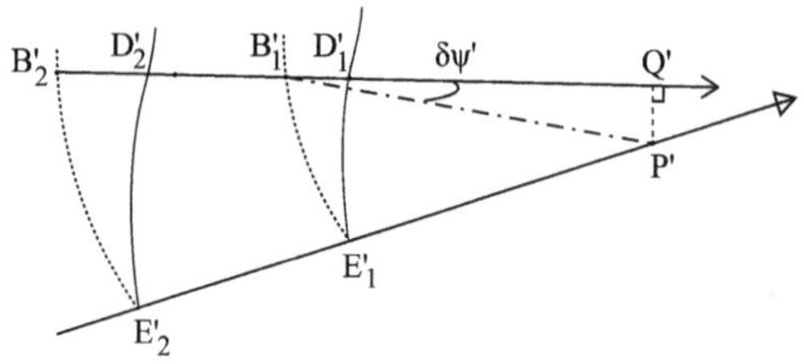

FIGURE 8.11 Effects on wave aberration of a skew aperture ray $B_2'D_2'B_1'D_1'Q'$ by change in radius of reference sphere with centre P′ of the sphere on the principal ray $E_2'E_1'P'$ remaining the same.

In practice, even with moderately large values of radii of the reference sphere, the magnitude of angular aberration $\delta\psi'$ is so small that $\Delta W' \ll W'$. It may be noted that if M′ is chosen as the centre of the reference spheres, $\Delta W' = 0$ for the specific aperture ray. However, for other rays of the pencil, the corresponding values of $\Delta W'$ depend on corresponding angular aberrations, and are given by relation (8.58).

In general, the aperture ray $B_2' D_2' B_1' D_1'$ is a skew ray, and it does not intersect the principal ray $E_2' E_1' P'$ (Figure 8.11). However, it is obvious that in this case the change in wave aberration $\Delta W'$ of the aperture ray for the change in radius of the reference sphere is also given by relation (8.58).

8.2.3.2.1 *Wave Aberration and Hamilton-Bruns' Eikonal*

In classical ray optical treatments of aberrations, one or the other types of eikonal, e.g. point eikonal, point angle eikonal, or angle eikonal, etc., have been used extensively. Usually, the origins in the object space or the image space are taken on the optical axis, and perpendiculars are drawn from these origins on the ray segments in the corresponding spaces. Often the paraxial conjugate points or the principal foci on the axis are taken as the origins. The origins need not necessarily lie on the optical axis. The chosen conjugate points on the principal ray can be taken as the origins. Let the conjugate points on the principal ray in the object and the image spaces be P and P′, and Q and Q′ are the feet of the perpendiculars from P and P′ on the aperture ray segments in the object space and the image space, respectively. In Figure 8.10, it may be seen that

$$E_1' P' = B_1' P' = \frac{B_1' Q'}{\cos\delta\psi'} \tag{8.59}$$

As the radius of the reference sphere $R' \to \infty$, E_1', $B_1' \to \infty$, and $\delta\psi' \to 0$. Therefore, it follows

$$B_1' P' \approx B_1' Q' = E_1' P' \tag{8.60}$$

The wavefront aberration, or wave aberration of the aperture ray, is seen to be equal to $\{[PP'] - [QQ']\}$ with respect to a reference sphere of an infinite radius of a curvature centred on P′ [90]. It is to be noted that $[QQ']$ is the Hamilton-Bruns' angle eikonal for origins at P and P′. When the image-forming pencil of rays originates from the point P in the object space, the corresponding expression for wave aberration reduces to $\{[PP'] - [PQ']\}$. In this case, $[PQ']$ is the point angle eikonal as defined in Section 7.1.2. The relations between the different Eikonals and the wavefront aberration are described with more detail in [91].

8.2.3.2.2 *Wave Aberration on the Exit Pupil*

From a ray optical point of view, the wavefront often takes a very complicated shape near the image point, so much so that it defies the development of any suitable analytical treatment for image quality assessment based on its deformation in shape from the desired spherical shape. However, a tractable analysis can be obtained when it is carried out in regions away from the focal region. Early studies on the ray optical treatment of optical aberrations were often carried out in the star space, where radius of the reference sphere is infinity.

However, it is now realized that the quality of imaging by an optical system is affected by two phenomena: aberrations and diffraction. It is shown in Chapter 17 that assessment of image quality in accordance with diffraction theory of image formation calls for a knowledge of the amplitude and phase of the wave at the exit pupil. Assuming amplitude to be uniform, the phase of the wave is directly related with the deformation of the geometrical wavefront expressed as wavefront aberration at the exit pupil. This is the primary reason why, in ray optical treatment of optical system analysis and design, the wave aberration at the exit pupil is taken into consideration.

8.2.4 Caustics and Aberrations

All measures of aberration discussed above have a common drawback, which is that they depend on certain choices that are subjective to some extent. For example, the value of all types of 'ray aberrations', namely, 'longitudinal', or 'transverse', or 'angular', depends on the choice of the image plane. Indeed, these aberrations change with a shift in axial location of the image plane. Similarly, the value of wave aberration is dependent on the choice of the reference sphere. It may be noted that the centre of the reference sphere changes with changes in axial location of the image plane. In practice, even determination of the best image plane calls for a knowledge of aberrations with respect to a set of axial locations of the image plane, so that the 'best' location for the image plane can be worked out on the basis of suitable criteria for minimization of aberrational effects.

In order to circumvent this problem, scientists have explored the use of 'caustics'. The word 'caustic' comes from the Greek 'καυστός', meaning burnt, via the Latin word 'causticus', meaning burning. 'The caustic surface can be thought of as the aberrated three-dimensional image of a point object produced by a lens' [92]. In general, at a point on a smooth surface, there is a single normal. The centres of curvature along different azimuths around the point lie on the normal. The sections corresponding to the azimuths, for which the radii of curvature are minimum and maximum, are called the principal sections, and the two centres of curvature corresponding to them are called the principal centres of curvature at that point on the surface. The two principal directions are necessarily perpendicular to each other. As the point on the surface is moved in the neighbourhood of the original point, the two loci of the two centres of curvature become two sheets that are called caustic surfaces, or simply caustics. Equivalently, because of aberrations introduced by the system, a homocentric bundle of rays originating from a point object is transformed into a heterocentric bundle in the image space, and the caustics can be visualized as the envelope of the system of rays near the image point. Note that all image-forming rays touch the caustics, leading to a concentration of light so that the caustic surface can be made readily visible.

In the jargon of differential geometry, the caustic surface is the evolute of the wavefront surface, and the wavefront surface is the involute of the caustic surface. In the case of aberration-free perfect imaging, the caustic surface reduces to a point—the image point. In the case of imaging of axial object points by axially symmetric imaging systems, one of the sheets of the caustic degenerates to a single line, the axis of revolution. Figure 8.12 illustrates the caustics for an axially symmetric image-forming wavefront.

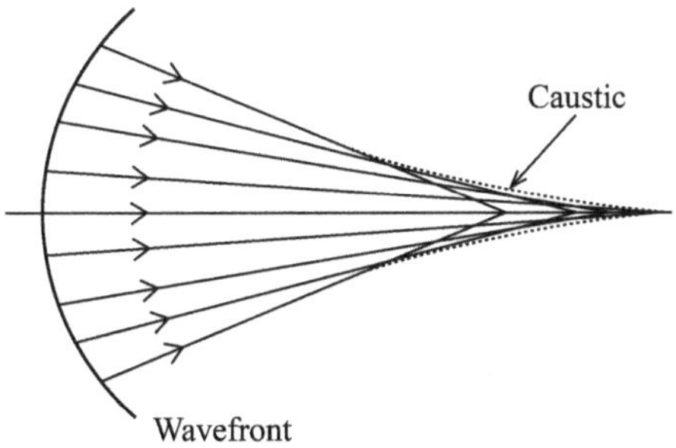

FIGURE 8.12(a) Ray geometry in formation of caustics of an axially symmetric wavefront.

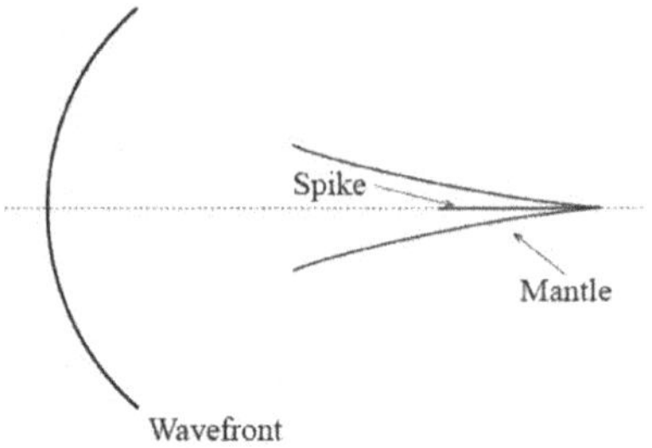

FIGURE 8.12(b) One of the caustic surfaces of an axially symmetric wavefront constitute the mantle, and the other caustic surface degenerates to a spike along the axis.

The origin of the caustics goes back to the early nineteenth century when Coddington [93] developed methods for calculating the principal radii of the curvature of a wavefront along the principal ray (see Section 7.4.2.1). 'Coddington equations are in fact a special case of the caustic surface equation' [94].

In connection with his researches in ophthalmology, for which he was awarded the Nobel prize in 1911, Allvar Gullstrand [95] studied these equations extensively, and perfected them on the basis of metric differential geometry. After about fifty years, Altrichter and Schäfer [96] suggested the use of computers in solving Gullstrand equations, and Kneisly III [97] completed the task. Among others, Herzberger [98] and Stavroudis [99] were chief proponents on the use of caustics in optical design [100–103]. In spite of its potential, the use of caustics in practical analysis and design of axisymmetric optical systems has so far been limited. However, investigations into their use in non-symmetric optical systems, e.g. in the analysis and synthesis of ophthalmic lenses like progressive addition lenses, are continuing [104].

8.2.5 Power Series Expansion of the Wave Aberration Function

Figure 8.13 shows the image space of an axially symmetric system imaging a point P with coordinates (ξ,η) on the object plane (not shown). The principal ray of the image-forming pencil from point P intersects the image plane at P′. The coordinates of the point P′ with respect to an origin at the centre E′, of the exit pupil, are $(\xi',\eta',\tilde{Z}')$. A section of the imaging wavefront passing through E′ is E′Q′. $E'Q_0'$ is a section of the reference sphere with P′, the chosen image point, as the centre, and E′P′ as the radius. $Q_0'A'P_1'$ is the image space segment of an aperture ray of the image-forming pencil originating from the object point P. The aperture ray intersects the reference sphere, the wavefront, and the exit pupil plane at Q_0', Q′,, and A′respectively. With respect to E′ as origin, the coordinates of the point Q_0' are (X',Y',Z'). As defined earlier, the wave aberration of this aperture ray is $\left[Q_0'\,Q'\right]=W$. It is obvious that the wave aberration is different for another aperture ray of the same pencil, because the points of intersection of the latter ray with the reference sphere and the wavefront are different. Therefore, in an image-forming pencil, wave aberration W is a function of the parameters specifying a ray. The image space ray is a straight line when the medium is homogeneous. This ray can be uniquely specified by the coordinates X′, Y′ of the point Q_0', and the coordinates (ξ,η) of the object point P, or the coordinates of the chosen image point (ξ',η'). The latter choice is more common in practice. Note that the point Q_0' lies on a prespecified reference sphere, and so only two coordinates are sufficient to specify it. Therefore, the functional form of W is

$$W \equiv W(\xi',\eta',X',Y') \tag{8.61}$$

In general, a polynomial W of the four variables can contain any combination of the variables, as well as their powers and products. But the symmetry properties, if any, of the system can be invoked to obtain definite guidelines on the nature of these combinations. Buchdahl [105] has studied extensively the effects of symmetries of optical systems on their aberrations.

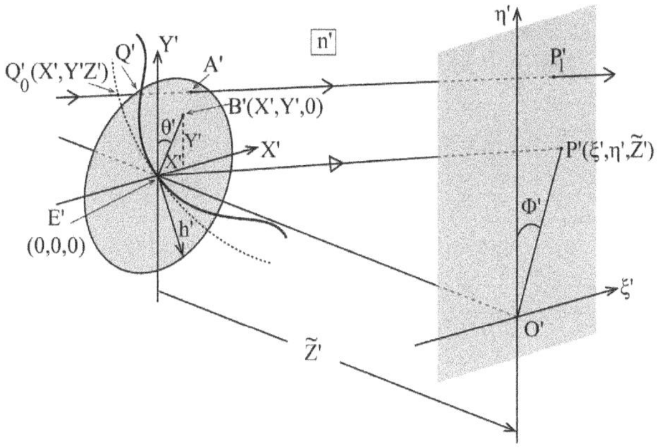

FIGURE 8.13 The principal ray E′P′ and an aperture ray $Q_0'Q'A'P_1'$ in the image space of an axially symmetric imaging system.

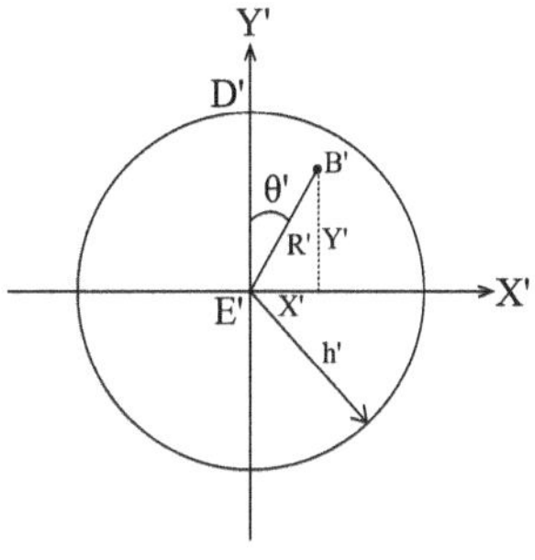

FIGURE 8.14 Polar coordinates for points on the exit pupil.

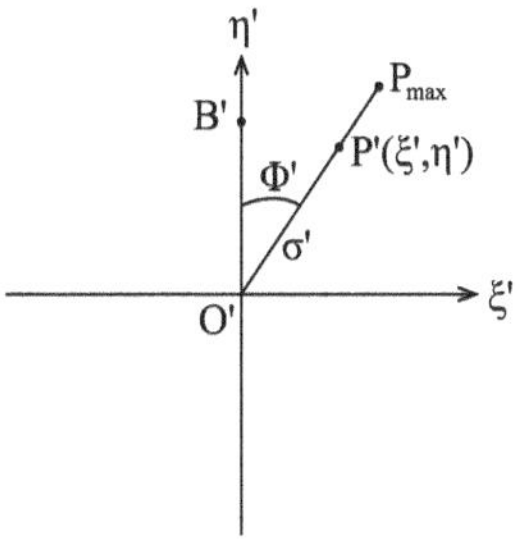

FIGURE 8.15 Polar coordinates of the image point.

We consider the case of optical systems that are rotationally symmetric around the optical axis. Further, it is assumed that the system consists of smooth surfaces. The treatment can be simplified by using polar coordinates for points on the exit pupil and the image plane. Figure 8.14 and Figure 8.15 show side views of the exit pupil and the image plane, respectively. The rectangular coordinates X', Y' of a point B' on the exit pupil is represented by R', θ' where $R' = E'B'$, and $\theta' = \angle D'E'B'$ (Figure 8.14). The coordinates are related by

$$X' = R'\sin\theta' \qquad Y' = R'\cos\theta' \tag{8.62}$$

Similarly, the rectangular coordinates ξ', η' of a point P′ on the image plane is represented by σ', φ' where $\sigma' = E'P'$, and $\varphi' = \angle B'O'P'$ (Figure 8.15). The coordinates are related by

$$\xi' = \sigma' \sin\varphi' \qquad \eta' = \sigma' \cos\varphi' \tag{8.63}$$

Note that the angles θ', φ' are taken with reference to the positive direction of the ordinate, contrary to the common usage of the positive direction of the abscissa in the definition of polar coordinates. This facilitates utilization of bilateral symmetry of optical systems around the meridian plane. In terms of these polar coordinates, aberration function W is represented as

$$W \equiv W(\xi', \eta', X', Y') \equiv W(\sigma', \varphi', R', \theta') \tag{8.64}$$

On account of this rotational symmetry, the wave aberration W must not change if there is a rigid rotation of the object plane, the entrance pupil, the exit pupil, and the image plane around the axis. If the angle of rotation around the axis is equal to $-\alpha$, the new angles are

$$\hat{\theta}' = \theta' + \alpha \qquad \hat{\varphi}' = \varphi' + \alpha \tag{8.65}$$

The new coordinates are $\hat{\xi}', \hat{\eta}', \hat{X}', \hat{Y}'$. They are related to the coordinates ξ', η', X', Y' by

$$\hat{\xi}'^2 + \hat{\eta}'^2 = \hat{\sigma}'^2 = \sigma'^2 = \xi'^2 + \eta'^2 \tag{8.66}$$

$$\hat{X}'^2 + \hat{Y}'^2 = \hat{R}'^2 = R'^2 = X'^2 + Y'^2 \tag{8.67}$$

Since the distances of the point on the exit pupil and the image point do not change by rigid rotation around the axis, $\hat{R}' = R'$, and $\hat{\sigma}' = \sigma'$.

From (8.65), we can write

$$\cos\left(\hat{\theta}' - \hat{\varphi}'\right) = \cos(\theta' - \varphi') \tag{8.68}$$

Multiplying the left side by $\hat{\sigma}'\hat{R}'$ and the right side by $\sigma'\hat{R}'$, it follows

$$\begin{aligned} &\left(\hat{\sigma}'\cos\hat{\varphi}'\right)\left(\hat{R}'\sin\hat{\theta}'\right) - \left(\hat{\sigma}'\sin\hat{\varphi}'\right)\left(\hat{R}'\cos\hat{\theta}'\right) \\ &= (\sigma'\cos\varphi')(R'\sin\theta') - (\sigma'\sin\varphi')(R'\cos\theta') \end{aligned} \tag{8.69}$$

(8.69) can be rewritten as

$$\hat{\eta}'Y' + \hat{\xi}'\hat{X}' = \eta'Y' + \xi'X' \tag{8.70}$$

From (8.66), (8.67), and (8.70), we get three rotationally invariant combinations $\left(\xi'^2 + \eta'^2\right)$, $\left(X'^2 + Y'^2\right)$, and $\left(\eta'Y' + \xi'X'\right)$. From four variables, it is possible to get only three independent invariants, and so all other invariant combinations must be functions of these three invariant combinations. For the sake of incorporating simplification in analysis of rotationally symmetric systems, we can put $\xi' = 0$ in the invariant combinations, so that they become $\eta'^2, \left(X'^2 + Y'^2\right), \left(\eta'Y'\right)$. This implies that only object points lying along the η'-axis are being considered. There is no loss of generality in the analysis, for in a rotationally symmetric system, by rotation around the optical axis, the η-axis can be made to superpose on a point located anywhere on the object/image plane. By change of variables, the three rotation invariant variables of the aberration function are η'^2, R'^2, and $\eta' R' \cos\theta'$. The radial component of the polar coordinates for points on the pupil can be normalized by the semi-diameter h' of the exit pupil, so that the coordinates (R', θ') can be expressed as (r', θ') where

$$r' = \left(\frac{R'}{h'}\right) \tag{8.71}$$

For points on the exit pupil, $0 \le r' \le 1$.

Similarly, a fractional parameter τ' defined as

$$\tau' = \left(\frac{\eta'}{\eta'_{max}}\right) \tag{8.72}$$

can be used to express the object/image height variable. The fractional height τ' lies within the range $(0,1)$. Note that both the variables r' and $\tau'$are dimensionless. Using these variables, the aberration function W takes the form

$$W\left(\tau'^2, r'^2, \tau' r' \cos\theta'\right) \tag{8.73}$$

In the following analysis, we shall be concerned only with the image space, and, therefore, for the sake of simplicity in notation, we omit the prime symbol associated with the image space parameters in most cases. Expression (8.73) will be written as

$$W\left(\tau^2, r^2, \tau r \cos\theta\right) \tag{8.74}$$

W can be expanded in a power series consisting of polynomials in the variables. The general term of the polynomial can be written as

$$\omega_{ijk}\left(\tau^2\right)^i\left(r^2\right)^j\left(\tau r \cos\theta\right)^k, \qquad i, j, k = 0, 1, 2, 3\ldots \tag{8.75}$$

The power series representing $W\left(\tau^2, r^2, \tau r \cos\theta\right)$ can be expressed as

$$W\left(\tau^2, r^2, \tau r \cos\theta\right) = \sum_i \sum_j \sum_k \omega_{ijk}\left(\tau^2\right)^i\left(r^2\right)^j\left(\tau r \cos\theta\right)^k \tag{8.76}$$

By regrouping the terms, (8.75) can be rewritten as

$$\omega_{ijk}\tau^{(2i+k)} r^{(2j+k)}\left(\cos\theta\right)^k = {}_pW_{mn}\tau^p r^m\left(\cos\theta\right)^n \tag{8.77}$$

where

$$p = 2i + k, \qquad m = 2j + k, \qquad n = k \tag{8.78}$$

In expression (8.75), the field and pupil terms are intermingled. Use of exponents p, m, and n in place of i, j, and k enables separation of these terms. This is highlighted by use of the notation ${}_pW_{mn}$ for the coefficients of the aberration polynomial terms. The prefix p is the degree of fractional field τ, the first suffix m is the power of the normalized radius r, and the second suffix n is the power of $(\cos\theta)$ in the particular aberration coefficient ${}_pW_{mn}$.

The wavefront aberration W can be expanded in a power series as

$$\begin{aligned} W(\tau^2, r^2, \tau r\cos\theta) = {}&{}_0W_{00} + {}_2W_{00}\tau^2 + {}_0W_{20}r^2 + {}_1W_{11}\tau r\cos\theta + {}_4W_{00}\tau^4 + \\ &{}_0W_{40}r^4 + {}_2W_{22}\tau^2r^2(\cos\theta)^2 + {}_2W_{20}\tau^2r^2 + {}_1W_{31}\tau r^3\cos\theta + {}_3W_{11}\tau^3r\cos\theta + \\ &{}_6W_{00}\tau^6 + {}_0W_{60}r^6 + {}_3W_{33}\tau^3r^3(\cos\theta)^3 + {}_4W_{20}\tau^4r^2 + {}_2W_{40}\tau^2r^4 + {}_5W_{11}\tau^5r\cos\theta + \\ &{}_4W_{22}\tau^4r^2(\cos\theta)^2 + {}_1W_{51}\tau r^5\cos\theta + {}_2W_{42}\tau^2r^4(\cos\theta)^2 + {}_3W_{31}\tau^3r^3\cos\theta + \\ &{}_8W_{00}\tau^8 + {}_0W_{80}r^8 + {}_4W_{44}\tau^4r^4(\cos\theta)^4 + \ldots.. \end{aligned} \tag{8.79}$$

The generation of the different polynomial terms in the above power series is illustrated in Table 8.1.

E′ is the centre of the exit pupil. By definition, the actual image-forming wavefront and the reference sphere touch each other at E′. Therefore, for $r = 0, W = 0$. Substituting $r = 0$ in the right-hand side of (8.79) we get

$$W = 0 = {}_0W_{00} + {}_2W_{00}\tau^2 + {}_4W_{00}\tau^4 + {}_6W_{00}\tau^6 + {}_8W_{00}\tau^8 + \ldots \tag{8.80}$$

The constant term ${}_0W_{00}$ is sometimes categorized as a 'piston' error to represent a fixed optical path difference between the reference sphere and the imaging wavefront for all rays of the image-forming pencil of rays.

The two terms ${}_0W_{20}r^2$ and ${}_1W_{11}\tau r\cos\theta$ correspond to a longitudinal shift of focus and transverse shift of focus, as discussed in sections 8.2.3.1.1 and 8.2.3.1.2. This implies that by a suitable choice of reference sphere, these two terms can be eliminated. Therefore, in a strict sense, they are not defects in imaging introduced by the imaging system. These terms are normally omitted in the polynomial expression for W, as they are not considered to be aberration terms. Based on these observations, the wave aberration polynomial for an axially symmetric optical system is expressed as

$$\begin{aligned} W = {}&\left[{}_0W_{40}r^4 + {}_2W_{22}\tau^2r^2(\cos\theta)^2 + {}_2W_{20}\tau^2r^2 + {}_1W_{31}\tau r^3\cos\theta + {}_3W_{11}\tau^3r\cos\theta\right] \\ &+ \left[{}_0W_{60}r^6 + {}_3W_{33}\tau^3r^3(\cos\theta)^3 + {}_4W_{20}\tau^4r^2 + {}_2W_{40}\tau^2r^4 + {}_5W_{11}\tau^5r\cos\theta\right. \\ &\left. + {}_4W_{22}\tau^4r^2(\cos\theta)^2 + {}_1W_{51}\tau r^5\cos\theta + {}_2W_{42}\tau^2r^4(\cos\theta)^2 + {}_3W_{31}\tau^3r^3\cos\theta\right] \\ &+ \left[{}_0W_{80}r^8 + {}_4W_{44}\tau^4r^4(\cos\theta)^4 + \ldots\right] + \cdots \end{aligned} \tag{8.81}$$

8.2.5.1 Aberrations of Various Orders

In the power series (8.81), the polynomial terms are shown in different groups; each group is bracketed. The first bracketed group contains five terms. In each of these terms, the sum Λ of the exponents of τ and r, i.e., $(p + m)$ is equal to 4. The next bracketed group contains nine terms, in each of which $\Lambda = (p + m)$ is equal to 6. Two terms of the third group where Λ is equal to 8 are shown next. Only three groups are shown in (8.81). As the power series is an infinite series, the infinite number of such groups can be formed. Each group of aberration polynomials constitutes aberrations of a specific 'order'. All aberration polynomial terms with the same value of (p + m) belong to the same order. Successive groups of aberration polynomials with $\Lambda = 4, 6, 8, \ldots$ are called primary aberrations, secondary aberrations, tertiary

TABLE 8.1

Generation of polynomial terms in power series expansion of the wave aberration function

i	j	k	p	m	n	Term
0	0	0	0	0	0	${}_0W_{00}$
1	0	0	2	0	0	${}_2W_{00}\tau^2$
0	1	0	0	2	0	${}_0W_{20}r^2$
0	0	1	1	1	1	${}_1W_{11}\tau r\cos\theta$
2	0	0	4	0	0	${}_4W_{00}\tau^4$
0	2	0	0	4	0	${}_0W_{40}r^4$
0	0	2	2	2	2	${}_2W_{22}\tau^2r^2(\cos\theta)^2$
1	1	0	2	2	0	${}_2W_{20}\tau^2r^2$
0	1	1	1	3	1	${}_1W_{31}\tau r^3\cos\theta$
1	0	1	3	1	1	${}_3W_{11}\tau^3r\cos\theta$
3	0	0	6	0	0	${}_6W_{00}\tau^6$
0	3	0	0	6	0	${}_0W_{60}r^6$
0	0	3	3	3	3	${}_3W_{33}\tau^3r^3(\cos\theta)^3$
2	1	0	4	2	0	${}_4W_{20}\tau^4r^2$
1	2	0	2	4	0	${}_2W_{40}\tau^2r^4$
2	0	1	5	1	1	${}_5W_{11}\tau^5r\cos\theta$
1	0	2	4	2	2	${}_4W_{22}\tau^4r^2(\cos\theta)^2$
0	2	1	1	5	1	${}_1W_{51}\tau r^5\cos\theta$
0	1	2	2	4	2	${}_2W_{42}\tau^2r^4(\cos\theta)^2$
1	1	1	3	3	1	${}_3W_{31}\tau^3r^3\cos\theta$
4	0	0	8	0	0	${}_8W_{00}\tau^8$
0	4	0	0	8	0	${}_0W_{80}r^8$
0	0	4	4	4	4	${}_4W_{44}\tau^4r^4(\cos\theta)^4$

aberrations, etc. Defining order by the value of Λ in the aberration polynomial, they are also called fourth order aberrations, sixth order aberrations, and eighth order aberrations, respectively. As ray aberration is scaled derivative of the wave aberration, the sum of the exponent of the field term and that of the pupil term of a ray aberration polynomial is $(\Lambda - 1)$, where Λ is the sum for the corresponding wave aberration

TABLE 8.2

Aberrations of various orders $[\Lambda = p + m]$

Order	Λ=4	Λ=6	Λ=8
	Primary Aberrations Seidel Aberrations	**Secondary Aberrations**	**Tertiary Aberrations**
Λ	Fourth order Aberrations	Sixth order Aberrations	Eighth order Aberrations
$(\Lambda - 1)$	Third order Aberrations	Fifth order Aberrations	Seventh order Aberrations
$[(\Lambda/2) - 1]$	First order Aberrations	Second order Aberrations	Third order Aberrations
Number of terms $N = \frac{1}{2}n(n+3)$ $n = (\Lambda/2)$	5	9	14

polynomial. Sometimes, $(\Lambda - 1)$ is used to specify the 'order' of aberrations, and consequently the groups are also called third order aberrations, fifth order aberrations, and seventh order aberrations. In older literature, the prevalent practice was to specify order by $\left(\frac{\Lambda}{2} - 1\right)$, so that the same groups used to be called first order aberrations, second order aberrations, third order aberrations, etc. [106]. In memory of Ludwig von Seidel who first worked out the mathematical relations to evaluate the primary aberrations of an axially symmetric lens system from the constructional parameters, the primary aberrations are also mentioned as Seidel aberrations. The various specifications for orders of aberrations are given in Table 8.2.

Table 8.3 presents the names of primary and secondary aberration terms. There are five primary aberration terms, namely primary spherical aberration, primary coma, primary astigmatism, primary curvature, and primary distortion. The adjective 'primary' needs to be implicitly or explicitly mentioned with each term to underscore the specific polynomial type. The number of secondary aberration terms is nine. Secondary spherical aberration, ${}_0W_{60}r^6$, is independent of field τ, and it is the counterpart of the primary spherical aberration, ${}_0W_{40}r^4$ in the fold of secondary aberrations. There is a term, ${}_2W_{40}\tau^2r^4$, which is proportional to τ^2, but it has quartic dependence on r, like primary spherical aberration. This is called oblique spherical aberration. Secondary linear coma, ${}_1W_{51}\tau r^5\cos\theta$, is the counterpart of primary coma, ${}_1W_{31}\tau r^3\cos\theta$. The addition of the word 'linear' in its name is somewhat redundant, however, it does facilitate emphasizing its characteristic difference with two more coma-like terms appearing in the secondary fold. They are ${}_3W_{31}\tau^3r^3\cos\theta$ and ${}_3W_{33}\tau^3r^3(\cos\theta)^3$. Like primary coma, each of them has a cubic dependence on aperture r, but, unlike primary coma they have a cubic dependence on field τ, so that they become secondary aberrations. They are often conjointly termed as 'elliptical coma', which obviously has no primary counterpart. At times, the two elliptical coma terms are distinguished as elliptical coma of type I and elliptical coma of type II, respectively. For secondary astigmatism, there are two terms. One of them, ${}_4W_{22}\tau^4r^2(\cos\theta)^2$, is the counterpart of the primary astigmatism term, ${}_2W_{22}\tau^2r^2(\cos\theta)^2$. There is one more aberration polynomial ${}_2W_{42}\tau^2r^4(\cos\theta)^2$ that demonstrates astigmatism-like characteristics due to its dependence on an even power of $\cos\theta$. This is often categorized as secondary astigmatism of type II, while the other astigmatism term, mentioned before, is categorized as secondary astigmatism of type I. The remaining two secondary aberration terms, namely secondary curvature, ${}_4W_{20}\tau^4r^2$, and secondary distortion, ${}_5W_{11}\tau^5r\cos\theta$, are obvious counterparts of primary curvature, ${}_2W_{20}\tau^2r^2$, and primary distortion, ${}_3W_{11}\tau^3r\cos\theta$. For higher order aberrations, each of them consists of terms akin to its lower orders, and a few new terms specific to it. No uniform notation on names of these terms exists. Table 8.4 gives the

TABLE 8.3

Nomenclature of primary and secondary aberration terms

Primary Aberrations	
${}_0W_{40}r^4$	Primary Spherical Aberration
${}_1W_{31}\tau r^3\cos\theta$	Primary Coma
${}_2W_{22}\tau^2r^2(\cos\theta)^2$	Primary Astigmatism
${}_2W_{20}\tau^2r^2$	Primary Curvature
${}_3W_{11}\tau^3r\cos\theta$	Primary Distortion
Secondary Aberrations	
${}_0W_{60}r^6$	Secondary Spherical Aberration
${}_2W_{40}\tau^2r^4$	Oblique Spherical Aberration
${}_1W_{51}\tau r^5\cos\theta$	Secondary Linear Coma
${}_3W_{31}\tau^3r^3\cos\theta + {}_3W_{33}\tau^3r^3(\cos\theta)^3$	Elliptical Coma
${}_4W_{22}\tau^4r^2(\cos\theta)^2 + {}_2W_{42}\tau^2r^4(\cos\theta)^2$	Secondary Astigmatism
${}_4W_{20}\tau^4r^2$	Secondary Curvature
${}_5W_{11}\tau^5r\cos\theta$	Secondary Distortion

TABLE 8.4

Tertiary aberrations

m	*n*=0	*n*=1	*n*=2	*n*=3	*n*=4
Tertiary Aberrations $p+m=8$					
1		${}_7W_{11}\tau^7r\cos\theta$			
2	${}_6W_{20}\tau^6r^2$		${}_6W_{22}\tau^6r^2(\cos\theta)^2$		
3		${}_5W_{31}\tau^5r^3\cos\theta$		${}_5W_{33}\tau^5r^3(\cos\theta)^3$	
4	${}_4W_{40}\tau^4r^4$		${}_4W_{42}\tau^4r^4(\cos\theta)^2$		${}_4W_{44}\tau^4r^4(\cos\theta)^4$
5		${}_3W_{51}\tau^3r^5\cos\theta$		${}_3W_{53}\tau^3r^5(\cos\theta)^3$	
6	${}_2W_{60}\tau^2r^6$		${}_2W_{62}\tau^2r^6(\cos\theta)^2$		
7		${}_1W_{71}\tau r^7\cos\theta$			
8	${}_0W_{80}r^8$				

TABLE 8.5
Quaternary aberrations

m	n=0	n=1	n=2	n=3	n=4	n=5
Quaternary Aberrations $p+m=10$						
1		${}_9W_{11}\tau^9 r\cos\theta$				
2	${}_8W_{20}\tau^8 r^2$		${}_8W_{22}\tau^8 r^2(\cos\theta)^2$			
3		${}_7W_{31}\tau^7 r^3\cos\theta$		${}_7W_{33}\tau^7 r^3(\cos\theta)^3$		
4	${}_6W_{40}\tau^6 r^4$		${}_6W_{42}\tau^6 r^4(\cos\theta)^2$		${}_6W_{44}\tau^6 r^4(\cos\theta)^4$	
5		${}_5W_{51}\tau^5 r^5\cos\theta$		${}_5W_{53}\tau^5 r^5(\cos\theta)^3$		${}_5W_{55}\tau^5 r^5(\cos\theta)^5$
6	${}_4W_{60}\tau^4 r^6$		${}_4W_{62}\tau^4 r^6(\cos\theta)^2$		${}_4W_{64}\tau^4 r^6(\cos\theta)^4$	
7		${}_3W_{71}\tau^3 r^7\cos\theta$		${}_3W_{73}\tau^3 r^7(\cos\theta)^3$		
8	${}_2W_{80}\tau^2 r^8$		${}_2W_{82}\tau^2 r^8(\cos\theta)^2$			
9		${}_1W_{91}\tau r^9\cos\theta$				
10	${}_0W_{10,0}r^{10}$					

composition of 14 tertiary aberration terms, and Table 8.5 gives the composition of 20 quaternary aberration terms.

8.2.5.2 Convergence of the Power Series of Aberrations

For all practical purposes, the power series of aberrations retains its utility if and only if the series is convergent, and if the rate of convergence is not unduly slow to make an analysis based on truncation of higher order terms ineffective. 'It is certain that it can only be valid for a restricted range of the variables, since discontinuities must occur for very large apertures or field angles' [107]. This type of discontinuity happens when a ray misses a surface, or it is totally internally reflected from a surface, etc. The general term for these phenomena is 'ray failure', and they are diagnosed during finite ray tracing [108]. As early as in 1921, Baker and Filon pointed out the possibility of success or failure of convergence of the power series on the choice of variables for the power series [109]. Stavroudis, prodded by Donald P. Feder of Eastman Kodak Laboratories, illustrated the problem of convergence of power series expansion of aberration function with a simple example [110]. Buchdahl, and later on, Forbes, reported their investigations on different aspects of convergence of the power series of aberration polynomials [111–119].

8.2.5.3 Types of Aberrations

A different classification of aberrations is often resorted to during the analysis and design of optical systems. The variation of aberration terms with a field is subsumed, and the aberration terms are broadly classified in three types. If the aberration term has complete symmetry around the optical axis, i.e. if it is independent of θ, so that $m = 0$, it is considered to be a 'spherical aberration type'. If the aberration term has only one plane of symmetry about the meridian plane, i.e. the aberration term involves only odd powers of $\cos\theta$, it is considered to be a 'coma type', and if the term has two planes of symmetry, the meridian and the sagittal planes, i.e. if the aberration term contains only even powers of $\cos\theta$, it is classified as an 'astigmatism type'.

For a specific fractional field τ, let the wave aberration be expressed as $W(x,y)$. The power series expansion of wave aberration W in terms of the corresponding polar coordinates (r,θ) are given in (8.81). This can be expressed in terms of the rectangular variables (x,y) as

$$\begin{aligned} W(x,y) = & \left[{}_0W_{40}\left(x^2+y^2\right)^2 +_2 W_{22}\tau^2y^2 +_2 W_{20}\tau^2\left(x^2+y^2\right) + \right. \\ & \left. {}_1W_{31}\tau\left(x^2+y^2\right)y + {}_3W_{11}\tau^3 y \right] + \left[{}_0W_{60}\left(x^2+y^2\right)^3 + {}_3W_{33}\tau^3y^3 + {}_4W_{20}\tau^4\left(x^2+y^2\right) + \right. \\ & W_{40}\tau^2\left(x^2+y^2\right)^2 + {}_5W_{11}\tau^5 y + {}_4W_{22}\tau^4y^2 + {}_1W_{51}\tau\left(x^2+y^2\right)^2 y + {}_2W_{42}\tau^2\left(x^2+y^2\right)y^2 + \\ & \left. {}_3W_{31}\tau^3\left(x^2+y^2\right)y \right] + \left[{}_0W_{80}\left(x^2+y^2\right)^4 + {}_4W_{44}\tau^4y^4 + \cdots \right] + \cdots \end{aligned} \tag{8.82}$$

The coma type component of wave aberration is given by $\frac{1}{2}\{W(x,y) - W(x,-y)\}$. On algebraic manipulation, we get

$$\begin{aligned} \frac{1}{2}\{W(x,y) - W(x,-y)\} = & {}_1W_{31}\tau r^3\cos\theta + {}_3W_{11}\tau^3 r\cos\theta + {}_3W_{33}\tau^3r^3(\cos\theta)^3 + \\ & {}_5W_{11}\tau^5 r\cos\theta + {}_1W_{51\tau r}{}^5\cos\theta + {}_3W_{31}\tau^3r^3\cos\theta + \cdots \end{aligned} \tag{8.83}$$

As expected, each term of the coma type component of wave aberrations contains a variable involving the odd power of $\cos\theta$. In the process, it includes the distortion terms along with the coma terms of all forms. Defining $\bar{W}(x,y)$ as

$$\bar{W}(x,y) = \frac{1}{2}\{W(x,y) + W(x,-y)\} \tag{8.84}$$

We note that the spherical aberration type of wave aberration is given by $\bar{W}\left(\sqrt{x^2+y^2},0\right)$. From (8.82) and (8.84), after algebraic manipulation we get

$$\bar{W}\left(\sqrt{x^2+y^2},0\right) =_0 W_{40}r^4 +_2 W_{20}\tau^2r^2 +_0 W_{60}r^6 +_4 W_{20}\tau^4r^2 +_2 W_{40}\tau^2r^4 + \ldots \tag{8.85}$$

Note that, besides spherical aberration of all orders and types, e.g., oblique spherical aberration, the spherical aberration type component of wave aberration, as defined above, includes curvatures of all orders. The astigmatism type component of the wave aberration is given by $\left[\bar{W}(x,y) - \bar{W}\left(\sqrt{x^2+y^2},0\right)\right]$. From (8.84), (8.85), and (8.82), it follows

$$\left[\bar{W}(x,y) - \bar{W}\left(\sqrt{x^2+y^2},0\right)\right] = {}_2W_{22}\tau^2r^2(\cos\theta)^2 + {}_4W_{22}\tau^4r^2(\cos\theta)^2 + {}_2W_{42}\tau^2r^4(\cos\theta)^2 + \ldots \tag{8.86}$$

Each term of the astigmatism type component of wave aberration contains a variable involving the even power of $\cos\theta$.

8.3 Transverse Ray Aberrations Corresponding to Selected Wave Aberration Polynomial Terms

From (8.11), we note that the transverse ray aberration components $\delta\xi'$ and $\delta\eta'$ corresponding to wave aberration $W(X',Y')$ are given by

$$\delta\xi' = -\frac{R'}{n'}\frac{\partial W(X',Y')}{\partial X'} \tag{8.87}$$

$$\delta\eta' = -\frac{R'}{n'}\frac{\partial W(X',Y')}{\partial Y'} \tag{8.88}$$

As done earlier in (8.71) for radial coordinate $r'\left[=(R'/h')\right]$, normalized coordinates (x',y') for points on the pupil can be defined as

$$x' = \frac{X'}{h'} \qquad y' = \frac{Y'}{h'} \tag{8.89}$$

Using the normalized coordinates, (8.87) and (8.88) reduce to

$$\delta\xi' = -\frac{1}{n'(h'/R')}\frac{\partial W(x',y')}{\partial x'} \tag{8.90}$$

$$\delta\eta' = -\frac{1}{n'(h'/R')}\frac{\partial W(x',y')}{\partial y'} \tag{8.91}$$

Omitting primes from the symbols, the transverse ray aberration components $\delta\xi$ and $\delta\eta$ of the aperture ray specified by normalized coordinates (x,y) on the pupil are given by

$$\delta\xi = -\frac{1}{n(h/R)}\frac{\partial W(x,y)}{\partial x} \tag{8.92}$$

$$\delta\eta = -\frac{1}{n(h/R)}\frac{\partial W(x,y)}{\partial y} \tag{8.93}$$

The power series expansion for the transverse ray aberration components $\delta\xi(x,y)$ and $\delta\eta(x,y)$ can be readily obtained from the above relations by differentiating the expression for $W(x,y)$ given in (8.82). Transverse ray aberrations corresponding to a few selected wave aberration polynomial terms are presented below.

8.3.1 Primary Spherical Aberration

In the power series expansion of wave aberration, the primary spherical aberration term is given by

$$W(r) \equiv W(x,y) = {}_0W_{40}r^4 = {}_0W_{40}\left(x^2+y^2\right)^2 \tag{8.94}$$

Incidentally, it may be noted that the name 'spherical aberration' for this aberration seems to be less appropriate than the name 'Öffnungsfehler' (aperture aberration) used in German literature on optics [120].

Using (8.92) and (8.93), the corresponding components of transverse ray aberration are

$$\delta\xi = -\frac{1}{n(h/R)} 4_0 W_{40} x\left(x^2 + y^2\right) \tag{8.95}$$

$$\delta\eta = -\frac{1}{n(h/R)} 4_0 W_{40} y\left(x^2 + y^2\right) \tag{8.96}$$

Therefore,

$$\delta\xi^2 + \delta\eta^2 = \left[\frac{1}{n(h/R)} 4_0 W_{40} r^3\right]^2 \tag{8.97}$$

In terms of ξ, η coordinates with origin at O′ on the image plane, let the coordinates of the image point P′ be $(0, \bar{\eta})$. Writing $\delta\xi = \xi - 0, \delta\eta = \eta - \bar{\eta}$ (8.97) reduces to

$$\xi^2 + (\eta - \bar{\eta})^2 = \left[\frac{1}{n(h/R)} 4_0 W_{40} r^3\right]^2 \tag{8.98}$$

This is the equation of a circle. The radius of the circle depends on r. The points of intersection of rays coming from a circular ring of radius r on the exit pupil form a circle of radius $\frac{1}{n(h/R)} 4_0 W_{40} r^3$ with its centre at $(0, \bar{\eta})$. Because of the occurrence of the term r^3 in the latter radius, the circles formed on the image plane by rays coming from equispaced rings on the exit pupil will not be equispaced.

Since this aberration is symmetric about the principal ray, it is sufficient to consider one of the two components in any section. Choosing the meridian section for which $x = 0$, we get

$$\delta\eta = \tilde{W}_{40} y^3 \tag{8.99}$$

where

$$\tilde{W}_{40} = -\frac{1}{n(h/R)} 4_0 W_{40} \tag{8.100}$$

Figure 8.16 illustrates the ray geometry on the meridian section when the optical axis is the principal ray, i.e., in the case of axial imagery. The corresponding y versus $\delta\eta$ relationship, as expressed by (8.99), is represented graphically in Figure 8.17.

8.3.1.1 Caustic Surface

In Figure 8.16, $A'O_1'$ is a ray in the meridional section of the image-forming pupil. With respect to the origin at O′, the coordinates of A′ and O_1' are $(0, y. - R)$, and $(0, \delta\eta, 0)$ respectively. Using (8.99), the equation of the ray $A'O_1'$ in η, ς coordinates is

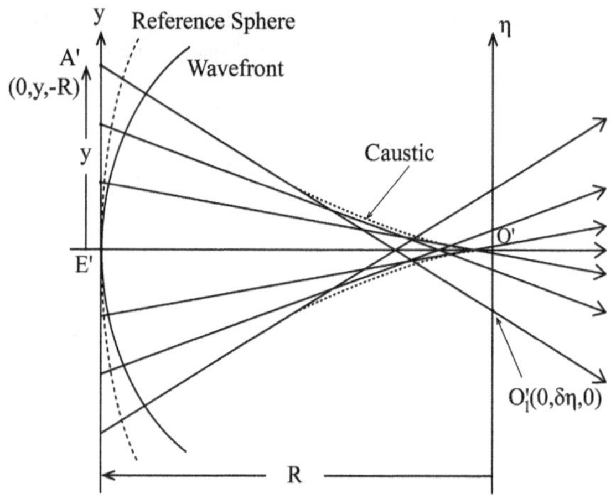

FIGURE 8.16 Ray geometry on the meridian section in case of axial imaging: the optical axis is the principal ray.

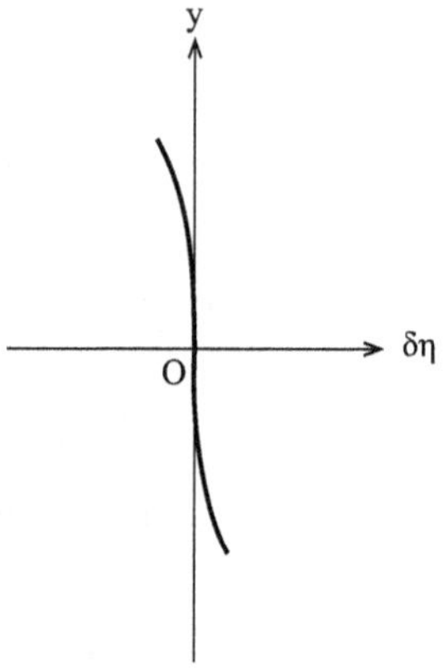

FIGURE 8.17 Cubic dependence of δη on y in presence of primary spherical aberration alone.

$$\eta = \tilde{W}_{40} y^3 - \frac{\left(y - \tilde{W}_{40} y^3\right)}{R} \varsigma \tag{8.101}$$

This equation represents a family of rays of the image-forming pupil in the meridian section. The parameter y defines a particular ray of the family. The envelope of this family can be determined by eliminating y between the equation (8.101), and the equation $\frac{d\eta}{dy} = 0$.

Differentiating (8.101), we get

$$y = \left[\frac{\varsigma}{3\tilde{W}_{40}(R+\varsigma)}\right]^{\frac{1}{2}} \tag{8.102}$$

From (8.101), it follows

$$\eta = -\frac{2}{3\sqrt{3}R}\left[\frac{\varsigma^{\frac{3}{2}}}{\left\{\tilde{W}_{40}(R+\varsigma)\right\}^{\frac{1}{2}}}\right] \tag{8.103}$$

Assuming $\varsigma \ll R$, (8.103) reduces to

$$\eta = -\frac{2}{3\sqrt{3\tilde{W}_{40}}}\left(\frac{\varsigma}{R}\right)^{\frac{3}{2}} \tag{8.104}$$

This is the equation of a semi-cubical parabola. Note that in the presence of a primary spherical aberration, the mantle of the caustic is a semi-cubical parabola, and the other part of the caustic reduces to a spike along the axis, as shown in Figure 8.12.

8.3.2 Primary Coma

The primary coma term in the power series expansion is given by

$$W(\tau, r, \theta) = {}_1W_{31}\tau r^3 \cos\theta \tag{8.105}$$

For a specific value of the fractional field τ, the primary coma can be expressed as

$$W(r, \theta) = W_{31} r^3 \cos\theta = W_{31} y\left(x^2 + y^2\right) \tag{8.106}$$

where $W_{31} \equiv {}_1W_{31}\tau$.

The corresponding components $\delta\xi, \delta\eta$ of the transverse ray aberration are

$$\delta\xi = -\frac{1}{n(h/R)} W_{31} r^2 \sin 2\theta \tag{8.107}$$

$$\delta\eta = -\frac{1}{n(h/R)} W_{31} r^2 (2 + \cos 2\theta) \tag{8.108}$$

From (8.106), it is seen that the primary coma term for the wave aberration is zero for any value of τ and r, when $\theta = \frac{\pi}{2}, \frac{3\pi}{2}$, i.e., at the sagittal plane. But, out of the two components of the transverse ray aberration, only $\delta\xi = 0$, and $\delta\eta \neq 0$.

The image point P′, the point of intersection of the principal ray with the image plane has the coordinates $(0, \bar{\eta})$, so that the coordinates of the point of intersection of a ray with the image plane are $\xi = \delta\xi, \eta = \bar{\eta} + \delta\eta$, where the transverse ray aberration components of the ray are $(\delta\xi, \delta\eta)$. From (8.107) and (8.108), we get

$$\xi^2 + \left[\eta - \left\{\bar{\eta} - \frac{2}{n(h/R)} W_{31} r^2\right\}\right]^2 = \left\{\frac{1}{n(h/R)} W_{31} r^2\right\}^2 \tag{8.109}$$

This is the equation of a circle with radius, $\frac{1}{n(h/R)} W_{31} r^2$ and the distance of the centre of the circle from the image point is $\left\{-\frac{2}{n(h/R)} W_{31} r^2\right\}$. This circle is the locus of the point of intersection of rays from a circular zone of radius r on the exit pupil. Note that both the distance of the location of the centre of the circle from the image point P′, and the radius of the circle tend to zero as $r \to 0$.

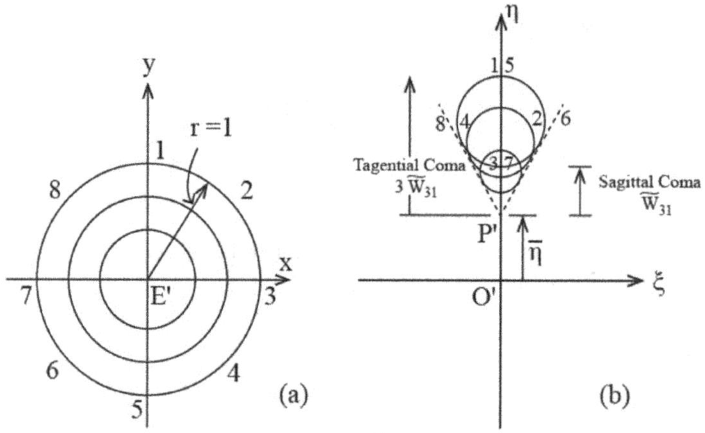

FIGURE 8.18 (a) Eight points marked at azimuths $\theta = (n-1)\frac{\pi}{4}, n = 1,2,\ldots,8$ on three concentric circular rings on the exit pupil (b) Loci of the points of intersection on the image plane of the eight rays passing through the points on each ring on the exit pupil are circles. Superposition of the latter circles constitutes the comatic image corresponding to an extraaxial point object.

Figure 8.18 explains the formation of the comatic flare with more detail. Side views of the exit pupil and the image plane at O′ are shown in Figure 8.18(a) and Figure 8.18(b), respectively. The chosen image point P′ has coordinates $(0, \bar{\eta})$ in the rectangular ξ, η coordinate system with the origin at O′. In Figure 8.18(a), points on the outer circular zone of the exit pupil, for which $r = 1$, are marked sequentially from m=1 to 8 at azimuths $\theta = (m-1)\frac{\pi}{4}$. The locus of the point of intersection on the image plane of the rays coming from the outer circular zone of the exit pupil is a circle shown in Figure 8.18(b). The centre of the circle has the coordinates $\{0, (\bar{\eta} - 2\tilde{W}_{31})\}$, and its radius is $\tilde{W}_{31}$, where

$$\tilde{W}_{31} = -\frac{1}{n(h/R)} W_{31} \tag{8.110}$$

Table 8.6 presents the values of $\delta\xi$ and $\delta\eta$ for rays coming from the points 1, 2, …, 8 on the circular edge of the exit pupil. The points on the image plane on the corresponding circle are also marked. Note that the intersection points on the image plane of rays coming from points on the semi-circle (containing the points 1, 2, 3, 4, 5) on the exit pupil form a complete circle. Indeed, intersection points on the image plane of the rays coming from points on the other part of the semi-circle (containing points 5, 6, 7, 8, 1) on the exit pupil also form the same circle. Circles of ray intersection points corresponding to rays from two other circular zones of the exit pupil are also shown in Figure 8.18(b). By symmetry, the two rays, coming from the edge of the sagittal section of the exit pupil, meet at a point (3,7) on the η-axis on the image plane. From Table 8.5, it is seen that the distance of this point of intersection from the image point P′ is $\tilde{W}_{31}$. This distance is known as the sagittal coma. Similarly, two rays coming from the edge of the meridian section of the exit pupil meet at the point (1,5) on the η-axis in the image plane. The distance of this common ray intersection point from the image point P′ is known as a tangential coma. It is equal to $3\tilde{W}_{31}$. As r changes from 0 to 1, the circles on the image plane overlap to form a diffuse pear-shaped patch as the image pattern corresponds to the point object. This patch looks like the tail of a comet, and so the image is called comatic image. The dotted lines that are symmetric with respect to the η-axis indicate the borders of this patch. The angle α made by the dotted line with the η-axis is equal to 30^0; it has the same value for all the circles. For example, for the last circle, the radius is $\tilde{W}_{31}$, and the distance of the centre from P′ is $2\tilde{W}_{31}$. Therefore, $\sin\alpha = \frac{\tilde{W}_{31}}{2\tilde{W}_{31}} = \frac{1}{2}$, and $\alpha = 30^0$. Note that Figure 8.18(b) does not convey an impression of ray density in the comatic patch. Detailed analysis shows that the ray density is maximum in the neighbourhood of the image point P′, and it decreases gradually at regions with increasing distance

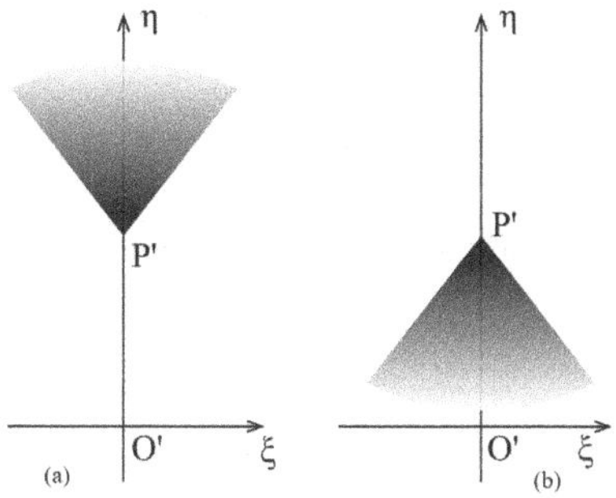

FIGURE 8.19 Ray density is maximum near the point P′ and reduces gradually at regions away from this point (a) Outer coma; (b) Inner coma.

TABLE 8.6

Values of $\delta\xi, \delta\eta$ for rays coming from edge $(r=1)$ of the circular exit pupil at selected values of θ. $\left[\tilde{W}_{31} = -\frac{1}{n(h/R)} W_{31}\right]$

Point	θ	$\delta\xi$	$\delta\eta$
1	0	0	$3\tilde{W}_{31}$
2	$\pi/4$	$\tilde{W}_{31}$	$2\tilde{W}_{31}$
3	$\pi/2$	0	$\tilde{W}_{31}$
4	$3\pi/4$	$-\tilde{W}_{31}$	$2\tilde{W}_{31}$
5	π	0	$3\tilde{W}_{31}$
6	$5\pi/4$	$\tilde{W}_{31}$	$2\tilde{W}_{31}$
7	$3\pi/2$	0	$\tilde{W}_{31}$
8	$7\pi/4$	$-\tilde{W}_{31}$	$2\tilde{W}_{31}$

from P′. Figure 8.19 illustrates this for the cases of (a) outer coma and (b) inner coma. The comatic patches extend towards or away from the origin of the image plane, depending on the sign of W_{31}.

8.3.3 Primary Astigmatism

The primary astigmatism term in the power series expansion of wave aberration is

$$W(\tau, r, \theta) = {}_2W_{22}\tau^2 r^2 (\cos\theta)^2 \tag{8.111}$$

Since the aberration is proportional to r^2, it is affected by a longitudinal shift of focus. Therefore, the defocus aberration term is considered together with astigmatism. So we consider the aberration term

$$W(\tau, r, \theta) = W_{20} r^2 + {}_2W_{22}\tau^2 r^2 (\cos\theta)^2 \tag{8.112}$$

For a specific value of the fractional field τ, we can write the aberration term as

$$W(r, \theta) = W_{20} r^2 + W_{22} r^2 (\cos\theta)^2 \tag{8.113}$$

Equivalently, in terms of rectangular coordinates (x, y),

$$W(x,y) = W_{20}x^2 + (W_{20} + W_{22})y^2 \tag{8.114}$$

The corresponding components $\delta\xi, \delta\eta$ of the transverse ray aberration are

$$\delta\xi = -\frac{1}{n(h/R)}.2W_{20}x = \tilde{W}_{20}x = \tilde{W}_{20}r\sin\theta \tag{8.115}$$

$$\delta\eta = -\frac{1}{n(h/R)}.2(W_{20} + W_{22})y = (\tilde{W}_{20} + \tilde{W}_{22})y = (\tilde{W}_{20} + \tilde{W}_{22})r\cos\theta \tag{8.116}$$

where

$$\tilde{W}_{20} = -\frac{2}{n(h/R)}W_{20} \quad \text{and} \quad \tilde{W}_{22} = -\frac{2}{n(h/R)}W_{22} = -\frac{2}{n(h/R)^2}W_{22}\tau^2 \tag{8.117}$$

From (8.115) – (8.117), it follows

$$\left\{\frac{\delta\xi}{\tilde{W}_{20}r}\right\}^2 + \left\{\frac{\delta\eta}{(\tilde{W}_{20} + \tilde{W}_{22})r}\right\}^2 = 1 \tag{8.118}$$

The equation (8.118) represents an ellipse with semi-axes $\tilde{W}_{20}r$ and $(\tilde{W}_{20} + \tilde{W}_{22})r$. The origin of the ellipse is the point of intersection of the principal ray with the transverse plane corresponding to defocus W_{20}. The shape of the ellipse is determined by relative values of $\tilde{W}_{20}$ and $\tilde{W}_{22}$. Five case are noteworthy. Figure 8.20 presents a rough sketch of the change in contour of an astigmatic wavefront in the focal region.

Case I. $W_{20} = 0$

We get $\tilde{W}_{20} = 0$, $\delta\xi = 0$ and $\delta\eta = \tilde{W}_{22}r\cos\theta$.

The locus of the point of intersection of rays emerging from points on the exit pupil with the image plane is a straight line along the η-axis. The point of intersection, P′ of the principal ray with the image plane at O′, is the midpoint of the line (Figure 8.20). All rays with values of r in the range (0, 1) will intersect the image plane within the range $(0, \bar{\eta} + \tilde{W}_{22})$ and $(0, \bar{\eta} - \tilde{W}_{22})$, so that the length of the line image is $2\tilde{W}_{22}$.

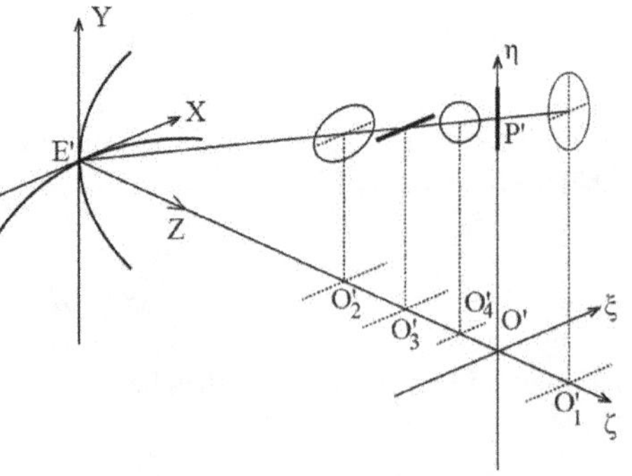

FIGURE 8.20 Change in contour of an astigmatic wavefront in the focal region (focal lines are shown thick to highlight them).

Case II. $W_{20} = -W_{22}$

This implies $\tilde{W}_{20} = -\tilde{W}_{22}$, so that $\delta\xi = -\tilde{W}_{22} r \sin\theta$, $\delta\eta = 0$.

On this defocus plane, the locus of intersection of rays emerging from points on the exit pupil is a straight line parallel to the ξ-axis. The length of the line image is $2\tilde{W}_{22}$. This is the same as for case I. The midpoint of the line is the point of intersection of the principal ray with this defocus plane. The latter is located at O'_3 on the axis (Figure 8.20) .

Case III. $W_{20} = -\frac{W_{22}}{2}$

It follows that $\tilde{W}_{20} = -\frac{\tilde{W}_{22}}{2}$, $\delta\xi = -\frac{\tilde{W}_{22}}{2} r \sin\theta$, $\delta\eta = -\frac{\tilde{W}_{22}}{2} r \cos\theta$. Therefore,

$$\delta\xi^2 + \delta\eta^2 = \left(\frac{1}{2}\tilde{W}_{22} r\right)^2 \tag{8.119}$$

This is the equation of a circle with radius $\frac{1}{2}\tilde{W}_{22} r$. The centre of the circle is the point of intersection of the principal ray with the transverse plane corresponding to a defocus $-\frac{W_{22}}{2}$. Note that the plane is midway between the two transverse planes of cases II and I. This defocus plane is located at O'_4 on the axis (Figure 8.20).

Case IV. $|W_{20} + W_{22}| < |W_{20}|$

Correspondingly, $|\tilde{W}_{20} + \tilde{W}_{22}| < |\tilde{W}_{20}|$. As a result, it is seen from (8.118) that the major axis of the contour ellipse lies parallel to the ξ-axis, and its minor axis is parallel to the η-axis.

Case V. $|W_{20} + W_{22}| > |W_{20}|$

In this case, the major axis of the contour ellipse lies parallel to the η-axis, and the minor axis is parallel to the ξ-axis.

Figure 8.20 illustrates the five cases as described above. Note that the line image S_1S_2 formed on the image plane at O' can be considered as the focus for rays from the x section of the pupil. It is called the sagittal astigmatic focal line, or simply the sagittal focal line. Similarly, the line image T_1T_2 formed on the image plane at O'_3 may be regarded as the focus for rays from the y section of the pupil. This is called the tangential astigmatic focal line, or the tangential focal line. It is important to note that the sagittal focal line is located in the meridional or tangential plane, and the tangential focal line is located in the sagittal plane (Figure 8.21). Also, per treatment based on theory of primary aberrations, the distance between the two focal lines is proportional to the square of the fractional field τ. The focal lines are at the centre of curvatures of the wavefront for the x and the y sections. They are, therefore, parts of the evolute of the wavefront.

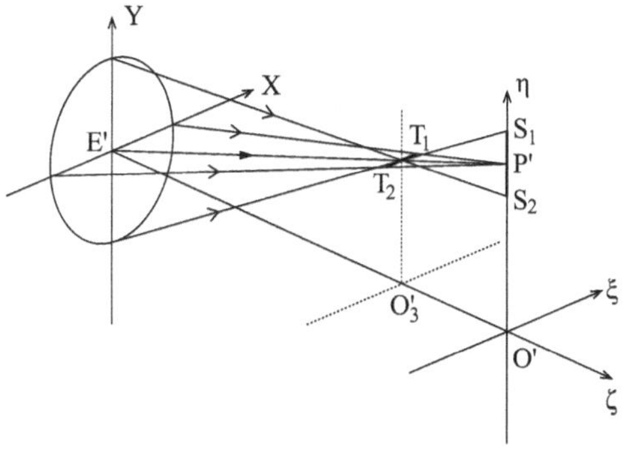

FIGURE 8.21 Sagittal focal line S_1S_2 on the meridian section. Tangential focal line T_1T_2 on the sagittal section.

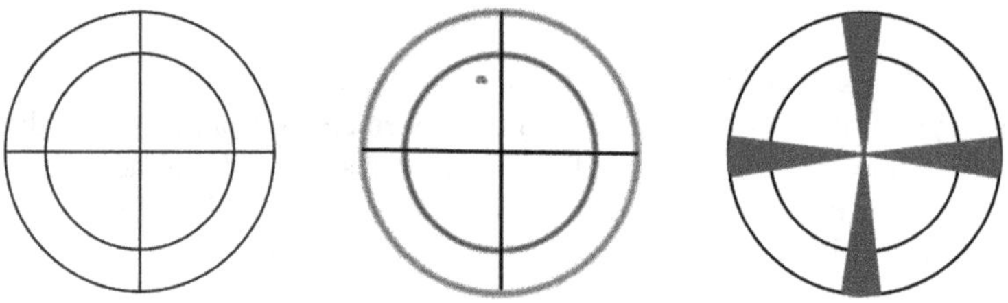

FIGURE 8.22 Image of a circular wheel with radial spokes. (a) object, (b) sagittal image, (c) tangential image.

The best image is usually taken to be located halfway between the tangential and sagittal image. As shown in case III above, all rays $(0 \le r \le 1)$ coming from the circular exit pupil pass through a circular patch of light of diameter $\tilde{W}_{22}$. This circular patch is also called 'the disc of least confusion'.

Figure 8.22 shows the images of a wheel with radial spokes. Note that the spokes (radial lines) are imaged sharply on the sagittal focal surface, and the circles are imaged sharply on the tangential focal surface. The corresponding images are called the sagittal image and the tangential image, respectively.

8.3.4 Primary Curvature

The primary astigmatism term in the power series expansion of wave aberration is

$$W(\tau, r, \theta) = {}_2W_{20}\tau^2 r^2 \tag{8.120}$$

Since the term contains an even function of pupil variables, it will be affected by a defocus or longitudinal shift of the image plane. Therefore, this aberration term is considered together with the defocus term as

$$W = W_{20}r^2 + {}_2W_{20}\tau^2 r^2 = \left(W_{20} + {}_2W_{20}\tau^2\right)\left(x^2 + y^2\right) \tag{8.121}$$

The components of transverse ray aberrations $\delta\xi, \delta\eta$ are given by

$$\delta\xi = -\frac{\partial W}{\partial x} = -\frac{1}{n(h/R)}\left(W_{20} + {}_2W_{20}\tau^2\right)2x \tag{8.122}$$

$$\delta\eta = -\frac{\partial W}{\partial y} = -\frac{1}{n(h/R)}\left(W_{20} + {}_2W_{20}\tau^2\right)2y \tag{8.123}$$

From the above equations, it is obvious that for any value of (x, y) or (r, θ)

$$\delta\xi = \delta\eta = 0, \text{ if } W_{20} = -{}_2W_{20}\tau^2 \tag{8.124}$$

This shows that all rays originating from fractional object height τ and passing through the pupils do not meet at a point on the chosen image plane. Nevertheless, all of them meet at a point on the transverse plane corresponding to defocus $W_{20} = -{}_2W_{20}\tau^2$. For a different object point with a different value of the fractional field τ, all rays originating from it meet at a point in the image space. But the latter image point is formed on a transverse plane that is different from the transverse plane of the earlier image point. The ς coordinate of the image point is determined by the value of defocus pertaining to the particular value of τ of the object point under consideration. (8.124) shows that it is proportional to the square of the fractional field parameter τ. Thus, the effect of the aberration called 'primary curvature' is to form a sharp image of a plane object on a curved surface.

The image so formed has a point-to-point correspondence with the object. The curved image surface $O'\tilde{P}'$ is tangential to the desired image plane on the axis (Figure 8.23). The image formed on the desired image plane $O'P'$ is not sharp; it loses its quality with increasing values of τ. In memory of Petzval, who first studied this phenomenon, this curvature of the image surface is called 'Petzval curvature'.

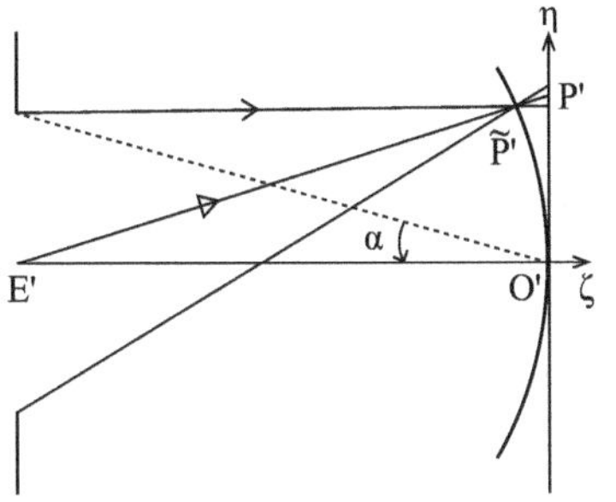

FIGURE 8.23 A sharp image $O'\tilde{P}'$ of a plane object OP on a curved surface. The object point P is imaged stigmatically at $\tilde{P}'$, and as a blur patch around the point P' of the desired plane image $O'P'$.

Astigmatism and field curvature are often grouped together as both of them are dependent on the square of the fractional field τ. Also, both of them are proportional to the normalized aperture term r, so their combined effect needs to be studied with a longitudinal shift of focus. Therefore, the combined aberration term becomes

$$W = W_{20}r^2 + {}_2W_{20}\tau^2 r^2 + {}_2W_{22}\tau^2 r^2 (\cos\theta)^2 \tag{8.125}$$

The components of transverse ray aberrations $\delta\xi, \delta\eta$ are

$$\delta\xi = -\frac{\partial W}{\partial x} = -\frac{1}{n(h/R)}\left(W_{20} + {}_2W_{20}\tau^2\right)2r\sin\theta \tag{8.126}$$

$$\delta\eta = -\frac{\partial W}{\partial y} = -\frac{1}{n(h/R)}\left[W_{20} + \left\{{}_2W_{20} + {}_2W_{22}\right\}\tau^2\right]2r\cos\theta \tag{8.127}$$

The relations (8.126) and (8.127) can be written succinctly as

$$\delta\xi = \left[\tilde{W}_{20} + {}_2\tilde{W}_{22}\tau^2\right]r\sin\theta \tag{8.128}$$

$$\delta\eta = \left[\tilde{W}_{20} + \left\{{}_2\tilde{W}_{20} + {}_2\tilde{W}_{22}\right\}\tau^2\right]r\cos\theta \tag{8.129}$$

where

$$\tilde{W}_{20} = -\frac{2}{n(h/R)}W_{20} \qquad {}_2\tilde{W}_{20} = -\frac{2}{n(h/R)}{}_2W_{20} \qquad {}_2\tilde{W}_{22} = -\frac{2}{n(h/R)}{}_2W_{22} \tag{8.130}$$

From (8.128) – (8.130), it can be seen that in the image-forming pencil of rays for the object/image point corresponding to fractional field τ, for all rays in the meridian plane (where $\theta = 0, \pi$), values of $\delta\xi$ and $\delta\eta$ become zero on the defocus plane specified by

$$W_{20} = -\left\{{}_2W_{20} + {}_2W_{22}\right\}\tau^2 \tag{8.131}$$

In the same pencil, for all rays in the sagittal section $\left(\text{where } \theta = \frac{\pi}{2}, \frac{3\pi}{2}\right)$, values of $\delta\xi$ and $\delta\eta$ become zero on the defocus plane, specified by

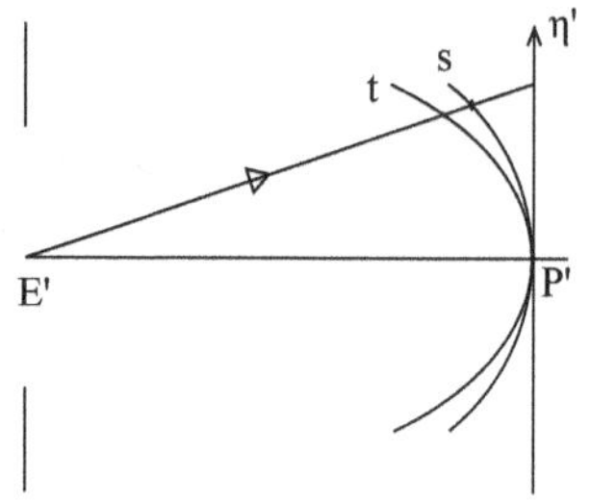

FIGURE 8.24 Meridional sections of the sagittal and tangential image surfaces in presence of both astigmatism and curvature.

$$W_{20} = -{}_2W_{20}\tau^2 \tag{8.132}$$

Therefore, if both the aberrations, namely primary curvature ${}_2W_{20}$ and primary astigmatism ${}_2W_{22}$ are nonzero, then the sagittal focal lines are formed on a curved image surface (Figure 8.24) instead of on the paraxial image plane, as shown in Figure 8.21 in the presence of primary astigmatism alone. The tangential or meridian focal lines are formed on another surface of greater or less curvature, depending on the signs and magnitudes of ${}_2W_{20}$ and ${}_2W_{22}$. Figure 8.24 presents a schematic diagram for the sagittal and tangential image surfaces. Note that, in general, the latter surfaces are different from the curved Petzval surface shown in Figure 8.23, since the latter implies ${}_2W_{22} = 0$.

8.3.5 Primary Distortion

In the presence of primary distortion alone, the aberration function W takes the form

$$W = {}_3W_{11}\tau^3 r\cos\theta = {}_3W_{11}\tau^3 y \tag{8.133}$$

The transverse ray aberration components $\delta\xi, \delta\eta$ are given by

$$\delta\xi = -\frac{1}{n(h/R)}\frac{\partial W}{\partial x} = 0 \tag{8.134}$$

$$\delta\eta = -\frac{1}{n(h/R)}\frac{\partial W}{\partial y} = -\frac{1}{n(h/R)}\,{}_3W_{11}\tau^3 \tag{8.135}$$

This shows that all rays of the image-forming pencil for the object/image fractional field τ meet at a point $\tilde{P}'$ on the η-axis (Figure 8.25). The point $\tilde{P}'$ is displaced from the desired image point P′ by an amount $\delta\eta$ that is proportional to τ^3. Therefore, it is not a simple change in linear magnification for all points of the object. The magnification of points lying further from the axis will be different from that of points lying nearer to the axis. The variation in magnification is proportional to τ^3. Figure 8.26(a)–(c) illustrates the resulting distortion in the image. The distortion occurring in the case of ${}_3W_{11} > 0$ is called barrel distortion, and the same in the case of ${}_3W_{11} < 0$ is called pincushion distortion. Figures 8.26(b) and 26(c) depict the two cases, respectively.

8.3.6 Mixed and Higher Order Aberration Terms

The earlier section enunciates the effects of each primary aberration alone on the location of ray intersection points around the desired image point. The analysis can be extended to the case of higher order aberration terms. Steward presented the ray intersection patterns for each of the nine secondary aberration terms [121]. In real optical imaging systems, they occur in combination with each other and with higher

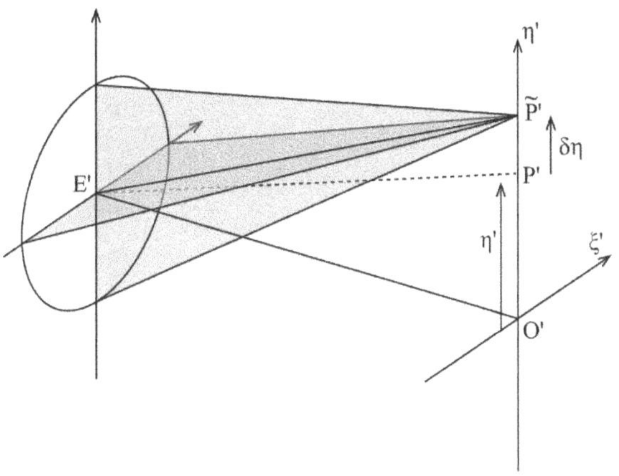

FIGURE 8.25 All rays of the image-forming pencil from the object point P (not shown) meet at the point $\tilde{P}'$ on the η'-axis, at a distance $\delta\eta$ from the desired image point P′ on the η'-axis.

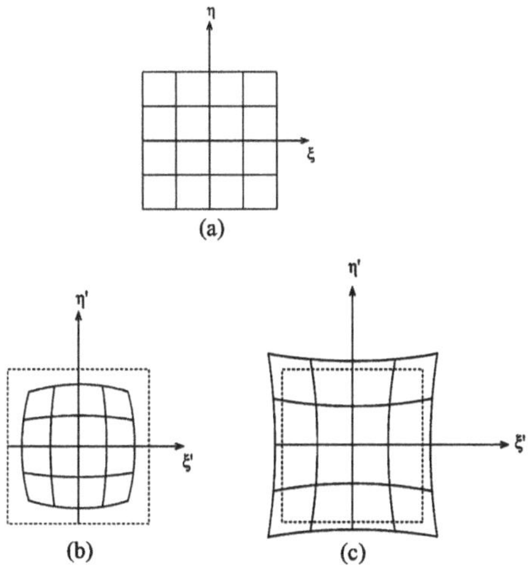

FIGURE 8.26 (a)–(c) A square object with grid lines. Image in the presence of (b) barrel distortion $\left({}_3W_{11} > 0\right)$. (c) pincushion distortion $\left({}_3W_{11} < 0\right)$.

order aberrations. The integrated effect on the location of the ray intersection points can, in principle, be worked out in the same manner as above. The combined effect of the aberrations can be visualized from different graphs, e.g. transverse ray aberration across the meridian or sagittal section, wavefront aberration graphs for the same sections, spot diagrams, etc. It is important to note that such graphs often incorporate the effects of all mixed and higher order aberrations for a specific value of fractional field τ. Of course, the sagittal and tangential image surfaces facilitate visualization of imaging of different field points around the image plane.

8.4 Longitudinal Aberrations

In practice, at times the effects of spherical aberration and astigmatism are analyzed as longitudinal aberrations. For example, the distance between the marginal focus and paraxial focus, defined earlier as 'longitudinal spherical aberration' is used. Similarly, the effect of astigmatism is analyzed by the axial distance between the two astigmatic focal lines. The relations between their longitudinal forms and corresponding transverse ray aberration and wave aberration forms are given below.

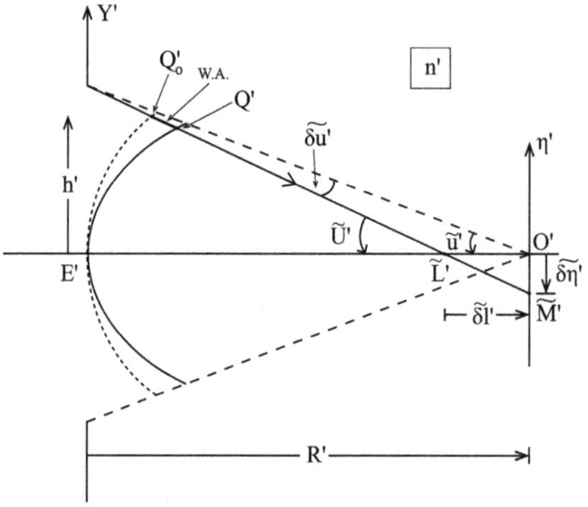

FIGURE 8.27(a) Longitudinal spherical aberration at margin of the pupil.

A. Longitudinal Spherical Aberration

Figure 8.27(a) shows the image space of an axially symmetric imaging system with its exit pupil and image plane located at E' and O' on the axis. $E'Q_0'$ is a reference sphere with its centre at O' and radius $E'O' = R'$. $E'Q'$ is the actual wavefront forming image of an axial object point O (not shown in figure). It passes through E'. The refractive index of the image space is n'. On account of rotational symmetry, the analysis is restricted to a single azimuth, and in the figure only the meridian section is shown. The point of intersection of the marginal ray $Q_0'Q'\tilde{M}'$ with the reference sphere is Q_0'. The coordinates of the point Q_0' with respect to E' as the origin is $(0, \tilde{Y}')$. Let $\tilde{Y}' = h'$. The marginal ray intersects the optical axis $E'O'$ at $\tilde{L}'$, and makes an angle $\angle Q_0'\tilde{L}'E' = \tilde{U}'$ with the axis. The paraxial angle $\angle Q_0'O'E' = \tilde{u}' = -(h'/R') = -(\tilde{Y}'/R')$. For a point with ordinate Y' on the reference sphere, the normalized coordinate $y' = (Y'/h') = (Y'/\tilde{Y}')$. By definition, for the marginal ray, the wavefront aberration $W = [Q_0'Q']$, the transverse ray aberration $= O'\tilde{M}' = \widetilde{\delta\eta}'$, the angular aberration $= \angle\tilde{L}'Q_0'O' = \widetilde{\delta\eta}'$, and the longitudinal spherical aberration $= \tilde{L}'O' = \widetilde{\delta l}'$.

Note that, by convention, the longitudinal aberration is taken as positive if the ray intersection point lies at the left of the paraxial image plane [122]. This is done to make longitudinal aberration positive corresponding to a positive value for wave aberration, and so the longitudinal aberration is measured by the distance from the marginal focus to the paraxial focus.

From (8.12) we get

$$\delta\eta' = -\frac{R'}{n'}\frac{\partial W}{\partial Y'} = -\frac{1}{n'(h'/R')}\frac{\partial W}{\partial y'} = \frac{1}{n'\tilde{u}'}\frac{\partial W}{\partial y'} \tag{8.136}$$

In the presence of primary spherical aberration $W = W_{40}r^4 = W_{40}y'^4$, so that

$$\delta\eta' = \frac{4}{n'\tilde{u}'}W_{40}y'^3 \tag{8.137}$$

For the marginal ray, $y' = 1$. So, its transverse ray aberration is given by

$$(\delta\eta')_{y'=1} = \widetilde{\delta\eta}' = \frac{4}{n'\tilde{u}'}W_{40} \tag{8.138}$$

Since in all cases of practical importance, $\widetilde{\delta u'} \ll \tilde{u}'$, we get $\tilde{U}' \cong \tilde{u}'$. Therefore, the longitudinal spherical aberration for the marginal ray is given by

$$(\delta l')_{y'=1} = \widetilde{\delta l'} = \frac{\widetilde{\delta \eta'}}{\tilde{u}'} = \frac{4}{n'\tilde{u}'^2} W_{40} \tag{8.139}$$

(8.139) can be rewritten as

$$W_{40} = \frac{1}{4} n'\tilde{u}'^2 \widetilde{\delta l'} \cong \frac{1}{4} n'h'^2 \left(\frac{1}{R' - \widetilde{\delta l'}} - \frac{1}{R'} \right) \tag{8.140}$$

The bracketed term on the right-hand side expresses the longitudinal spherical aberration as the difference in diopter between the marginal focus and the paraxial focus. It is represented by δD. Expressing W_{40} as δW, (8.140) can be expressed as

$$\delta W = \frac{1}{4} n'h'^2 \delta D \tag{8.141}$$

(8.141) gives the relation between the wave aberration for the marginal ray and the corresponding longitudinal spherical aberration expressed in diopter form.

Mouroulis and Macdonald [123] noted that casual inspection of expressions like (8.138) for transverse spherical aberration or (8.139) for longitudinal spherical aberration might lead to the wrong impression that the transverse spherical aberration is inversely proportional to the semi angular aperture u' of the system, or the longitudinal spherical aberration is inversely proportional to square of the semi angular aperture, $(u')^2$. In fact, to prevent the occurrence of such misconceptions, in our expressions (8.138) or (8.139), a 'curl' is used on top of the relevant parameters to emphasize that the quantities correspond to the marginal ray, and they cannot be taken as variables.

Figure 8.27(b) illustrates the case of an arbitrary ray, but not necessarily the marginal ray, of the image-forming pencil. The ray intersects the reference sphere at B'_0. Its coordinates are $(0, Y')$, with respect to the origin at E'. For this ray, we have $y' = (Y'/h')$, and $u' = -(Y'/R')$. From (8.137) we get

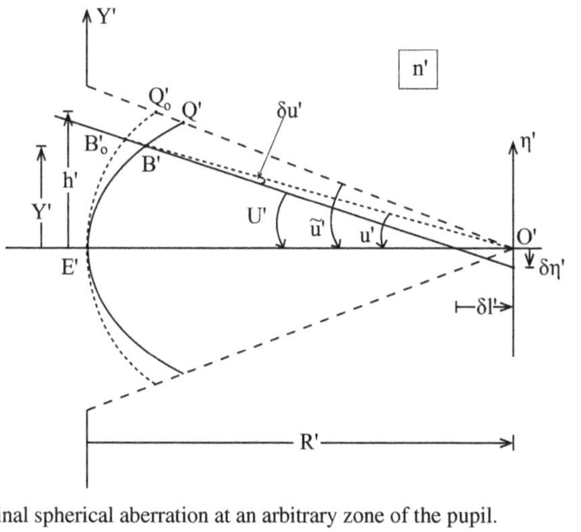

FIGURE 8.27(b) Longitudinal spherical aberration at an arbitrary zone of the pupil.

$$\delta\eta' = \frac{4W_{40}}{n'\tilde{u}'}(y')^3 = \frac{4W_{40}}{n'\tilde{u}'}\left(\frac{Y'}{h'}\right)^3 = \frac{4W_{40}}{n'\tilde{u}'}\left(\frac{Y'/R'}{h'/R'}\right)^3 = \frac{4W_{40}}{n'\tilde{u}'^4}(u')^3 \tag{8.142}$$

Since angular aberration $\delta u'$ for the aperture ray is much smaller than u', we have $U' \sim u'$. The longitudinal spherical aberration for this arbitrary image-forming ray is given by

$$\delta l' = \frac{\delta\eta'}{u'} \tag{8.143}$$

Substituting from (8.142) in (8.143), we get

$$\delta l' = \frac{4W_{40}}{n'\tilde{u}'^4}(u')^2 \tag{8.144}$$

The expressions (8.142) and (8.143) clearly show that the transverse spherical aberration and the longitudinal spherical aberration of an aperture ray of the image-forming pencil is proportional to the cube and the square of the semi aperture angle u'.

B. Longitudinal Astigmatism

The expression for primary astigmatism is

$$W = {}_2W_{22}\tau^2 r'^2 (\cos\theta)^2 \tag{8.145}$$

For a particular value of τ, W can be written as

$$W = W_{22} r'^2 (\cos\theta)^2 = W_{22} y'^2 \tag{8.146}$$

From (8.136) and (8.146), we get

$$\delta\eta' = \frac{1}{n'\tilde{u}'}\frac{\partial W}{\partial y'} = \frac{2}{n'\tilde{u}'}W_{22}y' \tag{8.147}$$

For the marginal ray with $y' = 1$,

$$\widetilde{\delta\eta'} = \frac{2}{n'\tilde{u}'}W_{22} \tag{8.148}$$

we have the expression for longitudinal astigmatism as

$$\widetilde{\delta l'} = \frac{2}{n'\tilde{u}'^2}W_{22} \tag{8.149}$$

This equation can be rewritten as

$$W_{22} = \frac{1}{2}n'\tilde{u}'^2\widetilde{\delta l'} \cong \frac{1}{2}n'h'^2\left(\frac{1}{R' - \widetilde{\delta l'}} - \frac{1}{R'}\right) \tag{8.150}$$

As before, expressing the bracketed term on the right-hand side by δD and denoting W_{22} by δW, it follows

$$\delta W = \frac{1}{4} n' h'^2 \delta D \tag{8.151}$$

Here, δD expresses the difference in diopters between the S and the T foci.

Let the point of intersection of an arbitrary ray of the image-forming pencil with the reference sphere be $(0, Y')$. If $u' = -(Y'/R')$, the longitudinal astigmatism $\delta l'$ is given by

$$\delta l' = \frac{\delta \eta'}{u'} \tag{8.152}$$

Substituting for $\delta\eta'$ from (8.147) in the above, we get the expression for longitudinal astigmatism for the chosen value of τ as

$$\delta l' = \frac{2}{n'\tilde{u}'} \frac{y'}{u'} W_{22} = \frac{2}{n'\tilde{u}'} \frac{Y'}{h'} \frac{R'}{Y'} W_{22} = \frac{2W_{22}}{n'(\tilde{u}')^2} \tag{8.153}$$

Note that the expression does not contain any quantity specific for the aperture ray. It shows that the longitudinal astigmatism is independent of the aperture.

8.5 Aplanatism and Isoplanatism

In Section 7.2.4.1, we noted the existence of 'aplanatic' conjugate points, which are perfect images of each other in the case of refraction by a spherical surface. The word 'aplanatic' is originally derived from the Greek word 'απλανητικός', that consists of 'a', meaning 'no' + planétos from the verb 'planaein' (≈ 'to wander'), i.e., 'no aberration'. 'Aplanatism' refers to the state of the optical imaging system that gives rise to aplanatic imaging.

Initially, there was no unanimity in the use of the term 'aplanatic' for describing the state of correction of aberration in optical imaging systems. Welford [124] notes that, in 1791, Blair [125] coined the term to mean improved correction of secondary spectrum in achromatic doublets. In 1827, Herschel [126] used the term to mean freedom from spherical aberration. In 1844, Hamilton [127] used a term 'direct aplanaticity' to mean freedom from spherical aberration, and another term 'oblique aplanaticity' to indicate freedom from linear coma. The use of the term aplanatism, as understood today, commenced from the seminal publications of Ernst Abbe [128–131]. The currently accepted meaning of 'aplanatism' in optics is stated explicitly by Welford in his publication. The term 'Aplanatism' is now

> thought of as applying to axisymmetric optical systems near the centre of the field; such a system is said to be aplanatic if it is free from spherical aberration, i.e. the axial image is aberration-less, and if there is no aberration varying linearly with field angle, i.e. no linear coma.
>
> [124]

The term 'isoplanatic' was first proposed by Staeble [130] in 1919 for axi-symmetric optical systems with residual spherical aberration, but zero linear coma. In the same year, it appeared in another independent report by Lihotzky [131] dealing with the same problem. Since the advent of analytical treatments based on Fourier series and transformations in optical image analysis and synthesis [132 – 134], the term 'isoplanatism' has a wider connotation based on stationarity of aberrations around an object/image point. Welford states explicitly that the term 'isoplanatism' can apply to any small region of the field of an optical system with any or no symmetry; the system is said to be isoplanatic in this region if the aberrations are stationary for small displacements of the object point, i.e. if the aberration has no component depending linearly on the object position. [124]

In this context, 'aplanatism' is a special case of 'isoplanatism' around an axial point in an axi-symmetric imaging system, where the spherical aberration is zero.

8.5.1 Coma-Type Component of Wave Aberration and Linear Coma

In Section 8.2.5.3, during our discussion on types of aberration in axi-symmetric optical imaging systems, it was noted that, for a specific value of field variable τ, aberration terms that only have a plane of symmetry around the meridian plane are classified as coma-type. Since the y-axis lies on the meridian plane, the coma-type component of the wave aberration $W(x,y)$ is given by $\left[\frac{1}{2}\{W(x,y)-W(x,-y)\}\right]$. Extending relation (8.83) to include terms up to quaternary aberrations, we have

$$\begin{aligned}W_{coma}(x,y) \equiv W_{coma}(r,\theta) = &\left[{}_1W_{31}\tau r^3\cos\theta + {}_1W_{51}\tau r^5\cos\theta + {}_1W_{71}\tau r^7\cos\theta + {}_1W_{91}\tau r^9\cos\theta + \ldots\right] \\ &+\left[\left({}_3W_{11}\tau^3 + {}_5W_{11}\tau^5 + {}_7W_{11}\tau^7 + {}_9W_{11}\tau^9 + \ldots\right)r\cos\theta\right] \\ &\left[\left\{\left({}_3W_{31}\tau^3 + {}_5W_{31}\tau^5 + {}_7W_{31}\tau^7 + \ldots\right)r^3\cos\theta + \left({}_3W_{51}\tau^3 + {}_5W_{51}\tau^5 + \ldots\right)\right.\right. \\ &r^5\cos\theta + \left({}_3W_{71}\tau^3\right)r^7\cos\theta\Big\} + \Big\{\left({}_3W_{33}\tau^3 + {}_5W_{33}\tau^5 + {}_7W_{33}\tau^7 + \ldots\right) \\ &r^3(\cos\theta)^3 + \left({}_3W_{53}\tau^3 + {}_5W_{53}\tau^5\right)r^5(\cos\theta)^3 + \left({}_3W_{73}\tau^3 + \ldots\right)r^7(\cos\theta)^3\Big\} \\ &\left.+\left\{\left({}_5W_{55}\tau^5 + \ldots\right)r^5(\cos\theta)^5 + \ldots\right\} + \ldots\right]\end{aligned} \tag{8.154}$$

In the above relation, W_{coma} is seen to be comprised of three groups of terms. Note that the degree in τ in none of the terms in any of the groups is an even number. The first group consists of linear coma terms where each term is proportional to τ. The second group consists of distortions of various orders. The third group consists of nonlinear coma terms of various orders. In the two latter groups, in each term the degree of τ is an odd number greater than one.

Abbe sine condition, aplanatism, and axial isoplanatism assume the size of the field to be small, so that aberration terms containing powers of τ higher than the first are ignored. The total linear coma, $W_{l.c.}$, is given by

$$W_{l.c.}(r,\theta) = {}_1W_{31}\tau r^3\cos\theta + {}_1W_{51}\tau r^5\cos\theta + {}_1W_{71}\tau r^7\cos\theta + {}_1W_{91}\tau r^9\cos\theta + \ldots \tag{8.155}$$

In terms of (x,y), it can be rewritten as

$$\begin{aligned}W_{l.c.}(x,y) &= {}_1W_{31}\tau\left(x^2+y^2\right)y + {}_1W_{51}\tau\left(x^2+y^2\right)^2 y + {}_1W_{71}\tau\left(x^2+y^2\right)^3 y + \ldots \\ &= \sum_{p=1} {}_1W_{(2p+1),1}\tau\left(x^2+y^2\right)^p y = \sum_{p=1} b_p\tau\left(x^2+y^2\right)^p y\end{aligned} \tag{8.156}$$

where $b_p = {}_1W_{(2p+1),1}$.

8.5.2 Total Linear Coma from Properties of Axial Pencil

Figure 8.28 shows the image space of an axi-symmetric imaging system with the exit pupil of E' on the axis. The radius of the exit pupil is h'. The vertex of the last refracting interface is A_k. The image plane is at O'. The principal ray $E'P'$ from the object point P (not shown) intersects the image plane at P'. Let $O'P' = \eta'$.

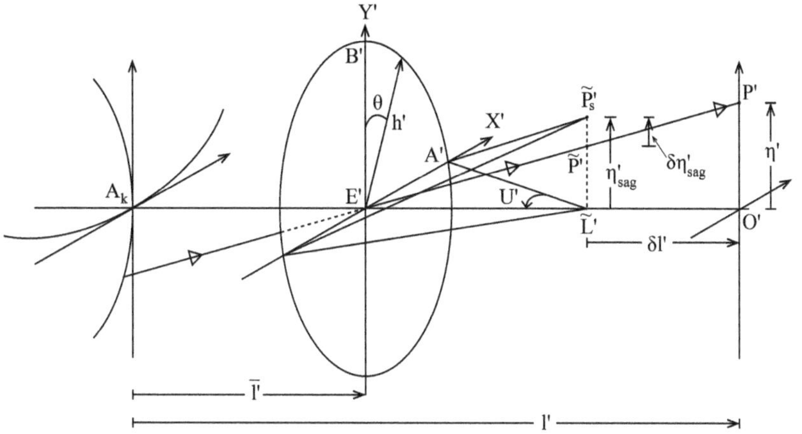

FIGURE 8.28 Total linear coma from properties of the axial pencils.

The distances $A_kO' = l'$, and $A_kE' = \bar{l}'$. In the sagittal section, the two marginal rays from the axial point O intersect the axis at $\tilde{L}'$. The semi angular aperture U′ of the axial pencil is large, and $\sin U' = -(h'/R')$. The longitudinal spherical aberration $L'O' = \delta l'$. The radius $R' = E'O' \cong E'\tilde{L}'$, since $\delta l' \ll R'$.

From the field point P (not shown) two marginal rays in the sagittal section intersect at $\tilde{P}'_s$. By symmetry, the point $\tilde{P}'_s$ lies on the meridional plane. Let the line $\tilde{P}'_s\tilde{L}'$ intersect E′P′ at $\tilde{P}'$. Since $\delta l'$ is small, $\tilde{P}'_s\tilde{L}'$ is perpendicular to the axis for all practical purposes.

Let $\tilde{P}'_s\tilde{L}' = \eta'_{sag}$, and $\tilde{P}'_sP' = \delta\eta'_{sag}$. From $\Delta E'\tilde{P}'\tilde{L}'$ and $\Delta EP'O'$ we get

$$\tilde{P}'\tilde{L}' = \frac{(l' - \bar{l}' - \delta l')}{(l' - \bar{l}')}\eta' \tag{8.157}$$

From the optical sine theorem described in Section 7.3.1.4.1, we can write for small values of η,

$$n\eta \sin U = n'\eta'_{sag} \sin U' \tag{8.158}$$

where U and U′ are the convergence angles of the real rays from the margin of the sagittal section of the object and the image spaces of refractive indies n and n′, respectively.

Using relation (8.12), the y component of transverse ray aberration due to linear coma is

$$\delta\eta' = -\frac{R'}{n'}\frac{\partial W_{l.c.}}{\partial Y'} = -\frac{R'}{n'h'}\frac{\partial W_{l.c.}}{\partial y'} \tag{8.159}$$

Introducing primes to indicate normalized coordinates (x', y') for points on the exit pupil, we obtain from (8.156),

$$W_{l.c.}(x', y') = \sum_{p=1} b_p \tau (x'^2 + y'^2)^p y' \tag{8.160}$$

From the last two equations, we get

$$\delta\eta' = -\frac{R'}{n'h'}\sum_{p=1} b_p \tau (x'^2 + y'^2)^{p-1}\{x'^2 + (2p+1)y'^2\} \tag{8.161}$$

For the point A′ with normalized coordinates $(1,0)$ at the margin of the sagittal section of the exit pupil, $\delta\eta'_{sag}$ is given by

$$\delta\eta'_{sag} = -\frac{R'}{n'h'}\sum_{p=1} b_p \tau \tag{8.162}$$

Note that the linear coma for the point B′ with normalized coordinates $(0,1)$ on the exit pupil is, by (8.160)

$$W_{l.c.}(0,1) = \sum_{p=1} b_p \tau \tag{8.163}$$

Therefore,

$$\delta\eta'_{sag} = -\frac{R'}{n'h'} W_{l.c.}(0,1) = \frac{1}{n'\sin U'} W_{l.c.}(0,1) \tag{8.164}$$

From Figure 8.28, it is seen that

$$\delta\eta'_{sag} = \tilde{P}'_s \tilde{P}' = \tilde{P}'_s \tilde{L}' - \tilde{P}'\tilde{L}' = \eta'_{sag} - \frac{(l' - \bar{l}' - \delta l')}{(l' - \bar{l}')}\eta' \tag{8.165}$$

Using (8.158), we get

$$\delta\eta'_{sag} = \frac{n\eta \sin U}{n'\sin U'} - \frac{(l' - \bar{l}' - \delta l')}{(l' - \bar{l}')}\eta' \tag{8.166}$$

Substituting from (8.164),

$$W_{l.c.}(0,1) = n'\sin U' \delta\eta'_{sag} = -H\left\{\frac{\sin U'}{u'}\frac{(l' - \bar{l}' - \delta l')}{(l' - \bar{l}')} - \frac{\sin U}{u}\right\} \tag{8.167}$$

where the paraxial invariant $H = nu\eta = n'u'\eta'$. Since all the components of the linear coma have the same azimuthal variation $\cos\theta$ in the pupil, for any zone with normalized radius r', the linear coma is given by

$$W_{l.c.}(r'\sin\theta, r'\cos\theta) = -H\cos\theta\left\{\frac{\sin U'}{u'}\frac{(l' - \bar{l}' - \delta l')}{(l' - \bar{l}')} - \frac{\sin U}{u}\right\} \tag{8.168}$$

'This remarkable expression gives the linear coma as a wavefront aberration in terms of the properties of the axial pencil alone, together with a knowledge of the position of the pupil', says Welford [135]. It is important to note that the equation (8.168), for the linear coma, refers strictly to the radius for which the finite ray of the axial pencil is traced. On the other hand, the relation holds good for very small field points to justify the neglect of terms higher than the first in τ in the general expression for coma. Further, in absence of spherical aberration, $\delta l' = 0$, and then, linear coma becomes independent of axial position of the pupil.

8.5.3 Offence against Sine Condition (OSC′)

For assessing the state of correction of an imaging system from linear coma, Conrady [136] proposed a dimensionless measure called 'offence against the sine condition (OSC′)'. It is defined by

$$\mathrm{OSC}' = \frac{\delta\eta'_{\mathrm{sag}}}{\left(\eta'_{\mathrm{sag}} - \delta\eta'_{\mathrm{sag}}\right)} \tag{8.169}$$

Using relations (8.165) – (8.166), it follows

$$\mathrm{OSC}' = \left(\frac{n\eta \sin U}{n' \sin U'}\right)\left(\frac{l' - \bar{l}'}{l' - \bar{l}' - \delta l'}\right)\frac{1}{\eta'} - 1 \tag{8.170}$$

Using the paraxial invariant $H = nu\eta = n'u'\eta'$, OSC′ can be expressed as

$$\mathrm{OSC}' = \left(\frac{\sin U}{u}\right)\left(\frac{u'}{\sin U'}\right)\left(\frac{l' - \bar{l}'}{l' - \bar{l}' - \delta l'}\right) - 1 \tag{8.171}$$

Paraxial ray trace for the marginal ray of the axial pencil usually commences with the convergence angle $u = \sin U$. In that case, the expression reduces to

$$\mathrm{OSC}' = \left(\frac{u'}{\sin U'}\right)\left(\frac{l' - \bar{l}'}{l' - \bar{l}' - \delta l'}\right) - 1 \tag{8.172}$$

Note that OSC′ is determined solely by the parameters of the axial pencil.

8.5.4 Staeble – Lihotzky Condition

From (8.171), by substituting $\mathrm{OSC}' = 0$, the condition for the absence of linear coma is obtained as

$$\left(\frac{\sin U'}{u'}\right) = \left(\frac{l' - \bar{l}'}{l' - \bar{l}' - \delta l'}\right)\left(\frac{\sin U}{u}\right) \tag{8.173}$$

It is important to note that not only primary linear coma but also linear coma of all higher orders is absent when this condition is satisfied. Since $\delta l' \neq 0$, i.e., spherical aberration is present, the system is not aplanatic, but it is isoplanatic over a small region around the axis on account of the absence of linear coma.

When spherical aberration is absent, $\delta l' = 0$, and equation (8.173) reduces to Abbe's sine condition

$$\frac{\sin U'}{u'} = \frac{\sin U}{u} \tag{8.174}$$

Reports on different aspects of Staeble – Lihotzky condition have appeared in publications by many investigators [137–142]. Investigations are continuing on the development of suitable relations for asymmetric systems [143–146].

8.6 Analytical Approach for Correction of Total Aberrations

Hristov reported an analytical approach for the correction of total aberrations in optical systems. Exact formulas for bilinear transformation from the object space to the image space are derived. Next, relative parameters in the matrix coefficients are introduced. By combining bilinear and matrix transformations, methods are developed for determining analytically the curvature and the thickness parameters required for the correction of total aberrations. Details of the method may be found in publications [147–151].

REFERENCES

1 L.N. Hazra, 'Journey to the Wonderland of Optical Aberrations: Problems and Pitfalls', Paper IT-OFTD-3, Proc. International Conf. on Optics & Optoelectronics, 12–15 Dec. 2005, IRDE, Dehradun, India.
2 A. Mark Smith, 'Alhacen on Image Formation and Distortion in mirrors, Book 6 of Alhacen's De Aspectibus, Vol. 2', *Trans. Amer. Phys. Soc.*, Vol. 98, Part I, Section 2 (2008) pp. 155–327.
3 A. Mark Smith, 'Alhacen on Refraction, Book 7 of Alhacen's De Aspectibus, Vol. 2', *Trans. Amer. Phys. Soc.*, Vol. 100, Part 3, Section 2 (2010) pp. 213–439.
4 J.H. Bridges, Ed., 'The 'Opus Majus' of Roger Bacon, Vol. II', Clarendon Press, Oxford (1847).
5 L. Nix and W. Schmidt, *Heronis Alexandrini opera quae supersunt omnia, Vol. II, Fasc. I, Mechanica et Catoptrica*, B. G. Teubner, Leipzig (1900).
6 J. Keplers, *Dioptrik*, D. Franke, Augsberg (1611); reprinted W. Englemann, Leipzig (1904).
7 R.B. Johnson, 'Historical perspectives on understanding optical aberrations', in *Lens Design*, Ed. W.J. Smith, CR 41, SPIE Press, Bellingham, Washington (1992).
8 R. Descartes, 'La Dioptrique' (1636), see *Discours de la method*, Flammarion, Paris (1966).
9 I. Newton, *Opticks*, Dover, New York (1952). [Originally published by Royal Society, London, in 1704].
10 J. Dollond, 'An account of some new experiments concerning the different refrangibility of light', *Phil. Trans.* Vol. 50 (1758) pp. 733–753.
11 A.C. Clairaut, 'Mémoire sur les moyens de perfectionner les lunettes d'approche', Mem. *de l'Acad., Paris* (1756) pp. 380–437, (1757) pp. 524–550, (1762) pp. 578–631.
12 J. le R. d'Alembert, *Œuvres completes*, Éditiones CNRS (2002).
13 A.C.S. Van Heel, 'Newton's work on Geometrical Optical Aberrations', *Nature*, Vol. 117 (1953) pp. 305–306.
14 J. Ramsden, 'A description of a new construction of eyeglasses for such telescopes as may be applied to mathematical instruments', *Phil. Trans.*, Vol. 73 (1783) pp. 94–99.
15 T. Young, 'On the theory of light and colour', *Phil. Trans.*, Vol. 103 (1802) pp. 12–48.
16 G.B. Airy, 'On the spherical aberration of eyepieces of telescopes', *Cambridge Phil. Trans.*, Vol. 2 (1827) pp. 227–252.
17 H. Coddington, *A System of Optics*, Simpkin and Marshall, London, Part I (1829) Part 2 (1830).
18 W.R. Hamilton, 'Theory of systems of rays', *Trans. Royal Irish Acad.*, Vol. 15 (1827) pp. 69–174; also, in *The Mathematical Papers of Sir William Rowan Hamilton*, Eds. A.W. Conway and J.L. Synge, Cambridge University Press (1931).
19 C.F. Gauss, 'Dioptrische Untersuchungen', *Abhandlungender Mathematischen Classe der Königlichen Gesellschaft der Wissenschaften zu Göttingen*, Vol. 1 (1838–1841) pp. 1–34.
20 L. Seidel, 'Zur Dioptrik. Über die Entwicklung der Glieder 3ter Ordnung, welche den Weg eines ausserhalb der Ebene der Axe gelegenen Lichtstrahles durch ein System brechender Medien bestimmen', *Astrnomische Nachrichten*, Vol. 43, No.1027 (1856) pp. 289–304; ibid., No. 1028, pp. 305–320; ibid., No.1029, 321. [An English translation of the combined papers 'About the third order expansion that describes the path of a light beam outside the plane of the axis through an optical system of refracting elements', Translator: R. Zehnder; personal communication, Jose Sasian (2010)].
21 A. Rakich and R. Wilson, 'Evidence supporting the primacy of Joseph Petzval in the discovery of aberration coefficients and their application to lens design,' *Proc. SPIE 6668, Novel Optical Systems Design and Optimization* X, (2007) 66680B1–13.
22 J. Sasián, 'Joseph Petzval lens design approach,' *Proc. SPIE 10590, International Optical Design Conference 2017, Denver* (2017) 10590171–15.
23 J. Petzval, *Bericht über die Ergebnisse einiger dioptrischen Untersuchungen*, Verlag von Conrad Adolph Hartleben, Pesth, (1843).

24 J. Petzval, 'Bericht über Dioptrische Untersuchungen', *Sitzungsberichte der mathem.-naturw. Classe der kaiserlichen Acad. Der Wiss.*, Vol. 26, Wien (1857) pp. 33–90.
25 J. Maxwell, 'On the general laws of optical instruments', *Quart. J. Pure and Appl. Math.*, Vol. 2 (1858) pp. 238–246.
26 Lord Rayleigh, 'Investigations in optics with special reference to the spectroscope', *Phil. Mag.*, Vol. 8 (1879) pp. 261–274, 403–411, 477–486.
27 E. Abbe, *Gesammelte Abhandlungen von Ernst Abbe*, G.Fischer, Jena (1904).
28 S. Czapski and O. Eppenstein, *Grundzügge der Theorie der Optischen Instrumente nach Abbe*, J.A.Barth, Leipzig (1924).
29 R.A. Sampson, 'On Gauss's Dioptrische Untersuchungen', *Proc. Lond. Math. Soc.*, Vol. 29 (1897) pp. 33–83.
30 R.A. Sampson, 'A New Treatment of Optical Aberrations', *Phil. Trans. Roy. Soc.* (London), series A, Vol. 212 (1913) pp. 149–183.
31 K. Schwarzschild, 'Untersuchungen zur geometrischen optic', *Abhandl. Königl. Ges. Wiss. Göttingen*, Vol. 4 (1905) Nos. 1–3.
32 A. Kerber, *Beiträge zur Dioptrik*, No.1, publ. by the author, Leipzig, (1895) pp. 36, 78, 82; ibid., No. 2, G. Fock, Leipzig, (1896) pp. 53, 63; ibid., No. 3, G. Fock, Leipzig (1897) p. 415; ibid., No. 4, G. Fock, Leipzig (1898) p. 415; ibid., No. 5, G. Fock, Leipzig (1899) pp. 344, 415.
33 T. Smith, 'On tracing rays through an optical system', *Proc. Phys. Soc. London*, Vol. 27, (1915) pp. 502–510; ibid., Vol. 30 (1917) pp. 221–233; ibid., Vol. 32 (1920) pp. 252–264; ibid., Vol. 33 (1921) pp. 174–178; ibid., Vol. 57 (1945) pp. 286–293.
34 A.E. Conrady, *Applied Optics and Optical Design, Part I*, Oxford University Press, London (1929).
35 A.E. Conrady, *Applied Optics and Optical Design, Part II*, Ed. R. Kingslake, Dover, New York (1960).
36 A. Kohlschütter, *Die Bildfehler fünfter ordnung optischer systeme*, Kaestner, Göttingen (1908).
37 J.P.C. Southall, *The Principles and Methods of Geometrical Optics*, Macmillan, New York (1913).
38 H.D. Taylor, A System of Applied Optics, Macmillan, London (1906).
39 É. Turrière, *Optique Industrielle*, Delagrave, Paris (1920).
40 C. Carathéodory, *Geometrische Optik*, Springer, Berlin (1937).
41 M. Berek, *Grundlagen der praktischen optik*, Walter de Gruyter, Berlin (1930).
42 G.C. Steward, *The Symmetrical Optical System*, Cambridge University Press, London (1928).
43 H. Boegehold, 'Zur Behandlung der Strahlenbegrenzung in 17. Und 18. Jahrhundert', *Central-Ztg. Opt. u. Mech.*, Vol. 49 (1928) pp. 94–95, 105–109.
44 M. Herzberger, 'Theory of image errors of the fifth order in rotationally symmetric systems', *J. Opt. Soc. Am.*, Vol. 29 (1939) pp. 395–406.
45 H.H. Hopkins, 'Monochromatic Lens Aberration Theory', *Phil. Mag., Ser.* 7, Article LXVI (1945) pp. 546–568.
46 F. Wachendorf, 'Bestimmung der Bildfehler 5 ordnung in zentierten optischen systemen', *Optik*, Vol. 5 (1949) pp. 80–85.
47 H.H. Hopkins, *Wave Theory of Aberrations*, Oxford University Press, London (1950).
48 H. Marx, 'Theorie der geometrisch-optischen bildfehler', in *Handbuch der Physik*, Ed. S. Flügge, Band XXIX, pp. 68–191, Springer-Verlag, Berlin (1967).
49 H.A. Buchdahl, *Optical Aberration Coefficients*, Oxford University Press, London, 1954.
50 Y. Matsui and K. Nariai, *Fundamentals of practical aberration theory*, World Scientific, Singapore, 1993; see also, in *Jap. J. Appl. Phys.*, Vol. 28, (1959) pp. 642–687; ibid, Vol. 29 (1960) pp. 184–256.
51 A. Cox, *A System of Optical Design*, Focal Press, London (1964).
52 J. Focke, 'Higher Order Aberration Theory', in *Progress in Optics*, Ed. E. Wolf, Vol. IV (1965) pp. 1–36.
53 J. Picht, *Grundlagen der geometrisch-optischen abbildung*, VEB Deutscher Verlag der Wissenschaften, Berlin (1955).
54 F. Zernike, Beugungstheorie der schneidenverfahrens und seiner verbesserten form, der Phasenkontrastmethode, *Physica*, 1, (1934) pp. 689–704.
55 J. Korringa, *Onderzoekingen op hat Gebied der Algebraische Optik*, Ph.D. thesis, Delft, 1942; also, in Physica, Vol. 8 (1941) p. 477.
56 B.R.A. Nijboer, 'The Diffraction Theory of Optical Aberrations', *Physica*, Vol. 10 (1943) pp. 679–692.
57 A.C.S. Van Heel, 'Calcul des aberrations du cinquième ordre at projets tenant compte de ces aberrations', in *La théorie des images optiques*, pp. 33–67, C.N. R.S., Paris (1949).

58 R.K. Luneburg, *Mathematical Theory of Optics*, University of California Press, Berkeley, California (1964).
59 A. Maréchal, 'Étude des effets combines de la diffraction et des aberrations géométriques sur l'image d'un point lumineux', *Rev. d'Optique*, Part I, Vol. 26 (1947) pp. 257–277; Part II, Vol. 27 (1948) pp. 73–92; Part III, Vol. 27 (1948) pp. 269–287.
60 T. Smith, 'Optical Calculations', in *A Dictionary of Applied Physics*, Ed. R. Glazebrook, Peter Smith, New York (1950).
61 R.E. Hopkins and R. Hanau, 'Aberration analysis and Third Order Theory', in *Military Standardization Handbook*, MIL-HDBK 141, Washington, D. C. (1962).
62 R. Kingslake, *Lens Design Fundamentals*, Academic, New York (1978).
63 A. Maréchal, E. Huygues, and P. Givaudon, 'Méthode de calcul des système optiques', in *Handbuch der Physik*, Ed. S. Flügge, Band XXIX, Springer Verlag, Berlin (1967).
64 M. Born and E. Wolf, *Principles of Optics*, Pergamon, Oxford (1980).
65 J. Braat, 'Polynomial expansion of severely aberrated wavefronts', *J. Opt. Soc. Am. A*, Vol. 4 (1987) pp. 643–650.
66 W. Brouwer, *Matrix Methods in Optical Instrument Design*, Benjamin, New York (1986).
67 H.A. Buchdahl, 'Invariant aberrations', *J. Opt. Soc. A*, Vol. 5 (1988) pp. 1957–1975.
68 Y.A. Klebtsov, 'Contribution to the theory of second order aberrations', *J. Opt. Technol.*, Vol. 61(1994) pp. 338–342.
69 H.A. Buchdahl, 'Spherical aberration coefficients of the second kind', *J. Opt. Soc. Am. A*, Vol. 13 (1996) pp. 1114–1116.
70 G.G. Slyusarev, *Aberration and Optical Design Theory*, Adam Hilger, Bristol (1984).
71 J. Flügge, *Das Photographische Objektiv*, Springer-Verlag, Wien (1955).
72 M. Rosete-Aguilar and J. Rayces, 'Renormalization of the Buchdahl-Rimmer third and fifth order geometric aberration coefficients to rms wave aberration function expressions', *J. Mod. Opt.*, Vol. 42 (1995) pp. 2435–2445.
73 M. Herzberger, Modern Geometrical Optics, Interscience, New York (1958).
74 J. Sasián, 'Theory of sixth-order wave aberrations', *Appl. Opt.*, Vol. 49 (2010) D69–93; errata Vol. 49 (2010) pp. 6502–6503.
75 W.T. Welford, Aberrations of optical systems, Adam Hilger, Bristol (1986).
76 P. Mouroulis and J. Macdonald, Geometrical Optics and Optical Design, Oxford Univ. Press, New York (1997).
77 V.N. Mahajan, Optical Imaging and Aberrations, Part I. Ray Geometrical Optics, SPIE Press, Bellingham, WA (1998).
78 J. Sasián, Introduction to Aberrations in Optical Imaging Systems, Cambridge University Press, Cambridge (2013).
79 A. Gullstrand, 'Die reelle optische Abbildung', *Svenska Vetensk. Handl.*, Vol. 41 (1906) pp. 1–119.
80 A. Gullstrand, 'Das allgemeine optisce Abbildungssystem', *Svenska Vetensk. Handl.*, Vol. 55 (1915) pp. 1–139.
81 A. Gullstrand, 'Optical Imagery' in Appendices to Part I of Helmholtz's *Treatise on Physiological Optics*, Ed. J.P.C. Southall, Dover, New York, 1962.
82 H.H. Hopkins, 'Image Formation by a general optical system. 1: General Theory', *Appl. Opt.*, Vol. 24 (1985) pp. 2491–2505.
83 H.H. Hopkins, 'Image Formation by a general optical system. 2: Computing methods', *Appl. Opt.*, Vol. 24 (1985) pp. 2506–2519.
84 Y. Wang and H.H. Hopkins, 'Ray-tracing and Aberration Formulae for a General Optical System', *J. Mod. Opt.*, Vol. 39 (1992) pp. 1897–1938.
85 W.T. Welford, Ref. 74, pp. 95–98.
86 B.R.A. Nijboer, 'The Diffraction Theory of optical aberrations Part I: General Discussion of the geometrical aberrations', *Physica*, Vol. 10 (1943) pp. 679–692.
87 D. Argentieri, Ottica Industriale, Ulrico Hoepli, Milan (1954) pp. 213–216.
88 J.L. Rayces, 'Exact relation between wave aberration and ray aberration', *Opt. Acta*, Vol. 11 (1964) pp. 85–88.
89 H.H. Hopkins, 'Calculation of the aberrations and image assessment for a general optical system', *Opt. Acta*, Vol. 28 (1981) pp. 667–714.
90 W.T. Welford, Ref. 74, p. 101.
91 A. Walther, *The Ray and Wave Theory of Lenses*, Cambridge Univ. Press, Cambridge (2006) pp. 59–68.

92 O.N. Stavroudis, 'Generalized ray tracing, caustic surfaces, generalized bending, and the construction of a novel merit function for optical design', *J. Opt. Soc. Am.*, Vol. 70 (1980) pp. 976–985.
93 H. Coddington, *A Treatise on the Reflexion and Refraction of Light*, Simpkin and Marshall, London (1829, 1830) Parts I and II.
94 D.L. Shealy, 'Caustic surface and the Coddington equations', *J. Opt. Soc. Am.*, Vol. 66 (1976) pp. 76–77.
95 A. Gullstrand, Ref. 80, pp. 15–20.
96 O. Altrichter and G. Schäfer, 'Herleitung der Gullstrandschen Grundgleischungen für Schiefe Strahlenbuschel aus den Hauptkrummungen der Wellenflasche', *Optik*, Vol. 13 (1956) pp. 241–253.
97 J.A. Kneisly III, 'Local curvatures of wavefronts in an optical system', *J. Opt. Soc. Am.* Vol. 54 (1964) pp. 229–235.
98 M. Herzberger, Ref. 75, p. 260.
99 O.N. Stavroudis, *The Optics of Rays, Wavefronts, and Caustics*, Academic, New York (1972) pp. 157–160.
100 A.M. Kassim and D.L. Shealy, 'Wavefront equation, caustics and wave aberration function of simple lenses and mirrors', *Appl. Opt.*, Vol. 27 (1988) pp. 516–522.
101 O.N. Stavroudis and R.S. Fronczek, 'Caustic surfaces and the structure of the geometric image', *J. Opt. Soc. Am.*, Vol. 66 (1976) pp. 795–800.
102 O.N. Stavroudis and R.C. Fronczek, 'Generalized ray tracing and the caustic surface', *Opt. Laser Technol.* Vol. 10, (1978) pp. 185–191.
103 O.N. Stavroudis, 'The k function in geometrical optics and its relationship to the archetypal wavefront and the caustic surface', *J. Opt. Soc. Am. A*, Vol. 12 (1995) pp. 1010–1016.
104 J.E.A. Landgrave and J.R. Moya-Cessa, 'Generalized Coddington equations in ophthalmic lens design', *J. Opt. Soc. Am. A*, Vol. 13 (1996) pp. 1637–1644.
105 H.A. Buchdahl, *An Introduction to Hamiltonian Optics*, Cambridge Univ. Press, Cambridge (1970).
106 H.H. Hopkins, Ref. 46, pp. 48–55.
107 W.T. Welford, Ref. 74, p. 141.
108 L.N. Hazra, 'Ray failures in finite ray tracing', *Appl. Opt.*, Vol. 24 (1985) pp. 4278–4280.
109 T.Y. Baker and L.N.G. Filon, 'On a theory of the second order longitudinal spherical aberration for a symmetrical optical system', *Phil. Trans. Roy. Soc. London A*, Vol. 221 (1921) pp. 29–71.
110 O.N. Stavroudis, Ref. 94, pp. 235–237.
111 H.A. Buchdahl, 'Power series of geometrical optics. I', *J. Opt. Soc. Am. A*, Vol. 1 (1984) pp. 952–957.
112 H.A. Buchdahl, 'Power series of geometrical optics. II', *J. Opt. Soc. Am. A*, Vol. 1 (1984) pp. 958–964.
113 H.A. Buchdahl, 'Power series of geometrical optics. III', *J. Opt. Soc. Am. A*, Vol. 2 (1985) pp. 847–851.
114 H.A. Buchdahl and G.W. Forbes, 'Power series of geometrical optics. IV', *J. Opt. Soc. Am. A*, Vol. 3 (1986) pp. 1142–1151.
115 H.A. Buchdahl, 'Power series of geometrical optics. V', *J. Opt. Soc. Am. A*, Vol. 3 (1986) pp. 1620–1628.
116 G. Forbes, 'Singularities of multivariate Lagrangian aberration functions', *J. Opt. Soc. Am. A*, Vol. 3 (1986) pp. 1370–1375.
117 G. Forbes, 'Extension of the convergence of Lagrangian aberration series', *J. Opt. Soc. Am. A*, Vol. 3 (1986) pp. 1376–1383.
118 G. Forbes, 'Acceleration of the convergence of multivariate aberration series', *J. Opt. Soc. Am. A*, Vol. 3 (1986) pp. 1384–1394.
119 H.A. Buchdahl, 'Spherical aberration coefficients of the second kind', *J. Opt. Soc. Am. A*, Vol. 13 (1996) pp. 1114–1116.
120 W.T. Welford, vide Ref. 74, p. 109.
121 G.C. Steward, *The Symmetrical Optical System*, Cambridge University Press, London (1958) pp. 39–41.
122 W.T. Welford, vide Ref. 74, p. 144.
123 P. Mouroulis and J. Macdonald, vide Ref. 75, pp. 218–219.
124 W.T. Welford, 'Aplanatism and Isoplanatism', Chapter VI in *Progress in Optics, Vol. XIII*, Ed. E. Wolf, North Holland, Amsterdam (1976) pp. 267–292.
125 R. Blair, 'Experiments and Observations on the Unequal Refrangibility of Light', *Trans. Roy. Soc., Edinburgh*, Vol. 3, Part II (1794) pp. 3–76.
126 J.F.W. Herschel, 'On the aberrations of compound lenses and object glasses', *Phil. Trans. Roy. Soc. London*, Vol. 111 (1821) pp. 222–267.
127 W.R. Hamilton, 'On the improvement of the double achromatic object glass', Paper presented to Royal Irish Academy, June 24, 1844, published as Article XXI in *The Mathematical Papers of Sir William Rowan Hamilton, Vol. I, Geometrical Optics*, Eds. A.W. Conway and J.L. Synge, (1931) pp. 387–460.

128 E. Abbe, 'A contribution to the Theory of the microscope, and the nature of microscopic vision', *Proceedings of the Bristol Naturalists' Society, Bristol, New Series*, Vol. I (1874) pp. 200–261; original German article in Schultze's Archiv für mikroscopische Anatomie, Vol. IX (1873) pp. 413–468.

129 E. Abbe 'Ueber die Bedingungen des Aplanatismus der Linsensysteme', *Chapter XI in Gesammelte Abhandlungen von Ernst Abbe*, Vol. I, Gustav Fischer, Jena (1904); also, in Sitzungsberichte der Jenaischen Gesellschaft für Medicin und Naturwissenschaft (1879) pp. 129–142.

130 F. Staeble, 'Isoplanatische Korrektion und Proportionalitätsbedingung', *Münchner Sitzungs Berichte* (1919) pp. 163–196.

131 E. Lihotzky, 'Verallgemeinerung der Abbeschen Sinusbedingung (als Bedingung für das Verschwinden der Koma in der unmittelbaren Nachbarschaft der Achse) für systeme mit nicht gehobener Längenaberration', *Wiener Sitzungs Berichte, mathematisch-naturwissenschaftliche Klasse* Vol. 128 (1919) pp. 85–90.

132 P.M. Duffieux, *L'Intégrale de Fourier et ses Applications à l'Optique*, privately published by Oberthur, Rennes (1946).

133 P.B. Fellgett and E.H. Linfoot, 'On the Assessment of Optical Images', *Phil. Trans. Roy. Soc. London, Series A*, Vol. 247 (1955) pp. 369–407.

134 J.W. Goodman, *Introduction to Fourier Optics*, McGraw-Hill, New York (1968).

135 W.T. Welford, vide Ref. 74, p. 174.

136 A.E. Conrady, vide Ref. 33, pp. 370–375.

137 H. Boegehold, 'Note on the Staeble and Lihotzky condition', *Trans. Opt. Soc. London*, Vol. 26 (1925) pp. 287–288.

138 M. Berek, Grundlagen der praktischen Optik', De Gruyter, Berlin (1930) pp. 71–80.

139 H.H. Hopkins, 'The optical sine condition', *Proc. Phys. Soc.* Vol. 58 (1946) pp. 92–99.

140 G.G. Slyusarev, 'Coma and departure from the sine condition' and 'generalized Staeble – Lihotzky equations', in Section 2.1.8 in *Aberration and Optical Design Theory*, [original Russian book title: *Metodi Rascheta Opticheskikh Sistem, Mashinostroenie*, Leningrad (1969)] Translator: J.H. Dixon, Adam Hilger, Bristol (1984) pp. 116–132.

141 H. Marx, 'Das von Schwarzschild eingeführte Seidel'sche Eikonal und die Staeble-Lihotzky'sche Isoplanasiebedingung,' *Optik*, Vol. 16 (1959) pp. 610–616.

142 M. Shibuya, 'Exact sine condition in the presence of spherical aberration', *Appl. Opt.*, Vol. 31 (1992) pp. 2206–2210.

143 W.T. Welford, 'The most general isoplanatism theorem', *Opt. Commun.*, Vol. 3 (1971) pp. 1–6.

144 H.H. Hopkins, vide Ref. 81.

145 C. Zhao and J.H. Burge, 'Conditions for correction of linear and quadratic field dependent aberrations in plane-symmetric optical systems', *J. Opt. Soc. Am. A* 19, pp. 2467–2472 (2002).

146 T.T. Elazhary, P. Zhou, C. Zhao, and J. H. Burge, 'Generalized sine condition', *Appl. Opt.*, Vol. 54, (2015) pp. 5037–5049.

147 B.A. Hristov, 'Exact Analytical Theory of Aberrations of Centered Optical Systems', *Opt. Rev.*, Vol. 20 (2013) pp. 395–419.

148 B.A. Hristov, 'Development of optical design algorithms on the base of the exact (all orders) geometrical aberration theory', *Proc. SPIE*, Vol. 8167 (2011) 81670E-1 to 81670E-12.

149 B.A. Hristov, 'Orthoscopic surfaces in the object space and image space of axi-symmetric optical systems', *Sov. J. Opt. Technol.*, Vol. 6 (1983) pp. 26–29.

150 B.A. Hristov, 'Sagittal curvature in the centered optical systems', *Sov. J. Opt. Technol.*, Vol. 2 (1972) pp. 21–24.

151 B.A. Hristov, 'Astigmatic function in the real optical systems' *Sov. J. Opt. Technol.*, Vol. 1 (1971) pp. 104–109.

9

Chromatic Aberrations

9.1 Introduction

The refractive index of any optical medium other than a vacuum is a function of frequency of the light travelling through it. Consequently, all paraxial and aberrational properties of an optical system are, in general, functions of frequency. Since the chromatic variation in paraxial characteristics of optical systems operating under broadband illumination causes degradation in the quality of imaging to the same extent as caused by monochromatic aberrations, it has become customary to consider these chromatic variations in paraxial characteristics to be aberrations. Indeed, 'the longitudinal chromatic aberration' and 'the transverse chromatic aberration' correspond to 'the chromatic variation in axial location of the image plane' and 'the chromatic variation in magnification', respectively, and reduction of these two defects to tolerable levels is a mandatory requirement for the practical usage of broadband optical imaging systems. The other types of chromatic aberration arise from the variation in monochromatic aberrations caused by changes in frequency of illumination. Usually, the latter chromatic aberrations are a small fraction of the monochromatic aberrations, and in practice, only in the case of high quality imaging systems does it become necessary to take proper account of them.

9.2 Dispersion of Optical Materials

In optical instruments, the optical materials used are mostly optical glass, optical plastics, or some crystalline materials [1]. Of course, the material has to be transparent over the spectral range of working illumination.

Figure 9.1 shows a schematic of the variation of the refractive index of an optical material with a wavelength over a long spectral range covering the UV-VIS-IR region of the electromagnetic radiation. Two dashed portions of the curve, one in the UV region and the other in the IR region, represent absorption bands. Note that there is a marked difference in the nature of variation in refractive index with wavelength across the absorption bands. The index rises significantly at each absorption band, and then begins to drop with increasing wavelength. This phenomenon is known as 'anomalous dispersion'. For most optical materials used in practical optical systems under the purview of this book, the absorption bands lie outside the spectral range of operation.

The refractive index n of an optical medium is equal to the ratio (c/v), where c and v are the speeds of light in the vacuum and the medium, respectively. For light of frequency f, let the vacuum wavelength be λ, so that $c = f\lambda$. In the medium of refractive index n, the wavelength of this light is (λ/n); its frequency f remains the same, so that $v = f(\lambda/n) = (c/n)$. Therefore, the refractive index n should ideally be considered as a function of the frequency f of light, and represented as $n(f)$. Nevertheless, traditionally the refractive index of optical materials is mentioned in terms of wavelength as $n(\lambda)$. It is important to note that λ refers to the vacuum wavelength of light.

Major optical glass manufacturers like SCHOTT, HOYA, OHARA, SUMITA, etc. provide values of refractive indices at a discrete set of wavelengths for each of their products. Typical wavelengths used in the description of the dispersion of optical glasses are well-known spectral lines (vide Table 8.1).

Note that the refractive index data of optical glasses given in datasheets are measured values of refractive indices in air [2]. Since the refractive index of air is very close to one, for most practical purposes those

DOI: 10.1201/9780429154812-9

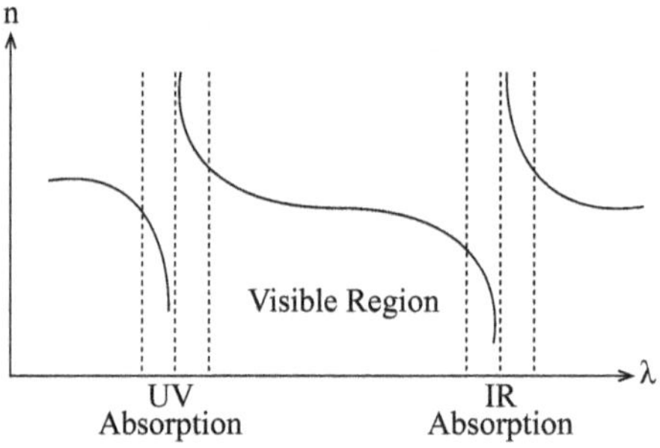

FIGURE 9.1 Anomalous dispersion of optical glass near the absorption bands in the infrared and ultraviolet regions. Normal dispersion in the visible region.

TABLE 9.1

Typical wavelengths used in description of dispersion of optical glasses

Approx. Wavelength (in nm)	Designation	Spectral line used
1013.98	t	IR mercury line
852.11	s	IR cesium line
706.52	r	Red Helium line
656.27	C	Red Hydrogen line
643.85	C′	Red Cadmium line
587.56	d	Yellow Helium line
546.07	e	Green Mercury line
486.13	F	Blue Hydrogen line
479.99	F′	Blue Cadmium line
435.83	g	Blue Mercury line
404.66	H	Violet Mercury line
365.01	i	Ultraviolet (UV) Mercury line

values may be considered to be refractive indices with respect to the vacuum. In demanding applications, however, the non-unity refractive index and dispersion of air are taken into account. A simple dispersion formula for air [3] is given as

$$n-1=a+\frac{b}{\lambda^2}+\frac{c}{\lambda^4} \tag{9.1}$$

In case of dry air, at standard pressure, the values of the coefficients a, b, and c are given for three values of temperature in Table 9.2. The values of λ are in nanometer (nm). Improved measurement procedures have led to relations that provide more accurate values for refractive indices for different types of air. Currently, the accepted dispersion relation [4–5] for standard air at temperature 15°C, pressure 101.325 kPa, and 0% humidity with 450 ppm CO_2 is

$$10^8\left(n_{as}-1\right)=\frac{k_1}{\left(k_0-\sigma^2\right)}+\frac{k_3}{\left(k_2-\sigma^2\right)} \tag{9.2}$$

TABLE 9.2

Values of constants in dispersion formula for air

t^0C	**a** $(\times 10^{-4})$	**b**	**c** $(\times 10^{4})$
0	2.87566	1.3412	3.777
15	2.72643	1.2288	3.555
20	2.58972	1.2259	2.576

where $\sigma\left(=\frac{1}{\lambda}\right)$ is the wavenumber in inverse micrometers and n_{as} is the refractive index at the vacuum wavelength λ. The constants are: $k_0 = 238.0185\ \mu m^{-2}$, $k_1 = 5792105\ \mu m^{-2}$, $k_2 = 57.362\ \mu m^{-2}$, and $k_3 = 167917\ \mu m^{-2}$. Other parameters of air remaining the same, in the presence of x_s ppm of CO_2, the refractive index n_{asx} of air is

$$(n_{asx} - 1) = (n_{as} - 1)\left[1 + 0.534 \times 10^{-6}(x_s - 450)\right] \tag{9.3}$$

From (9.2) and (9.3), we note that at vacuum wavelength 0.5 μm, the refractive index of air at temperature 15^0C, standard atmospheric pressure 101.325 kPa, 0% humidity, and 0 ppm CO_2 is 1.000278903.

9.2.1 Interpolation of Refractive Indices

Often, it becomes necessary to know the refractive index of an optical glass for unmeasured wavelengths. An analytical representation for $n(\lambda)$ is also necessary for computational purposes during the analysis or synthesis of optical systems. Search for the same commenced long back, and a number of empirical and semi-empirical formulae has been proposed. The wavelength λ is in μm (micron) in the following formulae. They may be classified as:

A. Cauchy – Schmidt – Conrady Dispersion Formulae [6–8]

$$\text{Cauchy:}\quad n(\lambda) = n_0 + \frac{a}{\lambda^2} + \frac{b}{\lambda^4} \tag{9.4}$$

$$\text{Schmidt:}\quad n(\lambda) = n_0 + \frac{a}{\lambda} + \frac{b}{\lambda^4} \tag{9.5}$$

$$\text{Conardy:}\quad n(\lambda) = n_0 + \frac{a}{\lambda} + \frac{b}{\lambda^{3.5}} \tag{9.6}$$

Three unknown terms, namely a, b, and n_0, are to be determined from experimentally measured values of refractive indices of the material at different wavelengths. Accuracy ~ 4D.

B. Cornu – Hartmann Dispersion Formulae [9–11]

$$\text{Cornu:}\quad n(\lambda) = n_0 + \frac{a}{(\lambda - \lambda_0)} \tag{9.7}$$

$$\text{Hartmann:}\quad n(\lambda) = n_0 + \frac{a}{(\lambda - \lambda_0)^b} \tag{9.8}$$

Constants a, b, n_0, and λ_0 are to be derived empirically from experimental results.

The exponent b varies between 0.5 and 2.0. For visible range, 'b' can be replaced by 1 or 1.2. Accuracy ~ 3 – 4 D.

C. Sellmeier Dispersion Formula [12]

$$\text{Sellmeier:}\quad n(\lambda)=\left[1+\sum_{j=1}^{N}\frac{a_j\lambda^2}{\lambda^2-\lambda_j^{\,2}}\right]^{\frac{1}{2}} \tag{9.9}$$

This is the most effective formula over a large wavelength range from UV to IR. The model, proposed in 1871, is based on the physical characteristics of normal and anomalous dispersion of optical materials. λ_j designates the centre of gravity of the j-th absorption band in the relevant spectral range. Accuracy depends on N and the choice of λ_j, $j=1,\ldots,N$. This formula is currently adopted by SCHOTT for characterization of dispersion of optical glass.

D. Helmholtz-Ketteler-Drude Dispersion Formula [13]

$$\text{Helmholtz-Ketteler-Drude:}\quad n(\lambda)=\left[\alpha_0+\sum_{j=1}^{N}\frac{a_j}{\lambda^2-\lambda_j^{\,2}}\right]^{\frac{1}{2}} \tag{9.10}$$

λ_j, $j=1,\ldots,N$ are the effective resonance wavelengths of the oscillators. Accuracy is expected to be comparable to the Sellmeier formula.

E. Herzberger Dispersion formulae [14–17]

$$\text{Herxberger(1942):}\quad n(\lambda)=n_0+n_1\lambda^2+\frac{n_2}{\lambda^2-\lambda_0^{\,2}}+\frac{n_3}{\left(\lambda^2-\lambda_0^{\,2}\right)^2} \tag{9.11}$$

$$\text{Herzberger(1959):}\quad n(\lambda)=n_0+n_1\lambda^2+\frac{n_2}{\lambda^2-0.028}+\frac{n_3}{\left(\lambda^2-0.028\right)^2} \tag{9.12}$$

In the Herzberger dispersion formulae, each glass is characterized by four coefficients, n_0, n_1, n_2, and n_3. λ_0 has a fixed value for all glasses. The near IR is taken care of by the coefficient n_1. The ultraviolet is taken care of by the coefficients n_2 and n_3, with an appropriate choice for λ_0. Initially, in 1942 Herzberger suggested $\lambda_0=0.187\,\mu m$, so that $\lambda_0^{\,2}$ is quoted as 0.035 in the corresponding formula. Later on, in 1959, he proposed $\lambda_0=0.168\,\mu m$, so that in the later formula $\lambda_0^{\,2}$ is quoted as 0.028. It turns out to be more appropriate for interpolation over the visible range.

F. Schott Dispersion Formula (old) [18–19]

$$\text{Schott(1966):}\quad n(\lambda)=\left[A_0+A_1\lambda^2+A_2\lambda^{-2}+A_3\lambda^{-4}+A_4\lambda^{-6}+A_5\lambda^{-8}\right]^{\frac{1}{2}} \tag{9.13}$$

This formula 'can be derived as a Laurent series expansion of finite order of the Sellmeier dispersion formula with an arbitrary number of effective resonance wavelengths. The λ^2 term comes from the IR absorption band and the λ^{-j} terms model the UV absorption bands' [20]. It was proposed by SCHOTT for characterization of dispersion of optical glasses, and it was widely accepted by glass manufacturers around the world. Indeed, some of them are still using it, whereas SCHOTT has adopted a different characterization scheme, as given below.

G. Schott Dispersion Formula (new) [20–21]

$$\text{Schott(1922):}\quad n(\lambda)=\left[1+\frac{B_1\lambda^2}{\left(\lambda^2-C_1\right)}+\frac{B_2\lambda^2}{\left(\lambda^2-C_2\right)}+\frac{B_3\lambda^2}{\left(\lambda^2-C_3\right)}\right]^{\frac{1}{2}} \tag{9.14}$$

In 1992, this Sellmeier dispersion formula with three effective resonance wavelengths was adopted as the standard dispersion formula for Schott optical glasses, and for each glass, the six coefficients are quoted in the datasheets. For example, for the SCHOTT glass NK7, the coefficients are:

$$B_1 = 1.03961212, B_2 = 0.231792344, B_3 = 1.01046945,$$

$$C_1 = 6.00069867\times 10^{-3}\,\mu m^2, C_2 = 2.00179144\times 10^{-2}\,\mu m^2, C_3 = 103.560653\,\mu m^2 \tag{9.15}$$

This equation is valid from UV to 2.3 μm in the IR. Accuracy is claimed to better than 5D over the visible range.

H. Buchdahl Dispersion Formula [22–26]

$$\text{Buchdahl:}\quad n(\omega) = n_0 + \sum_{j=1}^{N} \vartheta_j \omega^j \tag{9.16}$$

n_0 is the refractive index of the particular glass at selected base wavelength λ_0, and a chromatic coordinate ω is defined as

$$\omega = \frac{(\lambda - \lambda_0)}{1 + \alpha(\lambda - \lambda_0)} \tag{9.17}$$

where the value of α, a constant for all glasses, can be taken as 2.5 to model the dispersion of optical glasses over the visible range. The base wavelength is the wavelength normally used for the calculation of monochromatic aberrations; usually it is 0.5876 μm. The coefficients ϑ_j, called the dispersion coefficients, characterize the particular glass, and their values change from one glass to another. Note that the chromatic coordinate ω is a function of the wavelength, and is independent of the optical characteristics of the glass. Per the error analysis reported in reference [27], with a quadratic model, i.e., j = 2 in (9.16), accuracy is 4D over the visible range.

9.2.2 Abbe Number

The term 'dispersion' applies not only to the general phenomenon of variation of the refractive index of an optical material with wavelength and expressed by the function $n(\lambda)$, but also to the difference in refractive index of the material for two wavelengths, say λ_1 and λ_2. Dispersion of the material for the two wavelengths is specified as

$$\delta n = [\delta n]_{\lambda_1,\lambda_2} = n(\lambda_1) - n(\lambda_2) \tag{9.18}$$

Since the refractive index of normal glasses decreases with an increase in wavelength, the dispersion δn of an optical glass for two wavelengths λ_1 and λ_2, as defined above, is positive when $\lambda_1 < \lambda_2$.

At the wavelength λ, the power $K(\lambda)$ of a thin lens of refractive index $n(\lambda)$, and curvatures c_1 and c_2 is

$$K(\lambda) = [n(\lambda) - 1](c_1 - c_2) \tag{9.19}$$

The change in power of the lens with a change in wavelength from λ_1 to λ_2 is

$$\delta K = K(\lambda_1) - K(\lambda_2) = [n(\lambda_1) - n(\lambda_2)](c_1 - c_2) \tag{9.20}$$

Using (9.19), the above relation reduces to

$$\delta K = \frac{[n(\lambda_1) - n(\lambda_2)]}{[n(\lambda) - 1]} K(\lambda) = \frac{K(\lambda)}{V_\lambda} \tag{9.21}$$

where V_λ is given by

$$V_\lambda = \frac{[n(\lambda)-1]}{[n(\lambda_1)-n(\lambda_2)]} \tag{9.22}$$

This quantity is called the 'Abbe number' or 'Abbe value' of the optical material. Other names for the same are 'the reciprocal dispersive power', and 'the constringence'. Sometimes it is simply called 'V number' or 'V value', or simply as '$V_\#$'; in German literature, often the notation ν is used for the same. Optical glass manufacturers quote V values for two sets of wavelength combinations, λ, λ_1 and λ_2, over the visible range of wavelengths. The most commonly used V value in visible wavelength range is

$$V_d = \frac{n_d - 1}{n_F - n_C} \tag{9.23}$$

d, F and C are the three spectral lines, namely the yellow Helium line (0.58756 μm), the blue Hydrogen line (0.48613 μm), and the red Hydrogen line (0.65627 μm), respectively. They are usually chosen for visual instruments and systems. In order to cater to detectors with different spectral responses, another V value in visible wavelengths is also quoted by optical glass manufacturers. It is defined as

$$V_e = \frac{n_e - 1}{n_{F'} - n_{C'}} \tag{9.24}$$

e, F′ and C′ are the spectral lines, the green mercury line (0.54607 μm), the blue cadmium line (0.47999 μm), and the red cadmium line (0.64385 μm), respectively. Glass manufacturers provide refraction characteristics of available glasses as points on (n_d, V_d) and (n_e, V_e) maps. The latter are often called Abbe diagrams. The extreme ranges of n_d and V_d are (1.4,2.4) and (15,100), respectively. Figure 9.2 shows a current (n_d, V_d) Abbe diagram of the SCHOTT optical glasses. Note that the available glasses consist of a set of discreet points on the map. Therefore, even within the overall range, availability of optical glass with any arbitrary combination for n_d and V_d is not ensured.

9.2.3 Generic Types of Optical Glasses and Glass Codes

Optical glasses are broadly classified into two classes, crown and flint. Most crown glasses have $V_\# \geq 50$, and flint glasses have $V_\# \leq 50$. The designation 'crown' glasses owe its origin to the early methods of manufacture of such glasses. Spherical gobs used to be manufactured at the beginning, and these were subsequently blown up to resemble 'crowns'. The latter were then transformed into slabs by rotating the glass mass. The name 'flint' glasses originated in England, where the combination of ground 'flint' with lead oxide was used to obtain glasses that exhibited large chromatic dispersion.

Glass manufacturers use abbreviations or shortened technical names for different generic types of optical glasses. Often, the characteristic composition of the type is highlighted in its name. As an illustrative example, Table 9.3 presents a chart for abbreviations of the equivalent types of optical glasses marketed by three current glass manufacturers, namely SCOTT, HOYA, and OHARA. A small region of the (n_d, V_d) glass map (see Figure 9.2) is assigned for each glass type. All glasses with (n_d, V_d) characteristics lying within this region are identified by the abbreviation used for that generic type of glass. A prefix 'N' is used before the type abbreviation by SCHOTT optical glass company to indicate environmentally friendly alternatives to conventional glass types that contain lead and arsenic. Similarly, a prefix 'P' is used to indicate that the glass is suitable for precision moulding.

Besides the technical name for glass type, each optical glass is designated universally by a six-digit numerical code. The first three digits relate to the refractive index at the mid-wavelength (usually n_d), and the next three digits relate to the corresponding Abbe number. The glass code is given by

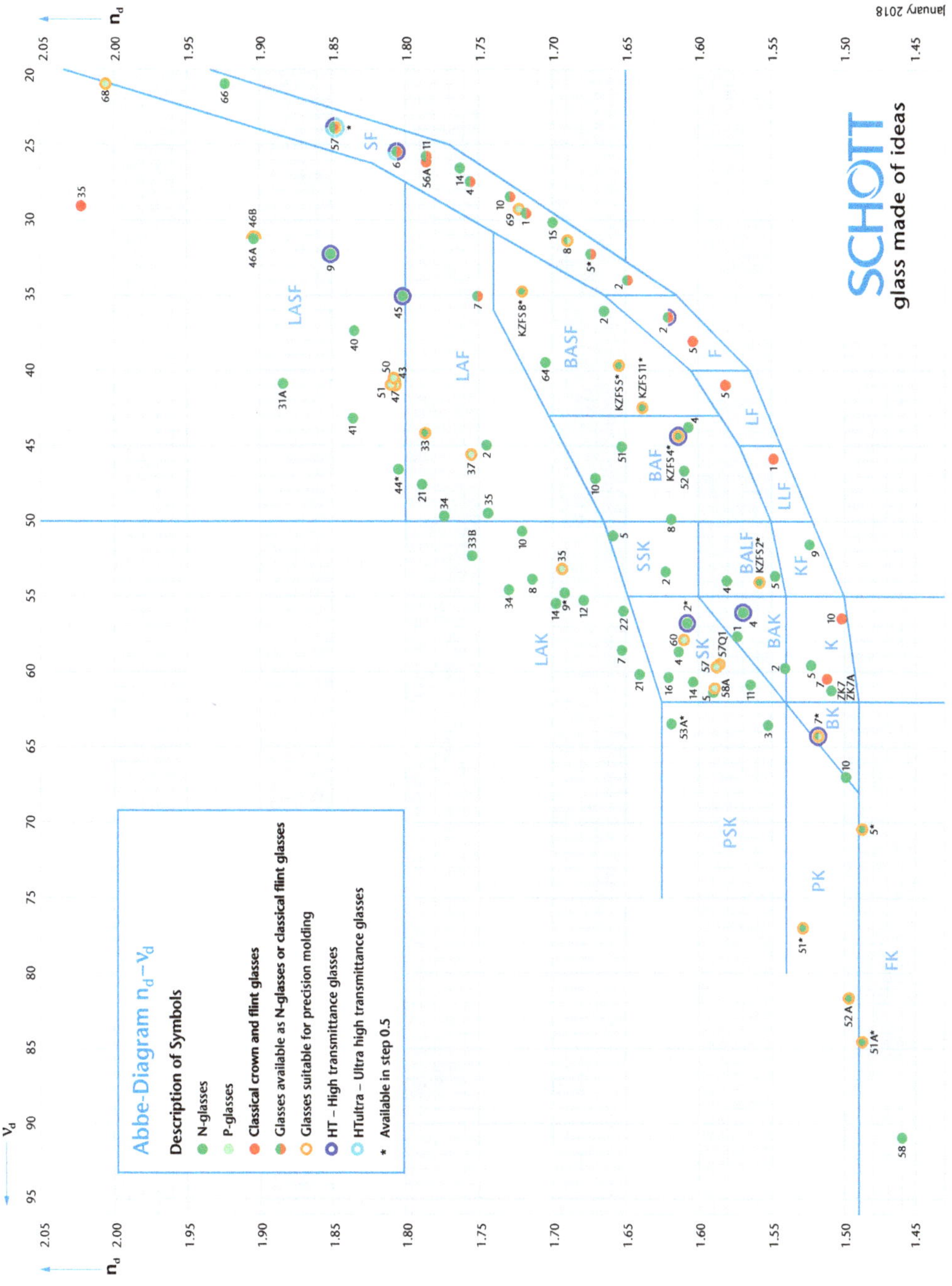

FIGURE 9.2 Abbe $(n_d - V_d)$ diagram of SCHOTT optical glasses [Courtesy: M/s SCHOTT Inc.].

TABLE 9.3

Generic types of common optical glasses and their abbreviated names

Generic Type	SCHOTT	HOYA	OHARA	Generic Type	SCHOTT	HOYA	OHARA
Fluorine Crown	FK	FC	FSL	Light Flint	LF	F	TIL
Phosphate Crown	PK	FCD	FPL	Extra Light Flint	LLF	FEL	TIL`
Dense Phosphate Crown	PSK	PCD	PHM	Barium Flint	BAF	BAF	BAH
Borosilicate Crown	BK	BSC	BSL	Flint	F	F	TIM
Crown	K	C	–	Dense Barium Flint	BASF	BAFD	BAH
Barium Crown	BAK	BAC	BAL	Dense Flint	SF	FD	TIH/ TIM
Dense Barium Crown	SK	BACD	BSM	Lanthanum Flint	LAF	LAF	LAH
Extra Dense Barium Crown	SSK	BACED	BSM	Dense Lanthanum Flint	LASF	TAF NBFD	LAH
Lanthanum Crown	LAK	LAC/TAC	LAL	Borate Flint/ Abnormal Dispersion Flint	KZFS	ADF	NBM

$$\text{Glass Code} \equiv NV \tag{9.25}$$

where each of N and V represent a three-digit number obtained from the values of n_d and V_d of the glass by relations given below.

$$N = \text{INT}\left[\{1000 \times (n_d - 1)\} + 0.5\right] \tag{9.26}$$

$$V = \text{INT}\left[10 \times V_d + 0.5\right] \tag{9.27}$$

$\text{INT}[x]$ represents the integer part of the real number x. Currently, SCHOTT is using a nine-digit code for optical glasses, where the first six digits are the same as above, and the last three digits relate to the density of the optical glass. Thus

$$\text{Nine-digit SCHOTT glass code} \equiv NVD \tag{9.28}$$

The three digits of D are obtained from the density ρ of the glass by using the relation

$$D = \text{INT}\left[100 \times \rho + 0.5\right] \tag{9.29}$$

Thus, the SCHOTT glass N-BK7 with $n_d = 1.51680$, $V_d = 64.17$, $\rho = 2.51\ \text{g/cm}^3$ has the six-digit international glass code 517642, and nine-digit SCHOTT glass code 517642251.

9.3 Paraxial Chromatism

The paraxial characteristics, e.g., the position and magnitude of paraxial images, of dioptric or catadioptric systems depend on the working wavelength of illumination. The term 'paraxial chromatism' refers to this variation.

9.3.1 A Single Thin Lens: Axial Colour and Lateral Colour

Figure 9.3 shows a thin lens L with curvatures c_1 and c_2 imaging a transverse object OP of height η. The axial distance of the object from the lens is LO = l. Let the refractive indices of the optical material of the lens at the three wavelengths d, F, and C be n_d, n_F, and n_C respectively. Corresponding powers K_d, K_F, and K_C of the thin lens at the three wavelengths are

$$K_d = (n_d - 1)(c_1 - c_2) \tag{9.30}$$

$$K_F = (n_F - 1)(c_1 - c_2) \tag{9.31}$$

$$K_C = (n_C - 1)(c_1 - c_2) \tag{9.32}$$

It follows that the chromatic variation in power for the change in wavelength from C to F is

$$(K_F - K_C) = (n_F - n_C)(c_1 - c_2) = \frac{K_d}{V_d} \tag{9.33}$$

The paraxial images of the axial object point O corresponding to wavelengths d, F, and C are formed at O'_d, O'_F, and O'_C. The corresponding image distances l'_d, l'_F, l'_C are given by

$$\frac{1}{l'_d} - \frac{1}{l} = K_d \tag{9.34}$$

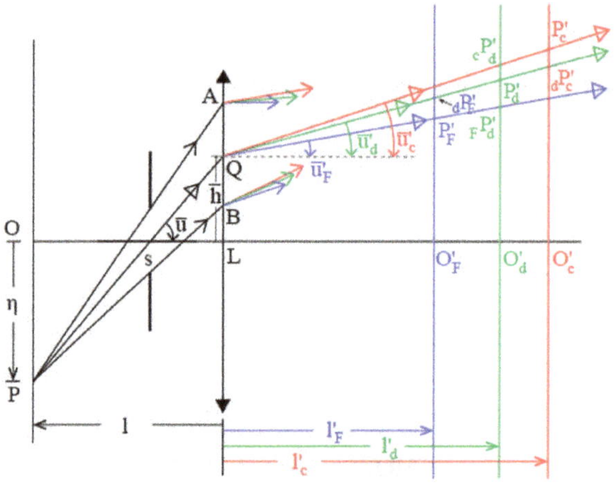

FIGURE 9.3 Paraxial chromatism of a thin lens with stop S away from the lens; object: OP = η; Images:

$$\left.\begin{array}{l} P'_F O'_F = \eta'_F,\ P'_C O'_C = \eta'_C,\ P'_d O'_d = \eta'_d \\ {}_F P'_d O'_d = \tilde{\eta}'_F,\ {}_C P'_d O'_d = \tilde{\eta}'_C \\ {}_d P'_F O'_F = \tilde{\eta}'_F,\ {}_d P'_C O'_C = \tilde{\eta}'_C \end{array}\right\}$$ Both axial colour and lateral colour present.

$$\frac{1}{l'_F} - \frac{1}{l} = K_F \tag{9.35}$$

$$\frac{1}{l'_C} - \frac{1}{l} = K_C \tag{9.36}$$

This chromatic variation of image distance is called longitudinal chromatic aberration, longitudinal colour, or axial colour. From (9.35) – (9.36), we find that the axial colour for a shift in wavelength from C to F is given by

$$\left(l'_F - l'_C\right) = -l'_F l'_C \left(K_F - K_C\right) \cong -l'^2_d \left(\frac{K_d}{V_d}\right) \tag{9.37}$$

For a very distant object, $l \rightarrow -\infty$, and $l'_d \rightarrow (1/K_d)$. Therefore, $\left(l'_F - l'_C\right) \rightarrow -\left(f_d/V_d\right)$, where the focal length f_d at d wavelength is equal to $(1/K_d)$. This implies that the axial colour of a single thin lens with a distant object is inversely proportional to the V number of its material.

> V is sometimes called the 'figure of merit' of a glass because the chromatic aberration becomes smaller and smaller when V grows. In case when the chromatic aberration cannot be corrected but is nevertheless objectionable (as for instance in ordinary eyepieces) we should therefore give preference to glasses of high V-value.
>
> [27]

The chromatic variation in power is also manifested as a chromatic variation in image height, or magnification. This variation in magnification is called transverse chromatic aberration, or oblique chromatic aberration, or simply transverse colour. In order to determine the images of the off-axial point P at different wavelengths, it is convenient to draw the line PL, extend it in the image space, and determine its points of intersection with the transverse image plane corresponding to that particular wavelength. In Figure 9.3, we note that the images of the off-axial point P at wavelengths F, d, and C are at P'_F, P'_d, and P'_C.

Let PAB be the polychromatic pencil of rays originating from the off-axial point P in the object space. Let the object height PO = η. The incident pencil is dispersed in the image space of the lens L. PSQ is the principal or chief ray of the off-axial pencil in the object space. On account of dispersion by the thin lens, the off-axial pencils corresponding to the different wavelengths are different; the corresponding chief rays are also different. Note that the chief ray of a particular wavelength in the image space passes through the paraxial image at that wavelength for the off-axial object point. Figure 9.3 shows three chief rays, namely QP'_d, QP'_C and QP'_F in the image space, and the transverse image planes $O'_dP'_d$, $O'_CP'_C$, and $O'_FP'_F$ corresponding to the object plane OP for d, C, and F wavelengths, respectively. Therefore, the paraxial images at d, C, and F wavelengths for the object PO are $P'_dO'_d\left(=\eta'_d\right)$, $P'_CO'_C\left(=\eta'_C\right)$, and $P'_FO'_F\left(=\eta'_F\right)$. The d-line principal ray QP'_d intersects the transverse image planes for C-line and F-line at the points ${}_dP'_C$ and ${}_dP'_F$. Let ${}_dP'_FO'_F = \hat{\eta}'_F$ and ${}_dP'_CO'_C = \hat{\eta}'_C$. The transverse image plane $P'_dO'_d$ corresponding to d line is intersected by the C line principal ray at ${}_CP'_d$, and by the F line principal ray at ${}_FP'_d$. Let ${}_FP'_dO'_d = \hat{\eta}'_F$, and ${}_CP'_dO'_d = \hat{\eta}'_C$.

From the similar triangles ΔLOP, $\Delta LO'_FP'_F$, $\Delta LO'_dP'_d$, and $\Delta LO'_CP'_C$ in Figure 9.3,

$$\frac{PO}{LO} = \frac{P'_FO'_F}{LO'_F} = \frac{P'_dO'_d}{LO'_d} = \frac{P'_CO'_C}{LO'_C}, \text{ i.e. } \quad \frac{\eta}{l} = \frac{\eta'_F}{l'_F} = \frac{\eta'_d}{l'_d} = \frac{\eta'_C}{l'_C} \tag{9.38}$$

We get

$$\left(\eta'_F - \eta'_C\right) = \eta'_d \frac{\left(l'_F - l'_C\right)}{l'_d} = -\eta'_d\, l'_d \left(\frac{K_d}{V_d}\right) \tag{9.39}$$

It is important to note that η'_F and η'_C are image heights at two axially separated image planes. So, the difference between them, as expressed above, is a combined effect of both transverse colour and axial colour. The two effects can be separated if the observation of transverse colour is made on a single image plane. For this, the d-line image plane is a convenient choice.

The convergence angle of the paraxial pupil ray in the object space is $\bar{u}$, and the same angle in the image space corresponding to F, d, and C wavelengths are $\bar{u}'_F$, $\bar{u}'_d$, and $\bar{u}'_C$, respectively. Let the height of the paraxial pupil ray on the lens plane be $\bar{h}$. Using a paraxial ray transfer equation for the d-line principal ray from the lens plane to the d-line image plane, we get

$$\eta'_d = \bar{h} + l'_d\, \bar{u}'_d \tag{9.40}$$

Substituting the above expression for η'_d in (9.39) we get

$$\left(\eta'_F - \eta'_C\right) = -l'_d\, \bar{h}\left(\frac{K_d}{V_d}\right) - l'^2_d\, \bar{u}'_d\left(\frac{K_d}{V_d}\right) \tag{9.41}$$

The paraxial refraction relation for the F-line principal ray at the thin lens is

$$\left(\bar{u}'_F - \bar{u}\right) = -\bar{h}\, K_F \tag{9.42}$$

Similarly, for the C-line principal ray we get

$$\left(\bar{u}'_C - \bar{u}\right) = -\bar{h}\, K_C \tag{9.43}$$

From (9.42) and (9.43), it follows

$$\left(\bar{u}'_F - \bar{u}'_C\right) = -\bar{h}\left(K_F - K_C\right) = -\bar{h}\left(\frac{K_d}{V_d}\right) \tag{9.44}$$

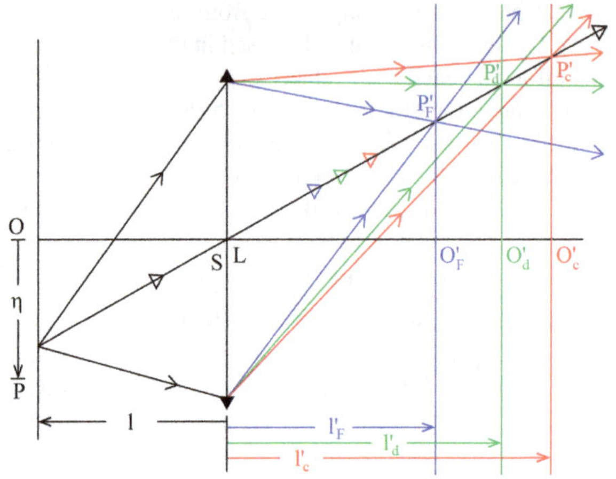

FIGURE 9.4 Paraxial chromatism of a thin lens with stop S on it. Only axial colour, no lateral colour.

The first term on the right-hand side of (9.41) can be written as

$$-l'_d\bar{h}\left(\frac{K_d}{V_d}\right)=l'_d\left(\bar{u}'_F-\bar{u}'_C\right)=\left(l'_d\bar{u}'_F+\bar{h}\right)-\left(l'_d\bar{u}'_C+\bar{h}\right)={}_F P'_dO'_d-{}_CP'_dO'_d=\left(\hat{\eta}'_F-\hat{\eta}'_C\right) \tag{9.45}$$

Similarly, using (9.37), the second term on the right-hand side of (9.41) can be expressed as

$$-l'^2_d\bar{u}'_d\left(\frac{K_d}{V_d}\right)=\left(l'_F-l'_C\right)\bar{u}'_d=\left(\bar{u}'_dl'_F+\bar{h}\right)-\left(\bar{u}'_dl'_C+\bar{h}\right)={}_d P'_FO'_F-{}_d P'_CO'_C=\left(\hat{\eta}'_F-\hat{\eta}'_C\right) \tag{9.46}$$

From (9.41), (9.45), and (9.46) it follows

$$\left(\eta'_F-\eta'_C\right)=\left(\tilde{\eta}'_F-\tilde{\eta}'_C\right)+\left(\hat{\eta}'_F-\hat{\eta}'_C\right) \tag{9.47}$$

The left-hand side of the above expression gives the lateral change in paraxial image heights for a change in wavelength from C to F; the image heights are measured on two different planes corresponding to the paraxial image planes for C and F wavelengths. When both the images are measured on the d-line image plane, the lateral colour on the observation plane corresponds to the first term on the right-hand side of (9.47). The second term on the right-hand side of (9.47) is the contribution from the axial colour that is added when the two image heights are measured on two axially separated planes [28–29].

When a stop is on the lens, the principal ray of the three wavelengths is not dispersed during its passage through the lens, so that

$$\bar{u}'_C=\bar{u}'_d=\bar{u}'_F=\bar{u} \tag{9.48}$$

Points ${}_CP'_d, P'_d,$ and ${}_FP'_d$ coincide, so that $\hat{\eta}'_F=\hat{\eta}'_C$, and the lateral colour is zero (Figure 9.4).

9.3.2 A Thin Doublet and Achromatic Doublets

A thin doublet implies two thin lenses in close contact. Let the curvatures of the refracting interfaces of the two lenses be c_1, c_2 and c_3, c_4, and let the refractive indices of the optical material of the two lenses be n_1 and n_2. Powers K_1 and K_2 of the two thin lenses at wavelength λ are

$$\left(K_\lambda\right)_1=\left[\left(n_\lambda\right)_1-1\right]\left(c_1-c_2\right)=\left[\left(n_\lambda\right)_1-1\right]\tilde{c}_1 \tag{9.49}$$

$$(K_\lambda)_2 = [(n_\lambda)_2 - 1](c_3 - c_4) = [(n_\lambda)_2 - 1]\tilde{c}_2 \tag{9.50}$$

where $\tilde{c}_1 = (c_1 - c_2)$, and $\tilde{c}_2 = (c_3 - c_4)$.

Power of the first thin lens at d, C, and F wavelengths are

$$(K_d)_1 = [(n_d)_1 - 1]\tilde{c}_1 \tag{9.51}$$

$$(K_C)_1 = [(n_C)_1 - 1]\tilde{c}_1 \tag{9.52}$$

$$(K_F)_1 = [(n_F)_1 - 1]\tilde{c}_1 \tag{9.53}$$

Power of the second thin lens at d, C, and F wavelengths are

$$(K_d)_2 = [(n_d)_2 - 1]\tilde{c}_2 \tag{9.54}$$

$$(K_C)_2 = [(n_C)_2 - 1]\tilde{c}_2 \tag{9.55}$$

$$(K_F)_2 = [(n_F)_2 - 1]\tilde{c}_2 \tag{9.56}$$

The change in power of the first thin lens for a change of working wavelength λ from C to F is

$$[(K_F)_1 - (K_C)_1] = [(n_F)_1 - (n_C)_1]\tilde{c}_1 = [(n_d)_1 - 1]\tilde{c}_1 \frac{[(n_F)_1 - (n_C)_1]}{[(n_d)_1 - 1]} = \frac{(K_d)_1}{(V_d)_1} \tag{9.57}$$

$(V_d)_1$ is the Abbe number for the material of the first thin lens for the wavelength combination d, C, and F. Similarly, the change in power of the second lens for the change in working wavelength from C to F is

$$[(K_F)_2 - (K_C)_2] = \frac{(K_d)_2}{(V_d)_2} \tag{9.58}$$

The total power K_d of the doublet at the wavelength d is

$$K_d = (K_d)_1 + (K_d)_2 \tag{9.59}$$

The change in power of the doublet for a change in wavelength from C to F is

$$(K_F - K_C) = \{(K_F)_1 - (K_C)_1\} + \{(K_F)_2 - (K_C)_2\} = \frac{(K_d)_1}{(V_d)_1} + \frac{(K_d)_2}{(V_d)_2} \tag{9.60}$$

Let the V values of optical materials for the two thin lenses of the doublet be $(V_d)_1$ and $(V_d)_2$. The required powers $\left(\breve{K}_d\right)_1$ and $\left(\breve{K}_d\right)_2$ for the thin lenses, to obtain a total power $\breve{K}_d$ for the doublet that has a prespecified amount of axial colour corresponding to $\left(\breve{K}_F - \breve{K}_C\right) = \delta K$, can be determined by solving the equations (9.59) and (9.60). The required powers $\left(\breve{K}_d\right)_1$ and $\left(\breve{K}_d\right)_2$ of the two thin lenses are

$$\left(\tilde{K}_d\right)_1 = \frac{(V_d)_1}{(V_d)_1 - (V_d)_2}\tilde{K}_d - \frac{(V_d)_1 (V_d)_2}{(V_d)_1 - (V_d)_2}\delta K \tag{9.61}$$

$$\left(\tilde{K}_d\right)_2 = \frac{(V_d)_2}{(V_d)_2 - (V_d)_1}\tilde{K}_d - \frac{(V_d)_1 (V_d)_2}{(V_d)_2 - (V_d)_1}\delta K \tag{9.62}$$

For achromatic doublets,

$$\delta K = \frac{\left(\tilde{K}_d\right)_1}{(V_d)_1} + \frac{\left(\tilde{K}_d\right)_2}{(V_d)_2} = 0 \tag{9.63}$$

'The condition for achromatism is then independent of the object distance, and we say the achromatism of a thin system is "stable" with regard to object distance' [30].

The expressions for required powers $\left(\tilde{K}_d\right)_1$ and $\left(\tilde{K}_d\right)_2$ of the individual thin lenses of an achromatic doublet reduce to

$$\left(\tilde{K}_d\right)_1 = \frac{(V_d)_1}{(V_d)_1 - (V_d)_2}\tilde{K}_d \tag{9.64}$$

$$\left(\tilde{K}_d\right)_2 = \frac{(V_d)_2}{(V_d)_2 - (V_d)_1}\tilde{K}_d \tag{9.65}$$

Achromatization of a thin doublet is possible only so long as V values of the optical materials of the two thin lenses are different. Therefore, it is necessary to use two different types of glass in the two lenses.

> The lens having the higher V has the same sign as the total focal length, whilst that having the lower V is of the opposite sign. The powers of the component lenses become smaller the greater the difference of the values of V, and smaller as the values of V themselves diminish.
>
> , [31]

9.3.2.1 Synthesis of a Thin Lens of a Given Power K_d and an Arbitrary $V_\#$

At times, it is required to have a single thin lens of given power K_d, and a prespecified amount of axial colour C_A. For a distant object, realization of this lens calls for an optical material of Abbe number $V_\# = -\frac{1}{K_d C_A}$. If no optical glass with this Abbe number $V_\#$ is available, the required lens may be synthesized by a thin doublet consisting of two thin lenses in close contact. The two lenses may be made from available optical glasses whose Abbe numbers are, say, $(V_d)_1$ and $(V_d)_2$. Obviously, the doublet is of the same power as that of the required thin lens, i.e., $K_d = (K_d)_1 + (K_d)_2$, where

$$(K_d)_1 = \left(\frac{V_1}{V_\#}\right)\left(\frac{V_\# - V_2}{V_1 - V_2}\right)K \tag{9.66}$$

$$(K_d)_2 = -\left(\frac{V_2}{V_\#}\right)\left(\frac{V_\# - V_1}{V_1 - V_2}\right)K \tag{9.67}$$

9.3.3 Secondary Spectrum and Relative Partial Dispersion

For an object distance l, the image distances l'_C and l'_F by a thin lens of power K_C and K_F at C and F wavelengths are

$$\frac{1}{l'_C} - \frac{1}{l} = K_C \tag{9.68}$$

$$\frac{1}{l'_F} - \frac{1}{l} = K_F \tag{9.69}$$

For a thin doublet lens achromatized at C and F wavelengths, $K_C = K_F$. Therefore, $l'_C = l'_F$. Let the power of the doublet lens at a third wavelength λ be K_λ. Using (9.49), (9.50), (9.53), and (9.56), we get

$$\begin{aligned}(K_\lambda - K_F) &= \left[(n_\lambda)_1 - (n_F)_1\right]\tilde{c}_1 + \left[(n_\lambda)_2 - (n_F)_2\right]\tilde{c}_2 \\ &= \frac{\left[(n_\lambda)_1 - (n_F)_1\right]}{\left[(n_F)_1 - (n_C)_1\right]}\frac{\left[(n_F)_1 - (n_C)_1\right]}{\left[(n_d)_1 - 1\right]}\left[(n_d)_1 - 1\right]\tilde{c}_1 \\ &\quad + \frac{\left[(n_\lambda)_2 - n_2(F)\right]}{\left[(n_F)_2 - (n_C)_2\right]}\frac{\left[(n_F)_2 - (n_C)_2\right]}{\left[(n_d)_2 - 1\right]}\left[(n_d)_2 - 1\right]\tilde{c}_2 = (P_{\lambda,F})_1\frac{(K_d)_1}{(V_d)_1} + (P_{\lambda,F})_2\frac{(K_d)_2}{(V_d)_2}\end{aligned} \tag{9.70}$$

where $(K_d)_1$ and $(K_d)_2$ are the powers of the first and the second thin lenses at wavelength d, and $(V_d)_1$ and $(V_d)_2$ are the V_d values of the optical materials of the two thin lenses. $(P_{\lambda,F})_{j=1,2}$ are dispersion characteristics of optical materials of the two thin lenses. $(P_{\lambda,F})$ is given by

$$(P_{\lambda,F}) = \frac{(n_\lambda - n_F)}{(n_F - n_C)} \tag{9.71}$$

Let l'_λ be the image distance at wavelength λ for the object distance l. Using the thin lens imaging equation,

$$\frac{1}{l'_\lambda} - \frac{1}{l} = K_\lambda \tag{9.72}$$

From (9.70) – (9.72),

$$(l'_\lambda - l'_F) = l'_\lambda l'_F (K_F - K_\lambda) \cong l'^2_d \left[(P_{\lambda,F})_1\frac{(K_d)_1}{(V_d)_1} + (P_{\lambda,F})_2\frac{(K_d)_2}{(V_d)_2}\right] \tag{9.73}$$

Figure 9.5 illustrates the paraxial chromatic effects in finite conjugate imaging by an achromatic doublet lens consisting of two thin lenses in close contact and a remote stop. Since the power of the doublet is equal at C and F wavelengths, the two images $P'_C O'_C$ and $P'_F O'_F$ are coincident. The images at other wavelengths have different axial locations, and they are of different lateral magnification. For example, at wavelength d, the image distance $l'_d \neq l'_F (= l'_C)$, and the image height $\eta'_d \neq \eta'_F (= \eta'_C)$. The axial distance $O'_d O'_F (= O'_d O'_C)$ is called the secondary spectrum at wavelength d, $(SS)_d$. Note that the value of the secondary spectrum is a function of the chosen wavelength [32–38].

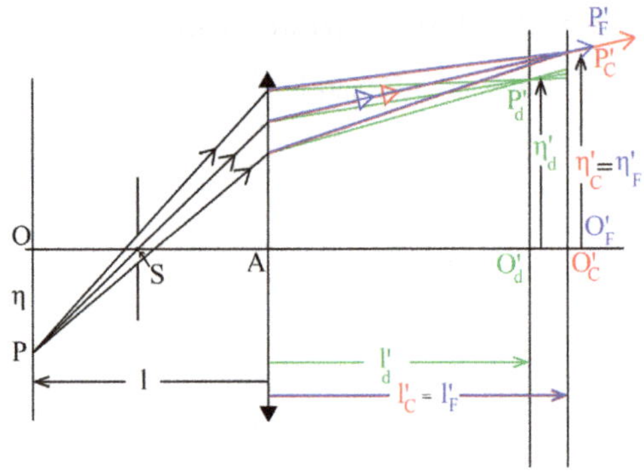

FIGURE 9.5 Imaging by a thin doublet (achromatized for C and F wavelengths) consisting of two thin lenses in contact. Object: η; Imag $\eta'_C = \eta'_F \neq \eta'_d$.

A general relative partial dispersion of an optical material is defined by

$$\left(P_{\lambda_1,\lambda_2}\right) = \frac{\left(n_{\lambda_1} - n_{\lambda_2}\right)}{\left(n_F - n_C\right)} \tag{9.74}$$

Let the power of the doublet lens at wavelengths λ_1 and λ_2 be K_{λ_1} and K_{λ_2}, respectively. The difference in power at the two wavelengths is given by

$$\begin{aligned}\left(K_{\lambda_1} - K_{\lambda_2}\right) &= \left[\left(n_{\lambda_1}\right)_1 - \left(n_{\lambda_2}\right)_1\right]\tilde{c}_1 + \left[\left(n_{\lambda_1}\right)_2 - \left(n_{\lambda_2}\right)_2\right]\tilde{c}_2 \\ &= \frac{\left[\left(n_{\lambda_1}\right)_1 - \left(n_{\lambda_2}\right)_1\right]}{\left[\left(n_F\right)_1 - \left(n_C\right)_1\right]} \frac{\left[\left(n_F\right)_1 - \left(n_C\right)_1\right]}{\left[\left(n_d\right)_1 - 1\right]}\left[\left(n_d\right)_1 - 1\right]\tilde{c}_1 \\ &+ \frac{\left[\left(n_{\lambda_1}\right)_2 - \left(n_{\lambda_2}\right)_2\right]}{\left[\left(n_F\right)_2 - \left(n_C\right)_2\right]} \frac{\left[\left(n_F\right)_2 - \left(n_C\right)_2\right]}{\left[\left(n_d\right)_2 - 1\right]}\left[\left(n_d\right)_2 - 1\right]\tilde{c}_2 = \left(P_{\lambda_1,\lambda_2}\right)_1 \frac{\left(K_d\right)_1}{\left(V_d\right)_1} + \left(P_{\lambda_1,\lambda_2}\right)_2 \frac{\left(K_d\right)_2}{\left(V_d\right)_2}\end{aligned} \tag{9.75}$$

Since, for an achromatic doublet,

$$\frac{\left(K_d\right)_1}{\left(V_d\right)_1} + \frac{\left(K_d\right)_2}{\left(V_d\right)_2} = 0 \tag{9.76}$$

It follows that

$$K_{\lambda_1} = K_{\lambda_2} \quad \text{if} \quad \left(P_{\lambda_1,\lambda_2}\right)_1 = \left(P_{\lambda_1,\lambda_2}\right)_2 \tag{9.77}$$

Optical glass manufacturers provide data of P_{λ_1,λ_2} for the selected combination of λ_1, λ_2. SCHOTT provides data for $P_{C,t}, P_{C,s}, P_{F,e}, P_{g,F}, P_{i,g}$. For the range of visible wavelengths, the glass manufacturers provide a $P_{g,F} - V_d$ diagram, where each available glass is represented as a point on the map. Figure 9.6 is a copy of the current $P_{g,F} - V_d$ diagram of SCHOTT optical glasses. It is significant to note that all normal glasses are clustered around a straight line. This line is called the normal line of optical glasses. The equation of the (g, F) normal line is

$$P_{g,F} = 0.6438 - 0.001682 V_d \tag{9.78}$$

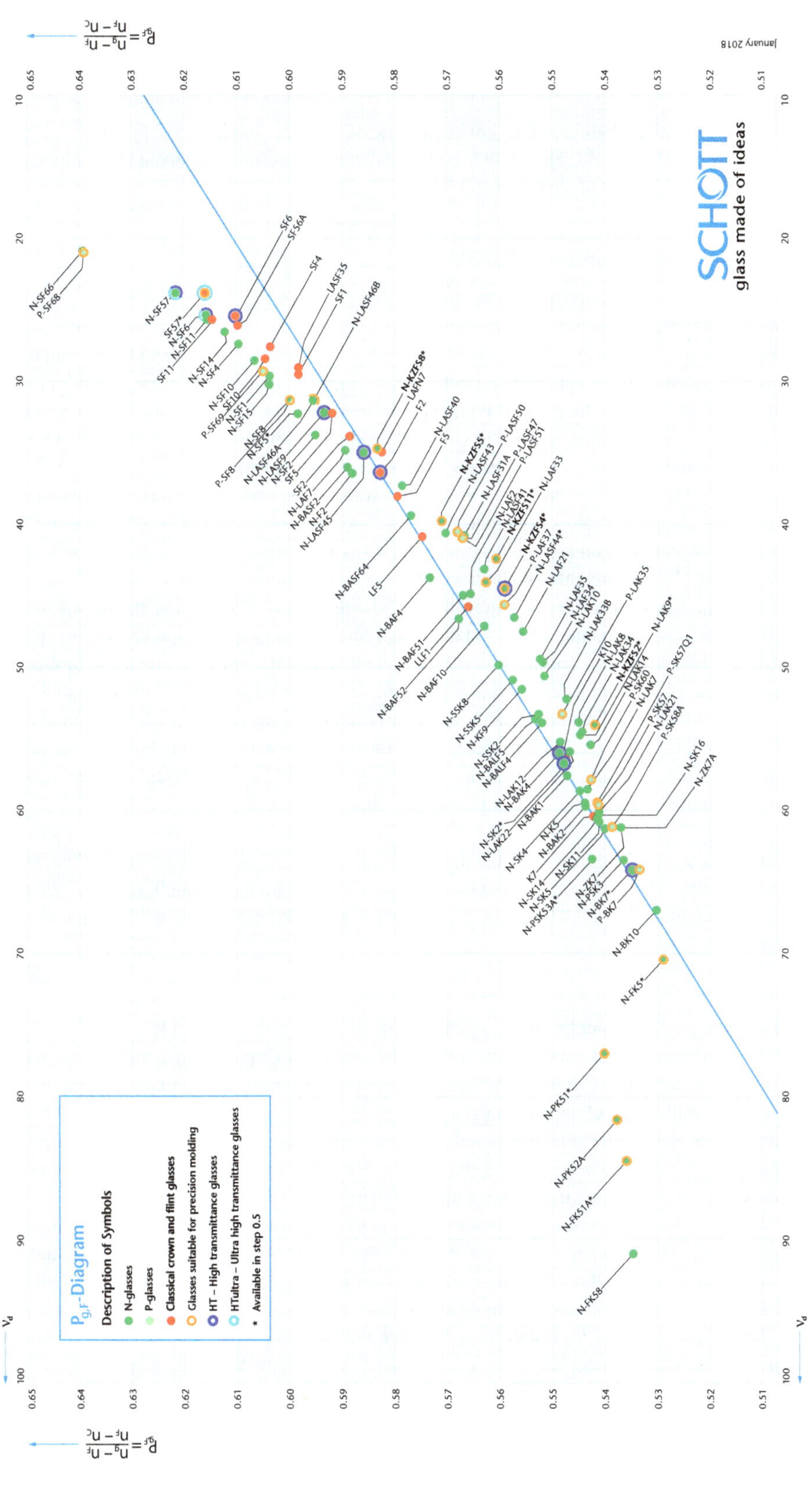

FIGURE 9.6 Abbe plot of relative partial dispersion $P_{g,F}$ values against V-value of Schott optical glasses [Courtesy: M/s SCHOTT Inc.].

For SCHOTT glasses [39], the position of the normal line is such that it passes through the points corresponding to glasses K7 $\left[V_d = 60.41, P_{g,F} = 0.5422\right]$ and F2 $\left[V_d = 36.37, P_{g,F} = 0.5828\right]$. The deviation $\Delta P_{g,F}$ of actual value of $P_{g,F}$ of a specific glass from that given by (9.78) is also quoted in optical glass datasheets.

This phenomenon of clustering normal glasses around a 'normal straight line' on the P – V diagram, such that no glasses with reasonably different V values have the same relative partial dispersion, gives rise to the effect called 'irrationality of dispersion' [40].

9.3.4 Apochromats and Superachromats

In the latter half 1870s, with a prodding by Fraunhofer, Ernst Abbe undertook the task of devising measures for the reduction of secondary spectrum that posed a serious bottleneck. In the process, he discovered the existence of a normal line in the P – V diagram of optical gasses, and found that the reduction or elimination of secondary spectrum in achromatic doublets calls for the use of optical materials that lie significantly away from the normal line. In collaboration with Otto Schott, he carried out extensive investigations at Zeiss company, and in the mid-1880s, a new secret glass named 'X' was used in new microscope objectives where secondary spectrum was significantly reduced [41–43]. In these achromatic doublets, the power is the same at three wavelengths, and the spherical aberration is also corrected at two wavelengths. Abbe christened them as 'apochromatic' lenses. These apochromatic and semi-apochromatic objectives conquered the microscope market for the firm of Zeiss. Around 1890, the secret unknown glass 'X' was revealed to be fluorite.

The limited availability of fluorites of required sizes, the problems of polishing the fluorites to optical quality, large thermal expansion, etc., restricted the use of fluorites for suppressing secondary spectrum in optical microscopy. Investigations continued in two different directions to overcome these problems. First came the development of non-crystalline 'Abnormal' optical glasses with (P,V) point lying away from the normal line. This was met with limited success. A second investigation was based on the conjecture that if the power of a thin lens system could be made equal at two wavelengths by using two thin lenses (of two different glasses) in close contact, the power of such a system could be made equal at three wavelengths by using three thin lenses (of three different glasses) in close contact.

> Mr. H. Dennis Taylor, best known as the designer of the 'Cooke' photographic objective, was the first to succeed in producing these remarkable triple objectives, which are now widely known as 'Photovisual' lenses because sharp photographs can be obtained at the visual focus without colour screens.
>
> [44–46]

Predesign of a thin three-lens apochromat is discussed in detail by Kingslake [47].

Figure 9.7 is a schematic illustration of the chromatic variation of power of a singlet, an achromatic doublet with the same power at C and F wavelengths, an apochromat or a three-color achromat with the same power at C, d, and F wavelengths. In optics literature, there is no unanimity on the nomenclature 'Apochromats'. Currently, the term is often used in the sense of 'a three colour achromat', which implies only the equality of power at three wavelengths. Abbe's definition of an apochromat is more restrictive, as mentioned earlier. Also, as described above, apochromats can be doublets or triplets.

Note that the equality of power at three wavelengths of the working range leads to a reduction of secondary spectrum over the range of the wavelength, but not complete absence. Indeed, the residual chromatic variation in power in three colour achromats over the wavelength range is called 'tertiary spectrum'. For reduction of tertiary spectrum, suggestions have been made to develop four colour achromats, which have the same power at four wavelengths over the working wavelength range. Proposals for hyperachromats, which have same power at five wavelengths over the working wavelength range, have also been made. Details of different proposals for apochromats and superachromats can be found in the references [48–77].

An alternative approach for tackling the problem of achromatization that led to some interesting observations is briefly described in the next section.

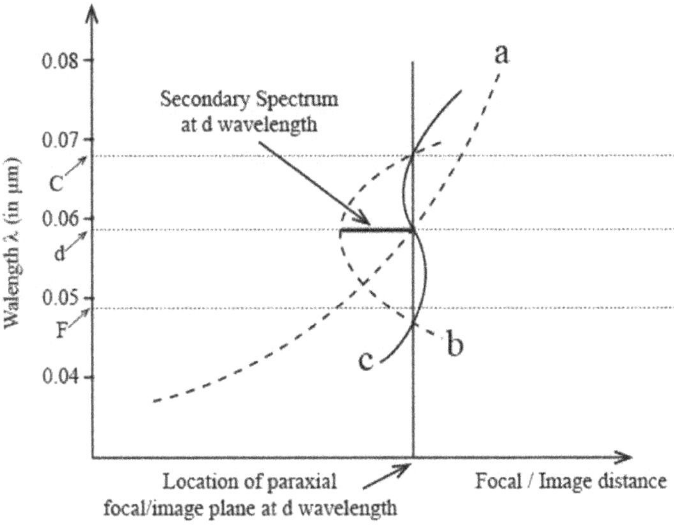

FIGURE 9.7 A schematic for the chromatic variation of focal length for lenses: (a) a singlet, (b) an achromatic doublet with the same power at C and F wavelengths, (c) a three-colour achromat with the same power at C. d and F wavelengths. [Arbitrary linear scale along the horizontal axis].

9.3.5 'Complete' or 'Total' Achromatization

'Complete' or 'Total' achromatism of an optical system implies a system that has the same paraxial characteristics at each wavelength within a wavelength range, say (λ_a, λ_b). Let us consider the simple case of a thin doublet consisting of two thin lenses in close contact. The refractive indices of optical materials of the two thin lenses are $n_1(\lambda)$ and $n_2(\lambda)$, and the differences in curvatures of the two surfaces of the two thin lenses are $\tilde{c}_1$ and $\tilde{c}_2$.

Let the powers of the doublet at any two wavelengths λ_1 and λ_2 within the range (λ_a, λ_b) be K_{λ_1} and K_{λ_2}, respectively. They are given by

$$K_{\lambda_1} = [n_1(\lambda_1) - 1]\tilde{c}_1 + [n_2(\lambda_1) - 1]\tilde{c}_2 \tag{9.79}$$

$$K_{\lambda_2} = [n_1(\lambda_2) - 1]\tilde{c}_1 + [n_2(\lambda_2) - 1]\tilde{c}_2 \tag{9.80}$$

It follows

$$K_{\lambda_1} - K_{\lambda_2} = [n_1(\lambda_1) - n_1(\lambda_2)]\tilde{c}_1 + [n_2(\lambda_1) - n_2(\lambda_2)]\tilde{c}_2 \tag{9.81}$$

When $K_{\lambda_1} = K_{\lambda_2}$

$$\frac{[n_1(\lambda_1) - n_1(\lambda_2)]}{[n_2(\lambda_1) - n_2(\lambda_2)]} = -\frac{\tilde{c}_2}{\tilde{c}_1} = C \tag{9.82}$$

The above relation can hold for any value of λ_1 and λ_2 within the range (λ_a, λ_b), if a relationship such as

$$n_1(\lambda) = Cn_2(\lambda) + \text{constant} \tag{9.83}$$

holds good for optical materials of the two lenses over the range (λ_a, λ_b). Available optical glasses do not satisfy this criterion to suitable tolerances. This fact is another manifestation of the 'irrationality of dispersion' described above [78].

9.3.5.1 Harting's Criterion

Power K_λ of a thin doublet at any wavelength λ is

$$K_\lambda = [n_1(\lambda) - 1]\tilde{c}_1 + [n_2(\lambda) - 1]\tilde{c}_2 \tag{9.84}$$

The rate of change of power with respect to the wavelength is minimum when $[dK_\lambda]/d\lambda = 0$. This implies

$$\frac{dn_1(\lambda)/d\lambda}{dn_2(\lambda)/d\lambda} = -\frac{\tilde{c}_2}{\tilde{c}_1} \tag{9.85}$$

For complete achromatism, i.e., the same power at all wavelengths within a finite spectral range, the ratio at the left-hand side of (9.85) must be a constant independent of λ. It follows that

$$\frac{d}{d\lambda}\left[\frac{dn_1(\lambda)/d\lambda}{dn_2(\lambda)/d\lambda}\right] = 0 \tag{9.86}$$

Equation (9.86) is a general criterion for the absence of secondary colour to obtain complete or total achromatism.

Let the dispersion relations for the optical materials of the two thin lenses of the doublet be expressed by Hartmann's dispersion formula (9.8) as

$$n_1(\lambda) = (n_0)_1 + \frac{a_1}{\{\lambda - (\lambda_0)_1\}^{b_1}} \tag{9.87}$$

$$n_2(\lambda) = (n_0)_2 + \frac{a_2}{\{\lambda - (\lambda_0)_2\}^{b_2}} \tag{9.88}$$

Using these formulae, we get from equation (9.85)

$$\frac{dn_1(\lambda)/d\lambda}{dn_2(\lambda)/d\lambda} = \frac{a_1 b_1 \{\lambda - (\lambda_0)_1\}^{-(b_1+1)}}{a_2 b_2 \{\lambda - (\lambda_0)_2\}^{-(b_2+1)}} = -\frac{\tilde{c}_2}{\tilde{c}_1} \tag{9.89}$$

In the above, $(n_0)_1, a_1, b_1, (\lambda_0)_1$ are the Hartmann dispersion constants for the optical material of the first thin lens of the doublet. Similarly, $(n_0)_2, a_2, b_2, (\lambda_0)_2$ are the Hartmann dispersion constants for the optical material of the second thin lens of the doublet. Using the general criterion (9.86) we get the required condition

$$(b_2 + 1)\{\lambda - (\lambda_0)_1\} = (b_1 + 1)\{\lambda - (\lambda_0)_2\} \tag{9.90}$$

This condition can be fulfilled if and only if

$$b_1 = b_2 \text{ and } (\lambda_0)_1 = (\lambda_0)_2 \tag{9.91}$$

This condition was first reported by Harting [79]. Subsequently, Moffitt published a derivation of the same [80]. Perrin [81] designated this condition for total achromatism as Harting's criterion. It is also mentioned in the book by Hardy and Perrin [82].

Assuming that the two glasses satisfy the condition (9.91), from (9.89) we get

$$\frac{\tilde{c}_1}{\tilde{c}_2} = -\frac{a_2}{a_1} \tag{9.92}$$

This implies that, for complete achromatism, the ratio of the net curvatures of the two thin lenses comprising the doublet are equal to the negative reciprocal of the ratio of the Hartmann dispersion coefficients of the two glasses. The required powers of the constituent thin lens elements of the thin doublet can be worked out as shown below.

Let K_1 and K_2 be powers of the two thin lenses constituting the thin doublet of power K. We have

$$\frac{K_1}{K_2} = \frac{\{n_1(\lambda)-1\}\tilde{c}_1}{\{n_2(\lambda)-1\}\tilde{c}_2} = -\frac{\{n_1(\lambda)-1\}a_2}{\{n_2(\lambda)-1\}a_1} = -\frac{\beta_2}{\beta_1} \tag{9.93}$$

where

$$\beta_2 = \frac{a_2}{n_2(\lambda)-1} \quad \text{and} \quad \beta_1 = \frac{a_1}{n_1(\lambda)-1} \tag{9.94}$$

Powers K_1 and K_2 are given by

$$K_1 = \left(\frac{\beta_2}{\beta_2-\beta_1}\right)K \tag{9.95}$$

$$K_2 = -\left(\frac{\beta_1}{\beta_2-\beta_1}\right)K \tag{9.96}$$

The powers are inversely proportional to $(\beta_2 - \beta_1)$. To minimize the powers of the elements, it is necessary to choose glasses so that absolute value of this difference is as large as possible.

The similarity of the relations (9.95) – (9.96) with the relations (9.64) – (9.65) using Abbe numbers of the glass elements is too apparent to overlook. Indeed, the relations (9.95) – (9.96) aim for complete achromatism over a spectral range, whereas the relations (9.64) – (9.65) look for equality of power at two wavelengths over the spectral range.

Alternative criteria for achromatism can be derived by using other dispersion formulae. Buchdahl [83] has deliberated in more detail on these aspects of achromatization in view of irrationality of dispersion of optical glasses.

9.3.6 A Separated Thin Lens Achromat (Dialyte)

For various reasons, such as overcoming difficulties in manufacturability or achieving better aberrational correction, at times it becomes necessary to use two thin lenses separated in air instead of the thin doublet (Figure 9.8). This lens structure is called 'air-spaced doublet'. In lens design jargon, it is also known as a 'dialyte' or 'dialyt'. The term 'dialyte' means 'parted' or 'separated', and it is derived from the Latin word 'dialyein' ['dia' (apart) + 'lyein' (loosen)]. In earlier times, the optical designers resorted to dialyte telescope objectives [84]

> in order to evade the then predominant difficulty of obtaining large disks of flint glass; for if the crown lens is placed in front, there is a considerable contraction of the cone of rays

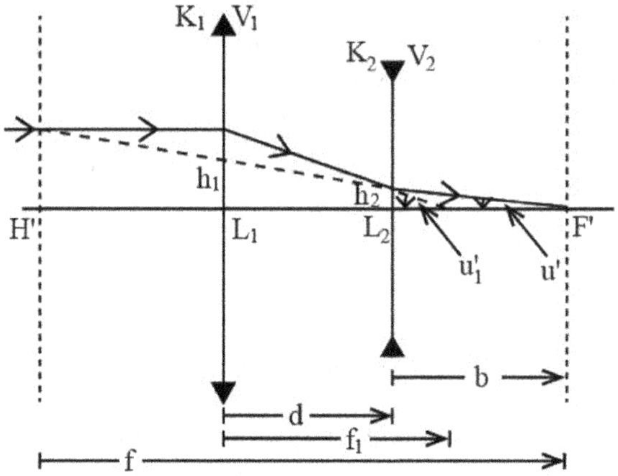

FIGURE 9.8 Imaging of an object at infinity by a dialyte lens consisting of two air-spaced thin lenses.

> before the flint glass is reached, and a smaller diameter of the latter is therefore sufficient. This reason for using separated lenses no longer exists; but the type has been revived in recent times.
>
> [85]

Figure 9.8 shows an axially symmetric system consisting of two thin lenses of power K_1 and K_2, separated by a distance d. The equivalent focal lengths of the two lenses are $f_1 (= 1/K_1)$ and $f_2 (= 1/K_2)$. The V numbers of the two lenses are V_1 and V_2. The heights of paraxial marginal ray from infinity at the two lenses are h_1 and h_2. The convergence angle of this ray in the image space is u', and the same in the intermediate space is u'_1. The equivalent focal length of the system is f, and its back focal length is b.

Using the paraxial ray transfer equation and following our sign convention, we get

$$h_2 = h_1 + du'_1 = h_1\left(1 - \frac{d}{f_1}\right) \tag{9.97}$$

It can be rewritten as

$$\frac{d}{f_1} = 1 - \frac{h_2}{h_1} = 1 - \frac{h_2 u'}{h_1 u'} = 1 - \frac{b}{f} \tag{9.98}$$

The ratio (d/f_1) is a dimensionless quantity, and may be represented by a single parameter $\tilde{k}$.

The sum of the contributions to longitudinal chromatic aberrations of the two thin lens elements when the system is achromatic must be zero, i.e.

$$h_1^2\left(\frac{K_1}{V_1}\right) + h_2^2\left(\frac{K_2}{V_2}\right) = 0 \tag{9.99}$$

Note that, since $h_2 \neq h_1$, and h_2 depends on $u'_1 (= u_1 - h_1 K_1)$, where u_1 is the convergence angle of the paraxial marginal ray from the axial object point, it is not possible to have 'stable' achromatization in the case of a dialyte; the condition for achromatism is different for different object distances. Figure 9.8 shows the case of an object at infinity, and the following treatment covers the same.

The total power K of the dialyte is

$$K = K_1 + K_2 + d\,K_1\,K_2 = \frac{1}{f} \tag{9.100}$$

where f is the equivalent focal length of the dialyte lens. Powers K_1 and K_2 of the two thin lenses for the achromatic dialyte lens can be determined, when the object is at infinity, from (9.98) – (9.100) in terms of back focal length b and equivalent focal length f as [86]

$$K_1 = \frac{V_1 b}{f(V_1 b - V_2 f)} \quad (9.101)$$

$$K_2 = -\frac{V_2 f}{b(V_1 b - V_2 f)} \quad (9.102)$$

The distance d between the two thin lenses is related to the focal lengths b and f of the dialyte, and the power K_1 of the first thin lens by

$$dK_1 = \left(1 - \frac{b}{f}\right) \quad (9.103)$$

Alternatively, the focal lengths of the two thin elements of the above case can be determined by the relations

$$f_1 = f\left[1 - \frac{V_2}{V_1(1-\tilde{k})}\right] \quad (9.104)$$

$$f_2 = f(1-\tilde{k})\left[1 - \frac{V_1(1-\tilde{k})}{V_2}\right] \quad (9.105)$$

The derivation does not take into account the variation in height h_2 at different wavelengths on account of dispersion of the incident ray by passage through the first lens. The resulting effects and the secondary spectrum of the dialyte have been discussed in [87]. A direct solution of the dialyte problem in the general case of a finite object distance is given in [88].

9.3.7 A One-Glass Achromat

It is obvious that it is impossible to obtain achromatism by using a single or any combination of thin lenses in close contact (called a 'monoplet' by Herzberger [88]) if the lenses are of the same optical glass. But it is possible to obtain an air-spaced achromat with only one type of glass. Substituting $V_1 = V_2$ in (9.104) and (9.105), the expressions for focal lengths $\hat{f}_1$ and $\hat{f}_2$ of the two separated thin lenses of a one-glass achromatic dialyte of focal length f are

$$\hat{f}_1 = \frac{\tilde{k}}{(\tilde{k}-1)} f \quad (9.106)$$

$$\hat{f}_2 = -\tilde{k}(\tilde{k}-1) f \quad (9.107)$$

where $\tilde{k} = (d/\hat{f}_1)$. Since d is necessarily positive, $\tilde{k}$ is of same sign as $\hat{f}_1$. From (9.106), it follows that $(\tilde{k}-1)$ must be of the same sign as f. This type of dialyte lens is commonly known as a Schupmann lens [89]. Details of one-glass Schupmann achromatic doublets are noted in [84].

One-glass achromats are used in Huygens eyepieces. The thin-lens layout of the latter can be found in a simple way [90–91]. The total power K of a system consisting of two thin lenses of power K_1 and K_2, separated axially by a distance d is

$$K = K_1 + K_2 - dK_1 K_2 \tag{9.108}$$

After differentiation, we get

$$\delta K = \delta K_1 + \delta K_2 - dK_1 \delta K_2 - dK_2 \delta K_1 \tag{9.109}$$

Let the differential power refer to the two wavelengths F-line and C-line, and the mean wavelength to the d-line. Using (9.33), we get, for the two thin lenses

$$\delta K_1 = \frac{(K_d)_1}{(V_d)_1} \qquad \delta K_2 = \frac{(K_d)_2}{(V_d)_2} \tag{9.110}$$

From (9.109), it follows

$$\delta K = K_F - K_C = \frac{(K_d)_1}{(V_d)_1} + \frac{(K_d)_2}{(V_d)_2} - d(K_d)_1 \frac{(K_d)_2}{(V_d)_2} - d(K_d)_2 \frac{(K_d)_1}{(V_d)_1} \tag{9.111}$$

In the above, $(V_d)_1$ and $(V_d)_2$ are the V numbers of the optical glasses of the first and the second thin lens, respectively. When the glasses are the same, $(V_d)_1 = (V_d)_2$, and (9.111) reduces to

$$(K_d)_1 + (K_d)_2 = 2d(K_d)_1 (K_d)_2 \tag{9.112}$$

Therefore, when the separation of the two thin lenses of the same glass is $\hat{d}$, given by

$$\hat{d} = \frac{(K_d)_1 + (K_d)_2}{2(K_d)_1 (K_d)_2} = \frac{1}{2}\left[(f_d)_1 + (f_d)_2\right] \tag{9.113}$$

we get $K_F = K_C$. Note that, for an object at infinity, this one-glass dialyte has the same equivalent focal length for the F- and C-line wavelengths. There is no chromatic variation in image height for these two wavelengths. However, chromatic variation of focus remains.

9.3.8 Secondary Spectrum Correction with Normal Glasses

It is well recognized that the correction of the secondary spectrum in achromatic doublets calls for the use of at least one abnormal dispersion glass. Nevertheless, search for alternatives for controlling secondary spectrum in optical systems has led to the exploration of liquid lenses [66, 68] and diffractive optical elements [92–93], among others.

Subsequently, it was realized that although it is not possible to eliminate the secondary spectrum in an imaging system consisting of a 'monoplet', it should be possible to control it in an optical system consisting of two or more air-spaced components. The basic reason for this is that in properly designed air-spaced multicomponent systems, the dispersion by one component can be compensated by the dispersion of the opposite sign by the other component. Note that the achromatization condition (9.99), when applied to air-spaced components, is incorrect, for it assumes h_2 is same for all wavelengths over the working wavelength range. In reality, the larger the separation between the two components, the larger the variation in height.

The first report on the correction of optical systems by normal glasses appeared in a patent by McCarthy in 1955. He noted:

> an optical system containing ordinary optical materials of at least two kinds can be substantially freed of secondary spectrum by constructing it of two components, one of which has

> chromatic overcorrection and the other of which has chromatic undercorrection and is placed at a distance from the first component. In the new system, the over-correction of the first component is balanced by the under-correction of the second to correct the system for primary colour and a choice of the proper separation between the components then results in a significant reduction and even complete elimination or over-correction of secondary spectrum.
> [94]

More than twenty years later, a comprehensive first order theory for correction of secondary spectrum was provided by Wynne [95–96]. Subsequently, few reports on application of this approach appeared in the literature [97–99].

9.3.9 A Thick Lens or a Compound Lens System

So long as one deals with imaging by a single thin lens or a monoplet, the principal or unit planes of all wavelengths lie on the lens itself. In the case of imaging by a thick lens, or of a lens system consisting of two or more air-spaced thin lenses, this is no longer true. Figure 9.9 is a schematic diagram of chromatic imaging by an axially symmetric system of an object at infinity. A_1 and A_k are the first and the last vertex of the lens. $P'_F\, O'_F$ and $P'_C\, O'_C$ are the transverse images at F and C wavelengths, respectively, of a transverse object at infinity. The first and the second principal or unit planes of the system corresponding to C and F wavelengths are located on the axis at H_C, H'_C and H_F, H'_F. Consequently, both the back focal length and the equivalent focal length of the system are different for C and F wavelengths, as marked on the diagram. In general, the cardinal points of the system suffer from chromatic aberrations [100].

Figure 9.10 shows the general ray configuration of a thick/composite lens system with zero axial colour at C and F wavelengths. It implies $O'_F \equiv O'_C$, so that $(\text{bfl})_F \equiv (\text{bfl})_C$. But $(\text{efl})_C \neq (\text{efl})_F$. The inequality $P'_F\, O'_F \neq P'_C\, O'_C$ implies that the system has residual lateral colour. Figure 9.11 shows the general ray

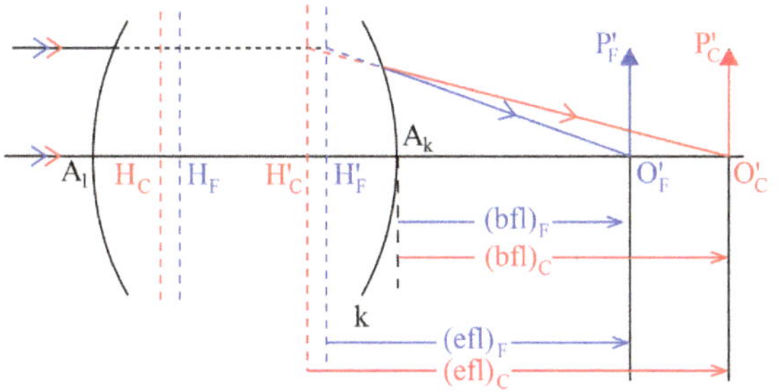

FIGURE 9.9 Imaging of an object at infinity by a thick/composite lens. First vertex: A_1; Last vertex: A_k. Principal planes H_F, H'_F, and transverse image $P'_F O'_F$ at wavelength F; Principal planes H_C, H'_C, and transverse image $P'_C O'_C$ at wavelength C.

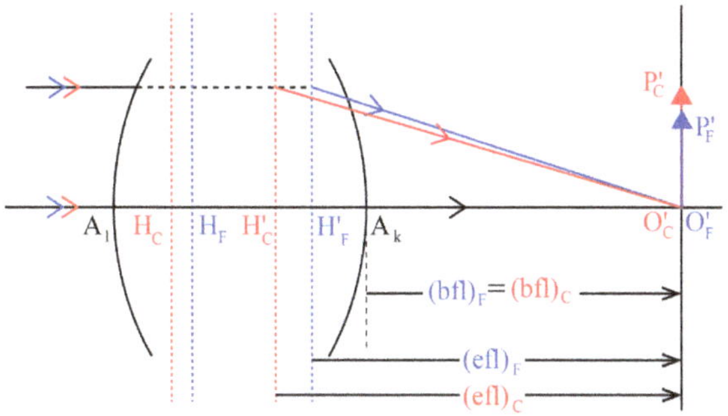

FIGURE 9.10 Zero axial colour at C and F wavelengths for a thick/composite lens implies $O'_C \equiv O'_F$, and, in general, $(\text{bfl})_C = (\text{bfl})_F, (\text{efl})_C \neq (\text{efl})_F$, and $P'_C O'_C \neq P'_F O'_F$ i.e. presence of residual lateral colour.

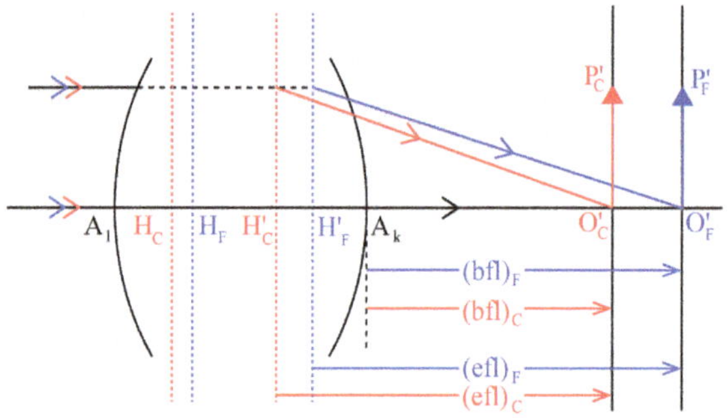

FIGURE 9.11 When $(efl)_C = (efl)_F$ for a thick/composite lens, lateral colour at C and F wavelengths is zero. But, in general, $H'_C \neq H'_F$ and $O'_C \neq O'_F$, so that residual axial colour $O'_C O'_F$ Is present.

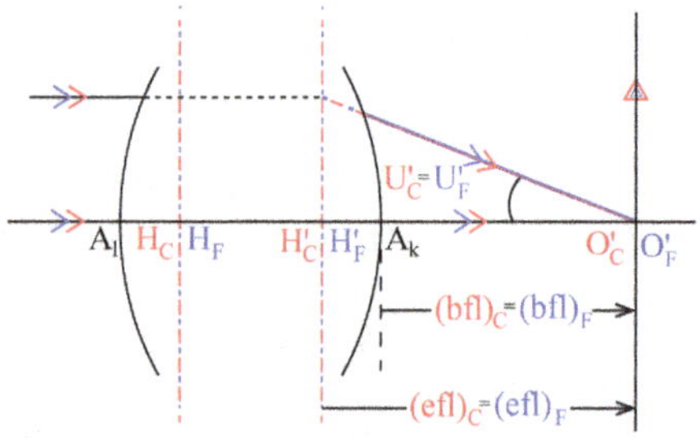

FIGURE 9.12 Paraxial achromatism of a thick/composite lens implying absence of both axial colour and lateral colour at C and F wavelengths is possible if both $(efl)_C = (efl)_F$ and $(bfl)_C = (bfl)_F$ simultaneously.

configuration for the system in the case of zero lateral colour at F and C wavelengths. In this case, $(efl)_F = (efl)_C$. But, $H'_C \neq H'_F$, and $O'_C \neq O'_F$. $O'_C\ O'_F$ is the residual axial colour.

Paraxial achromatism of a thick/composite lens implying absence of both axial colour and lateral colour at C and F wavelengths is possible if both conditions $(efl)_F = (efl)_C$, and $(bfl)_F \equiv (bfl)_C$ hold simultaneously. Figure 9.12 shows that this is possible if and only if both the axial images and the principal points corresponding to C and F wavelengths coincide.

Incidentally, it should be noted that Hristov extended his analytical approach for the correction of monochromatic aberrations to paraxial chromatic correction in axi-symmetric optical systems consisting of one or more thick lens elements [101].

9.4 Chromatism Beyond the Paraxial Domain

All chromatic effects and phenomena considered earlier in Section 9.3 are essentially based on changes in paraxial characteristics of imaging lenses caused by changes in working wavelength. It is obvious that the latter will cause concomitant changes in finite or real ray characteristics, e.g., spherical aberration, field curvature, and aberrations of all orders in the imaging configuration. In general, the tackling of finite or real ray chromatic effects is an involved problem. This case will be discussed in the next chapter in Section 10.2.

REFERENCES

1 N.J. Kriedl and J.L. Rood, 'Optical materials', in *Applied Optics and Optical Engineering, Vol. I*, Ed. R. Kingslake, Academic, New York (1965) pp. 153–200.

2 SCHOTT Technical information, Advanced Optics, 'TIE – 29 Refractive Index and Dispersion', April 2016, p.1.

3 W.F. Meggers and C.G. Peters, 'Measurements on the index of refraction of air for wavelengths from 2218 A to 9000 A', *Astrophysical J.*, Vol. 50 (1919) pp. 56–71. Also in, Sci. Papers of the Bureau of Standards, no. 327 (1918).
4 P.E. Ciddor, 'Refractive index of air: new equations for the visible and the near infrared', *Appl. Opt.*, Vol. 35 (1996) pp. 1566–1573.
5 J.A. Stone, Jr. and J.H. Zimmerman, 'Index of Refraction of Air', NIST, Feb. 2001. emtoolox.nist.gov/wavelength/Documentation.asp.
6 A.L. Cauchy, *Mémoire sur la dispersion de la lumière*, J.G. Calve, Prague (1836).
7 A.E. Conrady, 'On the chromatic correction of object glasses', *Mon. Not. Royal Astron. Soc.*, Vol. 64 (1904) pp. 182–188.
8 A.E. Conrady, *Applied Optics and Optical Design, Part II*, Ed. R. Kingslake, Dover, New York (1960) pp. 658–661.
9 J.F. Hartmann, 'Interpolationsformeln für das prismatische spectrum', *Publiationen des Astrophysikalischen Observatoriums Zu Potsdam*', *Vol.* 12 (1902).
10 A.C. Hardy and F.H. Perrin, *The Principles of optics*, McGraw-Hill, New York (1932).
11 H. Chrétien, Calcul des Combinaisons Optiques, Masson, Paris (1980); [First published in 1938).
12 W. Sellmeier, 'Zur Erklärung der abnormen Farenfölge im Spectrum einiger Substanzen', *Annalen der Physik und Chemie*, Vol. 219 (1871) pp. 272–282.
13 P. Drude, *The Theory of Optics*, Translated by C.R. Mann and R.A. Millikan, Longmans, Green, London (1902) pp. 382–399.
14 M. Herzberger, 'The Dispersion of Optical Glass', *J. Opt. Soc. Am.*, Vol. 32 (1942) pp. 70–77.
15 M. Herzberger and H. Jenkins, 'Color Correction in Optical Systems and Types of Glass', *J. Opt. Soc. Am.*, Vol. 39 (1949) pp. 984–989.
16 M. Herzberger, Modern Geometrical Optics, Interscience, New York (1958), p. 121.
17 M. Herzberger, 'Colour correction in optical systems and a new dispersion formula', *Opt. Acta*, Vol. 6 (1959) pp. 197–215.
18 R. Kingslake, *Lens Design Fundamentals*, Academic, New York (1978) p. 14.
19 W.J. Smith, *Modern Optical Engineering*, McGraw-Hill, New York (2000) p. 176.
20 F.T. Lentes, 'Refractive Index and Dispersion' in *The Properties of Optical Glass*, Eds. H. Bach and M. Neuroth, Springer, Berlin (1995) pp. 24–27.
21 SCHOTT Technical information, vide Ref. 2, p.5.
22 H.A. Buchdahl, *Optical Aberration Coefficients*, Dover, New York (1968) pp. 150–154. (First published by Oxford University Press in 1954) pp.150–151.
23 H.A. Buchdahl, *An Introduction to Hamiltonian Optics*, Cambridge University Press, London (1970) pp. 223–229.
24 P.N. Robb and R.I. Mercado, 'Calculation of refractive indices using Buchdahl's chromatic coordinates', *Appl. Opt.*, Vol. 22 (1983) pp. 1198–1215.
25 G.W. Forbes, 'Chromatic coordinates in aberration theory', *J. Opt. Soc. Am. A*, Vol. 1 (1984) pp. 344–349.
26 P.J. Reardon and R.A. Chipman, 'Buchdahl's glass dispersion coefficients calculated from Schott equation constants', *Appl. Opt.*, Vol. 28 (1989) pp. 3520–3523.
27 A.E. Conrady, *Applied Optics and Optical Design, Part I*, Oxford University Press, London (1929) p. 145.
28 P. Mouroulis and J. Macdonald, Geometrical Optics and Optical Design, Oxford Univ. Press, New York (1997), pp. 191–194.
29 A.E. Conrady, vide Ref. 27, p. 301.
30 R. Kingslake, vide Ref. 18, p.80.
31 A. Koenig, 'Theory of chromatic aberration', in *Geometrical Investigation of the formation of images in optical instruments*, Ed. M. von Rohr, (Forming Vol. I of *Die Theorie der optischen Instrumente, Bd.I* (Berlin, 1904), Translated by R. Kanthack, Dept. of Scientific and Industrial Research, His Majesty's Stationery Office, London (1920) p. 354.
32 S. Czapski, 'Das Sekundäre Spektrum', in *Grundzüge der Theorie der Optischen Instrumente nach Abbe*, Johann Ambrosius Barth, Leipzig (1904) pp. 174–180.
33 A. Koenig, 'Secondary Spectrum', vide Ref. 31, pp. 364–369.

34 J.P.C. Southall, 'Secondary Spectrum', in The Principles and Methods of Geometrical Optics, Macmillan, New York (1910) pp. 523–526.
35 T. Smith, 'Note on dispersion formulae and the secondary spectrum', *Trans. Opt. Soc. London*, Vol. 22 (1921) pp. 99–110.
36 M.J. Kidger, 'Secondary Spectrum', in Fundamental Optical Design, SPIE, Bellingham, Washington (2002) pp. 94–97.
37 A.E. Conrady, vide Ref. 27, pp. 155–159.
38 B.H. Walker, 'Understanding secondary colour', *Opt. Spectra*, June 1978, pp. 44–46.
39 SCHOTT Technical information, Advanced Optics, 'TIE – 29 Refractive Index and Dispersion', April 2016, p. 3.
40 W.T. Welford, *Aberrations of optical systems*, Adam Hilger, Bristol (1986) p. 200.
41 E. Abbe und O. Schott, Productionsverzeichniss *des glastechnischen Laboratoriums von SCHOTT und Genossen in Jena*: published as a prospectus in July, 1886 and re-printed in Gesammelte Abhandlungen von Ernst Abbe, Bd. II, Jena (1906), pp. 194–201.
42 E. Abbe: 'Ueber neue Mikroskope', in *Sitzungsberichte der Jenaischen Gesellschaft für Medizin und Naturwisseschaft*, Jena (1886) pp.107–128; reprinted in *Gesammelte Abhandlungen, Bd. I*, Jena (1904) pp. 450–472.
43 E. Abbe, 'On Improvements of the Microscope with the aid of new kinds of Optical glass', [Translated by H.A. Miers] *J. Roy. Micros. Soc.*, VI Part 2 (1886) pp. 316–321.
44 M. Herzberger, Modern Geometrical Optics, Interscience, New York (1958) p. 118.
45 A.E. Conrady, vide Ref. 27, p. 160.
46 H. Dennis Taylor, A System of Applied Optics, Macmillan, London (1906) p. 309.
47 R. Kingslake, vide Ref. 18, pp. 84–87.
48 N. v. d. W. Lessing, 'Selection of optical glasses in apochromats', *J. Opt. Soc. Am.*, Vol. 47 (1957) pp. 955–958.
49 N. v. d. W. Lessing, 'Further considerations on the selection of optical glasses in apochromats', *J. Opt. Soc. Am.*, Vol. 48 (1958) pp. 269–273.
50 R.E. Stephens, 'Selection of glasses for three-color achromats', *J. Opt. Soc. Am.*, Vol. 49 (1959) pp. 398–401.
51 R.E. Stephens, 'Four colour achromats and superachromats', *J. Opt. Soc. Am.*, Vol. 50 (1960) pp. 1016–1019.
52 M. Herzberger and N.R. McClure, 'The design of superachromatic lenses', *Appl. Opt.*, Vol. 2 (1963) pp. 553–560.
53 H. Drucks, 'Bemerkung zur Theorie der superachromats', *Optik*, Vol. 23 (1966) pp. 523–534.
54 M. Herzberger and H. Pulvermacher, 'Die farbfehlerkorrektion von multipletts', *Opt. Acta*, Vol. 17 (1970) pp. 349–361.
55 N. v. d. W. Lessing, 'Selection of optical glasses in superachromats', *Appl. Opt.*, Vol. 9 (1970) pp. 1655–1668.
56 N. v. d. W. Lessing, 'Further considerations on the selection of optical glasses in superachromats', *Appl. Opt.*, Vol. 9 (1970) pp. 2390–2391.
57 B.L. Nefedov, 'The design of apochromats made from two and three different glasses', *Sov. J. Opt. Tech.*, Vol. 40 (1973) pp. 46–57.
58 G.A. Mozharov, 'Graphical analysis method of choosing the glasses for the design of a four-lens three-color apochromat', *Sov. J. Opt. Tech.*, Vol. 44 (1977) pp. 146–148.
59 M.G. Shpyakin, 'Design of four colour thin apochromats', *Sov. J. Opt. Tech.*, Vol. 45 (1978) pp. 81–83.
60 M.G. Shpyakin, 'Calculation of the components of apochromats made from four types of glass for a broad spectral region', *Sov. J. Opt. Tech.*, Vol. 45 (1978) pp. 219–223.
61 G.A. Mozharov, 'Two component four-color apochromats', *Sov. J. Opt. Tech.*, Vol. 47 (1980) pp. 398–399.
62 R.W. Sinnott, 'An apochromatic triplet objective', *Sky and Telescope*, Vol. 62 (1981) pp. 376–381.
63 P.N. Robb, Selection of glasses', Proc. *1985 International Lens Design Conference, SPIE*, Vol. 554 (1985) pp. 60–75.
64 R.I. Mercado, 'The design of apochromatic optical systems', *Proc. 1985 International Lens Design Conference, SPIE*, Vol. 554 (1985) pp. 217–227.

65 R.I. Mercado, 'Designs of two-glass apochromats and superachromats', *Proc. International Lens Design Conference, SPIE*, Vol. 1354 (1990) pp. 262–272.
66 R.D. Sigler, 'Apochromatic correction using liquid lenses', *Appl. Opt.*, Vol. 16 (1990) pp 2451–2459.
67 R.I. Mercado, 'Correction of secondary and higher-order spectrum using special materials', Proc. *SPIE*, Vol. 1535 (1991) pp. 184–198.
68 R.D. Sigler, 'Designing apochromatic telescope objectives with liquid lenses' *Proc. SPIE*, Vol. 1535 (1991) pp. 89–112.
69 J. Maxwell, 'Tertiary spectrum manipulation in apochromats', *Pro. SPIE, 1990 International Les Design Conference*, Vol. 1354 (1991) pp. 408–416.
70 R.I. Mercado, 'Design of achromats and superachromats', *in Lens Design*, Vol. CR41, SPIE, Bellingham, Washington (1992) pp. 270–296.
71 T. Kryszczynski, 'Secondary spectrum aberrations of refractive optical systems in Refractometry', *in Proc. SPIE*, Vol. 2208 (1995) pp. 18–27.
72 P. Hariharan, 'Apochromatic lens combination, a novel design approach', *Opt. Laser Tech.*, Vol. 29 (1997) pp. 217–219.
73 P. Hariharan, 'Superachromatic lens combination', *Opt. Laser Tech.*, Vol. 31 (1999) pp. 115–118.
74 M.J. Kidger, 'Secondary spectrum and apochromats', chapter 5 in Intermediate Optical Design, SPIE, Bellingham, Massachusetts (2004) pp. 101–116.
75 A. Miks and J. Novak, 'Analysis and Synthesis of planachromats', *Appl. Opt.*, Vol. 49 (2010) pp. 3403–3410.
76 B.F.C. de Albuquerque, J. Sasian, F.L. de Sousa, and A.S. Montes, 'Method of glass selection for colour correction in optical system design', *Opt. Exp.*, Vol. 20 (2012) pp. 13592–13611.
77 J.R. Rogers, 'The importance of induced aberrations in the correction of secondary colour', *Adv. Opt. Techn.*, Vol. 2 (2013) pp. 41–51.
78 W.T. Welford, 'Optical Calculations and Optical Instruments, an Introduction', in *Handbuch der Physik, Band XXIX, Optische Instrumente*, Ed. S. Flügge, Springer, Berlin (1967) p. 22.
79 H. Harting, 'Zur Theorie des sekundaren Spektrums', *Zeitts. F. Instrumentenk.*, Vol. 31 (1911) pp. 72–79.
80 G.W. Moffitt, 'Complete achromatization of a two-piece lens', *Phys. Rev.*, Vol. 11 (1918) pp. 144–147
81 F.H. Perrin, 'A study of Harting's criterion for complete achromatism', *J. Opt. Soc. Am.*, Vol. 28 (1938) pp. 86–93.
82 A.C. Hardy and F.H. Perrin, The Principles of Optics, McGraw-Hill, New York (1932) pp. 115–116.
83 H.A. Buchdahl, 'Many colour correction of thin doublets', *Appl. Opt.*, Vol. 24 (1985) pp. 1878–1882.
84 P.L. Manly, *Unusual Telescopes*, Cambridge University Press, Cambridge (1999) p. 55.
85 A.E. Conrady, vide Ref. 27, p. 177.
86 W.J. Smith, vide Ref. 19, p. 411.
87 R. Kingslake and R.B. Johnson, *Lens Design Fundamentals*, SPIE Press, Bellingham, Massachusetts (2010) pp. 156–159.
88 M. Herzberger, vide Ref. 16, p. 86.
89 L. Schupmann, *Optical correcting device for refracting telescopes*, U. S. Patent, 6,20,978 (1899).
90 A.C. Hardy and F.H. Perrin, Ref. 77, pp. 111–112.
91 P. Mouroulis and J. Macdonald, vide Ref. 28, p. 195.
92 A.I. Tudorovskii, 'An objective with a phase plate', *Opt. Spectrosc.*, Vol. 6 (1959) pp. 126–133.
93 C.W. Chen, 'Application of diffractive optical elements in visible and infrared optical systems', in Lens Design, Ed. W. Smith, *SPIE Press*, Vol. CR 41, (1992) pp. 158–172.
94 E.L. McCarthy, *Optical System with Corrected Secondary Spectrum*, U. S. Patent 2,698,555A, (1955).
95 C.G. Wynne, 'Secondary spectrum correction with normal glasses', *Opt. Commun.*, Vol. 21 (1977) pp. 419–424.
96 C.G. Wynne, 'A comprehensive first order theory of chromatic aberration: secondary spectrum correction without special glasses', *Opt. Acta*, Vol. 25 (1978) pp. 627–636.
97 M. Rosete-Aguilar, 'Correction of secondary spectrum using normal glasses', *Proc. SPIE*, Vol. 2774 (1996) pp. 378–386.
98 M. Rosete-Aguilar, 'Application of the extended first-order chromatic theory to the correction of secondary spectrum', *Rev. Mex. Fis.*, Vol. 43 (1997) pp. 895–905.

99 R. Duplov, 'Apochromatic telescope without anomalous dispersion glasses', *Appl. Opt.*, Vol. 45, (2006) pp. 5164–5167.

100 G.G. Slyusarev, Aberration and Optical Design Theory, Adam Hilger, Bristol (1984), pp.211–213; First published in 1969 as *Metodi Rascheta Opticheskikh Sistem*, Mashinostroenie Press, Leningrad.

101 B.A. Hristov, 'Exact analytical design method for paraxial chromatic correction in axis-symmetric optical systems', *Proc. SPIE*, Vol. 7428 (2009) 74280R-1 to 74280R-12.

10

Finite or Total Aberrations from System Data by Ray Tracing

In an imaging lens system, aberrations of any type depend on two factors. First, aberrations are a function of the constructional parameters, e.g. curvatures, conic constants, aspheric coefficients and semi-diameters of the refracting/reflecting interfaces, axial thicknesses and optical materials of the individual lens elements constituting the system, and the inter lens separations, etc. of the lens system. In general, for monochromatic or quasi-monochromatic imaging systems, refractive indices of the constituent optical materials at the working wavelength are of concern. But, for systems operating under multi-wavelength or broadband illumination, the dispersion characteristics of the constituent optical materials need also to be given due consideration. Next, even for the same system, aberrations vary for different locations of the object-image conjugates, and for the size of the object/image. As enunciated earlier, any of the equivalent forms of aberrations—wave aberration, transverse or longitudinal ray aberration, angular aberration—are, in general, a function of the parameters defining the individual rays of the image-forming pencil. Aberrations of individual rays of an image-forming pencil are defined in terms of their respective deviations from a desired imaging characteristic. Often, the latter pertains to a reference ray that may or may not be a member of the actual image-forming pencil of rays.

In general, individual rays of an image-forming pencil, originating from an object point, follow different paths in the intermediate spaces of the lens system. Except for in trivial cases, the analytical expressions relating aberrations of individual rays, or the coefficients of the power series expansion of the aberration function, with the constructional parameters and conjugate parameters, are unwieldy, so much so that further analysis becomes practically impossible. Nevertheless, ray trace data, which contain the data of constructional parameters in implicit form, can act as an intermediary to obtain the much sought for aberration expressions that play useful roles in both analysis and synthesis of practical systems.

Paraxial ray tracing was discussed in Section 3.5.2; it involves sequential use of refraction equation and transfer equation for all refracting (reflecting) interfaces of the system from the object to the image. In the object space, the convergence angle of the incident ray and the height of the ray at the first surface need to be worked out separately with the help of the opening formula. Similarly, in the image space, closing formulae are used to determine the required image parameters. The logistics for the tracing of non-paraxial finite rays is identical. However, the practical implementation is computation-intensive. This is the case for primarily two reasons. First, the paraxial refraction formula (3.85) assumes the law of refraction as $ni = n'i'$, where i and i' are the angle of incidence and the angle of refraction, and n, n' are refractive indices of the two media. In real finite ray trace, the vector form of Snell's law, as given in equation (2.190), needs to be used. This leads to the set of equations (2.197) – (2.200) that are finite ray refraction equations for a spherical interface. The relations are more involved for non-spherical interfaces. Next, the paraxial transfer formula involves a transfer of the emergent ray to the tangent line at the vertex of the next refracting interface. However, in the case of finite ray trace, analytical formulae for transfer to the next interface is available only when the surface is of the second degree, e.g., a sphere, a conicoid, etc. In the case of other types of surfaces, such analytical formulae are not available, and determination of the point of incidence on that surface calls for iterative methods, which involve a substantial amount of numerical computation. Note that determination of the path of the light ray in part of the system having inhomogeneous medium, if any, involves the use of the vector form of the differential equation for light rays, as given in equation (2.167), and this is a numerical process that is computation intensive.

DOI: 10.1201/9780429154812-10

In the pre-digital computer era, these factors severely inhibited the practical use of finite ray tracing in the analysis and synthesis of optical systems. Easy availability of digital computers with phenomenal number crunching capability has brought about a radical change in the scenario, so much so that, in practice, the time required for the tracing of finite rays no longer poses any constraint for any reasonable analysis and synthesis of optical systems. Almost all software packages have built-in programs for tracing finite rays through large types of surfaces.

The basic principles of finite ray tracing, as briefly put forward above, are well-known. Nevertheless, fast, accurate, and efficient practical implementation with the required level of precision often calls for ingenuity. In the pre-digital computer period, the primary challenges were the need for accurate evaluation of trigonometric functions and the fast implementation of high precision arithmetic. A glimpse of different enterprises by intrepid investigators in optics to circumvent the problem can be obtained in references [1–7]. Since the incorporation of digital computers with phenomenal increase in number crunching ability, the capability of ray tracing has increased rapidly, and even ray tracing through non-symmetric, unconventional interfaces has become a routine affair. Many texts and references describe the ray tracing procedure in detail [7–34].

10.1 Evaluation of Total or Finite Wavefront Aberration (Monochromatic)

The term 'total' or 'finite' aberration implies the whole of the aberration introduced by the imaging system. It is a function of specific aperture and field coordinates. Each set of aperture and field coordinates corresponds to a particular ray of the image-forming pencil of rays. Therefore, different rays of the pencil have different values of aberration. All measures of aberration—wave aberration, angular aberration, longitudinal and transverse ray aberrations, as elucidated in Chapter 8—are 'total' or 'finite' aberrations.

Kerber [35] first worked out expressions for 'total' or 'finite' aberrations for longitudinal ray aberrations of axial pencils. Koenig and von Rohr [36] presented the 'Kerber difference formula' for evaluating total ray aberrations.. Later on, Conrady [37] gave expressions for 'total' wavefront aberrations of an axial pencil. Hopkins [38] extended the study to the case of meridional rays for an off-axial pencil. Welford [39] presented a review of these early works. A significant contribution for the accurate evaluation of the 'total' or 'finite' transverse ray aberration components is the so called 'Aldis Theorem'; Cox [40], who gave a simple extension of the theorem to include aspheric surfaces, first reported it. Welford [41], and recently Gross [42], underscored the importance of this theorem in the evaluation of total transverse ray aberration components of a system in terms of the transverse ray aberrations by the constituent refracting interfaces. In what follows, we shall deal with evaluation of total or finite wavefront aberration in axi-symmetric lens systems. Key features and concepts will be underscored, leaving aside the detailed algebraic derivations, for which we mostly refer to the original publications.

10.1.1 Wave Aberration by a Single Refracting Interface in Terms of Optical Path Difference (OPD)

In Figure 10.1, AP is a refracting interface between two media of refractive indices n and n′. $\tilde{Q}\tilde{V}$ and QV are any two wavefronts out of a family of geometrical wavefronts in the incident medium. Let this family originate from the same source point, say O. Note that, in general, the point O may not lie in the incident medium of the refracting interface, so the point is not shown in Figure 9.1. Similarly, let $\tilde{Q}'\tilde{V}'$ and Q′V′ be any two wavefronts in the corresponding family of geometrical wavefronts in the refraction medium. Let $\tilde{V}VAV'\tilde{V}'$ and $\tilde{Q}QPQ'\tilde{Q}'$ be two rays orthogonal to the families of wavefronts in the two spaces. Since both the incident and the refraction media are considered homogeneous and isotropic, asphericity of the geometrical wavefronts should preferably be represented in a form so that its value remains unaltered during propagation through a single medium. Consequently, it becomes possible to correlate directly the effect of refraction with the change in asphericity of the family of wavefronts upon refraction.

Let $\tilde{V}VAV'\tilde{V}'$ be considered the reference ray, and $\tilde{Q}QPQ'\tilde{Q}'$ be an arbitrary ray of the same pencil. The two rays intersect at the point S in the incident medium and at the point S′ in the refraction medium.

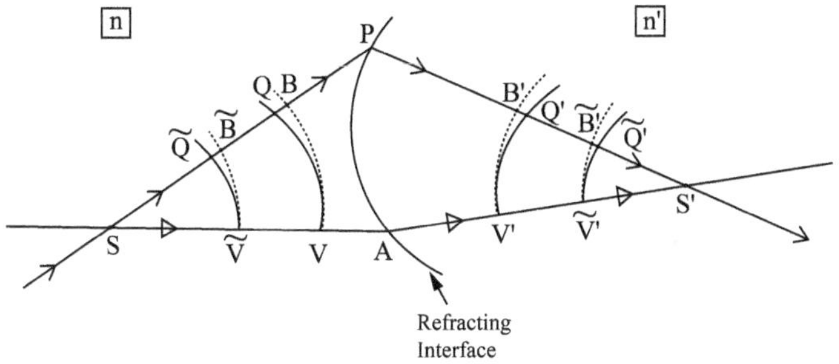

FIGURE 10.1 A refracting interface AP between two homogeneous media of refractive indices n and n'. $S\tilde{V}VAV'\tilde{V}'S'$ is the reference ray and $S\tilde{Q}QPQ'\tilde{Q}'S'$ is an arbitrary ray of the image-forming pencil.

Reference spheres $\tilde{B}\tilde{V}$ and BV are drawn with the centre at S and radii $S\tilde{V}$ and SV, respectively. It is seen that wave aberration W in the aperture ray is given by

$$W = [BQ] = [\tilde{B}\tilde{Q}] \tag{10.1}$$

Therefore, W is independent of the choice of radius of the reference sphere when the centre of the reference sphere is the point of intersection of the aperture ray with the reference ray. Similarly, in the refraction medium, the wave aberration W′ in the corresponding family of geometrical wavefronts in the refraction medium is given by

$$W' = [B'Q'] = [\tilde{B}'Q'] \tag{10.2}$$

By definition of wavefronts, we have

$$[VAV'] = [QPQ'] = [BPB'] - [BQ] + [B'Q'] = [BPB'] - W + W' \tag{10.3}$$

It follows that the wave aberration $\Delta W[= (W' - W)]$ introduced by the refraction at the single interface along the ray $\tilde{Q}QPQ'\tilde{Q}'$ is given by

$$\Delta W = (W' - W) = [VAV'] - [BPB'] = [SAS'] - [SPS'] \tag{10.4}$$

It can be rewritten as

$$\Delta W = \{[SA] - [SP]\} + \{[AS'] - [PS']\} = \{[AS'] - [PS']\} - \{[AS] - [PS]\} \tag{10.5}$$

It follows

$$\Delta W = \Delta\{[AS] - [PS]\} = \Delta\{n(AS - PS)\} \tag{10.6}$$

This is the basic relation for determining wave aberration introduced by a single refracting interface. Here, wave aberration is expressed as an optical path difference (OPD) [43]. For an optical system consisting of multiple refracting interfaces, the total wave aberration of a ray with respect to a reference ray can be obtained by summing up contributions from individual refracting interfaces.

10.1.2 Rays′ Own Focus and Invariant Foci of Skew Rays

Figure 10.2 shows the path of another ray $\tilde{Q}_1\tilde{B}_1Q_1B_1P_1B_1'Q_1'\tilde{B}_1'\tilde{Q}_1'$ of the same pencil. The ray is orthogonal to the wavefronts $\tilde{Q}\tilde{V}, QV, Q'V'$, and $\tilde{Q}'\tilde{V}'$ at $\tilde{Q}, Q, Q'$, and $\tilde{Q}'$ respectively. The ray intersects the reference

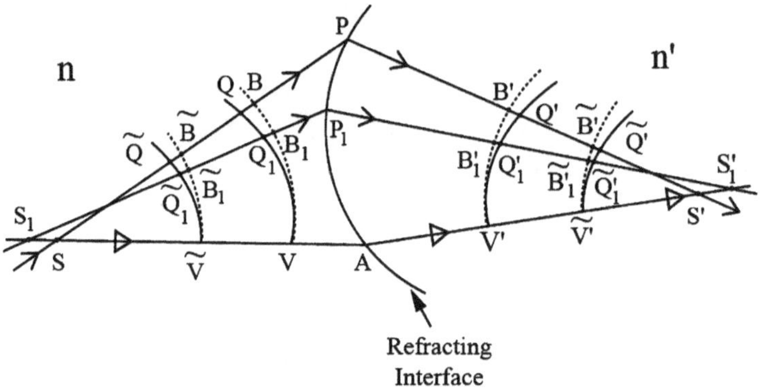

FIGURE 10.2 Path of another ray $S_1\tilde{Q}_1\tilde{B}_1Q_1B_1P_1B_1'Q_1'\tilde{B}_1'\tilde{Q}_1'S_1'$ of the same image-forming pencil as shown in Figure 10.1.

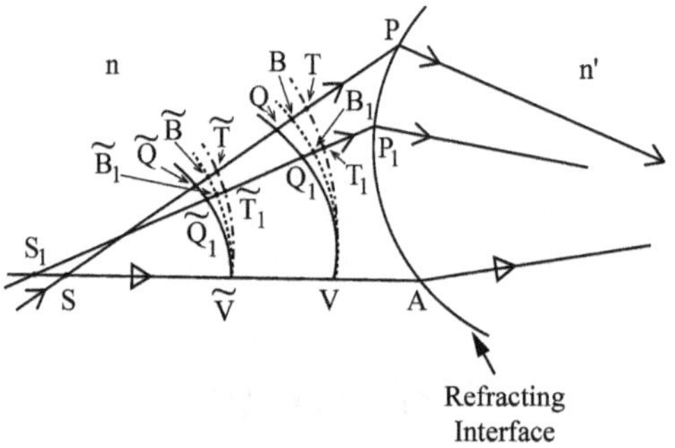

FIGURE 10.3 Reference spheres T_1V and $\tilde{T}_1\tilde{V}$ are concentric with centre at S_1. Reference spheres BB_1V and $\tilde{B}\tilde{B}_1\tilde{V}$ are concentric with centre at S.

spheres $\tilde{B}\tilde{V}, BV$ (both with centres at S), $B'V', \tilde{B}'\tilde{V}'$ (both with centres at S') at the points $\tilde{B}_1, B_1, B_1'$, and $\tilde{B}_1'$, respectively. It is obvious that

$$\tilde{B}_1\tilde{Q}_1 \neq B_1Q_1 \text{ and } Q_1' \neq \tilde{B}_1'\tilde{Q}_1' \tag{10.7}$$

Let the points of intersection of this ray with the reference ray be S_1 in the incident medium, and S_1' in the refraction medium. $\tilde{T}_1\tilde{V}$ and T_1V are two reference spheres drawn with centre at S_1 and radii $S_1\tilde{V}$, and S_1V respectively (Figure 10.3). The points of intersection of the ray $\tilde{Q}_1\tilde{B}_1Q_1B_1P_1$ with the latter reference spheres are $\tilde{T}_1$ and T_1. It is easily seen that $\tilde{T}_1\tilde{Q}_1 = T_1Q_1$. Thus, the aberration of this ray in the incident medium remains invariant with changes in the radius of the reference sphere if the centre of the reference spheres is taken at the point of intersection of the ray with the reference ray. Obviously, the same analysis applies to the part of the ray in the refraction medium. In terms of OPD, the change in wave aberration, ΔW_1 of this ray by the single refraction is

$$\Delta W_1 = \Delta\{[AS_1] - [PS_1]\} = \Delta\{n(AS_1 - PS_1)\} \tag{10.8}$$

In general, the points of intersection of different rays of a pencil with a reference ray are different. The expression for total wave aberration of a ray in an optical system consisting of multiple refracting interfaces as a summation of contribution by the individual interfaces calls for defining wave aberrations of the wavefronts in the different media with respect to the specific point of intersection of the ray with the reference ray in that medium. This point is called 'the ray's own focus' in that medium. In the final image space, aberrations of different rays of an image-forming pencil are measured with respect to a chosen image point. This calls for the addition of an aberration term corresponding to the shift of centre of the reference sphere from the ray's own focus in the image space to the chosen image point.

This treatment of wavefront aberration holds good for axial pencils and rays in the meridional section of an off axial pencil. It fails in the case of skew rays in off axial imagery, for skew rays do not intersect the reference ray that usually lies on the meridional plane in the case of axially symmetric systems. This problem was solved by H.H. Hopkins [44]. He called a point for which the aberration between two rays of a given pencil remains constant as the wave progresses an 'invariant focus' for the two rays. The midpoint of the line of shortest join between two skew rays can be used as a convenient invariant focus. In fact, any point lying on the straight line passing through the midpoints of equally inclined chords between the skew rays can serve the same purpose.

10.1.3 Pupil Exploration by Ray Tracing

In Section 4.1, the role of an aperture stop, and its images in the object space and image space, namely the entrance pupil and the exit pupil, in image formation was put forward. The discussion assumes the paraxial model of image formation so that the outer contour of the image-forming pencil from an axial point passes exactly through the periphery of the aperture stop in the intermediate space, wherever the stop is located, and the periphery of the entrance pupil in the object space, and the periphery of the exit pupil in the image space. In real optics, usually the pupil imagery is aberrated, so that both the entrance pupil and the exit pupil do not have sharp, curved lines for this periphery. In many cases, the real ray corresponding to the paraxial marginal ray from the axial object point may not pass through the system.

Further, in Section 4.2, the phenomenon of 'vignetting' was elucidated. It was shown that, except for the trivial case of imaging by a single lens element with a stop on it, the base of the actual pencil of rays forming image of an off-axial point is not the same as that of the axial pencil. Figures 4.5 and 4.6 illustrate the effect. It is evident that the effective pupil is different for different off-axial points. Note that the discussion of these effects in Chapter 4 assumes paraxial imagery. Obviously, pupil aberrations in real optical systems will modify the phenomenon substantially.

In classical ray-optical theories of aberrations, aberration associated with any aperture ray is described as a relative measure, i.e., the aberration of an individual ray is expressed with respect to that of a reference ray for which the aberration is assumed zero. The choice of the reference ray is to some extent arbitrary, and different treatments have taken recourse to different choices for the reference ray.

A convenient choice is to take the reference ray as the central ray of the image-forming pencil. Often the paraxial pupil ray (PPR) is chosen as the reference ray in ray optical treatment of aberrations. For axial imagery, this choice is okay because the optical axis is the pupil ray for axial object points.

However, this choice seems a bit weird for off axial imagery. Because of vignetting and pupil aberrations, the PPR is, in general, not the central ray of the image-forming pencil, and in many cases the PPR does not even belong to the real image-forming pencil. It will be seen later that the knowledge of wave aberration at the real exit pupil is required for correct evaluation of image quality assessment criteria according to diffraction theory of image formation.

It is now imperative to determine the real pupil for each field point. The determination of this effective pupil for each field point is called 'pupil exploration' [45]. The main functions of pupil exploration for a field point are to determine the effective pupil ray of the pencil of rays that forms the image of the point, and the rim rays that define the corresponding vignetted pupil. The methodology for the same has been described in references [46–49]. The technique for pupil exploration in general optical systems is described in [50]. A brief illustration for axially symmetric systems is given below.

Figure 10.4 shows the object space of an axially symmetric imaging lens system. The object plane is located at point O on the axis. The paraxial entrance pupil is located at point E on the axis. The paraxial pupil ray from the off-axial point Q is QE. Considering QE to be a real ray incident on the system, it is checked whether the ray passes through the system to the image space, or if it fails to pass. Next, the real ray QT_1 in the meridional plane, where ET_1 is a small quantity ε approximately equal to 1 per cent of the diameter of the paraxial entrance pupil, is traced to perform the test. The process is repeated for points T_m along ET_1. Note that $ET_m = m\varepsilon$, where $m = \pm 1, \pm 2, \ldots$, until the upper rim point T_U, and the lower rim point T_L are determined. Taking the Y value of the midpoint $\bar{T}$ of $T_U T_L$, a similar search is conducted along

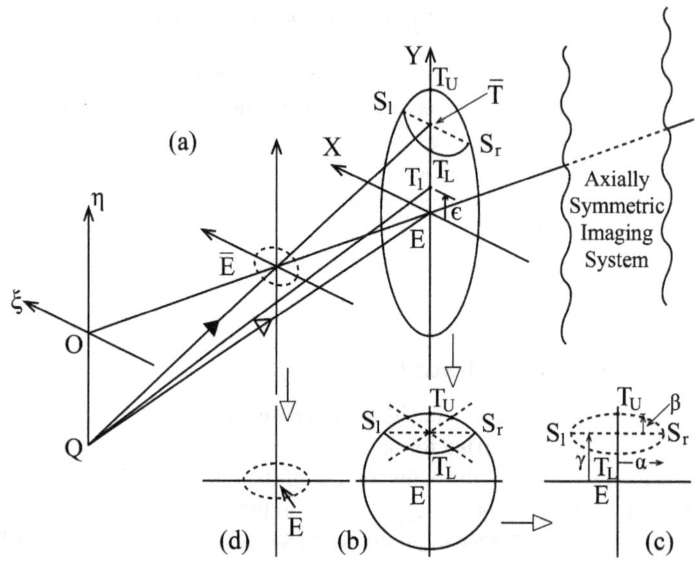

FIGURE 10.4 Pupil exploration for imaging of the point Q. (a) Isometric view of vignetted pupil at the paraxial entrance pupil plane (b) Side view of the eccentric vignetted pupil at the paraxial entrance pupil plane (c) Approximation of the vignetted pupil by an eccentric ellipse (d) Effective real pupil of elliptical contour centered at the point of intersection of the real pupil ray with the optical axis.

the X direction to determine the extreme side rim points. Similarly, the rim points of the effective pupil along the azimuths $\frac{\pi}{4}, \frac{3\pi}{4}$ are determined. A method of bisection can be used to increase the accuracy in the location of rim points by halving the separation of contiguous pass and fail rays until the desired accuracy is obtained.

The meridional ray $Q\bar{T}$ is the real pupil ray. $\bar{E}$, the point of intersection of the real pupil ray with the optical axis, is the axial location of the real entrance pupil for the imaging of the extra-axial object point Q. The projection of the vignetted pupil $T_U S_r T_L S_l$ on the transverse plane at $\bar{E}$ is the effective real entrance pupil. Often, an ellipse can approximate it. Figure 10.4 gives a rough illustration of the process.

With respect to the origin at E, the ellipse, approximating the intersection of the vignetted pencil of rays with the paraxial pupil plane, is represented by

$$\left(\frac{X}{\alpha}\right)^2 + \left(\frac{Y-\gamma}{\beta}\right)^2 = 1 \tag{10.9}$$

where α and β are the semi-major axis and the semi-minor axis of the ellipse, and γ is the vertical shift of the centre of the ellipse from the axial point E. Let the coordinates X, Y be normalized by h, the radius of the paraxial entrance pupil. More correctly h is the height of the point of intersection of the upper tangential edge ray from the axial point O with a reference pupil sphere with O as centre and OE as radius, as discussed in detail in Section 11.1. Equation (10.9) can be rewritten as

$$\left(\frac{x}{A}\right)^2 + \left(\frac{y-C}{B}\right)^2 = 1 \tag{10.10}$$

where

$$\alpha = Ah, \ \beta = Bh \text{ and } \gamma = Ch \tag{10.11}$$

Note that A, B, and C remain determinate, even when the pupil is at infinity.

The ellipse (10.10) is associated with the paraxial pupil sphere passing through E, the axial location of the paraxial pupil sphere. The real pupil sphere is considered to pass through the point $\bar{E}$, the point of intersection of the chosen chief ray of the actual image-forming pencil, corresponding to the extra-axial object point Q, with the optical axis. The ellipse associated with this pupil sphere has its centre at the pupil point $\bar{E}$. However, the lengths of its semi-axes are different from those of the ellipse represented by (10.10).

In terms of un-normalized (X,Y) coordinates, the semi-axes of the displaced ellipse on the paraxial pupil plane are Ah and Bh, whereas the ellipse at the point E has the axes as X_{max}, Y_{max}. Figure 10.5 illustrates the real pupil ray, as well as the edge rays in the sagittal and meridional sections, with the paraxial pupil ray shown in dashed lines. Using the properties of similar triangles in Figures 10.5(a) and 10.5(b), we get the relations

$$\frac{Ah}{X_{max}} = \frac{\bar{R}_0}{\bar{R}} = \frac{(B+C)h - \eta}{Y_{max} - \eta} = \frac{Ch - \eta}{-\eta} \tag{10.12}$$

where η = object height QO and radii of the paraxial and the real pupil spheres are QE= $\bar{R}_0$, and Q$\bar{E}$ = $\bar{R}$, respectively.

If R is the radius of the axial pupil sphere OE, and $\bar{N}_0$ is the Z direction cosine of the paraxial pupil ray QE from the extra-axial point Q, we have $R = \bar{N}_0\bar{R}_0$. From (9.12) the relation between the new and the old ellipses are obtained as

$$\left.\begin{aligned} X_{max} &= (Ah)\bar{N}_0\frac{\bar{R}}{R} \\ Y_{max} &= (Bh)\bar{N}_0\frac{\bar{R}}{R} \end{aligned}\right\} \tag{10.13}$$

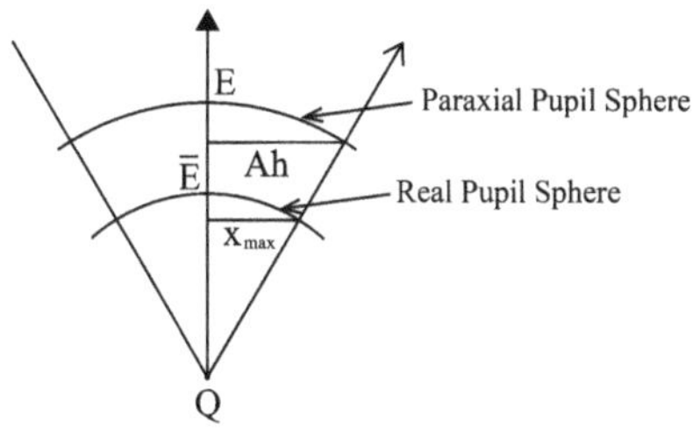

FIGURE 10.5(a) Sagittal section (Plan view) of the image-forming pencil of rays from the extra-axial point Q.

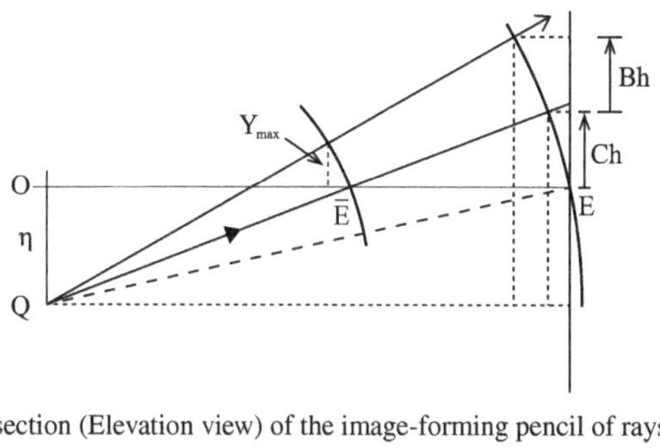

FIGURE 10.5(b) Meridional section (Elevation view) of the image-forming pencil of rays from the extra-axial point Q.

The new ellipse centered at $\bar{E}$ is given by

$$\left(\frac{X}{X_{max}}\right)^2 + \left(\frac{Y}{Y_{max}}\right)^2 = 1 \tag{10.14}$$

The division of X and Y values by their respective maximum values provides a scaling of the ellipse into a unit circle. It forms the basis of canonical pupil variables proposed by H.H. Hopkins (to be discussed in Chapter 11).

10.1.4 A Theorem of Equally Inclined Chords between Two Skew Lines

Equally inclined chords between two nonintersecting skew lines have found interesting applications in tackling problems in aberration theory. In Section 10.1.2, we found their effective use in defining wave aberrations of skew rays. In the following, we underscore a basic theorem of equally inclined chords that is effectively used in the computation of wave aberrations.

In Figure 10.6, two nonintersecting skew rays in a homogeneous medium are shown. Two different colours are used for the two rays to underscore that the rays are nonintersecting. The ray with direction cosines (L_0, M_0, N_0) passes through the point $P_0(X_0, Y_0, Z_0)$, and the ray with direction cosines (L, M, N) passes through the point $P(X, Y, Z)$. Let $P_0\underline{P}$ be a chord equally inclined to the two skew lines. Let the distance $\underline{P}P = e$. The coordinates of the point $\underline{P}$ are: $\{(X - Le), (Y - Me), (Z - Ne)\}$. The length e is thus positive if the displacement $\underline{P}P$ is in the same sense as that implied by the direction cosines (L, M, N). Since $P_0\underline{P}$ is equally inclined to the directions (L, M, N) and (L_0, M_0, N_0), we have

$$\Sigma L\{(X - Le) - X_0\} = \Sigma L_0\{X_0 - (X - Le)\} \tag{10.15}$$

It follows

$$e = \frac{\Sigma(L + L_0)(X - X_0)}{1 + \Sigma LL_0} \tag{10.16}$$

Note that this formula for the distance is independent of location of the origin and the orientation of the coordinate system employed.

10.1.5 Computation of Wave Aberration in an Axi-Symmetric System

In Section 8.3, the wave aberration of an aperture ray is defined to be equal to the optical path length along the aperture ray in the image space, from the point of intersection of the aperture ray with the reference sphere to the point of intersection of the aperture ray with the actual image-forming wavefront in

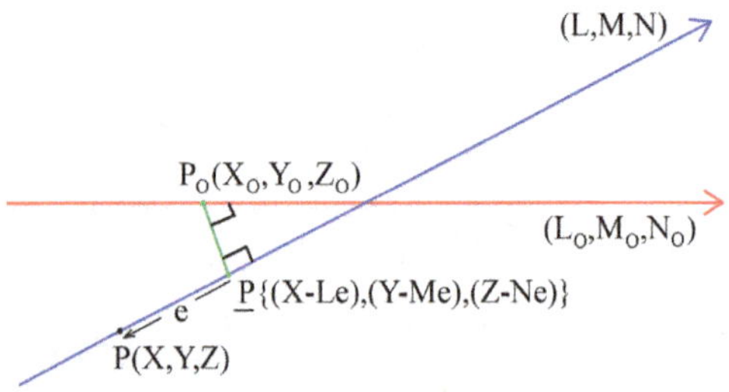

FIGURE 10.6 $P_0\underline{P}$ is an equally inclined chord between two skew lines, one passing through the point $P_0(X_0, Y_0, Z_0)$ and the other passing through the point $P(X, Y, Z)$.

the image space. The reference sphere has its centre at the point of intersection of the pupil ray with the chosen image plane, and its radius is equal to the distance from the centre of the exit pupil to this centre. Subsequently, it is also shown that this optical path length can be conveniently expressed as an OPD between two optical paths. One of the optical paths is from the object point to the centre of the exit pupil along the reference ray, i.e. the principal ray, and the other optical path is along the aperture ray from the object point to its point of intersection with the reference sphere. Hopkins [38, 44] and Welford [39, 51] underscored the problems and pitfalls in evaluation of the wave aberration by direct subtraction of the two optical paths. Some of these limitations can now be overcome by taking recourse to larger byte numbers in computer arithmetic. Nevertheless, straightforward subtraction is not an appropriate procedure for the purpose in view of the remaining pitfalls. In what follows, we present a procedure for the evaluation of wave aberration of axially symmetric systems. The formulae remain determinate for most practical situations, and provide accurate results. This sequence is adapted from the general procedure developed by the optics design group led by Hopkins [44, 52–53]. The key conceptual points highlighting merits of the approach are underscored below; for detailed derivation, we refer the original publications by Hopkins. For convenience, the notations are mostly identical with these publications. Cartesian sign convention is followed for the paraxial parameters.

Figure 10.7 shows an axially symmetric imaging lens system consisting of k refracting interfaces. It forms an image of the transverse object OQ. The paraxial image of OQ is O′Q′. For the sake of convenience, the diagram is shown in two dimensions. The analysis, however, is applicable to skew rays as well. The points $A_s, s = 1,\ldots,k$ are the vertices of the k refracting surfaces. The paraxial entrance pupil and the paraxial exit pupil are located at the points E and E′ on the axis. By pupil exploration, the effective real entrance pupil and real exit pupil are found to be located at the points $\bar{E}$ and $\bar{E}'$, respectively. $\bar{Q}\bar{E}\bar{P}_1\bar{P}_2\ldots\bar{P}_{k-1}\bar{P}_k\bar{E}'\bar{Q}'$ is the effective real pupil/principal/chief ray from the object point $\bar{Q}$. $\bar{Q}BP_1P_2\ldots P_{k-1}P_kB'$ is a general aperture ray of the image-forming pencil of rays from $\bar{Q}$. Note that the pupil/principal/chief ray is on the meridional/tangential plane, but the general aperture ray may be a skew ray. Points B and B′ are the points of intersection of the incident and emergent segments of the aperture ray respectively with the reference spheres $\bar{E}B$ and $\bar{E}'B'$, with centres at $\bar{Q}$ and $\bar{Q}'$, and radii $\bar{Q}\bar{E}$ and $\bar{Q}'\bar{E}'$.

The wavefront aberration W′ for the aperture ray $\bar{Q}BP_1P_2\ldots P_{k-1}P_kB'$ with respect to the reference pupil or principal ray $\bar{Q}\bar{E}\bar{P}_1\bar{P}_2\ldots\bar{P}_{k-1}\bar{P}_k\bar{E}'\bar{Q}'$ is given by the optical path difference (OPD)

$$W' = \left[\bar{E}\bar{P}_1\bar{P}_2\ldots\bar{P}_{k-1}\bar{P}_k\bar{E}'\right] - \left[BP_1P_2\ldots P_{k-1}P_kB'\right] \tag{10.17}$$

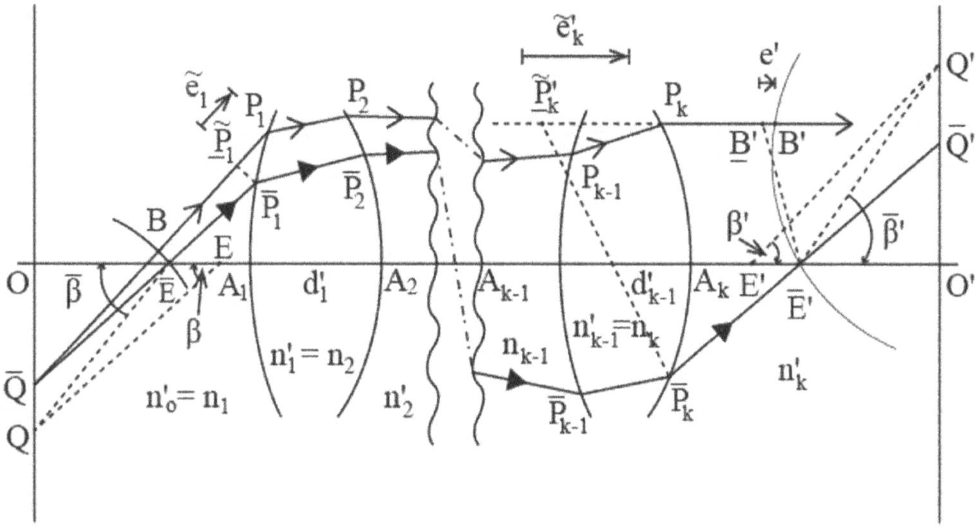

FIGURE 10.7 Calculation of the wavefront aberration W′.

The refractive indices of the different spaces of the optical system are

$$n'_0 = n_1, n'_1 = n_2, \ldots, n'_{k-1} = n_k, n'_k \tag{10.18}$$

Denoting by round brackets the geometrical path lengths, the lengths of the segments of the two rays are

$$\begin{array}{ll} D'_0 = (BP_1) & \bar{D}'_0 = (\bar{E}\bar{P}_1) \\ D'_1 = (P_1P_2) & \bar{D}'_1 = (\bar{P}_1\bar{P}_2) \\ \ldots\ldots\ldots & \ldots\ldots\ldots \\ D'_{k-1} = (P_{k-1}P_k) & \bar{D}'_{k-1} = (\bar{P}_{k-1}\bar{P}_k) \\ D'_k = (P_kB') & \bar{D}'_k = (\bar{P}_k\bar{E}') \end{array} \tag{10.19}$$

Note that the distance segments are directed lengths. Therefore, if the entrance pupil point $\bar{E}$ were in the interior of the system, both $D'_0 = (BP_1)$ and $\bar{D}'_0 = (\bar{E}\bar{P}_1)$ would be negative in sign. Similarly, if the exit pupil point $\bar{E}'$ were in the interior of the system, $D'_K = (P_kB')$ and $\bar{D}'_K = (\bar{P}_k\bar{E}')$ would be negative. Reflection is treated as a special case of refraction. So, after reflection at any surface, with light travelling from right to left, the distance segments D' and $\bar{D}'$ are of negative sign. As per usual convention, all refractive indices are taken to be negative. This continues for successive refractions until there is another reflection that makes light travel from left to right when D' and $\bar{D}'$ become positive, as do the refractive indices as well.

The direction cosines of the segments of the two rays are

$$\begin{array}{ll} BP_1 \rightarrow (L_1, M_1, N_1) & \bar{E}\bar{P}_1 \rightarrow (\bar{L}_1, \bar{M}_1, \bar{N}_1) \\ P_1P_2 \rightarrow (L_2, M_2, N_2) & P_1P_2 \rightarrow (\bar{L}_2, \bar{M}_2, \bar{N}_2) \\ \ldots\ldots\ldots & \ldots\ldots\ldots \\ P_{k-1}P_k \rightarrow (L_k, M_k, N_k) & \bar{P}_{k-1}\bar{P}_k \rightarrow (\bar{L}_k, \bar{M}_k, \bar{N}_k) \\ P_kB' \rightarrow (L'_k, M'_k, N'_k) & \bar{P}_k\bar{E}' \rightarrow (\bar{L}'_k, \bar{M}'_k, \bar{N}'_k) \end{array} \tag{10.20}$$

The coordinates of the points of intersection of the segments of the two rays with the refractive interfaces (with origins at the respective vertices) are denoted by

$$\begin{array}{ll} P_1 \rightarrow (X_1, Y_1, Z_1) & \bar{P}_1 \rightarrow (\bar{X}_1, \bar{Y}_1, \bar{Z}_1) \\ P_2 \rightarrow (X_2, Y_2, Z_2) & \bar{P}_2 \rightarrow (\bar{X}_2, \bar{Y}_2, \bar{Z}_2) \\ \ldots\ldots\ldots & \ldots\ldots\ldots \\ P_{k-1} \rightarrow (X_{k-1}, Y_{k-1}, Z_{k-1}) & \bar{P}_{k-1} \rightarrow (\bar{X}_{k-1}, \bar{Y}_{k-1}, \bar{Z}_{k-1}) \\ P_k \rightarrow (X_k, Y_k, Z_k) & \bar{P}_k \rightarrow (\bar{X}_k, \bar{Y}_k, \bar{Z}_k) \end{array} \tag{10.21}$$

With the above notations, the wavefront aberration (9.8) is given by

$$W' = \sum_{s=0}^{k} n'_s \bar{D}'_s - \sum_{s=0}^{k} n'_s D'_s \tag{10.22}$$

This may be rewritten as

$$W' = n_0'\left(\bar{D}_0' - D_0'\right) + \sum_{s=1}^{k-1} n_s'\bar{D}_s' - \sum_{s=1}^{k-1} n_s' D_s' + n_k'\left(\bar{D}_k' - D_k'\right) \tag{10.23}$$

Note that the summations, $\sum_{s=1}^{k-1} n_s'\bar{D}_s'$ and $\sum_{s=1}^{k-1} n_s' D_s'$ are the optical path lengths, $[\bar{P}_1\bar{P}_2 \ldots \bar{P}_{k-1}\bar{P}_k]$ and $[P_1P_2 \ldots P_{k-1}P_k]$ of the pupil ray and aperture ray, respectively, in the interior of the optical system. The first and the last term on the right-hand side correspond to optical path differences between the two rays in the initial object and final image spaces, respectively. The notation for axial distance between two vertices is $d_s' = (A_sA_{s+1})$. The equation (10.23) can be recast as

$$W' = n_0'\left(\bar{D}_0' - D_0'\right) + \bar{P} - P + n_k'\left(\bar{D}_k' - D_k'\right) \tag{10.24}$$

where

$$\bar{P} = \sum_{s=1}^{k-1} n_s'\left(\bar{D}_s' - d_s'\right) \tag{10.25}$$

$$P = \sum_{s=1}^{k-1} n_s'\left(D_s' - d_s'\right) \tag{10.26}$$

Numerical evaluation of the four terms in the right-hand side of equation (10.24) are described below. It is reiterated that the first term takes care of the ray paths in the initial object space from the object to the first refracting interface, and the last term does the same for the final image space from the last refracting interface to the image. The second and the third terms take care of the ray paths from the first refracting interface to the last refracting interface.

First term:
In the object space, a chord $\bar{P}_1\tilde{\underline{P}}_1$, equally inclined to the aperture ray $B\tilde{\underline{P}}_1P_1$ and the principal ray $\bar{E}\bar{P}_1$, and passing through $\bar{P}_1$, the point of intersection of the principal ray and the first refracting interface is drawn (figure 10.7). Since $\left(\bar{E}\bar{P}_1\right) = \left(B\tilde{\underline{P}}_1\right)$, we get

$$n_0'\left(\bar{D}_0' - D_0'\right) = n_0'\left\{\left(\bar{E}\bar{P}_1\right) - \left(BP_1\right)\right\} = -n_0'\left(\tilde{\underline{P}}_1P_1\right) = -n_1\tilde{e}_1 \tag{10.27}$$

where $\tilde{e}_1 = \left(\tilde{\underline{P}}_1P_1\right)$. In this symbol, a 'tilde' is used on 'ẽ' or $\tilde{\underline{P}}_1$ to underscore that the chord $\bar{P}_1\tilde{\underline{P}}_1$ used for this purpose is equally inclined to the aperture ray and the principal ray.

$\tilde{e}_1$ is calculated by using the theorem of equally inclined chords enunciated in Section 10.1.4. Substituting

$$(L, M, N) \leftarrow (L_1, M_1, N_1) \tag{10.28}$$

$$(L_0, M_0, N_0) \leftarrow (\bar{L}_1, \bar{M}_1, \bar{N}_1) \tag{10.29}$$

$$(X, Y, Z) \leftarrow (X_1, Y_1, Z_1) \tag{10.30}$$

$$(X_0, Y_0, Z_0) \leftarrow (\bar{X}_1, \bar{Y}_1, \bar{Z}_1) \tag{10.31}$$

in equation (10.16), we get the following expression for $\tilde{e}_1$.

$$\tilde{e}_1 = \frac{\Sigma\left(L_1 + \bar{L}_1\right)\left(X_1 - \bar{X}_1\right)}{1 + \Sigma L_1\bar{L}_1} \tag{10.32}$$

Last term:
In the image space, two chords $P_k\underline{\tilde{P}}'_k$ and $\bar{E}'\bar{B}'$, equally inclined to the aperture ray P_kB', and the principal ray $\bar{P}_k\bar{E}'$ are drawn. The chords pass through $\bar{P}_k$, the point of intersection of the principal ray and the last refracting interface, and $\bar{E}'$ respectively. Since $\left(\bar{P}_k\bar{E}'\right) = \left(\underline{\tilde{P}}'_k\bar{B}'\right)$, it follows

$$n'_k\left(\bar{D}'_k - D'_k\right) = n'_k\left(\bar{P}_k\bar{E}' - P_kB'\right) = n'_k\left(\bar{P}_k\bar{E}' - \underline{\tilde{P}}'_k\bar{B}' + \underline{\tilde{P}}'_kP_k - \underline{B}'B'\right) = n'_k\tilde{e}'_k - n'_ke' \tag{10.33}$$

where $\tilde{e}'_k = \left(\underline{\tilde{P}}'_kP_k\right)$ and $e' = \left(\underline{B}'B'\right)$.

Substituting

$$\left(L, M, N\right) \leftarrow \left(L'_k, M'_k, N'_k\right) \tag{10.34}$$

$$\left(L_0, M_0, N_0\right) \leftarrow \left(\bar{L}'_k, \bar{M}'_k, \bar{N}'_k\right) \tag{10.35}$$

$$\left(X, Y, Z\right) \leftarrow \left(X_k, Y_k, Z_k\right) \tag{10.36}$$

$$\left(X_0, Y_0, Z_0\right) \leftarrow \left(\bar{X}_k, \bar{Y}_k, \bar{Z}_k\right) \tag{10.37}$$

in equation (10.16), we get the following expression for $\tilde{e}'_k$.

$$\tilde{e}'_k = \frac{\Sigma\left(L'_k + \bar{L}'_k\right)\left(X_k - \bar{X}_k\right)}{1 + \Sigma L'_k\bar{L}'_k} \tag{10.38}$$

Evaluation of the remaining part, i.e., n'_ke', will be taken up after dealing with the second and third terms. This part corresponds to the focal shift from the invariant focus to the foal point $\bar{Q}'$

Third term:
A typical term of P, defined by (9.18), is given by $\left\{n'_s\left(D'_s - d'_s\right)\right\}$. Figure 10.8 illustrates the case for the $(s+1)$th medium. Two equally inclined chords, $A_s\underline{P}'_s$ and $A_{s+1}\underline{P}_{s+1}$ are drawn between the optical axis and the segment P_sP_{s+1} (extended, when necessary) of the skew aperture ray in the $(s+1)$th medium. Let $A_s\underline{P}_s$ be an equally inclined chord between the optical axis and the segment of the same aperture ray in the sth medium. We define

$$\underline{P}_sP_s = e_s \qquad \underline{P}'_sP_s = e'_s \qquad \underline{P}_{s+1}P_{s+1} = e_{s+1} \tag{10.39}$$

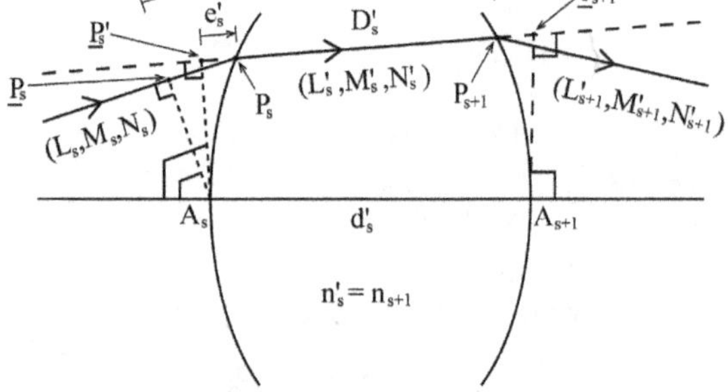

FIGURE 10.8 $A_S\underline{P}'_s$ and $A_{S+1}\underline{P}_{s+1}$ are two equally inclined chords between the axis and the skew ray P_SP_{S+1} (extended both ways).

Note that the symbol 'tilde' is now omitted from the top of e_s, e'_s, e_{s+1}, or $\underline{P}_s$, $\underline{P}'_s$, $\underline{P}_{s+1}$, for the relevant equally inclined chords are drawn between the aperture ray and the optical axis.

Substituting in equation (10.16) $L_0 = M_0 = 0, N_0 = 1$ for the optical axis, $X_0 = Y_0 = Z_0 = 0$ (considering A_s as the origin), $(L,M,N) \leftarrow (L_s, M_s, N_s)$ for the aperture ray in the sth medium, and $(X,Y,Z) \leftarrow (X_s, Y_s, Z_s)$ for the point P_s, we get

$$\underline{P}_s P_s = e_s = \frac{L_s X_s + M_s Y_s + (1+N_s) Z_s}{1+N_s} \tag{10.40}$$

Similarly, by substituting in equation (10.16), $L_0 = M_0 = 0, N_0 = 1$, $X_0 = Y_0 = Z_0 = 0$, and $(L,M,N) \leftarrow (L'_s, M'_s, N'_s)$ for the aperture ray in the $(s+1)$th medium, and $(X,Y,Z) \leftarrow (X_s, Y_s, Z_s)$ for the point P_s, we get

$$\underline{P}'_s P_s = e'_s = \frac{L'_s X_s + M'_s Y_s + (1+N'_s) Z_s}{1+N'_s} \tag{10.41}$$

Figure 10.9 shows the path of the skew aperture ray from its incidence on the first refracting interface at P_1 to its emergence from the kth refracting interface at P_k. The quantity P, defined in (10.26), is given by

$$\begin{aligned} P = [P_1 \ldots P_k] - [A_1 \ldots A_k] &= \sum_{s=1}^{k-1} n'_s (D'_s - d'_s) \\ &= n'_1 (D'_1 - d'_1) + n'_2 (D'_2 - d'_2) + \ldots + n'_{k-1} (D'_{k-1} - d'_{k-1}) \\ &= -n'_1 e'_1 + n'_1 e_2 - n'_2 e'_2 + n'_2 e_3 + \ldots - n'_{k-1} e'_{k-1} + n'_{k-1} e_k \\ &= -n'_1 e'_1 + (n_2 e_2 - n'_2 e'_2) + \ldots + (n_{k-1} e_{k-1} - n'_{k-1} e'_{k-1}) + n_k e_k \end{aligned} \tag{10.42}$$

P can be represented as

$$P = -n'_1 e'_1 - \sum_{s=2}^{k-1} (n'_s e'_s - n_s e_s) + n_k e_k \tag{10.43}$$

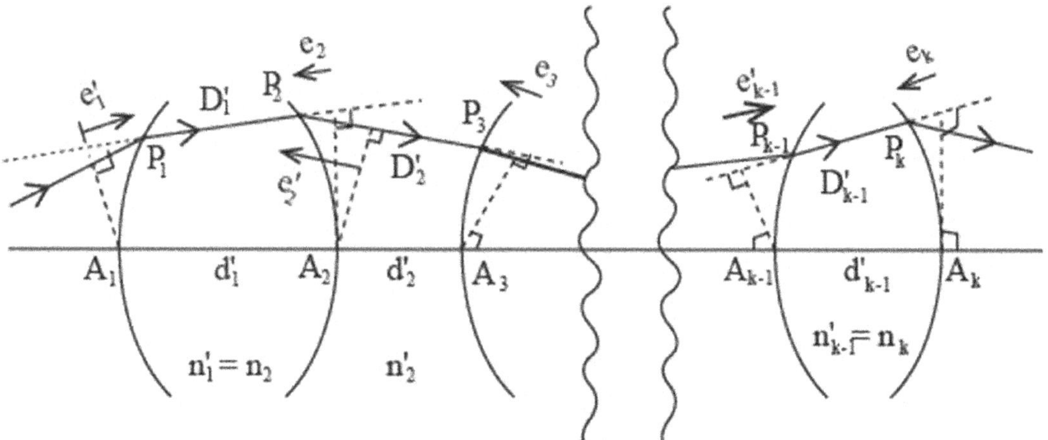

FIGURE 10.9 Passage of a skew aperture ray in an axially symmetric system consisting of k refracting interfaces.

Second Term:
The second term in equation (10.24) consists of $\bar{P}$, which is expressed in (10.25) as

$$\bar{P} = \sum_{s=1}^{k-1} n'_s \left(\bar{D}'_s - d'_s \right) \tag{10.44}$$

This equation can be recast in the following form by using corresponding quantities for the principal ray instead of the aperture ray in (10.43)

$$\bar{P} = -n'_1 \bar{e}'_s - \sum_{s=2}^{k-1} \left(n'_s \bar{e}'_s - n_s \bar{e}_s \right) + n_k \bar{e}_k \tag{10.45}$$

Evaluation of e′:
Figure 10.10 shows a skew aperture ray $\underline{\tilde{P}}'_k \bar{B}'B'M'$ and the principal ray $\bar{P}_k\bar{E}'\bar{Q}'$ in the image space. The principal ray lies on the meridional plane, and it intersects the image plane at $\bar{Q}'$. The latter is taken as the real image point corresponding to the object point $\bar{Q}$ (Figure 10.7). The height $\bar{Q}'O' = \eta'_k$. The paraxial image Q′ of the object point Q at the edge of the field is at height $Q'O' = \eta'$. The reference sphere $\bar{E}'B'$ has its centre at $\bar{Q}'$, and radius $\bar{E}'\bar{Q}' = \bar{R}'$, where $\bar{E}'$ is the centre of the real exit pupil, and it lies on the axis. The aperture ray intersects the reference sphere at B′. A chord equally inclined to the principal ray and the aperture ray is drawn through $\bar{E}'$. The chord intersects the aperture ray at the point $\underline{B}'$. The geometrical path length $\underline{B}'B' = e'$.

The direction cosines of the principal ray $P_k\bar{Q}'$ in the image space are $\left(0, \bar{M}'_k, \bar{N}'_k\right)$. The coordinates of the centre $\bar{Q}'$ of the reference sphere of radius $\bar{E}'\bar{Q}' = \bar{R}'$ is $\left(0, \bar{M}'_k\bar{R}', \bar{N}'_k\bar{R}'\right)$. With its origin at the real exit pupil point $\bar{E}'$, let the coordinates of points $\underline{B}'$ and B′ on the aperture ray be $\left(\underline{X}', \underline{Y}', \underline{Z}'\right)$ and (X, Y, Z), respectively. They are related by

$$X' = \underline{X}' + L'_k e' \tag{10.46}$$

$$Y' = \underline{Y}' + M'_k e' \tag{10.47}$$

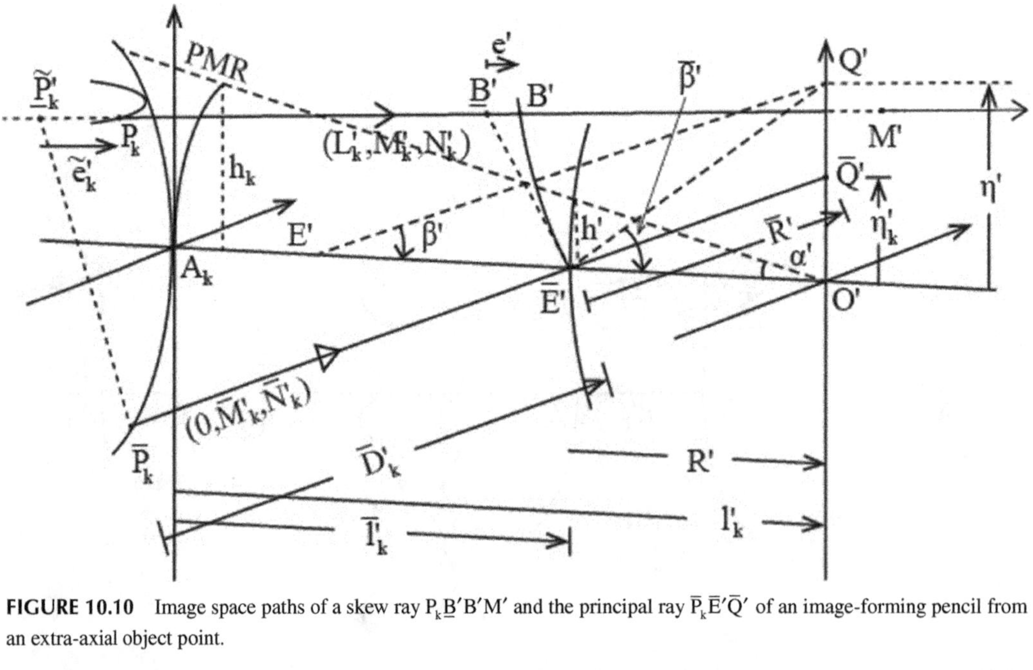

FIGURE 10.10 Image space paths of a skew ray $P_k\underline{B}'B'M'$ and the principal ray $\bar{P}_k\bar{E}'\bar{Q}'$ of an image-forming pencil from an extra-axial object point.

$$Z' = \underline{Z}' + N'_k e' \tag{10.48}$$

Equation of the reference sphere $\bar{E}'B'$ is

$$X'^2 + \left(Y' - \bar{M}'_k \bar{R}'\right)^2 + \left(Z' - \bar{N}'_k \bar{R}'\right)^2 = \bar{R}'^2 \tag{10.49}$$

Substituting from (10.46) – (10.48) in (10.49) we get

$$e'^2 - 2\left(\bar{R}'F\right)e' + \left(\bar{R}'J\right) = 0 \tag{10.50}$$

where

$$F = M'_k \bar{M}'_k + N'_k \bar{N}'_k - \frac{\underline{X}'L'_k + \underline{Y}'M'_k + \underline{Z}'N'_k}{\bar{R}'} \tag{10.51}$$

$$J = \frac{1}{\bar{R}'}\left(\underline{X}'^2 + \underline{Y}'^2 + \underline{Z}'^2\right) - 2\left(\underline{Y}'\bar{M}'_k + \underline{Z}'\bar{N}'_k\right) \tag{10.52}$$

The roots of (10.50) are

$$e' = \left(\bar{R}'F\right) \pm \sqrt{\left\{\left(\bar{R}'F\right)^2 - \left(\bar{R}'J\right)\right\}} \tag{10.53}$$

Note that, for the principal ray, the points $\bar{B}'$ and B' coincide with the point $\bar{E}'$, which is the origin of the rectangular coordinate system. Therefore, $\underline{X}' = \underline{Y}' = \underline{Z}' = 0$, and so, from (9.46), $J = 0$. In this special case, by definition, $e' = 0$. This implies that the negative sign in (9.50) needs to be chosen, so that

$$e' = \left(\bar{R}'F\right) - \sqrt{\left\{\left(\bar{R}'F\right)^2 - \left(\bar{R}'J\right)\right\}} \tag{10.54}$$

This relation can be simplified further to obtain

$$e' = \frac{\left(\bar{R}'J\right)}{\left(\bar{R}'F\right) + \sqrt{\left\{\left(\bar{R}'F\right)^2 - \left(\bar{R}'J\right)\right\}}} = \frac{J}{F + \sqrt{\left\{F^2 - \left(J/\bar{R}'\right)\right\}}} \tag{10.55}$$

Figure 10.11 shows the paraxial marginal ray (PMR), the paraxial pupil ray (PPR) $E'Q'$, and the real pupil ray $\bar{P}_k\bar{E}'\bar{Q}'$ in the image space. The convergence angles of the PMR and the PPR are α' and β', so that the paraxial angle variables in the image space are $u'_k = \sin\alpha'$ and $\bar{u}'_k = \tan\beta'$. Let $A_kE' = \hat{\bar{l}}'_k, A_k\bar{E}' = \bar{l}'_k$, and $A_kO' = l'_k$. The expression for paraxial invariant is

$$H' = n'_k \sin\alpha'\eta' = n'_k u'_k\left(\hat{\bar{l}}'_k - l'_k\right)\tan\beta' = n'_k\left[u'_k\hat{\bar{l}}'_k\bar{u}'_k - u'_k l'_k \bar{u}'_k\right] = n'_k\left[\bar{h}_k u'_k - h_k \bar{u}'_k\right] \tag{10.56}$$

On account of pupil aberrations and vignetting, the real pupil ray intersects the optical axis at $\bar{E}'$, which is not the same point as the paraxial exit pupil point E'. Effectively, this implies a shift of the stop from E' to $\bar{E}'$. Since the paraxial invariant remains unchanged by this shift, effectively the paraxial pupil ray is also changed from $E'Q'$ to $\bar{E}'Q'$. Let $\angle Q'\bar{E}'O' = \bar{\beta}'$. The convergence angle of the effective paraxial pupil ray is $\bar{u}'_k = \tan\bar{\beta}'$, and at the plane at $\bar{E}', \bar{h}_k = 0$, and $h_k = h'$. From (10.56) we get

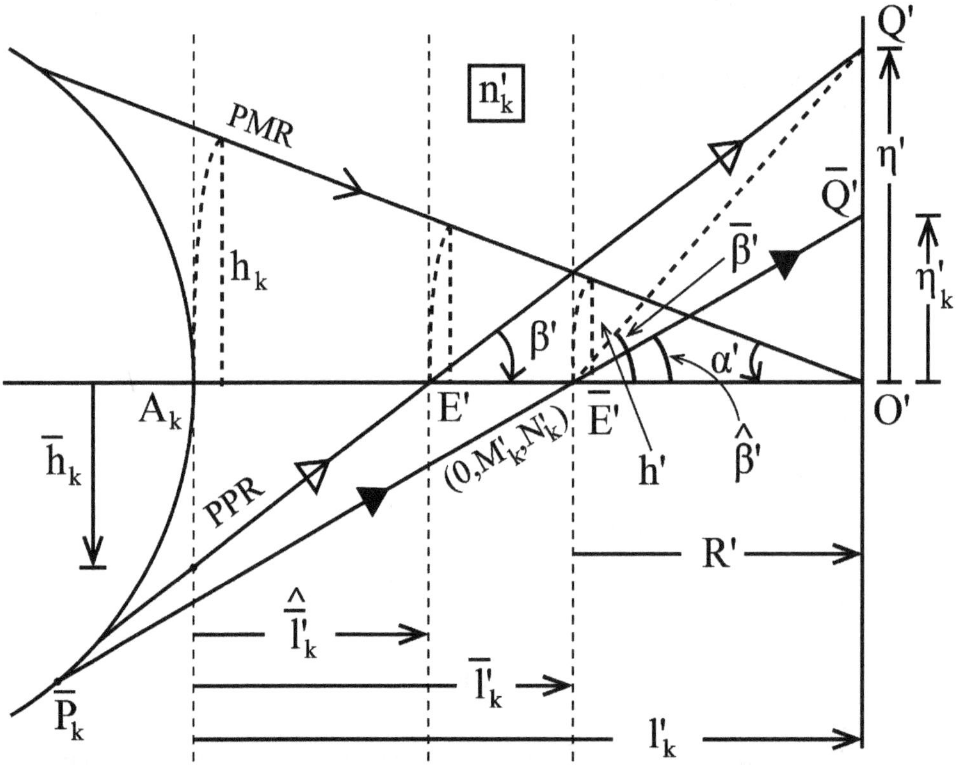

FIGURE 10.11 Image space paths of the paraxial marginal ray (PMR), the paraxial pupil ray (PPR), and the real pupil ray (with bold triangles) of an image-forming pencil from an extra-axial object point.

$$H' = -n'_k h' \bar{u}'_k \tag{10.57}$$

so that

$$h' = -\frac{H'}{n'_k \, \bar{u}'_k} \tag{10.58}$$

Using the above equation and the relation $h' = u'_k R'$, the radius $\underline{R}'$ can be written as

$$\bar{R}' = \frac{R'}{\bar{N}'_k} = \frac{h'}{\bar{N}'_k u'_k} = -\frac{H'}{n'_k u'_k \bar{u}'_k \bar{N}'_k} \tag{10.59}$$

From (10.55) we get the following relation for e'

$$e' = \frac{J}{F + \sqrt{\left\{F^2 + \left(n'_k u'_k \bar{u}'_k \bar{N}'_k / H'\right)J\right\}}} \tag{10.60}$$

Determinate forms for F and J in terms of image space parameters are given below.

$$F = \left(M'_k \bar{M}'_k + N'_k \bar{N}'_k\right) + \bar{N}'_k u'_k \left(\bar{M}'_k \underline{y}' + \bar{N}'_k \underline{z}'\right) \tag{10.61}$$

$$J = \bar{N}'_k u'_k \left[\underline{X}'_k \underline{x}' + \left(\underline{Y}'_k - \bar{Y}_k\right)\underline{y}' + \left(\underline{Z}'_k - \bar{Z}_k\right)\underline{z}'\right] - 2\left(h_k - u'_k \bar{Z}_k\right)\left(\bar{M}'_k \underline{y}' + \bar{N}'_k \underline{z}'\right) \tag{10.62}$$

where $(\underline{X}'_k, \underline{Y}'_k, \underline{Z}'_k)$ and $(0, \bar{Y}_k, \bar{Z}_k)$ are coordinates of the points $\underline{\tilde{P}}'_k$ and $\bar{P}_k$ respectively (Figure 10.10). The direction cosines of the aperture ray $\underline{\tilde{P}}'_k P_k \bar{B}' B' M'$ and the real principal ray $\bar{P}_k \bar{E}' \bar{Q}'$ are (L'_k, M'_k, N'_k) and $(0, \bar{M}'_k, \bar{N}'_k)$, respectively.

The coordinates of the points $P_k(X_k, Y_k, Z_k), \bar{P}_k(0, \bar{Y}_k, \bar{Z}_k)$ and the direction cosines of the aperture ray and the real pupil ray in the image space are obtained by finite ray trace. Using the relation (10.38) for $\tilde{e}'_k$, together with the direction cosines of the aperture ray in the image space, and the coordinates of P_k, we get

$$\underline{X}'_k = X_k - L'_k \tilde{e}'_k \qquad \underline{Y}'_k = Y_k - M'_k \tilde{e}'_k \qquad \underline{Z}'_k = Z_k - N'_k \tilde{e}'_k \tag{10.63}$$

$(\underline{x}', \underline{y}', \underline{z}')$ are the reduced coordinates

$$\underline{x}' = \frac{\underline{X}'}{h'} \qquad \underline{y}' = \frac{\underline{Y}'}{h'} \qquad \underline{z}' = \frac{\underline{Z}'}{h'} \tag{10.64}$$

where $(\underline{X}', \underline{Y}', \underline{Z}')$ are the coordinates of the point $\underline{B}'$. Denoting the distances $(\bar{P}_k \bar{E}') = (\underline{\tilde{P}}'_k \underline{B}')$ by $\bar{D}'_k$, and $\tilde{e}'_k = (\underline{\tilde{P}}'_k P_k)$, we get

$$\underline{X}' = X_k + L'_k(-\tilde{e}'_k + \bar{D}'_k) \tag{10.65}$$

$$\underline{Y}' = Y_k + M'_k(-\tilde{e}'_k + \bar{D}'_k) \tag{10.66}$$

Using the relation (10.58) for h', and the relation $\bar{D}'_k = (\bar{l}'_k - \bar{Z}_k)/\bar{N}'_k$, it follows

$$\underline{x}' = -\left(\frac{n'_k}{H'\bar{N}'_k}\right)\left\{\bar{u}'_k\left[\bar{N}'_k(X_k - L'_k\tilde{e}'_k) - L'_k\bar{Z}_k\right] + \bar{h}_k L'_k\right\} \tag{10.67}$$

$$\underline{y}' = -\left(\frac{n'_k}{H'\bar{N}'_k}\right)\left\{\bar{u}'_k\left[\bar{N}'_k(Y_k - M'_k\tilde{e}'_k) - M'_k\bar{Z}_k\right] + \bar{h}_k M'_k\right\} \tag{10.68}$$

The chord $\bar{E}'\underline{B}'$ is equally inclined to the ray directions (L'_k, M'_k, N'_k) and $(0, \bar{M}'_k, \bar{N}'_k)$. We use the relation (10.64) to obtain $\bar{z}'$. Considering points P_0, P (Figure 10.6) as points $\bar{E}', \underline{B}'$ (Figure 10.9), respectively, we substitute $(L_0, M_0, N_0) \leftarrow (0, \bar{M}'_k, \bar{N}'_k)$, $(X_0, Y_0, Z_0) \leftarrow (0,0,0)$, and $(L, M, N) \leftarrow (L'_k, M'_k, N'_k)$ with $(X, Y, Z) \leftarrow (\underline{X}', \underline{Y}', \underline{Z}')$. Since $\bar{E}'\underline{B}'$ is an equally inclined chord, the corresponding value of e in (10.16) must be zero. It follows

$$\Sigma(L'_k + \bar{L}'_k)\underline{X}' = L'_k\underline{X}' + (M'_k + \bar{M}'_k)\underline{Y}' + (N'_k + \bar{N}'_k)\underline{Z}' = 0 \tag{10.69}$$

since $\bar{L}'_k = 0$. On division by h', (10.69) yields the following relation for $\bar{z}'$

$$\bar{z}' = -\frac{L'_k\bar{x}' + (M'_k + \bar{M}'_k)\bar{y}'}{(N'_k + \bar{N}'_k)} \tag{10.70}$$

h' is the normalizing radius given by relation (10.58).

The convergence angle of the paraxial marginal ray in the final image space is u'_k, and its height on the k-th surface is h_k. In the case of axial imaging by a rotationally symmetric system, the paraxial exit pupil is taken as the effective exit pupil, and the paraxial principal ray calculations provide $(\bar{u}'_k, h_k)$, which can be used in the above calculations. For extra-axial imagery the effective values for $\bar{u}'_k, h_k$ are dependent on the fractional field parameter τ, and they are denoted by $\bar{u}'_k(\tau)$ and $h_k(\tau)$. The paraxial principal ray quantities may be considered as a special case when $\tau = 0$, and are denoted by $\bar{u}'_k(0)$ and $h_k(0)$. The corresponding quantities in the extra-axial case need to be determined specifically for each value of τ. In Figure 10.11,

$$\angle Q'E'O' = \beta', \qquad \angle Q'\bar{E}'O' = \bar{\beta}', \qquad \angle \bar{Q}'\bar{E}'O' = \hat{\beta}' \tag{10.71}$$

Let normalized heights H' and $\bar{H}'_0$ be defined as

$$H' = (n'_k \sin\alpha')\eta' \qquad \bar{H}'_0 = (n'_k \sin\alpha')\eta'_k \tag{10.72}$$

where $\eta' = Q'O'$, and $\eta'_k = \bar{Q}'O'$. As mentioned earlier, $\bar{u}'_k$ is given by

$$\bar{u}'_k(\tau) = \tan\bar{\beta}' = \frac{\eta'}{R'} = \left(\frac{\eta'}{\eta'_k}\right)\left(\frac{\eta'_k}{R'}\right) = \left(\frac{H'}{\bar{H}'_0}\right)\tan\hat{\beta}' \tag{10.73}$$

where

$$\tan\hat{\beta}' = \frac{\sin\hat{\beta}'}{\cos\hat{\beta}'} = \frac{-\bar{M}'_k}{\bar{N}'_k} \tag{10.74}$$

It follows

$$\bar{u}'_k(\tau) = -\left(\frac{H'}{\bar{H}'_0}\right)\frac{\bar{M}'_k}{\bar{N}'_k} \tag{10.75}$$

$\bar{h}_k$ is given by

$$\bar{h}_k(\tau) = \bar{l}'_k\bar{u}'_k(\tau) = \left(\frac{H'}{\bar{H}'_0}\right)\left\{\frac{-\bar{l}'_k\bar{M}'_k}{\bar{N}'_k}\right\} = \left(\frac{H'}{\bar{H}'_0}\right)\left\{\frac{\bar{N}'_k\bar{Y}_k - \bar{M}'_k\bar{Z}_k}{\bar{N}'_k}\right\} \tag{10.76}$$

In computation of e' by formula (10.60) for a specific τ, the value of $\bar{u}'_k(\tau)$ as given by (10.73) needs to be employed.

10.1.6 Computation of Transverse Ray Aberrations in an Axi-Symmetric System

Figure 10.12 illustrates the passages of the real pupil ray $\bar{P}_k\bar{Q}'$, and a skew aperture ray of the image-forming pencil, forming the image of an extra-axial object point Q (not shown), in the final image space. The paraxial image of the object point Q is at Q'. The point of intersection, $\bar{Q}'$, of the real pupil ray with the image plane at O' is considered to be the real image point. The skew aperture ray intersects the image plane at M'. With reference to the origin O' of a rectangular coordinate system, the coordinates of the points $\bar{Q}'$ and M' are $(0, \bar{\eta}'_k)$ and $(\delta\xi', \bar{\eta}'_k + \delta\eta')$. The components of transverse aberrations of the aperture ray are $\{\delta\xi', \delta\eta'\}$. Coordinates of the points P_k and $\bar{P}_k$ are (X_k, Y_k, Z_k) and $(0, \bar{Y}_k, \bar{Z}_k)$, respectively.

The equation of the line P_kM' is

$$\frac{\delta\xi' - X_k}{L'_k} = \frac{\bar{\eta}'_k + \delta\eta' - Y_k}{M'_k} = \frac{l'_k - Z_k}{N'_k} \tag{10.77}$$

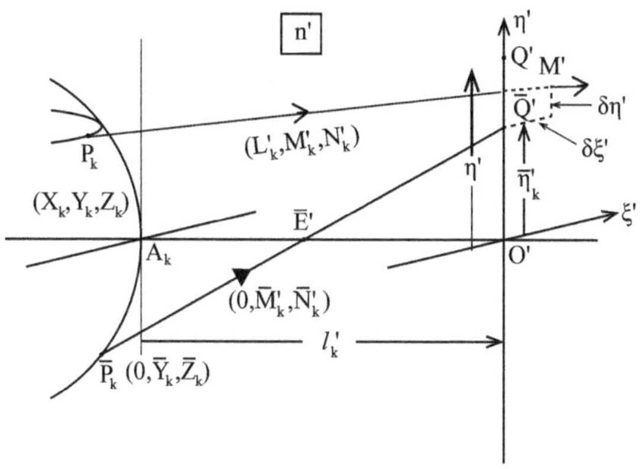

FIGURE 10.12 Transverse ray aberrations $\delta\xi', \delta\eta'$ of the aperture ray P_kM'.

Similarly, the equation of the line $\bar{P}_k\bar{Q}$ is

$$\frac{\bar{\eta}'_k - \bar{Y}_k}{\bar{M}'_k} = \frac{\bar{l}'_k - \bar{Z}_k}{\bar{N}'_k} \tag{10.78}$$

From (10.77) – (10.78), it follows

$$\delta\xi' = X_k + \frac{L'_k}{N'_k}(l'_k - Z_k) \tag{10.79}$$

$$\delta\eta' = Y_k + \frac{M'_k}{N'_k}(l'_k - Z_k) - \bar{\eta}'_k \tag{10.80}$$

where

$$\bar{\eta}'_k = \bar{Y}_k + \frac{\bar{M}'_k}{\bar{N}'_k}(l'_k - \bar{Z}'_k) \tag{10.81}$$

10.2 Measures for Nonparaxial Chromatism

Earlier, in Section 9.3, two measures for paraxial chromatism, namely axial colour or longitudinal chromatic aberration, and lateral colour or transverse chromatic aberration, are discussed. They essentially deal with variation in paraxial characteristics of optical imaging systems with changes in working wavelength. As such there is no ambiguity in their definition. It is not so in the case of nonparaxial chromatism. In the latter case, the chromatic aberration is defined as the difference in aberration between a second wavelength λ and a mean wavelength $\bar{\lambda}$. Measure for this chromatic aberration is obtained by two methods, which calculate slightly different quantities. 'The first, Conrady's differential technique, finds essentially the derivative at $\lambda = \bar{\lambda}$ of the wavefront aberration with respect to wavelength. In the second, the wavefront aberration for the second wavelength is found by separate ray tracing using the appropriate refractive indices' [52]. The two methods are applicable in both axial and extra-axial imagery for determining chromatic wavefront aberration. For the case of axial imagery, the multiple ray tracing approach has traditionally been used to determine the chromatic aberration as a longitudinal aberration, and the effect is called spherochromatism [54–56].

10.2.1 Spherochromatism

Figure 10.13 shows the imaging of an axial object point O by an axi-symmetric imaging lens system. O'_F, O'_C, and O'_d are the paraxial images of the point O at F, C, and d wavelengths, respectively. In the object space, OB is the common ray path for the marginal rays at F, C, and d wavelengths, as well as at all other wavelengths over the range. In the image space, the corresponding marginal rays at F, C, and d wavelengths intersect the optical axis at ${}_mO'_F$, ${}_mO'_C$, and ${}_mO'_d$, respectively. The latter points are the marginal foci at the corresponding wavelengths. Spherochromatism is defined as 'the chromatic variation of spherical aberration, and is expressed as the difference between the marginal spherical aberration in F and C light' [54].

Let

$$A_{k\,m}O'_F = L'_F \qquad A_{k\,m}O'_C = L'_C \qquad A_kO'_F = l'_F \qquad A_kO'_C = l'_C \tag{10.82}$$

Spherochromatism (SC) is expressed by the difference in longitudinal spherical aberration at F and C wavelengths

$$SC = (L'_F - l'_F) - (L'_C - l'_C) = O'_{F\,m}O'_F - O'_{C\,m}O'_C \tag{10.83}$$

Equivalently, spherochromatism can also be expressed as the difference in longitudinal forms of marginal chromatic aberration and of paraxial chromatic aberration

$$SC = (L'_F - L'_C) - (l'_F - l'_C) = {}_m O'_{C\,m}O'_F - O'_CO'_F \tag{10.84}$$

Note that the C and F wavelengths represent the upper and lower wavelengths of the range of wavelength for which spherochromatism is being considered. At an intermediate wavelength, say d, the paraxial focus O'_d and the marginal focus ${}_mO'_d$ are different, as shown in Figure 10.13.

Figures 10.14(a) and (b) illustrate two forms of presentation of the phenomenon of spherochromatism in imaging systems. From the single axial object point O, three finite rays of F, C, and d wavelengths are separately traced using the corresponding values for refractive indices for all optical materials. In the object space, the three rays follow the same path. The path is a straight line passing through the object point O and a specific point on the entrance pupil. All air spaces have the same refractive index of unity at all wavelengths. The process is repeated for the same object point O and a different point on the entrance pupil. Usually, three points on the entrance pupil with $r = 0.707, 0.866$, and 1.0 are chosen, and the corresponding axial images are located in the image space. The paraxial ray trace at each of the wavelengths from the same object point provides the location of axial focus for $r = 0$, i.e. the location of the paraxial images at the corresponding wavelengths. Figure 10.14(a) shows the zonal variation in locations of axial images for F, d, and C wavelengths separately. Figure 10.14(b) shows the variation in location of axial images over the wavelength range for paraxial, zonal, and marginal rays separately. Note

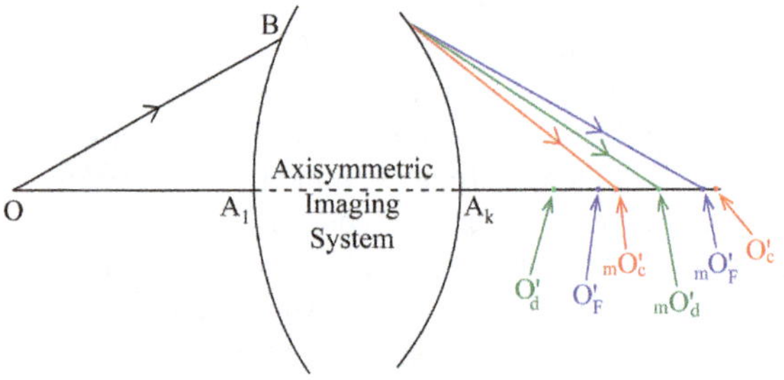

FIGURE 10.13 Axial object point O. At F, d, and C wavelengths, paraxial images are on the axis at the points O'_F, O'_d, and O'_C, and marginal images are at the points ${}_mO'_F$, ${}_mO'_d$, and ${}_mO'_C$, respectively.

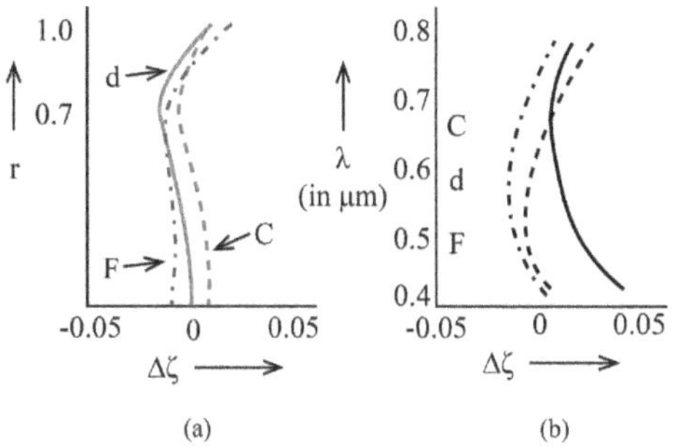

FIGURE 10.14 Representation of the phenomenon of spherochromatism in axially symmetric imaging systems (a) variation in location of axial foci corresponding to different zones at F, d, and C wavelengths (b) Chromatic variation in location of axial foci for paraxial, marginal, and axial rays. [$\Delta\zeta$ is in arbitrary relative scale].

that, in absence of spherochromatism, the curved lines for F, d, and C wavelengths in Figure 10.14(a) will be identical, but displaced along the abscissa, and similarly, the curved lines for the zonal, the marginal, and the paraxial rays in Figure 10.14(b) will be identical, but displaced along the abscissa.

Since spherochromatism is expressed as longitudinal chromatic aberration, the concept is highly appropriate for characterizing nonparaxial chromatism in axial imagery, but it has obvious limitations in characterizing it in the case of extra-axial imagery. The most widely used nonparaxial chromatism characterization formula appropriate for both axial and extra-axial imaging is expressed as a wavefront aberration. It was developed by Conrady [57–61]. The treatment given in the next section is adapted from that given by Hopkins in [52].

10.2.2 Conrady Chromatic Aberration Formula

Figure 10.15 shows the image space of an axi-symmetric imaging system forming the image of an extra-axial object point Q. $\bar{E}'_{\bar{\lambda}}B'_{\bar{\lambda}}$ is the emergent wavefront for the mean wavelength $\bar{\lambda}$ passing through the centre of the real exit pupil, and its principal ray $\bar{E}'_{\bar{\lambda}}\bar{Q}'_{\bar{\lambda}}$ intersects the image plane at $\bar{Q}'_{\bar{\lambda}}$. The reference sphere $\bar{E}'_{\bar{\lambda}}B'$ is of radius $\bar{E}'_{\bar{\lambda}}\bar{Q}'_{\bar{\lambda}}$, and its centre is at $\bar{Q}'_{\bar{\lambda}}$. $B'B'_{\bar{\lambda}}$ is a ray of the pencil of rays forming the image at $\bar{Q}'_{\bar{\lambda}}$ in light of wavelength $\bar{\lambda}$. It intersects the reference sphere and the wavefront corresponding to wavelength $\bar{\lambda}$ at B' and $B'_{\bar{\lambda}}$, respectively. The wavefront aberration $W'_{\bar{\lambda}}$ for this ray is given by

$$W'_{\bar{\lambda}} = \left[B'B'_{\bar{\lambda}}\right] \tag{10.85}$$

Let $\bar{E}'_{\lambda}B'_{\lambda}$ be the emergent wavefront for another wavelength λ passing through the point $\bar{E}'_{\bar{\lambda}}$. The ray for wavelength λ and passing through the point $\bar{E}'_{\bar{\lambda}}$ is different from $\bar{E}'_{\lambda}\bar{Q}'_{\lambda}$. Similarly, $B'B'_{\lambda}$ is a ray of wavelength λ and passes through the point B', and intersects the λ -wavefront at the point B'_{λ}. Thus, with reference to the reference sphere $\bar{E}'_{\bar{\lambda}}B'$, the wavefront aberration of the ray $B'B'_{\lambda}$ is given by

$$W'_{\lambda} = \left[B'B'_{\lambda}\right] \tag{10.86}$$

The chromatic aberration, defined as the change in wave aberration when the wavelength is varied from $\bar{\lambda}$ to λ is

$$\delta W'_{\lambda} = W'_{\lambda} - W'_{\bar{\lambda}} = \left[B'B'_{\lambda}\right] - \left[B'B'_{\bar{\lambda}}\right] \tag{10.87}$$

Let the ray $B'B'_{\bar{\lambda}}$ cut the λ -wavefront $\bar{E}'B'_{\lambda}$ at the point $\hat{B}'_{\lambda}$, and the angular chromatic aberration between the two rays $B'B'_{\lambda}$ and $B'B'_{\bar{\lambda}}$ be $\delta U'_{\lambda}$. We may write

$$\left(B'B'_{\lambda}\right) \cong \left(B'\hat{B}'_{\lambda}\right)\cos\delta U'_{\lambda} \cong \left(B'\hat{B}'_{\lambda}\right) \tag{10.88}$$

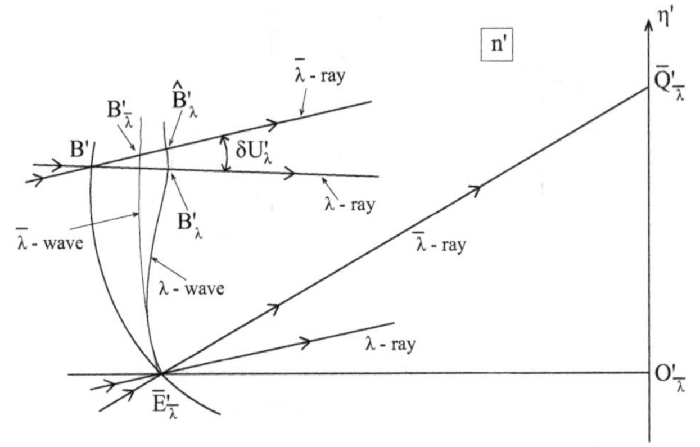

FIGURE 10.15 Image space parameters for Conrady chromatic aberration.

since the angle $\delta U'_{\lambda}$ is a small aberrational quantity tending to zero for most practical cases. By substitution, the equation (10.87) can be written as

$$\delta W'_{\lambda} = \left[B'\hat{B}'_{\lambda}\right] - \left[B'B'_{\bar{\lambda}}\right] = n'_{\lambda}\left(B'\hat{B}'_{\lambda}\right) - n'_{\bar{\lambda}}\left(B'B'_{\bar{\lambda}}\right) \tag{10.89}$$

Since both the lengths $B'\hat{B}'_{\lambda}$ and $B'B'_{\bar{\lambda}}$ are small, and $\left(n'_{\lambda} - n'_{\bar{\lambda}}\right)$ is also small, $\delta W'_{\lambda}$ can be approximated as

$$\delta W'_{\lambda} \cong n'_{\bar{\lambda}}\left(B'\hat{B}'_{\lambda}\right) - n'_{\bar{\lambda}}\left(B'B'_{\bar{\lambda}}\right) \cong n'_{\bar{\lambda}}\left(B'_{\bar{\lambda}}\hat{B}'_{\lambda}\right) \tag{10.90}$$

This shows that the chromatic aberration $\delta W'_{\lambda}$ is given, to a good approximation, as the optical distance between the λ -wavefront and the $\bar{\lambda}$-wavefront as measured along the $\bar{\lambda}$-ray, and provides the justification for using in all ray segments the geometrical path lengths of the $\bar{\lambda}$-ray in the Conrady chromatic aberration formula.

From (10.22) the wave aberration of the $\bar{\lambda}$-wavefront at B' is

$$(W')_{\bar{\lambda}} = \sum_{s=0}^{k}(n'_s)_{\bar{\lambda}}\bar{D}'_s - \sum_{s=0}^{k}(n'_s)_{\bar{\lambda}}D'_s = \sum_{s=0}^{k}(n'_s)_{\bar{\lambda}}\left[\bar{D}'_s - D'_s\right] \tag{10.91}$$

The distances $\bar{D}'_s$ and D'_s are the geometrical distances along the s-th segment of the principal ray passing through $\bar{E}'$ and the aperture ray passing through B', both rays being $\bar{\lambda}$-rays. At wavelength λ, the refractive indices of the different segments become

$$(n'_s)_{\lambda} = (n'_s)_{\bar{\lambda}} + \delta n'_s \tag{10.92}$$

and the geometrical distances $\bar{D}'_s$ and D'_s change to $\left(\bar{D}'_s + \delta\bar{D}'_s\right)$ and $\left(D'_s + \delta D'_s\right)$, respectively. The wave aberration of the λ -wavefront at B' is

$$(W')_{\lambda} = \sum_{s=0}^{k}\left\{(n'_s)_{\bar{\lambda}} + \delta n'_s\right\}\left[\bar{D}'_s + \delta\bar{D}'_s - D'_s - \delta D'_s\right] \tag{10.93}$$

The variation in wave aberration when wavelength changes from $\bar{\lambda}$ to λ is

$$\begin{aligned}\delta W'_{\lambda} &= (W')_{\lambda} - (W')_{\bar{\lambda}} \\ &= \sum_{s=0}^{k}\left(\delta n'_s\right)\left[\bar{D}'_s - D'_s\right] + \sum_{s=0}^{k}\left(n'_s\right)_{\bar{\lambda}}\left[\delta\bar{D}'_s - \delta D'_s\right] + \sum_{s=0}^{k}\left(\delta n'_s\right)\left[\delta\bar{D}'_s - \delta D'_s\right]\end{aligned} \tag{10.94}$$

By Fermat's principle, quantities like $\sum_{s=0}^{k}(n_s')_{\bar{\lambda}}\,\delta\bar{D}_s'$ and $\sum_{s=0}^{k}(n_s')_{\bar{\lambda}}\,\delta D_s'$ in the second term of the above are quantities of order 2 and above. The two quantities in the third term are obviously second order quantities. Neglecting quantities of order higher than the first, the Conrady chromatic aberration formula gives the first order chromatic change in wave aberration as

$$\delta W_\lambda' = (W')_\lambda - (W')_{\bar{\lambda}} \cong \sum_{s=0}^{k}(\delta n_s')\left[\bar{D}_s' - D_s'\right] \qquad (10.95)$$

Note that $(\delta n_s')$ is zero in any non-dispersive medium, e.g. air. The computation for the Conrady chromatic aberration can run concurrently with the computation of wave aberration as noted in Section 10.1.5; in the formulae it is only needed to replace n by δn.

10.2.3 Image Space Associated Rays in Conrady Chromatic Aberration Formula

A physical interpretation of the Conrady chromatic aberration formula (10.95) for the chromatic aberration $\delta W_\lambda'$ can be obtained from (10.90) and Figure 10.15. It is important to note that $\bar{E}_{\bar{\lambda}}'$ is the point where the principal ray of $\bar{\lambda}$ wavelength intersects the optical axis. Both the reference sphere, with centre at $\bar{Q}_{\bar{\lambda}}'$, and the chosen wavefront for λ wavelength pass through the same point $\bar{E}_{\bar{\lambda}}'$ on the axis. The image space associated principal ray for wavelength λ is shown in the figure. Again, the image space associated with the aperture ray of wavelength λ at the point B′ on the reference sphere is invoked for analyzing the chromatic effects. Note that in the object space, the two λ rays described above start from the same object point Q, like the $\bar{\lambda}$ rays, but the directions of the corresponding object space rays of the two wavelengths do not usually have the same direction. The Conrady formula gives a first order approximation to the chromatic aberration, and its interpretation calls for the use of the image space associated rays in defining chromatic aberration.

10.2.4 Evaluation of Exact Chromatic Aberration using Object Space Associated Rays

Figure 10.16 shows the image space of an axi-symmetric lens system forming image of an extra-axial point Q (not shown). In the object space, the principal rays of the two pencils of rays forming an image of the point Q at wavelengths λ and $\bar{\lambda}$ follow the same ray path, but, due to dispersion during passage through the system, in the image space the two rays follow separate paths. Similarly, an aperture ray of the λ-pencil and an aperture ray of the $\bar{\lambda}$-pencil follow the same path in the object space, but follow

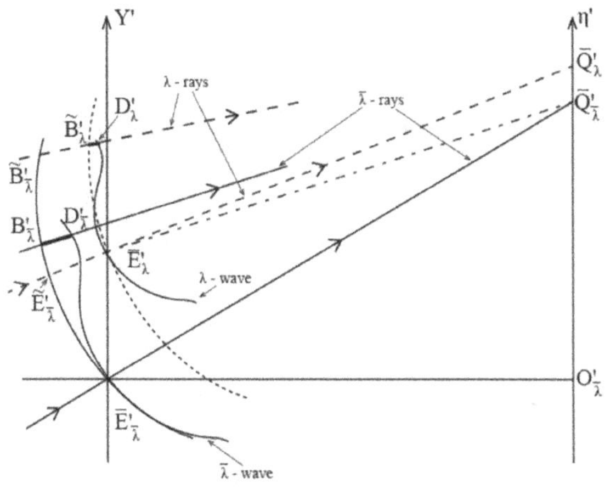

FIGURE 10.16 Image space geometry of object space associated principal rays and aperture rays at two wavelengths λ and $\bar{\lambda}$. Exact chromatic aberration $\delta W_\lambda' = W_\lambda' = W_\lambda' = \left[B_\lambda' D_\lambda'\right] - \left[B_{\bar{\lambda}}' D_{\bar{\lambda}}'\right]$.

different paths in the image space. So, in this treatment we are dealing with 'object space associated principal rays' and 'object space associated aperture rays'. A different form of object space associated rays is used by Agurok [62] for evaluating the chromatic aberration of real rays.

In the image space, the $\bar{\lambda}$-principal ray, $\bar{E}'_{\bar{\lambda}}Q'_{\bar{\lambda}}$, intersects the optical axis at $\bar{E}'_{\bar{\lambda}}$, the centre of the real exit pupil at wavelength $\bar{\lambda}$, and the image plane at $Q'_{\bar{\lambda}}$. $\bar{E}'_{\bar{\lambda}}B'_{\bar{\lambda}}$ is the $\bar{\lambda}$-reference sphere with $Q'_{\bar{\lambda}}$ as centre and $\bar{E}'_{\bar{\lambda}}Q'_{\bar{\lambda}}$ as radius, and $\bar{E}'_{\bar{\lambda}}D'_{\bar{\lambda}}$ is the $\bar{\lambda}$-wavefront passing through $\bar{E}'_{\bar{\lambda}}$. $B'_{\bar{\lambda}}D'_{\bar{\lambda}}$ is the $\bar{\lambda}$-aperture ray that intersects the $\bar{\lambda}$-reference sphere at $B'_{\bar{\lambda}}$, and the $\bar{\lambda}$-wavefront at $D'_{\bar{\lambda}}$. So, the $\bar{\lambda}$-aperture ray has aberration

$$W'_{\bar{\lambda}} = \left[B'_{\bar{\lambda}}D'_{\bar{\lambda}}\right] \tag{10.96}$$

The $\bar{E}'_{\lambda}\bar{Q}'_{\lambda}$ -principal ray, $\tilde{E}'_{\lambda}\bar{E}'_{\lambda}\bar{Q}'_{\lambda}$, intersects the $\bar{\lambda}$-reference sphere at $\tilde{E}'_{\lambda}$, the $\bar{\lambda}$-exit pupil plane at $\bar{E}'_{\lambda}$, and the chosen image plane at $\bar{Q}'_{\lambda}$. For the sake of comparison, a reference sphere $\bar{E}'_{\lambda}B'_{\lambda}$ is drawn with $Q'_{\bar{\lambda}}$ as centre and . as radius. Note that the centre of the λ -reference sphere is also $Q'_{\bar{\lambda}}$, and not $\bar{Q}'_{\lambda}$. $\bar{E}'_{\lambda}D'_{\lambda}$ is a λ-wavefront passing through the point $\bar{E}'_{\lambda}$. The λ -aperture ray, $\tilde{B}'_{\lambda}B'_{\lambda}D'_{\lambda}$, intersects the $\bar{\lambda}$-reference sphere at $\tilde{B}'_{\lambda}$, the λ -reference sphere at B'_{λ}, and the λ -wavefront at D'_{λ}. The λ -aperture ray has aberration

$$W'_{\lambda} = \left[B'_{\lambda}D'_{\lambda}\right] \tag{10.97}$$

The chromatic aberration for the object space associated rays, $\delta W'_{\lambda}$, is exactly given by

$$\delta W'_{\lambda} = W'_{\lambda} - W'_{\bar{\lambda}} = \left[B'_{\lambda}D'_{\lambda}\right] - \left[B'_{\lambda}D'_{\bar{\lambda}}\right] \tag{10.98}$$

without any approximation. The computation for W'_{λ} and $W'_{\bar{\lambda}}$ follows the same procedure as described in Section 10.1.5. Minor modification in the image space quantities is needed to take into account the focal shift from $\bar{Q}'_{\lambda}$ to $\bar{Q}'_{\bar{\lambda}}$. Necessary changes in computations are noted in ref. [52].

REFERENCES

1 A. Koenig and M. von Rohr, 'Computation of rays through a system of refracting surfaces', Chapter II in *Geometrical Investigation of the formation of images in optical instruments*, [original German book title: *Die Theorie der optischen Instrumente*, Ed. M. von Rohr Springer, Berlin (1904)] Translator, R. Kanthack, H. M. Stationery Office, Dept. of Scientific and Industrial Research (1920) pp. 35–82.

2 L. Silverstein, *Simplified Method of Tracing Rays through any Optical System of lenses, prisms and mirrors*, Longman, Green, London (1918).

3 T. Smith, 'Optical Calculations', in *Dictionary of Applied Physics*, Vol. IV, Ed., R. Glazebrook, Macmillan, London (1923) pp. 287–315.

4 T. Smith, 'On tracing rays through an optical system', *Proc. Phys. Soc.*, Vol. 27 (1915) 502–509; ibid., Vol. 30 (1918) pp. 221–233; ibid., Vol. 32 (1920) pp. 252–284; ibid., Vol. 33 (1921) pp. 174–178; ibid., Vol. 57 (1945) pp. 286–293.

5 A.E. Conrady, *Applied Optics and Optical Design*, Part I, Dover, New York (1957) [originally by Oxford University Press (1929)] pp. 3–71, 402–436.

6 G.G. Slyusarev, 'Raytracing through centred optical systems consisting of spherical surfaces', Chapter 1 in *Aberration and Optical Design Theory*, [original Russian book title: *Metodi Rascheta Opticheskikh Sistem, Mashinostroenie*, Leningrad (1969)] Translator: J.H. Dixon, Adam Hilger, Bristol (1984) pp. 1–47.

7 H. Chrétien, *Calcul des Combinaisons Optiques*, Masson, Paris (1980) [First published in 1938] pp. 687–725.

8 T.Y. Baker, 'Tracing skew rays through second degree surfaces', *Proc. Phys. Soc.*, Vol. 56 (1944) pp. 114–122.

9 D.P. Feder, 'Tracing of skew rays', *J. Res. Nat. Bur. Std.*, Vol. 45 (1950) pp. 61–63.

10 D.P. Feder, 'Optical calculations with automatic computing machinery', *J. Opt. Soc. Am.*, Vol. 41 (1951) pp. 630–635.

11 G. Black, 'Ray tracing on the Rochester University electronic computing machine', *Proc. Phys. Soc.* Vol. 67 B (1954) pp. 569–574.
12 J. Flügge, *Das Photographische Objektiv*, Springer, Wien (1955) pp. 16–30.
13 M. Herzberger, *Modern Geometrical Optics*, Interscience, New York (1958) pp. 3–68.
14 W. Weinstein, 'Literature survey on ray tracing', *Proc. Symp. Optical Design with digital computers, Technical optics section, Imperial College of Science and Technology, London*, June 5–7 (1956) pp. 15–26..
15 M. Born and E. Wolf, *The Principles of Optics*, Pergamon, Oxford (1984) [First edition in 1959] pp. 190–196.
16 P.W. Ford, 'New ray tracing scheme', *J. Opt. Soc. Am.*, Vol. 50 (1960) pp. 528–533.
17 G. Spencer and M.V.R.K. Murty, 'Generalized ray tracing procedure', *J. Opt. Soc. Am.*, Vol. 52 (1962) pp. 672–678.
18 R.E. Hopkins and R. Hanau, 'Fundamental methods of ray tracing', in *Military Standardization Handbook, MIL-HDBK-141, Optical Design*, U. S. Defense Supply Agency, Washington (1962) pp. 5–37.
19 A. Cox, Chapter 3, 'Ray Tracing' in *A System of Optical Design*, The Focal Press, London (1964) pp. 93–136.
20 J. Burcher, *Les combinaisons optiques: pratique des calculs*, Masson, Paris (1967) pp. 339–348.
21 A. Maréchal, E. Hugues, and P. Givaudon, 'Méthode de calcul des systems optiques', in Handbuch der Physik, Band XXIX, Optische Instrumente, Ed., S. Flügge, Springer, Berlin (1967) pp. 43–67.
22 H. Marx, 'Theorie des geometrisch-optischen Bildfehler', in Handbuch der Physik, Band XXIX, Optische Instrumente, Ed., S. Flügge, Springer, Berlin (1967) pp. 68–191.
23 R. Alpiar, 'Algebraic ray tracing on a digital computer', *Appl. Opt.*, Vol. 8 (1969) pp. 293–304.
24 R. Kingslake, *Lens Design Fundamentals*, Academic, New York (1978) pp. 19–47, 137–155.
25 D.C. O'Shea, *Elements of Modern Optical Design*, John Wiley, New York (1985) pp. 56–86, 177–184.
26 W.T. Welford, *Aberrations of Optical Systems*, Adam Hilger, Bristol (1986) pp. 45–78.
27 Y. Wang and H.H. Hopkins, 'Ray Tracing and Aberration Formulae for a general optical system', *J. Mod. Opt.*, Vol. 39 (1992) pp. 1897–1938.
28 D. Malacara and Z. Malacara, *Handbook of Lens Design*, Marcel Dekker, New York (1994) pp. 15–23, 55–78.
29 M.J. Kidger, *Fundamental Optical Design*, SPIE, Bellingham, Washington (2002) pp. 45–62.
30 H. Gross, *Handbook of Optical Systems,* Vol. I, Wiley, New York (2005) pp. 173.
31 E.L. Dereniak and T.D. Dereniak, '*Geometrical and Trigonometric Optics'*, Cambridge Univ. Press, Cambridge (2008) pp. 255–291, 328–346.
32 A. Romano, *Geometric Optics*, Birkhaüser, Boston (2010) pp. 17–20.
33 P.D. Lin, *Advanced Geometrical Optics*, Springer, Singapore (2017) pp. 29–69.
34 R. Flores-Hernández and A. Gómez-Vieyra, 'Ray Tracing' in *Handbook of Optical Engineering, Vol. I*, Fundamentals ad Basic Optical Instruments, Eds. D. Malacara-Hernández and B.J. Thompson, CRC Press, Boca Raton (2018) pp. 119–164.
35 A. Kerber, *Beiträge zur Dioptrik*, No. 1, (published by author) Leipzig (1895).
36 A. Koenig and M. von Rohr, Ref. 1, pp. 78–82.
37 A.E. Conrady, in *Dictionary of Applied Physics*, Vol. IV, Ed., R. Glazebrook, Macmillan, London (1923) pp. 212–217.
38 H.H. Hopkins, *Wave Theory of Aberrations*, Clarendon Press, Oxford (1950) pp. 142–147.
39 W.T. Welford, 'A new total aberration formula', *Opt. Acta*, Vol. 19 (1972) pp. 719–727.
40 A. Cox, vide Ref. 19, pp. 129–132.
41 W.T. Welford, Ref. 26, pp. 163–165.
42 H. Gross, 'Section 29.8.4. Aldis Theorem', in *Handbook of Optical Systems, Vol.3*, Wiley-VCH, Weinheim (2007) p. 57.
43 A.E. Conrady, *Applied Optics and Optical Design, Part II*, Dover, New York (1960) p. 587.
44 H.H. Hopkins, 'The Wave Aberration Associated with Skew rays', *Proc. Phys. Soc. B*, Vol. LXV (1952) pp. 934–942.
45 S.A. Comastri, 'Pupil exploration and calculation of vignetting', *Optik*, Vol. 85 (1990) pp. 173–176.
46 W.B. King, 'The Approximation of Vignetted Pupil Shape by an Ellipse', *Appl. Opt.*, Vol. 7 (1968) pp. 197–201.

47 J. Macdonald, 'Pupil exploration', Chapter 3 in *New Analytical and Numerical Techniques in Optical Design*, Ph. D. Thesis, University of Reading (1974).
48 I. Powell, 'Optical design and analysis program', *Appl. Opt.*, Vol. 17 (1978) pp. 3361–3367.
49 I. Powell, 'Pupil exploration and wave-front-polynomial fitting of optical systems', *Appl. Opt.*, Vol. 34 (1995) pp. 7986–7997.
50 Y. Wang and H.H. Hopkins, vide 27, pp. 1897–1938.
51 W.T. Welford, vide Ref. 26, pp. 162–171.
52 H.H. Hopkins, 'Calculation of the aberrations and image assessment for a general optical system', *Opt. Acta*, Vol. 28 (1981) pp. 667–714.
53 H.H. Hopkins, 'Image formation by a general optical system. 2: Computing methods', *Appl. Opt.*, Vol. 21 (1985) pp. 2506–2519.
54 R. Kingslake, vide Ref. 24, pp. 73–75.
55 W. Weinstein, 'Chromatic variation of spherical aberration', *Brit. J. Appl. Physics*, Vol. 1 (1950) pp. 67–73.
56 W.J. Smith, Modern Optical Engineering, Mc-Graw Hill, New York (2000) pp. 406–409.
57 A.E. Conrady, 'On the chromatic correction of object glasses', *M. N. Roy. Astron. Soc.*, Vol. 64 (1904) pp. 182–188.
58 A.E. Conrady, 'On the correction of object glasses (Second paper)', *M. N. Roy. Astron. Soc.*, Vol. 64 (1904) pp. 458–460.
59 D.P. Feder, 'Conrady's Chromatic Condition', *Res. Paper 2471, J. Res. Nat. Bur. Std.*, Vol. 52 (1954) pp. 43–49.
60 R. Kingslake, vide Ref. 24, pp. 93–98, 202–203.
61 W.T. Welford, vide Ref. 26, pp. 200–202.
62 A.B. Agurok, 'Chromatic aberration of real rays', *Sov. J. Opt. Technol.*, Vol. 46 (1979) pp. 711–715.

11

Hopkins' Canonical Coordinates and Variables in Aberration Theory

11.1 Introduction

The relations given in the earlier chapter take care of common numerical errors arising in the evaluation of aberrations. Nevertheless, it cannot be overlooked that these expressions, per se, become indeterminate when either/both of the object/image planes, or either/both of the entrance/exit pupils are at infinity or at a large distance from the system, e.g., in the case of camera or projection lenses, telescopic systems, or telecentric systems, etc. In each of these cases, special formulae need to be invoked. In Hopkins' treatment of image formation [1–4], this problem is overcome by specifying rays in the object and image spaces by using reduced coordinates for the object and image planes, and by defining suitable relative coordinates for the entrance and exit pupils.

The generalized set of extended paraxial coordinates, called 'canonical coordinates' by Hopkins, serves to extend the linear rules of paraxial optics to arbitrarily large fields and apertures. 'This leads to a consistent definition of pupils and oblique magnification for a real optical system. The failure of an optical system to meet these required relations leads to aberrations and is closely related to the establishment of isoplanatism' [5].

In order to take proper account of pupil aberrations and vignetting, the effective real pupil is determined by pupil exploration for each field point. It has already been enunciated in an earlier chapter how meaningful measures of aberration of the 'actual' wavefront are defined. Further insight on the correspondence between pencils of rays in the initial object space and the final image space is obtained by specifying rays in terms of reduced coordinates on the object/image plane and normalized variables for points on the pupil. Hopkins' variables for the pupils make use of the concept of a pupil sphere that has the object/image as the centre and the distance of the centre of the 'effective' entrance/exit pupil from the object/image point as radius. In a homogeneous space, a ray is specified in terms of the coordinates of two points. Usually, one of the two points is chosen to be the point of intersection of the ray with the object/image plane. The other point is taken to be the point of intersection of the ray with the entrance/exit pupil plane. In Hopkins' scheme, the former choice is retained, but a different point is chosen for the second point. The latter is the point of intersection of the ray with the reference sphere with the object/image as centre and the distance of the centre of the effective entrance/exit pupil as radius. The two rectangular (ξ, η) coordinates of the object/image point on the transverse object/image plane and the two rectangular (X,Y) coordinates of the point on the pupil sphere constitute the four parameters for specifying the ray. In Hopkins' canonical coordinates scheme, both the pupil and the object/image variables are normalized to obtain a new set of reduced variables. In the hyperspace of these variables, the linearization model applicable for paraxial optics can be extended to reasonably large fields and apertures. A consequence of this is that a regular mesh of rays on the entrance pupil from an object point gives rise to somewhat identical mesh of rays in the exit pupil in the image space. The difference between the two is usually small even in the case of moderately corrected systems, and this difference gets smaller with the reduction in residual aberrations, becoming zero in the case of aberration-free systems. A set of relations exists between pupil and image coordinates defined by Hopkins. These relations have a formal similarity with Hamilton's equations [6] in mechanics, and Hopkins was motivated thereby to call these coordinates 'canonical' [1].

DOI: 10.1201/9780429154812-11

Use of these canonical coordinates in the definition of wavefront aberration facilitates accurate definition and evaluation of diffraction-based image evaluation criteria for both axial and extra-axial imaging. We refer to the original publications [1–3] by Hopkins and his research group. In what follows, we underscore the key points that have direct bearing on the treatment of image formation and ray optical aberration theory for axially symmetric optical systems.

11.2 Canonical Coordinates: Axial Pencils

Figure 11.1 shows the imaging of an axial object point O by an axially symmetric system. E and E′ are the axial locations of the entrance pupil and the exit pupil. For the sake of simplicity, it is shown in two dimensions. The upper edge ray from O makes an angle α with the optical axis and cuts the reference sphere, with centre at O and radius $OE(= R)$, at B_U, which is at a height Y_{max} above the axis. The paraxial angle and height variables in the object space, and at the entrance pupil E, respectively are defined by

$$u = \sin\alpha \qquad h = Y_{max} \tag{11.1}$$

If the system is aberration free, the edge ray will emerge as $B'_U\, 0'$ where B'_U is at a height Y'_{max} above the axis, and $\angle B'_U O' E' = \alpha'$. The paraxial angle and height variables in the image space and at the exit pupil E′, respectively, are defined by

$$u' = \sin\alpha' \qquad h' = Y'_{max} \tag{11.2}$$

In the case of systems with residual aberrations, the emergent edge ray follows a path that is different from $B'_U 0'$. Nevertheless, $B'_U 0'$ is a close approximation to the actual emergent ray, and u′ and h′, as defined above, can be used to define $\sin\alpha'$ and Y'_{max}, respectively.

Figure 11.1 shows an arbitrary aperture ray OD from O. It intersects the entrance pupil sphere EDB_U at the point D with real space coordinates (X, Y). The reduced pupil coordinates of the point D are defined to be

$$x = (X / h) \qquad y = (Y / h) \tag{11.3}$$

Similarly, reduced pupil coordinates for the corresponding point $D'(X', Y')$ on the exit pupil are defined as

$$x' = (X' / h') \qquad y' = (Y' / h') \tag{11.4}$$

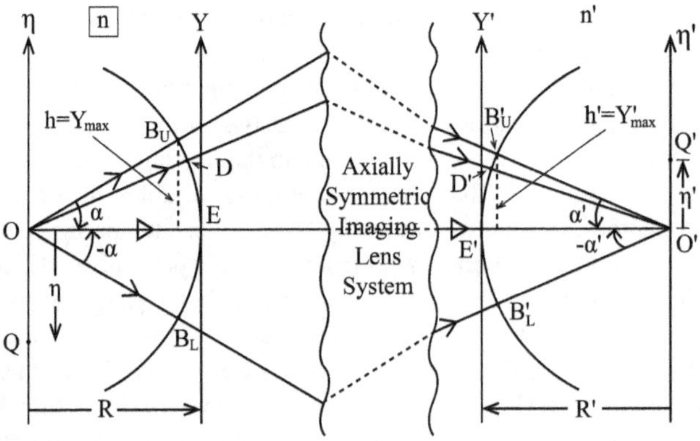

FIGURE 11.1 Canonical coordinates for an axial pencil.

In the absence of coma, a ray from an axial object point O, specified in the object space in terms of reduced pupil coordinates (x,y), will have its reduced exit pupil coordinates (x',y'), such that

$$x' = x \qquad y' = y \tag{11.5}$$

This is a consequence of perfect satisfaction of sine condition in the case of absence of coma. Even when a system is nominally corrected from coma, the relation (11.5) holds with small errors. Note that the relation (11.5) holds even when either or both the pupils are at infinity.

Let the real space coordinates of an object point Q and the corresponding image point Q′ be (ξ,η) and (ξ',η'), respectively. The corresponding reduced coordinates (G,H) and (G',H') are defined as

$$G = (n\sin\alpha)\xi \qquad H = (n\sin\alpha)\eta \tag{11.6}$$

$$G' = (n'\sin\alpha')\xi' \qquad H' = (n'\sin\alpha')\eta' \tag{11.7}$$

Since the paraxial magnification is

$$M = \frac{\xi'}{\xi} = \frac{\eta'}{\eta} = \frac{nu}{n'u'} = \frac{n\sin\alpha}{n'\sin\alpha'} \tag{11.8}$$

It follows

$$G' = G \qquad H' = H \tag{11.9}$$

If a ray from O has the components of transverse ray aberrations $\{\delta\xi',\delta\eta'\}$ on the image plane, the reduced forms of these aberrations are

$$\delta G' = (n'\sin\alpha')\delta\xi' \qquad \delta H' = (n'\sin\alpha')\delta\eta' \tag{11.10}$$

Note that the real space linear quantities on the object and image planes are normalized by multiplying them with the angular numerical aperture. In the case of object and/or image at infinity, the corresponding real space quantities tend to infinity, but the reduced quantities, as defined above, remain finite. They become the real space angular coordinates normalized by multiplication with the linear numerical aperture (nh) and/or $(n'h')$.

11.3 Canonical Coordinates: Extra-Axial Pencils

In the case of axial pencils of an axially symmetric imaging system, the upper edge ray $OB_UB'_UO'$ and the lower edge ray $OB_LB'_LO'$ are located symmetrically around the optical axis, and the optical axis is the effective pupil ray of the image-forming pencil of rays (Figure 11.1). Note that the direction cosine M_U of the upper edge ray is equal to $(-\sin\alpha)$, and the direction cosine M_L of the lower ray is

$$M_L = -M_U \tag{11.11}$$

The paraxial pupil ray and the real pupil ray are identical in the case of axial imagery. Since the direction cosine of the optical axis or the real pupil ray with respect to the η-axis is zero, the modulus of the difference in the direction cosines of the lower edge ray and the pupil ray is the same as the modulus of the difference in direction cosines of the upper edge ray and the pupil ray.

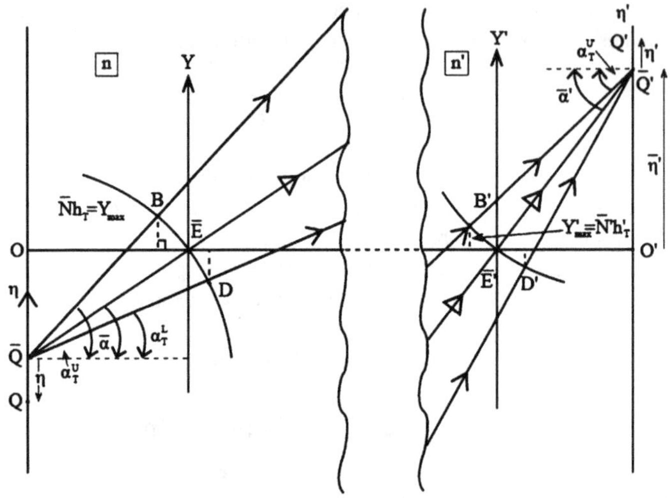

FIGURE 11.2 Canonical coordinates for an extra-axial pencil: meridian section.

In the case of extra-axial imagery in general, the paraxial pupil ray and the 'effective' real pupil ray are not identical. By definition, for any extra-axial object/image point, the axial locations of the paraxial pupils are the same as those for imaging the axial point on the same transverse plane. However, due to pupil aberrations and vignetting, the paraxial pupil ray, starting from the extra-axial object point and passing through the axial point of the paraxial entrance pupil, is no longer the central or chief ray of the image-forming pencil of rays. In Section 10.1.3, a numerical procedure, called 'pupil exploration', was described to determine the 'effective' real pupils for imaging the specific extra-axial point.

Figure 11.2 shows the meridian section in the object space and the image space of the pencil of rays forming image of an extra-axial object point $\bar{Q}$. The upper edge ray $\bar{Q}B$ and the lower edge ray $\bar{Q}D$ in the tangential section are determined by pupil exploration. The effective pupil ray of the real pencil of rays may be defined in different ways. After experimenting with different choices, Hopkins found that in both ray and wave optical treatment of image formation, a 'useful choice is to make the differences in the direction cosines with respect to the η-axis of the three rays to be equal' [7].

Let $\alpha_T^U, \bar{\alpha}, \alpha_T^L$ be the angles to the optical axis of the upper edge ray, the effective pupil ray, and the lower edge ray in the tangential section in the object space. Per the above choice for the angle $\bar{\alpha}$, we have

$$\left(\sin\alpha_T^U - \sin\bar{\alpha}\right) = -\left(\sin\alpha_T^L - \sin\bar{\alpha}\right) \quad (11.12)$$

Using the direction cosines $M_T^U = \left(-\sin\alpha_T^U\right)$, $\bar{M} = \left(-\sin\bar{\alpha}\right)$, $M_T^L = \left(-\sin\alpha_T^L\right)$, the above equation (11.12) can be written as

$$\left(M_T^U - \bar{M}\right) = -\left(M_T^L - \bar{M}\right) \quad (11.13)$$

This expression reduces to corresponding expression (11.11) in the axial case where $\bar{M} = 0$.

As defined above, the ray $\bar{Q}\bar{E}$ is the effective pupil ray. It intersects the optical axis at point $\bar{E}$. Note that, in general, on account of pupil aberration and vignetting, this ray does not pass through the centre of the physical aperture stop of the system. In the object space, $B\bar{E}D$ is the meridian section of a reference sphere drawn with $\bar{Q}$ as the centre and $\bar{Q}\bar{E}\left(=\bar{R}\right)$ as radius. The upper edge ray and the lower edge ray intersect the reference sphere at B and D, respectively. In Hopkins' canonical scheme, the effective pupil ray $\bar{Q}\bar{E}$ is considered to be the 'axis' ray of the extra-axial pencil of rays from the object point $\bar{Q}$. The tangential and sagittal differential rays along it are called tangential paraxial and sagittal paraxial rays, respectively. The latter rays can be traced by using suitably defined angle and height variables in the same manner as ordinary paraxial rays from an axial object point. Corresponding to equation (11.1) in the axial case, the angle variable u_T and the height variable h_T for the meridional section in the case of extra-axial imaging is defined as

$$u_T = \frac{\left(\sin\alpha_T^U - \sin\bar{\alpha}\right)}{\cos\bar{\alpha}} \qquad \bar{N}h_T = Y_{max} \tag{11.14}$$

where Y_{max} is the height of B above the optical axis. Note that the corresponding point for the lower edge ray is below the optical axis by the same distance. The formulae (11.14) are consistent for

$$Y_{max} = \bar{R}\left(\sin\alpha_T^U - \sin\bar{\alpha}\right) = \cos\bar{\alpha}\bar{R}u_T = \bar{N}h_T \tag{11.15}$$

where the usual paraxial relation $h_T = \bar{R}u_T$ is assumed to hold between the paraxial height and angle variables. It is obvious that the factors $(1/\cos\alpha)$ and $\bar{N}$ are occurring in the variables on account of obliquity of the pupil ray of the extra-axial pencil with respect to the optical axis.

If the upper edge ray has no transverse aberration and the generalized sine condition holds, the upper edge ray emerges as $B'\bar{Q}'$ in the image space with angle variable u'_T and height variable h'_T so that

$$u'_T = \frac{\left(\sin\alpha_T^{U'} - \sin\bar{\alpha}'\right)}{\cos\bar{\alpha}'} \qquad \bar{N}'h'_T = \bar{R}'\left(\sin\alpha_T^{U'} - \sin\bar{\alpha}'\right) = Y'_{max} \tag{11.16}$$

In the case of the presence of residual aberrations, $B'\bar{Q}'$ is not precisely the path of the emergent ray corresponding to the top edge ray $\bar{Q}B$ in the object space. Nevertheless, even for nominally corrected systems, $B'\bar{Q}'$ is a very close approximation to the path of the real emergent ray. The parameters of this approximate path are used for normalization purposes.

Figure 11.3 shows the sagittal section of imaging of the extra-axial point $\bar{Q}$ by the axially symmetric lens system. $\bar{Q}\bar{E}$ and $\bar{E}'\bar{Q}'$ are segments of the 'effective' real pupil ray from $\bar{Q}$ in the object space and the image space, respectively. In the object space, $\bar{Q}BG$ is a sagittal aperture ray from the extra-axial object point $\bar{Q}$. The sagittal ray lies on the plane that is perpendicular to the meridian plane and that contains the principal ray. $\bar{E}B$ is the arc of a circle that is the intersection of the plane Y = 0, with the reference sphere

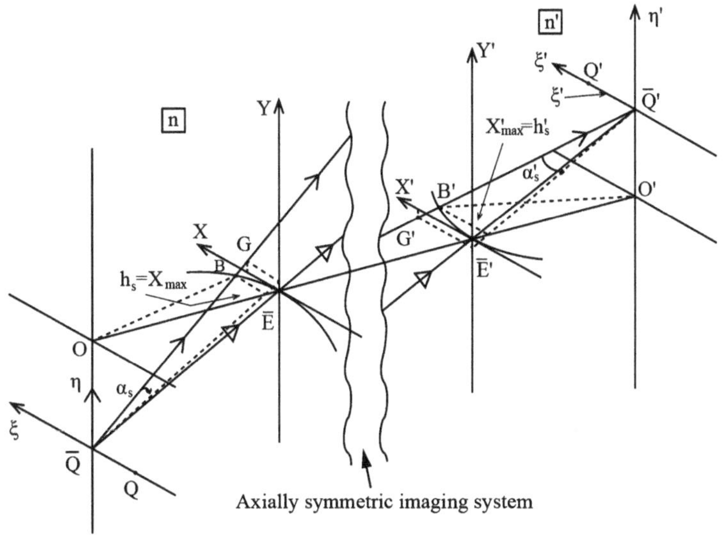

FIGURE 11.3 Canonical coordinates for an extra-axial pencil: sagittal section.

with centre at $\bar{Q}$ and radius $(\bar{Q}\bar{E})$. The sagittal rays from $\bar{Q}$ are rays $\bar{Q}$ that pass through the points on the arc $\bar{Q}B$; each of these points has $Y = 0$. Similarly, in the image space, $\bar{E}'B'$ is the arc of a circle that is the intersection of the plane $Y' = 0$ with the reference sphere with its centre at $\bar{Q}'$ and radius $(\bar{E}'\bar{Q}')$. The sagittal rays in the image space are comprised of rays that pass through a point on the arc $\bar{E}'B'$, and lie on the plane $\bar{E}'B'\bar{Q}'$ that is perpendicular to the meridional plane. It contains the image space segment $\bar{E}'\bar{Q}'$ of the principal ray

By pupil exploration, the edge ray $\bar{Q}BG$ in the sagittal section on one side of the meridional plane is determined. The aperture angle of this sagittal edge ray with the plane $X = 0$ is α_S. Note that, by symmetry, the aperture angle of the sagittal edge ray on the other side of the meridional plane (not shown in the diagram) is $-\alpha_S$. The angle and height variables of the sagittal paraxial ray are defined as

$$u_S = \sin\alpha_S \qquad h_S = X_{max} \tag{11.17}$$

where X_{max} is the X coordinate of the point B on the edge ray. It may be checked easily that they are consistent, as

$$X_{max} = \bar{R}\sin\alpha_S = \bar{R}u_S = h_S \tag{11.18}$$

The image space sagittal paraxial quantities are u'_s and h'_s. If the real sagittal edge ray has no transverse aberration and the isoplanatic condition holds, the real ray emerges exactly as $B'\bar{Q}'$, where

$$\sin\alpha'_S = u'_S \qquad X'_{max} = h'_S \tag{11.19}$$

In the case of an imaging system with residual aberrations, as before, the real sagittal ray follows a path that is different from $B'\bar{Q}'$. But, for nominally corrected systems, $B'\bar{Q}'$ is a good approximation for that path, so that sagittal paraxial ray parameters can be used for normalization purposes.

11.4 Reduced Pupil Variables

The reduced entrance pupil coordinates for a point on the entrance pupil with real space coordinates (X,Y) are defined as

$$x_S = \frac{X}{h_S} \qquad y_T = \frac{Y}{\bar{N}h_T} \tag{11.20}$$

Similarly, the reduced exit pupil coordinates for a point on the exit pupil with real space coordinates (X', Y') are defined as

$$x'_S = \frac{X'}{h'_S} \qquad y'_T = \frac{Y'}{\bar{N}'h'_T} \tag{11.21}$$

The periphery of the entrance pupil in real space coordinates is an ellipse

$$\left(\frac{X}{h_S}\right)^2 + \left(\frac{Y}{\bar{N}h_T}\right)^2 = 1 \tag{11.22}$$

In terms of reduced entrance pupil coordinates (x_s, y_T) it becomes a unit circle

$$x_S^2 + y_T^2 = 1 \tag{11.23}$$

If the isoplanatism condition holds in the neighbourhood of the ray passing through (x_S, y_T), its reduced exit pupil coordinates (x'_S, y'_T) are given by

$$x'_S = x_S \qquad y'_T = y_T \tag{11.24}$$

The periphery of the exit pupil in reduced coordinates is also a unit circle

$$x'^2_S + y'^2_T = 1 \tag{11.25}$$

In real space coordinates, the periphery of the exit pupil is the ellipse

$$\left(\frac{X'}{h'_S}\right)^2 + \left(\frac{Y'}{\bar{N}' h'_T}\right)^2 = 1 \tag{11.26}$$

It is to be noted that the semi-major and the semi-minor axes of the entrance pupil and the exit pupil are to be determined from the data obtained by iterative finite ray tracing, as described earlier. The quantities are obtained as

$$h_S = X_{max} \qquad \bar{N}h_T = Y_{max} \qquad h'_S = X'_{max} \qquad \bar{N}'h'_T = Y'_{max} \tag{11.27}$$

In general, the semi-axes of the elliptic periphery of the exit pupil are not the same as those of the entrance pupil, so that their eccentricities are different. But in reduced coordinates, both peripheries are unit circles centered on the axis.

11.5 Reduced Image Height and Fractional Distortion

In Figure 10.12, let the paraxial image of the extra-axial object point Q be Q', where $Q'O' = \eta'$. The real pupil ray intersects the η'-axis in the image plane at $\bar{Q}'$, where $\bar{Q}'O' = \bar{\eta}'_k$. The latter is given by relation (10.81). The reduced image heights are given by $H_0 = n'_k u'_k \eta' = n'u'\eta'$, and

$$\bar{H}'_0 = n'_k u'_k \bar{\eta}'_k = \frac{n'}{\bar{N}'_k}\left\{u'\left(\bar{Y}_k \bar{N}'_k - \bar{Z}_k \bar{M}'_k\right) + h_k \bar{M}'_k\right\} \tag{11.28}$$

where the substitutions $n' = n'_k$, and $u' = u'_k$ are made. Note that the reduced image heights remain determinate even when the image is at infinity.

The fractional geometrical distortion is given by

$$FD = \frac{\left(\bar{\eta}'_k - \eta'\right)}{\eta'} = \frac{\left(\bar{H}'_0 - H_0\right)}{H_0} \tag{11.29}$$

This expression also remains determinate when the image is at infinity.

11.6 Pupil Scale Ratios

In the case of imaging an axial point, the object space numerical aperture of the image-forming pencil of rays is

$$_{Obj}(N.A.)_{axial} = n \sin\alpha \tag{11.30}$$

where α is the angle that the edge ray along an azimuth makes with the axis (Figure 11.1). Note that in any section, e.g., the meridian section, the angle made by the lower edge ray with the axis is equal and opposite in sign to that of the upper edge ray. This characteristic is invoked when defining numerical aperture for the image-forming pencil from extra-axial object points.

Since, in general, the periphery of the entrance pupil and that of the exit pupil when imaging extra-axial points are not circles, the numerical aperture of the image-forming cone of rays is different in different azimuths. The object space numerical aperture of the upper edge ray in the meridian or tangential section of the image-forming pencil from the extra-axial object point is (Figure11.2)

$$_{\text{Obj}}(\text{N.A.})_T^U = n\left(\sin\alpha_T^U - \sin\bar{\alpha}\right) \tag{11.31}$$

From (11.12), we note that the object space numerical aperture of the lower edge ray in the meridian or tangential section of the image-forming pencil from the same extra-axial point is equal in magnitude and opposite in sign to that of the upper edge ray, exactly as for the axial pencil.

In the sagittal section (Figure 11.3), the object space numerical aperture is defined by

$$_{\text{Obj}}(\text{N.A.})_S = n\sin\alpha_S \tag{11.32}$$

Corresponding numerical apertures in the image space are defined by

$$_{\text{Img}}(\text{N.A.})_{\text{axial}} = n'\sin\alpha' \tag{11.33}$$

$$_{\text{Img}}(\text{N.A.})_T^U = n'\left(\sin\alpha_T^{U'} - \sin\bar{\alpha}'\right) \tag{11.34}$$

$$_{\text{Img}}(\text{N.A.})_S = n'\sin\alpha'_S \tag{11.35}$$

The tangential pupil scale ratios in the object space and in the image space are defined by

$$\rho_T = \frac{\left(\sin\alpha_T - \sin\bar{\alpha}\right)}{\sin\alpha} \qquad \rho'_T = \frac{\left(\sin\alpha_T' - \sin\bar{\alpha}'\right)}{\sin\alpha'} \tag{11.36}$$

where the superscript 'U' is omitted from the notations for angles $\alpha_T^U, \alpha_T^{U'}$.

Similarly, the sagittal pupil scale ratios in the object space and in the image space are defined by

$$\rho_S = \frac{\sin\alpha_S}{\sin\alpha} \qquad \rho'_S = \frac{\sin\alpha'_S}{\sin\alpha'} \tag{11.37}$$

Note that the tangential and the sagittal pupil scale ratios in the object and the image space are effectively the ratios of tangential and sagittal numerical apertures to the numerical aperture of the axial pencil in the corresponding space. If the object and/or the image is at infinity, the pupil scale ratios become the ratios of the appropriate linear apertures.

The pupil scale ratios have a photometric/radiometric significance. If the object is a Lambertian radiator, the ratio of the flux accepted from an extra-axial object point $\bar{Q}$ to the flux accepted from the axial point O is given by the product of the tangential and sagittal pupil scale ratios corresponding to that extra-axial object point, i.e.

$$\frac{F_{\bar{Q}}}{F_O} = \rho_S \rho_T \tag{11.38}$$

The corresponding relative illumination R on the image plane is

$$R = \rho'_S \rho'_T \tag{11.39}$$

This formula takes into account the effects of both the obliquity of the image-forming pupil and vignetting.

In the following, we provide the procedure and formulae for evaluation of the entrance pupil scale ratios and exit pupil scale ratios. Figures 11.4(a) and (b) show variables of the parabasal tangential marginal ray and of the parabasal sagittal marginal ray on the real principal ray $Q\bar{E}P_1$ (as the base ray) from the extra-axial object point Q in the object space. The direction cosines of the real principal ray in the object space are $(\bar{L}_1 \bar{M}_1 \bar{N}_1)$.

Equations (7.256) and (7.258) constitute the refraction and transfer formulae for s ray tracing. Similarly, equations (7.260) and (7.261) constitute the refraction and transfer formulae for t ray tracing. The initial variables for opening formulae for s ray and t ray tracing are given below.

In the object space $QP_1 = s_1 = t_1 = \left(\dfrac{l_1 - \bar{Z}_1}{\bar{N}_1} \right)$ and $Q\bar{E} = \bar{R}$. It follows

$$\left(u_S\right)_1 = \frac{X_{max}}{\bar{R}} \qquad \left(h_S\right)_1 = s_1 \left(u_S\right)_1 \tag{11.40}$$

$$\left(u_T\right)_1 = \frac{Y_{max}}{\bar{N}_1 \bar{R}} \qquad \left(h_T\right)_1 = t_1 \left(u_T\right)_1 \tag{11.41}$$

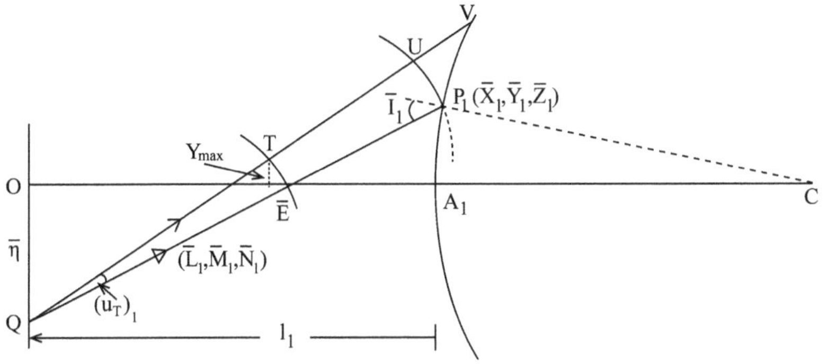

FIGURE 11.4(a) The parabasal tangential marginal ray on the real principal ray $Q\bar{E}P_1$ (as the base ray) from the extra-axial object point Q in the object space. (Elevation view).

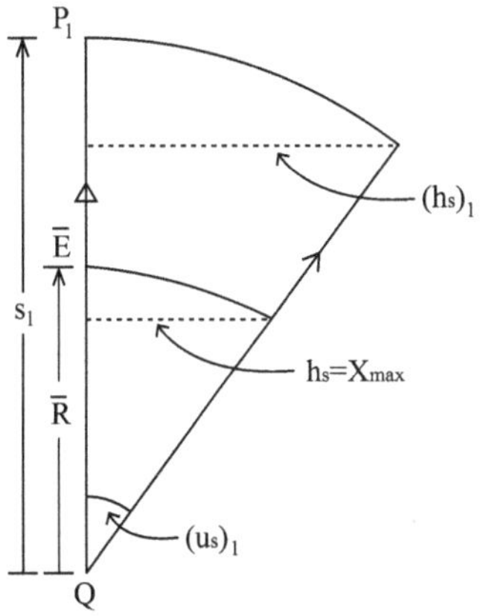

FIGURE 11.4(b) The parabasal sagittal marginal ray on the real principal ray $Q\bar{E}P_1$ (as the base ray) from the extra-axial object point Q in the object space. (Plan view).

11.6.1 Entrance Pupil Scale Ratios ρ_S and ρ_T

Using (11.17) and (10.13), we get

$$u_S = \sin\alpha_S = \frac{X_{max}}{\bar{R}} = A\bar{N}_0 \frac{h}{R} \tag{11.42}$$

$$u = \sin\alpha = \frac{h}{R} \tag{11.43}$$

$$\rho_S = \frac{\sin\alpha_S}{\sin\alpha} = \frac{u_S}{u} = A\bar{N}_0 \tag{11.44}$$

Using relations (11.14), (11.15), and (10.13), we get

$$u_T = \frac{\left(\sin\alpha_T^U - \sin\bar{\alpha}\right)}{\cos\bar{\alpha}} \tag{11.45}$$

$$Y_{max} = \left(\cos\bar{\alpha}\right)\bar{R}u_T = Bh\bar{N}_0 \frac{\bar{R}}{R} \tag{11.46}$$

$$\left(\sin\alpha_T^U - \sin\bar{\alpha}\right) = u_T \cos\bar{\alpha} = \frac{Y_{max}}{\bar{R}} = B\bar{N}_0 \frac{h}{R} \tag{11.47}$$

$$\rho_T = \frac{\left(\sin\alpha_T^U - \sin\bar{\alpha}\right)}{\sin\alpha} = B\bar{N}_0 \tag{11.48}$$

Note that $\bar{N}_0$ is the Z direction cosine of the paraxial pupil ray, and is given by

$$\bar{N}_0 = \frac{1}{\sqrt{1+\left(u_1\tau\right)^2}} \tag{11.49}$$

where u_1 is the convergence angle of the paraxial marginal ray from the axial object point in the object space, and τ is the fractional field of the extra-axial point under consideration.

Alternative derivations for ρ_S and ρ_T:

Let the edge ray of the axial pencil cut a reference sphere of centre O and radius $\bar{E}O$ at a height $\tilde{h}$. The latter may be considered the paraxial ray height at $\bar{E}$. The axial aperture angle is now given by

$$\sin\alpha = \frac{\tilde{h}}{\bar{N}_1\bar{R}} \tag{11.50}$$

From (11.42)

$$\sin\alpha_S = \frac{X_{max}}{\bar{R}} \tag{11.51}$$

so that

$$\rho_S = \frac{\sin\alpha_S}{\sin\alpha} = \bar{N}_1\left(\frac{X_{max}}{\tilde{h}}\right) \tag{11.52}$$

Similarly, from (11.47), we get

$$\left(\sin\alpha_T^U - \sin\bar{\alpha}\right) = \frac{Y_{max}}{\bar{R}} \tag{11.53}$$

giving

$$\rho_T = \frac{\left(\sin\alpha_T^U - \sin\bar{\alpha}\right)}{\sin\alpha} = \bar{N}_1\left(\frac{Y_{max}}{\tilde{h}}\right) \tag{11.54}$$

Relations (11.52) and (11.54) can be used to obtain paraxial approximations for ρ_S and ρ_T. In paraxial approximation

$$X_{max} \approx \tilde{h} \qquad Y_{max} \approx \bar{N}_1^2\tilde{h} \tag{11.55}$$

so that

$$\rho_S = \bar{N}_1 \qquad \rho_T = \bar{N}_1^3 \tag{11.56}$$

11.6.2 Exit Pupil Scale Ratios ρ'_S and ρ'_T

The entrance pupil scale ratio ρ_S is defined as

$$\rho_S = \frac{\sin\alpha_S}{\sin\alpha} = \frac{(n\sin\alpha_S)\bar{\eta}}{(n\sin\alpha)\bar{\eta}} = \frac{\bar{H}_S}{\bar{H}_0} \tag{11.57}$$

where $\bar{\eta} = \eta$, the paraxial object height since the object is distortion free.

The exit pupil scale ratio ρ'_S is defined as

$$\rho'_S = \frac{\sin\alpha'_S}{\sin\alpha'} = \frac{\left(n'\sin\alpha'_S\right)\bar{\eta}'}{(n'\sin\alpha')\bar{\eta}'} = \frac{\bar{H}'_S}{\bar{H}'_0} \tag{11.58}$$

where $\bar{\eta}'$ is the height of the point of intersection of the real pupil ray with the image plane from the optical axis. Using the property of sagittal invariant as shown in (7.269), we have $\bar{H}_S = \bar{H}'_S$. From (11.57)–(11.58), it follows

$$\frac{\rho_S}{\rho'_S} = \frac{\bar{H}'_0}{\bar{H}_0} = 1 + \left(\frac{\bar{H}'_0 - \bar{H}_0}{\bar{H}_0}\right) = 1 + \frac{\bar{\eta}' - \bar{\eta}}{\bar{\eta}} = 1 + FD \tag{11.59}$$

where FD is the fractional distortion. ρ'_S can be determined when ρ_S and FD are known by using the relation

$$\rho'_S = \frac{\rho_S}{1 + FD} \tag{11.60}$$

The exit pupil scale ratio ρ'_T is defined as

$$\rho'_T = \frac{\sin\alpha'_T - \sin\alpha'}{\sin\alpha'} = \frac{n'\left(\sin\alpha'_T - \sin\alpha'\right)\bar{\eta}'}{n'(\sin\alpha')\bar{\eta}'} = \frac{\bar{H}'_T}{\bar{H}'_0} \tag{11.61}$$

Using (11.16) we get

$$\bar{H}'_T = n'\bar{u}'_T\bar{\eta}' = n'\frac{\left(\sin\alpha_T^{U\prime} - \sin\bar{\alpha}'\right)}{\cos\bar{\alpha}'}\bar{\eta}' = n'\frac{\bar{N}'h'_T}{\bar{R}'}\bar{\eta}' \tag{11.62}$$

where $\bar{R}'$ is the radius $\bar{E}'\bar{Q}'$ of the exit pupil sphere (see Figure 11.2). The direction cosines of the pupil ray $\bar{E}'\bar{Q}'$ in the image space are $\left(0, \bar{M}', \bar{N}'\right)$. Since $\bar{\eta}' = \bar{R}'\bar{M}'$, (11.62) reduces to

$$\bar{H}'_T = n'\bar{N}'\bar{M}'h'_T \tag{11.63}$$

By t ray trace, the final k-th surface values $\left(h_T\right)_k$ and $\left(u_T\right)'_k$ are obtained. The paraxial tangential ray height at the exit pupil sphere is given by

$$h'_T = \left(h_T\right)_k \cos\bar{I}' - \left(u_T\right)'_k \bar{D}'_k \tag{11.64}$$

where $\bar{D}'_k$ is the distance from the point of intersection $\bar{P}_k$ of the pupil ray with the k-th surface to the exit pupil point $\bar{E}'$ on the axis. If the coordinates of the point $\bar{P}_k$ are $\left(0, \bar{Y}_k, \bar{Z}_k\right)$, $\bar{D}'_k$ is given by

$$\bar{D}'_k = \bar{P}_k\bar{E}' = \left(-\frac{\bar{Y}_k}{\bar{M}'}\right) \tag{11.65}$$

Using (11.64) and (11.65), equation (11.63) reduces to

$$\begin{aligned} \bar{H}'_T &= n'\bar{N}'\bar{M}'\left[\left(h_T\right)_k \cos\bar{I}' + \left(u_T\right)'_k \frac{\bar{Y}_k}{\bar{M}'}\right] \\ &= n'\bar{N}'\left\{\bar{M}'\left(h_T\right)_k \cos\bar{I}' + \bar{Y}_k\left(u_T\right)'_k\right\} \end{aligned} \tag{11.66}$$

Substituting the value of $\bar{H}'_T$ obtained from (11.66), ρ'_T can be evaluated by using relation (11.61).

Relations identical to (11.56) hold for paraxial approximations for ρ'_S and ρ'_T. The relations are

$$\rho'_S \approx \bar{N}'_k = \bar{N}' \qquad \rho'_T \approx \left(\bar{N}'_k\right)^3 = \left(\bar{N}'\right)^3 \tag{11.67}$$

where the Z direction cosine of the real pupil ray in the image space of an axially symmetric imaging system consisting of k refracting interfaces is $\bar{N}'_k = \bar{N}'$.

11.7 W_S and W_T from S and T Ray Traces

Figure 11.5 shows the final refracting interface $A_k\bar{P}_k$ of an axially symmetric imaging system consisting of k interfaces, with the real pupil ray $\bar{P}_k\bar{Q}'$ intersecting the paraxial image plane at $\bar{Q}'$. By S and T ray traces around the principal ray from the extra-axial object point, the sagittal and tangential foci S′ and T′

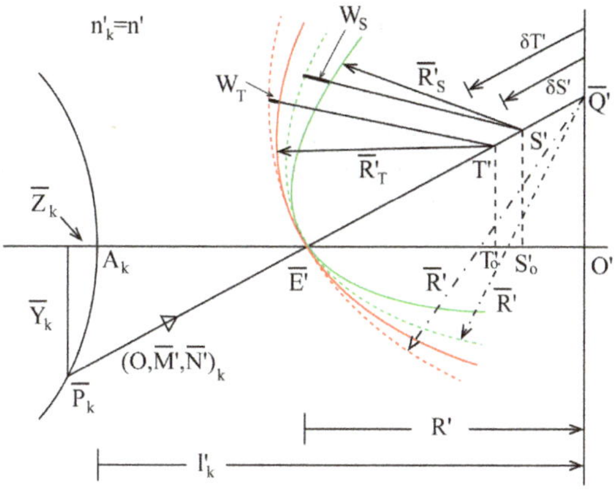

FIGURE 11.5 The sagittal focus S′ and the tangential focus T′ in the image space corresponding to the extra-axial object point Q (not shown). Physical interpretation of the effects of astigmatism in terms of wave aberrations W_S and W_T [Pseudo colour used for discrimination and better visualization].

in the image space are obtained (see Section 7.4.2.2). The customary measures for the sagittal and tangential focus errors are

$$\delta S' = (\bar{Q}'S') \qquad \delta T' = (\bar{Q}'T') \tag{11.68}$$

which are both of negative sign for the case shown in the diagram. From S ray trace, we get in the image space, with the usual notation,

$$\bar{P}_k S' = s'_k = \frac{(h_s)_k}{(u'_s)_k} \tag{11.69}$$

Similarly, from T ray trace, we get

$$\bar{P}_k T' = t'_k = \frac{(h_t \cos \bar{I}')_k}{(u'_t)_k} \tag{11.70}$$

Thus δS′ and δT′ can be calculated by using the formulae

$$\delta S' = \left[\frac{h_S}{u'_S} - \frac{l' - \bar{Z}}{\bar{N}'}\right]_k \tag{11.71}$$

$$\delta T' = \left[\frac{h_T \cos \bar{I}'}{u'_T} - \frac{l' - \bar{Z}}{\bar{N}'}\right]_k \tag{11.72}$$

The astigmatic focal difference δA′ is defined as

$$\delta A' = (S'T') = \delta T' - \delta S' \tag{11.73}$$

The case of negative δA′ is shown in the diagram.

In the small aperture approximation, the shape of the imaging wavefront through the exit pupil $\bar{E}'$ is such that along the tangential or meridian section it is a circle of radius $\bar{R}'_T$ and centre at T′, and along the sagittal section it is a circle of radius $\bar{R}'_S$ and centre at S′. For the case shown in the diagram, the wavefront lies in front of the reference sphere with its centre at $\bar{Q}'$ and radius $\bar{E}'\bar{Q}' = \bar{R}'$. Therefore, the wavefront aberrations in both the tangential and sagittal sections are positive. Indeed, with small aperture approximation, the wavefront aberration is positive along any azimuth in this case.

The focal difference terms, $\delta S', \delta T'$, and $\delta A'$, are of much help in easy visualization of astigmatism. But they cannot always be used as aberration measures. For example, if any of the image points $\bar{Q}'$, S′, or T′ be at infinity, the focal difference terms cannot act effectively as an aberration measure. To be useful directly in the aberration polynomial involving (x'_S, y'_T) as variables, the quantities

$$W_S = \frac{1}{2} n' (h'_S)^2 \left(\frac{1}{\bar{R}'_S} - \frac{1}{\bar{R}'} \right) \tag{11.74}$$

$$W_T = \frac{1}{2} n' (h'_T)^2 \left(\frac{1}{\bar{R}'_T} - \frac{1}{\bar{R}'} \right) \tag{11.75}$$

are, therefore, used in place of $\delta S'$ and $\delta T'$. In the above,

$$\bar{R}'_S = \bar{E}'S' \qquad \bar{R}'_T = \bar{E}'T' \qquad \bar{R}' = \bar{E}'\bar{Q}' \tag{11.76}$$

A physical interpretation of W_S and W_T is shown in Figure 11.5. Sections of the reference sphere centered at $\bar{Q}'$ and radius $\bar{R}' (= \bar{E}'\bar{Q}')$ are shown in Figure 11.5 along both the sagittal and the tangential azimuths by chained lines. The sagittal section of the paraxial wavefront along the principal ray is a circle with S′ as its centre and radius $\bar{R}'_S$ $(= \bar{E}'S')$. W_S is the optical path length along the paraxial marginal sagittal ray between the reference sphere and the sagittal section of the wavefront. Considering the pupil ray as a pseudo-optical axis, W_S is tantamount to change in wave aberration along the sagittal section for the marginal sagittal ray by longitudinal shift of focus from $\bar{Q}'$ to S′. Similarly, W_T is tantamount to change in wave aberration along the tangential or meridian section by longitudinal shift of focus from $\bar{Q}'$ to T′.

Retaining terms up to degree two in aperture in the expansion for wave aberration polynomial, $W(x'_S, y'_T)$ referred to $\bar{Q}'$ as focus, we get

$$W(x'_S, y'_T) = W_{20}(x'^2_S + y'^2_T) + W_{22} y'^2_T + O(3) \tag{11.77}$$

In the sagittal section, $y'_T = 0$, it reduces to

$$W(x'_S, 0) = W_{20} x'^2_S \tag{11.78}$$

For the marginal ray in the sagittal section, $x'_S = 1$, the wave aberration is

$$W(1,0) = W_{20} = W_S \tag{11.79}$$

as described above. Similarly, in the tangential or meridian section, $x'_S = 0$, we have

$$W(0, y'_T) = (W_{20} + W_{22}) y'^2_T \tag{11.80}$$

For the marginal ray in the tangential or meridian section, $y'_T = 1$, we get

$$W(0,1) = W_{20} + W_{22} = W_T \tag{11.81}$$

Noting from Figure 11.5 that

$$\bar{R}' - \bar{R}'_S = -\delta S' \text{ and } \bar{R}'\bar{N}'_k = R' \tag{11.82}$$

and using (11.71) for $\delta S'$, the expression (11.74) can be rewritten as

$$W_S = \frac{1}{2} n' \frac{(h'_S)^2}{\bar{R}'_S \bar{R}'} \left[\frac{l' - \bar{Z}}{\bar{N}'} - \frac{h_S}{u'_S} \right]_k \tag{11.83}$$

Let the paraxial ray height of the axial pencil at $\bar{E}'$ be $\tilde{h}'$, so that $R'u'_k = \tilde{h}'$. Equation (11.83) reduces to

$$W_S = \frac{h'_S}{\bar{R}'u'_k} \left[\frac{1}{2} n'u'_S u' \left\{ \frac{l' - \bar{Z}}{\bar{N}'} - \frac{h_S}{u'_S} \right\} \right]_k = \left(\frac{h'_S}{\tilde{h}'} \right) \bar{N}'_k \left[\frac{1}{2} n' \left\{ \frac{(h - u'\bar{Z})}{\bar{N}'} u'_S - h_S u' \right\} \right]_k \tag{11.84}$$

where $h_k = l'_k u'_k$ The first factor can be recast as

$$\left(\frac{h'_S}{\tilde{h}'} \right) \bar{N}'_k = \frac{(h'_S / \bar{R}')}{(\tilde{h}' / \bar{R}'\bar{N}'_k)} = \frac{(h'_S / \bar{R}')}{(\tilde{h}' / R')} = \frac{\sin \alpha_S{}'}{\sin \alpha'} = \rho'_S \tag{11.85}$$

Finally, a determinate expression for numerical computation of W_S is

$$W_S = \frac{1}{2} \rho'_S n'_k \left\{ \frac{(h - u'\bar{Z})}{\bar{N}'} u'_S - h_S u' \right\}_k \tag{11.86}$$

Similarly, using equation (11.72) for $\delta T'$, the expression (9.148) for W_T can be rewritten as

$$W_T = \frac{h'_T}{\bar{R}'u'_k} \left[\frac{1}{2} n'u'_T u' \left\{ \frac{l' - \bar{Z}}{\bar{N}'} - \frac{h_T \cos \bar{I}'}{u'_T} \right\} \right]_k = \left(\frac{h'_T}{\tilde{h}'} \right) \bar{N}'_k \left[\frac{1}{2} n' \left\{ \frac{(h - u'\bar{Z})}{\bar{N}'} u'_S - h_T \cos \bar{I}' u' \right\} \right]_k \tag{11.87}$$

Using equations (11.16) and (11.36) and the relation $\sin \alpha' = (\tilde{h}' / R') = (\tilde{h}' / \bar{R}'\bar{N}'_k)$, the factor before the bracketed term can be written as

$$\left(\frac{h'_T}{\tilde{h}'} \right) \bar{N}'_k = \frac{(h'_T / \bar{R}')}{(\tilde{h}' / \bar{R}'\bar{N}'_k)} = \frac{(h'_T / \bar{R}')}{(\tilde{h}' / R')} = \frac{1}{\bar{N}'_k} \frac{(\sin \alpha'_T - \sin \bar{\alpha}')}{\sin \alpha'} = \frac{\rho'_T}{\bar{N}'_k} \tag{11.88}$$

The determinate expression for W_T is

$$W_T = \frac{1}{2} n'_k \left(\frac{\rho'_T}{\bar{N}'_k} \right) \left\{ \frac{(h - u'\bar{Z})}{\bar{N}'} u'_T - h_T \cos \bar{I}' u' \right\}_k \tag{11.89}$$

The quantities W_S and W_T have the specific advantage that they directly give the coefficients W_{20} and W_{22} of the wave aberration polynomial when the aberration is expressed as a function of canonical pupil coordinates (x_S, y_T). Nevertheless, they are not useful in the visualization of the tangential and sagittal focal surfaces from an examination of the values of W_S and W_T at different field angles. It is important to note that when the s and t foci coincide, W_S and W_T are, in general, unequal. This is so because of the inequality in size of the real pupil along the tangential and the sagittal sections.

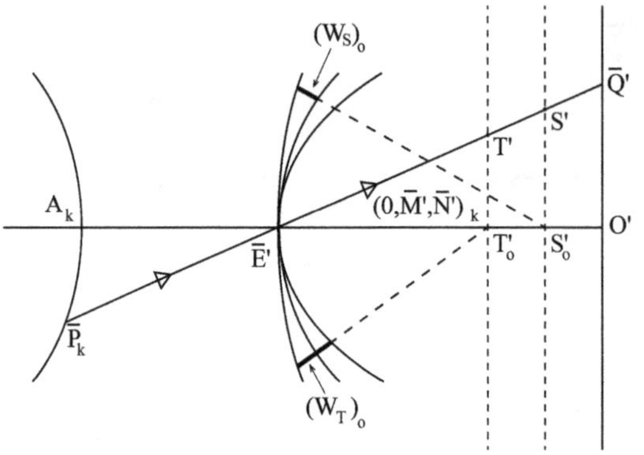

FIGURE 11.6 $(W_S)_0$ and $(W_T)_0$ are the amounts of axial defocus required to shift the paraxial image plane at O′ to transverse planes containing the points S′ and T′, respectively.

This problem can be circumvented by evaluating two related quantities $(W_S)_0$ and $(W_T)_0$. These are the amounts of axial defocus required to shift the paraxial image plane at O′ to planes containing the points S′ and T′, respectively. The axial defocus is considered as the change in wave aberration measured with respect to a reference sphere at $\bar{E}'$. Figure 11.6 illustrates $(W_S)_0$ and $(W_T)_0$ in the image space. $(W_S)_0$ and $(W_T)_0$ are defined as

$$(W_S)_0 = \frac{1}{2} n' (\tilde{h}')^2 \left(\frac{1}{\tilde{R}'_S} - \frac{1}{\tilde{R}'} \right) \tag{11.90}$$

$$(W_T)_0 = \frac{1}{2} n' (\tilde{h}')^2 \left(\frac{1}{\tilde{R}'_T} - \frac{1}{\tilde{R}'} \right) \tag{11.91}$$

where $\tilde{h}'$ is the height of the PMR (not shown in the figure), from the axial object point O, at a reference sphere through $\bar{E}'$ and centred at O′. The radii $\tilde{R}'$, $\tilde{R}'_S$, and $\tilde{R}'_T$ are given by

$$\bar{E}'O' = \tilde{R}' = \bar{N}'_k \bar{R}' \quad \bar{E}'S'_0 = \tilde{R}'_S = \bar{N}'_k \bar{R}'_S \quad \bar{E}'T'_0 = \tilde{R}'_T = \bar{N}'_k \bar{R}'_T \tag{11.92}$$

Substituting from (11.92) in (11.90) and (11.91), we get

$$(W_S)_0 = \left(\frac{\tilde{h}'}{h'_S} \right)^2 \frac{1}{\bar{N}'_k} \left\{ \frac{1}{2} n' (h'_S)^2 \left[\frac{1}{\bar{R}'_S} - \frac{1}{\bar{R}'} \right] \right\} \tag{11.93}$$

$$(W_T)_0 = \left(\frac{\tilde{h}'}{h'_T} \right)^2 \frac{1}{\bar{N}'_k} \left\{ \frac{1}{2} n' (h'_T)^2 \left[\frac{1}{\bar{R}'_T} - \frac{1}{\bar{R}'} \right] \right\} \tag{11.94}$$

The multiplying factor of the bracketed quantity in the right-hand side of equation (11.93) can be recast as

$$\left(\frac{\tilde{h}'}{h'_S} \right)^2 \frac{1}{\bar{N}'_k} = \frac{\bar{N}'_k \left(\tilde{h}' / \bar{N}'_k \bar{R}' \right)^2}{\left(h'_S / \bar{R}' \right)^2} = \frac{\bar{N}'_k \left(\tilde{h}' / \tilde{R}' \right)^2}{\left(\sin\alpha'_S \right)^2} = \frac{\bar{N}'_k \left(\sin\alpha' \right)^2}{\left(\sin\alpha'_S \right)^2} = \frac{\bar{N}'_k}{\left(\rho'_S \right)^2} \tag{11.95}$$

Similarly, the multiplying factor of the bracketed quantity in the right-hand side of equation (11.94) can be recast as

$$\left(\frac{\tilde{h}'}{h'_T}\right)^2 \frac{1}{\bar{N}'_k} = \frac{\bar{N}'_k\left(\tilde{h}'/\bar{N}'_k\bar{R}'\right)^2}{\left(h'_T/\bar{R}'\right)^2} = \frac{\bar{N}'_k\left(\tilde{h}'/\tilde{R}'\right)^2}{\left[\frac{\left(\sin\alpha'_T - \sin\bar{\alpha}'\right)}{\bar{N}'_k}\right]^2} = \bar{N}'^3_k\left(\frac{\sin\alpha'}{\left(\sin\alpha'_T - \sin\bar{\alpha}'\right)}\right)^2 = \frac{\bar{N}'^3_k}{\left(\rho'_S\right)^2} \quad (11.96)$$

From (11.74), (11.75), and (11.93) – (11.96), we get the relations between $\left[(W_S)_0, (W_T)_0\right]$ and $\left[W_S, W_T\right]$ as

$$\left.\begin{aligned} (W_S)_0 &= \frac{\bar{N}'_k}{\left(\rho'_S\right)^2} W_S \\ (W_T)_0 &= \frac{\bar{N}'^3_k}{\left(\rho'_S\right)^2} W_T \end{aligned}\right\} \quad (11.97)$$

Note that the above relations are determinate for any location of object/image and entrance/exit pupil of the system.

11.8 Local Sagittal and Tangential Invariants for Extra-Axial Images

Let Q be a point near to the extra-axial point $\bar{Q}$ (Figure 11.7). With its origin at $\bar{Q}$, the coordinates of the point Q are (ξ_S, η_T). Considering $\bar{Q}\bar{E}\bar{E}'\bar{Q}'$ as an effective axis for imaging in the neighbourhood of $\bar{Q}$, the 'paraxial' image of Q is defined to be the point Q′, where the ray emerging from Q and passing through point $\bar{E}'$ intersects the image plane. In general, this ray does not pass exactly through the point $\bar{E}$ in the object space, although the ray lies close to $\bar{E}$. Let the coordinates of the point Q′ be (ξ'_S, η'_T), with respect to the origin at $\bar{Q}'$ in the image space. Hopkins proved the existence of two paraxial invariants in the neighbourhood of the conjugates $\bar{Q}, \bar{Q}'$. The sagittal paraxial invariant is

$$\left(n\sin\alpha_S\right)\xi_S = \left(n'\sin\alpha'_S\right)\xi'_S \quad (11.98)$$

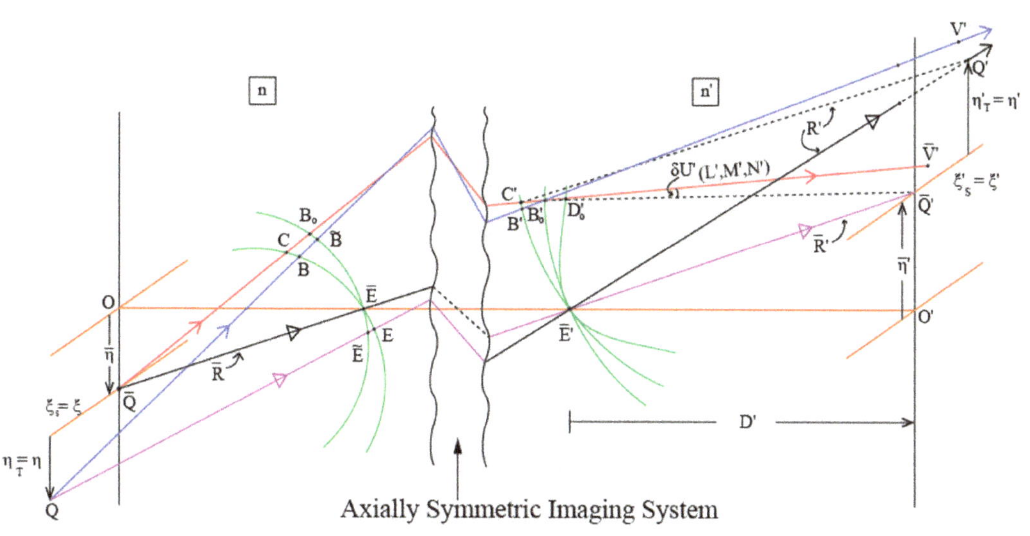

FIGURE 11.7 Local sagittal and tangential invariants for image space associated rays [Pseudo colour used for discrimination and better visualization].

and the tangential paraxial invariant is

$$\left\{n\left(\sin\alpha_T - \sin\bar{\alpha}\right)\right\}\eta_T = \left\{n'\left(\sin\alpha'_T - \sin\bar{\alpha}'\right)\eta'_T\right\} \tag{11.99}$$

where the angles $\alpha_S, \alpha'_S, \alpha_T, \alpha'_T, \bar{\alpha}, \bar{\alpha}'$ are the finite angles in the sagittal and meridian sections, as defined above.

Using simplified notations (ξ, η) for (ξ_S, η_T), and (ξ', η') for (ξ'_S, η'_T), by omitting the subscripts, the above invariants can be rewritten as

$$\left(n\sin\alpha_S\right)\xi = \left(n'\sin\alpha'_S\right)\xi' \tag{11.100}$$

$$\left\{n\left(\sin\alpha_T - \sin\bar{\alpha}\right)\right\}\eta = \left\{n'\left(\sin\alpha'_T - \sin\bar{\alpha}'\right)\right\}\eta' \tag{11.101}$$

We present below a derivation of these invariants. Our treatment is an adaptation of the initial derivation by Hopkins [8].

Let $\bar{Q}B_0B'_0\bar{V}'$ be an aperture ray from $\bar{Q}$ (Figure 11.7). It intersects the image plane at $\bar{V}'$, and the reference sphere $\bar{E}'B'_0$ at B'_0. The centre of the reference sphere is at $\bar{Q}'$, and its radius is $\bar{E}'\bar{Q}' = \bar{R}'$. We consider the aperture ray from Q to be the ray which passes through the point B'_0 in the image space. In the figure, that ray is seen to be $QB\tilde{B}B'B'_0V'$, which intersects the image plane at V'. Note that the reference pupil rays from the extra-axial points $\bar{Q}$ and Q pass through the same point $\bar{E}'$ in the image space, and the chosen aperture rays from the extra-axial points $\bar{Q}$ and Q pass through the same point B'_0 in the image space. In general, the object space segments of the pupil rays do not pass through any single point in the object space, and neither do the chosen aperture rays. Rays from Q, namely $Q\tilde{E}E\bar{E}'$ and $QB\tilde{B}B'B'_0$, are called 'image-space associated rays'.

$\bar{E}B_0$ and $\bar{E}'B'_0$ are sections of two reference spheres in the object space and the image space with centres at $\bar{Q}$ and $\bar{Q}'$, and radii $\bar{Q}\bar{E}$, $\bar{Q}'\bar{E}'$, respectively. $E\bar{E}BC$ and $\bar{E}'B'C'$ are sections of two reference spheres in the object space and the image space with centres at Q and Q', and radii $Q\bar{E}$, $Q'\bar{E}'$, respectively. Note that, in the object space, the aperture ray from $\bar{Q}$ intersects the reference sphere corresponding to $\bar{Q}$ at B_0, and the reference sphere for Q at the point C. The aperture ray from Q intersects the former reference sphere at $\tilde{B}$ and the latter reference sphere at B. The pupil ray from Q intersects the $\tilde{Q}-$ reference sphere at $\tilde{E}$, and the $Q-$ reference sphere at E. In the image space, at $\bar{E}'$ the two pupil rays are normal to their respective reference spheres centered at $\bar{Q}'$ and Q'. The aperture rays from Q and $\bar{Q}$ intersect the $Q'-$reference sphere at B' and C', respectively. By definition, both the aperture rays intersect the $\bar{Q}'-$reference sphere at B'_0. $\bar{E}'D'_0$ is the actual wavefront forming image of the point $\bar{Q}$ at $\bar{Q}'$.

By definition, we have

$$\bar{Q}\bar{E} = \bar{Q}B_0 \qquad QE = Q\bar{E} = QB \tag{11.102}$$

The wavefront aberration for the aperture ray $\bar{Q}B_0B'_0$ is given by

$$W_{\bar{Q}'} = \left[\bar{E}\ldots\bar{E}'\right] - \left[B_0\ldots B'_0\right] \tag{11.103}$$

The wavefront aberration for the aperture ray $QB\tilde{B}\ldots B'B'_0$ of the pencil forming the image of Q at Q' is given by

$$W_{Q'} = \left[E\ldots\bar{E}'\right] - \left[B\ldots B'\right] \tag{11.104}$$

From the above relations, we get

$$W_{Q'} - W_{\bar{Q}'} = \{[E\ldots\bar{E}'] - [\bar{E}\ldots\bar{E}']\} - \{[B\ldots B'] - [B_0\ldots B_0']\}$$
$$= \{-[\tilde{E}E] + [\tilde{E}\ldots\bar{E}'] - [\bar{E}\ldots\bar{E}']\} - \{[B\tilde{B}] + [\tilde{B}\ldots B_0'] - [B'B_0'] - [B_0\ldots B_0']\} \quad (11.105)$$

Both the angle $\angle\tilde{E}\bar{E}E$ and the distance $(E\bar{E})$ are of the same order as the object size $\bar{Q}Q\left(=\rho=\sqrt{\xi^2+\eta^2}\right)$, so that

$$(\tilde{E}E) = (E\bar{E})\tan\angle\tilde{E}\overline{EE} \sim O_\rho(1)O_\rho(1) = O_\rho(2) \quad (11.106)$$

By Fermat's principle, we have

$$[\tilde{E}E\ldots\bar{E}'] = [\bar{E}\ldots\bar{E}'] + O_\rho(2) \quad (11.107)$$

$$[\tilde{B}\ldots B'B_0'] = [B_0C'B_0'] + O_\rho(2) \quad (11.108)$$

The relation (11.105) reduces to

$$W_{Q'} - W_{\bar{Q}'} = [B'B_0'] - [B\tilde{B}] + O_\rho(2) \quad (11.109)$$

The distances CB_0 and $B\tilde{B}$ are related by

$$[CB_0] = [B\tilde{B}] + O_\rho(2) \quad (11.110)$$

Similarly,

$$[C'B_0'] = [B'B_0'] + O_\rho(2) \quad (11.111)$$

The equation (11.105) can thus be rewritten as

$$W_{Q'} - W_{\bar{Q}'} = [C'B_0'] - [CB_0] + O_\rho(2) \quad (11.112)$$

Note that CB_0 is the intercept of the aperture ray from $\bar{Q}$ lying between the reference spheres for Q and $\bar{Q}$. Similarly, $C'B_0'$ is the intercept of the aperture ray from $\bar{Q}$ lying between the reference spheres for Q' and $\bar{Q}'$. An approximate expression for $[C'B_0']$ in terms of $(X',Y'),(\xi',\eta'),\bar{R}'$, and n' is derived next.

Let B_0' have coordinates (X',Y',Z') with $\bar{E}'$ as origin. If $p'=(C'B_0')$, the coordinates of C' are

$$[(X'-L'p'),(Y'-M'p'),(Z'-N'p')] \quad (11.113)$$

where (L',M',N') are the direction cosines of the ray $C'B_0'\bar{V}'$. The reference sphere $\bar{E}'C'$ has its centre at Q' and its radius $\bar{E}'Q' = C'Q' = R'$. Coordinates of the point Q', with respect to the origin at $\bar{E}'$, are $(\xi', \bar{\eta}'+\eta', D')$ where $D' = \bar{E}'O'$. The equation of the reference sphere $\bar{E}'C'$ is

$$(C'Q')^2 = [(X'-L'p')-\xi']^2 + [(Y'-M'p')-(\bar{\eta}'+\eta')]^2 + [(Z'-N'p')-D']^2 = R'^2 \quad (11.114)$$

Note that

$$(\bar{E}'Q')^2 = R'^2 = (\xi')^2 + (\bar{\eta}'+\eta')^2 + (D')^2 \quad (11.115)$$

Therefore, (11.115) reduces to

$$(X' - L'p')^2 + (Y' - M'p')^2 + (Z' - N'p')^2 - 2(X' - L'p')\xi'$$
$$-2(Y' - M'p')(\bar{\eta}' + \eta') - 2(Z' - N'p')D' = 0 \tag{11.116}$$

Coordinates of the point $\bar{Q}'$ with respect to $\bar{E}'$ as origin are $(0, \bar{\eta}', D')$. The equation of the reference sphere $\bar{E}'B_0'$ with $\bar{Q}'$ as centre and $\bar{E}'\bar{Q}' = B_0'\bar{Q}_0' = \bar{R}'$ as radius is

$$(B_0'\bar{Q}')^2 = X'^2 + (Y' - \bar{\eta}')^2 + (Z' - D')^2 = \bar{R}'^2 \tag{11.117}$$

We have

$$(\bar{E}'\bar{Q}')^2 = \bar{R}'^2 = \bar{\eta}^2 + D'^2 \tag{11.118}$$

From the above two equations, we get

$$X'^2 + Y'^2 + Z'^2 - 2Y'\bar{\eta}' - 2Z'D' = 0 \tag{11.119}$$

Using (11.119), the equation (11.116) reduces to

$$p'^2 + 2\left[L'(\xi' - X') + M'(\bar{\eta}' + \eta' - Y') + N'(D' - Z')\right]p' - 2(X'\xi' + Y'\eta') = 0 \tag{11.120}$$

The square bracket may be seen to be the projection of $B_0'Q'$ on the ray $C'B_0'\bar{V}'$ with direction cosines (L', M', N'). For extended paraxial calculations, this projection can be approximated by $B_0'\bar{V}'$, which is equal to $\bar{R}'\cos\delta U'$, where $\delta U'$ is the angular aberration. (11.120) may thus be written as

$$p'^2 + 2(\bar{R}'\cos\delta U')p' - 2(X\xi + Y\eta) = 0 \tag{11.121}$$

from which we get

$$p' = \frac{(X'\xi' + Y'\eta')}{\bar{R}'\cos\delta U'}\left[1 + \left(\frac{p'}{2\bar{R}'\cos\delta U'}\right)\right]^{-1} \tag{11.122}$$

which is an exact relation satisfied by p'.

In Figure 11.7, $\bar{E}'D_0'$ is a section of the actual wavefront forming the image at $\bar{Q}'$. The aperture ray $C'B_0'D_0'\bar{V}'$ intersects the reference sphere $\bar{E}'B_0'$ at B_0', and the actual wavefront $\bar{E}'D_0'$ at D_0'. The wavefront aberration of the aperture ray is $[B_0'D_0']$. If the aberration of the same wavefront is referred to the reference sphere $\bar{E}'B'C'$, whose centre is at Q', the wavefront aberration is increased by the amount $[C'B_0'] = n'p'$. This change in wavefront aberration on account of the transverse shift of focus from $\bar{Q}'$ to Q' is denoted by $\delta W_T'$ so that

$$\delta W_T' = [C'B_0'] = n'p' \tag{11.123}$$

Using (11.122), we can write

$$\delta W_T' = n'\left\{\frac{(X'\xi' + Y'\eta')}{\bar{R}'\cos\delta U'}\right\}\left[1 + \left(\frac{\delta W_T'}{2n'\bar{R}'\cos\delta U'}\right)\right]^{-1} \tag{11.124}$$

Similarly, in the object space

$$\delta W_T = [CB_0] = np = n\left\{\frac{X\xi + Y\eta}{\bar{R}}\right\}\left[1 + \left(\frac{\delta W_T}{2n\bar{R}}\right)\right]^{-1} \tag{11.125}$$

The quantity $\cos\delta U$ does not appear in (11.125). This is because $\delta U = 0$ in the object space.

From the above, the first approximations for $\delta W_T'$ and δW_T are

$$\delta W_T' = [C'B_0'] = n'\left\{\frac{(X'\xi' + Y'\eta')}{\bar{R}'\cos\delta U'}\right\} + O_p(2) \tag{11.126}$$

$$\delta W_T = [CB_0] = np = n\left\{\frac{X\xi + Y\eta}{\bar{R}}\right\} + O_p(2) \tag{11.127}$$

In terms of the two expressions given above, (11.112) can be rewritten as

$$W_{Q'} - W_{\bar{Q}'} = n'\left\{\frac{(X'\xi' + Y'\eta')}{\bar{R}'\cos\delta U'}\right\} - n\left\{\frac{X\xi + Y\eta}{\bar{R}}\right\} + O_p(2) \tag{11.128}$$

Since the angular aberration $\delta U' = O(1)$ in the aperture variables (X', Y'), $\cos\delta U' = O(2)$. So, equation (11.128) reduces to

$$W_{Q'} - W_{\bar{Q}'} = n'\left\{\frac{(X'\xi' + Y'\eta')}{\bar{R}'}\right\} - n\left\{\frac{X\xi + Y\eta}{\bar{R}}\right\} + O(2) + O_p(2) \tag{11.129}$$

In the above, $(X, Y), (X', Y')$ are coordinates of B_0 and B_0' with respect to origins at $\bar{E}$ and $\bar{E}'$, respectively. $(\xi, \eta), (\xi', \eta')$ are coordinates of the points Q, Q' with respect to origins at $\bar{Q}$, and $\bar{Q}'$respectively. $\bar{R} = \bar{E}\bar{Q}$, $\bar{R}' = \bar{E}'\bar{Q}'$. The pupil variables X, X' may be normalized by paraxial ray heights of a close sagittal ray as

$$x_s = (X/h_s) \qquad x_S' = (X'/h_s') \tag{11.130}$$

where $h_s = X_{max}, h_S' = X_{max}'$ (Figure 11.2). We have

$$\sin\alpha_s = (h_s/\bar{R}) \qquad \sin\alpha_s' = (h_s'/\bar{R}') \tag{11.131}$$

The pupil variables Y, Y' are normalized by the product of the paraxial ray height of a close tangential ray and the Z direction cosine of the pupil ray as

$$y_T = (Y/\bar{N}h_T) \qquad y_T' = (Y'/\bar{N}'h_T') \tag{11.132}$$

where $\bar{N}h_T = Y_{max}$, and $\bar{N}'h_T' = Y_{max}'$ (Figure 11.2).

We have

$$(\sin\alpha_T - \sin\bar{\alpha}) = (\bar{N}h_T/\bar{R}) \qquad (\sin\alpha_T' - \sin\bar{\alpha}') = (\bar{N}'h_T'/\bar{R}') \tag{11.133}$$

For object/image elements in the neighbourhood of $\bar{Q}$ and $\bar{Q}'$, respectively, the reduced coordinates are defined as

$$G_S = n(\sin\alpha_S)\xi \qquad H_T = n(\sin\alpha_T - \sin\bar{\alpha})\eta \tag{11.134}$$

$$G'_S = n'(\sin\alpha'_S)\xi' \qquad H'_T = n'(\sin\alpha'_T - \sin\bar{\alpha}')\eta' \tag{11.135}$$

In terms of the reduced variables, the equation (11.129) can be written as

$$W_{Q'} - W_{\bar{Q}'} = (G'_S x'_S - G_S x_S) + (H'_T y'_T - H_T y_T) + O(2) + O_p(2) \tag{11.136}$$

In his extended paraxial analysis for extra-axial pencils, Hopkins [6] proved that for any aperture ray of the extra-axial pencil

$$x'_S = x_S + O(2) \qquad y'_T = y_T + O(2) \tag{11.137}$$

Substituting for (x'_S, y'_T) in (11.136), it follows

$$W_{Q'} - W_{\bar{Q}'} = (G'_S - G_S)x_S + (H'_T - H_T)y_T + O(2) + O_p(2) \tag{11.138}$$

Let

$$W_{\bar{Q}'}(x'_S, y'_T) = a_0 + a_1 x'_S + b_1 y'_T + O(2) \tag{11.139}$$

For $x'_S, y'_T \to 0, W_{\bar{Q}'}(x'_S, y'_T) \to 0$. Therefore, a_0 must be equal to zero. Since both the wavefront and the reference sphere are normal to the pupil ray at $\bar{E}'$,

$$W_{\bar{Q}'}(x'_S, y'_T) = W_{\bar{Q}'}(-x'_S, -y'_T) \tag{11.140}$$

This can hold for any value of x'_S, y'_T, if and only if $a_1 = b_1 = 0$. Thus, $W_{\bar{Q}'}(x'_S, y'_T) = O(2)$. Similarly, it can be shown that $W_{Q'}(x'_S, y'_T) = O(2)$. Thus, in the limit of small field size, (11.138) gives

$$(G'_S - G_S)x_S + (H'_T - H_T)y_T = 0 \tag{11.141}$$

for all values of x_S and y_T. This is possible only when

$$G'_S = G_S \qquad H'_T = H_T \tag{11.142}$$

From (11.134) and (11.135), it follows

$$(n'\sin\alpha'_S)\xi' = (n\sin\alpha_S)\xi \tag{11.143}$$

$$\{n'(\sin\alpha'_T - \sin\bar{\alpha}')\}\eta' = \{n(\sin\alpha_T - \sin\bar{\alpha})\}\eta \tag{11.144}$$

The 'paraxial' sagittal and tangential invariants in the neighbourhood of extra-axial points mentioned earlier in (11.100) and (11.101) are the same as (11.143) and (11.144). Note that there is a sagittal invariant for real rays akin to the paraxial one described above, but there is no real ray equivalent of the paraxial tangential invariant.

11.9 Reduced Coordinates on the Object/Image Plane and Local Magnifications

It may be recalled that the normal paraxial magnification M is derived from the Smith-Helmholtz-Lagrange invariant relationship

$$(n \sin\alpha)\eta = (n' \sin\alpha')\eta' \tag{11.145}$$

$$M = \frac{\eta'}{\eta} = \frac{(n \sin\alpha)}{(n' \sin\alpha')} \tag{11.146}$$

Similarly, the paraxial sagittal and tangential invariants in the neighbourhood of Q and Q′ lead to two local magnifications in the neighbourhood of these points (Figure 11.8). The local sagittal magnification M_S can be derived from (11.143) as

$$M_S = \frac{\xi'}{\xi} = \frac{(n \sin\alpha_S)}{(n' \sin\alpha')} \tag{11.147}$$

The local tangential magnification M_T follows from (11.144) as

$$M_T = \frac{\eta'}{\eta} = \frac{\{n(\sin\alpha_T - \sin\bar{\alpha})\}}{\{n'(\sin\alpha'_T - \sin\bar{\alpha}')\}} \tag{11.148}$$

The sagittal pupil scale ratios ρ_s and ρ'_s, defined by (11.37), provide a link between the paraxial sagittal magnification M_S with the ordinary paraxial magnification M by

$$M_S = \frac{\rho_s (n \sin\alpha)}{\rho'_s (n' \sin\alpha')} = \left(\frac{\rho_s}{\rho'_s}\right) M \tag{11.149}$$

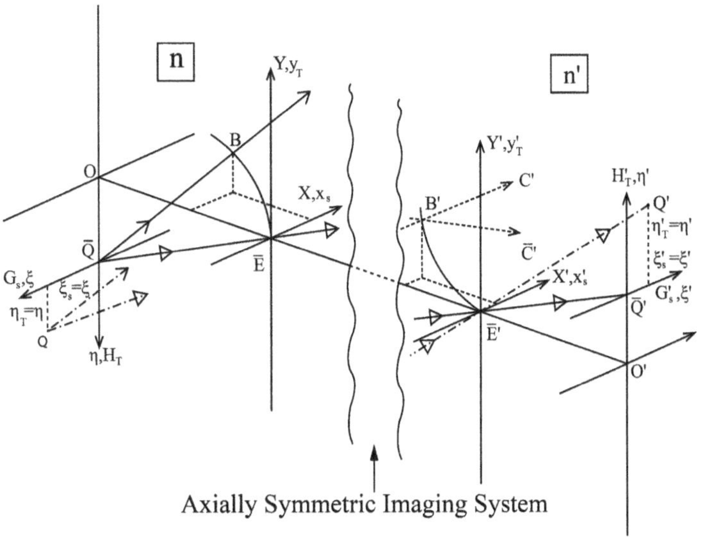

FIGURE 11.8 Reduced coordinates on the object/image plane and local magnifications.

Similarly, a link between the paraxial tangential magnification M_T with the ordinary paraxial magnification M is provided by the tangential pupil scale ratios ρ_T and ρ_T', defined by (11.36) as

$$M_T = \frac{\rho_T (n \sin \alpha)}{\rho_T' (n' \sin \alpha')} = \left(\frac{\rho_T}{\rho_T'} \right) M \tag{11.150}$$

In general, a small square located around $\bar{Q}$ will be imaged as a rectangle around $\bar{Q}'$. The aspect ratio of this rectangle in real space is (M_T / M_S), and

$$(M_T / M_S) = 1 \text{ when} (\rho_S / \rho_S') = (\rho_T / \rho_T') \tag{11.151}$$

Note that the local tangential and sagittal magnifications, as well as the object and image space pupil scale ratios, have specific values for a specific field point, so the aspect ratio of the rectangular image may be different at different values of the fractional field. The aspect ratio is the same over the whole field of view only when the imaging system is completely free from distortion, so that (11.151) holds good at any point over the field.

Reduced coordinates for points $Q(\xi, \eta)$ in the object plane, and $Q'(\xi', \eta')$ in the image plane are defined by

$$G_S = (n \sin \alpha_S) \xi \qquad H_T = \{ n (\sin \alpha_T - \sin \bar{\alpha}) \} \eta \tag{11.152}$$

$$G_S' = (n' \sin \alpha_S') \xi' \qquad H_T' = \{ n' (\sin \alpha_T' - \sin \bar{\alpha}') \} \eta' \tag{11.153}$$

From (11.143) and (11.144), it follows

$$G_S = G_S' \qquad H_T = H_T' \tag{11.154}$$

Note that in the hyperspace of these reduced coordinates, the image of a small square, located anywhere in the field, is always a square with aspect ratio unity.

Reduced values $\{ \delta G_S', \delta H_T' \}$ for the transverse aberrations $\{ \delta\xi', \delta\eta' \}$ are defined by

$$\delta G_S' = (n' \sin \alpha_S') \delta\xi' \qquad \delta H_T' = \{ n' (\sin \alpha_T' - \sin \bar{\alpha}') \} \delta\eta' \tag{11.155}$$

Note that reduced transverse aberrations remain determinate even when the image is at infinity.

11.10 Canonical Relations: Generalized Sine Condition

Figure 11.9 shows the image space of an axi-symmetric system with usual notations for the various parameters. It has been shown earlier in (8.11) – (8.12) that the components $\{ \delta\xi', \delta\eta' \}$ of the transverse aberration of an aperture ray $B'\bar{C}'$ of the pencil of rays forming an image at $\bar{Q}'$ is related to its wave aberration W at the point $B'(X', Y')$ by the relations

$$\delta\xi' = -\left(\frac{\bar{R}'}{n'} \right) \frac{\partial W}{\partial X'} \qquad \delta\eta' = -\left(\frac{\bar{R}'}{n'} \right) \frac{\partial W}{\partial Y'} \tag{11.156}$$

Using Hopkins' normalization scheme for points in the neighbourhood of the point $\bar{Q}'$ on the image plane, the reduced aberrations $\delta G_S'$ and $\delta H_T'$ are related to $\delta\xi'$ and $\delta\eta'$ by

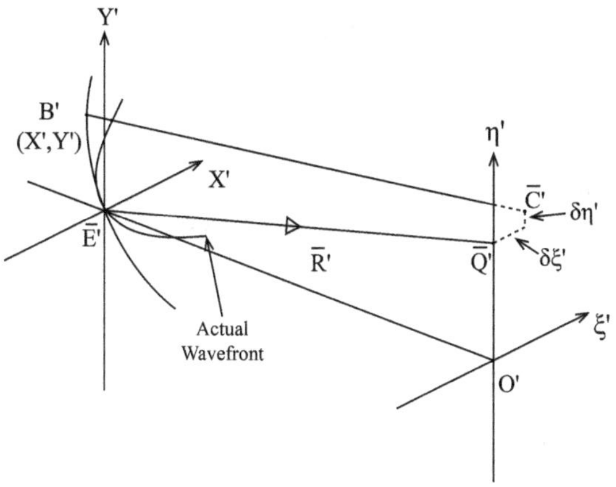

FIGURE 11.9 The image space of an axi-symmetric system with usual notations.

$$\delta G'_S = n'(\sin\alpha'_S)\delta\xi' = n'u'_S\delta\xi' = n'\left(\frac{h'_S}{\bar{R}'}\right)\delta\xi' \tag{11.157}$$

$$\delta H'_T = n'(\sin\alpha'_T - \sin\bar{\alpha}')\delta\eta' = n'\bar{N}'\left(\frac{h'_T}{\bar{R}'}\right)\delta\eta' \tag{11.158}$$

Hopkins' canonical variables (x'_S, y'_T) for a point on the exit pupil are related to the rectangular coordinates (X', Y') of the point by

$$x'_S = \frac{X'}{h'_S} \qquad y'_T = \frac{Y'}{\bar{N}'h'_T} \tag{11.159}$$

It follows

$$\frac{\partial W}{\partial x'_S} = h'_S\frac{\partial W}{\partial X'} \qquad \frac{\partial W}{\partial y'_T} = \bar{N}'h'_T\frac{\partial W}{\partial Y'} \tag{11.160}$$

On algebraic manipulation of the above relations, we get the first pair of canonical relations

$$\left.\begin{aligned} \delta G'_S &= -\frac{\partial W}{\partial x'_S} \\ \delta H'_T &= -\frac{\partial W}{\partial y'_T} \end{aligned}\right\} \tag{11.161}$$

From the discussions in Section 11.8, it is apparent that

$$G'_S = G_S + O_\rho(2) \qquad H'_T = H_T + O_\rho(2) \tag{11.162}$$

Substituting these values in (11.136) we get

$$W_{Q'} - W_{\bar{Q}'} = G_S(x'_S - x_S) + H_T(y'_T - y_T) + O_\rho(2) \tag{11.163}$$

If the points $\bar{Q}'$ and Q' are close, G'_S and H'_T will be small, and we take

$$G'_S \approx \delta G'_S \sim \delta G_S \qquad H'_T \approx \delta H'_T \sim \delta H_T \tag{11.164}$$

Expressing the small increment in aberration for the aperture ray through B'_0 when the object point is moved from $\bar{Q}'$ to Q' by $\delta W = W_{Q'} - W_{\bar{Q}'}$, the relation (11.163) may be written as

$$\delta W = \delta G_S (x'_S - x_S) + \delta H_T (y'_T - y_T) + O_\rho(2) \tag{11.165}$$

so that, in the limit $\rho \to 0$,

$$\left.\begin{aligned} \frac{\partial W}{\partial G_S} &= (x'_S - x_S) \\ \frac{\partial W}{\partial H_T} &= (y'_T - y_T) \end{aligned}\right\} \tag{11.166}$$

The above two relations constitute the second pair of canonical relations. The aberrations for rays through B'_0, of waves forming images in the neighbourhood of $\bar{Q}'$, is thus constant when

$$x'_S = x_S \qquad y'_T = y_T \tag{11.167}$$

This is the condition of isoplanatism, and it expresses the generalized form of sine condition. It is important to note that the satisfaction of this criterion for isoplanatism is irrespective of the magnitude of the geometrical distortion at $\bar{Q}'$.

In this treatment, the wavefront aberration W is taken as a function of the four reduced variables, and it is expressed as $W(x'_S, y'_T, G_S, H_T)$. The change in wavefront aberration when the object moves from $\bar{Q}$ to Q is

$$W(x'_S, y'_T, G_S, H_T) - W(x'_S, y'_T, 0, 0) = \left(\frac{\partial W}{\partial G_S}\right) G_S + \left(\frac{\partial W}{\partial H_T}\right) H_T \tag{11.168}$$

neglecting terms in degree 2 and above in G_S, H_T.

For the special case of axial imaging in the neighbourhood of O′, it is usual to suppress the subscripts S and T. For $G = 0$, the above relation reduces to

$$W(x', y'; 0, H) - W(x', y'; 0, 0) = \left(\frac{\partial W}{\partial H}\right) H = H(y' - y) \tag{11.169}$$

This gives the usual sine condition coma for the ray passing through reduced pupil variables (x', y') in the pencil forming image of the object point $(0, H)$.

A caveat: The notation for canonical coordinates for object/image planes is (G_S, H_T), and sometimes they are taken simply as (G, H) without the subscripts. This 'H' should not be confused with its more common usage as the paraxial Lagrange invariant.

REFERENCES

1 H.H. Hopkins, 'Canonical Pupil Coordinates in Geometrical and Diffraction Image Theory', *Proc. Conf. on Photographic and Spectroscopic Optics, Tokyo, Japan (1964), published in Japan J. Appl. Phys.*, Vol. 4, Supplement 1 (1965) pp. 31–35.
2 H.H. Hopkins and M.J. Yzuel, 'The Computation of Diffraction Patterns in the Presence of Aberrations', *Opt. Acta*, Vol. 17 (1970) pp. 157–182.
3 H.H. Hopkins, 'Canonical and Real-Space Coordinates Used in the Theory of Image Formation', in *Applied Optics and Optical Engineering, Vol. IX*, Eds. R.R. Shannon and J.C. Wyant, Academic, New York (1983) pp. 307–369.
4 H.H. Hopkins, 'Optical Design Calculations Using Canonical Coordinates', Parts I, II, IIIa and IIIb, University of Reading (1982), personal communication.
5 H.H. Hopkins, Ref. 3, p. 309.
6 H. Goldstein, *Classical Mechanics*, Addison-Wesley, Reading, Massachusetts (1965) p. 217.
7 H.H. Hopkins, Ref. 3, p. 316.
8 H.H. Hopkins, Ref. 4, Part II, pp. 87–96, 74–78.

12

Primary Aberrations from System Data

12.1 Introduction

Earlier, Chapter 10 elucidated how the total wavefront aberration of an axi-symmetric optical system could be evaluated from the data obtained by tracing suitable real rays through the system with the help of finite ray tracing formulae that make use of the system data. In the process, 'real' rays are involved, so that the evaluated wave front aberration is 'total'. The power series expansion for wave aberration of an axially symmetric optical system is an infinite series consisting of groups of terms of the same order. Rewriting expression (8.81) with the same notation

$$\begin{aligned} W = & \left[{}_0W_{40}r^4 + {}_2W_{22}\tau^2r^2(\cos\theta)^2 + {}_2W_{20}\tau^2r^2 + {}_1W_{31}\tau r^3\cos\theta + {}_3W_{11}\tau^3 r\cos\theta \right] \\ & + \left[{}_0W_{60}r^6 + {}_3W_{33}\tau^3r^3(\cos\theta)^3 + {}_4W_{20}\tau^4r^2 + {}_2W_{40}\tau^2r^4 + {}_5W_{11}\tau^5 r\cos\theta \right. \\ & \left. + {}_4W_{22}\tau^4r^2(\cos\theta)^2 + {}_1W_{51}\tau r^5\cos\theta + {}_2W_{42}\tau^2r^4 + {}_2W_{40}\tau^2r^4(\cos\theta)^2 + {}_3W_{31}\tau^3r^3\cos\theta \right] \\ & \left[{}_0W_{80}r^8 + {}_4W_{44}\tau^4r^4(\cos\theta)^4 + \cdots \right] + \cdots \end{aligned} \tag{12.1}$$

we note that the expression contains all terms of the first two orders, and a few terms of the next order.

The model of imaging used for the evaluation of 'total' wavefront aberration is such that all terms of the infinite series have been duly incorporated in the analysis. No approximation is involved in the determination of ray paths in the object space, in the image space, or in any of the intermediate spaces of the optical system. The exact role of each ray of the image-forming pencil, with regard to quantities related to imaging by the system, has been ascertained in the process. On the other hand, the paraxial model of imaging assumes $W = 0$, that holds in any system as $\tau, r \to 0$.

Looking at the form of the aberration series, it is obvious that, in general, as aperture and field of any system increase, higher order terms become predominant. At the lowest level, for very small values of aperture and field, the system performance can be effectively modelled by paraxial optics. It is expected that imaging characteristics of systems with small and not very large values of aperture and field can be modelled by retaining only the primary aberration terms. It is a grossly approximate model. Nevertheless, the model provides analytical and semi-analytical tools for handling the complex problems of imaging optics.

The wave aberration with only the primary aberration terms is represented by

$$W = \left[{}_0W_{40}r^4 + {}_2W_{22}\tau^2r^2(\cos\theta)^2 + {}_2W_{20}\tau^2r^2 + {}_1W_{31}\tau r^3\cos\theta + {}_3W_{11}\tau^3 r\cos\theta \right] \tag{12.2}$$

The corresponding components of transverse ray aberration are, with the same notations as used earlier,

$$\begin{aligned} \delta\xi &= -\frac{1}{n\sin\alpha}\frac{\partial W}{\partial x} \\ &= -\frac{1}{n\sin\alpha}\left[4\,{}_0W_{40}r^3\sin\theta + {}_1W_{31}\tau r^2\sin 2\theta + 2\,{}_2W_{20}\,\tau^2 r\sin\theta \right] \end{aligned} \tag{12.3}$$

DOI: 10.1201/9780429154812-12

$$\delta\eta = -\frac{1}{n\sin\alpha}\frac{\partial W}{\partial y}$$
$$= -\frac{1}{n\sin\alpha}\left[4\,{}_0W_{40}r^3\cos\theta + {}_1W_{31}\tau r^2\{2+\cos2\theta\} + 2\left({}_2W_{22} + {}_2W_{20}\right)\tau^2 r\cos\theta + {}_3W_{11}\tau^3\right] \quad (12.4)$$

As mentioned earlier, it is now said that the first breakthrough in this regard by developing formulae for determination of the different aberration coefficients from system data was made by Petzval. The earliest available treatment on evaluation of primary aberration coefficients from system data is that by Ludwig von Seidel [1].

Early approaches on derivation of primary aberrations of axi-symmetric systems from system data involved the use of modified trigonometrical ray tracing formula where the trigonometric functions were replaced by a series term where only terms relevant to third order analysis were retained. Later on, somewhat similar approaches were undertaken while using algebraic ray tracing methods [2–15]. From geometrical considerations, the contribution to primary aberrations by a single refracting interface in extra-axial imaging has been worked out by using the concept of auxiliary axis [15]. Appropriate measures for concatenating the ray aberration contributions by individual surfaces to obtain the ray aberration components for the whole system need to be undertaken. This has been facilitated by the use of matrix algebra [16–18].

On the other hand, Schwarzschild extended the concept of Hamilton-Bruns' eikonal for analyzing the problem of geometrical aberrations in optical imaging [19]. He introduced an eikonal, and called it the 'Seidel eikonal'. His treatment of the problem is analogous to a method he used in the calculations of orbital elements in celestial mechanics. The perturbation of the orbit has a direct analogy with aberrations in imaging optics. That is why the 'Schwarzschild eikonal' is also known as 'the perturbation eikonal of Schwarzschild'. Concatenation of the aberration contribution by individual surfaces for obtaining aberration of the composite system is more straightforward by this approach [20–23].

The approaches mentioned above are mostly directed towards primary ray aberrations and their numerical evaluation. Numerical evaluation of 'primary wavefront aberrations' was initiated by Conrady, as reported in [24], and Hopkins, in his landmark contribution 'Wave theory of aberrations' [25] presented a lucid description of the process of evaluation and use of primary wave aberrations. This treatment of primary wave aberrations is now widely accepted [26–32].

A significant common aspect of the different approaches adopted for the numerical evaluation of primary aberrations is that this evaluation can be accomplished using data of paraxial ray trace through the system. Each of the five primary aberrations of an optical system is expressed as a summation of contributions by the constituent refracting interfaces to corresponding primary aberrations. It is significant to note that these contributions are intrinsic to the surface, and are determined exclusively by the curvature of the refracting surface, refractive indices on both sides of the surface, and height and convergence angles of the paraxial marginal and pupil rays at the surface; primary aberration introduced by an earlier surface has no effect on the primary aberrations introduced by any of the surfaces following it. This is a characteristic that is unique to primary aberrations.

12.2 Validity of the Use of Paraxial Ray Parameters for Evaluating Surface Contribution to Primary Wavefront Aberration

This aspect has been deliberated upon by many investigators, including Welford [33]. The following is adapted from a nice elucidation of this somewhat tricky theme by Mouroulis and Macdonald [34]. Figure 12.1 shows the refraction of an axial wavefront EQ by an axially symmetric refracting surface AP. The refracted wavefront is E′Q′. Assume the surface to be an intermediate surface of an axi-symmetric imaging system. Let the paraxial object and the paraxial image for refraction by the interface be O and O′, respectively. The paraxial ray path is OPO′, and the corresponding real ray path through P is TPT′. In the object space, the real ray TBP and the paraxial ray OB_0P intersect the reference sphere EBB_0, with O as centre and OE as radius, at the points B and B_0, respectively. The corresponding points in the image space are B′ and B'_0.

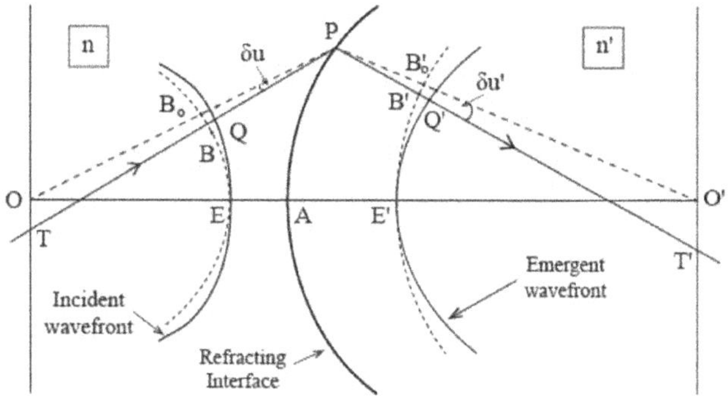

FIGURE 12.1 An axially symmetric refracting interface AP. Incident axial wavefront: EQ; Refracted axial wavefront: E′Q′; Paraxial ray path: OPO′; Corresponding real ray path: TPT′.

The wavefront aberrations of the incident and the emergent wavefronts are

$$W = [BQ] \qquad W' = [B'Q'] \tag{12.5}$$

Equating optical path lengths between the two wavefronts

$$[EAE'] = [QPQ'] = [BPB'] - [BQ] + [B'Q'] = [BPB'] - W + W' \tag{12.6}$$

The increment ΔW in aberration due to refraction by the surface is

$$\Delta W = W' - W = [EAE'] - [BPB'] = [EA] + [AE'] - [BP] - [PB'] \tag{12.7}$$

This can be rewritten as

$$\Delta W = \{[AE'] - [PB']\} - \{[AE] - [PB]\} = \Delta\{[AE] - [PB]\} \tag{12.8}$$

By definition, the paraxial ray $OB_0PB_0'O'$ is perpendicular to the reference spheres in the object and the image space at points B_0 and B_0', respectively. Therefore,

$$(PB_0) = (PB)\cos\delta u \qquad (PB_0') = (PB')\cos\delta u' \tag{12.9}$$

where δu and $\delta u'$ are the angular aberrations $\angle B_0PB$ and $\angle B_0'PB'$, respectively. Angular aberrations have the same aperture and field dependence as the transverse ray aberrations. Since wave aberration is at least of the fourth degree in aperture r and field τ, corresponding transverse ray aberration, as well as angular aberration, is at least of degree three in aperture r and field τ, i.e.

$$\delta u, \delta u' \approx O(3) \tag{12.10}$$

Since

$$\cos\delta u = 1 - \frac{(\delta u)^2}{2} + \ldots \tag{12.11}$$

(12.9) can be rewritten as

$$(PB) = (PB_0) + O(6) \qquad (PB') = (PB_0') + O(6) \tag{12.12}$$

From (12.8) it follows

$$\Delta W = \{[AE'] - [PB_0']\} - \{[AE] - [PB_0]\} + O(6)$$
$$= \{[AO'] - [PO']\} - \{[AO] - [PO]\} + O(6) \tag{12.13}$$

since

$$E'O' = B_0'O' \qquad EO = B_0O \tag{12.14}$$

Therefore, at primary aberration level,

$$\Delta W = \Delta\{[AO] - [PO]\} \tag{12.15}$$

This proves that the increment in primary wave aberration by refraction at a surface can be conveniently obtained from paraxial ray trace data. This is a direct consequence of the observation (12.12). A simple relation like (12.15) cannot be extended for aberrations of a higher order. Also, note that ΔW, given by (12.15), represents the contribution of this particular surface to the total primary aberration of the ray passing through P on the surface. The emergent wavefront E′Q′ can be considered an incident wavefront for the next surface, and so on. Ultimately, the primary aberration of the total system is obtained by summing up the contributions by individual surfaces.

This observation is called 'Summation theorem for primary aberrations'.

For the sake of simplicity, the proof is given for axial conjugates. The result, however, also holds good for extra-axial imaging.

12.3 Primary Aberrations and Seidel Aberrations

From (12.2), the expression for wave aberration containing only primary aberration terms is

$$W = \left[{}_0W_{40}r^4 + {}_1W_{31}\tau r^3\cos\theta + {}_2W_{22}\tau^2r^2(\cos\theta)^2 + {}_2W_{20}\tau^2r^2 + {}_3W_{11}\tau^3 r\cos\theta\right]$$
$$= \left[{}_0W_{40}\left(x^2+y^2\right)^2 + {}_1W_{31}\tau\left(x^2+y^2\right)y + {}_2W_{22}\tau^2y^2 + {}_2W_{20}\tau^2\left(x^2+y^2\right) + {}_3W_{11}\tau^3 y\right] \tag{12.16}$$

As mentioned earlier, the earliest available record for evaluating the primary aberration coefficients of an imaging lens system is the one by Seidel [9.62]. The Seidel notations for the primary aberration coefficients are different, and the relation between the two notations is given in Table 12.1. In terms of Seidel aberration coefficients, the equation (12.16) is written as

$$W = \frac{1}{8}S_I r^4 + \frac{1}{2}S_{II}\tau r^3\cos\theta + \frac{1}{2}S_{III}\tau^2r^2(\cos\theta)^2 + \frac{1}{4}\left(S_{III}+S_{IV}\right)\tau^2r^2 + \frac{1}{2}S_V\tau^3 r\cos\theta \tag{12.17}$$

Using (8.11) – (8.12), corresponding expressions for the transverse ray aberration components are

$$\delta\xi' = -\frac{R'}{n'}\frac{\partial W}{\partial X'} = -\frac{1}{n'(h'/R')}\frac{\partial W}{\partial x'} \tag{12.18}$$

$$\delta\eta' = -\frac{R'}{n'}\frac{\partial W}{\partial Y'} = -\frac{1}{n'(h'/R')}\frac{\partial W}{\partial y'} \tag{12.19}$$

TABLE 12.1

Relation between primary wave aberration coefficients and Seidel aberrations

Primary	Wave Aberration Coefficients	Seidel Aberrations
Spherical Aberration	${}_0W_{40} =$	$(S_I/8)$
Coma	${}_1W_{31} =$	$(S_{II}/2)$
Astigmatism	${}_2W_{22} =$	$(S_{III}/2)$
Curvature	${}_2W_{20} =$	$\{(S_{III}+S_{IV})/4\}$
Distortion	${}_3W_{11} =$	$(S_V/2)$

where $x' = (X'/h')$, and $Y' = (Y'/h')$. Omitting primes from symbols in the image space, we can write

$$\delta\xi = -\frac{1}{(n\sin\alpha)}\frac{\partial W}{\partial x} \qquad \delta\eta = -\frac{1}{(n\sin\alpha)}\frac{\partial W}{\partial y} \tag{12.20}$$

where $(n\sin\alpha)$ is the image space axial numerical aperture. From (12.16) it follows

$$\frac{\partial W}{\partial x} = 4_0W_{40}r^3\sin\theta +_1 W_{31}\tau r^2\sin 2\theta + 2_2W_{20}\tau^2 r\sin\theta \tag{12.21}$$

$$\frac{\partial W}{\partial y} = 4_0W_{40}r^3\cos\theta + {}_1W_{31}\tau r^2[2+\cos 2\theta] + 2({}_2W_{22} + {}_2W_{20})\tau^2 r\cos\theta + {}_3W_{11}\tau^3 \tag{12.22}$$

Using Seidel terms, equations (12.21) and (12.22) can be rewritten as

$$\frac{\partial W}{\partial x} = \frac{1}{2}S_I r^3\sin\theta + \frac{1}{2}S_{II}\tau r^2\sin 2\theta + \frac{1}{2}(S_{III}+S_{IV})\tau^2 r\sin\theta \tag{12.23}$$

$$\frac{\partial W}{\partial y} = \frac{1}{2}S_I r^3\cos\theta + \frac{1}{2}S_{II}\tau r^2[2+\cos 2\theta] + \frac{1}{2}(3S_{III}+S_{IV})\tau^2 r\cos\theta + \frac{1}{2}S_V\tau^3 \tag{12.24}$$

12.4 Seidel Aberrations in Terms of Paraxial Ray Trace Data

Optics literature abounds with a multitude of derivations of Seidel aberrations in terms of data obtained from paraxial ray trace through an axi-symmetric imaging system from the early days. In principle, the Seidel aberrations can be determined from the data of only one paraxial ray trace. However, with the passage of time, succinct expressions have emerged for the Seidel aberrations that make use of ray trace data of the paraxial marginal ray and paraxial pupil ray. Nevertheless, some of the expressions differ from each other to a multiplicative constant. In what follows, we adapt the treatment of Hopkins [35], as modified later by Welford [36].

12.4.1 Paraxial (Abbe's) Refraction Invariant

Figure 12.2 shows the refraction of a paraxial ray by a spherical interface VP between two optical media of refractive indices n and n′. The point V, lying on the optical axis, is the vertex of the spherical surface

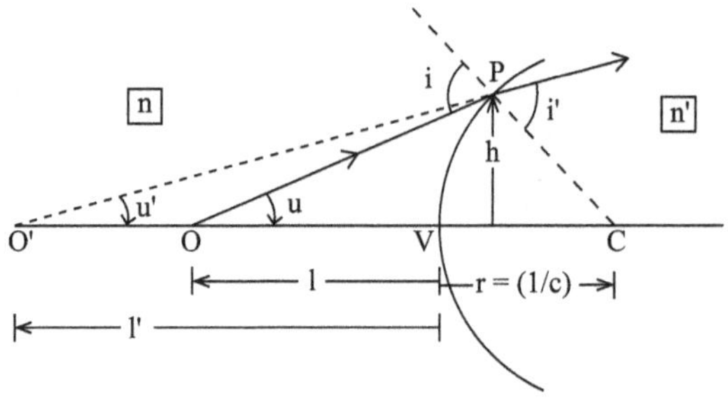

FIGURE 12.2 Paraxial (Abbe's) invariants at a refracting interface for the paraxial marginal ray; the invariant is A.

of curvature c and radius r, so that cr = 1. The object distance VO = l, and image distance VO′ = l′. The height of the paraxial ray on the interface is h, and the convergence angles of the incident ray and the refracted ray are u and u′, respectively.

From the paraxial conjugate equation (3.21) we have

$$\frac{n'}{l'} - \frac{n}{l} = \frac{(n' - n)}{r} \tag{12.25}$$

It can be rewritten as

$$n'\left(\frac{1}{l'} - \frac{1}{r}\right) = n\left(\frac{1}{l} - \frac{1}{r}\right) \tag{12.26}$$

This represents a quantity that remains invariant by this paraxial refraction. It is called the 'Zero invariant' [37]. In recognition of the contributions of Ernst Abbe by making significant use of this invariant relation in analytical aberration theory, it is also called 'Abbe's refraction invariant' [38]. Multiplying both sides of (9.210) by h, the invariant can be recast as

$$n'\left(\frac{h}{l'} - \frac{h}{r}\right) = n\left(\frac{h}{l} - \frac{h}{r}\right) \tag{12.27}$$

Since $h = -lu = -l'u'$, (12.27) reduces to

$$n'(hc + u') = n(hc + u) \tag{12.28}$$

The relation (12.28) is, indeed, the paraxial version of Snell's law, i.e. $ni = n'i'$.

Note that, unlike the well-known paraxial Lagrange invariant, Abbe's refraction invariant is not an invariant relationship that is valid for the whole system; it is valid only for the particular refracting interface, and the particular paraxial ray. For the paraxial marginal ray with parameters (nu, h) and the paraxial pupil ray with parameters $(n\bar{u}, \bar{h})$, the corresponding Abbe's refraction invariants A and $\bar{A}$ are given by

$$A = n(hc + u) = n'(hc + u') \tag{12.29}$$

$$\bar{A} = n(\bar{h}c + \bar{u}) = n'(\bar{h}c + \bar{u}') \tag{12.30}$$

12.4.2 Seidel Aberrations for Refraction by a Spherical Interface

A transverse object OB is imaged as O′B′ by a spherical refracting interface $V\bar{P}P$ with refractive indices n and n′ in the object space and the image space, respectively (Figure 12.3). OPO′ is the paraxial marginal ray (PMR), and $BE\bar{P}B'$ is the paraxial pupil ray (PPR). The spherical interface is of curvature $c(=1/r)$.

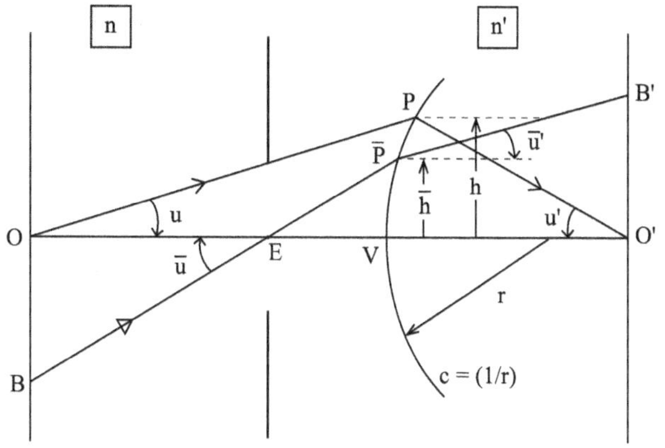

FIGURE 12.3 The paraxial image O′B′ of a transverse object OB by a spherical refracting interface $V\bar{P}P$. The paraxial marginal ray: OPO′; The paraxial pupil ray: $BE\bar{P}B'$.

The heights of the PMR and PPR on the interface are h and $\bar{h}$. The convergence angles of the PMR and the PPR in the object space are u and $\bar{u}$, and those in the image space are u′ and $\bar{u}'$. The Seidel aberrations S_I, S_{II}, S_{III}, S_{IV}, and S_V introduced in the wavefront by this refraction are:

$$\left.\begin{aligned}
S_I &= S_1 = -A^2 h\Delta\left(\frac{u}{n}\right)\\
S_{II} &= S_2 = -A\bar{A}h\Delta\left(\frac{u}{n}\right) = \frac{\bar{A}}{A}S_I\\
S_{III} &= S_3 = -\bar{A}^2 h\Delta\left(\frac{u}{n}\right) = \frac{\bar{A}}{A}S_{II}\\
S_{IV} &= S_4 = -H^2 c\Delta\left(\frac{1}{n}\right)\\
S_V &= S_5 = -\left[\frac{\bar{A}^3}{A}h\Delta\left(\frac{u}{n}\right) + \frac{\bar{A}}{A}H^2 c\Delta\left(\frac{1}{n}\right)\right] = \frac{\bar{A}}{A}\left[S_{III} + S_{IV}\right]
\end{aligned}\right\} \tag{12.31}$$

where A and $\bar{A}$ are given by (12.29) and (12.30), respectively. H is the paraxial Lagrange invariant, h is the height of the PMR on the surface, and the differences are:

$$\Delta\left(\frac{u}{n}\right) = \left[\frac{u'}{n'} - \frac{u}{n}\right] \qquad \Delta\left(\frac{1}{n}\right) = \left[\frac{1}{n'} - \frac{1}{n}\right] \tag{12.32}$$

The derivation of (12.31) is given by Welford [33] in detail, and the interested reader may consult his text. Note that the sign convention for paraxial angles used in these formulae are the opposite of the one used by Conrady, Hopkins, and many others earlier on [24–25]. However, sign conventions for all other parameters remaining the same, the new formulae can be obtained from the corresponding old formulae by substituting all paraxial angles with their negative value, e.g. 'u' with '−u'. When the axial object point O is at the centre of curvature of the spherical interface, A = 0; also, when O is very near this centre, A ≈ 0. In these cases, the expression for S_V, as given in (12.31), becomes indeterminate. To circumvent this problem, Welford derived an alternate expression for S_V

$$S_V = \bar{A}\left[\Delta\left(-h\bar{u}^2\right) - \Delta\left(\bar{h}^2 cu\right)\right] \tag{12.33}$$

where

$$\Delta\left(-h\bar{u}^2\right) = -h\left(\bar{u}'^2 - \bar{u}^2\right) \qquad \Delta\left(\bar{h}^2 cu\right) = \bar{h}^2 c\left(u' - u\right) \tag{12.34}$$

Expressing u and u′ in terms of A and A' respectively, the expression (12.33) for S_V can be rewritten as

$$S_V = -\bar{A}^3 h\Delta\left(\frac{1}{n^2}\right) + \bar{A}\bar{h}c\left(2\bar{A}h - A\bar{h}\right)\Delta\left(\frac{1}{n}\right) \tag{12.35}$$

Note that the expression for S_I does not contain any term pertaining to the PPR. S_I is entirely dependent on the parameters of the PMR that remains the same with change in axial location of the aperture stop/entrance pupil. Expressions for S_{II}, $S_{III,}$ and S_V do not only contain terms pertaining to PMR, but they also contain $\bar{A}$, the Abbe's refraction invariant for the PPR. This implies that they depend on the axial location of the pupil. The expression for S_{IV} is somewhat unique, for it does not contain any term pertaining to either the PMR or the PPR. S_{IV} is dependent only on the axial curvature of the refracting interface, the refractive indices on two sides of the interface, and the paraxial invariant H.

12.4.3 Seidel Aberrations in an Axi-Symmetric System Consisting of Multiple Refracting Interfaces

The summation theorem of primary aberrations, mentioned in Section 12.2, implies that the Seidel aberrations in an axi-symmetric system consisting of multiple refracting interfaces is given by the algebraic sum of the contributions by the individual refracting interfaces. In terms of the ray trace data of the PMR and the PPR, in an axisymmetric system consisting of k refracting interfaces, the five Seidel sums for the system is given by

$$\left.\begin{aligned}
S_I &= \sum_{j=1}^{k}\left(S_I\right)_j = -\sum_{j=1}^{k} A_j^2 h_j\left[\Delta\left(\frac{u}{n}\right)\right]_j \\
S_{II} &= \sum_{j=1}^{k}\left(S_{II}\right)_j = -\sum_{j=1}^{k} A_j\bar{A}_j h_j\left[\Delta\left(\frac{u}{n}\right)\right]_j \\
S_{III} &= \sum_{j=1}^{k}\left(S_{III}\right)_j = -\sum_{j=1}^{k} \bar{A}_j^2 h_j\left[\Delta\left(\frac{u}{n}\right)\right]_j \\
S_{IV} &= \sum_{j=1}^{k}\left(S_{IV}\right)_j = -H^2\sum_{j=1}^{k} c_j\left[\Delta\left(\frac{1}{n}\right)\right]_j \\
S_V &= \sum_{j=1}^{k}\left(S_V\right)_j = -\sum_{j=1}^{k}\left[\frac{\bar{A}_j^3}{A_j}h_j\left[\Delta\left(\frac{u}{n}\right)\right]_j + \frac{\bar{A}_j}{A_j}H^2 c_j\left[\Delta\left(\frac{1}{n}\right)\right]_j\right] = \sum_{j=1}^{k}\frac{\bar{A}_j}{A_j}\left[\left(S_{III}\right)_j + \left(S_{IV}\right)_j\right]
\end{aligned}\right\} \tag{12.36}$$

For any surface at which $A_j \to 0$, the alternate determinate forms for $\left(S_V\right)_j$, as given in (12.33) or (12.35), can be used.

12.4.4 Seidel Aberrations of a Plane Parallel Plate

For the sake of simplicity, we consider the special case of a plane parallel plate in air. An extension to the general case of non-air ambient medium is straightforward. Let the plane parallel plate be of thickness t, and the refractive index of the optical material of the plate be n. Let PQ_1Q_2R be the paraxial marginal ray through the plane parallel plate (Figure 12.4). Heights of the PMR at the two surfaces are h_1 and h_2. The convergence angle u of the incident ray is equal to the convergence angle u′ of the emergent ray from the plate. Since for the plane parallel plate $c_1 = c_2 = 0$, Abbe refraction invariants A_1 and A_2 for the PMR at the two surfaces are $A_1 = A_2 = u$. Similarly, Abbe refraction invariants for the paraxial pupil ray at the two surfaces are $\bar{A}_1 = \bar{A}_2 = \bar{u}$, where $\bar{u}$ is the angle of convergence of the PPR, or the angle of incidence of the PPR in the incidence space. The heights of the PMR at the two surfaces are related by

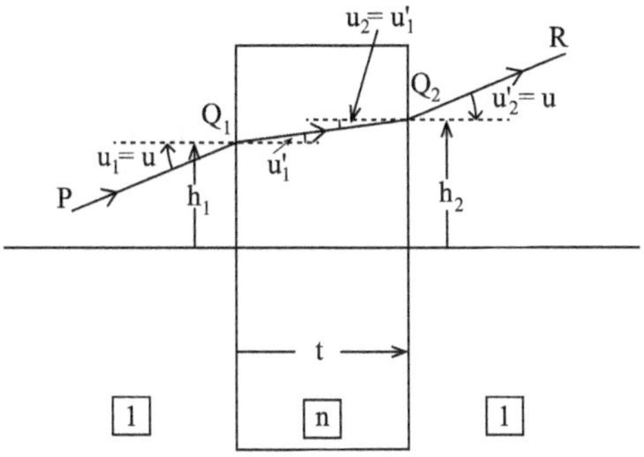

FIGURE 12.4 PQ_1Q_2R: A paraxial marginal ray through a plane parallel plate. $PQ_1 \parallel Q_2R$.

$$h_2 = h_1 + tu_1' = h_1 + t\frac{u}{n} \tag{12.37}$$

We have $nu_1' = u = nu_2$, so that

$$\left[\Delta\left(\frac{u}{n}\right)\right]_1 = \left[\frac{u_1'}{n} - u\right] = -\frac{n^2-1}{n^2}u \qquad \left[\Delta\left(\frac{u}{n}\right)\right]_2 = \left[u_2' - \frac{u_2}{n}\right] = \frac{n^2-1}{n^2}u \tag{12.38}$$

The Seidel sums S_I, S_I, S_{III}, S_{IV}, and S_V are given by

$$\left.\begin{aligned}
S_I &= -A_1^2 h_1\left[\Delta\left(\frac{u}{n}\right)\right]_1 - A_2^2 h_2\left[\Delta\left(\frac{u}{n}\right)\right]_2 = -u^3(h_2 - h_1)\frac{n^2-1}{n^2} = -\frac{n^2-1}{n^3}u^4 t\\
S_{II} &= \left(\frac{\bar{u}}{u}\right)S_I = -\frac{n^2-1}{n^3}u^3\bar{u}t\\
S_{III} &= \left(\frac{\bar{u}}{u}\right)S_{II} = -\frac{n^2-1}{n^3}u^2\bar{u}^2 t\\
S_{IV} &= 0\\
S_V &= \left(\frac{\bar{u}}{u}\right)S_{III} = -\frac{n^2-1}{n^3}u\bar{u}^3 t
\end{aligned}\right\} \tag{12.39}$$

It is important to note that the Petzval curvature term S_{IV} is zero, whatever the thickness of the plate is, or whatever the location of the object is with respect to the plate. All Seidel sums are individually zero when $u = 0$. Otherwise, S_I, S_{II}, S_{III}, and S_V are directly proportional to the thickness t of the plate. In high power oil immersion objectives, even a thin cover slip gives rise to a substantial amount of spherical aberration since u is large (as the numerical aperture is very high), and spherical aberration is proportional to the fourth power of u. In such applications, the microscope objectives need to be designed to possess residual aberrations for balancing the aberrations caused by the cover slips. On the other hand, microscope objectives, used in applications where there is no cover slip, are needed to be corrected from aberrations as best as possible.

The treatment of Seidel aberrations of plane parallel plates can be used for analyzing the Seidel aberrations contributed by a reflecting prism in optical systems. Different types of reflecting prisms are utilized for inverting or reverting the images. Usually, the incident and emergent faces of these prisms are plane, and the optical effects of such a prism is identical with that of a plane parallel plate, the thickness of the plate being determined from the tunnel diagram for the prism. Since for $u = 0$, each of the five Seidel sums is zero for any value of thickness t, the prisms are preferably used in a star space of an optical system to avoid the occurrence of unwanted aberrations.

12.4.5 Seidel Aberrations of a Spherical Mirror

Figure 12.5 shows the PMR emerging from an axial point object O and incident on a spherical mirror at the point Q that is at height h from the optical axis. After reflection, the reflected ray emerges as QO′. The curvature of the spherical mirror is $c(=1/r)$, and the convergence angles of the incident ray and the emergent ray are u and u′, respectively.

It was mentioned earlier that the paraxial refraction equation can be utilized for tackling the case of reflection by substituting $n' = -n$. So, from the paraxial refraction equation

$$n'u' - nu = -h(n' - n)c \tag{12.40}$$

we get

$$u' + u = -2hc \tag{12.41}$$

Therefore, it follows

$$\Delta\left(\frac{u}{n}\right) = -\frac{u'+u}{n} = \frac{2hc}{n} \qquad \Delta\left(\frac{1}{n}\right) = \frac{1}{n'} - \frac{1}{n} = -\frac{2}{n} \tag{12.42}$$

The Seidel coefficients S_I, S_{II}, S_{III}, S_{IV}, and S_V are given by

$$\left.\begin{aligned} S_I &= -\frac{2A^2h^2c}{n} \\ S_{II} &= \left(\frac{\bar{A}}{A}\right)S_I \\ S_{III} &= \left(\frac{\bar{A}}{A}\right)^2 S_I \\ S_{IV} &= \frac{2H^2c}{n} \\ S_V &= \left(\frac{\bar{A}}{A}\right)[S_{III} + S_{IV}] \end{aligned}\right\} \tag{12.43}$$

A major advantage for using reflecting surfaces is the complete absence of chromatic aberration of any order. There is an advantage, even in the case of monochromatic aberration, that is often overlooked. Note that in the case of reflection, the difference in refractive index between the emergent space and the

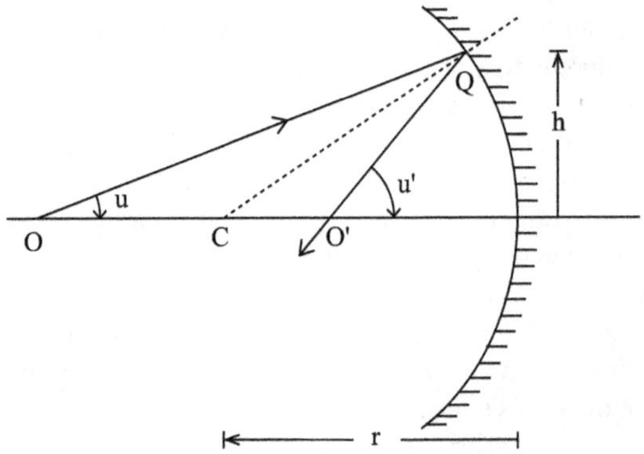

FIGURE 12.5 OQO′: A paraxial marginal ray by reflection in a single spherical mirror

incident space is 2.00 (assuming reflection in air) compared to 0.5 – 0.8 in the case of refraction. So, in the former case, a much smaller angle of incidence is required to get a given convergence. This leads to reduction in aberrations of all order.

A few special cases deserve attention:

Case I. Object at infinity

Since $u = 0$, $A = nhc$, $H = n(u\bar{h} - \bar{u}h) = -n\bar{u}h$. Equation (12.43) reduces to

$$\left.\begin{aligned} S_I &= -2nh^4c^2 \\ S_{II} &= -2nh^3c(\bar{h}c + \bar{u}) \\ S_{III} &= -2nh^2(\bar{h}c + \bar{u})^2 \\ S_{IV} &= \frac{2H^2c}{n} = 2n\bar{u}h^2c \\ S_V &= \frac{(\bar{h}c + \bar{u})}{hc}[S_{III} + S_{IV}] \end{aligned}\right\} \tag{12.44}$$

Case II. Stop is at the mirror

Since $\bar{h} = 0$, $\bar{A} = n\bar{u}$, $H = n(u\bar{h} - \bar{u}h) = -n\bar{u}h$ Equation (12.43) reduces to

$$\left.\begin{aligned} S_I &= -\frac{2A^2h^2c}{n} \\ S_{II} &= -2Ah^2\bar{u}c = \frac{2AhHc}{n} \\ S_{III} &= -\frac{2}{n}H^2c \\ S_{IV} &= \frac{2}{n}H^2c \\ S_V &= 0 \end{aligned}\right\} \tag{12.45}$$

In this case, if the object is at infinity, we have $u = 0, A = nhc$. The Seidel coefficients given by (12.45) reduce to

$$\left.\begin{aligned} S_I &= -2nh^4c^2 \\ S_{II} &= 2Hh^2c^2 \\ S_{III} &= -\frac{2}{n}H^2c \\ S_{IV} &= \frac{2}{n}H^2c \\ S_V &= 0 \end{aligned}\right\} \tag{12.46}$$

Case III. Stop at the centre of curvature

Since $\bar{h} = -r\bar{u}$, $\bar{A} = n(hc + \bar{u}) = 0$. From the general relation (12.43), we note that in this case $S_{II} = S_{III} = S_V = 0$, whatever the axial position of the object may be.

12.4.6 Seidel Aberrations of a Refracting Aspheric Interface

12.4.6.1 Mathematical Representation of an Aspheric Surface

The aspheric surfaces considered here are smooth surfaces of revolution about the optical axis. The plane and sphere are the simplest surfaces, and the different conicoid of revolution, e.g. paraboloid, ellipsoid, or hyperboloid, is also being considered. A plane, a sphere, or a conicoid may be 'figured' to have relatively small departures from the ideal geometric shape. A general aspheric surface that has no close relation to any simple geometric surface is also included in this treatment.

The central equations of a conicoid are of different forms for the different types, i.e., spheres, paraboloids, ellipsoids, etc. However, the vertex equations for all of them can be put in a form

$$X^2 + Y^2 + (1+Q)Z^2 - 2rZ = 0 \tag{12.47}$$

where the Z-axis is the axis of revolution, and it is along the optical axis of the system. The vertex radius of curvature is r, and the different values of Q give different surfaces, as indicated in Table 12.2. The corresponding plane curves, chosen to have the same vertex radius of curvature, r, are shown in Figure 12.6. In each figure, a section of the conicoid together with the circle of curvature at the vertex A is shown. The centre of the circle of curvature is at C_0. The general equation of the conicoid of revolution can be written in a form where Z is given as an explicit function of (X, Y) by

$$Z = \frac{c(X^2+Y^2)}{1+\sqrt{\{1-(1+Q)c^2(X^2+Y^2)\}}} \tag{12.48}$$

TABLE 12.2

Values of Q for different types of conicoid

$Q > 0$	Ellipsoid of revolution about the minor axis
$Q = 0$	Sphere
$-1 < Q < 0$	Ellipsoid of revolution about the major axis
$Q = -1$	Paraboloid
$Q < -1$	Hyperboloid

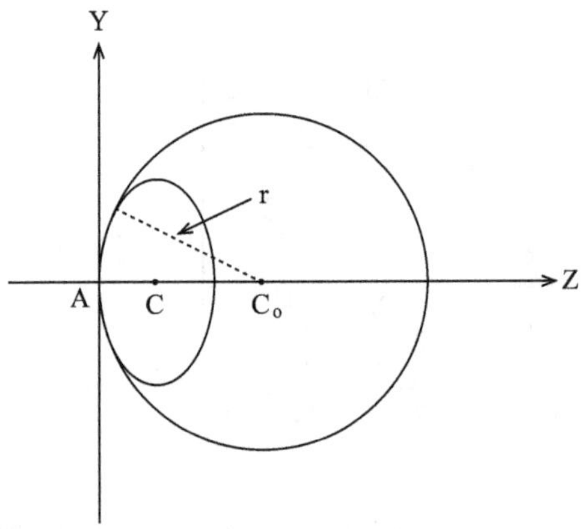

FIGURE 12.6(i) $Q > 0$. An ellipsoid of revolution about the minor axis. The vertex radius of the curvature at A is r.

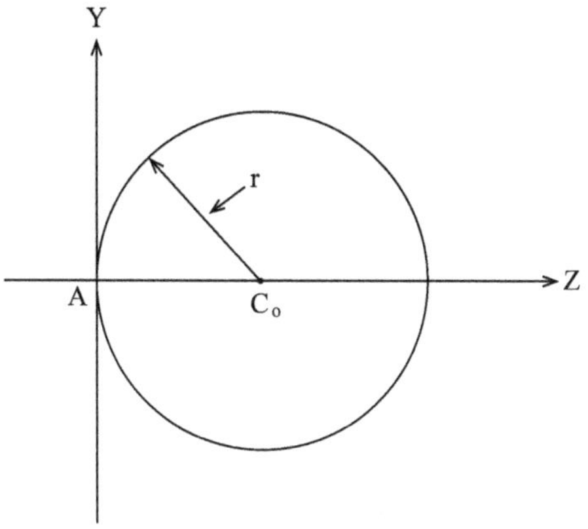

FIGURE 12.6(ii) Q = 0. A sphere of radius of curvature r.

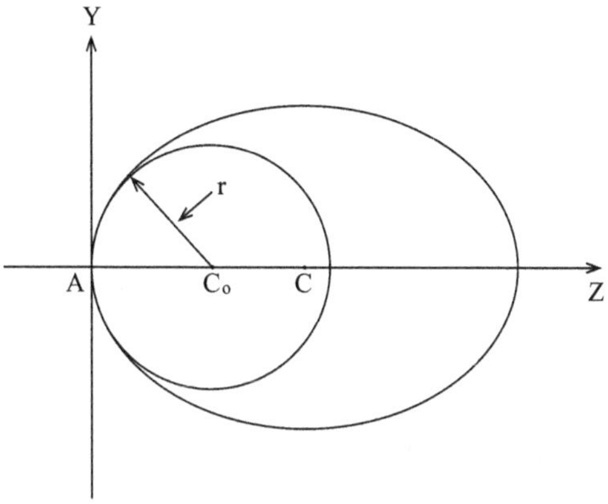

FIGURE 12.6(iii) $-1 < Q < 0$. An ellipsoid of revolution about the major axis. The vertex radius of the curvature at A is r.

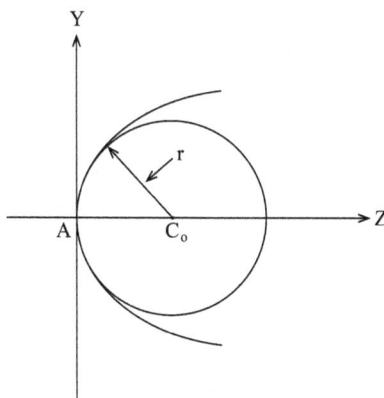

FIGURE 12.6(iv) $Q = -1$. A paraboloid of revolution about the axis. The vertex radius of curvature at A is r.

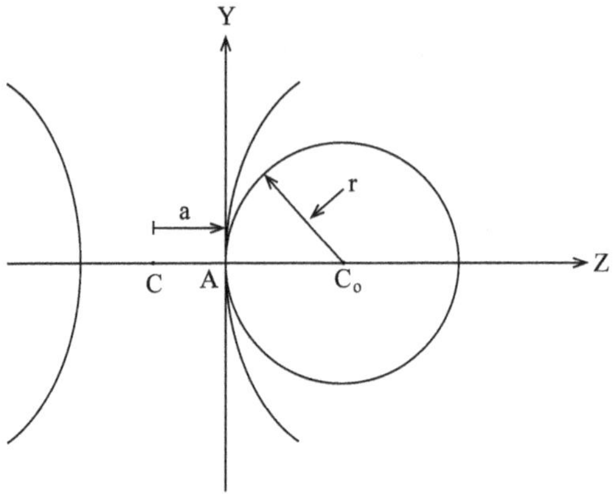

FIGURE 12.6(v) $Q < -1$. A hyperboloid of revolution about the axis. The vertex radius of curvature at A is r.

The general case of a 'figured' conicoid, where the conicoid is regarded as a 'parent surface' and 'figuring' terms are added, can be represented as

$$Z = \frac{c(X^2+Y^2)}{1+\sqrt{\{1-(1+Q)c^2(X^2+Y^2)\}}} + \sum_{j=2}^{J} a_{2j}(X^2+Y^2)^{2j} \tag{12.49}$$

where the highest power J of j usually does not exceed 6. Note that, since only those surfaces having symmetry about the axis are considered, in the 'figuring' terms, there is no term containing odd powers of (X^2+Y^2).

If $c = 0$, the equation (12.48) gives $Z = 0$. This is a plane. The only conicoid having a point of zero curvature is the degenerate case of a plane. In this case, (12.49) reduces to

$$Z = \sum_{j=2}^{J} a_{2j}(X^2+Y^2)^{2j} \tag{12.50}$$

This expression represents a general aspheric surface that is not close to any known quadratic surface.

12.4.6.2 Seidel Aberrations of the Smooth Aspheric Refracting Interface

For considering primary or Seidel aberrations, terms expressing shape of the surface depending on powers higher than the fourth in aperture need not be considered, since such terms produce only higher order aberrations. Also, the remaining aspheric equation can be separated into spherical and aspheric parts. The spherical parts give rise to the aberrations already discussed. The additional effect of the aspheric part remains to be considered.

Expanding the expression for Z in (12.49), and retaining terms up to degree four in aperture, we get the following expression for a figured conicoid

$$Z = \frac{1}{2}c(X^2+Y^2) + \left[\frac{1}{8}c^3(1+Q)+a_4\right](X^2+Y^2)^2 \tag{12.51}$$

For a perfect spherical surface of curvature c, $Q = a_4 = 0$, the expression is

$$Z = \frac{1}{2}c(X^2+Y^2) + \frac{1}{8}c^3(X^2+Y^2)^2 \tag{12.52}$$

Equation (12.51) can be rewritten as

$$
\begin{aligned}
Z &= \frac{1}{2}c\left(X^2+Y^2\right)+\left(\frac{1}{8}c^3+G\right)\left(X^2+Y^2\right)^2 \\
&= \frac{1}{2}c\left(X^2+Y^2\right)+\frac{1}{8}c^3\left(X^2+Y^2\right)^2+G\left(X^2+Y^2\right)^2
\end{aligned}
\tag{12.53}
$$

If the aperture stop, or any of the pupils is at the aspheric surface, the only effect of aspherizing is to change the spherical aberration, since all other Seidel aberrations depend on powers lower than the fourth in the aperture. The wavefront at the point (X,Y) on the pupil is advanced relative to the centre of the pupil by the optical path length $\delta_0 W_{40} = G\left(X^2+Y^2\right)^2(n'-n) = Gh^4\Delta(n)$. So, the Seidel coefficient $(S_I)_{asp}$ corresponding to the aspheric term G when the stop is at the surface is

$$
(S_I)_{asp} = 8Gh^4\Delta(n) \tag{12.54}
$$

If the stop or its conjugate is not at the aspheric surface, the Seidel coma, the astigmatism and the distortion of the refracting surface, which is being aspherized, are also changed, in addition to the change in Seidel spherical aberration. The changes in these oblique aberrations can be obtained by the use of a stop shift formula, discussed later.

12.5 Axial Colour and Lateral Colour as Primary Aberrations

'Axial colour' is defined as 'the chromatic variation in image distance', and 'lateral or transverse colour' is defined as 'the chromatic variation in image height or magnification'. In Section 9.3, 'axial colour' and 'lateral colour' are described as longitudinal chromatic aberration and transverse chromatic aberration, respectively. Since the magnitudes of these aberrations are of the order of primary aberrations, it is pertinent to seek expressions for the corresponding primary wavefront aberrations, or equivalently for the corresponding Seidel aberrations. The derivations given below are adapted from the treatments given in references [39–40].

Figure 12.7 shows refraction of polychromatic rays originating from an axial object point O (not shown), by an intermediate refracting surface AP in an axi-symmetric system. O_{λ_1} and O_{λ_2} are two paraxial images of the object point O at wavelengths λ_1 and λ_2, formed in the object space of the intermediate refracting surface AP by part of the imaging system preceding it. O'_{λ_1} and O'_{λ_2} are the images of O_{λ_1} and

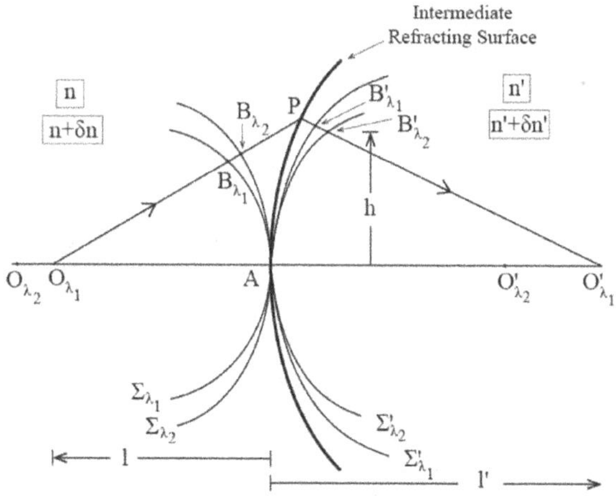

FIGURE 12.7 Change in axial colour by refraction at surface AP.

O_{λ_2}, respectively, formed by refraction at surface AP. At wavelengths λ_1 and λ_2, the refractive indices of the object space of the surface are n and $(n+\delta n)$ and those of the image space of the surface are n', and $(n'+\delta n')$ respectively. In the object space, the spherical wavefronts Σ_{λ_1} and Σ_{λ_2}, passing through the vertex A, have their centres at O_{λ_1} and O_{λ_2}. Similarly, in the image space, the spherical wavefronts Σ'_{λ_1} and Σ'_{λ_2}, passing through vertex A, have their centres at O_{λ_1} and O_{λ_2}. $O_{\lambda_1}PO'_{\lambda_1}$ is part of the paraxial marginal ray from O to O′ at wavelength λ_1. It intersects the spherical wavefronts $\Sigma_{\lambda_1}, \Sigma_{\lambda_2}, \Sigma'_{\lambda_1}$, and Σ'_{λ_2} at the points $B_{\lambda_1}, B_{\lambda_2}, B'_{\lambda_1}$, and B'_{λ_2} respectively. Let $AO_{\lambda_1} = l$ and $AO'_{\lambda_1} = l'$.

The wavefront aberration W, corresponding to axial colour $O_{\lambda_1}O_{\lambda_2}$ is

$$W = \left[B_{\lambda_1}B_{\lambda_2}\right] \tag{12.55}$$

Similarly, the wavefront aberration W′, corresponding to axial colour $O'_{\lambda_1}O'_{\lambda_2}$ is

$$W' = \left[B'_{\lambda_1}B'_{\lambda_2}\right] \tag{12.56}$$

The change in axial colour by the refraction at AP can be expressed equivalently by the change in wavefront aberration as

$$\delta_0 W_{20} = \Delta W = \left[B'_{\lambda_1}B'_{\lambda_2}\right] - \left[B_{\lambda_1}B_{\lambda_2}\right] = \left\{\left[PB_{\lambda_1}\right] - \left[PB'_{\lambda_1}\right]\right\} - \left\{\left[PB_{\lambda_2}\right] - \left[PB'_{\lambda_2}\right]\right\} \tag{12.57}$$

Figure 12.8 shows a spherical refracting interface AB of radius r with its centre at C. PBO is the paraxial marginal ray, and AP is a spherical wavefront of radius l, and with its centre at O. The PMR intersects the spherical wavefront at P, and the spherical refracting surface at B. The convergence angle of the PMR is u. $PD_1(=h_1)$ and $BD_2(=h_2)$ are two perpendiculars drawn from P and B on the optical axis ACO.

By geometry,

$$AD_2 \cong \frac{1}{2}h_2^2\frac{1}{r} \qquad AD_1 \cong \frac{1}{2}h_1^2\frac{1}{l} \tag{12.58}$$

Also, $(h_1 - h_2) = -PB\sin u \sim -PB.u$. For $u \to 0, h_1 = h_2 = h$. It follows

$$D_1D_2 = AD_2 - AD_1 = \frac{1}{2}h^2\left(\frac{1}{r} - \frac{1}{l}\right) \tag{12.59}$$

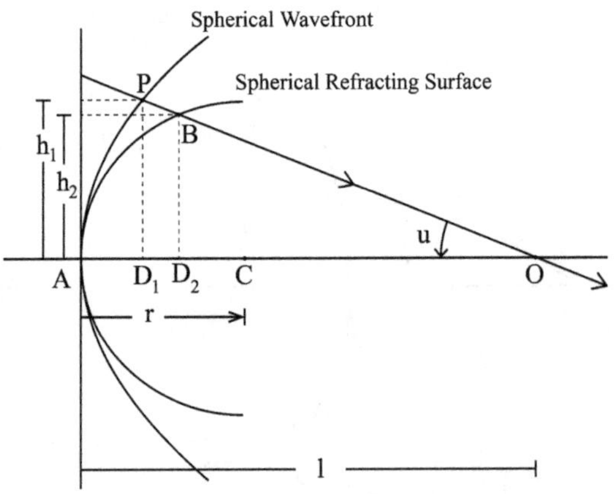

FIGURE 12.8 AB: a spherical refracting interface of radius r with centre at C. AP: a spherical wavefront of radius l and centre at O. PBO: a paraxial marginal ray.

$$PB = D_1D_2 \cos u \cong D_1D_2 = \frac{1}{2}h^2\left(c - \frac{1}{l}\right) \tag{12.60}$$

Using (12.60), we get

$$\left[PB_{\lambda_1}\right] = \frac{1}{2}nh^2\left(c - \frac{1}{l}\right) \qquad \left[PB'_{\lambda_1}\right] = \frac{1}{2}n'h^2\left(c - \frac{1}{l'}\right) \tag{12.61}$$

$$\left[PB_{\lambda_2}\right] = \frac{1}{2}(n + \delta n)h^2\left(c - \frac{1}{l}\right) \qquad \left[PB'_{\lambda_2}\right] = \frac{1}{2}(n' + \delta n')h^2\left(c - \frac{1}{l'}\right) \tag{12.62}$$

By substituting these values in (12.57), we get

$$\delta_0 W_{20} = \frac{1}{2}h^2\delta n'\left(c - \frac{1}{l'}\right) - \frac{1}{2}h^2\delta n\left(c - \frac{1}{l}\right) \tag{12.63}$$

The Abbe refraction invariant A for the PMR is

$$A = nh\left(c - \frac{1}{l}\right) = n'h\left(c - \frac{1}{l'}\right) \tag{12.64}$$

The primary wavefront aberration corresponding to axial colour introduced by refraction at the surface is given by

$$\delta_0 W_{20} = \frac{1}{2}Ah\Delta\left(\frac{\delta n}{n}\right) = \frac{1}{2}C_L = \frac{1}{2}C_I = \frac{1}{2}C_1 = \frac{1}{2}L \tag{12.65}$$

where $\Delta\left(\frac{\delta n}{n}\right) = \left(\frac{\delta n'}{n'}\right) - \left(\frac{\delta n}{n}\right)$. Different notations, e.g. C_L, C_I, C_1, or L etc., are used to denote Seidel longitudinal chromatic aberration.

Figure 12.9 shows the refraction at an intermediate surface of an axi-symmetric imaging system. The object and image spaces of this refracting surface are of refractive index n, n', respectively. The intermediate image $O_{\lambda_1}P_{\lambda_1}$ at wavelength λ_1 of the object OP (not shown), formed by the part of the

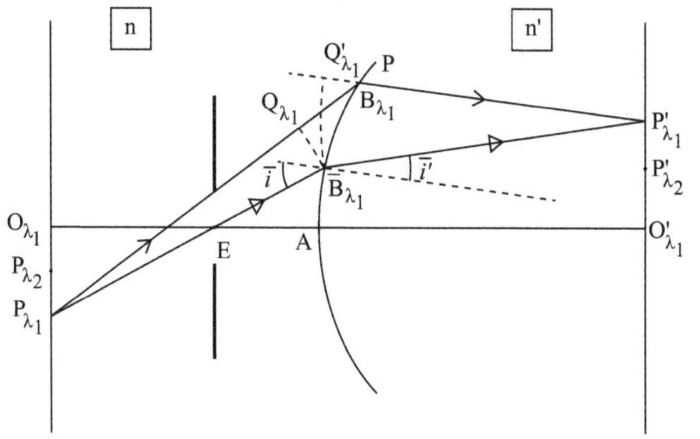

FIGURE 12.9 Change in lateral colour by refraction at surface AP.

system preceding the refracting surface at A, is the effective object for the refracting surface at A. The corresponding paraxial image by this refraction is $O'_{\lambda_1}P'_{\lambda_1}$. The paths of two paraxial rays from the extra-axial point P_{λ_1} are shown. These rays are the paraxial pupil ray $P_{\lambda_1}\bar{B}_{\lambda_1}P'_{\lambda_1}$, and an aperture ray $P_{\lambda_1}B_{\lambda_1}P'_{\lambda_1}$, which passes through the margin of the pupil. The paraxial height of the point B_{λ_1} with respect to the point $\bar{B}_{\lambda_1}$ is h, same as the height on the surface of the paraxial marginal ray from the axial point O_{λ_1}. $Q_{\lambda_1}\bar{B}_{\lambda_1}$ is a spherical wavefront in the object space; its centre is at P_{λ_1}. Similarly, $Q'_{\lambda_1}\bar{B}_{\lambda_1}$ is a spherical wavefront with its centre at P'_{λ_1} in the image space.

The wave aberration introduced in the aperture ray, with respect to the reference pupil ray by the refraction at the surface, is

$$W_{\lambda_1} = \left[P_{\lambda_1}\bar{B}_{\lambda_1}P'_{\lambda_1}\right] - \left[P_{\lambda_1}B_{\lambda_1}P'_{\lambda_1}\right] = n'\left(Q'_{\lambda_1}B_{\lambda_1}\right) - n\left(Q_{\lambda_1}B_{\lambda_1}\right) \tag{12.66}$$

If the angle of incidence and the angle of refraction of the PPR are $\bar{i}$ and $\bar{i}'$ respectively, the primary aberration part of W_{λ_1} is given by

$$W_{\lambda_1} \sim \left(n'h\bar{i}' - nh\bar{i}\right) = h\left(n'\bar{i}' - n\bar{i}\right) \tag{12.67}$$

For the same extra-axial object point P, the PPR at a second wavelength λ_2 is $P_{\lambda_2}E......P'_{\lambda_2}$. For the extra-axial point P, the lateral or transverse colour on the plane at O_{λ_1} is $P_{\lambda_1}P_{\lambda_2}$, and after refraction by the surface at A, the lateral or transverse colour at the image plane at O'_{λ_1} is $P'_{\lambda_1}P'_{\lambda_2}$. In order to determine the surface contribution to the change in lateral colour in terms of wavefront aberration, from (12.67) we determine

$$W_{\lambda_2} = W_{\lambda+\delta\lambda} = h\left\{\left(n' + \delta n'\right)\bar{i}' - \left(n + \delta n\right)\bar{i}\right\} \tag{12.68}$$

The primary wavefront aberration corresponding to transverse or lateral colour contributed by refraction at the surface is

$$\Delta_1 W_{11} = h\delta n'\bar{i}' - h\delta n\bar{i} = \left(n'\bar{i}'\right)h\left(\frac{\delta n'}{n'}\right) - \left(n\bar{i}\right)h\left(\frac{\delta n}{n}\right) = \bar{A}h\Delta\left(\frac{\delta n}{n}\right) = C_T = C_{II} = C_2 = T \tag{12.69}$$

where Abbe refraction invariant $\bar{A}$ for the pupil ray is

$$\bar{A} = n'\bar{i}' = n\bar{i} \tag{12.70}$$

Different notations C_T, C_{II}, or T, etc., are used to denote Seidel transverse chromatic aberration.

Like Seidel monochromatic aberrations, the total Seidel chromatic aberrations of an axi-symmetric imaging system consisting of multiple refracting interfaces is equal to the algebraic summation of contributions by the constituent surfaces. For an axi-symmetric system consisting of k refracting interfaces, the Seidel longitudinal and transverse chromatic aberrations are given by

$$\left.\begin{aligned} C_L = C_I = C_1 = L = \sum_{j=1}^{k} A_j h_j \left[\Delta\left(\frac{\delta n}{n}\right)\right]_j \\ C_T = C_{II} = C_2 = T = \sum_{j=1}^{k} \bar{A}_j h_j \left[\Delta\left(\frac{\delta n}{n}\right)\right]_j \end{aligned}\right\} \tag{12.71}$$

It is important to note that in the above derivation for axial and lateral colour, not all variations in h, A, and $\bar{A}$ are properly taken into account. Only the variation in refractive index with change in wavelength is duly taken into account. At wavelength λ, the refractive indices of the two media are n, n'. At wavelength

$(\lambda + \delta\lambda)$, the refractive indices of the two media are changed to $(n + \delta n),(n' + \delta n')$. Indeed, all paraxial parameters also change with variation in wavelength. In practice, the above treatment is justified by the argument that, in most cases, the neglected quantities are very small to affect the predictions by the treatment. A somewhat similar argument justifies the use of finite and not-so-small wavelength ranges in approximate expressions in the chromatic or spectral analysis of optical systems.

REFERENCES

1 L. Seidel, 'Zur Dioptrik. Über die Entwicklung der Glieder 3ter Ordnung, welche den Weg eines ausserhalb der Ebene der Axe gelegenen Lichtstrahles durch ein System brechender Medien bestimmen', *Astrnomische Nachrichten*, Vol. 43, (1856) No. 1027, pp. 289–304; ibid., No. 1028, pp. 305–320; ibid., No.1029, pp. 321–332.

2 S. Czapski, 'Die analytischen dioptrischen Theorien: die fünf Bildfehler Seidels', in *Grundzüge der Theorie der Optischen Instrumente nach Abbe*, Johann Ambrosius Barth, Leipzig (1904) pp. 149–152. [First published 1893].

3 A. Kerber, *Beiträge zur Dioptrik, Heft II*, Guster Fock, Leipzig (1896).

4 J. P. C. Southall, 'Seidel's theory of the spherical aberration of the third order', *Chapter VII in The Principles and Methods of Geometrical Optics Macmillan, New York* (1910) pp. 456–473.

5 A. Koenig and M. von Rohr, 'Seidel's theory of aberrations of the third order', in *Geometrical Investigation of the Formation of Images in Optical Instruments*, Ed. M. von Rohr, His Majesty's Stationery Office, London (1920) pp. 321–341.

6 T. Smith, 'Optical Calculations', in *Dictionary of Applied Physics*, Vol. IV, Ed., R. Glazebrook, Macmillan, London (1923) pp. 312–315.

7 G.C. Steward, The Symmetrical Optical System, Cambridge University Press (1928).

8 M. Berek, 'Die Seidel'sche Theorie der Aberrationen dritter Ordnung', Chapter 5 in Grundlagen der Praktischen Optik, Walter de Gruyter, Berlin (1930) pp. 41–71.

9 H.A. Buchdahl, 'Algebraic Theory of the Primary Aberrations of the Symmetrical Optical System', *J. Opt. Soc. Am.* Vol. 38 (1948) pp. 14–19.

10 D.P. Feder, 'Optical calculations with automatic computing machinery', *J. Opt. Soc. Am.*, Vol. 41 (1951) pp. 630–635.

11 J. Flügge, 'Die Theorie der Abbildungsfehler dritter ordnung', Chapter IV in Das Photographische Objektiv, Springer, Wien (1955) pp. 68–102.

12 J. Picht, *Grundlagen der geometrisch-optischen Abbildung*, VEB Deutscher Verlag der Wissenschaften, Berlin (1955) pp. 69–82.

13 F.D. Cruickshank and G.A. Hills, *The Third Order Aberration Coefficients of a refracting optical system*, University of Tasmania Press, Hobart, Australia (1960).

14 P. Baumeister, 'Aberration Analysis and Third Order Theory', Section 8 in *Military Standardization Handbook, MIL-HDBK-141, Optical Design*, U. S. Defense Supply Agency, Washington (1962) 8–1 to 8–21.

15 A.E. Conrady, *Applied Optics and Optical Design*, Part I, Dover, New York (1957) [originally by Oxford University Press (1929)], Section 27–28, pp. 52–55.

16 R.A. Sampson, 'A new treatment of optical aberrations', *Phil. Trans. Roy Soc. London A*, Vol. 212 (1913) pp.149–185.

17 W. Brouwer, 'The numerical calculation of the third order aberration coefficients', Chapter 9 in *Matrix methods in Optical Instrument Design*, W. A. Benjamin, New York (1964) pp. 121–135.

18 H. Chrétien, *Calcul des Combinaisons Optiques*, Masson, Paris (1980) [First published in 1938] pp. 687–725.

19 K. Schwarzschild, 'Untersuchungen zur geometrischen optic', *Abhandl. Königl. Ges. Wiss. Göttingen, Math-Phys. Kl.*, Vol. 4 (1905–1906) Nos. 1, 2, 3.

20 M. Herzberger, *Modern Geometrical Optics*, Interscience, New York (1958) pp. 299–379.

21 M. Born and E. Wolf, *Principles of Optics*, Pergamon, Oxford (1980) pp. 220–225. [First published 1959]

22 Y. Matsui and K. Nariai, Fundamentals of Practical Aberration Theory, World Scientific, Singapore (1993).

23 A. Romano, Geometric Optics, Birkhaüser, Boston (2010) pp. 201–217.

24 A.E. Conrady, *Applied Optics and Optical Design, Part II*, Dover, New York (1960), pp. 707–759.
25 H.H. Hopkins, *Wave Theory of Aberrations*, Clarendon Press, Oxford (1950), pp. 76–95.
26 G.G. Slyusarev, *Aberration and Optical Design Theory*, [original Russian book title: *Metodi Rascheta Opticheskikh Sistem, Mashinostroenie*, Leningrad (1969)] Translator: J.H. Dixon, Adam Hilger, Bristol (1984), pp. 48–244.
27 G.R. Rosendahl, 'A new derivation of third order aberration coefficients', *Appl. Opt.*, Vol. 6 (1967) pp. 765–771.
28 W.T. Welford, *Aberrations of the symmetrical optical system*, Academic, San Diego (1974).
29 W.T. Welford, 'Calculation of the Seidel aberrations', Chapter 8 in *Aberrations of optical systems*, Adam Hilger, Bristol (1986) pp. 130–161.
30 P. Mouroulis and J. Macdonald, *Geometrical Optics and Optical Design*, Oxford University Press, New York (1997) pp. 227–245.
31 V.N. Mahajan, 'Calculation of Primary Aberrations: Refracting Systems', in *Optical Imaging and Aberrations, Part I. Ray Geometrical Optics*, SPIE, Bellingham, Washington (1998) pp. 245–290.
32 J. Sasián, *Introduction to aberrations in optical imaging systems*, Cambridge University Press, Cambridge (2013).
33 W.T. Welford, *Aberrations of Optical Systems*, Adam Hilger, Bristol (1986), pp. 131–132.
34 P. Mouroulis and J. Macdonald, vide Ref. 9.90, pp. 228–229.
35 H.H. Hopkins, Chapter VI and Chapter VII in *Wave Theory of Aberrations*, Clarendon Press, Oxford (1950), pp. 66–95.
36 W.T. Welford, vide Chapter 8 'Calculation of the Seidel Aberrations', in *Aberrations of Optical Systems*, Adam Hilger, Bristol (1986), pp. 130–140.
37 P. Culmann, 'The Formation of Optical Images', in *Geometrical Investigation of the Formation of Images in Optical Instruments*, Ed. M. von Rohr, His Majesty's Stationery Office, London (1920) p. 133.
38 M. Born and E. Wolf, vide Ref. 9.81, p. 158.
39 W.T. Welford, 'Section 10.5 Expressions for the primary chromatic aberrations', in *Aberrations of Optical Systems*, Adam Hilger, Bristol (1986), pp. 202–206.
40 P. Mouroulis and J, Macdonald, vide Ref. 9.90, pp. 236–237.

13

Higher Order Aberrations in Practice

13.1 Evaluation of Aberrations of Higher Orders

Figure 13.1 presents a schematic diagram for illustrating the role of imaging models consisting of various orders of aberrations in the case of imaging an axial object point O by an axially symmetric system consisting of two refracting interfaces with their vertices located at A_1 and A_2 on the optical axis. $OP_1P_2O'_2$ is the finite ray path forming image of the point O at O'_2. Per the aberration free paraxial optics model, the image is formed at ${}_gO'_2$, and the paraxial ray path from the object to the image is $OP_{1\,g}P_{2\,g}O'_2$. According to an imaging model that also incorporates the primary aberrations, the image of O is at ${}_pO'_2$, and the approximate ray path is $OP_{1\,p}P_{2\,p}O'_2$. When the model also includes the secondary aberrations, the image is at ${}_sO'_2$, and the approximate ray path is $OP_{1\,s}P_{2\,s}O'_2$. Addition of the tertiary aberration effects in the model leads to the image at ${}_tO'_2$, and the approximate ray path $OP_{1\,t}P_{2\,t}O'_2$.

It is important to note that not only the final image, but also the height of the marginal ray on the second surface is different according to different models. Primary aberrations are a special case, where the OPD's can be evaluated from the paraxial ray paths. The validity of this simplification has been discussed in Section 12.2. For aberrations of a higher order, i.e. secondary aberrations or tertiary aberrations, etc., the orders of approximation are such that this simplification is no longer permitted. These aberrations consist of two components: intrinsic aberrations and induced aberrations. The former are named for the intrinsic nature of the surface, while the latter, which are also called extrinsic aberrations, arises out of interaction of the surface with the aberrations of the incident wavefront or rays.

'Image error theories of higher order have been developed on various foundations and with diverse methods, partially independent of another' [1]. Historically, Petzval seems to have dealt with higher order aberrations in connection with the design of his famous portrait lenses [2–3], but his derivations are not available in the published literature. As reported by von Rohr, early investigations were directed towards development of methods for determining longitudinal spherical aberration of higher order [4]. Conrady [5] and Bennett [6], among others, informed on these methods.

A general treatment for analyzing fifth order aberrations was initiated by Kohlschütter [7], who extended the study of primary aberrations as carried out by Schwarzschild [8] on the basis of the Seidel eikonal that he proposed. 'However, the formulas of the intrinsic coefficients as given by Kohlschütter are very complicated, and consequently the whole theory has not been used in practice, since that time' [1].

A different approach was adopted by T. Smith [9–10], who used the Hamilton-Bruns' angle eikonal in his studies on characteristics of aberration coefficients. This work was developed further by Korringa [11–12], Stephan [13], and van Heel [14]. Herzberger [15], using the point eikonal, referred to the object plane and the exit pupil for developing fifth order coefficients of image error for rotationally symmetric systems. Further extension of this study is presented in his treatise [16].

The widely used practical approach for treatment of higher order aberrations involves the use of trigonometric or algebraic ray tracing at the refracting interfaces. Marx developed a recursive computational method for the image error coefficients of arbitrary order [17]. Several ingenious attempts were made by many investigators to circumvent the formidable amount of numerical computation involved in evaluating the coefficients of higher order aberrations. Wachendorf [18] expanded the ray tracing equations to the fifth order in terms of direction tangents as a preliminary step to derive the fifth order aberration coefficients. G.W. Hopkins [19] extended the treatment to the seventh order, and called his method

DOI: 10.1201/9780429154812-13

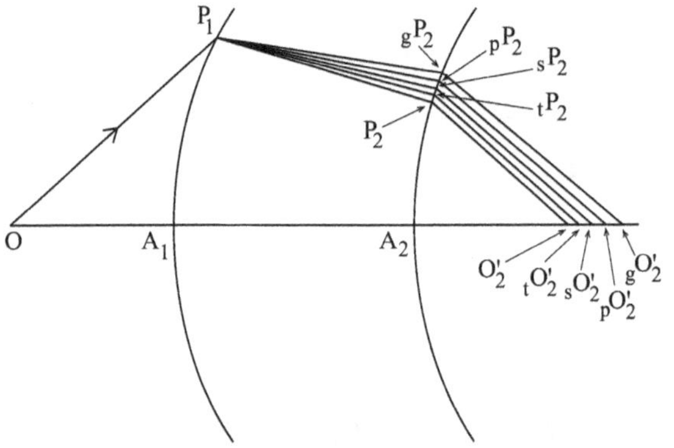

FIGURE 13.1 Schematic ray paths as per various models for imaging of an axial object point O by an axially symmetric system consisting of two refracting interfaces at A_1 and A_2 on the axis. Approximate ray paths as per (a) Gaussian (Paraxial) model: $OP_{1g}\,P_{2g}\,O'_2$ (b) Aberration models based on (i) Primary aberrations alone: $OP_{1p}\,P_{2p}\,O'_2$; (ii) Primary and secondary aberrations together: $OP_{1s}P_{2s}O'_2$; (iii) Primary, secondary, and tertiary aberrations together: $OP_{1t}\,P_{2t}\,O'_2$. Exact ray path incorporating all orders of aberration: $OP_1\,P_2\,O'_2$.

'proximate ray tracing' for the determination of the aberration coefficients. The approach followed by Cox [20] was slightly different by using expansion of direction cosines instead of the direction tangents.

By innovative algebraic manipulations, Buchdahl arrived at remarkably simple and reliable expressions for optical aberration coefficients of higher order [21–31]. For this purpose, he devised 'quasi-invariants', which reduce to the optical invariant in the paraxial limit. Cruickshank [32] demonstrated the use of the Buchdahl aberration coefficients in practical optical design. The thesis by Rimmer [33] presents expressions for optical aberration coefficients of Buchdahl in a form amenable for computer programming. Rimmer formulas were subsequently adjusted by Anderson [34] to make them conform with the conventions used by Feder [35]. This was done in connection with the developments in modular optical design by Stavroudis [36]. Brender Andersen extended Buchdahl's approach, and developed a method that is applicable also to systems without rotational symmetry [37], and he published Fortran codes for the computation of optical aberration coefficients of higher order in symmetric systems [38].

Bociort [39] reported on the use of computer algebra software in the derivation of analytic expressions for the higher order optical aberration coefficients. Later on, Bociort, Andersen, and Beckmann extended the study to the case of finite objects and object space telecentric systems [40]. Hoffmann sought explanation for higher order aberrations in terms of induced aberrations [41]. On a different note, Rosete-Aguilar and Rayces experimented with the renormalization of the Buchdahl – Rimmer third and fifth-order geometric aberration coefficients in order to identify the most important aberration in degradation of image quality [42–43]. Their analysis goes beyond the domain of ray optics and makes use of diffraction theory of image formation.

Sasián reported on the development of an 'ab initio' theory of sixth-order wave aberrations [44]. His book contains detailed formulae for the intrinsic and extrinsic aberration coefficients of the sixth order, and the connections between the Buchdahl – Rimmer coefficients and the wave aberration coefficients of his formulation [45].

13.2 A Special Treatment for Tackling Higher Order Aberrations in Systems With Moderate Aperture and Large Field

The Seidel theory of primary aberrations assumes the aperture and size of the field to be such that terms of degree higher than four can be neglected. In the case of axial imagery, there is no approximation involved with respect to τ, since this is zero. The approximation in τ has effects only as one considers extra-axial imaging, and it becomes more significant the greater the size of the field. It has two effects. First, the Seidel aberration coefficients do not correctly reflect the primary aberration types. Secondly, the significant effects of higher order aberrations are being ignored. The latter effects can be very significant, even at low relative apertures, where the size of the field is large. The usual approach for tackling aberrational analysis of these types of systems calls for the determination of a large number of aberration coefficients of higher order. Often, the appropriate highest order for the problem at hand is uncertain.

For systems with moderate to low relative aperture, but with large fields, Hopkins [46] suggested a different approach for taking into account the effect of higher order aberrations with significantly less computation. He proposed the tracing of an exact principal ray at the desired field angle with S and T ray fans, and determined aberration coefficients up to the fourth power of the aperture of the pencil surrounding the principal ray. The key points of this approach are underscored below.

The expression (8.81) for wave aberration polynomial shows the explicit form of all primary and secondary aberration terms, and only two tertiary aberration terms.

$$\begin{aligned} W = & \left[{}_0W_{40}r^4 + {}_2W_{22}\tau^2r^2(\cos\theta)^2 + {}_2W_{20}\tau^2r^2 + {}_1W_{31}\tau r^3\cos\theta + {}_3W_{11}\tau^3 r\cos\theta \right] \\ & + \left[{}_0W_{60}r^6 + {}_3W_{33}\tau^3r^3(\cos\theta)^3 + {}_4W_{20}\tau^4r^2 + {}_2W_{40}\tau^2r^4 + {}_5W_{11}\tau^5 r\cos\theta \right. \\ & \left. + {}_4W_{22}\tau^4r^2(\cos\theta)^2 + {}_1W_{51}\tau r^5\cos\theta + {}_2W_{42}\tau^2r^4(\cos\theta)^2 + {}_3W_{31}\tau^3r^3\cos\theta \right] \\ & \left[{}_0W_{80}r^8 + {}_4W_{44}\tau^4r^4(\cos\theta)^4 + \cdots \right] + \cdots \end{aligned} \tag{13.1}$$

All terms of tertiary and quaternary aberrations are given in Table 8.4 and Table 8.5, respectively. The wave aberration W is defined with respect to a reference sphere with its centre at the paraxial image. Two groups of terms are absent when the reference sphere is chosen to have its centre at the sagittal focus along the principal ray. For limitingly small aperture, i.e. $r \rightarrow 0$, but unrestricted value of τ, W reduces to

$$W = \left({}_3W_{11}\tau^3 + {}_5W_{11}\tau^5 + \ldots \right) r\cos\theta \tag{13.2}$$

This amounts to a lateral shift of the image point. The different terms in the bracket correspond to different orders of distortion. For a reference sphere centred on the principal ray, evidently

$$W_{11} = {}_3W_{11}\tau^3 + {}_5W_{11}\tau^5 + \ldots = 0 \tag{13.3}$$

The value of aberration for limitingly small aperture in the sagittal section, i.e. $\theta = \frac{\pi}{2}$, the terms remaining in W are

$$W = \left({}_2W_{20}\tau^2 + {}_4W_{20}\tau^4 + \ldots \right) r^2 \tag{13.4}$$

This amounts to a change of focus. The different terms in the bracket correspond to different orders of curvature. Therefore, if the centre of the reference sphere is located at the sagittal focus,

$$W_{20} = \left({}_2W_{20}\tau^2 + {}_4W_{20}\tau^4 + \ldots \right) = 0 \tag{13.5}$$

If the wave aberration W is measured with respect to a reference sphere with its centre at the sagittal focus, it follows that the expression (13.1) reduces to

$$\begin{aligned} W = & \left[{}_0W_{40}r^4 + {}_2W_{22}\tau^2r^2(\cos\theta)^2 + {}_1W_{31}\tau r^3\cos\theta \right] \\ & + \left[{}_0W_{60}r^6 + {}_3W_{33}\tau^3r^3(\cos\theta)^3 + {}_2W_{40}\tau^2r^4 + {}_4W_{22}\tau^4r^2(\cos\theta)^2 \right. \\ & \left. + {}_1W_{51}\tau r^5\cos\theta + {}_2W_{42}\tau^2r^4(\cos\theta)^2 + {}_3W_{31}\tau^3r^3\cos\theta \right] \\ & \left[{}_0W_{80}r^8 + {}_4W_{44}\tau^4r^4(\cos\theta)^4 + \cdots \right] + \cdots \end{aligned} \tag{13.6}$$

The above relation can be rewritten as

$$W = W_{22}r^2(\cos\theta)^2 + W_{31}r^3\cos\theta + W_{40}r^4 + W_{33}r^3(\cos\theta)^3 + W_{42}r^4(\cos\theta)^2 + W_{44}r^4(\cos\theta)^4 + \dots \quad (13.7)$$

where

$$\left.\begin{aligned} W_{22} &= {}_2W_{22}\tau^2 + {}_4W_{22}\tau^4 + \dots \\ W_{31} &= {}_1W_{31}\tau + {}_3W_{31}\tau^3 + \dots \\ W_{40} &= {}_0W_{40} + {}_2W_{40}\tau^2 + \dots \end{aligned}\right\} \quad (13.8)$$

and

$$\left.\begin{aligned} W_{33} &= {}_3W_{33}\tau^3 + {}_5W_{33}\tau^5 + \dots \\ W_{42} &= {}_2W_{42}\tau^2 + {}_4W_{42}\tau^4 + \dots \\ W_{44} &= {}_4W_{44}\tau^4 + {}_6W_{44}\tau^6 + \dots \\ &\dots \end{aligned}\right\} \quad (13.9)$$

Unlike the aberration coefficients in expression (13.6) where each of them has one prefix and two suffixes, each of the finite aberration coefficients in (13.7) has two suffixes and no prefix. The terms W_{22}, W_{31}, and W_{40} in (13.8) represent the primary aberration types – astigmatism, coma, and spherical aberration. The Seidel aberration coefficients for astigmatism, coma, and spherical aberration give only the first terms of each of the series in (13.8). The remaining terms in each series correspond to higher orders of the respective aberration types. The terms $W_{33}, W_{42}, W_{44}, \dots$ in (13.9) have no counterpart among the primary aberration types. The first terms in the series for W_{33} and W_{42} are secondary aberration terms, the elliptical coma of Type II and secondary astigmatism of Type II, respectively [see Table 8.3]. Note that secondary astigmatism of Type I and elliptical coma of type I are the second terms in the series for W_{22} and W_{31}, as given in (13.8). The second terms in the series for W_{33} and W_{42} are tertiary aberration terms, as is the first term in the series for W_{44}. The second term in the series for W_{44} is a quaternary aberration term. More finite aberration terms of the like of W_{33}, W_{42}, W_{44} can be found from higher order aberration terms. An analytical procedure for computing the finite aberration coefficients is described in detail by Weinstein [47–48].

Incidentally, it may be noted that Lu and Cao have recently developed a sixth order wave aberration theory appropriate for ultra-wide-angle optical systems, e.g. fish-eye lenses [49].

With this brief statement on the state-of-the-art in studies on higher order aberrations, we refer the interested author to the original works cited in the references for detailed information.

13.3 Evaluation of Wave Aberration Polynomial Coefficients From Finite Ray Trace Data

For rotationally symmetric systems, with usual notation, the wave aberration function W' for the ray coming from the object point at τ and passing through the point $(r,\theta)[\equiv(x,y)]$ on the exit pupil takes the form $W'(\tau^2, r^2, \tau r\cos\theta)$. From (8.76) – (8.78) we note that it can be expanded in a polynomial as

$$W'(\tau^2, r^2, \tau r\cos\theta) = \sum_i\sum_j\sum_k \omega_{ijk}(\tau^2)^i(r^2)^j(\tau r\cos\theta)^k = \sum_p\sum_m\sum_n {}_pW_{mn}\tau^m r^n(\cos\theta)^n \quad (13.10)$$

where $p = 2i + k$, $m = 2j + k$, and $n = k$. It is possible to investigate the properties of W' with respect to any one of the three variables, τ, r, $\cos\theta$, for fixed values of the remaining variables. In problems of optical system analysis and design, usually one needs to look into the characteristics for a set of specific values for τ. For a specific value of τ, the aberration function W' takes the form

$$W'(r,\theta) = \sum_m \sum_n W_{mn} r^m (\cos\theta)^n = \sum_m \sum_n W_{mn} r^{(m-n)} (r\cos\theta)^n \tag{13.11}$$

In rectangular coordinates, where $x = r\sin\theta$, $y = r\cos\theta$,

$$W'(x,y) = \sum_m \sum_n W_{mn} \left(x^2 + y^2\right)^{\frac{m-n}{2}} y^n \tag{13.12}$$

In the above, $W_{mn} = \sum_p {}_pW_{mn} \tau^p$. The indices m, n have the restriction $m \geq n$, and take on the values $m = 2,4,6,\ldots$ with $n = 0, 2, 4, \ldots$, and $m = 3, 5, 7, \ldots$ with $n = 1, 3, 5, \ldots$.

$W'(r,\theta)$ can be expressed as

$$W'(r,\theta) = \sum_{j=1} \sum_{k=0}^{j} \left[W_{2j,2k} r^{2j} (\cos\theta)^{2k} + W_{(2j+1),(2k+1)} r^{(2j+1)} (\cos\theta)^{(2k+!)} \right] \tag{13.13}$$

Equation (13.12) or (13.13) is a power series expansion of the aberration function W'. The polynomial on the right-hand side of the equation consists of an infinite number of terms. However, in almost all practical cases, the series is highly convergent, and a suitably truncated series can be used in practice. This is easier said than done. Not equal weightage is called for truncation in first and second suffixes, i.e. in m and n. Both the exclusion of required higher order terms and the inclusion of unwanted higher order terms have consequences that may cause the analysis to go haywire. Nevertheless, intelligent choices based on the nature of the system, coupled with necessary validatory checks, often provide a finite polynomial series for the aberration function, so much so that it can be used not only for rapid evaluation of image assessment criteria, but the aberration polynomial coefficients generated in the intermediate stage provide a scope for fast analysis of aberrational characteristics obviating the need for calculation of many aberration coefficients of higher orders.

Assuming (13.12) as a finite polynomial, the aberration coefficients W_{mn} may be found as a linear function of the aberrations $W_j \equiv W(x_j, y_j)$ along a suitable number of traced rays. Assume there are I number of W_{mp} coefficients in the series, and I number of suitably chosen rays are traced. The latter yields I number of linear equations. The j-th equation is

$$W_j = \sum_m \sum_n A(m,n;j) W_{mn} \tag{13.14}$$

where $A(m,n;j)$ is the coefficient of W_{mn} in the j-th equation, and it is a function of the exit pupil coordinates of the j-th ray. The set of I equations like (13.14) constitutes the matrix relation

$$\mathbf{W} = \mathbf{A}\tilde{\mathbf{W}} \tag{13.15}$$

where $\tilde{\mathbf{W}}$ and $\mathbf{W}$ are column vectors of the I polynomial coefficients and the wave aberrations of the I rays, respectively. $\mathbf{A}$ is a rectangular matrix whose elements depend on the exit pupil coordinates of the I rays.

Inversion of the matrix given by (13.15) gives

$$\tilde{\mathbf{W}} = \mathbf{A}^{-1}\mathbf{W} = \mathbf{B}\mathbf{W} \tag{13.16}$$

Individual equations of this matrix equation yield the polynomial coefficients as a linear combination of the aberrations of the I rays, as shown below:

$$W_{mn} = \sum_i B(m,n;i) W_i \tag{13.17}$$

This technique for determination of the aberration polynomial coefficients is not at all efficient from the point of view of numerical computation. Note that $W(x,y)$ in (13.12) calls for aberrations W at exit pupil points (x, y). In the process of ray tracing, the coordinates of ray intersection points with the exit pupil are not known beforehand. Usually, ray tracing commences with the coordinates of the object point and the coordinates of a point on the entrance pupil at which the object space ray intersects the entrance pupil. After the completion of ray tracing, the coordinates of the point of intersection of the emergent ray in the image space with the exit pupil is known. In general

$$(x,y)_{\text{entrance pupil}} \neq (x,y)_{\text{exit pupil}} \tag{13.18}$$

Therefore, the matrix inversion process noted in (13.16) needs to be carried out for every system separately for each selection of ray grid pattern.

The use of reduced pupil variables (x_S, y_T) in the scheme of canonical coordinates by Hopkins (see Chapter 11) obviates the need for this repetitive evaluation of the matrix **B**. The aberration function is expressed in terms of the reduced pupil variables as $W'(x'_S, y'_T)$, and it is given by

$$W'(x'_S, y'_T) = \sum_m \sum_n W_{mn} \left(x_S'^2 + y_T'^2 \right)^{\frac{m-n}{2}} y_T'^n \tag{13.19}$$

In this canonical scheme, the entrance pupil coordinates and the exit pupil coordinates of a ray are exactly the same for systems corrected from sine condition error. Even in the case of moderately corrected systems, $x'_S \cong x_S, y'_T \cong y_T$. If the difference between the two coordinates needs to be reduced, two different correction schemes have been explored by Singh [50] and Macdonald and Wang [51]. Whereas the approach by Singh involves evaluation of second derivatives of aberrations and is essentially an analytical approach, the semi-numerical approach of Macdonald and Wang is more straightforward from the point of view of practical implementation. The key point of both approaches is to determine the coordinates of the point of intersection on the entrance pupil of the exact ray (from the same object point) that yields the desired exit pupil coordinates for the corresponding ray in the image space. A finite ray traced from the object point through the corrected entrance pupil coordinates will intersect the exit pupil at the desired point (x'_T, y'_T).

Suppose the desired exit pupil coordinates are (x'_{Si}, y'_{T_i}). Take the initial values of the entrance pupil coordinates: $[x_S]_{(0)} = x'_{Si}, [y_T]_{(0)} = y'_{Ti}$. Trace the ray from the object point in the object space through $\{[x_S]_{(0)}, [y_T]_{(0)}\}$, and determine the exit pupil coordinates. Suppose the exit pupil coordinates so determined are $\left([x'_s]_{(0)}, [y'_T]_{(0)}\right)$.

Determine

$$[\Delta x'_S]_{(0)} = [x'_s]_{(0)} - x'_{Si}, [\Delta y'_T]_{(0)} = [y'_T]_{(0)} - y'_{Ti} \tag{13.20}$$

Take revised entrance pupil coordinates:

$$[x_S]_{(1)} = [x_S]_{(0)} - [\Delta x'_S]_{(0)}, [y_T]_{(1)} = [y_T]_{(0)} - [\Delta y'_T]_{(0)} \tag{13.21}$$

Trace a ray from the same object point through $\left\{[x_S]_{(1)},[y_T]_{(1)}\right\}$, and determine the exit pupil coordinates. The process may be repeated. For the k-th iteration, the correction for the entrance pupil coordinates is given by

$$[x_S]_{(k)} = [x_S]_{(k-1)} - \left[\Delta x'_S\right]_{(k-1)}, [y_T]_{(k)} = [y_T]_{(k-1)} - \left[\Delta y'_T\right]_{(k-1)} \tag{13.22}$$

The authors claimed a single iteration is normally sufficient to achieve one's goals. On our own, we have validated this claim in our routines.

The speed of computation of the polynomial coefficients increases manifold when the available information about the wavefront is utilized with each ray trace. During ray tracing, for a skew ray, along with the wave aberration $W'\left(x'_S, y'_T\right)$, given by (13.19), transverse ray aberrations are also determined. In Hopkins' canonical variables, the reduced transverse ray aberrations are

$$\delta G'_S\left(x'_S, y'_T\right) = -\frac{\partial W'\left(x'_S, y'_T\right)}{\partial x'_S} = -\sum_m \sum_n W_{mn} (m-n)\left(x'^2_S + y'^2_T\right)^{\left(\frac{m-n-2}{2}\right)} x'_S y'^n_T \tag{13.23}$$

$$\delta H'_T\left(x'_S, y'_T\right) = -\frac{\partial W'\left(x'_S, y'_T\right)}{\partial y'_T} = -\sum_m \sum_n W_{mn}\left(n x'^2_S + m y'^2_T\right)\left(x'^2_S + y'^2_T\right)^{\left(\frac{m-n-2}{2}\right)} y'^{n-1}_T \tag{13.24}$$

Therefore, for each skew ray trace, three linear equations can be set up for the unknown coefficients W_{mn} using (13.19), (13.23), and (13.24). $\delta G'_S\left(x'_S, y'_T\right)$ is identically zero for each meridional ray. Therefore, for each meridional ray trace, two linear equations can be set up. For an extra-axial pencil, two more equations can be set up.

$$W_S = W_{20} \tag{13.25}$$

$$W_T = W_{20} + W_{22} \tag{13.26}$$

These are already derived in Section 11.7. W_S and W_T are obtained in the course of tracing the paraxial sagittal ray and the paraxial tangential ray, respectively, about the pupil ray.

By tracing the suitably chosen five skew rays, Macdonald and Wang [51] determined the coefficients of a seventeen $(= 5 \times 3 + 2)$ term polynomial. Since different imaging systems have different combinations of aperture and field, different types of aberration polynomials are needed for specifying their aberration function. The polynomials vary both in the number of terms and also in the choice of terms. Singh [50] gave a list of three on-axis polynomials and eight extra-axial polynomials. For most optical imaging systems, one or the other type of these polynomials often serves the purpose of analytical specification of wave aberration function in practice.

In general, if N meridional and K skew rays, together with the principal ray and the S and T rays, are traced for an extra-axial object point, we can get $P(= 2N + 3K + 2)$ computed aberration values W_j, $j = 1,\ldots,P$. It is obvious that, for an axial object point $P = 2N$. The P aberration values W_j can be used to set up P simultaneous linear equations for the P unknown coefficients W_{mn}. The j-th equation has the form

$$W_j = \sum_m \sum_n A(m,n;j) W_{mn} \tag{13.27}$$

$A(m,n;j)$, the coefficient of W_{mn} is a function of the coordinates $\left(x'_S, y'_T\right)$, to which the j-th equation refers. The P simultaneous linear equations can be written in a matrix form

$$\mathbf{W} = \mathbf{A}\tilde{\mathbf{W}} \tag{13.28}$$

where $\tilde{\mathbf{W}}$ and $\mathbf{W}$ are column vectors of the P polynomial coefficients of the form W_{mn} and the P computed aberrations W_i, respectively. The vector $\tilde{\mathbf{W}}$ is related with the vector $\mathbf{W}$ by

$$\tilde{\mathbf{W}} = \mathbf{A}^{-1}\mathbf{W} = \mathbf{B}\mathbf{W} \tag{13.29}$$

The desired coefficient W_{mn} is obtained by multiplying a row of the matrix of $K(m,n;i)$ coefficients by the computed aberrations W_i, as shown below.

$$W_{mn} = \sum_{i=1}^{P} B(m,n;i)W_i \tag{13.30}$$

It is important and significant to note that, in this approach, elements of the matrix **B** need to be evaluated only once for a chosen set of rays on the pupil, and the chosen aberration polynomial. Elements of the matrix depend on the coordinates $\left(x'_S, y'_T\right)$. The same matrix can be used for determining the coefficients of an aberration polynomial of any lens system. They are called universal coefficients. Note that the total number of computed aberration data is equal to the number of coefficients of the aberration polynomial. In practice, the selection of ray grid needs to be based on other physical considerations. Choice of the specific coefficients in the aberration polynomial is also a tricky job, one that is primarily based on 'a priori' estimates on the presence of significant amounts of particular aberration types. These estimates may be based on the sizes of the aperture and field of the system. Also, practical implementation of the approach is facilitated by judicious utilization of symmetry considerations. Details of these aspects may be found in references [50–51].

For the sake of illustration, Figure 13.2(a) shows a ray sampling grid pattern used for determining the coefficients of an aberration polynomial in the case of axial imaging. On account of rotational symmetry, sampling along any one azimuth is sufficient. Usually, this is done along the y_T direction. The points of intersection with the canonical circular pupil of the two rays from the axial object point are shown. The four coefficients of an aberration polynomial of the form

$$W'(r) = W_{40}r^4 + W_{60}r^6 + W_{80}r^8 + W_{10.0}r^{10} \tag{13.31}$$

can be determined from the four aberration values, one wave aberration and one transverse ray aberration being obtained from each ray trace data. Similarly, Figure 13.2(b) shows a ray sampling grid pattern used for determining the coefficients of an aberration polynomial in the case of extra-axial imaging. On account of bilateral symmetry, sampling over a semicircle is adequate for this purpose. The points of intersection with the canonical circular pupil of four meridional rays and one skew ray from the extra-axial object point are shown in the figure. Note that from each meridional ray trace data, two values of aberration, one wave aberration and one transverse ray aberration, are obtained. From each sagittal ray trace data, three values of aberration, one wave aberration and two transverse ray aberration components, are obtained.

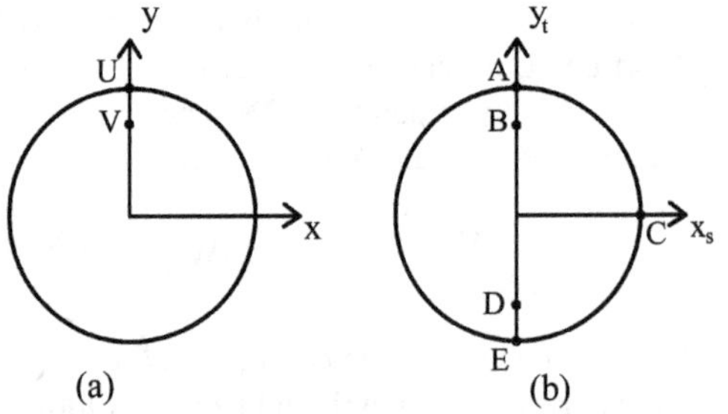

FIGURE 13.2 Footprints on the canonical circular pupil with typical ray sampling grids for (a) A 4-term aberration polynomial in the axial case: Location of points on the pupil of two rays from the extra-axial object point: (i) $U: x_S = 0, y_T = 1.0$, (ii) $V: x_S = 0, y_T = 0.707$; (b) A 13-term polynomial for an axial field point: Five rays: (i) $A: x_S = 0, y_T = 1.0$, (ii) $B: x_S = 0, y_T = 0.707$, (iii) $C: x_S = 1.0, y_T = 0$, (iv) $D: x_S = 0, y_T = -0.707$, (v) $E: x_S = 0, y_T = -1.0$.

Additionally, by parabasal S and T ray trace on the principal ray, two additional data for aberrations of the wavefront corresponding to the particular field point, W_S and W_T, are obtained. Singh [50] utilized a similar ray trace scheme for determining coefficients of the following thirteen term polynomial

$$W'(r,\theta) = W_{20}r^2 + W_{22}r^2(\cos\theta)^2 + W_{40}r^4 + W_{42}r^4(\cos\theta)^2 + W_{60}r^6 \\ + W_{62}r^6(\cos\theta)^2 + W_{80}r^8 + W_{10,0}r^{10} + W_{31}r^3\cos\theta \\ + W_{33}r^3(\cos\theta)^3 + W_{51}r^5\cos\theta + W_{71}r^7\cos\theta + W_{91}r^9\cos\theta \quad (13.32)$$

On a different note, Kidger [52] proposed the use of least squares fitting for calculation of the aberration polynomial. Typically, 25 to 30 rays are traced to determine the coefficients of a 15 term aberration polynomial, which in this case is described in terms of rectangular coordinates of the exit pupil. It is expected that the redundant information available in the least squares approach can be used in devising suitable methods to check the appropriateness of the choice of the aberration polynomial for the problem at hand. Macdonald and Wang made a significant observation in this context. They noted:

> the least-squares fit is better than the exact fit when the optical system is poorly corrected, whereas the exact fit should be used when the aberrations of the system are small. This is contrary to what might be expected since, for a poorly corrected system, high-order aberrations are typically large, so that more terms should be included in the aberration polynomial.
>
> [51]

Incidentally, in the related field of charged particle optics, similar approaches are often made to determine the coefficients of aberration polynomials [53–54]. An alternative semi-analytic approach is proposed by Kintner, who developed an orthogonal set of polynomials for a prespecified set of data points, and used the set of polynomials for generating the wavefront aberration polynomial [55].

REFERENCES

1 J. Focke, 'Higher order aberration theory', in *Progress in Optics, Vol. IV*, Ed. E. Wolf, North Holland, Amsterdam (1965) pp. 1–36.

2 J. Petzval, *Bericht über die Ergebnisse einiger dioptrischen Untersuchungen*, Verlag von Conrad Adolph Hartleben, Pesth, (1843). Reprinted 1975, Akadémiai Kiadó, Budapest.

3 J. Petzval, 'Bericht über Dioptrische Untersuchungen', *Sitzungsberichte der mathem.-naturw. Classe der kaiserlichen Acad. der Wiss.*, Vol. 26, Wien (1857) pp. 33–90.

4 A. Koenig and M. von Rohr, 'Theory of Spherical Aberration', Chapter V in *Geometrical Investigation of the formation of images in optical instruments*, [original German book title: *Die Theorie der optischen Instrumente*, Ed. M. von Rohr Springer, Berlin (1904)] Translator, R. Kanthack, H.M. Stationery Office, Dept. of Scientific and Industrial Research (1920) pp. 217–219, 235–239, 342.

5 A.E. Conrady, Ref. 5, Section [31].

6 H.F. Bennett, 'The reciprocal spherical aberration of an optical system including higher orders', *RP 466, Bureau of Standards, Journal of Research*, Vol. 9 (1932) pp. 187–225.

7 A. Kohlschütter, *Die Bildfehler fünfter ordnung optischer systeme*, Kaestner, Dissertation, Göttingen (1908).

8 K. Schwarzschild, 'Untersuchungen zur geometrischen optic', *Abhandl. Königl. Ges. Wiss. Göttingen, Math-Phys. Kl.*, Vol. 4 (1905–1906) Nos. 1, 2, 3.

9 T. Smith, 'The changes in aberrations when the object and stop are moved', *Trans. Opt. Soc. (London)* Vol. 23 (1922) pp. 311–321.

10 R.J. Pegis, 'The modern development of Hamiltonian optics', in Progress in Optics, Vol. I, Ed. E. Wolf, North Holland (Amsterdam).

11 J. Korringa, 'Classification of aberrations in rotationally symmetrical optical systems', *Physica*, Vol. 8 (1941) pp. 477–496.

12 J. Korringa, *Onderzoekingen op het Gebied der Algebraische Optiek*, Ph. D. Thesis, Delft (1942).

13 W.G. Stephan, *Practische Toepassingen op het Gebied der Algebraische Optica*, Ph. D. Thesis, Delft (1947).
14 A.C.S. van Heel, 'Calcul des aberrations du cinquième ordre et projets tenant compte de ces aberrations', in *La Théorie des Images Optiques*, Colloques internationaux 14–19 Oct 1946, C.N.R.S., Paris (1949).
15 M. Herzberger, 'Theory of image errors of the fifth order in rotationally symmetrical systems. I', *J. Opt. Soc. Am.*, Vol. 39 (1939) pp. 395–406.
16 M. Herzberger, *Modern Geometrical Optics*, Interscience, New York (1958), pp. 299–379.
17 H. Marx, 'Theorie des geometrisch-optischen Bildfehler', in Handbuch der Physik, Band XXIX, Optische Instrumente, Ed., S. Flügge, Springer, Berlin (1967) pp. 68–191.
18 F. Wachendorf, 'Bestimmung der Bildfehler 5. Ordnung in Zentrierten optischen systemen', *Optik*, Vol. 5 (1949) pp. 80–122.
19 G.W. Hopkins, 'Proximate ray tracing and optical aberration coefficients', *J. Opt. Soc. Am.*, Vol. 66, (1976) pp. 405–410.
20 A. Cox, Chapter 5, 'Fifth order aberrations' in *A System of Optical Design*, The Focal Press, London (1964) pp. 175–213.
21 H.A. Buchdahl, Optical Aberration Coefficients, Dover, New York (1968). [First published by Oxford University Press in 1954].
22 H.A. Buchdahl, 'Optical Aberration Coefficients. I. The Coefficient of Tertiary Spherical Aberration', *J. Opt. Soc. Am.*, Vol. 46 (1956) pp. 941–943.
23 H.A. Buchdahl, 'Optical Aberration Coefficients. II. The Tertiary Intrinsic Coefficients', *J. Opt. Soc. Am.*, Vol. 48 (1958) pp. 563–567.
24 H.A. Buchdahl, 'Optical Aberration Coefficients. III. The Computation of the Tertiary Coefficients', *J. Opt. Soc. Am.*, Vol. 48 (1958) pp. 747–756.
25 H.A. Buchdahl, 'Optical Aberration Coefficients. IV. The Coefficient of Quaternary Spherical Aberration', *J. Opt. Soc. Am.*, Vol. 48 (1958) pp. 757–759.
26 H.A. Buchdahl, 'Optical Aberration Coefficients. V. On the quality of predicted displacements', *J. Opt. Soc. Am.*, Vol. 49 (1959) pp. 1113–1121.
27 H.A. Buchdahl, 'Optical Aberration Coefficients. VI. On Computation Involving Coordinates Lying Partly in the Image Space', *J. Opt. Soc. Am.*, Vol. 50 (1960) pp. 534–539.
28 H.A. Buchdahl, 'Optical Aberration Coefficients. VII. The Primary, Secondary, and Tertiary Deformation and Retardation of the Wave Front', *J. Opt. Soc. Am.*, Vol. 50 (1960) pp. 539–544.
29 H.A. Buchdahl, 'Optical Aberration Coefficients. VIII. Coefficient of Spherical Aberration of Order Eleven', *J. Opt. Soc. Am.*, Vol. 50 (1960) pp. 678–683.
30 H.A. Buchdahl, 'Optical Aberration Coefficients. XII. Remarks Relating to Aberrations of any Order', *J. Opt. Soc. Am.*, Vol. 55 (1965) pp. 641–649.
31 H.A. Buchdahl, An Introduction to Hamiltonian Optics, Cambridge Univ. Press, London (1970).
32 F.D. Cruickshank and G.A. Hills, 'Use of Optical Aberration Coefficients in Optical Design', *J. Opt. Soc. Am.*, Vol. 50 (1960) pp. 379–387.
33 M.P. Rimmer, Optical Aberration Coefficients, M. Sc. Thesis, University of Rochester (1963).
34 D.W. Anderson, 'Control of fifth-order aberrations in modular optical design', *J. Opt. Soc. Am.*, Vol. 69 (1979) pp. 321–324.
35 D.P. Feder, 'Optical calculations with automatic computing machinery', *J. Opt. Soc. Am.*, Vol. 41 (1951) pp. 630–635.
36 O.N. Stavroudis, Chapter 7, 'The Fifth order' in *Modular Optical Design*, Springer, Berlin (1982) pp. 112–146.
37 T. Brender Andersen, 'Automatic computation of optical aberration coefficients', *Appl. Opt.*, Vol. 19 (1980) pp. 3800–3816.
38 T. Brender Andersen, 'Optical aberration coefficients: FORTRAN subroutines for symmetrical systems', *Appl. Opt.*, Vol. 20 (1981) pp.3263–3268.
39 F. Bociort, 'Computer Algebra Derivation of High-Order Optical Aberration Coefficients', ISSN 1381–1045, Technical Report no. 7, February 1995, Research Institute for Applications of Computer Algebra (RIACA), Amsterdam.
40 F. Bociort, T.B. Andersen, and L.H.J.F. Beckmann, 'High-order optical aberration coefficients: extension to finite objects and to telecentricity in object space', *Appl. Opt.*, Vol. 47 (2008) pp. 5691–5700.

41 J.M. Hoffman, Induced Aberrations in Optical Systems, Ph. D. Thesis, The University of Arizona (1993).
42 M. Rosete-Aguilar and J. Rayces, 'Renormalization of the Buchdahl-Rimmer third and fifth-order geometric aberration coefficients to rms wave aberration function expressions', *J. Mod. Opt.*, Vol. 42 (1995) pp. 2435–2445.
43 M. Rosete-Aguilar and J. Rayces, 'Renormalization of the Buchdahl-Rimmer coefficients of the reduced geometric aberration functions to rms spot size', *J. Mod. Opt.*, Vol. 43 (1996) pp. 685–692.
44 J. Sasián, 'Theory of sixth-order wave aberrations', *Appl. Opt.*, Vol. 49 (2010) D69 – D95. [vide corrigendum in Appl. Opt., Vol. 49 (2010) pp. 6502–6503].
45 J. Sasián, Chapter 14 'Sixth-order aberration coefficients' in *Introduction to Aberrations in Optical Imaging Systems*, Cambridge University Press, Cambridge (2013) pp. 187–204.
46 H.H. Hopkins, 'Higher order aberrations consequent upon oblique refraction at a spherical interface', in *La Théorie des Images Optiques*, Colloques internationaux 14–19 Oct 1946, C.N.R.S., Paris (1949).
47 W. Weinstein, 'Wave-front Aberrations of Oblique Pencils in a Symmetrical Optical System: Refraction and Transfer Formulae', *Proc. Phys. Soc. B*, Vol. 62 (1949) pp. 726–740. [vide corrigendum in Proc. Phys. Soc. B, Vol. 63 (1950) 221].
48 W. Weinstein, 'The Computation of Wave-Front Aberrations of Oblique Pencils in a Symmetrical Optical System', *Proc. Phys. Soc. B*. Vol. 63 (1950) pp. 709–723.
49 L. Lu and Y. Cao, 'Sixth-order wave aberration theory of ultrawide-angle optical systems', *Appl. Opt.*, Vol. 56 (2017) pp. 8570–8583.
50 R.N. Singh, 'A fast and simple method of determining the coefficients of the monochromatic aberration polynomial for image assessment and automatic optical design', *Opt. Acta*, Vol. 23 (1976) pp. 621–650.
51 J. Macdonald and Y. Wang, 'Sources of error in rapid fitting of wave-aberration polynomials', *Opt. Aca*, Vol. 33 (1986) pp. 621–636.
52 M.J. Kidger, 'The calculation of the optical transfer function using Gaussian quadrature', *Opt. Acta*, Vol. 25 (1978) pp. 665–680.
53 P.W. Hawkes, 'Aberrations' in *Handbook of Charged Particle Optics*, Ed. J. Orloff, CRC Press, Boca Raton, FL (2009).
54 M. Oral and B. Lencová, 'Calculation of aberration coefficients by raytracing', *Ultramicroscopy*, Vol. 109 (2009) pp. 1365–1373.
55 A.M. Plight, 'The calculation of the wavefront aberration polynomial', *Opt. Acta*, Vol. 27 (1980) pp. 717–721.

14

Thin Lens Aberrations

14.1 Primary Aberrations of Thin Lenses

The primary aberrations alone generally give an inadequate description of the aberrational correction of optical systems. Nevertheless, the reduction of primary aberrations to small values is a sine qua non, though not a sufficient condition, for achieving images of acceptable quality. The domain of the search space that can provide a system of good quality can, therefore, be significantly reduced. In systematic optical design, this is one of the key roles played by primary aberrations. However, the primary aberrations are fairly complicated functions of the constructional parameters of the system, so in general it is not possible to delineate the boundary of the effective multidimensional search space. At this point, the thin lens primary aberration theory becomes relevant in the analysis and design of optical systems.

In thin lens approximation for an optical system, the axial thickness of each lens element constituting the system is neglected in comparison with the radii of the two refracting interfaces, the semi-diameter of the aperture, or the focal length of that element, and the axial thicknesses of all lens elements are taken as zero. In multi-element systems, the axial separation between the elements is left unchanged. Obviously, in the final stages of detailed design, this approximation is of limited utility. However, the approximation is very useful in deriving the general principles of lens structures and in obtaining useful results in systems in which the field and the aperture are not too large.

Many sets of thin lens formulae have been proposed, differing mainly in choice of the variables used to specify the paraxial conjugates and in the 'shape' of the lens for a given power. Early works are referred to by Taylor [1] and von Rohr [2]. The different choices of the variables mentioned above lead to expressions of varying degree of complexity for primary aberrations of thin lenses [3–21]. However, the dimensionless, symmetrical variables originating from Coddington [3] are now widely used.

It is interesting to note that, though formulae for primary aberrations of thin lenses have been in existence for a long time, they were usually given for a thin lens in air until Wynne generalized the notion of a thin lens to mean a thin component of refractive index $(n_0 n)$ in a medium of index n, and gave the expressions for their primary aberrations. Hazra and Delisle [20] extended the study to the case of primary aberrations of a thin lens with different object and image space media. In what follows, the final expressions for the primary aberrations will be given. Details of derivation may be found in [20]. Note that the sign convention for the paraxial convergence angle used in this book follows that used in analytical geometry, and it is the opposite of what was used in our earlier publication.

It is also significant to note that Taylor [22] used an interesting thin lens model for approximating a single thick lens. In this model, a single thick lens is assumed to be made up of two thin lenses, each of which may be of plano-convex/plano-concave/convexo-plane/concavo-plane type, depending on the shape of the refracting interface, and separated by a plane parallel plate of thickness equal to the axial thickness of the single thick lens. Figure 14.1 illustrates the Taylor model applied to single thick lenses of different shapes. The four cases of approximation shown in (a) – (d) are:

(a) A biconvex thick lens ≈ a convexo-plane thin lens at vertex A_1, in contact with a plane parallel plate of the same material and thickness $t(= A_1A_2)$, in contact with a plano-convex thin lens at vertex A_2.

DOI: 10.1201/9780429154812-14

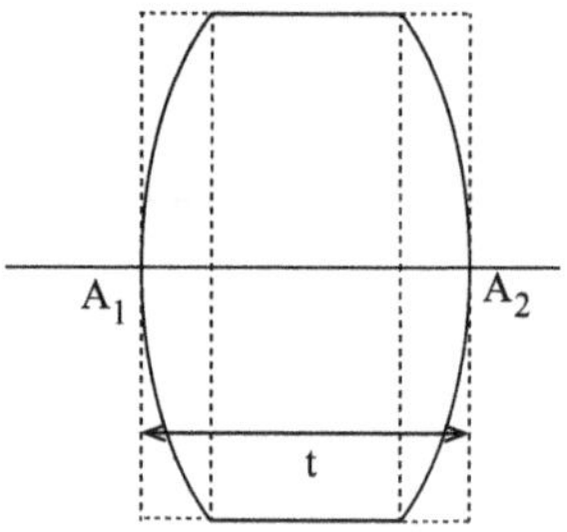

FIGURE 14.1(a) Taylor thin lens model of a single biconvex thick lens ≡ [a convexo-plane thin lens at A_1 + a plane parallel plate of thickness $t\left(= A_1A_2\right)$ + a plano-convex thin lens at A_2].

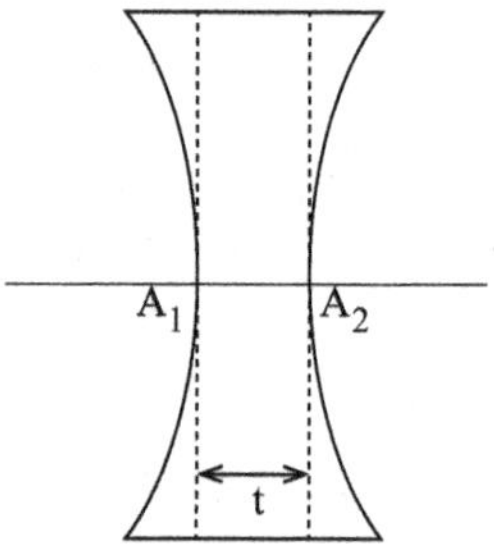

FIGURE 14.1(b) Taylor thin lens model of a single biconcave thick lens ≡ [a concavo-plane thin lens at A_1+ a plane parallel plate of thickness $t\left(= A_1A_2\right)$ + a plano-concave thin lens at A_2].

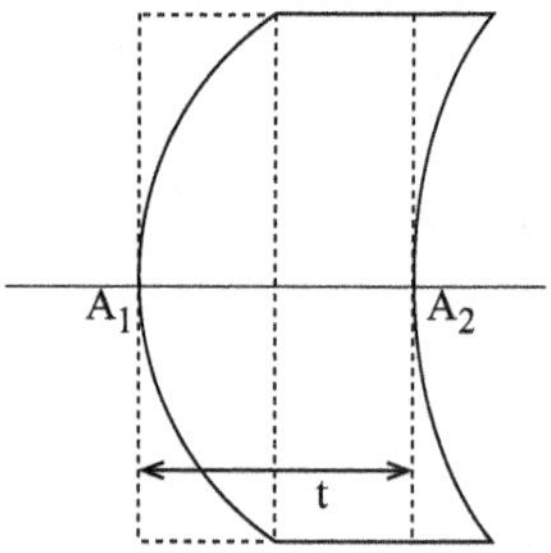

FIGURE 14.1(c) Taylor thin lens model of a single convexo-convex meniscus lens ≡ [a convexo-plane thin lens at A_1 + a plane parallel plate of thickness $t\left(= A_1A_2\right)$ + a plano-concave thin lens at A_2].

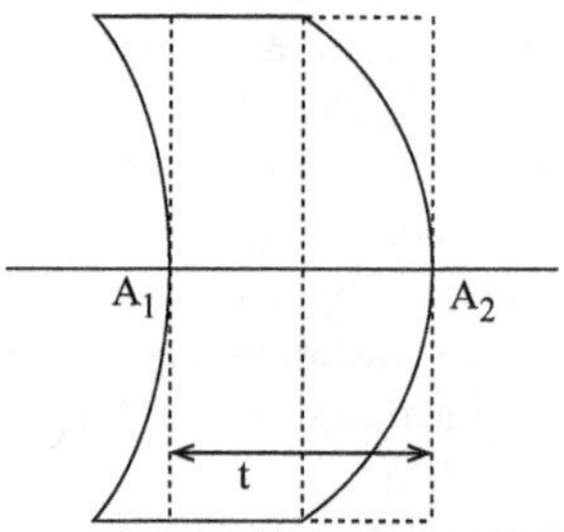

FIGURE 14.1(d) Taylor thin lens model of a single concavo-concave meniscus lens ≡ [a concavo-plane thin lens at A_1 + a plane parallel plate of thickness $t\left(= A_1A_2\right)$ + a plano-convex thin lens at A_2].

(b) A biconcave thick lens ≈ a concavo-plane thin lens at vertex A_1, in contact with a plane parallel plate of the same material and thickness $t(=A_1A_2)$, in contact with a plano-concave thin lens at vertex A_2.
(c) A thick convexo-convex meniscus lens ≈ a convexo-plane thin lens at vertex A_1, in contact with a plane parallel plate of the same material and thickness $t(=A_1A_2)$, in contact with a plano-concave thin lens at vertex A_2.
(d) A thick concavo-concave meniscus lens ≈ a concavo-plane thin lens at vertex A_1, in contact with a plane parallel plate of the same material and thickness $t(=A_1A_2)$, in contact with a plano-convex thin lens at vertex A_2.

Using these approximations, Taylor even obtained analytical forms for primary aberrations of multi-element optical systems. The values of primary aberrations of real systems obtained by his approach provide better approximations compared to conventional thin lens approximations.

14.2 Primary Aberrations of a Thin Lens (with Stop on It) in Object and Image Spaces of Unequal Refractive Index

Figure 14.2(a) shows a single lens of axial thickness d and refractive index μ. The curvatures of the first and the second refracting interfaces are c_1 and c_2, and the refractive indices of the object and image spaces are n_1 and n_2', respectively. The convergence angles of the paraxial marginal ray (PMR) and the paraxial pupil ray (PPR) in the object space and in the image space are u_1, u_2', and $\bar{u}_1, \bar{u}_2'$, respectively. The heights of the PMR on the consecutive refracting surfaces are h_1 and h_2. Using paraxial refraction and transfer equations consecutively for the PMR on the two consecutive surfaces of the lens, we get

$$\mu u_1' - n_1 u_1 = -h_1 (\mu - n_1) c_1 \tag{14.1}$$

$$u_2 = u_1' \tag{14.2}$$

$$h_2 = h_1 + d u_2 \tag{14.3}$$

$$n_2' u_2' - \mu u_2 = -h_2 (n_2' - \mu) c_2 \tag{14.4}$$

In thin lens approximation, $d \rightarrow 0$, so that $h_1 = h_2 = h$. The equivalent thin lens corresponding to the thick lens of Figure 14.2(a) is shown in Figure 14.2(b). For the thin lens, it follows

$$n_2' u_2' - n_1 u_1 = -hK \tag{14.5}$$

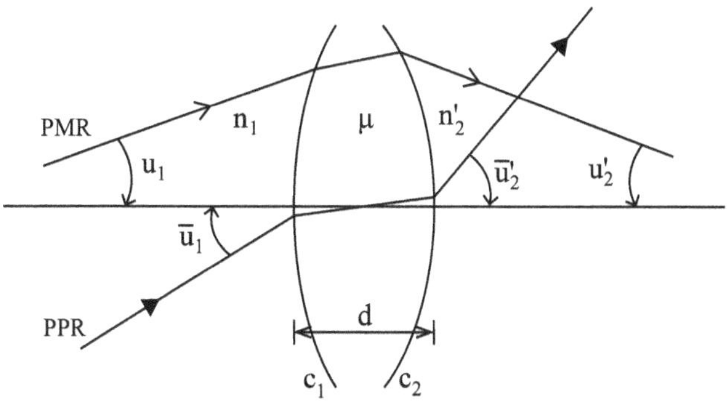

FIGURE 14.2(a) Paths of the paraxial marginal ray (PMR) and the paraxial pupil ray (PPR) through a single thick lens of axial thickness d and refractive index μ. The refractive indices of the object and image spaces are n_1 and n_2', and curvatures of the refracting interfaces are c_1 and c_2, respectively.

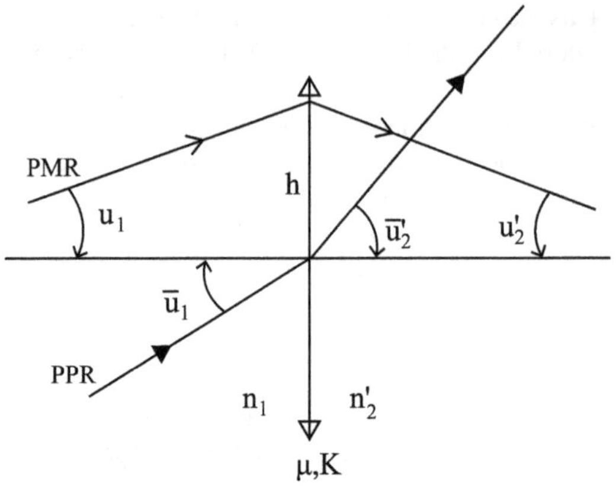

FIGURE 14.2(b) Equivalent thin lens for the singlet of Figure 14.2(a) assuming $d \rightarrow 0$. Stop is on the lens.

where K, the power of the thin lens, is given by

$$K = (\mu - n_1)c_1 - (\mu - n_2')c_2 \tag{14.6}$$

As mentioned before, the use of two composite parameters involving the variables c_1, c_2, u_1, and u_2' for the structure of the lens and its imaging configuration leads to simplified expressions for primary aberrations of a thin lens. The two parameters are called the 'Shape factor', and the 'Conjugate variable', respectively. The former is also known by other names, e.g. 'Bending' [6], 'Durchbeigung' [2], 'Coflexure' [1], and 'Cambrure' [7]. In our treatment, we follow the dimensionless, symmetrical variables suggested by Coddington and Hopkins for both the shape of the lens and the conjugate position.

The normalized shape variable X is defined by

$$X = \frac{(\mu - n_1)c_1 + (\mu - n_2')c_2}{K} \tag{14.7}$$

where power K is defined by (14.6).

From (14.6) and (14.7), we get

$$\left.\begin{aligned} c_1 &= \frac{K(X+1)}{2(\mu - n_1)} \\ c_2 &= \frac{K(X-1)}{2(\mu - n_2')} \end{aligned}\right\} \tag{14.8}$$

In the general case when $n_1 \neq n_2'$, a change ΔX in shape factor leads to different changes in the curvatures

$$\left.\begin{aligned} \Delta c_1 &= \frac{K}{2(\mu - n_1)}\Delta X \\ \Delta c_2 &= \frac{K}{2(\mu - n_2')}\Delta X \end{aligned}\right\} \tag{14.9}$$

but the power K remains unchanged. A normalized conjugate variable Y is defined as

$$Y = \frac{n_2' u_2' + n_1 u_1}{n_2' u_2' - n_1 u_1} = \frac{n_2' u_2' + n_1 u_1}{-hK} \tag{14.10}$$

The convergence angles in the object and image spaces are given by

$$\left.\begin{aligned} u_2' &= -\frac{hK}{2n_2'}(Y+1) \\ u_1 &= -\frac{hK}{2n_1}(Y-1) \end{aligned}\right\} \tag{14.11}$$

In terms of paraxial magnification $M = \left(n_1 u_1 / n_2' u_2' \right)$, Y can be expressed as

$$Y = \frac{1+M}{1-M} \tag{14.12}$$

The paraxial refraction invariant for the PMR at the first surface, A_1, is given by

$$A_1 = n_1 \left(hc_1 + u_1 \right) = \frac{1}{2} n_1 hK \left[\frac{X}{(\mu - n_1)} - \frac{Y}{n_1} + \frac{\mu}{n_1 (\mu - n_1)} \right] \tag{14.13}$$

For the PMR, the paraxial refraction invariant at the second surface is

$$A_2 = n_2' \left(hc_2 + u_2' \right) = \frac{1}{2} n_2' hK \left[\frac{X}{(\mu - n_2')} - \frac{Y}{n_2'} - \frac{\mu}{n_2' (\mu - n_2')} \right] \tag{14.14}$$

Since the stop is on the lens, $\bar{h} = 0$. The paraxial refraction invariants for the PPR, $\bar{A}_1$ and $\bar{A}_2$, at the first and the second surfaces are therefore given by

$$\bar{A}_1 = n_1 \left(\bar{h} c_1 + \bar{u}_1 \right) = n_1 \bar{u}_1 = -\frac{H}{h} \tag{14.15}$$

$$\bar{A}_2 = n_2' \left(\bar{h} c_2 + \bar{u}_2' \right) = n_2' \bar{u}_2' = -\frac{H}{h} \tag{14.16}$$

where H, the Smith-Helmholtz-Lagrange invariant is

$$H = n_1 \left(u_1 \bar{h} - \bar{u}_1 h \right) = -n_1 \bar{u}_1 h = n_2' \left(u_2' \bar{h} - \bar{u}_2' h \right) = -n_2' \bar{u}_2' h \tag{14.17}$$

The changes in ratio of the convergence angle of the PPR, with respect to the refractive index of the medium when the PPR undergoes refraction at the first and the second surfaces, are given by

$$\Delta_1 = \left[\Delta \left(\frac{u}{n} \right) \right]_1 = \frac{u_1'}{\mu} - \frac{u_1}{n_1} \tag{14.18}$$

$$\Delta_2 = \left[\Delta\left(\frac{u}{n}\right)\right]_2 = \frac{u_2'}{n_2'} - \frac{u_2}{\mu} \qquad (14.19)$$

Using (14.1) – (14.4), (14.8), and (14.11) for algebraic manipulation, (14.18) and (14.19) reduce to

$$\Delta_1 = -\frac{1}{2}hK\left[\frac{1}{n_1^2} + \frac{X}{\mu^2} - \frac{\mu^2 - n_1^2}{\mu^2 n_1^2}Y\right] \qquad (14.20)$$

$$\Delta_2 = -\frac{1}{2}hK\left[\frac{1}{(n_2')^2} - \frac{X}{\mu^2} + \frac{\mu^2 - (n_2')^2}{\mu^2 (n_2')^2}Y\right] \qquad (14.21)$$

Using the above relations, primary aberrations of the thin lens can be obtained. We quote the final results below. In order to simplify the notations, n_1 and n_2' are represented by n and n' in what follows.

A. Primary Spherical Aberration

The Seidel spherical aberration coefficient S_I of the thin lens is given by

$$S_I = -A_1^2 h\Delta_1 - A_2^2 h\Delta_2 \qquad (14.22)$$

Substituting for A_1 and A_2 from (14.13) and (14.14), and for Δ_1 and Δ_2 from (14.20) and (14.21), we get the following expression for S_I.

$$S_I = \frac{1}{8}h^4K^3\left(a_1X^3 + a_2Y^3 + a_3X^2Y + a_4XY^2 + a_5X^2 + a_6Y^2 + a_7XY + a_8X + a_9Y + a_{10}\right) \qquad (14.23)$$

where

$$a_1 = \frac{1}{\mu^2}\left[\frac{n^2}{(\mu - n)^2} - \frac{(n')^2}{(\mu - n')^2}\right] \qquad (14.24)$$

$$a_2 = -\frac{1}{\mu^2}\left[\frac{\mu^2 - n^2}{n^2} - \frac{\mu^2 - (n')^2}{(n')^2}\right] \qquad (14.25)$$

$$a_3 = -\frac{1}{\mu^2}\left[\frac{\mu + 3n}{\mu - n} - \frac{\mu + 3n'}{\mu - n'}\right] \qquad (14.26)$$

$$a_4 = \frac{1}{\mu^2}\left[\frac{2\mu + 3n}{n} - \frac{2\mu + 3n'}{n'}\right] \qquad (14.27)$$

$$a_5 = \frac{1}{\mu}\left[\frac{\mu + 2n}{(\mu - n)^2} + \frac{\mu + 2n'}{(\mu - n')^2}\right] \qquad (14.28)$$

$$a_6 = \frac{1}{\mu}\left[\frac{3\mu + 2n}{n^2} + \frac{3\mu + 2n'}{(n')^2}\right] \qquad (14.29)$$

$$a_7 = -\frac{4}{\mu}\left[\frac{\mu+n}{n(\mu-n)} + \frac{\mu+n'}{n'(\mu-n')}\right] \tag{14.30}$$

$$a_8 = \left[\frac{2\mu+n}{n(\mu-n)^2} - \frac{2\mu+n'}{n'(\mu-n')^2}\right] \tag{14.31}$$

$$a_9 = -\left[\frac{3\mu+n}{n^2(\mu-n)} - \frac{3\mu+n'}{(n')^2(\mu-n')}\right] \tag{14.32}$$

$$a_{10} = \mu^2\left[\frac{1}{n^2(\mu-n)^2} + \frac{1}{(n')^2(\mu-n')^2}\right] \tag{14.33}$$

A. Primary Coma

The Seidel coma coefficient S_{II} of the thin lens is given by

$$S_{II} = -A_1\bar{A}_1 h\Delta_1 - A_2\bar{A}_2 h\Delta_2 \tag{14.34}$$

Using (14.15) and (14.26), (14.34) reduces to

$$S_{II} = H\left(A_1\Delta_1 + A_2\Delta_2\right) \tag{14.35}$$

By substituting from (14.13), (14.14) for A_1, A_2, and from (14.20), (14.21) for Δ_1, Δ_2 in (14.35), we get, after algebraic manipulation, the following expression for S_{II}.

$$S_{II} = -\frac{1}{4}h^2K^2H\left(p_1X^2 + p_2Y^2 + p_3XY + p_4X + p_5Y + p_6\right) \tag{14.36}$$

where

$$p_1 = \frac{1}{\mu^2}\left[\frac{n}{\mu-n} - \frac{n'}{\mu-n'}\right] \tag{14.37}$$

$$p_2 = \frac{1}{\mu^2}\left[\frac{\mu^2-n^2}{n^2} - \frac{\mu^2-(n')^2}{(n')^2}\right] \tag{14.38}$$

$$p_3 = -\frac{1}{\mu^2}\left[\frac{\mu+2n}{n} - \frac{\mu+2n'}{n'}\right] \tag{14.39}$$

$$p_4 = \frac{1}{\mu}\left[\frac{\mu+n}{n(\mu-n)} + \frac{\mu+n'}{n'(\mu-n')}\right] \tag{14.40}$$

$$p_5 = -\frac{1}{\mu}\left[\frac{2\mu+n}{n^2} + \frac{2\mu+n'}{(n')^2}\right] \tag{14.41}$$

$$p_6 = \mu\left[\frac{1}{n^2(\mu - n)} - \frac{1}{(n')^2(\mu - n')}\right] \tag{14.42}$$

B. Primary Astigmatism

The Seidel astigmatism coefficient of the thin lens is given by

$$S_{III} = -\bar{A}_1^2 h\,\Delta_1 - \bar{A}_2^2 h\,\Delta_2 \tag{14.43}$$

Using equations (14.15), (14.16), (14.20), and (14.21) for $\bar{A}_1, \bar{A}_2, \Delta_1$, and Δ_2, respectively, we get from (14.43)

$$S_{III} = \frac{1}{2}H^2K\left[\left\{\frac{1}{n^2} + \frac{1}{(n')^2}\right\} - \left\{\frac{1}{n^2} - \frac{1}{(n')^2}\right\}Y\right] \tag{14.44}$$

C. Primary Curvature

The Seidel curvature coefficient S_{IV} of the thin lens is given by

$$S_{IV} = -H^2\left[c_1\left\{\Delta\left(\frac{1}{n}\right)\right\}_1 + c_2\left\{\Delta\left(\frac{1}{n}\right)\right\}_2\right] \tag{14.45}$$

Substituting from (14.8) for c_1 and c_2 in (14.48), we get

$$S_{IV} = \frac{H^2K}{2\mu}\left[\left(\frac{1}{n} + \frac{1}{n'}\right) + \left(\frac{1}{n} - \frac{1}{n'}\right)X\right] \tag{14.46}$$

D. Primary Distortion

From (12.35), we note that, with the usual notations, the Seidel aberration coefficient for primary distortion, S_V, arising at a refracting interface is given by

$$S_V = -\bar{A}^3 h\Delta\left(\frac{1}{n^2}\right) + \bar{A}\bar{h}c\left(2\bar{A}h - A\bar{h}\right)\Delta\left(\frac{1}{n}\right) \tag{14.47}$$

For a stop on the surface, $\bar{h} = 0$, and so (14.47) reduces to

$$S_V = -\bar{A}^3 h\Delta\left(\frac{1}{n^2}\right) \tag{14.48}$$

Therefore, for a thin lens with a stop on it, the Seidel coefficient for primary distortion is given by

$$S_V = -\bar{A}_1^3 h\left\{\Delta\left(\frac{1}{n^2}\right)\right\}_1 - \bar{A}_2^3 h\left\{\Delta\left(\frac{1}{n^2}\right)\right\}_2 \tag{14.49}$$

Since $\bar{A}_1 = \bar{A}_2 = -H/h$, the above equation reduces to

$$S_V = \frac{H^3}{h^2}\left[\frac{1}{n^2} - \frac{1}{(n')^2}\right] \tag{14.50}$$

E. Longitudinal Chromatic Aberration

The longitudinal chromatic aberration C_L of a thin lens with a stop on it is given by

$$C_L = A_1 h\left\{\Delta\left(\frac{\delta n}{n}\right)\right\}_1 + A_2 h\left\{\Delta\left(\frac{\delta n}{n}\right)\right\}_2 \tag{14.51}$$

where δn represents the absolute change in refractive index n over the working wavelength range. In our notation, δn, δμ, and *δn′* represent the corresponding changes in the object space, in the material of the thin lens, and in the image space, respectively.

Substituting the expressions for A_1 and A_2 from (14.13) and (14.14) in (14.51) we get

$$C_L = \frac{1}{2}h^2K\left[\left\{\frac{1}{\mu-n} + \frac{1}{\mu-n'}\right\}\delta\mu - \left\{\frac{\delta n}{n(\mu-n)} + \frac{\delta n'}{n'(\mu-n')}\right\}\mu \right.$$
$$\left. + \left\{\frac{\delta\mu}{\mu}\left(\frac{n}{\mu-n} - \frac{n'}{\mu-n'}\right) - \left(\frac{\delta n}{\mu-n} - \frac{\delta n'}{\mu-n'}\right)\right\}X + \left(\frac{\delta n}{n} - \frac{\delta n'}{n'}\right)Y\right] \tag{14.52}$$

F. Transverse Chromatic Aberration

The transverse chromatic aberration of a thin lens with a stop on it is given by

$$C_T = \bar{A}_1 h\left\{\Delta\left(\frac{\delta n}{n}\right)\right\}_1 + \bar{A}_2 h\left\{\Delta\left(\frac{\delta n}{n}\right)\right\}_2 \tag{14.53}$$

Since $\bar{A}_1 = \bar{A}_2 = -H/h$, the above equation reduces to

$$C_T = -H\left(\frac{\delta n'}{n'} - \frac{\delta n}{n}\right) \tag{14.54}$$

It is important to note that, in the general case of unequal object and image space media, primary aberrations of a thin lens with a stop on it have interesting characteristics. For example, the distortion coefficient S_V and the transverse chromatic aberration C_T are not zero. The Seidel coefficient for primary curvature varies linearly with the shape variable X, and the Seidel coefficient for primary astigmatism varies linearly with the conjugate variable Y. The Seidel coefficient C_L for longitudinal chromatic aberration varies linearly with both the shape variable X and the conjugate variable Y. S_{II}, the Seidel coefficient for primary coma, is a quadratic function of both the shape variable X and the conjugate variable Y. The Seidel coefficient for primary spherical aberration is a cubic function of both the shape variable X and the conjugate variable Y. We refer to publication [20] for details of physical interpretation of these variations.

14.3 Primary Aberrations of a Thin Lens (with Stop on It) With Equal Media in Object and Image Spaces

In this case, we have $n_2' = n' = n = n_1$. By substitution in equations (14.6) – (14.12), it follows

$$K = (\mu - n)(c_1 - c_2) \tag{14.55}$$

$$X = \frac{(c_1 + c_2)}{(c_1 - c_2)} \tag{14.56}$$

$$\left.\begin{aligned} c_1 &= \frac{K(X+1)}{2(\mu - n)} \\ c_2 &= \frac{K(X-1)}{2(\mu - n)} \end{aligned}\right\} \tag{14.57}$$

$$\Delta c_1 = \Delta c_2 = \frac{K}{2(\mu - n)}\Delta X \tag{14.58}$$

$$Y = \frac{u' + u}{u' - u} = \frac{u' + u}{-hK} \tag{14.59}$$

$$\left.\begin{aligned} u' &= -\frac{hK}{2n}(Y+1) \\ u &= -\frac{hK}{2n}(Y-1) \end{aligned}\right\} \tag{14.60}$$

Similarly, the expressions for A_1, A_2, Δ_1, and Δ_2 are obtained from the corresponding expressions for the general case by substituting $n_2' = n' = n = n_1$.

A. Primary Spherical Aberration

When $n = n'$, from (14.24) – (14.27) and (14.31) – (14.32), we note

$$a_1 = a_2 = a_3 = a_4 = a_8 = a_9 = 0 \tag{14.61}$$

The Seidel coefficient for primary spherical aberration $(S_I)_{n=n'}$ is given by

$$(S_I)_{n=n'} = \frac{1}{4}h^4K^3\left[\frac{\mu + 2n}{\mu(\mu - n)^2}X^2 + \frac{3\mu + 2n}{\mu n^2}Y^2 - \frac{4(\mu + n)}{\mu n(\mu - n)}XY + \frac{\mu^2}{n^2(\mu - n)^2}\right] \tag{14.62}$$

B. Primary Coma

When $n = n'$, from (14.37) – (14.39) and (14.42), we note

$$p_1 = p_2 = p_3 = p_6 = 0 \tag{14.63}$$

Therefore, the Seidel coefficient for primary coma $(S_{II})_{n=n'}$ of a thin lens when a stop is on the lens is given by

$$(S_{II})_{n=n'} = -\frac{1}{2}h^2K^2H\left[\frac{\mu+n}{\mu n(\mu-n)}X - \frac{2\mu+n}{\mu n^2}Y\right] \quad (14.64)$$

C. Primary Astigmatism

Substituting $n' = n$ in (14.44) we get the expression for the Seidel coefficient for primary astigmatism, $(S_{III})_{n=n'}$, of a thin lens when the stop is on the lens as

$$(S_{III})_{n=n'} = \frac{H^2K}{n^2} \quad (14.65)$$

D. Primary Curvature

Substituting $n' = n$ in (14.46), we get the expression for the Seidel coefficient for primary curvature, $(S_{IV})_{n=n'}$, of a thin lens as

$$(S_{IV})_{n=n'} = \frac{H^2K}{\mu n} \quad (14.66)$$

E. Primary Distortion

Substituting $n' = n$ in (14.50), we note that the Seidel coefficient for primary distortion, $(S_V)_{n=n'}$, of a thin lens is

$$(S_V)_{n=n'} = 0 \quad (14.67)$$

F. Longitudinal Chromatic Aberration

For the same medium in the object space and the image space, $n' = n$, and $\delta n' = \delta n$. Substituting these values in (14.52), we get the following expression for longitudinal chromatic aberration, $(C_L)_{n=n',\delta n'=\delta n}$ as

$$(C_L)_{n=n',\delta n=\delta n'} = h^2K\left[\frac{n\delta\mu - \delta n\mu}{n(\mu-n)}\right] \quad (14.68)$$

G. Transverse Chromatic Aberration

Substituting $n' = n$, and $\delta n' = \delta n$ in (14.54), we get the following expression for longitudinal chromatic aberration, $(C_T)_{n=n',\delta n=\delta n'}$ as

$$(C_T)_{n=n',\delta n=\delta n'} = 0 \quad (14.69)$$

14.4 Primary Aberrations of a Thin Lens (with Stop on It) in Air

For a thin lens in air, $n = n' = 1$, and $\delta n = \delta n' = 0$. The equations (14.55) – (14.60) reduce to

$$K = (\mu-1)(c_1 - c_2) \quad (14.70)$$

$$X = \frac{(c_1 + c_2)}{(c_1 - c_2)} \tag{14.71}$$

$$\left.\begin{aligned} c_1 &= \frac{K(X+1)}{2(\mu-1)} \\ c_2 &= \frac{K(X-1)}{2(\mu-1)} \end{aligned}\right\} \tag{14.72}$$

$$\Delta c_1 = \Delta c_2 = \frac{K}{2(\mu-1)}\Delta X \tag{14.73}$$

$$Y = \frac{u' + u}{u' - u} = \frac{u' + u}{-hK} \tag{14.74}$$

$$\left.\begin{aligned} u' &= -\frac{hK}{2}(Y+1) \\ u &= -\frac{hK}{2}(Y-1) \end{aligned}\right\} \tag{14.75}$$

Similarly, the values of A_1, A_2, Δ_1, and Δ_2 are modified with similar substitutions.

Finally, we get the following expressions for the seven Seidel aberration coefficients

$$(S_I)_{air} = \frac{1}{4}h^4K^3\left[\frac{\mu+2}{\mu(\mu-1)^2}\left\{X - \frac{2(\mu^2-1)}{(\mu+2)}Y\right\}^2 + \left\{\frac{\mu^2}{(\mu-1)^2} - \frac{\mu}{(\mu+2)}Y^2\right\}\right] \tag{14.76}$$

$$(S_{II})_{air} = -\frac{1}{2}h^2K^2H\left[\frac{\mu+1}{\mu(\mu-1)}X - \frac{2\mu+1}{\mu}Y\right] \tag{14.77}$$

$$(S_{III})_{air} = H^2K \tag{14.78}$$

$$(S_{IV})_{air} = \frac{H^2K}{\mu} \tag{14.79}$$

$$(S_V)_{air} = 0 \tag{14.80}$$

$$(C_L)_{air} = h^2K\left(\frac{\delta\mu}{\mu-1}\right) = h^2\frac{K}{V} \tag{14.81}$$

$$(C_T)_{air} = 0 \tag{14.82}$$

where $V = (\mu - 1)/\delta\mu$ is the Abbe number for the optical material of the lens.

Note that when the object space and the image space media are same, the dependence of the Seidel coefficients on the shape variable and the conjugate variable change significantly. The Seidel coefficients for both primary distortion and transverse colour become zero. The coefficients for Seidel astigmatism and Seidel curvature become independent of Y and X, respectively. The Seidel coefficient for axial colour also becomes independent of both X and Y. The dependence of the Seidel coefficient for primary spherical aberration on shape variable X and on conjugate variable Y is quadratic in each case. Similarly, the dependence of the Seidel coefficient for primary coma on X and on Y is linear in each case. Obviously, the nature of these dependences remains unaltered when the lens is in air, but the expressions are further simplified.

Figure 14.3 presents schematic diagrams illustrating the conjugate locations for various values of conjugate variable Y for a single thin lens in air, with a stop on the lens. Figure 14.3(a) corresponds to lenses of positive focal length, and Figure 14.3(b) corresponds to lenses of negative focal length.

Figure 14.4 shows the shapes of a single thin lens for different values of X for lenses of positive foal length and for lenses of negative focal length, respectively.

From (14.76) it is seen that $(S_I)_{air}$ is minimum when

$$X = \frac{2(\mu^2 - 1)}{\mu + 2} Y \tag{14.83}$$

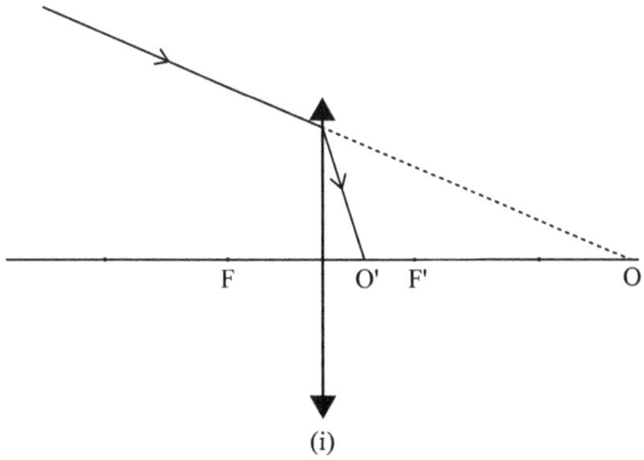

FIGURE 14.3(a) Axial locations of the conjugates O, O′ for a single thin lens of positive focal length for various values of conjugate variable Y; (i) Y > +1 (ii) Y = +1, (iii) Y = 0, (iv) Y = −1, and (v) Y < −1.

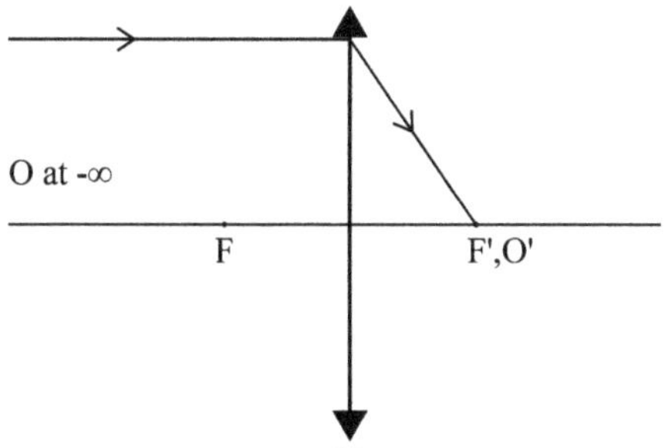

FIGURE 14.3(a) continued

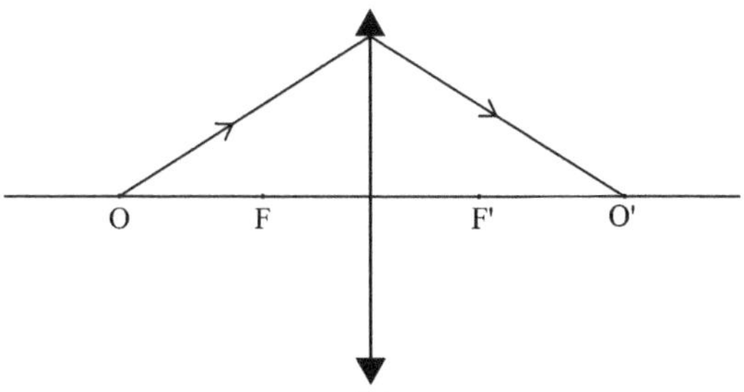

FIGURE 14.3(a) continued

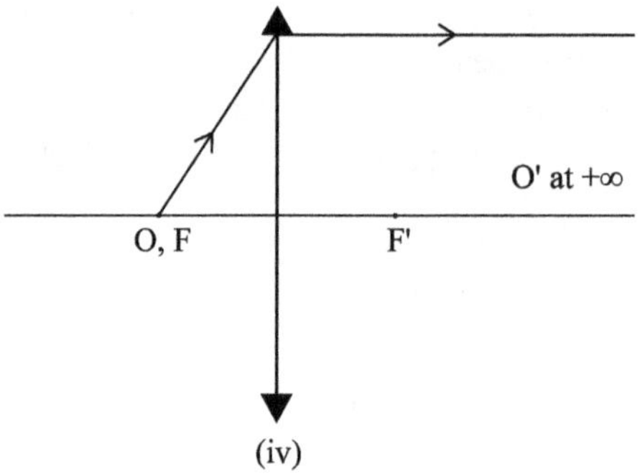

FIGURE 14.3(a) continued

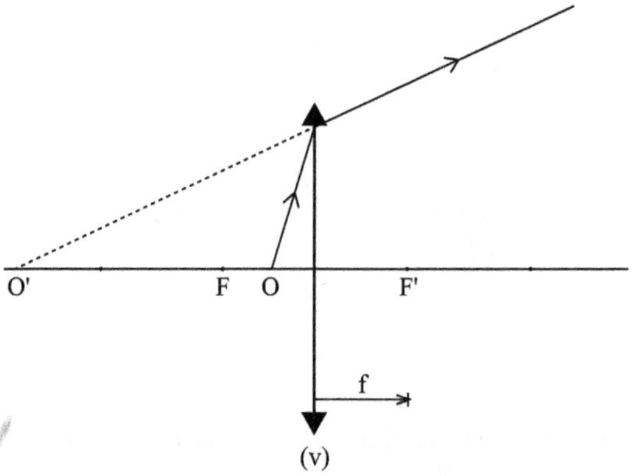

FIGURE 14.3(a) continued

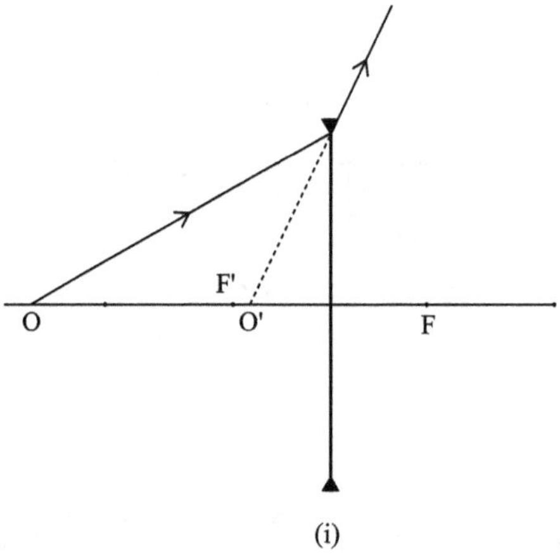

FIGURE 14.3(b) Axial locations of the conjugates O, O′ for a single thin lens of negative focal length for various values of conjugate variable Y; (i) Y > +1 (ii) Y = +1, (iii) Y = 0, (iv) Y = −1, and (v) Y < −1.

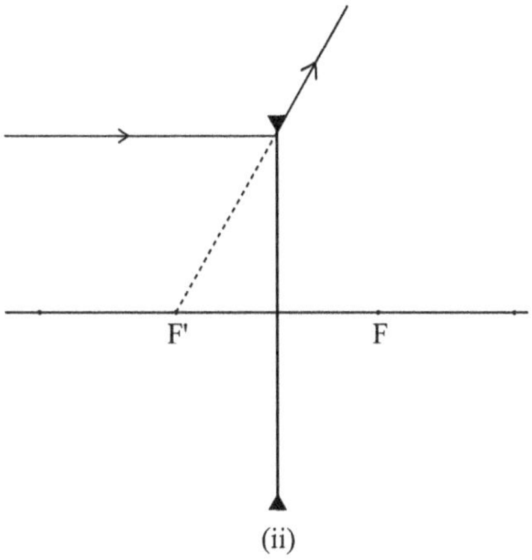

FIGURE 14.3(b) continued

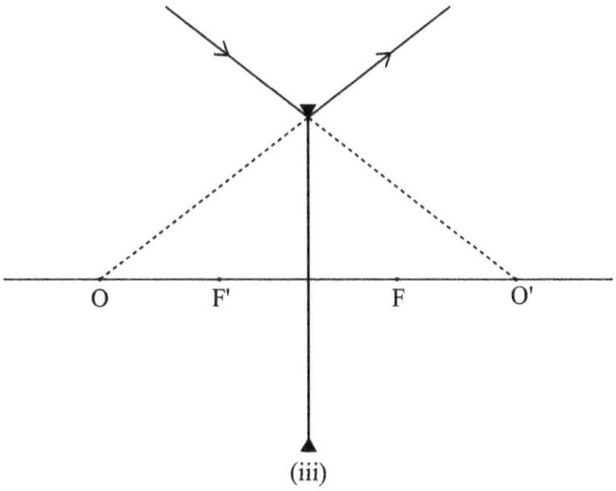

FIGURE 14.3(b) continued

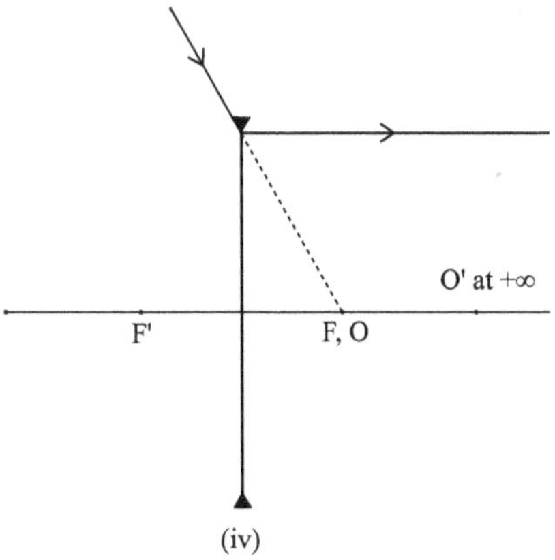

FIGURE 14.3(b) continued

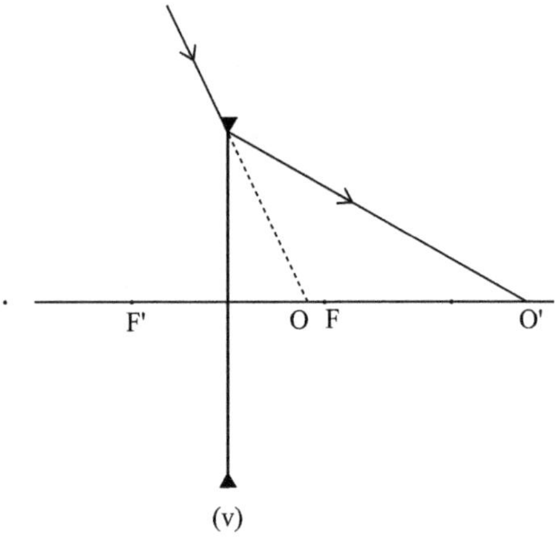

FIGURE 14.3(b) continued

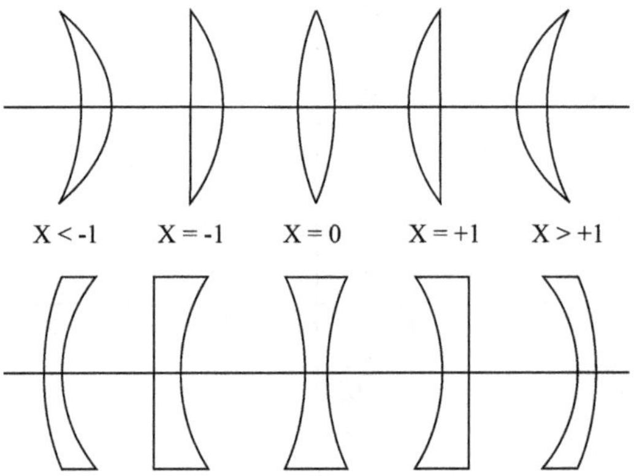

FIGURE 14.4 Shapes of a single thin lens for different values of the shape variable X. Lenses with (a) positive focal length; (b) negative focal length

The value of minimum Seidel spherical aberration for a thin lens of refractive index μ and with a stop on it is

$$\left[(S_I)_{air}\right]_{min} = \frac{1}{4}h^4K^3\left[\frac{\mu^2}{(\mu-1)^2} - \frac{\mu}{(\mu+2)}Y^2\right] \qquad (14.84)$$

Therefore, for $Y = 0$, i.e. when the object and the image are equidistant from the lens, the minimum $(S_I)_{air}$ is obtained when $X = 0$, and the minimum values of normalized $(S_I)_{air}$ for a lens of refractive index $\mu = 1.5, 1.7, 1.9, 2.4$, and 4.0 are 9, 5.9, 4.5, 2.9, and 1.8, respectively. Figure 14.5 shows the variation in values of normalized primary spherical aberration with shape factor X for different values of refractive index μ of a single thin lens for $Y = 0$. The curves in the figure show that, in general, for a thin lens with any given shape and conjugate variable, the Seidel spherical aberration reduces significantly with an increase in refractive index of the optical material of the lens. It is also apparent that for other values of Y, the shape variable X providing the minimum spherical aberration is given by (14.83). Since, for a given μ, $\left[(S_I)_{air}\right]_{min}$ is proportional to Y^2, and the shape variable X providing the minimum S_I is directly proportional to Y, this variation of $\left[(S_I)_{air}\right]_{min}$ with X can also be demonstrated in Figure 14.5, which shows an inverted parabola with its axis lying along the ordinate. The dashed line corresponds to $\mu = 1.5$, and the chain line corresponds to $\mu = 1.7$.

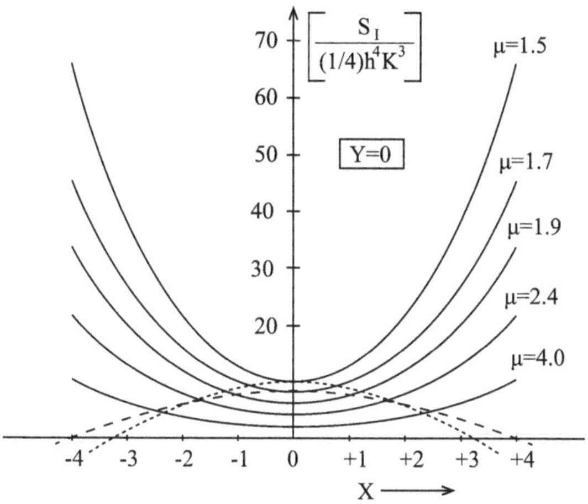

FIGURE 14.5 Normalized primary spherical aberration versus shape variable X for different values of refractive index μ of a single thin lens.

For a set of values of μ, figure 14.5 shows the variation of normalized Seidel spherical aberration with X for the case $Y = 0$. The variation of normalized Seidel spherical aberration with X for other values of Y can be obtained graphically by using the particular curve for $Y = 0$ for a specific μ in combination with the corresponding inverted parabola representing variation of $\left[(S_I)_{air}\right]_{min}$ with X. The graphical procedure is described next.

First, determine the value of X that provides the minimum Seidel spherical aberration for the new value of Y from equation (14.83). Then, determine the point of intersection of the ordinate at that value of X with the particular locus of minimum Seidel spherical aberration for the same value of μ. Then, the available parabola for $Y = 0$ corresponding to the same value of μ is translated without rotation until its vertex lies on the point of intersection as determined above. The resulting parabolic curve gives the variation of $(S_I)_{air}$ with X for the thin lens of refractive index μ at the required value of Y.

Note that for $Y = +1$, i.e. when the object is at infinity, the minimum Seidel spherical aberration is obtained with shape factor $X = 2(\mu^2 - 1)/(\mu + 2)$. It follows that for $\mu = 1.5$ and 1.6, $(c_1/c_2) = -6$ and -14, respectively, so that for infinitely distant objects the front surface should be much deeper than the rear surface. As a rule of thumb, an approximately plano convex lens with its convex side towards the infinity conjugate provides an image with a small residual Seidel spherical aberration.

From (14.77), it is seen that the Seidel coma is zero when

$$X = \frac{(2\mu + 1)(\mu - 1)}{(\mu + 1)} Y \tag{14.85}$$

For $Y = 0$, $X = 0$. It is significant to note that for the case of equal conjugates, an equi-convex/equi-concave lens gives zero Seidel coma, and minimum Seidel spherical aberration. For non-zero values of Y, no single value of X provides zero Seidel coma, and minimum Seidel spherical aberration. However, the values of X prescribed by equations (14.83) and (14.85) are numerically close, particularly for higher values of μ.

14.5 Structural Aberration Coefficients

The use of dimensionless shape and conjugate variables has led to expressions for Seidel coefficients of primary aberrations in a degenerate form enabling separation of the part arising out of intrinsic structure of the system, and the part occurring for the particular imaging configuration. Shack proposed normalized dimensionless aberration coefficients $\sigma_I, \sigma_{II}, \sigma_{III}, \sigma_{IV}, \sigma_V, \sigma_L$, and σ_T for a single thin lens with a stop on it, and named them structural aberration coefficients [23–24].

$$\left.\begin{aligned}
\sigma_I &= \frac{S_I}{\left(\frac{1}{4}h^4K^3\right)} = \left[\frac{\mu+2}{\mu(\mu-1)^2}\left\{X - \frac{2(\mu^2-1)}{(\mu+2)}Y\right\}^2 + \left\{\frac{\mu^2}{(\mu-1)^2} - \frac{\mu}{(\mu+2)}Y^2\right\}\right] \\
\sigma_{II} &= \frac{S_{II}}{\left(\frac{1}{2}h^2K^2H\right)} = \left[\frac{\mu+1}{\mu(\mu-1)}X - \frac{2\mu+1}{\mu}Y\right] \\
\sigma_{III} &= \frac{S_{III}}{H^2K} = 1 \\
\sigma_{IV} &= \frac{S_{IV}}{H^2K} = \frac{1}{\mu} \\
\sigma_V &= 0 \\
\sigma_L &= \frac{C_L}{h^2K} = \frac{1}{V} \\
\sigma_T &= 0
\end{aligned}\right\} \quad (14.86)$$

Geary [25] showed the use of these coefficients for depicting the relative contribution of third order coma and third order astigmatism as a simultaneous function of both field angle and f-number. Sasián [26] has described structural aberration coefficients arising in different cases of surfaces and multiple lenses, etc., and pointed out their practical usefulness.

14.6 Use of Thin Lens Aberration Theory in Structural Design of Lens Systems

The earliest use of thin lens aberration theory was in the design of achromatic, aplanatic doublet lenses undertaken in the latter half of eighteenth century by Clairaut [27] and D'Alembert [28]. About ninety years later, Mossotti [29] extended these studies, and gave a quintic equation for the ratio of powers of the two thin elements of the doublet. This quintic polynomial has three real roots; usually one of them is practically useful. Subsequently, Turrière [30] and Chrétien [31] called the quintic equation Clairaut – Mossotti's equation, and the aplanatic cemented doublets designed using this equation as Clairaut – Mossotti objectives. Smith [32] presented a general aberration theory of thin objectives constructed with three or more kinds of glasses. Smith and Milne [33] recalculated the telescope objectives of Steinheil and Voit. Boegehold [34] highlighted the contribution of Clairaut and D'Alembert in the design of telescope objectives. Conrady [35] published a systematic analytical procedure for the design of aplanatic doublets. Later on, comprehensive tables of power distributions for cemented doublets with a wide range of glass pairings were published by Brown and Smith [36], Slyusarev [37], and van Heel [38]. Subsequently, attempts were made to circumvent the problem of solving the quintic Clairaut – Mossotti equation by Steel [39–40]. For specific catalogue glass pairings, he reformulated the problem to reduce the difficult problem of tackling a quantic equation to solving a quadratic equation. In his formulation, the conjugate distances were determined for chosen power distribution. Sussman [41] proposed an approach that involved solving a quartic equation instead of the quintic equation. Hopkins and Rao [42] described a method based on normalized aberrations in which, for chosen values of coma and chromatic aberration, values of spherical aberration were determined for chosen glasses. Blandford [43] solved a generalized version of Mossotti's equation, and his algorithm produces, for precise values of spherical aberration and coma, available solutions for a set of preferred glasses where each glass is tried with every other available glass. The solutions are given in order of their closeness with the chromatic aberration target. In many applications, e.g. in laser based systems or in thermal imaging, correction of chromatic aberration becomes less important. Jamieson [44] pointed out the possibility of choosing suitable aplanatic doublets with reduced higher order aberrations.

Strict adherence to aberration targets at the thin lens design stage is not very meaningful from the point of view of practical design, and Khan and Macdonald [45] emphasized the need for floating aberration targets, and presented several useful nomographs for practical use. Szulc [46] presented equations that could be used for obtaining achromatic aplanatic doublets with reduced spherochromatism, by choosing a suitable second glass for a prespecified first glass. Banerjee and Hazra [47] formulated the problem of the design of doublet lenses with prespecified values of primary spherical aberration, central coma, and axial colour as a constrained multivariate nonlinear optimization problem, and they obtained desired solutions for a set of preferred catalogue optical glasses using a local optimization algorithm. Subsequently, improved results were obtained by using global optimization algorithms for both cemented and broken contact doublet lenses [48–50]. An interesting type of doublet called 'one radius doublet' was worked out by Van Heel [51] for infinity conjugate applications. The refracting interfaces of these cemented doublets are either plane or of a radius of the same absolute value. With suitable glasses, they can be made achromatic, but not perfectly aplanatic.

Thin lens aberration theory is also applied in the design of cemented triplets with prespecified targets for spherical aberration, central coma, and chromatic aberration. Various analytical and semi-numerical approaches are made to determine the required structural parameters for achieving the desired goals for aberrations [52–56]. Similar to the one radius doublets of Van Heel, Mikš [57] presented data for one radius triplets. Triplets in the form of a singlet in close contact with a cemented doublet can be made apochromatic. This structure was first suggested by Taylor as a photographic objective with reduced secondary spectrum over the visible range, and they 'are now widely known as 'Photovisual' lenses because sharp photographs can be obtained at the visual focus without colour screens' [58–59]. Chen et al. [60–61] presented algebraic methods for structural design of these apochromatic triplets in both cases of a singlet + a cemented doublet, and singlet + a broken contact doublet.

Thin lens aberration theory has also been extensively used in the structural design of optical systems consisting of two or more air-spaced thin components [62–84]. For two and three component lenses, practically useful lens structures are obtained analytically in a straightforward manner [62–67]. Banerjee and Hazra [68–69] used thin lens aberration theory in reformulating the problem of structural design of Cooke triplet lenses in a form that enabled the use of global optimization techniques. The complexity of the problem of the 'ab initio' structural design of multicomponent lens systems can be significantly reduced by suitable use of thin lens aberration theory, and a partial list of investigations is presented in references [70–84].

14.7 Transition from Thin Lens to Thick Lens and Vice Versa

For real applications, it is necessary to thicken the hypothetical thin lenses for which curvatures of the two refracting interfaces are known. It is apparent that the minimum allowable axial thickness is related to the diameter of the aperture of the lens. In general, any axial thickness larger than this minimum axial thickness is admissible, but, for obvious practical reasons, the upper limit of axial thickness should be as small as possible, unless there is any compelling reason, such as aberration reduction or special structural requirements.

Figure 14.6 shows the axial thickness d and edge thickness d_{edge} of a thick lens of positive focal length f and aperture diameter Ø. The radii of the first and the second refracting interfaces are r_1 and r_2. The sag of the first surface on the axis is s_1, and that of the second surface is s_2. The axial thickness of lens d is given by

$$d = d_{edge} + s_1 - s_2 \cong d_{edge} + \frac{Ø^2}{8}\left(\frac{1}{r_1} - \frac{1}{r_2}\right) \tag{14.87}$$

where $s_1 \cong \frac{Ø^2}{8r_1}$ and $s_2 \cong \frac{Ø^2}{8r_2}$. With the same degree of approximation, the expression for focal length of the lens of refractive index μ is

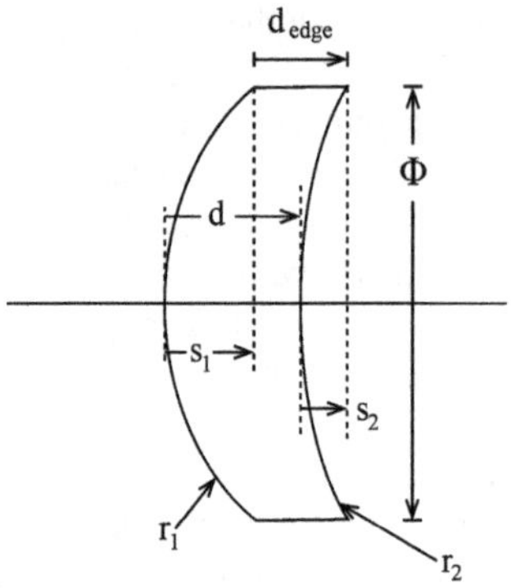

FIGURE 14.6 Axial thickness d and edge thickness d_{edge} of a thick lens of positive focal length f and aperture diameter Φ.

$$\frac{1}{f} = (\mu - 1)\left(\frac{1}{r_1} - \frac{1}{r_2}\right) \tag{14.88}$$

The expression for axial thickness d can be rewritten as

$$d \cong d_{edge} + \frac{\varnothing^2}{8(\mu - 1)f} \tag{14.89}$$

For lenses working in visible wavelengths $\mu \sim 1.5$, and (14.89) can be approximated as

$$d \cong d_{edge} + \frac{\varnothing^2}{4f} \tag{14.90}$$

The edge thickness, d_{edge}, has to be necessarily greater than the stipulated minimum values, $\left[\left(d_{edge}\right)_{min}\right]$.

For single thick lenses of negative focal lengths, the axial thickness d is determined from the condition

$$d \geq \left[\left(d_{axial}\right)_{min}\right] \tag{14.91}$$

Values of $\left[\left(d_{edge}\right)_{min}\right]$ and $\left[\left(d_{axial}\right)_{min}\right]$ are determined by stability and manufacturability of the lenses. Tables 14.1 and 14.2 give values of the same for lenses of different aperture diameter, as suggested by Slyusarev [83].

The power K of a single lens of thickness d is given by (3.158)

$$K = (\mu - 1)(c_1 - c_2) + \frac{(\mu - 1)^2}{\mu} dc_1c_2 \tag{14.92}$$

From the above relation, it is obvious that if μ, c_1, c_2 are kept unchanged, insertion of thickness in a thin lens leads to a change in power or focal length of the lens. Also, axial locations of the principal planes of the lens change. The required curvature and axial thickness for the lens to achieve the desired focal length may be obtained by scaling the system. In order to keep the equivalent focal length of the thick lens the same as that of the thin lens, it is necessary to use the scaled values of curvatures for the two surfaces, and the scaled value for axial thickness. It is also necessary to position the lens with respect to its preceding and succeeding components such that, if the axial distance of the preceding surface from the thin lens is

TABLE 14.1

Minimum edge thickness corresponding to different values of aperture diameter for a single thick lens element of positive focal length

Lens Aperture Diameter Φ (in mm)	Minimum edge thickness $\left[\left(d_{edge}\right)_{min}\right]$ (in mm)
$\Phi \leq 6$	0.6
$6 < \Phi \leq 10$	0.8
$10 < \Phi \leq 18$	1.0
$18 < \Phi \leq 30$	1.2
$30 < \Phi \leq 50$	1.6
$50 < \Phi \leq 80$	2.0
$80 < \Phi \leq 120$	2.5
$\Phi > 120$	3.0

TABLE 14.2

Minimum axial thickness for a single thick lens element of negative focal length

Lens Aperture Diameter Φ (in mm)	Minimum axial thickness $\left[\left(d_{axial}\right)_{min}\right]$ (in mm)
$\Phi \leq 50$	$0.08\ \Phi$ to $0.12\ \Phi$
$\Phi > 50$	$0.06\ \Phi$ to 0.1Φ

l, then the axial distance of the preceding surface from the first principal plane of the thick lens has to be l. Similarly, if l′ is the axial distance of the succeeding surface from the thin lens, then the axial distance of the succeeding surface from the second principal plane of the thick lens has to be l′. With a view to keep different optical characteristics of the thin lens and the corresponding thick lens the same, several approaches for the thickening of thin lenses have been proposed [83–94].

Note that although the equivalent focal length, and therefore some of the paraxial characteristics, can be kept the same, the thick system, as obtained above, does not automatically ensure that the path of paraxial rays through the thick lens remain unchanged from its thin counterpart. As a result, the primary aberrations of the thick lens often become significantly different from the thin lens primary aberrations. This is often cited as a major shortcoming of the practical use of the concept of thin lenses in practical use. However, it should be noted that use of generalized bending and suitable axial repositioning of the thick lens can circumvent this problem. Hopkins and Rao [83] suggested an iterative procedure for this purpose. In what follows, we present an analytical procedure that can significantly reduce the change in primary aberrations that occur after thickening. The method is adapted from a brief discussion of the topic by Blandford [89].

For the sake of generality, Figure 14.7(a) shows a part of an optical system where a thin lens, of refractive index μ, is located at an intermediate position. Its axial distance from the last vertex on the left is l, and its axial distance from the next vertex on the right is l′, so that $\hat{A}_j A_{j-1} = l$, and $\hat{A}_{j+1} A_{j+2} = l'$ The left and right vertices of the thin lens, $\hat{A}_j$ and $\hat{A}_{j+1}$, are superimposed on each other on the axis. Let the curvatures of the two refracting interfaces of the thin lens be c_1 and c_2, and let the convergence angle of the paraxial ray incident on the lens be u_1, and the convergence angle of the corresponding emergent paraxial ray be u_2'. Let the convergence angle of the paraxial ray inside the thin lens (not shown in figure) be $u_1' = u_2$, and the height of the paraxial ray on the thin lens be h.

Figure 14.7(b) shows the same part of the optical system after the thin lens is thickened. Power of the thin lens is

$$K = (\mu - 1)(c_1 - c_2) \tag{14.93}$$

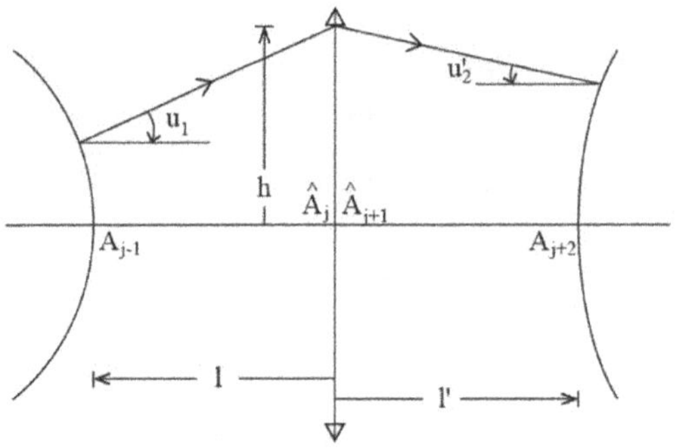

FIGURE 14.7(a) Path of a paraxial ray through a thin lens located between two refracting interfaces.

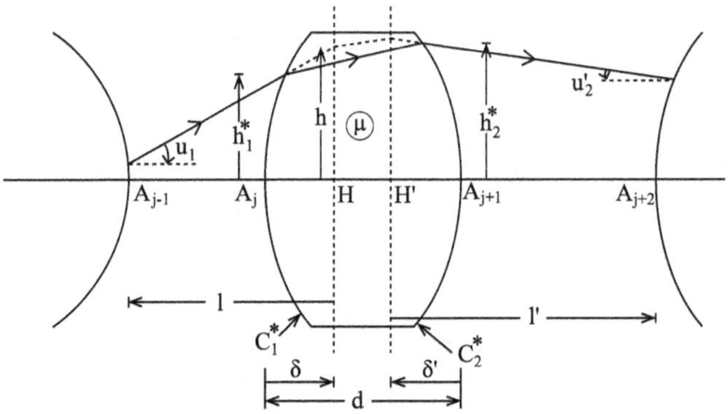

FIGURE 14.7(b) Changes in the path of the paraxial ray after the thin lens is thickened and shifted axially.

Let the thickness of the thick lens be d, and curvatures of the refracting interfaces of the thick lens be c_1^* and c_2^*. Power of the thick lens is

$$K = (\mu - 1)(c_1^* - c_2^*) + \frac{(\mu - 1)^2}{\mu} dc_1^* c_2^* \tag{14.94}$$

Let the height of the same paraxial ray be h_1^* on the first surface, and h_2^* on the second surface of the thick lens. For the thin lens, paraxial refraction and transfer equations yield

$$\left.\begin{array}{c} \mu u_2 - u_1 = \mu u_1' - u_1 = -h(\mu - 1)c_1 \\ h_2 = h_1 = h \\ u_2' - \mu u_2 = -h(1 - \mu)c_2 \end{array}\right\} \tag{14.95}$$

For the thick lens, the corresponding equations are

$$\left.\begin{array}{c} \mu u_2 - u_1 = \mu u_1' - u_1 = -h_1^*(\mu - 1)c_1^* \\ h_2^* = h_1^* + du_1' = h_1^* + du_2 \\ u_2' - \mu u_2 = -h_2^*(1 - \mu)c_2^* \end{array}\right\} \tag{14.96}$$

Comparing (14.95) and (14.96), it is noted that the convergence angles u_1, u_2, and u_2' remain unchanged for the transition from the thin lens to the thick lens if and only if the following holds.

$$\left.\begin{aligned} h_1^* c_1^* &= h c_1 \\ h_2^* c_2^* &= h c_2 \end{aligned}\right\} \quad (14.97)$$

For the powers of the two lenses to be equal, the right-hand sides of equations (14.93) and (14.94) need to be equal. This condition yields

$$(c_1 - c_2) = (c_1^* - c_2^*) + \frac{(\mu - 1)}{\mu} d c_1^* c_2^* \quad (14.98)$$

From (14.97) we get

$$\left.\begin{aligned} c_1^* &= \frac{h}{h_1^*} c_1 \\ c_2^* &= \frac{h}{h_2^*} c_2 \end{aligned}\right\} \quad (14.99)$$

From the third equation of (14.96) we get

$$(1 - \mu) c_2^* = -\frac{(u_2' - \mu u_2)}{h_2^*} \quad (14.100)$$

Substituting from (14.99) and (14.100) in (14.98), we get

$$h_1^* h_2^* (c_1 - c_2) = h c_1 h_2^* - h c_2 h_1^* + \frac{d}{\mu} h c_1 u_2' - d h c_1 u_2 \quad (14.101)$$

From the transfer equation in (14.96), we get $h_2^* = h_1^* + d u_2$. Substituting this value in the above equation, we obtain a quadratic equation in (h_1^*) as

$$(h_1^*)^2 + (d u_2 - h)(h_1^*) - \frac{d u_2' h c_1}{\mu (c_1 - c_2)} = 0 \quad (14.102)$$

For feasible cases, the equation has two real roots, out of which the one nearer to h is chosen.

With this solution for (h_1^*), the height of the paraxial ray is h at both the principal planes at H and H′. The locations of the principal planes are determined by

$$\left.\begin{aligned} A_j H &= \delta = \frac{(h - h_1^*)}{u_1} \\ A_{j+1} H' &= \delta' = \frac{(h - h_2^*)}{u_2'} \end{aligned}\right\} \quad (14.103)$$

The thick lens is to be positioned so that

$$\left.\begin{aligned} H A_{j-1} &= \hat{A}_j A_{j-1} = l \\ H' A_{j+2} &= \hat{A}_{j+1} A_{j+2} = l' \end{aligned}\right\} \quad (14.104)$$

Note that, if the paraxial marginal ray is chosen to remain unaltered after thickening, the primary spherical aberration will remain the same. But, since the path of the paraxial pupil ray is not constrained to remain the same, the other primary aberrations, depending on the parameters of the paraxial pupil ray, will change. A rule of thumb may help to keep all these changes small—choose either the paraxial marginal ray or the paraxial pupil ray as the paraxial ray, whose paths are to remain unaltered, by choosing the one that has a greater height on the thin lens. For keeping the paths of both the paraxial rays, one has to take recourse to effectively generalize bending suggested by Sutton [88], or the iterative procedure suggested by Hopkins and Rao [83].

14.8 Thin Lens Modelling of Diffractive Lenses

An important recent application of thin lens modelling is in the simulation of optical imaging characteristics of diffractive lenses. The 'lens' function of these optical elements is obtained primarily by the effect of diffraction in two-/three-dimensional structures instead of by refraction. These structures may be generated optically by interferometric or holographic techniques, or by computer controlled micro manipulating techniques, e.g. microlithography, laser plotting, diamond turning, etc. The structures generated by the latter techniques are also called 'kinoforms'.

The optical performance of a diffractive lens can be simulated by modelling it as a thin refractive lens of the same equivalent power as the diffractive lens. The refractive lens is assumed to be made of artificial material having an ultra-high refractive index [95–99]. The model is exact in the limit of the refractive index of the refractive lens tending to infinity. It has been observed by Hazra et al [100] that, in practice, any large value for the refractive index provides useful results [100]. This model enables the handling of diffractive lens elements in the same framework as refractive elements in the analysis and synthesis of optical systems.

REFERENCES

1 H.D. Taylor, *A System of Applied Optics*, Macmillan, London (1906).

2 M. von Rohr, *Geometrical Investigation of the formation of images in optical instruments*, [original German book title: *Die Theorie der optischen Instrumente*, Ed. M. von Rohr Springer, Berlin (1904)] Translator, R. Kanthack, His Majesty's Stationery Office, Dept. of Scientific and Industrial Research (1920), pp. 152–155, 230–234, 253–255, 270–271, 296–297.

3 H. Coddington, A Treatise on the Reflexion and Refraction of Light, Marshall, London (1829).

4 É. Turrière, Optique Industrielle, Delagrave, Paris (1920).

5 I.C. Gardner, 'Application of algebraic aberration equations to optical design', *Scientific Papers of the Bureau of Standards*, Vol. 22 (1926) pp. 73–203.

6 A.E. Conrady, *Applied Optics and Optical Design*, Part I, Dover, New York (1957) [originally by Oxford University Press (1929)] pp. 63–64, 320–328.

7 H. Chrétien, *Calcul des Combinaisons Optiques*, Masson, Paris (1980) [First published in 1938] pp. 176–178, 314, 400–401.

8 H.H. Hopkins, *Wave Theory of Aberrations*, Clarendon Press, Oxford (1950) pp. 119–140.

9 D. Argentieri, *Ottica Industrielle*, Ulrico Hoepli, Milan (1954) pp. 8–24.

10 B. Hăvelka, Geometrical Optics I, Czech Academy of Sciences, Prague (1955).

11 H.H. Emsley, *Aberrations of Thin Lenses*, Constable, London (1956).

12 M. Born and E. Wolf, *Principles of Optics*, Pergamon, Oxford (1980) pp. 220–225. [First published 1959] pp. 226–230.

13 C.G. Wynne, 'Thin-lens aberration theory', *Opt. Acta*, Vol. 8 (1961) pp. 255–265.

14 F.D. Cruickshank, *The thin lens and systems of thin lenses, Tracts in Geometrical Optics and Optical Design, No. 3*, The University of Tasmania, Hobart, Tasmania (1961).

15 W. Brouwer, Matrix methods in Optical Instruments Design, W. A. Benjamin, New York (1964) pp. 136–140.

16 J. Burcher, Les Combinaisons Optiques: Pratique des Calculs, Masson, Paris (1967) pp. 168–172, 199–200.

17 A.K. Ghosh and M. De, 'Elementary Studies on Aberrations of Lens Elements and Components. I. *Thin/Thick Lens – Axial Case, J. Opt. (India)*, Vol. 10 (1981) pp. 46–50.

18 G.G. Slyusarev, 'Third order aberrations in systems consisting of thin lenses', Chapter 3 in *Aberration and Optical Design Theory*, [original Russian book title: *Metodi Rascheta Opticheskikh Sistem, Mashinostroenie*, Leningrad (1969)] Translator: J. H. Dixon, Adam Hilger, Bristol (1984) pp. 245–276.

19 W.T. Welford, 'Thin Lens Aberrations', Chapter 12 in Ref. 9.26, pp. 226–239.

20 L.N. Hazra and C.A. Delisle, 'Primary aberrations of a thin lens with different object and image space media', *J. Opt. Soc. Am. A*, Vol. 15 (1998) pp. 945–953.

21 A. Mikš, 'Modification of the formulas for third-order aberration coefficients', *J. Opt. Soc. Am. A*, Vol. 19 (2002) pp. 1867–1871.

22 H.D. Taylor, 'Section II, The Theorem of Elements', in Ref. 9.146, pp. 20–24.

23 R.V. Shack, 'The use of normalization in the application of simple optical systems', *Proc. SPIE*, Vol. 54 (1974) pp. 155–162.

24 R.V. Shack, 'Structural aberration coefficients', in *OPTI 518, Introduction to aberrations*, Class Notes, College of Optical Sciences, Univ. Arizona (2010).

25 J.M. Geary, 'Coma and astigmatism versus field angle and f/number: a simple look', *Opt. Eng.*, Vol. 19 (1980) pp. 918–920.

26 J. Sasián, Chapter 11 'Structural aberration coefficients' in *Introduction to Aberrations in Optical Imaging Systems*, Cambridge University Press (2013) pp. 147–161.

27 A.C. Clairaut, 'Mémoires sur les moyens de perfectionner les lunette d'approche par l'usage d'objectifs composes de plusieurs matières différement réfrigentes', *Mém. Acad. R. Sci. Paris*, Vol. 1 (1763) pp. 380–437; Vol. 2 (1764) pp. 524–550; Vol. 3 (1769) pp. 578–631.

28 J. D'Alembert, 'Nouvelles recherches sur les verres optiques' *Mém. Acad. R. Sci. Paris*, Vol. 1 (1767) pp. 75–145; Vol. 2 (1768) pp. 63–105; Vol. 3 (1769) pp. 43–108.

29 O.F. Mossotti, Nuova Theoria Instrumenti Ottici, Casa Edifice Nistri, Pisa, Italy (1859) pp. 171–191.

30 E. Turrière, Optique Industrielle, Delagrave, Paris (1920) pp. 28–36.

31 H. Chrétien, 'Objectif de Clairaut- Mossotti', Section 42 in Ref. 7, pp. 419–422.

32 T. Smith, 'The aberration theory of thin objectives constructed with three or more kinds of glass', *Trans. Opt. Soc. London*, Vol. 22 (1921) pp. 111–121.

33 T. Smith and G. Milne, 'A recalculation of the telescope objectives of Steinheil and Voit', *Trans. Opt. Soc. London*, Vol. 22 (1921) pp. 122–126.

34 H. Boegehold, 'Die Leistung von Clairaut und d'Alembert für die Theorie des Fernrohrobjektiv', *Zeits. Instrum.*, Vol. 55 (1935) pp. 97–111.

35 A. E. Conrady, vide Chapters II, V, VI, VII in Ref. 6.

36 E. Brown and T. Smith, 'Systematic construction tables for thin cemented aplanatic lenses', *Philos. Trans. Roy. Soc. London. Ser. A*, Vol. 240 (1946) pp. 59–116.

37 G.G. Slyusarev, Tables for the Design of Two-lens Cemented Objectives, S I Vavilov State Optical Institute, Leningrad (1949).

38 A.C.S. van Heel, 'Paraxial and third order data of corrected doublets', *Opt. Acta*, Vol. 1 (1954) pp. 39–49.

39 W.H. Steel, 'On the choice of glasses for cemented achromatic aplanatic doublets', *Austral. J. Phys.*, Vol. 7 (1954) pp. 244–253.

40 W.H. Steel, 'A general equation for the choice of glass for cemented doublets', *Austral. J. Phys.*, Vol. 8 (1955) pp. 68–73.

41 M.H. Sussman, 'Cemented aplanatic doublets', *J. Opt. Soc. Am.*, Vol. 52 (1962) pp. 1185–1186.

42 H.H. Hopkins and V.V. Rao, Section 3 in 'The systematic design of two component objectives', *Opt. Acta*, Vol. 17 (1970) pp. 497–514.

43 B.A.F. Blandford, 'The selection of glass types for cemented doublets', in *The design of Optical Systems for Spatial Filtering*, Ph. D. Thesis, University of Reading, U. K.) pp. 137–144.

44 T.H. Jamieson, 'Lens systems: aplanatic, anastigmatic two and three element', *Appl. Opt.*, Vol. 15 (1976) pp. 2276–2282.

45 I.M. Khan and J. Macdonald, 'Cemented doublets. A method for rapid design', *Opt. Acta*, Vol. 29 (1982) pp. 807–822.

46 A. Szulc, 'Improved solution for the cemented doublet', *Appl. Opt.*, Vol. 35 (1996) pp. 3548–3558.

47 S. Banerjee and L.N. Hazra, 'Structural design of doublet lenses with prespecified targets', *Opt. Eng.*, Vol. 36 (1997) pp. 3111–3118.

48 S. Banerjee and L.N. Hazra, Simulated Annealing with constrained random walk in structural design of doublet lenses., *Opt. Eng.*, Vol. 37(1998) pp. 3260–3267.

49 S. Banerjee and L.N. Hazra, 'Structural design of broken contact doublets with prespecified aberration targets using genetic algorithm', *J. Mod. Opt.*, Vol. 49 (2002) pp. 1111–1123.
50 S. Banerjee and L.N. Hazra, 'Experiments with a genetic algorithm for structural design of cemented doublets with prespecified aberration targets', *Appl. Opt.*, Vol. 40 (2001) pp. 6265–6273.
51 A.C.S. Van Heel, 'One radius doublets', *Opt. Acta*, Vol. 2 (1955) pp. 29–35.
52 A.E. Conrady, Sec. 86 in *Applied Optics and Optical Design, Part II*, Dover, New York (1960) pp. 554–561.
53 R. Kingslake, *Lens Design Fundamentals*, Academic, New York (1978) pp. 173–178.
54 M.I. Khan, 'Cemented Triplets: A method for rapid design', *Opt. Acta*, Vol. 31 (1984) pp. 873–883.
55 C.H. Chen, S.G. Shiue and M.H. Lu, 'Method of solving cemented triplets with given primary aberrations', *J. Mod. Optics*, Vol. 44 (1997) pp. 753–763.
56 S. Chatterjee and L.N. Hazra, 'Structural design of cemented triplets by genetic algorithm', *Opt. Eng.*, Vol. 43 (2004) pp. 432–441.
57 A. Mikš, 'One radius triplets', *Appl. Opt.*, Vol. 41 (2002) pp. 1277–1281.
58 A.E. Conrady, *Applied Optics and Optical Design*, Part I, Dover, New York (1957) [originally by Oxford University Press (1929)] p. 160; vide also, Sec. 41, pp. 155–166.
59 A.E. Conrady, Sec. 88 in *Applied Optics and Optical Design, Part II*, Dover, New York (1960) pp. 578–584.
60 C.H. Chen, S.G. Shiue, and M.H. Lu, 'Method of solving a triplet comprising a singlet and a cemented doublet with given primary aberrations', *J. Mod. Opt.*, Vol. 44 (1997) pp. 1279–1291.
61 C.H. Chen, S.G. Shiue, and M.H. Lu, 'Method of solving triplets consisting of a singlet and air-spaced doublet with given primary aberrations', *J. Mod. Opt.*, Vol. 45 (1998) pp. 2063–2084.
62 H.H. Hopkins and V.V. Rao, Section 2 in 'The systematic design of two component objectives', *Opt. Acta*, Vol. 17 (1970) pp. 497–514.
63 R.E. Stephens, 'The Design of Triplet Anastigmat Lenses of the Taylor Type', *J. Opt. Soc. Am.*, Vol. 38 (1948) pp. 1032–1039. [vide corrections in ibid., Vol. 40 (1950) 407].
64 N.v.d.W. Lessing, 'Selection of Optical Glasses in Taylor Triplets (Special Method)', *J. Opt. Soc. Am.*, Vol. 48 (1958) pp. 558–562.
65 N.v.d.W. Lessing, 'Selection of Optical Glasses in Taylor Triplets (General Method)', *J. Opt. Soc. Am.*, Vol. 49 (1959) pp. 31–34.
66 R. Kingslake, 'Automatic Predesign of the Cooke Triplet Lens', *Proc. ICO Conf. on Optical Instruments and Techniques*, Imperial College, London (1961), Ed. K. J. Habell, Chapman and Hall, London (1962) pp. 107–120.
67 M. De, *Systematic Development of unsymmetrical triplets, J. Opt. (India)*, Vol. 29 (2000) pp. 131–138.
68 S. Banerjee and L. N. Hazra, 'Thin lens design of Cooke triplet lenses: application of a global optimization technique', *Proc. SPIE*, Vol. 3430 (1998) pp. 175–183.
69 S. Banerjee and L.N. Hazra, 'Genetic algorithm in structural design of Cooke triplet lenses', *Proc. SPIE*, Vol. 3737 (1999) pp. 172–179.
70 M. Berek, 'Synthese optischer Systeme auf der Grundlage der Theorie' Chapter 9 in *Grundlagen der praktischen Optik*, Walter de Gruyter, Berlin (1930) pp. 94–133.
71 I.C. Gardner, Chapter VII "Application of the algebraic aberration equations to optical design", Scientific Papers of the Bureau of Standards, Paper S550, National Bureau of Standards, USA, Vol. 22 (1926).
72 A.E. Conrady, Chapters XI, XVIII, XIX and XX in Applied Optics and Optical Design, Part II (in Italic), Dover, New York (1960), pp. 519–584, 760–826.
73 G.G. Slyusarev, 'Principal algebraic method equation' in Chapter 6 in *Aberration and Optical Design Theory*, [original Russian book title: *Metodi Rascheta Opticheskikh Sistem, Mashinostroenie*, Leningrad (1969)] Translator: J.H. Dixon, Adam Hilger, Bristol (1984) pp. 339–363.
74 D. Argentieri, Part I, 'Il progetto analitico dei systemi di lenti', in Ottica Industriale, Ulrico Hoepli, Milano, Italy (1954) pp. 3–175.
75 F.D. Cruickshank, *'Contribution des methods algébraiques au calcul des systemes optiques', Rev. d'Optique*, Vol. 35 (1956) pp. 292–299.
76 T.H. Jamieson, 'Thin-lens theory of zoom systems', *Opt. Acta*, Vol. 17 (1970) pp. 565–584.
77 R. Kingslake, Chapter 12 and 13 in *Lens Design Fundamentals*, Academic, New York (1978) pp. 233–295.

78 L.N. Hazra, 'Structural design of multicomponent lens systems', L.N. Hazra, *Appl. Opt*, Vol. 23 (1984) pp. 4440–4443.
79 L.N. Hazra and A.K. Samui, 'Design of individual components of a multicomponent lens system: use of a singlet', *Appl. Opt.*, Vol. 25 (1986) pp. 3721–3730.
80 M. De, Synthesis of Optical Systems – Structural Development and Design, *Part I, J. Opt.* (India), Vol. 27 (1998) pp. 95–104; Part II, *ibid.*, Vol. 28 (1999) pp. 41–52; Part III, *ibid.*, Vol. 29 (2000) pp. 1–4.
81 L.N. Hazra and S. Chatterjee, 'A Prophylactic Strategy for Global Synthesis in Lens Design', *Opt. Rev.*, Vol. 12 (2005) pp. 247–254.
82 S. Chatterjee and L.N. Hazra, 'Structural design of a lens component with prespecified aberration targets by evolutionary algorithm', *Proc. SPIE*, Vol. 6668 (2007) 66680 S-1 to S-12.
83 H.H. Hopkins and V.V. Rao, Section 4 in 'The systematic design of two component objectives', *Opt. Acta*, Vol. 17 (1970) pp. 497–514.
84 G.G. Slyusarev, *Aberration and Optical Design Theory*, [original Russian book title: *Metodi Rascheta Opticheskikh Sistem, Mashinostroenie*, Leningrad (1969)] Translator: J. H. Dixon, Adam Hilger, Bristol (1984) pp. 353–363.
85 M. Berek, 'Unwandlung diker Einzellinsen in Äquivalentlinsen der Dicke Null', in *Grundlagen der praktischen Optik*, Walter de Gruyter, Berlin (1930) pp. 86–89.
86 M. Herzberger, 'Replacing a thin lens by a thick lens', *J. Opt. Soc. Am.*, Vol. 34 (1944) pp. 114–115.
87 J.L. Rayces, 'Petzval sum of thick lenses', *J. Opt. Soc. Am.*, Vol. 45 (1955) p. 774.
88 L.E. Sutton, 'A method for localized variation of the paths of two paraxial rays', *Appl. Opt.*, Vol. 2 (1963) pp. 1275–1280.
89 B.A.F. Blandford, 'The design of components with finite thicknesses', in *The design of Optical Systems for Spatial Filtering,* Ph. D. Thesis, University of Reading, U. K.) pp. 148–151.
90 R. Kingslake, *Lens Design Fundamentals*, Academic, New York (1978) pp. 59–60.
91 G.W. Hopkins, Chapter 1, 'Basic Algorithms for Optical Engineering', in *Applied Optics and Optical Engineering, Vol. IX*, Eds. R. R. Shannon and J. C. Wyant, Academic, New York (1983) pp. 1–9.
92 W.T. Welford, *Aberrations of Optical Systems*, Adam Hilger, Bristol (1986) pp. 41–42.
93 P. Mouroulis and J. Macdonald, *Geometrical Optics and Optical Design*, Oxford University Press, New York (1997) pp. 272–280.
94 A. Mikš and P. Novák, 'Replacing thin lens by a thick lens', *Appl. Opt.*, Vol. 59 (2020) pp. 6327–6332.
95 W.C. Sweatt, 'Describing holographic elements as lenses', *J. Opt. Soc. Am.*, Vol. 67 (1977) pp. 803–808.
96 W.A. Kleinhans, 'Aberrations of curved zone plates and Fresnel lenses', *Appl. Opt.*, Vol. 16 (1977) pp. 1701–1704.
97 M.A. Gan, 'Third order aberrations and the fundamental parameters of axisymmetrical holographic elements', *Opt. Spectrosc.*, Vol. 47 (1979) pp. 419–422.
98 W.C. Sweatt, 'Mathematical equivalence between a holographic optical element and an ultra-high index lens', *J. Opt. Soc. Am.*, Vol. 69 (1979) pp. 486–487.
99 D.A. Buralli, G.M. Morris, and J.R. Rogers, 'Optical performance of holographic kinoforms', *Appl. Opt.*, Vol. 28 (1989) pp. 976–983.
100 L.N. Hazra, Y. Han, and C.A. Delisle, 'Kinoform lenses: Sweatt model and phase function', *Opt. Commun.*, Vol. 117 (1995) pp. 31–36.

15

Stop Shift, Pupil Aberrations, and Conjugate Shift

15.1 Axial Shift of the Aperture Stop

The role of the aperture stop in an optical imaging system was discussed in Chapter 4. In some optical systems, e.g. camera lenses, an exclusive iris diaphragm is used to control illumination in the image. This diaphragm is the aperture stop of the system. By changing the diameter of this diaphragm, the illumination in the image can be varied. In optical systems that do not have such built-in provisions for varying illumination in the image, the illumination in the image is effectively determined by the rim of one of the multitude of transverse openings of the elements comprising the system. This particular rim is the aperture stop in such a system. But it should be noted that, except for in trivial cases like a single lens with stop on it, no single rim in an optical system can act as the effective aperture stop for all axial positions of the object/image from $-\infty$ to $+\infty$. However, in general, the imaging systems are usually designed such that, over the working range of axial locations for object/image, the same rim acts as the aperture stop. It should also not be overlooked that for a given axial location of an aperture stop of fixed diameter, the angular aperture of the cone of rays forming image of the axial object point varies with changes in axial location of the object/image plane.

On the other hand, for a given conjugate location of the imaging system, any axial shift of the aperture stop of fixed diameter causes changes in the angular aperture of the cone of rays forming image of the axial object point. Indeed, to prevent any change in illumination in the axial image point with axial shift of the stop, it is necessary to ensure that the angular aperture of the cone of rays forming image of the axial object point remains unaltered. The latter can only be achieved by changing the diameter of the aperture stop such that at each axial location it fits the diameter of the originally chosen axial cone of rays.

In optics jargon, 'stop shift' in an imaging system implies an axial shift of the stop with concomitant change in the diameter of its aperture so that the cone of rays forming the axial image remains unaltered. However, the extra-axial imagery is altered significantly. Unlike the case of axial imaging by a regular cone of rays, each extra-axial point is imaged by a different oblique bundle of rays. The latter bundle is determined by the height of the extra-axial point, the axial location of the aperture stop, and the axial locations of other finite elements of the system. A conspicuous feature of the shape of this bundle is the phenomenon of vignetting described in Chapter 4 in paraxial approximation.

Figure 15.1 shows the imaging of an object OP by an axially symmetric imaging system. O′P′ is the chosen image plane (usually the paraxial image plane). The aperture stop is axially shifted from S_1 to S_2. Note the change in transverse size of the aperture stop so that the axial cone is just admitted. Σ_1, Σ_3, and Σ_2 are three wavefronts in the image space corresponding to the imaging of the extra-axial point P at P′. The wavefronts Σ_1, Σ_2 pass through the axial points at S_1, S_2, respectively. Part of the wavefront intercepted by the aperture stop at S_1 is shown by the chain lines, and part of the wavefront intercepted by the aperture stop at S_2 is shown by dashed lines. The two parts are distinctly different, and this is the cause for changes in extra-axial aberrations by an axial shift of the aperture stop.

15.1.1 The Eccentricity Parameter

Figures 15.2 and 15.3 show the object space of an axially symmetric lens system forming an image of a transverse object $OP = \eta$. In Figure 15.2, the aperture stop is on the first surface at A_1. The paraxial marginal ray (PMR) OD has a convergence angle u, and height h on the first surface. The paraxial pupil

DOI: 10.1201/9780429154812-15

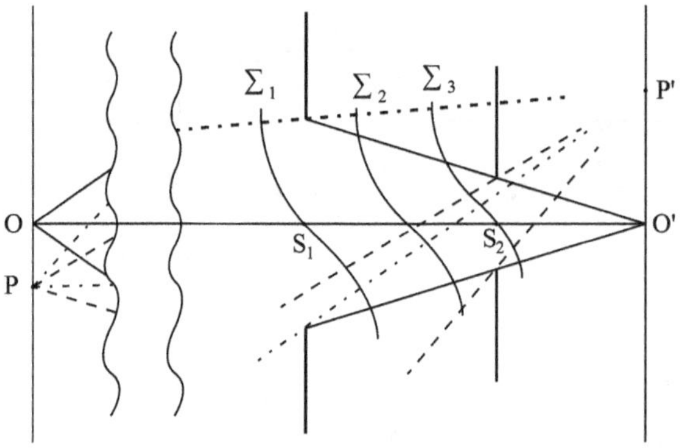

FIGURE 15.1 Σ_1, Σ_2, and Σ_3 are three arbitrary wavefronts of the pencil forming image of the extra-axial point P. An axial shift of the aperture stop from S_1 to S_2 allows different parts of the wavefronts in actual image formation. (Note that to keep the pencil forming image of the axial object point unchanged, axial shift is accompanied by proportional change in the aperture of the stop).

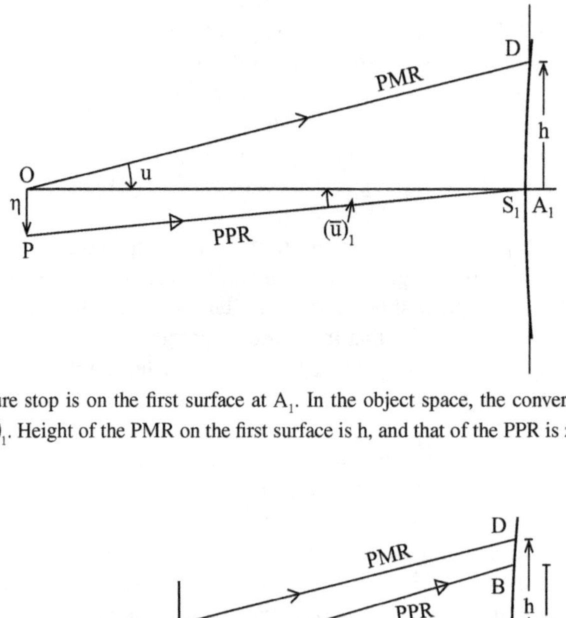

FIGURE 15.2 The aperture stop is on the first surface at A_1. In the object space, the convergence angle of the PMR is u, and that of the PPR is $(\bar{u})_1$. Height of the PMR on the first surface is h, and that of the PPR is zero.

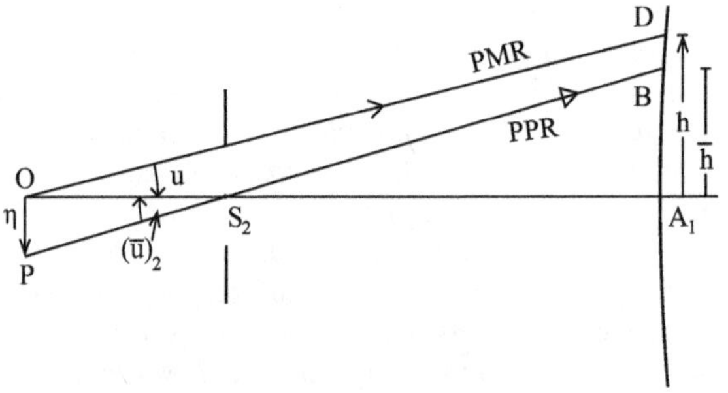

FIGURE 15.3 The aperture stop of the setup in Figure 15.2 is axially shifted to S_2. Its diameter is changed to keep the PMR unchanged. The convergence angle of the new PPR is $(\bar{u})_2$. Height of the PMR on the first surface is h, and that of the PPR is $\bar{h}$.

ray (PPR) PS_1 has a convergence angle $(\bar{u})_1$. Figure 15.3 shows the ray geometry when the aperture stop is axially shifted to S_2. The diameter of the aperture of the stop is changed to allow the PMR to remain unchanged. The PPR is changed from PS_1 to PS_2B. The convergence angle of the new PPR is $(\bar{u})_2$, and its height on the first surface is $\bar{h}$.

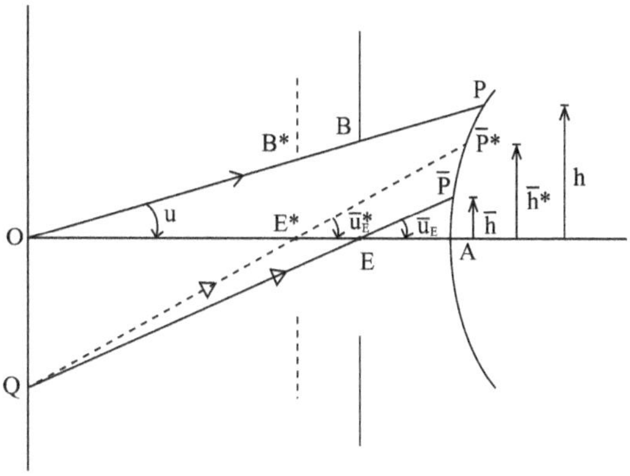

FIGURE 15.4 Axial shift of the aperture stop from E to E^*. Parameters of the PMR remain unchanged. Parameters of the PPR change from $(\bar{u}_E, h)$ to $(\bar{u}_E^*, \bar{h}^*)$. Stop radius changed from BE to B^*E^* for keeping H the same.

The eccentricity parameter E on the surface at A_1 is defined as

$$E = \frac{\bar{h}}{h} \tag{15.1}$$

Note that, at times [1], the eccentricity parameter is defined in a reduced form as $\left(\frac{1}{H}\right)\left(\frac{\bar{h}}{h}\right)$. However, we prefer the dimensionless form for eccentricity parameter.

For the sake of illustration, Figure 15.4 shows an axial shift of the aperture stop from E to E^* in the object space of an axi-symmetric imaging system. The paraxial marginal ray (PMR) OB^*BP remains unchanged so that the values of u and h, the convergence angle of the PMR, and its height on the surface at A are unaltered. The paraxial pupil ray (PPR) is $QE\bar{P}$ before the stop shift, and after the stop shift the paraxial pupil ray is $QE^*\bar{P}^*$. As a result, the parameters of the PPR are changed from $\bar{h}, \bar{u}_E$ to $\bar{h}^*, \bar{u}_E^*$. The corresponding change δE in eccentricity E on the first surface is given by

$$\delta E = \frac{\delta \bar{h}}{h} = \frac{\bar{h}^* - \bar{h}}{h} \tag{15.2}$$

15.1.2 Seidel Difference Formula

Figure 15.5 shows an aperture stop S in the intermediate space of refractive index $n_i' = n_{i+1}$ between two successive refracting interfaces with vertices at A_i and A_{i+1}. The convergence angle of the PMR is $u_i' = u_{i+1}$, and its height at the two surfaces are h_i and h_{i+1}. The convergence angle of the PPR is $\bar{u}_i' = \bar{u}_{i+1}$, and its heights at the two surfaces are $\bar{h}_i$ and $\bar{h}_{i+1}$. The height of the image in the intermediate space is $\eta_i' \equiv \eta_{i+1}$, and the axial distance of this image plane from A_i and A_{i+1} are l_i' and l_{i+1}, respectively. The distances $A_iA_{i+1} = d_i$, $A_iS = \bar{l}_i'$, and $A_{i+1}S = l_{i+1}$. We have

$$h_i = -l_i' u_i' \qquad h_{i+1} = h_i + d_i u_i' = -\left(l_i' - d_i\right) u_i' \tag{15.2}$$

$$\bar{h}_i = -\bar{l}_i' \bar{u}_i' \qquad \bar{h}_{i+1} = \bar{h}_i + d_i \bar{u}_i' = -\left(\bar{l}_i' - d_i\right) \bar{u}_i' \tag{15.3}$$

The difference in eccentricity at the $(i+1)$-th and i-th surfaces is

$$E_{i+1} - E_i = \frac{\bar{h}_{i+1}}{h_{i+1}} - \frac{\bar{h}_i}{h_i} = \frac{\bar{h}_{i+1} h_i - \bar{h}_i h_{i+1}}{h_{i+1} h_i} \tag{15.4}$$

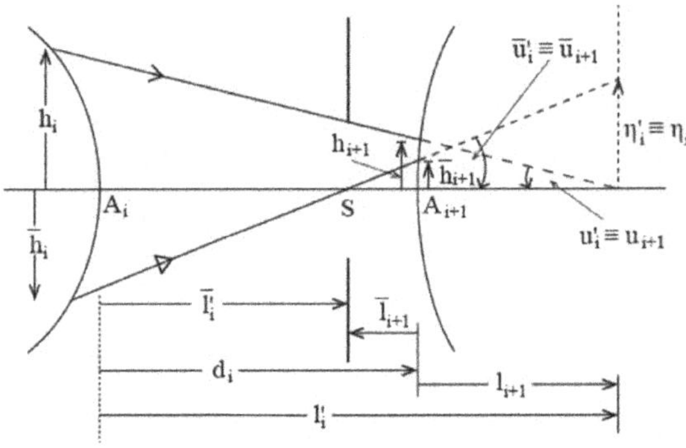

FIGURE 15.5 An aperture stop S in the intermediate space of refractive index $n'_i \equiv n_{i+1}$ between two successive refracting interfaces with vertices at A_i and A_{i+1}.

Using (15.2) – (15.3), (15.4) reduces to

$$E_{i+1} - E_i = -\frac{\bar{u}'_i \bar{u}'_i \left(l'_i - \bar{l}'_i \right) d_i}{h_{i+1} h_i} \tag{15.5}$$

The paraxial invariant H of the system is given by

$$H = n'_i u'_i \eta'_i = n'_i u'_i \left(l'_i - \bar{l}'_i \right) \bar{u}'_i \tag{15.6}$$

By substituting (15.6) in (15.5), it follows

$$E_{i+1} - E_i = -\frac{H d_i}{n'_i h_{i+1} h_i} \tag{15.7}$$

Thus, the increment in the eccentricity parameter from the i-th to the (i+1)-th surface is given by the product of the paraxial invariant and the axial separation between the two surfaces, divided by the product of the refractive index of the intermediate space and the heights of the PMR at the two surfaces. No term related to the PPR appears in the expression (15.7). This proves the Seidel difference formula:

> The difference in eccentricity parameter between two successive surfaces is independent of stop position.

If the stop position is known, the eccentricity parameters can be determined. Assume the axial position S of the stop with respect to the preceding and following surfaces at A_i and A_{i+1} are known (Figure 15.5). Let $SA_i = {}_S d_i$, and $SA_{i+1} = {}_S d_{i+1}$, and let the refractive index of the medium be $n_S \equiv n'_i \equiv n_{i+1}$. Let the heights of the PMR at the surfaces at A_i, S, A_{i+1} be h_i, h_S, h_{i+1}, respectively. At the stop surface, the height of the PPR, $\bar{h}_S = 0$. Therefore, $E_S = 0$. From (15.7), we can write

$$E_{i+1} - E_S = E_{i+1} = -\frac{H\, {}_S d_{i+1}}{n_s h_s h_{i+1}} \tag{15.8}$$

$$E_i - E_S = E_i = \frac{H\, {}_S d_i}{n_s h_s h_i} \tag{15.9}$$

The values of eccentricity at other surfaces can be found by working in both directions using the relation (15.7).

If the stop is shifted so that E_i changes to $E_i^* = E_i + \Delta E$, by Seidel difference formula, the new eccentricity E_{i+1}^* for the $(i+1)$-th surface is $E_{i+1}^* = E_{i+1} + \Delta E$. This implies that the change in eccentricity at a surface due to the stop shift is the same for all surfaces of the system.

The difference formula, described above, was developed by Seidel for determining the primary aberration coefficients from ray trace data of only the paraxial marginal ray using knowledge of axial location of aperture stop and the paraxial invariant H, obviating the need for tracing the paraxial pupil ray [2]. Given (nu, h) values for the paraxial marginal ray at a surface, the corresponding values for the paraxial pupil ray at the surface can be determined, as long as H for the system and the eccentricity E at the surface are known. The ray parameters for the PPR are:

$$\left.\begin{aligned} n\bar{u} &= \frac{nu\bar{h} - H}{h} = nuE - \frac{H}{h} = -\frac{H}{h}\left(1 - nuh\frac{E}{H}\right) \\ \bar{h} &= hE \end{aligned}\right\} \tag{15.10}$$

15.1.3 Stop-Shift Effects on Seidel Aberrations in Refraction by a Single Surface

Recalling the expressions for the monochromatic and chromatic Seidel aberration coefficients for a single refracting surface from (12.31), (12.65), and (12.69), we write

$$\left.\begin{aligned} S_I &= S_1 = -A^2 h\Delta\left(\frac{u}{n}\right) \\ S_{II} &= S_2 = -A\bar{A}h\Delta\left(\frac{u}{n}\right) = \frac{\bar{A}}{A}S_I \\ S_{III} &= S_3 = -\bar{A}^2 h\Delta\left(\frac{u}{n}\right) = \frac{\bar{A}}{A}S_{II} \\ S_{IV} &= S_4 = -H^2 c\Delta\left(\frac{1}{n}\right) \\ S_V &= S_5 = -\left[\frac{\bar{A}^3}{A}h\Delta\left(\frac{u}{n}\right) + \frac{\bar{A}}{A}H^2 c\Delta\left(\frac{1}{n}\right)\right] = \frac{\bar{A}}{A}\left[S_{III} + S_{IV}\right] \\ C_I &= C_1 = Ah\Delta\left(\frac{\delta n}{n}\right) \\ C_{II} &= C_2 = \bar{A}h\Delta\left(\frac{\delta n}{n}\right) \end{aligned}\right\} \tag{15.11}$$

The expression for the paraxial optical invariant H in terms of (nu, h) parameters for the PMR and the PPR at a surface is

$$H = n\left(u\bar{h} - \bar{u}h\right) \tag{15.12}$$

The Abbe invariant for the PMR, A, and for the PMR, $\bar{A}$ are given by

$$A = n\left(hc + u\right) \qquad \bar{A} = n\left(hc + \bar{u}\right) \tag{15.13}$$

From (15.12) – (15.13), we get

$$H = A\bar{h} - \bar{A}h \tag{15.14}$$

It follows

$$\delta H = A\delta\bar{h} + \bar{h}\delta A - \bar{A}\delta h - h\delta\bar{A} \tag{15.15}$$

By the stop shift, only the PPR is affected, and neither the PMR, nor the paraxial invariant, is affected. Therefore,

$$\delta H = \delta A = \delta h = 0 \tag{15.16}$$

Substituting these values in (15.15), we get

$$A\delta\bar{h} - h\delta\bar{A} = 0 \tag{15.17}$$

It follows

$$\delta\bar{A} = A\frac{\delta\bar{h}}{h} = A\delta\left(\frac{\bar{h}}{h}\right) = A\delta E \tag{15.18}$$

Let the new values of Seidel aberrations after a stop shift δE be represented by using an asterisk as an upper suffix to each of the aberration terms. The new value of $\bar{A}$ at the surface is $(\bar{A} + \delta\bar{A}) = (\bar{A} + A\delta E)$. Substituting this value for $\bar{A}$ in (15.11), and by simple algebraic manipulation, the new values of Seidel aberration coefficients after stop shift in the case of a single refracting interface are obtained as [3]

$$\left.\begin{aligned}
S_I^* &= S_I \\
S_{II}^* &= S_{II} + (\delta E)S_I \\
S_{III}^* &= S_{III} + 2(\delta E)S_{II} + (\delta E)^2 S_I \\
S_{IV}^* &= S_{IV} \\
S_V^* &= S_V + (\delta E)(3S_{III} + S_{IV}) + 3(\delta E)^2 S_{II} + (\delta E)^3 S_I \\
C_I^* &= C_I \\
C_{II}^* &= C_{II} + (\delta E)C_I
\end{aligned}\right\} \tag{15.19}$$

15.1.4 Stop-Shift Effects on Seidel Aberrations in an Axi-Symmetric System

Using the Seidel difference formula, the difference in eccentricity of two successive surfaces is independent of the stop position. This implies that a stop shift causing a change in eccentricity at a surface by δE, changes the eccentricity of all surfaces of the system by δE. Effects of this stop shift on the coefficients of Seidel aberrations of the axi-symmetric system consisting of k spherical surfaces are given below.

$$\left.\begin{aligned}
S_I^* &= \sum_{j=1}^{k}(S_I^*)_j = \sum_{j=1}^{k}(S_I)_j = S_I \\
S_{II}^* &= \sum_{j=1}^{k}(S_{II}^*)_j = \sum_{j=1}^{k}(S_{II})_j + (\delta E)\sum_{j=1}^{k}(S_I)_j = S_{II} + (\delta E)S_I \\
S_{III}^* &= \sum_{j=1}^{k}(S_{III}^*)_j = \sum_{j=1}^{k}(S_{III})_j + 2(\delta E)\sum_{j=1}^{k}(S_{II})_j + (\delta E)^2\sum_{j=1}^{k}(S_I)_j = S_{III} + 2(\delta E)S_{II} + (\delta E)^2 S_I \\
S_{IV}^* &= \sum_{j=1}^{k}(S_{IV}^*)_j = \sum_{j=1}^{k}(S_{IV})_j = S_{IV} \\
S_V^* &= \sum_{j=1}^{k}(S_V^*)_j = \sum_{j=1}^{k}(S_V)_j + (\delta E)\left[\sum_{j=1}^{k}\left\{3(S_{III})_j + (S_{IV})_j\right\}\right] + 3(\delta E)^2\sum_{j=1}^{k}(S_{II})_j + (\delta E)^3\sum_{j=1}^{k}(S_I)_j \\
&= S_V + (\delta E)[3S_{III} + S_{IV}] + 3(\delta E)^2 S_{II} + (\delta E)^3 S_I \\
C_I^* &= \sum_{j=1}^{k}(C_I^*)_j = \sum_{j=1}^{k}(C_I)_j = C_I \\
C_{II}^* &= \sum_{j=1}^{k}(C_{II}^*)_j = \sum_{j=1}^{k}(C_{II})_j + (\delta E)\sum_{j=1}^{k}(C_I)_j = C_{II} + (\delta E)C_I
\end{aligned}\right\} \tag{15.20}$$

where S_I^*, S_{II}^*, S_{III}^*, S_{IV}^*, S_V^*, C_I^*, C_{II}^* are the Seidel aberration coefficients for the whole axi-symmetric system after the stop shift, and S_I, S_{II}, S_{III}, S_{IV}, S_V, C_I, C_{II} are the Seidel aberration coefficients for the whole system before the stop shift.

15.1.5 Stop-Shift Effects on Seidel Aberrations in a Single Thin Lens

When the stop is on the thin lens, the aberrations are called 'central aberrations'. Each central Seidel aberration is denoted by a suffix 'c'. Let the stop be axially shifted to a position where the eccentricity $E = (\bar{h}/h)$. The heights of the PMR and the PPR at the new position of the stop are h and $\bar{h}$, respectively. Denoting Seidel aberration coefficients after the stop shift by an asterisk, we get

$$\left.\begin{aligned} S_I^* &= (S_I)_c \\ S_{II}^* &= (S_{II})_c + E(S_I)_c \\ S_{III}^* &= (S_{III})_c + 2E(S_{II})_c + E^2(S_I)_c \\ S_{IV}^* &= (S_{IV})_c \\ S_V^* &= E\{3(S_{III})_c + (S_{IV})_c\} + 3E^2(S_{II})_c + E^3(S_I)_c \\ C_I^* &= (C_I)_c \\ C_{II}^* &= E(C_I)_c \end{aligned}\right\} \qquad (15.21)$$

Note that $(S_V)_c = (C_T)_c = 0$, and so it is omitted from the expressions above. Spherical aberration S_I, curvature S_{IV}, and longitudinal chromatic aberration C_I are not affected by the stop shift. Central coma is a linear function of the shape variable X and of the conjugate variable Y. But S_{II}^* contains S_I, which is a quadratic function of both X and Y, so that the variation of S_{II}^* with X and Y becomes quadratic. Similarly, the variations of S_{III}^* and S_V^* with X and Y become quadratic.

15.1.6 Corollaries

From the discussions on the effects of stop shifts on Seidel aberrations of optical systems, many conclusions on aberrational characteristics of different combinations of lens structures may be drawn [4]. A few observations follow:

I. For a system with zero spherical aberration, a stop shift has no effect on coma.
II. For an aplanatic system where both spherical aberration and coma are zero, a stop shift has no effect on astigmatism.
III. A stop shift has no effect on monochromatic Seidel aberrations in an aplanatic system with a flat tangential field, i.e. when $\sum_{j=1}^{k}\{3(S_{III})_j + (S_{IV})_j\} = 0$.
IV. For a system with zero axial colour, a stop shift has no effect on lateral colour.
V. An optical system consisting of two separated components can be made achromatic if and only if each of the two components are achromatic. The restriction does not exist for systems with three or more separated components.
VI. An aplanatic system consisting of two separated components can be made free from distortion or astigmatism if the components are separately unaplanatic.
VII. No combination of thin lenses in contact can produce an anastigmatic system.

15.2 Pupil Aberrations

It has been discussed earlier that, in principle, it is not possible to obtain aberration-free imaging for more than a single set of object-image conjugate planes. Imaging at all other conjugate planes is affected by

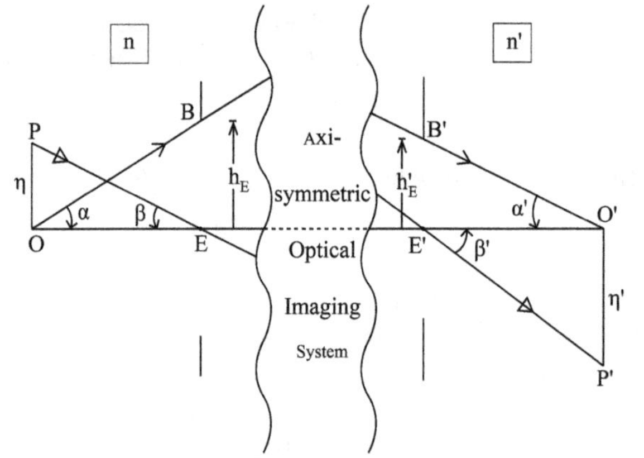

FIGURE 15.6 Role reversal: For imaging of object OP at O′P′, the entrance pupil and the exit pupil are at E and E′. For imaging of the entrance pupil at E as the exit pupil at E′, the object at O and the image at O′ are effectively the entrance pupil and the exit pupil, respectively.

aberrations of varying degree. In practice, optical imaging systems are usually designed to perform their primary function of forming an image of a plane object on an image plane. Concomitantly, the optical system images the aperture stop as the entrance pupil in the object space, and as the exit pupil in the image space. In general, both the entrance pupil and the exit pupil are aberrated images of each other. The pupil aberrations do not directly affect the quality of object-image imaging. But they do affect the illumination in the image, and in some applications they need to be taken due care of, either for removing severe non-uniformity in illumination in the image, or for enhancing illumination in outer fields which otherwise receive little illumination.

The study of pupil aberrations in axi-symmetric systems can be carried out by taking recourse to a role reversal for the object/image and entrance pupil/exit pupil [5]. In Figure 15.6, an object OP of height η is imaged by an axi-symmetric imaging system as O′P′ of height η'. The paraxial marginal ray OB…B′O′ makes an angle α with the optical axis in the object space, and an angle α' with the optical axis in the image space. The paraxial pupil ray PE…E′P′ makes an angle β with the optical axis in the object space, and an angle β' with the optical axis in the image space. Correspondingly, the paraxial angle variables in the object space are $u = \sin\alpha, \bar{u} = \tan\beta$, and those in the image space are $u' = \sin\alpha', \bar{u}' = \tan\beta'$. The semi-diameter of the entrance pupil is BE= h_E, and that of the exit pupil is $B'E' = h'_E$. As per our sign convention, u, η, h_E, h'_E are positive, and u', η', $\bar{u}$, $\bar{u}'$ are negative. The optical invariant H for object/image conjugate imaging is

$$H = nu\eta = n'u'\eta' \tag{15.22}$$

Alternatively, H is expressed as

$$H = -n\bar{u}h_E = -n'\bar{u}'h'_E \tag{15.23}$$

Note that, given u or h_E, the point P is to be taken above or below the axis in order for η or $\bar{u}$ to be of such sign as to make H positive in (15.22) and (15.23). Similarly, given u' or η', the point P′ is to be taken above or below the axis so that η' or $\bar{u}'$ will be of such sign as to make H positive in (15.22) and (15.23).

For pupil imaging, the entrance pupil/exit pupil is the object/image, and the object/image is the entrance pupil/exit pupil. Interchanging u and η with $\bar{u}$ and h_E, and u' and η' with $\bar{u}'$ and h'_E in (15.22) and (15.23), we get two equivalent expressions for $\bar{H}$, the paraxial invariant for imaging of the pupils

$$\bar{H} = n\bar{u}h_E = n'\bar{u}'h'_E = -H \tag{15.24}$$

$$\bar{H} = -nu\eta = -n'u'\eta' = -H \tag{15.25}$$

Note that $\bar{H}$ is of negative sign. The stop shift is expressed in terms of the eccentricity parameter $E = (\bar{h}/h)$, and the shift of object is expressed in terms of the eccentricity parameter $\bar{E} = (h/\bar{h})$. The paraxial Abbe invariant for the PMR at a refracting surface is $A = n(hc + u) = n'(hc + u')$, and the Abbe invariant for the PPR is $\bar{A} = n(\bar{h}c + \bar{u}) = n'(\bar{h}c + \bar{u}')$.

Interchanging the parameters of the paraxial marginal ray by the parameters of the paraxial pupil ray, and vice versa, the Seidel pupil aberration (monochromatic) terms for refraction at a spherical interface are obtained from (12.31) as

$$\left.\begin{aligned}
\bar{S}_I &= \bar{S}_1 = -\bar{A}^2\bar{h}\Delta\left(\frac{\bar{u}}{n}\right)\\
\bar{S}_{II} &= \bar{S}_2 = -\bar{A}A\bar{h}\Delta\left(\frac{\bar{u}}{n}\right) = \frac{A}{\bar{A}}\bar{S}_I\\
\bar{S}_{III} &= \bar{S}_3 = -A^2\bar{h}\Delta\left(\frac{\bar{u}}{n}\right) = \frac{A}{\bar{A}}\bar{S}_{II}\\
\bar{S}_{IV} &= \bar{S}_4 = -H^2c\Delta\left(\frac{1}{n}\right)\\
\bar{S}_V &= \bar{S}_5 = -\left[\frac{A^3}{\bar{A}}\bar{h}\Delta\left(\frac{\bar{u}}{n}\right) + \frac{A}{\bar{A}}H^2c\Delta\left(\frac{1}{n}\right)\right] = \frac{A}{\bar{A}}\left[\bar{S}_{III} + \bar{S}_{IV}\right]
\end{aligned}\right\} \tag{15.26}$$

In order to reiterate that in optics literature, either the Roman number or the equivalent Arabic number is used in the subscript of 'S' in the notations for the coefficients of Seidel aberrations; the equivalence is shown in equation (15.26). Using summation theorem for primary aberrations, the Seidel pupil aberration (monochromatic) terms for an axi-symmetric system consisting of k spherical refracting interfaces can be expressed as

$$\left.\begin{aligned}
\bar{S}_I &= \sum_{j=1}^{k}\left(\bar{S}_I\right)_j = -\sum_{j=1}^{k}\bar{A}_j^2\bar{h}_j\left[\Delta\left(\frac{\bar{u}}{n}\right)\right]_j\\
\bar{S}_{II} &= \sum_{j=1}^{k}\left(\bar{S}_{II}\right)_j = -\sum_{j=1}^{k}\bar{A}_jA_j\bar{h}_j\left[\Delta\left(\frac{\bar{u}}{n}\right)\right]_j\\
\bar{S}_{III} &= \sum_{j=1}^{k}\left(\bar{S}_{III}\right)_j = -\sum_{j=1}^{k}A_j^2\bar{h}_j\left[\Delta\left(\frac{\bar{u}}{n}\right)\right]_j\\
\bar{S}_{IV} &= \sum_{j=1}^{k}\left(\bar{S}_{IV}\right)_j = -H^2\sum_{j=1}^{k}c_j\left[\Delta\left(\frac{1}{n}\right)\right]_j\\
\bar{S}_V &= \sum_{j=1}^{k}\left(\bar{S}_V\right)_j = -\sum_{j=1}^{k}\left[\frac{A_j^3}{\bar{A}_j}\bar{h}_j\left[\Delta\left(\frac{\bar{u}}{n}\right)\right]_j + \frac{A_j}{\bar{A}_j}H^2c_j\left[\Delta\left(\frac{1}{n}\right)\right]_j\right] = \sum_{j=1}^{k}\frac{A_j}{\bar{A}_j}\left[\left(\bar{S}_{III}\right)_j + \left(\bar{S}_{IV}\right)_j\right]
\end{aligned}\right\} \tag{15.27}$$

Corresponding to (12.35), a determinate expression for $\bar{S}_V$ valid for all values of $\bar{A}$ is

$$\bar{S}_V = -A^3\bar{h}\Delta\left(\frac{1}{n^2}\right) + Ahc\left(2A\bar{h} - \bar{A}h\right)\Delta\left(\frac{1}{n}\right) \tag{15.28}$$

Similarly, the Seidel longitudinal chromatic aberration of the pupil, $\bar{C}_L$, and the Seidel transverse chromatic aberration of the pupil, $\bar{C}_T$, are obtained from (15.11) by interchanging A and $\bar{A}$, and replacing h by $\bar{h}$

$$\left.\begin{aligned}\overline{C}_L = \overline{C}_I = \overline{C}_1 = \overline{L} = \overline{A}\overline{h}\Delta\left(\frac{\delta n}{n}\right)\\ \overline{C}_T = \overline{C}_{II} = \overline{C}_2 = \overline{T} = A\overline{h}\Delta\left(\frac{\delta n}{n}\right)\end{aligned}\right\} \tag{15.29}$$

For an axi-symmetric system consisting of k spherical refracting interfaces, the Seidel longitudinal chromatic aberration of the pupil, $\overline{C}_L$, and the Seidel transverse chromatic aberration of the pupil, $\overline{C}_T$ are given by

$$\left.\begin{aligned}\overline{C}_L = \overline{C}_I = \overline{C}_1 = \overline{L} = \sum_{j=1}^{k}\overline{A}_j\overline{h}_j\left[\Delta\left(\frac{\delta n}{n}\right)\right]_j\\ \overline{C}_T = \overline{C}_{II} = \overline{C}_2 = \overline{T} = \sum_{j=1}^{k}A_j\overline{h}_j\left[\Delta\left(\frac{\delta n}{n}\right)\right]_j\end{aligned}\right\} \tag{15.30}$$

Note that the Seidel pupil aberration terms are denoted by putting a bar on the top of the corresponding notations for Seidel aberration terms for the imaging of the object. $\overline{S}_I \equiv \overline{S}_1$, the Seidel spherical aberration of the pupil is sometimes denoted by $S_{VI} \equiv S_6$, and regarded as a sixth Seidel aberration coefficient of object imagery. This term was contained in the expansion of the eikonal given by Schwarzschild [6], but he disregarded it in his discussion on primary aberrations of object imagery. Later on, Longhurst [7], among others, dealt with the importance of the correction of pupil spherical aberration in the design of eyepieces. Wynne [8] underscored the important role played by this term in studying the conjugate shift effects on primary aberrations. Similarly, $\overline{C}_I \equiv \overline{C}_1$, the Seidel longitudinal chromatic aberration of the pupil is sometimes denoted by $C_{III} \equiv C_3$ and regarded as the third Seidel chromatic aberration coefficient of the object imagery. S_{VI} and C_{III} for a single refracting surface are given by

$$S_{VI} = -\overline{A}^2\overline{h}\Delta\left(\frac{\overline{u}}{n}\right) \tag{15.31}$$

$$C_{III} = \overline{A}\overline{h}\Delta\left(\frac{\delta n}{n}\right) \tag{15.32}$$

Note that $\overline{S}_{VI} \equiv \overline{S}_6 = S_I$, and $\overline{C}_{III} \equiv \overline{C}_3 = C_I$.

15.2.1 Relation between Pupil Aberrations and Image Aberrations

For a single spherical refracting interface, the Seidel aberration coefficients for the pupil imagery and the object imagery are related by:

$$\left.\begin{aligned}\overline{S}_I = S_{VI} = E\left[S_V + \overline{A}\overline{h}Hc\Delta\left(\frac{1}{n}\right)\right]\\ \overline{S}_{II} = S_V - H\Delta\left(\overline{u}^2\right)\\ \overline{S}_{III} = S_{III} - H\Delta\left(u\overline{u}\right)\\ \overline{S}_{IV} = S_{IV}\\ \overline{S}_V = S_{II} - H\Delta\left(u^2\right)\\ \overline{S}_{VI} = S_I\end{aligned}\right\} \tag{15.33}$$

Note that the Seidel curvature term in the case of pupil imagery remains the same, as in the case of object imagery. The relations for axial and lateral colour for pupil imagery and object imagery are:

$$\left.\begin{aligned}\bar{C}_I &= C_{III} = C_I + \left(\bar{A}\bar{h} - Ah\right)\Delta\left(\frac{\delta n}{n}\right)\\ \bar{C}_{II} &= C_{II} + H\Delta\left(\frac{\delta n}{n}\right)\\ \bar{C}_{III} &= C_I\end{aligned}\right\} \tag{15.34}$$

(15.33) and (15.34) constitute a pair of important relations in classical aberration theory. We give below derivations of these formulae. The following set of twelve relations (i)-(xii), given by (15.35)-(15.46) facilitate these derivations.

i. $g = -hc \qquad \bar{g} = -\bar{h}c$ (15.35)

ii. $i = u + hc = (u - g) \qquad i' = u' + hc = (u' - g)$ (15.36)

iii. $\bar{i} = \bar{h}c + \bar{u} = (\bar{u} - \bar{g}) \qquad \bar{i}' = \bar{u}' + hc = (\bar{u}' - \bar{g})$ (15.37)

iv. $A = n(hc + u) = n'(hc + u') = ni = n'i' = n(u - g) = n'(u' - g')$ (15.38)

v. $\bar{A} = n(\bar{h}c + \bar{u}) = n'(\bar{h}c + \bar{u}') = n\bar{i} = n'\bar{i}' = n(\bar{u} - \bar{g}) = n'(\bar{u}' - \bar{g})$ (15.39)

vi. $H = n(u\bar{h} - \bar{u}h) = (A\bar{h} - \bar{A}h)$ (15.40)

vii. $A\Delta\left(\frac{1}{n}\right) = \Delta(i) = \Delta(u) \qquad \bar{A}\Delta\left(\frac{1}{n}\right) = \Delta(\bar{i}) = \Delta(\bar{u})$ (15.41)

viii. $A\Delta\left(\frac{1}{n^2}\right) = \Delta\left(\frac{u}{n}\right) - g\Delta\left(\frac{1}{n}\right) \qquad \bar{A}\Delta\left(\frac{1}{n^2}\right) = \Delta\left(\frac{\bar{u}}{n}\right) - \bar{g}\Delta\left(\frac{1}{n}\right)$ (15.42)

ix. $A\Delta\left(\frac{u}{n}\right) = \Delta(u^2) - g\Delta(u) = \Delta(ui) \qquad \bar{A}\Delta\left(\frac{\bar{u}}{n}\right) = \Delta(\bar{u}^2) - \bar{g}\Delta(\bar{u}) = \Delta(\bar{u}\bar{i})$ (15.43)

x. $A\Delta\left(\frac{\bar{u}}{n}\right) = \Delta(u\bar{u}) - g\Delta(\bar{u}) = \Delta(\bar{u}i) \qquad \bar{A}\Delta\left(\frac{u}{n}\right) = \Delta(u\bar{u}) - \bar{g}\Delta(u) = \Delta(u\bar{i})$ (15.44)

xi. $A\Delta\left(\frac{\bar{u}}{n}\right) - \bar{A}\Delta\left(\frac{u}{n}\right) = -cH\Delta\left(\frac{1}{n}\right)$ (15.45)

xii. $A^2\Delta\left(\frac{1}{n^2}\right) = \Delta(i^2) = \Delta(u^2) - 2g\Delta(u) \qquad \bar{A}^2\Delta\left(\frac{1}{n^2}\right) = \Delta(\bar{i}^2) = \Delta(\bar{u}^2) - 2\bar{g}\Delta(\bar{u})$ (15.46)

The derivation of the relations given in (15.33) and (15.34) follows.

I.
$$\begin{aligned}\bar{S}_I &= -\bar{A}^2\bar{h}\Delta\left(\frac{\bar{u}}{n}\right) = -E\bar{A}^2h\Delta\left(\frac{\bar{A}}{n^2} - \frac{\bar{h}c}{n}\right) = -E\left\{\frac{\bar{A}^3}{A}hA\Delta\left(\frac{1}{n^2}\right) - \bar{A}^2h\bar{h}c\Delta\left(\frac{1}{n}\right)\right\}\\ &= -E\left\{\frac{\bar{A}^3}{A}h\Delta\left(\frac{u}{n}\right) - \frac{\bar{A}^3}{A}hg\Delta\left(\frac{1}{n}\right) - \bar{A}^2h\bar{h}c\Delta\left(\frac{1}{n}\right)\right\}\\ &= -E\left\{\frac{\bar{A}^3}{A}h\Delta\left(\frac{u}{n}\right) - \frac{\bar{A}^2}{A}\left[-hc\Delta\left(\frac{1}{n}\right)\left(\bar{A}h - A\bar{h}\right)\right]\right\}\\ &= -E\left\{\frac{\bar{A}^3}{A}h\Delta\left(\frac{u}{n}\right) - \frac{\bar{A}}{A}Hch\bar{A}\Delta\left(\frac{1}{n}\right)\right\}\\ &= -E\left\{\frac{\bar{A}^3}{A}h\Delta\left(\frac{u}{n}\right) - \frac{\bar{A}}{A}Hc\left[\bar{h}A\Delta\left(\frac{1}{n}\right) - H\Delta\left(\frac{1}{n}\right)\right]\right\}\end{aligned}$$

$$= E\left\{-\left[\frac{\bar{A}^3}{A}h\Delta\left(\frac{u}{n}\right)+\frac{\bar{A}}{A}H^2c\Delta\left(\frac{1}{n}\right)\right]+\bar{A}\bar{h}Hc\Delta\left(\frac{1}{n}\right)\right\}$$

$$= E\left[S_V + \bar{A}\bar{h}Hc\Delta\left(\frac{1}{n}\right)\right] \tag{15.47}$$

II. $\bar{S}_{II} = -\bar{A}A\bar{h}\Delta\left(\frac{\bar{u}}{n}\right) = -\bar{A}A\bar{h}\Delta\left(\frac{\bar{A}}{n^2}-\frac{\bar{h}c}{n}\right)$

$$= -\bar{A}^2\left(\bar{A}h+H\right)\Delta\left(\frac{1}{n^2}\right)+\bar{A}\left(\bar{A}h+H\right)\bar{h}c\Delta\left(\frac{1}{n}\right)$$

$$= -\bar{A}^3h\Delta\left(\frac{1}{n^2}\right)-\bar{A}^2H\Delta\left(\frac{1}{n^2}\right)+\bar{A}^2h\bar{h}c\Delta\left(\frac{1}{n}\right)+\bar{A}H\bar{h}c\Delta\left(\frac{1}{n}\right) \tag{15.48}$$

$$\text{First Term} = -\bar{A}^3h\Delta\left(\frac{1}{n^2}\right) = -\frac{\bar{A}^3}{A}h\left\{A\Delta\left(\frac{1}{n^2}\right)\right\} = -\frac{\bar{A}^3}{A}h\left\{\Delta\left(\frac{u}{n}\right)-g\Delta\left(\frac{1}{n}\right)\right\}$$

$$= -\frac{\bar{A}^3}{A}h\Delta\left(\frac{u}{n}\right)+\frac{\bar{A}^2}{A}\left(A\bar{h}-H\right)\left(-hc\right)\Delta\left(\frac{1}{n}\right)$$

$$= -\frac{\bar{A}^3}{A}h\Delta\left(\frac{u}{n}\right)-\bar{A}^2\bar{h}hc\Delta\left(\frac{1}{n}\right)+\frac{\bar{A}^2}{A}hHc\Delta\left(\frac{1}{n}\right) \tag{15.49}$$

$$\text{Second Term} = -\bar{A}^2H\Delta\left(\frac{1}{n^2}\right) = -H\left(\frac{n'^2\bar{i}'^2}{n'^2}-\frac{n^2\bar{i}^2}{n^2}\right) = -H\Delta\left(\bar{i}^2\right)$$

$$= -H\Delta\left(\bar{u}^2\right)+2\bar{g}H\Delta\left(\bar{u}\right) = -H\Delta\left(\bar{u}^2\right)+2\bar{g}H\bar{A}\Delta\left(\frac{1}{n}\right)$$

$$= -H\Delta\left(\bar{u}^2\right)-2\bar{A}\bar{h}Hc\Delta\left(\frac{1}{n}\right) \tag{15.50}$$

Substituting from (15.49) and (15.50) in (15.48), we get

$$\bar{S}_{II} = -\frac{\bar{A}^3}{A}h\Delta\left(\frac{u}{n}\right)-\bar{A}^2\bar{h}hc\Delta\left(\frac{1}{n}\right)+\frac{\bar{A}^2}{A}hHc\Delta\left(\frac{1}{n}\right)-H\Delta\left(\bar{u}^2\right)$$

$$-2\bar{A}\bar{h}Hc\Delta\left(\frac{1}{n}\right)+\bar{A}^2h\bar{h}c\Delta\left(\frac{1}{n}\right)+\bar{A}H\bar{h}c\Delta\left(\frac{1}{n}\right)$$

$$= -\frac{\bar{A}^3}{A}h\Delta\left(\frac{u}{n}\right)+\frac{\bar{A}^2}{A}hHc\Delta\left(\frac{1}{n}\right)-\bar{A}\bar{h}Hc\Delta\left(\frac{1}{n}\right)-H\Delta\left(\bar{u}^2\right)$$

$$= -\frac{\bar{A}^3}{A}h\Delta\left(\frac{u}{n}\right)-\frac{\bar{A}}{A}\left\{\left(\bar{A}h-A\bar{h}\right)Hc\Delta\left(\frac{1}{n}\right)\right\}-H\Delta\left(\bar{u}^2\right)$$

$$= -\left\{\frac{\bar{A}^3}{A}h\Delta\left(\frac{u}{n}\right)+\frac{\bar{A}}{A}H^2c\Delta\left(\frac{1}{n}\right)\right\}-H\Delta\left(\bar{u}^2\right) \tag{15.51}$$

Therefore, we get

$$\bar{S}_{II} = S_V - H\Delta\left(\bar{u}^2\right) \tag{15.52}$$

III. $\bar{S}_{III} = -A^2\bar{h}\Delta\left(\frac{\bar{u}}{n}\right)$ and $S_{III} = -\bar{A}^2 h\Delta\left(\frac{u}{n}\right)$. Subtracting S_{III} from $\bar{S}_{III}$ we get

$$\begin{aligned}
\bar{S}_{III} - S_{III} &= -A\bar{h}A\Delta\left(\frac{\bar{u}}{n}\right) + \bar{A}h\bar{A}\Delta\left(\frac{u}{n}\right) \\
&= -A\bar{h}\left[\Delta(u\bar{u}) - g\Delta(\bar{u})\right] + \bar{A}h\left[\Delta(u\bar{u}) - \bar{g}\Delta(u)\right] \\
&= -H\Delta(u\bar{u}) + A\bar{h}(-hc)\Delta(\bar{u}) - \bar{A}c(-\bar{h}c)\Delta(u) \\
&= -H\Delta(u\bar{u}) - A\bar{A}h\bar{h}c\Delta\left(\frac{1}{n}\right) + A\bar{A}h\bar{h}c\Delta\left(\frac{1}{n}\right) = -H\Delta(u\bar{u})
\end{aligned} \tag{15.53}$$

Therefore, it follows

$$\bar{S}_{III} = S_{III} - H\Delta(u\bar{u}) \tag{15.54}$$

IV.

$$\begin{aligned}
S_{II} &= -\bar{A}Ah\Delta\left(\frac{u}{n}\right) = -\bar{A}Ah\Delta\left(\frac{A}{n^2} - \frac{hc}{n}\right) = -(\bar{A}h)\left\{A^2\Delta\left(\frac{1}{n^2}\right) - Ahc\Delta\left(\frac{1}{n}\right)\right\} \\
&= -(A\bar{h} - H)\left\{A^2\Delta\left(\frac{1}{n^2}\right) - Ahc\Delta\left(\frac{1}{n}\right)\right\} \\
&= -A^3\bar{h}\Delta\left(\frac{1}{n^2}\right) + A^2H\Delta\left(\frac{1}{n^2}\right) + A^2h\bar{h}c\Delta\left(\frac{1}{n}\right) - AhHc\Delta\left(\frac{1}{n}\right)
\end{aligned} \tag{15.55}$$

$$\begin{aligned}
\text{First Term} &= -\frac{A^3}{\bar{A}}\bar{h}\bar{A}\Delta\left(\frac{1}{n^2}\right) = -\frac{A^3}{\bar{A}}\bar{h}\left\{\Delta\left(\frac{\bar{u}}{n}\right) - \bar{g}\Delta\left(\frac{1}{n}\right)\right\} = -\frac{A^3}{\bar{A}}\bar{h}\Delta\left(\frac{\bar{u}}{n}\right) + \frac{A^2}{\bar{A}}(A\bar{h})\bar{g}\Delta\left(\frac{1}{n}\right) \\
&= -\frac{A^3}{\bar{A}}\bar{h}\Delta\left(\frac{\bar{u}}{n}\right) + \frac{A^2}{\bar{A}}(\bar{A}h + H)(-\bar{h}c)\Delta\left(\frac{1}{n}\right) \\
&= -\frac{A^3}{\bar{A}}\bar{h}\Delta\left(\frac{\bar{u}}{n}\right) - A^2h\bar{h}c\Delta\left(\frac{1}{n}\right) - \frac{A^2}{\bar{A}}\bar{h}Hc\Delta\left(\frac{1}{n}\right)
\end{aligned} \tag{15.56}$$

$$\begin{aligned}
\text{Second Term} &= A^2H\Delta\left(\frac{1}{n^2}\right) = H\left(\frac{n'^2 i'^2}{n'^2} - \frac{n^2 i^2}{n^2}\right) = H\Delta(i^2) = H\Delta(u^2) - 2gH\Delta(u) \\
&= H\Delta(u^2) + 2hcH\Delta(u) = H\Delta(u^2) + 2AhHc\Delta\left(\frac{1}{n}\right)
\end{aligned} \tag{15.57}$$

Substituting from (15.56) and (15.57) in (15.55), we get

$$\begin{aligned}
S_{II} &= -\frac{A^3}{\bar{A}}\bar{h}\Delta\left(\frac{\bar{u}}{n}\right) - A^2h\bar{h}c\Delta\left(\frac{1}{n}\right) - \frac{A^2}{\bar{A}}\bar{h}Hc\Delta\left(\frac{1}{n}\right) + H\Delta(u^2) + 2AhHc\Delta\left(\frac{1}{n}\right) \\
&\quad + A^2h\bar{h}c\Delta\left(\frac{1}{n}\right) - AhHc\Delta\left(\frac{1}{n}\right) \\
&= -\frac{A^3}{\bar{A}}\bar{h}\Delta\left(\frac{\bar{u}}{n}\right) - \frac{A^2}{\bar{A}}\bar{h}Hc\Delta\left(\frac{1}{n}\right) + H\Delta(u^2) + AhHc\Delta\left(\frac{1}{n}\right) \\
&= -\frac{A^3}{\bar{A}}\bar{h}\Delta\left(\frac{\bar{u}}{n}\right) - \frac{A}{\bar{A}}\left\{(A\bar{h} - \bar{A}h)Hc\Delta\left(\frac{1}{n}\right)\right\} + H\Delta(u^2) \\
&= -\left[\frac{A^3}{\bar{A}}\bar{h}\Delta\left(\frac{\bar{u}}{n}\right) + \frac{A}{\bar{A}}H^2c\Delta\left(\frac{1}{n}\right)\right] + H\Delta(u^2) = \bar{S}_V + H\Delta(u^2)
\end{aligned} \tag{15.58}$$

Therefore, it follows

$$\bar{S}_V = S_{II} - H\Delta\left(u^2\right) \tag{15.59}$$

V. By definition, $C_I = Ah\Delta\left(\frac{\delta n}{n}\right)$ and $\bar{C}_I = \bar{A}\bar{h}\Delta\left(\frac{\delta n}{n}\right)$

Therefore,

$$\bar{C}_I = C_I + \left(\bar{A}\bar{h} - Ah\right)\Delta\left(\frac{\delta n}{n}\right) \tag{15.60}$$

VI. By definition, $C_{II} = \bar{A}h\Delta\left(\frac{\delta n}{n}\right)$ and $\bar{C}_{II} = A\bar{h}\Delta\left(\frac{\delta n}{n}\right)$

By subtraction, we get

$$\bar{C}_{II} = C_{II} + \left(A\bar{h} - \bar{A}h\right)\Delta\left(\frac{\delta n}{n}\right) = C_{II} + H\Delta\left(\frac{\delta n}{n}\right) \tag{15.61}$$

For an axi-symmetric system consisting of k spherical refracting interfaces, the Seidel spherical aberration of the pupil is

$$\bar{S}_I = \sum_{j=1}^{k}\left(\bar{S}_I\right)_j = -\sum_{j=1}^{k}\bar{A}_j^2\bar{h}_j\left\{\Delta\left(\frac{\bar{u}}{n}\right)\right\}_j = S_{VI} = \sum_{j=1}^{k}\left(S_{VI}\right)_j = \sum_{j=1}^{k}E_j\left[\left(S_V\right)_j + H\bar{A}_jh_jc_j\left\{\Delta\left(\frac{1}{n}\right)\right\}_j\right] \tag{15.62}$$

Similarly, for the axi-symmetric system, the Seidel axial colour of the pupil is

$$\bar{C}_I = \sum_{j=1}^{k}\left(\bar{C}_I\right)_j = \sum_{j=1}^{k}\bar{A}_jh_j\left\{\Delta\left(\frac{\delta n}{n}\right)\right\}_j = C_{III} = \sum_{j=1}^{k}\left(C_{III}\right)_j = C_I + \sum_{j=1}^{k}\left(\bar{A}_j\bar{h}_j - A_jh_j\right)\left\{\Delta\left(\frac{\delta n}{n}\right)\right\}_j \tag{15.63}$$

where $C_I = \sum_{j=1}^{k}\left(C_I\right)_j$.

The remaining Seidel pupil aberration coefficients are related to the Seidel aberration coefficients for object imaging in an axi-symmetric system consisting of k spherical refracting surfaces by the relations given below.

$$\left.\begin{aligned}
\bar{S}_{II} &= \sum_{j=1}^{k}\left(\bar{S}_{II}\right)_j = \sum_{j=1}^{k}\left(S_V\right)_j - H\sum_{j=1}^{k}\left[\Delta\left(\bar{u}^2\right)\right]_j = S_V - H\left(\bar{u}_k'^2 - \bar{u}_1^{\,2}\right)\\
\bar{S}_{III} &= \sum_{j=1}^{k}\left(\bar{S}_{III}\right)_j = \sum_{j=1}^{k}\left(S_{III}\right)_j - H\sum_{j=1}^{k}\left[\Delta\left(u\bar{u}\right)\right]_j = S_{III} - H\left(u_k'\bar{u}_k' - u_1\bar{u}_1\right)\\
\bar{S}_{IV} &= \sum_{j=1}^{k}\left(\bar{S}_{IV}\right)_j = \sum_{j=1}^{k}\left(S_{IV}\right)_j = S_{IV}\\
\bar{S}_V &= \sum_{j=1}^{k}\left(\bar{S}_V\right)_j = \sum_{j=1}^{k}\left(S_{II}\right)_j - H\sum_{j=1}^{k}\left[\Delta\left(u^2\right)\right]_j = S_{II} - H\left(u_k'^2 - u_1^{\,2}\right)\\
\bar{S}_{VI} &= \sum_{j=1}^{k}\left(\bar{S}_{VI}\right)_j = \sum_{j=1}^{k}\left(S_I\right)_j = S_I
\end{aligned}\right\} \tag{15.64}$$

$$\left.\begin{aligned}\bar{C}_{II} &= \sum_{j=1}^{k}\left(\bar{C}_{II}\right)_j = \sum_{j=1}^{k}\left(C_{II}\right)_j + H\sum_{j=1}^{k}\left\{\Delta\left(\frac{\delta n}{n}\right)\right\}_j = C_{II} + H\left\{\left[\left(\frac{\delta n'_k}{n'_k}\right) - \left(\frac{\delta n_1}{n_1}\right)\right]\right\} \\ \bar{C}_{III} &= \sum_{j=1}^{k}\left(\bar{C}_{III}\right)_j = \sum_{j=1}^{k}\left(C_I\right)_j = C_I\end{aligned}\right\} \tag{15.64}$$

Note that while summation for the difference terms is carried out over the k surfaces, all intermediate terms are cancelled out, and only the terms pertaining to the final image space and the initial object space remain.

15.2.2 Effect of Stop Shift on Seidel Spherical Aberration of the Pupil

From (15.31), we rewrite the coefficient of the Seidel spherical aberration of the pupil S_{VI} by a single spherical refracting interface as

$$S_{VI} = -\bar{A}^2\bar{h}\Delta\left(\frac{\bar{u}}{n}\right) \tag{15.65}$$

After an axial shift of the stop by δE, the three parameters $\bar{A}, \bar{h}$, and $\bar{u}$ are changed to $\left(\bar{A}+\delta\bar{A}\right), \left(\bar{h}+\delta\bar{h}\right)$, and $\left(\bar{u}+\delta\bar{u}\right)$, respectively. From (15.18), we note

$$\delta\bar{h} = h\delta E \text{ and } \delta\bar{A} = A\delta E \tag{15.66}$$

Since H remains unchanged for any stop shift, we have

$$H = n\left(u\bar{h} - \bar{u}h\right) \qquad \delta H = n\left(u\delta\bar{h} - h\delta\bar{u}\right) = 0 \tag{15.67}$$

so that

$$\delta\bar{u} = \frac{\delta\bar{h}}{h}u = u\delta E \tag{15.68}$$

The new value of Seidel spherical aberration coefficient of the pupil S^*_{VI} is

$$S^*_{VI} = -\left(\bar{A}+\delta\bar{A}\right)^2\left(\bar{h}+\delta\bar{h}\right)\Delta\left(\frac{\bar{u}+u\delta E}{n}\right) \tag{15.69}$$

Using (15.66) and (15.68), it can be expressed as

$$S^*_{VI} = -\left(\bar{A}+A\delta E\right)^2\left(\bar{h}+h\delta E\right)\Delta\left(\frac{\bar{u}}{n}\right) - \left(\bar{A}+A\delta E\right)^2\left(\bar{h}+h\delta E\right)\delta E\Delta\left(\frac{u}{n}\right) \tag{15.70}$$

The first term on the right-hand side can be expanded to obtain

$$\begin{aligned}\text{First Term} = &-\bar{A}^2\bar{h}\Delta\left(\frac{\bar{u}}{n}\right) + (\delta E)\left\{-\bar{A}^2h\Delta\left(\frac{\bar{u}}{n}\right) - 2\bar{A}A\bar{h}\Delta\left(\frac{\bar{u}}{n}\right)\right\} \\ &+(\delta E)^2\left\{-A^2\bar{h}\Delta\left(\frac{\bar{u}}{n}\right) - 2\bar{A}Ah\Delta\left(\frac{\bar{u}}{n}\right)\right\} - (\delta E)^3 A^2h\Delta\left(\frac{\bar{u}}{n}\right)\end{aligned} \tag{15.71}$$

Expanding the second term on the right-hand side, we get

$$\text{Second Term} = (\delta E)\left\{-\bar{A}^2\bar{h}\Delta\left(\frac{u}{n}\right)\right\} + (\delta E)^2\left\{-\bar{A}^2 h\Delta\left(\frac{u}{n}\right) - 2\bar{A}A\bar{h}\Delta\left(\frac{u}{n}\right)\right\}$$
$$+(\delta E)^3\left\{-A^2\bar{h}\Delta\left(\frac{u}{n}\right) - 2\bar{A}Ah\Delta\left(\frac{u}{n}\right)\right\} + (\delta E)^4\left\{-A^2 h\Delta\left(\frac{u}{n}\right)\right\} \quad (15.72)$$

Adding (15.71) and (15.72), and using expressions for S_I, S_{II}, S_{III}, S_{VI}, and $\bar{S}_{II}$, $\bar{S}_{III}$, S^*_{VI} can be expressed as

$$S^*_{VI} = S_{VI} + (\delta E)\left\{2\bar{S}_{II} - \bar{A}^2 h\Delta\left(\frac{\bar{u}}{n}\right) - \bar{A}^2\bar{h}\Delta\left(\frac{u}{n}\right)\right\}$$
$$+(\delta E)^2\left\{S_{III} + \bar{S}_{III} - 2\bar{A}Ah\Delta\left(\frac{\bar{u}}{n}\right) - 2\bar{A}A\bar{h}\Delta\left(\frac{u}{n}\right)\right\}$$
$$+(\delta E)^3\left\{2S_{II} - A^2\bar{h}\Delta\left(\frac{u}{n}\right) - A^2 h\Delta\left(\frac{\bar{u}}{n}\right)\right\} + (\delta E)^4 S_I \quad (15.73)$$

The three bracketed quantities on the right-hand side of the above equation can be expressed solely in terms of coefficients of Seidel image aberrations by using relations given in (15.33).

$$\left\{2\bar{S}_{II} - \bar{A}^2 h\Delta\left(\frac{\bar{u}}{n}\right) - \bar{A}^2\bar{h}\Delta\left(\frac{u}{n}\right)\right\}$$
$$= 2\bar{S}_{II} - (\bar{A}\bar{h})\bar{A}\Delta\left(\frac{u}{n}\right) - (A\bar{h} - H)\bar{A}\Delta\left(\frac{\bar{u}}{n}\right)$$
$$= 2\bar{S}_{II} - (\bar{A}\bar{h})\left[\Delta(u\bar{u}) - \bar{g}\Delta(u)\right] - \bar{A}A\bar{h}\Delta\left(\frac{\bar{u}}{n}\right) + \bar{A}H\Delta\left(\frac{\bar{u}}{n}\right)$$
$$= 2\bar{S}_{II} - (\bar{A}\bar{h})\Delta(\bar{u}i) - (\bar{A}\bar{h})\left[g\Delta(\bar{u}) - \bar{g}\Delta(u)\right] + S_{II} + H\bar{A}\Delta\left(\frac{\bar{u}}{n}\right)$$
$$= 3\bar{S}_{II} - \bar{A}A\bar{h}\Delta\left(\frac{\bar{u}}{n}\right) - \bar{A}\bar{h}c\left[\bar{h}A\Delta\left(\frac{1}{n}\right) - h\bar{A}\Delta\left(\frac{1}{n}\right)\right] + H\left[\Delta(\bar{u}^2) - \bar{g}\Delta(\bar{u})\right]$$
$$= 4\bar{S}_{II} - H(\bar{h}c)\left[\bar{A}\Delta\left(\frac{1}{n}\right)\right] + H\Delta(\bar{u}^2) - H\bar{g}\Delta(\bar{u})$$
$$= 4\bar{S}_{II} + H\bar{g}\Delta(\bar{u}) + H\Delta(\bar{u}^2) - H\bar{g}\Delta(\bar{u})$$
$$= 4\bar{S}_{II} + H\Delta(\bar{u}^2)$$
$$= 4\left[S_V - H\Delta(\bar{u}^2)\right] + H\Delta(\bar{u}^2)$$
$$= 4S_V - 3H\Delta(\bar{u}^2) \quad (15.74)$$

$$\left\{S_{III} + \bar{S}_{III} - 2\bar{A}Ah\Delta\left(\frac{\bar{u}}{n}\right) - 2\bar{A}A\bar{h}\Delta\left(\frac{u}{n}\right)\right\}$$
$$= S_{III} + \bar{S}_{III} - 2A(A\bar{h} - H)\Delta\left(\frac{\bar{u}}{n}\right) - 2\bar{A}(\bar{A}h + H)\Delta\left(\frac{u}{n}\right)$$
$$= S_{III} + \bar{S}_{III} - 2A^2\bar{h}\Delta\left(\frac{\bar{u}}{n}\right) + 2HA\Delta\left(\frac{\bar{u}}{n}\right) - 2\bar{A}^2 h\Delta\left(\frac{u}{n}\right) - 2H\bar{A}\Delta\left(\frac{u}{n}\right)$$
$$= 3(S_{III} + \bar{S}_{III}) + 2H\left[A\Delta\left(\frac{\bar{u}}{n}\right) - \bar{A}\Delta\left(\frac{u}{n}\right)\right]$$

$$
\begin{aligned}
&= 3\left[S_{III} + S_{III} - H\Delta\left(u\bar{u}\right)\right] - 2H^2 c\Delta\left(\frac{1}{n}\right)\\
&= 6S_{III} + 2S_{IV} - 3H\Delta\left(u\bar{u}\right)
\end{aligned}
\tag{15.75}
$$

$$
\begin{aligned}
&\left\{2S_{II} - A^2 h\Delta\left(\frac{\bar{u}}{n}\right) - A^2\bar{h}\Delta\left(\frac{u}{n}\right)\right\}\\
&= 2S_{II} - Ah\left[A\Delta\left(\frac{\bar{u}}{n}\right)\right] - A\left(\bar{A}h + H\right)\Delta\left(\frac{u}{n}\right)\\
&= 2S_{II} - Ah\left[\bar{g}\Delta(u) + \Delta\left(u\bar{i}\right) - g\Delta\left(\bar{i}\right)\right] - A\bar{A}h\Delta\left(\frac{u}{n}\right) - H\left[A\Delta\left(\frac{u}{n}\right)\right]\\
&= 2S_{II} - Ah\left[\bar{g}\Delta(u) - g\Delta(\bar{u})\right] - Ah\left[\bar{A}\Delta\left(\frac{u}{n}\right)\right] - A\bar{A}h\Delta\left(\frac{u}{n}\right) - H\left[\Delta\left(u^2\right) - g\Delta(u)\right]\\
&= 4S_{II} - H\Delta\left(u^2\right) - A\bar{h}g\Delta(u) + Ahg\bar{A}\Delta\left(\frac{1}{n}\right) + Hg\Delta(u)\\
&= 4S_{II} - H\Delta\left(u^2\right) - A\bar{h}g\Delta(u) + \bar{A}hg\Delta(u) + Hg\Delta(u)\\
&= 4S_{II} - H\Delta\left(u^2\right) - Hg\Delta(u) + Hg\Delta(u)\\
&= 4S_{II} - H\Delta\left(u^2\right)
\end{aligned}
\tag{15.76}
$$

By substitution from (15.74) – (15.76) in (15.73), S_{VI}^*, the coefficient of Seidel spherical aberration of the pupil after the stop shift, is given in terms of the original coefficients of the Seidel aberrations by

$$
\begin{aligned}
S_{VI}^* = {} & S_{VI} + (\delta E)\{4S_V\} + (\delta E)^2\{6S_{III} + 2S_{IV}\} + (\delta E)^3\{4S_{II}\} + (\delta E)^4 S_I\\
& - H\left\{3(\delta E)\Delta\left(\bar{u}^2\right) + 3(\delta E)^2\Delta\left(u\bar{u}\right) + (\delta E)^3\Delta\left(u^2\right)\right\}
\end{aligned}
\tag{15.77}
$$

Note that the value of δE is same for all the refracting surfaces. Therefore, for an axi-symmetric imaging system consisting of k spherical refracting interfaces, the new value S_{VI}^* of the coefficient of the Seidel spherical aberration of the pupil of the system after the stop shift by δE is given by

$$
\begin{aligned}
S_{VI}^* &= \sum_{j=1}^{k}\left(S_{VI}^*\right)_j\\
&= \sum_{j=1}^{k}\left(S_{VI}\right)_j + (\delta E)\left\{4\sum_{j=1}^{k}\left(S_V\right)_j\right\} + (\delta E)^2\left\{6\sum_{j=1}^{k}\left(S_{III}\right)_j + 2\sum_{j=1}^{k}\left(S_{IV}\right)_j\right\} + (\delta E)^3\left\{4\sum_{j=1}^{r}\left(S_{II}\right)_j\right\}\\
&\quad + (\delta E)^4\left\{\sum_{j=1}^{k}\left(S_I\right)_j\right\} - H\left[3(\delta E)\left\{\bar{u}_k'^2 - \bar{u}_1^2\right\} + 3(\delta E)^2\left\{u_k'\bar{u}_k' - u_1\bar{u}_1\right\} + (\delta E)^3\left\{u_k'^2 - u_1^2\right\}\right]\\
&= S_{VI} + 4(\delta E)S_V + (\delta E)^2\left(6S_{III} + 2S_{IV}\right) + 4(\delta E)^3 S_{II} + (\delta E)^4 S_I\\
&\quad - H\left[3(\delta E)\left\{\bar{u}_k'^2 - \bar{u}_1^2\right\} + 3(\delta E)^2\left\{u_k'\bar{u}_k' - u_1\bar{u}_1\right\} + (\delta E)^3\left\{\bar{u}_k'^2 - u_1^2\right\}\right]
\end{aligned}
\tag{15.78}
$$

where S_I, S_{II}, S_{III}, S_{IV}, S_V, S_{VI} are the original Seidel aberration coefficients of the system before the stop shift. Similarly, u_1, u_k', and $\bar{u}_1, \bar{u}_k'$ are parameters of the PMR and PPR, respectively, before stop shift. Note that in the case of a system consisting of multiple refracting surfaces, only the quantities pertaining to the image space and the object space remain in the difference terms, since the quantities for the intermediate spaces cancel out.

15.2.3 Effect of Stop Shift on Seidel Longitudinal Chromatic Aberration of the Pupil

For object imagery, in equation (15.19) the coefficients of Seidel longitudinal chromatic aberration C_I^* and the Seidel transverse chromatic aberration C_{II}^* after the stop shift are expressed in terms of the original Seidel Chromatic aberration coefficients C_I and C_{II} before stop shift. For pupil imagery, the coefficient of the Seidel longitudinal chromatic aberration of the pupil after stop shift, C_{III}^*, can be expressed in terms of original Seidel chromatic aberration coefficients, C_I, C_{II} and C_{III}, before stop shift. The corresponding expression is derived below.

In an axi-symmetric imaging system, the coefficient of the Seidel longitudinal chromatic aberration of the pupil for a single refracting interface is given by

$$C_{III} = \bar{A}\bar{h}\Delta\left(\frac{\delta n}{n}\right) \tag{15.79}$$

After the stop shift by δE, the new values for $\bar{A}$ and $\bar{h}$ are $\left(\bar{A}+\delta\bar{A}\right)$ and $\left(\bar{h}+\delta\bar{h}\right)$, respectively, where $\delta\bar{A} = A\delta E$, and $\delta\bar{h} = h\delta E$. Therefore,

$$\begin{aligned}
C_{III}^* &= \left(\bar{A}+\delta\bar{A}\right)\left(\bar{h}+\delta\bar{h}\right)\Delta\left(\frac{\delta n}{n}\right) \\
&= \bar{A}\bar{h}\Delta\left(\frac{\delta n}{n}\right) + \bar{A}\left(\delta\bar{h}\right)\Delta\left(\frac{\delta n}{n}\right) + \left(\delta\bar{A}\right)\bar{h}\Delta\left(\frac{\delta n}{n}\right) + \left(\delta\bar{A}\right)\left(\delta\bar{h}\right)\Delta\left(\frac{\delta n}{n}\right) \\
&= C_{III} + (\delta E)C_{II} + (\delta E)A\bar{h}\Delta\left(\frac{\delta n}{n}\right) + (A\delta E)(h\delta E)\Delta\left(\frac{\delta n}{n}\right) \\
&= C_{III} + (\delta E)C_{II} + (\delta E)\bar{A}h\Delta\left(\frac{\delta n}{n}\right) + (\delta E)H\Delta\left(\frac{\delta n}{n}\right) + \left((\delta E)\right)^2 Ah\Delta\left(\frac{\delta n}{n}\right) \\
&= C_{III} + (\delta E)\{2C_{II}\} + (\delta E)^2 C_I + H\left\{(\delta E)\Delta\left(\frac{\delta n}{n}\right)\right\}
\end{aligned} \tag{15.80}$$

For an axi-symmetric imaging system consisting of k refracting interfaces, the coefficient of the Seidel longitudinal chromatic aberration of the pupil for the whole system after stop shift, C_{III}^*, is expressed in terms of original Seidel chromatic aberration coefficients, C_I, C_{II}, and C_{III}, before stop shift by

$$\begin{aligned}
C_{III}^* &= \sum_{j=1}^{k}\left(C_{III}^*\right)_j \\
&= \sum_{j=1}^{k}\left(C_{III}\right)_j + (\delta E)\left\{2\sum_{j=1}^{k}\left(C_{II}\right)_j\right\} + (\delta E)^2\left\{\sum_{j=1}^{k}\left(C_I\right)_j\right\} + H\left\{(\delta E)\Delta\left(\frac{\delta n}{n}\right)\right\} \\
&= C_{III} + 2(\delta E)C_{II} + (\delta E)^2 C_I + H(\delta E)\left\{\frac{\delta n_k'}{n_k'} - \frac{\delta n_1}{n_1}\right\}
\end{aligned} \tag{15.81}$$

Note that the relative dispersion characteristics of only the object space and the image space appear in the expression, since the same for all intermediate spaces cancels out.

15.2.4 A Few Well-Known Effects of Pupil Aberrations on Imaging of Objects [9–15]

a) Kidney bean effect: The presence of a large spherical aberration of the pupil in any one or more components operating in cascade in a multicomponent optical system that inhibits proper pupil matching and gives rise to undesirable nonuniform illumination on the image.
b) Loss of telecentricity in telecentric systems [10]: The presence of a large spherical aberration of the pupil that implies axial separation of effective pupils for different field points. Consequently, the

system loses telecentricity for all points except for the field point whose effective pupil matches the aperture stop.

c) Vignetting: The presence of a large spherical aberration of the pupil that causes vignetting for cones of light rays coming from extra-axial points, particularly for points lying at outer regions of the field.
d) Pupil walking [11]: The location of effective real entrance pupil changes both laterally and longitudinally in extra wide angle lenses, e.g. fish-eye lenses, in the presence of large pupil aberrations.
e) Slyusarev effect [12–13]: In wide angle or extra wide angle lenses, the relative illumination at large field points drops significantly following the cosine fourth formula, as mentioned earlier in Section 6.7, so much so that outer field points are imaged with practically negligible illumination. Introducing pupil coma to the system gives rise to larger exit pupils at larger values of field points, and the relative illumination at outer field points increases substantially to make them useful. The penalty is the concomitant large distortion of the image due to pupil coma.
f) Degradation of correction in adaptive optics systems [15]: Adaptive optics systems typically include an optical relay system that images the object of interest as well as the real pupil. The deformable mirrors in adaptive optics systems are often placed in a plane conjugate to the entrance pupil of the system formed by one or more relay systems to correct the wavefront error.
   In the presence of severe pupil aberrations, the image of the pupil formed in this conjugate suffers from displacement and distortion errors that vary with field angles. The deformable mirror located at a single position cannot take care of these errors, so much so that the AO system suffers from a type of anisoplanatism arising out of degradation in adaptive optics correction with variation of field.

15.3 Conjugate Shift

Conjugate shift implies an axial shift of the object plane with the concomitant axial shift of the image plane. On the basis of this shift, all optical imaging systems may be classified in three categories. In the first category falls the optical systems that need to perform quality imaging for a fixed conjugate location; in these systems, the axial location of the object plane, the image plane, and the imaging lens is fixed. Next, the second category consists of the optical imaging systems that, in addition to satisfying the objective of producing quality imaging for a specific conjugate location, as for the systems of the first category, are also required to produce images of tolerable quality over a specified neighbourhood of that conjugate location. Often, the axial location and the size of the stop remain the same when the conjugate location is varied in such systems. The third category includes all optical imaging systems that are decidedly required to be used for a wide conjugate range. The first two categories of systems are usually fixed focus systems, and the third category includes variable focus systems, e.g. zoom lenses. Conjugate shift effects on different aberrations are particularly required for the analysis and synthesis of optical imaging systems of the third category. A comprehensive treatment of this problem was undertaken by Smith [16]. His approach was based on eikonal theory. Pegis [17] demonstrated how Smith's approach leads to explicit stop shift and conjugate shift expressions for all orders of aberrations. Nevertheless, the inherent unwieldy nature of the treatment limited its use in practice, and it remains to be translated in a form that will be amenable for exploration in practice. However, simplified expressions for effects of stop shift and conjugate shift on primary aberrations were obtained and extensively used in practice [8, 18–20].

Figure 15.7 shows an axially symmetric spherical refracting interface at A on the optical axis. The original object $OQ(=\eta)$ is on a transverse plane at O on the axis. The stop is at E on the axis. OBP is the original paraxial marginal ray (PMR) from the axial point O. $QE\bar{P}$ is the paraxial pupil ray (PPR) from the extra-axial point Q. The convergence angles of the PMR and the PPR are u, and $\bar{u}$ respectively. The axial position of the object is changed from O to O^*. The axial location and transverse size of the stop

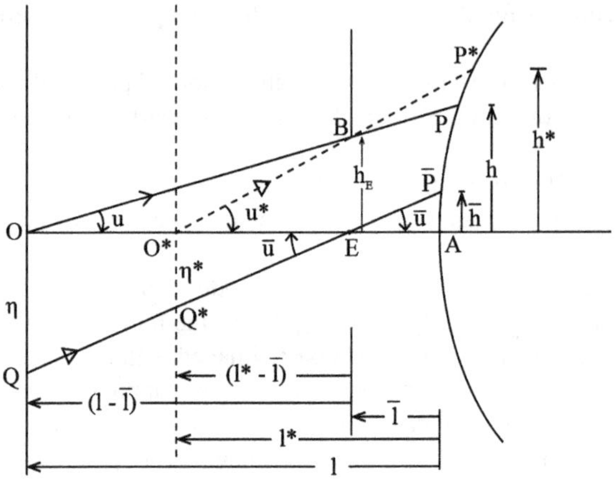

FIGURE 15.7 Axial shift of the object position from O to O*. Original PMR: OBP; new PMR: O*BP*; The pupil radius BE and the PPR QQ*E$\bar{P}$ are unchanged. Object height OQ is changed to O*Q* for keeping H the same.

is unaltered. The new paraxial marginal ray is O*BP*. The semi-diameter of the aperture stop is h_E. Let AO = l, AO* = l*, and AE = $\bar{l}$. It follows

$$EO = (l - \bar{l}), \qquad EO^* = (l^* - \bar{l}) \tag{15.82}$$

It is assumed that the paraxial optical invariant H is unchanged by the shift of the object plane. Let the new image height be η^*. From Figure 15.7, we get

$$\begin{aligned} H = nu\eta &= n\frac{h_E}{-(l-\bar{l})}\left[\bar{u}(l-\bar{l})\right] = -n\bar{u}h_E = -n\frac{O^*Q^*}{(l^*-\bar{l})}h_E \\ &= n\frac{h_E}{-(l^*-\bar{l})}\left(O^*Q^*\right) = nu^*\left(O^*Q^*\right) = nu^*\eta^* \end{aligned} \tag{15.83}$$

Therefore, $\eta^* = O^*Q^*$. This shows that the new position of the extra-axial image point is at Q*, where the original pupil ray cuts the new object plane. The incidence heights of the PPR, the original PMR, and the new PMR are $\bar{h}$, h, and h^* respectively. The amount by which the object is axially shifted may be represented by

$$\delta\bar{E} = \frac{\delta h}{\bar{h}} = \frac{h^* - h}{\bar{h}} \tag{15.84}$$

The changes in data of the PMR are

$$\delta h = \bar{h}\delta\bar{E} \tag{15.85}$$

H is represented by

$$H = A\bar{h} - \bar{A}h = n\left(u\bar{h} - \bar{u}h\right) \tag{15.86}$$

On differentiation, we get

$$\delta H = (\delta A)\bar{h} + A\left(\delta\bar{h}\right) - \left(\delta\bar{A}\right)h - \bar{A}(\delta h) = n\left[(\delta u)\bar{h} + u\left(\delta\bar{h}\right) - (\delta\bar{u})h - \bar{u}(\delta h)\right] \tag{15.87}$$

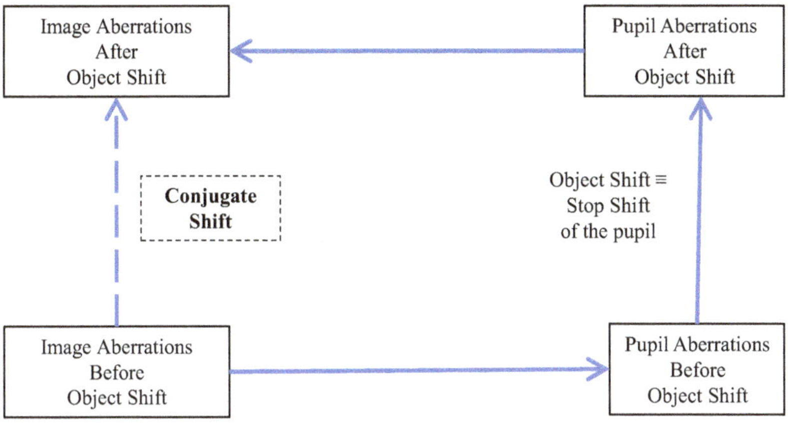

FIGURE 15.8 A flowchart for determining the effects of conjugate shift on Seidel aberrations in object imagery.

Since H and the PPR remain unchanged,

$$\delta H = \delta \bar{h} = \delta \bar{A} = 0 = \delta \bar{u} = \delta \bar{h} \tag{15.88}$$

Substituting from (15.88) in (15.87), we get

$$(\delta A)\bar{h} = \bar{A}(\delta h) \qquad (\delta u)\bar{h} = \bar{u}\delta h \tag{15.89}$$

which leads to the relations

$$\delta A = \bar{A}\frac{\delta h}{\bar{h}} = \bar{A}\delta\bar{E} \qquad \delta u = \bar{u}\frac{\delta h}{\bar{h}} = \bar{u}\delta\bar{E} \tag{15.90}$$

In Section 15.2, it was noted that, in the same imaging system, the case of pupil imagery involves a role reversal between object/image and entrance pupil/exit pupil (Figure 15.6). A comparison of Figure 15.4 and Figure 15.7 shows that an object shift is exactly equivalent to a stop shift in the imaging of the pupil.

The effects of conjugate shift on the coefficients of Seidel aberrations in object imagery are determined in a roundabout manner. First, the coefficients of Seidel pupil aberrations before object shift are expressed in terms of Seidel image aberrations before object shift (section 15.3.2). Next, the coefficients of Seidel pupil aberrations after object shift are expressed in terms of Seidel pupil aberrations before object shift, (i.e. stop shift in pupil imagery) (section 15.3.1). Then the Seidel image aberrations after conjugate shift are expressed in terms of Seidel pupil aberrations after object shift (section 15.3.3). Finally, the three expressions are combined sequentially to obtain composite expressions for the effects of conjugate shift on the coefficients of Seidel image aberrations (section 15.3.4). A flowchart is given in Figure 15.8 underscoring the operations for determining the effects of conjugate shift on Seidel aberrations in object imagery.

15.3.1 The Coefficients of Seidel Pupil Aberrations After Object Shift in Terms of the Coefficients of Seidel Pupil Aberrations Before Object Shift

The change in eccentricity parameter occurring in pupil imagery by an object shift is $\delta\bar{E}$, given by (15.84). The new values of the coefficients of Seidel pupil aberrations after the object shift (equivalently, stop shift for pupil imagery) are obtained from barred forms of equations (15.19), (15.77), and (15.80), and by interchanging $\{(\delta E), H, u, u'\}$ with $\{(\delta\bar{E}), \bar{H}, \bar{u}, \bar{u}'\}$. They are given by

$$\left.\begin{aligned}
\bar{S}_I^* &= \bar{S}_I \\
\bar{S}_{II}^* &= \bar{S}_{II} + (\delta\bar{E})\bar{S}_I \\
\bar{S}_{III}^* &= \bar{S}_{III} + 2(\delta\bar{E})\bar{S}_{II} + (\delta\bar{E})^2\bar{S}_I \\
\bar{S}_{IV}^* &= \bar{S}_{IV} \\
\bar{S}_V^* &= \bar{S}_V + (\delta\bar{E})\bar{S}_{IV} + 3(\delta\bar{E})\bar{S}_{III} + 3(\delta\bar{E})^2\bar{S}_{II} + (\delta\bar{E})^3\bar{S}_I \\
\bar{S}_{VI}^* &= \bar{S}_{VI} + 4(\delta\bar{E})\bar{S}_V + 2(\delta\bar{E})^2\bar{S}_{IV} + 6(\delta\bar{E})^2\bar{S}_{III} + 4(\delta\bar{E})^3\bar{S}_{II} + (\delta\bar{E})^4\bar{S}_I \\
&\quad -\bar{H}\left\{(\delta\bar{E})^3\Delta(\bar{u}^2) + 3(\delta\bar{E})^2\Delta(\bar{u}u) + 3(\delta\bar{E})\Delta(u^2)\right\} \\
\bar{C}_I^* &= \bar{C}_I \\
\bar{C}_{II}^* &= \bar{C}_{II} + (\delta\bar{E})\bar{C}_I \\
\bar{C}_{III}^* &= \bar{C}_{III} + 2(\delta\bar{E})\bar{C}_{II} + (\delta\bar{E})^2\bar{C}_I + \bar{H}\left[(\delta\bar{E})\left(\frac{\delta n}{n}\right)\right]
\end{aligned}\right\} \tag{15.91}$$

15.3.2 The Coefficients of Seidel Pupil Aberrations Before Object Shift in Terms of Coefficients of Seidel Image Aberrations Before Object Shift

The coefficients of Seidel pupil aberrations in terms of the coefficients of Seidel image aberrations before object shift are obtained from equations (15.33) and (15.34) as

$$\left.\begin{aligned}
\bar{S}_I &= S_{VI} \\
\bar{S}_{II} &= S_V - H\Delta(\bar{u}^2) \\
\bar{S}_{III} &= S_{III} - H\Delta(u\bar{u}) \\
\bar{S}_{IV} &= S_{IV} \\
\bar{S}_V &= S_{II} - H\Delta(u^2) \\
\bar{S}_{VI} &= S_I \\
\bar{C}_I &= C_{III} \\
\bar{C}_{II} &= C_{II} + H\Delta\left(\frac{\delta n}{n}\right) \\
\bar{C}_{III} &= C_I
\end{aligned}\right\} \tag{15.92}$$

Note that in the above equations, u, u', and $\bar{u}, \bar{u}'$ correspond with the convergence angles of the original PMR and PPR, respectively.

15.3.3 The Coefficients of Seidel Image Aberrations After Object Shift in Terms of Coefficients of Seidel Pupil Aberrations After Object Shift

After conjugate shift, the convergence angle of the new PMR is u^*. Using (15.90), u^* is given by

$$u^* = u + \delta u = u + \bar{u}(\delta\bar{E}) \tag{15.93}$$

Similarly, u'^*, the convergence angle of the new PMR after a refraction, is given by

$$u'^* = u' + \delta u' = u' + \bar{u}'(\delta\bar{E}) \tag{15.94}$$

Since the PPR remains unchanged for a conjugate shift, the convergence angles are unchanged, so that

$$\bar{u}^* = \bar{u} \qquad \bar{u}'^* = \bar{u}' \tag{15.95}$$

After the object shift, the coefficients of the Seidel image aberrations can be expressed in terms of the coefficients of Seidel pupil aberrations by using (15.33) – (15.34). The relations are given below.

$$\left.\begin{aligned}
S_I^* &= \bar{S}_{VI}^* \\
S_{II}^* &= \bar{S}_V^* + H\Delta\left(u^{*2}\right) = \bar{S}_V^* + H\Delta\left[\left(u + \bar{u}\left(\delta\bar{E}\right)\right)^2\right] \\
&= \bar{S}_V^* + H\left[\Delta\left(u^2\right) + 2\left(\delta\bar{E}\right)\Delta\left(u\bar{u}\right) + \left(\delta\bar{E}\right)^2\Delta\left(\bar{u}^2\right)\right] \\
S_{III}^* &= \bar{S}_{III}^* + H\Delta\left(u^*\bar{u}^*\right) = \bar{S}_{III}^* + H\Delta\left[\left(u + \bar{u}\left(\delta\bar{E}\right)\right)\bar{u}\right] = \bar{S}_{III}^* + H\left[\Delta\left(u\bar{u}\right) + \left(\delta\bar{E}\right)\Delta\left(\bar{u}^2\right)\right] \\
S_{IV}^* &= \bar{S}_{IV}^* \\
S_V^* &= \bar{S}_{II}^* + H\Delta\left(\bar{u}^{*2}\right) = \bar{S}_{II}^* + H\Delta\left(\bar{u}^2\right) \\
S_{VI}^* &= \bar{S}_I^* \\
C_I^* &= \bar{C}_{III}^* \\
C_{II}^* &= \bar{C}_{II}^* - H\Delta\left(\frac{\delta n}{n}\right) \\
C_{III}^* &= \bar{C}_I^*
\end{aligned}\right\} \tag{15.96}$$

15.3.4 Effects of Conjugate Shift on the Coefficients of Seidel Image Aberrations

Equation (15.96) expresses the coefficients of Seidel image aberrations after conjugate shift in terms of the coefficients of Seidel pupil aberrations after conjugate shift. The relations between the coefficients of Seidel pupil aberrations after conjugate shift with the corresponding coefficients before conjugate shift are given in equation (15.91). Using equation (15.91) the coefficients of Seidel image aberrations after conjugate shift can be expressed in terms of the coefficients of Seidel pupil aberrations before conjugate shift as

$$\begin{aligned}
S_I^* &= \bar{S}_{VI}^* \\
&= \bar{S}_{VI} + 4\left(\delta\bar{E}\right)\bar{S}_V + 2\left(\delta\bar{E}\right)^2\bar{S}_{IV} + 6\left(\delta\bar{E}\right)^2\bar{S}_{III} + 4\left(\delta\bar{E}\right)^3\bar{S}_{II} + \left(\delta\bar{E}\right)^4\bar{S}_I \\
&\quad -\bar{H}\left\{\left(\delta\bar{E}\right)^3\Delta\left(\bar{u}^2\right) + 3\left(\delta\bar{E}\right)^2\Delta\left(\bar{u}u\right) + 3\left(\delta\bar{E}\right)\Delta\left(u^2\right)\right\}
\end{aligned} \tag{15.97}$$

$$\begin{aligned}
S_{II}^* &= \bar{S}_V^* + H\left[\Delta\left(u^2\right) + 2\left(\delta\bar{E}\right)\Delta\left(u\bar{u}\right) + \left(\delta\bar{E}\right)^2\Delta\left(\bar{u}^2\right)\right] \\
&= \bar{S}_V + \left(\delta\bar{E}\right)\bar{S}_{IV} + 3\left(\delta\bar{E}\right)\bar{S}_{III} + 3\left(\delta\bar{E}\right)^2\bar{S}_{II} + \left(\delta\bar{E}\right)^3\bar{S}_I \\
&\quad + H\left[\Delta\left(u^2\right) + 2\left(\delta\bar{E}\right)\Delta\left(u\bar{u}\right) + \left(\delta\bar{E}\right)^2\Delta\left(\bar{u}^2\right)\right]
\end{aligned} \tag{15.98}$$

$$\begin{aligned}
S_{III}^* &= \bar{S}_{III}^* + H\Delta\left(u\bar{u}\right) + H\left(\delta\bar{E}\right)\Delta\left(\bar{u}^2\right) \\
&= \bar{S}_{III} + 2\left(\delta\bar{E}\right)\bar{S}_{II} + \left(\delta\bar{E}\right)^2\bar{S}_I + H\Delta\left(u\bar{u}\right) + H\left(\delta\bar{E}\right)\Delta\left(\bar{u}^2\right)
\end{aligned} \tag{15.99}$$

$$S_{IV}^* = S_{IV} \tag{15.100}$$

$$S_V^* = \bar{S}_{II}^* + H\Delta\left(\bar{u}^2\right) = \bar{S}_{II} + \left(\delta\bar{E}\right)\bar{S}_I + H\Delta\left(\bar{u}^2\right) \tag{15.101}$$

$$S_{VI}^* = \bar{S}_I^* = \bar{S}_I \tag{15.102}$$

$$C_I^* = \bar{C}_{III}^* = \bar{C}_{III} + 2(\delta\bar{E})\bar{C}_{II} + (\delta\bar{E})^2\bar{C}_I + \bar{H}\left[(\delta\bar{E})\Delta\left(\frac{\delta n}{n}\right)\right] \tag{15.103}$$

$$C_{II}^* = \bar{C}_{II}^* - H\Delta\left(\frac{\delta n}{n}\right) = \bar{C}_{II} + (\delta\bar{E})\bar{C}_I - H\Delta\left(\frac{\delta n}{n}\right) \tag{15.104}$$

$$C_{III}^* = \bar{C}_I^* = \bar{C}_I \tag{15.105}$$

The relations between the coefficients of Seidel pupil aberrations before object shift, with the coefficients of Seidel image aberrations before object shift are given in (15.92). Using (15.92) in equations (15.97) – (15.105) and noting that $\bar{H} = -H$, we get the conjugate shift equations as given below.

$$\left.\begin{aligned}
S_I^* &= S_I + (\delta\bar{E})\left[4S_{II} - H\Delta(u^2)\right] + (\delta\bar{E})^2\left[6S_{III} + 2S_{IV} - 3H\Delta(u\bar{u})\right] \\
&\quad + (\delta\bar{E})^3\left[4S_V - 3H\Delta(\bar{u}^2)\right] + (\delta\bar{E})^4 S_{VI} \\
S_{II}^* &= S_{II} + (\delta\bar{E})\left[3S_{III} + S_{IV} - H\Delta(u\bar{u})\right] + (\delta\bar{E})^2\left[3S_V - 2H\Delta(\bar{u}^2)\right] + (\delta\bar{E})^3 S_{VI} \\
S_{III}^* &= S_{III} + (\delta\bar{E})\left[2S_V - H\Delta(\bar{u}^2)\right] + (\delta\bar{E})^2 S_{VI} \\
S_{IV}^* &= S_{IV} \\
S_V^* &= S_V + (\delta\bar{E})S_{VI} \\
S_{VI}^* &= S_{VI} \\
C_I^* &= C_I + (\delta\bar{E})[2C_{II}] + (\delta\bar{E})^2 C_{III} - H(\delta\bar{E})\Delta\left(\frac{\delta n}{n}\right) \\
C_{II}^* &= C_{II} + (\delta\bar{E})C_{III} \\
C_{III}^* &= C_{III}
\end{aligned}\right\} \tag{15.106}$$

For an axi-symmetric system consisting of k refracting interfaces, the conjugate shift equations become

$$\left.\begin{aligned}
S_I^* &= S_I + (\delta\bar{E})[4S_{II}] + (\delta\bar{E})^2[6S_{III} + 2S_{IV}] + (\delta\bar{E})^3[4S_V] + (\delta\bar{E})^4 S_{VI} \\
&\quad - H\left[(\delta\bar{E})\{u_k'^2 - u_1^2\} + 3(\delta\bar{E})^2\{u_k'\bar{u}_k' - u_1\bar{u}_1\} + 3(\delta\bar{E})^3\{\bar{u}_k'^2 - \bar{u}_1^2\}\right] \\
S_{II}^* &= S_{II} + (\delta\bar{E})[3S_{III} + S_{IV}] + 3(\delta\bar{E})^2 S_V + (\delta\bar{E})^3 S_{VI} \\
&\quad - H\left[(\delta\bar{E})\{u_k'\bar{u}_k' - u_1\bar{u}_1\} + 2(\delta\bar{E})^2\{\bar{u}_k'^2 - \bar{u}_1^2\}\right] \\
S_{III}^* &= S_{III} + 2(\delta\bar{E})S_V + (\delta\bar{E})^2 S_{VI} - H\left[(\delta\bar{E})\{\bar{u}_k'^2 - \bar{u}_1^2\}\right] \\
S_{IV}^* &= S_{IV} \\
S_V^* &= S_V + (\delta\bar{E})S_{VI} \\
S_{VI}^* &= S_{VI} \\
C_I^* &= C_I + (\delta\bar{E})[2C_{II}] + (\delta\bar{E})^2 C_{III} - H(\delta\bar{E})\left\{\left(\frac{\delta n_k'}{n_k'}\right) - \left(\frac{\delta n_1}{n_1}\right)\right\} \\
C_{II}^* &= C_{II} + (\delta\bar{E})C_{III} \\
C_{III}^* &= C_{III}
\end{aligned}\right\} \tag{15.107}$$

where S_I, S_{II}, S_{III}, S_{IV}, S_V, S_{VI}, C_I, C_{II}, C_{III} are the coefficients of Seidel monochromatic and chromatic aberrations of the axi-symmetric imaging system before the conjugate shift, and each of them is a summation of the corresponding values for the constituent refracting interfaces. The starred quantities are the new values of the coefficients of the Seidel aberrations for the system after the conjugate shift represented by $(\delta\bar{E})$. Note that the difference of terms in different expressions involves only the parameters for the first and last spaces, since the contributions for all intermediate spaces cancels out.

15.3.5 The Bow–Sutton Conditions

Optical imaging systems tend to take advantage of the so-called 'symmetrical principle', where the parts of the optical system on the two sides of the aperture stop are symmetrical with respect to the stop. In cases where this symmetry is perfect, i.e. the two parts are exactly similar, and the object and image are of the same size and are at equal distances from the plane of the aperture stop, the system is called 'holosymmetrical', and in such an arrangement, coma S_{II}, distortion S_V, and lateral colour C_{II} are completely absent, i.e. $S_{II} = S_V = C_{II} = 0$. For other conjugate locations of the holosymmetrical system, there is an object shift by say, $\left(\delta\hat{\bar{E}}\right)$, and, by (15.107) the new values of coma $\hat{S}_V^*$ and lateral colour $\hat{C}_{II}^*$ are given by

$$\hat{S}_V^* = S_V + \left(\delta\hat{\bar{E}}\right)S_{VI} \tag{15.108}$$

$$\hat{C}_{II}^* = C_{II} + \left(\delta\hat{\bar{E}}\right)C_{III} \tag{15.109}$$

Since $S_V = C_{II} = 0$, for any value of $\left(\delta\hat{\bar{E}}\right)$, i.e. for any other conjugate location of the holosymmetrical system, we get

$$\left.\begin{aligned} \hat{S}_V^* &= 0, \text{ if and only if } S_{VI} = 0 \\ \hat{C}_{II}^* &= 0, \text{ if and only if } C_{III} = 0 \end{aligned}\right\} \tag{15.110}$$

The conditions (15.110) are known as the Bow–Sutton conditions [21–28]. Note that the expression for $\hat{S}_{II}^*$ is more involved, and no simple condition exists to ensure its absence for any conjugate location; however, in holosymmetrical systems, on account of intrinsic cancellation of a certain degree, the value of coma is, in general, less than that of unsymmetrical lens systems.

REFERENCES

1 W.T. Welford, *Aberrations of Optical Systems*, Adam Hilger, Bristol (1986) p. 81.

2 L. Seidel, 'Zur Dioptrik. Über die Entwicklung der Glieder 3ter Ordnung, welche den Weg eines ausserhalb der Ebene der Axe gelegenen Lichtstrahles durch ein System brechender Medien bestimmen', *Astronomische Nachrichten*, Vol. 43, (1856) No. 1027, pp. 289–304; ibid., No. 1028, pp. 305–320; ibid., No.1029, pp. 321–332.

3 H.H. Hopkins, *Wave Theory of Aberrations*, Clarendon Press, Oxford (1950) pp. 104–105.

4 P. Mouroulis and J. Macdonald, *Geometrical Optics and Optical Design*, Oxford University Press, New York (1997) p. 260.

5 H. Gross, 'Pupil aberrations', Section 29.5 in *Handbook of Optical Systems, Vol. 3*, Wiley, New York (2015) pp. 45–49.

6 K. Schwarzschild, 'Untersuchungen zur geometrischen optic', *Abhandl. Königl. Ges. Wiss. Göttingen*, Vol. 4 (1905) Nos. 1–3.

7 R.S. Longhurst, 'An investigation into the factors influencing the design of eyepieces', Thesis, Imperial College of Science and Technology, London.

8 C.G. Wynne, 'Primary Aberrations and Conjugate Change', *Proc. Phys. Soc.* (London), Vol. 65 B (1952) pp. 429–437.
9 J. Sasián, 'Pupil aberrations', Chapter 12 in *Introduction to Aberrations in Optical Imaging Systems*, Cambridge University Press, Cambridge (2013) pp. 162–172.
10 J. Sasián, 'Interpretation of pupil aberrations in imaging systems', *Proc. SPIE*, Vol. 6342 (2006) 634208-1–634208-4.
11 H.R. Fallah and J. Maxwell, 'Higher Order Pupil Aberrations in Wide Angle and Panoramic Optical Systems', Proc. SPIE, Vol. 2774, (1996) pp. 342–351.
12 G. Slussareff, 'L'Eclairement de l'image formée par les objectifs photographiques grand-angulaires', *J. Phys.* (U.S.S.R.) Vol. 4 (1941) pp. 537–545.
13 G. Slussareff, 'A Reply to Max Reiss', *J. Opt. Soc. Am.*, Vol. 36, No. 12, (1946), p. 707.
14 M.T. Chang and R.R. Shannon, 'Pupil aberrations in zoom systems', *Proc. SPIE*, Vol. 3129 (1997) pp. 205–216.
15 B.J. Baumann, Anisoplanatism in adaptive optics systems due to pupil aberrations', *Proc. SPIE*, Vol. 5903, Eds. R. K. Tyson and M. Lloyd-Hart (2005) 59030R-1 to 59030R-12.
16 T. Smith, 'The changes in aberrations when the object and stop are moved', *Trans. Opt. Soc.* (London), Vol. 23 (1921/1922) pp. 311–322.
17 R.J. Pegis, '§ 3. The Dependence of the Aberrations upon Object and Stop Position' in 'The Modern Development of Hamiltonian Optics', in *Progress in Optics Vol. I*, Ed. E. Wolf, North Holland, Amsterdam (1961) pp. 3–29.
18 M. Herzberger, *Modern Geometrical Optics*, Interscience, New York (1958) pp. 332–350.
19 K. Yamaji, 'Design of Zoom Lenses', in *Progress in Optics, Vol. VI*, Ed. E. Wolf, North-Holland, Amsterdam (1965) pp. 105–169.
20 H.H. Hopkins, 'An analytical technique for stable aberration correction in zoom systems', *Proc. SPIE*, Vol. 0399 (1983) pp. 100–133.
21 R.H. Bow, 'On Photographic Distortion', *Brit. Journal of Photography*, Vol. VIII (1861) pp. 417–419, 440–442.
22 T. Sutton, 'Distortion produced by lenses', *Phot. Notes*, Vol. VII (1862) No. 138, pp. 3–5.
23 A. Koenig and M. von Rohr, Chapter V in *Geometrical Investigation of the formation of images in optical instruments*, Ed. M. von Rohr [original German book title: *Die Theorie der Optischen Instrumente*, Ed. M. von Rohr Springer, Berlin (1904)] Translator, R. Kanthack, H.M. Stationery Office, Dept. of Scientific and Industrial Research (1920) pp. 244–247.
24 J.P.C. Southall, *The Principles and Methods of Geometrical Optics*, Macmillan, New York (1913) pp. 420–421.
25 A.E. Conrady, *Applied Optics and Optical Design, Part II*, Dover, New York (1960) pp. 491–493.
26 R. Kingslake, *Lens Design Fundamentals*, Academic, New York (1978) pp. 204–205.
27 J. Sasián, vide Ref. 8, p. 170.
28 D. Malacara-Hernández and Z. Malacara-Hernández, *Handbook of Optical Design*, CRC Press, Boca Raton (2013) pp. 151–152.

16

Role of Diffraction in Image Formation

16.1 *Raison d'Être* for 'Diffraction Theory of Image Formation'

Ray-optical theory is more or less adequate for the treatment of image formation by optical systems, so long as the residual aberrations in the imaging system are relatively large. This is because, in such cases, the predictions based on ray-optical theory are vindicated by experimental observations. However, in the case of optical systems with small values of residual aberrations, the ray-optical theory fails to provide correct interpretation for experimental observations [1]. Three of these observations that baffled scientists are noted below.

Observation I

Figure 16.1 shows the image space of an axially symmetric system with its exit pupil at E′ and image plane at O′ on the optical axis. In the case of an aberration-free system, the pencil of rays forming image of an axial object point O (not shown) converges at the axial image point O′. Equivalently, the imaging wavefronts are perfectly spherical with their centre at O′. In the case of a bright point in a dark background, it turns out that the intensity, defined by energy per unit area, is infinite at O′ and zero elsewhere. Experiments with optical imaging systems with smaller values of aberrations has shown that this conjecture cannot be true.

Observation II

Figure 16.2 shows the image space of an axially symmetric system with the exit pupil at E′ on the axis and a chosen image plane. It is assumed that the imaging system has residual spherical aberration. When the diameter of the exit pupil is A_+A_-, the marginal rays A_+a_+ and A_-a_- form a blur circle of diameter a_-a_+ on the chosen image plane. As the diameter of the exit pupil is gradually reduced from A_+A_- so that $A_+A_- > B_+B_- > C_+C_- > D_+D_-$, per ray-optical theory, the diameters of the corresponding blur circles should be such that $a_-a_+ > b_-b_+ > c_-c_+ > d_-d_+$. Practical observations show that this conjecture is valid up to a certain size of the exit pupil; any further reduction of diameter of the exit pupil leads to an increase, instead of a decrease, in the diameter of the blur circle. This observation was found to be the same for any value of spherical aberration.

Observation III

Figure 16.3 illustrates the third puzzling phenomenon. It shows the image space of an axially symmetric system with residual primary spherical aberration. The exit pupil is located at E′, and the paraxial image plane is at O'_p on the axis. The marginal rays intersect at O'_m on the axis. Rays from different zones of the exit pupil intersect the axis at different points between O'_m and O'_p, and the best focal plane appears to be the plane where the diameter of the blur circle is minimum. The transverse plane where the size of the blur is minimum is located at the point O'_{mb} on the axis. Per ray-optical theory, in the presence of primary spherical aberration, $O'_{mb}O'_p = \frac{3}{4}O'_mO'_p$. Experimental observations do not validate this conjecture. It is seen that the location of the best focal plane is midway between the paraxial and the marginal foci.

DOI: 10.1201/9780429154812-16

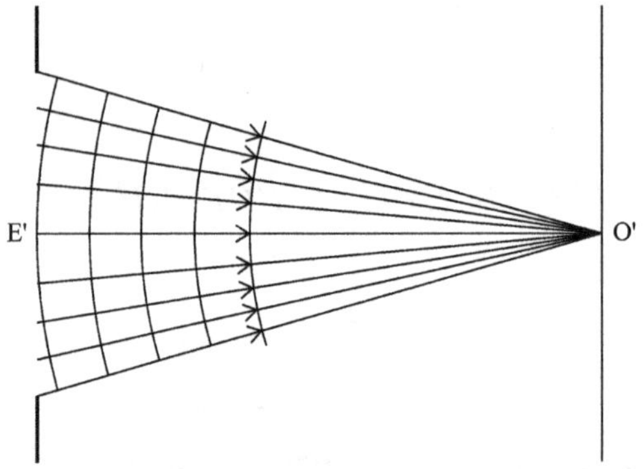

FIGURE 16.1 In the image space of an aberration free system, all rays originating from a point object O meet at the corresponding image point O′.

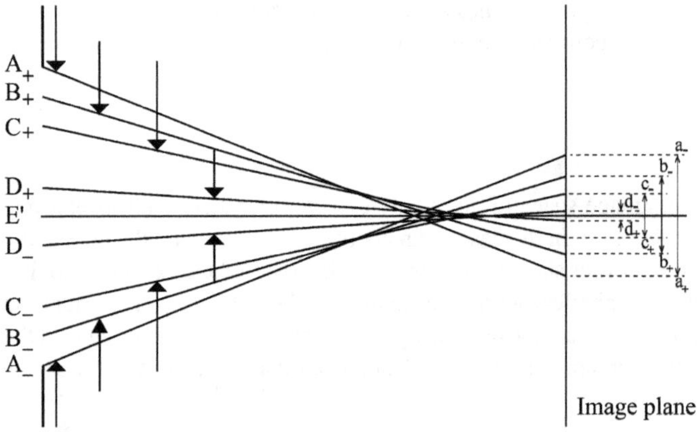

FIGURE 16.2 As the diameter of the image-forming pencil of rays from a point object is decreased, the diameter of the blur circle on a chosen image plane decreases up to a certain minimum size of the pencil, beyond which the diameter increases.

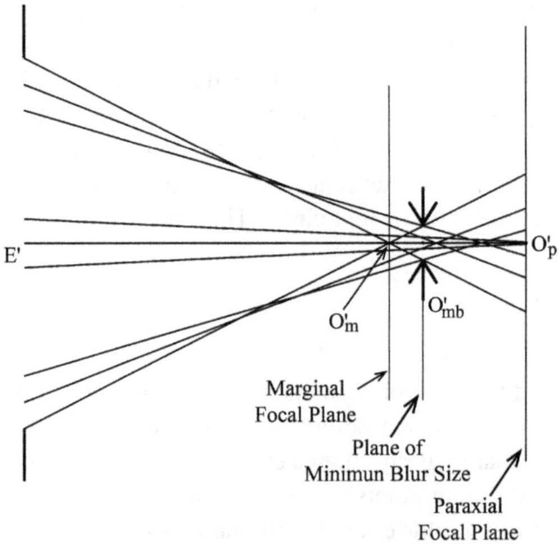

FIGURE 16.3 In the presence of primary spherical aberration, the axial location of the plane of minimum blur should be located at an axial distance from the paraxial focal plane in the direction of the marginal focus; as per ray optics, this distance is three-fourths of the distance of the marginal focus from the paraxial focus, however, experimental observations showed the factor to be $\frac{1}{2}$, not $\frac{3}{4}$.

In course of their search for the causes of these inconsistencies between theory and practice, the scientists noted that ray-optical theory did not take into account the effects of diffraction of the image-forming wave by the finite aperture of the imaging system. When the diffraction effects are properly incorporated in the theoretical analysis, the above-mentioned discrepancies no longer exist.

16.2 Diffraction Theory of the Point Spread Function

By some accounts, the first reference to diffraction effects appears in the work of Leonardo da Vinci in the fifteenth century [2]. However, the phenomenon of diffraction was first carefully observed and reported by Grimaldi in the seventeenth century. He coined the term diffraction, from the Latin diffringere [dis (apart) + frangere (to break)] meaning 'to break into pieces', referring to light breaking up into different directions. The first proposition of his book, entitled 'Physico-mathematics of Light, Colours and Rainbow' [3] is:

> Propositio I. Lumen propagator seu diffunditur non solum directè, refractè, ac reflexè, sed etiam alio quodam quarto modo, diffracte. [in Latin]

It means:

> Proposition 1. Light propagates or spreads not only directly, by refraction, and by reflection, but also by a somewhat different fourth mode, diffraction.

The prevalent theory of light used to explain the propagation of light in that period failed to explain the phenomenon of diffraction. It is likely that Huygens, one of the early proponents of wave theory of light, was unaware of the discoveries of Grimaldi, for there was no mention of the same by Huygens in his treatise [4]. Nevertheless, a heuristic assertion by Huygens on the propagation of wavefronts in connection with his investigations on wave theory of light played a monumental role in future investigations on the theory of diffraction. Huygens asserted that each point on a wavefront may be regarded as the centre of a secondary disturbance that gives rise to spherical wavelets, and the position of the wavefront at any later time is the envelope of all such wavelets. More than a century later, Fresnel showed that the phenomenon of diffraction can be explained by combining Huygens' principle with Young's principle of interference [5–7]. This leads to what is known as Huygens–Fresnel principle. Fresnel's treatment was put to rigorous mathematical testing by Kirchoff, giving rise to the Fresnel – Kirchoff diffraction formula [8]. Later on, the few internal inconsistencies in the theory were taken care of with the Rayleigh – Sommerfeld formulation of diffraction [9–11]. It has been reported that the two approaches yield essentially the same result when the aperture diameter is much greater than the wavelength of light [12–13]. Other approaches, e.g. boundary wave diffraction theory [14–18] or geometrical theory of diffraction [19], for tackling the phenomenon of diffraction have also been implemented; they seem to be particularly useful in special cases. A glimpse of the fascinating developments in diffraction theory may be obtained from the references [20–29].

For most problems in instrumental optics, the Huygens–Fresnel integral formulation provides sufficiently accurate and useful solutions. A brief description of the formulation is presented below. It is based on simple physical considerations of a heuristic nature.

16.2.1 The Huygens–Fresnel Principle

The Huygens–Fresnel principle provides the basic formula for calculating the effects produced by a monochromatic wave. In Figure 16.4, A is a surface over which the complex amplitude U_0 at each point such as P_0 is known. Let BPD be a wavefront of the secondary wave from the element dA of A at P_0. Since we are concerned with a diverging spherical wave, the complex amplitude at P is of the form

$$du_P = \frac{\alpha}{R}\exp\{-inkR\} \tag{16.1}$$

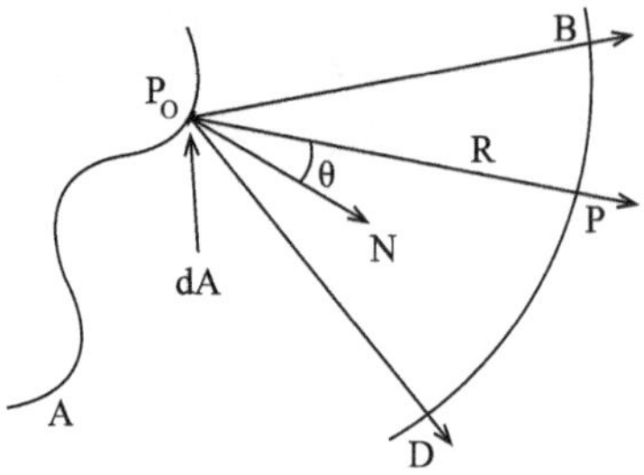

FIGURE 16.4 The Huygens-Fresnel integral.

where $R = (P_0P)$, n is the refractive index of the medium and $k = (2\pi / \lambda)$ and λ is the wavelength of the wave in vacuum. In the above, α is assumed to be the real amplitude over a spherical wavefront, centred on the point P_0 and of radius unity. Following the inverse square law of intensity, the term $\left(\frac{\alpha}{R}\right)$ represents the real amplitude at P. It is reasonable to suppose that α is proportional to the complex amplitude u_0 at P_0, and also to the size dA of the area element. It follows

$$\alpha = Cu_0 dA \tag{16.2}$$

where C is a constant. Let the angle between the line P_0P and the normal at P_0 be θ.

The resultant complex amplitude u_p at P is obtained by integrating contributions du_P from all elements of A as

$$u_P = C \iint_A u_0 \frac{\exp\{-inkR\}}{R} \chi(\theta) dA \tag{16.3}$$

where $\chi(\theta)$ is an obliquity factor. We must then take into account the observation that the amplitude of the secondary wave is maximum in the direction P_0N of the normal to dA, and the amplitude in other directions varies with θ. Slightly different values for the obliquity factor $\chi(\theta)$ are predicted by the Fresnel – Kirchoff and Rayleigh – Sommerfeld diffraction theories. In different cases, the dependence on θ is either $\cos\theta$ or $\frac{1}{2}(1+\cos\theta)$, which for $\theta = 20^0$, take on values 0.94 and 0.97, respectively. Therefore, (16.3) applies to cases with smaller angles of diffraction, where $\chi(\theta) \cong 1$, and (16.3) reduces to

$$u_P = C \iint_A u_0 \frac{\exp\{-inkR\}}{R} dA \tag{16.4}$$

The constant C can be determined by solving the integral for a case where u_p is known. The derivation for C given below follows a treatment of Hopkins [26].

Let us consider the value of u_p at a point on the axis of an aperture A, which is illuminated normally by a plane wave (Figure 16.5). O is the origin of a rectangular coordinate system with its Y, Z coordinate axes on the plane of the aperture, and the Z-axis is along the direction of propagation of the plane wave. The plane wave has a complex amplitude

$$u(z) = \alpha e^{-inkz} \tag{16.5}$$

so that, on the aperture

$$u_0 = u(0) = \alpha \tag{16.6}$$

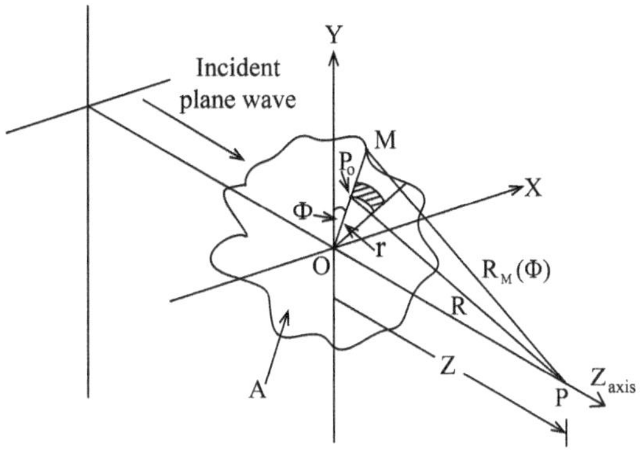

FIGURE 16.5 A plane wave incident normally on an irregular aperture in a transverse plane.

On the aperture, the element of area at P_0 is $dA = rdrd\phi$. The complex amplitude u_p at P is

$$u_p = u(z) = C \int_{\phi=0}^{2\pi} \int_{r=0}^{\rho(\phi)} \alpha \frac{e^{-inkR}}{R} rdrd\phi \tag{16.7}$$

where $\rho(\phi) = (OM)$ is the maximum radius of the aperture opening in the azimuth ϕ. Note that

$$R^2 = Z^2 + r^2 \tag{16.8}$$

so that $2RdR = 2rdr$, or $\frac{rdr}{R} = dR$. By changing variables from (r,ϕ) to (R,ϕ), (16.7) is rewritten as

$$u(z) = C\alpha \int_{\phi=0}^{2\pi} \int_{R=z}^{R_M(\phi)} e^{-inkR} dRd\phi \tag{16.9}$$

where $R_M(\phi) = (MP)$. By carrying out the integrations we get

$$u(z) = C\alpha \int_{\phi=0}^{2\pi} \left\{ \left[\frac{e^{-inkR}}{-ink} \right]_{R=z}^{R=R_M(\phi)} \right\} d\phi = \left(\frac{C}{ink} \right) \alpha e^{-inkz} \int_0^{2\pi} d\phi - \left(\frac{C}{ink} \right) \alpha \int_0^{2\pi} e^{-inkR_M(\phi)} d\phi \tag{16.10}$$

which reduces to

$$u(z) = \left(\frac{2\pi C}{ink} \right) \alpha e^{-inkz} - \left(\frac{C\alpha}{ink} \right) \int_0^{2\pi} e^{-inkR_M(\phi)} d\phi \tag{16.11}$$

Since, for an irregular shape of aperture, $R_M(\phi)$ varies by very many λ's over $\phi = 0$ to 2π, we shall have, for the real part of the integrand of the integral in (16.11), a cosine with very many oscillations (Figure 16.6), so that

$$\int_0^{2\pi} \cos\left\{ 2\pi \frac{n}{\lambda} R_M(\phi) \right\} d\phi \cong 0 \tag{16.12}$$

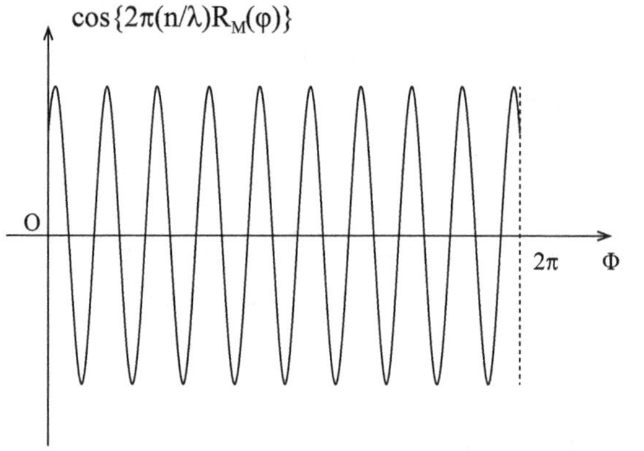

FIGURE 16.6 Oscillations with ϕ of the real part of the integrand of the integral in (16.11).

Similarly, the imaginary part of the integrand is also zero. We then have

$$u(z) = \left(\frac{2\pi C}{ink}\right)\alpha e^{-inkz} \tag{16.13}$$

which, to agree with the experiment must be the result given by geometrical ray optics, namely

$$u(z) = \alpha e^{-inkz} \tag{16.14}$$

From (16.13) and (16.14), we get

$$C\frac{2\pi}{ink} = C\frac{2\pi\lambda}{in2\pi} = 1 \text{ or } C = n\left(\frac{i}{\lambda}\right) \tag{16.15}$$

The complex amplitude u_p, given by (16.4) can be expressed as

$$u_p = n\left(\frac{i}{\lambda}\right)\iint_A u_0 \frac{e^{-inkR}}{R} dA \tag{16.16}$$

In air or in a vacuum, $n = 1$, and $C = \left(\frac{i}{\lambda}\right)$. Hence, in air or in a vacuum, (16.4) is written as

$$u_P = \left(\frac{i}{\lambda}\right)\iint_A u_0 \frac{\exp\{-ikR\}}{R} dA \tag{16.17}$$

16.2.2 Diffraction Image of a Point Object by an Aberration Free Axi-Symmetric Lens System

The following derivation is an adaptation from the techniques developed by Hopkins and his co-workers [27–28] for the computation of diffraction images in the course of his investigations on image quality assessment. Figure 16.7 shows the imaging of a general object point $\bar{Q}$ by an axially symmetric imaging system. The object plane, the entrance pupil, the exit pupil, and the image plane are located at O, $\bar{E}$, $\bar{E}'$ and O′ on the axis. The latter points are considered to be origins of a rectangular system of coordinates in the corresponding planes with (ξ,η), (X,Y), (X',Y'), and (ξ',η'), being parallel to a set of rectangular axes in the four planes, respectively. The third axis in the system of coordinates is taken along

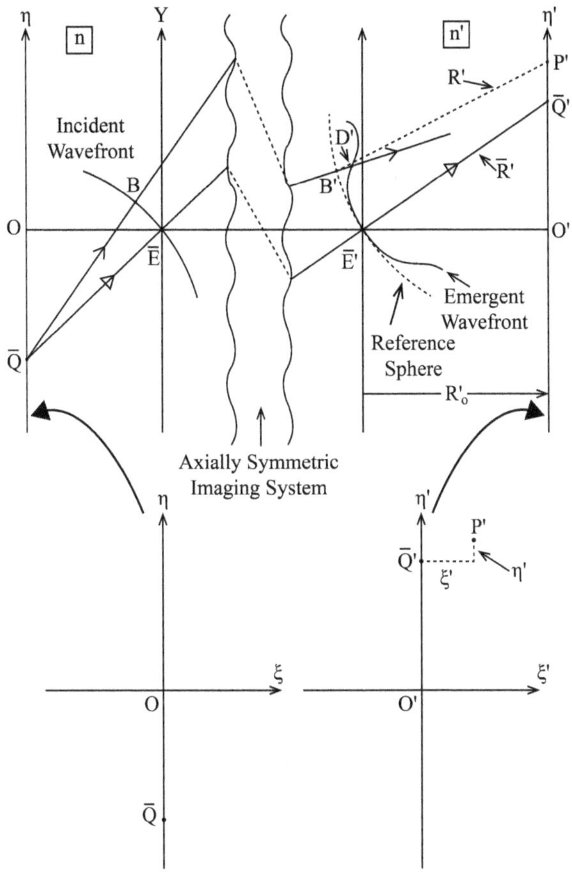

FIGURE 16.7 Imaging of a general object point $\bar{Q}$ by an axially symmetric imaging system; Top diagram: Elevation of the imaging system; Bottom diagrams: Side views of the object and the image planes.

the optical axis. For the sake of convenience, The top diagram in Figure 16.7 shows the elevation of the imaging diagram, and the two diagrams in the bottom of Figure 16.7 show the side views of the object plane and the image plane.

$\bar{Q}\bar{E}\bar{E}'\bar{Q}'$ is the principal or pupil ray of the image-forming pencil of rays, and $\bar{Q}BB'D'$ is another ray of the pencil. The entrance pupil sphere corresponding to the object point $\bar{Q}$ in the object space is $\bar{E}B$. Its centre is at $\bar{Q}$, and its radius is $\bar{Q}\bar{E}$. The spherical wave diverging from the object point $\bar{Q}$ is coincident with this sphere in the object space. In the image space, $\bar{E}'B'$, shown with dashed lines, is the exit pupil sphere. Its centre is the geometrical image point $\bar{Q}'$, and its radius is $\bar{Q}'\bar{E}'$. Note that the object point $\bar{Q}$ lies on the η-axis, and in an axially symmetric system, $\bar{Q}'$, the point of intersection of the pupil ray with the η' axis is taken as the geometrical image point.

If there is no aberration, the emergent wave is coincident with the exit pupil sphere. For this case, the distribution of complex amplitude over the reference sphere may be written as $\exp\{ikn'\bar{R}'\}/\bar{R}'$, where $k = (2\pi/\lambda)$, with λ the wavelength in a vacuum, n′ is the refractive index of the image space, and the radius of the exit pupil sphere is $\bar{R}' = (\bar{Q}'\bar{E}') = (\bar{Q}'B')$.

If the emergent wavefront $\bar{E}'D'$ is aberrated, as in Figure 16.7, the variation in complex amplitude over the exit pupil sphere is given by

$$F_0(X', Y') = \frac{\exp\{+ikn'\bar{R}'\}}{\bar{R}'} f(X', Y') \tag{16.18}$$

where

$$\begin{aligned} f(X',Y') &= \alpha(X',Y')\exp\{ikW(X',Y')\} && \text{within } A' \\ &= 0 && \text{outside } A' \end{aligned} \qquad (16.19)$$

is the pupil function. The factor $\alpha(X',Y')$ takes account of the variation of real amplitude over the pupil, and $W(X',Y') = n'(B'D')$ is the wavefront aberration for any point B' on the exit pupil sphere.

The complex amplitude at any point P' in the image plane in the neighbourhood of the point $\bar{Q}'$ is given by

$$F(\xi',\eta') = n'\left(\frac{i}{\lambda}\right)\iint_{A'} F_0(X',Y')\frac{\exp\{-ikn'R'\}}{R'}\,dA' \qquad (16.20)$$

where (ξ',η') are the rectangular coordinates of the point P' relative to the geometrical image point $\bar{Q}'$, the distance $R' = (B'P')$, and where dA' is an element on the exit pupil sphere.

Using (16.18), and noting that $dA' = dX'dY'/N'$, where (L', M', N') are the direction cosines of the straight line $B'\bar{Q}'$, (16.20) can be written as

$$F(\xi',\eta') = n'\left(\frac{i}{\lambda}\right)\iint_{A'} f(X',Y')\frac{\exp\{ikn'(\bar{R}'-R')\}}{\bar{R}'R'N'}\,dX'dY' \qquad (16.21)$$

The variation of $\{R'N'\}^{-1}$ during integration over the aperture A' causes a small variation in the real amplitudes of the secondary waves, to which the resultant amplitude is insensit'ive [29]. Ignoring this variation, and omitting the constant factor $\left\{n'\left(\frac{i}{\lambda}\right)/\bar{R}'R'N'\right\}$, the equation (16.21) reduces to

$$F(\xi',\eta') = \iint_{A'} f(X',Y')\exp\{ikn'(\bar{R}'-R')\}\,dX'dY' \qquad (16.22)$$

In the above equation, it is necessary to express the value of $(\bar{R}'-R')$ with high accuracy, since even an error of $\lambda/2$ in this quantity reverses the sign of the integrand for the point in 'view.

With reference to $\bar{E}'$ as the origin of a rectangular coordinate system, the coordinates of B', P' and $\bar{Q}'$ are (X', Y', Z'), $\{\xi', (\bar{\eta}'+\eta'), R'\}$, and $(0, \bar{\eta}', R'_0)$ respectively. Note that $\bar{E}O' = R'_0$.

R' and $\bar{R}'$ are given by

$$(R')^2 = (B'P')^2 = (X'-\xi')^2 + \{Y'-(\bar{\eta}'+\eta')\}^2 + (Z'-R'_0)^2 \qquad (16.23)$$

and

$$(\bar{R}')^2 = (\bar{E}'\bar{Q}')^2 = (B'\bar{Q}')^2 = X'^2 + (Y'-\bar{\eta}')^2 + (Z'-R'_0)^2 \qquad (16.24)$$

Subtracting (16.23) from (16.24) we get

$$(\bar{R}'-R') = \frac{2(X'\xi'+Y'\eta') - (\xi'^2+\eta'^2+2\bar{\eta}'\eta')}{(\bar{R}'+R')} \qquad (16.25)$$

Expressing the denominator on the right-hand side as

$$\left(\bar{R}'+R'\right)=2\bar{R}'-\left(\bar{R}'-R'\right) \tag{16.26}$$

(16.25) reduces to

$$\left(\bar{R}'-R'\right)=\left[\frac{\left(X'\xi'+Y'\eta'\right)}{\bar{R}'}-\frac{\left(\xi'^2+\eta'^2+2\bar{\eta}'\eta'\right)}{2\bar{R}'}\right]\left\{1-\frac{\left(\bar{R}'-R'\right)}{2\bar{R}'}\right\}^{-1} \tag{16.27}$$

With $\lambda=0.5\mu\text{m}$, and $\bar{R}'>10\text{mm}$, for any point P′ that has the maximum value of $\left(\bar{R}'-R'\right)\sim 100\lambda$, $\frac{\left(\bar{R}'-R'\right)}{2\bar{R}'}\leq 0.0025$. Omission of the last bracketed term in the above expression will give an error of 0.25λ in the value of 100λ for $\left(\bar{R}'-R'\right)$. If the maximum value of $\left(\bar{R}'-R'\right)\sim 10\lambda$, $\frac{\left(\bar{R}'-R'\right)}{2\bar{R}'}\leq 0.00025$. In this case, omission of the last bracketed term gives an error of 0.0025λ. When $\left(\bar{R}'-R'\right)\sim 1\lambda$, the error is reduced further to 0.000025λ.

In equation (16.22), the value of the integral becomes zero or very small if the exponential function executes many oscillations during the integration. The above analysis shows that omission of the last bracketed term causes small errors only when the large number of oscillations of the integrand has led to an insignificantly small resultant. In cases where the resultant has practically useful values, the error caused by omission of the last bracketed term is insignificantly small.

Therefore, omitting the last bracketed term in (16.27), we can write

$$kn'\left(\bar{R}'-R'\right)=\frac{2\pi n'}{\lambda}\left[\frac{\left(X'\xi'+Y'\eta'\right)}{\bar{R}'}-\frac{\left(\xi'^2+\eta'^2+2\bar{\eta}'\eta'\right)}{2\bar{R}'}\right] \tag{16.28}$$

From (16.22), it follows

$$F(\xi',\eta')=\exp\left\{-i\epsilon(\xi',\eta')\right\}\iint_{A'}f(X',Y')\exp\left\{i2\pi\left[\left(\frac{1}{\lambda}n'\frac{\xi'}{\bar{R}'}\right)X'+\left(\frac{1}{\lambda}n'\frac{\eta'}{\bar{R}'}\right)Y'\right]\right\}dX'dY' \tag{16.29}$$

where $\epsilon(\xi',\eta')$ is given by

$$\epsilon(\xi',\eta')=\left(\frac{2\pi n'}{\lambda}\right)\frac{\left(\xi'^2+\eta'^2+2\bar{\eta}'\eta'\right)}{2\bar{R}'} \tag{16.30}$$

Since the pupil function is zero outside A′, infinite limits can be used in the integration, and the above equation can be rewritten as

$$F(\xi',\eta')=\exp\left\{-i\epsilon(\xi',\eta')\right\}\iint_{+\infty}^{+\infty}f(X',Y')\exp\left\{i2\pi\left[\left(\frac{1}{\lambda}n'\frac{\xi'}{\bar{R}'}\right)X'+\left(\frac{1}{\lambda}n'\frac{\eta'}{\bar{R}'}\right)Y'\right]\right\}dX'dY' \tag{16.31}$$

The phase factor outside the integral is of modulus unity, and may be ignored in the case of incoherent image formation. However, in the cases of coherent and partially coherent image formation of extended objects, not only the phase term $\epsilon(\xi',\eta')$, but also the optical path differences along the pupil rays from adjacent object points may have a significant effect on the image [30–32].

In the case of incoherent image formation, the intensity $I(\xi',\eta')$ in points (ξ',η') of the diffraction pattern, apart from the multiplicative constant factors omitted earlier, is given by

$$I(\xi',\eta')=\left|F(\xi',\eta')\right|^2 \tag{16.32}$$

The complex amplitude distribution in the diffraction image of a point is called 'Amplitude Point Spread Function (APSF)'. Similarly, the intensity distribution in the diffraction image of a point is called 'Intensity Point Spread Function'. On account of larger occurrence of incoherent imaging systems in practice, often the latter is simply noted as 'Point Spread Function', or by its abbreviation PSF. The other names used for PSF include 'Impulse Response Function', 'Green's Function', 'Intensity/Amplitude distribution in the Far-field/Fraunhofer Diffraction Pattern', etc. In general, the PSF of an optical imaging system is unique for a specific pair of conjugate points, and it can adequately describe the performance of the optical system in its imaging. The effects of image motion, atmospheric turbulence, and other extraneous factors can be taken care of by suitably modifying the PSF.

With a change of variables from (ξ', η') to (p', q'), where

$$p' = \frac{1}{\lambda}\left(\frac{n'}{\bar{R}'}\right)\xi' \qquad q' = \frac{1}{\lambda}\left(\frac{n'}{\bar{R}'}\right)\eta' \tag{16.33}$$

and omitting the multiplying phase term, the expression for complex amplitude (16.31) reduces to

$$F(p', q') = \iint_{-\infty}^{+\infty} f(X', Y') \exp\{i2\pi(p'X' + q'Y')\} dX'dY' \tag{16.34}$$

The above relation shows that the complex amplitude at a point P′ in the diffraction pattern on the image plane normal to the optical axis is given exactly by the Fourier transform of the pupil function. It is important to note that this relation is valid if and only if the pupil function is expressed as a function of rectangular coordinates, (X', Y') of points B′on the reference sphere, often called the pupil sphere.

Although the equation (16.34) can be used for accurate evaluation of the complex amplitude and intensity at points in the diffraction pattern, it is obvious that indeterminacies arise in special cases when either the exit pupil or the image plan is at infinity. Also, in general, evaluation of the Fourier transform in (16.34) involves numerical integration over the irregular area A′. These problems can be circumvented by using canonical coordinates, as enunciated in Chapter 11.

From equation (11.21), we note that the canonical coordinates (x'_S, y'_T) of a point on the exit pupil sphere are related with its rectangular coordinates (X', Y') by the relations

$$x'_S = \frac{X'}{h_S} \qquad y'_T = \frac{Y'}{\bar{N}'h'_T} \tag{16.35}$$

The rectangular coordinates for the point P′ on the image plane with respect to the origin at $\bar{Q}'$ is (ξ', η'). The corresponding canonical coordinates (G'_S, H'_T) are given in equation (11.135) as

$$G'_S = n'\left(\frac{h_S}{\bar{R}'}\right)\xi' = n'(\sin\alpha'_S)\xi' \tag{16.36}$$

$$H'_T = n'\left(\frac{\bar{N}'h'_T}{\bar{R}'}\right)\eta' = n'(\sin\bar{\alpha}'_T - \sin\bar{\alpha}')\eta' \tag{16.37}$$

Figures 11.2 and 11.3 illustrate the different quantities of the above expressions. Using the above equations, we get

$$p'X' = \left(\frac{1}{\lambda}n'\frac{\xi'}{\bar{R}'}\right)X' = \frac{1}{\lambda}\left(n'\frac{h'_S}{\bar{R}'}\xi'\right)\frac{X'}{h'_S} = \frac{1}{\lambda}(n'\sin\alpha'_S\xi')x'_S = \frac{1}{\lambda}G'_S x'_S = u'_S x'_S \tag{16.38}$$

$$q'Y' = \left(\frac{1}{\lambda}n'\frac{\eta'}{\bar{R}'}\right)Y' = \frac{1}{\lambda}\left(n'\frac{\bar{N}'h'_T}{\bar{R}'}\eta'\right)\frac{Y'}{\bar{N}'h'_T} = \frac{1}{\lambda}\{n'(\sin\alpha'_T - \sin\bar{\alpha}')\eta'\}y'_T = \frac{1}{\lambda}H'_T y'_T = v'_T y'_T \tag{16.39}$$

where $u'_S = (G'_S / \lambda)$ and $v'_T = (H'_T / \lambda)$. It is convenient to use the wavelength of light in vacuum λ as the unit of length in the diffraction theory of image formation. Note that both the canonical coordinates (x'_S, y'_T) for a point on the pupil, and (u'_S, v'_T) for a point on the image plane are dimensionless. The corresponding quantities in the object space, i.e. (x_S, y_T) and (u_S, v_T) are similarly dimensionless.

Using these relations, and omitting the multiplicative phase term and the constant factors, the complex amplitude at a point P′ with canonical coordinates (u'_S, v'_T) in the diffraction pattern is given by the Fourier transform of the pupil function, and from (16.34), we get

$$F(u'_S, v'_T) = \iint_{-\infty}^{+\infty} f(x'_S, y'_T) \exp\{i2\pi(u'_S x'_S + v'_T y'_T)\} dx'_S dy'_T \tag{16.40}$$

The intensity at the point $P'(u'_S, v'_T)$ is, apart from the terms omitted earlier, given by

$$I(u'_S, v'_T) = |F(u'_S, v'_T)|^2 \tag{16.41}$$

It may be noted that in the above expression, the limits of integration formally put infinity to fit in the expression for Fourier transformation. As shown earlier, the pupil function is zero outside a finite region A′, so that the integration is to be carried out only over the finite region A′. The outer contour of this region in canonical coordinates (x'_S, y'_T) is a circle (see equation (11.25)) represented by

$$x'^2_S + y'^2_T = 1 \tag{16.42}$$

whereas, in rectangular coordinates (X', Y'), the contour is, in general, an ellipse (see equation (11.26)) represented by

$$\left(\frac{X'}{h'_S}\right)^2 + \left(\frac{Y'}{\bar{N}' h'_T}\right)^2 = 1 \tag{16.43}$$

It may be reiterated that the expressions for the complex amplitude and intensity as given by (16.40) and (16.41) hold good if and only if the pupil points are considered on the pupil sphere. If the points are taken on the exit pupil plane, instead of the pupil sphere, the Fourier transform relation does not strictly hold, and at a large field of view the errors may become large.

16.2.3 Physical Significance of the Omitted Phase Term

Since the modulus of the phase term is unity, it is not of much concern in practice in incoherent image formation, where the quantity of primary interest is intensity. In fact, the term is often neglected in treatments of incoherent image formation.. However, in cases of coherent and partially coherent image formation, the phase term cannot be neglected. The phase term $\epsilon(\xi', \eta')$ is given by

$$\epsilon(\xi', \eta') = \left(\frac{2\pi n'}{\lambda}\right) \frac{(\xi'^2 + \eta'^2 + 2\bar{\eta}'\eta')}{2\bar{R}'} \tag{16.44}$$

Figure 16.8 shows the image space of an extra-axial point by an axially symmetric system. For the sake of convenience, the diagram is given in two dimensions, although the formulae deal with all three dimensions. The exit pupil and the image plane are located at $\bar{E}'$ and O' on the axis. The distance $\bar{E}'O' = R'_0$. $\bar{E}'B'$ is the section of a pupil sphere with its centre at the geometrical image point $\bar{Q}'$ and radius $\bar{E}'\bar{Q}' = \bar{R}'$. With $\bar{E}'$ as its centre and $\bar{E}'\bar{Q}'$ as radius, a sphere is drawn. $W'\bar{Q}'$ is a section of this sphere, which is called the 'image sphere'. $\bar{T}'_1\bar{Q}'\bar{T}'_2$ is a line on the tangent to the image sphere at the point $\bar{Q}'$. The amplitude is

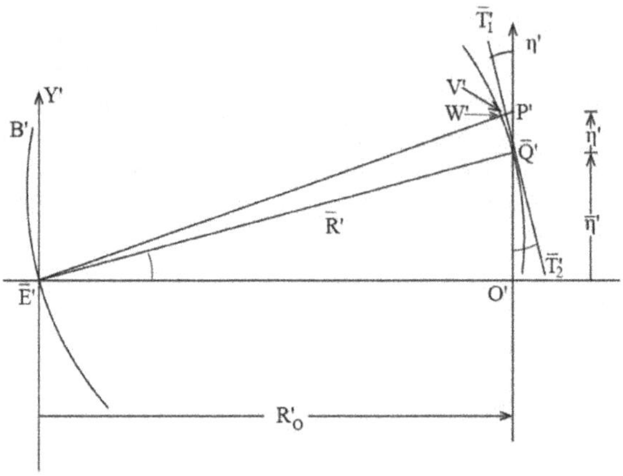

FIGURE 16.8 Confocal spheres in the image space of an axially symmetric system in case of imaging of an extra-axial point.

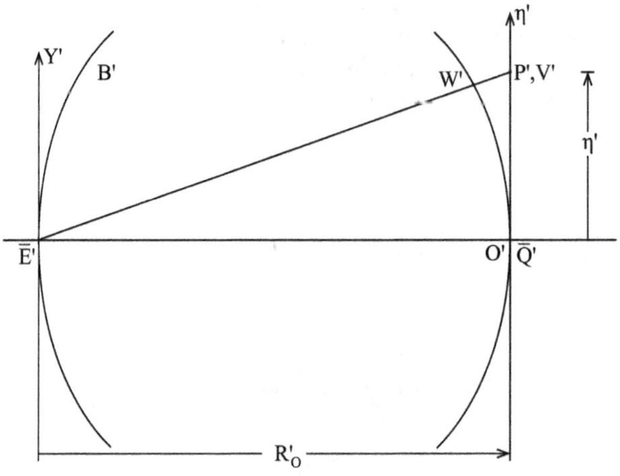

FIGURE 16.9 Confocal spheres for axial imaging in the image space of an axially symmetric system.

calculated at the point P′. The line $\bar{E}'P'$ intersects the image sphere and the tangent surface at points W′ and V′, respectively. In the diagram, the three marked angles are equal, i.e.

$$\angle\bar{Q}'\bar{E}'O' = \angle O'\bar{Q}'\bar{T}_2' = \angle P'\bar{Q}'\bar{T}_1' \tag{16.45}$$

For $\eta' \ll \bar{\eta}'$, the distance $V'P' \cong \frac{\bar{\eta}'}{\bar{R}'}\eta'$. Also, by geometry, $W'V' \cong \frac{\xi'^2 + \eta'^2}{2\bar{R}'}$. Therefore, we get

$$W'P' = \frac{\left(\xi'^2 + \eta'^2 + 2\bar{\eta}'\eta'\right)}{2\bar{R}'} \tag{16.46}$$

It is obvious that if the image sphere is taken as the image surface, the omitted phase term (16.44) becomes insignificant. On the other hand, even in the case of incoherent image formation, the complex amplitude for the pupil needs to be taken on the pupil sphere. The pupil sphere and the image sphere, as defined above, are confocal spheres between which the Fourier transform relationship is strictly valid. Figure 16.9 depicts the pupil sphere and the image sphere in the case of imaging of an axial point [33].

16.2.4 Anamorphic Stretching of PSF in Extra-Axial Case

For an axially symmetric imaging system with an axially symmetric pupil, the diffraction image of an axial point on the transverse image plane is symmetric around the axis. The subscripts S and T in the canonical coordinates become redundant, and one can use the following notations:

$$x'_S \to x'\left(=\frac{X'}{h'}\right) \qquad y'_T \to y'\left(=\frac{Y'}{h'}\right) \tag{16.47}$$

In the axial case, (x',y') are the canonical coordinates for a point on the exit pupil sphere; the rectangular coordinates of the point are (X',Y') with the origin at the location of the exit pupil on the optical axis; the maximum semi-diameter of the aperture of the exit pupil is h'. For points in the neighbourhood of the geometrical image point on the image plane, the simpler notations are:

$$G'_S \to G'\left[=n'(\sin\alpha')\xi'\right] \qquad H'_T \to H'\left[=n'(\sin\alpha')\eta'\right] \tag{16.48}$$

In the axial case, with a circular pupil, the maximum semi-angular aperture α' of the exit pupil is the same along any azimuth. (ξ',η') are rectangular coordinates of a point on the transverse image plane with the origin at the geometrical image point on the axis. In Chapter 11 it is shown that the canonical coordinates for a point with rectangular coordinates (ξ',η'), with respect to the origin at the geometrical image point $\bar{Q}'$, on the η-axis are given by (see equation (11.135)),

$$G'_S = n'\left(\sin\alpha'_S\right)\xi' \qquad H'_T = n'\left(\sin\alpha'_T - \sin\bar{\alpha}'\right)\eta' \tag{16.49}$$

On account of different normalizations in the axial and the extra-axial cases, the physical scale of an extra-axial diffraction pattern relative to that of the axial pattern is determined by the ratios

$$\left.\begin{aligned} \frac{G'_S}{G'} &= \frac{n'(\sin\alpha'_S)\xi'}{n'(\sin\alpha')\xi'} = \rho'_S \\ \frac{H'_T}{H'} &= \frac{n'(\sin\alpha'_T - \sin\bar{\alpha}')\eta'}{n'(\sin\alpha')\eta'} = \rho'_T \end{aligned}\right\} \tag{16.50}$$

where (G'_S,H'_T) and (G',H') are reduced coordinates of similar points in the neighbourhood of the extra-axial point $\bar{Q}'$ and the axial point O' respectively (Figure 16.7). ρ'_S and ρ'_T are the sagittal pupil scale ratio and the tangential pupil scale ratio, respectively (see equations (11.36) – (11.37)).

When the diffraction image at the extra-axial point is expressed as a function of $(u'_S,\ v'_S)$, it appears to be circular alike the diffraction image of the axial point. But in real pace coordinates (ξ',η') the pattern 'stretched' in the sagittal and the tangential directions by ρ'_S and ρ'_T, respectively, relative to the scale of the axial diffraction image. Since, in general, $\rho'_S \neq \rho'_T$, this stretching is called 'anamorphic stretching'. The actual diffraction pattern in terms of real rectangular coordinates (ξ',η') may be obtained from the pattern in terms of (u'_S,v'_T) by stretching the latter by the factors $(1/\rho'_S)$ and $(1/\rho'_T)$ in the sagittal and the tangential directions, respectively (Figure 16.10).

16.3 Airy Pattern

The complex amplitude F at a point $P'(u'_S,v'_T)$ in the diffraction image of an object point is given by equation (16.40). In polar coordinates representation, (16.40) becomes

$$F(p,\phi) = \iint_{-\infty}^{+\infty} f(r,\theta)\exp\left[i2\pi pr\cos(\theta-\phi)\right]rdrd\theta \tag{16.51}$$

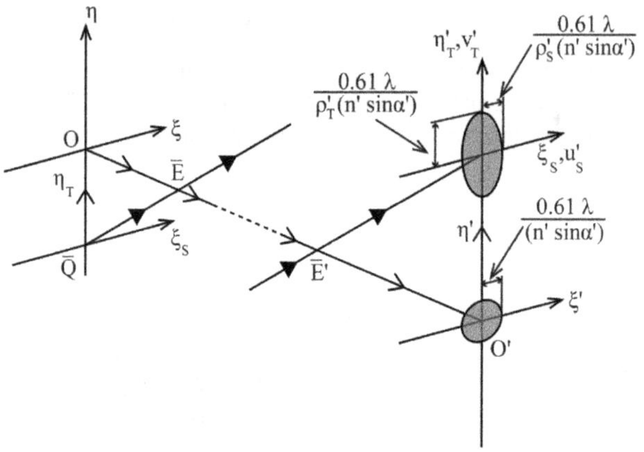

FIGURE 16.10 Anamorphic stretching of the PSF in extra-axial imaging.

The polar coordinates for a point with canonical coordinates (x'_S, y'_T) on the exit pupil are defined as

$$x'_S = r\sin\theta \qquad y'_T = r\cos\theta \tag{16.52}$$

Similarly, the polar coordinates for a point with canonical coordinates (u'_S, v'_T) on the image plane are defined as

$$u'_S = p\sin\phi \qquad v'_T = p\cos\phi \tag{16.53}$$

The representation of angles θ and ϕ follows the same convention as illustrated in Figure 8.14 and Figure 8.15. It should also be noted that the coordinates (u'_S, v'_T) or (p, ϕ), for a point in the diffraction pattern on the image plane, have the geometrical image point as the origin of the coordinate system.

In what follows, we omit the primes in image space quantities for the sake of simplicity, since we shall deal with only image space quantities. For an aberration-free system with unity amplitude over the exit pupil, the pupil function reduces to

$$\begin{aligned} f(r,\theta) \equiv f(r) &= 1, \qquad \text{for } 0 \le r \le 1 \\ &= 0, \qquad \text{for } r > 1 \end{aligned} \tag{16.54}$$

The corresponding complex amplitude at a point (p, ϕ) in the diffraction pattern is given by

$$F(p,\phi) = \int_0^1 \left\{ \int_0^{2\pi} \exp\left[i2\pi pr\cos(\theta - \phi)\right] d\theta \right\} r dr \tag{16.55}$$

Due to axial symmetry of the pupil function of the axially symmetric imaging system, it follows that $F(p,\phi)$ becomes the azimuth invariant, and it can be written as

$$F(p,\phi) \equiv F(p) = \int_0^1 \left\{ \int_0^{2\pi} \exp\left[i2\pi pr\cos(\theta - \phi)\right] d\theta \right\} r dr \tag{16.56}$$

The integral within the second brackets can be represented analytically by using Bessel functions [34–37]. The Bessel function of the first kind of order n and argument x, $J_n(x)$ has the integral representation

$$J_n(x) = \frac{i^{-n}}{2\pi} \int_0^{2\pi} \exp\left[i(n\alpha + x\cos\alpha)\right] d\alpha \tag{16.57}$$

For n = 0, (16.57) reduces to

$$J_0(x) = \frac{1}{2\pi} \int_0^{2\pi} \exp(ix\cos\alpha)\, d\alpha \tag{16.58}$$

Using the above relation, the equation (16.56) can be written as

$$F(p) = 2\pi \int_0^1 J_0(2\pi pr)\, r dr \tag{16.59}$$

The integration can be carried out by using the recurrence relation of Bessel functions

$$\frac{d}{dx}\left\{x^{n+1} J_{n+1}(x)\right\} = x^{n+1} J_n(x) \tag{16.60}$$

For n = 0, (16.60) reduces to

$$\frac{d}{dx}\left\{xJ_1(x)\right\} = xJ_0(x) \tag{16.61}$$

On integration, we get

$$\int_0^x \tilde{x} J_0(\tilde{x})\, d\tilde{x} = xJ_1(x) \tag{16.62}$$

Substituting $\tilde{x} = 2\pi pr$ and using (16.62), the equation (16.59) reduces to

$$F(p) = 2\pi \frac{1}{2\pi p}\frac{1}{2\pi p} \int_0^{2\pi p} \tilde{x} J_0(\tilde{x})\, d\tilde{x} = \pi \frac{2J_1(2\pi p)}{(2\pi p)} \tag{16.63}$$

The function $\left[2J_1(b)/b\right]$ is called the 'Besinc' or 'Jinc' function on account of its similarity with the $\left[\operatorname{sinc}(b) = \frac{\sin b}{b}\right]$ function.

The normalized complex amplitude $F_N(p)$, and normalized complex intensity $I_N(p)$ defined as

$$F_N(p) = \frac{F(p)}{F(0)} \qquad I_N(p) = \frac{I(p)}{I(0)} \tag{16.64}$$

are given by

$$F_N(p) = \frac{2J_1(2\pi p)}{(2\pi p)} \qquad I_N(p) = \left[\frac{2J_1(2\pi p)}{(2\pi p)}\right]^2 \tag{16.65}$$

However, for the sake of brevity, in what follows, the subscript is omitted in the notations for the normalized quantities, and the normalized amplitude and the normalized intensity are represented by $F(p)$ and $I(p)$, respectively.

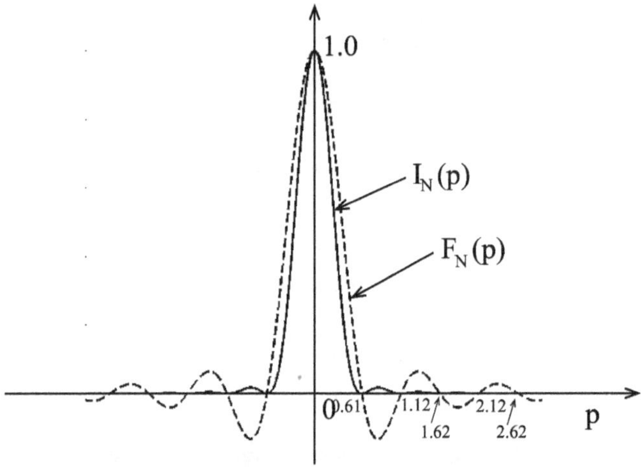

FIGURE 16.11 Normalized amplitude $F_N(p)$ and normalized intensity $I_N(p)$ for an Airy pupil.

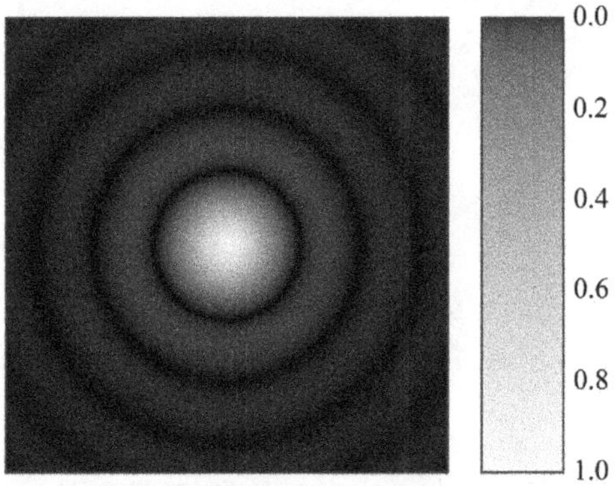

FIGURE 16.12 Diffraction image of an axial object point by an axially symmetric aberration-free imaging system.

Figure 16.11 shows variation of the normalized amplitude and the normalized intensity with the normalized distance p along an azimuth. Note that, on account of rotational symmetry in the aberration-free case, this variation is the same along any azimuth. Figure 16.12 shows the diffraction image of an axial point object formed by an axially symmetric aberration-free imaging system. It is seen that the principal maximum of the intensity distribution is at $p = 0$, where $I(0) = \left[\left\{\frac{2J_1(2\pi p)}{(2\pi p)}\right\}_{p=0}\right]^2 = 1$. For $p \neq 0, I(p) < I(0)$. But, $I(p)$ does not decrease monotonically with an increase in p, rather it oscillates with gradually diminishing amplitude. The minimum value of intensity is zero, and the intensity is zero for values of p given by roots of the equation

$$J_1(2\pi p) = 0 \; (p \neq 0) \tag{16.66}$$

The notation $p_0^m, m = 1,2,3,\ldots.$ is used to indicate the location of the m-th minimum of the Airy pattern. The values of $p_0^m, m = 1,2,3,\ldots.$ are given in Table 1.

The location of the secondary maxima can be obtained by making use of another form of the recurrence relation (16.60). It is given by

$$\frac{d}{dx}\{x^{-n}J_n(x)\} = -x^{-n}J_{n+1}(x) \tag{16.67}$$

For n = 1,

$$\frac{d}{dx}\left\{\frac{J_1(x)}{x}\right\} = -\frac{J_2(x)}{x} \tag{16.68}$$

The secondary maxima of $I(p)$, is given by the roots of the equation

$$J_2(2\pi p) = 0 \; (p \neq 0) \tag{16.69}$$

Historically, the intensity pattern in the diffraction image of an aberration-free circular aperture was first worked out by Airy in a somewhat different form in 1835 [38]. It may be noted that the first systematic application of the Bessel function in diffraction integrals was made by Lommel in 1870 [39].

In recognition of his contribution, this diffraction pattern is called the 'Airy pattern', and the corresponding circular aberration free unit amplitude pupil is often called the 'Airy pupil'. Table 16.1 gives a list of locations of the first few maxima and minima of the intensity distribution in the Airy pattern. Note that, with the increasing value of p, the separation between two successive minima or two successive maxima approaches the value 0.5.

16.3.1 Factor of Encircled Energy

At times, it becomes necessary to know what fraction of total energy in the diffraction pattern is contained within a specific central core area of the diffraction pattern. 'Encircled energy' is also termed 'total illuminance', or 'cumulative point spread function'. This function depicts the integrated behaviour of the point spread function. In comparison with the point spread function, it is a smooth function, and is easier to measure in practice.

The energy, $E(\omega)$ contained within a circle of radius ω and centred on the geometrical image point, i.e. the centre of the diffraction pattern, is given by

$$E(\omega) = \int_0^{2\pi}\int_0^{\omega} I(p)\,p\,dp\,d\phi = 2\pi\int_0^{\omega} I(p)\,p\,dp \tag{16.70}$$

The total energy in the diffraction pattern is

$$E(\infty) = \int_0^{2\pi}\int_0^{\infty} I(p)\,p\,dp\,d\phi = 2\pi\int_0^{\infty} \left[F(p)\right]^2 p\,dp \tag{16.71}$$

TABLE 16.1

Locations of the First Few Maxima and Minima in the Airy Pattern

p	Max/Min	I(p)
0	Max	1
0.6098	Min	0
0,8174	Max	0.0175
1.1166	Min	0
1.3396	Max	0.0042
1.6192	Min	0
1.8493	Max	0.0016
2.1205	Min	0
2.3549	Max	0.0008
2.6214	Min	0

By Rayleigh's theorem [40], we get

$$\int_0^\infty [F(p)]^2 \,pdp = \int_0^\infty |f(r)|^2 \,rdr \tag{16.72}$$

Substituting for $f(r)$ from (16.54), we get

$$E(\infty) = 2\pi\left(\frac{1}{2}\right) = \pi \tag{16.73}$$

The factor of encircled energy, $\varepsilon(\omega)$ is defined by the ratio of the energy inside a circle, concentric with the diffraction pattern, and of radius ω, with the total energy in the diffraction pattern, i.e.

$$\varepsilon(\omega) = \frac{E(\omega)}{E(\infty)} \tag{16.74}$$

Using (16.70) and (16.73), we can write

$$\varepsilon(\omega) = 2\int_0^\omega I(p)\,pdp = 2\int_0^\omega [F(p)]^2 \,pdp \tag{16.75}$$

From (16.63), F(p) is given by

$$F(p) = \pi\left[\frac{2J_1(2\pi p)}{2\pi p}\right] \tag{16.76}$$

Factor of encircled energy $\varepsilon(\omega)$ can be expressed as

$$\varepsilon(\omega) = 2\pi^2\int_0^\omega \left[\frac{2J_1(2\pi p)}{2\pi p}\right]^2 pdp = 2\int_0^{2\pi\omega} \frac{J_1^2(x)}{x}dx \tag{16.77}$$

For n = 0, (16.60) becomes

$$\frac{d}{dx}\{xJ_1(x)\} = xJ_0(x) \tag{16.78}$$

Carrying out the differentiation on the left side, and then multiplying both sides by $J_1(x)$, we get

$$J_1^2(x) + xJ_1(x)\frac{d}{dx}[J_1(x)] = xJ_0(x)J_1(x) \tag{16.79}$$

For n = 0, (16.67) becomes

$$\frac{d}{dx}[J_0(x)] = -J_1(x) \tag{16.80}$$

From (16.79) and (16.80) it follows

$$\frac{J_1^2(x)}{x} = J_0(x)J_1(x) - \frac{dJ_1(x)}{dx}J_1(x) = -\frac{1}{2}\frac{d}{dx}\left[J_0^2(x) + J_1^2(x)\right] \tag{16.81}$$

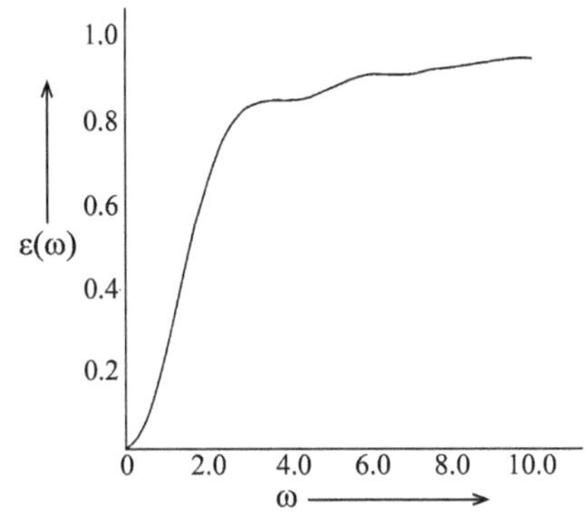

FIGURE 16.13 Variation of Factor of Encircled Energy (FEE) $\varepsilon(\omega)$ with ω.

On integration, we get

$$\int \frac{J_1^2(x)}{x} dx = -\frac{1}{2}\left[J_0^2(x) + J_1^2(x)\right] \tag{16.82}$$

(16.77) reduces to

$$\varepsilon(\omega) = 2\left[-\frac{1}{2}\left\{J_0^2(x) + J_1^2(x)\right\}\right]_0^{2\pi\omega} \tag{16.83}$$

Since $J_0(0) = 1$ and $J_1(0) = 0$, we get the following expression for the factor of encircled energy (FEE)

$$\varepsilon(\omega) = \left[1 - J_0^2(2\pi\omega) - J_1^2(2\pi\omega)\right] \tag{16.84}$$

This relation was first derived by Lord Rayleigh [41–42]. Figure 16.13 shows the variation of the FEE $\varepsilon(\omega)$ with ω. Note the monotonically increasing nature of $\varepsilon(\omega)$ with increasing values of ω.

Since the measurement of factor of encircled energy is relatively simpler than the measurement of the point spread function, designers and system analysts looked for rules of thumb that could enable the prediction of resolving power from measurements of the factor of encircled energy. Huber [42] undertook experiments in this regard, and observed that 'the resolving power could be related to the diameter of the central core of the image that contained 25 per cent of total light' [43]. In a latter publication, Hopkins et al. vindicated the observation, and concluded that the diameter of the central core containing 25–30 per cent of total light gives a heuristic estimate for resolving power in most practical systems.

Few significant studies on the computation of the factor of encircled energy and its practical measurement have been reported in the published literature [44–48]. Since the detectors used currently have a square area for receiving radiation, a similar concept called 'ensquared energy/power' has been developed for practical use [49].

16.4 Resolution and Resolving Power

It is seen above that the image of a point object, by a real optical imaging system of finite aperture, is not a point, but a non-uniform patch of light of finite, albeit small, size on the image plane. Equation (16.41) shows that, in general, the size, shape, and distribution of light intensity in the patch depend on the size of the pupil, the aberrations of the imaging system, non-uniformity, if any, in amplitude over the pupil, and the location of the object point with respect to the optical axis. Because the image of a point object is necessarily finite in size, it follows that when two adjacent points are below a certain minimum distance,

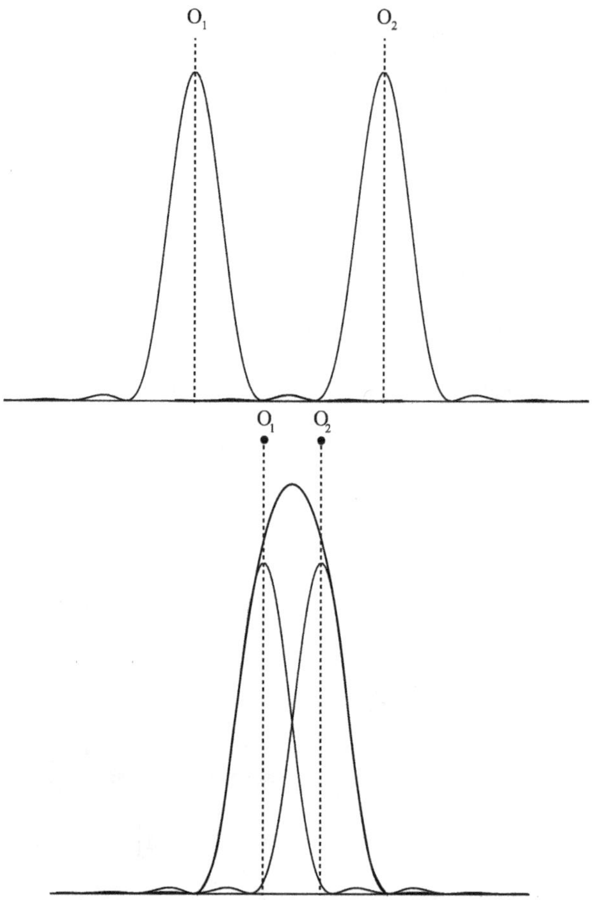

FIGURE 16.14 Top: Two clearly resolved points in the image Bottom: Two object points unresolved in the image. Intensity patterns of two close points add up to form an intensity pattern that looks like it is originating from a single object point.

the two image patches corresponding to the two object points coalesce into a single patch from which the existence of two separate image points can no longer be discerned, i.e. the two object points cannot be 'resolved' in the image. Figure 16.14(a) shows that the addition of the intensity patterns in two image patches corresponding to two object points gives rise to a composite intensity pattern from which the two separate points can be clearly resolved. When the two points lie much closer, they cannot be resolved from their composite intensity pattern (Figure 16.14(b)).

According to Webster's dictionary, in optics, 'the word "resolution" means the act, process or capability of distinguishing between two separate but adjacent objects, or sources of light or between two nearly equal wavelengths' [50]. The closely related term 'resolving power' implies the ability of an optical device to produce images of close objects such that the separate existence of the close objects is retained in the image as two close images, and can be used for further processing as required. However, this ability of an imaging system or device is manifested differently in the case of different classes of objects, e.g. two equally bright/dark points on a dark/bright background, two equally bright/dark lines on a dark/bright background, arrays of points or lines of equal/unequal brightness, etc. Obviously, the topic forms an integral component of 'image quality/performance assessment' in optical imaging systems. A detailed survey of the concept of resolution is given in reference [51]

The earliest scientific approach for tackling the problem of resolution was to seek an answer for the question: in an optical imaging system, what is the minimum separation between two bright object points in a dark background that can be resolved in the image? This topic is commonly referred to as 'Two-point resolution' in optics literature. An identical problem arises in the context of resolving two close spectral lines in spectroscopes, and so similar concepts of resolution have often been used in the fields of optical imagery and spectroscopy.

16.4.1 Two-Point Resolution

16.4.2 Rayleigh Criterion of Resolution

The well- known 'Rayleigh criterion of resolution' [52–53] stipulates:

> Two equal intensity points, placed symmetrically on two sides of the optical axis along any azimuth can be considered to be 'just resolved' when their separation is such that the principal maximum in the diffraction pattern of one of them falls on the first minimum in the diffraction pattern of the other one.

Figure 16.15 illustrates this limiting case. Note that for two object points separated by a distance Δ_{obj}, the distance Δ_{im} between the locations of the principal maxima in their diffraction patterns in the image is

$$\Delta_{\text{im}} = M\Delta_{\text{obj}} \tag{16.85}$$

where M is the transverse magnification between the object and the image. Note that for an object and/or image at infinity, the corresponding distances are to be represented in angular form, giving rise to the concept of 'angular resolution'.

From Table 16.1 we note that the distance of the first minimum of the diffraction pattern from the central maximum is $p_0^1 = 0.6098$. With reference to Figure 16.13, at the Rayleigh limit of resolution the central maxima of the two diffraction patterns are located at $p = +0.3049$, and $p = -0.3049$, and the corresponding intensity patterns $I_1(p)$ and $I_2(p)$ are

$$I_1(p) = \left[\frac{2J_1\{2\pi(p+0.3049)\}}{\{2\pi(p+0.3049)\}}\right]^2 \quad \text{and} \quad I_2(p) = \left[\frac{2J_1\{2\pi(p-0.3049)\}}{\{2\pi(p-0.3049)\}}\right]^2 \tag{16.86}$$

The composite intensity $C(p)$ along the azimuth p is

$$C(p) = I_1(p) + I_2(p) \tag{16.87}$$

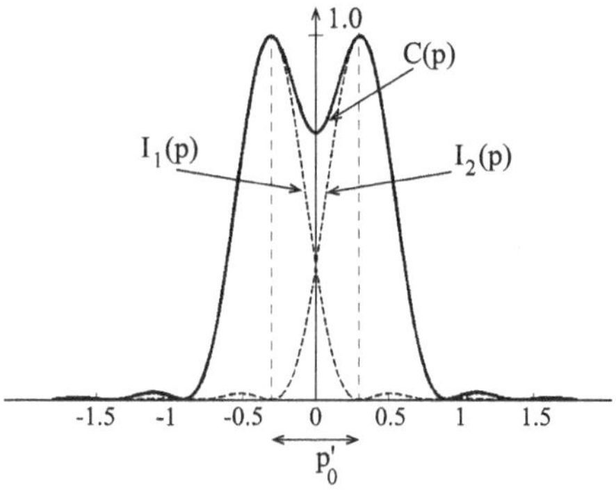

FIGURE 16.15 Intensities $I_1(p)$ and $I_2(p)$ in the images of two equal intensity bright points in a dark background. The points are placed symmetrically on two sides of the optical axis. The separation of the points is p'_0. The composite intensity $C(p) = I_1(p) + I_2(p)$.

Figure 16.15 illustrates the intensity distributions $I_1(p)$ and $I_2(p)$, and the composite intensity distributions $C(p)$ at the Rayleigh limit of resolution. Note that the dip in intensity on the axis at the midpoint of the separation between the two principal maxima is $C(p) = 0.81$.

The canonical variable p represents the normalized distance of a point, lying on the azimuth p, from the optical axis. It is given by

$$p = \left(\frac{n \sin\alpha}{\lambda}\right)\chi = \left(\frac{nh}{\lambda}\right)\tan\beta \tag{16.88}$$

where χ is the geometrical distance of the point from the axis. The semi-angular aperture of the axial image forming cone of light is α. The radius of the pupil is h, and the angle subtended by χ at the centre of the pupil is β. For small values of β, $\tan\beta \approx \beta$.

The distance of the first minimum of $I(p)$ from the optical axis is $p_0^1 = 0.61$. Let the corresponding geometrical distance be $\Delta\chi$ that subtends an angle $\Delta\beta$ at the centre of the pupil. Obviously, per the Rayleigh criterion, $\Delta\chi$ and $\Delta\beta$ are the least resolvable distance and the least resolvable angle, respectively. From (16.73), we get the following expressions for least resolvable distance/angle per the Rayleigh criterion of resolution.

$$\left.\begin{aligned} \Delta\chi &= 0.61\frac{\lambda}{n\sin\alpha} \\ \Delta\beta &= 1.22\frac{\lambda}{n\Phi} \end{aligned}\right\} \tag{16.89}$$

where the diameter ϕ of the pupil is equal to 2h. The first relation applies in the case of finite conjugate, and the next one is appropriate for an infinite conjugate. Note that, in general, $\Delta\chi$ and $\Delta\beta$ are not same for the object space and image space, and quantities specific to a space call for use of corresponding pupil parameters.

It is interesting to note the comments by Rayleigh himself on his famous criterion of resolution.

> This rule is convenient on account of its simplicity and it is sufficiently accurate in view of the necessary uncertainty as to what is meant by resolution. Perhaps in practice more favourable conditions are necessary to secure a resolution that would be thought satisfactory.
>
> [53]

Schuster [54] noted that the Rayleigh criterion of resolution was more suited for visual observation, and the central dip of the composite intensity distribution may not be appropriate for other detectors. He proposed a criterion of resolution that can be effectively used with any detection mechanism. The criterion is: Two adjacent points or lines are considered resolved when there is no overlap of their central lobes. This limit of resolution is double the Rayleigh limit of resolution. On the other hand, few proposed 'a generalized Rayleigh criterion' based solely on the magnitude of the dip in the middle of the composite intensity pattern, without any reference to the intensity distributions of the individual points or lines [55].

The comments of Sparrow are also noteworthy in this context:

> As originally proposed, the Rayleigh criterion was not intended as a measure of the actual limit of resolution, but rather as an index of the relative merit of different instruments.
>
> [56]

It should be noted that, at times, in optics literature, the Rayleigh criterion of resolution is called 'the Airy limit of resolution', presumably because the criterion is derived essentially by using the intensity distribution in the Airy pattern [55]. The well-known Dawes' resolution limit in astronomy, determined empirically by using a telescope on double stars of equal brightness, stipulates a lower value for the minimum resolvable angle compared to the Rayleigh criterion. It is given by $\Delta\beta = 1.02(\lambda / n\Phi)$ [56].

16.4.3 Sparrow Criterion of Resolution

The top diagram in the left of Figure 16.16 shows the variation of composite intensity C(p) with p for a large value of separation $\hat{p}$ between the maxima of the diffraction images of two equal intensity object points placed symmetrically on the optical axis along the azimuth p on the transverse object plane. The variation of C(p) versus p for a small value of $\hat{p}$ is shown in the top diagram in the right of Figure 16.16.

It should be noted that Figure 16.15, corresponding to the Rayleigh criterion of resolution, is a special case, where $\hat{p} = p_0^1$. In this case, the intensity curve has a minimum at $p = 0$, surrounded by two maxima at the locations of the central maxima of the two diffraction patterns. Note that the intensity curve is concave at $p = 0$. In the next case the intensity curve has a single maximum at $p = 0$, where the curve is convex. The transition from the first case to the second occurs at a certain value of $\hat{p}$, say $\tilde{p}$. If $\hat{p} > \tilde{p}$, the two bright maxima of C(p) are separated by a strip of lower intensity, as illustrated in the contour plots in the lower left diagram. The lower diagram on the right shows that this strip is no longer present when $\hat{p} \le \tilde{p}$. This distance $\tilde{p}$ is considered to be the limit of resolution in the Sparrow criterion.

The composite diffraction pattern C(p) is symmetric about the origin. All odd derivatives with respect to p vanish at $p = 0$, and the analytical statement of the Sparrow criterion of resolution [57] is

$$\frac{d^2}{dp^2} C(p,\tilde{p}) = 0 \text{ at } p = 0 \tag{16.90}$$

$C(p,\tilde{p})$ is given by

$$C(p,\tilde{p}) = \left[\frac{2J_1\left\{2\pi\left(p - \frac{\tilde{p}}{2}\right)\right\}}{\left\{2\pi\left(p - \frac{\tilde{p}}{2}\right)\right\}}\right]^2 + \left[\frac{2J_1\left\{2\pi\left(p + \frac{\tilde{p}}{2}\right)\right\}}{\left\{2\pi\left(p + \frac{\tilde{p}}{2}\right)\right\}}\right]^2 = I\left(p - \frac{\tilde{p}}{2}\right) + I\left(p + \frac{\tilde{p}}{2}\right) \tag{16.91}$$

In short notation, (16.90) can be rewritten as

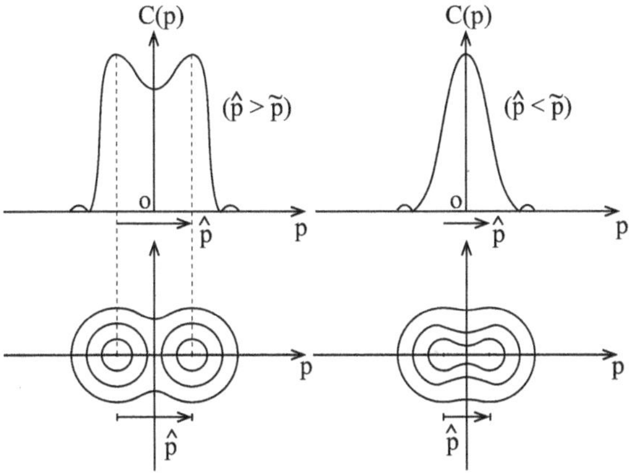

FIGURE 16.16 Diagrams for variation of composite intensity $C(p)$ with p in the top-left: $\hat{p} > \tilde{p}$, right: $\hat{p} < \tilde{p}$; corresponding contour lines of equal intensity on the image plane in the bottom-(left): $\hat{p} > \tilde{p}$, and (right): $\hat{p} < \tilde{p}$.

$$C''(0,\tilde{p}) = I''\left(0 - \frac{\tilde{p}}{2}\right) + I''\left(0 + \frac{\tilde{p}}{2}\right) = 0 \tag{16.92}$$

or,

$$C''(\tilde{p}) = I''\left(-\frac{\tilde{p}}{2}\right) + I''\left(\frac{\tilde{p}}{2}\right) = 0 \tag{16.93}$$

Note that $I(p)$ and $I''(p)$ are even functions of p, so that the Sparrow criterion of resolution may be reinterpreted as

$$I''\left(\frac{\tilde{p}}{2}\right) = 0 \tag{16.94}$$

This implies that 'the limit of resolution, $\tilde{p}$, is given by the separation of the two points of inflection of the intensity curve $I(p)$ of a single point with its diffraction maximum on the axis' [58]. Figure 16.17 shows $\tilde{p} = |C_1C_2|$, where C_1 and C_2 are the two points of inflection of the intensity curve.

Since $I(p) = [F(p)]^2$, we have

$$I'(p) = \frac{dI(p)}{dp} = 2F(p)F'(p) \tag{16.95}$$

and

$$I''(p) = 2[F'(p)]^2 + 2F(p)F''(p) \tag{16.96}$$

For $p = \frac{\tilde{p}}{2}$,

$$F\left(\frac{\tilde{p}}{2}\right) = \frac{2J_1(\pi\tilde{p})}{(\pi\tilde{p})} \qquad F'\left(\frac{\tilde{p}}{2}\right) = -4\pi\frac{J_2(\pi\tilde{p})}{(\pi\tilde{p})} \qquad F''\left(\frac{\tilde{p}}{2}\right) = -8\pi^2\left[\frac{J_2(\pi\tilde{p})}{(\pi\tilde{p})} - \frac{J_3(\pi\tilde{p})}{(\pi\tilde{p})}\right] \tag{16.97}$$

Using (16.94) and (16.97), the equation (16.96) leads to the following relation that can be used for determining $\tilde{p}$.

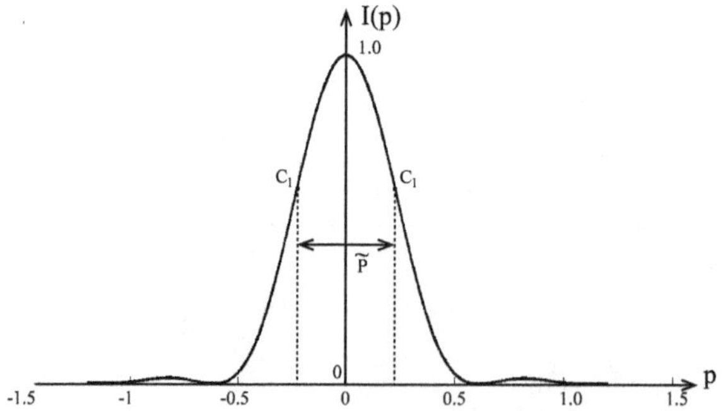

FIGURE 16.17 Sparrow limit of resolution between two equal intensity self-luminous points in a dark background is $\tilde{p} = 0.4736$. It is the distance between the two points of inflexion, C_1 and C_2, where $\frac{d^2\{I(p)\}}{dp^2} = 0$.

$$J_1(\pi\tilde{p})\left[J_2(\pi\tilde{p})-(\pi\tilde{p})J_3(\pi\tilde{p})\right]=(\pi\tilde{p})\left\{J_2(\pi\tilde{p})\right\}^2 \tag{16.98}$$

Evidently, no analytical solution of this nonlinear equation is available. However, the equation can be solved numerically to obtain $\tilde{p}=0.4736$. This value tallies well with the result given by Barakat [59].

The least resolvable distance/angle, per the Sparrow criterion of resolution in the case of incoherent illumination, is

$$\left.\begin{aligned}\Delta\chi &= 0.47\frac{\lambda}{n\sin\alpha}\\ \Delta\beta &= 0.94\frac{\lambda}{n\Phi}\end{aligned}\right\} \tag{16.99}$$

For an aberration-free optical imaging system, the least resolvable distance predicted by the Sparrow criterion of resolution is approximately 22 per cent lower than the same predicted by the Rayleigh criterion of resolution. Sparrow called his criterion of resolution 'the actual limit of resolution', and Osterberg [60] called it 'the physical limit of resolution'. Sparrow called his criterion of resolution an 'undulation condition'. It is relevant to note the remarks by Ramsay et al. on the Sparrow criterion of resolution:

> One conspicuous difficulty with Sparrow's criterion is that before the limit of resolution can be known, both the relative intensities of the lines to be resolved and the equation of the J curve (the composite intensity curve) must be known. In general this information is not available; and even if it were, the solution for …the least resolvable distance might not be possible in all cases.
>
> [55]

16.4.4 Dawes Criterion of Resolution

The limits of angular resolution $\Delta\beta$, stipulated by either the Rayleigh criterion (16.89) or the Sparrow criterion (16.99), are linearly proportional to λ, the wavelength of light (one radian is equal to 206265 arcseconds). Per the Rayleigh criterion, angular resolutions at three wavelengths corresponding to green ($\lambda = 0.555\mu m$), violet ($\lambda = 0.404\mu m$), and red ($\lambda = 0.650\mu m$) are:

$$\left.\begin{aligned}(\Delta\beta)_{green} &\cong \frac{140}{\varnothing}\text{arcsecond}\\ (\Delta\beta)_{violet} &\cong \frac{101.5}{\varnothing}\text{arcsecond}\\ (\Delta\beta)_{red} &\cong \frac{164}{\varnothing}\text{arcsecond}\end{aligned}\right\} \tag{16.100}$$

where aperture diameter $\varnothing$ is expressed in millimetres and one radian is taken as equal to 206265 arcseconds. Because the eye has maximum response in green light, usually $(140/\varnothing)$ is quoted as the Rayleigh limit of angular resolution in visual telescopes. For stars of equal brightness, the dip between the stars' maxima is 26 per cent and a human observer can conveniently resolve the two stars. It was experimentally found that even closer stars could be distinguished in practice. Based on a large number of experimental observations with small telescopes, Dawes [61–62] empirically determined the limit of angular resolution to be $(117/\Phi)$ arcseconds in the case of white double stars of equal brightness. His observations were found to hold good for seeing in telescopes of diameter about 150mm. His criterion implies a central dip of only 3.2 per cent between the intensity maxima.

16.4.5 Resolution in the Case of Two Points of Unequal Intensity

In the case of imaging of two points of unequal intensity by an axisymmetric aberration-free imaging system, obviously the Rayleigh criterion of resolution loses any significance. But the Sparrow criterion can still be used, so long as the inequality in intensity is not very large. The following procedure is proposed by Osterberg [60].

Suppose that the two dissimilar points of unequal intensity are located equidistant from the optical axis. The point $p = \ddot{p}$ at which $\frac{dC(p)}{dp} = 0$ is to be determined. The composite intensity pattern $C(p)$ is the weighted sum of the individual intensities by the two points. Note that unlike in the case of equal intensity, points $\ddot{p} \neq 0$ in general. The Sparrow limit of resolution of such particles is the separation $\dddot{p}$ between the points for which $\frac{d^2C(p)}{dp^2} = 0$ at $p = \ddot{p}$.

16.4.6 Resolution in the Case of Two Mutually Coherent Points

So far, we have considered the resolution of two self-luminous point objects in a dark background. Typical examples of such objects are astronomical stars in telescopy, or fluorescent point-like samples in microscopy. Since the light from two such points is practically independent of each other, the resultant intensity at any point in the wave field is a summation of individual intensities of light from the two points. Illumination of the light field in such cases is called 'incoherent illumination'. On the other hand, in the case of two non-self-luminous points, or two trans-illuminated points, the light from two such points is not independent of each other; consequently, the resultant intensity at any point in the wave field is equal to the squared modulus of the summation of the individual amplitudes of the light vibrations from the two object points. Illumination of the light field in such cases is called 'coherent illumination'. Incoherent and coherent illumination are two extreme conditions of illumination. In the vast majority of practical optical systems, the condition of illumination lies in between these two extreme cases. The illumination is then said to be 'partially coherent'. Also, in the presence of aberrations, the PSFs change significantly, and this change affects the resolving power of the system. The problem of resolution becomes too involved in such cases, and we refer to the references [63–70] for more details.

The Abbe/Rayleigh limit of resolution implicitly assumes incoherent illumination. If two mutually coherent object points, separated by the Rayleigh limit of resolution, are placed symmetrically on a transverse plane on two sides of the optical axis of an optical imaging system with an Airy pupil, the composite intensity distribution $I(p)$ in the image is given by

$$I(p) = |F(p)|^2 = |F_1(p) + F_2(p)|^2 \tag{16.100}$$

where

$$F_1(p) = \frac{2J_1[2\pi(p+0.3049)]}{[2\pi(p+0.3049)]} \quad \text{and} \quad F_2(p) = \frac{2J_1[2\pi(p-0.3049)]}{[2\pi(p-0.3049)]} \tag{16.101}$$

Figure 16.18 shows $F_1(p)$, $F_2(p)$, $F(p)$, and $I(p)$. Clearly, the composite intensity pattern does not resolve the two points, although they are separated by the Rayleigh limit of resolution. Compare this figure with Figure 16.15 illustrating resolution of the two object points separated by the Rayleigh limit of resolution when the object points are mutually incoherent.

The Sparrow criterion of resolution can be applied in this case to obtain a physical limit of resolution. Using the same notation, $\tilde{p}$, for the limit of resolution, the analytical statement for the Sparrow limit of resolution is similar to the equation (16.90) and is written as

$$\frac{d^2}{dp^2}C(p,\tilde{p}) = 0 \text{ at } p = 0 \tag{16.102}$$

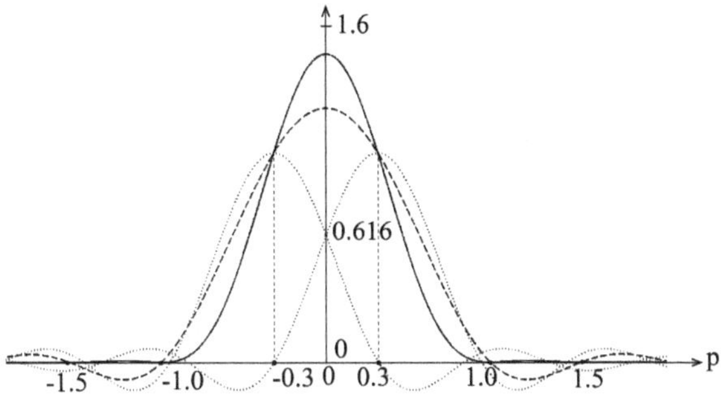

FIGURE 16.18 Image of two object points separated by the Rayleigh limit of resolution under coherent illumination. $F_1(p), F_2(p)$: Amplitude distributions of the two points (dotted lines). Composite amplitude distribution: $F(p) = F_1(p) + F_2(p)$ (dashed line). Composite intensity distribution: $I(p) = |F(p)|^2$ (solid line).

where $C(p, \tilde{p})$ is now given by a different expression as

$$C(p,\tilde{p}) = \left| F\left(p - \frac{\tilde{p}}{2}\right) + F\left(p + \frac{\tilde{p}}{2}\right) \right|^2 \tag{16.103}$$

Noting that

$$F(p) = \frac{2J_1(2\pi p)}{(2\pi p)} \qquad F'(p) = -\frac{4\pi J_2(2\pi p)}{(2\pi p)} \qquad F''(p) = -8\pi \left[\frac{J_2(2\pi p)}{(2\pi p)^2} - \frac{J_3(2\pi p)}{(2\pi p)} \right]^2 \tag{16.104}$$

From (16.102) – (16.103) we get

$$\begin{aligned} \frac{d^2}{dp^2} C(p,\tilde{p}) &= 2\left| F\left(p - \frac{\tilde{p}}{2}\right) + F\left(p + \frac{\tilde{p}}{2}\right) \right| \left\{ \left| F''\left(p - \frac{\tilde{p}}{2}\right) + F''\left(p + \frac{\tilde{p}}{2}\right) \right| \right\} \\ &\quad + 2\left[F'\left(p - \frac{\tilde{p}}{2}\right) + F'\left(p + \frac{\tilde{p}}{2}\right) \right]^2 \end{aligned} \tag{16.105}$$

From (16.104), it is seen that F(p) and $F''(p)$ are even functions of p, and $F'(p)$ is an odd function of p. Therefore, it follows

$$\frac{d^2}{dp^2} C(p,\tilde{p}) = 8F\left(p + \frac{\tilde{p}}{2}\right) F''\left(p + \frac{\tilde{p}}{2}\right) \tag{16.106}$$

When the Sparrow criterion (16.84) is satisfied, we get

$$F\left(\frac{\tilde{p}}{2}\right) F''\left(\frac{\tilde{p}}{2}\right) = 0 \tag{16.107}$$

Since $F\left(\frac{\tilde{p}}{2}\right) \neq 0$ for likely solutions of the equation $F''\left(\frac{\tilde{p}}{2}\right) = 0$, the relation (16.107) reduces to the equation $F''\left(\frac{\tilde{p}}{2}\right) = 0$. It is evident that $\tilde{p}$, the least resolvable distance according to the Sparrow criterion in the case of coherent illumination is equal to the distance between the two inflection points on the amplitude curve F(p) of a single point with its diffraction maximum on the axis. This was first noted by Luneburg [58]. It has been shown earlier, in Section 16.4.3, that the least resolvable distance per the Sparrow criterion in the case of incoherent illumination is equal to the distance between the two inflection points on the intensity curve I(p) of a single point with its diffraction maximum on the axis.

Using (16.104), by algebraic manipulation, we get the following nonlinear equation satisfied by $\tilde{p}$

$$(\pi\tilde{p})J_3(\pi\tilde{p}) = J_2(\pi\tilde{p}) \tag{16.108}$$

The above equation can be solved numerically to get a solution $\tilde{p} = 0.7321$. Therefore, the least resolvable distance/angle per the Sparrow criterion of resolution in the case of coherent illumination is

$$\left.\begin{aligned} \Delta\chi &= 0.73\frac{\lambda}{n\sin\alpha} \\ \Delta\beta &= 1.46\frac{\lambda}{n\Phi} \end{aligned}\right\} \tag{16.109}$$

Comparing (16.109) with (16.89) and (16.99), it is noted the least resolvable distance/angle per the Sparrow criterion of resolution in coherent illumination is larger than the Rayleigh limit by about 20 per cent. It is important to note that it is also larger than the least resolvable distance per Sparrow criterion in incoherent illumination by approximately 55 per cent. This observation prompted Luneburg to note: '… it will be more difficult to resolve coherent light sources than incoherent sources…' [58]. In general, this observation is somewhat misleading, for the case treated above assumes the two mutually coherent points to be of the same phase. In the case of non-zero phase difference between the two coherently illuminated points, the least resolvable distance/angle can be significantly different. This will be discussed in the next section.

16.4.7 Breaking the Diffraction Limit of Resolution

Lowering the least resolvable distance/angle in the image continues to pose a major challenge in the field of image formation. The Abbe/Rayleigh limit of resolution or the Sparrow limit of resolution are the ultimate limits of resolution that can be achieved by aberration-free imaging systems, so much so that such systems are euphemistically called 'diffraction limited systems'. Scientists and technologists are striving hard to overcome or to circumvent this so-called 'fundamental' limit of resolution by whatever means are available and seem appropriate for the purpose. These achievements are euphemistically called 'superresolution', 'hyperresolution', or 'ultraresolution', etc. In the 1960s, Kartashev [71] and Lukosz [72] carried out thought-provoking scientific analyses of superresolution on the basis of information theory. Further advances were undertaken by Zalevsky, Mendlovic, and Lohmann [73–74]. In the case of imaging accessible objects, for example in microscopy or nanoscopy, significant progress has already been made in biomedicine, in microfabrication, and in optical lithography [75–76]. We mention below a few of them.

16.4.7.1 Use of Phase-Shifting Mask

In Section 16.4.5, we noted that the least resolvable distance (as per the Sparrow criterion) $\tilde{p}$ in the case of coherent illumination is larger than that in case of incoherent illumination. It is important to note that our derivation of $\tilde{p}$ assumed the two coherently illuminated points to be in the same phase.

The least resolvable distance for the case of two coherently illuminated equi-phase points is not equal to that in the case of two coherently illuminated points with a phase difference between them.

For the sake of illustration, let us consider the imaging of two points, placed symmetrically around the optical axis on a transverse plane, by an axially symmetric optical imaging system in coherent illumination. The distance between the maxima of the corresponding diffraction patterns on the image plane is $\hat{p}$, and let the two points have a phase difference π. The composite amplitude distribution along the line joining the locations of the two diffraction maxima on the image plane is given by

$$F(p,\tilde{p})=\left\{\left[\frac{2J_1\left\{2\pi\left(p+\frac{\hat{p}}{2}\right)\right\}}{\left\{2\pi\left(p+\frac{\hat{p}}{2}\right)\right\}}\right]+e^{i\pi}\left[\frac{2J_1\left\{2\pi\left(p-\frac{\hat{p}}{2}\right)\right\}}{\left\{2\pi\left(p-\frac{\hat{p}}{2}\right)\right\}}\right]\right\}=\left\{\left[\frac{2J_1\left\{2\pi\left(p+\frac{\hat{p}}{2}\right)\right\}}{\left\{2\pi\left(p+\frac{\hat{p}}{2}\right)\right\}}\right]-\left[\frac{2J_1\left\{2\pi\left(p-\frac{\hat{p}}{2}\right)\right\}}{\left\{2\pi\left(p-\frac{\hat{p}}{2}\right)\right\}}\right]\right\} \quad (16.110)$$

Symmetry considerations show that $F(p,\hat{p})=0$ at $p=0$ for any value of $\hat{p}$. Consequently, the composite intensity $I(p,\hat{p})=\left|F(p,\hat{p})\right|^2=0$ at $p=0$ for any value of $\hat{p}$. However, destructive interference causes significant attrition in heights of the two peaks in the composite intensity pattern, setting thereby a practical limit on the least resolvable distance. Also note that the distance between the two peaks of the composite intensity pattern is not the same as the distance between the locations of the two individual diffraction maxima. Figure 16.19 illustrates the situation for three values of $\hat{p}=0.61, 0.47$ and 0.25.

It should be noted that a phase difference of any odd integral multiple of π produces the same effect. The required phase can be introduced by a thin plate of thickness t between one of the points where t is given by

$$t=j\frac{\lambda}{2(\mu-1)},\quad j=1,3,5,\ldots \quad (16.111)$$

where μ is the refractive index of the optical material of the thin plate at the working wavelength λ. The preparation of suitable phase shifting mask in the case of extended objects is an involved process, and so practical use of the phase shifting masks in high resolution lithography was initially hindered until the emergence of suitable techniques.

Following the pioneering use of these masks by Flanders et al. [77] and Shibuya [78], Levenson [79–80] has extensively used these phase shifting masks in IC fabrication by optical microlithography. Later developments are noted in references [81–82]. Another proposal for the use of phase shift masks in holographic storage was made by Hänsel and Polack [83].

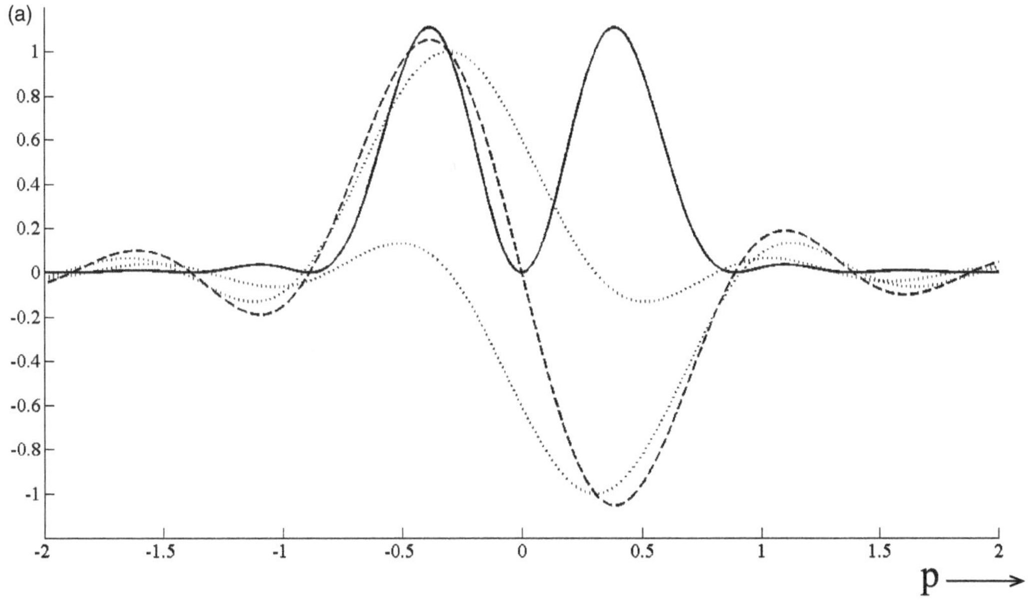

FIGURE 16.19 Image of two object points separated by a distance $\hat{p}$. The two points have a phase difference of π. $F_1(p), F_2(p)$: Amplitude distributions of the two points. Composite amplitude distribution: $F(p)=F_1(p)+F_2(p)$. Composite intensity distribution: $I(p)=\left|F(p)\right|^2$. (a) $\hat{p}=0.61$ (b) $\hat{p}=0.47$ (c) $\hat{p}=0.25$.

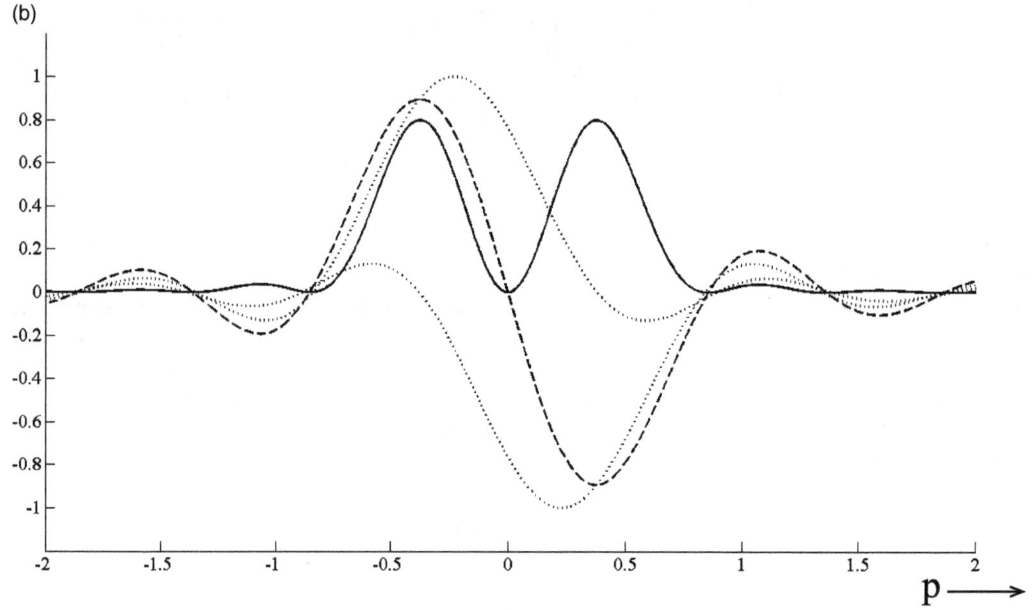

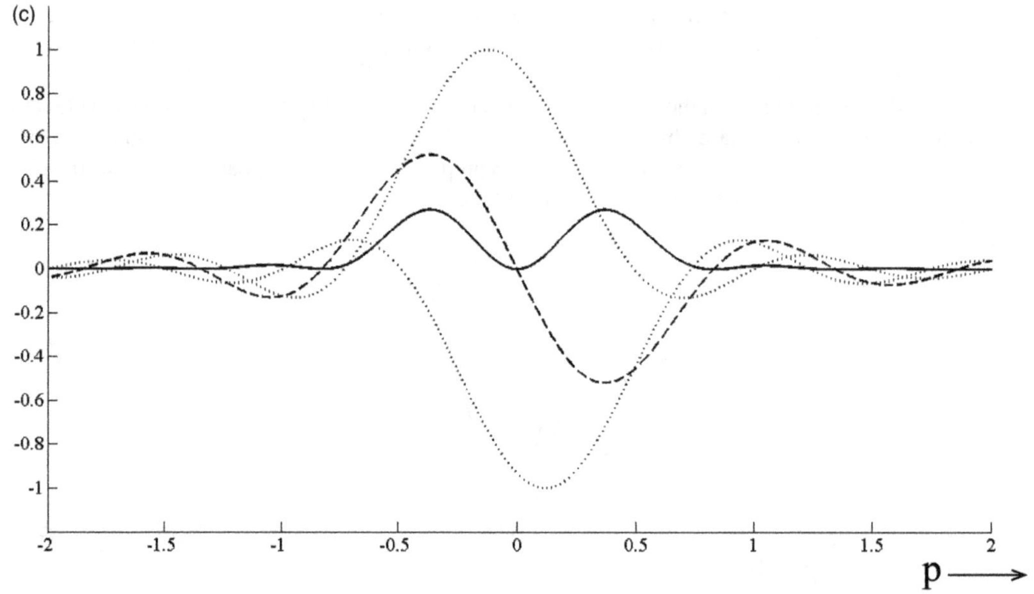

FIGURE 16.19 continued

16.4.7.2 Superresolution over a Restricted Field of View

It is obvious that an increase in the resolving power of imaging systems calls for a narrowing down of the central lobe of the point spread function. For an Airy pupil of given numerical aperture/diameter, the shape and size of the central lobe and the secondary lobes are fixed. In the case of an Airy pupil with central obscuration, the central lobe becomes narrower, thereby providing an increase in resolving power [84]. In the extreme case of an infinitesimally narrow ring aperture along the periphery of the pupil, the intensity point spread function becomes

$$I(p) = \left[J_0(2\pi p)\right]^2 \tag{16.112}$$

Since $J_0(x)$ has its first zero at x=2.4048, the first zero of $I(p)$ is at $p_0^1 = 0.38$. The least resolvable distance/angle according to the Rayleigh criterion of resolution is

$$\left.\begin{aligned} \Delta\chi &= 0.38\frac{\lambda}{(n\sin\alpha)} \\ \Delta\beta &= 0.76\frac{\lambda}{n\Phi} \end{aligned}\right\} \tag{16.113}$$

Applying the Sparrow criterion for incoherent illumination, we get, from (16.94), $\tilde{p} = 0.28$. Accordingly, the least resolvable distance/angle per the Sparrow criterion is

$$\left.\begin{aligned} \Delta\chi &= 0.28\frac{\lambda}{(n\sin\alpha)} \\ \Delta\beta &= 0.56\frac{\lambda}{n\Phi} \end{aligned}\right\} \tag{16.114}$$

Comparing (16.113) with (16.89), and (16.114) with (16.99), we note that, compared to an Airy pupil, the improvement of resolving power in the case of the narrow ring aperture on the periphery of the pupil is approximately 40 per cent. However, this improvement cannot be utilized in practice on account of the huge loss of intensity in the central lobe, and the concomitant rise in intensity in the neighbouring sidelobes.

Toraldo di Francia [85–87] proposed a way to circumvent the diffraction limit of resolution by aiming to overcome the limit over a restricted field of view by means of pupil plane filtering. He was the first to coin the term 'superresolution'. His goal was to achieve a narrow central lobe surrounded by an extended dark ring, surrounded by a zone of large intensity. The latter zone is excluded from the field of view, and the effective field of view is somewhat determined by the extent of the dark ring. The situation is illustrated in Figure 16.20. He explored the use of concentric annular rings on the exit pupil to achieve the desired goals. The point spread functions shown in Figure 16.20 show the normalized intensity, where the intensity at the centre of the diffraction pattern is shown as 1.0 for both the Airy pupil and the superresolving pupil. The absolute intensity at the central maximum of the diffraction pattern for the superresolving filter is less than the absolute intensity attained with an Airy pupil by many orders of magnitude, so much so that the filter cannot be used in practice. Investigations are continuing to explore suitable amplitude/phase/complex pupil

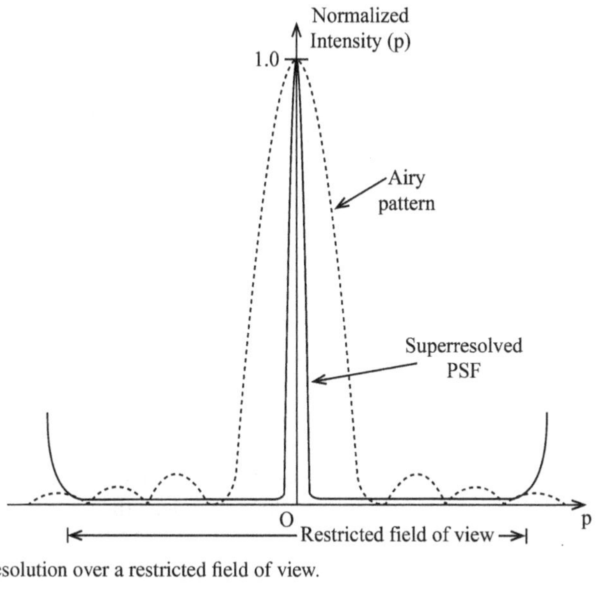

FIGURE 16.20 Superresolution over a restricted field of view.

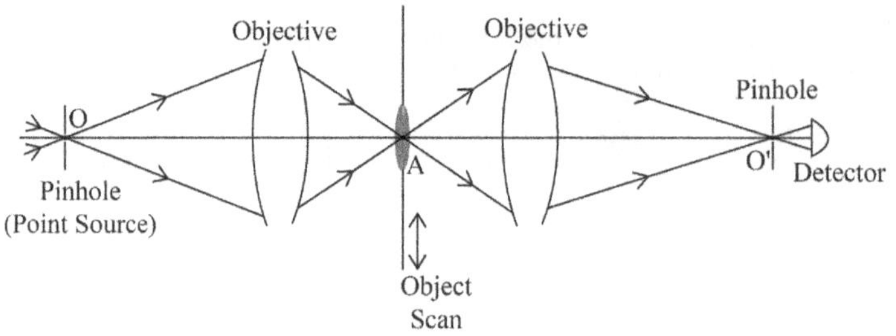

FIGURE 16.21 Schematic diagram of a confocal scanning microscope.

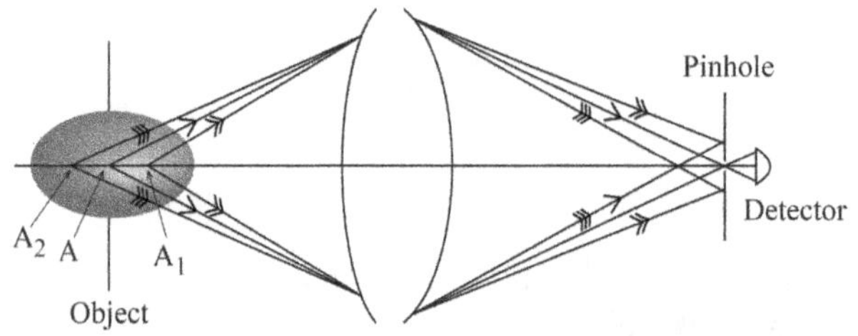

FIGURE 16.22 Better axial sectioning in confocal microscope. Rays from the object point A are allowed to reach the detector; rays from points like A_1 and A_2 are cut off by the pinhole.

plane filters for achieving useful superresolution over restricted fields of view by Boivin et al, Cox et al, Sales et al, and Hazra et al, among others. [88–98].

Earlier investigation by Bachl and Lukosz on superresolution imaging of a reduced object field is also noteworthy [99].

16.4.7.3 Confocal Scanning Microscopy

One of the basic reason for loss of contrast and resolution in conventional whole field microscopy is the overlap of in focus and out of focus images of neighbouring points on the desired image point. Confocal scanning microscopy minimizes this overlap by taking recourse to point-by-point scanning of the object in a transmission/reflection mode, and reconstructing the image from the scanned data. Figure 16.21 presents a schematic diagram of a confocal scanning microscope operating in transmission mode. Two objectives are in use. The first one images a point source O of light on a point A on the object. The second objective images the point A on a pinhole O′ in front of the detector. Assuming the size of the pinholes to be infinitesimally small, no in focus or out of focus light from any point other than the point A on the object reaches the detector. The source pinhole ensures that no off-axis point on the object is illuminated, and the pinhole in front of the detector ensures that light from other axial points like A_1 or A_2 can reach the detector. The latter effect is illustrated in Figure 16.22. Of course, in practice, the necessarily finite but small size of the pinholes causes departure from the ideal performance.

Minsky invented this scheme in the pre-laser days, and had a patent on his 'Microscopy Apparatus' [100]. But no significant use could be made lacking a suitable light source. Interest on the topic was revived after the advent of the laser, and several competing commercial products of 'Confocal laser scanning microscopes' are in routine use now. The basic theory and implementation of different schemes for confocal scanning are given in references [101–105].

Although the primary goal of confocal scanning is not 'breaking the diffraction limit of resolution', Minsky noted, 'This brings an extra premium because the diffraction patterns of both pinhole apertures are multiplied coherently: the central peak is sharpened and the resolution is increased' [105].

Assuming the two pinhole – objective arrangements to be identical, the amplitude point spread function is $\{F(p) \times F(p)\} = \{F(p)\}^2$. Consequently, the intensity point spread function is $\{F(p)\}^4$. Assuming the objective lenses to be aberration free, the intensity point spread function $I(p)$ is

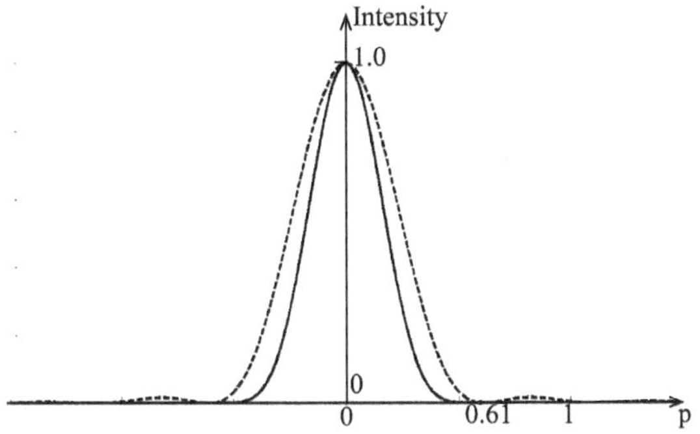

FIGURE 16.23 Dashed line: IPSF for a conventional diffraction limited microscope. $\text{IPSF} = \left[2J_1(2\pi p)/(2\pi p)\right]^2$. Solid line: IPSF for a diffraction limited confocal microscope. $\text{IPSF} = \left[2J_1(2\pi p)/(2\pi p)\right]^4$.

$$I(p) = \left[F(p)\right]^4 = \left[\frac{2J_1(2\pi p)}{(2\pi p)}\right]^4 \tag{16.115}$$

Compared to the intensity point spread function $\left\{2J_1(2\pi p)/(2\pi p)\right\}^2$ of a conventional microscope of the same numerical aperture, (16.115) shows that the point spread function in a confocal microscope is much narrower (Figure 16.23). Note that, per the Rayleigh criterion of resolution, there is no gain in resolution, since the location of the first zero, $p_0^1 = 0.61$ is same in both cases. But, per the Sparrow criterion, $\tilde{p} = 0.32$. The least resolvable distance in a confocal microscope using the Sparrow criterion is

$$\Delta\chi = 0.32\frac{\lambda}{n\sin\alpha} \tag{16.116}$$

Comparing this with the least resolvable distance in the case of incoherent illumination, as given by (16.99), we note that there is a gain of approximately 30 per cent in the case of objectives of the same numerical aperture. Current research in the field involves exploring the replacement of the pinhole by a detector array [106–107]. Each pixel of the detector array simultaneously gives a confocal image. The resultant four dimensional data set can be reconstructed to get images of improved resolution and signal to noise ratio

16.4.7.4 Near Field Superresolving Aperture Scanning

Alike diffractive optical elements, superresolution by near field aperture scanning is an example of prescient invention in the field of optics. In 1928, Synge [108] suggested a method 'for extending the microscopic resolution into the ultra-microscopic region'. On account of the non-availability of required technical support, the concept could not be implemented in practice at that time.

This approach for superresolution is based on the physical principle: In a measurement where the probing detector is less than one wavelength away from the specimen, the resolution is independent of the working wavelength, and it is determined by the size of the aperture of the detector. Current parlance says that the measurement is being undertaken in the 'near field'. The latter term should not be confused with the 'Fresnel region', which used to also be called the 'near field' in the past.

This principle was validated in 1972 at microwave frequencies ($\lambda = 3\text{cm}$) by Ash and Nicholls with a resolution of $\lambda/60$ [109]. Both Ash et al. [110] and Pohl et al. [111] reported the practical implementation of the concept in visible wavelengths in 1984. Incidentally, implementation of a scanning tunnelling microscope was also reported for similar purposes around the same time [112]. In both of these microscopes it is important to use a highly precise feedback mechanism to ensure the distance $d\,(d \ll \lambda)$ between the probe and specimen during scanning. Also, design of the probe and its tip plays a major role in the reliability and efficiency of this measurement [113]. Further details may be found in references [114–116].

REFERENCES

1 A.E. Conrady, 'Chapter III Physical Aspects of Optical Images' in *Applied Optics and Optical Design, Part one*, Dover, New York (1957) pp. 126–141.
2 M. Born and E. Wolf, *Principles of Optics*, Pergamon, Oxford (1980) p. 370.
3 F.M. Grimaldi, Physico mathesis de lumine, coloribus, et irde, aliisque annexis libri duo, Vittorio Bonati, Bologna, Italy (1665) pp. 1–11. [in Latin].
4 C. Huygens, *Traité de la lumière*, Leyden (1690). English Translation by S. P. Thomson, *Treatise on Light*, Macmillan, London (1912). [Note: Huygens published his *Traité* in 1690; however, in the preface to his book, Huygens states that in 1678 he first communicated his book to the French Royal Academy of Sciences.]
5 A. Fresnel, 'Note sur l'application du principe d'Huygens et de la théorie des interferences aux phénomènes de la réflexion et de la diffraction', (1818–1819) in *Œuvres complètes d'Augustin Fresnel*, Impériale, Paris (1916).
6 T. Young, 'The Bakerian Lecture. Experiments and Calculations relative to Physical Optics', *Phil. Trans. Roy. Soc. London*, Vol. 94 (1804) pp. 1–16.
7 T. Young, 'II. The Bakerian Lecture. On the Theory of Light and Colours', *Phil. Trans. Roy. Soc. London*, Vol. 92 (1802) pp. 12–48.
8 G. Kirchoff, 'Zur Theorie der Lichtstrahlen', *Ann. d. Physik.*, Vol. 254 (1883) pp. 663–695.
9 Lord Rayleigh, 'On the passage of waves through apertures in plane screens, and allied problems', *Phil. Mag.*, Vol. XLIII (1897) pp. 259–272; also in *Scientific Papers, Vol. IV*, Macmillan, University Press, Cambridge (1903) pp. 283–296.
10 A. Sommerfeld, 'Mathematische Theorie der Diffraction', *Math. Ann.*, Vol. 47 (1896) pp. 317–374. English Translation, *Mathematical Theory of Diffraction*, Birkhäuser, Springer Nature (2004).
11 A. Sommerfeld, *Optics, Vol. IV of Lectures on Theoretical Physics*, Academic, New York (1954).
12 E. Wolf and E.W. Marchand, 'Comparison of the Kirchhoff and the Rayleigh-Sommerfeld Theories of Diffraction at an Aperture', *J. Opt. Soc. Am.*, Vol. 54 (1964) pp. 587–594.
13 J.J. Stamnes, *Waves in Focal Regions*, Adam Hilger, Boston (1986) pp. 24–40.
14 G.A. Maggi, 'Sulla propagazione libera e perturbata della onde luminose in un mezzo isotropo', *Annai di Matematica Pura ed Applicata'*, Vol. 16 (1888) pp. 21–48.
15 A. Rubinowicz, 'Die Beugungswelle in der Kirchoffschen Theorie der Beugungsersceinungen', *Ann. d. Physik, Ser.* 4 Vol. 53 (1917) pp. 257–278.
16 S. Banerji, 'VIII. On the radiation of light from the boundaries of diffracting apertures', *Phil. Mag. Ser. 6*, Vol. 37 (1919) pp. 112–128.
17 G.N. Ramachandran, 'On the radiation from the boundary of diffracting apertures and obstacles', *Proc. Ind. Acad. Sci. A*, Vol. 20 (1945) pp. 165–176.
18 A. Rubinowicz, 'The Miyamoto – Wolf Diffraction Wave', in *Progress in Optics, Vol. IV*, Ed. E. Wolf, North – Holland, Amsterdam (1965) pp. 195–240.
19 J.B. Keller, 'Geometrical Theory of Diffraction', *J. Opt. Soc. Am.*, Vol. 52 (1962) pp. 116–130.
20 C.F. Meyer, *The Diffraction of Light, X rays and Material Particles*, The University Press, Chicago (1934).
21 F. Zernike, 'Diffraction and Optical Image Formation', *Proc. Phys. Soc.* (London), Vol. 61 (1948) pp. 158–164.
22 B.B. Baker and E.T. Copson, *The Mathematical Theory of Huygens' Principle*, Clarendon Press, Oxford (1950).
23 C.J. Bouwkamp, 'Diffraction Theory', *Rep. Progr. Phys.*, Vol. 17 (1954) pp. 35–100.
24 G. Toraldo di Francia, *La Diffrazione della Luce*, Edizioni Scientifiche Einaudi, Paolo Boringhieri, Torino (1958) [in Italian].
25 J.M. Cowley, *Diffraction Physics*, North Holland, Amsterdam (1971).
26 H.H. Hopkins, *Notes on Diffraction*, personal communication (1982).
27 H.H. Hopkins and M.J. Yzuel, 'The computation of diffraction patterns in the presence of aberrations', *Opt. Acta*, Vol. 17 (1970) pp. 157–182.
28 H.H. Hopkins, 'The Development of Image Evaluation Methods', *in Proc. SPIE*, Vol. 46, Image Assessment and Specification, Ed. D. Dutton, Rochester, New York (1974) pp. 2–18.
29 Lord Rayleigh, 'On images formed without reflection or refraction', *Phil. Mag., Fifth series*, Vol. XI (1881) pp. 214–218. Also, in *Scientific Papers by J. W. Strutt, Baron Rayleigh, Vol. I*, University Press, Cambridge (1899) pp. 513–517.

30 R.N. Singh, 'On image formation with coherent light', *Opt. Acta*, Vol. 23 (1976) pp. 597–606.
31 H.H. Hopkins, 'Image formation with coherent and partially coherent light', *Phot. Sc. and Eng.*, Vol. 21 (1977) pp. 114–123.
32 S.A. Rodionov, 'Diffraction in optical systems', *Opt. Spectrosc. (USSR)*, Vol. 46 (1979) pp. 434–438.
33 J.W. Goodman, *Introduction to Fourier Optics*, 3rd Edition, Viva Books, New Delhi (2007) pp. 73–74.
34 E.W. Wisstein, 'Bessel Function of the First Kind', mathworld.wolfram.com/BesselFunctionoftheFirstKind.html.
35 H.J. Weber and G.B. Arfken, 'Bessel Functions of the First Kind, $J_v(x)$', Section 12.1 in *Essential Mathematical Methods for Physicists,* Academic, Elsevier, San Diego, California (2004) pp. 589–611.
36 F.W.J. Oliver, 'Bessel Functions of Integer Order', §9.1 in Handbook of *Mathematical Functions with Formulas, Graphs and Mathematical Tables*, Eds. M. Abramowitz and I.A. Stegun, National Bureau of Standards, Applied Mathematics Series 55, US Govt. Printing Office, Washington (1972) pp. 355–434.
37 G.N. Watson, *A Treatise on the Theory of Bessel Functions*, Cambridge University Press, London (1966).
38 G.B. Airy, 'On the diffraction of an object-glass with circular aperture', *Trans. Camb. Philos. Soc.*, Vol. V, Part II, No. XII (1835) pp. 283–291.
39 E. Lommel, 'Ueber der Anwendung der Bessel'schen Functionen in der Theorie der Beugung', *Schlömilchs Zeitsch. f. Math. u. Phys.*, Vol. 15 (1870) pp. 141–169.
40 R. Bracewell, *The Fourier Transform and Its Applications*, McGraw-Hill, New York (1965) p. 250.
41 M. Born and E. Wolf, vide Ref. 2, pp. 397–398.
42 S. Huber, *Z. Instrumentenk.*, Vol. 63 (1943) pp. 333–341, 369–389.
43 R.E. Hopkins, H. Kerr, T. Lauroesch, and V. Carpenter, 'Measurements of Energy Distribution in Optical Images', Proceedings of the NBS Semi-Centennial Symposium on Optical Image Evaluation held during 18–20 October 1951, National Bureau of Standards Circular 526, Superintendent of Documents, U. S. Government Printing Office, Washington 25, D. C., Issued April 29,1954, pp. 183–198.
44 E. Wolf, 'Light distribution near focus in an error free diffraction image', *Proc. Roy. Soc. A*, Vol. 204 (1951) pp. 533–548.
45 J. Focke, 'Total illumination in an aberration free diffraction image'. *Opt. Acta*, Vol. 3 (1956) pp. 161–163.
46 G. Lansraux and G. Boivin, 'Numerical determination of the factor of encircled energy relative to a diffraction pattern of revolution', *Can. J. Phys.*, Vol. 36 (1958) pp. 1696–1709.
47 R. Barakat and A. Newman, 'Measurement of the total illuminance in a diffraction image. I. Point sources', *J. Opt. Soc. Am.*, Vol. 53 (1963) pp. 1365–1370.
48 R. Barakat and M.V. Morello, 'Computation of the total illuminance (encircled energy) of an optical system from the design data for rotationally symmetric aberrations', *J. Opt. Soc. Am.*, Vol. 54 (1964) pp. 235–240.
49 V.N. Mahajan, *Optical Imaging and Aberrations, Part II, Wave Diffraction Optics*, SPIE Press, Bellingham, Washington (2011).
50 *Webster's Encyclopedic Unabridged Dictionary of the English Language*, Gramercy, Random House, New York/Avenel (1996) p. 1221.
51 A.J. den Dekker and A. van den Bos, 'Resolution: a survey', *J. Opt. Soc. A. A*, Vol. 14 (1997) pp. 547–557.
52 Lord Rayleigh, 'Investigations in optics with special reference to the spectroscope', *Phil. Mag., Fifth series*, Vol. VIII (1879) pp. 261–274, 403–411, 477–486; Vol. IX (1880) pp. 40–55. Also, in *Scientific Papers by J. W. Strutt, Baron Rayleigh, Vol. I*, University Press, Cambridge (1899) pp. 415–459.
53 Lord Rayleigh, 'On the resolving power of telescopes I', *Phil. Mag., Fifth series*, Vol. X (1880) pp. 116–119. Also, in *Scientific Papers by J. W. Strutt, Baron Rayleigh, Vol. I*, University Press, Cambridge (1899) pp. 488–490.
54 A. Schuster, *An Introduction to the Theory of Optics*, Edward Arnold, London (1924) p. 158.
55 B.P. Ramsay, E.L. Cleveland, and O.T. Koppius, 'Criteria and the Intensity-Epoch slope', *J. Opt. Soc. Am.*, Vol. 31 (1941) pp. 26–33.
56 D.E. Stoltzmann, 'The Perfect Point Spread Function', in *Applied Optics and Optical Engineering, Vol. IX*, Eds. R. R. Shannon and J. C. Wyant, Academic, New York (1983) pp. 111–149.
57 C.M. Sparrow, 'On spectroscopic resolving power', *Astrophys. J.*, Vol. 44 (1916) pp. 76–86.

58 R.K. Luneburg, *Mathematical Theory of Optics*, University of California Press, Berkeley and Los Angeles (1964) pp. 344–348. [This book is a compilation of Lecture Notes by R. K. Luneburg at Brown University in the summer of 1944].
59 R. Barakat, 'Application of apodization to increase two-point resolution by the Sparrow Criterion. I. Coherent illumination', *J. Opt. Soc. Am.*, Vol. 52 (1962) pp. 276–283.
60 H. Osterberg, 'Microscope imagery and interpretations', *J. Opt. Soc. Am.*, Vol. 40 (1950) pp. 295–303.
61 H. Rutten and M. van Venrooij, Telescope Optics, Willmann-Bell, Richmond, Virginia (1988) pp. 206–208.
62 D.E. Stoltzmann, 'The Perfect Point Spread Function', Chapter 4 in *Applied Optics and Optical Engineering, Vol. IX,* Ed. R.R. Shannon and J.C. Wyant, Academic, New York (1983) pp. 111–148.
63 H.H. Hopkins, 'The concept of partial coherence in optics', *Proc. Roy. Soc. London A*, Vol. 208 (1951) pp. 263–277.
64 H.H. Hopkins, 'On the diffraction theory of optical images', *Proc. Roy. Soc. London A*, Vol. 217 (1953) pp. 408–432.
65 M. Born and E. Wolf, vide Ref. 2, pp. 491–535.
66 D.N. Grimes and B.J. Thomson, 'Two-point resolution with partially coherent light', *J. Opt. Soc. Am.*, Vol. 57 (1967) pp. 1330–1334.
67 B.J. Thomson, 'Image Formation with Partially Coherent Light', in *Progress in Optics, Vol. VII*, Ed. E. Wolf, North-Holland, Amsterdam (1969) pp. 170–230.
68 S.C. Som, 'Influence of partially coherent illumination and spherical aberration on microscopic resolution', *Opt. Acta*, Vol. 18 (1971) pp. 597–608.
69 M. De and A. Basuray, 'Two-point resolution in partially coherent light I. Ordinary microscopy, disc source', *Opt. Acta*, Vol. 19 (1972) pp. 307–318.
70 E.C. Kintner and R.M. Sillitto, 'Two-point resolution criteria in partially coherent imaging', *Opt. Acta*, Vol. 20 (1973) pp. 721–728.
71 A.I. Kartashev, 'Optical systems with enhanced resolving power', *Opt. Spectrosc.* Vol. 9 (1960) pp. 204–206.
72 W. Lukosz, 'Optical systems with resolving power exceeding the classical limit', *J. Opt. Soc. Am.*, Vol. 56 (1966) pp. 1463–1472.
73 Z. Zalevsky, D. Mendlovic, and A.W. Lohmann, 'Optical systems with improved resolving power', *in Progress in Optics*, Vol. XL, Ed. E. Wolf, North-Holland, Amsterdam, Elsevier (2000) pp. 271–341.
74 Z. Zalevsky and D. Mendlovic, *Optical Superresolution*, Springer, New York (2004).
75 F.M. Schellenberg, *Selected Papers on Resolution Enhancement Techniques in Optical Lithography, SPIE Milestone Series Vol. MS178,* SPIE Press, Bellingham, Washington (2004).
76 C. Cremer and B.R. Masters, 'Resolution enhancement techniques in microscopy', *Eur. Phys. Journal H*, Vol. 38 (2013) pp. 281–344.
77 D.C. Flanders, A.W. Hawryluk, and H.J. Smith, 'Spatial phase division – a new technique for exploring submicrometer-linewidth periodic and quasi-periodic patterns', *J. Vac. Sci. Tech.*, Vol. 18 (1979) pp. 1949–1952.
78 M. Shibuya, 'Projection Master for Transmitted Illumination', *Jap. Patent Appl.* No. Showa 57–62052 (April 1982), and Japanese Patent Showa 62–50811, Applied September 30, (1980) Issued October 27, 1987.
79 M.D. Levenson, N.S. Viswanathan, and R.A. Simpson, 'Improving resolution in photolithography with a phase-shifting mask', *IEEE Trans. Electron. Devices*, Vol. ED – 29 (1982) pp. 1828–1836.
80 M.D. Levenson, 'Wavefront engineering for photolithography', *Physics Today*, Vol. 46. No.7 (1993).
81 G.A. Cirino, R.D. Mansano, P. Verdonck, L. Cescato, and L.G. Nato, 'Diffractive phase – shift lithography photomask operating in proximity printing mode', *Opt. Exp.*, Vol. 18 (2010) pp. 16387–16404.
82 B.J. Lin, *Optical Lithography*, SPIE, Bellingham, Washington (2010) pp. 150–156.
83 H. Hänsel and W. Polack, 'Verfahren zur Herstellung einer Phasenmaske Amplitudenstruktur', DDR Patent #26 50 817, Issued November 17, 1977.
84 M. Born and E, Wolf, vide Ref. 2, pp. 416–418.
85 G. Toraldo di Francia, 'Nuovo pupille superresolvente', *Atti Fond. Giorgio Ronchi*, Vol. 7 (1952) pp. 366–372.
86 G. Toraldo di Francia, 'Super-gain antennas and optical resolving power', *Il Nuovo Cimento*, Vol. 9 Suppl. 3 (1952) pp. 426–438.
87 G. Toraldo di Francia, vide Ref. 24.
88 A. Boivin, *Theorie et calcul des Figures de Diffraction de Revolution*, Les Presses de l'Universite Laal, Quebec, and Gauthier – Villars, Paris (1964) [in French].

89 R. Boivin and A. Boivin, 'Optimized amplitude filtering for superresolution over a restricted Field-I. Achievement of maximum central irradiance under an energy constraint,' *Opt. Acta*, Vol. 27, (1980) pp. 587–610.
90 R. Boivin and A. Boivin, 'Optimized amplitude filtering for superresolution over a restricted field-II. Application of the impulse-generating filter', *Opt. Acta* Vol.27, (1980) pp. 1641–1670.
91 I.J. Cox, C.J.R. Sheppard, and T. Wilson, 'Reappraisal of arrays of concentric annuli as superresolving filters', *J. Opt. Soc. Am.*, Vol. 72 (1982) pp. 1287–1291.
92 T.R.M. Sales and G.M. Morris, 'Diffractive superresolution elements', *J. Opt. Soc. Am. A*, Vol. 14 (1997) pp. 1637–1646.
93 I. Leiserson, S.G. Lipson, and V. Sarafis, 'Superresolution in far-field imaging', *Opt. Lett.*, Vol. 25 (2000) pp. 209–211.
94 A. Ranfagni, D. Mugnai, and R. Ruggen, 'Beyond the diffraction limit: Super-resolving pupils', *J. Appl. Phys.*, Vol. 95 (2004) pp. 2217–2222.
95 D. Mugnai and A. Ranfagni, 'Further remarks on super-resolving pupils', *J. Appl. Phys.*, Vol. 102 (2007) 036103-1 to 3.
96 L.N. Hazra and N. Reza, 'Optimal design of Toraldo superresolving filters', *Proc. SPIE* 7787, 77870D (2010).
97 L.N. Hazra and N. Reza, 'Superresolution by pupil plane phase filtering', *Pramana* 75, (2010) pp. 855–867.
98 N. Reza and L. Hazra, 'Toraldo filters with concentric unequal annuli of fixed phase by evolutionary programming', *J. Opt. Soc. Am. A*, Vol. 30 (2013) pp. 189–195.
99 A. Bachl and W. Lukosz, 'Experiments on superresolution imaging of a reduced object field', *J. Opt. Soc. Am.*, Vol. 57 (1967) pp. 163–169.
100 M. Minsky, 'Microscopy Apparatus', U. S. Patent 3,013,467, Filed Nov. 7, 1957, Issued Dec. 19, 1961.
101 C.J.R. Sheppard and A. Choudhury, 'Image formation in the scanning microscope', *Opt. Acta*, Vol. 24 (1977) pp. 1051–1073.
102 T. Wilson and C.J.R. Sheppard, *Theory and Practice of Scanning Optical Microscopy*, Academic, London (1984).
103 S.G. Lipson, H. Lipson, and D.S. Tannhauser, *Optical Physics*, Cambridge University Press, Cambridge (1996) pp. 361–363.
104 C.J.R. Sheppard and D.M. Shotton, *Confocal Laser Scanning Microscopy*, Taylor & Francis, London (1997).
105 M. Minsky, 'Memoir on inventing the confocal scanning microscope', *SCANNING*, Vol. 10 (1988) pp. 128–138.
106 M. Castello, C.J.R. Sheppard, A. Diaspro, and G. Vicidomini, 'Image scanning microscopy with a quadrant detector', *Opt. Lett.*, Vol. 40 (2015) pp. 5355–5358.
107 C.J.R. Sheppard, 'Pixel reassignment in image scanning microscopy', *Plenary Talk, Int. Conf. on Optics and Electro-Optics (ICOL 2019)*, pp. 19–22 October 2019, Dehradun, India.
108 E.H. Synge, 'A suggested method for extending the microscopic resolution into the ultramicroscopic region', *Phil. Mag.*, Vol. 35 (1928) pp. 356–362.
109 E.A. Ash and G. Nicholls, 'Super-resolution aperture scanning microscope', *Nature*, Vol. 237 (1972) pp. 510–512.
110 A. Lewis, M. Isaacson, A. Harootunian, and A. Murray, 'Development of a 500 Å resolution light microscope: I – Light is efficiently transmitted through a $\lambda/6$ diameter aperture', *Ultramicroscopy*, Vol. 13 (1984) pp. 227–230.
111 D.W. Pohl, W. Denk, and M. Lanz, 'Optical Stethoscopy: image recording with resolution $\lambda / 20$', *Appl. Phys. Lett.*, Vol. 44 (1984) pp. 651–653.
112 G. Binnig and H. Rohrer, 'The scanning tunneling microscope', *Scientific American*, Vol. 253, No. 2 (Aug. 1985) pp. 50–58.
113 E. Betzig, J.K. Trautman, T.D. Harris, J.S. Weiner, and R.L. Kostelak, 'Breaking the diffraction barrier: optical microscopy on a nanometric scale, *Science*, Vol. 251 (1991) pp. 1468–1470.
114 E. Betzig and J.K. Trautman, 'Near-field optics: microscopy, spectroscopy, and surface modification beyond the diffraction limit', *Science*, Vol. 257 (1992) pp. 189–195.
115 D.W. Pohl and D. Courjon, Eds., *Near Field Optics*, Springer, Netherlands (1993).
116 M.A. Paesler and P.J. Moyer, *Near-Field Optics, Theory, Instrumentation, and Applications*, John Wiley, New York (1996).

17

Diffraction Images by Aberrated Optical Systems

In this chapter, we consider the combined effects of diffraction and residual aberrations of an imaging system on the quality of an image of a bright point object on a dark background.

17.1 Point Spread Function (PSF) for Aberrated Systems

The complex amplitude F at a point $P'(p,\phi)$ in the diffraction image of an object point is given by equation (16.51) as

$$F(p,\phi)=\iint_{-\infty}^{+\infty} f(r,\theta)\exp\left[i2\pi pr\cos(\theta-\phi)\right]rdrd\theta \tag{17.1}$$

where $f(r,\theta)$ is the complex amplitude over the reference sphere, and it is called the pupil function. It is given by

$$\begin{aligned} f(r,\theta) &= \alpha(r,\theta)\exp\left[ikW(r,\theta)\right] && \text{within } A \\ &= 0 && \text{outside } A \end{aligned} \tag{17.2}$$

The factor $\alpha(r,\theta)$ takes account of any variation of real amplitude over the wavefront, and $W(r,\theta)$ is the wavefront aberration for any point (r, θ) on the reference sphere. The symbol A represents the region of pupil over the reference sphere. Assuming uniform real amplitude over the wavefront, (17.1) reduces to

$$F(p,\phi)=\iint_{A} \exp\left[iS(r,\theta)\right]rdrd\theta \tag{17.3}$$

where

$$S(r,\theta)=2\pi\left\{W(r,\theta)+pr\cos(\theta-\phi)\right\} \tag{17.4}$$

Note that the wave aberration $W(r,\phi)$ is expressed in units of wavelength. The intensity PSF is given by

$$I(p,\phi)=\left|F(p,\phi)\right|^2 \tag{17.5}$$

The integral (17.3) with $W(r,\theta)$ representing a general aberration polynomial cannot be worked out analytically, except for in a few special cases.

DOI: 10.1201/9780429154812-17

17.1.1 PSF of Airy Pupil in Different Planes of Focus

For an Airy pupil, W = 0, and the intensity distribution in the PSF of the Airy pupil on the paraxial image plane is given by

$$I_N(p) = \left[\frac{2\,J_1(2\pi p)}{(2\pi p)}\right]^2 \tag{17.6}$$

The intensity distribution in the PSF of the Airy pupil on neighbouring image planes that are axially shifted from the paraxial image plane can be determined by incorporating a 'pseudo-aberration' term in W as

$$W = W_{20} r^2 \tag{17.7}$$

where W_{20} is commonly called the 'defocus' or 'defect of focus' term in the aberration polynomial, and it takes into account the effect of longitudinal shift of focus (see Section 8.2.3.1.1). In this special case, analytical expression for the corresponding integral was obtained by Lommel [1] in terms of special functions U_n and V_n introduced by him. In his memory, these functions are now called Lommel U_n and V_n functions.

Figures 17.1(a)–(i) show the variation of normalized intensity along an azimuth for the diffraction images of a point object in different planes of focus corresponding to W_{20} over (0, 0.25, 2.0).

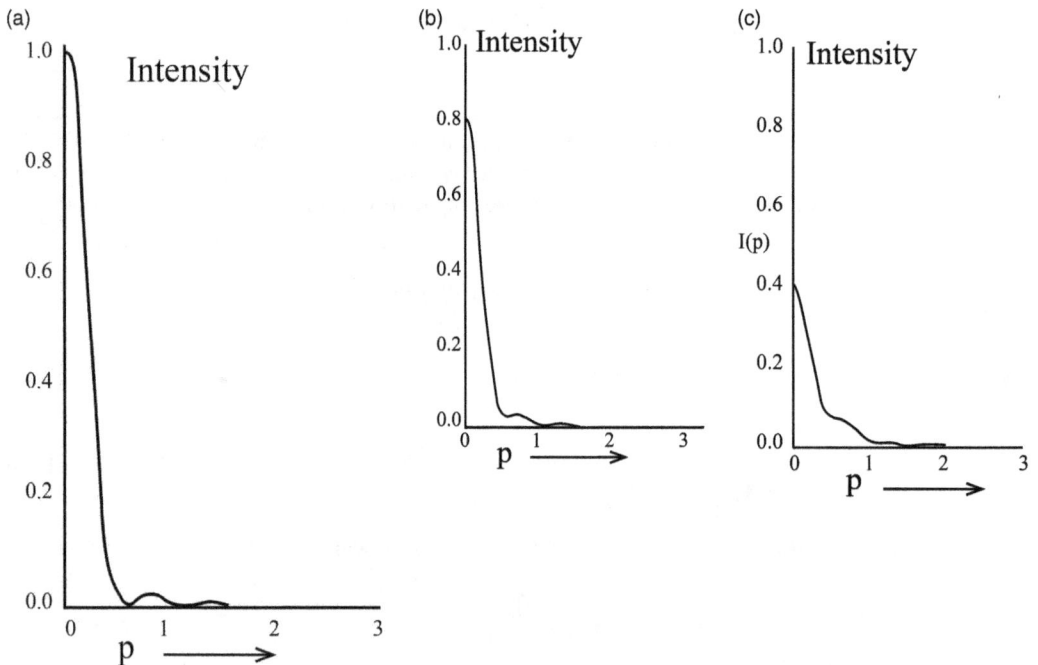

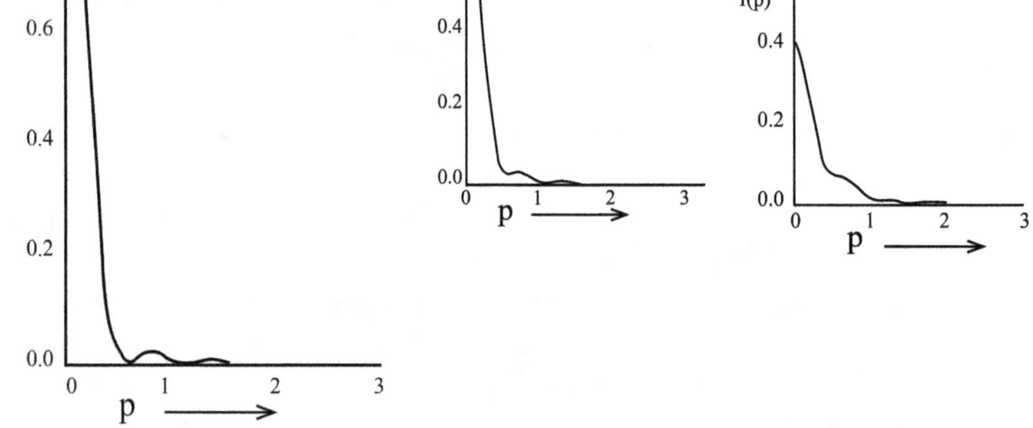

FIGURE 17.1 Distribution of normalized intensity along an azimuth for the diffraction images of a point object by an axially symmetric imaging system in different planes of focus corresponding to: (a) $W_{20} = 0$ (b) $W_{20} = 0.25$ (c) $W_{20} = 0.50$ (d) $W_{20} = 0.75$ (e) $W_{20} = 1.0$ (f) $W_{20} = 1.25$ (g) $W_{20} = 1.5$ (h) $W_{20} = 1.75$ (i) $W_{20} = 2.0$ (in units of wavelength).

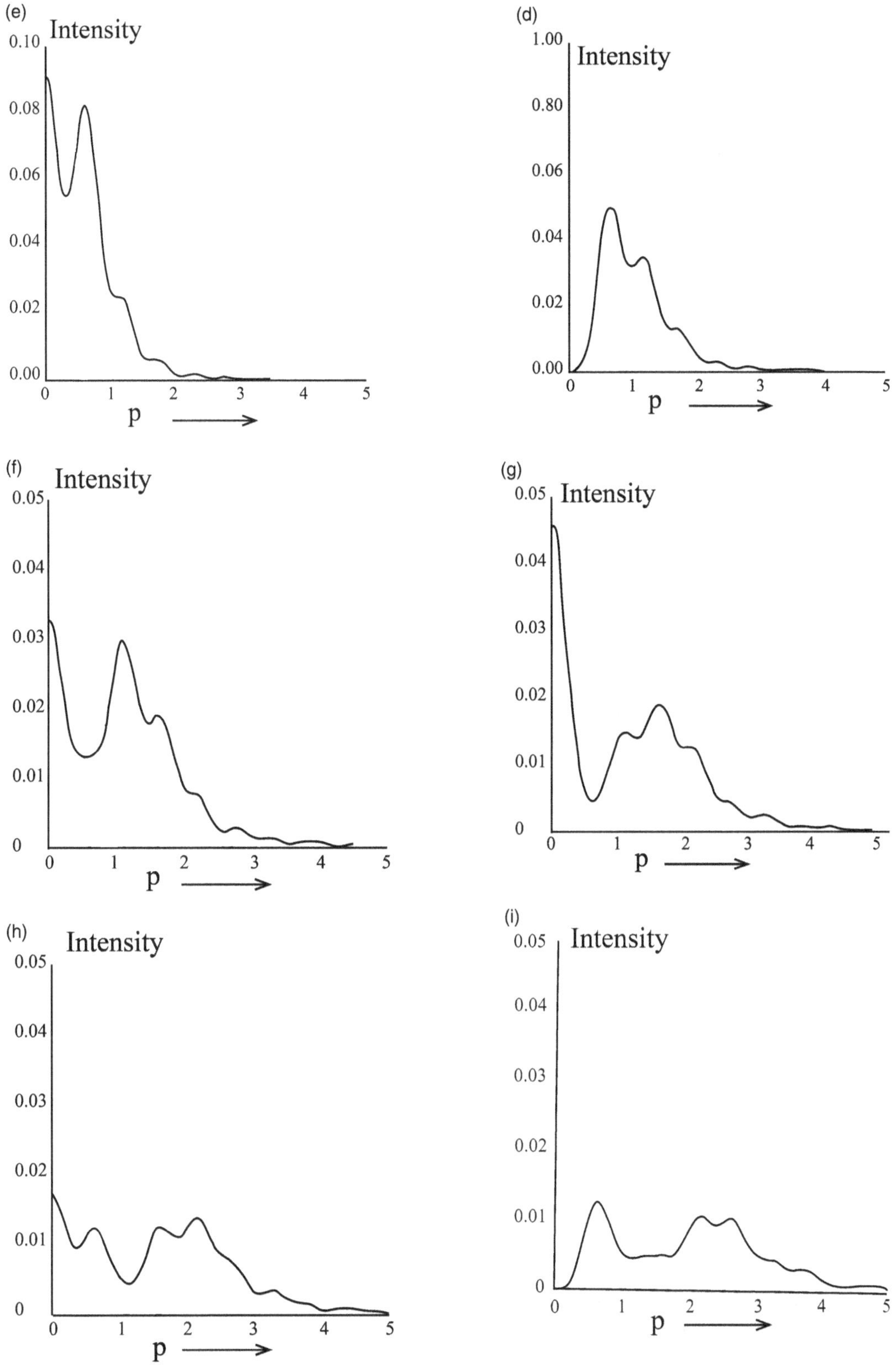

FIGURE 17.1 Continued

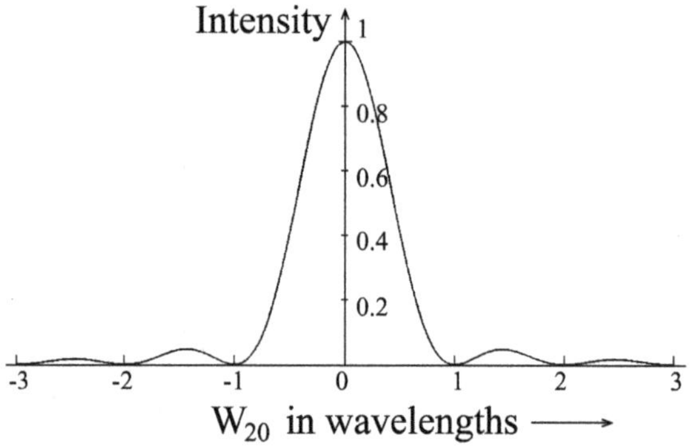

FIGURE 17.2 Distribution of the normalized intensity at the centre of the diffraction image of an axial point object by an axially symmetric aberration-free imaging system with shift of the focal plane.

17.1.2 Distribution of Intensity at the Centre of the PSF as a Function of Axial Position of the Focal Plane

The centre of the PSF corresponds with $p = 0$. Expressing the focal plane position in terms of W_{20}, the corresponding intensity $I(0, W_{20})$ at the centre for an Airy pupil is given by

$$I(0,W_{20}) = \left| \int_0^{2\pi} \int_0^1 \exp\left[ikW_{20}r^2\right] r\, dr\, d\theta \right|^2 \tag{17.8}$$

The normalized intensity as a function of W_{20} is expressed as

$$I_N(0,W_{20}) = \frac{I(0,W_{20})}{I(0,0)} = \left[\frac{\sin\left\{\frac{k}{2}W_{20}\right\}}{\left\{\frac{k}{2}W_{20}\right\}} \right]^2 \tag{17.9}$$

Figure 17.2 graphically illustrates the variation of the normalized intensity at the centre of the diffraction pattern with a shift of the focal plane. Note the regularly placed zeros in intensity on both sides of the best focus.

17.1.3 Determination of Intensity Distribution in and around Diffraction Images by Aberrated Systems

Notwithstanding the great difficulty in working out the diffraction integral in general cases, many significant attempts have been made from the early days to tackle the problem. Immediately after his seminal work on 'Airy pattern', G.B. Airy embarked upon the problem of determining 'the intensity of light in the neighbourhood of a caustic' [2]. Lord Rayleigh studied the influence of aberrations on the loss of intensity in the centre of the diffraction pattern [3]. Strehl studied the intensity along the principal ray [4]. Conrady initiated evaluation of the diffraction integral (16.53) by means of mechanical quadrature with a planimeter [5]. Further studies in the same line were carried out by Buxton and Martin [6–10]. The use of Simpson's rule in carrying out numerical integration of the diffraction integral was reported for the first time by Kingslake [11].

A different approach to tackling the problem was undertaken by Steward [12] and Picht [13]. They attempted series expansions for the intensity distribution in typical diffraction images. Often the procedure provided results as sought for, but, in general, the procedure leads to unwieldy expressions that are not amenable for computational purposes.

An extensive review of the research on this problem till the middle of the twentieth century is provided by Wolf [14]. A few years later, Black and Linfoot [15] used Simpson's rule for integration in the expression (17.3) for the case of primary spherical aberration. Hopkins [16] noted that a major disadvantage of using Simpson's rule for integration of highly oscillatory functions is the requirement of very close intervals to obtain reliably accurate results. He proposed an alternative method that involves dividing the area of integration into elements of area δA, defined by the circles of radii $r = \overline{r} \pm \Delta r$ and azimuths $\theta = \overline{\theta} \pm \Delta\theta$, where $(\overline{r}, \overline{\theta})$ are the coordinates of the mid-point of δA [17]. The expression (17.3) may then be written as

$$F(p,\phi) = \sum_{\overline{r}} \sum_{\overline{\theta}} \int_{\overline{\theta}-\Delta\theta}^{\overline{\theta}+\Delta\theta} \int_{\overline{r}-\Delta r}^{\overline{r}+\Delta r} \exp\left[iS(r,\theta)\right] r dr d\theta \tag{17.10}$$

For each of the small mesh elements δA, the phase function $S(r,\theta)$ can be approximated by neglecting the terms in degree two and above in expansion of $S(r,\theta)$ about the point $(\overline{r}, \overline{\theta})$, i.e.

$$S(r,\theta) \cong S(\overline{r},\overline{\theta}) + (r-\overline{r}) + \left\{\frac{\partial}{\partial r} S(r,\theta)\right\}_{\substack{r=\overline{r}\\ \theta=\overline{\theta}}} + \left\{\frac{\partial}{\partial \theta} S(r,\theta)\right\}_{\substack{r=\overline{r}\\ \theta=\overline{\theta}}} \tag{17.11}$$

Subsequently, further modifications of the method were suggested with a goal to achieve the same degree of accuracy by using a lesser number of mesh elements, thereby reducing the computational load for evaluation of the PSF [18–20]. In a later publication, Hopkins laid down the basic justification for the use of this approach in numerical integration of highly oscillatory functions [21].

A common characteristic of the expressions for $F(p,\phi)$ or $I(p,\phi)$ given above is that they provide the computational results as sought for, but they are not convenient for use in further analysis. A suitable technique for the purpose was proposed by De and Hazra [22], who used the average value of S over a single mesh element as the fixed value for S over the mesh element δA, i.e.

$$S(r,\theta) \cong \tilde{S} = \left[\int_{\overline{\theta}-\Delta\theta}^{\overline{\theta}+\Delta\theta} \int_{\overline{r}-\Delta r}^{\overline{r}+\Delta r} S(r.\theta) r \, dr \, d\theta\right] / \left[\int_{\overline{\theta}-\Delta\theta}^{\overline{\theta}+\Delta\theta} \int_{\overline{r}-\Delta r}^{\overline{r}+\Delta r} r \, dr \, d\theta\right] \tag{17.12}$$

For obtaining better accuracy with lower number of mesh elements, a further modification involving the standard deviation of S from the average value in each element δA is suggested by Hazra [23]. Incidentally, these latter techniques also eliminated the tricky problem of numerical differentiation of the function S at the centres of mesh elements.

Figure 17.3 shows the distribution of axial intensity for three values of primary spherical aberration: (a) $W_{40} = 1\lambda$, (b) $W_{40} = 2\lambda$, and (c) $W_{40} = 3\lambda$. Note the axial shift of the peak intensity in the diffraction patterns. The intensity distributions in the corresponding diffraction images on the transverse planes of best focus are given in Figure 17.4. For each case, the influence of focal shift on the intensity distribution is shown for two other transverse planes, equidistant from the plane of best focus and lying on opposite sides of the plane of best focus, in Figure 17.5. Figure 17.6 shows the intensity distribution on the Gaussian/paraxial focal plane along the azimuth $\phi = 0$ for five values of primary coma: (a) $W_{31} = 0.16\lambda$, (b) $W_{31} = 0.32\lambda$, (c) $W_{31} = 0.63\lambda$, (d) $W_{31} = 1.26\lambda$, and (e) $W_{31} = 1.89\lambda$. Note the gradual fall of peak intensity, and the characteristic shifts of the centres of the diffraction pattern on the transverse plane from the location of the geometric image. For three values of primary coma, namely, (a) $W_{31} = 0.63\lambda$, (b) $W_{31} = 1.26\lambda$, and (c) $W_{31} = 1.89\lambda$, Figure 17.7 shows the influence of focal shift on the intensity distribution at two other transverse planes, equidistant from the Gaussian/paraxial image plane, and lying on opposite sides of the plane.

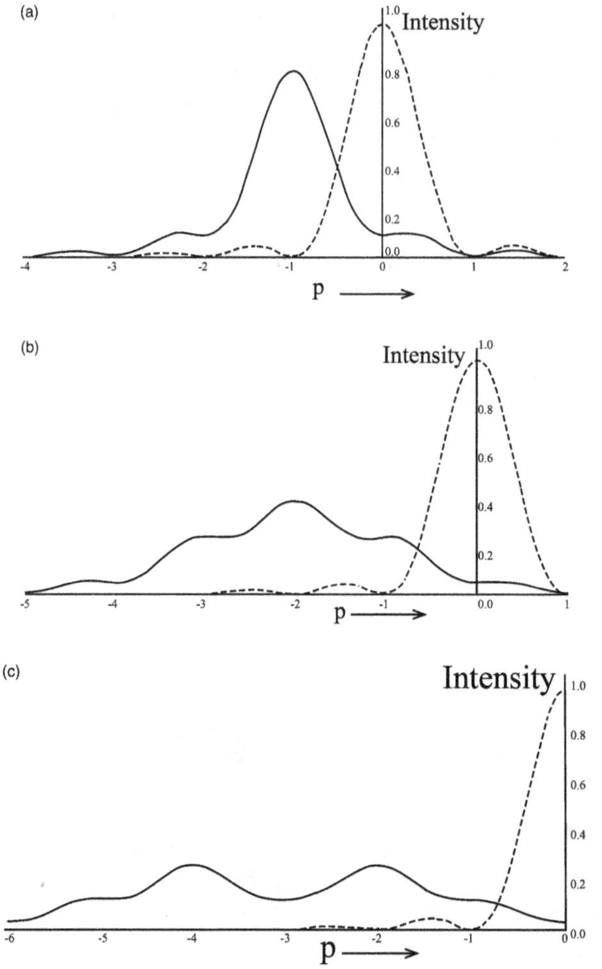

FIGURE 17.3 Distribution of the normalized axial intensity in the diffraction image of an axial point object by an axially symmetric imaging system with primary spherical aberration: (a) $W_{40} = 1\lambda$, (b) $W_{40} = 2\lambda$, and (c) $W_{40} = 3\lambda$.

Note: Dashed line: Airy intensity pattern.

The integration to be carried out for determining the complex amplitude can be represented as a two-dimensional Fourier transform of the pupil function. The availability of fast Fourier transform (FFT) methods, e.g., the Cooley-Tukey algorithm or any of its many variants, opened up new possibilities for numerical evaluation of the complex amplitude [24–25]. However, the basic integration scheme employed in FFT methods is usually a crude quadrature technique that is likely to influence the final results significantly. Also, the algorithm provides the values of $F(u)$ at a prespecified set of u over a wide range that is compatible with the technique. It is not very convenient for the evaluation of $F(u)$ at an arbitrary value u, or for a set of arbitrarily chosen values of u. Barakat provides a brief review of the quadrature techniques and different FFT algorithms in diffraction integrals of incoherent imagery [26]. A few modified algorithms for alleviating some of the above-mentioned problems of PSF evaluation, etc. are reported in the literature [27–29].

17.1.4 Spot Diagrams [30–39]

It is enunciated above that the quality of image in an imaging system is affected by both residual aberrations in the system, and diffraction of the image-forming wavefront by finite aperture of the imaging system.

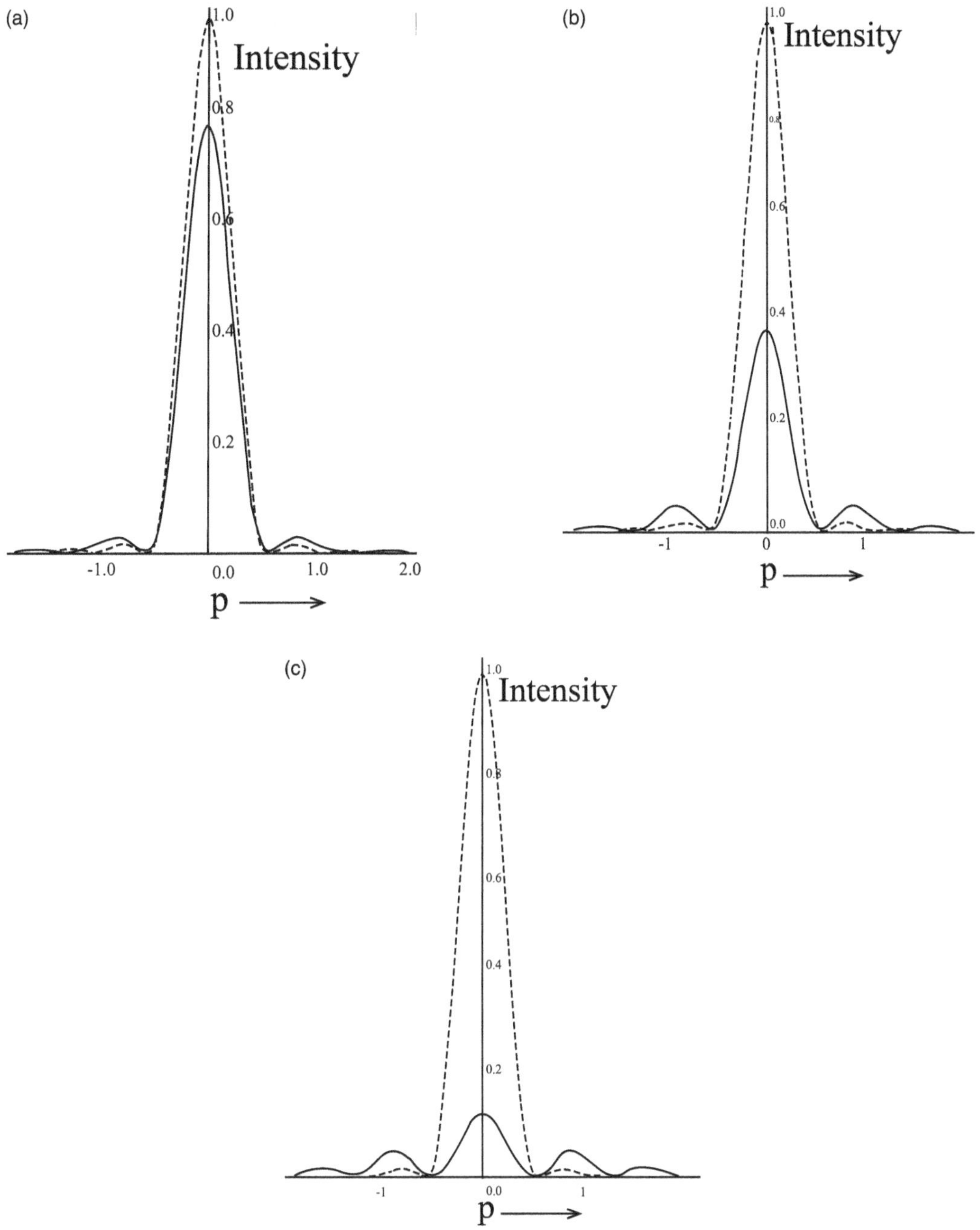

FIGURE 17.4 Distribution of normalized intensity along any azimuth on the transverse plane of best focus in the diffraction image of an axial point object by an axially symmetric imaging system with primary spherical aberration: (a) $W_{40} = 1\lambda$, (b) $W_{40} = 2\lambda$, and (c) $W_{40} = 3\lambda$.

Note: Dashed line: Airy intensity pattern.

Nevertheless, in imaging systems with residual aberrations that are not small, aberrations play the predominant role in image formation, so much so that, for many practical purposes, a rough estimate of the image quality can be obtained by ignoring the diffraction effects. From an object point, a set of rays are traced through the system, and the points of intersection of these rays with a chosen image plane are determined. When the number of ray intersection points is sufficiently large, the diagram containing

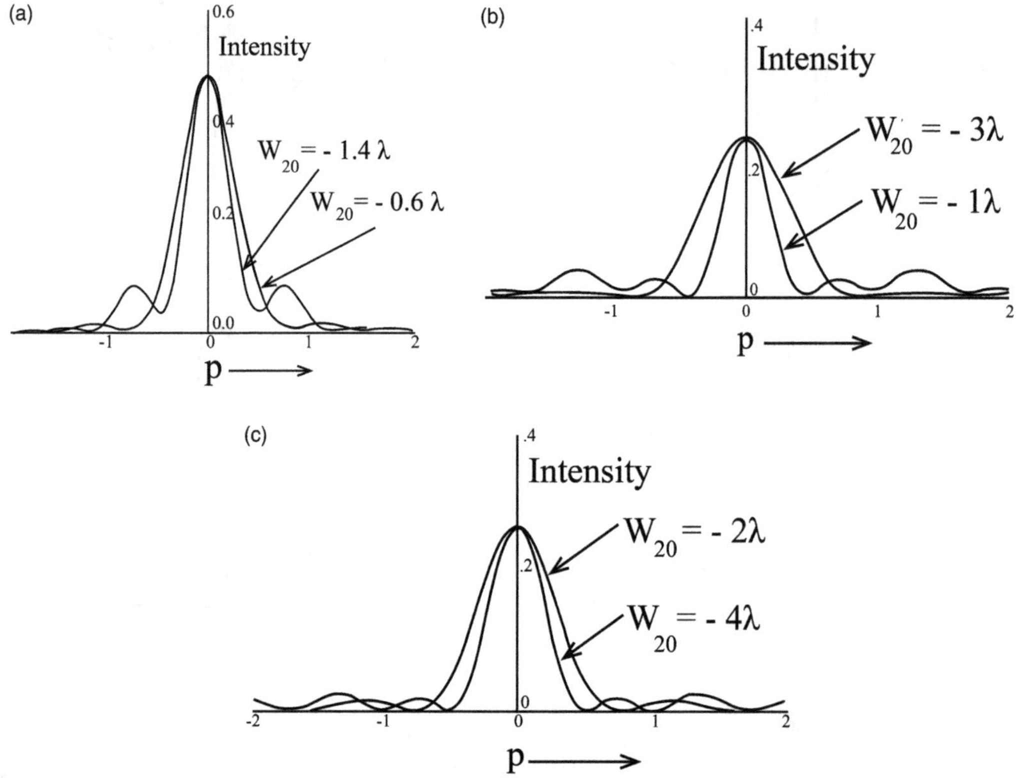

FIGURE 17.5 For each case of Figure 17.4, distribution of normalized intensity along any azimuth on two other transverse planes, equidistant from the plane of best focus, and lying on opposite sides of the plane of best focus: (a) $W_{40} = 1\lambda$, (b) $W_{40} = 2\lambda$, and (c) $W_{40} = 3\lambda$.

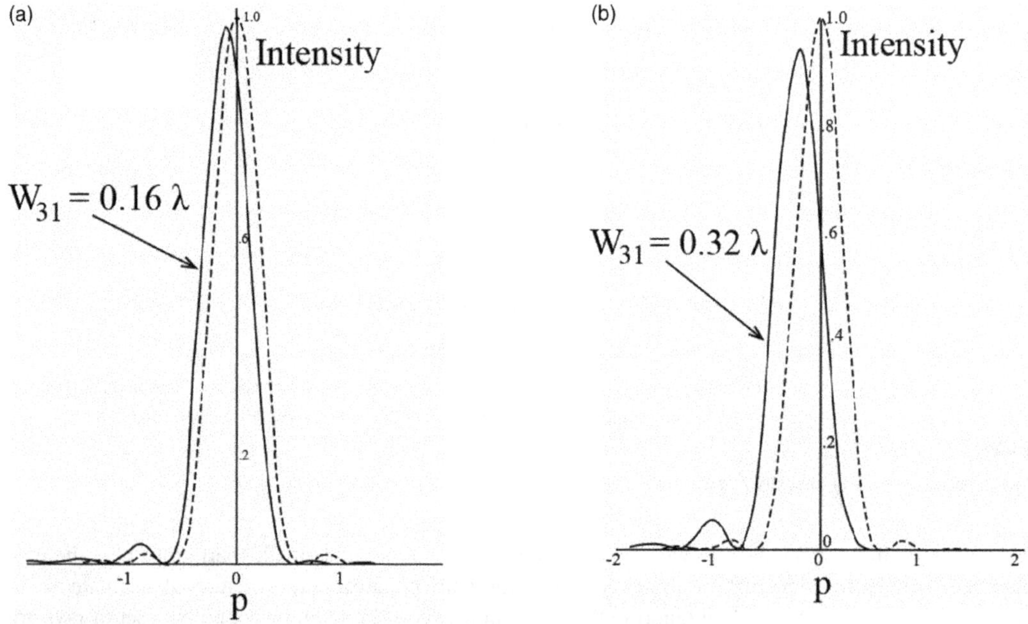

FIGURE 17.6 Distribution of normalized intensity on the Gaussian/paraxial focal plane along the azimuth $\phi = 0$ in the presence of primary coma: (a) $W_{31} = 0.16\lambda$ (b) $W_{31} = 0.32\lambda$ (c) $W_{31} = 0.63\lambda$ (d) $W_{31} = 1.26\lambda$ and (e) $W_{31} = 1.89\lambda$.

Note: Dashed line: Airy intensity pattern.

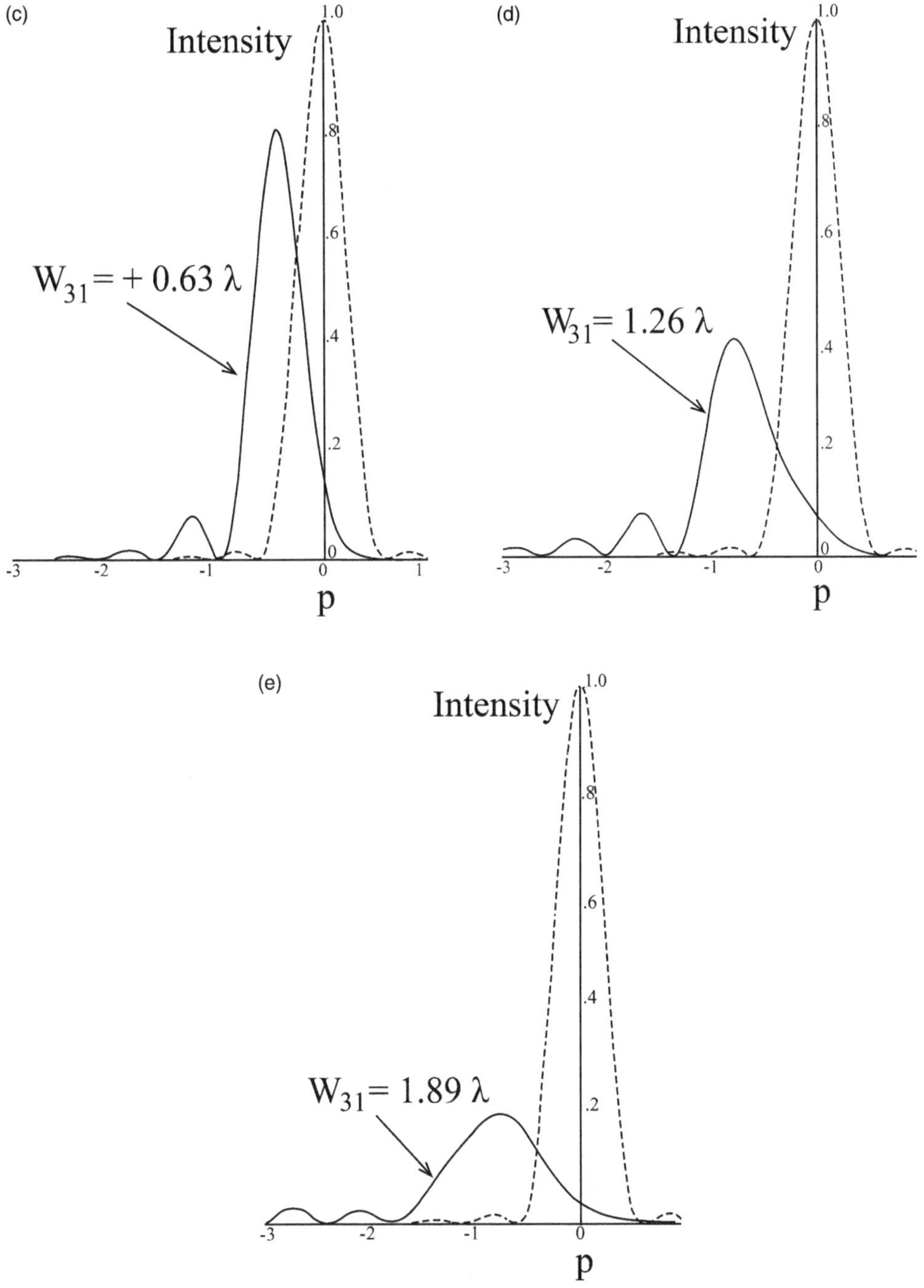

FIGURE 17.6 Continued

the ray intersection points gives an impression of the quality of the image of the particular point object. This diagram is called the 'spot diagram'. Obviously, the spot diagram is dependent on the choice of the set of rays chosen for the display. Nevertheless, with a large number of rays evenly distributed over the pupil on a rectangular or polar grid, the spot diagram contains a definite signature of the composite ray aberrational effects on the structure of the image. When the rays are distributed equally over the

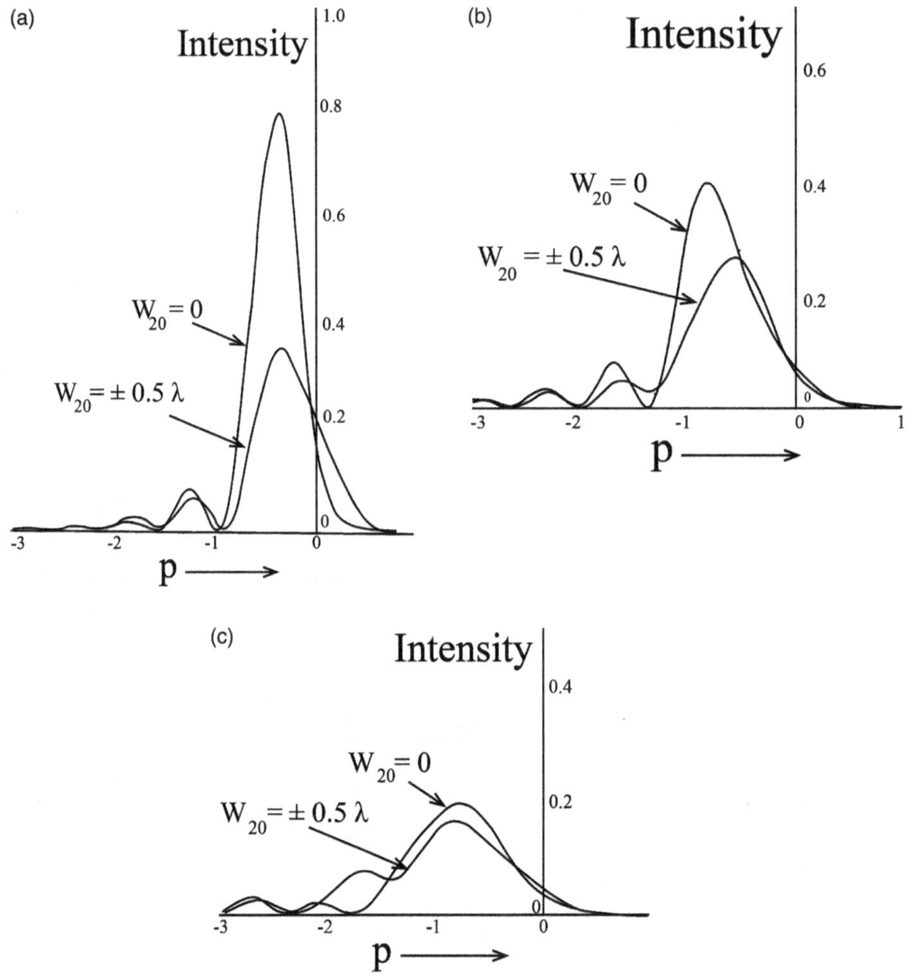

FIGURE 17.7 Effect of focal shift on the intensity distribution at two other transverse planes, equidistant from the Gaussian/paraxial image plane, and lying on opposite sides of the plane in the presence of primary coma: (a) $W_{31} = 0.63\lambda$ (b) $W_{31} = 1.26\lambda$ and (c) $W_{31} = 1.89\lambda$.

exit pupil, the resulting spot diagram indicates the distribution of illuminance in the image of the point. Chromatic effects in the image can be studied by independently tracing rays in multiple wavelengths, and by assigning different colours for the ray intersection points corresponding to the different wavelengths. In practice, coefficients of the transverse ray aberration polynomial pertaining to the specific object point are first determined by tracing a small set of rays. Next, the polynomial is used to determine the ray intersection points for a large grid of rays. For axially symmetric systems, bilateral symmetry about the meridional plane can be utilized for reducing the computation time significantly.

17.2 Aberration Tolerances

In optical imaging systems, the tolerances on image quality evaluation/assessment pertinent to the particular application are of practical importance. Similarly, for non-imaging systems, the tolerances on performance criteria vis-à-vis end goals of the system become important. Except for in the trivial cases, no direct link of the performance criteria with constructional parameters of the optical system is available.

Aberrations of a system provide the needed link or interface between these two levels because aberrations for any configuration can be worked out from a given set of constructional parameters of an optical system. The latter involves the use of well-established ray tracing methods. Determination of an image performance criterion from a given set of aberrations is a more involved numerical procedure. Therefore, at the stage of preliminary design of optical systems, control of aberrations constitutes the central task. This involves bringing down aberrations to tolerable levels. In practice, optical systems can be classified into three different categories in accordance with the magnitude of aberration that can be tolerated for providing the desired image quality in that particular application. In the first category belong the optical systems where diffraction plays an insignificant role in comparison to ray aberrations in determining the quality of the image. All diffraction limited systems where diffraction, in comparison to ray aberrations, plays the pivotal role in image quality, constitute the other extreme category. The intermediate category is constituted by optical systems for which the assessment of image quality calls for a proper reckoning of both diffraction and ray aberration. It is obvious that no uniform aberration tolerance can be applicable for all cases.

17.2.1 Rayleigh Quarter-Wavelength Rule

Studies on aberration tolerances commenced with diffraction limited systems. Strictly speaking, image quality in the latter systems is solely determined by diffraction effects occurring due to the finite size of the aperture of the system, and aberration plays no role the process. Obviously, an ideal diffraction limited system is to be free from any aberration. For example, in an aberration-free imaging system of circular aperture, the PSF is the Airy pattern characterized by a bright central lobe surrounded by a series of rings of gradually diminishing brightness. It is now known that with a gradual increase of aberrations, the effects on the PSF are somewhat similar for any combination of aberration types. The effects are a decrease in intensity in the centre of the diffraction pattern, and an increase in intensity in the outer lobes. Further, the half width of the central lobe remains the same up to a certain level of aberration.

Lord Rayleigh pioneered a systematic study on the influence of aberrations in imaging. At that time, it was still not feasible to evaluate the PSF of aberrated systems. Of the three characteristics of PSF of aberrated systems, Lord Rayleigh could work out the central intensity in the diffraction patterns produced by aberrated systems [40]. He noted that, in the presence of primary spherical aberration $W_{40} = \frac{\lambda}{8}$, the central intensity in the diffraction pattern is 0.9464 times the central intensity in the case of an aberration-free system. Further increase of primary spherical aberration to $\frac{\lambda}{4}$ and $\frac{\lambda}{2}$ causes further reduction to values 0.8003 and 0.3947, respectively. On the basis of this observation, Rayleigh made the comment: 'aberration begins to be decidedly prejudicial when the wave-surface deviates from its proper place by about quarter of a wavelength' [40]. Nine years later, Rayleigh reaffirmed his observation, and noted again:

> In general, we may say that aberration is unimportant, when it nowhere (or at any rate over a relatively small area only) exceeds a small fraction of the wavelength (λ). Thus in estimating the intensity at a focal point, where, in the absence of aberration, all the secondary waves would have exactly the same phase, we see that an aberration nowhere exceeding $\frac{1}{4}\lambda$ can have but little effect.
>
> [41]

In the same article, he made a conclusive statement:

> aberration begins to be distinctly mischievous when it amounts to about a quarter period, i.e. when the wave-surface deviates at each end by a quarter wave-length.... The general

> conclusion is that an aberration between the centre and circumference of a quarter period has but little effect upon the intensity at the central point of the image.
>
> [41]

This is the origin of the well-known 'Rayleigh quarter wavelength rule'. It may be stated analytically as: The imaging performance of an optical system is effectively diffraction limited so long as

$$\max\left|W(r,\theta)\right| \leq \frac{\lambda}{4} \tag{17.13}$$

Often it is also stated that an optical imaging system is effectively diffraction limited if the wavefront at the exit pupil is contained inside two concentric spheres with their common centre at the geometric image point, and their radii differing by $(\lambda/4)$.

The 'Rayleigh quarter wavelength rule' is specified in terms of maximum deviation of the wave surface from the reference sphere, and does not mention any other restriction on the overall shape of the wave surface. This implies that the rule does not impose any restriction on the coefficients of the individual aberrations composing the wave surface. Consequently, it is no longer feasible to assign unique tolerances for the individual aberration coefficients. An additional image evaluation criterion becomes necessary for fixing the latter tolerances [42].

As mentioned earlier, Rayleigh's observation on the decrease of central intensity in the aberrated diffraction pattern was based on primary spherical aberration alone, and he noted that when the wave-surface deviates at each end by a quarter wavelength, the central intensity decreases by 20 per cent from the unaberrated case. Rayleigh hypothesized this as the maximum fall that can be tolerated in a 'diffraction limited system'. Instead of the smooth variation in the case of primary spherical aberration where the aberration varies monotonically from zero at centre to $\frac{\lambda}{4}$ at the edge of the exit pupil, it is indeed possible to have wavefronts that undergo several oscillations in wavefront from centre to the edge without violating the $\frac{\lambda}{4}$ rule, and Barakat showed that in some cases the central intensity falls below 0.8 of the unaberrated case [43]. However, van den Bos reported that some anomaly in the representation of the aberration function used by Barakat led to incorrect results [44]. In any case, a caveat on unreserved use of the Rayleigh $\frac{\lambda}{4}$ rule is called for in general. Conrady formulated theoretically and applied in practice tolerances on focus error and on aberrations using the Rayleigh $\frac{\lambda}{4}$ rule as the criterion for permissible wavefront errors [45].

17.2.2 Strehl Criterion

Towards the end of the nineteenth century, Strehl [4, 46] noted that the most significant effect of the presence of small aberration on the diffraction image of a point source was to reduce the intensity at the principal maximum of the diffraction pattern, i. e. at the 'diffraction focus', with more light appearing concomitantly in the outer parts of the pattern. He proposed to use the ratio of this reduced intensity at the diffraction focus, $I_{ab}(0)$, with the intensity at the diffraction focus in absence of any aberration, $I_0(0)$, as a measure of the influence of aberration on image quality. Strehl designated this measure as 'Definitionshelligkeit'.

> The literal translation of this word, namely 'definition brightness' is just as ambiguous in English as the original word is in German. It is sometimes rendered in English as the 'Strehl definition' but it seems preferable to use the phrase 'the Strehl intensity ratio', or simply 'Strehl ratio'.

[47]

Wetherell notes that the word 'Definition' is used here in the sense 'distinctness of outline or detail' [48]. The Strehl ratio, like the Rayleigh criterion, is appropriate for highly corrected systems.

Analytically, the Strehl criterion is stated as: A real optical imaging system with residual aberrations can be considered to be 'diffraction limited' if the Strehl ratio does not fall below 0.8, i.e., if

$$\frac{I_{ab}(0,0)}{I_0(0,0)} \geq 0.8 \tag{17.14}$$

17.2.3 Strehl Ratio in Terms of Variance of Wave Aberration

Within its range of applicability, the practical usefulness of the Strehl ratio was greatly extended by Maréchal who showed that the loss of relative intensity at the diffraction focus is related to the variance of the wave aberration [49–50]. Assuming uniform real amplitude over the unit circular pupil, the expression for complex amplitude $F_{ab}(0,0)$ at the diffraction focus is obtained from (17.1) as

$$F_{ab}(0,0) = \int_0^{2\pi}\int_0^1 e^{ikW(r,\theta)} r \, dr \, d\theta \tag{17.15}$$

The mean value of aberration, $\bar{W}$, over the pupil is given by

$$\bar{W} = \left\{\int_0^{2\pi}\int_0^1 W(r,\theta) r dr d\theta\right\} / \left\{\int_0^{2\pi}\int_0^1 r dr d\theta\right\} = \frac{1}{\pi}\int_0^{2\pi}\int_0^1 W r dr d\theta \tag{17.16}$$

In the above, $W(r,\theta)$ is represented by the shortened notation W. In terms of this mean value of aberration, $F_{ab}(0,0)$ can be expressed as

$$F_{ab}(0,0) = e^{ik\bar{W}} \int_0^{2\pi}\int_0^1 e^{ik(W-\bar{W})} r \, dr \, d\theta \tag{17.17}$$

Assuming the variation of $(W - \bar{W})$ over the pupil to be small enough for neglecting terms with powers greater than 2 of $\left[k(W - \bar{W})\right]$ in the expansion of $e^{ik(W-\bar{W})}$, $F_{ab}(0,0)$ reduces to

$$F_{ab}(0,0) = \pi e^{ik\bar{W}} \left[1 - \frac{1}{2}k^2\left(\overline{W^2} - \bar{W}^2\right)\right] \tag{17.18}$$

where

$$\overline{W^2} = \left\{\int_0^{2\pi}\int_0^1 W^2 r \, dr \, d\theta\right\} / \left\{\int_0^{2\pi}\int_0^1 r \, dr \, d\theta\right\} = \frac{1}{\pi}\int_0^{2\pi}\int_0^1 W^2 r \, dr \, d\theta \tag{17.19}$$

The Strehl intensity ratio can be expressed as

$$\text{S.R.} = \frac{I_{ab}(0,0)}{I_0(0,0)} = \frac{|F_{ab}(0,0)|^2}{|F_0(0,0)|^2} \cong \left\{1 - \frac{2\pi^2}{\lambda^2}E\right\}^2 = \left\{1 - \frac{k^2E}{2}\right\}^2 = (\text{S.R.})_1 \tag{17.20}$$

where $k = \frac{2\pi}{\lambda}$ and variance E is given by

$$E = \frac{1}{\pi}\int_0^{2\pi}\int_0^1 W^2 r\, dr d\theta - \left\{\frac{1}{\pi}\int_0^{2\pi}\int_0^1 W r\, dr d\theta\right\}^2 \tag{17.21}$$

E is the variance of $W(r,\theta)$ over the exit pupil. Variance E is the 'mean-square deviation' of the wavefront from the spherical form. It can also be expressed as $E = (\Delta W)^2$, where ΔW is the standard deviation, or the root mean square deviation of the wavefront from its spherical shape. E is expressed as

$$E = (\Delta W)^2 = \int_0^{2\pi}\int_0^1 (W - \bar{W})^2 r\, dr\, d\theta \Big/ \int_0^{2\pi}\int_0^1 r\, dr\, d\theta \tag{17.22}$$

This expression reduces to (17.21).

Neglecting terms of degree higher than 2 in E, expression (17.20) reduces to

$$\text{S.R.} \cong 1 - \left(\frac{2\pi}{\lambda}\right)^2 E = 1 - k^2E = (\text{S.R.})_2 \tag{17.23}$$

Although it is an approximation of the Maréchal expression, its use 'has the advantage that it is more directly related to the extremal properties of Zernike's circle polynomials' [51].

Maréchal's relation between the Strehl ratio and the variance of wave aberration holds good for imaging systems with small values of aberration. Extensive numerical computations have shown that the necessary condition to be satisfied is that the Strehl ratio of the system should exceed 0.5.

For systems with $\text{S.R.} \geq 0.5$, use of relation (17.20) correctly predicts an increase in Strehl ratio for a decrease in variance E. But in the case of systems with $\text{S.R.} < 0.5$, this may no longer be true [52]. An empirical relation of the form

$$\text{S.R.} \cong e^{-k^2E} = (\text{S.R.})_3 \tag{17.24}$$

is found to remain valid over a wider range, providing useful results so long as the Strehl ratio of the system is larger than 0.3 [53].

By algebraic manipulations, an interrelationship among the three approximations for the Strehl ratio given in (17.20), (17.23), and (17.24) is

$$(\text{S.R.})_3 = (\text{S.R.})_2 + \frac{1}{2}(k^2E)^2 = (\text{S.R.})_1 + \frac{1}{4}k^4E^2 \tag{17.25}$$

It follows that

$$(\text{S.R.})_3 > (\text{S.R.})_2 > (\text{S.R.})_1 \tag{17.26}$$

Numerical computations show that, for moderate values of aberrations, $(\text{S.R.})_1$ and $(\text{S.R.})_2$ underestimates the correct value of the Strehl ratio, whereas $(\text{S.R.})_3$ overestimates the correct value.

17.2.3.1 Use of Local Variance of Wave Aberration

The accuracy and range of validity can be significantly increased by making use of local averages and local variances of wave aberration. It involves dividing the total area of the pupil into subintervals, and to use the local values of the average and variance of wave aberration. Obviously, the accuracy of computation of Strehl ratio by using these values increases with the increase in the number of subintervals. A brief mathematical analysis is given below. For the sake of simplicity, the treatment given below is restricted to azimuth invariant rotationally symmetric aberration function $W(r)$. An extension to the case of aberration function $W(r,\theta)$ is straightforward.

From (17.15) the complex amplitude $F_{ab}(0,0)$ at the diffraction focus is given by

$$F_{ab}(0,0) = \int_0^{2\pi}\int_0^1 e^{ikW(r,\theta)} r\,dr\,d\theta \tag{17.27}$$

Using the change of variables $t = r^2$, this relation can be rewritten as

$$F_{ab}(0,0) = \pi\int_0^1 e^{ikW(t)}dt \tag{17.28}$$

Let the interval $(0,1)$ for t be divided in N equal subintervals of width 2ϵ, so that $2\epsilon N = 1$. The p-th subinterval in the range $\{(t_p - \epsilon),(t_p + \epsilon)\}$ has its midpoint $t_p = \frac{(2p-1)}{2N}$. In each subinterval, the averages are

$$\bar{W}_p = \frac{1}{2\epsilon}\int_{t_p-\epsilon}^{t_p+\epsilon} W(t)dt \qquad \overline{W_p^2} = \frac{1}{2\epsilon}\int_{t_p-\epsilon}^{t_p+\epsilon} [W(t)]^2 dt \tag{17.29}$$

Using the above, equation (17.28) can be written as

$$F_{ab}(0,0) = \pi\sum_{p=1}^{N} e^{ik\bar{W}_p}\int_{t_p-\epsilon}^{t_p+\epsilon} e^{ik[W-\bar{W}_p]}dt \tag{17.30}$$

Using the approximation

$$e^{ik[W-\bar{W}_p]} \cong \left\{1 + ik(W-\bar{W}_p) - \frac{1}{2}[k(W-\bar{W}_p)]^2\right\} \tag{17.31}$$

(17.30) reduces to

$$F_{ab}(0,0) = \pi 2\epsilon\sum_{p=1}^{N} e^{ik\bar{W}_p}\left[1 - \frac{1}{2}\left(\overline{W_p^2} - \bar{W}_p^2\right)\right] \tag{17.32}$$

$$\text{We have } F_0(0,0) = \pi 2\epsilon N = \pi \tag{17.33}$$

The normalized complex amplitude at the diffraction focus is

$$F_N(0,0) = \frac{F_{ab}(0,0)}{F_0(0,0)} = 2\epsilon\sum_{p=1}^{N} e^{ik\bar{W}_p}\left[1 - \frac{1}{2}k^2\left(\overline{W_p^2} - \bar{W}_p^2\right)\right] \tag{17.34}$$

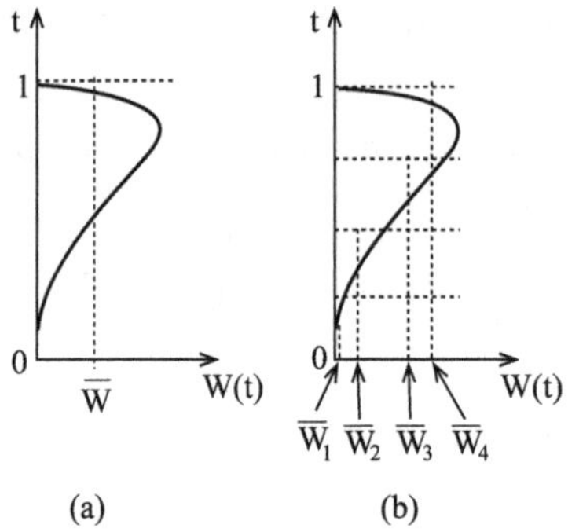

FIGURE 17.8 (a) Average of the aberration function over $(0,t)$ is $\bar{W}$ (Maréchal case) (b) Local averages $\bar{W}_n, n = 1,2,3,4$ over the four subzones of equal width.

This can be rewritten as

$$F_N(0,0) = 2\epsilon \sum_{p=1}^{N} e^{ik\bar{W}_p}\left[1 - \frac{1}{2}k^2 E_p\right] \tag{17.35}$$

where E_p is the p-th local variance of the aberration function, and is given by

$$E_p = \left(\overline{W_p^2} - \bar{W}_p^2\right) \tag{17.36}$$

Using (17.34), the Strehl ratio can be expressed as

$$\text{S.R.} = I_N(0,0) = |F_N(0,0)|^2 = 4\epsilon^2 \sum_{p=1}^{N}\sum_{q=1}^{N}\left[1 - \frac{1}{2}k^2 E_p\right]\left[1 - \frac{1}{2}k^2 E_q\right]\cos\left\{k\left(\bar{W}_p - \bar{W}_q\right)\right\} \tag{17.37}$$

This expression for Strehl Ratio in terms of local variances reduces to the original Maréchal expression (17.23) when $N = 1$. Figure 17.8 illustrates the difference between the average over the whole domain, and the local averages over subzones. Similar differences exist between the overall variance as used in Maréchal treatment, and the proposed local variances over subzones. Obviously, the accuracy in computation of the Strehl ratio and the range of validity of the expression in correlating variance with the Strehl ratio increases with increasing values of N. For large values of N, an approximate relation can be obtained by retaining the local average values of aberration and neglecting the local variances in the above expression

$$\text{S.R.} = 4\epsilon^2 \sum_{p=1}^{N}\sum_{q=1}^{N}\cos\left\{k\left(\bar{W}_p - \bar{W}_q\right)\right\} \tag{17.38}$$

The use of these local characteristics of wave aberration in computation of the Strehl ratio is illustrated in reference [23].

17.2.3.2 Tolerance on Variance of Wave Aberration in Highly Corrected Systems

For highly corrected systems, the tolerance criterion is usually

$$\text{S.R.} \geq 0.8 \tag{17.39}$$

Using (17.20), the tolerance criterion may be expressed in terms of variance E of wave aberration as

$$E \le \frac{\lambda^2}{187} \tag{17.40}$$

This implies that the tolerable value of root mean square deviation ΔW of the wavefront is given by

$$\Delta W \le \frac{\lambda}{13.67} \sim \frac{\lambda}{14} \tag{17.41}$$

It is important to note that the tolerable value of variance of wave aberration by imposition of the Strehl criterion of tolerance, as formulated by Maréchal, is based on the overall deformation or deviation of the wavefront from the spherical shape. This deviation or deformation is an integrated or combined effect of all terms of the aberration polynomial describing the wavefront.

Let us first consider a few special cases.

17.2.3.3 Tolerance on Axial Shift of Focus in Aberration-Free Systems

It has been shown in Section 8.2.3.1.1 that a longitudinal or axial shift of the focal/image plane can be taken into account as wavefront aberration $W_{20}r^2$, where W_{20} is called 'the defect of focus' aberration term. It is approximately related with $\Delta\zeta$, the axial shift of the focal/image plane by

$$W_{20} = \frac{1}{2} n'(\sin u')^2 \Delta\zeta \tag{17.42}$$

where n' is the refractive index, and u' is the semi-angular aperture in the image space. It follows

$$\bar{W} = \frac{1}{\pi}\int_0^{2\pi}\int_0^1 W_{20} r^2 r \, dr \, d\theta = \frac{W_{20}}{2} \tag{17.43}$$

$$\overline{W^2} = \frac{1}{\pi}\int_0^{2\pi}\int_0^1 \left(W_{20} r^2\right)^2 r \, dr \, d\theta = \frac{W_{20}^2}{3} \tag{17.44}$$

$$E = (\Delta W)^2 = \overline{W^2} - \bar{W}^2 = \frac{W_{20}^2}{3} - \frac{W_{20}^2}{4} = \frac{W_{20}^2}{12} \tag{17.45}$$

Using (17.40), it is seen that, per the Strehl criterion, the tolerance on the axial shift of focus is

$$|W_{20}| \le \frac{\lambda}{3.95} \sim \frac{\lambda}{4} \tag{17.46}$$

In fact, this was the case considered by Rayleigh while stipulating his famous 'quarter wavelength rule'.

For an object at infinity, $(n' \sin u') = \frac{1}{2F_{\#}}$. Assuming image space refractive index $n' = 1$, from (17.42) and (17.46), we get

$$\Delta\zeta = 2\lambda F_{\#}^2 \tag{17.47}$$

Table 17.1 shows the values of $\pm\Delta\zeta$ corresponding to selected values of F number for working wavelength 0.5 μm. This shows the unavoidable uncertainty associated with the measurement of the focal length of

TABLE 17.1

Values of $\Delta\zeta$ at Different F-number for a Diffraction Limited Imaging System (Working wavelength $\lambda = 0.5\mu m$)

F-number	$\pm\Delta\zeta$ (in mm)
1	0.001
5	0.025
10	0.100
15	0.225
20	0.400
30	0.900
50	2.500

diffraction-limited axially symmetric imaging systems. In practice, the uncertainty will be higher for aberrated systems.

17.2.3.4 Tolerance on Primary Spherical Aberration

In the presence of primary spherical aberration alone, the aberration function is expressed as

$$W(r) = W_{40}r^4 \tag{17.48}$$

Values of $\bar{W}$ and $\overline{W^2}$ are given by

$$\bar{W} = \frac{W_{40}}{3} \qquad \overline{W^2} = \frac{W_{40}^{\ 2}}{5} \tag{17.49}$$

so that variance is

$$E = (\Delta W)^2 = \overline{W^2} - \bar{W}^2 = \frac{4}{45}W_{40}^{\ 2} \tag{17.50}$$

Using (17.40), we note that, to satisfy the Strehl criterion, the tolerance on primary spherical aberration is

$$|W_{40}| \le \frac{\lambda}{4.08} \sim \frac{\lambda}{4} \tag{17.51}$$

17.3 Aberration Balancing

It is important to note that the tight tolerance on primary spherical aberration, as shown above, corresponds to imaging on the paraxial or Gaussian image plane. It is a common observation that the quality of image is often better on transverse planes that are slightly away from the Gaussian image plane. The location of the best image plane in the presence of primary spherical aberration can be explored by considering an aberration function $W(r)$ that incorporates the effects of both primary spherical aberration and 'defocusing' simultaneously. $W(r)$ can be expressed as

$$W(r) = W_{20}r^2 + W_{40}r^4 \tag{17.52}$$

$\bar{W}$ and $\overline{W^2}$ are given by

$$\bar{W} = \frac{1}{\pi}\int_0^{2\pi}\int_0^1 \left[W_{20}r^2 + W_{40}r^4\right] r dr d\theta = \frac{W_{20}}{2} + \frac{W_{40}}{3} \tag{17.53}$$

$$\overline{W^2} = \frac{1}{\pi}\int_0^{2\pi}\int_0^1 \left[W_{20}r^2 + W_{40}r^4\right]^2 r dr d\theta = \frac{W_{20}^{\ 2}}{3} + \frac{W_{40}^{\ 2}}{5} + \frac{W_{20}W_{40}}{2} \tag{17.54}$$

Variance, or the mean square wavefront deformation E, is given by

$$E = \left(\Delta W\right)^2 = \overline{W^2} - \bar{W}^2 = \left[\frac{W_{20}^{\ 2}}{12} + \frac{4}{45}W_{40}^{\ 2} + \frac{W_{20}W_{40}}{6}\right] \tag{17.55}$$

The relationship between W_{20} and W_{40} that leads to a minimum value of E for a given value of W_{40} can be found from the minimization condition

$$\frac{dE}{dW_{20}} = \frac{d\left[\left(\Delta W\right)^2\right]}{dW_{20}} = 0 \tag{17.57}$$

From (17.55) and (17.56), it follows

$$W_{20} = -W_{40} \tag{17.57}$$

The above relation implies that for a specific value of W_{40}, the defocused plane specified by $W_{20} = -W_{40}$ provides the minimum value of variance that in turn leads to the maximum value of Strehl ratio. This phenomenon is called 'Aberration balancing' [54–56]. Substituting from (17.55), in (17.53) we obtain

$$E = \left[\frac{1}{12} + \frac{4}{45} - \frac{1}{6}\right]W_{40}^{\ 2} = \frac{1}{180}W_{40}^{\ 2} \tag{17.58}$$

With defect of focus given by (17.57), the tolerable amount of W_{40} per the Strehl criterion (17.39) can be worked out from the Maréchal variance criterion (17.40) by using (17.58) as

$$W_{40}^{\ 2} \le \frac{180}{187}\lambda^2 \tag{17.59}$$

The tolerable value for W_{40} is approximately

$$\left|W_{40}\right| \le 0.98\lambda \sim \lambda \tag{17.60}$$

It should be noted that on the optimum focal plane specified by $W_{20} = -W_{40}$, the Strehl criterion can be satisfied by a relaxed tolerance for primary spherical aberration. Whereas on the paraxial focal plane, satisfaction of the Strehl criterion calls for a strict tolerance $\frac{\lambda}{4}$ for primary spherical aberration, the tolerance is relaxed by approximately four times on the optimum focal plane.

Figure 17.4(a)–(c) presents the point spread function at best focus, i.e. at the plane corresponding to $W_{20} = -W_{40}$ in the presence of three values of primary spherical aberration (a) $W_{40} = 1\lambda$, (b) $W_{40} = 2\lambda$, and (c) $W_{40} = 3\lambda$. That the Strehl intensity is significantly higher at the best focal plane compared to transverse planes on two opposite sides of the best focal plane can be clearly seen from the curves in Figure 17.5(a)–(c).

17.3.1 Tolerance on Secondary Spherical Aberration with Optimum Values for Primary Spherical Aberration and Defect of Focus

When secondary spherical aberration is considered singly, the aberration function is

$$W(r) = W_{60} r^6 \tag{17.61}$$

$\bar{W}$ and $\overline{W^2}$ of this function are given by

$$\bar{W} = \frac{1}{\pi}\int_0^{2\pi}\int_0^1 W_{60} r^6 r dr d\theta = \frac{W_{60}}{4} \tag{17.62}$$

$$\overline{W^2} = \frac{1}{\pi}\int_0^{2\pi}\int_0^1 \left(W_{60} r^6\right)^2 r dr d\theta = \frac{W_{60}^{\ 2}}{7} \tag{17.63}$$

So that variance E is

$$E = (\Delta W)^2 = \frac{9}{112}(W_{60})^2 \tag{17.64}$$

For satisfying Strehl criterion, variance E has to satisfy the condition (17.40), so that

$$\frac{9}{112}(W_{60})^2 \le \frac{\lambda^2}{187} \tag{17.65}$$

The tolerance on secondary spherical aberration, when considered singly, is

$$|W_{60}| \le \frac{\lambda}{3.87} \sim 0.258\lambda \tag{17.66}$$

This tolerance can be relaxed significantly in the concomitant presence of optimum values for primary spherical aberration and defect of focus. When the three aberrations are present together, the aberration function is of the form

$$W(r) = W_{20} r^2 + W_{40} r^4 + W_{60} r^6 \tag{17.67}$$

$\bar{W}$ and $\overline{W^2}$ of this function are given by

$$\bar{W} = \frac{1}{\pi}\int_0^{2\pi}\int_0^1 \left[W_{20} r^2 + W_{40} r^4 + W_{60} r^6\right] r\, dr\, d\theta = \frac{W_{20}}{2} + \frac{W_{40}}{3} + \frac{W_{60}}{4} \tag{17.68}$$

$$\overline{W^2} = \frac{1}{\pi}\int_0^{2\pi}\int_0^1 \left[W_{20} r^2 + W_{40} r^4 + W_{60} r^6\right]^2 r dr d\theta$$

$$= \frac{W_{20}^{\ 2}}{3} + \frac{W_{40}^{\ 2}}{5} + \frac{W_{60}^{\ 2}}{7} + \frac{W_{20}W_{40}}{2} + \frac{W_{40}W_{60}}{3} + \frac{2}{5}W_{60}W_{20} \tag{17.69}$$

Variance E is expressed as

$$E = \overline{W^2} - (\bar{W})^2 = \frac{W_{20}{}^2}{12} + \frac{4}{45} W_{40}{}^2 + \frac{9}{112} W_{60}{}^2 + \frac{1}{6} W_{20} W_{40} + \frac{1}{6} W_{40} W_{60} + \frac{3}{20} W_{60} W_{20} \tag{17.70}$$

For any specific value of W_{60}, the optimum values for W_{40} and W_{20} that minimize E can be obtained from the conditions

$$\frac{\partial E}{\partial W_{40}} = \frac{\partial E}{\partial W_{20}} = 0 \tag{17.71}$$

Using (17.70), the conditions (17.71) yield

$$\left.\begin{aligned} \frac{8}{45} W_{40} + \frac{1}{6} W_{20} + \frac{1}{6} W_{60} = 0 \\ \frac{1}{6} W_{40} + \frac{1}{6} W_{20} + \frac{3}{20} W_{60} = 0 \end{aligned}\right\} \tag{17.72}$$

Let

$$\alpha = \left(\frac{W_{40}}{W_{60}}\right) \quad \text{and} \quad \beta = \left(\frac{W_{20}}{W_{60}}\right) \tag{17.73}$$

(17.72) can be rewritten as

$$\left.\begin{aligned} 16\alpha + 15\beta + 15 = 0 \\ 10\alpha + 10\beta + 9 = 0 \end{aligned}\right\} \tag{17.74}$$

Solving (17.74), we get

$$\alpha = -\frac{3}{2} \qquad \beta = \frac{3}{5} \tag{17.75}$$

Therefore, for any specific value of W_{60}, the optimum values for W_{40} and W_{20} that minimize E are

$$\left.\begin{aligned} W_{40} = -\frac{3}{2} W_{60} \\ W_{20} = \frac{3}{5} W_{60} \end{aligned}\right\} \tag{17.76}$$

Substituting (17.76) in (17.70), the variance E reduces to $E = \frac{1}{2800} W_{60}{}^2$. For satisfying the Strehl criterion, variance E has to satisfy the condition (17.40), so that

$$\frac{1}{2800} W_{60}{}^2 \le \frac{\lambda^2}{187} \tag{17.77}$$

So, the tolerance on secondary spherical aberration is

$$|W_{60}| \leq 3.87\lambda \tag{17.78}$$

in the presence of balanced values of primary spherical aberration $W_{40} = -\frac{3}{2}W_{60}$ and defect of focus $W_{20} = \frac{3}{5}W_{60}$. It is important to note that the tolerance on secondary spherical aberration can be significantly relaxed when the optimum balancing of primary spherical aberration and the defect of focus are utilized.

17.3.2 Tolerance on Primary Coma with Optimum Value for Transverse Shift of Focus

In the presence of primary coma alone, the aberration function is of the form

$$W(r,\theta) = W_{31}r^3 \cos\theta \tag{17.79}$$

It is seen that

$$\bar{W} = 0, \overline{W^2} = \frac{W_{31}^2}{8}, \text{ and Variance } E = \overline{W^2} - (\bar{W})^2 = \frac{W_{31}^2}{8} \tag{17.80}$$

The Strehl criterion is satisfied when variance E satisfies the condition (17.40), so that

$$\frac{W_{31}^2}{8} \leq \frac{\lambda^2}{187} \tag{17.81}$$

The corresponding tolerance on W_{31} is

$$|W_{31}| \leq \frac{\lambda}{4.83} \sim 0.2\lambda \tag{17.82}$$

Unlike the symmetric aberrations, no relaxation in the tolerance for primary coma can be obtained by incorporation of longitudinal shift of focus. This can also be inferred from the curves in Figure 17.7(a)–(c). However, significant relaxation can be obtained by incorporating a balanced amount of transverse shift of focus. This should be obvious from the curves in Figure 17.6. Note that in Figure 17.6(a), $W_{31} = 0.16\lambda$. The Strehl intensity, i.e. the intensity at $p = 0$, is about 0.87. But the peak intensity of the diffraction pattern is much higher, about 0.97 at $p = -0.13$. The latter implies a transverse shift of focus. In the case of Figure 17.6(b), $W_{31} = 0.32\lambda$. Note that the Strehl intensity at $p = 0$ falls to 0.62, but with a larger transverse shift of focus, the peak intensity in the diffraction pattern is about 0.92. For the next case, with $W_{31} = 0.63\lambda$, the Strehl intensity at $p = 0$ is as low as 0.2, but with further transverse shift of focus, the peak intensity is 0.8.

It has been shown earlier that the transverse shift of focus can be taken account of as an aberration term of the form $W_{11} r\cos\theta$. In the case of presence of the latter in combination with primary coma, the aberration function takes the form

$$W(r,\theta) = W_{11} r\cos\theta + W_{31}r^3 \cos\theta \tag{17.83}$$

$\bar{W}$ and $\overline{W^2}$ are given by

$$\bar{W} = 0 \qquad \overline{W^2} = \frac{W_{11}^2}{4} + \frac{W_{31}^2}{8} + \frac{W_{11}W_{31}}{3} \tag{17.84}$$

Variance E is expressed as

$$E = (\Delta W)^2 = \overline{W^2} - (\bar{W})^2 = \frac{W_{11}{}^2}{4} + \frac{W_{31}{}^2}{8} + \frac{W_{11}W_{31}}{3} \tag{17.85}$$

For a specific value of W_{31}, the optimum value for W_{11} that minimizes the variance E can be obtained from the condition

$$\frac{\partial E}{\partial W_{11}} = 0 \tag{17.86}$$

From (17.85) and (17.86), the optimum value of W_{11} is

$$W_{11} = -\frac{2}{3} W_{31} \tag{17.87}$$

Substituting this value of W_{11} in (17.85), the expression for variance E reduces to

$$E = \frac{W_{31}{}^2}{72} \tag{17.88}$$

In accordance with condition (17.40) on variance E for satisfying the Strehl criterion, we get

$$\frac{W_{31}{}^2}{72} \le \frac{\lambda^2}{187} \tag{17.89}$$

Therefore, in the presence of the optimum value of transverse shift of focus, the tolerance on the primary coma is

$$|W_{31}| \le 0.62\lambda \tag{17.90}$$

The Strehl tolerance (17.82) on primary coma, when considered alone, is significantly relaxed when the optimum value for the transverse shift of focus is utilized simultaneously.

17.3.3 Tolerance on Primary Astigmatism with Optimum Value for Defect of Focus

In the presence of primary astigmatism, the aberration function takes the form

$$W(r,\theta) = W_{22} r^2 (\cos\theta)^2 \tag{17.91}$$

$\bar{W}$ and $\overline{W^2}$ are given by

$$\bar{W} = \frac{W_{22}}{4} \qquad \overline{W^2} = \frac{W_{22}{}^2}{8} \tag{17.92}$$

so that the variance E becomes

$$E = \overline{W^2} - (\bar{W})^2 = \frac{W_{22}{}^2}{16} \tag{17.93}$$

In order to satisfy the Strehl criterion, the variance E needs to satisfy the inequality so that

$$\frac{W_{22}^{\ 2}}{16} \le \frac{\lambda^2}{187} \tag{17.94}$$

In the presence of the single aberration term, primary astigmatism, the tolerance on W_{22} is expressed as

$$|W_{22}| \le 0.29\lambda \tag{17.95}$$

This tolerance can be somewhat relaxed in defocused planes lying in the neighbourhood of the Gaussian focal plane. The aberration term encompassing the two aberrations together is represented by

$$W(r,\theta) = W_{20}r^2 + W_{22}r^2(\cos\theta)^2 \tag{17.96}$$

$\bar{W}$ and $\overline{W^2}$ corresponding to the above aberration function are

$$\bar{W} = \frac{1}{2}\left[W_{20} + \frac{W_{22}}{2}\right] \qquad \overline{W^2} = \frac{W_{20}^{\ 2}}{3} + \frac{W_{22}^{\ 2}}{8} + \frac{W_{20}W_{22}}{3} \tag{17.97}$$

Variance of this aberration function is

$$E = (\Delta W)^2 = \overline{W^2} - (\bar{W})^2 = \frac{W_{20}^{\ 2}}{12} + \frac{W_{22}^{\ 2}}{16} + \frac{W_{20}W_{22}}{12} \tag{17.98}$$

The optimum value for W_{20} minimizing the variance E for a specific value of W_{22} can be obtained from the condition $\frac{\partial E}{\partial W_{20}} = 0$. Using (17.98) we get

$$\frac{\partial E}{\partial W_{20}} = \frac{W_{20}}{6} + \frac{W_{22}}{12} = 0 \tag{17.99}$$

i.e., the optimum value for W_{20} is given by

$$W_{20} = -\frac{1}{2}W_{22} \tag{17.100}$$

Substituting this value of W_{20} in (17.98), the minimum value of variance E is $\{W_{22}^{\ 2}/24\}$. The Strehl criterion is satisfied if the inequality (17.40) holds good. This implies the following condition

$$\frac{W_{22}^{\ 2}}{24} \le \frac{\lambda^2}{187} \tag{17.101}$$

Therefore, in the presence of optimal value for defocusing $W_{20} = -(W_{22}/2)$, the tolerance on primary astigmatism becomes

$$W_{22} \le 0.35\lambda \tag{17.102}$$

Higher order aberrations in lens systems typically depends on the basic structure of the system. Changes in curvatures or thicknesses do not usually modify the higher order aberrations by large amounts, whereas the primary aberrations are relatively more susceptible to these changes. In practical lens design,

TABLE 17.2

Strehl Tolerances for Highly Corrected Systems with Different Types of Residual Aberrations

Aberration	Aberration Polynomial $W(r)$	Tolerances on Aberration Coefficients W_{mn} and Aberration balancing conditions
Defocus	$W_{20}r^2$	$\lvert W_{20}\rvert \le 0.25\lambda$
Primary Spherical Aberration	$W_{40}r^4$	$\lvert W_{40}\rvert \le 0.25\lambda$
Primary Spherical Aberration (with choice of best focal plane)	$W_{20}r^2 + W_{40}r^4$	$\lvert W_{40}\rvert \le \lambda$ $W_{20} = -W_{40}$
Secondary Spherical Aberration	$W_{60}r^6$	$W_{60} \le 0.26\lambda$
Secondary Spherical Aberration (with choice of best focal plane and of primary spherical aberration)	$W_{20}r^2 + W_{40}r^4 + W_{60}r^6$	$\lvert W_{60}\rvert \le 3.87\lambda$ $W_{40} = -1.5\ W_{60}$ $W_{20} = 0.60\ W_{60}$
Primary Coma	$W_{31}r^3 \cos\theta$	$\lvert W_{31}\rvert \le 0.2\ \lambda$
Primary Coma (with choice of transverse shift of focus)	$W_{11}r\cos\theta + W_{31}r^3 \cos\theta$	$\lvert W_{31}\rvert \le 0.62\lambda$ $W_{11} = -(2/3)W_{31}$
Secondary Linear Coma (with choice of lateral shift of focus and of primary coma)	$W_{11}r\cos\theta + W_{31}r^3\cos\theta + W_{51}r^5\cos\theta$	$\lvert W_{51}\rvert \le 2.53\ \lambda$ $W_{11} = 0.30\ W_{51}$ $W_{31} = -1.20\ W_{51}$
Combined Coma type aberration terms	$W_{11}r\cos\theta + W_{31}r^3\cos\theta + W_{33}r^3(\cos\theta)^3$	$\lvert W_{31}\rvert \le 0.78\lambda$ $W_{11} = -0.17\ W_{31}$ $W_{33} = -W_{31}$
Primary Astigmatism	$W_{22}r^2(\cos\theta)^2$	$\lvert W_{22}\rvert \le 0.29\lambda$
Primary Astigmatism (with choice of defocus)	$W_{20}r^2 + W_{22}r^2(\cos\theta)^2$	$\lvert W_{22}\rvert \le 0.35\lambda$ $W_{20} = -0.5\ W_{22}$
Combined Astigmatism type aberration terms	$W_{20}r^2 + W_{22}r^4(\cos\theta)^2 + W_{44}r^4(\cos\theta)^4$	$\lvert W_{42}\rvert \le 1.77\lambda$ $W_{20} = -0.125\ W_{42}$ $W_{44} = -W_{42}$

aberration balancing often facilitates the tackling of intransigent aberrations. Table 17.2 gives a succinct presentation of Strehl Tolerances for highly corrected systems with different types of residual aberrations. Note the relaxation in tolerances when balanced aberrations of similar types are present simultaneously.

17.3.4 Aberration Balancing and Tolerances on a FEE-Based Criterion

Balancing aberrations and aberration tolerances, as enunciated in the earlier sections, is based on maximization of the Strehl ratio, or, equivalently, minimization of the variance of wave aberration.

As Maréchal mentions explicitly, this approach is useful only in the case of highly corrected systems with a small amount of residual aberrations. For practical systems that can work even with somewhat large amounts of residual aberrations, diffraction plays a minor role in settling image quality, so much so that often the diffraction effects can be ignored and image quality evaluation and tolerancing can be done solely on the basis of ray optical aberrations. In between these two limiting cases of optical imaging systems lie a large number of systems that call for the balancing of aberrations and subsequent tolerancing of aberrations on the basis of image evaluation criteria that retains validity in this intermediate domain. Effective central illumination, an image evaluation criterion based on 'Factor of Encircled Energy' (FEE) is useful in this context.

Effective central illumination, $C(\omega)$, within a circle of radius ω and centre at the chosen image point in the diffraction image of a point object is defined by [57]

$$C(\omega) = E(\omega)/\pi\omega^2 \tag{17.103}$$

where $E(\omega)$ is the energy within a circle of radius ω with its centre at the chosen image point in the diffraction pattern on the image plane corresponding to a point object. It is obvious that

$$\lim_{\omega \to 0} C(\omega) = I(0) \tag{17.104}$$

Denoting the aberration-free quantities by subscript '0' and the aberrated quantities by subscript 'ab',

$$\mathcal{C}_{ab}(\omega) = \frac{E_{ab}(\omega)}{\pi\omega^2} \quad \text{and} \quad \mathcal{C}_0(\omega) = \frac{E_0(\omega)}{\pi\omega^2} \tag{17.105}$$

'Relative Effective Central Illumination' (RECI) is defined by

$$\mathcal{C}_R(\omega) = \frac{\mathcal{C}_{ab}(\omega)}{\mathcal{C}_0(\omega)} = \frac{E_{ab}(\omega)}{E_0(\omega)} = \frac{\varepsilon_{ab}(\omega)}{\varepsilon_0(\omega)} \tag{17.106}$$

where the factor of encircled energy, FEE, for the aberrated and unaberrated systems are given by

$$\varepsilon_{ab}(\omega) = \frac{E_{ab}(\omega)}{E_{ab}(\infty)} \text{ and } \varepsilon_0(\omega) = \frac{E_0(\omega)}{E_0(\infty)} \tag{17.107}$$

Note that, by Parseval's theorem

$$E_{ab}(\infty) = E_0(\infty) = \pi \tag{17.108}$$

(17.106) may be rewritten as

$$\mathcal{C}_R(\omega) = 1 - \frac{\Delta\varepsilon(\omega)}{\varepsilon_0(\omega)} \tag{17.109}$$

where

$$\Delta\varepsilon(\omega) = \varepsilon_0(\omega) - \varepsilon_{ab}(\omega) \tag{17.110}$$

Again, from (17.106) we can write

$$\lim_{\omega\to 0} \mathcal{C}_R(\omega) = \frac{\lim_{\omega\to 0} \mathcal{C}_{ab}(\omega)}{\lim_{\omega\to 0} \mathcal{C}_0(\omega)} = \frac{I_{ab}(0)}{I_0(0)} \tag{17.111}$$

The right-hand side of the above expression is the Strehl ratio, which is expressed as

$$\text{S.R.} = \frac{I_{ab}(0)}{I_0(0)} = 1 - \frac{\Delta I_{ab}(0)}{I_0(0)} \tag{17.112}$$

Comparing (17.109) with (17.111), we note that $\left[\Delta\varepsilon(\omega)/\varepsilon_0(\omega)\right]$ represents the relative loss in RECI, $\mathcal{C}_R(\omega)$, and it reduces to the loss in the Strehl definition in the limit of $\omega \to 0$.

A general criterion for an imaging system to obtain satisfactory performance can be stipulated on the basis of RECI or FEE as

$$\frac{\Delta\varepsilon(\omega)}{\varepsilon_0(\omega)} \le 0.2 \tag{17.113}$$

for a prespecified ω. This criterion reduces to the Strehl criterion in the limit $\omega \to 0$. The value of ω to be chosen in a particular case may be of the order of the least resolvable distance expected from the system.

Treatment for aberration balance and tolerancing can be carried out in the same manner as done in earlier sections for Strehl tolerances based on Maréchal balancing. As an illustrative example, we present tolerances on secondary spherical aberration with optimum values for primary spherical aberration and defect of focus on the basis of criterion (17.113). Figure 17.9 shows the variation of balancing factors $\alpha\left(=\left\{\frac{W_{20}}{W_{60}}\right\}\right)$ and $\beta\left(=\left\{\frac{W_{40}}{W_{60}}\right\}\right)$, with changes in the value of ω. Figure 17.10 shows the variation of tolerances in secondary spherical aberration, $\left|W_{60}\right|_{max}$, with changes in value of ω, when the corresponding optimal values for W_{40} and W_{20} are maintained. Note that the values of α, β, *and* $\left|W_{60}\right|_{max}$ tend to Maréchal values based on Strehl tolerances as $\omega \to 0$. Similarly, in the presence of primary spherical aberration alone, the optimum values for the balancing factor $\gamma\left(=\frac{W_{20}}{W_{40}}\right)$ corresponding to different values of ω are given in Figure 17.11. Details of tolerancing on the criterion based on RECI can be found in our publication [58].

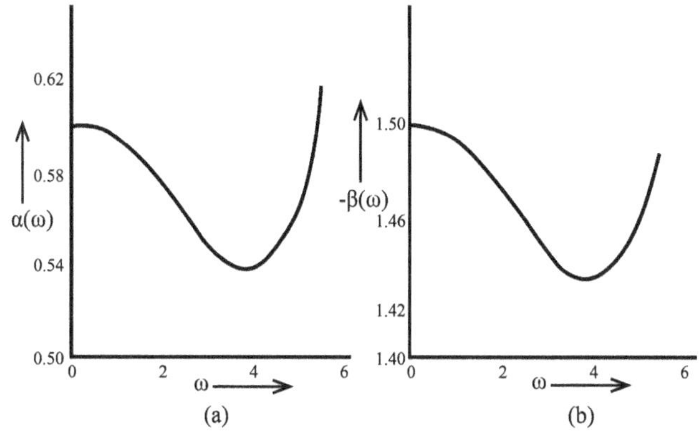

FIGURE 17.9 Aberration balancing parameters $\alpha(\omega)$ and $\beta(\omega)$ versus ω. (a) $\alpha = \frac{W_{20}}{W_{40}}$ (b) $\beta = \frac{W_{40}}{W_{60}}$.

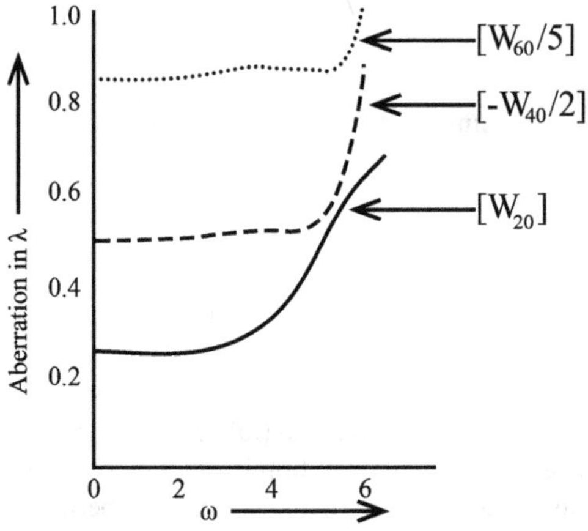

FIGURE 17.10 Variation in tolerable values of W_{60} in the presence of balanced values of W_{20} and W_{40} with changes in ω, the radius of the circle within which the FEE is maximized.

Note: Solid line: W_{20} dashed line: $W_{40} / 2$ chain line: $W_{60} / 5$.

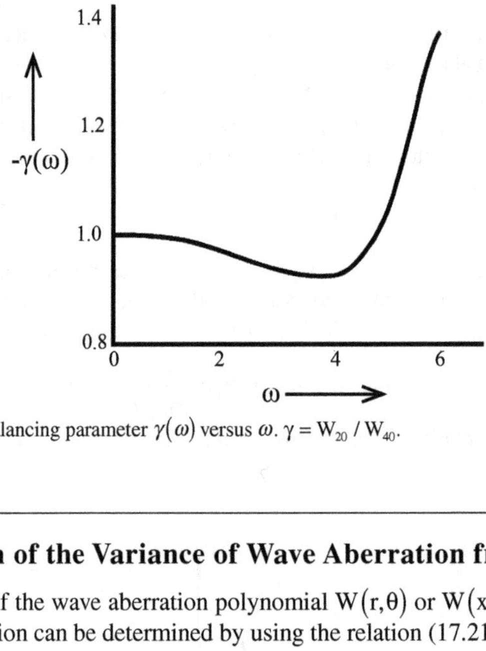

FIGURE 17.11 Aberration balancing parameter $\gamma(\omega)$ versus ω. $\gamma = W_{20} / W_{40}$.

17.4 Fast Evaluation of the Variance of Wave Aberration from Ray Trace Data

It was shown earlier that, if the wave aberration polynomial $W(r,\theta)$ or $W(x,y)$ is known, then the variance E of the wave aberration can be determined by using the relation (17.21) as

$$E = \frac{1}{\pi}\int_0^{2\pi}\int_0^1 W^2 \, r \, dr \, d\theta - \left\{\frac{1}{\pi}\int_0^{2\pi}\int_0^1 W \, r \, dr \, d\theta\right\}^2 \tag{17.114}$$

This expression assumes a circular exit pupil. In general, on account of pupil aberrations and vignetting effects, the effective exit pupil is no longer circular. However, the required result may be obtained by suitably modifying the above expression in each case. This problem can be circumvented by taking recourse to canonical coordinates (x'_S, y'_T) for the pupil, proposed by Hopkins (see Chapter 11). The exit pupil remains a unit circle, and so the above relation does not require any change. Other advantages of using the canonical coordinates are mentioned subsequently.

Assume the polar form for the aberration polynomial as

$$W(r,\theta)=\sum_m\sum_n W_{mn} r^m (\cos\theta)^n \tag{17.115}$$

where

$$r\cos\theta = y'_T \qquad r\sin\theta = x'_S \tag{17.116}$$

Substituting (17.115) in (17.114), we get

$$E=\frac{1}{\pi}\int_0^{2\pi}\int_0^1 \sum_m\sum_n\sum_p\sum_q W_{mn}W_{pq} r^{m+p}(\cos\theta)^{n+q}\, r dr d\theta$$

$$-\left\{\frac{1}{\pi}\int_0^{2\pi}\int_0^1 \sum_m\sum_n W_{mn} r^m (\cos\theta)^n\, r dr d\theta\right\}^2 \tag{17.117}$$

Defining

$$H(m,n)=\frac{1}{\pi}\int_0^{2\pi}\int_0^1 r^m(\cos\theta)^n\, r dr d\theta \tag{17.118}$$

It can be seen that

$$\begin{aligned} H(m,n) &= \frac{2}{m+2}\,\frac{n-1}{n}\,\frac{n-3}{n-2}\cdots\frac{3}{4}\,\frac{1}{2} && \text{for n even} \\ &= 0, && \text{for n odd} \end{aligned} \tag{17.119}$$

The variance E can be written as

$$E=\sum_m\sum_n\sum_p\sum_q H(m+p,n+q)W_{mn}W_{pq}$$

$$-\sum_m\sum_n\sum_p\sum_q H(m,n)H(p,q)W_{mn}W_{pq} \tag{17.120}$$

In Section 13.3 we dealt with a method for fast evaluation of the wave aberration polynomial coefficients from finite ray trace data. The latter involves the wave aberration and transverse ray aberrations for each finite ray traced through the system, and the S and T ray trace data. The total number P of aberration data W_j obtained thereby is equal to the number of aberration coefficients W_{mn} in the aberration polynomial. Equation (13.30) shows that each aberration coefficient W_{mn} is expressed by the relation

$$W_{mn}=\sum_{i=1}^{P} B(m,n;i)W_i \tag{17.121}$$

Using (17.121), (17.120) can be expressed as

$$E=\sum_i\sum_j\sum_m\sum_n\sum_p\sum_q\left[H(m+p,n+q)-H(m,n)H(p,q)\right]B(m,n;i)B(p,q;j)W_iW_j \tag{17.122}$$

The above relation for variance is actually a quadratic combination of the computed aberrations, as shown below

$$E = \sum_i \sum_j P(i,j) W_i W_j \tag{17.123}$$

The coefficients $P(i,j)$ are given by

$$P(i,j) = \sum_m \sum_n \sum_p \sum_q \left[H(m+p,n+q) - H(m,n)H(p,q) \right] B(m,n;i) B(p,q;j) \tag{17.124}$$

The coefficients $P(i,j)$ are unique for the particular combination of the aberration polynomial and the corresponding ray grid pattern for the computed aberrations, and are not related with any characteristics of the imaging system. They are universally valid for any lens system, and this obviates the need to solve the equation for each case.

It should be noted that the variance E can also be directly evaluated from the wave aberration values for R rays from the object point passing through R chosen points on the exit pupil. For each of these rays, the necessary sine error correction to determine the exact entrance pupil coordinates should be incorporated by the method outlined in Section 13.3. The expression for variance takes the quadratic for

$$E = \sum_{i=1}^{R} \sum_{j=1}^{R} P_0(i,j) W_i W_j \tag{17.125}$$

where

$$\left.\begin{aligned} P_0(i,j) &= \frac{1}{R} - \frac{1}{R^2}, \quad & i = j \\ &= -\frac{1}{R^2}, \quad & i \neq j \end{aligned}\right\} \tag{17.126}$$

Incidentally, many other image evaluation parameters, e.g. the integrated radius of gyration $\mathcal{R}$ of the geometrical PSF, can be expressed in terms of computed aberrations of a few selected rays as

$$\mathcal{R} = \sum_i \sum_j \tilde{P}(i,j) W_i W_j \tag{17.127}$$

where $\tilde{P}(i,j)$ are a set of universal coefficients that need to be evaluated once, and only once, for a standard ray grid pattern, and that can be used in case of any system.

17.5 Zernike Circle Polynomials

In our treatment of optical aberrations so far, we have used the expansion of the aberration function $W(x,y)$ or $W(r,\theta)$ in a power series. In tackling problems where integration over the unit circle is to be carried out, it is obvious that a prudent approach is to expand the aberration function W in terms of a complete set of polynomials that are orthogonal in the interior of the unit circle. One such set was introduced by Zernike [59–61]. Out of the infinite number of the complete set of polynomials in the interior of the unit circle involving the two variables r and θ, the set of orthogonal polynomials proposed by Zernike has certain unique properties of invariance. The most notable among them is the property of rotational invariance so that any Zernike polynomial $Z_n^m(r,\theta)$ can be represented as a product of the form

$$Z_n^m(r,\theta) = R_n^m(r) G(\theta) \tag{17.128}$$

where $G(\theta)$ is a continuous function that repeats itself every 2π radians, and rotating the coordinate system by an angle α does not change the form of the polynomial, i.e.

$$G(\theta+\alpha)=G(\theta)G(\alpha) \tag{17.129}$$

The set of trigonometric functions

$$G(\theta)=e^{\pm im\theta} \tag{17.130}$$

where m is any positive integer or zero, meets the requirement.

Zernike circle polynomials are given by

$$Z_n^m(r,\theta)=R_n^m(r)e^{\pm im\theta} \tag{17.131}$$

Indices n and m are integers ≥ 0, while $n \geq m$ and $(n-m)$ is even. $R_n^m(r)$ is a polynomial in r, containing the terms, $r^m, r^{m+2}, \ldots, r^{n-2}, r^n$. These polynomials are called radial polynomials, and they are closely related to Jacobi's polynomials (terminating hypergeometric series) [62]. The polynomials $R_n^m(r)$ are even in r, if n or m is even; they are odd in r, if n or m is an odd number. They are given by [62]

$$R_n^m(r)=(-1)^{\frac{(n-m)}{2}}\begin{pmatrix}\frac{n+m}{2}\\ m\end{pmatrix}r^m \quad F\left(\frac{(n+m+2)}{2},-\frac{(n-m)}{2},(m+1),r^2\right) \tag{17.132}$$

where F is the hypergeometric function. In terms of Jacobi's polynomials, the above expression can be written as

$$R_n^m(r)=(-1)^{\frac{(n-m)}{2}}\begin{pmatrix}\frac{n+m}{2}\\ m\end{pmatrix}r^m G_{\frac{(n-m)}{2}}\left(m+1,m+1,r^2\right)$$

$$=\frac{r^{-m}}{\left(\frac{n-m}{2}\right)!}\left(\frac{d}{d\left(r^2\right)}\right)^{\frac{n-m}{2}}\left\{\left(r^2\right)^{\frac{n+m}{2}}\left(r^2-1\right)^{\frac{n-m}{2}}\right\} \tag{17.133}$$

where $G_j(p,q,x)$ are Jacobi's polynomials. This can be further simplified to obtain an explicit formula for $R_n^m(r)$ as

$$R_n^m(r)=\sum_{s=0}^{(n-m)/2}\frac{(n-s)!}{s!\left(\frac{n+m}{2}-s\right)!\left(\frac{n-m}{2}-s\right)!}r^{n-2s} \tag{17.134}$$

The normalization has been chosen so that for any integer value for n, $R_n^m(1)=1$.

An aberration function $W(\tau,r,\theta)$ corresponding to a field point represented by the fractional field τ, and a point on the exit pupil represented in polar coordinates as (r,θ) can be expanded in terms of a complete set of orthonormal Zernike circle polynomials as

$$W(\tau,r,\theta)=\sum_{n=0}^{\infty}\sum_{m=0}^{n}a_{nm}(\tau)Z_n^m(r,\theta) \tag{17.135}$$

For the sake of simplicity, the notations

$$W(r,\theta)\equiv W(\tau,r,\theta) \text{ and } c_{nm}\equiv a_{nm}(\tau) \tag{17.136}$$

are used to represent the parameters for a specific field point τ. It follows

$$W(r,\theta)=\sum_{n=0}^{\infty}\sum_{m=0}^{n}c_{nm}Z_n^m(r,\theta) \tag{17.137}$$

where c_{nm} are the Zernike polynomial expansion coefficients. n and m are non-negative integers, while $n \geq m$, and $(n-m)$ is even.

In an axially symmetric optical system, on account of lateral symmetry about the meridional plane, the coefficients of $\sin m\theta$ are zero. So, the Zernike circle polynomials to be considered are of the form

$$Z_n^m(r,\theta)=R_n^m(r)\cos m\theta \tag{17.138}$$

In order to study orthogonality properties of these polynomials, let us consider the integral

$$\int_0^{2\pi}\int_0^1 Z_n^m(r,\theta)Z_{n'}^{m'}(r,\theta)r\,dr\,d\theta=\left[\int_0^1 R_n^m(r)R_{n'}^{m'}(r)\,dr\right]\left[\int_0^{2\pi}\cos m\theta\cos m'\theta\,d\theta\right] \tag{17.139}$$

We note that

$$\left.\begin{aligned}\int_0^{2\pi}\cos m\theta\cos m'\theta\,d\theta &= 0, \quad \text{for } m\neq m'\\ &=\pi, \quad \text{for } m=m'\geq 1\\ &=2\pi, \quad \text{for } m=m'=0\end{aligned}\right\} \tag{17.140}$$

This implies that the orthogonality condition for the angular functions is

$$\int_0^{2\pi}\cos m\theta\cos m'\theta\,d\theta=\pi(1+\delta_{m0})\delta_{mm'} \tag{17.141}$$

where δ represents Kronecker delta, so that

$$\left.\begin{aligned}\delta_{mm'}&=1, \quad for\ m=m'\\ &=0, \quad for\ m\neq m'\end{aligned}\right\} \text{ and } \left.\begin{aligned}\delta_{m0}&=1, \quad for\ m=0\\ &=0, \quad for\ m\neq 0\end{aligned}\right\} \tag{17.142}$$

From (17.139), it follows that the orthogonality condition for the radial polynomials needs to consider only the case $m=m'$, since whenever $m\neq m'$, the integral (17.140) becomes zero. The orthogonality property of the radial polynomials is given by

$$\int_0^1 R_n^m(r)\,R_{n'}^m(r)r\,dr=\frac{\delta_{nn'}}{2(n+1)} \tag{17.143}$$

In the above, we quote the result. Derivation of this result is quite involved, and interested readers may look into the thesis of Nijboer [63] for more information. The orthogonality condition of Zernike polynomials is

$$\int_0^{2\pi}\int_0^1 Z_n^m(r,\theta)Z_{n'}^m(r,\theta)r\,dr\,d\theta=\int_0^1 R_n^m(r)R_{n'}^m(r)r\,dr\int_0^{2\pi}(\cos m\theta)^2\,d\theta=\pi\frac{(1+\delta_{m0})\delta_{nn'}}{2(n+1)} \tag{17.144}$$

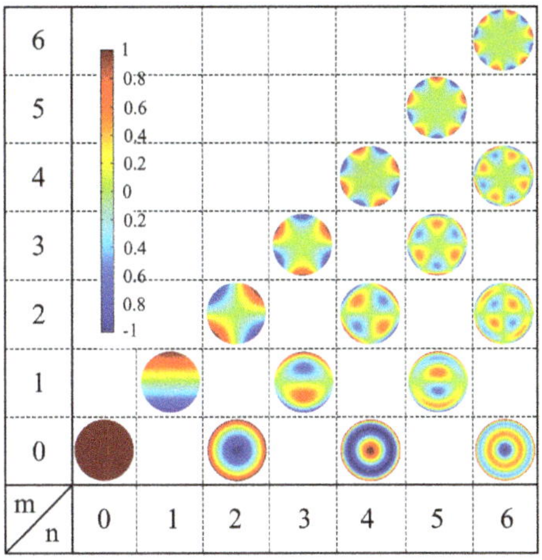

FIGURE 17.12 Zernike circle polynomials $Z_n^m(r,\theta)$ for radial order n and azimuthal order m for $0 \le (n,m) \le 6$ over the circular pupil with colour coding of amplitude.

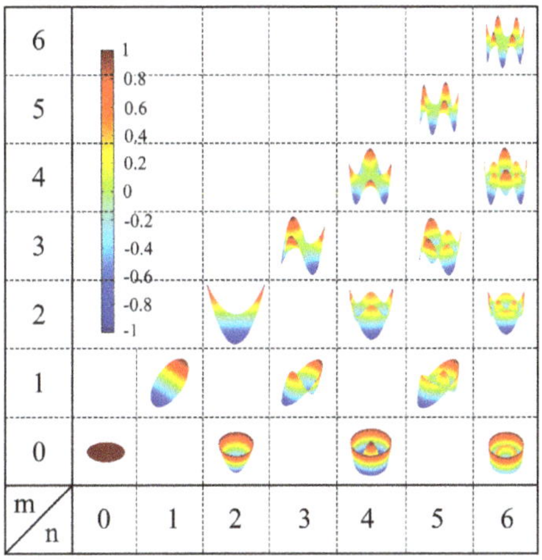

FIGURE 17.13 3-D surface profiles of the Zernike polynomials of Figure 17.12.

Table 17.3 gives the explicit forms of the radial polynomials $R_n^m(r)$ for the first few indices, $m \le 6$, $n \le 6$. Figure 17.12 shows the Zernike circle polynomials $Z_n^m(r,\theta)$ for radial order n and azimuthal order m for $0 \le (n,m) \le 6$ over the circular pupil with colour coded amplitude. Figure 17.13 shows the corresponding 3-D profiles. The 16 Zernike aberrations for axially symmetric systems usually follow the nomenclature given in Table 17.4.

Note that one of the Zernike polynomials, $Z_0^0(r,\theta) = R_0^0(r) = 1$. From the orthogonality property of the Zernike polynomials, it follows that the mean value $\overline{Z_n^m}$ of any other Zernike polynomial with index $n \ne 0$ is identically zero, i.e.

$$\overline{Z_n^m} = \left\{\int_0^{2\pi}\int_0^1 Z_n^m(r,\theta)\, r\, dr\, d\theta\right\} / \left\{\int_0^{2\pi}\int_0^1 r\, dr\, d\theta\right\} = 0 \quad (17.145)$$

and $\overline{Z_0^0} = 1$.

TABLE 17.3

The radial polynomials $R_n^m(r)$ for $m \le 6, n \le 6$

n / m	0	1	2	3	4	5	6
0	1		$2r^2-1$		$6r^4-6r^2+1$		$20r^6-30r^4+12r^2-1$
1		r		$3r^3-2r$		$10r^5-12r^3+3r$	
2			r^2		$4r^4-3r^2$		$15r^6-20r^4+6r^3$
3				r^3		$5r^5-4r^3$	
4					r^4		$6r^6-5r^4$
5						r^5	
6							r^6

TABLE 17.4

Nomenclature for the Zernike Aberrations Corresponding to the Zernike Polynomials $Z_n^m(r)$ for $m \le 6, n \le 6$

Notation	Name of Zernike Aberration
Z_0^0	Piston
Z_2^0	Defocus
Z_4^0	Balanced Primary Spherical Aberration
Z_6^0	Balanced Secondary Spherical Aberration
Z_1^1	Tilt
Z_3^1	Balanced Primary Coma
Z_5^1	Balanced Secondary Coma
Z_2^2	Primary Astigmatism
Z_4^2	Balanced Secondary Astigmatism
Z_6^2	Balanced Tertiary Astigmatism
Z_3^3	Elliptical Coma/ Trefoil (Secondary)
Z_5^3	Balanced Elliptical Coma/ Trefoil (Tertiary)
Z_4^4	Tetrafoil (Tertiary)
Z_6^4	Balanced Quaternary Tetrafoil
Z_5^5	Pentafoil
Z_6^6	Hexafoil

The mean value of the square $\overline{\left(Z_n^m\right)^2}$ of any Zernike polynomial is

$$\overline{\left(Z_n^m\right)^2} = \left\{\int_0^{2\pi}\int_0^1 \left(Z_n^m(r,\theta)\right)^2 r\,dr\,d\theta\right\} / \left\{\int_0^{2\pi}\int_0^1 r\,dr\,d\theta\right\} = \frac{1}{2(n+1)} \tag{17.146}$$

Using the above relations, the mean $\bar{W}$, the mean square $\overline{W^2}$, and the variance E of aberration function $W(r,\theta)$ (described in (17.137)) can be expressed in terms of the coefficients c_{nm} of the Zernike polynomial expansion as

$$\left.\begin{aligned} \bar{W} &= c_{00} \\ \overline{W^2} &= c_{00}^{\ 2} + \sum_{n=1}^{\infty}\sum_{m=0}^{n} \frac{c_{nm}^{\ 2}}{2(n+1)} \\ E = \overline{W^2} - \left(\bar{W}\right)^2 &= \sum_{n=1}^{\infty}\sum_{m=0}^{n} \frac{c_{nm}^{\ 2}}{2(n+1)} \end{aligned}\right\} \tag{17.147}$$

Note that the variance is given by the weighted sum of the squares of the coefficients of the Zernike polynomial expansion of the aberration function. Recalling from (17.20) and (17.23) the expression for the Strehl ratio in terms of variance of wave aberration, we get

$$\text{Strehl Ratio} = \left\{1 - \frac{k^2E}{2}\right\}^2 \cong 1 - k^2E \tag{17.148}$$

From the above relations, it is significant to note that each term in the Zernike polynomial expansion of an aberration function makes a unique contribution to the variance of the aberration function, and thereby to the Strehl ratio in the image. Addition or subtraction of terms in the expansion affects the final result to the extent of unique contribution by the particular terms. This is a consequence of the orthogonality of the Zernike aberration polynomials. Each Zernike polynomial represents a combination of terms of the power series expansion. This combination is not arbitrary; the constituent power series terms are perfectly balanced to provide minimum variance across the pupil. For example, $R_6^0(r)$ represents a secondary spherical aberration term that is balanced with primary spherical aberration and defocus terms to minimize its variance. Convenient relations between coefficients of the power series expansion and those of the Zernike polynomials have been worked out by many investigators. Zernike circle polynomials are being increasingly used in allied areas of adaptive optics, ophthalmology, opto-mechanical design, optical metrology, etc. A partial list of significant investigations on these polynomials and their applications is given in references [64–105].

It should be noted that Zernike polynomials were originally developed as orthogonal polynomials in the interior of a unit circle with uniform amplitude. For the interior of a different shape and/or nonuniform amplitude, the polynomials will be different. They need to be developed from appropriate orthogonality conditions. For annular pupils, several investigators have worked out Zernike annular polynomials [55, 75, 102–104]. For general aperture shapes, corresponding orthogonal polynomials can be derived by Gram–Schmidt orthogonalization of Zernike polynomials [105]. The case of Gaussian amplitude variation over the pupil has been treated in references [106–107].

A disadvantage of the use of Zernike polynomial expansion for aberration function should be noted. 'The coefficients of single aberrations in the expansion of aberration will change completely, if the radius of the exit pupil is varied, because the variable r in $R_n^m(r)$ must naturally vary from 0 to 1' [63].

Using Zernike circle polynomials, Nijboer [63] and Nienhuis [108] developed semi-analytical formulae for calculating intensity distributions in the image plane of axially symmetric optical systems with small amounts of residual aberrations. Their approach for evaluation is commonly known as the Nijboer-Zernike theory. The following relation plays a key role in development of formulae in their theory.

$$\int_0^1 R_n^m(r) J_m(vr) r dr = (-1)^{\frac{n-m}{2}} \frac{J_{n+1}(v)}{v} \tag{17.149}$$

Also, the original Nijboer-Zernike theory is applicable only in the case of uniform amplitude over the exit pupil. In the recent past, an improved semi-analytical approach for calculation of the intensity distributions in the presence of reasonably larger values of residual aberrations and defocusing is proposed. The method also permits a nonuniform amplitude over the pupil, and makes use of a new Bessel series representation of diffraction integrals. The approach is called the 'Extended Nijboer Zernike Theory' [109–111].

In classical aberration theory, aberration is expressed as a function of the coordinates of a point on the exit pupil, and of the coordinates of a point on the image plane. Formally, the aberration function is expressed in double power series expansion in terms of coordinates of the pupil, and those of the image. Similarly, to leverage the advantages of Zernike polynomials, a double Zernike polynomial expansion is also proposed [112–114].

17.6 Role of Fresnel Number in Imaging/Focusing

The Fresnel number is a dimensionless characteristic of the imaging/focusing geometry of an electromagnetic wave in the image space of an optical system. It is defined as

$$\text{Fresnel number } N_F = \frac{a^2}{\lambda z} \tag{17.150}$$

where a is the characteristic size, e.g. the radius of a circular aperture, λ is the working wavelength, and z is the distance of the observation plane from the aperture. Recalling Fresnel zone construction in diffraction theory, it may be noted that the Fresnel number, as defined above, is the number of half-period zones on the diffracting aperture, counted from the centre to the edge of the aperture, as seen from the centre of the plane of observation. Across the inner and outer circles of any half-period zone, a phase transition by π takes place.

Equivalently, the Fresnel number is the difference in half wavelengths, between the optical path along the edge of the aperture to the observation point and the perpendicular distance of the observation point from the aperture.

For a typical focusing lens in conventional visible optics, let $a = 10\text{mm}, f = 100\text{mm}$,. and $\lambda = 0.5\mu\text{m}$. As per (17.121), the Fresnel number $N_F = 2000$. In the majority of optical systems in practice, the Fresnel number $N_F \gg 1$. On the other hand, the imaging geometry of several laser-based optical systems leads to small Fresnel numbers—in some cases even less than unity. For example, let us consider a long focal length infrared laser system with the parameters: $a = 5\text{mm}$, $f = 3\text{m}$, and $\lambda = 10.6\mu\text{m}$. The Fresnel number for this system is $N_F = 0.79$ [115–119].

A direct consequence of having a small Fresnel number in imaging/focusing geometry is that the centre of curvature of the phase front in the exit pupil and the point of maximum intensity may well be separated axially. This separation of the two points is called 'focal shift'. The phenomenon is observed in cases of hard-edged uniform amplitude pupils as well as Gaussian pupils. The commonly used scalar diffraction theory has an underlying assumption that is equivalent to large Fresnel number in imaging geometry, and, therefore cannot predict this focal shift. Due care is necessary in predicting the position of maximum intensity along the principal ray [120–121]. Modified forms of Fresnel numbers are also suggested for different types of pupils and ray geometry [117, 122–123].

17.7 Imaging/Focusing in Optical Systems with Large Numerical Aperture

Our discussions so far on the image of a bright point on a dark background, i.e. the point spread function by an optical imaging system, is based on the following approximations: (a) convergence angles are small, (b) polarization effects are ignored, and (c) the exit pupil is large in size compared to the working wavelength. When the numerical aperture of the imaging system is large, obviously the first two assumptions do not hold good. In the latter case, the amplitude of the imaging wavefront becomes non-uniform, even if the amplitude of the incident wavefront is perfectly uniform. The polarization state of the incident wavefront gets altered significantly. These two effects affect the diffraction pattern significantly, even in the absence of aberrations of any form. Obviously, residual aberrations will also influence the diffraction pattern further. Models of varying degrees of complexity have been called upon by different investigators. At one extreme are a few models that are simple modifications of the PSF evaluation techniques based on scalar diffraction theory. On the other hand, vector diffraction theory is invoked to obtain detailed and more accurate description of the diffraction image. Most important works in the field may be found in references [124–141].

REFERENCES

1 E. Lommel, 'Die Beugungserscheinungen einer kreisrunden Oeffnung und eines kretsrunden Schirmchens theoretisch und experimentell bearbeitet', *Bayerisch. Akad. d. Wiss.*, Vol. 15 (1884) pp. 233–337. English Translation by G. Bekefi and G.A. Woonton, 'Theoretical and Experimental Investigations of Diffraction Phenomena at a circular aperture and obstacle', Defence Research Board, Ottawa, Canada. [Courtesy, A. Boivin, Univ. Laval, Quebec, Canada]

2 G.B. Airy, 'On the intensity of light in the neighbourhood of a caustic', *Trans. Camb. Phil. Soc.*, Vol. 6, Part III, No. XVII, (1838) pp. 379–402.

3 Lord Rayleigh, 'Investigations in optics, with special reference to the spectroscope. §4. Influence of Aberration', *Philos. Mag.*, Vol. 8 (1879) pp. 403–411; also in *Scientific Papers, Vol. I*, University Press, Cambridge (1899) pp. 428–436.

4 K. Strehl, *Theorie des Fernrohres auf Grund der Beugung des Lichtes*, Johann Ambrosius Barth, Leipzig (1894).

5 A.E. Conrady, 'Star-Discs', *Mon. Not. Roy. Astron. Soc.*, Vol. 79 (1919) pp. 575–593.

6 A. Buxton, 'Star-Discs', *Mon. Not. Roy. Astron. Soc.*, Vol. 81 (1921) pp. 547–566.

7 L.C. Martin, 'A physical study of spherical aberration', *Trans. Opt. Soc. London*, Vol. 23 (1921) pp. 63–92.

8 L.C. Martin, 'Star Discs (3), The Effect of Coma', *Mon. Not. Roy. Astron. Soc.*, Vol. 82 (1922) pp. 310–318.

9 L.C. Martin, 'A physical study of coma', *Trans. Opt. Soc. London*, Vol.24 (1922–23) pp. 1–9.

10 A. Buxton, 'Note on the Effect of Astigmatism on Star-Discs', *Mon. Not. Roy. Astron. Soc.*, Vol. 83 (1923) pp. 475–480.

11 R. Kingslake, 'The diffraction structure of the elementary coma image', *Proc. Phys. Soc.*, Vol. 61 (1948) pp. 147–158.

12 G.C. Steward, 'Chapter VI. Diffraction Patterns Associated with the Symmetrical Optical System', in *The Symmetrical Optical System*, University Press, Cambridge, (1928) pp. 72–88.

13 J. Picht, *Optische Abbildung, Einführung in die Wellen und Beugungstheorie Optischer Systems*, Viewg, Braunschweig (1931).

14 E. Wolf, 'The diffraction theory of aberrations', *Rep. Progr. Phys. London*, Vol. 14 (1951) pp. 95–120.

15 G. Black and E.H. Linfoot, 'Spherical aberration and the information content of optical images', *Proc. Roy. Soc. London, Series A*, Vol. 239 (1957) pp. 522–540.

16 H.H. Hopkins, 'The numerical evaluation of the frequency response of optical systems', *Proc. Phys. Soc. London, B*, Vol. 52 (1957) pp. 1002–1005.

17 H.H. Hopkins and M.J. Yzuel, 'The computation of diffraction patterns in the presence of aberrations', *Opt. Acta*, Vol. 17 (1970) pp. 157–182.

18 M.J. Yzuel and F.J. Arlegui, 'A study on the computation accuracy of the aberrational diffraction images', *Opt. Acta*, Vol. 27 (1980) pp. 549–562.

19 J.J. Stamnes, B. Spjelkavik, and H.M. Pedersen, 'Evaluation of diffraction integrals using local phase and amplitude approximations', *Opt. Acta*, Vol. 30 (1983) pp. 207–222.

20 J.J. Stamnes, *Waves in Focal Regions*, Adam Hilger, Bristol (1986) pp. 60–85.

21 H.H. Hopkins, 'Numerical Evaluation of a class of double integrals of oscillatory functions', *IMA J. Num. Analysis*, Vol. 9, No. 1 (1989) pp. –80.

22 M. De and L.N. Hazra, 'Walsh functions in problems of optical imagery', *Opt. Acta*, Vol. 24 (1977) pp. 221–234.
23 L.N. Hazra, 'Numerical evaluation of a class of integrals for image assessment', *Appl. Opt.*, Vol. 27 (1988) pp. 3464–3467.
24 J.W. Cooley and I.W. Tukey, 'An algorithm for the machine calculation of complex Fourier series', *Math. Comput.*, Vol. 19 (1965) pp. 297–301.
25 E.O. Brigham, *The Fast Fourier Transform*, Prentice-Hall, Englewood Cliffs, New Jersey (1974).
26 R. Barakat, 'The Calculation of Integrals Encountered in Optical Diffraction Theory', in *The Computer in Optical Research: Methods and Applications*, Ed. B.R. Frieden, Springer, Berlin (1980) pp. 55–80.
27 A.E. Siegman, 'Quasi Fast Hankel Transform', *Opt. Lett.*, Vol. 1 (1977) pp. 13–15.
28 S. Szapiel, 'Point Spread Function Computation: Quasi-Digital Method', *J. Opt. Soc. Am. A*, Vol. 2 (1985) pp. 3–5.
29 S. Szapiel, "Point-Spread Function Computation: Analytic End Correction in the Quasi-Digital Method," *J. Opt. Soc. Am. A*, Vol. 4 (1987) pp. 625–628.
30 D.G. Hawkins and E.H. Linfoot, 'An improved type of Schmidt camera', *Mon. Not. Roy. Astron. Soc.*, Vol. 105 (1945) pp. 334–344.
31 M. Herzberger, 'Light distribution in the optical image', *J. Opt. Soc. Am.*, Vol. 37 (1947) pp. 485–493.
32 O. Stavroudis and D.P. Feder, 'Automatic computation of spot diagrams', *J. Opt. Soc. Am.*, Vol. 44 (1954) pp. 163–170.
33 E.H. Linfoot, *Recent Advances in Optics*, Clarendon, Oxford (1955) pp. 200–206, 224–227.
34 F.A. Lucy, 'Image Quality Criteria derived from skew traces', *J. Opt. Soc. Am.*, Vol. 46 (1956) pp. 699–706.
35 R.E. Hopkins and R. Hanau, in *Military Standardization Handbook, MIL-HDBK-141, Optical Design*, Defense Supply Agency, Washington (1962) 8–1 to 8–3.
36 K. Miyamoto, 'Image evaluation by spot diagram using a computer', *Appl. Opt.*, Vol. 2 (1963) pp. 1247–1250.
37 O.N. Stavroudis and L.E. Sutton, *Spot Diagrams for the Prediction of Lens Performance from Design Data*, U. S. Department of Commerce, National Bureau of Standards Monograph 93 (1965).
38 W.J. Smith, *Modern Optical Engineering*, Mc-Graw Hill, New York (2000) pp. 360–361.
39 D. Malacara-Hernández and Z. Malacara-Hernández, *Handbook of Optical Design*, CRC Press, Boca Raton, Florida (2013) pp. 211–216.
40 Lord Rayleigh, 'Investigations in optics with special reference to the spectroscope. §4. Influence of Aberrations', *Phil. Mag., Fifth series*, Vol. VIII (1879) pp. 403–411. Also, in *Scientific Papers by J. W. Strutt, Baron Rayleigh, Vol. I*, University Press, Cambridge (1899) pp. 428–436.
41 Lord Rayleigh, 'Wave Theory of Light. §13, Influence of Aberrations. Optical Power of Instruments', *Encyclopædia Britannica*, Vol. XXIV (1888) pp. 435–437. Also, in *Scientific Papers by J. W. Strutt, Baron Rayleigh, Vol. III*, University Press, Cambridge (1902) pp. 100–106.
42 U. Bose, 'On Optical Tolerances – I, Rayleigh $\lambda / 4$ criterion', *Bull. Opt. Soc. India*, Vol. 2 (1968) pp. 47–51.
43 R. Barakat, 'Rayleigh Wavefront Criterion'. *J. Opt. Soc. Am.*, Vol. 55 (1965) pp. 572–573.
44 A. van den Bos, 'Rayleigh wave-front criterion: comment', *J. Opt. Soc. Am. A*, Vol. 16 (1999) pp. 2307–2309.
45 A.E. Conrady, 'Optical tolerances' in Applied Optics and Optical Design, Part one, Dover, New York (1957), pp. 136–139, 198, 393–395, 433.
46 K. Strehl, 'Ueber Luftschlieren und Zonenfehler', *Zeitschrift für Instrumentenkunde*, Vol. XXII (1902) pp. 213–217.
47 H.H. Hopkins, 'The development of image evaluation methods', *Proc. SPIE*, Vol. 46, 'Image Assessment & Specification', Ed., D. Dutton, (1974) pp. 2–18.
48 W.B. Wetherell, 'The Calculation of Image Quality', Chapter 6 in *Applied Optics and Optical Engineering, Vol. VIII*, Eds., R.R. Shannon and J.C. Wyant, Academic, New York (1980) pp. 171–315.
49 A. Maréchal, 'Étude des effets combinés de la diffraction et des aberrations géométriques sur l'image d'un point lumineux', *Rev. d'Opt. Théorique Instrum.* Vol. 26 (1947) pp. 257–277.
50 A. Maréchal and M. Françon, *Diffraction, Structure des Images*, Masson, Paris (1970) p. 107.
51 M. Born and E. Wolf, *Principles of Optics*, 6th ed., Pergamon, Oxford (1984) pp. 463–464. [First edition, 1959].

52 W.B. King, 'Dependence of the Strehl Ratio on the magnitude of the variance of the wave aberration', *J. Opt. Soc. Am.*, Vol. 58 (1968) pp. 655–661.
53 V.N. Mahajan, 'Strehl ratio for primary aberrations in terms of their aberration variance,' *J. Opt. Soc. Am.* Vol. 73 (1983) pp. 860–861.
54 E.L. O'Neill, *Introduction to Statistical Optics*, Addison-Wesley, Reading, Massachusetts (1963) pp. 58–68.
55 B. Tatian, 'Aberration balancing in rotationally symmetric lenses', *J. Opt. Soc. Am.*, Vol. 64 (1974) pp. 1083–1091.
56 G. Martial, 'Strehl ratio and aberration balancing', *J. Opt. Soc. Am. A*, Vol. 8 (1991) pp. 164–176.
57 M. De, L.N. Hazra, and P. Sengupta, 'Apodization of telescopes working in a turbulent medium', *Opt. Acta*, Vol. 22 (1975) pp. 125–139.
58 M. De, L.N. Hazra, and P.K. Purkait, 'Walsh functions in lens optimization I. FEE-based criterion', *Opt. Acta*, Vol. 25 (1978) pp. 573–584.
59 F. Zernike, 'Diffraction theory of knife-edge test and its improved form, the phase contrast method', *Mon. Not. R. Astron. Soc.*, Vol. 94, (1934) pp. 377–384.
60 F. Zernike, Beugungstheorie des Schneidenverfahrens und seiner verbesserten Form, der Phasenkontrastmethode, *Physica* Vol.1, (1934) pp. 689–794.
61 F. Zernike, 'The diffraction theory of aberrations', in *Proc. NBS semi centennial symposium on Optical Image Evaluation*, 18–20 Oct. 1951, U.S. Dept. of Commerce, Nat. Bur. Standards Circular No. 526, Washington (1954) pp. 1–8.
62 R. Courant and D. Hilbert, *Methods of Mathematical Physics, Vol. I*, Interscience, New York (1953) pp. 90–91. [Original German edition by Springer, Berlin (1937)].
63 B.R.A. Nijboer, *The Diffraction Theory of Aberrations*, Ph. D. Thesis, Univ. Groningen (1942) pp. 19–24.
64 B.R.A. Nijboer, 'The Diffraction Theory of Optical Aberrations. Part I: General Discussion of the Geometrical Aberrations', Physica, Vol. 10 (1943) pp. 679–692.
65 B.R.A. Nijboer, 'The Diffraction Theory of Optical Aberrations. Part II. Diffraction pattern in the presence of small aberrations', Physica, Vol. 13 (1947) pp. 605–620.
66 A.B. Bhatia and E. Wolf, 'On the circle polynomials of Zernike and related orthogonal sets,' *Proc. Cambridge Philos. Soc.*, Vol. 50, (1954) pp. 40–48.
67 E.H. Linfoot, *Recent Advances in Optics*, Clarendon, Oxford (1958) pp. 48–58.
68 M. Born and E. Wolf, *Principles of Optics*, 6th ed., Pergamon, Oxford, pp. 464–480. [First ed. 1959].
69 E.L. O'Neill, *Introduction to Statistical Optics*, Addison-Wesley, Reading, Massachusetts (1963) pp. 58–68.
70 S.N. Bezdidko, 'The use of Zernike polynomials in optics,' *Sov. J. Opt. Tech.*, Vol. 41 (1974), pp. 425–429.
71 E.C. Kintner, 'Some comments on the use of Zernike polynomials in optics,' *Opt. Commun.* Vol. 18 (1976), pp. 235–237.
72 E.C. Kintner, 'A recurrence relation for calculating the Zernike polynomials,' *Opt. Acta*, Vol. 23 (1976), pp. 499–500.
73 E.C. Kintner, 'On the mathematical properties of the Zernike polynomials', *Opt. Acta*, Vol. 23 (1976) pp. 679–680.
74 W.J. Tango, 'The circle polynomials of Zernike and their application in optics,' *Appl. Phys.* Vol. 13 (1977) pp. 327–332.
75 R. Barakat, 'Optimum balanced wave-front aberrations for radially symmetric amplitude distributions: generalizations of Zernike polynomials', *J. Opt. Soc. Am.*, Vol. 70 (1980) pp. 739–742.
76 J.Y. Wang and D.E. Silva, 'Wave-front interpretation with Zernike polynomials', *Appl. Opt.* Vol. 19 (1980) pp. 1510–1518.
77 V.N. Mahajan, 'Zernike annular polynomials for imaging systems with annular pupils', *J. Opt. Soc. Am.*, Vol.71 (1981) pp. 75–78, 1408.
78 R.K. Tyson, 'Conversion of Zernike aberration coefficients to Seidel and higher order power-series aberration coefficients,' *Opt. Lett.*, Vol. 7, (1982) pp. 262–264.
79 G. Conforti, 'Zernike aberration coefficients from Seidel and higher-order power series coefficients,' *Opt. Lett.*, Vol. 8, (1983) pp. 407–408.
80 C.J. Kim and R.R. Shannon, 'Catalog of Zernike polynomials', in *Applied Optics and Optical Engineering,* Vol. X, R.R. Shannon and J.C. Wyant, Eds. Academic, New York, N.Y. (1987) pp. 193–221.

81 B. Swantner, 'Comparison of Zernike polynomial sets in commercial software,' *Opt & Phot. News*, (September 1992), pp. 42–43.
82 R.K. Tyson, 'Using Zernike polynomials', *Opt. & Phot. News* (December1992) p. 3.
83 J.R. Rogers, 'Zernike polynomials', *Opt. & Phot. News*, (August 1993) pp. 2–3.
84 V.N. Mahajan, 'Zernike Circle Polynomials and Optical Aberrations of Systems with Circular Pupils', in Engineering Laboratory Notes Section, Ed., R.R. Shannon, *Supplement to Appl. Opt.*, Vol. 33(1994) pp. 8121–8124.
85 V.N. Mahajan, *Optical Imaging and Aberrations. Part I*, SPIE Press, Bellingham, Washington (1998) pp. 163–169.
86 J. Schwiegerling, 'Scaling Zernike expansion coefficients to different pupil sizes', *J. Opt. Soc. Am. A*, Vol. 19 (2002) pp. 1937–1945.
87 C.E. Campbell, 'Matrix method to find a new set of Zernike coefficients from an original set when the aperture radius is changed', *J. Opt. Soc. Am. A*, Vol. 20 (2003) pp. 209–217.
88 G. Dai, 'Scaling Zernike expansion coefficients to smaller pupil sizes: a simpler formula,' *J. Opt. Soc. Am. A*, Vol. 23 (2006) pp. 539–547.
89 H. Shu, L. Luo, G. Han, and J.L. Coatrieux, "General method to derive the relationship between two sets of Zernike coefficients corresponding to different aperture sizes," *J. Opt. Soc. Am. A*, Vol. 23 (2006) pp. 1960–1966.
90 A.J.E.M. Janssen and P. Dirksen, 'Concise formula for the Zernike coefficients of scaled pupils,' *J. Microlith., Microfab., Microsyst.*, Vol. 5 (2006) 030501.
91 V.N. Mahajan, 'Zernike coefficients of a scaled pupil', *Appl. Opt.*, Vol. 49 (2010) pp. 4374–5377.
92 V. Lakshminarayanan and A. Fleck, 'Zernike polynomials: A Guide', *Journal of Modern Optics*, Vol. 58 (2011) pp. 545–561.
93 R.J. Noll, 'Zernike polynomials and atmospheric turbulence,' *J. Opt. Soc. Am.* 66 (1976) p. 207–211.
94 J.C. Wyant and K. Creath, 'Basic Wavefront Aberration Theory for Optical Metrology', Chapter 1 in *Applied Optics and Optical Engineering, Vol. XI*, Academic. Boston (1992) pp. 28–38.
95 V. Genberg, G. Michels, and K. Doyle, 'Orthogonality of Zernike Polynomials', *Proc. SPIE* Vol. 4771 (2002) pp. 276–286.
96 L.N. Thibos, R.A. Applegate, J.T. Schwiegerling, and R. Webb, 'Standards for reporting the optical aberrations of eyes', *J. Refractive Surgery*, Vol. 18 (2002) S652–S660.
97 H. Gross, Ed., *Handbook of Optical Systems, Vol. 2*, Wiley, Weinheim (2005) pp. 212–216.
98 V.N. Mahajan, 'Zernike Polynomials and Wavefront Fitting', Chapter 13 in *Optical Shop Testing*, Ed. D. Malacara, 3rd Edition, John Wiley, New Jersey (2007) pp. 498–546.
99 A.J.E.M. Janssen and P. Dirksen, 'Computing Zernike polynomials of arbitrary degree using the discrete Fourier transform,' *J. Europ. Opt. Soc. Rap. Public.*, Vol. 2, (2007) 07012.
100 A.J.E.M. Janssen, S. van Haver, P. Dirksen, and J.J.M. Braat, "Zernike representation and Strehl ratio of optical systems with variable numerical aperture," *J. Mod. Opt.*, Vol. 55, (2008) pp. 1127–1157.
101 E. Alizadeh, S.M. Lyons, J.M. Castle, and A. Prasad, 'Measuring systematic changes in invasive cancer cell shape using Zernike moments', *Integrative Biology, Sept.* (2016). DOI: 10.1039/c6ib00100a.
102 V.N. Mahajan, 'Zernike annular polynomials for imaging systems with annular pupils,' *J. Opt. Soc. Am.*, Vol. 71, pp. 75–85, 1408 (1981); *J. Opt. Soc. Am. A*, Vol.1 (1984) p. 685.
103 V.N. Mahajan, 'Zernike annular polynomials and optical aberrations of systems with annular pupils', *Appl. Opt.* 33, (1994) pp. 8125–8127.
104 G. Dai and V.N. Mahajan, "Zernike annular polynomials and atmospheric turbulence," *J. Opt. Soc. Am. A*, Vol. 24 (2007) pp. 139–155.
105 W. Swantner and W.W. Chow, 'Gram–Schmidt orthogonalization of Zernike polynomials for general aperture shapes,' *Appl. Opt.*, Vol. 33 (1994) pp. 1832–1837.
106 V.N. Mahajan, 'Uniform versus Gaussian beams: a comparison of the effects of diffraction, obscuration, and aberrations,' *J. Opt. Soc. Am. A*, Vol. 3 (1986) pp. 470–485.
107 V.N. Mahajan, 'Zernike-Gauss polynomials for optical systems with Gaussian pupils,' *Appl. Opt.* Vol. 34 (1995) pp. 8057–8059.
108 K. Nienhuis, *On the influence of diffraction on image formation in the presence of aberrations*, Ph.D. thesis (University of Groningen, Groningen, The Netherlands (1948).
109 A.J.E.M. Janssen, 'Extended Nijboer–Zernike approach for the computation of optical point-spread functions', *J. Opt. Soc. Am. A*, Vol. 19 (2002) pp. 849–857.

110 J.J.M. Braat, P. Dirksen, and A.J.E.M. Janssen, 'Assessment of an extended Nijboer–Zernike approach for the computation of optical point-spread functions,' *J. Opt. Soc. Am. A*, Vol. 19 (2002), pp. 858–870.
111 A.J. Janssen, J.J. Braat, and P. Dirksen, 'On the computation of the Nijboer-Zernike aberration integrals at arbitrary depth of focus', *J. Mod. Opt.*, Vol. 51 (2004) pp. 687–703.
112 W. Kwee and J.J.M. Braat, 'Double Zernike expansion of the optical aberration function,' *Pure Appl. Opt.* 2, (1993) pp. 21–32.
113 I. Agurok, 'Double expansion of wavefront deformation in Zernike polynomials over the pupil and field-of-view of optical systems: lens design, testing, and alignment,' *Proc. SPIE*, Vol. 3430 (1998) pp. 80–87.
114 J.J.M. Braat and A.J.E.M. Janssen, 'Double Zernike expansion of the optical aberration function from its power series expansion', Vol. 30, *J. Opt. Soc. Am. A* (2013) pp. 1213–1222.
115 Y. Li and E. Wolf, 'Three-dimensional intensity distribution near the focus in systems of different Fresnel numbers', *J. Opt. Soc. Am. A*, Vol. 1 (1984) pp. 801–808.
116 J.J. Stamnes, *Waves in Focal Regions*, Adam Hilger, Bristol (1986) pp. 240–416.
117 A.E. Siegman, *Lasers*, University Science Books, Mill Valley, California (1986) pp. 675–679.
118 H. Kogelnik and T. LI, 'Laser beams and resonators', *Appl. Opt.*, Vol. 5 (1966) pp. 1550–1567.
119 Y. Li, 'Dependence of the focal shift on Fresnel number and f number', *J. Opt. Soc. Am.*, Vol. 72 (1982) pp. 770–774.
120 S. Szapiel, 'Maréchal intensity formula for small Fresnel-number systems', *Optics Letters*, Vol. 8 (1983) pp. 327–329.
121 C.J.R. Sheppard, 'Imaging in optical systems of finite Fresnel number', *J. Opt. Soc. Am. A*, Vol. 3 (1986) pp. 1428–1432.
122 M. Martínez-Corral, C.J. Zapata-Rodríguez, P. Andrés, and E. Silvestre, 'Effective Fresnel-number concept for evaluating the relative focal shift in focused beams', *J. Opt. Soc. Am. A*, Vol. 15 (1998) pp. 449–455.
123 Y. Yao, J. Zhang, Y. Zhang, Q. Bi, and J. Zhu, 'Off-axis Fresnel numbers in laser systems', *High Power Laser Science and Engineering*, Vol. 2 (2014), pp. 1–6. doi:10.1017/hpl.2014.22
124 V.S. Ignatowsky, 'Diffraction by a lens having arbitrary opening', *Trans. Opt. Inst. Petrograd I, paper IV* (1919). (in Russian).
125 V.S. Ignatowsky, 'Diffraction by a parabolic mirror having arbitrary opening', *Trans. Opt. Inst. Petrograd, I, paper V* (1920). (in Russian).
126 H.H. Hopkins, 'The Airy disc formula for systems of high relative aperture', *Proc. Phys. Soc.*, Vol. 55 (1943) pp. 116–128.
127 H.H. Hopkins, 'A note on the polarization of a plane polarized wave after transmission through a system of centered refracting surfaces, and some effects at the focus', *Proc. Phys. Soc. London*, Vol. 56 (1944) pp. 48–51.
128 B. Richards, 'Diffraction in Systems of High Relative Aperture', in *Astronomical Optics and Related Subjects*, Z. Kopal, Ed. (North-Holland, Amsterdam, 1955), pp. 352–359.
129 B. Richards and E. Wolf, 'Electromagnetic diffraction in optical systems I. An integral representation of the image field,' *Proc. R. Soc. London, Ser. A* 253, (1959) pp. 349–357.
130 B. Richards and E. Wolf, 'Electromagnetic diffraction in optical systems II. Structure of the image field in an aplanatic system,' *Proc. R. Soc. London, Ser. A* 253, (1959) pp. 358–379.
131 R. Barakat, 'The intensity distribution and total illumination of aberration-free diffraction images', *Chapter III in Progress in Optics*, Vol. I, North-Holland, Amsterdam (1961) pp. 68–108.
132 A. Boivin and E. Wolf, 'Electromagnetic field in the neighbourhood of the focus of a coherent beam', *Phys. Rev.*, Vol. 138, (1965) B1561–B1565.
133 A. Boivin, J. Dow, and E. Wolf, 'Energy flow in the neighbourhood of the focus of a coherent beam', *J. Opt. Soc. Am* 57, (1967) pp. 1171–1175.
134 C.J.R. Sheppard and T. Wilson, 'The image of a single point in microscopes of large numerical aperture', *Proc. R. Soc. London Ser. A* 379, (1982) pp. 145–158.
135 J.J. Stamnes, *Waves in Focal Regions*, Adam Hilger, Bristol (1986) pp. 455–481.
136 C.J.R. Sheppard and H.J. Mathews, 'Imaging in high aperture optical systems', *J. Opt. Soc. Am. A*, Vol. 4 (1987) pp. 1354–1360.
137 C.J.R. Sheppard and P. Török, 'Electromagnetic field in the focal region of an electric dipole wave', *Optik*, Vol. 104, (1997) pp. 175–177.
138 M. Gu, *Advanced Optical Imaging Theory*, Springer, Berlin (2000). pp. 143–176.

139 J.J.M. Braat, S. van Haver, A.J.E.M. Janssen, and P. Dirksen, 'Assessment of Optical Systems by Means of Point-spread Functions', in *Progress in Optics, Vol. 51*, E. Wolf, Ed. (Elsevier B.V., 2008), Chap. 6, pp. 349–468.
140 R. de Bruin, H.P. Urbach, and S.F. Pereira, 'On focused fields with maximum electric field components and images of electric dipoles', *Opt. Expr.*, Vol. 19 (2011) pp. 9157–9171.
141 J. Braat and P. Török, *Imaging Optics*, Cambridge Univ. Press, Cambridge (2019).

18

System Theoretic Viewpoint in Optical Image Formation

18.1 Quality Assessment of Imaging of Extended Objects: A Brief Historical Background

In two previous chapters, we were mostly concerned with the imaging of point objects, more specifically with the imaging of one or two bright points in a dark background. In the early days of optical instruments, such as telescopes and microscopes, the quality of imaging used to be assessed on the basis of the quality of image of a point object. The Rayleigh criterion and the Strehl criterion of image quality is of great use in the study of image quality for highly corrected optical systems that are near diffraction limited in performance. Maréchal's treatment of the optimum balancing of residual aberrations and tolerancing of aberrations based on the Strehl criterion is also applicable for near diffraction limited systems.

> They are, however, unsuitable for general use for two reasons. First, and more fundamentally, because the diffraction image of a point source in the presence of larger aberrations no longer has a pronounced central maximum surrounded by weak diffraction rings; it can become very complicated in form. Second, it is not easy to appreciate the influence of a reduced Strehl ratio for the image of a single point on the image of an extended object.
>
> [1]

By the end of the 1850s, Foucault, now famous for his knife edge test, suggested that a striped pattern would be a better target than a point object for assessing the quality of imaging of extended objects by an optical system [2]. About fourteen years later, Abbe reported his momentous observations on microscope imagery by considering the imaging of a coherently illuminated diffraction grating [3]. Rayleigh extended the studies by Abbe to 'the case of a grating or a row of points perforated in an opaque screen and illuminated by plane waves of light' [4]. Rayleigh also considered the case of self-luminous or incoherently illuminated object points. The studies of Abbe and Rayleigh were primarily concerned with determining the 'limit of resolution' (termed by them as 'resolving power') in the different cases. It was observed that a factor of 2 was involved in the case of incoherent imagery compared to the coherent case. It is noteworthy that Rayleigh made use of Fourier analysis in his derivation. The 'limit of resolution' corresponds to what is known as 'the cut-off frequency' in modern terminology.

The concept of spatial frequency has no overt connection with Rayleigh's work. Frequency analysis was first implemented for audio signals, because of the nature of human hearing.

> Its extension to electric circuit analysis is also natural, since electric circuits were used often to reproduce sound and since they involved time sequences of events. The extension of frequency analysis to optical imagery began in the late 1920s and early 1930s with the development of two technologies: (1) the recording of sound on film for talking pictures and (2) the sequential scanning of images for transmission over wire (photo telegraphy) and over radio waves (television).
>
> [5]

DOI: 10.1201/9780429154812-18

Reports of corresponding works appeared in publications by Mackenzie [6] and by Mertz and Gray [7]. Concepts akin to transfer function in linear systems were pursued and reported in two publications by Frieser [8] and Wright [9].

In the early 1940s, Duffieux laid the foundations of what is now known as 'Fourier optics' through a series of publications culminating in his book on the topic [10]. Around the same time, Luneburg derived the limits of resolution in both cases of coherently and incoherently illuminated objects by using Fourier analysis [11]. Several authors explored the use of system theoretic and communication theoretic viewpoints in the quality assessment of optical images formed on sensors, like television displays and photographic films, etc. [12–16].

Frequency analysis of imaging systems, in general, is applicable to images formed by any procedure, and it provides a systematic approach for quantitative assessment of the quality of final images in comparison to the original object. Image formation may consist of several processes in cascade, e.g. forming an optical image of the object by lens on a detector/detector array, processing this intermediate image for transmission/record, etc., electronically or chemically, and finally using electro-optic/optical procedure for display of the image. This is a typical approach adopted in television engineering and cinematography. In order to incorporate the optical imaging process in the whole system analysis, many concepts were borrowed from electrical communication engineering. The process is by no means straightforward, for the concepts utilized in electrical communication need to be properly assimilated with the intricacies of optical image formation. Significant strides were undertaken by the pioneers in the field to accomplish the task, and ultimately this led to the development of a system theoretic approach that enabled the assessment of the quality of intermediate or final images in an image-forming system consisting of multiple stages. In this enterprise, Hopkins, along with his coworkers, played a preeminent role in the 1950s by laying the foundations of frequency analysis of images formed by optical lens/mirror systems [17–32]. Other notable contributors during this period were Steel [33–34], O'Neill [35–36], Shack [37], Parrent and Drane [38], Linfoot [39], Black and Linfoot [40], Lohmann [41], and Maréchal and Françon [42].

The evaluation of the actual transfer function of a real optical system needs to take into account the effects of both aberration and diffraction. Except for the special case of aberration-free diffraction limited systems, no analytical solution to the problem is available, and it is necessary to take recourse to numerical procedures for evaluating the transfer function. The latter procedures involve a huge amount of numerical computation to obtain reasonably accurate results. In order to circumvent this problem, Hopkins and his co-workers embarked upon a geometrical optics approximation to the transfer function [21–23, 27, 29–30]. More or less concurrently, Miyamoto [43–46] explored this possibility of obtaining a relatively simple expression by using geometrical optic approximation for the transfer function, obviating the need for extensive numerical computation. It was found that the geometrical optics approximation for transfer function retains its validity for very small values of spatial frequency. Since, in the case of images formed by non-diffraction limited systems with somewhat large values of residual aberrations (e.g. camera lenses), the range of spatial frequency of interest is also small, it was contemplated that the geometrical transfer function can play a useful role in practice. However, attempts by few other investigators [47–48] to extend its scope of application met limited success. Recently, Mahajan and his coworkers [49–51] have questioned the utility of calculating the geometrical transfer function. Indeed, with the ready availability of fast digital computers with phenomenally increasing number crunching ability, it is now practically feasible to evaluate the diffraction OTF's within a reasonable time, so much so that the use of grossly approximate concepts like geometrical OTF is getting somewhat obscure in the practical analysis and synthesis of optical systems.

Point spread function (PSF) is the basic criterion for quality assessment of optical imaging systems with a small field. For extended object imagery, several criteria, including line spread function (LSF), edge spread function (ESF), bar spread function (BSF), etc., are used in practice. Each of these criterion has its advantages and disadvantages, and its niche application area. These functions are related with the optical transfer function, so that any of them can be derived from the OTF, and vice versa. In the 1860s, several investigations were carried out on evaluation of the OTF from the design data, and also on suitable methods for measuring the OTF of imaging lenses [52–64]. Towards the end of this decade, investigations were carried out on the use of the Fast Fourier Transform (FFT) algorithm in OTF computations [65–70]. It was found that this approach was not often satisfactory in the computation of image quality evaluation parameters to cater to the specific needs of analysis of optical imaging systems. The search for alternative methods for fast and accurate calculation of the OTF continued in the next decade [71–75]. At this time, most in-house and commercial software for analysis and design had provisions for computing the optical transfer function from design data. However, the predictions for OTF for the same system often did not tally with each other. Hopkins and Dutton [76] reported an interlaboratory comparison of MTF measurements and computations on a large wide-angle lens and pointed out the variations in MTF values in off-axial imagery. Macdonald [77] carried out a comparison of results of various OTF calculation programs and underscored the possible remedies for preventing this divergence of results. It is important to note that, even two decades after the publication of Macdonald's paper, reports appeared on the uncertainties in MTF computational results and the need for lens design code standardization to remove the uncertainties [78–80]. Similar problems in experimental measurements of the OTF of imaging lenses were also reported by others [81–85]. An approach for standardization in this regard was initiated in different countries, e.g. the United Kingdom, Germany, Japan, the United States of America, France, Holland, and others [86–95]. After many deliberations among the stakeholders, the International Standards Organization (ISO) started releasing a set of standards in this field from the mid-1990s. These standards are supposed to be reviewed every five years [96–100]. Several books on the topic of optical transfer function cover developments in the field [101–103]. A tome containing selected papers on the foundations and theory of optical transfer function has been compiled by Baker [104] and published by the SPIE under its Milestone series.

18.2 System Theoretic Concepts in Optical Image Formation

Operations of most systems, including communication systems, can be modelled as a two port system, where output $o(t)$ of the system is represented by a system operator Θ, operating on the input $i(t)$ [105–107] (Figure 18.1).

In the case of an optical imaging system, usually the input is a two-dimensional object $B(u,v)$, and the output is a two-dimensional image $B'(u',v')$ (Figure 18.2).

Note that, for the sake of analytical convenience, actual rectangular coordinates (ξ, η) and (ξ', η') for a pair of conjugate points in object and image planes, respectively, are suitably scaled to obtain the coordinates (u, v) and (u',v') for them. By using invariance relations in optical imaging theory, the scaling is done so that $u' = u$, $v' = v$ gives the position of the geometrical image of the point (u,v).

In most general terms, the object or image functions $B(u,v)$ or $B'(u',v')$ represent any property that characterizes the object or the image under the given conditions of illumination and imaging. Although Figure 18.2 represents the operation of an 'optical' imaging system, it should be obvious that an extension of the same model in the case of images produced by non-optical means, e.g. in the case of electro-optical imaging, electronic imaging, chemical imaging, computational imaging, acoustic imaging, etc., is straightforward.

In general, behaviour of most real systems, including the 'Optical Imaging Systems', is nonlinear. Since handling nonlinear systems is not easy, scientists and technologists attempt to make approximate models of these systems so that the systems can be treated as linear systems, facilitating significantly

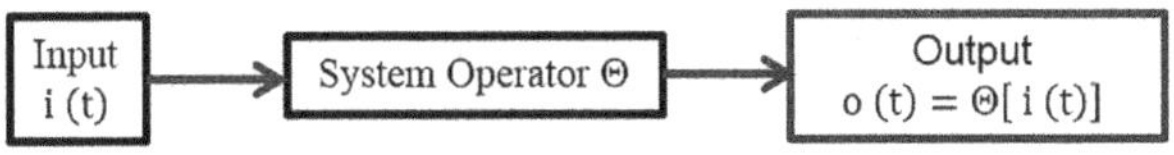

FIGURE 18.1 A two-port system with system operator Θ.

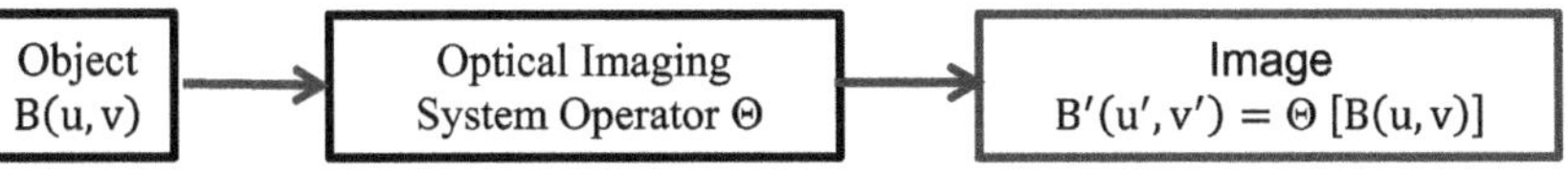

FIGURE 18.2 An optical imaging system with operator Θ.

analytical or semi-analytical handling of them; however, often the range and scope of applicability of these approximate models are severely restricted [108–113].

18.2.1 Linearity and Principle of Superposition

An optical imaging system is 'linear' when it satisfies the principle of superposition. Figure 18.3 shows three images, corresponding to three different objects, produced by an imaging system characterized by the operator Θ. In Figure 18.3(a), the system produces an image $B_1'(u',v')$ for an object $B_1(u,v)$, and in Figure 18.3(b) the system produces an image $B_1'(u',v')$ for an object $B_2(u,v)$. Thus we get

$$B_1'(u',v') = \Theta\left[B_1(u,v)\right] \tag{18.1}$$

$$B_2'(u',v') = \Theta\left[B_2(u,v)\right] \tag{18.2}$$

The imaging system is linear if and only if the following relation holds good.

$$\alpha B_1'(u',v') + \beta B_2'(u',v') = \Theta\left[\alpha B_1(u,v) + \beta B_2(u,v)\right] \tag{18.3}$$

This relation expresses mathematically the principle of superposition. Figure 18.3(c) illustrates imaging for the special case $\alpha = \beta = 1$, where the relation (18.3) holds good. It is obvious that (18.3) cannot hold good for arbitrary values for α and β. The domain for values of α and β, over which the relation (18.3) holds good, sets the 'dynamic range' over which the optical imaging system satisfies the principle of superposition, and the system is linear. Outside this range, the system exhibits nonlinear characteristics [114].

18.2.2 Space Invariance and Isoplanatism

Figure 18.4(a) shows the image $B'(u',v')$, corresponding to an object $B(u,v)$, produced by an optical imaging system characterized by the system operator Θ. For the sake of illustration, the object is chosen to be centred on the optical axis; assume the image is also centred on the axis. Thus

$$B'(u',v') = \Theta\left[B(u,v)\right] \tag{18.4}$$

Figure 18.4(b) shows the image when the same object is displaced, and is centred at the point (u_0, v_0). The imaging system is said to be space invariant if and only if the same image is obtained on the image plane, but the image is now laterally shifted and centred on an image point in consonance with the object point (u_0, v_0).

Mathematically, an imaging system is said to be space invariant if and only if the following relation holds good.

$$B'(u'-u_0, v'-v_0) = \Theta\left[B(u-u_0, v-v_0)\right] \tag{18.5}$$

This, indeed, is a stringent condition that is hardly satisfied in real optical systems. Since aberrations in an optical imaging system vary for different locations of the field point, the quality of image is bound to vary in different field locations. This implies that a lateral shift of an object does produce a lateral shift of the image, but the latter image is not exactly identical with the earlier image. In practice, optical imaging systems are basically space variant, and they may be considered space invariant in the immediate neighbourhood of a field point, so long as the variation in aberration with field location can be neglected. This assumption of non-variation in aberration at a field point may be justified over a small region, surrounding

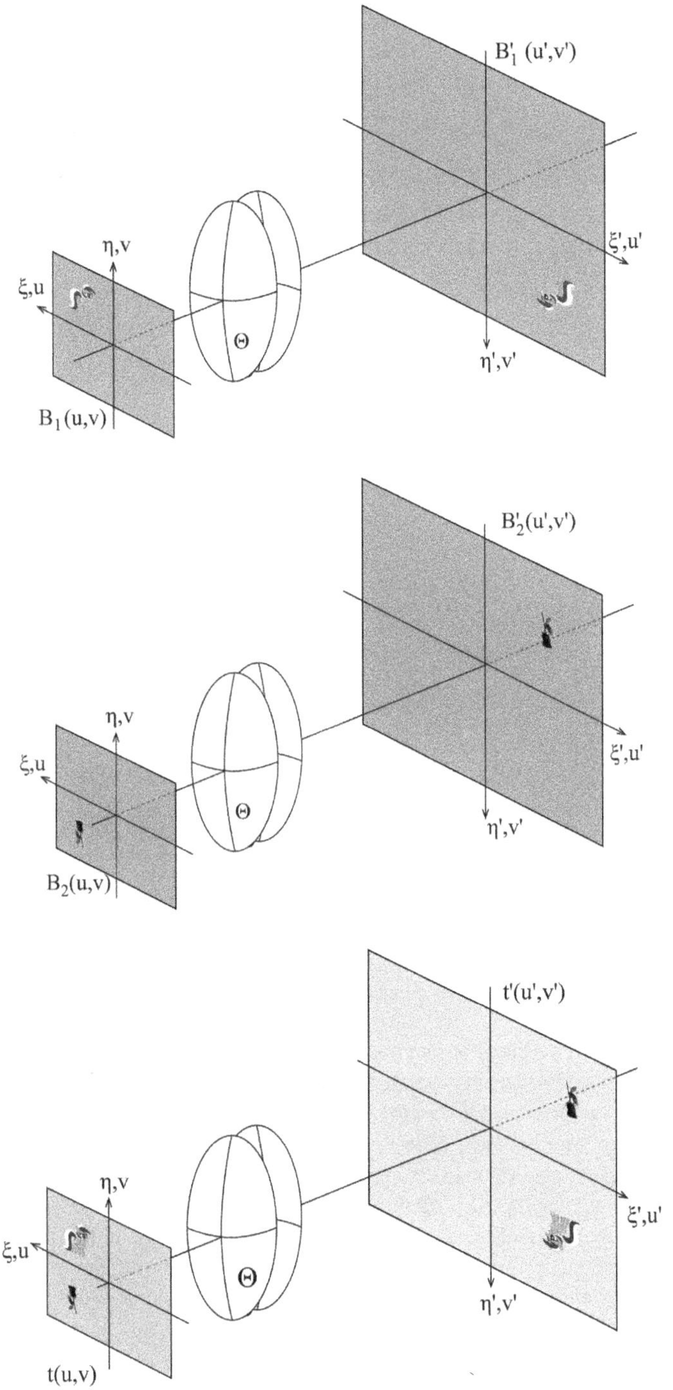

FIGURE 18.3 An optical imaging system satisfying the principle of superposition. (a) Image of an object $B_1(u,v)$ alone: $B_1'(u',v') = \Theta[B_1(u,v)]$ (b) Image of an object $B_2(u,v)$ alone: $B_2'(u',v') = \Theta[B_2(u,v)]$ (c) Image when the objects are present together: $t(u',v') = \Theta[t(u,v)]$ where $t(u',v') = [B_1'(u',v') + B_2'(u',v')]$ and $t(u,v) = [B_1(u,v) + B_2(u,v)]$.

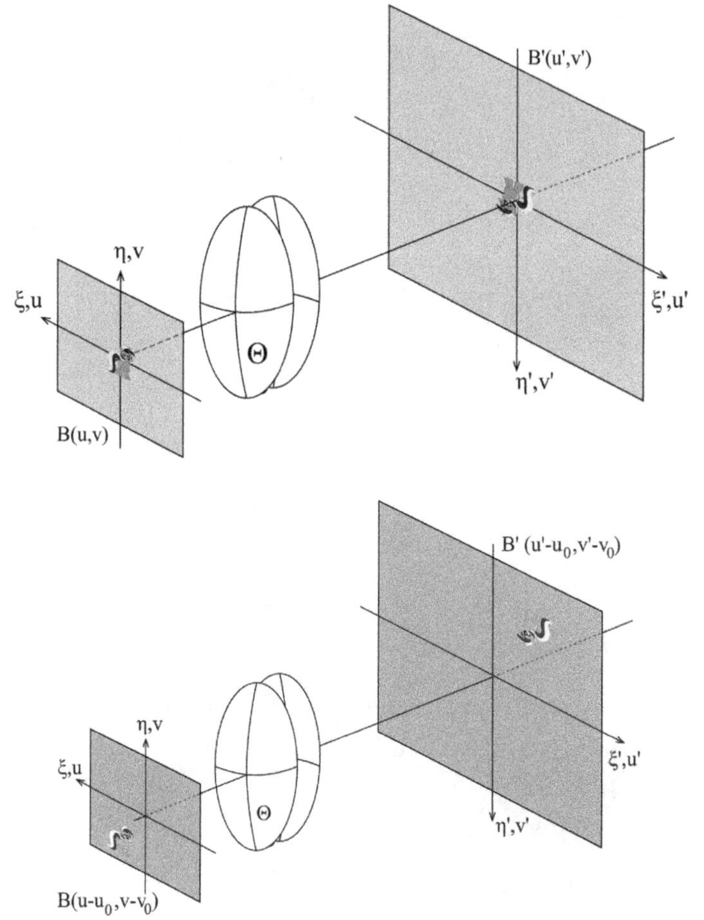

FIGURE 18.4 A space invariant or isoplanatic optical imaging system (a) Image of an object placed on the origin: $B'(u',v') = \Theta[B(u,v)]$ (b) Image when the object is shifted to the point (u_0, v_0): $B'(u' - u_0, v' - v_0) = \Theta[B(u - u_0, v - v_0)]$.

the aberrated image of the field point on the image plane, within which the intensity in the image of the field point differs from zero. For an axial point, the latter condition is called 'aplanatism', and at off-axial points the general term is 'isoplanatism' [115]. Different aspects of the application of Fourier analysis to space variant imaging systems have been discussed in reference [116]. Practical analysis of optical imaging systems invokes a concept of 'local' space invariance in different parts of the field. For axially symmetric optical imaging systems, it is assumed that the variation of aberrations with variation in field location is smooth, and image quality assessment is carried out at three field locations, namely, on-axis, at 0.707 field, and at the edge of the field. Sometimes this analysis is extended to five points on the field. Of course, for demanding applications calling for full field detail, the number of field points may be much more [117].

18.2.3 Image of an Object by a Linear Space Invariant Imaging System

Canonical coordinates, enunciated in earlier chapters, for points on the object $[u_S, v_T]$, image $[u'_S, v'_T]$, entrance pupil $[x_S, y_T]$, and exit pupil $[x'_S, y'_T]$ facilitate the application of frequency response techniques for both axial and extra-axial imaging. In what follows, for the sake of simplicity, the subscripts are usually omitted from notations of the coordinates, except in some cases where full notation is used to prevent the unwanted occurrence of any ambiguity. However, primes are retained to indicate image space quantities.

A two-dimensional object $B(u,v)$ can be represented as

$$B(u,v) = \iint_{-\infty}^{+\infty} B(u_0,v_0)\delta(u-u_0,v-v_0)du_0dv_0 \tag{18.6}$$

where $\delta(u,v)$ is the Dirac delta function. For a linear imaging system satisfying the principle of superposition, the image $B'(u',v')$ is given by

$$B'(u',v') = \iint_{-\infty}^{+\infty} Bu_0v_0\Theta\left[\delta(u-u_0,v-v_0)\right]du_0dv_0 \tag{18.7}$$

In general, the response at point (u',v') on the image plane for a point object at (u_0,v_0) is

$$\Theta\left[\delta(u-u_0,v-v_0)\right] = G(u',v';u_0,v_0) \tag{18.8}$$

For a space invariant system, the response $G(u',v';u_0,v_0)$ reduces to

$$G(u',v';u_0,v_0) = G(u'-u_0,v'-v_0) \tag{18.9}$$

From (18.7), we get

$$B'(u',v') = \iint_{-\infty}^{+\infty} B(u_0,v_0)G(u'-u_0,v'-v_0)du_0dv_0 \tag{18.10}$$

Changing the current variables from (u_0,v_0) to (u,v), the equation (18.10) may be rewritten as

$$B'(u',v') = \iint_{-\infty}^{+\infty} B(u,v)G(u'-u,v'-v)dudv \tag{18.10}$$

The image $B'(u',v')$ is given by the convolution of the object $B(u,v)$ with the point spread function $G(u,v)$. Denoting the convolution by an asterisk, (18.10) can be written succinctly as

$$B'(u',v') = B'(u,v) = B(u,v)*G(u,v) \tag{18.11}$$

$G(u,v)$ is the response on the image plane corresponding to a bright point object against a dark background. The point object is located on the axis. In parlance of system theory, it is a unit impulse input, $\delta(u,v)$, and the corresponding output is unit impulse response, $G(u,v)$, that is called the point spread function in optics.

18.3 Fourier Analysis [118–122]

One of the key reasons for application of transforms, e.g. Fourier transforms, in the analysis of linear systems stems from the above relation (18.11) that shows that the output is given by a convolution of the input with the unit impulse response function of the system. The Fourier transform of the output, however, is equal to the product of the Fourier transforms of the input and of the unit impulse response. Multiplication is much less numerically intensive than convolution; also the former is more amenable for analytical manipulation.

Fourier Transforms $b(s,t), g(s,t)$, and $b'(s,t)$ of $B(u,v), G(u',v')$, and $B'(u',v')$ respectively are given by

$$\left.\begin{aligned} b(s,t) &= \iint_{-\infty}^{+\infty} B(u,v) e^{-i2\pi(us+vt)} dudv \\ g(s,t) &= \iint_{-\infty}^{+\infty} G(u',v') e^{-i2\pi(u's+v't)du'dv'} \\ b'(s,t) &= \iint_{-\infty}^{+\infty} B'(u',v') e^{-i2\pi(u's+v't)du'dv'} \end{aligned}\right\} \quad (18.12)$$

where (s,t) are spatial frequencies. In the Fourier domain, we have the following multiplicative relationship among the Fourier component of the image in terms of the Fourier component of the object and the Fourier component of the point spread function.

$$b'(s,t) = b(s,t) \times g(s,t) \quad (18.13)$$

$g(s,t)$ is the 'frequency response' of the imaging system at the frequency pair (s,t) for the two-dimensional object $B(u,v)$, whose image is $B'(u',v')$.

The inverse Fourier transforms, given below, express the object, the image, and the point spread function in terms of their corresponding Fourier components.

$$\left.\begin{aligned} B(u,v) &= \iint_{-\infty}^{+\infty} b(s,t) e^{i2\pi(us+vt)} dsdt \\ G(u',v') &= \iint_{-\infty}^{+\infty} g(s,t) e^{i2\pi(u's+v't)} dsdt \\ B'(u',v') &= \iint_{-\infty}^{+\infty} b'(s,t) e^{i2\pi(u's+v't)} dsdt \end{aligned}\right\} \quad (18.14)$$

18.3.1 Alternative Routes to Determine Image of an Object

Two alternative routes are available for determining the image of an object.

Route I. The whole operation is in direct function space. Relation (18.11) shows that the image is obtained from a convolution of the object with the point spread function. A schematic is given in Figure 18.5.

Route II. The operation consists of three steps. The first step involves a Fourier transformation of the object to determine its frequency spectrum. The next step is to obtain the frequency spectrum of the image by multiplying the object frequency spectrum by the frequency transfer function of the system. The last step involves an inverse Fourier transformation of the image frequency spectrum to obtain the image. The steps are illustrated below in Figure 18.6.

18.3.2 Physical Interpretation of the Kernel of Fourier Transform

Let us consider the real part of the kernel of the Fourier transform in (18.14)

$$\text{Real part of the kernel} = \cos\left[2\pi(us + vt)\right] \quad (18.15)$$

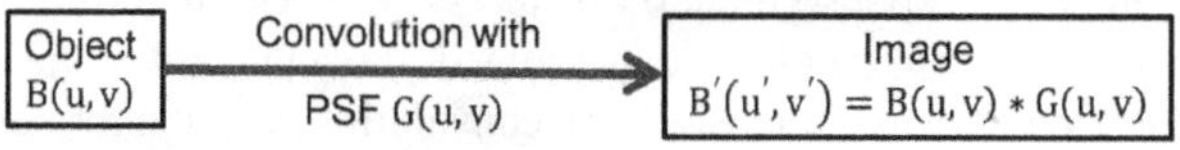

FIGURE 18.5 Image from object by convolution operation in direct space.

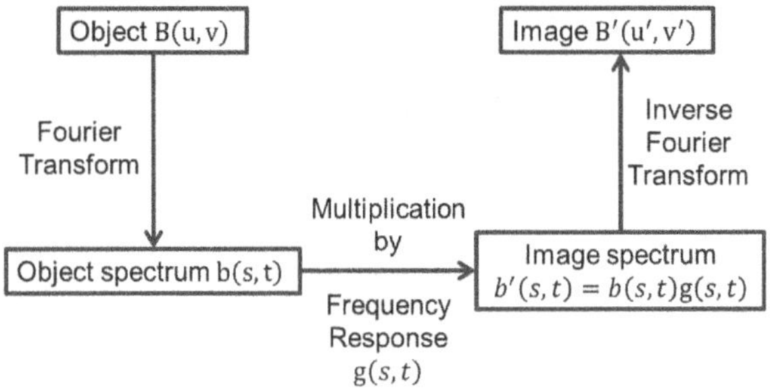

FIGURE 18.6 Image from object by using multiplication in frequency space.

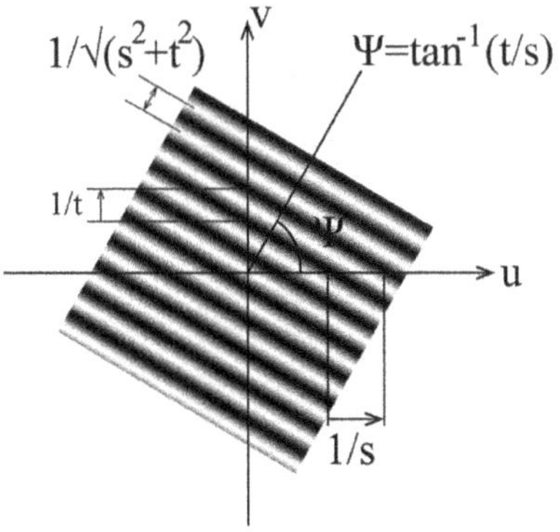

FIGURE 18.7 A two-dimensional cosinusoidal grating in (u, v) space with its axis along the direction given by the azimuth $\psi\{= \tan^{-1}(t/s)\}$ is effectively a one-dimensional grating with its axis along the direction of ψ and with lines of constant intensity perpendicular to this direction.

Assume that we are considering object/image/point spread function as a distribution of intensity that is a non-negative quantity. To ensure that (18.15) is non-negative, let us consider a function $P(u,v)$ defined by

$$P(u,v) = \frac{1}{2}\left[1 + \cos\{2\pi(us + vt)\}\right] \tag{18.16}$$

Figure 18.7 shows a contour diagram of the function $P(u,v)$ in the two-dimensional (u,v) space for the frequency pair (s,t). Note that the function represents a cosine distribution of intensity along the direction given by $\psi\{= \tan^{-1}(t/s)\}$, and has lines of constant intensity perpendicular to this direction. If new rectangular axes $[\tilde{u},\tilde{v}]$ are chosen by rotating the $[u,v]$ axes by angle ψ, then the intensity distribution has a cosinusoidal variation along the direction of $\tilde{u}$, and is constant along the direction of $\tilde{v}$. Effectively, the two-dimensional cosinusoidal grating represented by (18.16) is a one dimensional grating of period $S\left\{=\left[1/\left(s^2+t^2\right)^{\frac{1}{2}}\right]\right\}$ in $[\tilde{u},\tilde{v}]$ coordinate system, and it is represented as

$$P(\tilde{u}) = \frac{1}{2}\left[1 + \cos\{2\pi\tilde{u}S\}\right] \tag{18.17}$$

To emphasize the direction of the period of the grating, it may be represented by the notation, $P(\tilde{u},\Psi)$.

The first relation in (18.14) expresses a two-dimensional object $B(u,v)$ as a summation ('integration') of its Fourier frequency components $b(s,t)$. It follows that an arbitrary two-dimensional object $B(u,v)$ may be considered to be consisted of one-dimensional sinusoidal gratings $P(\tilde{u},\Psi)$ of different periods

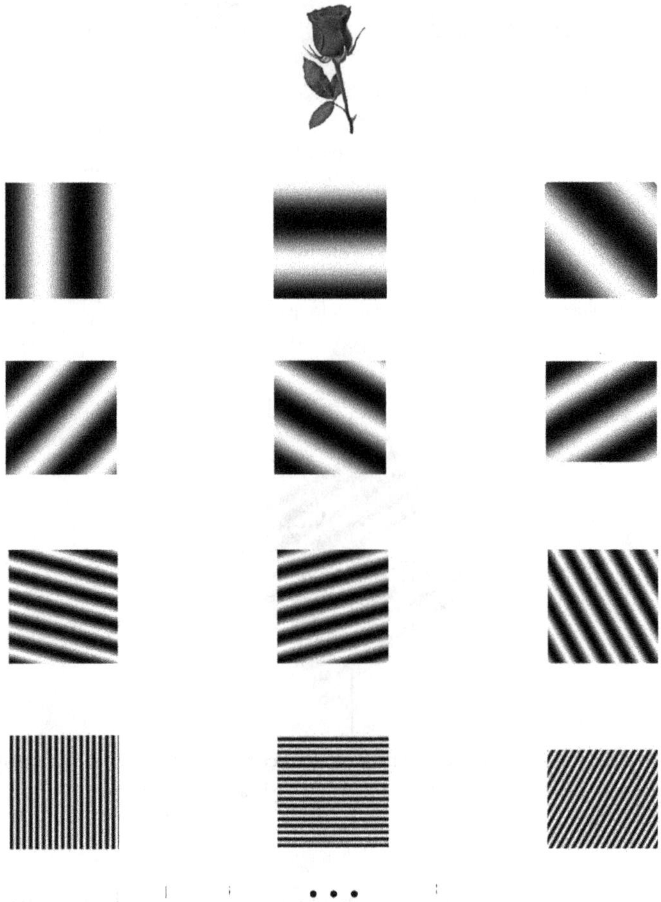

FIGURE 18.8 An object on a two-dimensional plane and its constituent gratings.

and lying in different azimuths, each grating having its characteristic modulation index $b(S)$ specific to the object $B(u,v)$. The latter one-dimensional gratings of different periods are the basic building blocks of any two-dimensional object. Two such objects vary by the number of gratings, the values of b(S) for the various constituent gratings and the azimuths along which the gratings are placed. Figure 18.8 shows a two-dimensional object, and some of its constituent gratings.

> This has an important practical consequence in that the imaging properties of any system may be studied using slits or gratings in different azimuths as test objects and image-scanning screens, with a resulting photometric gain by a factor which may be as large as 10^4.
>
> [32]

This is why the imaging characteristics of an axially symmetric system are described by means of curves of a single variable for a number of azimuths. On account of axial symmetry, for on axis imaging, only one curve suffices, but for each off axial point of interest, at least two curves corresponding to two mutually perpendicular azimuths, namely the tangential and the sagittal directions, are used.

18.3.3 Reduced Period and Reduced Spatial Frequency

On account of rotational symmetry around the optical axis, the images of one-dimensional structures, e.g. a line or a grating lying near the axis, are independent of the azimuth. Let us take the case $\psi = 0$, and consider a one-dimensional periodic object lying in the direction of the η-axis, and having a real space period $\hat{\xi}$ The corresponding real space spatial frequency is $\hat{R} = (1/\hat{\xi})$. The reduced period $\hat{u}$ and reduced spatial frequency $\hat{s}(=1/\hat{u})$ are given by

$$\left.\begin{aligned} \hat{u} &= \left(\frac{n\sin\alpha}{\lambda}\right)\hat{\xi} \\ \hat{s} &= \frac{1}{\hat{u}} = \left(\frac{\lambda}{n\sin\alpha}\right)\hat{R} \end{aligned}\right\} \tag{18.18}$$

The reduced period $\hat{u}$ and reduced spatial frequency $\hat{s}$ are dimensionless quantities, whereas the real space period $\hat{\xi}$ and real space spatial frequency $\hat{R}$ have the units of length, and cycles or line pairs per unit length, respectively. A caveat is in order. In some engineering disciplines, the real space spatial frequency is defined in terms of 'lines' per unit length, instead of 'line pairs' per unit length.

The real space period and real space spatial frequency corresponding to $\hat{u}$ and $\hat{s}$ are given by the converses of equation (18.18), as given below

$$\left.\begin{aligned} \hat{\xi} &= \left(\frac{\lambda}{n\sin\alpha}\right)\hat{u} \\ \hat{R} &= \left(\frac{n\sin\alpha}{\lambda}\right)\hat{s} \end{aligned}\right\} \tag{18.19}$$

Similarly, a one-dimensional periodic image lying in the direction of the η'-axis, and having a real space period $\hat{\xi}'$, has a real space spatial frequency $\hat{R}' = (1/\hat{\xi}')$. The reduced period and spatial frequency are given by

$$\left.\begin{aligned} \hat{u}' &= \left(\frac{n'\sin\alpha'}{\lambda}\right)\hat{\xi}' \\ \hat{s}' &= \frac{1}{\hat{u}'} = \left(\frac{\lambda}{n'\sin\alpha'}\right)\hat{R}' \end{aligned}\right\} \tag{18.20}$$

The real space values corresponding to reduced values $\hat{u}'$ and $\hat{s}'$ are given by the converses of equation (18.20).

For one-dimensional periodic objects located at an extra-axial point on the η-axis, with lines in the real-space azimuth $\tilde{\psi}$, and relative to the η-axis and period $\hat{\hat{\xi}}$, the reduced period and reduced spatial frequency are no longer independent of the azimuth $\tilde{\psi}$, and they also depend on the tangential and sagittal numerical apertures pertaining to the particular extra-axial point (see Sections 11.6.1 and 11.6.2). In what follows, we quote the final results, and the interested reader may look in reference [118] for derivation.

The reduced period of a grating of real-space period $\hat{\hat{\xi}}$ in the real space azimuth $\tilde{\psi}$ is given by

$$\hat{u} = \left[\left(\frac{\cos\tilde{\psi}}{\rho_S}\right)^2 + \left(\frac{\sin\tilde{\psi}}{\rho_T}\right)^2\right]^{-\frac{1}{2}} \left(\frac{n\sin\alpha}{\lambda}\right)\hat{\hat{\xi}} \tag{18.21}$$

The real space frequency $\hat{R}\left(=1/\hat{\hat{\xi}}\right)$, and the reduced spatial frequency $\hat{s}(=1/\hat{u})$ are related by

$$\hat{s} = \left[\left(\frac{\cos\tilde{\psi}}{\rho_S}\right)^2 + \left(\frac{\sin\tilde{\psi}}{\rho_T}\right)^2\right]^{\frac{1}{2}} \left(\frac{\lambda}{n\sin\alpha}\right)\hat{R} \tag{18.22}$$

For the case $\tilde{\psi} = 0$, the grating lines are parallel to the η-axis, so we write $\hat{\xi}$ in place of, $\hat{\hat{\xi}}$ and (18.21) – (18.22) reduce to

$$\left.\begin{aligned} \hat{u} &= \rho_S \left(\frac{n\sin\alpha}{\lambda}\right)\hat{\xi} \\ \hat{s} &= \frac{1}{\rho_S}\left(\frac{\lambda}{n\sin\alpha}\right)\hat{R} \end{aligned}\right\} \tag{18.23}$$

For the case $\tilde{\psi} = \frac{\pi}{2}$, the grating lines are perpendicular to the η-axis, and the period of the grating lies along the η-axis, so that we write $\hat{\eta}$ to represent the grating period $\hat{\hat{\xi}}$ and relations (18.21) – (18.22) reduce to

$$\left.\begin{aligned} \hat{u} &= \rho_T \left(\frac{n\sin\alpha}{\lambda}\right)\hat{\eta} \\ \hat{s} &= \frac{1}{\rho_T}\left(\frac{\lambda}{n\sin\alpha}\right)\hat{R} \end{aligned}\right\} \tag{18.24}$$

The relation (18.23) shows that the sagittal numerical aperture, $\rho_S(n\sin\alpha)$, alone, is involved in the determination of reduced period $\hat{u}$ along the ξ direction, which is in the azimuth perpendicular to the grating lines in this case. Similarly, it is noted from (18.24) that the tangential numerical aperture, $\rho_T(n\sin\alpha)$, alone, determines the reduced period along the η direction, which is in the azimuth perpendicular to the grating lines in this case.

The following cases should be noted:

(i) for the case $\tilde{\psi} = 0$, the relation (18.23) reduces to (18.18) only when $\rho_S = 1$.
(ii) for the case $\tilde{\psi} = \frac{\pi}{2}$, the relation (18.24) reduces to (18.18) only when $\rho_T = 1$.
(iii) in general, relations (18.21) and (18.22) for the extra-axial case, reduce to relation (18.18) for the axial case if and only if $\rho_S = \rho_T = 1$.

18.3.4 Line Spread Function

The intensity distribution in the image of an infinitesimally narrow line of unit strength lying along the axis $u = 0$ in the object plane is called the line spread function. For a linear system satisfying the principle of superposition, the line spread function is the superposition of the images of the different points along the line object. The point spread function in the images of the different points along the line object is markedly dependent on the aberrations of the system, and the imaging system is, in general, space variant. However, in all useful systems, around a given point in the image plane, the aberration does not change significantly over a region of the order of several times the whole width of the line spread function. Consequently, the image in the immediate neighbourhood of the point will not be affected by those parts of the object whose images fall outside this region, since their response has already fallen to zero. Under the circumstances, a concept of 'local invariance' is invoked to obtain a simplified model of optical imaging, and the line spread function (LSF) is expressed in terms of point spread function (PSF).

A line object along the v-axis is represented as

$$B(u,v) = \delta(u) \tag{18.25}$$

The corresponding image is

$$B'(u',v') = \int_{-\infty}^{+\infty}\left[\int_{-\infty}^{+\infty}\delta(u)G(u'-u,v'-v)du\right]dv = \int_{-\infty}^{+\infty}G(u',v')dv' \tag{18.26}$$

The LSF is this image. It is a one-dimensional function, and is related with the PSF by

$$L(u') = \int_{-\infty}^{+\infty}G(u',v')dv' \tag{18.27}$$

The limits of the integration are nominally taken as $\pm\infty$. In practice, the range of integration is over a small domain of V'. This is so because the PSF $G(u',v')$ is centered on the point $u'=0, v'=0$, and the value of $G(u',v')$ falls rapidly, so much so that $G(u',v')$ effectively becomes zero when $\sqrt{u'^2+v'^2}$ exceeds a few times the whole width of the LSF.

18.3.5 Image of a One-Dimensional Object

Let us consider an arbitrary one-dimensional object. It is expressed as

$$B(u,v) = B(u) \tag{18.28}$$

Its image is given by

$$B'(u',v') = \int_{-\infty}^{+\infty}B(u)\left[\int_{-\infty}^{+\infty}G(u'-u,v'-v)dv\right]du \tag{18.29}$$

Using (18.27), we get

$$\int_{-\infty}^{+\infty}G(u'-u,v'-v)dv = L(u'-u) \tag{18.30}$$

It follows that the image of the one-dimensional object is one-dimensional

$$B'(u',v') = \int_{-\infty}^{+\infty}B(u)L(u'-u)du = B'(u') \tag{18.31}$$

It is to be noted that the image of a one-dimensional object is given by the convolution of the one dimensional object and the LSF.

18.3.6 Optical Transfer Function (OTF), Modulation Transfer Function (MTF), and Phase Transfer Function (PTF)

The relation (18.31) expresses the one-dimensional image as a convolution of the one-dimensional object $B(u)$ and the line spread function $L(u)$. The Fourier spectra of the image $b'(s)$ is equal to the product of the Fourier spectra of the object $b(s)$ and the same of the line spread function $l(s)$. Thus

$$b'(s) = b(s)\,l(s) \tag{18.32}$$

where

$$\left.\begin{aligned} b'(s) &= \int_{-\infty}^{+\infty} B'(u')e^{-i2\pi u's}du' \\ b(s) &= \int_{-\infty}^{+\infty} B(u)e^{-i2\pi us}du \\ l(s) &= \int_{-\infty}^{+\infty} L(u')e^{-i2\pi u's}du' \end{aligned}\right\} \tag{18.33}$$

The normalized form $D(s)$ of the spectrum $l(s)$ is given by

$$D(s) = \frac{l(s)}{l(0)} = T(s)\exp\left[i\theta(s)\right] \tag{18.34}$$

where

$$l(0) = \int_{-\infty}^{+\infty} L(u')du' \tag{18.35}$$

Use of this normalized form $D(s)$ for the spectrum of LSF instead of $l(s)$ eliminates unimportant multiplying factors in subsequent treatment. Therefore, to obtain the spectrum of the image, we use

$$b'(s) = b(s)\ D(s) \tag{18.36}$$

Let us consider a one-dimensional cosinusoidal grating lying in the direction of the u-axis, and having a reduced spatial frequency $\bar{s}$ as an object. The grating $B(u)$ is represented by

$$B(u) = \alpha + \beta\cos(2\pi u\bar{s}) \tag{18.37}$$

In exponential notation, it can be expressed as

$$B(u) = \alpha + \frac{\beta}{2}\left[e^{i2\pi u\bar{s}} + e^{-i2\pi u\bar{s}}\right] \tag{18.38}$$

Its spectrum $b(s)$ is given by

$$b(s) = \alpha\delta(s) + \frac{\beta}{2}\left[\delta(s-\bar{s}) + \delta(s+\bar{s})\right] \tag{18.39}$$

The spectrum $b'(s)$ of the image is

$$b'(s) = \alpha\delta(s)D(s) + \frac{\beta}{2}D(s)\left[\delta(s-\bar{s}) + \delta(s+\bar{s})\right] \tag{18.40}$$

The image $B'(u')$ is given by the inverse Fourier transform of the image spectrum $b'(s)$.

$$B'(u') = \alpha\int_{-\infty}^{+\infty}\delta(s)\ D(s)\ e^{i2\pi u's}ds + \frac{\beta}{2}\int_{-\infty}^{+\infty}\left[\delta(s-\bar{s}) + \delta(s+\bar{s})\right]D(s)e^{i2\pi u's}ds \tag{18.41}$$

On integration, (18.41) reduces to

$$B'(u') = \alpha D(0) + \frac{\beta}{2} D(\bar{s}) e^{i2\pi u'\bar{s}} + \frac{\beta}{2} D(-\bar{s}) e^{-i2\pi u'\bar{s}} \tag{18.42}$$

Using the relations

$$\left.\begin{aligned} D(0) &= 1 \\ D(\bar{s}) &= T(\bar{s}) e^{i\theta(\bar{s})} \\ D(-\bar{s}) = D^*(\bar{s}) &= T(\bar{s}) e^{-i\theta(\bar{s})} \end{aligned}\right\} \tag{18.43}$$

we get the image

$$\begin{aligned} B'(u') &= \alpha + \frac{\beta}{2} T(\bar{s}) \left[e^{i\{2\pi u'\bar{s} + \theta(\bar{s})\}} + e^{-i\{2\pi u'\bar{s} + \theta(\bar{s})\}} \right] \\ &= \alpha + \beta T(\bar{s}) \cos\{2\pi u'\bar{s} + \theta(\bar{s})\} \end{aligned} \tag{18.44}$$

Note that the image of the one-dimensional grating (18.37) produced by a linear space-invariant imaging system is a one-dimensional grating of the same spatial frequency $\bar{s}$. The modulation of the object grating has changed from β to $\{\beta T(\bar{s})\}$. $\theta(\bar{s})$ is the spatial phase shift, and it shifts the component of the image of frequency $\bar{s}$ by an amount $\delta(u) = -\theta(\bar{s}) / 2\pi\bar{s}$ relative to the geometrical image.

If $\theta(s)$ is directly proportional to s, the different components of the image of different frequency are shifted by the same amount, and this gives rise to a spatial shift of the image. On the other hand, if $\theta(s)$ has a component that depends nonlinearly on s, different spatial frequency components of the image are shifted differently, giving rise to phase distortion and consequent blurring of the image [123–125].

In general, a linear space-invariant imaging system is characterized by the functions noted in relation (18.34)

$$D(s) = T(s) e^{i\theta(s)} \tag{18.45}$$

The three functions are called

$$\left.\begin{aligned} D(s)&: \text{Optical Transfer Function (OTF)} \\ T(s)&: \text{Modulation Transfer Function (MTF)} \\ \theta(s)&: \text{Phase Transfer Function (PTF)} \end{aligned}\right\} \tag{18.46}$$

Figure 18.9 illustrates roles of $T(s)$ and $\theta(s)$ in formation of the image of a one-dimensional cosinusoidal grating [32]. Note that our choice of suitable normalization of $l(s)$ ensures that an object $B(u) = \alpha$ is imaged as $B'(u') = \alpha$. However, this model does not take into account any loss of light by absorption, scattering, etc. during the process of image formation. It should also be noted that the complex exponential functions are the 'eigenfunctions' of linear invariant systems, since they retain their original form (except for a multiplicative complex constant) in the output. The transfer function of the system describes the continuum of eigenvalues of the system.

The above treatment is given in terms of reduced variables, so that it is applicable to both axial and extra-axial imagery, and for objects lying on any azimuth ψ. For each case, the formulae necessary for transition from the reduced to the real variables are given in Section 18.3.3.

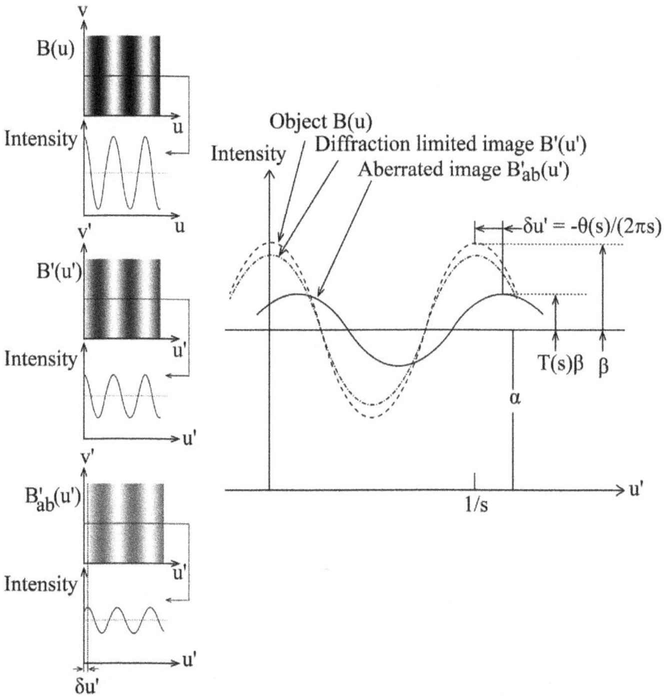

FIGURE 18.9 Roles of Modulation Transfer Function T(s) and Phase Transfer Function θ(s) in formation of the image of a one-dimensional cosinusoidal grating.

18.3.7 Effects of Coherence in Illumination on Extended Object Imagery [126–132]

So far, the exact property of the object or the image that characterizes them in the functions B or B′ has not been clearly spelt out, except in mentioning that this property should satisfy the conditions for linear shift-invariant imagery. Indeed, this property is decided by the level of coherence, if any, between the waves originating from different points of the object for forming the image. Below, we give a heuristic description of the effects of coherence on image formation.

If the different waves originating from the points on the object are mutually uncorrelated, then over a reasonable time of response, no sustained interference effects exist, the intensity point spread functions add up, and the object/image is considered linear in 'intensity' for the sake of linear system theoretic analysis of optical imaging. On the other hand, if the different points on the object are illuminated such that the waves originating from the different points are fully correlated with each other, the amplitude point spread functions add up, and the object/image is to be considered linear in 'amplitude' for the same purpose. Note that intensity is the squared modulus of the amplitude [126].

For the intermediate case of partial correlation of the waves originating from different object points and taking part in image formation, the system is neither linear in intensity nor in amplitude. In this situation, the imaging systems are linear in mutual intensity. Details of the concept of mutual intensity may be found in references [127–132]. In what follows, we restrict our treatment to the two extreme cases of full coherence and no coherence, i.e. we shall deal with the two special cases, coherent image formation and incoherent image formation.

18.4 Abbe Theory of Coherent Image Formation [133–136]

In the course of his pathbreaking investigations on improving the quality of microscopic imagery, Abbe came out with a different interpretation for the formation of images in microscopes, and explained it as a

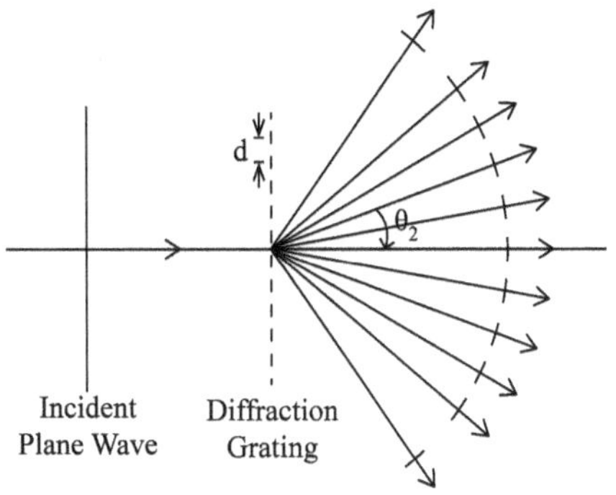

FIGURE 18.10 Diffraction of a normally incident plane wave by an amplitude grating.

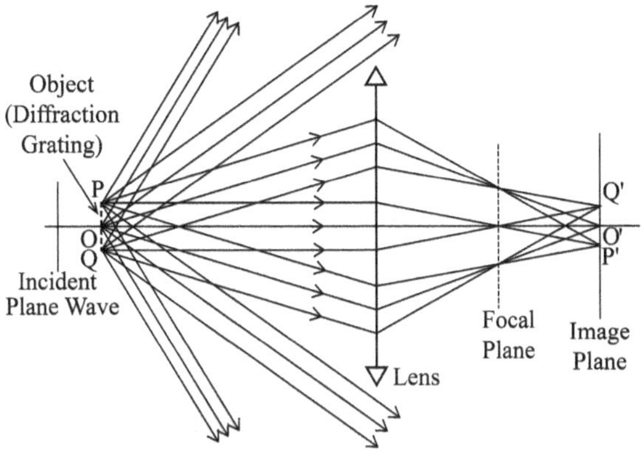

FIGURE 18.11 Two stages in formation of an image as per the Abbe theory of coherent imagery.

two-step process. Figure 18.10 shows that a series of plane waves in different directions emerges from a diffraction grating when a plane wave is incident on it. In the case of normal incidence, the directions of the emergent plane waves are given by

$$d\sin\theta_n = n\lambda, \quad n = 0, \pm 1, \pm 2, \pm 3, \ldots \tag{18.47}$$

where d is the period of the grating, and λ is the working wavelength.

Figure 18.11 shows a diffraction grating as an object. For the sake of simplicity in drawing, the lens used for imaging is considered a thin lens. The observations, however, are applicable for any real lens system. Out of the multitude of plane waves emerging from the diffraction grating, a limited number passes through the aperture of the imaging lens and comes to their respective focal spots on the focal plane. Had the size of the lens aperture been larger, a greater number of plane waves would have passed through the imaging lens, giving rise to more focal spots on the focal plane. This is the first stage in the Abbe theory of coherent image formation. Note that each focal spot corresponds with a particular plane wave of the multitude of plane waves originating from the diffraction grating and going in different directions. The waves converging on the different focal spots become diverging after passing through the aerial focal spots. The latter waves interfere with each other in the common region on the image plane and form an image. The image will be identical (within a scale factor that takes into account the magnification and sign) with the object, i.e. the diffraction grating, if and only if all plane waves originating from the object pass through the lens unhindered and take part in image formation. The journey of the waves from the aerial focal spots to the final image constitutes the second stage of the Abbe theory of coherent image formation. Imaging a diffraction grating of a different period gives rise to a different set of focal points. In general, the analysis of the imaging of any arbitrary object can be carried out in terms of Fourier components of the object. Each of these Fourier components turns out to be a cosinusoidal grating.

Indeed, the Abbe treatment of coherent image formation is the precursor of 'Fourier optics'.

The physical interpretation of coherent image formation by Abbe facilitates understanding the limit of resolution in microscopic imagery. Moreover, it lays down the foundations of what is now called 'analogue optical signal processing'.

18.5 Transfer Function, Point Spread Function, and the Pupil Function

In the diffraction theory of image formation enunciated in the last two chapters, canonical coordinates are used for the object/image and the entrance/exit pupil. An important advantage of this is that the same formulae can be used for both extra-axial imaging and axial imaging. The pertinent features are reiterated in the context of frequency response in optical imaging.

Figure 18.12 shows the image space of an axially symmetric optical system. The effective exit pupil is located at $\bar{E}'$, and the image plane is located at O′. $\bar{E}'\bar{Q}'$ is the pupil ray of a pencil forming the geometrical image at $\bar{Q}'$ of an object point $\bar{Q}$ lying on the meridional plane (object space not shown). P′ is a point on the image plane, and it has canonical coordinates (u'_S, v'_T) relative to $\bar{Q}'$ as origin. The real space coordinates of the point P′ relative to $\bar{Q}'$ as origin are (ξ', η'), The v'_T axis lies in the meridional plane containing the optical axis, as well as the points Q and $\bar{Q}'$. B′ is a point on the exit pupil sphere with its centre at $\bar{Q}'$, and radius $\bar{R}'_0 = (\bar{E}'\bar{Q}')$. The canonical and real space coordinates of the point B′ are (x'_S, y'_T) and (X', Y'), respectively.

The relations between (u'_S, v'_T) and (ξ', η') are:

$$\left.\begin{aligned} u'_S &= \frac{1}{\lambda} G'_S = \frac{1}{\lambda}\rho'_S\left[n'(\sin\alpha')\xi'\right] \\ v'_T &= \frac{1}{\lambda} H'_T = \frac{1}{\lambda}\rho'_T\left[n'(\sin\alpha')\eta'\right] \end{aligned}\right\} \tag{18.48}$$

The details of pupil scale ratios ρ'_S and ρ'_T were given in Section 11.6.

The relations between (x'_S, y'_T) and (X', Y') are:

$$\left.\begin{aligned} x'_S &= \frac{X'}{h'_S} \\ y'_T &= \frac{Y'}{\bar{N}'h'_T} \end{aligned}\right\} \tag{18.49}$$

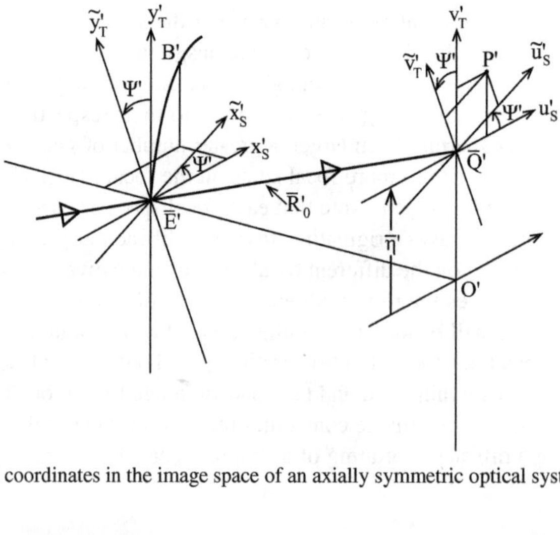

FIGURE 18.12 Canonical coordinates in the image space of an axially symmetric optical system.

$\bar{N}'$ is the direction cosine of the effective pupil ray $\bar{E}'\bar{Q}'$ with respect to the optical axis, and $(\bar{N}'h'_T)$ is the height of the upper tangential ray at the pupil sphere above the optical axis. The perpendicular distance of the point of intersection of the sagittal edge ray with the exit pupil sphere from the optical axis is h'_S. Reduced pupil variables were discussed with more detail in Section 11.4.

For the image at $\bar{Q}'$, the pupil function is denoted by

$$f(X',Y') = f(x'_S,y'_T) = \tau(x'_S,y'_T)\exp\left[i2\pi W(x'_S,y'_T)\right] \tag{18.50}$$

where $\tau(x'_S,y'_T)$ is the real amplitude, and $W(x'_S,y'_T)$ is the wavefront aberration at B'. The pupil function expressed by (18.50) describes the variation of complex amplitude over the pupil sphere as a function of the canonical coordinates (x'_S,y'_T). Note that the pupil function in reduced coordinates (x'_S,y'_T) has an outer periphery that is a unit circle given by $x_S'^2 + y_T'^2 = 1$. Outside this unit circle, for any value of (x'_S,y'_T), the value of pupil function is zero.

In Section 16.2.2, it was shown that the complex amplitude at a point P' on the image plane is given by

$$U_{P'} = \exp\{-i\epsilon(\xi',\eta')\}F(u'_S,v'_T) \tag{18.51}$$

where

$$F(u'_S,v'_T) = \iint_{-\infty}^{+\infty} f(x'_S,y'_T)\exp\{i2\pi(u'_Sx'_S + v'_Ty'_T)\}dx'_Sdy'_T \tag{18.52}$$

is the Fourier transform of the pupil function $f(x'_S,y'_T)$. $F(u'_S,v'_T)$ is called the amplitude point spread function (APSF). The inverse Fourier transform gives the pupil function in terms of the APSF.

$$f(x'_S,y'_T) = \iint_{-\infty}^{+\infty} F(u'_S,v'_T)\exp\{-i2\pi(u'_Sx'_S + v'_Ty'_T)\}du'_Sdv'_T \tag{18.53}$$

It is shown in Section 16.2.3 that $\epsilon(\xi',\eta') = 0$, if the point P' is taken not on the image plane, but on the confocal image sphere (with its centre at $\bar{E}'$ and radius $\bar{E}'\bar{Q}'$). It should be realized that the existence of relation (18.52) expressing the amplitude distribution in the image of a point object as a Fourier transform of the corresponding pupil function is one of the main causes for growth and development of Fourier optics. Comments by Hopkins are noteworthy in this regard: 'The Laplace transforms also admit of simple convolution-type theorems and could also be used for the analysis of image formation, as is now common in the corresponding treatment of circuit theory. The formula for the response would, however, lack the simplicity' [32], as available by using Fourier transform.

18.5.1 Amplitude Transfer Function (ATF) in Imaging Systems Using Coherent Illumination

An imaging system under coherent illumination is linear in amplitude. The amplitudeimage $A'(u'_S,v'_T)$ of a two-dimensional object with amplitude distribution $A(u_S,v_T)$ is given by

$$A'(u'_S,v'_T) = \iint_{-\infty}^{+\infty} A(u_S,v_T)F(u'_S - u_S,v'_T - v_T)du_Sdv_T \tag{18.54}$$

where $F(u'_S,v'_T)$ is the amplitude point spread function (APSF). The two-dimensional Fourier spectrum of the image $A'(u'_S,v'_T)$ is

$$a'(s,t) = a(s,t)f(s.t) \tag{18.55}$$

The three Fourier spectra are given by

$$a'(s,t) = \iint_{-\infty}^{+\infty} A'(u'_S, v'_T) \exp\{-i2\pi(u'_S s + v'_T t)\} du'_S dv'_T \tag{18.56}$$

$$a(s,t) = \iint_{-\infty}^{+\infty} A(u_S, v_T) \exp\{-i2\pi(u_S s + v_T t)\} du_S dv_T \tag{18.57}$$

$$f(s,t) = \iint_{-\infty}^{+\infty} F(u'_S, v'_T) \exp\{-i2\pi(u'_S s + v'_T t)\} du'_S dv'_T \tag{18.58}$$

In the above, $f(s,t)$ is the unnormalized two-dimensional amplitude transfer function (ATF). The normalized form is

$$d(s,t) = \frac{f(s,t)}{f(0,0)} \tag{18.59}$$

with

$$C = f(0,0) = \iint_{-\infty}^{+\infty} F(u'_S, v'_T) du'_S dv'_T \tag{18.60}$$

we write

$$A'(u'_S, v'_T) = \frac{1}{C} \iint_{-\infty}^{+\infty} A(u_S, v_T) F(u'_S - u_S, v'_T - v_T) du_S dv_T \tag{18.61}$$

So, we have

$$a'(s,t) = a(s,t) d(s.t) \tag{18.62}$$

With this normalized form, the image of $A(u_S, v_T) = 1$ is

$$A'(u'_S, v'_T) = \frac{1}{C} \iint_{-\infty}^{+\infty} F(u'_S - u_S, v'_T - v_T) du_S dv_T = \frac{1}{C} \iint_{-\infty}^{+\infty} F(\hat{u}_S, \hat{v}_T) d\hat{u}_S d\hat{v}_T = 1 \tag{18.63}$$

A one-dimensional object imaged in the azimuth $\psi = 0$ (Figure 18.12), and having amplitude distribution $A(u_S)$, gives an amplitude distribution in the image

$$A'(u'_S) = \int_{-\infty}^{+\infty} A(u_S) F(u'_S - u_S) du_S \tag{18.64}$$

where $F(u'_S)$ is the amplitude line spread function. If the line lies in the meridian plane, its image lies along the v'_T axis, and the amplitude line spread function is given by

$$F(u'_S) = \int_{-\infty}^{+\infty} F(u'_S, v'_T) dv'_T \tag{18.65}$$

where $F(u'_S, v'_T)$ is the amplitude point spread function. The Fourier spectrum of the one-dimensional image $A'(u'_S)$ is

$$a'(s) = a(s) f(s) \tag{18.66}$$

where $f(s)$ is the unnormalized one-dimensional amplitude transfer function, and, using (18.65), it is expressed as

$$f(s) = \int_{-\infty}^{+\infty} F(u'_S)\exp[-i2\pi u'_S s]du'_S = \iint_{-\infty}^{+\infty} F(u'_S, v'_T)\exp[-i2\pi u'_S s]du'_S dv'_T \tag{18.67}$$

From (18.58) and (18.67), it follows

$$f(s) = f(s,0) \text{ and } f(0) = f(0,0) \tag{18.68}$$

The normalized one-dimensional amplitude transfer function is

$$d(s) = \frac{f(s)}{f(0)} = \frac{f(s,0)}{f(0,0)} = d(s,0) \tag{18.69}$$

where $d(s,t)$ is the normalized two-dimensional amplitude transfer function. Like the expression (18.63) for the two-dimensional case, we replace the relation (18.64) by

$$A'(u'_S) = \frac{1}{C}\int_{-\infty}^{+\infty} A(u_S)F(u'_S - u_S)du_S \tag{18.70}$$

to obtain

$$a'(s) = a(s)d(s) \tag{18.71}$$

Using this normalized form for the one-dimensional amplitude transfer function, we note that the image of a one-dimensional object $B(u_S) = 1$ is given by $B'(u'_S) = 1$.

Using (18.53), (18.58), and (18.59) – (18.60), the normalized two-dimensional amplitude transfer function is expressed in terms of the pupil function given by (18.50) as

$$d(s,t) = \frac{1}{C}f(s,t) = \frac{1}{C}f(x'_S, y'_T) \tag{18.72}$$

where $C = f(0,0)$. The normalized one-dimensional amplitude transfer function for structures imaged in the azimuth $\psi = 0$ is given by

$$d(s) = d(s,0) = \frac{1}{C}f(s,0) = \frac{1}{C}f(x'_S, 0) \tag{18.73}$$

18.5.1.1 ATF in Coherent Diffraction Limited Imaging Systems

From (18.50), the pupil function $f(x'_S, y'_T)$ is given by

$$\begin{aligned} f(x'_S, y'_T) &= \tau(x'_S, y'_T)\exp[i2\pi W(x'_S, y'_T)], & (x'^2_S + y^2_T) \le 1 \\ &= 0 & (x'^2_S + y'^2_T) > 1 \end{aligned} \tag{18.74}$$

In a diffraction limited imaging system,

$$W(x'_S, y'_T) = 0 \tag{18.75}$$

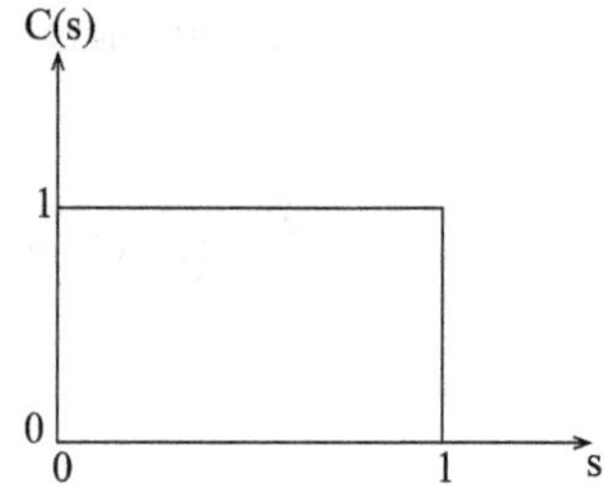

FIGURE 18.13 Normalized one-dimensional amplitude transfer function for a diffraction limited axially symmetric imaging system.

and the real amplitude function is usually taken as unity, so that the pupil function reduces to

$$\begin{aligned} f(x', y_T') &= 1, \quad (x_S'^2 + y_S'^2) \le 1 \\ &= 0, \quad (x_S'^2 + y_S'^2) > 1 \end{aligned} \tag{18.76}$$

From the above relation, we get $f(0,0)=1$. Therefore, $d(s,t)=f(s,t)$. From (18.72) and (18.76), we note that the normalized two-dimensional amplitude transfer function is

$$\begin{aligned} d(s,t) &= 1, \quad (s^2 + t^2) \le 1 \\ &= 0, \quad (s^2 + t^2) > 1 \end{aligned} \tag{18.77}$$

The normalized one-dimensional amplitude transfer function for structures imaged in the azimuth $\psi = 0$ is

$$\begin{aligned} d(s) = d(s,0) = f(s,0) &= 1, \quad s \le 1 \\ &= 0, \quad s > 1 \end{aligned} \tag{18.78}$$

Figure 18.13 shows the normalized one-dimensional amplitude transfer function for an axially symmetric imaging system.

In the case of amplitude transfer function, the cut-off in reduced spatial frequency is $s = 1$. From the relationship between real space period and reduced spatial frequency given above in Section 18.3.3, $s = 1$ corresponds to the least resolvable distance $\Delta\xi'$, given by

$$\Delta\xi' = \frac{\lambda}{n' \sin\alpha'} \tag{18.79}$$

18.5.1.2 Effects of Residual Aberrations on ATF in Coherent Systems

In the relation (18.74), it is seen that the residual aberration $W(x_S', y_T')$ gives rise to a phase term in the pupil function. Assuming $\tau(x_S', y_T') = 1$, the normalized two-dimensional amplitude transfer function is expressed by

$$\begin{aligned} d(s,t) &= \exp[i2\pi W(s,t)], \quad (s^2 + t^2) \le 1 \\ &= 0, \quad\quad\quad\quad\quad\quad\;\; (s^2 + t^2) > 1 \end{aligned} \tag{18.80}$$

The normalized one-dimensional amplitude transfer function for structures imaged in the azimuth $\psi = 0$ is

$$\begin{aligned} d(s) = d(s,0) &= \exp[i2\pi W(s,0)], \quad s \le 1 \\ &= 0, \quad\quad\quad\quad\quad\quad\;\; s > 1 \end{aligned} \tag{18.81}$$

18.5.2 Optical Transfer Function (OTF) in Imaging Systems Using Incoherent Illumination

An imaging system operating under incoherent illumination is linear in intensity. So the transfer function in this case is intensity transfer function. However, in view of widespread occurrence of such optical imaging systems, the 'intensity transfer function' is commonly called the 'Optical Transfer Function (OTF)'. Similarly, unless otherwise specified, the term 'point spread function (PSF)' implies the intensity point spread function.

The intensity point spread function is the squared modulus of the amplitude point spread function

$$G(u'_S, v'_T) = \left|F(u'_S, v'_T)\right|^2 \tag{18.82}$$

The image $B'(u'_S, v'_T)$ of a two-dimensional object $B(u_S, v_T)$ is given by

$$B'(u'_S, v'_T) = \iint_{-\infty}^{+\infty} B(u_S, v_T) G(u'_S - u_S, v'_T - v_T) du_S dv_T \tag{18.83}$$

The two-dimensional Fourier spectrum of $B'(u'_S, v'_T)$, $B(u_S, v_T)$, and $G(u'_T, v'_T)$ are $b'(s,t), b(s,t)$, and $g(s,t)$, respectively. They are related by

$$b'(s,t) = b(s,t) g(s,t) \tag{18.84}$$

The unnormalized two-dimensional intensity transfer function, $g(s,t)$, is the inverse Fourier transform of the intensity point spread function, $G(u'_S, v'_T)$, i.e.

$$g(s,t) = \iint_{-\infty}^{+\infty} G(u'_S, v'_T) \exp\{-i2\pi(u'_S s + v'_T t)\} du'_S dv'_T \tag{18.85}$$

The normalized form for the transfer function, $D(s,t)$ is obtained as

$$D(s,t) = \frac{g(s,t)}{g(0,0)} \tag{18.86}$$

Denoting $g(0,0)$ by Q, i.e.

$$Q = g(0,0) = \iint_{-\infty}^{+\infty} G(u'_S, v'_T) du'_S dv'_T \tag{18.87}$$

By Rayleigh's (Plancherel's) theorem [137], the above relation can be expressed also as

$$Q = \iint_{-\infty}^{+\infty} \left|F(u'_S, v'_T)\right|^2 du'_S dv'_T = \iint_{-\infty}^{+\infty} \left|f(x'_S, y'_T)\right|^2 dx'_S dy'_T \tag{18.88}$$

In place of (18.83) we write

$$B'(u'_S, v'_T) = \frac{1}{Q} \iint_{-\infty}^{+\infty} B(u_S, v_T) G(u'_S - u_S, v'_T - v_T) du_S dv_T \tag{18.89}$$

so that

$$b'(s,t) = b(s,t)D(s,t) \tag{18.90}$$

Using the normalized transfer function, the image of the object $B(u_S, v_T) = 1$ becomes

$$B'(u'_S, v'_T) = \frac{1}{Q}\int\int_{-\infty}^{+\infty} G(u'_S - u_S, v'_T - v_T)du_S dv_T = 1 \tag{18.91}$$

An incoherent line object on the meridional plane is represented by $B(u_S, v_T) = \delta(u_S)$. The line is lying along the v_T axis. The corresponding image is given by

$$B'(u'_S, v'_T) = \int_{-\infty}^{+\infty}\left[\int_{-\infty}^{+\infty}\delta(u_S)G(u'_S - u_S, v'_T - v_T)du_S\right]dv_T = \int_{-\infty}^{+\infty} G(u'_S, v'_T)dv'_T = G(u'_S) \tag{18.92}$$

$G(u'_S)$ is the line spread function for a line object with intensity distribution $\delta(u_S)$. The intensity distribution in the image of a one-dimensional object with intensity distribution $B(u_S)$ is given by

$$B'(u'_S) = \int_{-\infty}^{+\infty} B(u_S)G(u'_S - u_S)du_S \tag{18.93}$$

The Fourier spectrum $b'(s)$ of $B'(u'_S)$ is equal to the product of the Fourier spectra $b(s)$ and $g(s)$ of $B(u_S)$ and $G(u'_S)$, respectively,

$$b'(s) = b(s)g(s) \tag{18.94}$$

Note that $g(s)$ is the unnormalized intensity transfer function, and it is represented by

$$g(s) = \int_{-\infty}^{+\infty} G(u'_S)\exp\{-i2\pi u'_S\}du'_S = \int\int_{-\infty}^{+\infty} G(u'_S, v'_T)\exp\{-i2\pi u'_S s\}du'_S dv'_T \tag{18.95}$$

From (18.85) and (18.95), we note

$$g(s) = g(s,0) \text{ and } g(0) = g(0,0) \tag{18.96}$$

A normalized form for the one-dimensional intensity transfer function may be defined by

$$D(s) = \frac{g(s)}{g(0)} = \frac{g(s,0)}{g(0.0)} = D(s,0) \tag{18.97}$$

Note that the equal intensity lines of the one-dimensional object/image of the above relation are assumed to be lying parallel to the v_T / v'_T axis.

Writing the following relation in place of relation (18.93)

$$B'(u'_S) = \frac{1}{Q}\int_{-\infty}^{+\infty} B(u_S)G(u'_S - u_S)du_S \tag{18.98}$$

it follows

$$b'(s) = b(s)D(s) \tag{18.99}$$

and the image of $B(u_S) = 1$ is $B'(u'_S) = 1$.

Using autocorrelation, theorem [138] and the relations (18.82), (18.85), (18.86), and (18.88), we get the following expression for the two-dimensional intensity transfer function/optical transfer function (OTF) in terms of the pupil function

$$D(s,t) = \frac{1}{Q}\int\int_{-\infty}^{+\infty} f\left(x'_S + \frac{s}{2}, y'_T + \frac{t}{2}\right) f^*\left(x'_S - \frac{s}{2}, y'_T - \frac{t}{2}\right) dx'_S dy'_T \tag{18.100}$$

with the normalizing factor Q given by (18.88). The asterisk symbol indicates a complex conjugate.

For one-dimensional objects with constant intensity along the η_T direction, i.e. objects that are imaged in the azimuth $\psi = 0$, the one-dimensional transfer function is given by

$$D(s) = \frac{1}{Q}\int\int_{-\infty}^{+\infty} f\left(x'_S + \frac{s}{2}, y'_T\right) f^*\left(x'_S - \frac{s}{2}, y'_T\right) dx'_S dy'_T \tag{18.101}$$

Extra-axial one-dimensional objects, e.g. lines or gratings with real space azimuth ψ_0, relative to the η_T-axis, have in general reduced azimuth $\psi \neq \psi_0$. This is because the pupil scale ratios, ρ_S and ρ_T, used in scaling from real space to canonical coordinates are, in general, not equal. In object space, they are related by

$$\tan\psi = \frac{\rho_S}{\rho_T}\tan\psi_0 \tag{18.102}$$

Similarly, in the image plane the two angles are related by

$$\tan\psi' = \frac{\rho'_S}{\rho'_S}\tan\psi'_0 \tag{18.103}$$

In general, $\bar{\psi}'_0 \neq \psi'$, and $\psi_0 \neq \psi'_0$. But the reduced azimuths for the line on the object plane and its image in the image plane are equal, i.e. $\psi' = \psi$, since reduced coordinates of the image (u'_S, v'_T) and those of the object (u_S, v_T) are equal $[u'_S = u_S, v'_T = v_T]$. The real space and reduced azimuths in the object plane and the image plane are illustrated in Figure 18.14.

Let us consider a one-dimensional structure in the reduced azimuth ψ. Figure 18.12 shows a set of rotated coordinates $(\tilde{u}'_S, \tilde{v}'_T)$ on the image plane, and a set of rotated coordinates $(\tilde{x}'_S, \tilde{y}'_T)$ on the exit pupil plane. Both rotations are by angle ψ' from the meridional plane in the anticlockwise direction. The equal intensity lines of the one-dimensional structure lie parallel to the $\tilde{v}'_T$-axis. Since $\psi' = \psi$, the equal intensity lines in the object space lie parallel to the $\tilde{v}_T$-axis in the object plane.

Let $\tilde{G}(\tilde{u}'_S, \tilde{v}'_T)$ be the intensity point spread function referred to the rotated axes. Then

$$\tilde{G}(\tilde{u}'_S, \tilde{v}'_T) = \left|\tilde{F}(\tilde{u}'_S, \tilde{v}'_T)\right|^2 \tag{18.104}$$

where $\tilde{F}(\tilde{u}'_S, \tilde{v}'_T)$ is the amplitude point spread function $F(u'_S, v'_T)$ expressed as a function of $(\tilde{u}'_S, \tilde{v}'_T)$. A point P′, with coordinates $(\tilde{u}'_S, \tilde{v}'_T)$ in the rotated system has coordinates

$$\left.\begin{aligned} u'_S &= \tilde{u}'_S\cos\psi' - \tilde{v}'_T\sin\psi' \\ v'_T &= \tilde{v}'_T\cos\psi' + \tilde{u}'_S\sin\psi' \end{aligned}\right\} \tag{18.105}$$

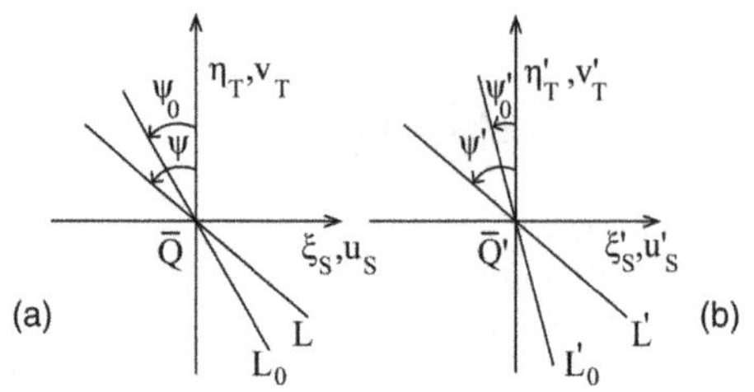

FIGURE 18.14 (a) Object plane, (b) Image plane. Image of a line passing through the extra-axial point $\bar{Q}$ in object plane passes through $\bar{Q}'$ in the image plane. Real azimuth of the line L_0 is ψ_0 in object plane, and that of its image L'_0 in image plane is ψ'_0 with respect to coordinates (ξ_S,η_T) and (ξ'_S,η'_T), respectively. In reduced coordinates (u_S,v_T) and (u'_S,v'_T), the line becomes L and L′ in the object and the image plane, respectively. The reduced azimuth of the line is Ψ in the object plane, and ψ' in the image plane. In general, $\psi' = \psi, \psi'_0 \neq \psi_0$.

in the unrotated system. Therefore, we write

$$\tilde{F}(\tilde{u}'_S,\tilde{v}'_T) = F\{(\tilde{u}'_S\cos\psi' - \tilde{v}'_T\sin\psi'),(\tilde{v}'_T\cos\psi' + \tilde{u}'_S\sin\psi')\} \tag{18.106}$$

Using (18.52), we get

$$\tilde{F}(\tilde{u}'_S,\tilde{v}'_T) = \int\int_{-\infty}^{+\infty} f(x'_S,y'_T)\exp\left\{i2\pi\left[\begin{matrix}(\tilde{u}'_S\cos\psi' - \tilde{v}'_T\sin\psi')x'_S\\ +(\tilde{v}'_T\cos\psi' + \tilde{u}'_S\sin\psi')y'_T\end{matrix}\right]\right\}dx'_S dy'_T \tag{18.107}$$

where $f(x'_S,y'_T)$ is the pupil function referred to the unrotated axes. Note that point B has the coordinates (x'_S,y'_T) in the unrotated axes system, and it has the coordinates $(\tilde{x}'_S,\tilde{y}'_T)$ in the rotated system. They are related by

$$\left.\begin{aligned} x'_S &= \tilde{x}'_S\cos\psi' - \tilde{y}'_T\sin\psi' \\ y'_T &= \tilde{x}'_T\cos\psi' + \tilde{x}'_S\sin\psi' \end{aligned}\right\} \tag{18.108}$$

By substitution from (18.108), it follows

$$(\tilde{u}'_S\cos\psi' - \tilde{v}'_T\sin\psi')x'_S + (\tilde{v}'_T\cos\psi' + \tilde{u}'_S\sin\psi')y'_T = (\tilde{u}'_S\tilde{x}'_S + \tilde{v}'_T\tilde{y}'_T) \tag{18.109}$$

Using this relation and changing variables from (x'_S,y'_T) to $(\tilde{x}'_S,\tilde{y}'_T)$, the relation (18.107) reduces to

$$\tilde{F}(\tilde{u}'_S,\tilde{v}'_T) = \int\int_{-\infty}^{+\infty} \tilde{f}(\tilde{x}'_S,\tilde{y}'_T)\exp\{i2\pi(\tilde{u}'_S\tilde{x}' + \tilde{v}'_T\tilde{y}'_T)\}d\tilde{x}'_S d\tilde{y}'_T \tag{18.110}$$

where $\tilde{f}(\tilde{x}'_S,\tilde{y}'_T)$ is the pupil function referred to the rotated axes $(\tilde{x}'_S,\tilde{y}'_T)$. It is important to note that the amplitude point spread function $\tilde{F}(\tilde{u}'_S,\tilde{v}'_T)$ expressed in terms of rotated coordinates $(\tilde{u}'_S,\tilde{v}'_T)$ is the Fourier transform of the pupil function $\tilde{f}(\tilde{x}'_S,\tilde{y}'_T)$ expressed in terms of rotated axes $(\tilde{x}'_S,\tilde{y}'_T)$.

Relation (18.52) expresses the amplitude point spread function $F(u'_S,v'_T)$ in terms of pupil function $f(x'_S,y'_T)$ when both of them are referred to unrotated axes (u'_S,v'_T) and (x'_S,y'_T), respectively, i.e. when $\psi' = \psi = 0$. The relation (18.110) expresses the amplitude point spread function $F(\tilde{u}'_S,\tilde{v}'_T)$ in terms of the pupil function $f(\tilde{x}'_S,\tilde{y}'_T)$ when both of them are referred to rotated axes $(\tilde{u}'_S,\tilde{v}'_T)$ and $(\tilde{x}'_S,\tilde{y}'_T)$, respectively. Similarity in form of the two expressions, namely (19.52) and (18.110) should be noted. This implies that results obtained for line structures in the azimuth $\psi = 0$ may be written in the same form to apply to any other azimuth.

In (18.92), we showed that the line spread function of a line object placed on an extra-axial object point Q along the v_T axis is given by

$$G(u'_S) = \int_{-\infty}^{+\infty} G(u'_S, v'_T) dv'_T \tag{18.111}$$

For the sake of clarity, in what follows, we use a different notation, $L(u'_S)$, for the line spread function. It is iterated that $L(u'_S) = G(u'_S)$. We have

$$G(u'_S, v'_T) = \left|F(u'_S, v'_T)\right|^2 = F(u'_S, v'_T) F^*(u'_S, v'_T) \tag{18.112}$$

where * denotes the complex conjugate.

Using (18.52) in (18.112), the line spread function, $L(u'_S)$, corresponding to the case $\psi = 0$, can be expressed as

$$\begin{aligned} L(u'_S) = \int_{-\infty}^{+\infty} &\left\{ \iint_{-\infty}^{+\infty} f(x'_{S_1}, y'_{T_1}) \exp\left[i2\pi\left(u'_S x'_{S_1} + v'_T y'_{T_1}\right)\right] dx'_{S_1} dy'_{T_1} \right\} \\ &\times \left\{ \iint_{-\infty}^{+\infty} f(x'_{S_2}, y'_{T_2}) \exp\left[i2\pi\left(u'_S x'_{S_2} + v'_T y'_{T_2}\right)\right] dx'_{S_2} dy'_{T_2} \right\}^* dv'_T \end{aligned} \tag{18.113}$$

This may be rewritten as

$$\begin{aligned} L(u'_S) = \int_{y'_{T_1}=-\infty}^{+\infty} &\left\{ \int_{x'_{S_1}=-\infty}^{+\infty} f(x'_{S_1}, y'_{T_1}) \exp\left[i2\pi u'_S x'_{S_1}\right] dx'_{S_1} \right\} \int_{y'_{T_2}=-\infty}^{+\infty} \left\{ \int_{x'_{S_2}=-\infty}^{+\infty} f(x'_{S_2}, y'_{T_2}) \exp\left[i2\pi u'_S x'_{S_2}\right] dx'_{S_2} \right\}^* \\ &\times \left\{ \int_{v'_T=-\infty}^{+\infty} \exp\left[i2\pi\left(y'_{T_1} - y'_{T_2}\right) v'_T\right] dv'_T \right\} dy'_{T_1} dy'_{T_2} \end{aligned} \tag{18.114}$$

Now the integration over v'_T yields

$$\int_{v'_T=-\infty}^{+\infty} \exp\left[i2\pi\left(y'_{T_1} - y'_{T_2}\right) v'_T\right] dv'_T = \delta(y'_{T1} - y'_{T2}) \tag{18.115}$$

Substituting this in (18.114) and carrying out the integration over y'_{T2}, we get

$$L(u'_S) = \int_{-\infty}^{+\infty} \left\{ \int_{-\infty}^{+\infty} f(x'_{S_1}, y'_{T_1}) \exp\left[i2\pi u'_S x'_{S_1}\right] dx'_{S_1} \right\} \left\{ \int_{-\infty}^{+\infty} f(x'_{S_2}, y'_{T_1}) \exp\left[i2\pi u'_S x'_{S_2}\right] dx'_{S_2} \right\}^* dy'_{T_1} \tag{18.116}$$

Combining the products of the integrand and omitting the superfluous subscripts, the line spread function for a line imaged in the azimuth $\psi = 0$ is given by

$$L(u'_S) = \int_{-\infty}^{+\infty} \left| \int_{-\infty}^{+\infty} f(x'_S, y'_T) \exp(i2\pi u'_S x'_S) \, dx'_S \right|^2 dy'_T \tag{18.117}$$

For a line placed in the azimuth ψ in an extra-axial region, the line spread function will be of the same form as (18.117), and it can be expressed in terms of the rotated pupil function as

$$\tilde{L}(\tilde{u}'_S) = \int_{-\infty}^{+\infty} \left| \int_{-\infty}^{+\infty} \tilde{f}(\tilde{x}'_S, \tilde{y}'_T) \exp(i2\pi \tilde{u}'_S \tilde{x}'_S) d\tilde{x}'_S \right|^2 d\tilde{y}'_T \tag{18.118}$$

The OTF for a one-dimensional structure in the azimuth ψ, $D(s;\psi)$ is given by the normalized inverse Fourier transform of the intensity line spread function, $\tilde{L}(\tilde{u}'_S)$ as

$$D(s;\psi) = \frac{1}{Q} \int_{-\infty}^{+\infty} \tilde{L}(\tilde{u}'_S) \exp\{-i2\pi s \tilde{u}'_S\} d\tilde{u}'_S \tag{18.119}$$

Now, the line spread function $\tilde{L}(\tilde{u}'_S)$ can be expressed in terms of the point spread function $\tilde{G}(\tilde{u}'_S, \tilde{v}'_T)$ as

$$\tilde{L}(\tilde{u}'_S) = \int_{-\infty}^{+\infty} \tilde{G}(\tilde{u}'_S, \tilde{v}'_T) d\tilde{v}'_T \tag{18.120}$$

The equation (18.119) may be written as

$$D(s;\psi) = \frac{1}{Q} \int\int_{-\infty}^{+\infty} \tilde{G}(\tilde{u}'_S, \tilde{v}'_T) \exp\{-i2\pi s \tilde{u}'_S\} d\tilde{u}'_S d\tilde{v}'_T \tag{18.121}$$

We note from (18.105) that

$$\tilde{u}'_S = u'_S \cos\psi' + v'_T \sin\psi' \tag{18.122}$$

Changing variables from $(\tilde{u}'_S, \tilde{v}'_T)$ to (u'_S, v'_T), (18.121) is written as

$$D(s;\psi) = \frac{1}{Q} \int\int_{-\infty}^{+\infty} G(u'_S, v'_T) \exp\{-i2\pi[(s\cos\psi)u'_S + (s\sin\psi)v'_T]\} du'_S dv'_T \tag{18.123}$$

$G(u'_S, v'_T)$ is the intensity PSF referred to the unrotated axes. Note that the normalizing factor Q is

$$Q = \int\int_{-\infty}^{+\infty} \tilde{G}(\tilde{u}'_S, \tilde{v}'_T) d\tilde{u}'_S d\tilde{v}'_T = \int\int_{-\infty}^{+\infty} G(u'_S, v'_T) du'_S dv'_T \tag{18.124}$$

and it is same as the normalizing factor given in (18.87) – (18.88).

From (18.85), (18.86), and (18.123) it is obvious that the one-dimensional optical transfer function for structures imaged in the azimuth ψ is related with the two-dimensional transfer function referred to the unrotated axes by the relation

$$D(s;\psi) = D(s\cos\psi, s\sin\psi) \tag{18.125}$$

From (18.100), it follows

$$D(s;\psi) = \frac{1}{Q} \int\int_{-\infty}^{+\infty} f\left(x'_S + \frac{s\cos\psi}{2}, y'_T + \frac{s\sin\psi}{2}\right) f^*\left(x'_S - \frac{s\cos\psi}{2}, y'_T - \frac{s\sin\psi}{2}\right) dx'_S dy'_T \tag{18.126}$$

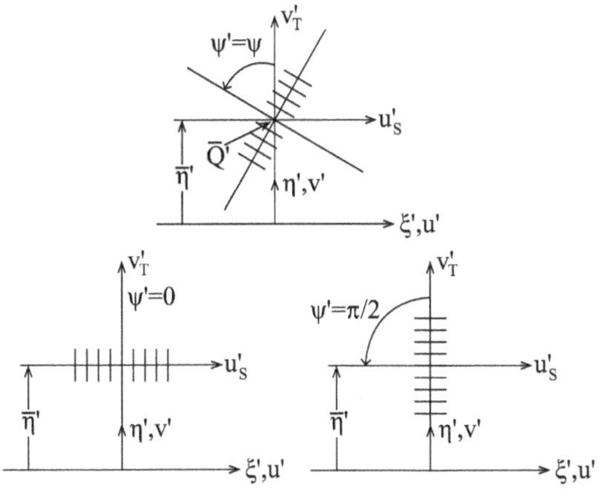

FIGURE 18.15 Orientations in the image of a one-dimensional periodic object/grating placed at an extra-axial location. Top: rotated at azimuth ψ; Bottom right: Tangential: $\psi = \frac{\pi}{2}$; Bottom left: Sagittal: $\psi = 0$.

This relation reduces to (18.101) in the special case $\psi = 0$. The latter is called sagittal OTF. The other special case is tangential OTF where $\psi = \frac{\pi}{2}$. The top diagram in Figure 18.15 shows the image of a one-dimensional periodic grating/object placed at the extra-axial point $\bar{Q}$ on the object plane. The grating/object is placed at an azimuth ψ in reduced coordinates (u_S, v_T). This means the equal intensity lines of the grating/object are inclined relative to the v_T axis at an angle ψ in the anticlockwise direction. The geometrical image of the point $\bar{Q}$ is at $\bar{Q}'$, and in reduced coordinates (u'_S, v'_T), the azimuth of the equal intensity lines in the image is $\psi' = \psi$. When the equal intensity lines are perpendicular to the v'_T axis, i.e. parallel with the u'_S axis, $\psi = \frac{\pi}{2}$, and the orientation is called 'tangential' (Bottom right diagram in Figure 18.15). The orientation is said to be 'sagittal' when the equal intensity lines are parallel with the v'_T axis, i.e. perpendicular to the u'_S axis, so that $\psi = 0$ (Bottom left diagram in Figure 18.15).

Denoting the rotated axes $(\tilde{u}'_S, \tilde{v}'_T)$ by (u', v'), and the rotated pupil coordinates $(\tilde{x}'_S, \tilde{y}'_T)$ by (x', y'), relation (18.110) for the amplitude point spread function can be rewritten as

$$F_\psi(u', v') = \int_{-\infty}^{+\infty}\int f_\psi(x', y') \exp\{i2\pi(u'x' + v'y')\} dx'dy' \tag{18.127}$$

Note that $f_\psi(x', y')$ is the rotated pupil function denoted previously by $\tilde{f}(\tilde{x}'_S, \tilde{y}'_T)$.

Since $x'_S = x_S, y'_T = y_T$ for a specific ray in a nominally isoplanatic system, a further simplification in notation may be obtained by omitting the primes and writing $x = x', y = y'$. Thus, (18.127) can be written as

$$F_\psi(u', v') = \int_{-\infty}^{+\infty}\int f_\psi(x, y) \exp\{i2\pi(u'x + v'y)\} dxdy \tag{18.128}$$

Using (18.108), the rotated pupil function $f_\psi(x, y)$ can be written in terms of the ordinary unrotated pupil function referred to the axes (x'_S, y'_T) as

$$f_\psi(x, y) = f(x\cos\psi - y\sin\psi, y\cos\psi + x\sin\psi) \tag{18.129}$$

Using the simplified notation, as above, the intensity line spread function for a line in the azimuth ψ, given by equation (18.118), can be written as

$$L(u';\psi)=\int_{y=-\infty}^{+\infty}\left|\int_{x=-\infty}^{+\infty} f_{\psi}(x,y)\exp(i2\pi u'x)dx\right|^2 dy \tag{18.130}$$

where $f_{\psi}(x,y)$ is the rotated pupil function. Substituting from (18.129), we get the following expression for line spread function for a line in the azimuth ψ in terms of unrotated pupil function $f(x,y)$ referred to the original axes

$$L(u';\psi)=\int_{-\infty}^{+\infty}\left|\int_{-\infty}^{+\infty} f\{(x\cos\psi-y\sin\psi),(y\cos\psi+x\sin\psi)\}\exp(i2\pi u'x)dx\right|^2 dy \tag{18.131}$$

With the simplified notation, the equation (18.126) expressing the one-dimensional transfer function for the azimuth ψ in terms of the unrotated pupil function may be rewritten as

$$D(s,\psi)=\frac{1}{Q}\int\int_{-\infty}^{+\infty} f\left(x+\frac{s\cos\psi}{2},y+\frac{s\sin\psi}{2}\right)f^*\left(x-\frac{s\cos\psi}{2},y-\frac{s\sin\psi}{2}\right)dxdy \tag{18.132}$$

where the normalizing factor Q is given by

$$Q=\int\int_{-\infty}^{+\infty}|f(x,y)|^2\,dxdy \tag{18.133}$$

Expressions for the two special cases of importance are given next in terms of the simplified notation.

The sagittal OTF:

$$D(s;0)=\frac{1}{Q}\int\int_{-\infty}^{+\infty} f\left(x+\frac{s}{2},y\right)f^*\left(x-\frac{s}{2},y\right)dxdy \tag{18.134}$$

The tangential OTF:

$$D\left(s;\frac{\pi}{2}\right)=\frac{1}{Q}\int\int_{-\infty}^{+\infty} f\left(x,y+\frac{t}{2}\right)f^*\left(x,y-\frac{t}{2}\right)dxdy \tag{18.135}$$

18.5.2.1 OTF in Incoherent Diffraction Limited Imaging Systems

Using simplified notation $x=x'_S, y=y'_T$, the pupil function in an axially symmetric system with a circular pupil is represented by

$$\begin{aligned} f(x,y)&=\tau(x,y)\exp[ikW(x,y)] \quad &(x^2+y^2)\le 1,\\ &=0, &(x^2+y^2)>1 \end{aligned} \tag{18.136}$$

For a diffraction limited system, $W(x,y)=0$. Assuming uniform amplitude over the pupil, we put $\tau(x,y)=1$, and the pupil function of the diffraction limited Airy pupil is

$$\begin{aligned} f(x,y)&=1, \quad (x^2+y^2)\le 1\\ &=0, \quad (x^2+y^2)>1 \end{aligned} \tag{18.137}$$

The frequency response formula (18.132) for OTF reduces to

$$D(s;\psi)=\frac{1}{Q}\iint_{-\infty}^{+\infty} f\left(x+\frac{s\cos\psi}{2}, y+\frac{s\sin\psi}{2}\right) f^*\left(x-\frac{s\cos\psi}{2}, y-\frac{s\sin\psi}{2}\right)dxdy=\frac{1}{Q}\iint_S dxdy \tag{18.138}$$

where S is the region of overlap of the two relatively displaced pupils, centred on the points $\left(\pm\frac{s\cos\psi}{2}, \pm\frac{s\sin\psi}{2}\right)$, respectively. Note that the pupils are unit circles.

For this pupil, the normalizing factor Q given in (18.107) becomes

$$Q=\iint_{-\infty}^{+\infty}|f(x,y)|^2 dxdy=\iint_A dxdy=\int_0^{2\pi}\int_0^1 r dr d\theta=\pi \tag{18.139}$$

A is the domain of the pupil that is a circular aperture of the unit radius. Figure 18.16 illustrates the domain of the pupil A, the displaced pupils A_1, A_2, and the region of integration S for the case $\psi=0$. From (18.138) and (18.139), we note that the normalized one-dimensional intensity transfer function, or OTF, $D(s;\psi)$, along the azimuth ψ is given by

$$D(s;\psi)=\frac{1}{Q}\iint_S dxdy=\frac{1}{\pi}[2\beta-\sin 2\beta] \tag{18.140}$$

where

$$\beta=\cos^{-1}\left|\frac{s}{2}\right| \tag{18.141}$$

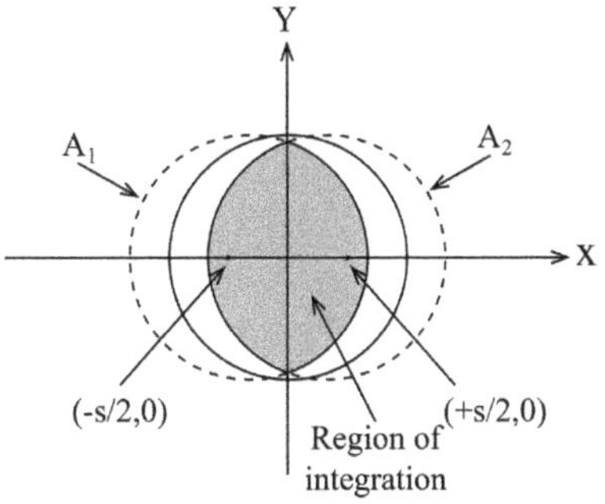

FIGURE 18.16 The unit circle pupil function A, and the region of integration in the frequency response integral for the case $\psi = 0$.

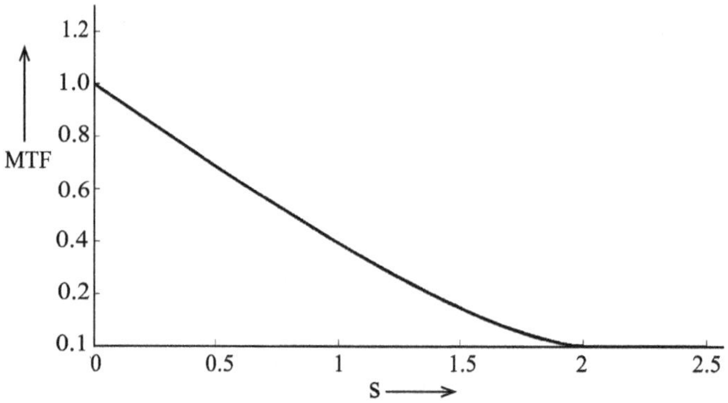

FIGURE 18.17 Modulation Transfer Function (MTF) of a diffraction limited axially symmetric imaging system with circular pupil.

Further, for diffraction limited systems with Airy pupil, on account of symmetry, $D(s)$ is same for any azimuth,

$$D(s) = D(s,0) = D\left(s, \frac{\pi}{2}\right) = D(s, \psi) \tag{18.142}$$

Often, a special notation, $D_0(s)$, is used for the MTF of diffraction limited systems. Figure 18.17 shows the modulation transfer function (MTF) for diffraction limited systems operating in incoherent illumination. Note that the phase transfer function (PTF) is zero for any value of frequency s in this case.

A simple approximation to this diffraction limited MTF was reported by Hufnagel [139]. In our notation, it can be represented by

$$D(s) \cong 1 - 0.625s + 0.015625s^4 \tag{18.143}$$

For any value of s over $0 \le s \le 2$, the accuracy is ± 0.003.

The cut-off frequency is $s_{cutoff} = 2$. From (18.3.3), the period of the highest resolved spatial frequency is

$$\Delta\xi'^{incoherent}_{min} = \frac{0.5\lambda}{n' \sin\alpha'} \tag{18.144}$$

This is the classical diffraction limit for the least resolvable distance in systems operating under incoherent illumination.

In Section 18.5.1.1, it is seen that the cut off frequency in the case of imaging systems operating under coherent illumination is $s^{coherent}_{cutoff} = 1$. The least resolvable distance corresponding to this reduced spatial frequency is

$$\Delta\xi'^{coherent}_{min} = \frac{\lambda}{n' \sin\alpha'} \tag{18.145}$$

In an infinite conjugate imaging system with the object at infinity, $F_{\#} = [1/(2\sin\alpha')]$.

Assuming $n' = 1$, and $\lambda = 0.5\mu m$, the highest resolvable real space frequency in the image under coherent illumination is

$$R^{coherent}_{max} = \frac{1}{\Delta\xi'^{coherent}_{min}} = \frac{n' \sin\alpha'}{\lambda} = \left(\frac{1000}{F_{\#}}\right) \text{cycles/mm} \tag{18.146}$$

With the same system configuration, the highest resolvable real space frequency in the image under incoherent illumination is given by

$$R_{max}^{incoherent} = \frac{1}{\Delta\xi'^{incoherent}_{min}} = \frac{n'\sin\alpha'}{0.5\lambda} = \left(\frac{2000}{F_{\#}}\right) \text{cycles / mm} \tag{18.147}$$

From (18.146) and (18.147), it is seen that $R_{max}^{incoherent}$ is twice the value of $R_{max}^{coherent}$. It would, however, be wrong to conclude that the resolving power of an optical imaging system operating under incoherent illumination is double that of an optical imaging system operating under coherent illumination.

> A major flaw in the above argument lies in the direct comparison of the cut off frequencies in the two cases. Actually, the two are not directly comparable, since the cutoff of the amplitude transfer function determines the maximum frequency component of the image amplitude while the cutoff of the optical transfer function determines the maximum frequency component of the image intensity.
>
> [140]

18.5.2.2 Effects of Residual Aberrations on OTF in Incoherent Systems

For an imaging system with residual aberration $W(x,y)$, the pupil function is

$$\left.\begin{aligned} f(x,y) &= \exp\left[ikW(x,y)\right], \quad && \left(x^2+y^2\right)\le 1 \\ &= 0, && \left(x^2+y^2\right) > 1 \end{aligned}\right\} \tag{18.148}$$

assuming a uniform real amplitude of value unity over the pupil.

Note that in reduced coordinates (x,y), elucidated earlier, the pupil is a unit circle, whatever the actual shape of the pupil may be. From (18.132), the normalized one-dimensional OTF along the azimuth ψ is given by

$$D(s;\psi) = \frac{1}{Q}\iint_S \exp\left[ikW\left(x+\frac{s\cos\psi}{2}, y+\frac{s\sin\psi}{2}\right)\right]\times\exp\left[-ikW\left(x-\frac{s\cos\psi}{2}, y-\frac{s\sin\psi}{2}\right)\right]dxdy \tag{18.149}$$

where S is the region of overlap of the two sheared unit circles.

From (18.139), it is seen that $Q = \pi$. The relation (18.149) may be rewritten as

$$\begin{aligned} D(s;\psi) &= \frac{1}{\pi}\iint_S \exp\left\{ik\left[W\left(x+\frac{s\cos\psi}{2}, y+\frac{s\sin\psi}{2}\right) - W\left(x-\frac{s\cos\psi}{2}, y-\frac{s\sin\psi}{2}\right)\right]\right\}dxdy \\ &= \frac{1}{\pi}\iint_S \exp\left\{iksV(x,y;s,\psi)\right\}dxdy \end{aligned} \tag{18.150}$$

where $V(x,y;s,\psi)$ is called the wave aberration difference function. It is defined by

$$V(x,y;s,\psi) = \frac{1}{s}\left[W\left(x+\frac{s\cos\psi}{2}, y+\frac{s\sin\psi}{2}\right) - W\left(x-\frac{s\cos\psi}{2}, y-\frac{s\sin\psi}{2}\right)\right] \tag{18.151}$$

Let us recall the statement of Schwartz's inequality [141–144]. For any two functions $f_1(x)$ and $f_2(x)$, the following inequality holds good.

$$\left|\int_a^b f_1(x) f_2(x)\,dx\right|^2 \le \int_a^b \left[f_1(x)\right]^2 dx \int_a^b \left[f_2(x)\right]^2 dx \qquad (18.152)$$

Extending the relation to two dimensions, we get

$$\left|\iint_{\wp} f_1(x,y)\, f_2(x,y)\,dxdy\right|^2 \le \iint_{\wp} \left|f_1(x,y)\right|^2 dxdy \iint_{\wp} \left|f_2(x,y)\right|^2 dxdy \qquad (18.153)$$

Note that limits of integration in the definite integrals are the same for all integrals in (18.152) and (18.153).

Let

$$\left.\begin{aligned} f_1(x,y) &= \exp\left[ikW\left(x+\frac{s\cos\psi}{2}, y+\frac{s\sin\psi}{2}\right)\right], && \left(x+\frac{s\cos\psi}{2}\right)^2+\left(y+\frac{s\sin\psi}{2}\right)^2 \le 1 \\ &= 0, && \left(x+\frac{s\cos\psi}{2}\right)^2+\left(y+\frac{s\sin\psi}{2}\right)^2 > 1 \end{aligned}\right\} \qquad (18.154)$$

and

$$\left.\begin{aligned} f_2(x,y) &= \exp\left[-ikW\left(x-\frac{s\cos\psi}{2}, y-\frac{s\sin\psi}{2}\right)\right], && \text{for } \left(x-\frac{s\cos\psi}{2}\right)^2+\left(y-\frac{s\sin\psi}{2}\right)^2 \le 1 \\ &= 0, && \text{for } \left(x-\frac{s\cos\psi}{2}\right)^2+\left(y-\frac{s\sin\psi}{2}\right)^2 > 1 \end{aligned}\right\} \qquad (18.155)$$

Substituting for $f_1(x,y)$ and $f_2(x,y)$ from (18.154) and (18.155), the left-hand side of the relation (18.153) becomes

$$\begin{aligned} &\left|\iint_S \exp\left\{ik\left[W\left(x+\frac{s\cos\psi}{2}, y+\frac{s\sin\psi}{2}\right) - W\left(x-\frac{s\cos\psi}{2}, y-\frac{s\sin\psi}{2}\right)\right]\right\} dxdy\right|^2 \\ &= \left[\pi\, D(s;\psi)\right]^2 \end{aligned} \qquad (18.156)$$

where S is the common area of overlap of the two mutually displaced unit circles with centres at $\left(\pm\frac{s\cos\psi}{2}, \pm\frac{s\sin\psi}{2}\right)$.

Similarly, by these substitutions, the right-hand side of the relation (18.153) becomes

$$\left|\iint_S dxdy\right|^2 = \left[\pi D_0(s)\right]^2 \qquad (18.157)$$

$D_0(s)$ is the normalized OTF for an aberration-free diffraction limited imaging system.

By Schwartz's inequality, expressed by (18.153), we get from the above relations

$$D(s;\psi) \le D_0(s) \qquad (18.158)$$

Therefore, in any optical imaging system with residual aberrations, the OTF for any frequency at any azimuth is less than the OTF for the same frequency of a diffraction limited aberration-free system, except for $s = 0$ and $s = 2$. At these two frequencies, aberrations have no effect on the corresponding OTF values. At $s = 0$, the two displaced unit circles overlap, and from (18.156) and (18.157), we get

$$D(0;\psi) = \frac{1}{\pi}\iint_A dxdy = D_0(0) = 1 \tag{18.159}$$

where A is the area of the unit circle. In general, by using Schwartz's inequality, it can be shown that, for any circular pupil $f(x,y)$, $D(s;\psi)$ cannot be greater than unity.

At $s = 2$, there is no overlap of the displaced unit circles, so that

$$D(2;\psi) = D_0(2) = 0 \tag{18.160}$$

Also, it can also be seen that the slope at $s = 0$ for the OTF curve is independent of the amount of aberration in the system

$$\left[\frac{d}{ds}D(s,\psi)\right]_{s=0} = \left[\frac{d}{ds}D_0(s)\right]_{s=0} = -\frac{2}{\pi} \tag{18.161}$$

'This result provides an explanation of why the coarser structure in an object may be discerned even in very poorly corrected systems' [32].

The value of the integral (18.150) is determined by the range of variation of the argument of the integrand. When the aberration is large, the aberration difference function is also large, so that the OTF is only appreciably different from zero for smaller values of s. Therefore, finer details are not revealed in the image with good contrast. Imaging of finer details with good contrast necessitates smaller values of aberrations of the system.

18.5.2.3 Effects of Defocusing on OTF in Diffraction Limited Systems

In Chapter 8 it was discussed that in the case of a not-too-large aperture or field, 'defocusing' or 'longitudinal shift of focus' can be expressed as an aberration term

$$W(x,y) = W_{20}\left(x^2 + y^2\right) \tag{18.162}$$

where the 'defocusing aberration' coefficient W_{20} is related with the axial shift $\delta z'$ of the image plane by

$$W_{20} = \frac{1}{2}n'(\sin\alpha')^2\,\delta z' \tag{18.163}$$

The pupil function is

$$\left.\begin{aligned} f(x,y) &= exp\left[ikW_{20}\left(x^2+y^2\right)\right], & \left(x^2+y^2\right) \le 1 \\ &= 0, & \left(x^2+y^2\right) > 1 \end{aligned}\right\} \tag{18.164}$$

On account of rotational symmetry of the pupil function, the normalized one-dimensional OTF $D(s;\psi)$ is independent of ψ, so that

$$D(s) = D(s;\psi) = D(s,0) = \frac{1}{\pi}\iint_S \exp\left[ik\left\{W\left(x+\frac{s}{2},y\right) - W\left(x-\frac{s}{2}\right)\right\}\right]dxdy \tag{18.165}$$

Using (18.162), for 'defocusing aberration', (18.165) reduces to

$$D(s) = \frac{1}{\pi}\iint_S \exp[i\gamma x]\,dx\,dy \tag{18.166}$$

where

$$\gamma = kW_{20}2|s| = \frac{4\pi W_{20}}{\lambda}|s| \tag{18.167}$$

A semi-analytical formula for evaluating the integral was given by Hopkins [21]. It is given by

$$\begin{aligned} D(s) = {} & \frac{4}{\pi\gamma}\cos\left(\frac{1}{2}\gamma|s|\right)\left\{\beta J_1(\gamma) + \sum_{n=1}^{\infty}(-1)^{n+1}\frac{\sin(2n\beta)}{2n}\left[J_{2n-1}(\gamma) - J_{2n+1}(\gamma)\right]\right\} \\ & -\frac{4}{\pi\gamma}\sin\left(\frac{1}{2}\gamma|s|\right)\left\{\sum_{n=0}^{\infty}(-1)^{n}\frac{\sin(2n+1)\beta}{2n+1}\left[J_{2n}(\gamma) - J_{2(n+1)}(\gamma)\right]\right\} \end{aligned} \tag{18.168}$$

where $\beta = \cos^{-1}\left|\frac{s}{2}\right|$, and $J_m(p)$ is a Bessel function of the first kind, order m, and argument p. The two series involving the difference of Bessel functions of different orders are slowly converging. Because of symmetry of the system, $D(s) = D(-s)$, and so the modulus of s, $|s|$ is used in the above relation. Figure 18.18 shows the changes in MTF of a diffraction limited lens at different amounts of defocusing. The curves, labelled by integers $m = 0,1,2,\ldots$etc., correspond to defocusing by amounts $(m\lambda/\pi)$ introduced between the edge and the centre of the aperture. Note that, on account of symmetry, the OTF is real for all values of s, and the value of OTF is either positive, zero, or negative. A negative value of OTF at a frequency implies 'contrast reversal' at that frequency compared to the contrast at a frequency where the value of OTF is positive. This corresponds to a value of π for the PTF at that frequency.

Stokseth [48] gave the following empirical formula in place of the semi-analytical formula (18.168).

$$\left.\begin{aligned} D(s) &= 2\left(1 - 0.69s + 0.0076s^2 + 0.043s^3\right)\frac{J_1\{\gamma(1-0.5s)\}}{\{\gamma(1-0.5s)\}}, && \text{for } 0 \le s < 2 \\ &= 0, && \text{for } s \ge 2 \end{aligned}\right\} \tag{18.169}$$

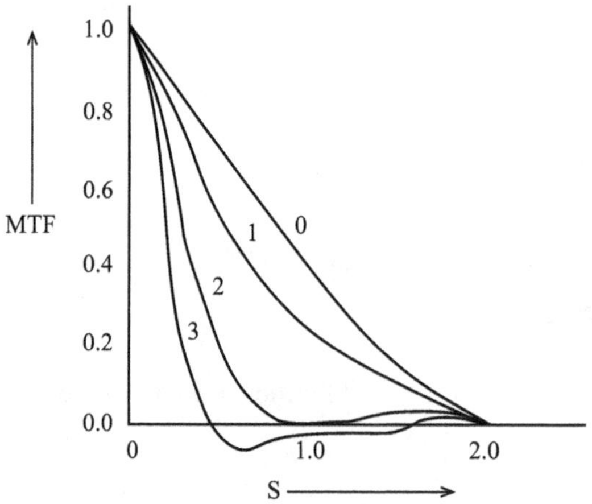

FIGURE 18.18 Effect of defocusing on the MTF of a diffraction limited axially symmetric imaging system with circular pupil.

where $\gamma = \frac{4\pi}{\lambda} W_{20} s$. This formula gives accurate results for values of $|W_{20}| \geq 5\lambda$. The accuracy becomes poorer with the decrease in value of defocusing.

In the mid-1950s, Steel [145] proposed an empirical formula for estimating MTF in the case of small defocusing. About thirty years later, Shannon [146] revived interest on the formula. In our notation, the empirical expression for the MTF $D(s)$ on a plane axially shifted by $\delta z'$ can be expressed as

$$D(s) = D_0(s) \times DMF(s). \tag{18.170}$$

where $D_0(s)$ is the MTF for the system in the absence of aberrations, i.e. the MTF for the diffraction limited system, and $DMF(s)$ is a multiplying factor for defocusing.

For a circular pupil with radius unity, from (18.140) we get

$$D_0(s) = \frac{1}{\pi}[2\beta - \sin 2\beta] \quad \text{where } \beta = \cos^{-1}\left|\frac{s}{2}\right| \tag{18.171}$$

The defocus multiplying factor $DMF(s)$ is given by

$$DMF(s) = \frac{2J_1(2\pi\chi)}{(2\pi\chi)} \tag{18.172}$$

where

$$\chi = s(2-s)\frac{W_{20}}{\lambda} \tag{18.173}$$

From (18.163), we get

$$W_{20} = \frac{1}{2} n'(\sin\alpha')^2 \delta z' \tag{18.174}$$

Assuming $n' = 1$, and using the relation $F_{\#} \cong 1/[2(N.A.)]$, W_{20} can also be represented by

$$W_{20} = \frac{\delta z'}{8(F_{\#})^2} \tag{18.175}$$

Using (18.170) for predicting MTF on defocused image planes, yield an accuracy of 0.017 over the range of defocus from zero to 2.2 waves (0.6 wave, rms). Values of MTF for aberration-free diffraction limited lens with circular aperture are given with six decimal place accuracy for a range of defocusing, as presented by Levi and Austing [62]. Incidentally, in the same paper, they also published values of LSF on the plane of geometrical focus to six decimal place accuracy for an axially symmetric diffraction limited lens with circular pupil and the line passing through the axis in the object plane.

18.5.2.4 OTF in Incoherent Imaging Systems with Residual Aberrations

The influence of residual aberrations on the OTF of an optical imaging system is, in general, a reduction in value of the modulation or contrast transfer function at all frequencies. In the presence of any asymmetrical aberration, there may also be a phase transfer function leading to a phase shift of the particular

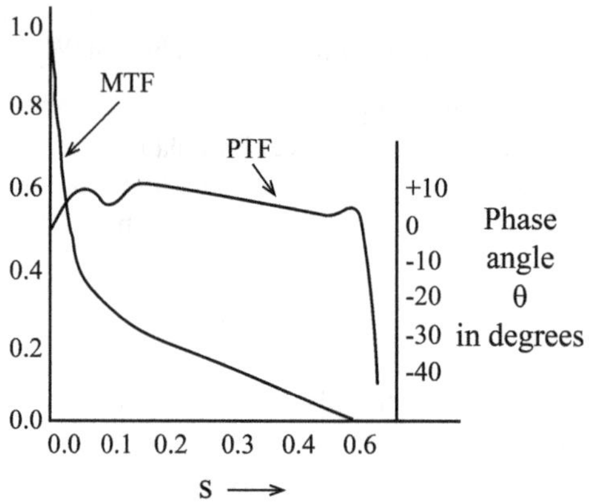

FIGURE 18.19 Modulation Transfer Function (MTF) and Phase Transfer Function (PTF) of a 90mm F/5.6 lens at *35⁰* field.

frequency components. Extensive studies on effects of spherical aberration, coma, and astigmatism on the OTF have been carried out, and we refer to the classic publications in this regard [18.22–18.33, 18.57–18.60].

In practice, the residual aberrations of any system consist of a composite of aberrations of different types and orders. For each field point, the aberration polynomial has to be determined by tracing rays through the system from the object to the image plane under consideration. If the aberration function determined thereby is a symmetrical function, the pupil function becomes symmetrical, and the OTF becomes real, so that the phase transfer function $\theta(s)$ is either zero, or π. In cases where the aberration function becomes asymmetrical, it gives rise to other values for phase transfer function. However, the effect of a linear phase term is a shift of the total image, and this in no way affects the quality of the image. However, the presence of a nonlinear component in the phase term affects the quality of the image significantly because the mutual superposition of different frequency components in the image gives rise to loss of sharpness and clarity in the image. Therefore, in general, a good MTF in an imaging system by itself is no guarantee for good quality of the image, unless the value of the PTF is within tolerable limits. As an illustrative example, Figure 18.19 shows the MTF and the PTF for imaging by a 90mm F/5.6 lens at 35^0 field [147].

Rosenhauer et al. studied the relations between the axial aberrations of photographic lenses and their optical transfer function [148]. In particular, they made significant observations on the effects of stopping down on OTF in cases of overcorrected and undercorrected lenses.

Shannon [149–150] gave an empirical formula for average MTF $D(s)$. In our notation, the formula can be expressed as

$$D(s) = D_0(s) \times Ab(s) \tag{18.176}$$

where $D_0(s)$ is the MTF of aberration-free diffraction limited system and $Ab(s)$ is a multiplying factor depending upon the root mean square (rms) wavefront aberration, W_{rms}. Shannon formula cover the range of aberrations from zero to almost 0.5 waves rms error. From (18.140), for a circular aperture with radius unity, $D_0(s)$ is given by

$$D_0(s) = \frac{1}{\pi}[2\beta - \sin 2\beta] \quad \text{where } \beta = \cos^{-1}\left|\frac{s}{2}\right| \tag{18.177}$$

The multiplying factor $Ab(s)$ is taken as

$$Ab(s) = 1 - \left[\left(\frac{W_{rms}}{0.18}\right)^2 \left\{1 - 4\left(\left|\frac{s}{2}\right| - 0.5\right)^2\right\}\right] \tag{18.178}$$

Note that, for $W_{rms} = 0.18$, at $s = 0,1,2$, values of $Ab(s) = 1, 0, 1$, respectively. Formula (18.176) provides a reasonable level of prediction for small aberrations, but these are average values with the possibility of large variations in some cases.

18.5.2.5 Effects of Nonuniform Real Amplitude in Pupil Function on OTF

We recall the expression (18.136) for the pupil function of an axially symmetric system with a circular pupil in terms of reduced or canonical coordinates

$$\left.\begin{aligned} f(x,y) &= \tau(x,y)\exp\left[ikW(x,y)\right], \quad &(x^2+y^2)\le 1 \\ &= 0, &(x^2+y^2)>1 \end{aligned}\right\} \tag{18.179}$$

While considering the effects of aberrations alone on the OTF in the earlier sections, we assumed that the pupil is of uniform real amplitude of value unity, i.e.

$$\left.\begin{aligned} \tau(x,y) &= 1, \quad (x^2+y^2)\le 1 \\ &= 0, \quad (x^2+y^2)>1 \end{aligned}\right\} \tag{18.180}$$

Now, we shall first consider the case of an aberration-free diffraction limited system with nonuniform amplitude over the pupil. The pupil function is then given by

$$\left.\begin{aligned} f(x,y) &= \tau(x,y), \quad &(x^2+y^2)\le 1 \\ &= 0, &(x^2+y^2)>1 \end{aligned}\right\} \tag{18.181}$$

Since the pupil function has no imaginary term, the phase transfer function will be either zero or π for any value of spatial frequency s. The OTF, or more specifically, the MTF is given by

$$\tilde{D}(s;\psi) = \frac{1}{\tilde{Q}}\iint_S \tau\left(x+\frac{s\cos\psi}{2}, y+\frac{s\sin\psi}{2}\right)\tau\left(x-\frac{s\cos\psi}{2}, y-\frac{s\sin\psi}{2}\right)dxdy \tag{18.182}$$

where the normalizing factor $\tilde{Q}$ is given by

$$\tilde{Q} = \tilde{D}(0;\psi) = \iint_S \left[\tau(x,y)\right]^2 dxdy \tag{18.183}$$

By Schwartz's inequality (18.153), we get

$$\begin{aligned} \tilde{I}(s;\psi) &= \left|\iint_S \tau\left(x+\frac{s\cos\psi}{2}, y+\frac{s\sin\psi}{2}\right)\tau\left(x-\frac{s\cos\psi}{2}, y-\frac{s\sin\psi}{2}\right)dxdy\right|^2 \\ &\le \iint_S \left|\tau\left(x+\frac{s\cos\psi}{2}, y+\frac{s\sin\psi}{2}\right)\right|^2 dxdy \iint_S \left|\tau\left(x-\frac{s\cos\psi}{2}, y-\frac{s\sin\psi}{2}\right)\right|^2 dxdy \end{aligned} \tag{18.184}$$

Let the real amplitude of the pupil be

$$\left.\begin{aligned} \tau(x,y) &\le 1, \quad (x^2+y^2)\le 1 \\ &= 0, \quad (x^2+y^2)>1 \end{aligned}\right\} \tag{18.185}$$

For this pupil, the right-hand side of (18.184) follows the inequality

$$\iint_S \left| \tau\left(x + \frac{s\cos\psi}{2}, y + \frac{s\sin\psi}{2} \right) \right|^2 dxdy \iint_S \tau\left(x - \frac{s\cos\psi}{2}, y - \frac{s\sin\psi}{2} \right) dxdy$$
$$\le \left[\iint_S dxdy \right]^2 = I_0(s) \tag{18.186}$$

From (18.184) and (18.186)

$$\tilde{I}(s;\psi) \le I_0(s) \text{ or, } \left[\frac{\tilde{I}(s;\psi)}{I_0(s)} \right] = i < 1 \tag{18.187}$$

When $\tau(x,y)$ is equal to unity, $\tilde{Q}$, as given in (18.183), reduces to

$$Q = \iint_S dxdy = \pi \tag{18.188}$$

$Q = \pi$ is used as a normalizing factor in the calculation of OTF for aberration-free or aberrated imaging systems.

It is significant to note that, by suitable choice of the function $\tau(x,y)$, $\tilde{Q}$ can be made smaller than Q. Therefore, $\left[\tilde{Q}/Q\right] = q < 1$. From (18.187) – (18.188), we get

$$\frac{\tilde{I}(s;\psi)}{\tilde{Q}} \frac{1}{\frac{I_0(s)}{Q}} = \frac{\tilde{D}(s;\psi)}{D_0(s)} \tag{18.189}$$

From the above, it is seen that

$$\left. \begin{aligned} \tilde{D}(s;\psi) > D_0(s), \quad \text{if} \frac{i}{q} > 1 \\ \tilde{D}(s;\psi) < D_0(s), \quad \text{if} \frac{i}{q} < 1 \end{aligned} \right\} \tag{18.190}$$

This implies that, at some frequencies, the normalized MTF of an imaging system with nonuniform amplitude over the pupil can be made higher than the MTF of a diffraction limited system. This is a key reason why mixed filters involving both phase and amplitude are pursued in the tailoring of frequency response characteristics, although use of amplitude filters causes loss of energy in the output.

18.5.2.6 Apodization and Inverse Apodization

The point spread function of a diffraction limited optical imaging system consists of a central circular bright lobe surrounded by rings of gradually diminishing brightness. Per the Rayleigh criterion of resolution, two bright points of equal intensity in a dark background can be resolved when the peaks of their central lobes in the diffraction image are separated by a distance equal to or greater than the radius of the central lobe. But when the brightness of one point is much less than that of the other, the diffraction image of the faint object gets submerged under the secondary or tertiary bright rings of the diffraction image of the bright point. Consequently, the two points cannot be resolved. A remedy is sought via tailoring of the

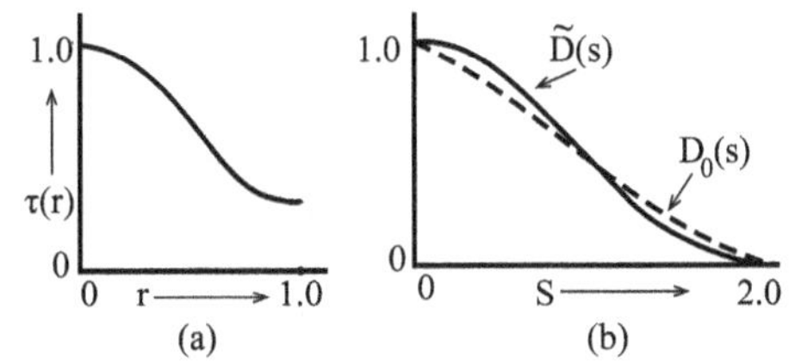

FIGURE 18.20 (a) An amplitude filter with amplitude tapering from centre to edge (b) Corresponding MTF $\tilde{D}(s)$. MTF of a diffraction limited system: $D_0(s)$.

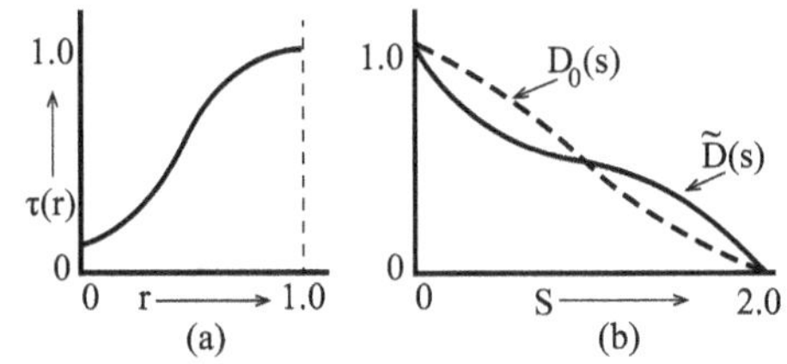

FIGURE 18.21 (a) An amplitude filter with amplitude tapering from edge to centre (b) Corresponding MTF $\tilde{D}(s)$. MTF of a diffraction limited system:$D_0(s)$.

point spread function of the imaging system by using amplitude filters so that the bulges in the secondary or tertiary rings in the diffraction image no longer exist, and the central lobe expands with gradually tapering intensity. As a result, the faint diffraction image of the second point located under the secondary or tertiary rings of the original diffraction pattern stand out distinctly along with the diffraction image of the neighbouring bright point. Common practical examples of this situation occur in astronomy where one has to resolve a faint star in the immediate neighbourhood of a bright star, or in spectroscopy where one has to resolve a very weak spectral line next to a bright spectral line.

Etymologically, the word 'apodization' is derived from the Greek words 'α', meaning 'to take away' and 'ποδοσ' ('padas'), meaning 'feet'. Typically, the amplitude filters that taper the amplitude from the centre to the edge of the pupil result in higher contrast for the coarse detail. Figure 18.20(a) shows the amplitude distribution along an azimuth in a typical axially symmetric amplitude filter of the type mentioned. The corresponding MTF $\tilde{D}(s)$ is shown in Figure 18.20(b) along with the MTF $D_0(s)$ of a diffraction limited system. Note that $\tilde{D}(s) > D_0(s)$ over a lower frequency passband. But $\tilde{D}(s) < D_0(s)$ over the remaining frequency band until $s = 2$, where MTF for both filters become zero.

On the other hand, Figure 18.21(a) shows the amplitude distribution along an azimuth for an amplitude filter in which the amplitude tapers from the edge to the centre. Figure 18.21(b) shows the MTF corresponding to this filter, and also the MTF of a diffraction limited system. Significantly, in this case, $\tilde{D}(s) < D_0(s)$ over the lower frequency passband, and $\tilde{D}(s) > D_0(s)$ over the remaining frequency band until $s = 2$, where the MTF for both filters becomes zero. In the earlier literature, sometimes this effect was called 'superresolution', but, nowadays the term 'superresolution' is used to imply resolution beyond the diffraction limit (see Section 16.4.7). In spite of the gain in MTF over a high frequency passband, the limit of resolution achieved by using the amplitude filter is the diffraction limit specified by $s = 2$. Therefore, it is more appropriate to call this effect 'inverse apodization' to highlight its differences with the conventional 'apodization', as defined earlier [151–160]. Another growing tendency is to use the term 'apodization' for any sort of amplitude filtering on the pupil of an imaging system.

An interesting possibility for improving the quality of image formation in imaging systems with residual aberrations lies in the use of suitable amplitude filters. Also, tailoring the PSF or the OTF of imaging systems by using mixed filters consisting of both amplitude filter and phase filter on the pupil of an image-forming system has attracted the attention of researchers. Details of these are beyond our scope, and we give a partial list of important publications in this area for the interested reader [161–168].

18.6 Aberration Tolerances based on OTF

In Sections 17.2 and 17.3, two different approaches for aberration tolerances are presented. The first one is a comprehensive treatment developed by Maréchal; it is based on the Strehl criterion, which is, in practice equivalent to the Rayleigh criterion of aberration. Maréchal's treatment is suitable for highly corrected systems, e.g. microscope objectives. But, in a large number of practical applications, optical systems that have much larger aberrations are capable of producing good quality images of objects in which the detail is coarse in comparison with the limit of resolution of the system. In many of these systems, the effect of diffraction on the quality of imaging is not significant compared to the effect of residual aberrations of the system, so much so that ray optical analysis is usually sufficient in the tolerancing of aberrations. However, a large number of optical systems belongs to the intermediate domain, where aberrations are not small, and the effects of diffraction are also significant. With a view to develop appropriate aberration tolerances for these systems, a second approach based on the factor of encircled energy in the point spread function was explored by investigators. An outline of this approach was presented in the last chapter.

An alternative route for suitable aberration tolerances based on the concept of the OTF was developed by Hopkins [169].

18.6.1 The Wave Aberration Difference Function

From (18.150), we note that in the case of a pupil represented by a unit circle, the normalized one-dimensional OTF along the azimuth ψ is given by

$$D(s;\psi)\frac{1}{\pi}\iint_{S}\exp\{iksV(x,y;s,\psi)\}dxdy \tag{18.191}$$

where $V(x,y;s,\psi)$ is the wave aberration difference function, expressed as

$$V(x,y;s,\psi)=\frac{1}{s}\left[W\left(x+\frac{s\cos\psi}{2},y+\frac{s\sin\psi}{2}\right)-W\left(x-\frac{s\cos\psi}{2},y-\frac{s\sin\psi}{2}\right)\right] \tag{18.192}$$

In the absence of aberrations, $W(x,y)=0$, and the corresponding OTF is

$$D_0(s;\psi)=\frac{1}{\pi}\iint_{S}dxdy=\frac{S}{\pi} \tag{18.193}$$

S is the area of the region of integration, i.e. the common area of overlap of the two pupils whose centres are displaced to the points $\left(\pm\frac{s\cos\psi}{2},\pm\frac{s\sin\psi}{2}\right)$. $D_0(s;\psi)$ is necessarily real, so there is no phase shift in the absence of aberrations. The relative frequency response of the system at a particular frequency and orientation, $M(s,\psi)\exp\{i\theta(s;\psi)\}$, is defied to be the ratio of the frequency response of the system in the presence of aberration to the response obtained for the same system in the absence of aberration at the same frequency and orientation, i.e.

$$M(s,\psi)\exp\{i\theta(s;\psi)\}=\left[D(s;\psi)/D_0(s,\psi)\right] \tag{18.194}$$

In the above

$$D(s;\psi)=T(s;\psi)\exp\{i\theta(s;\psi)\}\text{ and }D_0(s;\psi)=T_0(s;\psi) \tag{18.195}$$

where s and ψ represent the frequency and the azimuth respectively. $T(s;\psi)$ and $T_0(s;\psi)$ are the contrast values of the same system in the presence and in the absence of aberrations, respectively. $M(s,\psi)$ is equal to the ratio $[T(s;\psi)/T_0(s;\psi)]$. By Schwartz's inequality, the numerical value of $M(s,\psi)$ is always less than unity.

The mean value of the wave aberration difference function $V(x,y;s,\psi)$ over the region S is

$$\overline{V(x,y;s,\psi)} = \frac{1}{S}\iint_S \exp\{iksV(x,y;s,\psi)\}dxdy \tag{18.196}$$

Using the above relations, we get

$$M(s,\psi)\exp\{i\theta(s;\psi)\} = \exp\left\{iks\overline{V(x,y;s,\psi)}\right\}\frac{1}{S}\iint_S \exp\left\{iks\left[V(x,y;s,\psi) - \overline{V(x,y;s,\psi)}\right]\right\}dxdy \tag{18.197}$$

Using shortened notations V and $\bar{V}$ for $V(x,y;s,\psi)$ and $\overline{V(x,y;s,\psi)}$, respectively, (18.160) may be written as

$$M(s,\psi)\exp\{i\theta(s;\psi)\} = \exp\{iks\bar{V}\}\frac{1}{S}\iint_S \exp\{iks[V-\bar{V}]\}dxdy \tag{18.198}$$

The value of $M(s,\psi)$ is equal to the modulus of the integral on the right-hand side of the above equation. Expanding the integrand and retaining terms up to degree 2 in $[V-\bar{V}]$, we get

$$\begin{aligned} M(s,\psi)\exp\{i\theta(s;\psi)\} &\cong \exp\{iks\bar{V}\}\frac{1}{S}\iint_S\left\{1 + iks(V-\bar{V}) - \frac{1}{2}k^2s^2(V-\bar{V})^2\right\}dxdy \\ &= \exp\{iks\bar{V}\}\left\{1 - \frac{2\pi^2s^2}{\lambda^2}K(s,\psi)\right\} \end{aligned} \tag{18.199}$$

where

$$K(s,\psi) = \frac{1}{S}\iint_S V^2dxdy - \left\{\frac{1}{S}\iint_S Vdxdy\right\}^2 \tag{18.200}$$

represents the variance of the wave aberration difference function $V(x,y;s,\psi)$ over the region S. Separating the modulus and argument in (18.199), we get

$$\theta(s;\psi) = iks\overline{V(x,y;s,\psi)} \tag{18.201}$$

$$M(s,\psi) = 1 - \frac{2\pi^2s^2}{\lambda^2}K(s,\psi) \tag{18.202}$$

for the phase shift and relative contrast modulation, respectively.

18.6.2 Aberration Tolerances based on the Variance of Wave Aberration Difference Function

A criterion for the permissible aberration in an optical system can be formulated by the condition

$$M(s,\psi) \geq 0.8 \tag{18.203}$$

It should be noted that the criterion relates to a single frequency s at the orientation ψ. Using (18.202), the tolerance condition may be reformulated as

$$K(s,\psi) \leq \frac{0.1\lambda^2}{\pi^2 s^2} \tag{18.204}$$

When (18.204) is satisfied, the contrast of the image of the component of normalized spatial frequency s in orientation ψ will be not less than 0.80 times the contrast, which would be obtained in the image for the same frequency and orientation in the absence of aberration.

(18.200) gives $K(s,\psi)$, the variance of the wave aberration difference function $V(x,y;s,\psi)$ over the region S as a homogeneous positive quadratic function of the coefficients of the wave aberration polynomial. $K(s,\psi)$ can be minimized with respect to any aberration coefficient W_{mn} by satisfying

$$\frac{\partial K(s,\psi)}{\partial W_{mn}} = 0 \tag{18.205}$$

A set of linear equations is obtained involving the aberration coefficients W_{mn}. This set leads to unique values of the ratios of the aberration coefficients of various orders, including the coefficient of defocus as an aberration, giving a maximum response for a given spatial frequency and orientation.

Using (18.204) – (18.205) as the optimum focal plane, balancing conditions for optimum aberration correction and tolerance on aberration for a given frequency and orientation may be determined. These vary with frequency. Detailed investigations on this were carried out by Hopkins [169]. Results of these investigations reveal that over the spatial frequency range $0.2 \leq s \leq 1.4$, the aberration tolerances and balancing conditions tally fairly well with the Maréchal aberration tolerances and balancing conditions based on the Strehl criterion. For values of spatial frequency over the range $1.4 < s < 2.0$, the OTF for aberration-free systems is very small, so much so that results obtained by the imposition of Hopkins' spatial frequency criterion over this range are prone to numerical errors, and in any case, they are not of much practical importance.

However, two important conclusions may be drawn from these studies. First,

> even when the range of spatial frequencies of interest only extends as far as $s = 0.20$ the degree of correction of aberration required to give 80% of the maximum obtainable contrast at the upper limit will be sufficient to give correspondingly good contrast for the whole frequency range up to the theoretical bandwidth limit $s = 2$
>
> [32]

The second conclusion relates to the case when very good image quality is required, but only for low values of normalized spatial frequency, e.g. for values of s less than 0.1 or 0.2. In such cases, the tolerance theory based on OTF is of great practical value, for the permissible aberrations may then be considerably in excess of the Rayleigh quarter wavelength limit, and the Maréchal treatment is no longer meaningful. In the case of a third category of imaging systems, response for the lower frequencies may be allowed to fall, provided an enhanced response is obtained for the higher frequency range. No useful formulation for the treatment of tolerances exists for such cases. The treatment of aberration tolerancing based on the factor of encircled energy, as developed in Section 17.3.4, may be explored for this purpose.

We quote below the formulae for aberration balancing and aberration tolerances for the spatial frequency range $s < 0.20$ in a few select cases. Details may be found in the publication [169].

A. Depth of focus: $W(r) = W_{20}r^2$ (18.206)

$$(W_{20})_{max} = \pm\frac{0.10\lambda}{s} \tag{18.207}$$

B. Primary spherical aberration with balanced defocusing:

$$W(r) = W_{20}r^2 + W_{40}r^4 \tag{18.208}$$

$$\left.\begin{aligned}(W_{40})_{max} &= \pm\left(\frac{0.212}{s} + 0.33\right)\lambda \\ (W_{20} / W_{40}) &= -(1.33 - 1.10s + 0.70s^2)\end{aligned}\right\} \tag{18.209}$$

C. Secondary spherical aberration with balanced primary spherical aberration and defocusing:

$$W(r) = W_{20}r^2 + W_{40}r^4 + W_{60}r^6 \tag{18.210}$$

$$\left.\begin{aligned}(W_{60})_{max} &= \pm\left(\frac{0.58}{s} + 1.50s\right) \\ (W_{40} / W_{60}) &= -(1.80 - 1.50s + 1.50s^2) \\ (W_{20} / W_{60}) &= +(0.90 - 1.40s + 2.10s^2)\end{aligned}\right\} \tag{18.211}$$

D. Primary coma: $W(r,\phi) = W_{31}r^3 \cos\phi$ (18.212)

$$\left.\begin{aligned}\text{Azimuth } \psi &= \frac{\pi}{2} \\ W_{31} &= \pm\left[\frac{0.142}{s} + 0.16\right]\lambda \\ \theta(s) &= \mp[0.89 + 0.24s]\end{aligned}\right\} \tag{18.213}$$

$$\left.\begin{aligned}\text{Azimuth } \psi &= 0 \\ W_{31} &= \pm\left[\frac{0.246}{s} + 0.19\right]\lambda \\ \theta(s) &= 0\end{aligned}\right\} \tag{18.214}$$

E. Secondary coma with balanced primary coma:

$$W(r,\phi) = W_{31}r^3 \cos\phi + W_{51}r^5 \cos\phi \tag{18.215}$$

$$\left.\begin{aligned} &\text{Azimuth } \psi = \frac{\pi}{2} \\ &W_{51} = \pm\left[\frac{0.37}{s} + 0.83\right]\lambda \\ &(W_{31}/W_{51}) = -\left[1.49 - 1.26s + 0.85s^2\right] \\ &\theta(s) = \mp[1.16 + 0.35s] \end{aligned}\right\} \tag{18.216}$$

$$\left.\begin{aligned} &\text{Azimuth } \psi = 0 \\ &W_{51} = \pm\left[\frac{0.64}{s} + 0.96\right]\lambda \\ &(W_{31}/W_{51}) = -\left[1.49 - 0.94s + 0.41s^2\right] \\ &\theta(s) = 0 \end{aligned}\right\} \tag{18.217}$$

18.7 Fast Evaluation of Variance of the Wave Aberration Difference Function from Finite Ray Trace Data

The wave aberration difference function $V(x,y;s,\psi)$ is defined in (18.151) by

$$V(x,y;s,\psi) = \frac{1}{s}\left[W\left(x + \frac{s\cos\psi}{2}, y + \frac{s\sin\psi}{2}\right) - W\left(x - \frac{s\cos\psi}{2}, y - \frac{s\sin\psi}{2}\right)\right] \tag{18.218}$$

where $x \equiv x'_S$ and $y \equiv y'_T$. The reduced spatial frequency is s, and the azimuth along which the MTF is being evaluated is ψ. The variance $K(s,\psi)$ of the wave aberration difference function over the overlapping region S of the sheared exit pupils with centres at $\left(\frac{s\cos\psi}{2}, \frac{s\sin\psi}{2}\right)$ and $\left(-\frac{s\cos\psi}{2}, -\frac{s\sin\psi}{2}\right)$ is given by

$$K(s,\psi) = \frac{1}{S}\iint_S \left[V(x,y;s,\psi)\right]^2 dxdy - \left\{\frac{1}{S}\iint_S V(x,y;s,\psi)dxdy\right\}^2 \tag{18.219}$$

From (13.12), we get the polynomial expression for $W(x,y)$ as

$$W(x,y) = \sum_m \sum_n W_{mn}\left(x^2 + y^2\right)^{\frac{m-n}{2}} y^n \tag{18.220}$$

For the sake of simplicity, the prime symbol over W is omitted in the above expression. Using the above relation in (18.218), the wave aberration difference function can be written as

$$V(x,y;s,\psi) = \sum_m \sum_n W_{mn} h(m,n;s,\psi) \tag{18.221}$$

where

$$h(m,n;s,\psi)=\frac{1}{s}\left[\begin{array}{l}\left\{\left(x+\frac{s\cos\psi}{2}\right)^2+\left(y+\frac{s\sin\psi}{2}\right)^2\right\}^{\frac{m-n}{2}}\left(y+\frac{s\sin\psi}{2}\right)^n\\-\left\{\left(x-\frac{s\cos\psi}{2}\right)^2+\left(y-\frac{s\sin\psi}{2}\right)^2\right\}^{\frac{m-n}{2}}\left(y-\frac{s\sin\psi}{2}\right)^n\end{array}\right] \quad (18.222)$$

Let two sets of coefficients be defined as

$$H(m,n;s,\psi)=\frac{1}{S}\iint_S h(m,n;s,\psi)dS \quad (18.223)$$

$$J(m,n;p,q;s,\psi)=\frac{1}{S}\iint_S h(m,n;s,\psi)h(p,q;s,\psi)dS \quad (18.224)$$

Expressing W_{mn} in terms of P computed aberrations W_i, $i=1,\ldots,P$, as in (17.121) or (13.30), we have

$$W_{mn}=\sum_{i=1}^{P}B(m,n;i)W_i \quad (18.225)$$

Using relations (18.221) – (18.225), the relation (18.219) for variance $K(s,\psi)$ of the wave aberration difference function can be expressed in a quadratic form as

$$K(s,\psi)=\sum_i\sum_j P(i,j;s,\psi)W_iW_j \quad (18.226)$$

where

$$P(i,j;s,\psi)=\sum_m\sum_n\sum_p\sum_q\left[J(m,n;p,q;s,\psi)-H(m,n;s,\psi)H(p,q;s,\psi)\right]K(m,n;i)K(p,q;j) \quad (18.227)$$

Note that, for a prespecified set of values for (s,ψ), and for a set of prespecified points on the exit pupil, the coefficients $P(i,j;s,\psi)$ are needed to be calculated once, and only once for this combination. The number of prespecified points on the exit pupil is related with the aberration polynomial to be used for representing the aberration. Different systems require different types of aberration polynomials for proper representation of their aberrations. Therefore, different sets of coefficients catering to different choices for (s,ψ) and aberration polynomials may be calculated once, and only once, for possible use in different systems. Variance of the wave aberration difference function can be obtained by substituting the computed aberration values in the quadratic relation (18.226). This approach obviates the need for equation solving or extensive numerical computation in individual cases.

18.8 Through-Focus MTF

In Section 18.5.2.3, the effect of defocusing on the MTF of diffraction limited system was studied. In the case of imaging by an optical system with residual aberrations, determination of the best focal/image plane poses a tricky problem. In practice, the problem is tackled by determining the MTF over

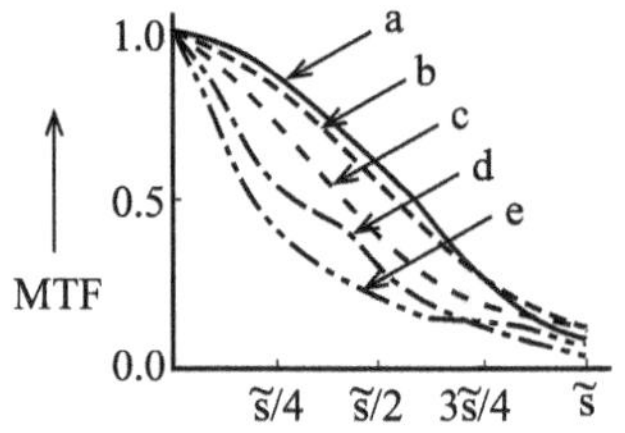

FIGURE 18.22 MTFs over $(0,\tilde{s})$ (where $\tilde{s} < 0.10$) at image planes corresponding to (a) $W_{20} = 0$, (b) $W_{20} = +m\lambda$ (c) $W_{20} = -m\lambda$ (d) $W_{20} = +2m\lambda$ (e) $W_{20} = -2m\lambda$.

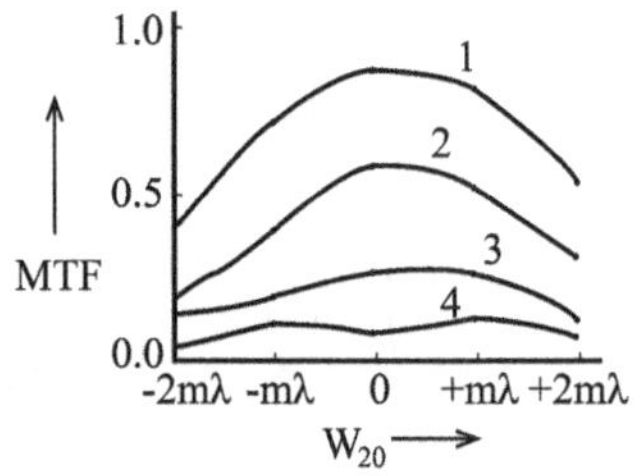

FIGURE 18.23 Through Focus MTFs over shift of image plane corresponding to W_{20} over $(-2\,m\lambda, +2\,m\lambda)$ for spatial frequencies: $\text{Curve 1} - \frac{\tilde{s}}{4}, \text{Curve 2} - \frac{\tilde{s}}{2}, \text{Curve 3} - \frac{3\tilde{s}}{4}, \text{Curve 4} - \tilde{s}$.

the bandwidth of interest at a set of neighbouring planes on both sides of the paraxial or Gaussian image plane (Figure 18.22). From these MTF curves, the variation in frequency response with axial shift of image plane may be determined at one or more selected frequencies (Figure 18.23).

The latter may be used to determine the optimum location of the focal/image plane.

Similarly, through-focus MTF curves at the specific frequency of primary interest in the system can be plotted showing the variation of the MTF with changes in the axial location of the image plane. This may be repeated for different field positions. In systems with residual aberrations, the different curves corresponding to different field positions do not overlap. The relative shapes and positions of these curves vary considerably from lens to lens. These curves play a useful role in determining the effective depth of focus over the whole field, and in setting up related tolerances [170].

18.9 Interrelationship between PSF, LSF, ESF, BSF, and OTF

The basic element in optical image formation is the image of a bright point in a dark background. This is called the point spread function (PSF). In real optical imaging systems, the PSF is space variant, so that the PSF at the edge of the field is not the same as that on the axis, and neither of them is the same as the PSF at other regions of the field. For practical purposes, the changes in the PSF in the neighbourhood of a point is neglected, so that within this small neighbourhood, the PSF may be considered as space invariant and, therefore, the image of a spatially incoherent object lying exclusively within this domain may be regarded as the superposition of the weighted intensity distributions of the PSF from all the points of the object. Imaging quality of an extended object needs to be specified for a few selected field points.

In principle, all the image evaluation parameters, namely the point spread function (PSF), the line spread function (LSF), the edge spread function (ESF), the bar spread function, and the optical transfer function (OTF), are equivalent, and so alternative procedures are available for the practical measurement of OTF. The same observation holds in the case of measurement of any of the image evaluation criteria. Choice of a particular approach in a specific case depends upon several factors, e.g. photometric considerations, types of detectors utilized, working wavelengths, availability of high resolution targets, and opto-mechanical setups, etc. A glimpse at investigations carried out on this important topic may be obtained from references [171–199]. Incidentally, it should be noted that, compared to the other image quality assessment criteria, the OTF possesses a unique capability of seamless integration with similar quality evaluation parameters of constituent components of the overall optical system, other than the component involved in optical imaging.

It is important to realize that, on account of the space variant nature of the PSF, the images of the same spatially incoherent extended object are not the same when the location and orientation of the extended object vary over the field. As already enunciated in earlier sections, the use of reduced or canonical coordinates for the entrance/exit pupil and the object/image considerably facilitates the treatment of extra-axial imagery by enabling a unified treatment, irrespective of location and orientation of the object.

In Section 18.5.2, the relation between the pupil function and the amplitude point spread function (APSF), the intensity point spread function (PSF), the line spread function (LSF), and the optical transfer function (OTF), and interrelation between the latter image quality evaluation parameters was dealt with. The relations are applicable for both axial and extra-axial imagery. Figure 18.12 illustrates the image space, including the exit pupil and the image plane for the general case.

In the following, we discuss two relationships that deserve special mention.

18.9.1 Relation between PSF, LSF, and OTF for Circularly Symmetric Pupil Function

Since the pupil function is circularly symmetric, the APSF, the PSF, the LSF, and the OTF are also circularly symmetric. Omitting the subscripts in coordinates, the point spread function is denoted by $G(u',v')$. Shifting to polar coordinates (w',ϕ), where

$$\left.\begin{aligned} u' &= w'\sin\phi \\ v' &= w'\cos\phi \end{aligned}\right\} \tag{18.228}$$

we write

$$G(u',v') \equiv G(w',\phi) = G(w') \tag{18.229}$$

since the PSF is rotationally symmetric.

The line spread function corresponding to a line object along the v-axis is given by

$$L(u') = \int_{-\infty}^{+\infty} G(u',v')dv' = 2\int_{0}^{+\infty} G(u',v')dv' = 2\int_{|u'|}^{+\infty} \frac{G(w')w'dw'}{\sqrt{w'^2 - u'^2}} \tag{18.230}$$

This relation shows that the line spread function $L(u')$ is equal to the Abel transform of the circularly symmetric point spread function $G(w')$ [200].

The two-dimensional transfer function (unnormalized) can be written from (18.85) as

$$D(s,t) = \int\int_{-\infty}^{+\infty} G(u',v')\exp\{i2\pi(u's + v't)\}du'dv' \tag{18.231}$$

In the circularly symmetric case, by converting to polar coordinates, the OTF can be represented as

$$D(s,t) \equiv D(\tilde{s},\theta) = D(\tilde{s}) \tag{18.232}$$

Note that both ϕ and θ are measured with respect to the meridian plane so that

$$\left.\begin{aligned} s &= \tilde{s}\sin\theta \\ t &= \tilde{s}\cos\theta \end{aligned}\right\} \tag{18.233}$$

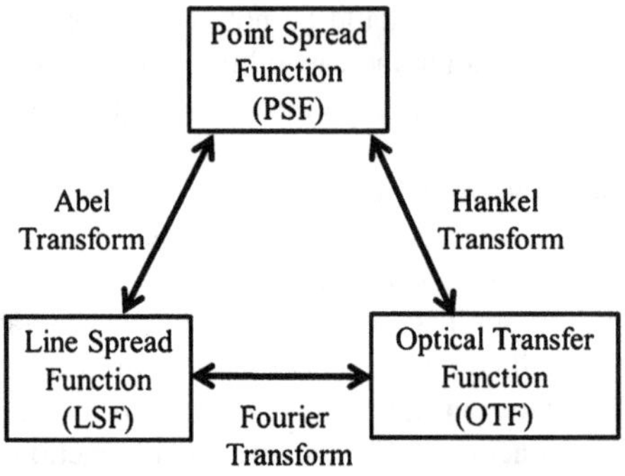

FIGURE 18.24 Interrelationship between the PSF, the LSF, and the OTF in an imaging system with circularly symmetric pupil function.

Following similar mathematical manipulations as in Section 16.3, we get the following expression for the unnormalized OTF

$$D(\tilde{s}) = \int_0^{\infty} G(w') J_0(2\pi w'\tilde{s}) dw' \tag{18.234}$$

Thus, the OTF is given by the 'Fourier-Bessel' or 'Hankel' transform of order zero of the circularly symmetric PSF $G(w')$ [201].

Alternatively, using (18.95), the OTF can be obtained by a one-dimensional Fourier Transform of the line spread function as

$$D(\tilde{s}) = \int_{-\infty}^{+\infty} L(u') \exp\{-i2\pi u'\tilde{s}\} du' \tag{18.235}$$

The interrelationship between the PSF, the LSF, and the OTF in an optical imaging system with circularly symmetric pupil function is illustrated with the help of a triangular flowchart in Figure 18.24.

18.9.2 Relation between ESF, LSF, and OTF

Figure 18.12 shows the image space of an optical imaging system. The geometrical image of an extra-axial object point $\bar{Q}$ (not shown in the figure) is at $\bar{Q}'$. For the sake of simplification in notation, in what follows, we omit the subscripts and represent the coordinates (u'_S, v'_T) and $(\tilde{u}'_S, \tilde{v}'_T)$ by (u', v') and $(\tilde{u}', \tilde{v}')$, respectively. Corresponding changes in the object space are implied.

An edge along the $\tilde{V}$ axis passing through the point $\bar{Q}$ on the object plane is represented by

$$\left.\begin{aligned} B(\tilde{u},\tilde{v}) = H(\tilde{u}) &= 0, \quad \tilde{u} < 0 \\ &= 1, \quad \tilde{u} > 0 \end{aligned}\right\} \tag{18.236}$$

Its image is

$$B(\tilde{u}',\tilde{v}') = \int_{-\infty}^{+\infty} H(\tilde{u}) \left[\int_{-\infty}^{+\infty} G(\tilde{u}' - \tilde{u}, \tilde{v}' - \tilde{v}) d\tilde{v} \right] d\tilde{u} \tag{18.237}$$

where $G(\tilde{u}',\tilde{v}')$ is the point spread function. The bracketed quantity inside the integrand can be written in terms of line spread function $L(\tilde{u}')$ as

$$\left[\int_{-\infty}^{+\infty} G(\tilde{u}'-\tilde{u},\tilde{v}'-\tilde{v})d\tilde{v}\right] = L(\tilde{u}'-\tilde{u}) \tag{18.238}$$

Using (18.236) and substituting from (18.238), equation (18.237) reduces to

$$B(\tilde{u}',\tilde{v}') = \int_{-\infty}^{+\infty} H(\tilde{u})L(\tilde{u}'-\tilde{u})d\tilde{u} = \int_{0}^{\infty} L(\tilde{u}'-\tilde{u})d\tilde{u} = H(\tilde{u}') \tag{18.239}$$

where the 'edge spread function (ESF)', also called the 'edge response function (ERF)' is given by

$$H(\tilde{u}') = \int_{-\infty}^{\tilde{u}'} L(\chi)d\chi \tag{18.240}$$

A substitution $\chi = (\tilde{u}'-\tilde{u})$ has been made above. From (18.240), it follows that the edge gradient $\Gamma(\tilde{u}')$ at any point $\tilde{u}'$ in the image is

$$\Gamma(\tilde{u}') = \frac{dH(\tilde{u}')}{d\tilde{u}'} = L(\tilde{u}') \tag{18.241}$$

It shows that the edge gradient at any point $\tilde{u}'$ in the edge spread function is equal to the intensity at $\tilde{u}'$ in the line spread function. However, it was pointed out by Hopkins and Zalar [201] that the strict equality between the two, as given in (18.231), cannot be used in practice, for different normalizations are used for the LSF and the ESF, and so there is a numerical factor between the gradient of the normalized edge spread function and the normalized line spread function.

Using the relation (18.118), the unnormalized line spread function $L(\tilde{u}')$ corresponding to the edge along the $\tilde{V}$ axis in the object space can be expressed as

$$L(\tilde{u}') = \int_{\tilde{y}'} \left| \int_{\tilde{x}'} f(\tilde{x}',\tilde{y}')\exp(i2\pi\tilde{u}'\tilde{x}')d\tilde{x}' \right|^2 d\tilde{y}' \tag{18.242}$$

The limits of integration in the above expression extend over the whole region of the exit pupil. The geometrical image of the edge along the line $\tilde{u}=0$ lies along the line $\tilde{u}'=0$ on the image plane. The normalization factor used for the LSF is the intensity at $\tilde{u}'=0$ of the LSF produced by the optical system in the absence of any aberration and focus error. Denoting the latter intensity by $L_0(0)$, from (18.242) we get

$$L_0(0) = \int_{\tilde{y}'} \left| \int_{\tilde{x}'} f(\tilde{x}',\tilde{y}')d\tilde{x}' \right|^2 d\tilde{y}' \tag{18.243}$$

The normalized LSF, $\hat{L}(\tilde{u}')$, is defined by

$$\hat{L}(\tilde{u}') = \frac{L(\tilde{u}')}{L_0(0)} = \frac{1}{L_0(0)} \int_{\tilde{y}'} \left| \int_{\tilde{x}'} f(\tilde{x}',\tilde{y}')exp(i2\pi\tilde{u}'\tilde{x}')d\tilde{x}' \right|^2 d\tilde{y}' \tag{18.244}$$

From the above equations, by using the Schwartz's inequality relation, it can be seen that

$$\left.\begin{aligned} L(\tilde{u}') &< L_0(0) \\ L_0(\tilde{u}') &< L_0(0) \end{aligned}\right\} \tag{18.245}$$

The Strehl ratio deals with the fall in peak intensity in the PSF of an aberrated system in comparison with the peak intensity in the PSF of the same system in absence of aberrations. Maréchal developed aberration tolerances and conditions for the optimum balancing of aberrations based on the Strehl criterion, which stipulates a Strehl ratio ≥ 0.8 for tolerable diffraction limited systems. This topic has been discussed in detail in Section 17.2. For extended object imagery, a similar criterion may be obtained on the basis of the fall in peak intensity in the LSF in the presence of aberrations. This criterion states that the tolerable diffraction limited systems should have:

$$\hat{L}(0) = \left[L(0)/L_0(0)\right] \geq 0.8 \tag{18.246}$$

The ratio $\hat{L}(0)$ is known as the Struve ratio [203]. Aberration tolerances and the optimum balancing of aberrations on the basis of the Struve criterion have been derived for selected aberration polynomials by Hopkins and Zalar [202] and Mahajan [204]. The reported numerical results for the tolerances have small but significant differences with the tolerance values obtained by Maréchal on the basis of the Strehl criterion.

The pupil function $f(\tilde{x}',\tilde{y}')$ is given by

$$f(\tilde{x}',\tilde{y}') = \tau(\tilde{x}',\tilde{y}') \quad \exp\{ikW(\tilde{x}',\tilde{y}')\} \tag{18.247}$$

where $\tau(\tilde{x}',\tilde{y}')$ is the variation of real amplitude, and $W(\tilde{x}',\tilde{y}')$ is the wavefront aberration. For the case of an aberration-free pupil of uniform amplitude, $W(\tilde{x}',\tilde{y}') = 0$ and $\tau(\tilde{x}',\tilde{y}') = 1$. For a circular aperture of unit radius, (18.243) reduces to

$$L_0(0) = \int_{-1}^{+1} \left| \int_{-\sqrt{(1-y^2)}}^{+\sqrt{(1-y^2)}} d\tilde{x}' \right|^2 d\tilde{y}' = \frac{16}{3} \tag{18.248}$$

For the edge spread function (ESF), the normalization factor is the constant intensity at points very far from the geometrical image of the edge. From (18.232), this intensity is given by

$$H(\infty) = \int_{-\infty}^{+\infty} L(\chi)d\chi = \int_{-\infty}^{+\infty} L(\tilde{u}')d\tilde{u}' \tag{18.249}$$

Considering the LSF as an incoherent superposition of the images of the points along the line $-\infty \leq \tilde{v} \leq +\infty$, we get the LSF in terms of the PSF as

$$L(\tilde{u}') = \int_{-\infty}^{+\infty} G(\tilde{u}',\tilde{v}')d\tilde{v}' \tag{18.250}$$

From (18.240), it follows

$$H(\infty) = \int\int_{-\infty}^{+\infty} G(\tilde{u}',\tilde{v}')d\tilde{u}'d\tilde{v}' \tag{18.251}$$

It is shown in (18.88) that, by Plancherel's theorem, also known as Rayleigh's theorem, the above relation reduces to

$$H(\infty)=\int\int_{-\infty}^{+\infty}\left|F(\tilde{u}',\tilde{v}')\right|^2 d\tilde{u}'d\tilde{v}'=\int\int_{-\infty}^{+\infty}\left|f(\tilde{x}',\tilde{y}')\right|^2 d\tilde{x}'d\tilde{y}'=\tilde{Q} \tag{18.252}$$

In the case of a pupil with uniform real amplitude, we take $\tau(\tilde{x}',\tilde{y}')=1$, so that $\left|f(\tilde{x}',\tilde{y}')\right|=1$. The normalizing factor $H(\infty)$ reduces to

$$H(\infty)=Q=\iint_A d\tilde{x}'d\tilde{y}' \tag{18.253}$$

where A is the domain of the exit pupil.

The normalized form of ESF is defined by

$$\hat{H}(\tilde{u}')=\frac{H(\tilde{u}')}{H(\infty)}=\frac{H(\tilde{u}')}{H(\infty)} \tag{18.254}$$

In the case of an exit pupil of circular aperture of unit radius, it follows

$$H(\infty)=Q=\iint_A d\tilde{x}'d\tilde{y}'=\int_0^{2\pi}\int_0^1 r dr d\theta=\pi \tag{18.255}$$

and the corresponding normalized ESF is

$$\hat{H}(\tilde{u}')=\frac{1}{\pi}H(\tilde{u}') \tag{18.256}$$

From (18.243) – (18.244), the normalized line spread function, in the case of an exit pupil with uniform real amplitude and a circular aperture of unit radius, is given by

$$\hat{L}(\tilde{u}')=\frac{L(\tilde{u}')}{L_0(0)}=\frac{3}{16}L(\tilde{u}') \tag{18.257}$$

Using (18.241) and (18.256) – (18.257), we get the relationship between the normalized edge gradient and the normalized LSF as

$$\hat{\Gamma}(\tilde{u}')=\frac{d\hat{H}(\tilde{u}')}{d\tilde{u}'}=\frac{1}{\pi}\frac{dH(\tilde{u}')}{d\tilde{u}'}=\frac{1}{\pi}L(\tilde{u}')=\left(\frac{16}{3\pi}\right)\hat{L}(\tilde{u}') \tag{18.258}$$

or, the normalized LSF is related with normalized edge gradient by

$$\hat{L}(\tilde{u}')=\left(\frac{3\pi}{16}\right)\hat{\Gamma}(\tilde{u}') \tag{18.259}$$

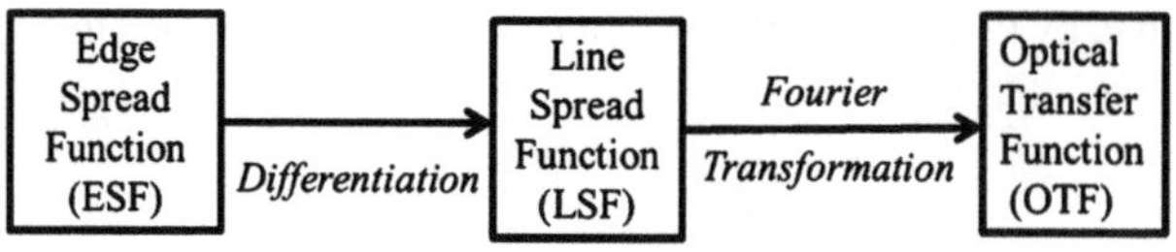

FIGURE 18.25 Flowchart for OTF from ESF via LSF.

From (18.119), we note that the OTF for a one-dimensional structure in the azimuth ψ, $D(s;\psi)$ is given by the normalized inverse Fourier transform of the intensity line spread function, $\tilde{L}(\tilde{u}'_S)$ as

$$D(s;\psi) = \frac{1}{Q}\int_{-\infty}^{+\infty} \tilde{L}(\tilde{u}'_S)\exp\{-i2\pi s\tilde{u}'_S\}d\tilde{u}'_S \tag{18.260}$$

In the case of an exit pupil of uniform real amplitude over a circular aperture of radius unity, it is shown in (18.255) that $Q = \pi$. Therefore, (18.260) reduces to

$$D(s;\psi) = \frac{1}{\pi}\int_{-\infty}^{+\infty} \tilde{L}(\tilde{u}'_S)\exp\{-i2\pi s\tilde{u}'_S\}d\tilde{u}'_S \tag{18.261}$$

Determination of OTF from ESF via LSF is illustrated with the help of a flowchart in Figure 18.25.

18.9.3 BSF and OTF

Bar spread function (BSF) is the response of the optical imaging system to bar targets as objects. On account of the ease and convenience of their use, for image quality assessment of optical systems they are widely used in practice. A particular bar target consists of a finite set of alternating bright and dark bars of equal width. The contrast modulation in a target is usually unity. The inverse of the centre to centre spacing of two bright or dark bars is termed 'frequency' of that particular bar target. Obviously, such a pattern consisting of a finite number of bars cannot be characterized by a single frequency. A test target consists of a finite number of such bar targets consisting of harmonics of the coarsest frequency. For each frequency, the bar target consists of two identical targets in mutually perpendicular orientations. The orientations are the same for each frequency.

Currently, systems operating in visible wavelengths are usually characterized by three bar targets, commonly called 'USAF 1951 resolution target' [205]. Occasionally, two bar targets, also known as Cobb chart (BS1613), and five bar targets, called NBS 1963 target (BS 4657), are also used. All these targets allow direct observation of response to discrete values for spacing between the bars, or for a discrete number of 'bar frequencies'. The problem can be circumvented by using sector targets; the commonly available ones consist of 36 pairs of bright and dark sectors [206]. The 'frequency' in a sector target changes continually along the radius. However, the sector target is also a binary target like the other linear targets. Infrared or Thermal imaging systems are often characterized by four bar targets [207].

In principle, a Fourier representation of the three bar patterns provides a link between the contrast transfer in the bar target images and the MTF. Several attempts have been made to explore this link for determination of MTF from simple measurements. Details may be found in references [208–210].

18.10 Effects of Anamorphic Imagery in the Off-Axis Region on OTF Analysis

In general, axially symmetric imaging systems produce anamorphic, also called anamorphotic, imagery in the extra-axial region. This means that a small square placed on the axis is imaged as a square. But

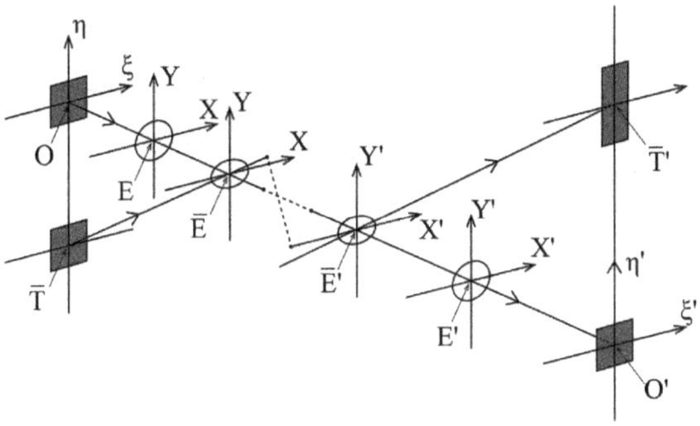

FIGURE 18.26 Anamorphic imaging in the extra-axial region of an axially symmetric system.

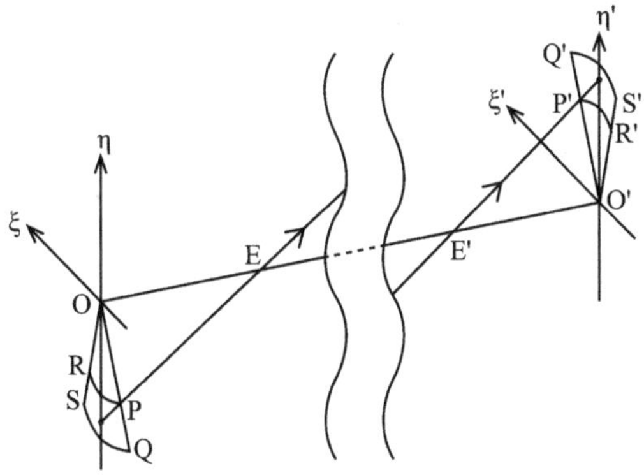

FIGURE 18.27 Imaging of the segment of a circular annulus by an axially symmetric system.

the same square object, when placed in the extra-axial region, is imaged as a rectangle. Figure 18.26 illustrates this phenomenon, where the square object placed at O is imaged as a square at O′, but when the square object is placed at $\bar{T}$, it is imaged as a rectangle at $\bar{T}'$, the geometrical image of $\bar{T}$. Note that the line along the diagonal in the square object makes an angle 45^0 with the meridian plane, but its image, the diagonal in the rectangular image does not make the same angle with the meridian plane. This is a result of the inequality of different magnifications in extra-axial imagery. The case of a perfect imaging system is an exception.

Figure 18.27 shows an extra-axial object RPQS that is imaged by an axially symmetric imaging system as R′P′Q′S′. The object is a segment of an annulus centred on O, and the axial symmetry of the imaging system causes the object to be imaged as a segment of an annulus subtending the same angle at O′ as the object segment at O, i.e. $\angle R'O'P' = \angle ROP$. Let

$$OR = OP = \zeta \text{ and } O'R' = O'P' = \zeta' \tag{18.262}$$

The finite magnification M_0 is given by

$$M_0 = \frac{\zeta'}{\zeta} \tag{18.263}$$

Also, the magnification of the arc RP is

$$\frac{R'P'}{RP} = \left(\frac{\zeta'}{\zeta}\right)\left(\frac{\angle R'O'P'}{\angle ROP}\right) = \left(\frac{\zeta'}{\zeta}\right) = M_0 \tag{18.264}$$

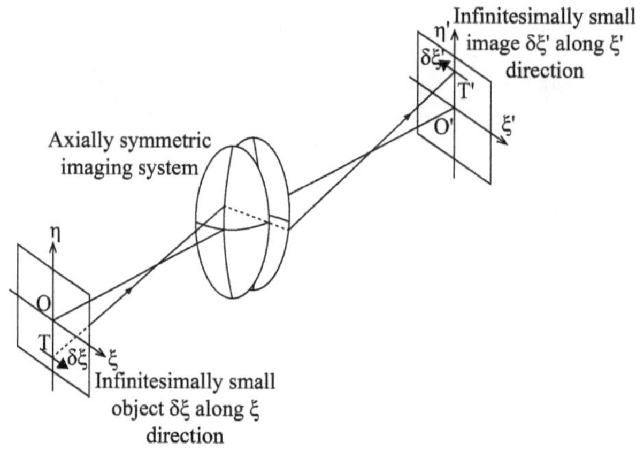

FIGURE 18.28 Sagittal magnification around an extra-axial point.

Figure 18.28 shows an infinitesimally small object $\delta\xi$ along the ξ direction. It is located at T on the meridian section at a distance $OT = \eta$ from the axis. By symmetry, the image is an infinitesimally small element $\delta\xi'$ along the ξ' direction. The image is located at T' on the meridian section at a distance $O'T' = \eta'$ from the axis. The sagittal magnification M_S is

$$M_S = \lim_{\delta\xi \to 0} \left\{ \frac{\delta\xi'}{\delta\xi} \right\} \tag{18.265}$$

Since the arc magnification M_0 remains the same in the limit of $(RP) \to 0$, it follows

$$M_S = M_0 \tag{18.266}$$

It should be noted that, in general, $M_S (= M_0)$ varies with object height η, and neither M_0 nor M_S is equal to the paraxial or Gaussian magnification M. Only in the case of distortion free imaging systems does $M_S = M_0 = M$. In the presence of distortion, let η' and η'_G be the real and the Gaussian or paraxial image heights of an object of height η in the meridian section. We have

$$M_0 = \frac{\eta'}{\eta} \text{ and } M = \frac{\eta'_G}{\eta} \tag{18.267}$$

Defining fractional distortion D by

$$D = \frac{\eta' - \eta'_G}{\eta'_G} \tag{18.268}$$

we get the relation between the sagittal magnification and the Gaussian magnification as

$$M_S = M_0 = M(1 + D) \tag{18.269}$$

The tangential magnification is similarly defined as

$$M_T = \lim_{\delta\eta \to 0} \left\{ \frac{\delta\eta'}{\delta\eta} \right\} \tag{18.270}$$

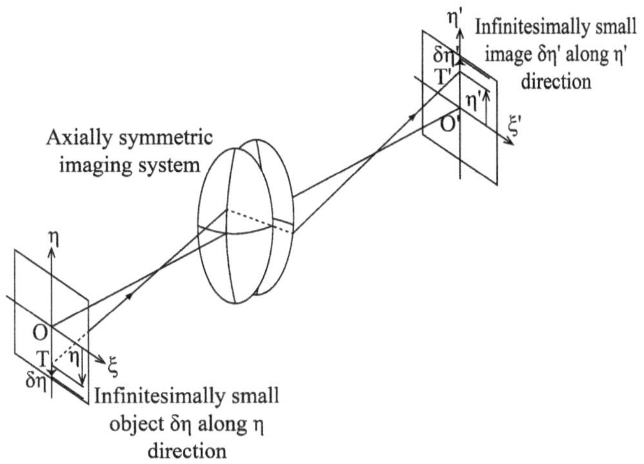

FIGURE 18.29 Tangential magnification around an extra-axial point.

Figure 18.29 shows the imaging of an infinitesimally small object located at the point T, and lying along the η direction.

The variation in magnification with change in object height η may be expressed as

$$M_0 = (M_0)_\eta = \frac{\eta'}{\eta} \qquad (M_0)_{\eta+\delta\eta} = \frac{\eta' + \delta\eta'}{\eta + \delta\eta} \tag{18.271}$$

Expanding by Taylor series, we get

$$(M_0)_{\eta+\delta\eta} = M_0 + \delta\eta \frac{\partial M_0}{\partial \eta} + O(\delta\eta^2) \tag{18.272}$$

where $O(\delta\eta^p)$ indicates all terms involving powers of degree p and above in $\delta\eta$.

From (18.271) – (18.272), it follows

$$\frac{\delta\eta'}{\delta\eta} = M_0 + \eta \frac{\partial M_0}{\partial \eta} + O(\delta\eta) \tag{18.273}$$

Therefore, the relation (18.270) reduces to

$$M_T = \lim_{\delta\eta \to 0} \left\{ \frac{\delta\eta'}{\delta\eta} \right\} = M_0 + \eta \frac{\partial M_0}{\partial \eta} \tag{18.274}$$

Thus, an infinitesimally small, square, extra-axial object is imaged as a rectangle with sagittal and tangential magnifications M_0 and $\left(M_0 + \eta \frac{\partial M_0}{\partial \eta} \right)$, respectively. It is shown in (18.269) that in the presence of fractional distortion D, the sagittal magnification M_S is equal to $M(1+D)$ where M is the paraxial magnification. In the presence of primary distortion ${}_3W_{11}\tau^3 r\cos\theta$, it can be shown that the tangential magnification $M_T = M(1+3D)$. If the orientation of an object grating of N cycles/mm is changed from

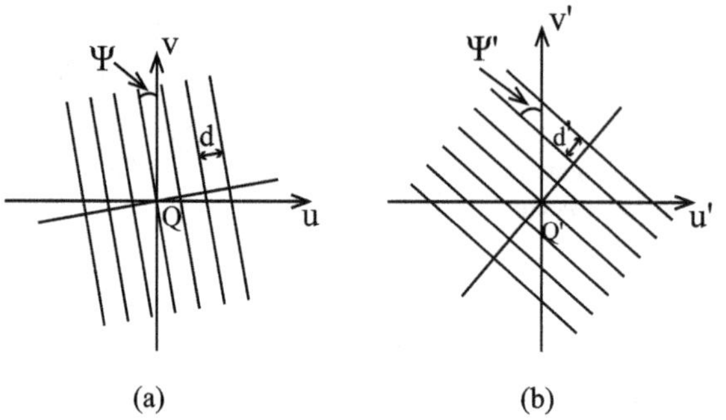

FIGURE 18.30 (a) A small sinusoidal grating of period d placed at an extra-axial point Q. Equal intensity lines of the grating make an angle ψ with the meridian section. (b) Image of the grating around the point Q′, the geometrical image of Q. The period and orientation of the image grating are d′ and ψ'.

tangential to sagittal azimuth, i.e. the angle ψ between the direction of the lines of equal intensity and the meridian section is changed from zero to $(\pi/2)$, the corresponding images will have frequencies N_S' and N_T', such that

$$\frac{N_S'}{N_T'} = \frac{M_T}{M_S} = \frac{1+3D}{1+D} = 1 + \frac{2D}{1+D} \tag{18.275}$$

Thus, in the presence of 5 per cent fractional distortion (primary), the spatial frequency in the image of an object grating changes by approximately 10 per cent when the grating is rotated from the meridional to the sagittal section.

Figure 18.30(a) shows a sinusoidal grating oriented at angle ψ with the meridional plane and placed in the extra-axial region of an axially symmetric optical imaging system. The grating is of period d. In general, both the period d′ and the orientation ψ' of its image are different from those of the object, i.e. $d' \neq d$, and $\psi' \neq \psi$. The image of the object grating is shown in Figure 18.30(b).

The period of the object grating is d. Let its period along the ξ and the η directions be $\delta\xi$ and $\delta\eta$, respectively. Equal intensity lines of the object grating make an angle ψ with the meridian section. For the object grating, we can write

$$\sin\psi = \frac{d}{\delta\eta} \quad \cos\psi = \frac{d}{\delta\xi} \quad \tan\psi = \frac{\delta\xi}{\delta\eta} \tag{18.276}$$

Corresponding parameters for the image grating are denoted by primes, and the above relations take the forms

$$\sin\psi' = \frac{d'}{\delta\eta'} \quad \cos\psi' = \frac{d'}{\delta\xi'} \quad \tan\psi' = \frac{\delta\xi'}{\delta\eta'} \tag{18.277}$$

Using the relations for sagittal and tangential magnifications at the extra-axial region as expressed by

$$M_S = \frac{\delta\xi'}{\delta\xi} \qquad M_T = \frac{\delta\eta'}{\delta\eta} \tag{18.278}$$

and after algebraic manipulations, the period and the orientation of the image grating can be expressed in terms of the period and the orientation of the object grating as [118]

$$\left.\begin{aligned} d' &= d\left\{\left(\frac{\cos\psi}{M_S}\right)^2 + \left(\frac{\sin\psi}{M_T}\right)^2\right\}^{-\frac{1}{2}} \\ \tan\psi' &= \left(\frac{M_S}{M_T}\right)\tan\psi \end{aligned}\right\} \quad (18.279)$$

In both numerical evaluation and experimental measurement of OTF in extra-axial imagery of axially symmetric systems, overlooking this effect gives rise to serious errors.

18.11 Transfer Function in Cascaded Optical Systems

One of the primary reasons for the popularity of OTF or MTF as an image quality assessment criterion in modern optical or electro-optical systems is that the concept of transfer function can be used to quantify the quality of the overall system in terms of transfer function of the constituent sub-systems. In most cases, the OTF of the overall system is given by the product of the OTFs of the subsystems. Subsystems usually consist of one or more optical imaging systems in conjunction with optomechanical and/or optoelectronic/photonic systems. So long as the role of each subsystem is to function as an independent imaging system that transforms the irradiance distribution in an object into the irradiance distribution in an image, the procedure of multiplication of transfer functions of the subsystems to obtain transfer function for the overall system remains valid.

Problems arise in the case of cascaded or concatenated optical imaging systems. More often than not, the transfer function of the overall optical imaging system can be obtained by a multiplication of the transfer function of its constituent optical imaging subsystems. This depends on the degree of coherence of illumination used in imaging. DeVelis and Parrent noted:

> we can multiply the transfer functions of the individual elements to obtain the system transfer function under only two circumstances: (1) If the system is coherent and we wish to work in complex amplitudes. (2) If, regardless of the coherence, we are willing to analyse the system in terms of the mutual intensity.
>
> [211]

For incoherent illumination, the OTF of the overall imaging system cannot be obtained by multiplying the OTFs of the optical imaging subsystems. In the case of coherent illumination, the ATF (or, the pupil function) of the overall system is given by the product of the ATFs (or, the pupil functions) of the subsystems [212]. These observations may be validated by assuming a hypothetical two-stage imaging relay system, where the first stage has a residual aberration W and the residual aberration of the second stage is –W. As standalone systems, each stage of the relay system has an OTF that is lower than the diffraction limited OTF, and their product is more so. But, as a concatenated system, the relay as a whole is diffraction limited, since the aberration of the second stage cancels out the aberration of the first stage. Obviously, the OTF of the relay is much higher than what is obtained by multiplying OTFs of the two stages. Note that these observations assume that the intermediate image(s) of the concatenated systems are aerial. In a two-stage relay with conjugate aberrations in the two stages, if the intermediate image produced by the first stage falls on a diffuser, and the image is reimaged by the second stage, the final image is no longer diffraction limited, and the OTF of the two-stage relay is equal to the product of the OTFs of the constituent stages [213].

18.12 Image Evaluation Parameters in Case of Polychromatic Illumination

All image evaluation parameters considered so far assume monochromatic illumination. For practical purposes, they are useful even for quasi-monochromatic illumination, in which case the bandwidth of the illumination is assumed to be infinitesimally small. In the case of incoherent imaging and polychromatic illumination with finite bandwidth, often the polychromatic versions of the image evaluation parameters can be obtained by incoherent superposition of the weighted monochromatic parameters over the working passband of illumination. The weighting factors take care of the variation of the system response with the wavelength while undertaking the specific function for image quality assessment. The spectral composition of the illuminating radiation, the spectral transmittance of the imaging system, and the spectral response of the detector for recording the image are the three major factors that determine the weighting factor.

18.12.1 Polychromatic PSF

In the general case of extra-axial imagery, we use the same reduced coordinates, namely (u'_S, v'_T) for points on the image plane. The origin of the corresponding coordinate system is at $\bar{Q}'$, the geometrical image point of an extra-axial object point at $\bar{Q}$, which has coordinates (0,0) with respect to $\bar{Q}$ as origin, and $(0,\eta)$ with respect to the axial object point O on the optical axis (see Figure 8.12). As mentioned earlier, image analysis based on points located on the meridional plane is sufficient on account of bilateral symmetry available with axially symmetric systems.

The monochromatic intensity point spread function at wavelength λ is denoted by $G_\lambda(u'_S, v'_T)$. It is the same as $G(u'_S, v'_T)$ in relation (18.82). A subscript λ is used now to underscore the monochromatic nature of the quantity. The polychromatic point spread function (PPSF) is defined by

$$G_{poly}(u'_S, v'_T) = \int_\lambda P(\lambda) G_\lambda(u'_S, v'_T) d\lambda \Big/ \int_\lambda P(\lambda) d\lambda \tag{18.280}$$

where the weighting factor $P(\lambda)$ is given by

$$P(\lambda) = E(\lambda) M(\lambda) D(\lambda) \tag{18.281}$$

$E(\lambda), M(\lambda)$, and $D(\lambda)$ are the normalized spectral distribution in the object radiance, the normalized spectral transmittance of the imaging lens system, and the normalized spectral response of the detector, respectively. For practical purposes, in the case of illumination with continuous spectra over a wavelength range (λ_1, λ_2), the two integrations in (18.280) are replaced by summations over a closely spaced, discrete set of wavelengths. (18.280) reduces to

$$G_{poly}(u'_S, v'_T) = \frac{\sum_\lambda P(\lambda) G_\lambda(u'_S, v'_T)}{\sum_\lambda P(\lambda)} \tag{18.282}$$

For each wavelength, the PSF is to be determined by independent ray tracing. It is obvious that the polychromatic PSF can be significantly different for different sources with different $E(\lambda)$, and different detectors with different $D(\lambda)$. Reports of few studies on different aspects of polychromatic PSF have appeared in the literature [214–220].

18.12.2 Polychromatic OTF

Several investigators have reported on the polychromatic OTF [221–229]. In the same way as the polychromatic PSF is formulated in terms of monochromatic PSFs, the polychromatic OTF is also formulated in terms of monochromatic OTFs. However, the concept of polychromatic OTF remains valid if and only if the following conditions are satisfied:

(i) The two general conditions—the principle of superposition and the property of space invariance—that need to be satisfied for applying the concept of monochromatic OTF are also required for using the concept of polychromatic OTF.

(ii) The object needs to have constant spectral composition, both spatially and temporally.
(iii) Across the detector area the spectral sensitivity should be uniform.
(iv) The local magnifications, including the sagittal magnification, M_S, and the tangential magnification, M_T, do not vary significantly with wavelength.

These requirements have been explicitly mentioned by Barnden [222–223], and Takeda [224] noted some interesting consequences in image quality assessment when the concept of polychromatic OTF is put in practice, e.g. two optical systems suffering from different amounts of chromatic aberration may lead to the same polychromatic OTFs for the two systems. We put below the theory of polychromatic OTF. It follows the treatment of Barnden [223].

Let a one-dimensional grating $B(\tilde{\xi})$ be placed at the object point $\bar{Q}$ (in the extra-axial region) with the equal intensity lines parallel to $\tilde{\eta}$-axis. The geometrical image of the point $\bar{Q}$ is $\bar{Q}'$, and the image of the one-dimensional grating is a one-dimensional grating at $\bar{Q}'$ with its equal intensity lines along the $\tilde{\eta}'$ direction (Figure 18.12).

The Fourier transform $b(R)$ of the object grating $B(\tilde{\xi})$ is

$$b(R) = \int_{-\infty}^{+\infty} B(\tilde{\xi}) e^{-i2\pi R\tilde{\xi}} d\tilde{\xi} \tag{18.283}$$

where R is the spatial frequency. Let $b(R)$ be written in terms of amplitude and phase as

$$b(R) = \beta(R) e^{i\varphi(R)} \tag{18.284}$$

The object $B(\tilde{\xi})$ can be written as

$$B(\tilde{\xi}) = \int_{-\infty}^{+\infty} b(R) e^{i2\pi R\tilde{\xi}} dR = \int_{-\infty}^{+\infty} \beta(R) e^{i\left[2\pi R\tilde{\xi} + \varphi(R)\right]} \tag{18.285}$$

Let the intensity of the object grating along the $\tilde{\xi}$-axis for the wavelength λ be

$$B_\lambda(\tilde{\xi}) = E(\lambda) B(\tilde{\xi}) \tag{18.286}$$

where $E(\lambda)$ is the spectral composition of the source illuminating the grating. $B_\lambda(\tilde{\xi})$ is called the monochromatic object grating. The polychromatic object grating is defined by

$$B_{Poly}(\tilde{\xi}) = \int_\lambda B_\lambda(\xi) d\lambda = \int_\lambda E(\lambda) d\lambda \int_{-\infty}^{+\infty} \beta(R) e^{i\left[2\pi R\tilde{\xi} + \varphi(R)\right]} dR \tag{18.287}$$

Equation (18.279) shows that, at any wavelength, the periods d and d' of the object grating and the corresponding image grating are related by $d' = md$, where m is given by

$$m = \left\{\left(\frac{\cos\psi}{M_S}\right)^2 + \left(\frac{\sin\psi}{M_T}\right)^2\right\}^{-\frac{1}{2}} \tag{18.288}$$

The corresponding spatial frequencies are related by $R = mR'$.

The Fourier spectrum of the image of the monochromatic object grating can be represented by

$$b'_\lambda(R') = E(\lambda) g_\lambda(R') b(mR') \tag{18.289}$$

where $g_\lambda(R')$ is the inverse Fourier transform of the line spread function $L_\lambda(\tilde{\xi}')$. It is given by

$$g_\lambda(R') = \int_{-\infty}^{+\infty} L_\lambda(\tilde{\xi}') e^{-i2\pi R'\tilde{\xi}'} d\tilde{\xi}' \tag{18.290}$$

The monochromatic image grating can be represented by

$$B_\lambda(\tilde{\xi}') = E(\lambda) \int_{-\infty}^{+\infty} b(mR') g_\lambda(R') \exp\{i2\pi R'\tilde{\xi}'\} dR' \tag{18.291}$$

Following the usual definition of monochromatic OTF, denoted by $D_\lambda(R')$, we have

$$D_\lambda(R') = \frac{g_\lambda(R')}{g_\lambda(0)} = T_\lambda(R') \exp\{i\theta_\lambda(R')\} \tag{18.292}$$

where $T_\lambda(R')$ and $\theta_\lambda(R')$ represent the monochromatic MTF and the monochromatic PTF, respectively. Equation (18.281) can be rewritten as

$$B_\lambda(\tilde{\xi}') = E(\lambda) \int_{-\infty}^{+\infty} b(mR') g_\lambda(0) D_\lambda(R') \exp\{i2\pi R'\tilde{\xi}'\} dR' \tag{18.293}$$

Let $S(\lambda)$ be the spectral response of the detector. The recorded intensity of the polychromatic image grating is given by

$$B_{Poly}(\tilde{\xi}') = \int_\lambda S(\lambda) B_\lambda(\tilde{\xi}') d\lambda \tag{18.294}$$

By substituting for $B_\lambda(\tilde{\xi}')$ from (18.293), the above relation takes the form

$$B_{Poly}(\tilde{\xi}') = \int_{-\infty}^{+\infty} \beta(mR') \exp\{i[2\pi R'\tilde{\xi}' + \varphi'(mR')]\} \left\{ \int_\lambda E(\lambda) S(\lambda) g_\lambda(0) D_\lambda(R') d\lambda \right\} dR' \tag{18.295}$$

A polychromatic OTF, $D_{Poly}(R')$, is defined by

$$D_{Poly}(R') = \frac{\int_\lambda E(\lambda) S(\lambda) g_\lambda(0) D_\lambda(R') d\lambda}{\int_\lambda E(\lambda) S(\lambda) g_\lambda(0) d\lambda} \tag{18.296}$$

The above definition gives the correctly normalized form for polychromatic OTF, since, by definition, $D_\lambda(0) = 1$ for any value of λ, and so the relation (18.296) gives $D_{Poly}(0) = 1$. The polychromatic OTF, $D_{Poly}(R')$, can be expressed in terms of polychromatic MTF, $T_{Poly}(R')$, and polychromatic PTF, $\theta_{Poly}(R')$, as

$$D_{Poly}(R') = T_{Poly}(R') \exp[i\theta_{Poly}(R')] \tag{18.297}$$

Representing the normalizing factor in (18.296) with W, we get

$$\int_\lambda E(\lambda) S(\lambda) g_\lambda(0) D_\lambda(R') d\lambda = W T_{Poly}(R') \exp[i\theta_{Poly}(R')] \tag{18.298}$$

Substituting from the above relation in (18.295), intensity in the polychromatic grating can be expressed by

$$B_{Poly}(\tilde{\xi}') = W\int_{-\infty}^{+\infty}\beta(mR')T_{Poly}(R')\exp\left\{i\left[2\pi R'\tilde{\xi}' + \varphi'(mR') + \theta_{Poly}(R')\right]\right\}dR' \tag{18.299}$$

The contrast in the polychromatic image grating is

$$C'_{Poly}(R') = \frac{\beta(mR')}{\beta(0)}\frac{T_{Poly}(R')}{T_{Poly}(0)} = \frac{\beta(R)}{\beta(0)}T_{Poly}(R') \tag{18.300}$$

since $R = mR'$ and $T_{Poly}(0) = 1$. From (18.285), the contrast in the object grating is

$$C(R) = \frac{\beta(R)}{\beta(0)} \tag{18.301}$$

Comparing the last two expressions, it is seen that, as in the case of monochromatic MTF, the polychromatic MTF is given by the ratio of the image contrast to object contrast.

From (18.292) and (18.298), it is seen that

$$WT_{Poly}(R')\left\{\cos\theta_{Poly}(R') + i\sin\theta_{Poly}(R')\right\}$$

$$= \int_{\lambda}E(\lambda)S(\lambda)g_{\lambda}(0)T_{\lambda}(R')\cos\theta_{\lambda}(R')d\lambda + i\int_{\lambda}E(\lambda)S(\lambda)g_{\lambda}(0)T_{\lambda}(R')\sin\theta_{\lambda}(R')d\lambda \tag{18.302}$$

It follows that the polychromatic PTF is given by

$$\theta_{Poly}(R') = \tan^{-1}\left[\frac{\int_{\lambda}^{\epsilon}E(\lambda)S(\lambda)g_{\lambda}(0)T_{\lambda}(R')\sin\theta_{\lambda}(R')d\lambda}{\int_{\lambda}^{\epsilon}E(\lambda)S(\lambda)g_{\lambda}(0)T_{\lambda}(R')\cos\theta_{\lambda}(R')d\lambda}\right] \tag{18.303}$$

Taking the design wavelength $\bar{\lambda}$ as a reference wavelength, the factor $g_{\lambda}(0)$ may be conveniently replaced by a factor K_{λ}, defined as

$$K_{\lambda} = \frac{g_{\lambda}(0)}{g_{\bar{\lambda}}(0)} \tag{18.304}$$

For practical purposes, the integrals in the definition of polychromatic OTF in (18.286) may be replaced by summations as

$$D_{Poly}(R') = \frac{\sum_{\lambda}E(\lambda)S(\lambda)K_{\lambda}D_{\lambda}(R')}{\sum_{\lambda}E(\lambda)S(\lambda)K_{\lambda}} \tag{18.305}$$

If the summation is carried out over a sufficiently large number of wavelengths within the bandwidth of illumination, this expression provides useful results for both polychromatic MTF and polychromatic PTF. Levi [230] published polychromatic MTF values for the aberration-free lens for a series of combinations of six commonly used light sources and twelve common detectors.

Incidentally, it should be noted that the Japan Optical Engineering Research Association (JOERA) carried out an extensive investigation on the most reasonable number of mesh divisions of the pupil to get monochromatic MTF, and the most economical sampling method of spectral wavelengths to calculate white light MTF concluded that about 400 mesh divisions and seven sampling wavelengths are sufficient for evaluation of white light MTF [231–232].

18.13 Information Theoretic Concepts in Image Evaluation

In the most general terms, the purpose of an optical system is to give information about the object. But, the purposes are so varied that no unique image quality assessment criterion can fill the bill in all cases. However, the large variety of optical systems may be categorized in two broad classes.

The first category comprises the vast majority of optical systems where information about the object is needed to be readily available, without any extensive post-processing of the available image. Obviously, in such cases, the image should resemble the object as closely as possible, so that it is treated as a faithful reproduction of the object. Any optical difference between the image and the object, other than any difference in scale, is considered a defect in the image. An appropriate theory of image quality assessment calls for the development of a quantitative estimate for the similarity between the object and the image, and this estimate provides a measure of the information displayed in the image.

The second category consists of optical systems where ready availability of information about the object is not the prime requirement, and various image interpretation procedures are resorted to for extracting maximum information available in the image. Obviously, the resemblance of the available image with the object is not a basic requirement in this case, and the theory of image quality assessment needs to be based on the amount of implicit information contained in the image, and on the likelihood of this information being made explicit by the subsequent interpretation process.

Fellgett and Linfoot [233–238] carried out extensive studies on various aspects of this problem. For the first category of systems, they worked out image quality assessment criteria that are based on the Fourier optical analysis of image formation. For tackling the second category of systems, they combined this treatment with the concept of information introduced by Shannon [239–240] in the context of communication in the presence of noise. Subsequently, information theory has emerged as a major discipline in statistical communication theory [241].

For the first category of systems, three criteria are proposed for the assessment of image quality. These provide quantitative measures for similarity between the object and the image. In what follows, we use canonical variables and coordinates for object/image points, and the same notations as used in this chapter. The canonical coordinates of an object point and its geometrical image point are the same. An extended object $B(u,v)$ is imaged as $B'(u,v)$. The intensity point spread function of the imaging system is $G(u,v)$. In the Fourier domain, the object spectrum and the image spectrum are $b(s,t)$ and $b'(s,t)$, respectively, and the OTF of the imaging system is $g(s,t)$, which is equal to zero outside the area $\mathcal{F}$ on the frequency space. $\mathcal{F}$ is determined by the bounds on frequencies s and t. It is assumed that the intensity distributions in the object and the image are normalized, so that

$$\iint_{\infty} B'(u,v)\,du\,dv = \iint_{\infty} B(u,v)\,du\,dv \tag{18.306}$$

The three image assessment criteria are defined below in both the direct space and the frequency space.

I. Relative structural content

$$T = \left[\iint_{\infty} B'^2 du\,dv \Big/ \iint_{\infty} B^2 du\,dv\right] = \left[\iint_{\mathcal{F}} |g|^2 |b|^2 ds\,dt \Big/ \iint_{\mathcal{F}} |b|^2 ds\,dt\right] \tag{18.307}$$

II. Fidelity

$$\Phi = \left[1 - \left\{\iint_{\infty} (B' - B)^2 \, dudv \Big/ \iint_{\infty} B^2 dudv\right\}\right]$$
$$= \left[1 - \left\{\iint_{\mathcal{F}} (|1 - g|)^2 |b|^2 \, dsdt \Big/ \iint_{\mathcal{F}} |b|^2 \, dsdt\right\}\right] \tag{18.308}$$

III. Correlation quality

$$Q = \left[\frac{\iint_{\infty} BB' dudv}{\iint_{\infty} B^2 dudv}\right] = \left[\frac{\iint_{\mathcal{F}} g|b|^2 \, dsdt}{\iint_{\mathcal{F}} |b|^2 \, dsdt}\right] \tag{18.309}$$

Note that the three criteria are related by

$$Q = \frac{1}{2}(T + \Phi) \tag{18.310}$$

For tackling the second category of imaging systems, Fellgett and Linfoot deduced that the statistical mean information (S.M.I.) content per unit area of image is

$$\text{S.M.I.} = \frac{1}{f^2} \iint_{\mathcal{F}} \log \left\{1 + \frac{|gg_r|^2 \overline{|b(s,t)|^2}}{|gg_r|^2 \overline{|\nu|^2} + \overline{|\nu'|^2}}\right\}^{\frac{1}{2}} dsdt \tag{18.311}$$

where g and g_r are the OTFs of the lens system and the receiver, respectively. $\overline{|b|^2} \equiv \overline{b(s,t)^2}$ is the statistical mean of the spectral power of the intensity distribution on the object plane. $\overline{|\nu|^2}$ and $\overline{|\nu'|^2}$ are the means of the spectral powers of the noise $n(u,v)$ in the object and $n'(u,v)$ in the image, respectively. The Fourier transforms of $n(u,v)$ and $n'(u,v)$ are

$$\nu(s,t) = \iint_{\infty} n(u,v) e^{-i2\pi(us+vt)} dudv$$
$$\nu'(s,t) = \iint_{\infty} n'(u,v) e^{-i2\pi(us+vt)} dudv \tag{18.312}$$

In the case of object and image in the presence of noise, the image quality assessment criteria need to be redefined by

I. Relative structural content

$$T = \left[\iint_{\infty} B'^2 dudv \Big/ \iint_{\infty} B^2 dudv\right] = \left[\iint_{\mathcal{F}} g^2 \overline{|b|^2} dsdt \Big/ \iint_{\mathcal{F}} \overline{|b|^2} dsdt\right] \tag{18.314}$$

II. Fidelity

$$\Phi = \left[1 - \left\{\iint_{\infty} (B' - B)^2 \, dudv \Big/ \iint_{\infty} B^2 dudv\right\}\right]$$
$$= \left[1 - \left\{\iint_{\mathcal{F}} (|1 - g|)^2 \overline{|b|^2} dsdt \Big/ \iint_{\mathcal{F}} \overline{|b|^2} dsdt\right\}\right] \tag{18.315}$$

III. Correlation quality

$$Q = \left[\iint_{\infty} \langle BB' \rangle dudv \Big/ \iint_{\infty} \langle B^2 \rangle dudv \right] = \left[\iint_{\mathcal{F}} g\overline{|b|^2} dsdt \Big/ \iint_{\mathcal{F}} \overline{|b|^2} dsdt \right] \tag{18.316}$$

Note that the effects of the receiver can be incorporated in the treatment by replacing g by (gg_r) in the above relations.

Relative structural content 'T may be considered to express a statistical mean information content in the case where the noise of the object plane can be neglected and the image details are almost completely smothered in the noise in the image plane' [242]. Also, Fidelity $\Phi \leq 1$, and consequently the fidelity defect, $(1-\Phi)$, correspond with the normalized distance in function space between the functions $B'(u,v)$ and $B(u,v)$. Lastly, it is significant to note that, if $|b|^2$ is constant, the correlation quality becomes equal to the Strehl ratio, to a multiplying constant [243].

REFERENCES

1 H.H. Hopkins, 'Introductory: Modern methods of image assessment', *Proc. SPIE* Vol. 274 (1981) pp. 2–11.

2 L. Foucault, 'Mémoir sur la construction des télescopes en verre argenté', *Ann. de L'Observatoire imp. de Paris, Memoirs*, Vol. 5 (1859) pp. 197–237.

3 E. Abbe, 'Beiträge zur Theorie des Mikroskops und der Mikroskopischen Wahrnemung', *Arch. f. Mikrosk. Anat.*, Vol. 9 (1873) pp. 413–468. English Translation by H.E. Fripp, 'A contribution to the theory of the microscope, and the nature of microscopic vision', *Proc. Bristol Naturalists' Society*, Vol. 1 (1876) pp. 200–272.

4 Lord Rayleigh (J.W. Strutt), 'On the theory of optical images, with special reference to the microscope', *Philos. Mag.*, Vol. 42 (1896) pp. 167–195. Also in, *Scientific papers of Lord Rayleigh, Vol. IV*, Cambridge Univ. Press, London (1899) pp. 235–260.

5 W.B. Wetherell, 'The Calculation of Image Quality', Chapter 6 in *Applied Optics and Optical Engineering, Vol. VIII*, Eds. R.R. Shannon and J.C. Wyant, Academic, New York (1980) pp. 171–315.

6 D. Mackenzie, 'The effects of the recording slit width on high frequency response', *Trans. Soc. Motion Pict. Eng.*, Vol. 12 (1928) pp. 730–747.

7 P. Mertz and F. Gray, 'A theory of scanning and its relation to the characteristics of the transmitted signal in telephotography and television', *Bell Syst. Tech. J.*, Vol. 13 (1934) pp. 464–515.

8 H. Frieser, 'Ueber das Auflosungsvermogen von Photographischen Schichten', *Kino-Technik*, Vol. 17 (1935) 167–172. English Translation by D.A. Sinclair, 'On the resolving power of photographic emulsions'; Nuclear Science Abstracts, TT-484; available with U.S. Atomic Energy Commission, Office of Technical Information Extension, Oak Ridge, Tennesse.

9 W.D. Wright, 'Television Optics', *Reports on Progress in Physics, London*, Vol. 5 (1938) pp. 203–209.

10 P.M. Duffieux, *L'Intégral de Fourier et ses Applications à l'optique*, Chez l'Auteur, Rennes, Société Anonyme des Imprimeries Oberthur (1946); also, by Masson, Paris (1970). English Translation, *The Fourier Transform and Its Applications to Optics second edition*, John Wiley, New York (1983). Duffieux is recorded to have the following publications (not available to the author) (i) *Ann. Phys.*, Vol. 14 (1940) pp. 302–338. (ii) *Bull. Soc. Sci. Bretagne*, Vol. 17 (1940) pp. 107–114. (iii) *Ann. Phys.*, Vol. 17 (1942) p. 209.

11 R.K. Luneburg, *Mathematical Theory of Optics*, University of California Press, Berkeley (1964) pp. 354–359. Originally published as mimeographed lecture notes, Brown University, Providence, Rhode Island (1944).

12 E.W.H. Selwyn, 'The photographic and visual Resolving Power of lenses', *Photogr. J., Section B*, Vol. 88 (1948) pp. 6–12.

13 O.H. Schade, 'Electro-optical characteristics of Television systems. Introduction and part I: characteristics of vision and visual systems', *RCA. Rev.*, Vol. 9 (1948) pp. 5–37

14 O.H. Schade, 'Electro-optical characteristics of television systems, part II: electro-optical specifications for television systems', *RCA Rev.*, Vol. 9 (1948) pp. 245–286.
15 O.H. Schade, 'Electro-optical characteristics of television systems, part III: electro-optical characteristics of camera systems', *RCA Rev.*, Vol. 9 (1948) pp. 490–530.
16 O.H. Schade, 'Electro-optical characteristics of television systems, part IV: correlation and evaluation of electro-optical characteristics of imaging systems', *RCA Rev.*, Vol. 9 (1948) pp. 653–686.
17 P. Elias, D.S. Grey, and D.Z. Robinson, 'Fourier treatment of optical processes', *J. Opt. Soc. Am.*, Vol. 42 (1952) pp. 127–134.
18 P. Elias, 'Optics and communication theory', *J. Opt. Soc. Am.*, Vol. 43 (1953) pp. 229–232.
19 J.W. Coltman, 'The specification of imaging properties by response to a sine wave input', *J. Opt. Soc. Am.*, Vol. 44 (1954) pp. 468–471.
20 H.H. Hopkins, 'Problems of Image Assessment and Optical Criteria for Lens Design', *Proc. Symposium on Optical Design with Digital Computers, Technical Optics Section, Imperial College of Science and Technology*, London, 5–7 June, 1954, pp. 67–77.
21 H.H. Hopkins, 'The frequency response of a defocussed optical system', *Proc. Roy. Soc. London, Ser. A*, Vol. 231 (1955) pp. 91–103.
22 M. De, 'The Influence of Astigmatism on the Response Function of an Optical System, *Proc. Roy. Soc. London Ser. A*. Vol. 233 (1955) pp. 91–104.
23 H.H. Hopkins, 'The frequency response of optical systems', *Proc. Phys. Soc. B*, Vol. 69 (1956) pp. 562–576.
24 A.M. Goodbody, 'The influence of spherical aberration on the response function of an optical system', *Proc. Phys. Soc. (London) Ser. B*, Vol. 72 (1958) pp. 411–424.
25 H.H. Hopkins, 'The numerical evaluation of the frequency response of optical systems', *Proc. Phys. Soc. (London) Ser. B*, Vol. 70 (1957) pp. 1002–1005.
26 H.H. Hopkins, 'The aberration permissible in optical systems', *Proc. Phys. Soc. (London) Ser. B*, Vol. 70 (1957) pp. 449–470.
27 H.H. Hopkins, 'Geometrical-optical treatment of frequency response', *Proc. Phys. Soc. (London) Ser. B*, Vol. 70 (1957) pp. 1162–1172.
28 M. De and B.K. Nath, 'Response of optical systems suffering from primary coma', *Optik*, Vol. 19 (1958) pp. 739–750. (vide comments by Barakat et al in reference [59])
29 N.S. Bromilow, 'Geometrical-optical calculation of frequency response for systems with spherical Aberration', *Proc. Phys. Soc. (London) Ser. B*, Vol. 71 (1958) pp. 231–237.
30 A.S. Marathay, 'Geometrical-optical calculation of frequency response for systems with coma', *Proc. Phys. Soc. (London) Ser. B*, Vol. 74 (1959) pp. 721–730.
31 A.M. Goodbody, 'The influence of coma on the response function of an optical system', *Proc. Phys. Soc. (London) Ser. B*, Vol. 75 (1960) pp. 677–688.
32 H.H. Hopkins, 'The application of frequency response techniques in optics', *21st Thomas Young Oration, Proc. Phys. Soc. (London) Ser. B*, Vol. 79 (1962) pp. 889–919.
33 W.H. Steel, 'A study of the combined effects of aberration and a central obscuration of the pupil on contrast of an optical image. Application to microscope objectives using mirrors', *Rev. d'Optique*, Vol. 32 (1953) pp. 4, 143, 269.
34 W.H. Steel, 'The defocussed images of sinusoidal gratings', *Opt. Acta*, Vol. 3 (1956) pp. 65–74.
35 E.L. O'Neill, 'Transfer function for an annular aperture', *J. Opt. Soc. Am.*, Vol. 46 (1956) pp. 285–288. Errata, in *J. Opt. Soc. Am.*, ibid., p. 1096.
36 E.L. O'Neill, *Introduction to Statistical Optics*, Addison-Wesley, Reading, Massachusetts (1963) pp. 58–68.
37 R.V. Shack, 'Outline of practical characteristics of an image-forming system', *J. Opt. Soc. Am.*, Vol. 46 (1956) pp. 755–757.
38 G.B. Parrent and C.J. Drane, 'The effect of defocussing and third order spherical aberration on the transfer function of a two-dimensional optical system', *Opt. Acta*, Vol. 3 (1956) pp. 195–197.
39 E.H. Linfoot, 'Transmission factors and optical design', *J. Opt. Soc. Am.*, Vol. 46 (1956) pp. 740–752.
40 G. Black and E.H. Linfoot, 'Spherical aberration and the information content of optical images', *Proc. R. Soc. London. A*, Vol. 239 (1957) pp. 522–540.

41 A. Lohmann, 'Übertrager Theorie Der Optischen Abbildung, Angewandt Auf Die Apodisation', Ergebnisse Der Internationalen Konferenz Für Wissenschaftliche Photographie, Köln, 24–27 September 1956; also in Photographische Korrespondenz, Vol. 94 (1958) pp. 620–631.
42 A. Maréchal and M. Françon, *Diffraction: Structure des Images*, Éditions de la Revue d'Optique Théorique et Instrumentale (1960).
43 K. Miyamoto, 'On a comparison between wave optics and geometrical optics by using Fourier analysis, I. general theory', *J. Opt. Soc. Am.*, Vol. 48 (1958) pp. 57–63.
44 K. Miyamoto, 'Comparison between wave optics and geometrical optics using Fourier analysis. II. Astigmatism, coma, spherical aberration', *J. Opt. Soc. Am.*, Vol. 48 (1958) pp. 567–575.
45 K. Miyamoto, 'On a comparison between wave optics and geometrical optics by using Fourier analysis. III. Image evaluation by spot diagram', *J. Opt. Soc. Am.*, Vol. 49 (1959) pp. 35–40.
46 K. Miyamoto, 'Wave optics and geometrical optics in optical design', in *Progress in Optics, Vol. I*, Ed. E. Wolf, North-Holland, Amsterdam (1961) pp. 31–66.
47 E.H. Linfoot, 'Convoluted spot diagrams and the quality evaluation of photographic images', *Opt. Acta*, Vol. 9 (1962) pp. 81–100.
48 P.A. Stokseth, 'Properties of a defocussed optical system', *J. Opt. Soc. Am.*, Vol. 59 (1969) pp. 1314–1321.
49 V.N. Mahajan, *Optical Imaging and Aberrations, Part II.*, SPIE Press, Bellingham, Washington, 2nd Edition (2011) pp. 41–44.
50 J.A. Diaz and V.N. Mahajan, 'Diffraction and geometrical transfer functions: calculation time comparison', *Proc. SPIE*, Vol. 10735 (2017) 103750D.
51 J.A. Diaz and V.N. Mahajan, 'Geometrical optical transfer function: is it worth calculating?', *Appl. Opt.*, Vol. 56 (2017) pp. 7998–8004.
52 R. Barakat, 'Computation of the transfer function of an optical system from the design data for rotationally symmetric aberrations. Part I. Theory', *J. Opt. Soc. Am.*, Vol. 52 (1962) pp. 985–991.
53 R. Barakat and M.V. Morello, 'Computation of the transfer function of an optical system from the design data for rotationally symmetric aberrations. Part II. Programming and numerical results', *J. Opt. Soc. Am.*, Vol. 52 (1962) pp. 992–997.
54 F.D. Smith, 'Optical image evaluation and the transfer function', *Appl. Opt.*, Vol. 2 (1963) pp. 335–350.
55 W. Brouwer, E.L. O'Neill, and A. Walther, 'The role of Eikonal and matrix methods in contrast transfer calculations', *Appl. Opt.*, Vol. 2 (1963) pp. 1239–1246.
56 R. Barakat and D. Lev, 'Transfer functions and total illuminance of high numerical aperture systems obeying the sine condition', *J. Opt. Soc. Am.*, Vol. 53 (1963) pp. 324–332.
57 R. Barakat, 'Numerical results concerning the transfer functions and total illuminance for optimum balanced fifth-order spherical aberration', *J. Opt. Soc. Am.*, Vol. 54 (1964) pp. 38–44.
58 R. Barakat and A. Houston, 'Transfer function of an annular aperture in the presence of spherical aberration', *J. Opt. Soc. Am.*, Vol. 55 (1965) pp. 538–541.
59 R. Barakat and A. Houston, 'Transfer function of an optical system in the presence of off-axis aberrations', *J. Opt. Soc. Am.*, Vol. 55 (1965) pp. 1142–1148.
60 J.B. DeVelis, 'Effect of symmetric and asymmetric aberrations on the phase of the transfer function', *J. Opt. Soc. Am.*, Vol. 55 (1965) pp. 1632–1638.
61 R. Barakat, 'Determination of the optical transfer function directly from the edge spread function', *J. Opt. Soc. Am.*, Vol. 55 (1965) pp. 1217–1221.
62 L. Levi and R.H. Austing, 'Tables of the modulation transfer function of a defocussed perfect system', *Appl. Opt.*, Vol. 7 (1968) pp. 967–974. Errata in ibid., p.2258.
63 K. Murata, 'Instruments for the Measuring of Optical Transfer Functions', in *Progress in Optics, Vol. V*, Ed. E. Wolf, North-Holland, Amsterdam (1966) pp. 199–245.
64 K. Rosenhauer and K.J. Rosenbruch, 'The measurement of the optical transfer function of lenses', *in Reports on Progress in Physics*, Vol. XXX (1967) Part I, pp. 1–25.
65 S.H. Lerman, W.A. Minnick, M.P. Rimmer, and R.R. Shannon, 'New method of computing the optical transfer function', *Spring Meeting of the OSA, J. Opt. Soc. Am.*, Vol. 57 (1967) pp. 566.
66 S.H. Lerman, 'Application of the fast Fourier transform to the calculation of the optical transfer function', *Proc. SPIE* Vol. 13 (1969) pp. 51–70.

67 J.W. Cooley and J.W. Tukey, 'An algorithm for the machine calculation of complex Fourier series', *Math. Computation*, Vol. 19 (1965) pp. 297–301.
68 E.O. Brigham, *The Fast Fourier Transform*, Prentice-Hall, Englewood Cliffs, New Jersey (1974).
69 D. Heshmaty-Manesh and S.C. Tam, 'Optical transfer function calculation by Winograd's fast Fourier transform', *Applied Optics*, Vol. 21 (1982) pp. 3273–3277.
70 J.W. Cooley and J.W. Tukey, 'On the origin and publication of the FFT paper', *Current Contents*, Vol. 33 No. 51–52, Dec. 20–27 (1993) pp. 8–9.
71 J. Macdonald, 'The calculation of the optical transfer function', *Opt. Acta*, Vol. 18 (1971) pp. 269–290.
72 E.C. Kintner and R.M. Sillitto, 'A new 'analytic' method for computing the optical transfer function', *Opt. Acta*, Vol. 23 (1976) pp. 607–619.
73 M.J. Kidger, 'The calculation of the optical transfer function using Gaussian quadrature', *Opt. Acta*, Vol. 25 (1978) pp. 665–680.
74 A.M. Plight, 'The rapid calculation of the optical transfer function for on-axis systems using the orthogonal properties of Tchebycheff polynomials', *Opt. Acta*, Vol. 25 (1978) pp. 849–860.
75 P.K. Purkait, L.N. Hazra, and M. De, 'Walsh functions in lens optimization II. Evaluation of the diffraction –based OTF for on-axis imagery', *Opt. Acta*, Vol. 28 (1981) pp. 389–396.
76 R.E. Hopkins and D. Dutton, 'Interlaboratory comparison of MTF measurements and computations on a large wide-angle lens', *Opt. Acta*, Vol. 18 (1971) pp. 105–121.
77 J. Macdonald, 'Comparison of results of various OTF calculation programs', *Opt. Engng.*, Vol. 14 (1975) pp. 166–168.
78 R.B. Johnson and W. Swantner, 'MTF computational uncertainties', *OE Reports, #104*, August (1992).
79 M.P. Rimmer, 'MTF computational issues', *OE Reports, #109*. January (1993).
80 D.W. Griffin, 'Lens design code standardization', *OSA Proc. Int. Opt. Design Conf.*, Vol. 22, Ed. G.W. Forbes (1994) pp. 28–31.
81 Ir. J.A.J. van Leunan, 'Problems of OTF standardisation for non-perfect imaging systems', *Proc. SPIE*, Vol. 46 (1974) pp. 81–85.
82 P. Kuttner, 'Interlaboratory comparisons of MTF measurements and calculations and OTF standards in Germany', *Opt. Engng.*, Vol. 14 (1975) pp. 151–156.
83 A.C. Marchant, E.A. Ironside, J.F. Attryde, and T.L. Williams, 'The reproducibility of MTF measurements', *Opt. Acta*, Vol. 22 (1975) pp. 249–264.
84 P. Kuttner, 'Interlaboratory comparisons of MTF measurements and calculations', *Opt. Acta*, Vol. 22 (1975) pp. 265–275.
85 T. Nitou and T. Ose, 'Interlaboratory comparison of MTF measurements in Japan', *Proc. SPIE*, Vol. 98 (1976) pp. 21–27.
86 A.C. Marchant, 'Accuracy in image evaluation: setting up an OTF standards laboratory', *Opt. Acta*, Vol. 18 (1971) pp. 133–137.
87 A.C. Marchant, 'Progress in the United Kingdom towards objective standards for air camera lenses', *Proc. SPIE*, Vol. 46 (1974) pp. 67–72.
88 P. Kuttner, 'OTF Standards in Germany', *Proc. SPIE*, Vol. 46 (1974) pp. 59–66.
89 P. Kuttner, 'Review of German standards on image assessment', *Opt. Engng.*, Vol. 17 (1978) pp. 90–93.
90 T. Ose and K. Murata, 'Standards of OTF in Japan', *Proc. SPIE*, Vol. 46 (1974) pp. 73–80.
91 T. Nakamura, Y. Sekine, T. Nito, and T. Ose, 'OTF standardization in Japan', *Proc. SPIE*, Vol. 98 (1976) pp. 120–124.
92 J. Simon, 'Measurement of the MTF in France', *Proc. SPIE*, Vol. 98 (1976) pp. 125–131.
93 Ir. J.A.J. van Launen, 'OTF (Optical Transfer Function) standardization' (in Holland), *Proc. SPIE*, Vol. 98 (1976) pp. 132–135.
94 C.L. Norton, G.C. Brock, and R. Welch, 'Optical and modulation transfer functions', *Photogrammetric Engineering and Remote Sensing*, Vol. 43 (1977) pp. 613–636.
95 *ANSI Guide to Optical Transfer Function Measurement and Reporting, ANSI PH 3.57:78 (R1989)*, American National Standards Institute, New York. Superseded by *Optics and Optical Instruments – Optical Transfer Function – Principles and Procedures of Measurement NAPM IT3.612:1997*, National Association of Photographic Manufacturers, Inc., USA.

96 'Optics and photonics–Optical transfer function–Definitions and mathematical Relationships', ISO standard 9334:2012. (Last reviewed and confirmed in 2017).
97 'Optics and photonics–Optical transfer function–Principles and procedures of measurement', ISO standard 9335:2012. (Last reviewed and confirmed in 2017).
98 'Optics and optical instruments–accuracy of optical transfer function (OTF) measurement', ISO standard 11421:1997. (Last reviewed and confirmed in 2017)
99 'Photography – Electronic still picture imaging – Resolution and spatial frequency responses', ISO standard 12233:2017.
100 'Optics and photonics – Optical transfer function – Principles of measurement of modulation transfer function (MTF) of sampled imaging systems', ISO standard 15529:2010. (Last reviewed and confirmed in 2015).
101 C.S. Williams and O.A. Becklund, *Introduction to the Optical Transfer Function*, John Wiley, New York (1989).
102 T.L. Williams, *The Optical Transfer Function of Imaging Systems*, Institute of Physics, Bristol (1999).
103 G.D. Boreman, *Modulation Transfer Function in Optical Electro-Optical Systems*, Vol. TT52, SPIE, Bellingham, Washington (2001).
104 L.R. Baker, *Selected Papers on Optical Transfer Function: Foundation and Theory*, SPIE Milestone Series, Vol. MS 59, (1992).
105 B.P. Lathi, Signals, *Systems and Communications*, Wiley, New York (1965).
106 G.R. Cooper and C.D. McGillem, *Methods of Signal and System Analysis*, Holt, Rinehartand Winston, New York (1967).
107 A.B. Carlson, *Communication Systems*, McGraw-Hill Kogakusha, Tokyo (1968).
108 J.W. Goodman, *Introduction to Fourier Optics, First Edition*, McGraw Hill, New York (1968).
109 J.C. Dainty and R. Shaw, *Image Science,* Academic Press, London (1974).
110 J.D. Gaskill, *Linear Systems, Fourier Transforms, and Optics*, Wiley, New York, NY (1978).
111 G.O. Reynolds, J.B. DeVelis, G.B. Parrent, and B.J. Thompson, *The New Physical Optics Notebook: Tutorials in Fourier Optics*, SPIE, Bellingham, Washington, and AIP, New York (1989).
112 E.G. Steward, *Fourier Optics: An Introduction*, Dover, New York (2004).
113 O.K. Ersoy, *Diffraction, Fourier Optics and Imaging*, Wiley-Interscience, New Jersey (2007).
114 E. Ingelstam, 'Attempts to treat nonlinear imaging devices', Invited paper, ICO conference on 'Photographic and Spectroscopic Optics', Tokyo and Kyoto, September 1964; abridged version published by Institutet För Optisk Forskning, Kungl. Tekniska Högskolan, Stockholm 70, Sweden.
115 W.T. Welford, 'Aplanatism and isoplanatism', in *Progress in Optics, Vol. XIII*, Ed. E. Wolf, North-Holland, Amsterdam (1976) pp. 267–292.
116 A.W. Lohmann and D.P. Paris, 'Space-variant image formation', *J. Opt. Soc. Am.*, Vol. 55 (1965) pp. 1007–1013.
117 B. Dube, R. Cicala, A. Closz, and J.P. Rolland, 'How good is your lens? Assessing performance with MTF full-field displays', *Appl. Opt.*, Vol. 56 (2017) pp. 5661–5667.
118 H.H. Hopkins, 'Canonical and real-space coordinates used in the theory of image formation', Chapter 8 *in Applied Optics and Optical Engineering*, Vol. IX, Ed. R. R. Shannon and J. C. Wyant, Academic, New York (1983) pp. 307–369.
119 A. Papoulis, *The Fourier Integral and Its Applications*, McGraw-Hill, New York (1962).
120 E.H. Linfoot, *Fourier Methods in Optical Image Evaluation*, Focal, London (1964).
121 R. Bracewell, *The Fourier Transform and Its Applications*, McGraw-Hill, New York (1965).
122 A. Papoulis, *Systems and Transforms with Applications in Optics*, McGraw Hill, New York, NY (1968).
123 R.V. Shack, 'On the significance of the phase transfer function, *Proc. SPIE*, Vol. 46 (1974) pp. 39–43.
124 R.E. Hufnagel, 'Significance of phase of optical transfer functions', *J. Opt. Soc. Am.*, Vol. 58 (1968) pp. 1505–1506.
125 H.H. Hopkins, 'Image shift, phase distortion and the optical transfer function', *Opt. Acta*, Vol. 31 (1984) pp. 345–368.
126. M. Françon, *Diffraction: Coherence in Optics*, Pergamon, Oxford (1966) [Translated from the French book by B. Jeffrey]
127 H.H. Hopkins, 'The concept of partial coherence in optics', *Proc. Roy. Soc. A*, Vol. 208 (1951) pp. 263–277.
128 H.H. Hopkins, 'On the diffraction theory of optical images', *Proc. Roy. Soc. A*, Vol. 217 (1953) pp. 408–432.

129 R.E. Kinzly, 'Investigations of the influence of the degree of coherence upon images of edge objects', *J. Opt. Soc. Am.*, Vol. 55 (1965) pp. 1002–1007.
130 M. Born and E. Wolf, *Principles of Optics*, Pergamon, New York (1959) Chapter X.
131 M.J. Beran and G.B. Parrent, *Theory of Partial Coherence*, Prentice-Hall, New Jersey (1964).
132 P.S. Considine, 'Effects of coherence on imaging systems', *J. Opt. Soc. Am.*, Vol. 56 (1966) pp. 1001–1009.
133 E. Abbe, 'Beiträge zur Theorie des Mikroskops und der mikroskopischen Wahrnehmung', in M. Schultze's Archiv für mikroskopische Anatomie, Vol. IX (1873) pp. 413–468. Also, in *Gesammelte Abhandlungen von Ernst Abbe, Erster Band: Abhandlungen über die Theorie des Mikroskops*, Gustav Fischer, Jena (1904) pp. 45–100. English Translation by H.E. Fripp, 'A Contribution to the Theory of the Microscope, and the nature of Microscopic Vision', *Proc. Bristol Naturalists' Society, New Series*, Vol. I (1876) pp. 200–272.
134 H. von Helmholtz, 'On the limits of the optical capacity of the microscope', *Proc. Bristol Naturalists' Society, New Series*, Vol. I (1876) pp. 407–440. [English Translation of German article in Poggendorff's Annalen (1874)]
135 Ref. 130, Sixth Edition (1984) pp. 419–424.
136 'The Abbe Theory of Imaging', Chapter 21 in *Handbook of Optical Systems, Vol. 2*, Ed. H. Gross, Wiley-VCH (2005) pp. 239–282.
137 R. Bracewell, Ref. 121, pp. 112–113.
138 R. Bracewell, Ref. 121, pp. 115–117.
139 R.E. Hufnagel, 'Simple approximation to diffraction-limited MTF', *Appl. Opt.*, Vol. 10 (1971) pp. 2547–2548.
140 J.W. Goodman, *Introduction to Fourier Optics, Third Edition*, Viva Books, New Delhi and Roberts & Company, Colorado (2007) p. 154.
141 E.L. O'Neill, Ref. 36, pp. 86–87.
142 R. Bracewell, Ref. 121, p. 159.
143 A. Papoulis, Ref. 122, p.177.
144 J.W. Goodman, Ref. 140, pp. 149–141.
145 W.H. Steel, 'The defocussed image of sinusoidal gratings', *Opt. Acta*, Vol. 3 (1956) pp. 65–74.
146 R.R. Shannon, 'A useful optical engineering approximation', *Optics and Photonics News*, Vol. 5(4) (1994) p. 34.
147 J. Macdonald, Personal communication.
148 K. Rosenhauer, K.-J. Rosenbruch, and F.A. Sunder-Plaβmann, 'The relation between the axial aberrations of Photographic lenses and their optical transfer function', *Appl. Opt.*, Vol. 5 (1966) pp. 415–420.
149 R.R. Shannon, 'Aberrations and their effects on images', *Proc. SPIE*, Vol. 531(1985) pp. 27–48.
150 R.R. Shannon, 'Some Recent Advances in the Specification and Assessment of Optical Images', in Proc. 1969 International Commission for Optics Meeting, published as *Optical Instruments and Techniques*, Oriel Press, England (1970) pp. 331–345.
151 R. Straubel, VIII. Intern. Kongr, f. Photographie, Dresden (1931); Pieter Zeeman Verhandelingen, Martnus Nijhoff, den Haag (1935) p. 302.
152 R.K. Luneburg, Ref. 11, pp. 349–358.
153 R. Barakat, 'Solution of the Luneberg apodization problems', *J. Opt. Soc. Am.*, Vol. 52 (1962) pp. 264–275.
154 J.E. Wilkins, Jr., 'Luneberg apodization problems', *J. Opt. Soc. Am.*, Vol. 53 (1963) pp. 420–424.
155 P. Jacquinot and B. Roizen-Dossier, 'Apodisation', *Chapter II in Progress in Optics*, Vol. III, Ed., E. Wolf, North-Holland, Amsterdam (1964) pp, 29–186.
156 L.N. Hazra, Maximization of Strehl's ratio in the farfield pattern of a telescope under partially coherent illumination, *J. Opt (India).*, Vol. 4, (1975) pp. 51–54.
157 L.N. Hazra and A. Banerjee, 'Application of Walsh functions in generation of optimum apodizers', *J. Opt.(India)*, Vol. 5 (1976) pp. 19–26.
158 L.N. Hazra, 'A new class of amplitude filters', *Opt. Commun.*, Vol. 21 (1977) pp. 232–236.
159 J.E. Wilkins, Jr., 'Apodization for maximum Strehl ratio and specified Rayleigh limit of resolution', *I, J. Opt. Soc. Am.*, Vol. 67 (1977) pp. 1027–1030; II, *J. Opt. Soc. Am.*, Vol. 69 (1979) pp. 1526–1530.
160 J.A. Macdonald, 'Apodization and frequency response with incoherent light', *Proc. Phys. Soc. London*, Vol. 72 (1958) pp. 749–756.

161 A. Lohmann, 'Übertragertheorie der optischen abbildung, angewandt aufdie apodisation', sonderdruk aus *Wissenschaftliche Photographie*, Egebnisse der Int. Konf. für Wissenschaftliche Photographie, Köln, 24–27 Sept., 1956, Verlag Dr. Othmar Helwich, Darmstadt (1958) pp. 620–631.
162 A. Boivin, 'Théorie des pupilles circulaires operant une modulation radiale simultanée de la phase et de l'amplitude', *Chapter IV in Théorie et calcul des figures de diffraction de revolution, A. Boivin, Univ. Laval, Quebec, Gauthier-Villars, Paris* (1964).
163 S.C. Biswas and A. Boivin, 'Influence of spherical aberration on the performance of optimum apodizers', *Opt. Acta*, Vol. 23 (1976) pp. 569–588.
164 M.J. Yzuel and F. Calvo, 'A study of the possibility of image optimization by apodization filters in optical systems with residual aberrations', *Opt. Acta*, Vol. 26 (1979) pp. 1397–1406.
165 L.N. Hazra, P.K. Purkait, and M. De, 'Apodization of aberrated pupils', *Can. J. Phys.*, Vol. 57 (1979) pp. 1340–1346.
166. J.P. Mills and B.J. Thompson, 'Effect of aberrations and apodization on the performance of coherent optical systems. I. The amplitude impulse response II. Imaging', *J. Opt. Soc. Am. A*, Vol. 3 (1986) pp. 694–703, 704–716.
167 P. Bernat-Molina, J. Castejón-Mochón, A. Bradley, and N. Lopez-Gill, 'Focus correction in an apodized system with spherical aberration', *J. Opt. Soc. Am. A*, Vol. 32 (2015) pp. 1556–1563.
168 A.N.K. Reddy, M. Hashemi, and S.N. Khonina, 'Apodization of two-dimensional pupils with aberrations', *Pramana-J. Phys.*, Vol. 90 (2018) pp. 77–85.
169 H.H. Hopkins, 'The aberration permissible in optical systems', *Proc. Phys. Soc. London, Sec. B*, Vol. 70 (1957) pp. 449–470.
170 C.S. Williams and O.A. Becklund, Ref. 101, pp. 250–255.
171 A. Cox, 'Image Evaluation by Edge Gradients', in Proceedings of the NBS Semi-Centennial Symposium on Optical Image Evaluation held during 18–20 October 1951, National Bureau of Standards Circular 526, Superintendent of Documents, U. S. Government Printing Office, Washington 25, D. C., Issued April 29,1954, pp. 267–273.
172 R.V. Shack, 'A Proposed Approach to Image Evaluation', in Proceedings of the NBS Semi-Centennial Symposium on Optical Image Evaluation held during 18–20 October 1951, National Bureau of Standards Circular 526, Superintendent of Documents, U. S. Government Printing Office, Washington 25, D. C., Issued April 29, 1954, pp. 275–286.
173 W.H. Steel, 'Calcul de la repartition de la lumière dans l'image d'une ligne', *Rev. Opt.*, Vol. 31 (1952) pp. 334–340.
174 W. Weinstein, 'Light distribution in the image of an incoherently illuminated edge', *J. Opt. Soc. Am.*, Vol. 44 (1954) pp. 610–615.
175 R.L. Lamberts, G.C. Higgins, and R.N. Wolfe, 'Measurement and analysis of the distribution of energy in optical images', *J. Opt. Soc. Am.*, Vol. 48 (1958) pp. 487–490.
176 R.L. Lamberts, 'Relationship between the sine wave response and the distribution of energy in the optical image of a line', *J. Opt. Soc. Am.*, Vol. 48 (1958) pp. 490–495.
177 F. Scott, R.M. Scott, and R.V. Shack, 'The use of edge gradients in determining modulation transfer functions', *Photographic Science and Engineering*, Vol. 7 (1963) pp. 345–349.
178 R. Barakat and A. Houston, 'Line spread function and cumulative line spread function for systems with rotational symmetry', *J. Opt. Soc. Am.*, Vol. 54 (1964) pp. 768–773.
179 E.W. Marchand, 'Derivation of the point spread function from the line spread function', *J. Opt. Soc. Am.*, Vol. 54 (1964) pp. 915–919.
180 E.W. Marchand, 'From the line to point spread function: the general case', *J. Opt. Soc. Am.*, Vol. 55 (1965) pp. 352–355
181 B. Tatian, 'Method for obtaining the transfer function from the edge response function', *J. Opt. Soc. Am.*, Vol. 55 (1965) pp. 1014–1019.
182 R. Barakat and A. Houston, 'Line spread and edge spread functions in presence of off-axis aberrations', *J. Opt. Soc. Am.* 55 (1965) pp. 1132–1135.
183 R. Barakat, 'Determination of the optical transfer function directly from the edge spread function', *J. Opt. Soc. Am.*, Vol. 55 (1965) pp. 1217–1221.
184 S.H. Lerman and R.R. Shannon, 'Two-Dimensional Image Prediction and Enhancement', *Proc. SPIE*, Vol. 10 (1967) pp. 79–84.

185 M.E. Rabedeau, 'Effect of truncation of line spread and edge response functions on the computed optical transfer function', *J. Opt. Soc. Am.*, Vol. 59 (1969) pp. 1309–1311.

186 B. Tatian, 'Asymptotic expansions for correcting truncation error in transfer function calculation', *J. Opt. Soc. Am.* Vol. 61 (1971) pp. 1214–1224.

187 P.L. Smith, 'New techniques for estimating the MTF of an imaging system from its edge response', *Appl. Opt.*, Vol. 11 (1972) pp. 1424–1425.

188 B. Tatian, 'Comment on: New techniques for estimating the MTF of an imaging system from its edge response', *Appl. Opt.*, Vol. 11 (1972) p. 2975.

189 P.L. Smith, 'Author's reply to Comment on: New techniques for estimating the MTF of an imaging system from its edge response', *Appl. Opt.*, Vol. 11 (1972) p. 2975.

190 J.C. Dainty and R. Shaw, *Image Science*, Academic, London (1974).

191 H.H. Hopkins, 'The Development of Image Evaluation Methods', *Proc. SPIE*, Vol. 46 (1974) pp. 2–

192 J. Macdonald, 'New analytical and numerical techniques in optical design', Ph. D. Thesis, University of Reading, U.K. (1974).

193 A. Basuray, S.K. Sarkar, and M. De, 'A new method for edge image analysis', *Opt. Acta*, Vol. 26 (1979) pp. 349–356.

194 S. Szapiel, *Diffraction-based Image Assessment in Optical Design*, Wydawnictwa Politechniki Warszawskiej, Warszawa (1986).

195 R. Chmelik, 'Focusing and the optical transfer function in a rotationally symmetric optical system', *Appl. Opt.*, Vol. 33 (1994) pp. 3702–3704.

196 G.D. Boreman and S. Yang, *'Modulation transfer function measurement using three- and four-bar targets', Appl. Opt.*, Vol. 34 (1995) pp. 8050–8052.

197 M.L. Calvo, M. Chevalier, V. Lakshminarayanan, and P.K. Mondal, 'Resolution criteria and modulation transfer function (MTF)/ line spread function (LSF) relationship in diffraction limited systems', *J. Opt. (India)*, Vol. 25 (1996) pp. 1–21.

198 A. Manzanares, M.L. Calvo, M. Chevalier, and V. Lakshminarayanan, 'Line spread function formulation proposed by W. H. Steel: a revision', *Appl. Opt.*, Vol. 36 (1997) pp. 4362–4366.

199 R. Shaw, 'A century of image quality', *Proceedings of the Conference on Image Processing, Image Quality and Image Capture Systems (PICS-99)*, Savannah, Georgia, USA, April 25–28 1999. IS&T- The Society for Imaging Science and Technology (1999) pp. 221–224.

200 R. Bracewell, Ref. 121, pp. 262, 296.

201 J.W. Goodman, Ref. 140, pp. 10–12.

202 H.H. Hopkins and B. Zalar, 'Aberration tolerances based on the line spread function', *J. Mod. Opt.*, Vol. 34 (1987) pp. 371–406.

203 W. Struve, 'Beitrag zur Theorie der Diffraction an Fernröhren', *Annalen der Physik*, Vol. 253, Issue 13 (1882) pp. 1008–1016.

204 V.N. Mahajan, Ref. 49, pp. 62, 196.

205 H. Osterberg, 'Evaluation of Phase Optical Tests', Section 26 in MIL-HDBK-141, Military Standardization Handbook: Optical Design, Defence Supply Agency, Washington (1962) 26–1 to 26–10.

206 M. Shah, S. Sarkar, and A. Basuray, 'Evaluation of the optical transfer function from an incoherent image of a sector disc', *Opt. Acta*, Vol. 33 (1986) pp. 1287–1293.

207 R. Hughes, 'Sensor model requirements for TAWS/IRTSS operation', M.S. Thesis, Naval Postgraduate School, Monterey, California, p. 8.

208 J.W. Coltman, 'The specification of imaging properties by response to a sine wave input', *J. Opt. Soc. Am.*, Vol. 44 (1954) pp. 468–471.

209 R.L. Lucke. 'Deriving the Coltman correction for transforming the bar transfer function to the optical transfer function (or the contrast transfer function to the modulation transfer function)', *Appl. Opt.*, Vol. 37 (1998) pp. 7248–7252.

210 G.D. Boreman, Modulation Transfer Function in Optical and Electro-Optical Systems, SPIE Press, Bellingham, Massachusetts, Washington (2001).

211 J. DeVelis and G.B. Parrent, *Jr., Transfer Function for Cascaded Optical Systems, J. Opt. Soc. Am.*, Vol. 57 (1967) pp. 1486–1490.

212 R.E. Swing, 'The case for the pupil function', *Proc. SPIE*, Vol. 46 (1974) pp. 104–113.

213 T.L. Williams, The Optical Transfer Function of Imaging Systems, Institute of Physics, Bristol (1999) pp. 33–35.

214 W.S. Kovach, 'Energy distribution in the PSF for an arbitrary passband', *Appl. Opt.*, Vol. 13 (1974) pp. 1769–1771.
215 M.J. Yzuel and J. Santamaria, 'Polychromatic optical image. Diffraction limited system and influence of the longitudinal chromatic aberration', *Opt. Acta*, Vol. 22 (1975) pp. 673–690.
216 M.J. Yzuel and J. Bescós, 'Polychromatic off axis optical image. Influence of the transverse chromatic aberration', *Opt. Acta*, Vol. 22 (1975) pp. 913–931.
217 M.J. Yzuel and J. Bescós, 'Colour in the polychromatic off-axis point-spread function in the presence of primary coma and astigmatism', *Opt. Acta*, Vol. 23 (1976) pp. 529–548.
218 I.V. Peĭsakhson and T.A. Cherevko, 'Numerical methods of estimating the image quality for optical systems operating in a wide spectral region', *Sov. J. Opt. Technol.*, Vol. 54 (1987) pp. 310–315.
219 H.S. Dhadwal and J. Hantgan, 'Genealized point spread function for a diffraction limited aberration free imaging system with polychromatic illumination, *Opt. Engng.*, Vol. 28 (1989) pp. 1237–1240.
220 V.N. Mahajan, Ref. 49, pp. 205–208.
221 L. Levi, 'Detector response and perfect lens MTF in polychromatic light', *Appl. Opt.*, Vol. 8 (1969) pp. 607–616.
222 R. Barnden, 'Calculation of axial polychromatic optical transfer function', *Opt. Acta*, Vol. 21 (1974) pp. 981–1003.
223 R. Barnden, 'Extra-axial polychromatic optical transfer function', *Opt. Acta*, Vol. 23 (1976) pp. 1–24.
224 M. Takeda, 'Chromatic aberration matching of the polychromatic optical transfer function', *Appl. Opt.*, Vol. 20 (1981) pp. 684–687.
225 D. Yu. Gal'pern, 'Frequency contrast characteristic of optical systems possessing magnification chromatism', *Sov. J. Opt. Technol.*, Vol. 45 (1978) pp. 205–209.
226 M. Subbarao, 'Optical transfer function of a diffraction-limited system for polychromatic illumination', *Appl. Opt.*, Vol. 29 (1990) pp. 554–558.
227 I.V. Peĭsakhson, 'Image contrast in optical systems possessing magnification chromatism', *J. Opt. Technol.* Vol. 64 (1997) pp. 217–220.
228 I.V. Peĭsakhson, 'Estimating the effect of chromatic aberrations on the contrast of an optical image', Vol. 64 (1997) pp. 221–223.
229 V.N. Mahajan, Ref. 49, pp. 208–209.
230 L. Levi, 'Detector response and perfect-lens-MTF in polychromatic light', *Appl. Opt.*, Vol. 8 (1969) pp. 607–616.
231 S. Minami and R. Ogawa, 'Improvements in the calculation of white light modulation transfer function (MTF) in Japan Optical Engineering Research Association (JOERA)', *Proc. SPIE*, Vol. 237, International Lens Design Conference (1980) pp. 164–173.
232 R. Ogawa and S. Tachihara, 'Calculation of white light optical transfer function (OTF) in the existence of lateral chromatic aberration', *Proc. SPIE*, Vol. 237, International Lens Design Conference (1980) pp. 173–181.
233 P.B. Fellgett and E.H. Linfoot, 'On the assessment of optical images', *Proc. Roy. Soc. London*, Vol. 247 (1955) pp. 369–407.
234 E.H. Linfoot, 'Information theory and optical images', *J. Opt. Soc. Am.*, Vol. 45 (1955) pp. 808–819.
235 E.H. Linfoot, *Recent Advances in Optics*, Clarendon, Oxford (1955).
236 E.H. Linfoot, 'Transmission factors and optical design', *J. Opt. Soc. Am.*, Vol. 46 (1956) pp. 740–752.
237 E.H. Linfoot, 'Image quality and optical resolution', *Opt. Acta*, Vol. 4 (1957) pp. 12–16.
238 E.H. Linfoot, 'Quality evaluations of optical systems', *Opt. Acta*, Vol. 5 (1958) pp. 1–14.
239 C.E. Shannon, 'A mathematical theory of communication', *Bell Syst. Tech. Jour.*, Vol. XXVII (1948) pp. 379–423, 623–656.
240 C.E. Shannon, 'Communication in the presence of noise', *Proc. Inst. Radio Engrs.*, Vol. 37 (1949) pp. 10–21; also, *in Proc. IEEE*, Vol. 86 (1998) pp. 447–457.
241 D. Middleton, An Introduction to Statistical Communication Theory, Mc-Graw Hill, New York (1960).
242 K. Miyamoto, 'Wave optics and geometrical optics in optical design', Chapter II in Progress in Optics, Vol. I, Ed. E. Wolf, North-Holland, Amsterdam (1961) pp. 60–62.
243 A. K. Ghosh, 'Image Evaluation Techniques: 1. Different physical parameters and their numerical evaluation', J. Opt. (Ind.), Vol. 1 (1972) 1–6.

19

Basics of Lens Design

Etymologically, the word 'lens' in optics is derived from 'lēns', the Latin word for 'lentil', a commonly eaten kind of legume. The similarity in the shape of the biconvex lens, used in early days for focusing, with that of the lentil prompted adoption of this nomenclature for a glass lens in optics. Over the years, 'lens' has become a generic name for any device used for manipulating the shape of a wavefront or a beam, or of the centric characteristics of a pencil of rays over any subdomain of electromagnetic or similar waves where the concepts of visible light are applicable. However, in what follows, our discussion is restricted to imaging lenses that operate in the visible wavelengths, and the neighbouring infrared and ultraviolet regions. For practical purposes of analysis and design, a mirror is often treated as a special case of a lens with incident and emergent media having refractive indices of opposite sign.

A single piece of glass or any transparent material that is rotationally symmetric around an axis containing centres of the two refracting spherical interfaces of the piece is a typical example of a 'lens'. In general, any one of the interfaces—or both of them—can be of any shape that is symmetric around the axis. Strictly speaking, it is a single *lens element*. The same word 'lens' is used to describe the imaging system of most optical imaging systems, e.g. 'camera lens', 'projection lens', etc. Although not specified, any of the latter lenses implies a *lens system* that mostly consist of several lens elements. At times, in a multi-element lens system, the multiple elements lie in axially separated clusters of lens elements, where in a single cluster the lens elements lie in close proximity in comparison to inter-cluster axial separations. In the jargon of optical engineering, these clusters are often called *lens components*. Figure 19.1 shows an imaging lens system that consists of four lens components. From the object side, the first lens component consists of four lens elements, and the second, third, the fourth component each consist of two lens elements, so that the lens system consists of ten (= 4+2+2+2) lens elements. However, there are occasional variations in this usage of the terms, and so, for correct understanding of the particular use of the word 'lens' in a specific case, it is prudent to be guided by the context.

Traditionally, optical imaging elements composed of homogeneous refracting materials used to be called lenses. But, at present the term 'lens' is used in an extended sense to indicate any optical device that forms the image of an object. The device may consist not only of optical elements composed of homogeneous or inhomogeneous optical material, but also of reflecting optical elements and diffractive optical elements. The use of terms like 'mirror lens', 'diffractive lens', 'gradient index (GRIN) lens' is now in vogue. In practice, optimal design of these elements pertaining to specific requirements of different applications poses a major challenge in optical system design, except for trivial cases. Fortunately, in most cases, the design of these lenses can be accommodated within the framework of the process of traditional lens design.

19.1 Statement of the Problem of Lens Design

The problem of lens design involves determining an appropriate lens system that can cater to the imaging requirements of an end-user. It is a typical inverse problem, as demonstrated in Figure 19.2. Commonly, the user requirements are mentioned in terms of a few combinations from the following list:

- Object/Image size (Linear or Angular)
- Magnification (Transverse/Longitudinal/Angular)
- Object/Image location and/or object-to-image throw
- Nature of object: self-luminous/trans-illuminated/aerial

DOI: 10.1201/9780429154812-19

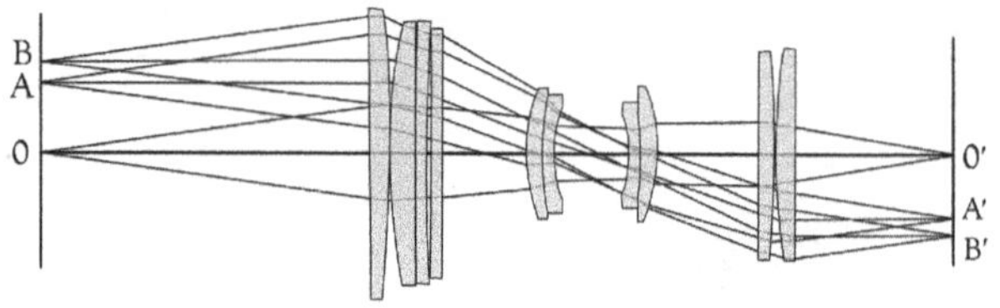

FIGURE 19.1 A lens 'system' forming image O′A′B′ of an object OAB. It consists of four lens 'components' and ten lens 'elements'.

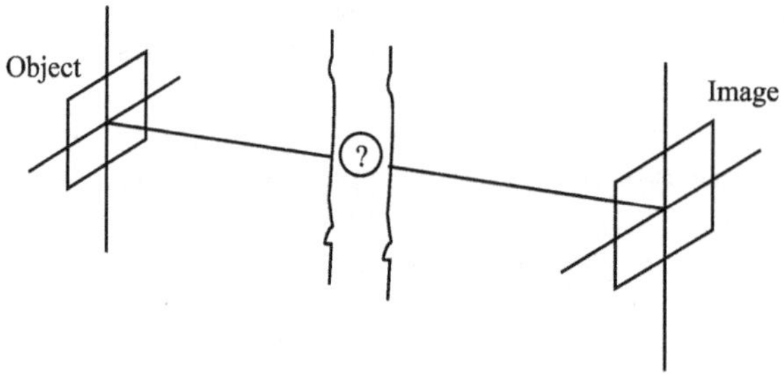

FIGURE 19.2 Lens design is typically an inverse problem.

- Image brightness
- Resolution
- Sharpness
- Wavelength range of illumination (Broadband/Monochromatic/Narrowband)
- Detector: Eye/CCD/Film etc.
- Optical material specifications

19.2 Lens Design Methodology

It is the task of the lens designer to sort out the issue of conflicting requirements, if any, in consultation with the end-user, and to set out the optical specifications for the required lens system. A list of lens specifications for conventional applications is given below:

1. Gaussian/paraxial specifications, e.g., focal length, F-no./numerical aperture, paraxial invariant, etc.
2. One or more of the image quality assessment criteria based on spot diagram/point spread function (PSF)/optical transfer function (OTF)/edge spread function (ESF)/bar spread function (BSF), etc.
3. Image brightness/illumination
4. Depth of field/focus

Concurrently, the lens designer needs to keep in view the following constraints:

a) Physical realizability
b) Manufacturability
c) Material availability
d) Cost limitation
e) Available time
f) Ergonomic and environmental considerations

Some of these constraints are problem specific; they need to be ascertained by the lens designer in close consultation with the user of the design. In general, the lens designer should have knowledge of the optical fabrication and testing technology in order to avoid prescribing unrealistic designs or designs that call for unnecessary complexities in fabrication and testing processes [1–5]. Kingslake deliberated in detail on these issues, and also on the interrelationship between lens design and optical system design in his treatises [6–8]. Welford gave a broad overview on the topic of optical design in the broader context of instrument design [9]

At the outset, the prospective system variables available to the lens designer consist of the following:

1. Number of lens elements
2. Shapes of the refracting interfaces, usually specified by their axial curvatures, conic constants, and aspheric coefficients
3. Thicknesses of the lens elements and inter-element separations
4. Stop position
5. Optical materials of the lens elements

However, in practice, not all of them are always available as variables; some of them are assigned fixed values. This may occur for the whole process of design in some cases. In order to follow a semi-analytical approach in the process of design, often a strategy of dynamic allocation in variables is adopted in practice by releasing more variables at later stages.

Figure 19.3 presents an outline of the flowchart for lens design methodology adopted in practice.

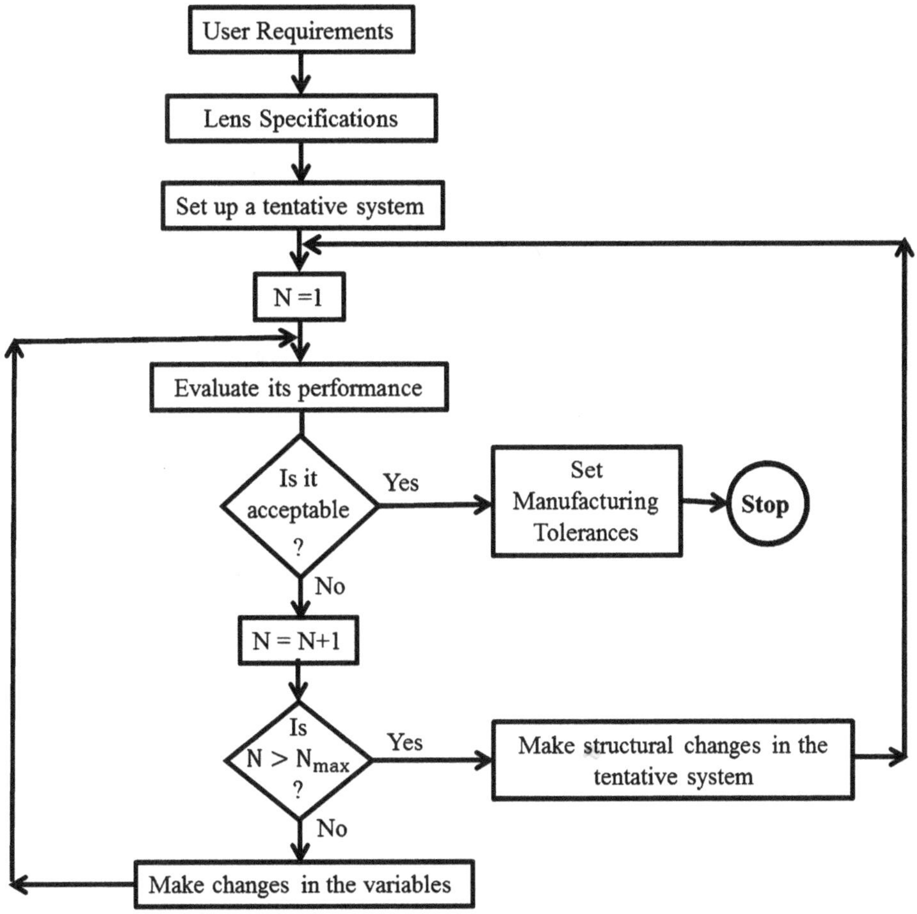

FIGURE 19.3 A flowchart for lens design methodology.

The operations in the flowchart and basic requirements for their effective implementation are detailed below.

Step I. Conversion of user requirements to a set of lens specifications calls for a good understanding of optical instruments, components, and devices, and the of role and characteristics of the underlying lens systems performing the optical tasks. Many excellent books and articles have dealt with the theory and practice of optical instruments from the inception of scientific investigations on the topic. A partial list of the well-known books on the topic is provided in references [10–41].

Step II. Setting up a tentative system from the set of required lens specifications and the set of constraints for the particular case is a tricky job. Success of a design venture depends to a large extent on proper accomplishment of this task. In practice, for a job akin to any of the well-known class of optical systems, significant cue or help may be obtained from

- Company file of similar products
- Available information of a competing product, if any
- Patent literature
- Books on lens design with lens database and with description of lens structures
- Lens data bank of many commercial optical design software

A few well-known books are noted in references [42–48].

On the other hand, for designing entirely new type or class of optical systems, or for designing a conventional system using unconventional devices, components, or elements, it becomes imperative to take recourse to a strategy for 'ab initio' design. The latter calls for judicious use of the theories of paraxial optics, thin lens aberrations, primary and higher order aberration, eikonals, photometry, and radiometry, etc. Different aspects of these theories have been enunciated in earlier chapters of this book.

This step is also known as 'structural design' of lens systems. The overall structure of the lens system, characterized by the number of components, the number of elements in the individual components, the powers of the components and elements, the axial thicknesses of the lens elements, the separation between the elements, the refractive index and dispersion of optical material of the individual lens elements, the maximum diameters of the individual elements, the axial location and diameter of the aperture stop, the curvatures and other surface characteristics (if any) of the interfaces of the tentative system, etc., are worked upon during the structural design phase of the lens system.

For systems operating in infrared or ultraviolet wavelengths, the choice of optical material for the lens components is limited. But, in visible wavelengths, the availability of a relatively larger number of optical glasses with varying refraction and dispersion characteristics provides a wide variety of options for tackling monochromatic and chromatic aberrations. However, the choice of glass for the individual lens elements in a multi-element lens system poses a veritable problem in practice. Recently, environmental considerations have reduced the number of options in glass choice significantly, and tackling the problem of glass choice has become somewhat easier. It should also be noted that the structural design of lens systems is primarily based on ray optical theory.

There are subtle differences among the methods adopted at this stage by various schools of optical design. Essentially, all of them operate with a 'top down strategy' to achieve the design goals with as few components as possible, and with as few elements as possible for each component. Concurrently, it is mandatory to comply with the prespecified (often implied) constraints for the specific design. A glimpse at the significant investigations and reports in this field over the last hundred years may be obtained from references [15, 49–133].

The tentative lens system obtained as above is taken as a starting design. An iterative procedure is adopted for gradually attaining the required specifications by incorporating small changes in available degrees of freedom at each iteration. The strategy to incorporate 'small' changes in degrees of freedom in each iteration is adopted for tackling the inherent nonlinearity in relations between the specifications and the degrees of freedom.

Step III. Performance Evaluation

The 'performance' of a lens developed from a new design is assessed in terms of its ability to provide the required specifications. The latter quantities are of three types. The first type is paraxial in nature, the second type is related to image quality, and the third type deals with violation, if any, of prespecified and inherent constraints of the system. It need not be overemphasized that determination of proper specifications and their tolerances has a very important role in ensuring feasible solutions in practice [134–135].

No explicit analytical relationship between the image evaluation parameters like PSF, OTF, ESF, etc. and the degrees of freedom consisting of constructional parameters of the lens system can be obtained in general. In practice, the journey from the degrees of freedom to the image evaluation parameters is a two stage process. The first stage involves determining the ray and/or wave aberrations from the constructional parameters using ray tracing. In the next stage, the desired image evaluation parameters are evaluated from the aberration values with the help of oscillatory integrals. In cases where the spot diagrams are used as image evaluation parameters, they are usually calculated by using interpolation techniques from a set of aberrations for each field point. Reduction of aberrations is a common goal for most imaging lenses, and so, usually, at the initial stages of design, a set of transverse and/or wave aberrations at chosen field and aperture points are used as measures of image quality. However, at a later stage, when the design is near completion, the required image evaluation parameters are calculated, and fine tuning in values of the degrees of freedom may be incorporated.

Step IV. Lens System Performance Tolerancing

The dialog box 'Is it O.K.?' is an acceptance – rejection test carried out on the tentative lens over the course of design. A set of performance measures for the tentative system, including the paraxial characteristics, the image evaluation parameters, and the compliance with constraints of the system, was evaluated in step III. Depending on the intended application, it is required to work out a set of tolerance criteria for each of the performance measures. When performance of a tentative lens system is in conformity with these tolerance criteria, the system is tentatively accepted, and the manufacturing tolerances for the system are worked out in the next step. A brief description of the manufacturing tolerances is given in Section 19.8.

Step V. The dialog box 'Is $N > N_{max}$?' checks whether the number N of the iteration procedure, which aims at gradually driving the lens system towards achieving the desired performance by incorporating small changes in the degrees of freedom, is less than or greater than N_{max}, a preassigned large integer number. If $N \leq N_{max}$, the iterative procedure continues with the running lens structure, and moves to Step VI. If not, the procedure moves to sub-step V(a), where a new lens structure is taken up as a tentative design for the lens design problem and the iterative procedure is repeated.

Step VI. Next we move on to the implementation of a systematic, semi-empirical, or ad hoc optimization procedure to change the values of the available degrees of freedom to achieve the design goals with the tentative structure adopted for tackling the problem. Like other iterative semi-numerical algorithms, usually the iterative procedure is continued until a stagnation occurs in the procedure, or until a prespecified finite number of times, whichever is earlier. The next chapter deals this topic in detail.

19.3 Different Approaches for Lens Design

Like many other areas of engineering design, the ready availability of computers with phenomenal data handling ability has led to the emergence of two different schools of thought in tackling the problem of lens design. Their opinions are briefly put forward below:

I. Lens design can be completely stated in explicit mathematical terms, and hence a computer can be expected to carry out the total lens design procedure.
II. Lens design inevitably requires qualitative judgments and compromises to be made, and hence the computer should be regarded as a tool capable of presenting the designer with provisional solutions.

The former opinion is mostly held by machine intelligence enthusiasts. The vast majority of practicing lens designers subscribes to the second view. This is because, in practice, there can hardly ever be a perfect system, and lens designers have to make a choice out of the multitude of alternative systems with varying amounts and types of defects. It seems the debate will continue in the foreseeable future, and the field will undergo unfettered progress in the coming days with the emergence of lens systems which are currently beyond imagination. A publication by Dilworth and Shafer [136] in 2013 on this issue is a significant record of contemporary state-of-the-art lens design.

19.4 Tackling Aberrations in Lens Design

Aberration of an arbitrary ray originating from a field point and passing through a specific point on the aperture is a function of the available degrees of freedom in the imaging lens system. In general, this functional relationship is highly nonlinear, so much so that a direct analytical treatment for controlling aberration by changing the values of the degrees of freedom, such as the curvatures of the interfaces, thicknesses and optical materials of the lens elements, air separations, etc., may be ruled out, except for in trivial cases. Since the advent of fast digital computers, the problem of lens design is being tackled as a problem of nonlinear optimization in a constrained multivariate hyperspace. However, for obtaining a useful starting design, particularly in situations where the lens system to be designed is not conventional, or when any of the components or elements to be utilized in the system are unconventional, designers have to take recourse to aberration theory. At times, 'ab initio' design may unfold novel structure for the lens system. Aberration theory also provides a broad understanding of the roles of different components and elements in the operation of the lens system. On the basis of systematic analysis of the characteristics of the various types of axi-symmetric lens systems and their components and elements, a few observations are, in general, very useful in tackling the intricate problem of controlling aberrations. Kidger has recorded many of them [137]. Some of the key observations are listed below.

19.4.1 Structural Symmetry of the Components on the Two Sides of the Aperture Stop

'A large and important class of .... objectives derive great advantage .... from "the symmetrical principle", which consists in placing two similar lenses or lens combinations symmetrically with reference to the centre of the limiting diaphragm of the system' [138]. In this article, Conrady put forward the virtues of symmetry in imaging geometry on the two sides of the aperture stop vis-à-vis correction of aberrations in the context of photographic objectives in particular, but his treatment is valid for any type of imaging lens with moderate to large aperture and field. Until now, this structural symmetry or near-symmetry about the aperture stop has been conspicuous in most imaging lens systems of moderate to large aperture and field [139–140].

An imaging lens is called 'holosymmetric' if the two components of the lens on the two sides of the aperture stop are exactly similar and are of the same scale, and if the object and image are also of equal size and at equal distances from the plane of the aperture stop. In this case, full benefits of the symmetrical principle can be harnessed. In particular, distortion, coma, and transverse chromatic aberration are completely absent. For a holosymmetric system operating at unequal conjugate distances, absence of distortion necessitates elimination of pupil spherical aberration for each component of the system. Similarly, in this case, absence of transverse chromatic aberration in imaging calls for achromatic pupil imaging for each component. Except for in extreme cases of high aperture ratio and large separation, there is very little coma for any conjugate position in holosymmetric systems.

There are imaging systems in which the two components on the two sides of the aperture stop are exactly similar, but are made to a different scale. These are called hemisymmetrical systems. In such systems, distortion is completely absent for a relation of object to image in the same proportion as that between the two components. For other conjugates, the absence of distortion calls for elimination of pupil spherical aberration for the individual components. In this case, coma is generally reduced, but not eliminated. Usually, for one particular conjugate location, coma is eliminated completely. Conversely, for distant objects, the required scaling factor for the front and the back components to eliminate coma needs to be worked out analytically. For the elimination of transverse chromatic aberration in imagery, it is required to obtain achromatism for pupil imaging for each component.

> Symmetrical systems also offer great advantages ... as to spherical and chromatic correction of the axial pencils. If this has been established for the single components for a bundle of parallel rays coming from the direction of the diaphragm, it is usually practically perfect for the combined system under all conditions under which it is likely to be employed. Naturally the highest degree of benefit is obtained when the single components are themselves reasonably free from coma, for then we obtain practically the full advantage of the symmetrical type even with strongly hemisymmetrical combinations.
>
> [138]

On a macro level, the departure from symmetry in the two components on both sides of the pupil gives rise to inequality of power between the two components. Shifting the power between the subcomponents or elements provides a convenient tool in the arsenal of the lens designer during the phase of structural design of many asymmetrical systems [141]. In general, the pursuit of symmetry constitutes a major task in designing wide angle reflective systems [142]. Sasian [143] proposed two parameters for quantifying power distribution and lens symmetry in the search for the optimum structure of lenses.

19.4.2 Axial Shift of the Aperture Stop

No convenient formula for the effect of axial shift of stop on total or finite ray aberration has yet been derived. However, analytical formulae for the effects of axial shift of stop on primary and secondary aberrations are available. These provide deep insights for controlling aberrations in lens systems. Few observations on aberrational characteristics of different combinations of lens structures have already been underscored in Section 15.1.6.

19.4.3 Controlled Vignetting

The phenomenon of vignetting is discussed in Section 4.2 as per the paraxial model of imaging. The actual effect of vignetting in real optical systems is determined by ray tracing during pupil exploration for each extra-axial point (see Section 10.1.3). Since vignetting causes less illumination in the extra-axial image compared to the axial image, usually the designers make attempts to reduce vignetting as much as possible. However, in extreme cases, vignetting can be used to cut down the badly aberrated rim rays for improving the quality of extra-axial imagery. Sometimes a separate vignetting stop located at a convenient position is used for the purpose.

19.4.4 Use of Thin Lens Approximations

Approximating a multicomponent lens system consisting of many components and elements by a single thin lens element is an absurd proposition. But, for many practical purposes, an appropriate thin lens model can play a useful role in the analysis and synthesis of a multicomponent lens system. In this model, only the individual lens elements or components are assumed thin by neglecting their thicknesses, but no such assumption is made regarding the separations between the elements or components. This model facilitates analytical treatment of the initial structural layout of even complex and demanding lens systems. Thin lens aberration theory, elucidated in Chapter 14, plays a pivotal role in this venture in combination with the theories of stop shift and conjugate shift, as presented in Chapter 15. A few rules of thumb on the advantages of using higher index glasses, on the proper location of aspheric and/or diffractive surfaces, on splitting or bending, on the use of ghost surfaces in tackling chromatic aberration, etc., emerge directly from this approach. It is imperative that any thin lens design be converted into a real thick lens design by introducing appropriate thicknesses to the elements (see Section 14.7). Further 'fine-tuning' of the design parameters is often required for tackling the changes in aberrations caused by thickening.

19.4.5 D-number and Aperture Utilization Ratio

At the initial stage of 'ab initio' structural design, each component of a multicomponent lens system is specified in terms of three paraxial parameters: its power K, and the heights h and $\bar{h}$ of the paraxial marginal ray and the paraxial principal ray, respectively. The next stage in design involves determining the detailed structure for each component. The latter implies the number of elements in the component, and the curvatures, thickness, optical material, etc. for the individual elements comprising the component. Obviously, the suitability of a particular structure out of the multitude of alternative structures available for the component may be based on criteria that are likely to reduce the possibility of occurrence of large amounts of higher order aberrations that are often uncontrollable.

In order to ascertain practical feasibility of the detailed structural design of individual components, two dimensionless parameters are proposed by Hopkins [81, 92]. The first parameter is called D-number (written in short as $D_{\#}$). It is a characteristic of the component as a whole, and is defined as

$$D_{\#} = \frac{1}{\Phi K} = \frac{f}{\Phi} \tag{19.1}$$

where K $(=1/f)$ is power of the component, and Φ is the aperture diameter of the component. In order to permit the full aperture pencil from the edge of the object to pass through unobstructed, at this stage, Φ is usually taken as (Figure 19.4)

$$\Phi = 2\left(|h| + |\bar{h}|\right) \tag{19.2}$$

Note that $D_{\#} \le F_{\#}$, where $F_{\#}$ is the relative aperture F-number of the component. $D_{\#}$ is equal to $F_{\#}$ if and only if the aperture stop or any of the pupils are located at the component. For each surface of the elements of the component, a measure of the fraction of the geometric sphere required to pass the full aperture pencil from the edge of the object is the aperture utilization ratio (AUR) R, which is given by

$$R = \frac{\Phi c}{2} = \frac{\bar{c}}{2D_{\#}} \tag{19.3}$$

where the normalized surface curvature $\bar{c}$ of the surface is equal to (c/K). From the point of view of higher-order aberrations, the suitability of a solution for the component is indicated by the value of aperture utilization ratio corresponding to $|\bar{c}_{max}|$. It is given by

$$R_{max} = |\bar{c}_{max}| / 2D_{\#} \tag{19.4}$$

Except for in special cases like an aplanatic surface, generally a value of R_{max} approaching unity indicates that the solution is not suitable because of the likelihood of the occurrence of a large amount

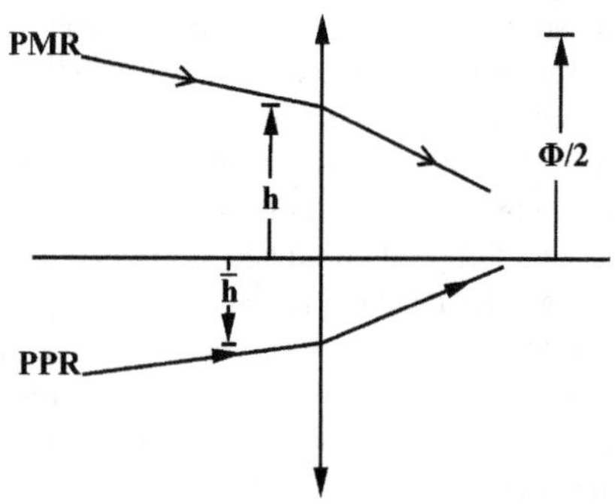

FIGURE 19.4 Maximum diameter Φ of a component that permits the full aperture pencil from the edge of the object pass unhindered.

of unmanageable higher-order aberrations. However, the acceptable value for aperture utilization ratio R depends on the type of the system and is decided by the designer based on prior experience or on the basis of a few tests for the problem at hand.

19.5 Classification of Lens Systems

There are different schemes for classification of lens system for the purpose of analysis and design. The classification may be done on the basis of similarity in either one or more imaging functions or in operational or structural characteristics, e.g.

- Lens systems of zero power and non-zero power
- Lens systems of small field of view and large field of view
- Lens systems with entrance pupil/exit pupil/both at infinity or at a finite distance
- Lens systems based on different combinations of relative aperture, field of view and location of aperture stop
- Lens systems based on working wavelength or operating wavelength range
- Lens systems constituted of dioptric, or catoptric or catadioptric elements

The list is not exhaustive; many other classification schemes exist in practice. Details of a few schemes and their significance from the point of view of analysis and design are given below.

19.5.1 Afocal Lenses

In terms of axial locations of object and image, lens systems with non-zero power can have three possible combinations: (i) object at infinity, image at a finite distance, (ii) object at a finite distance, image at infinity, (iii) both object and the image at finite distances.

Lens systems of zero power form image of an object at infinity also at infinity. These systems are also called 'afocal' or 'telescopic' systems. It should be noted that afocal systems form the image of an object at a finite distance also at a finite distance. Significantly, afocal systems are systems with constant transverse magnification. They have been discussed in detail in Section 3.11. Proper definition of angular magnification in the case of an afocal system operating at infinite conjugates is discussed in Section 4.8.

19.5.2 Telephoto Lenses and Wide-Angle Lenses

In terms of an angular field of view, optical imaging lenses may have a field of view as small as a fraction of a second of arc in the case of some telescope or microscope objectives to as large as near 180 degrees in the case of fish-eye lenses. The field of view of all other lenses lies in between the two. In the case of a specific photographic camera, the size of the detector (a film, or a CCD, or a CMOS array) recording the image is fixed. In order to form on the detector an image of an object with large magnification, it becomes necessary to use an imaging lens with a large equivalent focal length and a small field of view. This type of lens is called a telephoto lens. On the other hand, it becomes necessary to use a lens with small equivalent focal length if an object with a large field of view is to be recorded on the same detector. The latter lenses are called 'wide angle', 'retro focus', or 'reversed telephoto' lenses. All of them share the common feature of a small equivalent focal length, but they differ slightly in their structures [44–45].

As per common usage in photography, a 'standard' lens has an equivalent focal length that is approximately equal to the diagonal of the detector format. A 'wide angle' lens is one whose equivalent focal length is significantly less than the diagonal of the detector format. Since the field of view of such lenses is greater than that of the standard lens, they are called 'wide angle' lenses. A 'long focus' lens is one whose equivalent focal length exceeds that of the standard lens significantly. Consequently, the field of view of these lenses is also much smaller than that of the standard lens. However, they are seldom called 'narrow angle' lenses. It is more common to call these lenses 'telephoto' lenses in view of the wide spread use of this particular type of design to obtain long equivalent focal length compared to the back focal length in reflex cameras.

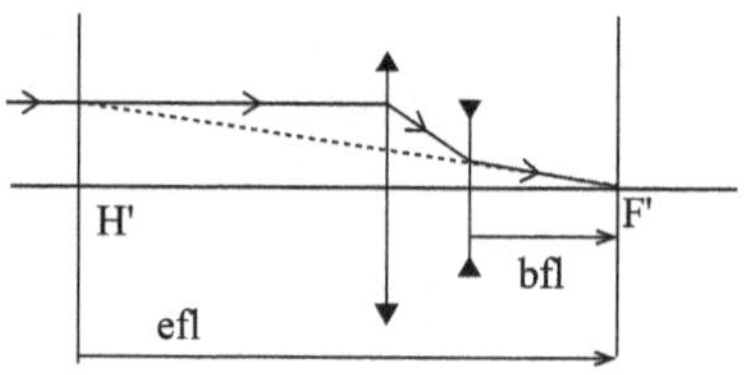

FIGURE 19.5 A telephoto lens. The effective focal length is larger than the back focal length.

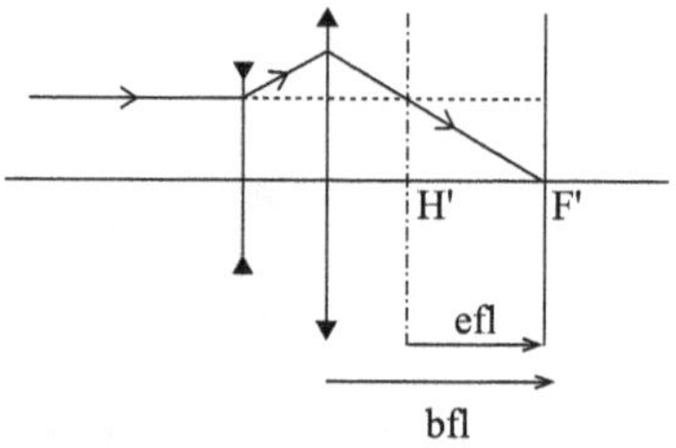

FIGURE 19.6 A wide angle or reversed telephoto lens: The effective focal length is smaller than the back focal length.

The axial distance of the detector plane from the vertex of the first interface in the front of a lens system is called the 'length' of the system. For a single component lens system, the equivalent focal length is approximately equal to the 'length' of the imaging system. An equivalent focal length larger than the length of the system can be obtained by using a two component system, where the first component is of positive power, and the second component is of negative power (Figure 19.5). In photographic optics, the ratio of the length of the system to the equivalent focal length of the system characterizes the 'telephoto effect' of the imaging system. The second characteristic of the telephoto lens is its 'back focal length', i.e. the axial distance of the vertex of the last interface from the detector. Obviously, a telephoto lens can be made more compact by using a lower value of telephoto ratio. On the other hand, admissible minimum and maximum values for the back focal length are usually determined by optomechanical considerations in a reflex camera system. By using refractive lenses with spherical surfaces, it is difficult to obtain a telephoto ratio lower than 0.65. A much lower value for the ratio can be obtained by using either catoptric or catadioptric imaging lenses. The latter are also called 'mirror lenses' [44].

By using an opposite distribution of power between the two components, i.e. by using a two component system consisting of a front component of negative power, separated from a second component of positive power, it is possible to obtain an imaging lens system that has an equivalent focal length much smaller than the back focal length (Figure 19.6). These lenses are called 'reversed telephoto' or 'retro focus' lenses.

19.5.3 Telecentric Lenses

In general, the paraxial principal ray in the image space corresponding to any extra-axial object point is inclined to the optical axis. As a result, the heights of the points of intersection of this ray with transverse planes on the image plane and the neighbouring in-focus and out-of-focus planes are not the same. In real systems, this point of intersection corresponds with the centre of the blur circle, or more specifically the brightest point in the blur patch corresponding to an extra-axial object point. Thus any measurement based on the height of this point above the axis is susceptible to choice of the image plane. However, if the aperture stop is located at the front focus of the imaging system, the entrance pupil is coincident on it, but the exit pupil lies at infinity. This implies that the paraxial principal ray corresponding to any extra-axial object point lies parallel to the axis, and, therefore, the height remains unchanged with small changes in axial location of the image plane. This is called an 'image space telecentric' system. Similarly, an 'object space telecentric' system that can take care of uncertainties in axial location of the object plane can be obtained by placing an aperture stop on the back focus of the imaging system. 'Both space telecentricity'

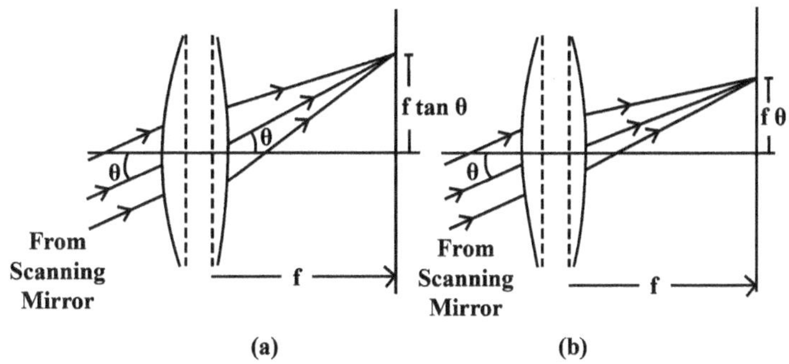

FIGURE 19.7 Pre-objective scanning by (a) an aberration-free lens of focal length f. Field height is $(f\tan\theta)$ at field angle θ; (b) An $f \cdot \theta$ scan lens of focal length f. Field height is $(f \times \theta)$ at field angle θ.

ensues by using a two component imaging system, where the back focus of the first component is coincident with the front focus of the second component, and the aperture stop of the system is placed at the common focal point. Telecentric stops are described with more detail with the help of illustrations in Section 4.11.

19.5.4 Scanning Lenses

Two types of scanning are used in optical scanning systems. In post-objective scanning, the objective lens works on the axis, so that any lens corrected for good axial imaging can be used. But the material to be scanned has to be on a prespecified curved surface—a great disadvantage for general use in practice. More often than not this problem is solved by taking recourse to pre-objective scanning. Practical implementation calls for flat field objective lenses. Also, proper synchronization in scanning demands that the scanning spot must move along the scan line at a uniform velocity. This is possible if and only if the displacement of the spot is linearly proportional to the field angle θ. Note that in the case of an aberration-free lens of equivalent focal length f, the corresponding displacement is $(f\tan\theta)$, as shown in Figure 19.7(a). For correct synchronized scanning, the displacement needs to be $(f \cdot \theta)$, as shown in Figure 19.7(b). The required fractional distortion, $[\tan\theta - \theta]$ for values of field angle θ is shown in Figure 19.8. A lens that has this amount of pincushion distortion can provide the required synchronization. This type of lens is known as $f \cdot \theta$ scan lens. Further requirements for image space telecentricity, image plane clearance, etc. in specific cases are noted by Westcott [144].

19.5.5 Lenses with Working Wavelength beyond the Visible Range

The availability of transparent materials, such as different types of optical glasses in the visible range of wavelengths, played a key role in the early growth and development of refractive lens elements of the optical quality, and consequently led to the emergence of a wide range of refractive optical lenses over the years. The latter has been facilitated by simultaneous development of high quality reflective optical elements. However, the situation is not similar in other wavelength ranges in the electromagnetic spectrum that fall under the domain of optics.

The Terahertz (THZ) wavelength range falls in the far infrared band, which is above the microwave band in frequency, and below the mid infrared. Typically, the wavelength range is between 3 mm to 30 μm, and the corresponding frequency range is between 0.1 THz to 10 THz. THz radiation is nonionizing due to low photon energy, and THz waves do not penetrate into metals. But they can penetrate into organic materials such as skin, plastics, cloth, or paper products.

In the THz range of wavelengths, refracting materials with low absorption are few. THz lenses are currently available, and they are used in some applications [145]. These lenses are made of crystalline material like HRFZ-Si (High Resistivity Float Zone Silicon), or Teflon, i.e. PTFE (Polytetrafluoroethylene), or other plastic materials like PE (Polyethylene) or TPX (Polymethylpentene). Often they have high

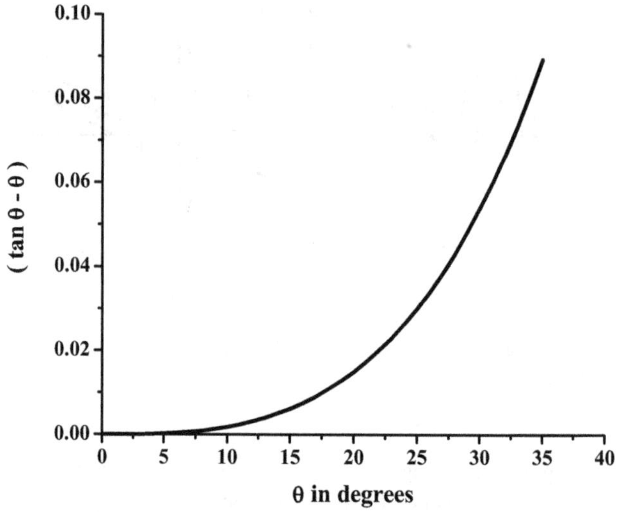

FIGURE 19.8 Variation of $(\tan\theta - \theta)$ with θ.

dispersion and non-negligible absorption at working wavelengths, and so reflecting systems are preferred for focusing or imaging.

The infrared wavelength band just exceeding the wavelength of visible light is known as Near IR (NIR) or Short Wave IR (SWIR). Many optical glasses retain useful transmission characteristics over this range, and objects are usually trans-illuminated, and so modalities of lens design remain practically the same as lenses operating in visible wavelengths. However, the situation vis-à-vis the availability of optical materials is tight in the Mid Wave IR (MWIR) and Long Wave IR (LWIR) wavelength ranges of (3 μm – 5 μm) and (8 μm – 12 μm), respectively. MWIR and LWIR are commonly known as thermal wavebands. Refracting lenses at these ranges use germanium, silicon, zinc sulphide, zinc selenide, and fluorides like magnesium fluoride, calcium fluoride, and barium fluoride. The objects of the imaging systems at MWIR and LWIR emit the radiation used in image formation. In general, special considerations, e.g. athermalization, elimination of narcissus effect, etc., are needed in the treatment of corresponding imaging systems [146–148]. The higher refractive indices of the more commonly used optical material in the latter lenses, and also the higher value of working wavelength, opens up new possibilities of fabrication and design of these IR lenses by using aspheric and diffractive/binary elements. For broadband operations, reflecting systems are often called upon.

Beyond the short wavelength region of visible waveband lie the ultraviolet (UV) waves with wavelengths shorter than 400 nm. The different spectral ranges in this region are characterized as Near UV (NUV: 400 nm – 300 nm), Middle UV (MUV: 300 nm – 200 nm), Far UV (FUV: 200nm – 40 nm), and Extreme UV (EUV: 40 nm – 5 nm). At times, a different classification is adopted for the first two ranges as (UVA: 400 nm – 320 nm), (UVB: 320 nm – 280 nm), and (UVC: 280 nm – 200 nm). Many optical materials that are transparent at visible wavelengths become absorbing in the UV region. For the majority of optical glasses, the 'cut-off' in transmission at short wavelengths falls in the (350 nm – 300 nm) spectral region. Transparency in the UV region is obtained only with few optical materials like fused silica (SiO_2), quartz, and fluorides, such as Calcium Fluoride (CaF_2), Magnesium Fluoride (MgF_2), Lithium Fluoride (Li F), etc. Scattering effects are very strong in the UV region. High surface quality and material homogeneity are necessary to obtain good quality performance in optical systems operating in this region. Because electromagnetic waves with wavelengths of the Far UV and Extreme UV region are strongly absorbed in air, the optical systems operating at these wavelengths need to operate in a vacuum, and consequently the spectral region (200 nm – 5 nm) comprising both Far UV and Extreme UV is also called Vacuum UV (VUV).

Electromagnetic (e.m.) waves with still shorter wavelengths in the range (5 nm – 0.01 nm) are called X rays. The spectral region is subdivided between Soft X Rays (SXR: 5 nm – 0.1 nm) and Hard X Rays (HRX: 0.1 nm – 0.01nm).

It should be noted that photon energy becomes increasingly larger with the decrease in wavelength of e.m. waves. The photon energy E = hν, where h is Planck's constant, and is equal to $4.135667516 \times 10^{-15}$ eV.sec, and ν, the frequency of the e.m. wave is related with its vacuum wavelength λ by $\nu\lambda = c = 3\times10^{8}\,m/sec$, the speed of light in vacuum. A rule of thumb for the calculation of photon energy E of e.m. wave of wavelength λ is

$$E(\text{in eV}) = \frac{1240.7}{\lambda(\text{in mm})} \cong \frac{1234.5}{\lambda(\text{in mm})} \tag{19.5}$$

by using 'count to five' approximation. It follows that for hard X rays with wavelengths in the range (0.1 nm – 0.01 nm), the photon energies lie in the range (12.5 keV – 125 keV), and for soft X rays with wavelength in the range (5 nm – 0.1 nm) the photon energies lie in the range (250 eV – 12.5 keV). For EUV with wavelengths in the range (40 nm – 5 nm), the photon energies lie in the range (30 eV – 250 eV). It may be seen that for e.m. waves with lower photon energies, in the visible region and in the ultraviolet region with wavelengths larger than 200 nm, and waves with higher photon energies in the hard X Ray region, many materials become transparent, and it is not necessary to utilize vacuum isolation techniques in the hard X Ray region, and in the UV region with wavelengths above 200nm. By contrast, soft X rays and EUV waves are easily absorbed in air. Typically, the EUV optics and SXR optics are identical, whereas HXR optics have unique characteristics.

The complex refractive index $n(\nu)$ at a frequency ν in the EUV and the X Ray region is usually expressed by

$$n(\nu) = 1 - \delta(\nu) + i\beta(\nu) \tag{19.6}$$

where $\delta(\nu)$ is the dispersive part and yields phase shift, and $\beta(\nu)$ represents the absorption or attenuation. Typically, $\delta(\nu)$ lies in the range $(10^{-5}, 10^{-6})$, and $\beta(\nu)$ varies by several orders of magnitude in the range $(10^{-7}, 10^{-11})$. It is obvious that reflective components akin to visible optics are practically impossible, since the reflection coefficient is very small for any angle of incidence. However, since the refractive index of material at these wavelengths is lower than one, albeit near to it, there is a critical angle of incidence i_c at which there is minimal refraction, and the incident beam passes along the vacuum/air-material interface. For any angle of incidence above this, the incident ray is reflected back totally. Assuming $\beta \approx 0$ for the sake of simplicity, the critical angle of incidence is

$$\sin i_c = 1 - \delta \tag{19.7}$$

It is seen that the critical angle of incidence i_c is nearly equal to 90^0, and its complementary angle $\theta_c = 90^0 - i_c$ is the angle that the incident ray makes with the interface. From (19.7) we get

$$\cos\theta_c = 1 - \delta \tag{19.8}$$

For practical purposes, the rays with angles of incidence lying in the range $(0, \theta_c)$ are said to have grazing incidence on the interface. Since these rays are totally reflected back in the vacuum/air by the interface, the phenomenon is called 'total external reflection'. This phenomenon is effectively utilized in the development of useful components for focusing in synchrotron optics, X Ray astronomy, etc., or for the collection of X Rays by polycapillary optics. Multilayer interference coatings are also used to obtain reflective optics at these wavelengths. Bragg reflection off flat or bent crystals are also utilized to obtain useful components. Compound refractive lenses using a series of many refractive lenses in cascade are also being explored. Note that the individual refractive lenses used in compound refractive lenses for focusing X Rays are bi-concave in shape because the refractive index of the lens material is less than the refractive index of the air/vacuum. Diffractive elements like Fresnel zone plates play useful and significant roles in accomplishing optical functions in the domain of hard X Rays and in the frontier region of Gamma rays [149–153].

19.5.6 Unconventional Lenses and Lenses using Unconventional Optical Elements

Conventional lenses are axially symmetric optical systems consisting of one or more axially symmetric lens elements with spherical interfaces. The elements are placed in air/vacuum, and are made of transparent optical materials, e.g. glass, plastic, etc., that are homogeneous. Mirrors are considered to be the special case of a refractive lens element with refractive indices on the two sides of it being of opposite sign. Departure from any or more characteristics of the conventional optical elements gives rise to unconventional lens elements and systems. Conventionally, a lens is meant to satisfy a single imaging configuration, as shown in Figure 19.1. In some practical applications, it becomes necessary to use a lens system that can provide multiple imaging configuration by incorporating small changes in conventional lens elements comprising the lens, or by using unconventional lens elements. These lens systems are also categorized as unconventional lenses. A basic reason for this categorization stems from the fact that often analysis and design of the unconventional lens systems call for special treatment under the broad framework of analysis and design of conventional lens systems. A partial list of unconventional lenses is given below. More details on them may be found in the enclosed references.

- Axially symmetric aspheric lenses [Section 12.4.6.1; 154–156]
- Anamorphic, cylindrical and toric lenses [157–162]
- Fresnel lenses, nonimaging concentrators and collimators [163–167]
- Freeform surfaces and lenses [168–179]
- Luneburg lenses [Section 7.2.3.3]
- Rod lenses [180–182]
- Gradient index lenses [183–185]
- Varifocal and Zoom lenses [186–190]
- Ophthalmic and progressive ophthalmic lenses [191–193]
- Systems using variable focus lens elements [194–200]
- Diffractive and binary lenses [201–205]
- Geodesic lenses [206–207]

19.6 A Broad Classification of Lenses based on Aperture, Field of View, Axial Location of Aperture Stop, Image Quality, and Generic Type of Lens Elements

Figure 19.9 gives a broad classification of lens systems underscoring the degree of complexity in their analysis and synthesis. At the first level, the lenses are classified on the basis of generic types of lens elements used in the system. Lenses using one or more unconventional optical elements need special considerations for their analysis and synthesis, and so they belong to a separate class. Note that most members of this class have unique characteristics that call for special attention.

Lenses using conventional refracting or reflecting optical elements, e.g. lenses and mirrors, are further subdivided into two classes on the basis of location of their aperture stop with reference to the lens. Lenses having 'remote' stops need to have larger aperture diameters for exclusive passage of off-axial pencils of rays, and special considerations need to be undertaken for tackling undesirable vignetting effects. These lenses are treated apart from other conventional lens systems where the aperture stop lies 'close' to the lens. Lenses belonging to the latter class are further classified in three classes on the basis of image quality. In one class belong all lenses that provide 'excellent' image quality for axial objects. These lenses may be further classified on the basis of the size of their relative aperture. In this category, microscope objectives with large relative aperture are a class apart from telescope objectives with small relative aperture. The next class includes all lenses that provide 'excellent' image quality both on-axis and off-axis. Currently design and fabrication of these high resolution diffraction limited lenses with demanding size of field pose a veritable challenge to the optical engineers [208–210]. Most other lenses catering to the vast range of optical imaging in practice are supposed to provide 'good' quality images over an extended field of view, and they constitute the next class. In Figure 19.9, the grey boxes underneath each

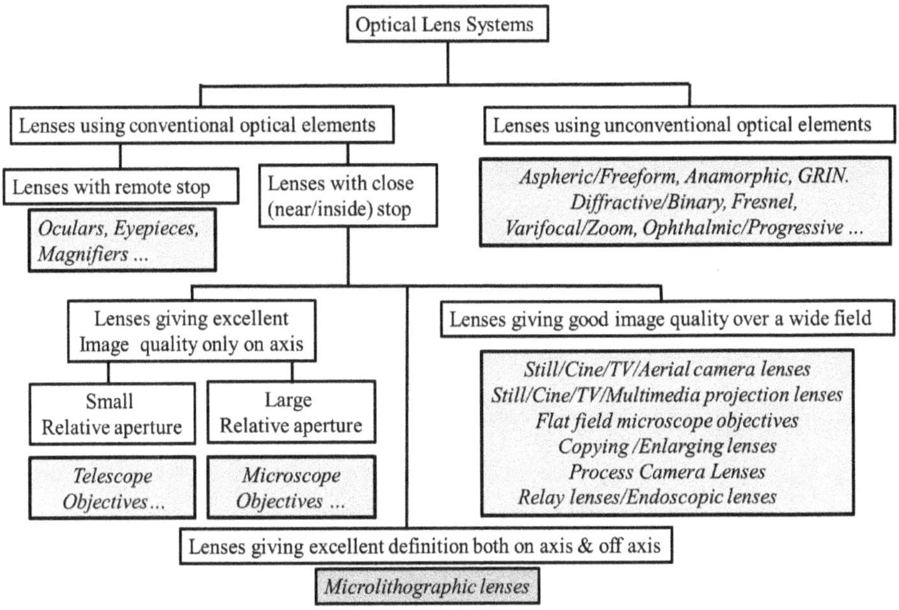

FIGURE 19.9 A broad classification of optical lens systems.

lens class contain the names of representative members of the class. Often, similar principles of analysis and synthesis turn out to be appropriate for members of each class, in spite of large variation in their practical uses [211–212].

19.7 Well-Known Lens Structures in Infinity Conjugate Systems

As a rule of thumb, any object distance or image distance larger than about ten times the equivalent focal length of a lens may be considered as an infinite distance for practical purposes, and the corresponding imaging configuration is called infinity conjugate imaging. In the fields of photographic optics and projection optics, applications call for lenses of different combinations of field of view and relative aperture. For small values of field of view in combination with large values of relative aperture, e.g. F-number, often a simple lens structure involving a single lens component with one or more lens elements serves the purpose. As the value of field of view increases and/or the value of the F-number decreases, these lenses of simple structure fail to provide good quality imaging, and it becomes necessary to use lenses with a greater number of components and elements. Use of more components opens up the possibility of using different structures. It is significant to note that there is a distinct structural similarity between the multitude of designs that provide images of good quality for certain combinations of relative aperture and field of view, so much so that lens systems of certain distinctive structures have become known by names associated with either the company marketing the lens, e.g. 'Cooke triplet', or by names given to them by the company first marketing them, e.g. 'Tessar', 'Hypergon', etc. Some of these lens structures also have names associated with the designer who first embarked upon that particular structure, e.g. 'Petzval', 'Double Gauss', etc. The evolution of these lens systems is a fascinating development in the field of optical engineering. A glimpse of these developments may be obtained from a partial list of available review articles written by pioneers in the field [213–223].

Figure 19.10 gives a collection of lens designs as a function of lens speed and semi field of view. They provide good quality imaging in broadband visible light. Note that the boundaries of the different lens types are not strict, and they are drawn to provide a rough estimate on the region on the map where that particular lens type is optimum. In this analysis, it is assumed that the lens elements used in these structures have spherical refracting interfaces. Structural similarity in members of a particular type

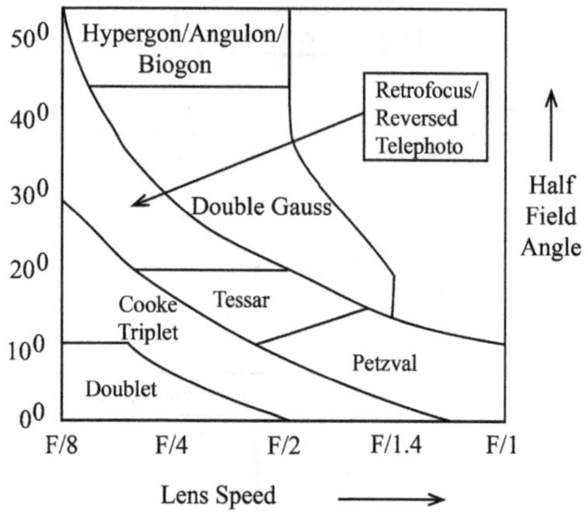

FIGURE 19.10 A collection of lens types plotted as a function of lens speed and semi field of view.

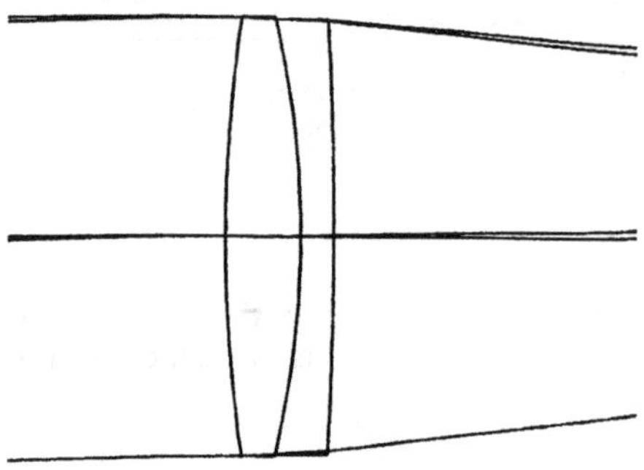

FIGURE 19.11 A cemented Fraunhofer doublet; crown/positive power lens in front.

implies similarity in the number of components, in the distribution of power in the components, and in the overall layout of the components, allowing for minor changes in detailed composition of the individual components. Obviously, use of aspheric or diffractive surfaces will modify the map significantly. A brief description of the types shown in Figure 19.10 is given below:

- Doublets: a lens system/component consisting of two single lens elements in close contact is called a doublet. There are two classes of doublets. In one class, the two lens elements are cemented to form a lens component that can be handled as a single element, and they are called cemented doublets. When the two close lens elements are not cemented, there is a small air gap between the two elements, and these 'split' or 'broken contact' doublets constitute the other class. In the case of achromatic doublets, the two elements of the doublets are of opposite power, and usually the positive element is made of crown glass, and the negative element is made of flint glass. Figure 19.11 shows a cemented doublet with a positive power/crown lens element in the front. This is called a 'Fraunhofer doublet'. When the lens element of negative power (flint) of a cemented doublet is in the front, it is called a 'Steinheil doublet' (Figure 19.12). Both Fraunhofer and Steinheil doublets have their split/broken contact versions. A split/broken contact Fraunhofer doublet is shown in Figure 19.13. In the case of a split achromatic doublet with positive power crown lens in the front, another useful solution emerges where both surfaces of the negative power (flint) element are convex to the front element. Figure 19.14 shows the structure of this type of doublet that is known as a 'Gauss doublet'. A significant characteristic of a Gauss doublet is that its spherical aberration is nearly insensitive to variation in the wavelength of the working illumination.
- Cooke Triplet: A 'Cooke Triplet' lens consists of three air-spaced single lens elements with a stop near the middle lens. Figure 19.15 shows a Cooke triplet where the middle lens is of negative power, and both outer lens elements are of positive power. In some versions of triplet, the power

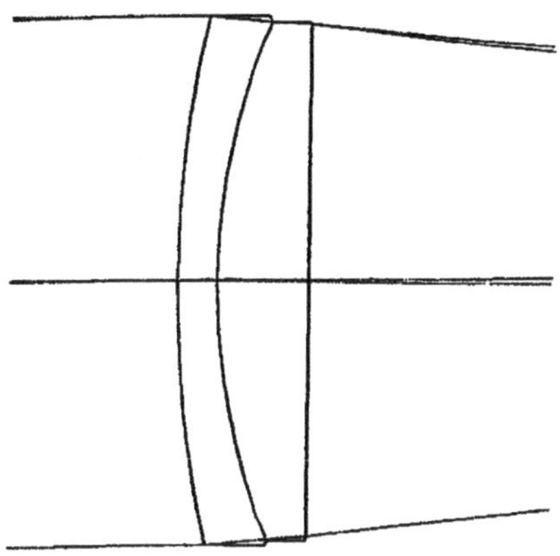

FIGURE 19.12 A cemented Steinheil doublet; flint/negative power lens in front.

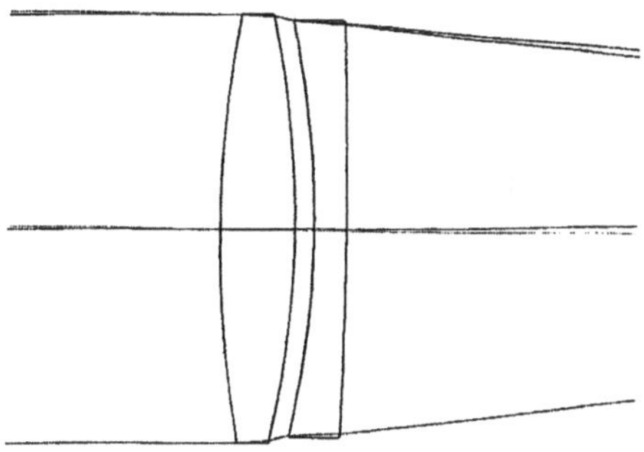

FIGURE 19.13 A split/broken contact Fraunhofer doublet; crown/positive power lens in front.

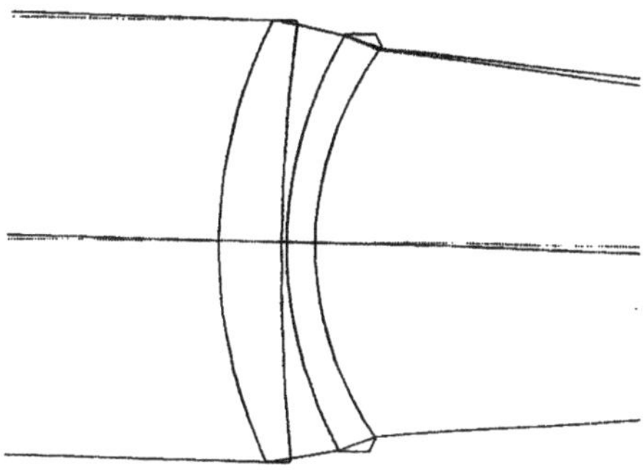

FIGURE 19.14 A split/broken contact Gauss doublet; crown/positive power lens in front–second type.

distribution may be opposite. This lens was invented in 1894 by H. Dennis Taylor, optical manager of M/s T. Cooke and Sons Ltd., York, and manufactured by M/s Taylor, Taylor and Hobson Ltd., Leicester, as M/s Cooke of York did not wish to manufacture photographic lenses. However, the triplet lenses were to be known as 'Cooke triplets' out of respect for Taylor's employer. Dennis Taylor claimed to have worked out the triplet entirely by algebraic formulae developed by himself [224].

- Tessar: A 'Tessar' lens is an air-spaced three component lens. It appears as a modification of the 'Cooke Triplet' lens, with the last component being changed by a cemented doublet (Figure 19.16). However, the origin of the lens type is different. It was invented by Paul Rudolph of M/s Zeiss in 1999 in the course of his attempts to improve Zeiss anastigmats. 'Lenses of the Tessar type have been made by the millions, by Zeiss and every other manufacturer, and they are still being produced as an excellent lens of intermediate aperture' [225].

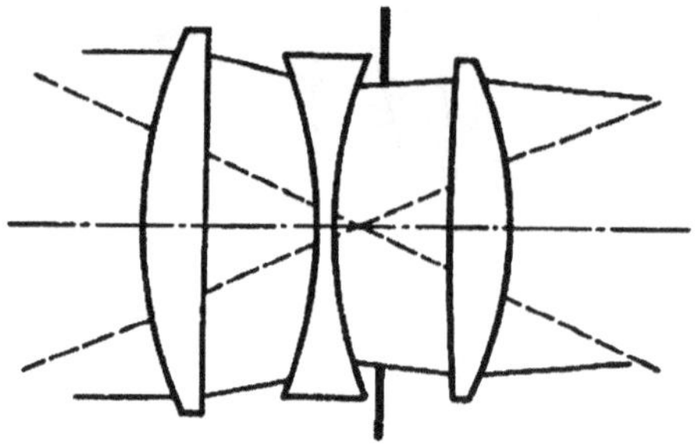

FIGURE 19.15 A Cooke Triplet lens.

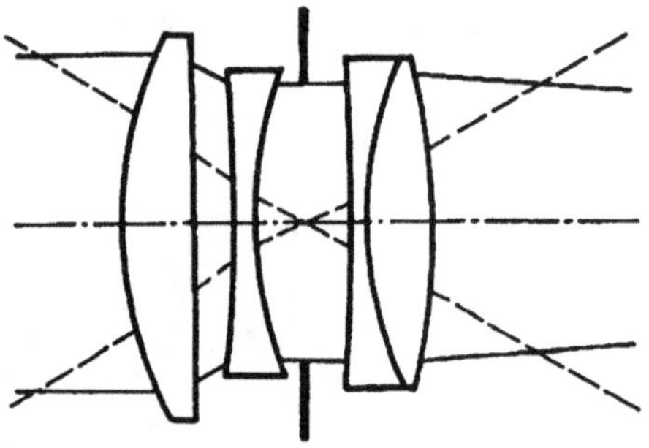

FIGURE 19.16 A Tessar lens.

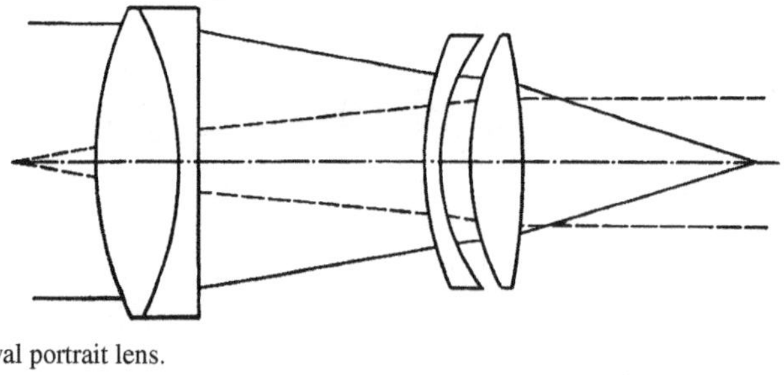

FIGURE 19.17 A Petzval portrait lens.

- Petzval: This type of lens first appeared as a fast portrait lens in 1840 as a landmark in the field of photography, closely on the heels of the emergence of the daguerreotype process. The original lens, as shown in Figure 19.17, embodies an air-spaced two component system consisting of a cemented doublet in the front with a broken contact doublet behind it. In spite of several modifications, Petzval lenses consisting of two air-spaced doublets have wide applications including major ones in projection systems.
- Retrofocus: Retrofocus lenses are also known as reversed telephoto lenses. They consist of a negative lens component in front of a positive lens component. This type of system is favourable for both a high relative aperture and a wide angular field. In these systems, the equivalent focal length can be made significantly smaller than the back focal length. The principle is illustrated in Section 19.5.2. Figure 19.18 presents the schematic diagram of a retrofocus objective.
- Double Gauss: In the course of his attempts to develop good quality anastigmats, in 1896, Paul Rudolph explored the use of Gauss doublets, mentioned earlier, in symmetric arrangement on both sides of the aperture stop. Incidentally, it may be mentioned that the word 'Anastigmat' is a double negative. As mentioned earlier, the word 'a-stigmatism' means 'no point', or failure to form a point

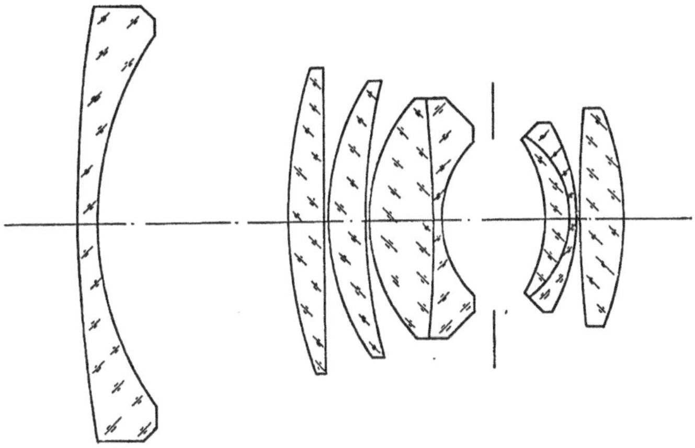

FIGURE 19.18 A Retrofocus lens.

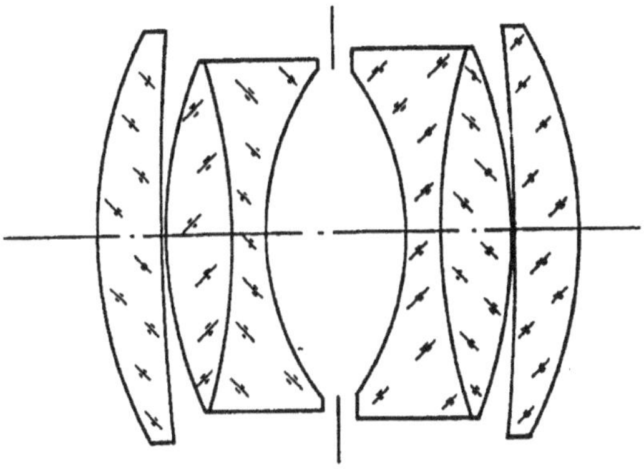

FIGURE 19.19 A Double-Gauss lens.

image. 'An-a-stigmatism' means a 'no-no-point', a negation of the earlier effect. This implies a good definition in the image. Rudolph had to incorporate major innovations, e.g. thickening of the negative meniscus, reduction of the distance between the two elements in the Gauss doublet, and, last but not the least, inserting a buried surface in the negative element. The resulting symmetric arrangement is shown in Figure 19.19, and it is known as the 'Planar'. Figure 19.19 presents a schematic of the 'Zeiss Planar' lens. Later on, in the 1920s, H.W. Lee of the Taylor Hobson Company improved the performance of this lens by relinquishing strict symmetry around the pupil, and this opened the floodgate of a whole range of 'Unsymmetrical Double Gauss' lenses, by a large number of companies with different trade names, many of which still hold the fort in a wide range of applications requiring lower F-number and higher field of view.

- Hypergon/Angulon/Biogon: Figure 19.20 shows the schematic of the Goerz Hypergon lens that has anastigmatic field out to $\pm 67^0$. With two deep meniscus lens elements, placed symmetrically on both sides of the aperture stop, it has a ball-like exterior form. However, because of lack of correction of spherical and chromatic aberrations, the aperture is limited to F/20. Also, there is a drastic fall-off in illumination towards the extreme limits of the wide field. Remedies to these defects were found by insertion of two components, with even two or three elements in one of them, in place of the single meniscus on each side of the aperture stop. The lenses so designed have back focal length so short that they cannot be used in SLR cameras, but they have found applications in rangefinder cameras. On account of symmetrical structure, the lenses are practically free from distortion, and they have also become standard lenses in aerial photography and photogrammetry. The

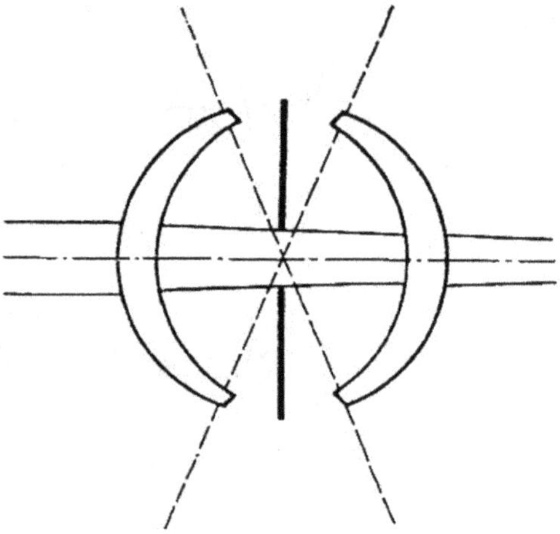

FIGURE 19.20 A Goerz Hypergon lens.

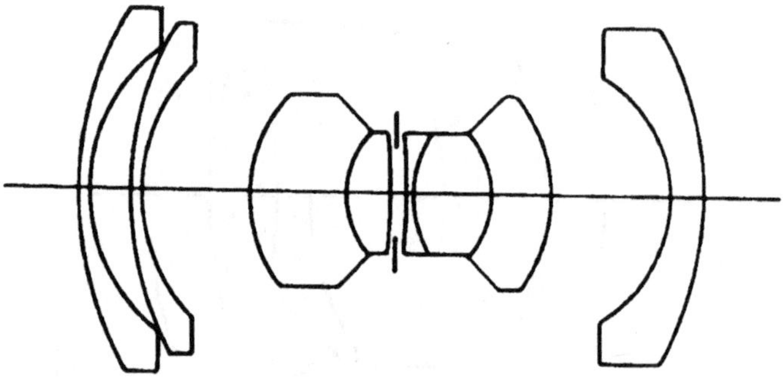

FIGURE 19.21 A Zeiss Biogon lens.

well-known lenses of this type are the Wild Aviogon, the Schneider Super Angulon, and the Zeiss Biogon. A schematic diagram of the Zeiss Biogon lens is shown in Figure 19.21. The Biogon lenses are said to be used in the famous Hasselblad cameras in NASA lunar landings in Apollo missions. The latter camera is popularly called the moon camera [226].

Note that in some figures, the upper marginal ray path and the lower marginal ray path for the axial pencil are shown by solid lines, and the upper pupil ray path and the lower pupil ray path for the extra-axial point at the edge of the field are shown by dashed lines. In Figure 19.15, Figure 19.16, Figure 19.17, and Figure 19.21, they are clearly discernible. However, in Figures 19.11, 19.12, 19.13, and 19.14, since the F-number of the system is large, and the field of view is very small, the marginal and pupil ray paths look nearly superposed near the lens. However, no ray paths are shown in Figure 19.18, Figure 19.19, and Figure 19.21.

19.8 Manufacturing Tolerances

The design of a lens system, as described so far, ensures that its imaging performance satisfies the stipulated image quality criteria. No manufacturing error of any kind is taken into account in the evaluation of image quality. It is obvious that manufacturing errors in the constituent lens elements and in their assembly in composing the system would affect the performance of the system. More often than not, the consequences of these errors are a significant degradation of the image quality from the level estimated for the designed system. It is evident that this degradation can be lessened by reducing the manufacturing errors. But the latter can only be achieved with a concomitant increase in the cost of manufacture. In practice, a trade-off is sought by setting up a set of manufacturing tolerances for each constructional parameter of the constituent optical elements and components, and also for dimensions of the optomechanical components used in assembly of the optical elements and components in the system. The goal is to ensure that the fabricated product yields the required level of optical performance at a reasonable cost [227–235].

> Tolerance analysis uses one or more methods of sensitivity, inverse sensitivity, worst case, standard deviation or Monte Carlo in the process of assigning tolerance limits and characterizing their effect on system performance. Sensitivity and inverse sensitivity consider performance changes for individual errors in system construction; worst case, standard deviation and Monte Carlo methods attempt to model the accumulated effects of multiple errors occurring in the fabrication process.
>
> [231]

In 'sensitivity analysis' based tolerancing method, prespecified or default tolerance limits are assigned for each constructional parameter by turn, and corresponding changes in system performance is evaluated in each case. If the changes are 'tolerable', the tolerances are set for the constructional parameters. In the case of 'inverse sensitivity' based tolerancing, for each constructional parameter, the permissible error in its value corresponding to a prespecified or default change in system performance is determined, and the tolerance for that parameter is set on the basis of this error value.

The second category of tolerancing methods makes use of different models based on different assumptions on the probabilistic nature of the occurrence of errors in the constructional parameters. In the 'worst case' model, it is assumed that the performance change is statistically independent, such that the change in performance of a system with several errors occurring simultaneously can be modelled by properly combining the changes in performance caused by each error occurring individually. It is further assumed that the maximum performance change of the system in the case of simultaneous occurrence of several errors occurs at the extremes of the tolerance range set for individual parameters. Although the assumptions are simplistic, and are often unrealistic, the result acts as a useful guide for actual system performance.

In the 'standard deviation' model, the probability of the system performance achieving a prespecified level can be directly related to the standard deviation of the performance changes computed in a sensitivity analysis or targeted performance changes in an inverse sensitivity analysis. The standard deviation σ in the performance of an optical system with I constructional parameters is expressed as

$$\sigma = \sqrt{\left(\frac{\Delta f_{x_1}}{\Delta x_1}\right)^2 \left(\sigma_{x_1}\right)^2 + \ldots + \left(\frac{\Delta f_{x_i}}{\Delta x_i}\right)^2 \left(\sigma_{x_i}\right)^2 + \ldots + \left(\frac{\Delta f_{x_I}}{\Delta x_I}\right)^2 \left(\sigma_{x_I}\right)^2} \tag{19.9}$$

where Δf_{x_i} is the performance change in the system when the constructional parameter x_i is changed to its tolerance limit $x_i + \Delta x_i$, and σ_{x_i} is the standard deviation of the errors in the construction parameter x_i. When probability density function of the errors in a constructional parameter is known, the standard deviation of its errors is expressed by $\sigma_{x_i} = C\Delta x_i$, where C is a constant determined by the probability density function. The required standard deviation σ in performance of the system can be obtained by suitably distributing the tolerances Δx_i, $i = 1,\ldots,I$.

The Monte Carlo model for tolerancing is implemented via 'simulation' of a series of random sampling of states for performance evaluation. In practice, random states are simulated by generating random numbers that determine which constructional parameters are to be perturbed. Any number (decided by a random number) of constructional parameters may be perturbed concurrently, and each perturbation can be by any amount lying within its tolerance limit (constrained by probability density function for that parameter) and is decided by another random number. The performance of the system is evaluated for each perturbed state. The mean value of performance and the standard deviation is calculated after each iteration. On account of the inherent stochastic nature of the simulation, 'Monte Carlo' simulation unfolds any hidden effect in the presence of multiple perturbations. Often the method is used in combination with other tolerancing techniques.

Elements of the sensitivity matrix can be obtained by using numerical differentiation. The accuracy of computational results may be increased by using central differences instead of forward differences. A major problem in tolerancing arises from specification of the performance requirements in a form that can be evaluated rapidly with sufficient accuracy. One of the earliest approaches was a scheme utilized by Hopkins and Tiziani in a treatment of lens centring errors [236]. Later on, Rimmer extended the approach in formulation of a few other performance indices [237–238]. More details on techniques of tolerance analysis may be found in references [239–241].

REFERENCES

1 F. Twyman, *Prism and Lens Making*, Adam Hilger, Bristol (1952).
2 D.F. Horne, *Optical Production Technology*, Adam Hilger, Bristol (1977).
3 B.K. Johnson, *Optics and Optical Instruments*, Dover Publications, New York (1960).
4 D. Malacara, *Optical Shop Testing*, Wiley, New York (1978).
5 W. Zschommler, *Feinoptik-Glasbearbeitung (Werkkünde für den Feinoptiker)*, Carl Hanser, Munich (1963). English Translation *Precision Optical Glassworking* by G.K. Sachdev and J. Maxwell, SPIE Volume 472, Macmillan, London (1984).
6 R. Kingslake, 'Lens Design', in Chapter 1 in *Applied Optics an d Optical Engineering,* Vol. III, Ed. R. Kingslake, Academic Press, New York (1965).
7 R. Kingslake, *Lens Design Fundamentals*, First Edition, Academic Press, New York (1978). Updated Second Edition by R. Kingslake and R. Barry Johnson, SPIE Press, Bellingham, Washington (2010).
8 R. Kingslake, *Optical System Design*, Academic Press, New York (1983).
9 W.T. Welford, 'Optical Design', Chapter 4 in *A Guide to Instrument Design*, produced by Scientific Instrument Manufacturers' Association of Great Britain(SIMA) and The British Scientific Instrument Research Association (SIRA), Taylor & Francis, London (1963).
10 S. Czapski and O. Eppenstein, *Grundzüge der Theorie der Optische Instrumente nach Abbe*, Johann Ambrosius Barth, Leipzig (1904).
11 M. von Rohr, Ed., *Die Bilderzeugung in optischen Instrumenten vom Standpunkte der geometrischen Optik*, [*Die Theorie der optischen instrumente I*], Julius Springer, Berlin (1904) English Translation *Geometrical Investigation of the Formation of Images in Optical Instruments,* by R. Kanthack, His Majesty's Stationery Office, Department of Scientific and Industrial Research, London (1920).
12 E.T. Whittaker, *The Theory of Optical Instruments*, Second Edition, Hafner Publishing, New York (1907). [First Published 1907].
13 J.P.C. Southall, *The Principles and Methods of Geometrical Optics, especially as applied to the Theory of Optical Instruments*, Macmillan, New York (1910).
14 A. Gleichen, *Die Theorie der modernen optischen instrumente*, F. Enke, Stuttgart (1911); English translation *The Theory of Modern Optical Instruments* by H.H. Emsley and W. Swaine, His Majesty's Stationery Office, Department of Scientific and Industrial Research, London (1918).
15 È. Turrière, *Optique Industrielle*, Librairie Delagrave, Paris (1920).
16 D.H. Jacobs, *Fundamentals of Optical Engineering*, McGraw-Hill, New York (1943).
17 G.S. Monk and W.H. McCorkle, Eds., *Optical Instrumentation*, McGraw-Hill, New York (1954).
18 J. Flügge, *Das Photographische Objektiv*, Springer, Wien (1955).
19 G.A. Boutry, *Optique Instrumentale*, Masson et Cie, Paris (1946). English Translation *Instrumental Optics* by R. Auerbach, Hilger and Watts, London (1961).
20 L.C. Martin, *Technical Optics*, Vol. I and Vol. II, Pitman, London (1948, 1950).
21 K.J. Habell and A. Cox, *Engineering Optics*, Pitman and Sons, London (1953).
22 W.T. Welford, 'Optical Calculations and Optical Instruments, an Introduction' in *Handbook der Physik*, Band XXIX, Optische Instrumente, Springer-Verlag, Berlin (1967).
23 'Design of optical instruments', W. Brouwer and A. Walther, Chapter 11 in *Advanced Optical Instruments*, Ed., A.C.S. van Heel, North-Holland, Amsterdam (1967) pp. 571–631.
24 R. Kingslake, Ed., *Applied Optics and Optical Engineering*, Vol. 4, Optical Instruments, Part 1, Vol. 5, Optical Instruments, Part 2, Academic Press, New York (1967, 1969).

25 P. Bousquet, *Spectroscopy and Its Instrumentation*, Adam Hilger, London (1971). Translated from original French title 'Spectroscopie Instrumentale' by K.M. Greenland.
26 B.N. Begunov, N.P. Zakaznov, S.I. Kiryushin, and V.I. Kuzichev, *Optical Instrumentation: Theory and Design*, Mir Publishers, Moscow (1988), Translated from the original Russian edition of 1981 by M. Edelev.
27 R.E. Hopkins, 'The components in the basic optical systems', Chapter 3 in *Methods of Experimental Physics*, Vol. 25, Ed. D. Malacara, Academic Press (1988) pp. 59–118.
28 D.S. Goodman, 'Basic optical instruments', Chapter 4 in *Methods of Experimental Physics*, Vol. 25, Ed. D. Malacara, Academic Press (1988) pp. 119–237.
29 S.F. Ray, *Applied Photographic Optics*, Focal Press, London (1988).
30 Y. Matsui, Introduction to Imaging Optics, JOEM, Japan Optomechatronics Association (1988). [in Japanese]
31 M. Leśniewski, *Optical System Design*, Warsaw University of Technology, Warsaw (1990). [in Polish, *projektowanie układów optycznych*, Wydawnictwa Politechniki Warszawskiej, Waszawa (1990)]
32 J.R. Rogers, 'Optical Lenses', Chapter 3 in *Handbook of microwave and optical components*, Vol. 3 Optical components, Ed. K. Chang, John Wiley, New York (1990) pp. 102–147.
33 M. Bass, Editor in Chief, *Handbook of Optics*, Second Edition, Vol. II, Optical Society of America, Part 2 Optical Instruments, Chapters 15–23, McGraw-Hill, New York (1995).
34 M.J. Riedl, *Optical Design Fundamentals for Infrared Systems*, Second Edition, Tutorial Text in Optical Engineering, Vol. TT 48, SPIE Press, Bellingham, Washington (2001) [First published in 1995].
35 R.R. Shannon, *The Art and Science of Optical Design*, Cambridge University Press, Cambridge (1997).
36 B.H. Walker, *Optical Engineering Fundamentals*, Prentice-Hall, New Delhi (2003). [First published by SPIE as Tutorial Text in Optical Engineering, Vol. TT30 in 1998]
37 A. Nussbaum, *Optical System Design*, Prentice Hall, New Jersey (1998).
38 R.E. Fischer, B. Tadic-Galeb, and P.R. Yoder, *Optical System Design*, McGraw-Hill, New York (2008).
39 J.P. Goure, Ed., *Optics in Instruments*, ISTE and Wiley, London (2011).
40 D. Malacara-Hernández and B.J. Thompson, Eds., *Fundamentals and Basic Optical Instruments*, Vol. 1, CRC Press, Taylor and Francis, Boca Raton (2018).
41 D. Malacara-Hernández and B.J. Thompson, Eds., *Advanced Optical Instruments and Techniques, Vol. 2*, CRC Press, Taylor and Francis, Boca Raton (2018).
42 A. Cox, *A System of Optical Design*, The Focal Press, London (1964).
43 L. Levi, *Applied Optics: A Guide to Optical System Design*, Vol. I, John Wiley, New York (1968).
44 S.F. Ray, *The Lens in Action*, Focal Press, London (1976).
45 S.F. Ray, *The Lens and All Its Jobs*, Focal Press, London (1977).
46 M. Laikin, *Lens Design*, Fourth Edition, CRC Press, Taylor & Francis, Boca Raton (2007). [First published in 1991].
47 W.J. Smith, *Modern Lens Design: A Resource Manual*, Mc-Graw Hill, New York (1992).
48 I. Livshits, D. Dilworth, S. Bezdid'ko, G.L. Golovanevsky, V.G. Golovanevsky, and E.D. Golovanevsky, 'All the world's lenses: a database of over 20,000 lenses from the patent literature', *Proc. SPIE*, Vol. 2537 (1995) pp. 124–128.
49 I.C. Gardner, *Application of the algebraic aberration equations to optical design*, Scientific Papers of the Bureau of Standards, No. 550 [Part of Vol. 22], U.S. Government Printing Office, Washington (1927).
50 M. Berek, *Grundlagen der praktischen Optik: Analyse und Synthese optischer Systeme*, Walter de Gruyter, Berlin (1930).
51 H. Chrétien, *Calcul des Combinaisons Optiques*, Fifth Edition, Masson, Paris (1980). [First published in 1938].
52 E.D. Brown and T. Smith, 'Systematic constructional tables for thin cemented aplanatic lenses', *Philos. Trans. Roy. Soc. London, Ser. A*, Vol. 240, (1946) pp. 59–116.
53 R.E. Stephens, 'The design of triplet anastigmat lenses of the Taylor type', *J. Opt. Soc. Am.* Vol. 38, pp. 1032–1039 (1948). [Errata in *J. Opt. Soc. Am.* Vol. 40 (1950) pp. 406–407]
54 D. Argentieri, *Ottica Industriale*, Second Edition, Editore Ulrico Hoepli, Milano (1954). [in Italian]
55 W.H. Steel, 'A general equation for the choice of glass for cemented doublets', *Aust. J. Phys.* Vol. 8, (1955) pp. 68–73.
56 F.D. Cruickshank, 'Contribution des methods algébraiques au calcul des systems optiques' *Revue d'Optique* Vol. 35, (1956) pp. 292–299.

57 N.v.d.W. Lessing, 'Selection of optical glasses in apochromats', *J. Opt. Soc. Am.* Vol. 47 (1957) pp. 955–958.
58 F.D. Cruickshank, 'The design of photographic objectives of the triplet family I. The design of the triplet type 111 objective', *Aust. J. Phys.* Vol. 11, (1958) pp. 41–54.
59 N.v.d.W. Lessing, 'Further considerations on the selection of optical glasses in apochromats', *J. Opt. Soc. Am.*, Vol. 48 (1958) pp. 269–273.
60 N.v.d.W. Lessing, 'Selection of optical glasses in Taylor triplets (special method)', *J. Opt. Soc. Am.* Vol. 48, (1958) pp. 558–562.
61 M. Herzberger, *Modern Geometrical* Optics, Interscience, New York (1958).
62 R.E. Stephens, 'Selection of glasses of three color achromats'. *J. Opt. Soc. Am.*, Vol. 49 (1959) pp. 398–401.
63 N.v.d.W. Lessing, 'Selection of optical glasses in Taylor triplets (General Method)', *J. Opt. Soc. Am.* Vol. 49, (1959) pp. 31–34.
64 N.v.d.W. Lessing, 'Selection of optical glasses in Taylor triplets with residual longitudinal chromatic aberration', *J. Opt. Soc. Am.* 49, (1959) pp. 872–874.
65 F.D. Cruickshank, 'The design of photographic objectives of the triplet family II. The initial design of compound triplet systems', *Aust. J. Phys.* Vol. 13, (1960) pp. 27–42
66 A.E. Conrady, *Applied Optics and Optical Design*, Part II, Dover, New York (1960).
67 C.G. Wynne, 'Thin-lens aberration theory', *Opt. Acta* 8, (1961) pp. 255–265.
68 R. Kingslake, 'Automatic Predesign of the Cooke Triplet Lens', *Proc. Conf. on Optical Instruments and Techniques London 1961*, Ed. K. J. Habell, Chapman and Hall, London (1962) pp. 107–120.
69 H.H. Hopkins, 'The Gaussian Optics of Multi-Lens Systems', *Proc. Conf. on Optical Instruments and Techniques London 1961*, Ed. K.J. Habell, Chapman and Hall, London (1962) pp. 133–159.
70 R.E. Hopkins, 'Method of lens design', Section 9, and 'An application of the method of lens design', Chapter 10, in *Optical Design, Military Standardization Handbook, MIL-HDBK-141*, U. S. Department of Defense (1962).
71 M.H. Sussman, 'Cemented aplanatic doublets', *J. Opt. Soc. Am.* Vol. 52 (1962) pp. 1185–1186.
72 W.J. Smith, *Modern Optical Engineering: The Design of Optical systems*, Third Edition, McGraw-Hill, New York (2000). [First published in 1966].
73 J. Burcher, *Les Combinaisons Optiques, Pratique des Calculs*, Masson & Cie, Paris (1967).
74 H.H. Hopkins and V.V. Rao, 'The systematic design of two-component objectives', *Opt. Acta* Vol. 17, (1970) pp. 497–514.
75 R.E. Kingslake, Chapters 6, 9, 11–15 in Ref. 7.
76 B.H. Walker, 'The 20% rule: selecting optical glass', *Optical Spectra*, December (1978) pp. 42–45.
77 J. Macdonald, Lecture 16 and Lecture 17, 'Optical Design', Lecture Notes, Third European Optics Summer School,!6–27 July 1979, University of Reading, Berkshire, U.K.
78 M. Leśniewski and W. Magdziarz, 'GABAR – an interactive tool for dimensional predesign of arbitrary optical systems', *Optik*, Vol. 75 (1979) pp. 135–137.
79 D. Shafer, 'Simple method for designing lenses', *Proc. SPIE*, Vol. 237, 1980 International Lens Design Conference (1980) pp. 234–241.
80 D. Shafer, 'Optical design with only two surfaces', *Proc. SPIE*, Vol. 237, 1980 International Lens Design Conference (1980) pp. 256–261.
81 M.I. Khan and J. Macdonald, 'Cemented doublets. A method for rapid design', *Opt. Acta* Vol. 29, (1982) pp. 807–822.
82 O.N. Stavroudis, *Modular Optical Design*, Springer, Berlin (1982).
83 K. D. Sharma and S. V. RamaGopal, 'Design of achromatic doublets: evaluation of the double graph technique', *Appl. Opt.*, Vol. 22 (1983) 497–500.
84 L.N. Hazra, 'Structural design of multicomponent lens systems', *Appl. Opt.*, Vol. 23 (1984) pp. 4440–4443.
85 G.G. Slyussarev, *Aberration and Optical Design Theory*, Adam Hilger, Bristol (1984). Translated by J.H. Dixon from *Metodi Rascheta Opticheskikh Sistem* (in Russian), Mashinostroenie Press, Leningrad (1969).
86 M.I. Khan, 'Cemented triplets. A method for rapid design', *Opt. Acta*, Vol. 31 (1984) pp. 873–883.
87 M.J. Kidger, 'Glass selection in optical design', *Proc. SPIE*, Vol. 531, Geometrical Optics, (1985) pp. 178–186.

88 P N. Robb, 'Selection of optical glasses. I: Two materials', *Appl. Opt.*, Vol. 24 (1985) pp. 1864–1877.
89 D.C. O'Shea, 'The Design Process', Chapter 11 in *Elements of Modern Optical Design*, John Wiley, New York (1985).
90 D.C. Sinclair, 'First order optical system layout', *Proc. SPIE*, Vol. 531 (1985) pp. 11–26.
91 R.G. Bingham and M.J. Kidger, 'The principles of aberration-corrected optical systems', *Proc. SPIE*, Vol. 554, 1985 International Lens Design Conference (1986) pp. 88–94.
92 L.N. Hazra and A.K. Samui, 'Design of the individual components of a multicomponent lens system: use of a singlet', *Appl. Opt.*, Vol. 25, (1986) pp. 3721–3730.
93 D. Shafer, 'Optical design and the relaxation response', *Proc. SPIE*, Vol. 766 (1987) pp. 91–99.
94 D.R. Shafer, 'The triplet: an embarrassment of riches', *Opt. Eng.*, Vol. 27 (1988) pp. 1035–1038.
95 H. Rutten and M.v. Venrooij, *Telescope Optics: Evaluation and Design*, Willmann-Bell, Inc., Richmond, Virginia (1988).
96 D. Korsch, *Reflective Optics*, Academic. Boston (1991).
97 S. Zhang and R. Shannon, 'Lens design using minimum number of glasses', *Proc. SPIE*, Vol. 2263 (1992) pp. 2–9.
98 D. Malacara and Z. Malacara, *Handbook of Lens Design*, Marcel Dekker, New York (1994).
99 W.J. Smith, 'Techniques of First-order Layout', Chapter 32 in *OSA Handbook of Optics, Vol. I*, M. Bass, Editor in Chief, McGraw-Hill, New York (1995) 32.1 to 32.16.
100 P. Mouroulis and J. Macdonald, *Geometrical Optics and Optical Design*, Oxford University Press, New York (1997).
101 S. Banerjee and L.N. Hazra, 'Structural design of doublet lenses with prespecified targets', *Opt. Eng.*, Vol. 36, (1997) pp. 3111–3118.
102 C. Chen, S. Shiue, and M. Lu, 'Method of solving cemented triplets with given primary aberration', *J. Mod. Opt.*, Vol. 44, (1997) pp. 753–761.
103 C. Chen, S. Shiue, and M. Lu, 'Method of solving a triplet comprising a singlet and a cemented doublet with given primary aberrations', *J. Mod. Opt.*, Vol. 44, (1997) pp. 1279–1291.
104 W.J. Smith, *Practical Optical System Layout and use of stock lenses*, McGraw-Hill, New York (1997).
105 S. Banerjee and L.N. Hazra, 'Simulated Annealing with constrained random walk in structural design of doublet lenses', *Opt. Eng.*, Vol. 37, No.12, (1998) pp. 3260–3267.
106 S. Banerjee and L.N. Hazra, 'Experiments with GSA techniques in structural design of doublet lenses', *Proc. SPIE*, Vol. 3482, (1998) pp. 126–134.
107 G.H. Smith, *Practical Computer-Aided Lens Design*, Willman-Bell, Richmond, VA (1998).
108 S. Banerjee and L.N. Hazra, 'Thin lens design of Cooke triplet lenses: application of a global optimization technique', *Proc. SPIE*, Vol. 3430, (1998) pp. 175–183.
109 S. Banerjee and L.N. Hazra, 'Genetic algorithm in structural design of Cooke triplet lenses', *Proc. SPIE*, Vol. 3737, (1999) pp. 172–179.
110 W.J. Smith, *Modern Optical Engineering: The Design of Optical Systems*, Third Edition, McGraw-Hill, New York (2000) [First published in 1966].
111 J. Tesar. 'Using small glass catalogs', *Opt. Eng.*, Vol. 39 (2000) pp. 1816–1821.
112 S. Banerjee and L.N. Hazra, 'Experiments with a genetic algorithm for structural design of cemented doublets with prespecified aberration targets', *Appl. Opt.*, Vol. 40, (2001) pp. 6265–6273.
113 J.M. Geary, *Introduction to Lens Design with Practical ZEMAX Examples*, Willman-Bell, Richmond, VA (2002).
114 S. Banerjee and L.N. Hazra, 'Structural design of broken contact doublets with prespecified aberration targets using genetic algorithm', *J. Mod. Opt.*, Vol. 49, (2002) pp. 1111–1123.
115 M.J. Kidger, *Fundamental Optical Design*, SPIE Press, Bellingham, Washington (2002).
116 D.N. Frolov, R.M. Raguzin, and V.A. Zverev, 'Optical system of a modern microscope', *J. Opt. Technol.*, Vol. 69 (2002) pp. 610–613.
117 M.J. Kidger, *Intermediate Optical Design*, SPIE Press, Bellingham, Washington (2004).
118 S. Chatterjee and L.N. Hazra, Structural design of cemented triplets by genetic algorithm, *Opt. Eng.*, Vol. 43, (2004) pp. 432–441.
119 C.L. Tien, W.S. Sun, C.C. Sun, and C.H. Lin, 'Optimization design of the split doublet using the shape factors of the third order aberrations for a thick lens', *J. Mod. Opt.*, Vol. 51 (2004) pp. 31–47.
120 A.P. Grammatin and E.V. Kolesnik, 'Calculation for optical systems, using the method of simulation modelling', *J. Opt. Technol.*, Vol. 71 (2004) pp. 211–213.

121 S. Chatterjee and L.N. Hazra, Optimal design of individual components of multicomponent optical systems, Proceedings International Conference on Optics and Optoelectronics (ICOL) CD-ROM Paper OP-ODFT-3, Instruments Research & Development Establishment, Dehradun (2005).
122 T. Kryszczyński and M. Leśniewski, 'Method of the initial optical design and its realization', *Proc. SPIE*, Vol. 5954 (2005) 595411-1 to 595411–12.
123 S. Chatterjee and L.N. Hazra, Structural design of a lens component with pre-specified aberration targets by evolutionary algorithm, *Proc. SPIE*, Vol. 6668,0S1–12 (2007).
124 W.S. Sun, C.H. Chu, and C.L. Tien, 'Well-chosen method for an optical design of doublet lens design', *Opt. Exp.*, Vol. 17 (2009) pp. 1414–1428.
125 I. Livshits and V. Vasilyev, 'Q and A tutorial on optical design', *Adv. Opt. Technol.*, Vol. 2 (2013) pp. 31–39.
126 D.C. Dilworth, *SYNOPSYS Supplement to Joseph M. Geary's Introduction to Lens Design*, William-Bell, Richmond, VA (2013).
127 C. Velzel, *A Course in Lens Design*, Springer, Dordrecht (2014).
128 I. Livshits and D.C. Dilworth, 'Practical tutorial: A simple strategy to start a pinhole lens design', *Adv. Opt. Technol.*, Vol. 4 (2015) pp. 413–427
129 H. Sun, *Lens Design: A Practical Guide*, CRC Press, Taylor & Francis, Boca Raton (2017).
130 D.C. O'Shea and J.L. Bentley, *Designing Optics Using CODEV*, Vol. PM 292, SPIE Press, Bellingham, Massachusetts (2018).
131 Z. Hou, M. Nikolic, P. Benitez, and F. Bociort, 'SMS2D designs as starting points for lens optimization', *Opt. Exp.*, Vol. 26 (2018) pp. 32463–32474.
132 J. Sasián, *Introduction to Lens Design*, Cambridge University Press, Cambridge (2019).
133 S.N. Bezdid'ko and A.F. Shirankov, 'Structural and size synthesis of the initial designs of optical systems', *J. Opt. Technol.*, Vol. 86 (2019) pp. 544–550.
134 R.E. Parks, 'Optical component specifications', *Proc. SPIE*, Vol. 237 (1980) pp. 455–463.
135 R.R. Shannon, 'Optical Specifications', Chapter 35 in *OSA Handbook of Optics, Vol. I*, M. Bass, Editor in Chief, McGraw-Hill, New York (1995) 35.1 to 35.12.
136 D.C. Dilworth and D. Shafer, 'Man versus machine: a lens design challenge', *Proc. SPIE*, Vol. 8841 (2013) 88410G-1 to 88410G-18.
137 M.J. Kidger, 'Principles of Lens Design', Chapter 7 in Ref. 115, pp. 139–166.
138 A.E. Conrady, Ref. 66, pp. 791–801.
139 R. Kingslake, Chapter 12–14 in Ref. 7, pp. 233–334.
140 W.E. Woeltche, 'Structure and image forming properties of asymmetrical wide angle lenses for 35mm photography', *Appl. Opt.*, Vol. 7 (1968) pp. 343–351.
141 E. Glatzel, 'New lenses for microlithography', *Proc. SPIE*, Vol. 237 (1980) pp. 310–320.
142 R. Abel and M. R. Hatch, 'The pursuit of symmetry in wide-angle reflective optical designs' *Proc. SPIE*, Vol. 237 (1980) pp. 271–280.
143 J.M. Sasian and M.R. Descour, 'Power distribution and symmetry in lens systems', *Opt. Engng.*, Vol. 37 (1998) pp. 1001–1007.
144 M. Westcott, 'Sorting out the scan lens puzzle', *Photonics Spectra*, (May 1984) pp. 91–93.
145 K.–E. Peiponen, A, Zeitler, and M. Kuwata-Gonokami, *Terahertz Spectroscopy and Imaging*, Springer, Berlin (2013).
146 J.M. Llyod, *Thermal Imaging Systems*, Plenum Press, New York (1982).
147 W.L. Wolfe, *Introduction to Infrared System Design*, SPIE, Bellingham, Washington (1996).
148 M.J. Riedl, *Optical Design Fundamentals for Infrared Systems*, SPIE, Bellingham, Washington (2006).
149 A.G. Michette, *Optical Systems for Soft X-Rays*, Plenum, New York (1986).
150 L.N. Hazra, Y. Han, and C. Delisle, 'Stigmatic Imaging by zone plates: a generalized treatment', *J. Opt. Soc. Am. A*, Vol. 10 (1993) pp. 69–74.
151 L.N. Hazra, Y. Han, and C. Delisle, 'Imaging by zone plates: axial stigmatism at a particular order', *J. Opt. Soc. Am. A*, Vol. 11 (1994) pp. 2750–2754.
152 D. Attwood, *Soft X-Rays and Extreme Ultraviolet Radiation*, Cambridge University Press, Cambridge (1999).
153 A.G. Michette and S.J. Pfauntsch, *X-Ray Sources and Imaging: Basics and Applications*, Wiley-VCH, Berlin (2020).

154 G.G. Slyusarev, 'Aspheric surfaces: A. Optical systems with an axis of symmetry', in Chapter 9 of *Aberration and Optical Design Theory*, [original Russian book title: *Metodi Rascheta Opticheskikh Sistem, Mashinostroenie*, Leningrad (1969)] Translator: J.H. Dixon, Adam Hilger, Bristol (1984) pp. 515–570.
155 R.K. Luneburg, *Mathematical Theory of Optics*, University of California Press, Berkeley (1964) pp. 139–151. [originally as mimeographed notes published by Brown University in 1944]
156 M. Born and E. Wolf, *Principles of Optics*, Pergamon, Oxford (1980) pp. 197–202.
157 J.C. Burfoot, 'Third-order aberrations of 'doubly symmetric' systems', *Proc. Phys. Soc.* Vol. LXVII, Ser. B (1954) pp. 523–528.
158 C.G. Wynne, 'The primary aberrations of anamorphotic lens systems', *Proc. Phys. Soc.* Vol. LXVII, Ser. B (1954) pp. 529–537.
159 G.G. Slyusarev, 'Aspheric surfaces: B. Anamorphotic systems', in Chapter 9 of *Aberration and Optical Design Theory*, [original Russian book title: *Metodi Rascheta Opticheskikh Sistem, Mashinostroenie*, Leningrad (1969)] Translator: J.H. Dixon, Adam Hilger, Bristol (1984) pp. 571–594.
160 W.T. Welford, *Aberrations of Optical Systems*, Adam Hilger, Bristol (1986) pp. 210–214.
161 W.J. Smith, Ref. 110, pp. 287–291.
162 V.N. Mahajan, 'Anamorphic Systems', Chapter 13 in *Optical Imaging and Aberrations III*, SPIE, Bellingham, Washington (2013) pp. 349–367.
163 O.E. Miller, J.H. McLeod, and W.T. Sherwood, 'Thin sheet plastic Fresnel lenses of high aperture', *J. Opt. Soc. Am.*, Vol. 41 (1951) pp. 807–815.
164 W.T. Welford and R. Winston, *The Optics of Nonimaging Concentrators – Light and Solar Energy*, Academic, New York (1978).
165 D. Malacara-Hernandez and Z. Malacara-Hernandez, *Handbook of Lens Design*, Marcel Dekker, New York (1994) pp. 372–373.
166 R. Leutz and A. Suzuki, *Nonimaging Fresnel Lenses: Design and Performance of Solar Concentrators*, Springer, Berlin (2001).
167 J. Chavez, *Introduction to Nonimaging Optics*, CRC Press, Boca Raton (2008).
168 'Freeform optics involve optical designs with at least one freeform surface, which, according to ISO 17450-1:2011, has no translational or rotational symmetry about axes normal to mean plane'. (The Center for freeform optics, USA)
169 J.C. Miñano, P. Benitez, and A. Santamaria, 'Freeform optics for illumination', *Opt. Rev.*, Vol. 16 (2009) pp. 99–102.
170 K. Fuerschbach, J.P. Rolland, and K.P. Thomson, 'A new family of optical systems employing φ-polynomial surfaces', *Opt. Exp.*, Vol. 19 (2011) pp. 21919–21928.
171 G.W. Forbes, 'Characterizing the shape of freeform optics', *Opt. Exp.*, Vol. 20 (2012) pp. 2483–2499.
172 I. Kaya, K.P. Thomson, and J.P. Rolland, 'Comparative assessment of freeform polynomials as optical surface descriptions', *Opt. Exp.*, Vol. 20 (2012) pp. 22683–22691.
173 K.P. Thompson and J.P. Rolland, 'Freeform optical surfaces', *Opt. Phot. News*, Vol. 23(6), (2012) pp. 30–35.
174 K.P. Thompson, P. Benítez, and J.P. Rolland, 'Freeform optical surfaces: report from OSA's first incubator meeting', *Opt. Phot. News*, Vol. 23(9), (2012) pp. 32–37.
175 F. Duerr, P. Benítez, J.C. Miñano, Y. Meuret, and H. Thienpont, 'Analytic free-form lens design in 3D coupling three ray sets using two lens surfaces', *Opt. Express* 20(10), (2012) pp. 10839–10846.
176 F. Fang, Y. Cheng, and X. Zhang, 'Design of freeform optics', *Adv. Opt. Techn.* Vol. 2 (2013) pp. 445–453.
177 C. Menke and G.W. Forbes, 'Optical design with orthogonal representation of rotationally symmetric and freeform aspheres', *Adv. Opt. Techn.* Vol. 2 (2013) pp. 97–109.
178 F. Duerr, Y. Meuret, and H. Thienpont, 'Potential benefits of free-form optics in on-axis imaging applications with high aspect ratio', *Opt. Express* 21(25), (2013) pp. 31072–31081.
179 C. Liu and H. Gross, 'Numerical optimization strategy for multi-lens imaging systems containing freeform surfaces', *Appl. Opt.*, Vol. 57 (2018) pp. 5758–5768.
180 H.H. Hopkins, 'Optical Principles of the Endoscope', in *Endoscopy*, Ed. G. Berci, Appleton Century Crofits, New York (1976) pp. 3–26.
181 S.J. Dobson and J. Ribeiro, 'The primary aberration characteristics of thin-lens models of common relay systems', *Meas. Sci. Technol.*, Vol. 5 (1994) pp. 32–36.

182 D, Cheng, Y. Wang, L. Yu, and X. Liu, 'Optical design and evaluation of a 4 mm cost-effective ultra-high-definition arthroscope', *Biomed. Opt. Express*, Vol. 5 (2014) pp. 2697–2714.
183 E.W. Marchand, *Gradient Index Optics*, Academic, New York (1978).
184 D.T. Moore, 'Gradient index optics: a review', *Appl. Opt.*, Vol. 19 (1980) pp. 1035–1038.
185 F. Bociort, *Imaging properties of gradient index lenses*, Köster, Berlin (1994).
186 R. Kingslake, 'The development of the zoom lens', *J. SMPTE*, Vol. 69 (1960) pp. 534–544.
187 W.J. Smith, Ref. 110, pp. 291–296.
188 A.D. Clark, *Zoom Lenses*, Adam Hilger, London (1973).
189 K. Yamaji, 'Design of zoom lenses', Chapter IV in *Progress in Optics, Vol. VI*, Ed. E. Wolf, North-Holland, Amsterdam (1967) pp. 107–170.
190 L.N. Hazra and S. Pal, 'A novel approach for structural synthesis of zoom systems', *Proc. SPIE*, Vol. 7786 (2010) 778607–1 to 778607–11.
191 D.A. Atchison, 'Spectacle lens design: a review', *Appl. Opt.*, Vol. 31 (1992) pp. 3579–3585.
192 G.H. Guilino, 'Design philosophy for progressive addition lenses', *Appl. Opt.*, Vol. 32 (1993) pp. 111–117.
193 L. Li, T.W. Raasch, and A.Y. Yi, 'Simulation and measurement of optical aberrations of injection molded progressive addition lenses', *Appl. Opt.*, Vol. 52 (2013) pp. 6022–6029.
194 S. Kuiper and B.H.W. Hendricks, 'Variable-focus liquid lens for miniature cameras', *Appl. Phys. Lett.*, Vol. 85 (2004) pp. 1128–1130.
195 B. Berge, 'Liquid lens technology: principle of electrowetting based lenses and applications to imaging', Proc. 18th IEEE conference on MEMS (2005) pp. 227–230.
196 H. W. Ren and S. T. Wu, 'Variable focus liquid lens', *Opt. Exp.*, Vol. 13 (2007) 5931–5936.
197 S. Reichelt and H. Zappe, 'Design of spherically corrected, achromatic variable-focus liquid lenses', *Opt. Exp.*, Vol. 15 (2007) pp. 14146–14154.
198 R. Peng, J. Chen, and S. Zhuang, 'Electrowetting-actuated zoom lens with spherical-interface liquid lenses', *J. Opt. Soc. Am. A*, Vol. 25 (2008) pp. 2644–2650.
199 P. Valley, D.L. Mathine, M.R. Dodge, J. Schwiegerling, G. Payman, and N. Peyghambarian, 'Tunable focus flat liquid-crystal diffractive lens', *Opt. Lett.*, Vol. 35 (2010) pp. 336–338.
200 R. Mizutani, S. Pal, L. Hazra, and Y. Otani, 'Non-moving component zoom lens', 73rd Autumn Meeting, The Japan Society of Applied Physics, pp. 11–14 September 2012, Ehime University, Matsuyama, Japan (2012) [in Japanese].
201 D.A. Buralli and G.M. Morris, 'Design of a wide field diffractive landscape lens', *Appl. Opt.*, Vol. 28 (1989) pp. 3950–3959.
202 E. Noponen, A. Vasara, J. Turunen, J.M. Miller, and R. Taghizadeh, 'Synthetic diffractive optics in the resonance domain', *J. Opt. Soc. Am. A*, Vol. 9 (1992) pp. 1206–1213.
203 L. N. Hazra and C. A. Delisle, 'Higher order kinoform lenses: diffraction efficiency and aberrational properties', *Opt. Eng.*, Vol. 36 (1997) 1500–1507.
204 L.N. Hazra, Y. Han, and C.A. Delisle, 'Kinoform lens: Sweatt model and phase function', *Opt. Commun.* Vol. 117 (1995) pp. 31–36.
205 D.C. O'Shea, T.J. Suleski, A.D. Kathman, and D.W. Prather, *Diffractive Optics*, SPIE, Bellingham, Washington (2004).
206 G.C. Righini, V. Russo, S. Sottini, and G. Toraldo di Francia, 'Geodesic lenses for guided optical waves', *Appl. Opt.*, Vol. 12 (1973) pp. 1477–1481.
207 P.J.R. Laybourn and G.C, Righini, 'New design of thin-film lens', *Electronics Lett.*, Vol. 22 (1986) pp. 343–345.
208 T. Matsuyama, Y. Ohmura, and D.M. Williamson, 'The lithographic lens: its history and evolution', *Proc. SPIE*, Vol. 6154 (2006) 615403-1 to 615403-14.
209 R.I. Gordon and M.P. Rimmer, (Lens) Design for (Chip) Manufacture: Lens tolerancing based on linewidth calculations in hyper-NA immersion lithography systems', *Proc. SPIE*, Vol. 6154 (2006) 61540K-1 to 61540K-12.
210 A. Dodoc, 'Towards the global optimum in lithographic lens design', *Proc. Int. Opt. Design Conf.*, IODC 2010, Paper IWD-3 (2010).
211 J. Flügge, 'Systematik der photographischen Objektive', in *Das Photographische Objektiv, Springer, Wien* (1955) pp. 165–199.

212 G. Franke, 'The Classification of optical systems', Part C in *Physical Optics in Photography*, The Focal Press, London (1966) [translated by K. S. Ankersmit from *Photographische Optik*, Akademische Verlagsgesellschaft, Frankfurt am Main (1964)] pp. 126–161.
213 W. Taylor and H.W. Lee, 'The development of the photographic lens' *Proc. Phys. Soc. London*, Vol. 47 Part 3 (1930) pp. 502–518.
214 R. Kingslake, 'The design of wide-aperture photographic objectives', *J. Appl. Phys.*, Vol. 11 (1940) pp. 56–69.
215 H.W. Lee, 'New lens systems', *Reports on Progress in Physics*, Vol. VII (1940) pp. 130–149.
216 C.G. Wynne, 'New lens systems', *Reports on Progress in Physics*, Vol. XIX (1956) pp. 298–325.
217 E. Glatzel, 'New developments in photographic objectives', in *Optical Instruments and Techniques 1969*, Ed. J. Home Dickson, Proc. ICO Conf., Univ. Reading (1969) pp. 407–428.
218 W. Wöltche, 'New development and trends in photographic optics at Zeiss', *Brit. J. Photography*, (25 Jan. 1980) pp. 76–79.
219 W. Wöltche, 'New development and trends in photographic optics at Zeiss Part 2', *Brit. J. Photography*, (1 Feb. 1980) pp. 94–99.
220 W. Wöltche, 'Optical system design with reference to the evolution of the double Gauss lens', 1980 Int. Lens Design Conf., *Proc. SPIE*, Vol. 237 (1980) pp. 202–215.
221 J. Hoogland, 'Systematics of photographic lens types', 1980 Int. Lens Design Conf., *Proc. SPIE*, Vol. 237 (1980) pp. 216–221.
222 W. Mandler, 'Design of basic double Gauss lenses', 1980 Int. Lens Design Conf., *Proc. SPIE*, Vol. 237 (1980) pp. 222–232.
223 E. Glatzel, 'New lenses for microlithography', 1980 Int. Lens Design Conf., *Proc. SPIE*, Vol. 237 (1980) pp. 310–320.
224 H.D. Taylor, *A system of Applied Optics*, Macmillan, London (1906).
225 R. Kingslake, *A History of the Photographic Lens*, Academic, New York (1989) p. 87.
226 J. Kammerer, 'The moon camera and its lenses', *Opt. Eng.*, Vol.11 (1972) G73–G78.
227 F.D. Cruickshank, 'A system of transfer coefficients for use in the design of lens systems: IV. The estimation of the tolerances permissible in the production of a lens system', *Proc. Phys. Soc. (London)*, Vol. 57 (1945) pp. 426–429.
228 P.C. Foote and R.A. Woodson, 'Lens Design and Tolerance Analysis Methods and Results', *J. Opt. Soc. Am.*, Vol. 38 (1948) pp. 590–599.
229 D.G. Koch, 'A statistical approach to lens tolerancing', *Proc. SPIE*, Vol. 147 (1978) pp. 71–82.
230 W.J. Smith, 'Fundamentals of establishing an optical tolerance budget', *Proc. SPIE*, Vol. 531 (1985) pp. 196–204.
231 P.O. McLaughlin, 'A primer on tolerance analysis', *Design Notes*, Summer 1991, Vol. 2, No. 3, Sinclair Optics, Fairport, N.Y.
232 R.R. Shannon, 'Tolerancing Techniques', Chapter 36 in *OSA Handbook of Optics, Vol. I*, M. Bass, Editor in Chief, McGraw-Hill, New York (1995) 36.1 to 36.12.
233 M.J. Kidger, 'Design for Manufacturability', Chapter 11 in Ref. 117, pp. 215–222.
234 J.L. Bentley, 'Where do optical tolerances come from?', *Proc. SPIE*, Vol. 10315 (2005) 1031501-1 to 1031501-3.
235 R.E. Fischer, B. Tadic-Galeb, and P.R. Yoder, *Optical System Design*, Second Edition, McGraw-Hill, New York (2008) pp. 347–388.
236 H.H. Hopkins and H.J. Tiziani, 'A theoretical and experimental study of lens centring errors and their influence on optical image quality', *Brit. J. Appl. Phys.*, Vol. 17 (1966) pp. 33–54.
237 M. Rimmer, 'Analysis of perturbed lens systems', *Appl. Opt.*, Vol. 9 (1970) pp. 533–537.
238 M.P. Rimmer, 'A Tolerancing procedure based on Modulation Transfer Function (MTF)', *Proc. SPIE*, Vol. 147 (1978) pp. 66–70.
239 R.R. Shannon, 'Tolerance Analysis', Chapter 6 in '*The Art and Science of Optical Design*', Cambridge University Press, Cambridge (1997) pp. 356–387.
240 M.I. Kaufman, B.B. Light, R.M. Malone, M.K. Gregory and D. K. Frayer, 'Technique for analysing lens manufacturing data with optical design applications', *Proc. SPIE*, Vol. 9573 (2015) 95730M-1 to 95730M-18.
241 J. Sasián, 'Lens Tolerancing', Chapter 10 in Ref. 132, pp. 110–125.

20

Lens Design Optimization

In common parlance, the word 'Optimization' means an action or process for making something as good or effective as possible. Merriam-Webster dictionary gives an elaborate description, and defines 'Optimization' as an act, process, or methodology of making something (such as a design, a system, or a decision) as fully perfect, functional, or effective as possible. From the early days of scientific lens design, its methodology has essentially followed the flowchart given in Figure 19.3, which is elucidated in Section 19.2. It is obvious that the flowchart is drawn to optimize the process of lens design. However, phenomenal changes in the detailed form or nature of the different boxes for image evaluation, decision making, tolerancing, etc., and the determination of required changes in variables at each iteration and in the modes of implementation of these boxes, have taken place over the years. During the last seventy years, techniques for practical implementation of lens design optimization have been continually improved upon in order to cater to the exigencies of complex optical systems. Significantly enough, these changes in the field of lens design are coterminous with the metamorphosis that the classical field of optics is undergoing during the same period. In the area of lens design, the primary impetus has been provided by the growing availability of digital computers with continually increasing number crunching and data processing abilities. Over the years, optical designers attempted to gain maximum leverage out of the process by gradually setting higher goals. In the next section, we provide a brief history of this fascinating transition from 'manual lens design' to 'automatic lens design'. A more appropriate nomenclature for the latter is 'Computer-aided lens design'. The spatiotemporal developments in this field during the second half of the twentieth century will be underscored in the next section.

20.1 Optimization of Lens Design: From Dream to Reality

The optimization of lens design involves intensive numerical computation. The magnitude of the latter increases exponentially with the increasing complexity of lens systems. In the pre-digital computer days, lens designers showed great ingenuity in developing approximate analytical or semi-analytical models that could reduce the need for extensive numerical computation in order to arrive at working solutions. Of course, the optimality of such solutions was often questionable. On the other hand, approximations inherent in these models lose their validity with an increase in the number of components for multicomponent optical systems. Consequently, optical lens systems of the pre-digital computer era typically used to be constituted of at most six or seven lens elements with spherical interfaces. Incidentally, it should also be noted that the other major reason for this constraint on the upper limit of lens elements to be used in practical systems was the non-availability in this period of vacuum coating techniques for anti-reflection coatings on glass-air interfaces.

It should not be overlooked that the lens designers were the earliest users of all computational tools or devices, including 'higher' figure log tables, slide rules, mechanical calculating machines, electric calculators of various forms, stored program computers, etc., developed for expediting 'arithmetic' computation. In order to harness maximum gain in computational speed, the designers often used to update the formulae used for ray tracing. Major changes took place with the arrival of digital computing machinery, which prompted the transition from the use of logarithmic and trigonometric formulae for ray tracing to deterministic formulae restricted to basic arithmetic computations. A glimpse of the highlights in development of ray tracing methods during the first sixty years of the twentieth century may be obtained from the references [1–23].

DOI: 10.1201/9780429154812-20

The first systematic investigation on lens design optimization was undertaken at Harvard University in the early 1950s by J.G. Baker and his co-workers under the tutelage of the U.S. Air Force through a contract with the Perkin-Elmer Corporation. Results of this research appeared in a series of technical reports entitled 'The Utilization of Automatic Calculating Machinery in the Field of Optical Design' during 1951–1955. However, these documents were declassified only in 1959 [24]. The investigators primarily utilized the optimum gradient method or the steepest descent method for optimization. This method was proposed originally by Cauchy in 1847 [25]. Baker observed that, after the first few iterations, the rate of convergence of the method was too slow, making it impracticable. Their attempts to improve convergence of the method met limited success. Around the same time, at the University of Pennsylvania, Rosen et al. [26–27], supported by the Department of the Army, Ordnance Corps, Frankford Arsenal, Philadelphia, explored the use of an IBM card-programmed calculator in order to improve the state of correction of a triplet objective using the method of least-squares, invented by Legendre [28].

The use of electronic digital computers in lens design was initiated in the United Kingdom at the University of Manchester by Black [29–30] under a project sponsored by the British Scientific Instrument Association. Like Baker, Black used a 'merit function' that he called 'performance number'. His experience with the method of steepest descent in lens design optimization was not satisfactory, and he embarked upon a modified form of gradient method in order to circumvent the problem of computing derivatives. He used a variable-by-variable method where the performance number was successively minimized with respect to one variable at a time. The method was slow in execution and it used to suffer from stagnation problems. Black attempted to accelerate the procedure by taking recourse to 'block' and 'group' operations. These are essentially modifications of the variable-by-variable methods where more than one variable is changed simultaneously in the case of a 'block' operation, and a 'group' operation is an extrapolation after a set of changes providing improvements in the performance function is already identified.

At the Institute of Optics, University of Rochester, R.E. Hopkins and his associates started exploring the suitable use of computers in lens design optimization from the outset. The earliest publication by the group is on automatic correction of third-order aberrations by a card-programmed calculator [31]. The next publication reports on use of the same technique for the design of telescope doublets by the same calculator [32]. Thorough investigations on different aspects of the two basic design problems were carried out, and new algorithms were developed. Also, the research group would duly upgrade their computer programs and design approaches with the availability of more powerful digital computers [33–43]. The method of optimization developed by Hopkins and his co-workers motivated many others to pursue similar approaches in subsequent years, so much so that even twenty years later, reports appear with the results of experiments with constrained optimization using Spencer's method [44].

In the early 1950s, Havliček [45–46], from the J. Stefan Institute at Ljubljana, Slovenia, reported on the fine correction of optical systems. Wachendorf [47] deliberated on the conditions to be satisfied for an optical system to be optimized, and also on the problem of defining suitable merit function. In the latter half of the 1950s, Meiron published three papers [48–50] from Israel on his investigations on three different techniques for optimization of lens design. The first one dealt with the steepest-descent method, and the second was the standard least-squares method. The third technique was essentially a variable-by-variable method. Later on, after shifting to the Perkin-Elmer Corporation in Connecticut, USA, Meiron reported his investigations on the use of the damped least-squares technique for lens design optimization [51]. Girard presented a doctoral thesis entitled 'Calcul Automatique en Optique Géométrique' at l'Université de Paris in 1957 [52].

Feder, one of the pioneers in the field, started his investigations on automatic lens design methods when the SEAC (Standards Eastern Automatic Computer) computer designed and built at National Bureau of Standards, Washington became available to him in 1954. He reported on his exploration of available techniques, primarily the method of steepest descent, the linearization methods, and the least-squares method, applied to a suitably defined merit function [53–54]. All his attempts encountered poor convergence. Therefore, after shifting to Eastman Kodak in 1957, he modified the steepest descent method to form a new approach, entitled the conjugate gradient method. The latter method was developed by Hestenes and Stiefel for solving a set of linear equations [55], and Feder extended it to the nonlinear case. Some experiments with it were reported in [56]. Quoting Curry [57], Feder made the important observation that the convergence of a gradient method depends upon the choice of metric in the space of the independent variables. In his benchmark paper on 'automatic optical design' in 1963, Feder noted that they were still pursuing both the conjugate gradient method and the damped least-squares method as they were not sure which one will be more useful in general [58]. Later on, Feder seems to have been convinced about the more effectiveness of the damped least-squares approach in lens design problems [59–60].

C.G. Wynne of Wray (Optical Works) Ltd., Kent, UK, clearly identified the reasons for the slow convergence of the steepest descent method for the type of equations encountered in lens design optimization problems, and to overcome the problem he devised a method, that he called 'Successive Linear Approximation with Maximum Steps (SLAMS)' [61–62]. In fact, this method is somewhat similar to the method proposed by Levenberg for solving nonlinear least-squares problems in applied mathematics [63]. A similar technique was also proposed by Girard [52] for tackling the lens design optimization problem. These three methods seem to have been developed independently. Interestingly, the technique was developed independently for the fourth time in a different context by Marquardt [64]. Shortly after his initial works with SLAMS, Wynne shifted to Imperial College, London and continued his investigations on computer-aided lens [65–68] design optimization. Notable members of his research group who continued his works in this field are P.M.J.H. Wormell and M.J. Kidger. Wormell reported the developments in successive versions of the lens design optimization programs at Imperial College [69]. Finkler extended the treatment for design of optical systems with extended depth of focus [70]. Kidger developed a highly successful commercial software package entitled 'SIGMA'. In 1993, he published a review article on different aspects of the DLS optimization method in lens design [78].

In 1961, Glatzel from Carl Zeiss, Oberkochen, Germany published the mathematical basis of an adaptive correction method for lens design [72]. The method was in many ways similar to the initial attempts by R.E. Hopkins and his co-workers at the Institute of Optics, Rochester [31–33, 38]. Subsequently, the two approaches diverged significantly. In Glatzel's adaptive method, target values are set for a small number of selected aberrations, and for others, generally higher order aberrations, limits are set. The mathematical basis of the method requires that the number of controlled aberrations should not exceed the number of free parameters. Note that the initial target values are rarely the final desired values. The targets are reset in the course of optimization in accordance with the state of correction. The diagnostics provided by the method in the course of optimization facilitate overcoming the stagnation and slow convergence of merit function-based approaches [73–74]. Rayces et al. reported on the development of a computer optimization program that experimented with both Glatzel's approach and Spencer's approach [75]. Later on, Maxwell et al. proposed a modification of Glatzel's adaptive approach by incorporating a quadratic approximation to the individual aberration variations [76–77]. It should not be overlooked that adaptive approaches have the inherent potential to circumvent the critical problem of being restricted in the search space to a local minimum of the merit function.

In 1963, Fialovszky from Budapest, Hungary published a differential method for the fine correction of optical systems. This method had similarities with Spencer's method, and formulae were presented for correcting one set of errors, while simultaneously another set of errors could be put below prespecified tolerance limits [72].

From Lincoln Laboratory, MIT, Grey proposed a different approach to tackle the problem of ill-conditioning and nonlinearity that gives rise to slow convergence and stagnation in lens design optimization [79–82]. He appreciated the significant role played by the classical aberration theory in the treatment of optical systems. Nevertheless, he identified two basic shortcomings in this theory that plagued lens design optimization at the final stages of design. Firstly, in general, it is not possible to manipulate an individual aberration coefficient in a manner that leaves the other aberration coefficients unperturbed. Next, each individual aberration coefficient is a function of all constructional parameters of the system, so that, in general, change in any parameter causes changes in all aberration coefficients.

In order to tackle these problems, Grey introduced a new concept of orthogonal aberration theory. According to this theory, aberration coefficients are orthogonal only in reference to a specific defect function, say ψ, for the lens design. Therefore, the defect function ψ is minimized when the orthogonal aberration coefficients, say G_i, are individually set equal to zero. Note that the tricky problem of balancing aberrations does not arise in this case.

Each aberration coefficient G_i is a function of the constructional parameters x_i, $i = 1,\ldots,N$. The coordinate base X may be replaced by a multidimensional variable U with coordinates u_i. Each u_i is a function of the constructional parameters, and there are as many u_i as there are x_i. The coordinate base U is an orthogonal coordinate base if there is a one-to-one correspondence between the u_i and the G_i. Grey calls aberration theory orthonormal if it is orthogonal, has an orthogonal coordinate base, and if $\frac{\partial G_i}{\partial u_j} = \delta_{ij}$, where

$$\left.\begin{aligned} \delta_{ij} &= 1, \quad i = j \\ &= 0, \quad i \neq j \end{aligned}\right\} \tag{20.1}$$

According to Grey:

> The specifications laid down for an orthonormal aberration theory are so broad that they cannot be fulfilled precisely. They can be fulfilled approximately, but only to the extent that the second derivatives of the optical aberration with respect to construction parameters can be ignored. Their use is thereby limited to differential correction of a lens. Within any region of parameter space sufficiently small that the second derivatives of the aberration can be ignored, an orthonormal aberration theory can be constructed and used to make differential corrections. Within this region, the transformation from the X coordinate base to the U coordinate base can be linear.
>
> [79]

The transformation from the X to U coordinate basis is implemented by the original Gram-Schmidt Orthonormalization method, or any of its modified forms [83]. A detailed description of the merit function in Grey's lens design program is given by Seppala [84].

At Los Alamos Scientific Laboratory, New Mexico, investigations on lens design optimization were initiated by Brixner from the outset, and the procedure has been continually modified over approximately three decades by him in collaboration with his associates [85–96]. One of the unique features of their approach is to use a merit function based on 'many-ray image-spot size and position data' obtained by ray tracing. For optimization, they claimed to have used an increment-vector damping technique developed by T.C. Doyle. This technique is a refinement of the Levenberg damping technique. Brixner identified the root causes of stagnation and slow convergence in the optimization procedure and suggested remedies for accelerating convergence [93].

The use of diffraction-based criteria of image quality as merit/error function in automatic lens design was initiated by H.H. Hopkins of Imperial College, UK in 1966 [97]. His method expresses the defect in image quality at a field point in quadratic form, with its vector elements consisting of wave and transverse ray aberrations of a selected number of rays from a particular field point. Use of canonical coordinates for points on the entrance/exit pupil transforms the entrance/exit pupils to unit circles (see Chapter 11). Further, the use of canonical variables for object/image enables a universal kernel of the quadratic form of the image defect. The kernel is universal for any lens system in the sense that it needs to be evaluated once, and only once, for a prespecified set of rays from a particular fractional field point. Whereas the merit/error function used in other approaches is typically the sum of squares of a set of weighted values of transverse ray aberrations or wave aberrations, Hopkins introduced the variance of wave aberration (which is directly linked with the Strehl ratio) as a merit function for highly corrected systems, and the variance of the wave aberration difference function (which is directly linked with the optical transfer function) as a merit function for moderately corrected systems. Lenskii, Meiron, King, Offner, Rosenbruch, Gostick, Hazra, and Szapiel, among others, have deliberated on the different aspects of this enterprise, particularly on extending the range of validity of the expressions used for variances [98–105]. In 1998, Hopkins and Kadkly published an extensive article on the Reading University optical design program (ROD) explaining the details of their merit function, as well as the damping factors and mode of boundary condition control by penalty functions [106].

During the initial period of experimentation with the damped least-squares (DLS) method for lens design optimization, the choice of a numerical value for the damping factor was somewhat arbitrary, and was based on the observation that a large value for the damping factor leads to small parameter changes, and vice versa. Often, the damping factor was chosen by trial and error method. However, it was duly realized that the rate of convergence depends critically on the choice of damping factor.

Some investigations were also directed to develop techniques for obviating the need for an explicit damping factor. K.B. O'Brien from IIT Research Institute, Chicago reported a lens design code 'DIDOC' [107]. The acronym stands for 'Desired Image Distribution using Orthogonal Constraints'. It involves an orthogonal transformation of the normal equations for constraining the solution to a hypercube in the transformed space, and eliminates the need for a damping factor. O'Brien used a technique developed by Herzberger for tackling the nonlinear least-squares problem [108–109]. H. Brunner from Ernst Leitz, Wetzlar, Germany used differential ray tracing formulae to obtain the terms of the Taylor series expansion of the merit function to the quadratic form, thereby obtaining an accuracy higher than the normal least-squares procedure. The single steps in the iterative process could be used without damping, and Brunner reported a rapid convergence [110]. Basu and Hazra observed that incorporation of a line search procedure provides good convergence even without any damping of the least-squares procedure [111].

D.R. Buchele from NASA, Cleveland, Ohio reported a technique for automatic determination of the damping factor by use of homogeneous second derivatives of aberrations [112]. In order to minimize the machine time of computation of the second derivatives, D.C. Dilworth used the differences in the first derivatives between successive iterations to obtain approximate second derivatives of aberrations with respect to each parameter [113–114]. He called this technique a pseudo-second-derivative (PSD) method. Note that, in these treatments, the contribution of second order mixed derivatives is assumed to be negligibly small compared to that of the homogeneous derivatives. Observations by Grey [79] and Wynne and Wormell [66] on the magnitudes and signs of these derivatives seem to support the validity of this assumption. A. Faggiano proposed that the convergence of the method could be improved further by scaling the additions to the diagonal matrix by an additional multiplicative damping factor, which starts with a very low value, and increases as necessary to allow the system to develop as fast as possible [115].

Another approach for accelerating convergence was proposed by P.N. Robb, who suggested to scale the change vector recursively after each iteration, increasing the scale factor exponentially until the optical figure of merit passes through a minimum [116]. E.D. Huber proposed utilizing the approximate second derivative information to predict the values of the first derivatives, and even the merit function after altering the design in each iteration; the basic idea is to circumvent the time-consuming calculation of these quantities at each iteration. Huber called his approach an extrapolated-least-squares (ELS) optimization method, and presented results in support of improvements obtained by the ELS method over the conventional DLS method [117–118].

From Osaka University, Japan, T. Suzuki, Y. Ichioka, S. Yonezawa, and M. Koyama proposed an interesting formulation of the lens design optimization problem in terms of 'floating' aberration targets. They introduced the concept of 'target domains' in place of 'target values'. A relaxation method is used to solve the system of linearized inequalities, and iterative cycles are used to find the solutions to the system of nonlinear inequalities [119–120]. H. Ooki from Nikon Corporation, Tokyo, Japan proposed a method for lens design optimization by using singular value decomposition, and called it a 'Rank Down Method (RDM)' [121]. Ooki observed that an appropriate choice of the rank down constant enables the designer to reflect his design goals, and to accelerate the convergence. In the first half of the 1990s, K. Tanaka and his co-workers at Canon Inc., Japan, revisited the problem of DLS optimization, and came out with new observations on the tricky problems of fixing damping factors and the handling of boundary conditions [122–128].

In 1969, Volosov and Zeno of Leningrad Institute of Motion Picture Engineers, Leningrad, USSR, reported on their implementation of the DLS method for lens design optimization [129]. They claimed to achieve better convergence and a more uniform variation of the parameters by changing the metrics of the variables so that all the variables are normalized to unity at each step. At about the same time, A.P. Grammatin wrote an exhaustive article on computer aided lens design optimization. The article appeared as Chapter 7 of the seminal book by G.G. Slyussarev [130]. The important role of line search at each iteration was underscored in this article, along with a few possible modifications in implementation of

Newton's method or the least-squares method. S.N. Bezdidko proposed the use of orthogonal Zernike polynomials in the expression for merit function, e.g. the variance of wave aberration, to expand the local boundedness of the solution [131]. Gan et al. has described in detail the state-of-the-art optical design software developed at S.I. Vavilov State Optical Institute, St. Petersburg, Russia [132]. In another publication in 1994, Gan provides a brief historical development of the field in Russia [133].

During the second half of the twentieth century, investigators in different parts of the world carried out many other experiments in the course of their developments in automatic or semi-automatic optical designs. The author of this work has limited access to the corresponding publications which occasionally appeared as solitary publications in open literature. However, we note below a few of these publications that came to our notice. In 1979, J.R. de F. Moneo and I. Juvells of the University of Barcelona, Spain reported on their experiments on the use of a new type of merit function formed from real ray trace data in DLS optimization [134]. N.v.d.W. Lessing of Council for Scientific & Industrial Research, Pretoria, South Africa proposed an iterative method in 1980 for the automatic correction of aberrations of optical systems. The method does not require any merit function, and uses a sort of adaptive approach for the correction of targeted aberrations by changes in the available parameters [135]. J. Kross from Optisches Institut der Technischen Universität, Berlin (West) reported in 1990 the principles of optimization in lens design developed at their institute. The key distinguishing features of their approach are insistence and utilization of linearization and dynamic adaption to actual conditions [136]. F.A. Aurin of Carl Zeiss, Oberkochen, Germany revisited the problem of optimization of the Seidel image errors in a multicomponent air-spaced lens system by bending of thin lens components. Aurin reported in 1990 different schemes for obtaining a set of useful local minima as solutions to the problem [137].

20.2 Mathematical Preliminaries for Numerical Optimization

Mathematically, lens design can be formulated as a problem of nonlinear optimization in a constrained multivariate hyperspace. Mathematical preliminaries for understanding different approaches adopted in the field of lens design optimization are given below. Details may be found in the reference books on mathematical optimization [138–143].

20.2.1 Newton–Raphson Technique for Solving a Nonlinear Algebraic Equation [144]

Solving a nonlinear algebraic equation of a single variable x, say

$$g(x) = 0 \tag{20.2}$$

means determining the value/values α of x that satisfy the equation, i.e. $g(\alpha) = 0$. Assume the function $g(x)$ to be smooth and continuously differentiable over a range (a, b) within which there is a single root α. Let an initial estimate for the root be $x^{(0)}$. We write the unknown root α as $\alpha = x^{(0)} + h^{(0)}$, where the error in locating the root is represented by $h^{(0)}$. Using Taylor expansion, we write

$$\begin{aligned} g(\alpha) = g\left(x^{(0)} + h^{(0)}\right) &= g\left(x^{(0)}\right) + h^{(0)} \left\{\frac{d[g(x)]}{dx}\right\}_{x=x^{(0)}} + \frac{1}{2}\left[h^{(0)}\right]^2 \left\{\frac{d^2[g(x)]}{dx^2}\right\}_{x=x^{(0)}} + \dots \\ &= g\left(x^{(0)}\right) + h^{(0)} g'\left(x^{(0)}\right) + \frac{1}{2}\left[h^{(0)}\right]^2 g''\left(x^{(0)}\right) + \dots = 0 \end{aligned} \tag{20.3}$$

Neglecting terms in degree 2 and above, an estimate for $h^{(0)}$ may be obtained as

$$h^{(0)} \cong -\frac{g\left(x^{(0)}\right)}{g'\left(x^{(0)}\right)} \tag{20.4}$$

Using this approximation for $h^{(0)}$, an improved estimate $x^{(1)}$ for the root α is given by

$$x^{(1)} = x^{(0)} + h^{(0)} = x^{(0)} - \frac{g\left(x^{(0)}\right)}{g'\left(x^{(0)}\right)} \tag{20.5}$$

A recursion relation follows as

$$x^{(k+1)} = x^{(k)} + h^{(k)} = x^{(k)} - \frac{g\left(x^{(k)}\right)}{g'\left(x^{(k)}\right)} \qquad k = 1,2,\ldots,K \tag{20.6}$$

For smooth and continuously differentiable functions, the root α may be determined to the desired level of accuracy by suitable number K of iterations. This numerical method for solving a nonlinear equation is known as the Newton–Raphson technique.

20.2.2 Stationary Points of a Univariate Function

Stationary points, such as the minima, the maxima, or the saddle points, of a univariate function $f(x)$ satisfy the equation

$$f'(x) = \frac{d\left[f(x)\right]}{dx} = 0 \tag{20.7}$$

Representing the function $f'(x)$ by $g(x)$, we note that the equation (20.7) is identical with equation (20.2). The problem of determining the stationary points of any univariate function $f(x)$ reduces to the problem of solving roots α of the equation $g(x) = f'(x) = 0$. The type of the stationary point can be ascertained from the second derivative $f''(\alpha)$ of the function at the root. The stationary point is (i) a minimum if $f''(\alpha) > 0$, (ii) a maximum if $f''(\alpha) < 0$, and (iii) a saddle point if $f''(\alpha) = 0$.

20.2.3 Multivariate Minimization

Unconstrained multivariate minimization constitutes the core problem in nonlinear optimization theory. There is no single method that has emerged as a panacea in the field, and a host of alternative approaches leading to different methods is continually being experimented with in search of treatments that can be relied upon in specific areas. Comparative studies on several methods that are typically used in problems related to optical design have also appeared [145–147]. A monograph was published in 1971 on the computer-aided optimization techniques prevailing in the field of lens design [148]. In what follows, we lay down the basics of mathematical optimization methods that have played significant roles in the non-linear optimization of lens design.

20.2.3.1 Basic Notations

Let $f(x_1, x_2, \ldots, x_n)$ be a function of n number of independent variables $x_1, x_2, \ldots, x_n$. It is convenient to gather the variables together in a vector

$$\mathbf{x} = \begin{bmatrix} x_1 \\ x_2 \\ \vdots \\ x_n \end{bmatrix} \tag{20.8}$$

and write the multivariate function succinctly as $f(\mathbf{x})$. Often $\mathbf{x}$ is referred to as a point in the hyperspace of variables. Strictly speaking, $\mathbf{x}$ is the position vector of the point in the multidimensional hyperspace. Determining stationary points of the function $f(\mathbf{x})$ is the core task in optimization theory, and so in the corresponding literature this function is referred to as the 'objective' function. The usual statement of the task is

$$\text{minimize } f(\mathbf{x}) \qquad \mathbf{x} \in \mathbb{R}^n \tag{20.9}$$

where $\mathbb{R}^n$ represents the linear space of all n-dimensional vectors. The symbol $\in$, as in '$\mathbf{x} \in \mathbb{R}^n$' means $\mathbf{x}$ is a point in the n-dimensional space. Note that, in general, any maximization problem can be simply recast as a minimization problem as

$$\max_{\mathbf{x}} f(\mathbf{x}) = -\min_{\mathbf{x}}\left[-f(\mathbf{x})\right] \tag{20.10}$$

The n first partial derivatives of the objective function $f(\mathbf{x})$ can be represented in vector form as the gradient vector

$$\mathbf{g}(\mathbf{x}) = \begin{bmatrix} \partial f(\mathbf{x}) / \partial x_1 \\ \partial f(\mathbf{x}) / \partial x_2 \\ \vdots \\ \partial f(\mathbf{x}) / \partial x_n \end{bmatrix} \tag{20.11}$$

Defining the vector operator ∇ as

$$\nabla = \begin{bmatrix} \partial / \partial x_1 \\ \partial / \partial x_2 \\ \vdots \\ \partial / \partial x_n \end{bmatrix} \tag{20.12}$$

(20.11) reduces to

$$\mathbf{g}(\mathbf{x}) = \nabla f(\mathbf{x}) \tag{20.13}$$

The n^2 second partial derivatives of $f(\mathbf{x})$ are conveniently represented as a matrix $\mathbf{G}(\mathbf{x})$ known as the Hessian matrix. It is given by

$$\mathbf{G}(\mathbf{x}) = \begin{bmatrix} \partial^2 f(\mathbf{x}) / \partial x_1^2 & \ldots & \partial^2 f(\mathbf{x}) / \partial x_1 \partial x_n \\ \vdots & \ldots & \vdots \\ \partial^2 f(\mathbf{x}) / \partial x_n \partial x_1 & \ldots & \partial^2 f(\mathbf{x}) / \partial x_n^2 \end{bmatrix} \tag{20.14}$$

Defining a matrix operator ∇^2 as $\nabla\nabla^T$ where the superscript T represents a transpose, we get

$$\nabla^2 = \begin{bmatrix} \partial^2 / \partial x_1^2 & \ldots & \partial^2 / \partial x_1 \partial x_n \\ \vdots & \ldots & \vdots \\ \partial^2 / \partial x_n \partial x_1 & \ldots & \partial^2 / \partial x_n^2 \end{bmatrix} \tag{20.15}$$

$\mathbf{G}(\mathbf{x})$ can, therefore, be written succinctly as

$$\mathbf{G}(\mathbf{x}) = \nabla^2 f(\mathbf{x}) \tag{20.16}$$

A large class of methods of multivariate minimization used for lens design optimization is an adaptation of iterative approach utilized in solving nonlinear equation by techniques like Newton–Raphson method. Of course, the latter technique needs to be modified for use in the case of function of multiple variables. In general, the k-th iteration involves computation of a search vector $[\mathbf{p}(\mathbf{x})]^{(k)}$. The latter is used to obtain an improved estimate $\mathbf{x}^{(k+1)}$ for location of the minimum for $f(\mathbf{x})$ from its current estimate $\mathbf{x}^{(k)}$ by using a relation

$$\mathbf{x}^{(k+1)} = \mathbf{x}^{(k)} + \alpha^{(k)}[\mathbf{p}(\mathbf{x})]^{(k)} \tag{20.17}$$

where $\alpha^{(k)}$ is usually determined by a search technique to be explained later. Usually, a strategy is adopted to obtain a stable method so that, for any k,

$$f\left(\mathbf{x}^{(k+1)}\right) < f\left(\mathbf{x}^{(k)}\right) \tag{20.18}$$

This implies that the objective function is monotonically decreasing from one iteration to the next.

20.2.3.2 The Method of Steepest Descent [25]

The relation (20.18) can be used to determine the search vector $[\mathbf{p}(\mathbf{x})]^{(k)}$. By Taylor series expansion of $f\left(\mathbf{x}^{(k+1)}\right)$ we may write for $\alpha^{(k)} \to 0$,

$$f\left(\mathbf{x}^{(k+1)}\right) = f\left(\mathbf{x}^{(k)} + \alpha^{(k)}[\mathbf{p}(\mathbf{x})]^{(k)}\right) \cong f\left(\mathbf{x}^{(k)}\right) + \alpha^{(k)}\left\{[\mathbf{g}(\mathbf{x})]^{(k)}\right\}^{T}[\mathbf{p}(\mathbf{x})]^{(k)} \tag{20.19}$$

where $\mathbf{g}(\mathbf{x})$ is the gradient vector given in (20.11). In order to satisfy the condition (20.18) with $\alpha^{(k)} > 0$, it is obvious that the following relation must be satisfied

$$\left\{[\mathbf{g}(\mathbf{x})]^{(k)}\right\}^{T}[\mathbf{p}(\mathbf{x})]^{(k)} < 0 \tag{20.20}$$

This relation is called the 'descent criterion'.

Assuming an angle θ between the gradient vector $\mathbf{g}(\mathbf{x})$ and the search vector $\mathbf{p}(\mathbf{x})$, (20.19) reduces to

$$f\left(\mathbf{x}^{(k+1)}\right) - f\left(\mathbf{x}^{(k)}\right) \cong \alpha^{(k)}\left|[\mathbf{g}(\mathbf{x})]^{(k)}\right|\left|[\mathbf{p}(\mathbf{x})]^{(k)}\right|\cos\theta \tag{20.21}$$

For a specific combination of $\alpha^{(k)}$, $\left|[\mathbf{g}(\mathbf{x})]^{(k)}\right|$, and $\left|[\mathbf{p}(\mathbf{x})]^{(k)}\right|$, the right-hand side of the above equation is most negative when $\theta = \pi$. Therefore, the greatest reduction in function value in a particular iteration is obtained when

$$[\mathbf{p}(\mathbf{x})]^{(k)} = -[\mathbf{g}(\mathbf{x})]^{(k)} \tag{20.22}$$

It is obvious that the negative gradient direction satisfies the descent criterion (20.20) if $[\mathbf{g}(\mathbf{x})]^{(k)} \neq 0$. The direction of negative gradient is called the steepest descent direction, and its use in iteration (20.17) is called 'the method of steepest descent'. Figure 20.1 shows the operation of the method in the case of

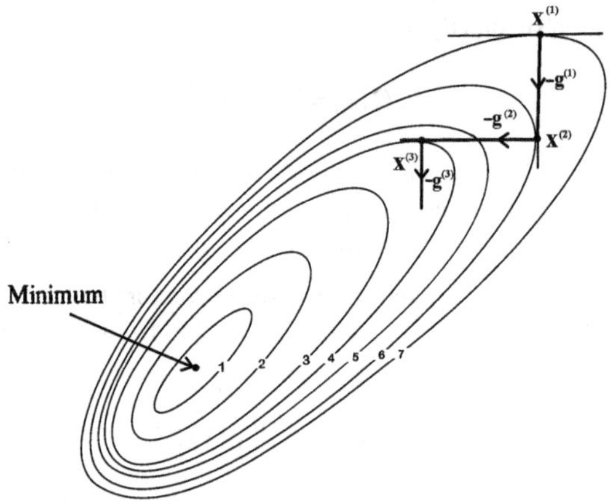

FIGURE 20.1 Method of steepest descent for locating a minimum for a function of two variables.

a typical quadratic function. It is seen that the iteration generates a sequence of points which follows a zig-zag path to the minimum. In each iteration, the function is locally decreasing at the fastest rate until it reaches a minimum. Location of this minimum in each iteration corresponds to the value of $\alpha^{(k)}$, usually determined by a line search technique. Although it is a stable method, the rate of convergence is found to be very slow in problems of lens design optimization.

20.2.3.3 Newton's Method

If the function $f(\mathbf{x})$ has a minimum at the point $\boldsymbol{\alpha}$, we have

$$\mathbf{g}(\boldsymbol{\alpha}) = 0 \tag{20.17}$$

Let an initial estimate for the point $\boldsymbol{\alpha}$ be $\mathbf{x}^{(0)}$. Defining the error in the initial estimate by $\mathbf{p}^{(0)}$, we have

$$\boldsymbol{\alpha} = \mathbf{x}^{(0)} + \mathbf{p}^{(0)} \tag{20.18}$$

By Taylor series expansion, we can write

$$\mathbf{g}(\boldsymbol{\alpha}) = \mathbf{g}\left\{\mathbf{x}^{(0)} + \mathbf{p}^{(0)}\right\} = \left\{\mathbf{g}(\mathbf{x})\right\}_{\mathbf{x}=\mathbf{x}^{(0)}} + \left\{\mathbf{G}(\mathbf{x})\right\}_{\mathbf{x}=\mathbf{x}^{(0)}} \mathbf{p}^{(0)} + O(2) = 0 \tag{20.19}$$

Using the notations

$$\mathbf{g}^{(0)} = \left\{\mathbf{g}(\mathbf{x})\right\}_{\mathbf{x}=\mathbf{x}^{(0)}} \quad \text{and} \quad \mathbf{G}^{(0)} = \left\{\mathbf{G}(\mathbf{x})\right\}_{\mathbf{x}=\mathbf{x}^{(0)}} \tag{20.20}$$

we get from (20.19)

$$\mathbf{p}^{(0)} \cong -\left[\mathbf{G}^{(0)}\right]^{-1} \mathbf{g}^{(0)} \tag{20.21}$$

An improved estimate, $\mathbf{x}^{(1)}$ for the root $\boldsymbol{\alpha}$ is

$$\mathbf{x}^{(1)} = \mathbf{x}^{(0)} + \mathbf{p}^{(0)} = \mathbf{x}^{(0)} - \left[\mathbf{G}^{(0)}\right]^{-1} \mathbf{g}^{(0)} \tag{20.22}$$

For quadratic functions, the inequality signs in (20.21) become an equality sign, since higher order derivative terms are zero. However, for non-quadratic functions, the following recursive relation may be used to determine the minimum by a finite number of iterations. This is known as Newton's method.

A recursive relation for improved values of the minimum $\boldsymbol{\alpha}$ is given by

$$\left.\begin{aligned} \mathbf{p}^{(k)} &= -\left[\mathbf{G}^{(k)}\right]^{-1}\mathbf{g}^{(k)} \\ \mathbf{x}^{(k+1)} &= \mathbf{x}^{(k)} + \mathbf{p}^{(k)} \end{aligned}\right\}, \quad k = 0,1,2,\ldots.\mathrm{K} \tag{20.23}$$

Alternatively, the relation is also expressed as

$$\left.\begin{aligned} \mathbf{G}^{(k)}\mathbf{p}^{(k)} &= -\mathbf{g}^{(k)} \\ \mathbf{x}^{(k+1)} &= \mathbf{x}^{(k)} + \mathbf{p}^{(k)} \end{aligned}\right\}, \quad k = 0,1,2,\ldots.\mathrm{K} \tag{20.24}$$

Note that a practical implementation of Newton's method requires, at each stage of iteration, determination of the Hessian matrix whose elements are the second order derivatives of the objective function. Several attempts are undertaken to circumvent this numerically intensive task. These attempts may be categorized into two distinct classes: the conjugate gradient methods and the quasi-Newton methods, also known as variable metric methods. The well-known 'Davidon-Fletcher-Powell formula' belongs to the latter category [142–143]

20.2.4 Nonlinear Least-Squares [28, 149]

The approach elucidated in the earlier section is effective for sufficiently smooth objective functions. However, in the case of objective functions of special forms, it is often possible to devise methods that are more efficient in practice. An important special form that has been most widely used in practice is when an objective function, say $\Psi(\mathbf{x})$ of independent variables $\mathrm{x}_{\mathrm{j},}\mathrm{j} = 1,2,\ldots,\mathrm{n}$ is a sum of the squares of m nonlinear functions $\mathrm{f}_{\mathrm{i}}(\mathbf{x})$, $i = 1, 2,\ldots, m$, where each of the latter functions is a function of the independent variables $\mathrm{x}_{\mathrm{j},}$ $\mathrm{j} = 1,,2,\ldots,\mathrm{n}$, i.e.

$$\psi(\mathbf{x}) = \sum_{\mathrm{i}=1}^{\mathrm{m}}\left[\mathrm{f}_{\mathrm{i}}(\mathbf{x})\right]^2 \tag{20.25}$$

The minimization of functions like $\psi(\mathbf{x})$ is called 'nonlinear least-squares' in optimization literature. The functions $\mathrm{f}_{\mathrm{i}}(\mathbf{x})$, $i = 1, 2,\ldots, m$ can be written succinctly in vector form as

$$\mathbf{f}(\mathbf{x}) = \begin{bmatrix} \mathrm{f}_1(\mathbf{x}) \\ \mathrm{f}_2(\mathbf{x}) \\ \vdots \\ \mathrm{f}_{\mathrm{m}}(\mathbf{x}) \end{bmatrix} \tag{20.26}$$

The objective function $\psi(\mathbf{x})$ can be written as

$$\psi(\mathbf{x}) = \mathbf{f}^T(\mathbf{x})\mathbf{f}(\mathbf{x}) \tag{20.27}$$

The first partial derivative of the objective function $\psi(\mathbf{x})$, as given in (20.25) with respect to the independent variable x_{j}, is given by

$$\frac{\partial\psi(\mathbf{x})}{\partial\mathrm{x}_{\mathrm{j}}} = 2\sum_{i=1}^{m}\mathrm{f}_{\mathrm{i}}(\mathbf{x})\frac{\partial\mathrm{f}_{\mathrm{i}}(\mathbf{x})}{\partial\mathrm{x}_{\mathrm{j}}} = 2\sum_{i=1}^{m}\mathrm{f}_{\mathrm{i}}\frac{\partial\mathrm{f}_{\mathrm{i}}}{\partial\mathrm{x}_{\mathrm{j}}} \tag{20.28}$$

The gradient vector $\mathbf{g}(\mathbf{x})$ is represented by

$$\mathbf{g}(\mathbf{x}) = \begin{bmatrix} \partial\Psi(\mathbf{x})/\partial x_1 \\ \partial\Psi(\mathbf{x})/\partial x_2 \\ \cdots \\ \partial\Psi(\mathbf{x})/\partial x_n \end{bmatrix} = 2\begin{bmatrix} \partial \mathrm{f}_1(\mathbf{x})/\partial \mathrm{x}_1 & \cdots & \cdots & \partial \mathrm{f}_\mathrm{m}(\mathbf{x})/\partial \mathrm{x}_1 \\ \vdots & \cdots & \cdots & \vdots \\ \partial \mathrm{f}_1(\mathbf{x})/\partial \mathrm{x}_\mathrm{n} & \cdots & \cdots & \partial \mathrm{f}_\mathrm{m}(\mathbf{x})/\partial \mathrm{x}_n \end{bmatrix} \begin{bmatrix} f_1(\mathbf{x}) \\ \vdots \\ \vdots \\ f_m(\mathbf{x}) \end{bmatrix} = 2\mathbf{J}^T(\mathbf{x})\mathrm{f}(\mathbf{x}) \quad (20.29)$$

where the $(\mathrm{m}\times\mathrm{n})$ Jacobian matrix $\mathbf{J}(\mathbf{x})$ is defined by

$$\mathbf{J}(\mathbf{x}) = \begin{bmatrix} \partial \mathrm{f}_1(\mathbf{x})/\partial \mathrm{x}_1 & \cdots & \partial \mathrm{f}_1(\mathbf{x})/\partial \mathrm{x}_\mathrm{n} \\ \vdots & \cdots & \vdots \\ \vdots & \cdots & \vdots \\ \partial \mathrm{f}_\mathrm{m}(\mathbf{x})/\partial \mathrm{x}_1 & \cdots & \partial \mathrm{f}_\mathrm{m}(\mathbf{x})/\partial \mathrm{x}_n \end{bmatrix} = \frac{\partial(\mathrm{f}_1,\mathrm{f}_2,\ldots\ldots,\mathrm{f}_\mathrm{m})}{\partial(\mathrm{x}_1,\mathrm{x}_2,\ldots,\mathrm{x}_\mathrm{n})} \quad (20.30)$$

The $(\mathrm{n}\times\mathrm{n})$ Hessian matrix $\mathbf{G}(\mathbf{x})$ of $\psi(\mathbf{x})$ is represented as

$$\mathbf{G}(\mathbf{x}) = \nabla^2\psi(\mathbf{x}) = \begin{bmatrix} \partial^2\psi(\mathbf{x})/\partial \mathrm{x}_1{}^2 & \cdots & \partial^2\psi(\mathbf{x})/\partial \mathrm{x}_1\partial \mathrm{x}_\mathrm{n} \\ \vdots & \vdots & \vdots \\ \partial^2\psi(\mathbf{x})/\partial \mathrm{x}_\mathrm{n}\partial \mathrm{x}_1 & \cdots & \partial^2\psi(\mathbf{x})/\partial \mathrm{x}_\mathrm{n}{}^2 \end{bmatrix} \quad (20.31)$$

G_{kj}, the kj-element of the Hessian matrix $\mathbf{G}(\mathbf{x})$, may be obtained by differentiating (20.28) with respect to x_k. It is given by

$$\mathrm{G}_{\mathrm{kj}} = \frac{\partial^2\psi(\mathbf{x})}{\partial \mathrm{x}_\mathrm{k}\partial \mathrm{x}_\mathrm{j}} = \frac{\partial}{\partial \mathrm{x}_\mathrm{k}}\left\{\frac{\partial\psi(\mathbf{x})}{\partial \mathrm{x}_\mathrm{j}}\right\} = 2\sum_{\mathrm{i}=1}^{\mathrm{m}}\left\{\frac{\partial \mathrm{f}_\mathrm{i}}{\partial \mathrm{x}_\mathrm{k}}\frac{\partial \mathrm{f}_\mathrm{i}}{\partial \mathrm{x}_\mathrm{j}} + \mathrm{f}_\mathrm{i}\frac{\partial^2 \mathrm{f}_\mathrm{i}}{\partial \mathrm{x}_\mathrm{k}\partial \mathrm{x}_\mathrm{j}}\right\} \quad (20.32)$$

Let the $(\mathrm{n}\times\mathrm{n})$ Hessian matrix of $\mathrm{f}_\mathrm{i}(\mathbf{x})$ be $\mathbf{T}_\mathrm{i}(\mathbf{x})$. It is expressed as

$$\mathbf{T}_i(\mathbf{x}) = \nabla^2\mathrm{f}_i(\mathbf{x}) = \begin{bmatrix} \partial^2\mathrm{f}_i(\mathbf{x})/\partial \mathrm{x}_1{}^2 & \cdots & \partial^2\mathrm{f}_\mathrm{i}(\mathbf{x})/\partial \mathrm{x}_1\partial \mathrm{x}_\mathrm{n} \\ \vdots & \vdots & \vdots \\ \partial^2\mathrm{f}_i(\mathbf{x})/\partial \mathrm{x}_\mathrm{n}\partial \mathrm{x}_1 & \cdots & \partial^2\mathrm{f}_\mathrm{i}(\mathbf{x})/\partial \mathrm{x}_\mathrm{n}{}^2 \end{bmatrix} \quad (20.33)$$

Note that there are m number of Hessian matrices $\mathbf{T}_\mathrm{i}(\mathbf{x})$, $\mathrm{i}=1,2,\ldots,\mathrm{m}$. From the relations (20.31) – (20.33), the complete Hessian matrix $\mathbf{G}(\mathbf{x})$ of $\psi(\mathbf{x})$ can be written as

$$\mathbf{G}(\mathbf{x}) = 2\mathbf{J}^T(\mathrm{x})\mathbf{J}(\mathrm{x}) + 2\sum_{\mathrm{i}=1}^{\mathrm{m}}\mathrm{f}_\mathrm{i}(\mathbf{x})\mathbf{T}_\mathrm{i}(\mathbf{x}) \quad (20.34)$$

The formation of relation (20.34) may be clarified with the help of the following observations. The $(\mathrm{m}\times\mathrm{n})$ Jacobian matrix has already been defined in (20.30). Note that $\left[\mathbf{J}^T(\mathrm{x})\mathbf{J}(\mathrm{x})\right]$, the first part of $\mathbf{G}(\mathbf{x})$, is an $(\mathrm{n}\times\mathrm{n})$ matrix

$$[\mathbf{J}^T(\mathbf{x})\mathbf{J}(\mathbf{x})] = \begin{bmatrix} \partial f_1(\mathbf{x})/\partial x_1 & \dots & \dots & \partial f_m(\mathbf{x})/\partial x_1 \\ \vdots & \dots & \dots & \vdots \\ \partial f_1(\mathbf{x})/\partial x_n & \dots & \dots & \partial f_m(\mathbf{x})/\partial x_n \end{bmatrix} \begin{bmatrix} \partial f_1(\mathbf{x})/\partial x_1 & \dots & \partial f_1(\mathbf{x})/\partial x_n \\ \vdots & \dots & \vdots \\ \vdots & \dots & \vdots \\ \partial f_m(\mathbf{x})/\partial x_1 & \dots & \partial f_m(\mathbf{x})/\partial x_n \end{bmatrix}$$

$$= \begin{bmatrix} \sum_{i-1}^{m} \frac{\partial f_i(\mathbf{x})}{\partial x_1}\frac{\partial f_i(\mathbf{x})}{\partial x_1} & \dots & \sum_{i=1}^{m} \frac{\partial f_i(\mathbf{x})}{\partial x_1}\frac{\partial f_i(\mathbf{x})}{\partial x_n} \\ \vdots & \dots & \vdots \\ \sum_{i=1}^{m} \frac{\partial f_i(\mathbf{x})}{\partial x_n}\frac{\partial f_i(\mathbf{x})}{\partial x_1} & \dots & \sum_{i=1}^{m} \frac{\partial f_i(\mathbf{x})}{\partial x_n}\frac{\partial f_i(\mathbf{x})}{\partial x_n} \end{bmatrix} \tag{20.35}$$

The second part of $\mathbf{G}(\mathbf{x})$ may be conveniently represented by defining an $(n \times n)$ matrix $\mathbf{S}(\mathbf{x})$ as given below

$$\mathbf{S}(\mathbf{x}) = \sum_{i=1}^{m} f_i(\mathbf{x})\mathbf{T}_i(\mathbf{x}) = \begin{bmatrix} \sum_{i=1}^{m} f_i(\mathbf{x})\frac{\partial^2 f_i(\mathbf{x})}{\partial x_1 \partial x_1} & \dots & \sum_{i=1}^{m} f_i(\mathbf{x})\frac{\partial^2 f_i(\mathbf{x})}{\partial x_1 \partial x_n} \\ \vdots & \dots & \vdots \\ \sum_{i=1}^{m} f_i(\mathbf{x})\frac{\partial^2 f_i(\mathbf{x})}{\partial x_n \partial x_1} & \dots & \sum_{i=1}^{m} f_i(\mathbf{x})\frac{\partial^2 f_i(\mathbf{x})}{\partial x_n \partial x_n} \end{bmatrix} \tag{20.35}$$

Thus, the Hessian matrix $\mathbf{G}(\mathbf{x})$ can be expressed as

$$\mathbf{G}(\mathbf{x}) = 2\mathbf{J}^T(\mathbf{x})\mathbf{J}(\mathbf{x}) + 2\mathbf{S}(\mathbf{x}) \tag{20.36}$$

Omitting the argument, (20.36) can be written as

$$\mathbf{G} = 2\mathbf{J}^T\mathbf{J} + 2\mathbf{S} \tag{20.37}$$

Using Newton's method described in an earlier section, we recall the recursive relation (20.24) for determining the minimum of the function $\psi(\mathbf{x})$

$$\left.\begin{aligned} \mathbf{G}^{(k)}\mathbf{p}^{(k)} &= -\mathbf{g}^{(k)} \\ \mathbf{x}^{(k+1)} &= \mathbf{x}^{(k)} + \mathbf{p}^{(k)} \end{aligned}\right\}, \quad k = 0,1,2,\dots K \tag{20.38}$$

Substituting for $\mathbf{g}$ and $\mathbf{G}$ from (20.29) and (20.36), respectively, the recursive relation reduces to

$$\left.\begin{aligned} \left\{\left\{[\mathbf{J}(\mathbf{x})]^{(k)}\right\}^T\left\{[\mathbf{J}(\mathbf{x})]^{(k)}\right\} + [\mathbf{S}(\mathbf{x})]^{(k)}\right\}\mathbf{p}^{(k)} &= -\left\{[\mathbf{J}(\mathbf{x})]^{(k)}\right\}^T[\mathbf{f}(\mathbf{x})]^{(k)} \\ \mathbf{x}^{(k+1)} &= \mathbf{x}^{(k)} + \mathbf{p}^{(k)} \end{aligned}\right\}, \quad k = 0,1,2,\dots,K \tag{20.39}$$

The matrix equation in the recursive relation does represent a set of n linear equations in the variables $p_i^{(k)}$, $i = 1,2,\dots,n$. At each iteration, this set of linear equations is to be solved to determine the variables $p_i^{(k)}$, $i = 1,2,\dots,n$. This set of linear equations are called 'normal equations with total Hessian matrix'. When Newton's method is applied to the sum-of-squares type function $\psi(\mathbf{x})$, in each iteration, the Hessian matrix $\left[\mathbf{G}(\mathbf{x})\right]^{(k)}$ is expressed as a sum of two terms. The first term, $\left\{\left[\mathbf{J}(\mathbf{x})\right]^{(k)}\right\}^T\left\{\left[\mathbf{J}(\mathbf{x})\right]^{(k)}\right\}$, involves evaluation of $(m\times n)$ number of first derivatives, $\left[\frac{\partial f_i(\mathbf{x})}{\partial x_j}\right]$, $i=1,2,\ldots,m,\ j=1,2,\ldots,n$. The second term, $\left[\mathbf{S}(\mathbf{x})\right]^{(k)}$, involves evaluation of $\frac{1}{2}mn(n+1)$ second derivatives of the form $\left[\frac{\partial^2 f_i(\mathbf{x})}{\partial x_j \partial x_k}\right]$, $i=1,2,\ldots,m,\ j=1,2,\ldots,n,\ k=1,2,\ldots,n$, even allowing for symmetry. This implies that in a typical case where $m=80$, $n=30$, in each iteration determination of the Hessian matrix $\left[\mathbf{G}(\mathbf{x})\right]^{(k)}$ involves evaluation of $2400\{=m\times n=(80\times 30)\}$ first derivatives, and $72000\{=mn^2=(80\times 30^2)\}$ second derivatives. However, in practice, the number of the second derivatives to be evaluated may be reduced by invoking symmetry. In the present case, it would be necessary to evaluate $2400\{=m\times n=(80\times 30)\}$ homogeneous second derivatives and $34800\left\{=\frac{1}{2}mn(n-1)=\frac{1}{2}\times 80\times 30\times(30-1)\right\}$ mixed second derivatives. The relative importance of the two terms of the Hessian matrix $\left[\mathbf{G}(\mathbf{x})\right]^{(k)}$ in determination of the change vector $\mathbf{p}^{(k)}$ depends on the relative magnitude of the two types of derivatives and the signs of the different derivatives of the same type. Although arguments are given at times in favour of neglect of the second order derivatives, in general, it is indeed very difficult to make judicious predictions in this regard.

Two broad classes of specialized algorithms exist for tackling the situation. One of them ignores the term $\left[\mathbf{S}(\mathbf{x})\right]^{(k)}$ completely. The resulting method is called 'The Gauss–Newton method'. The corresponding algorithms are called 'small residual algorithms'. Some methods tend to approximate the term $\left[\mathbf{S}(\mathbf{x})\right]^{(k)}$ in some way. The corresponding algorithms are called 'large residual algorithms'. 'Damped least-squares', the most widely used technique, emerged as a modification of the 'Gauss–Newton method'. Current versions of algorithms for damped least-squares in lens design optimization are designed to minimize the effect of ignoring $\left[\mathbf{S}(\mathbf{x})\right]^{(k)}$ by partially incorporating the effects of homogeneous second derivatives.

20.2.4.1 The Gauss–Newton Method

Neglecting the term $\left[\mathbf{S}(\mathbf{x})\right]^{(k)}$ containing the second derivative terms, the normal equations (20.39) become

$$\left\{\left\{\left[\mathbf{J}(\mathbf{x})\right]^{(k)}\right\}^T\left\{\left[\mathbf{J}(\mathbf{x})\right]^{(k)}\right\}\right\}\mathbf{p}^{(k)}=-\left\{\left[\mathbf{J}(\mathbf{x})\right]^{(k)}\right\}^T\left[\mathbf{f}(\mathbf{x})\right]^{(k)} \tag{20.40}$$

The Gauss–Newton method involves solving these normal equations to obtain $\mathbf{p}^{(k)}$ in each iteration and then to use it in the updating relation

$$\mathbf{x}^{(k+1)}=\mathbf{x}^{(k)}+\mathbf{p}^{(k)} \tag{20.41}$$

The convergence properties of the Gauss–Newton method can be greatly enhanced by the single expedient of incorporating a line search at each iteration [150–153]. This involves replacing the updating relation (20.41) by

$$\mathbf{x}^{(k+1)}=\mathbf{x}^{(k)}+\alpha^{(k)}\mathbf{p}^{(k)} \tag{20.42}$$

where

$$\alpha^{(k)}=\min_{\alpha^{(k)}}\left\{\Psi\left[\mathbf{x}^{(k)}+\alpha^{(k)}\mathbf{p}^{(k)}\right]\right\} \tag{20.43}$$

Obviously, an exact line search cannot be implemented in a finite number of steps. However, an approximate line search is adequate for most practical purposes. In the field of lens design optimization, Slyussarev and his co-workers, notably Grammatin, proposed this technique as an alternative to the DLS method [130]. Basu and Hazra carried out a thorough investigation, and concluded that the higher rate of convergence of the Gauss–Newton method augmented by a suitable line search procedure may be leveraged by normally adopting it for iterative optimization purposes in lens design; in the case of failure of the normal course due to singularity of the coefficient matrix or for similar reasons, a switchover to the DLS or some other procedure can be called upon to overcome the problem [154]. It has been observed that practical implementation of such hybrid approaches holds the key to solving complicated nonlinear optimization problems [155].

20.2.4.2 The Levenberg–Marquardt Method

A major cause of failure of the Gauss–Newton method in practice is singularity in the coefficient matrix $\left\{\left\{[\mathbf{J}(\mathbf{x})]^{(k)}\right\}^T\left\{[\mathbf{J}(\mathbf{x})]^{(k)}\right\}\right\}$. The Levenberg–Marquardt method provides an effective remedy by modifying the normal equations (20.40) in an iteration to

$$\left\{\left\{[\mathbf{J}(\mathbf{x})]^{(k)}\right\}^T\left\{[\mathbf{J}(\mathbf{x})]^{(k)}\right\}+\boldsymbol{\mu}^{(k)}\mathbf{I}\right\}\mathbf{p}^{(k)}=-\left\{[\mathbf{J}(\mathbf{x})]^{(k)}\right\}^T[\mathbf{f}(\mathbf{x})]^{(k)} \tag{20.44}$$

where $\boldsymbol{\mu}^{(k)} \geq 0$ is a scalar quantity, and I is an identity matrix of order n.

When $\mu^{(k)}=0$, the Levenberg–Marquardt method reduces to the Gauss–Newton method. As $\boldsymbol{\mu}^{(k)} \to \infty$, the effect of the term $\boldsymbol{\mu}^{(k)}\mathrm{I}$ increasingly dominates over that of the term $\left\{\left\{[\mathbf{J}(\mathbf{x})]^{(k)}\right\}^T\left\{[\mathbf{J}(\mathbf{x})]^{(k)}\right\}\right\}$, so much so that $\mathbf{p}^{(k)} \to -\left[\mu^{(k)}\right]^{-1}\left\{[\mathbf{J}(\mathbf{x})]^{(k)}\right\}^T[\mathbf{f}(\mathbf{x})]^{(k)}$. This represents an infinitesimal step in the direction of the steepest descent. Between these two extremes, both $\left\|\mathbf{p}^{(k)}\right\|$ and the angle between the search vector $\mathbf{p}^{(k)}$ and the gradient vector $[\mathbf{g}(\mathbf{x})]^{(k)}$ decrease monotonically as $\boldsymbol{\mu}^{(k)}$ increases [152]. The quantity $\left\|\mathbf{p}^{(k)}\right\|$ indicates Euclidean norm or 2-norm of the vector $\mathbf{p}^{(k)}$. In the original method devised by Levenberg [63], $\boldsymbol{\mu}^{(k)}$ is to be chosen so that

$$\mu^{(k)}=\min_{\mu^{(k)}}\left\{\Psi\left[\mathbf{x}^{(k)}+\mathbf{p}^{(k)}\right]\right\} \tag{20.45}$$

Although the method is closely related to the line search technique, it was inefficient because for each value of $\mu^{(k)}$, the system of linear equations (20.44) has to be solved for $\mathbf{p}^{(k)}$ in the same iteration. Marquardt [64], and later on Fletcher [156] and Powell [157], suggested different means of overcoming these problems. In common parlance, the factor $\boldsymbol{\mu}^{(k)}$ is called the damping factor, as incorporation of this factor in the treatment conspicuously 'dampens', i.e. reduces, the magnitude of the change vector $\mathbf{p}^{(k)}$. In the field of lens design optimization, several heuristic approaches for damping have been experimented with, e.g. additive damping, multiplicative damping, preferred number damping, etc.

The methods mentioned above for the treatment of small residual problems in nonlinear least-squares can also work for large residual problems, but the rate of convergence may be unacceptably slow for many applications. In this context, the Gill – Murray method uses the singular value decomposition of the Jacobian matrix essentially to make a decision between Newton and Gauss–Newton iterations [158]. Matsui and Tanaka [126] have also deliberated on the use of a combination of eigenvalues and singular value decomposition for DLS problems.

All iterative minimization schemes based on Newton's method guarantee a monotonic decrease of the objective function in successive iterations. It has been observed that enforcing strict monotonicity of the function values can, at times, considerably slow the rate of convergence of the minimization process, especially in the presence of narrow curved valleys [159]. Investigations on nonmonotone Levenberg–Marquardt algorithms have shown their ability to accelerate the iteration progress, particularly in the case where the objective function is ill-conditioned [160–161]. Preliminary investigations with nonmonotone Levenberg–Marquardt algorithms in lens design optimization produced encouraging results [162]. Figure 20.2 presents a broad classification of nonlinear optimization methods, most of which have been experimented with for lens design optimization.

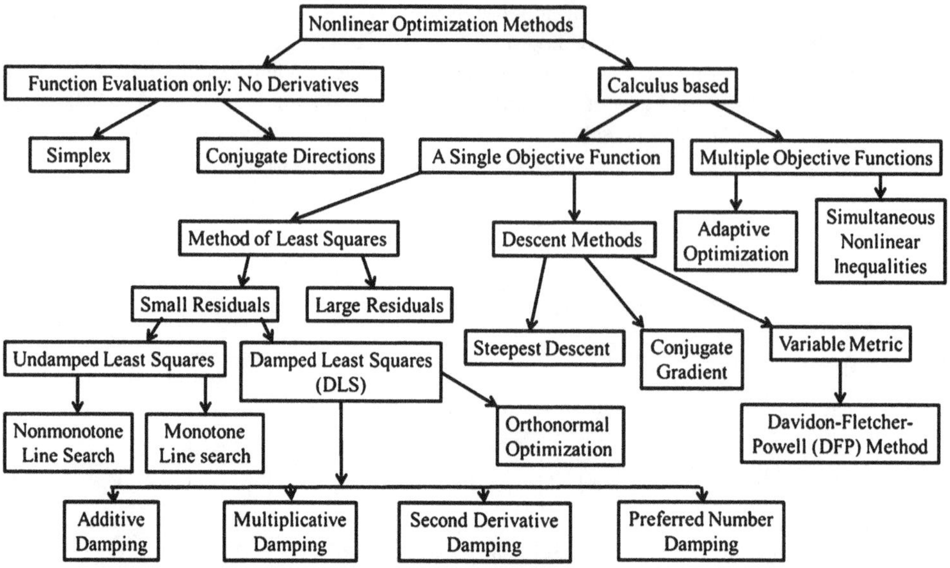

FIGURE 20.2 A broad classification of the nonlinear optimization methods used in lens design.

20.2.5 Handling of Constraints

In the optimization methods described above, the independent variables **x** are allowed to take any finite value, and so they are called methods for unconstrained optimization. In many practical problems, it is known beforehand that certain points in the multivariate hyperspace are not acceptable as solutions. Restraints on the use of values of **x** are formulated as 'constraints', and the search for optimum solutions in such cases constitutes the core task of a 'constrained optimization' problem that is usually stated as:

$$\left.\begin{array}{c} \min\limits_{\mathbf{x}} f(\mathbf{x}) \quad \mathrm{x} \in \mathbb{R}^n \\ \text{subject to} \\ c_s(\mathbf{x}) = 0, \quad s = 1,\ldots,S \\ \text{and} \\ c_t(\mathbf{x}) \geq 0, \quad t = 1,\ldots,T \end{array}\right\} \tag{20.46}$$

The constraints $c_s(\mathbf{x}) = 0,\ s = 1,\ldots,S$ are called equality constraints, and the constraints $c_t(\mathbf{x}) \geq 0,\ t = 1,\ldots,T$ are called inequality constraints. $(S+T)$ is usually less than n, the number of independent variables. A point satisfying all the constraints is called a feasible point, and the set of all feasible points constitutes the feasible region. All other points in $\mathbb{R}^n$, the space of the n independent variables are infeasible points. Note that a relation like $c(\mathbf{x}) = 0$ represents a hypersurface in the n-dimensional space. The feasible region satisfying the equality constraints consists of points lying in the common intersection of the hypersurfaces $c_s(\mathbf{x}) = 0$. For each inequality constraint, the feasible region lies on one side of the hypersurface. In the presence of both equality and inequality constraints, the overall feasible region is to satisfy all the constraints simultaneously.

Traditionally, minimization problems in the presence of equality constraints have been tackled by using the method of Lagrange multipliers [163]. It is assumed that both the objective function $f(\mathbf{x})$ and the equality constraint functions $c_s(\mathbf{x}),\ s = 1,\ldots,S$ have continuous first partial derivatives. A brief statement of constrained optimization by using the Lagrange multiplier method follows.

The objective function $f(\mathbf{x})$ and the constraint functions $c_s(\mathbf{x})$ are combined by using undetermined multipliers to form the Lagrange function (also known as the Lagrangian) defined by

$$\mathcal{L}(x_1,\ldots,x_n;\lambda_1,\ldots,\lambda_S)=f(x_1,\ldots,x_n)+\sum_{s=1}^{S}\lambda_s c_s(x_1,\ldots,x_n) \tag{20.47}$$

where λ_s, $s=1,\ldots,S$ are known as Lagrange multipliers. The constrained optimization problem reduces effectively to an unconstrained optimization problem involving the $(n+S)$ variables $(x_1,\ldots,x_n;\lambda_1,\ldots,\lambda_S)$. The $(n+S)$ equations to be solved are:

$$\left.\begin{aligned}
&\frac{\partial f}{\partial x_1}+\lambda_1\frac{\partial c_1}{\partial x_1}+\lambda_2\frac{\partial c_2}{\partial x_1}+\ldots+\lambda_S\frac{\partial c_S}{\partial x_1}=0\\
&\frac{\partial f}{\partial x_2}+\lambda_1\frac{\partial c_1}{\partial x_2}+\lambda_2\frac{\partial c_2}{\partial x_2}+\ldots+\lambda_S\frac{\partial c_S}{\partial x_2}=0\\
&\qquad\qquad\ldots\quad\ldots\\
&\qquad\qquad\ldots\quad\ldots\\
&\frac{\partial f}{\partial x_n}+\lambda_1\frac{\partial c_1}{\partial x_n}+\lambda_2\frac{\partial c_2}{\partial x_n}+\ldots+\lambda_S\frac{\partial c_S}{\partial x_n}=0\\
&\qquad c_1(x_1,x_2,\ldots,x_n)=0\\
&\qquad c_2(x_1,x_2,\ldots,x_n)=0\\
&\qquad\qquad\ldots\quad\ldots\\
&\qquad\qquad\ldots\quad\ldots\\
&\qquad c_S(x_1,x_2,\ldots,x_n)=0
\end{aligned}\right\} \tag{20.48}$$

The inequality constraints of the form

$$c_t(x_1,x_2,\ldots,x_n)\geq 0,\quad t=1,\ldots,T \tag{20.49}$$

can be replaced by equalities to satisfy the formulation. This may be implemented by introducing the so called 'slack variables' $x_{n+1},x_{n+2},\ldots,x_{n+T}$ defined by

$$c_t(\mathbf{x})=x_{n+t}^{\ 2}\geq 0,\quad t=1,\ldots,T \tag{20.50}$$

so that these constraints become

$$c_t(\mathbf{x})-x_{n+t}^{\ 2}=0,\quad t=1,\ldots,T \tag{20.51}$$

In practice, constraints can be of two types, linear and nonlinear. While linear constraints can be dealt with as shown above in a straightforward manner, nonlinear constraints often call for special treatments. One of the common approaches is to use linearization procedure as applied to objective functions. Few augmented Lagrangian methods are now used in the field of optimization to tackle non-linear constraints.

Alternative approaches for constrained minimization are the penalty and barrier function methods. The basic approach is to replace the search for the minimum in the constrained minimization problem with the successive minimization of a sequence of functions $\varnothing^{(k)}(\mathbf{x})$ formed by suitably combining the objective function and the constraint functions. It can be seen that the minima $\left\{\mathbf{x}^{(k)}\right\}^*$ of the functions $\varnothing^{(k)}(\mathbf{x})$ tend to the constrained minimum $\mathbf{x}^*$ of the objective function $f(\mathbf{x})$ as the number of iterations $k\to\infty$.

Barrier function methods are suitable for inequality constraints only. The functions $\varnothing^{(k)}(\mathbf{x})$ are defined by

$$\varnothing^{(k)}(\mathbf{x})=f(\mathbf{x})+\rho^{(k)}\sum_{t=1}^{T}\varphi\{c_t(\mathbf{x})\} \tag{20.52}$$

where $\rho^{(k)}$ is a positive scalar, and the remainder of the second term is the barrier function for the T inequality constraints. Each constraint has its individual barrier function $\varphi\{c_t(\mathbf{x})\}$. Common examples of barrier functions are:

(a) the inverse barrier function defined by

$$\varphi\{c_t(\mathbf{x})\} = [c_t(\mathbf{x})]^{-1} \tag{20.53}$$

(b) the log barrier function defined by

$$\varphi\{c_t(\mathbf{x})\} = -\ln\{c_t(\mathbf{x})\} \tag{20.54}$$

Note that both these functions have the property that $\varphi\{c_t(\mathbf{x})\} \to \infty$ as $c_t(\mathbf{x}) \to 0$. Also, the effect of the barrier term is delayed nearer to the boundary as $\rho^{(k)} \to 0$. This effect may be varied in different iterations by suitably choosing the value of $\rho^{(k)}$ in different iterations.

As there is no way of ensuring that the minimum satisfies all the equality constraints, as it is, the barrier function method is not suitable for equality constraints. However, approaches based on suitable penalty functions instead of the barrier functions can be developed to handle both equality and inequality constraints. In this case, the procedure is such that the modification of the objective function is invoked only when the constraints are violated, and the effect sharply increases with even a slight increase in violations.

As before, the basic approach is to replace the search for the minimum in the constrained minimization problem with the successive minimization of a sequence of functions $\varnothing^{(k)}(\mathbf{x})$ formed by suitably combining the objective function with both the equality and the inequality constraint functions. The sequence of functions being minimized in successive iterations is now defined by

$$\varnothing^{(k)}(\mathbf{x}) = f(\mathbf{x}) + \rho^{(s)} \sum_{s=1}^{S} \Phi\{c_s(\mathbf{x})\} + \rho^{(k)} \sum_{t=1}^{T} \Phi\{c_t(\mathbf{x})\} \tag{20.55}$$

where Φ are the penalty functions. For equality constraints, the penalty functions may be defined by

$$\Phi\{c_s(\mathbf{x})\} = c_s^2(\mathrm{x}) \tag{20.56}$$

and for inequality constraints, the penalty functions may be defined by

$$\Phi\{c_t(\mathbf{x})\} = [\min\{0, c_t(\mathbf{x})\}]^2 \tag{20.57}$$

More appropriate penalty functions are often used in specific cases. Use of the penalty function approach in lens design optimization problems will be discussed in more detail in a subsequent section.

20.3 Damped Least-Squares (DLS) Method in Lens Design Optimization

Most lens design optimization programs in practice use some form of the damped least-squares method. In this section we describe the details of the 'Computer Optimization of Lens Design' (COLD) package developed by us at the University of Calcutta for in-house use. Mostly, it is an adaptation of the Reading University optical design program (ROD) reported by Hopkins and Kadkly [106], with a few modifications. In what follows, most notations are similar to those used by Hopkins.

20.3.1 Degrees of Freedom

Any constructional parameter describing the lens system is a potential degree of freedom. However, in practice, not all constructional parameters of a lens system are variable, and so, in a lens system, constructional parameters that can be varied are called the degrees of freedom in design of that lens system. A tentative list of possible degrees of freedom in an axi-symmetric lens system in air is given below:

- Number of lens components in the system
- Number of lens elements in each lens component
- Axial curvatures of all refracting/reflecting interfaces
- Conic constants of all refracting/reflecting interfaces
- Figuring coefficients of all refracting/reflecting interfaces
- Axial thicknesses of all lens elements
- Air separations between successive lens elements/components
- Refractive index of optical material of each element
- Dispersion of optical material of each element
- Axial location of the aperture stop
- Semi-diameter of each refracting/reflecting interface

In practice, not all degrees of freedom of a class of lens systems are always available as degrees of freedom in practical design. Usually, degrees of freedom in practical design of a specific lens system are comprised of a list consisting of selected items from the above parameters.

20.3.2 Formation of the Objective Function

In the field of mathematical optimization, the 'objective function' is also known as the 'performance index'. However, in the field of lens design optimization, commonly used terms for the function are the 'defect function' or the 'error function', and also, euphemistically, the 'merit function', although the goal of optimization is to minimize this function.

In practice, the problem of lens design is essentially a problem of multi-objective optimization where each of these objectives can be described as a function of the degrees of freedom. The typical objective function of the lens design optimization problem consists of a suitable combination of these functions, corresponding to different objectives. Each of the constituent functions is defined such that its value is ≥ 0 at any feasible point in the hyperspace of variables, and a smaller value of the function indicates a higher level of attainment of the corresponding objective. The objective function, or defect function Ψ, of the overall problem is usually given by a weighted sum of these functions, where each weighting factor is ≥ 0, so that $\Psi \geq 0$ at any feasible point in the hyperspace of the degrees of freedom.

One of the most important objectives of a lens design is the satisfaction of certain criterion of image quality. The latter is prescribed in accordance with the requirements of practical application of the lens system. Extended object imagery is, in general, space variant, and so quality of image varies from one field point to the other. The determination of the quality of an image at a specific field point calls for a knowledge of the actual wavefront that is forming the image point. A perfectly spherical wavefront with its centre at the object point is transformed by the imaging lens system into an aspherical wavefront that forms the corresponding image. The actual image corresponding to an object point is, in general, not a point, giving rise to degradation of image quality. The asphericity of the image-forming wavefront is characterized by the wave aberration function and/or the transverse ray aberration function. In aberration theory, as described in an earlier chapter, this aberration function is expressed in a power series. For practical purposes, it is assumed that the function is smooth and continuous over the range of field and aperture under view. Further, assuming the power series to be sufficiently convergent, a truncated power series with a finite number of polynomial terms is used to express the aberration function. The coefficients of the polynomial terms in the aberration function are determined by tracing a set of rays from the object point through different points on the aperture. Note that the aberration polynomial to be used by truncating the power series is different for the axial point and the extra-axial points. Also, the choice of polynomial terms in the truncated series depends to a large extent on the magnitude of relative aperture and the field of the particular system. Again, the selection of ray fan, i.e. selection of points on the pupil through which rays from the field point under view are to be traced, is a tricky issue. For the case of axial imagery, the choice is not so difficult, but in the case of extra-axial imagery, it is not straightforward, and often, one has to take recourse to the trial-and-error method, or an empirical approach based on experience. For axially symmetric systems, considerable simplification may be obtained by invoking the symmetry considerations. Suitable methods for determining the aberration polynomials, catering to the needs of lens design optimization are described in Section 13.3. Forbes has proposed the use of Gaussian quadrature in ray selection [164].

The aspects of image quality considered above are solely based on ray optical theory of optical imaging. In fact, at each field point, in conjunction with aberrations, the degradation of image quality is also affected by diffraction effects arising out of the finite size of aperture of the imaging system. The basic requirements for system performance functions for image quality were spelt out clearly by Spencer: they should be 'small enough to avoid redundancy, large enough to define reasonably well the imagery over the whole field, and simple enough to be calculated rapidly' [40].

One of the simplest choices of defect function Υ related to image quality is the sum of weighted aberrations (wave/ray) $W_{\tau,i}$ at a selected number of points i on the pupil from a prespecified number of fractional field points τ:

$$\Upsilon = \sum_{\tau}\sum_{i}\left[\omega_{\tau,i}W_{\tau,i}\right]^2 \tag{20.58}$$

where $\omega_{\tau,i}$ are the weighting factors. The choice of number and location of the points i over the pupil is, however, arbitrary, and the selection of suitable values for the weighting factors $\omega_{\tau,i}$ poses a difficult problem. Nevertheless, the choice of defect function of the form given in (20.58) is widespread, since it directly fits in the scheme of nonlinear least-squares, and, ultimately, in the DLS method for lens design optimization. Another major drawback of this defect function is that it does not take into account the effect of diffraction in image formation.

In order to use diffraction-based criteria of image quality in lens design optimization, H.H. Hopkins [97] proposed the use of variance E of the wave aberration function for diffraction limited systems, and the variance $K(s,\psi)$ of the wave aberration difference function in lens design optimization. The two variances are related with the Strehl ratio, and the MTF at spatial frequency s and azimuth ψ, respectively (see Sections 17.2.3 and 18.6.1). Both defect functions E and $K(s,\psi)$ are expressed in a positive definite quadratic form as

$$\left.\begin{aligned} E &= \sum_{i}\sum_{j}P(i,j)W_iW_j \\ K(s,\psi) &= \sum_{i}\sum_{j}P(i,j;s,\psi)W_iW_j \end{aligned}\right\} \tag{20.59}$$

It may be recalled that both $P(i,j)$ and $P(i,j;s,\psi)$ are universal coefficients to be used in any system. $P(i,j)$ need to be evaluated once, and only once, for a particular combination of the chosen aberration polynomial and the corresponding ray grid pattern for the computed aberrations. Similarly, for a specific frequency and orientation (s,ψ), $P(i,j;s,\psi)$ need to be evaluated once, and only once, for a particular combination of the chosen aberration polynomial and the corresponding ray grid pattern for the computed aberrations. Incidentally, this approach knocks out the empirical assignment of the weighting factors $\omega_{\tau,i}$, as required in (20.58).

In what follows, we shall deal with only the variance of wave aberration E as the component of point image quality in defect function. For axially symmetric systems, usually the image quality is considered at three fractional field points where $\tau = 0, 0.707$, and 1.0. In demanding applications, image quality at more field points may be incorporated in the treatment. In the case of broadband systems, for each field point, the mean square value of the Conrady chromatic aberration $\partial W'_{Chr}$ for the rays of the pencil from the field point used for determining coefficients of the aberration polynomial, constitute the next component of image quality in the error function. The third component of the defect function takes into account aberrations of image shape on the chosen image plane by incorporating the square of the difference of the fractional distortion D from its target value D_0 for each of the extraaxial field points.

The part of the defect function involving optical image quality, $\Psi_{optical}$ takes the form

$$\Psi_{\text{optical}} = \sum_{\tau} \omega_{\tau}^{2} \sum_{i} \sum_{j} P(i,j) W_i W_j + \sum_{\tau} \omega_{\tau,\text{Chr}}^{2} \sum_{i} \left(\partial W'_{\tau,\text{Chr}} \right)_i^2 + \sum_{\substack{\tau \\ \tau \neq 0}} \omega_{\tau,\text{Dist}}^{2} \left(D - D_0 \right)_{\tau}^{2} \tag{20.60}$$

where $\omega_{\tau}, \omega_{\tau,\text{Chr}}$, and $\omega_{\tau,\text{Dist}}$ are the weighting factors. From Section 17.4, we note that the variance E can also be directly evaluated from the wave aberration values for R rays from the object point passing through R chosen points on the exit pupil. Recalling (17.125) and (17.126), we note that the expression for variance then takes the quadratic form

$$E = \sum_{i=1}^{R} \sum_{j=1}^{R} P_0(i,j) W_i W_j \tag{20.61}$$

where

$$\left.\begin{aligned} P_0(i,j) &= \frac{1}{R} - \frac{1}{R^2}, \quad i = j \\ &= -\frac{1}{R^2}, \quad i \neq j \end{aligned}\right\} \tag{20.62}$$

In cases where simply the weighted sums of squares of a set of aberrations, as in (20.58), are used in optical defect function, it may be noted that $P(i,j)$ in (20.60) would have $P(i,j) = 0$, for $j \neq i$.

Boundary conditions of the lens design optimization problem consist of two types. The first group deals with the Gaussian characteristics, e.g. position and size of the image, position of the entrance and/or the exit pupil, etc. They are tackled by setting target values G_0 for each of the appropriate paraxial quantities, and by considering $(G - G_0)_k$ to be pseudo-aberrations. Weighted sum of squares of these pseudo-aberrations are added to image quality defect function Ψ_{optical}. The second group of boundary conditions is based on considerations for manufacturability, availability of glass types, etc. They are tackled by suitably defined penalty functions P_h. The augmented defect function Ψ used in optimization is given by

$$\Psi = \Psi_{\text{optical}} + \sum_{k} \omega_{k,G}^{2} \left(G - G_0 \right)_k^2 + \sum_{h} P_h^2 \tag{20.63}$$

where $\omega_{k,G}$ is the weighting factor for the k-th Gaussian characteristic. Details on implementation of the control of boundary conditions will be given in a subsequent section.

It is apparent that the problem of lens design is essentially a problem of multi-objective optimization. The objective function Ψ, as formulated in (20.63) above, is a form commonly adopted in lens design optimization. The objectives included in (20.63) are not exhaustive, and other objectives may also be included, if it is necessary. An important practical aspect that deserves inclusion is 'tolerance desensitization'. It is obvious that every constructional variable will be subjected to manufacturing tolerances. The loosening of tight tolerances goes a long way in ensuring performance and cost optimization. Grey [165] stressed the importance of incorporating tolerance sensitivity in optimization procedure. Dobson and Cox [166] reported a procedure for incorporating 'tolerance desensitization' as a part of the defect function.

20.3.3 Least-Squares Optimization with Total Hessian Matrix

It may be recalled from discussions in Sections 20.2.3 and 20.2.4 that when Newton's method is applied to sum-of-squares type objective functions, in each iteration, the Hessian matrix is expressed as a sum of two terms. The first term involves evaluation of only first derivatives, and the second term involves evaluation of the second derivatives of the constituent functions of the sum-of-squares type objective function (see equation (20.39)). This is one of the primary reasons for widespread use of the method of least-squares, or a modified version of the same in nonlinear optimization, for in practice, so called 'small residual algorithms' often provide useful solutions. In these algorithms, the contributions of the second derivatives in the Hessian matrix are neglected, thereby obviating the need for evaluation of a large number of second order derivatives.

However, the arguments often put forward in justifying the neglect of contributions by the second derivative terms are not cogent enough, and the real reason for neglecting these terms is the difficulty in handling so many additional computations in each iteration. Indeed, it will be seen in what follows that incorporation of these effects, even if only partially, can facilitate the process of optimization. Before embarking on a description of this procedure, it may be interesting to look into an investigation by Basu and Hazra on least-squares optimization of four different types of lenses with the help of a large residual algorithm where the total Hessian matrix is used in obtaining coefficients of the normal equations [167]. Interested readers may note that in all four cases the method provides a fast convergence at the initial stages of the optimization run. But, after the initial stages of iteration, the speed of convergence is reduced significantly, and this can hardly justify the large amount of additional computation required for evaluation of the total Hessian in each iteration. Since, in many cases, the cost of numerical computation is no longer beyond reach, this option may be explored in extreme cases. Maybe some sort of matrix updating procedure can facilitate the process.

20.3.4 General Form of the Damping Factor Λ_r

In the DLS optimization procedure described here, the damping factor Λ_r for the r-th parameter, p_r is given by the product of three factors as

$$\Lambda_r = \lambda \lambda_r \hat{\lambda}_r \tag{20.64}$$

In the above λ is a 'global damping factor', common for all variable parameters. Its role is to limit the step length in the parameter space. λ_r and $\hat{\lambda}_r$ are differential damping factors that have specific values for the r-th parameter, p_r. The nonlinearity of the r-th parameter is taken care of by λ_r by taking into consideration effects of the homogeneous second derivatives of the aberrations with respect to the parameter p_r, and so λ_r is called a 'second-derivative damping factor'.

The values of $\hat{\lambda}_r$, $r = 1,2,\ldots,n$ are set once before commencement of the optimization procedure, and it is kept unchanged in all iterations. The values of λ and λ_r are determined in each iteration. Determination of these factors will be described in subsequent sections.

20.3.5 The Scaling Damping Factor $\hat{\lambda}_r$

The constructional parameters of a lens system are of different types. They have different dimensions, and usually very different numerical values. Therefore, a reasonably sized change in one parameter type will have a very different numerical value from that for another type. In order to take into account this disparity and eliminate its unwanted effects on the optimization process, a scaling damping factor $\hat{\lambda}_r$ is utilized.

Let the change in a constructional parameter p_r be denoted by x_r, and for each parameter type, a maximum desirable change, $\hat{x}$ per iteration is chosen. The scaling damping factor $\hat{\lambda}_r$ is the reciprocal of $\hat{x}$, i.e. $\hat{\lambda}_r = (1/\hat{x})$. Typical values for $\hat{x}$ used for different types of parameters are:

Curvatures	$\hat{x} = \delta c = .01_{mm}^{-1}$
Axial separations	$\hat{x} = \delta d = 1.0_{mm}$
Refractive indices	$\hat{x} = \delta\mu = 0.05$
Dispersions	$\hat{x} = \delta(\delta\mu) = 0.005$

These values are appropriate for systems operating in visible wavelengths. For infrared systems, a smaller value, say, $\hat{x} = \delta c = 0.001 mm^{-1}$ may be used for curvatures. Note that the values given above are to be used for all parameters of the given type.

20.3.6 Truncated Defect Function

A major reason for the complexity of lens design accrues from the inherent nonlinear behaviour of the different aberrations with respect to changes in constructional parameters. The problem is compounded by the variation in nonlinearity of the different parameters. The second derivative damping factor λ_r is a differential damping factor that is designed to take into account this nonlinear behaviour on the basis of approximate estimates for the effects of the second derivative terms on the defect function.

From (20.60) and (20.63), the defect function Ψ_0 at the start of an iteration can be written as

$$\Psi_0 = \sum_{\tau}\omega_\tau^2\sum_i\sum_j P(i,j)W_iW_j + \sum_{\tau}\omega_{\tau,\mathrm{Chr}}^2\sum_i\left(\partial W'_{\tau,\mathrm{Chr}}\right)_i^2 + \sum_{\substack{\tau\\ \tau\neq 0}}\omega_{\tau,\mathrm{Dist}}^2\left(D-D_0\right)_\tau^2 + \sum_k \omega_{k,G}\left(G-G_0\right)_k^2 + \sum_h P_h^2 \tag{20.65}$$

Let $p_1, p_2, \ldots, p_N$ be the constructional parameters, and changes in these parameters be expressed by $x_1(=\delta p_1), x_2(=\delta p_2), \ldots, x_N(=\delta p_N)$, respectively. After these changes, the value of the defect function will be

$$\begin{aligned}
\Psi = {} & \sum_{\tau}\omega_\tau^2\sum_i\sum_j P(i,j)\left\{W_i + \sum_{n=1}^{N}\frac{\partial W_i}{\partial p_n}x_n + \frac{1}{2}\sum_{n=1}^{N}\sum_{m=1}^{N}\frac{\partial^2 W_i}{\partial p_n\partial p_m}x_nx_m + O\left(x^3\right)\right\}\\
& \times\left\{W_j + \sum_{v=1}^{N}\frac{\partial W_j}{\partial p_v}x_v + \frac{1}{2}\sum_{v=1}^{N}\sum_{\mu=1}^{N}\frac{\partial^2 W_j}{\partial p_v\partial p_\mu}x_vx_\mu + O\left(x^3\right)\right\} + \sum_{\tau}\omega_{\tau,\mathrm{Chr}}^2\sum_i\left\{\left(\partial W'_{\tau,\mathrm{Chr}}\right)_i\right.\\
& \left. + \sum_{n=1}^{N}\frac{\partial\left(\partial W'_{\tau,\mathrm{Chr}}\right)_i}{\partial p_n}x_n + \frac{1}{2}\sum_{n=1}^{N}\sum_{m=1}^{N}\frac{\partial^2\left(\partial W'_{\tau,\mathrm{Chr}}\right)_i}{\partial p_n\partial p_m}x_nx_m + O\left(x^3\right)\right\}\\
& \times\left\{\left(\partial W'_{\tau,\mathrm{Chr}}\right)_i + \sum_{v=1}^{N}\frac{\partial\left(\partial W'_{\tau,\mathrm{Chr}}\right)_i}{\partial p_v}x_v + \frac{1}{2}\sum_{v=1}^{N}\sum_{\mu=1}^{N}\frac{\partial^2\left(\partial W'_{\tau,\mathrm{Chr}}\right)_i}{\partial p_v\partial p_{v\mu}}x_vx_\mu + O\left(x^3\right)\right\}\\
& + \sum_{\substack{\tau\\ \tau\neq 0}}\omega_{\tau,\mathrm{Dist}}^2\left\{\left(D-D_0\right)_\tau + \sum_{n=1}^{N}\frac{\partial D_\tau}{\partial p_n}x_n + \frac{1}{2}\sum_{n=1}^{N}\sum_{m=1}^{N}\frac{\partial^2 D_\tau}{\partial p_n\partial p_m}x_nx_m + O\left(x^3\right)\right\}\\
& \times\left\{\left(D-D_0\right)_\tau + \sum_{v=1}^{N}\frac{\partial D_\tau}{\partial p_v}x_v + \frac{1}{2}\sum_{v=1}^{N}\sum_{\mu=1}^{N}\frac{\partial^2 D_\tau}{\partial p_v\partial p_\mu}x_vx_\mu + O\left(x^3\right)\right\}\\
& + \sum_k\omega_{k,G}^2\left\{\left(G-G_0\right)_k + \sum_{n=1}^{N}\frac{\partial G_k}{\partial p_n}x_n + \frac{1}{2}\sum_{n=1}^{N}\sum_{m=1}^{N}\frac{\partial^2 G_k}{\partial p_n\partial p_m}x_nx_m + O\left(x^3\right)\right\}\\
& \times\left\{\left(G-G_0\right)_k + \sum_{v=1}^{N}\frac{\partial G_k}{\partial p_v}x_v + \frac{1}{2}\sum_{n=1}^{N}\sum_{\mu=1}^{N}\frac{\partial^2 G_k}{\partial p_v\partial p_\mu}x_vx_\mu + O\left(x^3\right)\right\}\\
& + \sum_h\left\{P_h + \sum_{n=1}^{N}\frac{\partial P_h}{\partial p_n}x_n + \frac{1}{2}\sum_{n=1}^{N}\sum_{m=1}^{N}\frac{\partial^2 P_h}{\partial p_n\partial p_m}x_nx_m + O\left(x^3\right)\right\}\\
& \times\left\{P_h + \sum_{v=1}^{N}\frac{\partial P_h}{\partial p_v}x_v + \frac{1}{2}\sum_{v=1}^{N}\sum_{\mu=1}^{N}\frac{\partial^2 P_h}{\partial p_v\partial p_\mu}x_vx_\mu + O\left(x^3\right)\right\}
\end{aligned} \tag{20.66}$$

Note that the second derivatives of Gaussian characteristics with respect to the constructional parameters, $\frac{\partial^2 G_k}{\partial p_\nu \partial p_\mu}$, are all zero, and so they will be omitted in what follows. By algebraic manipulations, (20.66) can be simplified in the following form.

$$\Psi = \Psi_0 + \Psi_1 + \Psi_2 + O(x^3) \tag{20.67}$$

where Ψ_0 is the original value of the defect function given by (20.65). $O(x^3)$ represents terms in total degree ≥ 3 in the parameter changes. Ψ_1 and Ψ_2 are represented by

$$\Psi_1 = \sum_{n=1}^{N} A_n x_n \quad \text{and} \quad \Psi_2 = \sum_{n=1}^{N}\sum_{m=1}^{N} B_{mn} x_n x_m \tag{20.68}$$

The coefficients A_n and B_{mn} are given by

$$A_n = \left\{ \begin{aligned} &\sum_\tau \omega_\tau^2 \sum_i \sum_j P(i,j)\left(W_i \frac{\partial W_j}{\partial p_n} + W_j \frac{\partial W_i}{\partial p_n} \right) + \sum_\tau \omega_{\tau,\mathrm{Chr}}^2 \sum_i 2\left(\partial W'_{\tau,\mathrm{Chr}}\right)_i \frac{\partial\left(\partial W'_{\tau,\mathrm{Chr}}\right)_i}{\partial p_n} \\ &+ \sum_{\substack{\tau \\ \tau\neq 0}} \omega_{\tau,\mathrm{Dist}}^2 \left[2\left(D - D_0\right)_\tau \frac{\partial D_\tau}{\partial p_n} \right] + \sum_k \omega_{k,G}\left[2\left(G - G_0\right)_k \frac{\partial G_k}{\partial p_n} \right] + \sum_h \left[2P_h \frac{\partial P_h}{\partial p_n} \right] \end{aligned} \right\} \tag{20.69}$$

$$B_{mn} = \left[\begin{aligned} &\sum_\tau \omega_\tau^2 \sum_i \sum_j P(i,j)\left[\frac{\partial W_i}{\partial p_n}\frac{\partial W_j}{\partial p_m} + \frac{1}{2}\left(W_i \frac{\partial^2 W_j}{\partial p_n \partial p_m} + W_j \frac{\partial^2 W_i}{\partial p_n \partial p_m} \right) \right] \\ &+ \sum_\tau \omega_{\tau,\mathrm{Chr}}^2 \sum_i \left[\frac{\partial\left(\partial W'_{\tau,\mathrm{Chr}}\right)_i}{\partial p_n}\frac{\partial\left(\partial W'_{\tau,\mathrm{Chr}}\right)_i}{\partial p_m} + \left(\partial W'_{\tau,\mathrm{Chr}}\right)_i \frac{\partial^2\left(\partial W'_{\tau,\mathrm{Chr}}\right)_i}{\partial p_n \partial p_m} \right] \\ &+ \sum_{\substack{\tau \\ \tau\neq 0}} \omega_{\tau,\mathrm{Dist}}^2 \left[\frac{\partial D_\tau}{\partial p_n}\frac{\partial D_\tau}{\partial p_m} + \left(D - D_0\right)_\tau \frac{\partial^2 D_\tau}{\partial p_n \partial p_m} \right] \\ &+ \sum_k \omega_{k,G}\left(\frac{\partial G_k}{\partial p_n}\frac{\partial G_k}{\partial p_m} \right) + \sum_h \left(\frac{\partial P_h}{\partial p_n}\frac{\partial P_h}{\partial p_m} + P_h \frac{\partial^2 P_h}{\partial p_n \partial p_m} \right) \end{aligned} \right] \tag{20.70}$$

Equation (20.67) for the defect function can be rewritten as

$$\Psi = \Psi_0 + \sum_{n=1}^{N} A_n x_n + \sum_{n=1}^{N}\sum_{m=1}^{N} B_{mn} x_n x_m + O(3) \tag{20.71}$$

Note that the coefficient A_n involves only the first derivatives of the aberrations and boundary constraints with respect to the constructional parameter p_n. But the coefficient B_{mn} involves not only the first derivatives of the aberrations and boundary constraints with respect to the constructional parameter p_n and p_m, but also the second derivatives of the aberrations and boundary constraints with respect to the constructional parameter p_n and p_m. In a case with N variables and M number of aberrations, pseudo-aberrations, and boundary conditions, determination of Ψ calls for evaluation of (MN) first derivatives and (MN^2) second derivatives. However, the number of second derivatives to be evaluated can be reduced by invoking symmetry, since $(\partial^2 W / \partial p_m \partial p_n) = (\partial^2 W / \partial p_n \partial p_m)$. In practice, (MN) number of homogeneous second derivatives, and $\left[\frac{1}{2} MN(N-1)\right]$ number of mixed second derivatives have to be evaluated.

It has already been mentioned in Section 20.2.4 that the practical difficulty in calculating the large number of second derivatives prompted the practitioners of optimization to use the 'small residuals' approximation, where all the second derivatives are omitted.

Retaining only terms containing the first derivatives, and excluding all terms containing the second derivatives, the coefficient B_{mn}, given by (20.70), reduces to

$$\hat{B}_{mn} = \left\{ \begin{array}{l} \sum_{\tau} \omega_{\tau}^2 \sum_{i} \sum_{j} P(i,j) \left[\frac{\partial W_i}{\partial p_n} \frac{\partial W_j}{\partial p_m} \right] + \sum_{\tau} \omega_{\tau,Chr}^2 \sum_{i} \left[\frac{\partial \left(\partial W'_{\tau,Chr} \right)_i}{\partial p_n} \frac{\partial \left(\partial W'_{\tau,Chr} \right)_i}{\partial p_m} \right] \\ + \sum_{\substack{\tau \\ \tau \neq 0}} \omega_{\tau,Dist}^2 \left[\frac{\partial D_{\tau}}{\partial p_n} \frac{\partial D_{\tau}}{\partial p_m} \right] + \sum_{k} \omega_{k,G} \left(\frac{\partial G_k}{\partial p_n} \frac{\partial G_k}{\partial p_m} \right) + \sum_{h} \left(\frac{\partial P_h}{\partial p_n} \frac{\partial P_h}{\partial p_m} \right) \end{array} \right\} \tag{20.72}$$

When all terms of $O(3)$ and the second degree terms depending on the second-order derivatives are excluded, the 'truncated defect function' $\hat{\Psi}$ takes the form

$$\hat{\Psi} = \Psi_0 + \sum_{n=1}^{N} A_n x_n + \sum_{n=1}^{N} \sum_{m=1}^{N} \hat{B}_{mn} x_n x_m \tag{20.73}$$

where $\hat{B}_{mn}$ is given by (20.72) and A_n is the same as given in (20.69). The original defect function is Ψ_0. It is expressed in equation (20.65). By neglecting the higher order terms, the DLS optimization procedure effectively optimizes the truncated defect function $\hat{\Psi}$.

20.3.7 Second-Derivative Damping Factor λ_r

The approach for optimizing with the truncated defect function is effectively a 'small residuals' formulation of the optimization problem. All terms containing the second derivatives of aberrations are summarily excluded from the approach. However, it may be noted that the number of homogeneous second derivatives is the same as the number of first derivatives, and with a little additional computation, the homogeneous second derivatives can be obtained from the first derivatives, which are being determined in each iteration. These homogeneous derivatives are utilized to obtain the second derivative damping factors λ_r. They can at least ensure that those parameters with larger values of $\frac{\partial^2 W}{\partial p_n^2}$ are damped down to smaller changes compared with other parameters.

After a change $x_r (= \delta p_r)$ in a single variable p_r, the value of the defect function, including all terms involving second derivatives and excluding higher order terms in $O(3)$, will be given by

$$\Psi = \Psi_0 + A_r x_r + B_{rr} x_r^2 \tag{20.74}$$

where $|x_r|$ will be limited to values $\leq \hat{x}_r$ by the parameter scaling factor $\hat{\lambda}_r$. Coefficients A_r and B_{rr} are obtained from (20.69) and (20.70) by restricting to terms $m = n = r$

$$A_r = \left\{ \begin{array}{l} \sum_{\tau} \omega_{\tau}^2 \sum_{i} \sum_{j} P(i,j) \left(W_i \frac{\partial W_j}{\partial p_r} + W_j \frac{\partial W_i}{\partial p_r} \right) + \sum_{\tau} \omega_{\tau,Chr}^2 \sum_{i} 2 \left(\partial W'_{\tau,Chr} \right)_i \frac{\partial \left(\partial W'_{\tau,Chr} \right)_i}{\partial p_r} \\ + \sum_{\substack{\tau \\ \tau \neq 0}} \omega_{\tau,Dist}^2 \left[2 \left(D - D_0 \right)_{\tau} \frac{\partial D_{\tau}}{\partial p_r} \right] + \sum_{k} \omega_{k,G} \left[2 \left(G - G_0 \right)_k \frac{\partial G_k}{\partial p_r} \right] + \sum_{h} \left[2 P_h \frac{\partial P_h}{\partial p_r} \right] \end{array} \right\} \tag{20.75}$$

$$B_{rr} = \left\{ \begin{aligned} & \sum_\tau \omega_\tau^2 \sum_i \sum_j P(i,j) \left[\frac{\partial W_i}{\partial p_r} \frac{\partial W_j}{\partial p_r} + \frac{1}{2} \left(W_i \frac{\partial^2 W_j}{\partial p_r^2} + W_j \frac{\partial^2 W_i}{\partial p_r^2} \right) \right] \\ & + \sum_\tau \omega_{\tau,Chr}^2 \sum_i \left[\frac{\partial (\partial W'_{\tau,Chr})_i}{\partial p_r} \frac{\partial (\partial W'_{\tau,Chr})_i}{\partial p_r} + (\partial W'_{\tau,Chr})_i \frac{\partial^2 (\partial W'_{\tau,Chr})_i}{\partial p_r^2} \right] \\ & + \sum_{\substack{\tau \\ \tau \neq 0}} \omega_{\tau,Dist}^2 \left[\frac{\partial D_\tau}{\partial p_r} \frac{\partial D_\tau}{\partial p_r} + (D - D_0)_\tau \frac{\partial^2 D_\tau}{\partial p_r^2} \right] \\ & + \sum_k \omega_{k,G} \left(\frac{\partial G_k}{\partial p_r} \frac{\partial G_k}{\partial p_r} \right) + \sum_h \left(\frac{\partial P_h}{\partial p_r} \frac{\partial P_h}{\partial p_r} + P_h \frac{\partial^2 P_h}{\partial p_r^2} \right) \end{aligned} \right\} \tag{20.76}$$

Substituting for A_r from (20.75) and for B_{rr} from (20.76) in (20.74), and then separating out the change $\delta\Psi_1$ in Ψ produced by the terms containing first derivatives alone, and the change $\delta\Psi_2$ in Ψ produced by the remaining terms containing second derivatives, we get

$$\Psi = \Psi_0 + \delta\Psi_1 + \delta\Psi_2 \tag{20.77}$$

where

$$\begin{aligned} \delta\Psi_1 = & \left\{ \begin{aligned} & \sum_\tau \omega_\tau^2 \sum_i \sum_j P(i,j) \left(W_i \frac{\partial W_j}{\partial p_r} + W_j \frac{\partial W_i}{\partial p_r} \right) + \sum_\tau \omega_{\tau,Chr}^2 \sum_i 2 (\partial W'_{\tau,Chr})_i \frac{\partial (\partial W'_{\tau,Chr})_i}{\partial p_r} \\ & + \sum_{\substack{\tau \\ \tau \neq 0}} \omega_{\tau,Dist}^2 \left[2 (D - D_0)_\tau \frac{\partial D_\tau}{\partial p_r} \right] + \sum_k \omega_{k,G} \left[2 (G - G_0)_k \frac{\partial G_k}{\partial p_r} \right] + \sum_h \left[2 P_h \frac{\partial P_h}{\partial p_r} \right] \end{aligned} \right\} \hat{x}_r \\ & + \left\{ \begin{aligned} & \sum_\tau \omega_\tau^2 \sum_i \sum_j P(i,j) \left[\frac{\partial W_i}{\partial p_r} \frac{\partial W_j}{\partial p_r} \right] + \sum_\tau \omega_{\tau,Chr}^2 \sum_i \left[\frac{\partial (\partial W'_{\tau,Chr})_i}{\partial p_r} \right]^2 \\ & + \sum_{\substack{\tau \\ \tau \neq 0}} \omega_{\tau,Dist}^2 \left[\frac{\partial D_\tau}{\partial p_r} \right]^2 + \sum_k \omega_{k,G} \left(\frac{\partial G_k}{\partial p_r} \right)^2 + \sum_h \left(\frac{\partial P_h}{\partial p_r} \right)^2 \end{aligned} \right\} \hat{x}_r^2 \end{aligned} \tag{20.78}$$

$$\begin{aligned} \delta\Psi_2 = \Bigg\{ & \sum_\tau \omega_\tau^2 \sum_i \sum_j P(i,j) \left[\frac{1}{2} \left(W_i \frac{\partial^2 W_j}{\partial p_r^2} + W_j \frac{\partial^2 W_i}{\partial p_r^2} \right) \right] \\ & + \sum_\tau \omega_{\tau,Chr}^2 \sum_i \left[(\partial W'_{\tau,Chr})_i \frac{\partial^2 (\partial W'_{\tau,Chr})_i}{\partial p_r^2} \right] \\ & + \sum_{\substack{\tau \\ \tau \neq 0}} \omega_{\tau,Dist}^2 \left[(D - D_0)_\tau \frac{\partial^2 D_\tau}{\partial p_r^2} \right] + \sum_h \left(P_h \frac{\partial^2 P_h}{\partial p_r^2} \right) \Bigg\} \hat{x}_r^2 \end{aligned} \tag{20.79}$$

using the largest expected value for x_r, i.e. when $x_r = \hat{x}_r$.

The second derivative damping factor λ_r for the change x_r in the r-th variable is defined in terms of the ratio of the contribution $\delta\Psi_2$ to the change in Ψ of the second derivatives with respect to the variable p_r to the whole contribution $\delta\Psi_1$ of the first derivatives to this change as

$$\lambda_r = 1 + \left| \frac{\delta\Psi_2}{\delta\Psi_1} \right|_r \tag{20.80}$$

Note that $\lambda_r = 0$ implies $\Lambda_r = 0$, so that there is no damping on the variable p_r. In order to avoid the inadvertent occurrence of such situations, a constant number is added to the ratio. On the other hand $\lambda_r \to \infty$ when $\delta\Psi_1 \to 0$. The latter implies the first derivatives of the different parameters with respect to this particular variable produce no contribution to the change in defect function. The physical significance of this is that the particular variable is effectively frozen. It is significant to note that second derivative damping factor defined by (20.80) is determined by the influence on the predicted value of Ψ of the omission of these second-derivative terms, and not simply by the magnitude of these derivatives. Of course, definition of the second-derivative damping factor may be modified in different ways.

20.3.8 Normal Equations

Neglecting all second-degree terms depending on the second order aberration derivatives, and all terms of degree three and above as represented by $O(3)$, the 'truncated defect function' $\hat{\Psi}$ is expressed in equation (20.73) by

$$\hat{\Psi} = \Psi_0 + \sum_{n=1}^{N} A_n x_n + \sum_{n=1}^{N}\sum_{m=1}^{N} \hat{B}_{mn} x_n x_m \tag{20.81}$$

where x_n, $n = 1,2,\ldots,N$ represent changes in the N variables. The expressions for the coefficients A_n and $\hat{B}_{mn}$ are given in (20.69) and (20.72), respectively. Ψ_0, on the right-hand side of equation (20.81), is the 'true defect function' at start of the iteration before changes. The expression for Ψ_0 is given in (20.65).

The condition for minimization of the truncated defect function $\hat{\Psi}$ with respect to a change x_r in variable p_r is

$$\frac{\partial\hat{\Psi}}{\partial x_r} = A_r + \sum_{m=1}^{N} \hat{B}_{mr} x_m + \sum_{n=1}^{N} \hat{B}_{rn} x_n = 0 \tag{20.82}$$

Noting from (20.72) that $\hat{B}_{nr} = \hat{B}_{rn}$, and replacing m by n in the first summation on the right-hand side of (20.82) we get the normal equations in the undamped case as

$$\frac{\partial\hat{\Psi}}{\partial x_r} = \sum_{n=1}^{N} b_{rn} x_n + a_r = 0, \quad r = 1,2,\ldots,N \tag{20.83}$$

where the substitutions $b_{rn} = 2\hat{B}_{rn}$ and $a_r = A_r$ have been made.

In damped least-squares optimization, an 'augmented defect function' $\tilde{\Psi}$ is minimized. $\tilde{\Psi}$ is defined by

$$\tilde{\Psi} = \hat{\Psi} + \sum_{n=1}^{N} (\Lambda_n x_n)^2 \tag{20.84}$$

where $\hat{\Psi}$ is the truncated defect function, and Λ_n is the overall damping factor on x_n. The general form for Λ_n is given in (20.64). The normal equations for minimizing the augmented defect function $\tilde{\Psi}$ are given by

$$\frac{\partial\tilde{\Psi}}{\partial x_r} = \frac{\partial\hat{\Psi}}{\partial x_r} + \frac{\partial}{\partial x_r}\left[\sum_{n=1}^{N} (\Lambda_n x_n)^2\right] = 0, \quad r = 1,2,\ldots,N \tag{20.85}$$

They may be simplified to obtain

$$\sum_{n=1}^{N} b_{rn} x_n + a_r + 2\Lambda_r^2 x_r = 0, \quad r = 1,2,\ldots,N \tag{20.86}$$

Comparing with (20.83), it may be noted that in the case of damped least-squares, $2\Lambda_r^2$ is added to the diagonal coefficients of the undamped least-squares.

If $\check{x}_1, \check{x}_2,\ldots, \check{x}_N$ are the roots of the DLS normal equations (20.86), the minimum value of the augmented defect function is given by

$$\tilde{\Psi}(p_1+\check{x}_1,p_2+\check{x}_2,\ldots,p_N+\check{x}_N)=\hat{\Psi}(p_1+\check{x}_1,p_2+\check{x}_2,\ldots,p_N+\check{x}_N)+\sum_{n=1}^{N}(\Lambda_n\check{x}_n)^2 \tag{20.87}$$

Note that

$$\tilde{\Psi}(p_1+\check{x}_1,p_2+\check{x}_2,\ldots,p_N+\check{x}_N)\le\tilde{\Psi}(p_1,p_2,\ldots,p_N)\equiv\tilde{\Psi}_0=\Psi_0 \tag{20.88}$$

The equality sign in the above expression holds good only when the original system is already at the minimum of the true defect function Ψ_0, so that $\check{x}_1=\check{x}_2=\ldots=\check{x}_N=0$. Indeed, before changes, both the augmented defect function and the truncated defect function reduce to the true defect function Ψ_0.

Otherwise, as shown in (20.88), the minimum value of the augmented defect function will always be less than Ψ_0. From (20.87) it may also be noted that the minimum value of the truncated defect function will always be less than the value of the augmented defect function at the minimum, which is less than the true defect function before changes. The difference between the truncated defect function $\hat{\Psi}$ and the true defect function Ψ becomes increasingly small as the changes in the variables become smaller. Changes along the direction $(\check{x}_1,\check{x}_2,\ldots,\check{x}_N)$ in the hyperspace of variables $p_1,p_2,\ldots,p_N$ will, therefore, initially give a lower value of the defect function, but with larger step lengths, the effect of second derivatives will be pronounced to vitiate the predictions based on the truncated form of the defect function. The role of the second derivative damping factor λ_r, discussed in the last section, is to use larger damping factors for those variables whose second derivatives have larger effects on the defect function compared to other variables for obtaining a more favourable direction in the hyperspace of variables, so that a larger step length along the direction can be utilized.

20.3.9 Global Damping Factor λ

Using (20.64), the augmented defect function $\tilde{\Psi}$ defined by (20.84) can be expressed as

$$\tilde{\Psi}=\hat{\Psi}+\sum_{n=1}^{N}\Lambda_n^2x_n^2=\hat{\Psi}+\sum_{n=1}^{N}\lambda^2\left(\lambda_n\hat{\lambda}_n\right)^2x_n^2 \tag{20.89}$$

The roles of the two components, namely the scaling damping factor $\hat{\lambda}_n$ and the second derivative damping factor λ_n, of the overall damping factor Λ_n have been enunciated in the earlier sections. Now we will describe the role of the global damping factor λ and its determination. It has been noted above that the roots $\left(\check{x}_1,\check{x}_2,\ldots,\check{x}_N\right)$ of the normal equations (20.86) represent changes $(\delta p_1,\delta p_2,\ldots,\delta p_N)$ in the variables $(p_1,p_2,\ldots,p_N)$. In the N-dimensional hyperspace of the variables, the original system is located at the point $(p_1,p_2,\ldots,p_N)$, and the changed system is located at the point $\left(p_1+\check{x}_1,p_2+\check{x}_2,\ldots,p_N+\check{x}_N\right)$, so that the roots $\left(\check{x}_1,\check{x}_2,\ldots,\check{x}_N\right)$ represent a direction in the hyperspace. Let us define a relative step length s along this direction so that $s=0$ represents the original system, and $s=1$ represents the DLS solution. Defect functions of the systems represented by points lying on this direction may be explored by considering changes $\left(s\check{x}_1, s\check{x}_2, \ldots, s\check{x}_N\right)$ in the variables. Obviously, as the value of s increases from zero by small amounts, defect functions of the corresponding systems go on decreasing until a minimum is reached for a certain value of s, say $\overline{s}$. For $s > \overline{s}$, the effect of second derivatives will be pronounced, and the defect function increases from the minimum value. In each iteration cycle, the value of $\overline{s}$ can be determined by utilizing a line search procedure. The global damping factor for the $(k+1)$-th iteration, $\lambda^{(k+1)}$, is obtained from the global damping factor for the k-th iteration, $\lambda^{(k)}$, by dividing the latter by $\overline{s}^{(k)}$, i.e.

$$\lambda^{(k+1)} \leftarrow \left\{\lambda^{(k)} / \overline{s}^{(k)}\right\}, k = 1,2\ldots \tag{20.90}$$

Note that the global damping factors from the second iteration onwards are generated by the above relation. For the first iteration, an initial value $\lambda^{(0)}$ for the global damping factor is to be given as part of the input.

With this backdrop in view, the following procedure is adopted for determining the global damping factor.

Step I. An initial value $\lambda^{(0)}$, say 1.0 is taken as the global damping factor. A factor $\beta = 1.3$ or 1.4 is chosen. Specify the scaling damping factor $\hat{\lambda}_r$ for all the variables.

Step II. Calculate the value of the defect function $\Psi_0^{(0)}$ for the start system with variables $\left(p_1^{(0)}, p_2^{(0)}, \ldots, p_N^{(0)}\right)$. Determine the second-derivative damping factors $\lambda_r^{(0)}$ for all variables, and the overall damping factor $\Lambda_r^{(0)} = \lambda\hat{\lambda}_r\lambda_r^{(0)}$ for all variables. Solve the DLS normal equations for the values $(\check{x}_1^{(0)}, \check{x}_2^{(0)}, \ldots, \check{x}_N^{(0)})$.

Step III. Calculate using ray tracing the value of the true defect function $\Psi_1^{(0)}$ for the system with changed variables with reduced step length $s = 1$, i.e. with the set of variables $(p_1 + \check{x}_1^{(0)}, p_2 + \check{x}_2^{(0)}, \ldots, p_N + \check{x}_N^{(0)})$.

Step IV. If $\Psi_1^{(0)} > \Psi_0^{(0)}$, it implies that the chosen initial value $\lambda^{(0)}$ for the global damping factor is too small. The first iteration is repeated with $\lambda^{(0)}$ replaced by $\left(\beta\lambda^{(0)}\right)$. This procedure is continued till $\Psi_1 < \Psi_0$, and this is then taken as the first iteration. The corresponding change vector is noted as $\left(\check{x}_1^{(1)}, \check{x}_2^{(1)}, \ldots, \check{x}_N^{(1)}\right)$. The superscript indicates that the change vector is obtained as roots of the DLS normal equations in the first iteration. The true defect function of the starting system in the first iteration is $\Psi_0^{(1)} \equiv \Psi_0^{(0)}$, and the true defect function of the system with variables $(p_1 + \check{x}_1^{(1)}, p_2 + \check{x}_2^{(1)}, \ldots, p_N + \check{x}_N^{(1)})$ is represented by $\Psi_1^{(1)}$. A special case may occur if the chosen initial value $\lambda^{(0)}$ turns out to be too small, and system aberrations are very large, so much so that ray tracing fails on account of 'ray failures' and the value of $\Psi_1^{(0)}$ cannot be calculated. In such cases, $\lambda^{(0)}$ is to be replaced by $\left(\overline{\beta}\lambda^{(0)}\right)$, where $\overline{\beta} \gg 1.0$. The value of $\overline{\beta}$ can be determined by a few trial runs. Note that the global damping factor $\lambda^{(1)}$ for the first iteration is the value of the same used to obtain the defect function $\Psi_1 < \Psi_0$.

Step V. $\Psi_1^{(1)}$ is obtained using a step length $s = 1$ in the direction of the vector $\left(s\check{x}_1^{(1)}, s\check{x}_2^{(1)}, \ldots, s\check{x}_N^{(1)}\right)$. Put $s = \beta$, and determine by ray tracing the true defect function $\Psi_2^{(1)}$ of the resulting system with variables $\left(p_1 + \beta\check{x}_1^{(1)}, p_2 + \beta\check{x}_2^{(1)}, \ldots, p_N + \beta\check{x}_N^{(1)}\right)$.

Step VI. If $\Psi_2^{(1)} < \Psi_1^{(1)}$, put $s = \beta^2$, and calculate using ray tracing the true defect function $\Psi_3^{(1)}$ of the system with variables $(p_1 + \beta^2\check{x}_1^{(1)}, p_2 + \beta^2\check{x}_2^{(1)}, \ldots, p_N + \beta^2\check{x}_N^{(1)})$. Using

$$s_\nu = \beta^{\nu-1}, \quad \nu = 1,2,3,\ldots \tag{20.91}$$

the process is continued with system variables $(p_1 + s_v \check{x}_1^{(1)}, p_2 + s_v \check{x}_2^{(1)}, \ldots, p_N + s_v \check{x}_N^{(1)})$ until $\Psi_v^{(1)} > \Psi_{v-1}^{(1)}$.

Step VII. A quadratic in $(s - s_v)$ is fitted to the defect function values $(\Psi_v, \Psi_{v-1}, \Psi_{v-2})$, corresponding to the relative step lengths (s_v, s_{v-1}, s_{v-2}).

$$\Psi = \Psi_v + (s - s_v)A + (s - s_v)^2 B \tag{20.92}$$

From the three points corresponding to step lengths s_v, s_{v-1}, s_{v-2}, we get

$$\Psi = \Psi_v \tag{20.93}$$

$$\Psi = \Psi_{v-1} + (s_{v-1} - s_v)A + (s_{v-1} - s_v)^2 B \tag{20.94}$$

$$\Psi = \Psi_{v-2} + (s_{v-2} - s_v)A + (s_{v-2} - s_v)^2 B \tag{20.95}$$

From the three equations (20.93) – (20.95) we get

$$\left.\begin{aligned} A &= -\left(\frac{1}{s_{v-2} - s_{v-1}}\right)\left[\left(\frac{s_{v-1} - s_v}{s_{v-2} - s_v}\right)(\Psi_{v-2} - \Psi_{½}) - \left(\frac{s_{v-2} - s_v}{s_{v-1} - s_v}\right)(\Psi_{v-1} - \Psi_v)\right] \\ B &= \left(\frac{1}{s_{v-2} - s_{v-1}}\right)\left[\left(\frac{1}{s_{v-2} - s_v}\right)(\Psi_{v-2} - \Psi_{½}) - \left(\frac{1}{s_{v-1} - s_v}\right)(\Psi_{v-1} - \Psi_v)\right] \end{aligned}\right\} \tag{20.96}$$

The equation (20.92) can be rewritten as

$$\Psi = \Psi_v + B\left\{s - \left[s_v - \frac{A}{2B}\right]\right\}^2 - \left(\frac{A^2}{4B}\right) \tag{20.97}$$

From the above relation, it is apparent that the value of $s = \overline{s}$ at which Ψ is a minimum is given by

$$\overline{s} = \left[s_v - \frac{A}{2B}\right] \tag{20.98}$$

The corresponding minimum value of the defect function is

$$\Psi_{min} = \Psi_v - \left(\frac{A^2}{4B}\right) \tag{20.99}$$

Using the values of A and B from (20.96) in (20.98) we get

$$\overline{s} = s_v + \frac{\left[(s_{v-1} - s_v)^2(\Psi_{v-2} - \Psi_{½}) - (s_{v-2} - s_v)^2(\Psi_{v-1} - \Psi_v)\right]}{2\left[(s_{v-1} - s_v)(\Psi_{v-2} - \Psi_v) - (s_{v-2} - s_v)(\Psi_{v-1} - \Psi_v)\right]} \tag{20.100}$$

Using $s_v = \beta^{v-1}$ when v is a positive integer greater than 2, the relative step length $\overline{s}$ at which the minimum of the true defect function occurs is given by

$$\bar{s} = \frac{(\beta+1)\beta^{(v-3)}\left[\Psi_v - (\beta^2+1)\Psi_{v-1} + \beta^2\Psi_{v-2}\right]}{2\left[\Psi_v - (\beta+1)\Psi_{v-1} + \beta\Psi_{v-2}\right]} \tag{20.101}$$

A special case occurs when $v = 2$. In this case, $s_{v-2} = s_0 = 0, \Psi_{v-2} = \Psi_0; s_{v-1} = s_1 = 1, \Psi_{v-1} = \Psi_1;$ $s_v = s_2 = \beta, \Psi_v = \Psi_2$. Substituting these values in (20.100), the expression for $\bar{s}$ yielding the minimum of the true defect function is given by

$$\bar{s} = \frac{\left[\Psi_2 - \beta^2\Psi_1 + (\beta^2 - 1)\Psi_0\right]}{2\left[\Psi_2 - (\beta+1)\Psi_1 + \beta\Psi_0\right]} \tag{20.102}$$

Step VIII. For commencing the next iteration, the value of global damping factor λ is to be changed by $(\lambda / \bar{s})$. It is already noted in (20.90) that from one iteration to the next, the global damping factor is to be changed by dividing itself by the corresponding value of $\bar{s}$ obtained in the earlier iteration.

20.3.10 Control of Gaussian Parameters

If Gaussian parameters like equivalent focal length, magnification, back focal length, or location of the exit pupil, etc. are required to have certain target values, the method of Lagrange multipliers, described in Section 20.2.5, may be used. However, in our program, we have included the weighted violations of such Gaussian targets in the defect function and treated them as pseudo-aberrations. The control on Gaussian parameters for the overall system can be implemented via absolute or relative control of the heights of the paraxial marginal ray and/or the paraxial pupil ray on the last refracting surface, and their angles of convergence in the final image space. Axial locations, etc. of the images and/or pupils in the intermediate spaces may also be similarly controlled by fixing appropriate Gaussian parameters for the requirements. The target G_0 for a parameter may be specified by the initial system value for the parameter or by a prespecified value as system input data.

20.3.11 Control of Boundary Conditions

A type of boundary condition encountered in practical lens design involve structures of the elements/components of the system. Commonly, these are the edge thicknesses for convex lens elements and convex air separations, and the axial thicknesses for concave lens elements and concave air separations. The other type of boundary conditions are imposed by the finite range of refractive index and Abbe number provided by the available glass types. Both types of boundary conditions are controlled by suitable penalty functions defined exclusively for each of them. These penalty functions are included in the defect function of the DLS optimization procedure. They are, therefore, needed to be defined so that they are not only continuous, but they also have continuous first and second derivatives with respect to the constructional variables.

20.3.11.1 Edge and Centre Thickness Control

For each thickness to be controlled, a maximum value t_{max} and a minimum value t_{min} are specified as inputs. If t is the current value of the thickness, a penalty function *P* is defined by

$$P = (\alpha f + \beta)^{\gamma} \tag{20.103}$$

where

$$f = \left\{\frac{2t - (t_{max} + t_{min})}{t_{max} - t_{min}}\right\}^2 \tag{20.104}$$

The power γ is chosen to be a suitably large number, say 50 or 60. α is a weighting factor, and β is a small number chosen to avoid the quantity $(\alpha f + \beta)$ becoming inconveniently small when the value of f is zero. For most practical cases, values like $\alpha = 1.0, \quad \beta = 0.01$ are satisfactory. It may be checked that, so long as t lies within the range $t_{min} < t < t_{max}$, the values of f lie within the range $(0,1)$, and the corresponding values of P lie within the range $(0,1.817)$. When $t = t_{max}$ or t_{min}, $f = 1.0$, and $P = 1.817$. But, with the slight increase of t from t_{max}, or decrease of t from t_{min}, the value of P increases sharply, e.g. when $f = 1.1, P \sim 525$. The sharp boundary can be made sharper with larger values of the power γ.

20.3.11.2 Control of Boundary Conditions Imposed by Available Glass Types

A discrete number of glass types are usually available as variables in lens design. Figure 9.2 shows the Abbe diagram for available Schott glasses over the visible wavelengths. Each glass is represented by a point on this $(n_d - V_d)$ map. Note that the cluster of available glasses lies within a region that may be approximately delineated by a closed boundary on the map.

For the sake of optimization, available glass type is assumed to be a continuous variable lying within this closed boundary. The closed boundary may be defined with the help of a combination of an appropriate number L of piecewise continuous curved or straight lines. Violation of the boundary is tackled by using penalty functions of the form

$$P = \sum_{l=1}^{L} (\alpha f_l + \beta)^{\gamma} \tag{20.105}$$

where each f_l is defined by an equation representing the l-th part of the boundary. Roles of the constants (α, β, γ) are the same as discussed in the earlier section. Typical boundaries in practice are triangles, rectangular polygons, or re-entrant polygons, with one or more of their sides being straight or curved lines. The interested reader may look for details in [106].

20.4 Evaluation of Aberration Derivatives

Practical implementation of the DLS optimization procedure involves determination of (MN) number of first derivatives and an equal number of homogeneous second derivatives of each of the M number of aberrations, pseudo-aberrations, and penalty functions with respect to the N number of constructional variables. Also, determination of manufacturing tolerances for the constructional variables requires the derivatives of the aberrations with respect to the constructional variables. The topic of differential analysis originated from the work of Seidel [168]. Notable contributions in the field have been made by Kerber [169], Schwarzschild [170], Nušl [171], and Smith [172]. But these studies, for the most part, were restricted to the analysis of transverse aberrations of meridional rays. Herzberger [173] extended the analysis to the case of transverse aberration of skew rays.

The application of differential correction techniques to lens design seems to have been taken up seriously by M'Aulay and Cruickshank in Tasmania, Australia during the second world war, as per their reports during 1942–1943 in 'Secret Papers' of the Optical Munitions Panel, Australia [174–175]. In 1943, Stempel published an account of differential changes at a single surface in connection with the design of the Huygenian eyepiece [176]. The first description of the differential method adopted by M'Aulay and Cruickshank for adjusting the aberration of a lens system appeared in open literature in 1945 [177]. They considered the effect of refraction on a meridional ray caused by a small perturbation of a constructional variable. The new ray was specified with respect to the original ray by quasi-paraxial type coordinates. A system of transfer coefficients was developed to specify the changes in the path of a ray in the final image space resulting from small changes in surface curvature, refractive index, and axial separation of successive surfaces, made within the system. These coefficients provided the link between the perturbation in a variable with the tangential ray aberrations. In a series of publications, Cruickshank described different aspects of these transfer coefficients [178–180]. M'Aulay discussed an approximate method that can be used for fast execution of this transfer [181]. Few other reports of investigations in this problem appeared around this time [182–184].

Buchdahl published his formulation of optical aberration coefficients, and in Part III of his book, he dealt with the first and higher derivatives of aberration coefficients [185]. The formulae provided for derivatives of the aberration coefficients are exact. However, they cannot be used directly in the framework of optical system assessment that utilizes the canonical variables and coordinates for points on the object/image plane and entrance/exit pupil (as described in Chapter 11). Indeed, the aberration function in the Hopkins scheme is defined in terms of reduced variables and coordinates for points on the pupil and the image, whereas the aberration function in the Buchdahl scheme is defined as a function in terms of points on the Gaussian pupil and image. Therefore, it is not straightforward to convert aberration coefficients of one scheme into those of the other. Several reports have appeared on use of identical merit functions in lens design following the Buchdahl aberration scheme. Andersen claims to have simplified some of Buchdahl's derivative formulae [186–194]. On a different note, Bhattacharya and Hazra reported the use of Gaussian brackets to obtain succinct expressions for the derivatives of paraxial parameters and Seidel aberration coefficients [195].

Feder reported differentiation of algebraic ray tracing equations to obtain the derivatives. In his scheme, aberration of a ray was described in terms of a 'characteristic function' H which, per his definition, is equivalent to the wave aberration only when the latter was measured on a reference sphere with an infinite radius [196]. Obviously, this is a special case. In a later publication [197], he removed the approximations in his earlier results, and gave exact formulae for skew rays. Few other reports dealing with formulae for the aberration derivatives with varying degrees of approximation appeared [198–201].

H.H. Hopkins started investigations on aberration derivatives after the publication of M'Aulay and Cruickshank's papers. He noted that considerable simplification would result by the use of Fermat's principle of optical path. A direct consequence of this observation is that any first order change in wavefront aberration of a ray due to a change in the constructional variable of a lens system may be transferred to the image space. Blaschke used this concept in developing a procedure for the differential correction of optical systems [202]. On the same basis, Suzuki and Uwoki presented a technique for adjusting wavefront aberrations with changes in variables [203].

The first order derivative of the i-th aberration W_i with respect to the j-th construction parameter p_j may be construed as a combination of three components as shown below [204]:

$$\frac{\partial W_i}{\partial p_j} = (\text{local contribution}) + (\text{induced contribution}) + (\text{focal shift term}) \tag{20.106}$$

The formulae for aberration derivatives were updated significantly by researchers in the group of H.H. Hopkins over several decades, and reported in Ph.D. theses published by members of his research group [205–209]. Some of them were meant for removing approximations made in earlier versions, but, notably, a few of them arose out of new conceptual developments in the treatment of the aberrations and pseudo-aberrations. Initially, it was assumed that the contribution of the last two terms in relation (20.106) is insignificant, and in practice, only the first term on the right-hand side is to be taken into consideration. But this conjecture turned out to be wrong. In fact, the aberrations and pseudo-aberrations used in formulation of the defect function in lens design optimization belong to two different classes. Correct interpretation of members of one class, e.g. fractional distortion, Conrady $(d - D)$ chromatic correction, etc., calls for use of image space associated rays, whereas members of the other class can be properly interpreted only by using object space associated rays. For the former case, the assumption of neglecting the contribution of the last two terms in (20.103) holds good, but for most other cases of monochromatic and chromatic aberrations, the object space associated rays are to be considered, and that calls for taking into consideration all three contributions. The local contribution, as noted above, denotes the change in aberration from the neighbourhood of the perturbed surface, and it solely depends on the type of the variable being perturbed. The induced contribution denotes the change in aberration caused by successive surfaces; it is calculated with the help of a set of 4×4 transfer matrix for the finite ray at each successive surface. The third contribution, in the form of focal shift terms, is required to refer the resultant derivative to the new reference sphere in the case of wave aberration, or the new pupil ray in the case of transverse ray aberration.

Notwithstanding the significant developments in reliable formulae for analytical derivatives of aberrations and pseudo-aberrations, their practical utilization in lens design optimization is somewhat restricted. This may have been caused by the general non-availability of the formulae in open literature. Also, the increasing availability of fast computing machinery with phenomenal number-crunching ability has reduced the time required for evaluating the derivative by numerical differentiation by several orders of magnitude so that many users have lost the motivation to undertake the formidable task of proper understanding and coding of these formulae. A brief description for calculating the derivatives by numerical differentiation is given below.

The first order derivative of the i-th aberration $W_i(p_1,\ldots,p_j,\ldots,p_N)$ with respect to the j-th constructional variable can be expressed as [210–212]

$$\frac{\partial W_i(p_1,\ldots,p_j,\ldots,p_N)}{\partial p_j} = \lim_{\Delta p_j \to 0} \frac{W_i\{p_1,\ldots,(p_j+\Delta p_j),\ldots,p_N\} - W_i(p_1,\ldots,p_j,\ldots,p_N)}{\Delta p_j} \tag{20.107}$$

For the sake of conciseness, we express $W_i(p_1,\ldots,p_j,\ldots,p_N)$ by $W_i(p_j)$. Equation (20.107) takes the form

$$\frac{\partial W_i(p_j)}{\partial p_j} = \lim_{\Delta p_j \to 0} \frac{W_i(p_j+\Delta p_j) - W_i(p_j)}{\Delta p_j} \tag{20.108}$$

Expanding the function $W_i(p_j)$ by a Taylor series about the point $p_j = p_{j,0}$ we get

$$W_i(p_{j,0}+\Delta p_j) = W_i(p_{j,0}) + \Delta p_j\left[\frac{\partial W_i}{\partial p_j}\right]_{p_j=p_{j,0}} + \frac{1}{2}(\Sigma p_j)^2\left[\frac{\partial^2 W_i}{\partial p_j^2}\right]_{p_j=p_{j,0}} + \frac{1}{6}(\Delta p_j)^3\left[\frac{\partial^3 W_i}{\partial p_j^3}\right]_{p_j=p_{j,0}} + \ldots \tag{20.109}$$

$$W_i(p_{j,0}-\Delta p_j) = W_i(p_{j,0}) - \Delta p_j\left[\frac{\partial W_i}{\partial p_j}\right]_{p_j=p_{j,0}} + \frac{1}{2}(\Delta p_j)^2\left[\frac{\partial^2 W_i}{\partial p_j^2}\right]_{p_j=p_{j,0}} - \frac{1}{6}(\Delta p_j)^3\left[\frac{\partial^3 W_i}{\partial p_j^3}\right]_{p_j=p_{j,0}} + \ldots \tag{20.110}$$

From (20.109) and (20.110), it follows

$$\left[\frac{\partial W_i}{\partial p_j}\right]_{p_j=p_{j,0}} = \frac{W_i(p_{j,0}+\Delta p_j) - W_i(p_{j,0})}{\Delta p_j} + O(1) \tag{20.111}$$

$$\left[\frac{\partial W_i}{\partial p_j}\right]_{p_j=p_{j,0}} = \frac{W_i(p_{j,0}+\Delta p_j) - W_i(p_{j,0}-\Delta p_j)}{2\Delta p_j} + O(2) \tag{20.112}$$

$$\left[\frac{\partial^2 W_i}{\partial p_j^2}\right]_{p_j=p_{j,0}} = \frac{\left[W_i(p_{j,0}+\Delta p_j) - 2W_i(p_{j,0}) + W_i(p_{j,0}-\Delta p_j)\right]}{(\Delta p_j)^2} + O(2) \tag{20.113}$$

Equations (20.111) and (20.112) are the forward difference formula and the central difference formula for the first derivatives. Equation (20.113) is the formula for calculating the homogeneous second derivatives.

For determining the first derivative, the forward difference formula is the fastest, and it requires tracing only one additional ray along with the base ray in each case. For calculating the first derivative by the central difference formula, for each derivative, two additional rays are needed to be traced in each case. The accuracy in the latter case is greater by one degree of magnitude. The aberrational data for the two additional rays (in the case of central difference) can be used in combination with the data for the base ray to evaluate the homogeneous second derivative by formula (20.113).

From (20.107) – (20.108), it is evident that the increment in the variable to be used in numerical differentiation needs to be small. The small increments will cause small changes in aberrations. On the other hand, a very small increment gives rise to changes in aberrations that are of the order of numerical errors. Therefore, the optimum range for the increment is linked with the word length of the computer. Jamieson proposed a rule of thumb: 'if the computer word is P decimal digits long, then the derivative increment is $10^{-(P/2)}$and is constant for all variables' [213]. Matsui and Tanaka reported a method to estimate the adequacy of increments of variables for evaluating aberration derivatives in least-squares optimization problems [128]. Their method involves determining a series of eigenvalues of a squared Jacobian matrix for a prespecified set of constraints. The large amount of additional computations necessitated in the process has restricted the practical use of this approach.

A thorough investigation covering a large number of systems was undertaken by Basu and Hazra to develop a suitable rule of thumb for increments in construction variables in computation of first order aberration derivatives, homogeneous second derivatives, and a few selected mixed second derivatives [214–215]. It was noted that a double precision arithmetic using 8-byte word with fifteen decimal digits of precision is usually sufficient for the computation of aberrations. Increment Δp in a variable p should normally be given as a relative increment $(\Delta p)_{rel} = rp$, where the factor $r = 10^{-\nu}$, with ν being a positive integer. When the variable $p \rightarrow 0$, the relative increment cannot be used, and an absolute increment $(\Delta p)_{abs} = \epsilon = 10^{-\alpha}$ may be used for any variable. α is a small positive integer. As a rule of thumb, ν or α may be taken as an integer over the range $2 \leq \nu \leq 4$; this entails a 1 per cent to 0.01 per cent change in the variable in the case of relative increment, and an absolute value in the range $(0.01, 0.0001)$ when the value of the variable is zero, or its absolute value tends to zero.

REFERENCES:

1 A. Koenig and M. von Rohr, 'Computation of Rays through a System of Refracting Surfaces', Chapter II in *Geometrical Investigation of the formation of images in optical instruments*, [original German book title: *Die Theorie der optischen Instrumente*, Ed. M. von Rohr Springer, Berlin (1904)] Translator, R. Kanthack, H.M. Stationery Office, Dept. of Scientific and Industrial Research (1920) pp. 35–82.

2 T. Smith, 'On tracing rays through an optical system', *Proc. Phys. Soc. London*, Vol. 27 (1915) pp. 502–509.

3 T. Smith, 'On tracing rays through an optical system (2nd paper)', *Proc. Phys. Soc. London*, Vol. 30 (1918) pp. 221–233.

4 L. Silberstein, *Simplified Method of Tracing Rays through Any Optical System*, Longmans, Green, London (1918).

5 T. Smith, 'On tracing rays through an optical system (3rd paper)', *Proc. Phys. Soc. London*, Vol. 32 (1920) pp. 252–264.

6 T. Smith, 'On tracing rays through an optical system (4th paper)', *Proc. Phys. Soc. London*, Vol. 33 (1921) pp. 174–178.

7 M. Herzberger, 'Über die Durchrechnung von Strahlen durch optische Systeme', *Zeitschrift für Physik*, Vol. 43 (1927) pp. 750–768.

8 G.G. Slyussarev, *Methods of Calculating Optical Systems*, 1st edn., Gostekhizdat, Leningrad (1937). [in Russian '*Metodi Rascheta Opticheskikh Sistem*']

9 H. Chrétien, 'Calculs Numériques', Chapter XI in *Calcul des Combinaisons Optiques*, Masson, Paris (1980) pp. 687–725 [First published in 1938].

10 A.E. Conrady, 'Numerical Calculations' in *Applied optics and Optical Design, Part I*, Oxford University Press, London (1929) pp. 10–18, 46–60; 'Trigonometrical tracing of oblique pencils', ibid., pp. 402–436.

11 T.Y. Baker, 'Tracing skew rays through second degree surface', *Proc. Phys. Soc. London*, Vol. 56 (1944) pp. 114–122.

12 J.G. Baker, '*Design and Development of an Automatically Focussing 40 inch f 5.0 Distortionless Telephoto and Related lenses for High- altitude Aerial Reconnaissance*', Technical Report, N.D.R.C., Section 16.1, Optical Instruments (1944). [Baker used Mark I calculator for ray tracing. It required 120 secs to trace one ray through one spherical surface.]
13 T. Smith, 'On tracing rays through an optical system (5th paper)', *Proc. Phys. Soc. London*, Vol. 57 (1945) pp. 286–293.
14 D.P. Feder, 'Optical calculations with automatic computing machinery', *J. Opt. Soc. Am.*, Vol 41 (1951) pp. 630–635. [Feder used both the commercially available IBM Card Programmed Electronic Calculator, and the SEAC (Standards Eastern Automatic Computer) machine that was a thousand times faster than the Mark I calculator.]
15 W. Allen and R.H. Stark, 'Ray tracing using the IBM card programmed electronic calculator', *J. Opt. Soc. Am.* Vol. 41 (1951) pp. 636–640.
16 M. Herzberger, '*Some Remarks on Ray Tracing*', *J. Opt. Soc. Am.*, Vol. 41 (1951) pp. 805–807.
17 W. Weinstein, 'Iterative ray tracing', *Proc. Phys. Soc., Section B*, Vol. 65 (1952) pp. 731–735.
18 H. Marx, 'Linearisierung der Durchrechnungsformeln für Windschiefe Strahlen', *Opt. Acta*, Vol. 1 (1954) pp. 127–140.
19 G. Black, 'Ray tracing in the Manchester University Electronic Computing Machine', *Proc. Phys. Soc., Section B*, Vol. 67 (1954) pp. 569–574.
20 J. Flügge, 'B. Die Durchrechnungsverfahren für achsnahe und achsferne Strahlen' in Chapter II, Das Photographische Objektiv', Springer, Wien (1955) pp. 16–30.
21 W. Weinstein, 'Literature survey on ray tracing', Proc. Symp. Optical Design with Digital Computers, Technical Optics Section, Imperial College of Science & Technology, 5–7 June (1956) pp. 15–26.
22 M. Herzberger, 'Automatic ray tracing', *J. Opt. Soc. Am.*, Vol. 47 (1957) pp. 736–739.
23 G.H. Spencer and M.V.R.K. Murty, 'General ray tracing procedure', *J. Opt. Soc. Am.*, Vol. 52 (1962) pp. 672–678.
24 J.G. Baker, '*The Utilization of Automatic Calculating Machinery in the field of Optical Design*', Technical Reports 1–12. And a Supplement, prepared by Perkin-Elmer Corp., under Air Force Contract AF33 (038) – 10836, issued at intervals during 1951–1955 (Declassified in 1959).
25 L.A. Cauchy, 'Méthode générale pour la resolution des systems d'équations simultanées', *Compte Rendu à l'Académie des Sciences, Paris*, Vol. 25 (1847) pp. 536–538.
26 S. Rosen and C. Eldert, 'Least-squares method for optical correction', *J. Opt. Soc. Am.*, Vol. 44 (1954) pp. 250–252.
27 S. Rosen and A.M. Chung, 'Application of the least-squares method', *J. Opt. Soc. Am.*, Vol. 46 (1956) pp. 223–226.
28 A.M. Legendre, 'Sur la Méthode des moindres quarrés', in *Nouvelles methods pour la determination des orbites des comètes*, Firmin-Didot, Paris (1805) pp. 72–80.
29 G. Black, 'Use of electronic digital computers in optical design', *Nature*, Vol. 175 (1955) pp. 164–165.
30 G. Black, 'On the automatic design of optical systems', *Proc. Phys. Soc., Section B*, Vol. 68 (1955) pp. 729–736.
31 R.E. Hopkins, C.A. McCarthy, and R. Walters, 'Automatic correction of third-order aberrations', *J. Opt. Soc. Am.*, Vol. 45 (1955) pp. 363–365.
32 R.E. Hopkins and T. Lauroesch, 'Automatic design of telescope doublets', *J. Opt. Soc. Am.*, Vol. 45 (1955) pp. 992–994.
33 C.A. McCarthy, 'A note on the automatic correction of third-order aberrations', *J. Opt. Soc. Am.*, Vol. 45 (1955) pp. 1087–1088.
34 H.D. Korones and R.E. Hopkins, 'Some effects of glass choice in telescope doublets', Vol. 40 (1959) pp. 869–871.
35 W.P. Hennessy and G.H. Spencer, 'Automatic correction of first and third order aberrations', *J. Opt. Soc. Am.*, Vol. 50 (1960) p. 494.
36 R.E. Hopkins, 'Third-order and fifth-order analysis of the triplet', *J. Opt. Soc. Am.*, Vol. 52 (1962) pp. 389–394.
37 R.E. Hopkins, 'Optical Design on Large Computers', in *Proceedings of the Conference on Optical Instruments and Techniques London 1961*, Ed., K. J. Habell, Chapman and Hall, London (1962) pp. 65–78.

38 R.E. Hopkins and G. Spencer, 'Creative thinking and computing machines in optical design', *J. Opt. Soc. Am.*, Vol. 52 (1962) pp. 172–176.
39 R.E. Hopkins, 'Re-evaluation of the problem of optical design', *J. Opt. Soc. Am.*, Vol. 52 (1962) pp. 1218–1222.
40 G.H. Spencer, 'A flexible automatic lens correction procedure', *Applied Optics*, Vol. 2 (1963) pp. 1257–1264.
41 R.E. Hopkins, 'A series of lenses designed for optimum performance', in Proc. Conf. Photographic and Spectroscopic Optics 1964, *Jap. J. Applied Physics*, Vol. 4, Suppl. 1 (1965) pp. 60–67.
42 R.E. Hopkins and H.A. Unvala, in *Proceedings of the conference 'Lens Design with Large Computers'*, Ed. W. L. Hyde, Institute of Optics, Rochester (1967).
43 T.R. Sloan and R.E. Hopkins, 'Design of double gauss systems using aspherics', *Appl. Opt.*, Vol. 6 (1967) pp. 1911–1915.
44 J.L. Rayces and L. Lebich, 'Experiments on constrained optimization with Spencer's method', *Opt. Eng.*, Vol. 27 (1988) pp. 1031-1034.
45 F.I. Havliček, 'Zur Feinkorrektion optischer Systeme', *Optik*, Vol. 9 (1952) p. 333.
46 F.I. Havliček, *'Über die Verwendung von Differenzen der Seidelschen Koeffizienten bei der Korrektur von optischen Systemen'*, *Optik*, Vol. 10 (1953) p. 475.
47 F. Wachendorf, 'Die Bestimmung eines optimalen Linesnsystems', *Optik*, Vol. 12 (1955) pp. 329–359.
48 J. Meiron and H.M. Loebenstein, 'Automatic correction of residual aberrations', *J. Opt. Soc. Am.*, Vol. 47 (1957) pp. 1104–1109.
49 J. Meiron, 'Automatic lens design by the least squares method', *J. Opt. Soc. Am.* Vol. 49 (1959) pp. 293–298.
50 J. Meiron and G. Volinez, 'Parabolic approximation method for automatic lens design', *J. Opt. Soc. Am.*, Vol. 50 (1960) pp. 207–211.
51 J. Meiron 'Damped least-squares method for automatic lensdesign, *J. Opt. Soc. Am.*, Vol. 55 (1965) pp. 1105–1109.
52 A. Girard, 'Calcul Automatique en Optique Géométrique', *Rev. d'Optique*, Vol. 37 (1958) pp. 225–241, 397–424.
53 D.P. Feder, 'Automatic lens design methods', *J. Opt. Soc. Am.*, Vol. 47 (1957) pp. 902–912.
54 D.P. Feder, 'Calculation of the optical merit function and its derivatives with respect to the system parameters', *J. Opt. Soc. Am.*, Vol. 47 (1957) pp. 913–925.
55 M.R. Hestenes and E.L. Stiefel, 'Methods of conjugate gradients for solving linear systems', *J. Res. Nat. Bur. Standards, Section B*, Vol. 49 (1952) pp. 409–432.
56 D.P. Feder, 'Automatic lens design with a high-speed computer', *J. Opt. Soc. Am.*, Vol. 52 (1962) pp. 177–183.
57 H.B. Curry, 'The method of steepest descent for non-linear minimization problems', *Quart. Appl. Math.*, Vol. 2 (1944) pp. 258–261.
58 D.P. Feder, 'Automatic optical design', *Appl. Opt.*, Vol. 2 (1963) pp. 1209–1226.
59 D.P. Feder, in *Recent Advances in Optimization Techniques*, Eds. A. Lavi and T.P. Vogl, Wiley, New York (1966) p. 6.
60 D.P. Feder, 'Optical design with automatic computers', *Appl. Opt.*, Vol. 11 (1972) pp. 53–58.
61 C.G. Wynne, 'Lens design by electronic digital computer: I', *Proc. Phys. Soc.*, Vol. 73 (1959) pp. 777–787.
62 M. Nunn and C.G. Wynne, 'Lens design by electronic digital computer: II', Vol. 74 (1959) pp. 316–329.
63 K. Levenberg, 'A method for the solution of certain nonlinear problems in least squares', *Quart. Appl. Math.*, Vol. 2 (1944) pp. 164–168.
64 D.W. Marquardt, 'An algorithm for least squares estimation of nonlinear parameters', *J. Soc. Ind. Appl. Math.*, Vol. 11 (1963) pp. 431–441.
65 C.G. Wynne, 'The Relevance of Aberration Theory to Computing Machine Methods', in *Proceedings of the conference on Optical Instruments and Techniques London 1961*, Ed. K. J. Habell, Chapman and Hall, London (1961) pp. 79–94.
66 C.G. Wynne and P.M.J.H. Wormell, 'Lens design by computer', *Appl. Opt.*, Vol. 2 (1963) pp. 1233–1238.
67 C.G. Wynne, 'Some examples of lens designing by computer', in Proc. Conf. Photographic and Spectroscopic Optics 1964, *Jap. J. Appl. Physics*, Vol. 4, Suppl. 1 (1965) pp. 81–85.
68 M.J. Kidger and C.G. Wynne, 'Experiments with lens optimization procedures', *Opt. Acta*, Vol. 14 (1967) pp. 279–288.

69 P.M.J.H. Wormell, 'Version 14, a program for the optimization of lens design', *Opt. Acta*, Vol. 25 (1978) pp. 637–654.
70 R. Finkler, 'The design of optical systems with extended depth of focus', *Opt. Acta*, Vol. 25 (1978) pp. 655–663.
71 M.J. Kidger, 'Use of the Levenberg-Marquardt (damped least-squares) optimization method in lens design', *Opt. Eng.*, Vol. 32 (1993) pp. 1731–1739.
72 E. Glatzel, 'Ein neues Verfahren zur automatischen Korrektur optischer Systeme mit electronischen Rechenmaschinen', *Optik*, Vol. 18 (1961) pp. 577–580.
73 E. Glatzel and R. Wilson, 'Adaptive automatic correction in optical design', *Appl. Opt.*, Vol. 7 (1968) pp. 265–276.
74 J. Rayces, 'Ten years of lens design with Glatzel's adaptive method', *Proc. SPIE*, Vol. 237 (1980) pp. 75–84.
75 J. Rayces and L. Lebich, 'Ray code: an aberration coefficient oriented lens design and optimization program', *Proc. SPIE*, Vol. 766 (1987) pp. 230–245.
76 J. Maxwell and C.C. Hull, 'Multidimensional quadratic extrapolation method for the correction of aberrations in lens systems', *Appl. Opt.*, Vol. 31 (1992) pp. 2234–2240.
77 C.C. Hull and J. Maxwell, 'A quadratic adaptive method for aberration correction', *J. Mod. Opt.*, Vol. 42 (1995) pp. 1213–1229.
78 L. Fialovszky, 'Anwendung einer Differentialmethode und der Ausgleichungrechnung zur Feinkorrektion optischer Systeme', *J. Opt. Soc. Am.*, Vol. 53 (1963) pp. 807–811.
79 D.S. Grey, 'Aberration theories for semiautomatic lens design by electronic computers. I. Preliminary remarks, *J. Opt. Soc. Am.*, Vol. 53 (1963) pp. 672–676.
80 D.S. Grey, 'Aberration theories for semiautomatic lens design by electronic computers. II. A specific computer program', *J. Opt. Soc. Am.*, Vol. 53 (1963) pp. 677–680.
81 D.S. Grey, in *Recent Advances in Optimization Techniques*, Ed. A. Lavi and T.P. Vogl, Wiley, New York (1966) p. 69.
82 D.S. Grey, 'Recent Developments in Orthonormalization of Parameter Space', in *Proc. Conf. Lens design with Large Computers*, Ed. W.L. Hyde, Rochester, New York (1967).
83 L.W. Cornwell, R.J. Pegis, A.K. Rigler, and T.P. Vogl, 'Grey's method for nonlinear optimization', *J. Opt. Soc. Am.*, Vol. 63 (1973) pp. 576–581.
84 L.G. Seppala, 'Optical interpretation of the merit function in Grey's lens design program', *Appl. Opt.*, Vol. 13 (1974) pp. 671–678.
85 B. Brixner, 'Automatic lens design for nonexperts', *Appl. Opt.*, Vol. 2 (1963) pp. 1281–1286.
86 B. Brixner, 'Faster LASL lens design program', *Appl. Opt.*, Vol. 11 (1973) pp. 2703–2708.
87 T.C. Doyle, Nonlinear least squares optimization of a continuous N-parameter system, LASL Report LA-DC-72–1018, Los Alamos, New Mexico (1972).
88 T.C. Doyle, Automatic lens design by nonlinear least squares optimization of a continuous N-parameter system, LASL Report LA-DC-73–518, Los Alamos, New Mexico (1973).
89 M. Klein and B. Brixner, 'CLIMB: a new Los Alamos Scientific Laboratory lens design code', *Appl. Opt.*, Vol. 15 (1975) pp. 2583–2587.
90 B. Brixner, 'Lens design merit functions: rms image spot size and rms optical path difference'. *Appl. Opt.*, Vol. 17 (1978) pp. 715–716.
91 B. Brixner, 'Lens design that simultaneously optimizes image spots, optical path difference (OPD), Diffraction modulation transfer function (DMTF) and Seidel aberrations, *Proc. SPIE*, Vol. 190 (1979) pp. 2–4.
92 B. Brixner, 'Lens Design and local minima' *Appl. Opt.*, Vol. 20 (1981) pp. 384–387.
93 B. Brixner, 'Accelerating convergence in automatic lens design', *Appl. Opt.*, Vol. 20 (1981) pp. 2452–2456.
94 B. Brixner, 'A nearly ideal lens optimization procedure', *Proc. SPIE*, Vol. 554 (1985) pp. 52–57.
95 B. Brixner and M.M. Klein, 'Optimization experiments with a double Gauss lens', *Opt. Eng.*, Vol. 27 (1988) pp. 420–423.
96 B. Brixner and M.M. Klein, 'Optimization to eliminate two lens elements that have undesirable shapes', *Opt. Eng.*, Vol. 27 (1988) pp. 1027–1030.
97 H.H. Hopkins, 'The use of diffraction-based criteria of image quality in automatic optical design', *Opt. Acta*, Vol. 13 (1966) pp. 343–369.

98 A.V. Lenskii, 'Optical Transfer Function in the region of low spatial frequencies as a quality criterion in the automatic correction of optical systems', *Opt. Spectrosc.*, Vol. 24 (1968) pp. 229–231.
99 J. Meiron, 'The use of merit functions based on wavefront aberrations in automatic lens design', *Appl. Opt.*, Vol. 7 (1968) pp. 667–672.
100 W.B. King, 'Use of the Modulation Transfer Function (MTF) as an Aberration-Balancing Merit Function in Automatic Lens design', in *Optical Instruments and Techniques* (Proc. ICO Conference, July 1969, Reading University, U. K.), Ed. J. Home Dickson, Oriel Press (1970) pp. 359–366.
101 A. Offner, 'Extended Range Diffraction-Based Merit Function for Least Squares Type Optimization', in *Optical Instruments and Techniques* (Proc. ICO Conference, July 1969, Reading University, U. K.), Ed. J. Home Dickson, Oriel Press (1970) pp. 367–374.
102 K.-J. Rosenbruch, 'Use of OTF-based criteria in automatic optical design', *Opt. Acta*, Vol.22 (1975) pp. 291–300.
103 R.W. Gostick, 'OTF-based optimization criteria for automatic optical design', *Optical and Quantum Electronics*, Vol. 8 (1976) pp. 31–37.
104 L.N. Hazra, 'Extended range diffraction-based merit function for lens design optimization', *Proc. SPIE*, Vol. 656 (1986) pp. 20–25.
105 S. Szapiel, *Diffraction-based Image Assessment in Optical Design*, Institute of Design of Precise and Optical Instruments, Warsaw Technical University, Warsaw (1986).
106 H.H. Hopkins and A. Kadkly, 'Control of damping factors and boundary conditions in optimization programs', *J. Mod. Opt.*, Vol.35 (1988) pp. 49–74.
107 K.B. O'Brien, 'Automatic optical design of desired image distributions using orthogonal constraints', *J. Opt. Soc. Am.*, Vol. 54 (1964) pp. 1252–1255.
108 M. Herzberger and R. Morris, 'A contribution to the method of least squares', *Quart. Appl. Math.*, Vol. 5 (1947) pp. 354–357.
109 M. Herzberger, 'The normal equation of the method of least squares and their solution', *Quart. Appl. Math.*, Vol. 7 (1949) pp. 217–223.
110 H. Brunner, Automatisches Korrigieren unter Berücksichtigung der zweiten Ableitungen der Güteunktion', *Opt. Acta*, Vol. 18 (1971) pp. 743–758.
111 J. Basu and L. Hazra, 'Role of line search in least-squares optimization of lens design', *Opt. Eng.*, Vol. 33 (1994) pp. 4060–4066.
112 D.R. Buchele, 'Damping factor for the least-squares method of optical design', *Appl. Opt.*, Vol. 7 (1968) pp. 2433–2436.
113 D.C. Dilworth, 'Pseudo-second-derivative matrix and its application to automatic lens design', *Appl. Opt.*, Vol. 17 (1978) pp. 3372–3375.
114 D.C. Dilworth, 'Improved Convergence with the pseudo-second-derivative (PSD) Optimization Method', *Proc. SPIE*, Vol. 399 (1983) pp. 159–165.
115 A. Faggiano, 'Automatic lens design with pseudo-second-derivative matrix: a contribution', *Appl. Opt.*, Vol. 19 (1980) pp. 4226–4229.
116 P.N. Robb, 'Accelerating convergence in automatic lens design', *Appl. Opt.*, Vol. 18 (1979) pp. 4191–4194.
117 E.D. Huber, 'Extrapolated Least-Squares optimization: a new approach to least-squares optimization in lens design', *Appl. Opt.*, Vol. 21 (1982) pp. 1705–1707.
118 E.D. Huber 'Extrapolated least-squares optimization in optical design', *J. Opt. Soc. Am. A*, Vol. 2 (1983) pp. 544–554.
119 T. Suzuki, Y. Ichioka, S. Yonezawa, and M. Koyama, 'Automatic lens design', *Jap. J. Appl. Physics*, Vol. 4, Suppl. 1 (1995) pp. 74–80.
120 T. Suzuki and S. Yonezawa, 'System of simultaneous nonlinear inequalities and automatic lens-design method', *J. Opt. Soc. Am.*, Vol. 56 (1966) pp. 677–683.
121 H. Ooki, 'Rank down method for automatic lens design', *Proc. SPIE*, Vol. 1354 (1990) pp. 171–176.
122 K. Tanaka, 'Linearization of nonlinear automatic lens design problems', in Proc. Conf. Photographic and Spectroscopic Optics 1964, *Appl. Opt.*, Vol. 29 (1990) p. 4537.
123 K. Tanaka, 'A formal linearization procedure of constrained nonlinear automatic lens design problems I. Formulation by means of penalty function method', *J. Opt. (Paris)*, Vol. 21 (1990) pp. 241–245.
124 H. Matsui and K. Tanaka, 'Formulation of a constrained automatic lens design problem by means of the augmented Lagrangian function', *J. Opt. (Paris)*, Vol. 24 (1993) pp. 173–175.

125 H. Matsui and K. Tanaka, 'Formulation of a nonlinear automatic lens design problem with boundary conditions by means of the Kuhn-Tacker's optimality Condition', *J. Opt. (Paris)*, Vol. 24 (1993) pp. 11–14.
126 H. Matsui and K. Tanaka, 'Solution of the damped least-squares problem by using a combination of eigenvalue and singular value decomposition', *Appl. Opt.*, Vol. 31 (1992) pp. 2241–2243.
127 H. Matsui and K. Tanaka, 'Determination method of an initial damping factor in the damped-least-squares problem', *Appl. Opt.*, Vol. 33 (1994) pp. 2411–2418. [Errata, *Appl. Opt.*, Vol. 34 (1995) 40].
128 H. Matsui and K. Tanaka, 'Method for estimating numerical adequacy of derivative increments of variables in the damped least-squares automatic lens design problem', *Appl. Opt.*, Vol. 34 (1995) pp. 642–647.
129 D.S. Volosov and N.V. Zeno, 'Automatic correction of the aberrations of complex optical systems', *Appl. Opt.*, Vol. 8 (1969) pp. 289–292.
130 A.P. Grammatin, 'Automatic Methods of Designing Optical Systems', Chapter 7 in *Aberration and Optical Design Theory*, G.G. Slyussarev [original Russian book title: *Metodi Rascheta Opticheskikh Sistem, Mashinostroenie*, Leningrad (1969)] Translator: J.H. Dixon, Adam Hilger, Bristol (1984) pp. 379–463.
131 S.N. Bezdidko, 'Optimization of optical systems using orthogonal polynomials', *Opt. Spectrosc.* Vol. 48 (1980) pp. 670–671.
132 M.A. Gan, D.D. Zhdanov, V.V. Novoselskiy, S.I. Ustinov, A.O. Fedorov, and I.S. Potyemin, 'DEMOS: State-of-the-art application software for design, evaluation, and modeling of optical systems', *Opt. Eng.*, Vol. 31 (1992) pp. 696–700.
133 M.A Gan, 'Automation of optical-system design', *J. Opt. Technol.*, Vol. 61 (1994) pp. 566–572.
134 J.R. De, F. Moneo, and I. Juvells, 'Correccion Automatica de Sistemas Opticos Mediante Marchas Exactas de Rayos'. *Parte I. Optica Pura y Aplicada*, Vol. 12 (1979) pp. 91–96; Parte II. Ibid., 97–104. [in Spanish]
135 N.v.d.w. Lessing, 'Method for the automatic correction of the aberrations of optical systems', *Appl. Opt.*, Vol. 19 (1980) pp. 487–488.
136 J. Kross, 'Principles of optimization in lens design developed at the Institute of Optics in Berlin (West)', *Proc. SPIE*, Vol. 1354 (1990) pp. 165–170.
137 F.A. Aurin, 'Optimization of the Seidel image errors by bending of lenses using a 4th degree merit function', *Proc. SPIE*, Vol. 1354 (1990) pp. 180–185.
138 M.J. Box, D. Davies, and W.H. Swann, *Non-Linear Optimization Techniques*, ICI Monograph No. 5, Oliver and Boyd, Edinburgh (1969).
139 W. Murray, Ed., *Numerical Methods for Unconstrained Optimization*, Academic, London (1972).
140 P.R. Adby and M.A.H. Dempster, *Introduction to Optimization Methods*, Chapman and Hall, London (1974).
141 D.M. Greig, *Optimisation*, Longman, London (1980).
142 L.E. Scales, *Introduction to Non-Linear Optimization*, Macmillan, London (1985).
143 R. Fletcher, *Practical Methods of Optimization*, John Wiley, Chichester (1987).
144 F.B. Hildebrand, *Introduction to Numerical Analysis*, Tata McGraw-Hill, New Delhi (1984) pp. 575–578.
145 A.K. Rigler and R.J. Pegis, 'Optimization Methods in Optics', in *The Computer in Optical Research: Methods and Applications*, Ed. B.R. Frieden, Springer, Berlin (1980) pp. 211–268.
146 L.W. Cornwell and A.K. Rigler, 'Comparison of four nonlinear optimization methods', *Appl. Opt.*, Vol. 11 (1972) pp. 1659–1660. [voir Errata, ibid., p. 2264]
147 T. Takahashi, 'An experimental analysis of optimization algorithms using a model function', *Optik, Band,* Vol. 35 (1972) pp. 101–115.
148 T.H. Jamieson, *Optimization Techniques in Lens Design*, Adam Hilger, London (1971).
149 C.L. Lawson and R.J. Hanson, *Solving Least Squares Problems*, Prentice-Hall, Englewood Cliffs, NJ (1974).
150 H.O. Hartley, 'The modified Gauss-Newton method for the fitting of nonlinear regression functions by least squares', *Technometrics*, Vol. 3 (1961) pp. 269–280.
151 R.N. Madison, 'A procedure for nonlinear least squares refinement in adverse practical conditions', *J. Assoc. Comput. Mach.*, Vol. 13 (1966) pp. 124–134.
152 J. Kowalik and M.R. Osborne, *Methods for Unconstrained Optimization Problems*, Elsevier, New York (1968).

153 Y. Bard, 'Comparison of gradient methods for the solution of nonlinear parameter estimation problems', *SIAM J. Numer. Anal.*, Vol. 7 (1970) pp. 157–186.

154 J. Basu and L.N. Hazra, 'Experiments with three versions of least squares optimization of lens design', Digest, XXVI Optical Society of India Symposium, Regional Engineering College, Warangal, 4–7 February (2000) p. 43.

155 L. Nazareth, 'Some recent approaches to solving large residual nonlinear least squares problems', *SIAM Review*, Vol. 22 (1980) pp. 1–11.

156 R. Fletcher, 'A modified Marquardt subroutine for nonlinear least squares', Technical Report No. AERE-R-6799, Atomic Energy Research Establishment, Harwell, United Kingdom (1971).

157 M.J.D. Powell, 'A Hybrid Method for Nonlinear Equations', in *Numerical Methods for Nonlinear Algebraic Equations*, Ed. P. Rabinowitz, Gordon and Breach, London (1970).

158 P.E. Gill and W. Murray, 'Algorithms for the solution of nonlinear least squares problem', *SIAM J. Numer. Anal.*, Vol. 15 (1978) pp. 977–992.

159 L. Grippo, F. Lampariello, and S. Lucidi, 'A nonmonotone line search technique for Newton's method', *SIAM J. Numer. Anal.*, Vol. 23 (1986) pp. 707–716.

160 M.C. Ferris and S. Lucidi, 'Nonmonotone stabilization methods for nonlinear equations', *J. Optimization Theory and Applications*, Vol. 81 (1994) pp. 53–71.

161 J.Z. Zhang and L.H. Chen, 'Nonmonotone Levenberg-Marquardt algorithms and their convergence analysis', *J. Optimization Theory and Applications*, Vol. 92 (1997) pp. 393–418.

162 S. Chatterjee, J. Basu, and L.N. Hazra, Non-monotone Levenberg-Marquardt algorithms in least squares optimization in lens design, Digest, XXVI Optical Society of India symposium, Regional Engineering College, Warangal,4–6 February (2000) p. 37.

163 W. Kaplan, *Advanced Calculus*, Addison-Wesley, Reading, Massachusetts (1959) pp. 128–129.

164 G.W. Forbes, 'Optical system assessment for design: numerical ray tracing in the Gaussian pupil', *J. Opt. Soc. Am. A*, Vol. 5 (1988) pp. 1943–1956. Errata: ibid., Vol. 6 (1989) p. 1123.

165 D.S. Grey, 'Tolerance sensitivity and optimization', *Appl. Opt.*, Vol. 9 (1970) pp. 523–526.

166 S.J. Dobson and A. Cox, 'Automatic desensitization of optical systems to manufacturing errors', *Meas. Sci. Technol.*, Vol. 6 (1995) pp. 1056–1058.

167 J. Basu and L.N. Hazra, 'Total hessian in least squares optimization of lens design', *J. Opt. (India)*, Vol. 29 (2000) pp. 95–104.

168 L. Seidel, 'Trigonometrische Formeln für den allgemeinen Fall der Brechung des Lichtes an Centrierten Sphärischen Flächen', *Münchener Sitzungsber.*, Vol. 2 (1866) pp. 263–283.

169 A. Kerber, *Beiträge zur Dioptrik, Heft 1,* Leipzig 1895, Selbstverlag, 14 S.

170 K. Schwarzschild, *Über differenzen Formeln zur Durchrechnung optischer Systeme*, Göttingen Nachr. (1907) pp. 531–570.

171 Fr. Nušl, *'Über allgemeine differenzen Formeln', Bull. Intern. De l'Acad. Des Sci. de Bohême*, Vol. 12 (1907) pp. 84–116.

172 T. Smith, 'The general form of the Smith-Helmholtz equation', *Trans. Opt. Soc. London*, Vol. 31 (1929–30) pp. 241–248.

173 M. Herzberger, *'Über die Durchrechnung von Strahlen durch optische Systeme', Zeits. F. Physik*, Vol. 43 (1927) pp. 750–768.

174 A.L. M'Aulay, *Secret Papers*, Optical Munitions Panel, Australia, November (1942) and July (1943).

175 F.D. Cruickshank, *Secret Papers*, Optical Munitions Panel, Australia, November (1942) and May (1943).

176 W.M. Stempel, 'An empirical approach to lens designs. The Huygens eyepiece', *J. Opt. Soc. Am.*, Vol. 33 (1943) pp. 278–292.

177 A.L. M'Aulay and F.D. Cruickshank, 'A differential method of adjusting the aberrations of a lens system', *Proc. Phys. Soc. (London)*, Vol. 57 (1945) pp. 302–310.

178 F.D. Cruickshank, 'A system of transfer coefficients for use in the design of lens systems: I. The general theory of the transfer coefficients', *Proc. Phys. Soc. (London)*, Vol. 57 (1945) pp. 350–361; 'A system of transfer coefficients for use in the design of lens systems: II. A second-order correction term', ibid., pp. 362–367; 'A system of transfer coefficients for use in the design of lens systems: III. The contribution to the image aberrations made by the individual surfaces of a lens system', ibid., pp. 419–425; 'A system of transfer coefficients for use in the design of lens systems: IV. The estimation of the tolerances permissible in the production of a lens system', ibid., pp. 426–429; 'A system of transfer coefficients for

use in the design of lens systems: V. Transfer coefficients for the astigmatism at small aperture and finite obliquity', ibid., pp. 430–435.

179 F.D. Cruickshank, 'A system of transfer coefficients for use in the design of lens systems: VI. The chromatic variations of tangential aberrations', *Proc. Phys. Soc. (London)*, Vol. 58 (1946) pp. 296–302.

180 F.D. Cruickshank, 'The paraxial differential transfer coefficients of a lens system', *J. Opt. Soc. Am.*, Vol. 36 (1946) pp. 13–19.

181 A.L. M'Aulay, 'A transfer method for deriving the effect on the image formed by an optical system from ray changes produced at a given surface', *Proc. Phys. Soc. (London)*, Vol. 57 (1945) pp. 435–440.

182 T. Smith, 'Variational formulae in optics', *Proc. Phys. Soc.*, Vol. 57 (1945) pp. 558–564.

183 W.M. Stempel, 'A differential adjustment method of refining optical systems', *J. Opt. Soc. Am.*, Vol. 38 (1948) pp. 935–953.

184 A.E. Glancy, 'Differential correction of an optical system', *J. Opt. Soc. Am.*, Vol. 41 (1951) pp. 389–396.

185 H.A. Buchdahl, 'Part III. The Adjustment of the Monochromatic and Chromatic Aberration Coefficients of the Symmetrical Optical System by Changes in the Constitution of the System' in *Optical Aberration Coefficients*, Dover, New York (1968). [First published by Oxford University Press in 1954] pp. 192–296.

186 F.D. Cruickshank and G.A. Hills, 'Use of optical aberration coefficients in optical design', *J. Opt. Soc. Am.*, Vol. 50 (1960) pp. 379–387.

187 C.J. Woodruff, 'A comparison, using orthogonal coefficients, of two forms of aberration balancing', *Opt. Acta*, Vol. 22 (1975) pp. 933–941.

188 C.J. Woodruff, 'The wavefront aberration difference function: a comparative study', *Opt. Acta*, Vol. 23 (1976) pp. 773–784.

189 P.N. Robb, 'Analytic merit function based on Buchdahl's aberration coefficients', *J. Opt. Soc. Am.*, Vol. 66 (1976) pp. 1037–1041.

190 P.N. Robb, 'Lens design using optical aberration coefficients', *Proc. SPIE*, Vol. 237 (1980) pp. 109–118.

191 T.B. Andersen, 'Optical aberration functions: derivatives with respect to axial distances for symmetric systems', *Appl. Opt.*, Vol. 21 (1982) pp. 1817–1823.

192 T.B. Andersen, 'Optical aberration functions: chromatic aberrations and derivatives with respect to refractive indices for symmetric systems', *Appl. Opt.*, Vol. 21 (1982) pp. 4040–4044.

193 S.C. Tam, G.D.W. Lewis, S. Doric, and D. Heshmaty-Manesh, 'Diffraction analysis of rotationally symmetric optical systems using computer-generated aberration polynomials', *Appl. Opt.* Vol. 22, (1983) pp. 1181–1187.

194 T.B. Andersen, 'Optical aberration functions: derivatives with respect to surface parameters for symmetrical systems', *Appl. Opt.*, Vol. 24 (1985) pp. 1122–1129.

195 K. Bhattacharya and L.N. Hazra, 'Analytical derivatives for optical system analysis: use of Gaussian brackets', *J. Opt. (India)*, Vol. 18, (1989) pp. 57–67.

196 D.P. Feder, 'Calculation of an optical merit function and its derivatives with respect to the system parameters', *J. Opt. Soc. Am.*, Vol. 47 (1957) pp. 913–925.

197 D.P. Feder, 'Differentiation of ray-tracing equations with respect to construction parameters of rotationally symmetric optics', *J. Opt. Soc. Am.*, Vol. 58 (1968) pp. 1494–1505.

198 S.H. Brewer, 'Surface-contribution algorithms for analysis and optimization', *J. Opt. Soc. Am.*, Vol. 66 (1976) pp. 8–13.

199 B. Tatian, 'Comment on "Surface-contribution algorithms for analysis and optimization"', *J. Opt. Soc. Am.*, Vol. 66 (1976) pp. 628–630.

200 S.H. Brewer, 'Author's reply to comments by Tatian', Vol. 66 (1976) p. 630.

201 I.P. Agurok and M.A. Duinovskij, 'Analytical calculation of the derivatives of lateral and wave aberrations with respect to the structural parameters of optical systems', *Sov. J. Opt. Technol.*, Vol. 57 (1990) pp. 292–294.

202 W.S.S. Blaschke, 'A procedure for the differential correction of optical systems allowing large parameter changes', *Opt. Acta*, Vol. 3 (1956) pp. 10–23.

203 T. Suzuki and L. Uwoki, 'Differential method for adjusting the wave-front aberrations of a lens system', *J. Opt. Soc. Am.*, Vol. 49 (1959) pp. 402–404.

204 J. Macdonald and Y. Wang, 'Analytical derivatives in damped-least squares optimisation', *Proc. SPIE*, Vol. 655 (1986) pp. 49–53.

205 W.B. King, *New Techniques for the evaluation and design of optical systems*, Ph. D. Thesis, University of London (1966).
206 J. Macdonald, *New Analytical and Numerical Techniques in Optical Design*, Ph. D. Thesis, University of Reading, United Kingdom (1974).
207 D.R.J. Campbell, Centring errors in optical systems, Ph. D. thesis, University of Reading, United Kingdom (1976).
208 K. Youern, *Analytical Derivatives of the Aberrations and Gaussian Properties of Optical Systems with respect to the Constructional Parameters*, Ph. D. Thesis, University of Reading, United Kingdom (1983).
209 Y. Wang, Analytical Derivatives and other techniques for improving the effectiveness of automatic optical design, Ph. D. thesis, University of Reading, United Kingdom (1986).
210 C.E. Gerald and P.O. Wheatley, *Applied Numerical Analysis*, Addison-Wesley, Reading, Massachusetts (1994) pp. 310–328.
211 N.I. Danilina, N.S. Dubrovskaya, O.P. Kvasha, and G.L. Smirnov, *Computational Mathematics*, [Translated from Russian by I. Aleksanova], Mir Publishers, Moscow (1988) pp. 374–378.
212 E.A. Volkov, *Numerical Methods*, Mir Publishers, Moscow (1986), [Original Russian edition published in 1962] pp. 53–62.
213 T.H. Jamieson, Ref. 148, pp. 60–61.
214 J. Basu and L.N. Hazra, 'Numerical evaluation of aberration derivatives', *J. Opt.(India)*, Vol. 25, No.1 (1996) pp. 37–60.
215 J. Basu and L.N. Hazra, 'A note on mixed second order aberration derivatives', *J. Opt.(India)*, Vol. 26, No. 3 (1997) pp. 157–159.

21

Towards Global Synthesis of Optical Systems

21.1 Local Optimization and Global Optimization: The Curse of Dimensionality

Figure 21.1 shows multiple minima of a function $f(x)$ in the range $a \leq x \leq b$. Each of them is a minimum in its immediate neighbourhood, so that these minima are called the local minima. Out of them, the minimum that has the lowest value of the function is called the global minimum. In Figure 21.1, the global minimum is located at $x = x_g$. All methods for numerical optimization elucidated so far require an initial estimate for the minimum. In the case of unimodal functions with one single minimum in a given range, these methods will locate the unique minimum from any initial estimate lying within the range. For multimodal functions with multiple minima in a given range, all these optimization procedures locate the minimum that lies in the neighbourhood of the initial estimate, i.e. the local minimum. A global or quasi-global minimum can be obtained by using a local optimization algorithm if and only if the initial estimate lies in the neighbourhood of the global or quasi-global minimum.

For a univariate function, the search for the global minimum over a given range may be carried out by function evaluations at a reasonable number of sampling points over the range. But for multivariate functions, it becomes intractable with an increase in the number of variables. Assume 100 samples in each dimension of an N variable system are adequate for the purpose, and each sample evaluation of the function requires 1μ second, the total time T required for function evaluation is $T = 100^N \mu\text{sec}$. Table 21.1 gives the value of T for a few values of N. It is obvious that, in general, the sampling approach is an infeasible proposition in practice. This problem belongs to the class of NP-hard problems, for which no algorithm is known to give an exact solution within polynomial time [1]. In optimization literature, this phenomenon is called 'the curse of dimensionality' [2]. It has been pointed out that a practical remedy to this otherwise intractable problem lies in devising approximate-solution algorithms whose running time T is proportional to small powers of T, say $T = N^p$, where p is a small number. An attempt to develop an algorithm for $p = 3$ is reported in [3].

The complexity of the problem may be appreciated by visualizing a two-dimensional landscape in a mountain range with a multitude of peaks and valleys, where a blind person is seeking either the highest peak or the lowest valley with only a stick at hand. A few studies on the merit function landscape related to relatively simple lens systems have appeared [4–7]. They illustrate the raison d'être of the different approaches adopted for global optimization in this particular field. The problem of global optimization in lens design has been critically put forward in a few publications [8–11]. In general, activities on global optimization were bolstered by two publications of Dixon and Szegö in the 1970s [12–13].

On a different note, ready availability of optimization software has facilitated experimentation in the field with novel ideas. At least two publications are promoting the development of open source algorithms for both global and local optimization [14–15]. Hopefully, they will lead to a proliferation of new ideas and concepts in this fascinating area.

21.2 Deterministic Methods for Global/Quasi-Global Optimization

As there is no analytical method for determining the globally or quasi-globally optimum design for a lens system, except for in trivial cases, several empirical procedures have been suggested to overcome

DOI: 10.1201/9780429154812-21

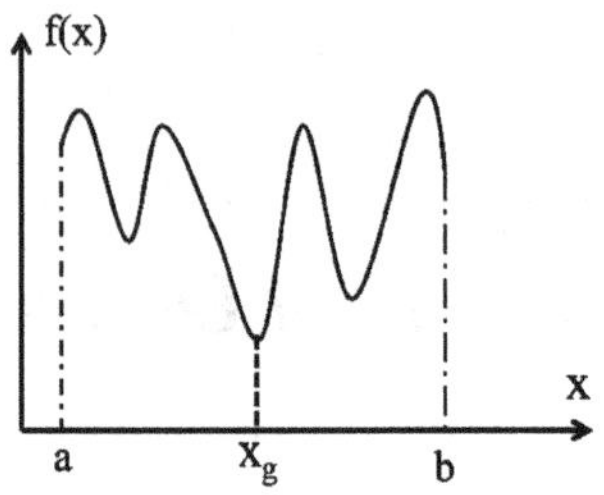

FIGURE 21.1 A univariate function $f(x)$ with multiple minima in a range $(a \le x \le b)$.

TABLE 21.1
Time to Perform a Grid Search in N Dimensions

N	Time
1	100 microsecond
2	10 millisecond
5	2.78 hours
10	3.17×10^{6} years
15	3.17×10^{16} years
20	3.17×10^{26} years

the limitation imposed by the quality of the local minimum obtained from the choice of an initial system [16]. In two dimensions, most of these procedures may be derived from the practical experience of hiking in the mountains with a specific goal. However, in higher dimensions, it is difficult to visualize the complexity of the design space for making useful conjectures. A few of these procedures are described in the following.

21.2.1 Adaptive/Dynamic Defect Function

The defect function used in the widely used local optimization algorithms in lens design consists of weighted sums of squares of several image quality parameters and several boundary conditions expressed as pseudo aberrations or penalty functions. Changes in the weighting factors, or in the image quality parameter, etc. give rise to a different defect function. The variation in the value of the defect function with the number of iterations is unique for a specific function, and often varies significantly for a different defect function. During optimization run with one defect function, when the system reaches near the local minimum, the reduction in defect function with further iterations becomes insignificant. At this time, an escape from the local minimum is often possible by switching to a different defect function [17]. A new minimum is sought by using the local optimization algorithm. The process may be repeated a number of times in pursuance of 'better' minima.

21.2.2 Partitioning of the Design Space

It is common conjecture that the local minima of the merit function in the hyperspace of construction variables are not spread out uniformly in the design space. Some of the minima may be located as a cluster in certain regions, some of them may be located wide apart, and there are certain regions in the design space where the presence of any minimum is nearly impossible. The latter may be identified from physical considerations on the nature of the individual variables, and such regions can be excluded from the search domain. Based on observations made by a sample survey, the remaining feasible space may be partitioned into a finite number of domains. In each of these domains, the local minimum is sought with the help of a local optimization algorithm. The 'best' minimum obtained thereby is adjudged a 'quasi-global' minimum. A big catch of this approach is the determination of a proper sampling procedure. Even in the case of a moderately complex lens system, the number of variables is often not so small as to justify use of a moderate number of samples. Nevertheless, even when using a crude sampling in this domain-partitioning approach, useful results are obtained in practice [18–21].

In design of lens systems, the number of lens elements/components is an important degree of freedom. A great flexibility in the design process can be obtained by dynamic addition or subtraction of surfaces and lenses during the optimization process. A 'sequential cluster algorithm' catering to this problem is reported [22–23].

21.2.3 The Escape Function and the 'Blow-Up/Settle-Down' Method

The 'escape function (EF)' or the 'blow-up/settle down (BUSD)' method is in fact a special case of the heuristic method proposed in Section 21.2.1 . When the exact (or an approximate) location $(\tilde{x}_1, \tilde{x}_2, \ldots, \tilde{x}_n)$ of a local minimum of the defect function Ψ is known from using a local optimization algorithm like the DLS method, a modified defect function Ψ_{mod} is formed by adding a term Ψ_{Esc} to the original defect function Ψ. Therefore, we get

$$\Psi_{mod} = \Psi + \Psi_{Esc} \tag{21.1}$$

where Ψ_{Esc} is given by the square of an escape function f_{Esc} defined by

$$f_{Esc} = \sqrt{H} \exp\left\{ -\frac{1}{2W^2} \sum_{j=1}^{N} \alpha_j \left(x_j - \tilde{x}_j \right)^2 \right\} \tag{21.2}$$

The parameters H and W take positive values and determine the power and range of influence of the escape function; α_j is a scale factor for the j-th variable. This form for the escape function was first proposed by Isshiki et al. [24–28]. The role of the escape function is to increase the value of the defect function near the local minimum. When optimization is recommenced with the local optimization algorithm, the design is forced to escape the local minimum, which no longer exists in the modified defect function, and seek a new local minimum. A series of local minima can be determined by setting up the escape function at each local minimum, providing a choice for the most suitable one. Smooth functioning of this approach depends to a large extent on proper setting up of the two parameters W and H for each minimum. However, Isshiki observed that the choice is not so critical for obtaining useful results in large variable systems. Shafer proposed the use of large changes in design variables along the DLS direction to force the design out of a local minimum and to settle down in another valley with a different minimum. He called his algorithm the 'blow-up/settle-down (BUSD)' method [11].

21.2.4 Using Saddle Points of Defect Function in Design Space

The gradient of the defect function vanishes concurrently at all minima, all maxima, and all points of inflection or saddle points of the function in the feasible design space. Consequently, all three types of stationary or critical points play distinct roles in the topography of the defect function landscape in the multidimensional hyperspace. Indeed, critical points constitute distinct networks characterizing the defect function in the design space. In this context, the important role of saddle points can be appreciated by visualizing a mountain pass, the central point of which is an effective saddle point. Optimal passages between various valleys or ravines in the mountainous landscapes are provided by the mountain passes. Obviously, two local minima are located on opposite sides of a mountain pass, and the journey from the minimum in one valley to the minimum of a neighbouring valley is optimum when undertaken through the mountain pass linking the two valleys.

An important characteristic of critical points for which the Hessian matrix of the defect function has a non-zero determinant is the number of negative eigenvalues of the Hessian matrix at the critical point.

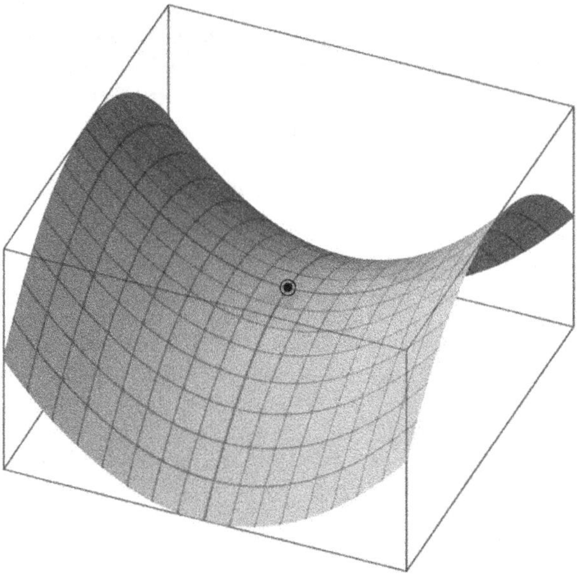

FIGURE 21.2 A saddle point S on a surface $z = f(x, y)$.

This number is called the Morse, or the Hessian index [29]. Along the direction defined by the eigenvector corresponding to the negative eigenvalue, the defect function is a maximum at the critical point. Intuitively, a two-dimensional saddle point is a minimum along a certain direction, and a maximum along the perpendicular direction (Figure 21.2). Critical points in the n-dimensional case have a set of mutually orthogonal directions. Along some of these directions, the critical points are a minimum, and along the other directions, called the downward directions, the critical points are maxima. For a particular critical point, the Morse index is the number of downward directions. Therefore, for minima and maxima, the Morse index is 0 and n, respectively, and saddle points have a Morse index between 1 and $(n-1)$. For a saddle point in the two-dimensional case, the Morse index is 1. In general, a saddle point of Morse index 1 in an n-dimensional case is a maximum along one direction, which can be visualized as the downward direction of a two-dimensional saddle point. Along each of the remaining $(n-1)$ orthogonal directions, the saddle point of Morse index 1 is a minimum, akin to the upward direction in the two-dimensional case.

If a local minimum is known, a new local minimum can be found by detecting saddle points of Morse index 1 in the vicinity of the known minimum. A local minimum in the neighbouring region can be obtained by the use of a local optimization algorithm on the other side of each of these saddle points. The process can be repeated in search of 'better' minima. This possibility of utilizing the saddle points for circumventing the limitations of local optimization procedures was identified in the early days of systematic investigations in global optimization [30–31]. Bociort et al. [5–6, 32] initiated investigations on the possibility of global optimization in optical system design by using saddle point detection in the vicinity of local minima. Later on, it was found that, without a priori information about them, detecting saddle points of Morse index 1 is computationally very expensive. They devised a new approach called 'saddle point construction' by inserting null lenses. Experiments with this technique provided quite encouraging results [33–37].

21.2.5 Using Parallel Plates in Starting Design

In the early days of automatic lens design by local optimization methods, Brixner [38] proposed a radically different approach. His method consisted of using as starting design a set of plane parallel glass plates, or a set of plates where one or two elements are given power and the rest are flat plates. During the process of optimization, the plates acquire power and the system evolves into a design that is a local optimum.

It is obvious that the method is highly sensitive to the number of glass plates, the thicknesses of the plates, the separations between them, the axial locations of the plates with respect to the object/image, etc.

For systems where the type of optimum structure is often known, use of non-specific structure for the sake of obtaining optimum solution, as above, can hardly be justified. However, for new types of systems this approach may provide useful solutions by repeating the process with different initial conditions. The latter would give rise to different solutions, with the possibility of a few good solutions.

Shafer commented:

> The reason this approach can yield high performance designs is that, because most of the surfaces start flat, the process has an excellent chance of stopping at a local minimum at which the surfaces have relatively long radii, compared to those at other minima. A local minimum with long radii generally will have much smaller higher-order aberrations, and therefore better performance, than a minimum that has any strong radii.
>
> [11]

21.3 Stochastic Global Optimization Methods

The 'Crisis of Dimensionality' associated with the problem of global optimization of optical lens design, discussed in Section 21.1, rules out the possibility of tackling the problem in a deterministic manner within any reasonable time frame, except for in trivial cases. The 'best' solution obtained by any of the heuristic methods discussed in the last section can at best be called a 'quasi-global' solution. An alternative option to tackle this problem is to take recourse to stochastic methods. As in many other fields of science and technology, investigators are exploring different stochastic algorithms for tackling the problem of global optimization of lens design. A glimpse of the major approaches that have been experimented with in lens design is given below.

21.3.1 Simulated Annealing

Simulated annealing, also called statistical cooling or Monte Carlo annealing can be regarded as a Monte Carlo improvement method. The method of simulated annealing can be traced back to Metropolis, who attempted to simulate the behaviour of an ensemble of atoms in equilibrium at a given temperature [39]. Significant improvements in the solution of combinatorial optimization problems were achieved through the simulation of 'annealing' using the Metropolis algorithm [40–43]. The algorithm was successful in tackling the problem of global optimization of multimodal functions of continuous variables [43–48]. A brief description of the algorithm is given below.

Let $f(\mathbf{x})$ be a function to be minimized. $f(\mathbf{x})$ is defined over an n-dimensional continuous space consisting of the variables $x \equiv (x_1, x_2, \ldots, x_n)$. Each variable x_j has its individual range defined by the finite continuous interval $a_j < x_j < b_j$. The simulated annealing algorithm proceeds iteratively.

Starting from a given initial point $\mathbf{x}^{(0)}$, it tends to the global minimum of the cost function through a succession of points $\mathbf{x}^{(1)}, \mathbf{x}^{(2)}, \ldots, \mathbf{x}^{(k)}, \mathbf{x}^{(k+1)}, \ldots$. At each iteration, a new candidate point is generated by making random step $\Delta\mathbf{x}^{(k)}$ from the current point $\mathbf{x}^{(k)}$. The value of the defect function is evaluated at the point $\left[\mathbf{x}^{(k)} + \Delta\mathbf{x}^{(k)}\right]$. Let the change in defect function be Δf. The candidate point is accepted or rejected according to the probability p given by

$$\left.\begin{aligned} p &= 1, && \text{for } \Delta f \le 0 \\ p &= \exp\left[-\frac{\Delta f}{T}\right], && \text{for } \Delta f > 0 \end{aligned}\right\} \tag{21.3}$$

This criterion is known as the Metropolis criterion. By analogy with statistical mechanics, the control parameter T is called 'temperature'. T is initially set at a high value, and slowly decreased. The random walk is undertaken at each value of T. Note that at large values of T, the acceptance probability of an uphill move is much larger than the same at lower values of T. The process is terminated at a small value of T when the acceptance probability of an uphill move becomes so small that no further uphill move is practically accepted. This implies that the point has reached near the global minimum.

In order to accelerate convergence of the procedure by preventing the tendency to leave the neighbourhood of the global minimum, Bohachevsky et al. proposed a modified form of acceptance probability given by

$$\left.\begin{aligned} p &= 1, && \text{for } \Delta f \le 0 \\ p &= \exp\left[-f^{-g}\frac{\Delta f}{T}\right], && \text{for } \Delta f > 0 \end{aligned}\right\} \tag{21.4}$$

where g is an arbitrary positive number. They called their method a 'generalized simulated annealing' (GSA) technique that is expected to deliver the global minimum by a lower number of iterations. Note that a too large value of g implies a significant departure from the basics of the Monte Carlo algorithm. Of course, for $g = 0$, (21.4) reduces to the Metropolis form of the Monte Carlo simulated annealing method. Generalized simulated annealing methods have been successfully used in lens design optimization problems [49–53]. Forbes et al. reported a heuristic approach that can be used as a guide to develop step generation and temperature control methods for improving efficiency of the simulated annealing algorithm [54–56]. They called their modified technique an 'adaptive simulated annealing' method. Hazra et al. reported the use of the GSA technique for global optimization at the structural design stage of lens design [57–58].

21.3.2 Evolutionary Computation

The origin of the field of evolutionary computation may be traced back to the investigations in late 1950s and early 1960s on the simulation of genetic systems by digital computers for developing novel procedures for optimization, machine learning, etc. [59–62]. By the mid-1960s, three main forms of evolutionary algorithms, 'Evolutionary Programming' (EP) [63], 'Genetic Algorithm' (GA) [64], and 'Evolution Strategies' (ES) [65–66], had emerged. Although all three approaches concern evolutionary procedures, there are subtle differences in approach and emphasis on detail.

Over the next 25 years, each of the three branches developed quite independently of each other, resulting in unique parallel developments. From the late 1980s, interaction among the various evolutionary algorithm communities increased significantly [67–68], so much so that it led to a consensus for a new name encompassing all related activities in this area—'Evolutionary Computation' [69]. However, often the term 'Genetic Algorithm (GA)' is used to represent the application of any form of evolutionary computation in engineering design [70–71].

A brief description of the basic features of GA in the context of the global optimization of lens design is given below.

(i) GA is an iterative optimization procedure. Instead of working with a single solution in each iteration, a GA works with a number of solutions (collectively known as a population) in each iteration.
(ii) The initial population may be generated randomly, or any problem-specific knowledge may be incorporated to initialize the population. Each individual of the population represents a point in the feasible solution space of the variables. An individual has a chromosomal representation, where a chromosome is a string of symbols. It is usually, but not necessarily, a bit string, i.e. a string of binary digits.
(iii) Three different genetic operators, namely, reproduction, crossover, and mutation, are applied sequentially to update the current population.
(iv) During the reproduction phase, trials or opportunities are allocated to individuals in proportion to their relative fitness values. Different types of selection procedures exist, e.g. roulette-wheel or tournament selection, etc.

(iv) Implementation of simple GA uses single point crossover, where the two mating chromosomes are each cut once at a randomly chosen crossover site, and sections after the cuts are swapped.
(v) Mutation is applied to each child after crossover. It randomly alters each gene with a very small probability.

Success of the GA depends to a large extent on the choice of population size, proper selection of probabilities of the genetic operators, retaining genetic diversity through iterations, etc. Further details may be obtained from the publications on the application of GA for global optimization in lens design [72–87]. A review presents a list of important publications in the field up to 2018 [88].

21.3.3 Neural Networks, Deep Learning, and Fuzzy Logic

Artificial neural networks (ANN), usually called 'neural networks' (NN) or connectionist systems, are computing systems inspired by the biological neural networks that constitute animal's brains. The NN, per se, is not an algorithm, but a computational structure to which several algorithms, e.g. backpropagation, etc., can be applied. The graded response neural network model, proposed by Hopfield [89–91], has been used to tackle the well-known combinatorial optimization problem, called the travelling salesman problem [92–93]. Weller [94] explored the possibility of utilizing NN in lens design. His observation was that the use of NN in lens optimization problems was going to be limited, for the NNs do not appear to be very useful for numerically sensitive problems. However, he noted that the application of neural nets to the problem of lens selection for starting design would be more successful. This observation is vindicated by recent research on the topic based on 'deep learning'. Deep learning is a machine learning method based on artificial neural networks with representation learning by methods like back propagation [95–96]. Côte et al. [97–98] and Yang et al. [99] reported their exploration of deep learning methods for starting designs in lens design optimization.

On the other hand, Macdonald et al. [100] mentioned that the mapping of an optical design optimization problem onto a neural net, in general, is difficult. However, he demonstrated the use of neural networks in solving a lens design optimization problem that cannot be tackled by other methods in a straightforward manner. The particular problem tackled by him using NN is essentially a combinatorial optimization problem where the chromatic aberration in a twenty element zoom lens system is minimized by selection of glass for each lens element from a finite set of prespecified optical glass types. The ability of neural networks to work with conflicting or incomplete, even fuzzy, data sets is currently being harnessed to tackle many challenging problems.

Although the origin of 'Fuzzy Logic' is sometimes traced back to the introduction of multiple-valued logic by J. Lukasiewicz in 1920 [101], it is widely accepted that 'Fuzzy Logic' emerged from the theory of fuzzy sets introduced by L. Zadeh in his seminal publication in 1965 [102].

Along with 'probabilistic reasoning', 'neural computing', and 'genetic algorithms', 'fuzzy logic' has emerged as a very important item in the arsenal of the practitioners of soft computing [103].

> Soft computing replaces the traditional time-consuming and complex techniques of hard computing with more intelligent processing techniques. The key aspect for moving from hard to soft computing is the observation that the computational effort required by conventional approaches which makes in many cases the problem almost infeasible, is a cost paid to gain a precision that in many applications is not really needed or, at least, can be relaxed without a significant effect on the solution.
>
> [104]

Hybrid techniques combining the pertinent useful features of the different algorithms are being experimented with to tackle complex problems. 'Fuzzy Logic Adaptive Genetic Algorithm' (FLAGA) and 'Adaptive-Neural-Network based Fuzzy Inference System' (ANFIS) are two examples which were utilized in problems of lens design optimization [105–108].

21.3.4 Particle Swarm Optimization

'Particle Swarm Optimization' (PSO) is a heuristic technique of search methodology, based on the notion of collaborative behaviour or swarming in groups of biological organisms, e.g. birds, bees, etc. The PSO is similar to techniques of evolutionary programming in the sense that both approaches are based on a random initial population, and the members share information with their neighbours among the population to perform the search using a combination of deterministic and probabilistic rules. However, instead of using the three classical genetic operators—selection, crossover, and mutation—each candidate solution termed 'particle' in PSO adjusts its trajectory in the objective space according to its own motion and the motion of its companions. The absence of the genetic operators in PSO makes it relatively easier to implement in practice. The algorithm was developed by Kennedy and Eberhart [109–110]. It was used for tackling problems of lens design optimization first by Qin [111–112]. Subsequently, a few other reports on the use of the PSO in tackling other problems of optical design appeared [113–114].

In basic PSO, a swarm system consisting of p particles is considered. Every position of each particle in the hyperspace of n variables is a candidate solution. In the search space, the particles change their position according to the law of inertia in such a manner that each individual particle tries to reach its most optimal position, and the swarm as a whole converges towards its most optimal position. The system is initialized with a population of random candidate solutions in the search space. The particles update their velocity and position in random direction in each iteration. Each particle remembers the position of the best fitness value achieved by it so far, and the associated solution is called 'pbest'; the solution with highest fitness value across all particles called 'gbest'. At the start of kth iteration, their coordinates in the design space are specified as:

$$\text{the 'pbest' for the ith particle } \overset{\text{pbest}}{x_{i,j}^{(1,k-1)}}, \quad j = 1,\ldots,n \tag{21.5}$$

$$\text{the 'gbest' for the swarm} \equiv \overset{\text{gbest}}{x_{j}^{(1,k-1)}}, \quad j = 1,\ldots,n \tag{21.6}$$

At the end of each iteration, 'pbest' and/or 'gbest' are updated if higher fitness value is achieved. The velocity and position are multidimensional vectors, and updates of them are governed by

$$v_{i,j}^{(k+1)} = \omega v_{i,j}^{k} + \Delta v_{i,j}^{k} \tag{21.7}$$

$$x_{i,j}^{(k+1)} = x_{i,j}^{k} + \Delta x_{i,j}^{k} \tag{21.8}$$

where

$$\Delta x_{i,j}^{k} = v_{i,j}^{(k+1)} \Delta k = v_{i,j}^{(k+1)} \tag{21.9}$$

$$\Delta v_{i,j}^{k} = c_1 r_{1,j} \left[\overset{\text{pbest}}{x_{i,j}^{(1,k-1)}} - x_{i,j}^{k} \right] + c_2 r_{2,j} \left[\overset{\text{pbest}}{x_{j}^{(1,k-1)}} - x_{i,j}^{k} \right] \tag{21.10}$$

Note that the updated velocity $v_{i,j}^{(k+1)}$ has three components: the inertial component $\{\omega v_{i,j}^{k}\}$, the cognitive component or the memory of the particle $\left\{ c_1 r_{1,j} \left[\overset{pbest}{x_{i,j}^{(1,k-1)}} - x_{i,j}^{k} \right] \right\}$, and the social component $\left\{ c_2 r_{2,j} \left[\overset{pbest}{x_{j}^{(1,k-1)}} - x_{i,j}^{k} \right] \right\}$. The three parameters lie ideally in the ranges $0 \le \omega \le 1.2$ and $0 \le c_1, c_2 \le 2$. $r_{1,j}, r_{2,j}$

are two uniformly distributed random numbers in the range $(0,1)$. For each j, they are regenerated for each velocity update.

More details on the PSO algorithm used in a related problem in optics may be found in publication [115].

21.4 Global Optimization by Nature-Inspired and Bio-Inspired Algorithms

In literature on optimization, the methods for global optimization enunciated in Sections 21.2 and 21.3, are said to be based on meta-heuristics.

> A meta-heuristic refers to a master strategy that guides and modifies other heuristics to produce solutions beyond those that are normally generated in a quest for local optimality. The heuristics guided by such a meta-strategy may be high level procedures or may embody nothing more than a description of available moves for transforming one solution into another, together with an associated evaluation rule.
>
> [116]

In this context, the numerical procedures for local optimization, described in Sections 20.2 and 20.3, are called simple heuristic procedures.

Generally, for global optimization, the adopted meta-heuristics are based on the iterative improvement of either a population of solutions, as in evolutionary algorithms, swarm-based algorithms, or a single solution, as in simulated annealing or tabu search. The meta-heuristics are mostly developed on the basis of critical observations of natural and biological events or phenomena. The corresponding algorithms are called nature-inspired or bio-inspired algorithms [117–120]. The two major categories of these algorithms are based on simulation of (i) the phenomenon of evolution, and (ii) the collective behaviour of swarms. Algorithms belonging to the first category are 'procedure based', as explained in Section 21.3.2, and those belonging to the second category are 'equation based', as described in Section 21.3.4. It is estimated that currently there are more than one hundred different nature-inspired algorithms and variants [119]. A partial list of some of the well-known algorithms is given below.

A. Iterative updating of single solution by random walk:
   - Generalized Simulated Annealing (GSA)
   - Adaptive Simulated Annealing (ASA)
   - Tabu Search (TS)

B. Procedure-based iterative algorithms related to evolution:
   - Genetic Algorithm (GA)
   - Genetic Programming (GP)
   - Evolution Strategies (ES)
   - Differential Evolution (DE)
   - Messy GA (MGA)
   - Structured GA (STGA)
   - Selfish Gene Algorithm (SGA)
   - Niching Algorithm (NA)
   - Immunity Genetic Algorithm (IGA)
   - Covariance Matrix Adaptation Evolution Strategy (CMAES)
   - Paddy Field Algorithm (PFA)
   - Scatter Search Algorithm (SSA)
   - Mind Evolutionary Computation (MEC)

C. Equation-based iterative algorithms related with swarm intelligence:
   - Particle Swarm Optimization (PSO)
   - Ant Colony Optimization (ACO)
   - Cuckoo Search Algorithm (CSA)
   - Artificial Bee Colony Algorithm (ABC)

- Fish Swarm Algorithm (FSA)
- Firefly Algorithm (FFA)
- Bat Algorithm (BA)
- PS2O Algorithm (modified PSO)
- Invasive Weed Colony Optimization (IWC)
- Flower Pollination Algorithm (FPA)
- Bacterial Foraging Algorithm (BFA)
- Artificial Immune System Algorithm (AIS)
- Group Search Optimizer (GSO)
- Shuffled Frog Leaping Algorithm (SFLA)
- Intelligent Water Drops Algorithm (IWD)
- Biogeography Based Optimization (BBO)
- Gravitational Search Algorithm (GSA)
- Black Hole Algorithm (BHA)
- Krill Herd Algorithm (KHA)
- Charged Particle System Algorithm (CPS)
- Eagle Strategy (EAS)

Hybrid algorithms combining two or more algorithms from the above are excluded from the list.

Results of 'Black-Box Optimization Benchmarking (BBOB2009)' as decided in 2009, of 31 real parameter optimization algorithms (from the above list) applied to 24 noiseless benchmark functions are presented in [121]. Investigations were restricted to 40 dimensions, as searching in larger dimensions and multimodal functions turned out to be more difficult. The choice of the best algorithm depends remarkably on the available budget of function evaluations.

Few algorithms from the above list have yet been experimented with for lens design optimization. Except for the applications already noted in Section 21.3, reports have appeared on applications of firefly algorithm [122], and of mind evolutionary computation (MEC) [123–124] in lens design optimization. MEC is an interesting algorithm where operations of 'similartaxis' and 'dissimilation' are employed instead of crossover and mutation operators in GA's. Similartaxis is defined as a process by which an individual becomes a winner by competing within the group. In MEC, the operator 'similartaxis' undertakes the local search, whereas the other operator 'dissimilation' undertakes the global search.

Houllier and Lépine [125] reported a comparative study of five optimization algorithms, the Particle Swarm Optimization, the Gravity Search Algorithm, the Cuckoo Search, the Covariance Matrix Adaptation Evolution Strategy, and the Nelder & Mead Simplex Search, for conventional and freeform optical design. They assessed the performance of these search algorithms using the Black Box Optimization Benchmarking Suite (BBOB2009). Simultaneously, they compared the performance of the five algorithms in optimizing one rotationally symmetric lens system, and a system using freeform optics. The performance of the algorithms was compared with that of a commercial optical design software. One of their concluding remarks: 'one should not expect a given search algorithm to outperform all the others for every problem and that it would be beneficial to select the correct algorithm for each problem'. This tallies well with the 'No free lunch theorems for optimization' put forward formally by Wolpert and Macready [126] two decades ago.

> The NFL theorem states that if an algorithm A can outperform another algorithm B for finding the optima of some objective functions, then there are some other functions on which B will outperform A. In other words, both A and B can perform equally well over all these functions if their performance is averaged over all possible problems or functions.
>
> [119]

Nevertheless, common observation of practitioners is that some algorithms seem to perform better than others in most problems. Some recent studies seemed to indicate that free lunches may exist, especially for multi-objective optimization [127], co-evolution [128], or continuous optimization [129]. The current status of the NFL theorems and their implications for metaheuristic optimization has been reviewed by Joyce and Hermann [130].

The emergence of these nature-inspired algorithms has opened up new frontiers in optical design. However, effective utilization of the algorithms depends on sorting out few basic issues. In practical problem solving, one is always concerned with the actual individual performance of solving a particular class of problems, not the averaged performance over all problems. Consequently, benchmarking with a finite set of algorithms and a finite set of functions is not very useful in many cases. Useful benchmarking should be tagged with specific characteristics of the particular class of problem. Next, each of these algorithms has some algorithm-dependent parameters, and the values of these parameters can affect the performance of the algorithm under consideration. However, it is not yet clear how to tune these parameters to achieve the best performance. A trial-and-error approach becomes necessary for each class of application. Proper theoretical analysis for understanding the stability, convergence, and robustness of the algorithms remains to be properly undertaken. Finally, even a moderately complex optical design problem involves a large number of variables. Therefore, the 'large scale' scalability of an algorithm needs to be ascertained, since an algorithm that works well for small scale problems is not automatically guaranteed to perform well in practically acceptable time scale in large scale problems.

21.5 A Prophylactic Approach for Global Synthesis

Notwithstanding the great achievements in global optimization of multimodal functions, as enunciated in the earlier sections, the scope for direct use of these methods in tackling the problem of optimization of optical design, in general, is somewhat limited. For simple systems with small numbers of variables, the methods can provide the desired solutions. As complexity in optical systems increases, the total number of variables and constraints tends to increase exponentially, so much so that it becomes practically impossible to obtain the desired solution within a reasonable time frame. Taking recourse to parallel computing may provide some relaxation, but the basic problem of large scale scalability still persists.

In order to circumvent this problem, a prophylactic approach for 'global synthesis' is undertaken. Note the change in the nature of the search. In this approach, the optimal solution is synthesized by using global search procedures for obtaining required characteristics of the individual components, and also for determining the composition of these components. Suitable prophylactic measures are adopted, wherever possible. This strategy of design makes use of the principle of paraxial optics, thin lens aberration theory, global optimization procedures, and a local optimization procedure, as discussed in Section 20.4. The strategy is being upgraded as and when new results are available. It has been reported earlier in a few publications [131–133]

The first step in this approach involves determination of the required paraxial characteristics and Petzval curvature (in the case of extended field imagery) to satisfy the problem specifications. This involves understanding the optics requirements on the basis of user-designer interaction. Step two involves working out the structural layout of one or more lens systems that may cater to the required paraxial specifications for achieving the imaging goals. At this stage, the lens system is assumed to consist of one or more thin components. Determining structural layout implies determination of the number of thin lens components, their individual powers, inter-lens separations, axial location of the stop, the diameters of the components, etc. (Figure 21.3). In the case of 'ab initio' design, these are determined analytically by using theories of paraxial optics. Often, multiple layouts provide alternative solutions. The basic flow diagram pertaining to this strategy is given in Figure 21.4, which displays the pertinent features. The basic strategy is to reduce any formidable design problem with large degrees of freedom into a set of sub-problems with considerably lower degrees of freedom. These sub-problems can be conveniently tackled within a reasonable time budget.

Step three in the design procedure involves determination of optimal values for central aberrations—the primary spherical aberration, the primary coma (with stop located on the thin component), and the longitudinal chromatic aberration—for each of the components to achieve a prespecified set of values for primary aberrations of the overall system. This task can be formulated as an optimization problem. It was

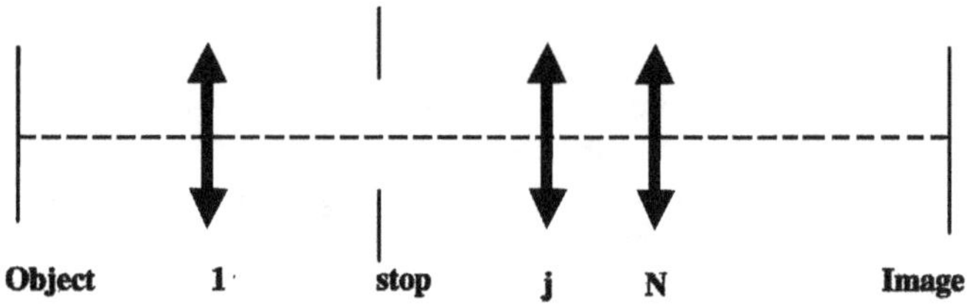

FIGURE 21.3 Structural layout of an N component thin lens system.

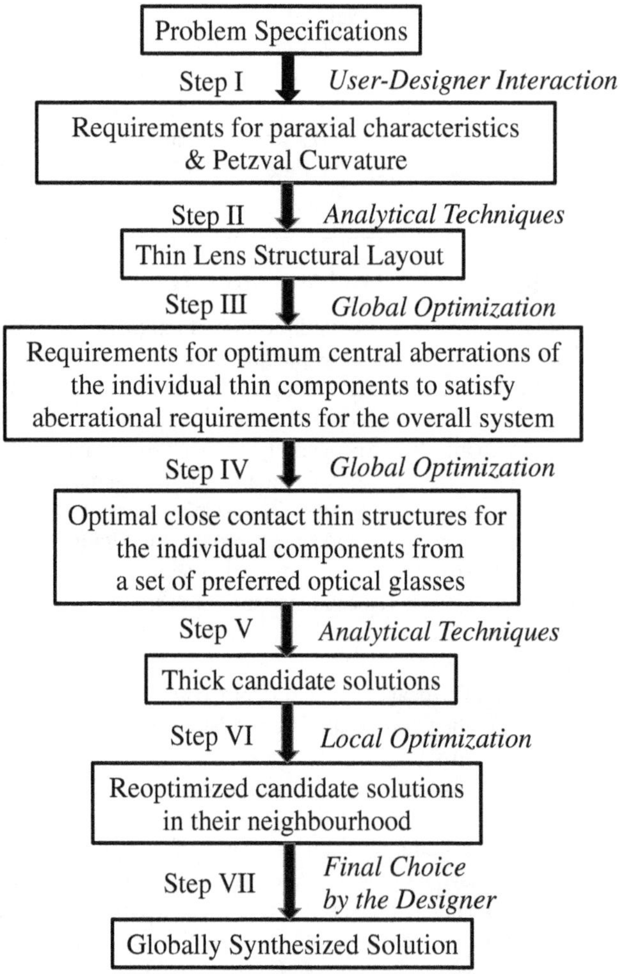

FIGURE 21.4 Basic flowchart for prophylactic lens design.

tackled initially by DLS optimization [134]. Subsequently, global optimization methods like simulated annealing and genetic algorithm were utilized for better results [58, 135].

Step four involves determination of the structure of one or more thin elements that can provide the required central aberrations and paraxial characteristics of each of the constituent components of the system. Selection of glass for each of the elements is taken care of by the optimization procedure implemented for determining the structure of the elements of the thin components. The optimal glass for each element is determined from a set of prespecified preferred optical glasses during implementation of the optimization procedure. For each component, search for optimum structure commences from

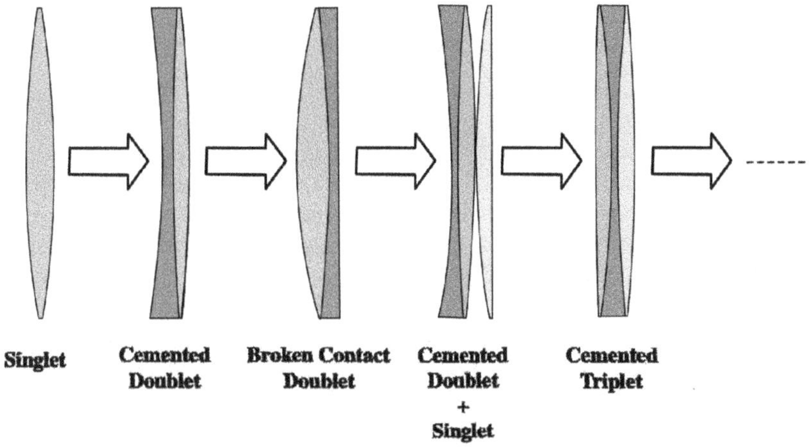

FIGURE 21.5 Synthesis of a component by structures of increasing complexity.

the simplest structure, i.e. the thin singlet to thin multi-element structures in close contact, such as the cemented doublet, the broken contact doublet, the cemented triplet, etc., and combination of different thin elements, like a singlet and a doublet, or two doublets in close contact (Figure 21.5). The search for optimum singlet can be undertaken analytically [136]. Initially, the DLS algorithm was used in the search for the optimum doublet [137]. Later on, generalized simulated annealing was used in search of a global solution [57]. It was soon realized that strict adherence to aberration targets at the thin lens design stage is not very meaningful, the aberration targets for the components and elements should be allowed to 'float', and the resulting set of quasi-global optima may be retained for further scrutiny at a later stage. This idea prompted utilization of population updating type global optimization algorithms, and a series of investigations on the use of genetic algorithms for implementing step three and step four of this approach was undertaken [79–80, 82–83, 138–139]. Incidentally, the use of one or more aspheric or diffractive surfaces may be explored during implementation of steps three and four.

At the end of step four, a few candidate thin lens solutions emerge. Step five involves the thickening of these solutions. It is obvious that both the paraxial characteristics and the aberrations will change significantly when the thin elements are thickened. But, suitable alterations in curvatures and axial locations of the elements can annul the changes in primary aberrations and in paraxial characteristics. Suitable analytical procedures for thickening have been discussed in Section 14.7.

Each candidate solution obtained after step five is reoptimized in its neighbourhood with the help of a local optimization algorithm, e.g. the DLS optimization procedure described in Section 20.4. (Step six). In step seven, the globally synthesized solution for the given problem is identified from the set of available options from step six. This is based on considerations like cost, weight, ease of manufacture, ergonomics, optomechanical compatibility, etc.

It may be noted that our approach is restricted to primary aberrations. It is possible to extend the analysis to aberrations of higher order, e.g. the secondary aberrations. But the presence of both intrinsic and extrinsic aberrations in the case of secondary aberrations will not make the thin lens treatment meaningful for most practical purposes. Rather, prophylactic steps for preventing the occurrence of large amounts of secondary aberrations are undertaken at the two stages of implementation of global optimization. In step three, specific upper limits for central aberrations of the components for the acceptable solutions are set. In step four, while composing structures of the individual elements, upper limits for aperture utilization ratio (see Section 19.4.5) are set for each refracting surface. Finite ray aberrations are properly accounted for during implementation of final local optimization with defect function using diffraction-based image evaluation parameters.

It should be noted that the meta-heuristics adopted in this prophylactic search for 'quasi-global' solutions in optimization of lens systems can circumvent the problem of 'large scale' scalability in practical applications.

21.6 Multi-Objective Optimization and Pareto-Optimality

The approach for lens design optimization enunciated so far involves optimization of a single objective function characterizing quality of the optical system. This function, commonly known as the defect/error/merit function in the field of optical design, is a function of the constructional variables of the system. It is estimated that lower the value of the objective function, the better the quality of the system.

The single objective function in lens design is formed by the weighted summation of different functions representing different objectives of the optical system. The latter objectives refer to the quality of imaging at different field points, both chromatic and monochromatic, and also to the satisfaction of different equality and inequality constraints on the paraxial characteristics of the imaging system and also on the feasibility of construction variables (for a representative example see Section 20.3.2). In reality, the problem of lens design optimization is one of multi-objective (MO) optimization. The weighting factors for the different objectives in the single objective (SO) function are to be decided at the outset of the optimization procedure. In optimization theory, this process of converting an MO optimization problem into an SO optimization problem is called 'scalarization', since the objective vector consisting of multiple objectives is transformed into a single scalar function by this process.

It is obvious that, in this approach, not only will the optimal solution be different when the weighting factors for the multiple objectives are changed, but the corresponding values for the multiple objectives will also be changed. On account of the inherent nonlinear relationship between the objectives and the construction variables, the choice of appropriate weighting factors poses a veritable problem in practice. Therefore, the designer is constrained by his intuition on relative weights on the objectives, and often tries to obtain a desired optimum solution by taking recourse to a trial-and-error approach. A remedy is sought via taking recourse to multi-objective optimization [140–145].

A general MO problem requiring the optimization of m objectives may be formulated as:

$$\text{Minimize } \boldsymbol{\Psi} \equiv \boldsymbol{\Psi}(\mathbf{x}) = \left[\Psi_1(\mathbf{x}), \Psi_2(\mathbf{x}), \ldots, \Psi_m(\mathbf{x})\right]^T \tag{21.11}$$

where each objective function $\Psi_j(\mathbf{x}), (j = 1, \ldots, m)$ of the objective vector $\boldsymbol{\Psi}$ is sought to be minimized. The constructional variable vector (in optimization terminology 'decision vector') $\mathbf{x}$ is represented by

$$\mathbf{x} = (x_1, x_2, \ldots, x_n)^T \in \Omega \tag{21.12}$$

where Ω is the hyperspace of construction variables. The space spanned by the objective vectors is called the objective space.

The utopian solution $\mathbf{x}_{uto}$ is the solution that minimizes all objectives simultaneously. It may be represented by

$$\mathbf{x}_{uto} \in \Omega : \forall \mathbf{x} \in \Omega, \quad \Psi_j(\mathbf{x}_{uto}) \le \Psi_j(\mathbf{x}) \quad \text{for } j \in \{1, 2, \ldots, m\} \tag{21.13}$$

where $\in$ means 'is an element of', ':' means 'such that', $\forall$ means 'for all'. For $m = 1$, the MO problem reduces to an SO problem. The utopian solution is the global optimum. For $m \ge 1$, the utopian solution does not exist in general, since the individual objective functions $\Psi_j(\mathbf{x}), (j = 1, \ldots, m)$ are typically conflicting. Instead of a single solution, a set of solutions representing different trade-offs among the individual objectives are available.

The characteristic features of dominance of one solution over the other for developing systematic treatment of multi-objective optimization was unfolded by an Italian named Vilfredo Pareto in a different context in the field of political economy [146]. He introduced two new concepts, which are now called Pareto-dominance and Pareto-optimality [147]. In the context of minimization, a solution vector $\mathbf{u} = [u_1, u_2, \ldots, u_n]^T$ is said to 'Pareto-dominate' another solution vector $\mathbf{v} = [v_1, v_2, \ldots, v_n]^T$, if and only if,

$$\left.\begin{array}{ll}\forall j \in (1, m), & \Psi_j(\mathbf{u}) \leq \Psi_j(\mathbf{v}) \\ \text{and } \exists j \in (1, m): & \Psi_j(\mathbf{u}) < \Psi_j(\mathbf{v})\end{array}\right\} \tag{21.14}$$

where '$\exists$' means 'there exists'. The relation (21.14) implies that $\boldsymbol{\Psi}(\mathbf{u})$ is better than $\boldsymbol{\Psi}(\mathbf{v})$ for all the constituent objectives, and there is one objective function for which $\boldsymbol{\Psi}(\mathbf{u})$ is strictly better than $\boldsymbol{\Psi}(\mathbf{v})$. This concept is utilized to compare and rank the solution vectors in the context of multi-objective optimization.

A solution $\mathbf{x}^*$ is said to be 'Pareto optimal' if and only if there does not exist another solution that dominates it. From (21.14) it follows that no change in the Pareto optimal solution $\mathbf{x}^*$ can improve any one of the objectives without adversely affecting one other objective. The objective vector $\boldsymbol{\Psi}(\mathbf{x}^*)$ corresponding to the Pareto optimal solution $\mathbf{x}^*$ is called a Pareto dominant vector, or non-inferior or non-dominated vector. On account of the inherent conflict among at least some of the objectives of the multi-objective problem, there would be several Pareto optimal solutions. These solutions constitute a Pareto optimal set. The corresponding objective vectors are said to be on the Pareto front or Pareto frontier in the objective space.

A knowledge of the Pareto fronts pertaining to real life problems of multi-objective optimization facilitates informed decision making by providing ready visualization of the scope for the attainment of conflicting objectives [148]. No analytical technique for evaluating the Pareto front exists. For relatively simpler problems, the MO problem can be transformed into an SO problem by using one of the scalarization methods, e.g. the weighted aggregation approach, and an estimate of the Pareto front can be obtained by carrying out SO optimization with appropriate combinations of weighting factors. In general, several population-based meta-heuristic algorithms have recently emerged to tackle this problem [149]. Recently, Albuquerque et al. have initiated the use of Pareto fronts in a multi-objective approach developed for optical design [150–152].

21.7 Optical Design and Practice of Medicine

It is instructive to look into the analogy between the process of optical design and the practice of medicine [153]. Table 21.2 underscores the major points of this analogy. In the field of medicine, the role of

TABLE 21.2

Analogy between the Practice of Medicine and Optical Design

Medicine	Optical Design
1. Anatomy	a. Ray Tracing
	b. Basic Laws of Ray Optics
2. Diagnosis	a. Gaussian Optics
	b. Seidel Optics
	c. Finite Ray (Wave) Aberrations
	d. Spot Diagram
	e. PointSpreadFunction; LineSpreadFunction; EdgeSpreadFunction; Bar Spread Function; OpticalTransferFunction
3. Therapy	a. Symmetry Considerations
	b. Power distribution
	c. Axial location of Stop
	d. Addition/Subtraction of Elements/Components
	e. …..
4. Prophylaxis	a. Use of quasi-globally optimum starting design
	b. Choice of suitable candidate solutions for components and elements with low probability of occurrence of large higher order aberrations

suitable prophylactic measures is well-established, not only for preventing occurrence of diseases, but also for facilitating therapeutic measures in the treatment of the patient. Similarly, efficiency of optimization algorithms for lens design can be significantly improved, if suitable prophylactic measures are properly taken into account in formulation of the metaheuristics utilized in development of these algorithms.

REFERENCES

1 M.R. Garey and D.S. Johnson, *Computers and Intractability: A Guide to the Theory of NP-Completeness*, W.H. Freeman, San Francisco (1979).
2 P.E. Gill, W. Murray, and M.H. Wright, *Practical Optimization*, Academic, New York (1981).
3 T.G. Kuper and T.I. Harris, 'A new look at global optimization for optical design', *Phot. Spectra*, January 1992, pp. 151–160.
4 D. Sturlesi and D.C. O'Shea, 'The search for global minimum in optical design', *Proc. SPIE*, Vol. 1168 (1989) pp. 92–106.
5 E. van Driel, F. Bociort, and A. Serebriakov, 'Topography of the merit function landscape in optical system design', *Proc. SPIE*, Vol. 5249 (2004) pp. 353–363.
6 F. Bociort, E. van Driel, and A. Serebriakov, 'Networks of local minima in optical system optimization', *Opt. Lett.*, Vol. 29 (2004) pp. 189–191.
7 I. Agurok, 'Multi-extremum optimization in lens design: navigation through merit function valleys maze', *arXiv preprint arXiv.1907.08676* (2019) pp. 1–24.
8 M.E. Harrigan, 'New approach to the optimization of lens systems', *Proc. SPIE*, Vol. 237 (1980) pp. 66–74.
9 G. Forbes and A.E.W. Jones, 'Global optimization in lens design', *Opt. Phot. News*, March 1992, pp. 23–29.
10 T.G. Kuper and T.I. Harris, 'Global optimization for lens sesign: an emerging technology', *Proc. SPIE*, Vol. 1780 (1993) pp. 14–28.
11 D. Shafer, 'Global optimization in optical design', *Computers in Phys.*, Vol. 8 (1994) pp. 188–195.
12 L.C.W. Dixon and G.P. Szegö, Eds., *Towards Global Optimization*, North-Holland, Amsterdam (1975).
13 L.C.W. Dixon and G.P. Szegö, Eds., *Towards Global Optimization 2*, North-Holland, Amsterdam (1978).
14 A. Yabe, *Optimization in Lens Design*, SPIE SPOTLIGHT, Vol. SL 36, Bellingham, Washington (2018).
15 C.C. Reichert, T. Gruhonjic, and A. Herkommer, 'Development of an open source algorithm for optical system design, combining genetic and local optimization', *Opt. Eng.*, Vol. 59 (2020) 055111–1 to 055111–13.
16 F. Bociort, A. Serebriakov, and J. Braat, 'Local optimization strategies to escape from poor local minima' *Proc. SPIE*, Vol. 4832 (2002) pp. 218–225.
17 A. Serebriakov, F. Bociort, and J. Braat, 'Finding new local minima by switching merit functions in optical system optimization', *Opt. Eng.*, Vol. 44 (2005) 100501–1 to 100501–3.
18 D. Sturlesi and D.C. O'Shea, 'Future of global optimization in optical design', *Proc. SPIE*, Vol. 1354 (1990) pp. 54–68.
19 D. Sturlesi and D.C. O'Shea, 'Global view of optical design space', *Opt. Eng.*, Vol. 30 (1991) pp. 207–218.
20 C.C. Meewella and D.Q. Mayne, 'Efficient domain partitioning algorithms for global optimization of rational and Lipschitz continuous functions', *J. Optimization Theory and Applic.*, Vol. 61 (1989) pp. 247–270.
21 S. Banerjee and L.N. Hazra, 'Experiments with GSA techniques in structural design of doublet lenses', *Proc. SPIE*, Vol. 3482 (1998) pp. 126–134.
22 G. Elsner, 'A new sequential cluster algorithm for optical lens design', *J. Optimization Theory and Applic.*, Vol. 59 (1988) pp. 165–172.
23 M. Walk and J. Niklaus, 'Some remarks on computer-aided design of optical lens systems', *J. Optimization Theory and Applic.*, Vol. 59 (1988) pp. 173–181.
24 M. Isshiki, H. Ono, and S. Nakadate, 'Lens design: an attempt to use 'escape function' as a tool in global optimization', *Opt. Rev.*, Vol. 2 (1995) pp. 47–51.

25 M. Isshiki, H. Ono, K. Hiraga, J. Ishikawa, and S. Nakadate, 'Lens design: global optimization with escape function', *Opt. Rev.*, Vol. 2 (1995) pp. 463–470.

26 M. Isshiki, 'Global optimization with escape function', *Pro. SPIE*, Vol. 3482 (1998) pp. 104–109.

27 M. Isshiki, L. Gardner, and G. Groot Gregory, 'Automated control of manufacturing sensitivity during optimization', *Proc. SPIE*, Vol. 5249 (2004) pp. 343–352.

28 M. Isshiki, D.C. Sinclair, and S. Kaneko, 'Lens design: global optimization of both performance and tolerance sensitivity', *Proc. SPIE*, Vol. 6342 (2007) 63420N-1 to 63420N-10.

29 J. Hart, 'Morse Theory for Implicit Surface Modeling', in Mathematical Visualization, Eds. H. -C. Hege and K. Polthier, Springer, Berlin (1998) p. 257.

30 G. Treccani, L. Trabattoni, and G.P. Szegö, 'A Numerical Method for the Isolation of Minima', in *Minimisation Algorithms, Mathematical Theories and Computer Results*, Ed. G.P. Szegö, Academic, New York (1972) pp. 239–255.

31 D. Corles, 'On regions of attraction as a basis for global minimization', in Ref. 12, pp. 55–95.

32 A. Serebriakov, 'Optimization and analysis of deep UV imaging systems', Ph. D. Thesis, TU Delft, Netherlands (2005).

33 F. Bociort and M. van Turnhout, 'Finding new local minima in lens design landscapes by constructing saddle points', *Opt. Eng.*, Vol. 48 (2009) 063001.

34 I. Livshits, Z. Hou, P. van Grol, Y. Shao, M.V. Turnhout, P. Urbach, and F. Bociort, 'Using saddle points for challenging optical design tasks', *Proc. SPIE*, Vol. 9192 (2014) 919204-1 to 919204-8.

35 M. van Turnhout, P. van Grol, F. Bociort, and H.P. Urbach, 'Obtaining new local minima in lens design by constructing saddle points', *Opt. Exp.*, Vol. 23 (2015) pp. 6679–6691.

36 Z. Hou, I. Livshits, and F. Bociort, 'One dimensional searches for finding new lens design solutions efficiently', *Appl. Opt.*, Vol. 55 (2016) pp. 10449–10456.

37 Z. Hou, I. Livshits, and F. Bociort, 'Practical use of saddle-point construction in lens design', *Proc. SPIE*, Vol. 10690 (2018) 1069007-1 to 1069007-10.

38 B. Brixner, 'Automatic lens design for nonexperts', *Appl. Opt.*, Vol. 2 (1963) pp. 1281–1286.

39 N. Metropolis, A.W. Rosenbluth, M.N. Rosenbluth, and A.H. Teller, 'Equation of state calculations by fast computing machines', *J. Chem. Phys.*, Vol. 21 (1953) pp. 1087–1092.

40 S. Kirkpatrick, C.D. Gelatt, Jr., and M.P. Vecchi, 'Optimization by simulated annealing', *Science*, Vol. 220 (1983) pp. 671–679.

41 E. Bonomi and J.–L. Lutton, 'The N-city travelling salesman problem: statistical mechanics and the metropolis algorithm', *SIAM Review*, Vol. 26 (1984) pp. 551–568.

42 V. Černy, 'Thermodynamical approach to the traveling salesman problem: an efficient simulation algorithm', *J. Opt. Theor. Appln.*, Vol. 45 (1985) pp. 41–51.

43 E.H. Aarts and P.J.M. van Laarhoven, 'Statistical cooling: a general approach to combinatorial optimization problems', *Philips J. Res.*, Vol. 40 (1985) pp. 193–226.

44 D. Vanderbilt and S.G. Louie, 'A Monte Carlo simulated annealing approach to optimization over continuous variables', *J. Comput. Phys.*, Vol. 56 (1984) pp. 259–271.

45 I.O. Bohachevsky, M.E. Johnson, and M.L. Stein, 'Generalized simulated annealing for function optimization', *Technometrics*, Vol. 28 (1986) pp. 209–217.

46 A. Corana, M. Marchesi, C. Martini, and S. Ridella, 'Minimizing multimodal functions of continuous variables with the "simulated annealing" algorithm', *ACM Trans. Math. Software*, Vol. 13 (1987) pp. 262–280.

47 P.J.M. van Laarhoven and E.H.L. Aarts, *Simulated Annealing: Theory and Applications*, D. Reidel, Holland (1987).

48 A. Dekkers and E. Aarts, 'Global optimization and simulated annealing', *Mathematical Programming*, Vol. 50 (1991) pp. 367–393.

49 I.O. Bohachevsky, V.K. Viswanathan, and G. Woodfin, 'An "intelligent" optical design program', *Proc. SPIE*, Vol. 485 (1984) pp. 104–112.

50 V.K. Viswanathan, I.O. Bohachevsky, and T.P. Cotter, 'An attempt to develop an "intelligent" lens design program, *Proc. SPIE*, Vol. 554 (1985) pp. 10–17.

51 G.K. Hearn, 'Design optimization using generalized simulated annealing', *Proc. SPIE*, Vol. 818 (1987) pp. 258–264.

52 S.W. Weller, 'Simulated annealing: what good is it?' *Proc. SPIE*, Vol. 818 (1987) pp. 265–274.

53 G.K. Hearn, 'Practical use of generalized simulated annealing optimization on microcomputers', *Proc. SPIE*, Vol. 1354 (1990) pp. 186–197.
54 G.W. Forbes and A.E.W. Jones, 'Towards global optimization with adaptive simulated annealing', *Proc. SPIE*, Vol. 1354 (1990) pp. 144–153.
55 A.E.W. Jones, *Global Optimization in Lens Design*, Ph. D. Thesis, University of Rochester, New York (1992).
56 A.E.W. Jones and G.W. Forbes, 'An adaptive simulated annealing algorithm for global optimization over continuous variables', *J. Global Optim.*, Vol. 6 (1995) pp. 1–37.
57 S. Banerjee and L.N. Hazra, 'Simulated annealing with constrained random walk for structural design of doublet lenses', *Opt. Eng.*, Vol. 37 (1998) pp. 3260–3267.
58 S. Banerjee and L.N. Hazra, 'Thin lens design of Cooke triplet lenses: application of a global optimization technique', *Proc. SPIE*, Vol. 3430 (1998) pp. 175–183.
59 A.S. Fraser, 'Simulation of genetic systems by automatic digital computers', *Aust. J. Biol. Sci.*, Vol. 10 (1957) pp. 484–499.
60 R.M. Friedberg, 'A learning machine, part I', *IBM J.*, Vol. 2 (1958) pp. 2–13.
61 R.N. Friedberg, B. Dunham, and J.H. North, 'A learning machine, part II', *IBM J.*, Vol. 3 (1959) pp. 282–287.
62 H.J. Bremermann, 'Optimization through Evolution and Recombination', in *Self Organizing Systems*, Eds. M.C. Yavits, G.T. Jacobi, and G.D. Goldstein, Spartan, Washington, D.C. (1962) pp. 93–106.
63 L.J. Fogel, A.J. Owens and M.J. Walsh, 'Artificial Intelligence through simulated evolution', Wiley, New York (1966) pp. 93–106.
64 J.H. Holland, 'Genetic algorithms and the optimal allocation of trials', *SIAM J. Comput.*, Vol. 2 (1973) pp. 88–105.
65 I. Rechenberg, *Evolutionsstrategie: Optimierung technischer Systeme nach Prinzipien der Biologischen Evolution*, Frommann-Holzborg, Stuttgart (1973).
66 H.P. Schwefel, *Evolutionsstrategie und numerische Optimierung*, Ph. D. Thesis, Technische Universität, Berlin (1975).
67 L. Davis, Ed., *Genetic Algorithms and Simulated Annealing*, Pitman, London (1987).
68 D.E. Goldberg, *Genetic Algorithms in Search, Optimization and Machine Learning*, Addison-Wesley, Reading, Massachusetts (1989).
69 Z. Michalewicz, *Genetic Algorithms + Data Structures = Evolution Programs*, Springer, Berlin (1992).
70 M. Mitchell, *An Introduction to Genetic Algorithms*, The MIT Press, Cambridge, Massachusetts (1996).
71 M. Gen and R. Cheng, *Genetic Algorithms and Engineering Design*, John Wiley, New York (1997).
72 E. Betensky, 'Postmodern lens design', *Opt. Eng.*, Vol. 32 (1993) pp. 1750–1756.
73 D.C. van Leijenhorst, C.B. Lucasius, and J.M. Thijssen, 'Optical design with the aid of a genetic algorithm', *BioSystems*, Vol. 37 (1996) pp. 177–187.
74 X. Chen and K. Yamamoto, 'An experiment in genetic optimization in lens design', *J. Mod. Opt*, Vol. 44 (1997) pp. 1693–1702.
75 I. Ono, S. Kobayashi, and K. Yoshida, 'Global and multi-objective optimization for lens design by real coded genetic algorithms', *Proc. SPIE*, Vol. 3482 (1998) pp. 110–121.
76 L. Hazra and S. Banerjee, 'Genetic algorithm in structural design of Cooke triplet lenses', *Proc. SPIE*, Vol. 3737 (1999) pp. 172–179.
77 K.E. Moore, 'Algorithm for global optimization of optical systems based on genetic competition', *Proc. SPIE*, Vol. 3780 (1999) pp. 40–47.
78 D. Vasiljević, 'Optimization of the Cooke triplet with the various evolution strategies and the damped least squares', *Proc. SPIE*, Vol. 3780 (1999) pp. 1–9.
79 S. Banerjee and L.N. Hazra, Experiments with a genetic algorithm for structural design of cemented doublets with prespecified aberration targets, *Appl. Opt.*, Vol. 40 (2001) pp. 6265–6273.
80 S. Banerjee and L.N. Hazra, Structural design of broken contact doublets with prespecified aberration targets using genetic algorithm, *J. Mod. Opt.*, Vol. 49 (2002) pp. 1111–1123.
81 D. Vasiljević, Classical and Evolutionary algorithms in the optimization of optical systems, Springer, New York. [Originally published by Kluwer Academic Publishers in 2002].
82 S. Chatterjee and L.N. Hazra, *Structural design of cemented triplets by genetic algorithm, Opt. Eng.*, Vol. 43, (2004) pp. 432–441.

83 S. Chatterjee and L.N. Hazra, Optimum values for genetic operators in evolutionary optimization of structural design of cemented doublets, *J. Opt. (India)*, Vol. 33 (2004) pp. 109–118.
84 S. Thibault, C. Gagné, J. Beaulieu, and M. Parizeau, 'Evolutionary algorithms applied to lens design: case study and analysis', *Proc. SPIE*, Vol. 5962 (2005) 596209.
85 Y.C. Fang and J. Macdonald, 'Optimizing chromatic aberration calibration using a novel genetic algorithm', *J. Mod. Opt.*, Vol. 53 (2006) pp. 1411–1427.
86 Y.C. Fang, C.M. Tsai, J. Macdonald, and Y.C. Pai, 'Eliminating chromatic aberration in Gauss-type lens design using a novel genetic algorithm', *Appl. Opt.*, Vol. 46 (2007) pp. 2401–2410.
87 C. Gagné, J. Beaulieu, M. Parizeau, and S. Thibault, 'Human-competitive lens system design with evolution strategies', *Appl. Soft Comput.*, Vol. 8 (2008) pp. 1439–1452.
88 K. Höschel and V. Lakshminarayanan, 'Genetic algorithms for lens design: a review', *J. Opt. (India)*, Vol. 48 (2019) pp. 134–144.
89 J.J. Hopfield, 'Neural networks and physical systems with emergent collective computational abilities', *Proc. Natl. Acad. Sci., USA*, Vol. 79 (1982) pp. 2554–2556.
90 J.J. Hopfield, 'Neurons with graded response have collective computational properties like those of two-state neurons', *Proc. Natl. Acad. Sci. USA*, Vol. 81 (1984) pp. 3088–3092.
91 J.J. Hopfield and D.W. Tank, 'Neural computation of decision in optimization problems', *Biol. Cybern.*, Vol. 52 (1985) pp. 141–152.
92 G.V. Wilson, 'On the stability of the travelling salesman problem algorithm of Hopfield and Tank', *Biol. Cybern.*, Vol 58 (1988) pp. 63–70.
93 S.U. Hegde, J.L. Sweet, and W.B. Levy, 'Determination of parameters in a Hopfield-Tank computational network', *Proc. IEEE 2nd Annual Conf. on Neural Nets, II* (1988) pp. 291–298.
94 S.W. Weller, 'Neural network optimization, components and design selection', *Proc. SPIE*, Vol. 1354 (1990) pp. 371–378.
95 Y. Le Cun, Y. Bengio, and G. Hinton, 'Deep learning', *Nature* Vol. 521 No. 7553 (2015) pp. 436–444.
96 J. Patterson and A. Gibson, *Deep Learning: A Practitioner's Approach,* O'Reilly Media, Sebastopol, California (2017).
97 G. Côté, J.-F. Lalonde, and S. Thibault, 'Toward Training a Deep Neural Network to Optimize Lens Designs', in *Frontiers in Optics / Laser Science*, (Optical Society of America, 2018), p. JW4A.28.
98 G. Côté, J.-F. Lalonde, and S. Thibault, 'Extrapolating from lens design databases using deep learning', *Opt. Exp.*, Vol. 27 (2019) pp. 28279–28292.
99 T. Yang, D. Cheng, and Y. Wang, 'Direct generation of starting points for freeform off-axis three-mirror imaging system design using neural network based deep learning', *Opt. Exp.*, Vol. 27 (2019) pp. 17228–17238.
100 J. Macdonald, A.J. Breese, and N.L. Hanbury, 'Optimization of a lens design using a neural network', *Proc. SPIE*, Vol. 1965 (1993) pp. 431–442.
101 J. Lukasiewicz, 'O Logice Trójwartościowej', *Ruch Filozoficzny*, Vol. 5 (1920) pp. 170–171. [English Translation 'On Three Valued Logic', in *Polish Logic 1920 – 1939*, Ed. S. McCall, Clarendon, Oxford (1967) pp. 16–18.]
102 L.A Zadeh, 'Fuzzy sets', *Information Control*, Vol. 8 (1965) pp. 338–353.
103 L.A. Zadeh, 'Fuzzy logic, neural networks and soft computing', *Commun. ACM*, Vol. 37 (1994) pp. 77–84.
104 I.P. Cabrera, P. Cordero, and M. Ojeda-Aciego, 'Fuzzy Logic, Soft Computing, and Applications', in *Bio-Inspired Systems: Computational and Ambient Intelligence.*, Eds. J. Cabestany, F. Sandoval, A. Prieto, J.M. Corchado, Lecture Notes in Computer Science, Vol 5517, Springer, Berlin (2009) pp. 236–244.
105 I.P. Agurok, A.A. Kostrzewski, and T.P. Jannson, 'Fuzzy-logic adaptive genetic algorithm (FLAGA) in optical design', *Proc. SPIE*, Vol. 4787 (2002) pp. 179–185.
106 J.-H. Sun and B.-R. Hsueh, 'Optical design and multivariate optimization with fuzzy method for miniature zoom optics', *Opt. Lasers Eng.*, Vol. 49 (2011) pp. 962–971.
107 J.–S.R. Jang, 'ANFIS: adaptive-neural-network based fuzzy inference systems', *IEEE Trans. Syst. Man Cybern.*, Vol. 23 (1993) pp. 665–685.
108 D. Petković, N.T. Pavlović, S. Shamshirband, M.L. Mat Kiah, N. Badrul Amur, and M.Y. Idna Idris, 'Adaptive neuro-fuzzy estimation of optimal lens system parameters', *Opt. Lasers Eng.*, Vol. 55 (2014) pp. 84–93.

109 J. Kennedy and R.C. Eberhart, 'Particle swarm optimization', *Proc. 4th Int. Conf. Neural Networks, Perth, Western Australia*, IEEE Service Center, Piscataway, New Jersey (1995) pp. 1942–1948.
110 R.C. Eberhart and Y. Shi, 'Particle Swarm Optimization: Developments, Application and Resources', in *Proc. Congress on Evolutionary Computation 2000, Seoul, Korea*, IEEE Service Center, Piscataway, New Jersey (2001) pp. 81–86.
111 H. Qin, 'Particle swarm optimization applied to automatic lens design', *Opt. Commun.*, Vol. 284 (2011) pp. 2763–2766.
112 H. Qin, 'Aberration correction of a single aspheric lens with particle swarm algorithm', *Opt. Commun.*, Vol. 285 (2012) pp. 2996–3000.
113 C. Menke, 'Application of particle swarm optimization to the automatic design of optical systems', *Proc. SPIE*, Vol. 10690 (2018) 106901A.
114 D. Guo, L. Yin, and G. Yuan, 'New automatic design method based on combination of particle swarm optimization and least squares', *Opt. Exp.*, Vol. 27 (2019) pp. 17027–17040.
115 S. Mukhopadhyay and L.N. Hazra, Section 3C on 'Multiobjective optimization using PSO' in 'Pareto Optimality between width of central lobe and sidelobe intensity in the farfield pattern of lossless phase only filters for enhancement of transverse resolution', *Appl. Opt.*, Vol. 54 (2015) pp. 9205–9212.
116 F. Glover and M. Laguna, *Tabu Search*, Kluwer Academic Publishers, Boston (1997) p. 17.
117 M. Wahde, *Biologically Inspired Optimization Methods: An Introduction*, WIT Press, Southampton (2008).
118 S. Binitha and S. Siva Sathya, 'A survey of bio inspired optimization algorithms', *Int. J. Soft Computing and Engineering*, Vol. 2 (2012) pp. 137–151.
119 X.–S. Yang, 'Nature-Inspired optimization algorithms: challenges and open problems', *J. Computational Science*, Article 101104 (2020) pp. 1–15.
120 J. Kennedy, R.C. Eberhart, and Y. Shi, *Swarm intelligence*, Academic, London (2001).
121 N. Hansen, A. Auger, R. Ros, S. Finck, and P. Pošík, 'Comparing results of 31 algorithms from the black-box optimization benchmarking BBOB-2009', *Proceedings of the 12th Annual Conference Companion on Genetic and Evolutionary Computation* (2010). pp. 1689–1696.
122 S. Shamshirband, D. Petković, N.T. Pavlović, S. Ch, T.A. Altameem, and A. Gani, 'Support vector machine firefly algorithm based optimization of lens systems', *Appl. Opt.*, Vol. 54 (2015) pp. 37–45.
123 X. Zhou and C. Sun, 'Convergence of MEC in bounded and continuous search space', *2004 IEEE Int. Conf. Systems, Man and Cybernetics, IEEE Explore* (2004) pp. 1412–1419.
124 M. Sakharov, T. Houllier, and T. Lépine, 'Mind Evolutionary Computation Co-Algorithm for Optimizing Optical Systems', in *Proc. Fourth Int. Scientific Conference on "intelligent Information Technologies for Industry" (IITI'19)*, Eds. S. Kovalev, V. Tarassov, V. Snasel, and A. Sukhanov, Springer Nature, Switzerland AG (2020) pp. 476–486.
125 T. Houllier and T. Lépine, 'Comparing optimization algorithms for conventional and freeform optical design', *Opt. Exp.*, Vol. 27 (2019) pp. 18940–18957.
126 D.H. Wolpert and W.G. Macready, 'No free lunch theorems for optimization', *IEEE Trans. Evol. Comput.* Vol. 1 (1997) pp. 67–82.
127 D. Corne and J. Knowles, 'Some multiobjective optimizers are better than others', *Evol. Comput.*, Vol. 4 (2003) pp. 2506–2512.
128 D.H. Wolpert and W.G. Macready, 'Coevolutionary free lunches', *IEEE Transactions Evol. Comput.*, Vol. 9 (2005) pp. 721–735.
129 A. Auger and O. Teytaud, 'Continuous lunches are free plus the design of optimal optimization algorithms', *Algorithmica*, Vol.57 (2010) pp. 121–146.
130 T. Joyce and J.M. Hermann, 'A Review of No Free Lunch Theorems, and their Implications for Metaheuristic Optimisation', in *Nature-Inspired Algorithms and Applied Optimization*, Ed. X. –S. Yang, Springer, Cham, Switzerland (2018) pp. 27–52.
131 L.N. Hazra, 'Changing Frontiers in Optical design', in Eds. J. Joseph, A. Sharma and V. K. Rastogi, *Perspectives in Modern Optics and Instrumentation*, Anita Publications, New Delhi, (2002) pp. 30–35.
132 L. N. Hazra, 'Towards global synthesis of lens systems', *DST-SERC Highlights*, Published by Department of Science & Technology, Govt. of India, (2003).
133 L.N. Hazra and S. Chatterjee, 'A prophylactic strategy for global synthesis in lens design', *Optical Review*, Vol.12, No.3 (2005) pp. 247–254.

134 L.N. Hazra, 'Structural design of multicomponent lens systems', *Appl. Opt* 23 (1984) pp. 4440–4443.

135 L.N. Hazra and S. Banerjee, 'Genetic algorithm in structural design of Cooke triplet lenses', *Proc. SPIE*, Vol. 3737 (1999) pp. 172–179.

136 L.N. Hazra and A.K. Samui, 'Design of individual components of a multicomponent lens system: use of a singlet', *Appl. Opt.*, Vol. 25 (1986) pp. 3721–3730.

137 S. Banerjee and L.N. Hazra, 'Structural design of doublet lenses with prespecified targets', *Opt. Eng.*, Vol. 36, No.11 (1997) pp. 3111–3118.

138 S. Banerjee and L.N. Hazra, 'Genetic algorithm for lens optimisation at structural design phase', in *Optics and Optoelectronics: Theory, Devices and Application, Vol. I*, Eds. O.P. Nijhawan et al., Narosa Publishing, New Delhi, (1999) pp. 499–504.

139 S. Chatterjee and L.N. Hazra, Structural design of a lens component with pre-specified aberration targets by evolutionary algorithm, *Proc. SPIE*, Vol.6668 (2007). 66680S-1 to 66680S-12.

140 D.E. Goldberg, 'Multiobjective optimization' in Ref. 68, pp. 197–201.

141 K. Deb, *Multi-Objective Optimization using Evolutionary Algorithms*, Jon Wiley, Chichester (2001).

142 A. Konak, D.W. Coit, and A. e Smith, 'Multiobjective optimization using genetic algorithms: a tutorial', *Reliability Eng, and Systems Safety*, Vol. 91 (2006) pp. 992–1007.

143 J. Branke, K. Deb, K. Miethinen, and B. Słowiński, Eds., *Multiobjective Optimization: Interactive and Evolutionary Approaches*, Springer, Berlin (2008).

144 N. Gunantara, 'A review of multi-objective Optimization: Methods and its applications', *Cogent Engineering*, Vol. 5 (2018) 1502242, pp. 1–16.

145 C.A. Coello Coello, 'Multi-objective Optimization', in *Handbook of Heuristics*, Eds. R. Marti, P. Panos and M. G. C. Resende, Springer Int. Publ. AG (2018) pp. 1–28.

146 V. Pareto, *Cours D'Economie Politique, Vol. I and II*, F. Rouge, Lausanne (1896–97).

147 Y. Censor, 'Pareto optimality in multiobjective problems', *Appl. Math. Optim.*, Vol. 4 (1977) pp. 41–59.

148 L.N. Hazra and S. Mukhopadhyay, 'Pareto Optimality between Far-field Parameters of Lossless Phase-only Filters', in *Advances in Optical Sciences and Engineering, Springer Proceedings in Physics,* 194 (2017) pp. 79–87.

149 P Ngatchou, A. Zarei, and M.A. El-Sharkawi, 'Pareto Multi-Objective Optimization', *Proc. 13th Int. Conf. Intelligent Systems Application to Power Systems, Arlington, Virginia,* IEEE Explore (2006).

150 B.F.C. de Albuquerque, L.-Y. Liao, A.S. Montes, F.L. de Sousa, and J. Sasian, 'A multi-objective approach in the optimization of optical systems taking into account tolerancing', *Proc. SPIE*, Vol. 8131 (2011) 813105-1 to 813105-9.

151 B.F.C. de Albuquerque, J. Sasian, F.L. de Sousa, and A.S. Montes, 'Method of glass selection for color correction in optical system design', *Opt. Exp.*, Vol. 20 (2012) pp. 13592–13611.

152 B.F.C. de Albuquerque, L.-Y. Liao, A.S. Montes, 'Multi-objective approach for the automatic design of optical systems', *Opt. Exp.*, Vol. 24 (2016) pp. 6619–6643.

153 M. Herzberger, 'Geometrical Optics', Chapter 2 in *Handbook of Physics*, Eds. E.U. Condon and H. Odishaw, McGraw-Hill, New York (1958) 6–20 to 6–46.

Epilogue

Looking back over the last 21 chapters, we note that most topics of ray optics and wave optics that form the building blocks of optical system analysis have been covered. Further, many topics, including the types and characteristics of different types of lens systems, lens design methodology, the mathematical background of constrained nonlinear optimization, deterministic and stochastic approaches for global optimization—which constitute the basic requisites for understanding—and the implementation and innovation in synthesis of optical systems, are presented with necessary detail. In most cases, the topics are initiated with a brief historical review for underscoring the fast-changing scenario in the area compounded by rapid technological developments in the recent past.

With proliferation of novel and challenging types of optical systems, it is quite natural that investigators will come forward with new proposals and suggestions for facilitating the analysis and synthesis of these systems. Over the years, across the globe, developments have been phenomenal. From the available publications, in general, some of the accounts turn out to be perfunctory, but some others are really thought-provoking. Attempts have been made in this treatise to include as many of the latter as possible. Nevertheless, some omissions have been inevitable, if for no other reason but to keep the treatise within manageable size. We note below a few of the significant contributions which could not be accommodated in this treatise.

Diapoint aberration theory is a different way for tackling the three-dimensional image quality in the immediate neighbourhood of the image plane of axisymmetric optical imaging systems. Although the theory has been described in detail by Herzberger in his book and in other publications [19.61], practical utilization of the method remains limited. Uses of Lie algebraic theory in geometrical optics has provided new analytical formulations. It appears that the treatment of complex systems is facilitated significantly if the concise and modular formulations are properly utilized [5.66–5.68]. Two other significant omissions are the polarization aberration theory and the modal aberration theory. Details on these two topics may be found in the book by Sasian [8.78].

Obviously, analysis and synthesis of nonimaging optics calls for a somewhat different approach, albeit most of the basic principles of imaging optics, discussed earlier, hold good in the case of nonimaging optics, also [19.163]. Currently, simultaneous multiple surface (SMS) design method is being experimented with in both imaging and nonimaging applications. Details may be found in references [19.131, 19.167].

Index

D

E

H

I

O

P

Q

R

S

T

For Product Safety Concerns and Information please contact our EU
representative GPSR@taylorandfrancis.com
Taylor & Francis Verlag GmbH, Kaufingerstraße 24, 80331 München, Germany

www.ingramcontent.com/pod-product-compliance
Lightning Source LLC
Chambersburg PA
CBHW080338220326
41598CB00030B/4534
* 9 7 8 1 0 3 2 0 3 7 0 8 0 *